Handbook of Randomized Computing

COMBINATORIAL OPTIMIZATION

VOLUME 9

Through monographs and contributed works the objective of the series is to publish state of the art expository research covering all topics in the field of combinatorial optimization. In addition, the series will include books which are suitable for graduate level courses in computer science, engineering, business, applied mathematics, and operations research.

Combinatorial (or discrete) optimization problems arise in various applications, including communications network design, VLSI design, machine vision, airline crew scheduling, corporate planning, computer-aided design and manufacturing, database query design, cellular telephone frequency assignment, constraint directed reasoning, and computational biology. The topics of the books will cover complexity analysis and algorithm design (parallel and serial), computational experiments and applications in science and engineering.

The titles published in this series are listed at the end of this volume.

Handbook of
Randomized Computing
Volume I

edited by

Sanguthevar Rajasekaran
University of Florida

Panos M. Pardalos
University of Florida

John H. Reif
Duke University

and

José Rolim
University of Geneva

SPRINGER SCIENCE+BUSINESS MEDIA, B.V.

A C.I.P. Catalogue record for this book is available from the Library of Congress.

ISBN 978-1-4613-4886-3   ISBN 978-1-4615-0013-1 (eBook)
DOI 10.1007/978-1-4615-0013-1

Printed on acid-free paper

Contents

Volume II

The technique of randomization has been employed to solve numerous problems of computing both sequentially and in parallel. Examples of randomized algorithms that are asymptotically better than their deterministic counterparts in solving various fundamental problems abound. Randomized algorithms have the advantages of simplicity and better performance both in theory and often in practice. This book is a collection of articles written by renowned experts in the area of randomized parallel computing.

A brief introduction to randomized algorithms

In the analysis of algorithms, at least three different measures of performance can be used: the best case, the worst case, and the average case. Often, the average case run time of an algorithm is much smaller than the worst case. For instance, the worst case run time of Hoare's quicksort is $O(n^2)$, whereas its average case run time is only $O(n \log n)$. The average case analysis is conducted with an assumption on the input space. The assumption made to arrive at the $O(n \log n)$ average run time for quicksort is that each input permutation is equally likely. Clearly, any average case analysis is only as good as how valid the assumption made on the input space is.

Randomized algorithms achieve superior performances without making any assumptions on the inputs by making coin flips within the algorithm. Any analysis done of randomized algorithms will be valid for all possible inputs.

A randomized algorithm can be thought of as one wherein certain decisions are made based on the outcomes of coin-flips made in the algorithm. A randomized algorithm with one possible sequence of outcomes for the coin flips can be considered as being different from the same algorithm with a different sequence of outcomes for the coin flips. Thus a randomized algorithm can be conceived of as a family of algorithms. For a given input, some of the algorithms in this family might have a 'poor performance'. We must ensure that the number of such bad algorithms in the family is only a small fraction of the total number of algorithms. If for *any* input we can find at least $(1 - \epsilon)$ (ϵ being very close to 0) portion of algorithms in the family that will have a 'good performance' on that input, then clearly, a random algorithm in the family will have a 'good performance' on any input with probability $\geq (1 - \epsilon)$. In this case we say that this family of algorithms (or this randomized algorithm) has a 'good performance' with probability at least $(1 - \epsilon)$. We refer to ϵ as the error probability. Notice that this probability is independent of the input distribution. The phrase 'good performance' can be interpreted in a number of ways. It could mean that the algorithm outputs the correct answer or that its run time is small. We can thus think of different types of randomized algorithms depending on how this phrase is interpreted. A *Las Vegas* algorithm is a randomized algorithm that always outputs the correct answer but whose run time is a random variable (possibly with a small mean). In contrast, a *Monte Carlo* algorithm is a randomized algorithm which runs for a predetermined amount of time but whose output may be incorrect occasionally.

Two simple examples

We now give two simple examples of randomized algorithms. The first is a Las Vegas algorithm and the second is a Monte Carlo algorithm.

[Repeated Element]. Let $a[\ ]$ be an array of n elements wherein there are $\frac{n}{2}$ distinct elements and $\frac{n}{2}$ copies of another element. The problem is to identify the repeated element.

It is easy to see that any deterministic algorithm that solves this problem will take at least $\frac{n}{2} + 2$ time in the worst case. An adversary, who has perfect knowledge about the algorithm and who is in charge of selecting the input, can ensure that the first $\frac{n}{2} + 1$ elements examined by the algorithm are all distinct. As a result, the algorithm will be forced to examine more than $\frac{n}{2} + 1$ elements.

A simple $O(n)$ time deterministic algorithm for this problem partitions the array into $\lceil \frac{n}{3} \rceil$ parts with three elements in each part (excepting possibly for one part). It then searches the individual parts for the repeated element. Clearly, at least one of the parts will have at least two copies of the repeated element. This algorithm runs in time $\Theta(n)$.

A simple and elegant Las Vegas algorithm that takes only $O(\log n)$ time *with high probability* can be devised as follows. (By *high probability* we mean a probability that is $\geq (1 - n^{-\alpha})$ for any fixed α). There are many *stages* in the algorithm. In any stage, two random numbers i and j are picked from the range $[1, n]$. These numbers are picked independently with replacement. After picking i and j check if $i \neq j$ and $a[i] = a[j]$. If so, the repeated element has been found. If not, the next stage is entered. The stages are repeated as many times as it takes to come up with the correct answer.

The run time of this algorithm can be shown to be $O(\log n)$ with high probability. The probability of finding the repeated element in any given stage is $P = \frac{n/2(n/2-1)}{n^2}$, which is $\geq \frac{1}{5}$ for all $n \geq 10$. Thus, the probability that the algorithm does not find the repeated element in the first $c\alpha \log n$ (c is a constant to be fixed) stages is

$$< (4/5)^{c\alpha \log n} = n^{-c\alpha \log (5/4)}$$

which will be $< n^{-\alpha}$ if we pick $c \geq \frac{1}{\log (5/4)}$.

In other words, the algorithm takes no more than $\frac{1}{\log (5/4)}\alpha \log n$ stages with probability $\geq 1 - n^{-\alpha}$. Since each stage takes $O(1)$ time, the run time of the algorithm is $O(\log n)$ with high   □

[Approximate Median]. The second example we consider has an array $a[\ ]$ of n numbers as input. The problem is to identify an approximate median of the n numbers. Assume, without loss of generality, that the array numbers are distinct.

For any set X of distinct numbers, define the *rank* of any element y in X as $|\{x \in X | x < y\}| + 1$. I.e., the rank of y is one plus the number of elements in X that are less than y. We want to find an element of $a[\ ]$ whose rank in $a[\ ]$ is in the interval $\left(\left(\frac{1}{2} - \delta\right) n, \left(\frac{1}{2} + \delta\right) n\right)$, where δ is a specified constant.

The technique of random sampling can be used to develop a Monte Carlo algorithm for this problem. Let S be a random sample of $a[\ ]$ with $|S| = s$. Each element of S is picked independently, and uniformly randomly from $a[\ ]$ with replacement. We expect that the median of S will have a rank close to $\frac{n}{2}$ in $a[\ ]$. In fact the Monte Carlo algorithm picks a random sample S of size $\frac{2(1-2\delta)}{\delta^2}\alpha \log n$ from $a[\ ]$, finds the median of S, and outputs this sample median.

It can be shown (we skip the proof) that the output of the algorithm is correct with probability $\geq (1 - n^{-\alpha})$.   □

Parallel Machine Models

Parallel computing has become popular with the rapid advances that have been made in the VLSI technology. If we use P processor to solve a problem, then there is a potential of reducing the (sequential) run time by a factor of up to P. If S is the best known sequential run time and if T is the parallel run time using P processors, then $PT \geq S$. Otherwise we can simulate the parallel algorithm using a single processor and get a run time better than S (which will be a contradiction). We refer to PT as the *work done* by the parallel algorithm. Any parallel algorithm for which $PT = O(S)$ will be referred to as an *optimal work* algorithm.

In sequential computing the Random Access Machine (RAM) has been widely accepted as a valid model of computing. On the other hand, there exist a large number of parallel computing models, partly due to the fact that differing parallel architectures can be conceived of and have been built in practice.

In any parallel machine we can still think of each processor as a RAM. Variations among different architectures arise from the ways in which they implement interprocessor communications. Parallel models can be categorized broadly into *parallel comparison trees, shared memory models,* and *fixed connection machines.*

A parallel comparison tree is analogous to the sequential comparison (or decision) tree. It is typically employed for the study of comparison problems such as sorting, selection, merging, etc. An algorithm under this model is represented as a tree. The computation starts at the root. P pairs of input keys are compared in parallel at the root (P being the number of processors). Based on the outcomes of these comparisons, the computation branches to an appropriate child of the root. Each node of the tree corresponds to comparison of P pairs of input keys. The computation terminates at a leaf node which has enough information to output the correct answer. Thus there is a tree corresponding to every input size. For a given instance of the problem, a branch in the tree is traversed. The worst case run time is proportional to the depth of the tree. This model takes into account only the comparison operations performed.

A shared memory model (also called the Parallel Random Access Machine (PRAM)) is a collection of RAMs working in synchrony where communication takes place with the help of a common block of shared memory. If, for example, processor i wants to communicate with processor j it can do so by writing a message in memory cell j which can then be read by processor j.

More than one processor may want to access the same cell at the same time for either reading from or writing into. Depending on how these conflicts are resolved, a PRAM can further be classified into three. In an Exclusive Read and Exclusive Write (EREW) PRAM neither concurrent reads nor concurrent writes are allowed. In a Concurrent Read and Exclusive Write (CREW) PRAM concurrent reads are permitted but not concurrent writes. Finally, a Concurrent Read and Concurrent Write (CRCW) PRAM allows both concurrent reads and concurrent writes. A mechanism for handling write conflicts is needed for a CRCW PRAM, since the processors trying to write at the same time in the same cell can possibly have different data to write and we should determine which data gets written. This is not a problem in the case of concurrent reads since the data read by different processors will be the same. In a Common-CRCW PRAM, concurrent writes are permissible only if the processors trying to access the same cell at the same time have the same data to write. In an Arbitrary-CRCW PRAM, if more than one processor tries to write in the same cell at the same time, an arbitrary one of them succeeds. In a Priority-CRCW PRAM, processors have assigned priorities. Write conflicts are resolved using these priorities.

A directed graph can be used to represent a fixed connection machine (or a fixed connection network). The nodes of this graph correspond to processing elements and the edges correspond to communication links. If two processors are connected by an edge, they can communicate in a unit step. Two processors not connected by an edge can communicate by sending a message along a path that connects the two processors. Each processor in a fixed connection machine is a RAM. Examples of fixed connection machines include the mesh, the hypercube, the star graph, etc.

A mesh is an $n \times n$ square grid whose nodes are processors and whose edges are communication links. The *diameter* of a mesh is $2n - 2$. (*Diameter* of a graph is defined to be the maximum of the shortest distance between any two nodes in the graph.) Diameter of a fixed connection machine is often a lower bound on the solution time of any nontrivial problem on the machine. The *degree* of a fixed connection network should be as small as possible for it to be physically realizable. (The *degree* of a fixed connection machine is defined to be the maximum number of neighbors for any node.) The degree of a mesh is four.

A hypercube of dimension n has 2^n nodes. Any node in a hypercube can be denoted as an n-bit binary number. Let x and y be the binary representations of any two nodes in a hypercube. Then, these two nodes will be connected by an edge if and only if the hamming distance between x and y is one, i.e., x and y differ in exactly one bit position. Thus, the degree of a hypercube with 2^n nodes is n. The diameter of a 2^n-noded hypercube can also be seen to be n. Butterfly, CCC, de Bruijn, etc. are networks that are very closely related to a hypercube.

The articles in this book span all of the above models of computing.

Intended use of the book

This book is meant for use by researchers, developers, educators, and students in the area of randomized computing. Since randomization has found applications in a wide variety of domains, the ideas and techniques illustrated in this book should prove useful to a very wide audience.

This book can also be used as a text in a graduate course dealing with randomized computing. Graduate students who plan to conduct research in this area will find this book especially invaluable. The book can also be used as a supplement to any course on algorithms, complexity, or parallel computing.

Acknowledgements

We are very thankful to the authors who have written these chapters under a very tight schedule. We thank the staff of Kluwer Academic Publishers, in particular, John Martindale and Sharon Donovan. We also gratefully acknowledge the partial support from the National Science Foundation through grant EIA-9872507.

Sanguthevar Rajasekaran
Panos Pardalos
John H. Reif
Jóse Rolim
January 2001

Other Reading

A partial list of books that deal with randomization is given below. This by no means is an exhaustive list of all the books in the area.

1. N. Alon and J. H. Spencer, *The Probabilistic Method*, Wiley-Interscience Publication, 1992.

2. B. Bollobas, *Random Graphs*, Academic Press, 1985.

3. T.H. Cormen, C. Leiserson and R.L. Rivest, *Introduction to Algorithms*, MIT Press, Cambridge, MA, 1991.

4. E. Horowitz, S. Sahni, and S. Rajasekaran, *Computer Algorithms*, (W. H. Freeman Press, 1998).

5. J. Já Já, *An Introduction to Parallel Algorithms*, Addison-Wesley Publishers, 1992.

6. T. Leighton, *Introduction to Parallel Algorithms and Architectures: Arrays Trees–Hypercube*, Morgan-Kaufmann Publishers, 1992.

7. R. Motwani and P. Raghavan, *Randomized Algorithms*, Cambridge University Press, 1995.

8. K. Mulmuley, *Computational Geometry: An Introduction Through Randomized Algorithms*, Prentice-Hall, 1994.

9. P.M. Pardalos and S. Rajasekaran, editors, *Advances in Randomized Parallel Computing*, Kluwer Academic Publishers, 1999.

10. P.M. Pardalos, S. Rajasekaran, and J. Rolim, editors, *Randomization Methods in Algorithm Design*, DIMACS Series in Discrete Mathematics and Theoretical Computer Science, Volume 43, American Mathematical Society, 1999.

11. J.H. Reif, editor, *Synthesis of Parallel Algorithms*, Morgan-Kaufmann Publishers, 1992.

Pankaj K. Agarwal, Earl J. McLean Jr. Professor of Computer Science, joined Duke University in 1989 after completing his PhD in computer science from Courant Institute of Mathematical Sciences, New York. He also spent the academic year 1989-90 at DIMACS, New Jersey. Dr. Agarwal's research interests include computational geometry, geographic information systems, spatial databases, molecular biology, computer graphics and visualization, and robotics. He is the author/co-author of three books, he has edited one book, and he has published over 160 technical papers in various journals and conference proceedings. He is a recipient of a National Young Investigator award, Alfred P. Sloan Fellowship, and Bass Fellowship. He currently serves on the editorial boards of five journals; he was the program chair of Sixteenth Aannual Symposium on Computational Geometry; and he regularly serves on program committees of international conferences.

Robert Bohlin received the M.Sc. degree in Automation Engineering from Chalmers University of Technology, Göteborg, Sweden, in 1997, and is currently working toward the Ph.D. degree in Industrial Mathematics at the same university. His research interest is path planning for robots.

Bogdan S. Chlebus is on the faculty of Institute of Informatics at the Warsaw University in Warsaw, Poland. His primary research interests are in distributed and parallel computing. Recently he has concentrated on the issues of communication in networks, fault-tolerant computing, and efficient use of randomization in algorithms.

Anne Condon's research interests are in the areas of probabilistic and interactive complexity classes, design and analysis of algorithms for computationally intractable problems, and DNA computing. Condon received a B.Sc. degree from University College, Cork, Ireland in 1982 and a Ph.D. from the University of Washington in 1987. Her thesis, a study of game-theoretic complexity classes, won an ACM Distinguished Dissertation award and was published by the MIT Press. Condon received a National Young Investigator Award in 1992. She has over 40 research publications in refereed conferences and journals. Condon is currently a Professor in the Department of Computer Science at U. Wisconsin, and was on the faculty of the Computer Sciences Department at U. Wisconsin between 1987 and 1999.

Xiaotie Deng got his B. Sci. at Tsinghua University, Beijing, China, in 1982, and his M. Sci. at Chinese Academy of Sciences, Beijing, China in 1984, his Ph.D. at Stanford University, California, USA, in 1989. After finishing PhD, he received an International Research Fellowship from Natural Science and Engineering Council of Canada to do postdoctoral research at Simon Fraser University, at British Columbia, Canada. In 1991, he joined York University, Toronto, as an assistant professor, and then associate professor. In 1997, he joined City University of Hong Kong, Hong Kong, China. The research work of Dr. Deng includes online algorithms, approximation algorithms, algorithmic aspect of game theory and management science. He is currently interested in applying expertise in algorithms and complexity to study internet searching and financial optimization.

Josep Díaz is Full Professor at the Universitat Politècnica de Catalunya. His main topics of interest are algorithms and combinatorics.

Devdatt Dubhashi received his B.Tech. from IIT Delhi and his Ph.D. from Cornell University. He worked at the Max-Planck-Institut for Informatik, Germany and at BRICS (Basic Research in Computer Science, a centre of the Danish National Research Foundation) before returning to IIT Delhi as Assitant Professor. Currently, he is with the Department of Computing, Chalmers University, Sweden. His main research interests are in probabilistic techniques for the design and analysis of algorithms.

Michael Goodrich received his Ph.D. in Computer Science from Purdue University in 1987. He is currently a professor in the Department of Computer Science at Johns Hopkins University, and director of the Johns Hopkins Center for Algorithm Engineering. He is an editor for the *International Journal of Computational Geometry & Applications*, *Journal of Computational and System Sciences*, and *Journal of Graph Algorithms and Applications*. He has published extensively in the areas of algorithm engineering and computational geometry, and his research is supported by the National Science Foundation, the Army Research Office, and the Defense Advanced Research Projects Agency.

Juraj Hromkovic was born in 1958 , Bratislava, Czechoslovakia. He received Dr. rer. nat. degree from Comenius University , Bratislava in 1982. Futher degrees: PhD at Comenius University in 1986, habilitation at Comenius University in 1989, DrSc degree (the highest scientific title in some Eastern European countries) in 1990. In 1989 he was an associative professor at the Comenius University. From 1990 to 1994 he was a visiting professor at the University of Paderborn in Germany, and from 1994 to 1997 he was an associative professor at the University of Kiel in Germany. In 1994 he got the title professor from the president of the Slovak republic. Recently he is a full professor at the Technological University (RWTH) of Aachen. He has published two books and more than 100 scientific papers. His main topics of research interest were computational complexity (especially the study of the power of nondeterminsm and randomization), formal languages and automata theory, communication complexity, parallelism and communication in networks, and algorithmics for hard problems.

David Karger (A.B. 1989, Harvard University, Ph.D. 1994, Stanford University) is a Harold E. Edgerton Associate Professor of Computer Science and a member of the Laboratory for Computer Science at the Massachusetts Institute of Technology. His research interests include algorithms and information retrieval. Professor Karger's work in algorithms has focused on applications of randomization to network optimization problems. His dissertation on the topic was awarded the ACM 1994 Doctoral Dissertation Award and the 1997 Tucker Prize. He has published more than 30 technical articles and two book chapters and has served on program committees for the Symposium on Discrete Algorithms and the Symposium on the Foundations of Computer Science. Professor Karger's work on information retrieval includes the co-development of the Scatter/Gather information retrieval system at Xerox PARC, which suggested several novel approaches for efficiently retrieving information from massive corpora and presenting it effectively to users. He has received two patents related to his work on this project.

Lydia Kavraki is an assistant professor of Computer Science at Rice University since 1996. She completed her B.S. degree in Computer Science in Greece and obtained her M.S. and Ph.D. degrees in Computer Science from Stanford University in 1992 and 1995 respectively. Kavraki also holds a joint appointment with the Department of Bioengineering at Rice University. Kavraki's research investigates algorithms and system architectures for solving geometric problems arising in the physical world. She is particularly interested in problems in the areas of motion planning, assembly sequencing, manufacturing, and applications in computational chemistry (pharmaceutical drug design) and medicine (robot-assisted surgery). Kavraki received the National Science Foundation CAREER award (Career Development Award) in 1997 and a Sloan Fellowship in 2000. She has authored more than fifty scientific papers published in refereed journals and conferences and has given more than sixty talks on her research work in the US and Europe. Kavraki was a co-editor of the book

"Robotics: The Algorithmic Perspective" published in 1998 and an invited speaker at the Frontiers of Engineering Symposium of the National Academy of Engineering in 1998. She has served in the program committees of several conferences (ICRA, IROS, IJCAI, AAAI, WAFR, SoCG) and co-chaired the 1998 International Workshop on Algorithmic Foundations of Robotics.

Vladik Kreinovich received his M.S. in Mathematics and Computer Science from Leningrad University, Russia, in 1974, and Ph.D. from the Institute of Mathematics, Soviet Academy of Sciences, Novosibirsk, in 1979. In 1975-80, he worked with the Soviet Academy of Sciences, in particular, in 1978-80, with the Special Astrophysical Observatory (representation and processing of uncertainty in radioastronomy). In 1982-89, he worked on error estimation and intelligent information processing for the National Institute for Electrical Measuring Instruments, Russia. In 1989, he was a Visiting Scholar at Stanford University (Artificial Intelligence). Since 1990, he is with Department of Computer Science, University of Texas at El Paso. Also, served as an invited professor in Paris (University of Paris VI), Hong Kong, etc. Main interests: representation and processing of uncertainty, especially interval computations and intelligent control. Member of the editorial board of the international journal "Reliable Computing" (formerly, "Interval Computations") and several other journals. Published 4 books and more than 400 papers.

Danny Krizanc received his BSc from University of Toronto in 1983 and his PhD from Harvard University in 1988. After graduating, he spent one year at the Centruum voor Wiskunde en Informatica in Amsterdam as a research scientist. Between 1989 and 1992, he was an Assistant Professor at The University of Rochester, Rochester, New York. In 1992 he joined Carleton University in Ottawa, Canada becoming an Associate Professor in 1995. Currently he holds an Associate Professor position at Wesleyan University in Middletown, Connecticut. His research interests include practical and theoretical aspects of parallel computing, distributed computing and networking.

Peter Bro Miltersen completed his PhD at the University of Aarhus in 1993. Afterwards he spent 3 years at the Universities of Warwick and Toronto doing postdoctoral research. He is currently an associate professor at the University of Aarhus and a member of the board of the BRICS international PhD school.

Michael Mitzenmacher graduated summa cum laude in Mathematics and Computer Science from Harvard in 1991. After a year in England studying mathematics at Cambridge University on a Churchill Fellowship, he earned his Ph.D. in computer science at U.C. Berkeley in 1996. He worked at Digital Systems Research Center until 1999, when he began his current position at Harvard as an Assistant Professor of Computer Science. His research interests include load balancing, error-correcting and erasure codes, compression, stochastic bin-packing problems, and on-line algorithms.

Sotiris Nikoletseas is currently a Senior Researcher and Director of Research Unit 1 ("Foundations of Computer Science, Relevant Technologies and Applications") at the Computer Technology Institute (CTI), Patras, Greece and also an External Lecturer at the Computer Engineering and Informatics Department of Patras University, Greece. His research interests include Probabilistic Techniques and Random Graphs, Average Case Analysis of Graph Algorithms and Randomized Algorithms, Fundamental Issues in Parallel and Distributed Computing with an emphasis in Mobile Computing, Approximate Solutions to Computationally Hard Problems. He has published several scientific articles in major international conferences and journals and has co-authored (with Paul Spirakis) a book on Probabilistic Techniques. He has been invited speaker in important international scientific events and Universities. He has been a referee for major journals and important international conferences. He has also served in the Program and Organizing Committees of International Conferences and Workshops. He has directed or participated in many European Union funded R&D projects.

Jordi Petit is assistant professor at the Universitat Politècnica de Catalunya. His topics of interest are design, analysis and implementation of algorithms and combinatorial optimization problems on random graphs.

Sanguthevar Rajasekaran received his Ph.D. degree in Computer Science from Harvard University in 1988. Currently he is a Professor in the Department of Computer and Information Science and Engineering at the University of Florida. His research interests include Parallel Algorithms, Randomized Computing, Data Mining, Computational Biology, Combinatorial Optimization, Parsing, and Image Processing. Dr. Rajasekaran has published more than 100 papers in journals, conferences, and books. He is a co-author of the texts *Computer Algorithms* and *Computer Algorithms/C++* both published by W.H. Freeman Press. He has edited several books on algorithms. He serves on the editorial boards of IEEE Transactions on Compurers, Journal of Parallel and Distributed Computing, and Journal of Interconnection Networks. His research is funded by the National Science Foundation and the Environment Protection Agency.

Abhiram Ranade received a B.Tech degree in Electrical Engineering from IIT Bombay in 1981 and a doctorate in Computer Science from Yale University in 1989. He was was an Assistant Professor of Electrical Engineering and Computer Science at the University of California, Berkeley during 1988-95. Since 1995 he is on the faculty of IIT Bombay, currently as a Professor of

Computer Science and Engineering. His research concerns theoretical computer science, parallel algorithms and high performance computing. He has served on program committees of leading conferences such as STOC (Symposium on the Foundations of Computer Science), SPAA (Symposium on Parallel Algorithms and Architectures), and Europar. He is currently an associate editor of IEEE Transactions on Parallel and Distributed Systems.

Andréa Werneck Richa graduated cum laude in Computer Science from the Federal University of Rio de Janeiro (UFRJ), Brazil, in 1989. After receiving an M.S. in Computer Science in 1992 from the graduate school in Engineering (COPPE) at UFRJ, she joined the Algorithms, Combinatorics and Optimization Ph.D. program in the School of Computer Science at Carnegie Mellon University. She earned her M.S. and Ph.D. degrees in this program in 1995 and 1998, respectively. Prof. Richa joined the Department of Computer Science and Engineering at Arizona State University as an Assistant Professor in 1998. Her research interests focus primarily on network algorithms (for distributed, wireless, or mobile scenarios), resource allocation, and parallel network architectures; and on randomized and approximation algorithms, and combinatorial optimization in general.

Dana Ron is a lecturer in the Department of Electrical Engineering Systems, at Tel Aviv University, Israel. She received her B.A., M.Sc., and Ph.D. from the Hebrew University in 1987, 1989, and 1995, respectively. She was an NSF postdoctoral fellow at MIT's Laboratory for Computer Science, 1995-1997, and a Bunting fellow at MIT and Radcliffe, 1997-1998. She joined the faculty of Electrical Engineering - Systems in the fall of 1998.

Sandeep Sen obtained his PhD from Duke University in 1989 and spent a year at AT&T Bell Laboratories (Murray Hill) thereafter. Since 1991 he has been a faculty in the department of Computer Science and Engineering in IIT Delhi, India where he is currently a Professor. He was a Visiting faculty on Sabbatical in UNC Chapel hill in 1998-99. His research interests are primarily in Algorithms and Theoretical Computer science including Computational Geometry, Randomization techniques and Computation in memory hierarchy.

Maria Serna is associated professor in the Software Department at the Universitat Politècnica de Catalunya. Her main topic of interest is the design an analysis of algorithms for graph problems.

Ramesh K. Sitaraman received his B. Tech. in electrical engineering from the Indian Institute of Technology, Madras. He obtained his Ph.D. in computer science from Princeton University in 1993. Dr. Sitaraman is currently at Akamai Technologies, a company devoted to efficient and reliable delivery of content on the Internet. He is also an Associate Professor of Computer Science at the University of Massachusetts at Amherst, where he is a member of the Theoretical Computer Science group, and co-directs the Theoretical Aspects of Parallel and Distributed Systems (TAPADS) Laboratory. Dr. Sitaraman's research focuses on fundamental theoretical issues in the design and use of parallel and distributed systems, and communication networks. His specific interests include communication protocols, fault tolerance, and algorithms for delivering web content and streaming media in the internet. He is a recipient of an NSF CAREER Award and a Lilly Fellowship. He is a member of the IEEE and the ACM.

Paul Spirakis, born in 1955, obtained his PhD from Harvard University, USA, in 1982. He worked as a post doc at Harvard and as a faculty member at the Courant Institute. Then he became a full Professor in the University of Patras Greece in 1990. He is also now the Director of the Computer Technology Institute of Greece. He won the top prize of the Greek Math Association in 1973. He is a Distinguished Visiting Scientist of Max Planck Informatik. His research interests include probabilistic algorithms, parallel and distributed algorithms and network protocols, exact analysis of algorithms, algorithms for mobile computing and experimental algorithms. Also, Complexity of problems. He has published extensively in the top journals and conferences of the field. He has published two books with Cambridge University Press and four books in Greek, all in the above fields. He is a member of the Council of EATCS, the Greek representative in the European Union for Information Society, and a consultant of the Greek State and the EU, also of the major Greek firms in Informatics. He is usually a member of the scientific program or steering commitees of the major conferences of his field, and a member of ACM, EATCS, Math. Society of America and the Greek Computer Society.

Anand Srivastav is a Professor of Mathematics at the University of Kiel in Germany. He received the doctoral degree from the University of Muenster in Germany in 1988 with a thesis in functional analysis. ¿From 1988 - 1993 he has been an Assitant Professor at the Research Institute of Discrete Mathematics of the University of Bonn. Research stays led him to the IMA, Minneapolis, Courant Institute, Yale University, Free University of Berlin and Humboldt University of Berlin. In 1996 he received the habilitation degree in computer science from the Free University of Berlin. His research interest are in combinatorial optimization, algorithms, randomization and derandomization, discrepancy theory and applications of discrete optimization to real-world problems.

Roberto Tamassia received his Ph.D. in Electrical and Computer Engineering from the University of Illinois at Urbana-Champaign in 1988. He is currently a professor in the Department of Computer Science at Brown University. He is an editor for *Computational Geometry: Theory and Applications* and the *Journal of Graph Algorithms and Applications*, and he has previously served on the editorial board of IEEE *Transactions on Computers*. He has published extensively in the areas of algorithm design and analysis, and graph drawing, and his research is supported by the National Science Foundation, the Army Research Office, and the Defense Advanced Research Projects Agency.

Raúl Trejo is a professor at Instituto Tecnológico y de Estudios Superiores de Monterrey (ITESM) at Estado de México, in México, where he has taught since 1994. He obtained his B.S. degree in México and is currently a candidate for Ph. D. at the University of Texas at el Paso. At ITESM, Trejo has participated in several educative projects regarding the integrated teaching of the introductory CS and Math curricula. He is a member of AAAI. His major fields of interest are reliable computing and functional programming. His personal interests include fiction writing, music and collectibles.

Santosh Vempala is an Assistant Professor of Mathematics at MIT. He got his PhD at Carnegie Mellon in 1997. He was a Miller fellow at Berkeley during the year 1998-1999 and has been at MIT since. His research interests are in geometry, randomness and combinatorics, especially in the context of polynomial-time algorithms. Once a year, he teaches a graduate course called "An eye for elegance".

Chapter 1

RANDOM SAMPLING: SORTING AND SELECTION

Danny Krizanc

Dept. of Math. and Comp. Sci.

Wesleyan University

dkrizanc@wesleyan.edu

Sanguthevar Rajasekaran

Dept. of Comp. and Info. Sci. and Engg.

University of Florida

raj@cise.ufl.edu

Abstract   Random sampling techniques have played a vital role in the design of sorting
and selection algorithms for numerous models of computing. In this article we
provide a summary of sorting and selection algorithms that have been devised
using random sampling. Models of computations treated include the parallel
comparison tree, the PRAM, the mesh, the mesh with fixed, reconfigurable, and
optical buses, the hypercube family, and parallel disk systems.

1.   INTRODUCTION

Comparison problems such as sorting and selection have been studied by
researchers extensively owing to their paramount importance. Given a sequence
of n keys the problem of sorting is to rearrange this sequence in nondecreasing
order. The selection problem takes as input a sequence of n keys and an integer
i $(1 \leq i \leq n)$. The problem is to identify the ith smallest key of the sequence.

Optimal (comparison based) sequential RAM algorithms are known for
sorting and selection. Sorting algorithms such as mergesort, heapsort, etc. run
in time $O(n \log n)$ time in the worst case (see e.g., [3, 30]). The selection
algorithm of Blum et al. [15] runs in linear time.

Optimal or near-optimal algorithms for sorting and selection have been
developed for numerous other models of computing as well. The technique of
random sampling has been successfully employed in many of these algorithms.

1

S. Rajasekaran et al (eds.), Handbook of Randomized Computing, Volume 1, pp, 1–21.
© 2001 *Kluwer Academic Publishers.*

In this article we provide a survey of some of these algorithms given for parallel models of computing.

Notation. Throughout this article we let n denote the input size and p denote the number of processors available.

1.1   AN INTRODUCTION TO MODELS OF COMPUTING

In this section we give a brief introduction to the models of computing considered in this article.

1.1.1   The Parallel Comparison Tree.
The Parallel Comparison Tree (PCT) model [84] is the natural generalization of the sequential comparison tree model [3] to the parallel setting. The basic operation available to processors is the comparison of two keys. With p processors, p comparisons may be performed simultaneously in one step. Depending on which of the 2^p possible results is attained, the next set of p comparisons is chosen. The computation ends when sufficient information is discovered about the relationships of the keys to specify the solution to the given problem. The deterministic complexity of a problem in this model is the number of steps required for the worst case input or the minimum depth of a tree solving the problem, as a function of the size of the input sequence and the number of processors used.

We note that in this model we do not consider any of the overheads, such as processor communication, memory accesses, etc., associated with performing the comparisons and making the appropriate deductions from the results of the comparisons. However, in cases where the cost of comparisons dominates the computation, upper bounds in this model can often be translated into upper bounds in more restricted models and in all cases where algorithms base their decisions solely upon comparisons, lower bounds in this model translate to lower bounds in these other models.

The model is easily extended to allow random computations. In the randomized PCT model, at each step we introduce a probability distribution over the choice of which p comparisons are to be performed. In this case, the complexity is the expected number of steps required on the worst case input.

1.1.2   The Parallel Random Access Machine.
The Parallel Random Access Machine (PRAM) is the natural generalization of the RAM model to the parallel setting. In it, p synchronous processors, each identical to a RAM, communicate through the use of a shared memory. There are three main variants of the PRAM depending on what restrictions are placed on concurrent access to the same memory cell in the shared memory. The Exclusive Read Exclusive Write (EREW) PRAM does not allow any simultaneous access to the same memory cell. The Concurrent Read Exclusive Write (CREW) PRAM

allows concurrent reads to take place but does not allow concurrent writes. The Concurrent Read Concurrent Write (CRCW) PRAM allows both concurrent reads and concurrent writes. There are three standard varieties of CRCW PRAM depending on the interpretation of a concurrent write operation. In the common CRCW PRAM, it is required that concurrent writes to a cell are all writing the same value. In the arbitrary CRCW PRAM model an arbitrary value among those being written is chosen. In the priority CRCW PRAM the value written by the lowest indexed processor is the result of the concurrent write. It is easy to see that the CRCW PRAM is stronger than the CREW PRAM which in turn is stronger than the EREW PRAM. It is easy to show that they are all three logarithmically related.

1.1.3   The Mesh. A mesh is a $\sqrt{p} \times \sqrt{p}$ square grid where there is a processor at each grid point. Every processor is connected to its four or less neighbors through bidirectional links. Each processor can communicate with all of its neighbors in one unit of time.

1.1.4   Mesh with Buses. Two variants of the mesh assume the existence of electrical communication buses: 1) the mesh connected computer with fixed buses (denoted as M_f), and 2) the mesh with reconfigurable buses (denoted as M_r).

In M_f each row and each column has an associated broadcast bus. A bus can be used to broadcast a message in every time step. It is assumed that the message broadcast along a bus can be read by all the processors connected to this bus in the same time unit.

M_r also employs buses but these buses are reconfigurable. Reconfigurability of the buses is achieved as follows. Each processor has (at most) four switches, one for each of its neighbors. This switch can be dynamically set on or off. If a switch of a processor is on, it means that the processor is connected to the corresponding neighbor. Depending on how the switches of the processors are set, we can get several disjoint buses. For instance we can form a row bus by setting all the switches along the row on. Each bus functions similar to the buses of M_f, i.e., a message can be broadcast in any bus at every time step and this message can be read by all the processors connected to the bus in the same time step.

In M_f and M_r we assume the existence of electrical buses. Optical technology can be employed to realize these buses in which case we get meshes with optical buses. Several such models have been investigated in the literature.

1.1.5   The Hypercube. A hypercube of dimension ℓ has $p = 2^\ell$ nodes and $\ell 2^{\ell-1}$ edges. Each node in an ℓ-dimensional hypercube can be labelled with an ℓ-bit binary number. Nodes x and y in a hypercube will be connected by a bidirectional link if and only if x and y (considered as binary numbers) differ in exactly one bit position. Thus there are exactly ℓ edges going out of (and coming into) any vertex.

If a hypercube processor can communicate with only one neighbor at any time step, this version of the hypercube will be called the *sequential model*. If a processor can communicate with all its neighbors in a time step then this variant of the hypercube is called the *parallel model*.

A multitude of other important networks are related to the hypercube. Among the more important (constant degree) members of the hypercube family are the Cube Connected Cycle (CCC), the Butterfly and the de Bruijn family of networks. For definitions of these networks see [46].

1.1.6   The Star Graph.   Let $s_1 s_2 \ldots s_n$ be a permutation of n symbols. For $1 < j \leq n$, we define $SWAP_j(s_1 s_2 \ldots s_n) = s_j s_2 \ldots s_{j-1} s_{j+1} \ldots s_n$. An n-star graph is a graph $S_n = (V, E)$ with $|V| = n!$ nodes, where $V = \{s_1 s_2 \ldots s_n | s_1 s_2 \ldots s_n$ is a permutation of n different symbols$\}$,   and $E = \{(u, v) | v = SWAP_j(u)$ for some $j, 1 < j \leq n\}$.

1.1.7   Parallel Disk Systems.   With the widening gap between processor speeds and disk access speeds, the I/O bottleneck has become critical. Parallel Disk Systems (PDS) have been introduced to alleviate this bottleneck [85]. In this model there are D distinct and independent disk drives. The disks can simultaneously transmit a block of data. A block consists of B records. If M is the internal memory size, then one usually requires that $M \geq 2DB$. While analyzing algorithms developed for this model, one typically computes the number of I/O operations needed for the algorithm. Local computations are neglected since the time for I/O is much more than the time for local computations.

2.   RANDOM SAMPLING

Random sampling has been employed in the development of numerous sorting and selection algorithms. One of the early papers that dealt with sampling was due to Frazer and McKellar [25]. They proposed the following sorting algorithm which can be thought of as a generalization of the quicksort algorithm [29]: 1) Sample $o(n)$ keys from the input and use any (possibly nonoptimal) algorithm to sort them; 2) Use these sample keys to partition the input into subsequences; and 3) Sort each subsequence independently. Sorting algorithms for several models of computing have been designed using this technique.

Random sampling has also dominated the arena of selection algorithms. As an example, Floyd and Rivest [24] proposed the following scheme for selection:

1) Randomly pick $o(n)$ keys from the input and identify two keys ℓ_1 and ℓ_2 from this sample such that the element to be selected has a value in the range $[\ell_1, \ell_2]$ and not many input keys are in the range $[\ell_1, \ell_2]$; 2) Delete all the input keys that are outside the range $[\ell_1, \ell_2]$; and 3) Perform an appropriate selection from out of the remaining keys. The number of comparisons made by this algorithm to identify the ith smallest key is $n + \min\{i, n - i\} + o(n)$ with high probability. The proof of this fact and related sampling bounds are generally encapsulated in a "sampling lemma."

2.1   A SAMPLING LEMMA

Several sampling lemmas have been proven in the literature. One of the basic lemmas deals with the following sampling process. Let $S = \{k_1, k_2, \ldots, k_s\}$ be a random sample from a sequence X of n numbers. Let the sorted order of S be $k'_1, k'_2, \ldots, k'_s$. If r_i is the rank of k'_i in X, many algorithms benefit from a high probability confidence interval for r_i. (The rank of any element k in X is the number of elements $\leq k$ in X.) A proof of the following Lemma can be found in [74].

Lemma 2..1 *For every* α, *Prob.* $\left(|r_i - i\frac{n}{s}| > \sqrt{3\alpha}\frac{n}{\sqrt{s}}\sqrt{\log n} \right) < n^{-\alpha}$.

Notation. We say a randomized algorithm has a resource bound of $\widetilde{O}(f(n))$ if there exists a constant c such that the amount of resource used is no more than $c\alpha f(n)$ **on any input** of size n with probability $\geq (1 - n^{-\alpha})$ (for any $\alpha > 0$). In an analogous manner, we could also define the functions $\widetilde{o}(.)$, $\widetilde{\Omega}(.)$, etc.

2.2   ORGANIZATION OF THIS PAPER

The rest of this article is organized as follows. Sections 3 and 4 are devoted to sorting and selection problems, respectively. In Section 5 we provide some concluding remarks.

3.   SAMPLING BASED SORTING

3.1   A GENERAL THEME

The following is an idea introduced by Frazer and McKellar [25] that has been implemented over a variety of models.

Algorithm I

Step 1. Pick a random sample of n^ϵ (for some constant $\epsilon < 1$) input keys.

Step 2. Sort this sample (using any nonoptimal algorithm).

Step 3. Partition the input using the sorted sample as splitter keys.

Step 4. Sort each part separately in parallel.

It is easy to show using lemma 2..1 that the splitter keys very evenly distribute the keys so that in the last step the work done by each processor is approximately the same, with high probability.

3.2   THE PCT

One of the classical results in parallel sorting is Batcher's algorithm [10]. This algorithm is based on the idea of bitonic sorting and was proposed for the hypercube and hence can be run on stronger models such as any of the PRAMs and the PCT as well. Batcher's algorithm runs in $O(\log^2 n)$ time when sorting n keys using n processors. Followed by this, a very nearly optimal algorithm was given by Preparata [62]. Preparata's algorithm used $n \log n$ processors and took $O(\log n)$ time. Finding a logarithmic time optimal parallel algorithm for sorting remained an open problem for a long time in spite of numerous attempts.

Finally in 1981, Reischuk was able to design a randomized logarithmic time optimal algorithm for the CREW PRAM [78] which implies the result for the randomized PCT. At around the same time Ajtai, Komlós, and Szemerédi announced their sorting network of depth $O(\log n)$ [4]. This established the upper bound for sorting on the PCT for the case $p \leq n$. This was extended by Alon, Azar and Vishkin [7] to the case $p > n$. Taken together their results show the complexity of sorting n keys on a p processor PCT is $\Theta(\log n / \log(1 + p/n))$. The matching lower bound for randomized or deterministic sorting was first shown by Alon and Azar [6]. A significantly simpler proof of the same result was provided by Boppana [16].

3.3   THE PRAM

As was stated above, Reischuk was the first to design an optimal logarithmic time a randomized CREW PRAM algorithm for sorting [78]. His algorithm may be derived from Algorithm I. The AKS sorting circuit [4] implies the existence of a deterministic EREW PRAM algorithm running in logarithmic time. However the size of the circuit was $O(n \log n)$ and also the underlying constant in the time bound was enormous. Leighton subsequently was able to reduce the circuit size to $O(n)$ using the technique of columnsort [45]. Though several attempts have been made to improve the constant in the time bound, the algorithm of [4] remains a result of only theoretical interest.

In 1987 Cole presented an optimal logarithmic time EREW PRAM algorithm for sorting, the constant in the time bound being reasonably small [18]. In the same paper, a sub-logarithmic time algorithm for sorting on the CRCW PRAM is also given. This algorithm uses $n(\log n)^\epsilon$ processors, the run time being $O(\frac{\log n}{\log \log \log n})$. Here ϵ is any constant > 0. The lower bound result of Beame and Hastad states that any CRCW PRAM sorting algorithm will have to take $\Omega(\frac{\log n}{\log \log n})$ time in the worst case as long as the processor bound is only a polynomial in the input size n [11]. Rajasekaran and Reif [73] were able to obtain a randomized algorithm for sorting on the CRCW PRAM that runs in time $\tilde{O}(\frac{\log n}{\log \log n})$, the processor bound being $n(\log n)^\epsilon$, for any fixed $\epsilon > 0$. This algorithm is also processor-optimal, i.e., to achieve the same time bound the processor bound can not be decreased any further.

3.4   THE MESH

The first asymptotically optimal sorting algorithm for the mesh was given by Thompson and Kung [83]. Their algorithm can sort n numbers on a $\sqrt{n} \times \sqrt{n}$ mesh in $O(\sqrt{n})$ time. Since the diameter of an n-node mesh is $2\sqrt{n} - 2$, [83]'s algorithm is clearly optimal. Thompson and Kung's algorithm is based on the idea of odd-even merging. Since a mesh has a large diameter, it is imperative to have not only asymptotically optimal algorithms but also they should have small underlying constants in their time bounds. Often times, the challenge in designing mesh algorithms lies in reducing the constants in time bounds.

Subsequent to Thompson and Kung's algorithm, Schnorr and Shamir gave a $3\sqrt{n} + o(\sqrt{n})$ time algorithm [79]. They also proved a lower bound of $3\sqrt{n} - o(\sqrt{n})$ for sorting. However, both the upper bound and the lower bound were derived under the assumption of no queueing. Ma, Sen, and Scherson [49] gave a near optimal algorithm for a related model. Kaklamanis et al. presented a very interesting algorithm for sorting with a run time of $2.5\sqrt{n} + \tilde{o}(\sqrt{n})$ [35]. This algorithm was randomized and used queues of size $O(1)$. The underlying idea here is the same as that of Algorithm I. Kaklamanis and Krizanc later improved this time bound to $2\sqrt{n} + \tilde{o}(\sqrt{n})$ [34].

The idea of using $O(1)$ sized queues has been successfully employed to design better deterministic sorting algorithms as well. Kunde has presented a $2.5\sqrt{n} + o(\sqrt{n})$ step algorithm [43]; Nigam and Sahni have given a $(2 + \epsilon)\sqrt{n} + o(\sqrt{n})$ time algorithm (for any fixed $\epsilon > 0$) [56]; Also Kaufmann, Sibeyn, and Torsten have offered a $2\sqrt{n} + o(\sqrt{n})$ time algorithm [36]. The third algorithm closely resembles the one given by [34] and Algorithm I.

The problem of $k - k$ sorting is to sort a mesh where k elements are input at each node. The bisection lower bound for this problem is $\frac{k\sqrt{n}}{2}$. For example, if we have to interchange data from one half of the mesh with data from the other half, $\frac{k\sqrt{n}}{2}$ routing steps will be needed. A very nearly optimal randomized algorithm for $k - k$ sorting is given in [65]. Kunde [44] has matched this result with a deterministic algorithm.

3.5   MESHES WITH BUSES

For a mesh with fixed buses, it is easy to design a logarithmic time algorithm for sorting n numbers using a polynomial (in n) number of processors (see e.g., [66]). However, if the mesh is of size $\sqrt{n} \times \sqrt{n}$, then the bisection lower bound for sorting will be $\Omega(\sqrt{n})$. The same lower bound holds for a mesh with a reconfigurable bus system also. In general, we can obtain impressive speedups on M_r and M_f if the number of processors used is much more than the input size.

When the input size n is the same as that of the network size, sorting can be done using a randomized algorithm on M_r in time that is only $\tilde{o}(\sqrt{n})$ more than the time needed for packet routing under the same settings as has been proven in [72]. This randomized algorithm is also similar to Algorithm I. In [41], Krizanc, Rajasekaran, and Shende show that on M_f also, sorting can be done in time that is nearly the same as the time needed for packet routing. The best known algorithm for packet routing on M_r takes time $\frac{17}{18}\sqrt{n} + \tilde{o}(\sqrt{n})$ [17]. For M_f, the best known packet routing time is $0.79\sqrt{n} + \tilde{o}(\sqrt{n})$ [80]. Therefore, sorting can be done on M_r in time $\frac{17}{18}\sqrt{n} + \tilde{o}(\sqrt{n})$ and on M_f in time $0.79\sqrt{n} + \tilde{o}(\sqrt{n})$.

An interesting feature of M_r is that sorting can be done on it in time $O(1)$ using a quadratic number of processors. In contrast, sorting can not be done in $O(1)$ time even on the CRCW PRAM, given only a polynomial number of processors [11]. A constant time algorithm using n^3 processors appears in [86]. The processor bound was improved to n^2 in independent works [33], [48], [54], [57].

3.6   THE HYPERCUBE

Batcher's algorithm runs in $O(\log^2 n)$ time on an n-node hypercube [10]. This algorithm uses the technique of bitonic sorting. Odd-even merge sorting can also be employed on the hypercube to obtain the same time bound. Nassimi and Sahni [55] gave an elegant $O(\log n)$ time algorithm for sorting which uses $n^{1+\epsilon}$ processors (for any fixed $\epsilon > 0$). This algorithm, known as *sparse enumeration sort*, has found numerous applications in the design of other sorting algorithms on various interconnection networks. A variant of Algorithm I was employed by Reif and Valiant to derive an optimal randomized algorithm for sorting on the CCC [77]. The best known deterministic algorithm for sorting on the hypercube (or any variant) is due to Cypher and Plaxton and it takes $O(\log n \log \log n)$ time [21]. This algorithm makes use of the technique of deterministic sampling and the underlying constant in the time bound is rather large. An excellent description of this algorithm can be found in [46]. Hsu and Wei [31] have recently presented an $O(dn^2 \log d)$ time algorithm for sorting on the $d^n = N$ node de Bruijn network. If $d = 2$, their algorithm runs in time $2 \log^2 N$ steps.

3.7   THE STAR GRAPH

Menn and Somani [51] employed an algorithm similar to that of Schnorr and Shamir [79] to show that sorting can be done on a star graph with $n!$ nodes in $O(n^3 \log n)$ time. Rajasekaran and Wei [76] have offered a randomized algorithm with a time bound of $\widetilde{O}(n^3)$. This algorithm is based on a randomized selection algorithm that they derive. A summary of this algorithm follows:

There are n phases in the algorithm. A star graph with $n!$ nodes is denoted as S_n. In the first phase they perform a selection of n uniformly distributed keys and as a consequence route each key to the correct sub-star graph S_{n-1} it belongs to. In the second phase, sorting is local to each S_{n-1}. At the end of second phase each key will be in its correct S_{n-2}. In general, at the end of the ℓth phase, each key will be in its right $S_{n-\ell}$ (for $1 \le \ell \le n - 1$). Selection in each phase takes $\widetilde{O}(n^2)$ time. Making use of these selected keys, every input key figures out the $S_{n-\ell}$ it belongs to in $O(n^2)$ time. The keys are routed to the correct $S_{n-\ell}$'s in $\widetilde{O}(n)$ time. Thus each phase takes $\widetilde{O}(n^2)$ time, accounting for a total of $\widetilde{O}(n^3)$ time.

The above approach differs from Algorithm I. However, random sampling is used in the selection algorithm of [76].

3.8   PARALLEL DISK SYSTEMS

The problem of disk sorting was first studied by Aggarwal and Vitter in their fundamental paper [2]. In the model they considered, each I/O operation results in the transfer of D blocks each block having B records. A more realistic model was envisioned in [85]. Several asymptotically optimal algorithms have been given for sorting on this model. Nodine and Vitter's optimal algorithm [58] involves solving certain matching problems. Aggarwal and Plaxton's optimal algorithm [1] is based on the Sharesort algorithm of Cypher and Plaxton. Vitter and Shriver gave an optimal randomized algorithm for disk sorting [85]. All these results are highly nontrivial and theoretically interesting. However, the underlying constants in their time bounds are high.

In practice the simple disk-striped mergesort (DSM) is used [9], even though it is not asymptotically optimal. DSM has the advantages of simplicity and a small constant. Data accesses made by DSM is such that at any I/O operation, the same portions of the D disks are accessed. This has the effect of having a single disk which can transfer DB records in a single I/O operation. An $\frac{M}{DB}$-way mergesort is employed by this algorithm. To start with, initial runs are formed in one pass through the data. At the end the disk has N/M runs each of length M. Next, $\frac{M}{DB}$ runs are merged at a time. Blocks of any run are uniformly striped across the disks so that in future they can be accessed in parallel utilizing the full bandwidth. Each phase of merging involves one pass through the data. There are $\frac{\log(N/M)}{\log(M/DB)}$ phases and hence the total number of passes made by DSM is $\frac{\log(N/M)}{\log(M/DB)}$. In other words, the total number of I/O read operations performed by the algorithm is $\frac{N}{DB}\left(1 + \frac{\log(N/M)}{\log(M/DB)}\right)$. The constant here is just 1.

A known lower bound on the number of passes for parallel disk sorting is $\Omega\left(\frac{\log(N/B)}{\log(M/B)}\right)$. Here N is the input size, M is the core memory size, and B is the block size. If one assumes that N is a polynomial in M and that B is small (which are readily satisfied in practice), the lower bound simply yields $\Omega(1)$ passes. A number of optimal algorithms that make only $O(1)$ passes have been proposed in the literature. So, the challenge in the design of parallel disk sorting algorithms is in reducing this constant. If $M = 2DB$, the number of passes made by DSM is $1 + \log(N/M)$, which indeed can be very high.

Recently, much work has been done that deals with the practical aspects of parallel disk systems. Pai, Schaffer, and Varman [59] analyzed the average case performance of a simple merging algorithm, employing an approximate model of average case inputs. Barve, Grove, and Vitter [9] have presented a simple randomized algorithm (SRM) and analyzed its performance. The analysis involves the solution of certain occupancy problems. The expected number R_{SRM} of I/O read operations made by their algorithm is such that

$$R_{SRM} \leq \frac{N}{DB}\left[1 + \frac{\ln(N/M)}{\ln kD}\frac{\ln D}{k\ln\ln D}\left(1 + \frac{\ln\ln\ln D}{\ln\ln D} + \frac{1+\ln k}{\ln\ln D} + O(1)\right)\right] \quad (1.1)$$

The algorithm merges $R = kD$ runs at a time, for some integer k. When $R = \Omega(D\log D)$, the expected performance of their algorithm is optimal. However, in this case, the internal memory needed is $\Omega(BD\log D)$. They have also compared SRM with DSM through simulations and shown that SRM performs better than DSM.

In a recent work, Rajasekaran [67] has presented a simple algorithm (called (ℓ, m)-merge sort (LMM)) that is asymptotically optimal (under the assumptions that N is a polynomial in M and B is small) and the underlying constant is small. LMM is as simple as the DSM. LMM makes less number of passes through the data than DSM when D is large. Recent implementation results [60] [71] indicate that LMM is competitive in practice.

4.   SELECTION ALGORITHMS

The sequential selection algorithm of Blum et. al. works as follows: 1) Partition the input of n numbers into groups with 5 elements in each group;

2) Find the median of each group; 3) Recursively compute the median M of the group medians; 4) Partition the input into two using M as the splitter key. Part I has all the input keys $\leq M$ and Part II has the remaining keys. Identify the part that has the key to be selected and recursively perform an appropriate selection in this part.

One can easily show that the above algorithm runs in time $O(n)$. This is a good example of how deterministic sampling can be employed. A variant of the above has been used in all the deterministic parallel algorithms for selection.

Likewise, random sampling has been effectively applied to derive optimal or near optimal selection algorithms in various parallel models. A summary of such an algorithm is given below. To begin with all the input keys are *alive*. We are interested in selecting the ith smallest key.

Algorithm II

Step 1. Sample a set S of $o(n)$ keys at random from the collection X of alive keys.

Step 2. Sort the set S.

Step 3. Identify two keys l_1 and l_2 in S whose ranks in S are $i\frac{s}{n} - \delta$ and $i\frac{s}{n} + \delta$ respectively, δ being a 'small' integer.
(* Using lemma 2..1 or a variant it is easy to show the rank of l_1 in X is $< i$, and the rank of l_2 in X is $> i$, with high probability. *)

Step 4. Eliminate all the keys in X which are either $< l_1$ or $> l_2$.

Step 5. Repeat Steps 1 through 4 until the number of alive keys is 'small'.

Step 6. Finally, concentrate and sort the alive keys.

Step 7. Perform an appropriate selection on the alive keys.

Next we enumerate known parallel selection algorithms on various models and show how the above theme has been used repeatedly.

4.1   THE PCT

Valiant [84] showed a deterministic lower bound for selection on the PCT. A deterministic upper bound was shown by Azar and Pippenger [8] (building on the work of Ajtai et al. [5]). Their results together show that deterministic selection from a sequence of n keys using p processors requires $\Theta(n/p + \log(\log n/\log(2 + p/n)))$ steps.

Meggido [50] and independently Reischuk [78] showed that the above lower bound could be "beaten" using randomization by providing an optimal randomized PCT algorithm for selection that runs in $\Theta(n/p + 1)$ steps. Both of their algorithms are implementations of Algorithm II on the randomized PCT.

The results above show that there is a significant gap between the randomized and deterministic parallel complexity of selection in the PCT model. A partial explanation of this phenomenon is given in [37] where a tight tradeoff between the amount of randomness used by a randomized PCT for selection and its performance, measured by the time it requires to complete its computation with a given failure probability, is shown.

4.2   THE PRAM

A straight forward implementation of Algorithm II on any of the PRAMs will yield an optimal randomized $\tilde{O}(\log n)$ time parallel algorithm for selection. On the CRCW PRAM, a similar algorithm can be used to solve the problem of finding the maximum of n given numbers in $\tilde{O}(1)$ time using n processors [75]. Cole used the idea of deterministic sampling to design an $O(\log n \log^* n)$ time $\frac{n}{\log n \log^* n}$ processor EREW PRAM algorithm [19]. The time bound of this algorithm has been improved to $O(\log n)$ using deterministic sampling as well as algorithms for approximate prefix computation [26].

4.3   THE MESH

The problem of selection on the mesh where the number of processors is equal to the input size has been studied by many researchers. The best known algorithm is due to Condon and Narayanan [20]. This randomized algorithm has a run time of $1.15\sqrt{n} + \tilde{o}(\sqrt{n})$ and is similar to Algorithm II. The best known deterministic algorithm has a run time of $1.44\sqrt{n} + o(\sqrt{n})$ [40]. Krizanc and Narayanan provide optimal (to within an additive term) $\sqrt{n} + o(\sqrt{n})$ step algorithms for certain special cases of selection, e.g., the maximum [38]. For the case $n > p$ Krizanc and Narayanan [39], present a deterministic algorithm with a run time of $O(\min\{p \log \frac{n}{p}, \max\{\frac{n}{p^{2/3}}, \sqrt{p}\}\})$. Rajasekaran, Chen, and Yooseph [70] have presented both deterministic and randomized algorithms for selection when $n > p$. Their deterministic algorithm has a run time of $O(\frac{n}{p} \log \log p + \sqrt{p} \log n)$ and the randomized algorithm resembles Algorithm II and runs in time $\tilde{O}((\frac{n}{p} + \sqrt{p}) \log \log p)$. A new deterministic selection scheme has been proposed in [70]. The idea is to employ the sequential algorithm of Blum et. al. [15] with some crucial modifications.

A summary of the selection scheme of [70] is given below since it can be applied to any interconnection network to obtain good performance. To begin with, each one of the p processors has $\frac{n}{p}$ keys.

Algorithm III

$N := n$

Step 0. *if* $\log(n/p)$ is $\leq \log\log p$ *then*

    sort the elements at each node

 else

    partition the keys at each node into $\log p$
    equal parts such that keys in one part will
    be $\leq$ keys in parts to the right.

repeat

  Step 1. In parallel find the median of keys at each
  node. Let M_q be the median and N_q be the number
  of remaining keys at node q, $1 \leq q \leq p$.

  Step 2. Find the weighted median of $M_1, M_2, \ldots, M_p$
  where key M_q has a weight of N_q, $1 \leq q \leq p$. Let
  M be the weighted median.

  Step 3. Count the rank r_M of M from
  out of all the remaining keys.

  Step 4. *if* $i \leq r_M$ *then*

    eliminate all the remaining keys that are $> M$

   else

    eliminate all the remaining keys that are $\leq M$.

  Step 5. Compute E, the number of keys eliminated.

  if $i > r_M$ *then* $i := i - E$; $N := N - E$.

until $N \leq c$, c being a constant.

Output the ith smallest key from out of the remaining keys.

When the above algorithm is implemented on the mesh the resulting time bound is $O(\frac{n}{p} \log\log p + \sqrt{p} \log n)$.

4.4 MESHES WITH BUSES

Here we consider the problem of selection when $n = p$. On a mesh with reconfigurable buses, a lower bound of $\Omega(\log\log n)$ applies for comparison based deterministic selection, since selection even on the parallel comparison tree model has the same lower bound. ElGindy and Wegrowicz [23] applied an algorithm similar to that of [53] and showed that selection can be done on a p-node M_r in $O(\log^2 p)$ time. Followed by this, Doctor and Krizanc [22] presented a very simple randomized algorithm (similar to Algorithm II) that achieves the same time bound with high probability. This time bound was improved to $O(\log p)$ by Hao, McKenzie, and Stout [27]. Using an algorithm similar to that of Algorithm II and some other crucial properties of

M_r, Rajasekaran [64, 69] gave an $O(\log \log p \log^* p)$ expected time randomized algorithm.

On the other hand, $\Omega(p^{1/6})$ is a lower bound for selection on M_f [42]. A very nearly optimal algorithm has been given in [42]. An optimal randomized algorithm can be found in [64, 69].

4.5   THE HYPERCUBE

A plethora of algorithms have been proposed for selection on the hypercube (both for the case $p = n$ and the case $p < n$). For the case $p = n$, an optimal $\tilde{O}(\log n)$ time randomized algorithm has been given in [77] and [63]. The algorithm in [77] is for sorting and hence can be applied for selection as well. On the other hand, the algorithm given in [63] is very simple. [63]'s algorithm has been implemented on CM-2 and empirical results are promising [70]. The best known deterministic algorithm is due to Berthomé et. al. [13] and has a run time of $O(\log n \log^* n)$.

For the case of $p < n$ on the sequential model, [61]'s deterministic algorithm runs in time $O(\frac{n}{p} \log \log p + \log^2 p \log(\frac{n}{p}))$ whereas the randomized algorithm of [63] has a run time of $\tilde{O}(\frac{n}{p} \log \log p + \log p \log \log p)$. A lower bound for this problem is $\frac{n}{p} \log \log p + \log p$. On the weak parallel model, [61]'s deterministic algorithm has a run time of $O(\frac{n}{p} + \log p \log \log p)$ and the randomized algorithm of [63] has a run time of $\tilde{O}(\frac{n}{p} + \log p)$. A lower bound for selection on this model is $\frac{n}{p} + \log p$. All of these algorithms use the technique of sampling (either deterministic or randomized). A slightly better deterministic algorithm can be obtained using Algorithm III as has been shown in [70]. The run time is $O(\frac{n}{p} \log \log p + \log^2 p \log \log p)$. If a better sorting algorithm is discovered for the hypercube, this time bound will improve further.

4.6   THE STAR GRAPH

The only known selection algorithm on the star graph is due to Rajasekaran and Wei [76]. This randomized algorithm runs in time $\tilde{O}(n^2)$ on an $n!$-node star graph. Within the same asymptotic time bound, this algorithm can perform n different selections. A sorting algorithm with a run time of $\tilde{O}(n^3)$ follows from this algorithm and is discussed in section 3.7.

4.7   PARALLEL DISK SYSTEMS

In [68] two algorithms have been presented for selection on the PDS model. The first algorithm is randomized and the second algorithm is deterministic. The number of parallel I/O read operations needed for either is $O\left(\frac{N}{DB}\right)$, where N is the number of input keys, D is the number of disks, and B is the block size.

Thus the algorithms are asymptotically optimal. Due to the small underlying constants, the algorithms have the potential of being practical as well. The randomized algorithm is based the general theme given above.

5. CONCLUSIONS

In this article we have surveyed known parallel algorithms for sorting and selection on various models of computing. We have also identified some very commonly used techniques for the design of such algorithms.

References

[1] A. Aggarwal and C. G. Plaxton, Optimal Parallel Sorting in Multi-Level Storage, *Proc. Fifth Annual ACM Symposium on Discrete Algorithms*, 1994, pp. 659-668.

[2] A. Aggarwal and J. S. Vitter, The Input/Output Complexity of Sorting and Related Problems, *Communications of the ACM* 31(9), 1988, pp. 1116-1127.

[3] A.V. Aho, J.E. Hopcroft, and J.D. Ullman, *The Design and Analysis of Computer Algorithms*, Addison-Wesley Publishing Company, 1974.

[4] M. Ajtai, J. Komlós and E. Szemerédi, An $O(n \log n)$ Sorting Network, *Proc. 15th ACM Symposium on Theory of Computing*, 1983, pp. 1-9.

[5] M. Ajtai, J. Komlós, W. Steiger, and E. Szemerédi, Optimal Parallel Selection Has Complexity $O(\log \log n)$, *Journal of Computer and System Science*, 38, 1989, pp. 125-133.

[6] N. Alon and Y. Azar, The Average Complexity of Deterministic and Randomized Parallel Comparison Sorting Algorithms, *SIAM Journal of Computing*, 17, 1988, pp. 1178-1192.

[7] N. Alon, Y. Azar and U. Vishkin, Tight Complexity Bounds for Parallel Comparison Sorting, *Proc. of 29th IEEE Symposium on Foundations o Computer Science*, 1986, pp. 502-510.

[8] Y. Azar and N. Pippenger, Parallel Selection, *Discrete Applied Mathematics*, 27, 1990, pp. 49-58.

[9] R. Barve, E. F. Grove, and J. S. Vitter, Simple Randomized Mergesort on Parallel Disks, *Parallel Computing* 23(4-5), 1997, pp. 601-631.

[10] K.E. Batcher, Sorting Networks and their Applications, *Proc. 1968 Spring Joint Computer Conference*, vol. 32, AFIPS Press, 1968, pp. 307-314.

[11] P. Beame and J. Hastad, Optimal Bounds for Decision Problems on the CRCW PRAM, *Proc. 19th ACM Symposium on Theory Of Computing*, 1987, pp. 83-93.

[12] Y. Ben-Asher, D. Peleg, R. Ramaswami, and A. Schuster, The Power of Reconfiguration, *Journal of Parallel and Distributed Computing*, 1991, pp. 139-153.

[13] P. Berthomé, A. Ferreira, B.M. Maggs, S. Perennes, and C.G. Plaxton, Sorting-Based Selection Algorithms for Hypercubic Networks, *Proc. International Parallel Processing Symposium*, 1993, pp. 89-95.

[14] G.E. Blelloch, C.E. Leiserson, B.M. Maggs, C.G. Plaxton, S.J. Smith, M. Zagha, A Comparison of Sorting Algorithms for the Connection Machine CM-2, *Proc. 3rd Annual ACM Symposium on Parallel Algorithms and Architectures*, 1991, pp. 3-16.

[15] M. Blum, R. Floyd, V.R. Pratt, R. Rivest, and R. Tarjan, Time Bounds for Selection, *Journal of Computer and System Science*, 7(4), 1972, pp. 448-461.

[16] R. Boppana, The Average-Case Parallel Complexity of Sorting, *Information Processing Letter*, 33, 1989, pp. 145-146.

[17] J.C. Cogolludo and S. Rajasekaran, Permutation Routing on Reconfigurable Meshes, *Proc. Fourth International Symposium on Algorithms and Computation*, Springer-Verlag Lecture Notes in Computer Science 762, 1993, pp. 157-166.

[18] R. Cole, Parallel Merge Sort, *SIAM Journal on Computing*, vol. 17, no. 4, 1988, pp. 770-785.

[19] R. Cole, An Optimally Efficient Selection Algorithm, *Information Processing Letters*, 26, 1988, pp. 295-299.

[20] A. Condon and L. Narayanan, Upper and Lower Bounds for Selection on the Mesh, *Algorithmica*, 30, 1998, pp. 1-30.

[21] R.E. Cypher and C.G. Plaxton, Deterministic Sorting in Nearly Logarithmic Time on the Hypercube and Related Computers, *Proc. ACM Symposium on Theory of Computing*, 1990, pp. 193-203.

[22] D.P. Doctor and D. Krizanc, Three Algorithms for Selection on the Reconfigurable Mesh, Technical Report TR-219, School of Computer Science, Carleton University, February 1993.

[23] H. ElGindy and P. Wegrowicz, Selection on the Reconfigurable Mesh, *Proc. International Conference on Parallel Processing*, 1991, Vol. III, pp. 26-33.

[24] R.W. Floyd and R.L. Rivest, Expected Time Bounds for Selection, *Communications of the ACM*, 18(3), 1975, pp. 165-172.

[25] W.D. Frazer and A.C. McKellar, Samplesort: A Sampling Approach to Minimal Storage Tree Sorting, *Journal of the ACM*, 17(3), 1970, pp. 496-507.

[26] T. Hagerup and R. Raman, An Optimal Parallel Algorithm for Selection, *Proc. 5th Annual ACM Symposium on Parallel Algorithms and Architectures*, 1993, pp. 346-355.

[27] E. Hao, P.D. McKenzie and Q.F. Stout, Selection on the Reconfigurable Mesh, *Proc. Frontiers of Massively Parallel Computation*, 1992, pp. 38-45.

[28] W.L. Hightower, J.F. Prins, J.H. Reif, Implementation of Randomized Sorting on Large Parallel Machines, *Proc. 4th Annual ACM Symposium on Parallel Algorithms and Architectures*, 1992, pp. 158-167.

[29] C.A.R. Hoare, Quicksort, *The Computer Journal*, 5, 1962, pp. 10-15.

[30] E. Horowitz, S. Sahni, and S. Rajasekaran, *Computer Algorithms*, W.H. Freeman Press, 1998.

[31] D.F. Hsu, D.S.L. Wei, Permutation Routing and Sorting on Directed de Bruijn Networks, Technical Report, University of Aizu, Japan, 1994.

[32] J. Jang, H. Park, and V.K. Prasanna, A Fast Algorithm for Computing Histograms on a Reconfigurable Mesh, *Proc. Frontiers of Massively Parallel Computing*, 1992, pp. 244-251.

[33] J. Jang and V.K. Prasanna, An Optimal Sorting Algorithm on Reconfigurable Mesh, *Proc. International Parallel Processing Symposium*, 1992, pp. 130-137.

[34] C. Kaklamanis and D. Krizanc, Optimal Sorting on Mesh-Connected Processor Arrays, *Proc. 4th Annual ACM Symposium on Parallel Algorithms and Architectures*, 1992, pp. 50-59.

[35] C. Kaklamanis, D. Krizanc, L. Narayanan, and Th. Tsantilas, Randomized Sorting and Selection on Mesh Connected Processor Arrays, *Proc. 3rd Annual ACM Symposium on Parallel Algorithms and Architectures*, 1991, pp. 17-28.

[36] M. Kaufmann, S. Torsten, and J. Sibeyn, Derandomizing Algorithms for Routing and Sorting on Meshes, *Proc. 5th Annual ACM-SIAM Symposium on Discrete Algorithms*, 1994, pp. 669-679.

[37] D. Krizanc, Time-Randomness Tradeoffs in Parallel Computation, *Journal of Algorithms*, 20, 1996, pp. 1-19.

[38] D. Krizanc, and L. Narayanan, Optimal Algorithms for Selection on a Mesh-Connected Processor Array, *Proc. IEEE Symposium on Parallel and Distributed Processing*, 1992, pp. 70-76.

[39] D. Krizanc and L. Narayanan, Multi-packet Selection on a Mesh-Connected Processor Array, *Proc. International Parallel Processing Symposium*, 1992, pp. 602-605.

[40] D. Krizanc, L. Narayanan and R. Raman, Fast Deterministic Selection on a Mesh-Connected Processor Array, *Algorithmica*, 15, 1996, pp. 319-332.

[41] D. Krizanc, S. Rajasekaran, and S. Shende, A Comparison of Meshes with Static Buses and Unidirectional Wrap-Arounds, *Parallel Processing Letters*, 3, 1993, pp. 119-114.

[42] V.K.P. Kumar and C.S. Raghavendra, Array Processor with Multiple Broadcasting, *Journal of Parallel and Distributed Computing*, 4, 1987, pp. 173-190.

[43] M. Kunde, Concentrated Regular Data Streams on Grids: Sorting and Routing Near to the Bisection Bound, *Proc. IEEE Symposium on Foundations of Computer Science*, 1991, pp. 141-150.

[44] M. Kunde, Block Gossiping on Grids and Tori: Sorting and Routing Match the Bisection Bound Deterministically, *Proc. European Symposium on Algorithms*, 1993, pp. 272-283.

[45] T. Leighton, Tight Bounds on the Complexity of Parallel Sorting, *IEEE Transactions on Computers*, C34(4), 1985, pp. 344-354.

[46] T. Leighton, *Introduction to Parallel Algorithms and Architectures: Arrays–Trees–Hypercube*, Morgan-Kaufmann Publishers, 1992.

[47] H. Li and Q. Stout, *Reconfigurable Massively Parallel Computers*, Prentice-Hall Publishers, 1991.

[48] R. Lin, S. Olariu, J. Schwing, and J. Zhang, A VLSI-optimal Constant Time Sorting on Reconfigurable Mesh, *Proc. European Workshop on Parallel Computing*, 1992, pp. 16-27.

[49] Y. Ma, S. Sen, and D. Scherson, The Distance Bound for Sorting on Mesh Connected Processor Arrays is Tight, *Proc. IEEE Symposium on Foundations of Computer Science*, 1986, pp. 255-263.

[50] N. Meggido, Parallel Algorithms for Finding the Maximum and the Median Almost Surely in Constant Time, Technical Report, School of Computer Science, Carnegie-Mellon University, Pittsburg, PA, Oct. 1982.

[51] A. Menn and A.K. Somani, An Efficient Sorting Algorithm for the Star Graph Interconnection Network, *Proc. International Conference on Parallel Processing*, 1990, vol. 3, pp. 1-8.

[52] O. Menzilcioglu, H.T. Kung, and S.W. Song, Comprehensive Evaluation of a Two-Dimensional Configurable Array, *Proc. 19th Symposium on Fault Tolerant Computing*, 1989, pp. 93-100.

[53] J.I. Munro and M.S. Paterson, Selection and Sorting with Limited Storage, *Theoretical Computer Science*, 12, 1980, pp. 315-323.

[54] K. Nakano, D. Peleg, and A. Schuster, Constant-time Sorting on a Reconfigurable Mesh, Manuscript, 1992.

[55] D. Nassimi and S. Sahni, Parallel Permutation and Sorting Algorithms and a New Generalized Connection Network, *Journal of the ACM*, 29(3), 1982, pp. 642-667.

[56] M. Nigam and S. Sahni, Sorting n^2 Numbers on $n \times n$ Meshes, *Proc. International Parallel Processing Symposium*, 1993, pp. 73-78.

[57] M. Nigam and S. Sahni, Sorting n Numbers on $n \times n$ Reconfigurable Meshes with Buses, *Proc. International Parallel Processing Symposium*, 1993, pp. 174-181.

[58] M. H. Nodine and J. S. Vitter, Large Scale Sorting in Parallel Memories, *Proc. Third Annual ACM Symposium on Parallel Algorithms and Architectures*, 1990, pp. 29-39.

[59] V. S. Pai, A. A. Schaffer, and P. J. Varman, Markov Analysis of Multiple-Disk Prefetching Strategies for External Merging, *Theoretical Computer Science*, 128(2), 1994, pp. 211-239.

[60] M. D. Pearson, Fast Out-of-Core Sorting on Parallel Disk Systems, Technical Report PCS-TR99-351, Dartmouth College, Computer Science, Hanover, NH, June 1999, ftp://ftp.cs.dartmouth.edu/TR/TR99-351.ps.Z.

[61] C.G. Plaxton, Efficient Computation on Sparse Interconnection Networks, Ph. D. Thesis, Department of Computer Science, Stanford University, 1989.

[62] F. Preparata, New Parallel Sorting Schemes, *IEEE Transactions on Computers*, C27(7), 1978, pp. 669–673.

[63] S. Rajasekaran, Randomized Parallel Selection, *Proc. Symposium on Foundations of Software Technology and Theoretical Computer Science*, 1990, pp. 215-224.

[64] S. Rajasekaran, Mesh Connected Computers with Fixed and Reconfigurable Buses: Packet Routing, Sorting, and Selection, *Proc. First Annual European Symposium on Algorithms*, Springer-Verlag Lecture Notes in Computer Science 726, 1993, pp. 309-320.

[65] S. Rajasekaran, $k - k$ Routing, $k - k$ Sorting, and Cut Through Routing on the Mesh, *Journal of Algorithms* 19, 1995, pp. 361-382.

[66] S. Rajasekaran, Mesh Connected Computers with Fixed and Reconfigurable Buses: Packet Routing and Sorting, *IEEE Transactions on Computers*, 45(5), 1996, pp. 529-539.

[67] S. Rajasekaran, A Framework For Simple Sorting Algorithms On Parallel Disk Systems, *Proc. 10th Annual ACM Symposium on Parallel Algorithms and Architectures*, 1998, pp. 88-97.

[68] S. Rajasekaran, Selection Algorithms for the Parallel Disk Systems, *Proc. International Conference on High Performance Computing*, 1998, pp. 103-110.

[69] S. Rajasekaran, Selection on Mesh Connected Computers with Fixed and Reconfigurable Buses, *Journal of Algorithms*, 29, 1998, pp. 68-81.

[70] S. Rajasekaran, W. Chen, and S. Yooseph, Unifying Themes for Parallel Selection, *Proc. Fourth International Symposium on Algorithms and Computation*, Springer-Verlag Lecture Notes in Computer Science 834, 1994, pp. 92-100.

[71] S. Rajasekaran and X. Jin, A Practical Realization of Parallel Disks, to appear in *Proc. International Workshop on High Performance Scientific and Engineering Computing with Applications*, 2000.

[72] S. Rajasekaran and T. McKendall, Permutation Routing and Sorting on the Reconfigurable Mesh, Technical Report MS-CIS-92-36, Department of Computer and Information Science, University of Pennsylvania, May 1992.

[73] S. Rajasekaran and J.H. Reif, Optimal and Sub-Logarithmic Time Randomized Parallel Sorting Algorithms, *SIAM Journal on Computing*, 18(3), 1989, pp. 594-607.

[74] S. Rajasekaran and J.H. Reif, Derivation of Randomized Sorting and Selection Algorithms, in *Parallel Algorithm Derivation and Program Transformation*, Edited by R. Paige, J.H. Reif, and R. Wachter, Kluwer Academic Publishers, 1993, pp. 187-205.

[75] S. Rajasekaran, and S. Sen, Random Sampling Techniques and Parallel Algorithms Design, in *Synthesis of Parallel Algorithms*, Editor: J.H. Reif, Morgan-Kaufman Publishers, 1993, pp. 411-451.

[76] S. Rajasekaran and D.S.L. Wei, Selection, Routing, and Sorting on the Star Graph, *Proc. International Parallel Processing Symposium*, 1993, pp. 661-665.

[77] J.H. Reif and L.G. Valiant, A Logarithmic Time Sort for Linear Size Networks, *Journal of the ACM*, 34(1), 1987, pp. 60-76.

[78] R. Reischuk, Probabilistic Parallel Algorithms for Sorting and Selection, *SIAM Journal of Computing*, 14(2), 1985, pp. 396-409.

[79] C. Schnorr and A. Shamir, An Optimal Sorting Algorithm for Mesh-Connected Computers, *Proc. ACM Symposium on Theory of Computing*, 1986, pp. 255-263.

[80] J.F. Sibeyn, M. Kaufmann, and R. Raman, Randomized Routing on Meshes with Buses, *Proc. European Symposium on Algorithms*, 1993, pp. 333-344.

[81] T.M. Stricker, Supporting the Hypercube Programming Model on Mesh Architectures (A Fast Sorter for iWarp Tori), *Proc. 4th Annual ACM Symposium on Parallel Algorithms and Architectures*, 1992, pp. 148-157.

[82] X. Thibault, D. Comte, and P. Siron, A Reconfigurable Optical Interconnection Network for Highly Parallel Architecture, *Proc. Symposium on the Frontiers of Massively Parallel Computation*, 1988, pp. 437-442.

[83] C.D. Thompson and H.T. Kung, Sorting on a Mesh Connected Parallel Computer, *Communications of the ACM*, 20(4), 1977, pp. 263-271.

[84] L. G. Valiant, Parallelism in Comparison Problems, *SIAM Journal of Computing*, 4, 1975, pp. 348-355.

[85] J. S. Vitter and E. A. M. Shriver, Algorithms for Parallel Memory I: Two-Level Memories, *Algorithmica* 12(2-3), 1994, pp. 110-147.

[86] B.F. Wang, G.H. Chen, and F.C. Lin, Constant Time Sorting on a Processor Array with a Reconfigurable Bus System, *Information Processing Letters*, 34(4), 1990, pp. 187-192.

Chapter 2

SIMPLIFIED ANALYSES OF RANDOMIZED ALGORITHMS FOR SEARCHING, SORTING, AND SELECTION

Michael T. Goodrich [†]

Dept. of Computer Science, Johns Hopkins Univ., Baltimore, MD 21218.

goodrich@cs.jhu.edu

Roberto Tamassia[‡]

Dept. of Computer Science, Brown Univ., Providence, RI 02912.

rt@cs.brown.edu

Abstract   We describe simplified analyses of well-known randomized algorithms for searching, sorting, and selection. The proofs we include are quite simple and can easily be made a part of a Freshman-Sophomore Introduction to Data Structures (CS2) course and a Junior-Senior level course on the design and analysis of data structures and algorithms (CS7/DS&A). We show that using randomization in data structures and algorithms is safe and can be used to significantly simplify efficient solutions to various computational problems.

1.   INTRODUCTION

We live with probabilities all the time, and we easily dismiss as "impossible" events with very low probabilities. For example, the proba-

*This work was announced in preliminary form in the Proceedings of the 13th SIGCSE Technical Symposium on Computer Science Education, 1999, 53–57.

[†]The work of this author was supported in part by the U.S. Army Research Office under grant DAAH04–96–1–0013, and by the National Science Foundation under grant CCR–9625289.

[‡]The work of this author was supported in part by the U.S. Army Research Office under grant DAAH04–96–1–0013, and by the National Science Foundation under grants CCR–9732327 and CDA–9703080.

S. Rajasekaran et al (eds.), Handbook of Randomized Computing, Volume 1, pp. 23–34.
© 2001 *Kluwer Academic Publishers.*

bility of a U.S. presidential election being decided by a single vote is estimated at 1 in 10 million*. The probability of being killed by a bolt of lightning in any given year is estimated at 1 in 2.5 million†. And, in spite of Hollywood's preoccupation with it, the probability that a large meteorite will impact the earth in any given year is about 1 in 100 thousand‡. Because the probabilities of these events are so low, we safely assume they will not occur in our lifetime.

Why is it then that computer scientists have historically preferred *deterministic* computations over randomized computations? Deterministic algorithms certainly have the benefit of provable correctness claims and often have good time bounds that hold even for worst-case inputs. But as soon as an algorithm is actually implemented in a program P, we must again deal with probabilistic events, such as the following:

- P contains a bug,
- we provide an input to P in an unexpected form,
- our computer crashes for no apparent reason,
- P's environment assumptions are no longer valid.

Since we are already living with bad computer events such as these, whose probabilities are arguably much higher than the bad "real-world" events listed in the previous paragraph, we should be willing to accept probabilistic algorithms as well. In fact, fast randomized algorithms are typically easier to program than fast deterministic algorithms. Thus, using a randomized algorithm may actually be safer than using a deterministic algorithm, for it is likely to reduce the probability that a program solving a given problem contains a bug.

1.1 SIMPLIFYING ANALYSES

In this paper we describe several well-known randomized algorithms but provide new, simplified analyses of their asymptotic performance bounds. In fact, our proofs use only the most elementary of probabilistic facts. We contrast this approach with traditional "average-case" analyses by showing that the analyses for randomized algorithms need not make any restrictive assumptions about the forms of possible inputs. Specifically, we describe how randomization can easily be incorporated into discussions of each of the following standard algorithmic topics:

- searching in dictionaries,

---

*wizard.ucr.edu/polmeth/working_papers97/gelma97b.html
†www.nassauredcross.org/sumstorm/thunder2.htm
‡newton.dep.anl.gov/newton/askasci/1995/astron/AST63.HTM

- sorting,
- selection.

We discuss each of these topics in the following sections.

2.   SEARCHING IN DICTIONARIES

A *dictionary* is a data structure that stores key-value pairs, called items, and supports search, insertion and deletion operations. We consider *ordered dictionaries* where an ordering is defined over the keys. An interesting alternative to balanced binary search trees for efficiently realizing the ordered dictionary abstract data type (ADT) is the *skip list* [3–7]. This structure makes random choices in arranging items in such a way that searches and updates take $O(\log n)$ time *on average*, where n is the number of items in the dictionary. Interestingly, the notion of average time used here does not depend on any probability distribution defined on the keys in the input. Instead, the running time is averaged over all possible outcomes of random choices used when inserting items in the dictionary.

2.1   SKIP LISTS

A *skip list* S for an ordered dictionary D consists of a series of sequences $\{S_0, S_1, \ldots, S_h\}$. Each sequence S_i stores a subset of the items of D sorted by nondecreasing key plus items with two special keys, denoted $-\infty$ and $+\infty$, where $-\infty$ is smaller than every possible key that can be inserted in D and $+\infty$ is larger than every possible key that can be inserted in D. In addition, the sequences in S satisfy the following:

- Sequence S_0 contains every item of dictionary D (plus the special items with keys $-\infty$ and $+\infty$).
- For $i = 1, \ldots, h-1$, sequence S_i contains (in addition to $-\infty$ and $+\infty$) a randomly generated subset of the items in sequence S_{i-1}.
- Sequence S_h contains only $-\infty$ and $+\infty$.

An example of a skip list is shown in Figure 2.1. It is customary to visualize a skip list S with sequence S_0 at the bottom and sequences $S_1, \ldots, S_{h-1}$ above it. Also, we refer to h as the *height* of skip list S.

Intuitively, the sequences are set up so that S_{i+1} contains more or less every other item in S_i. As we shall see in the details of the insertion method, the items in S_{i+1} are chosen at random from the items in S_i by picking each item from S_i to also be in S_{i+1} with probability $1/2$. That is, in essence, we "flip a coin" for each item in S_i and place that item in S_{i+1} if the coin comes up "heads." Thus, we expect S_1 to have

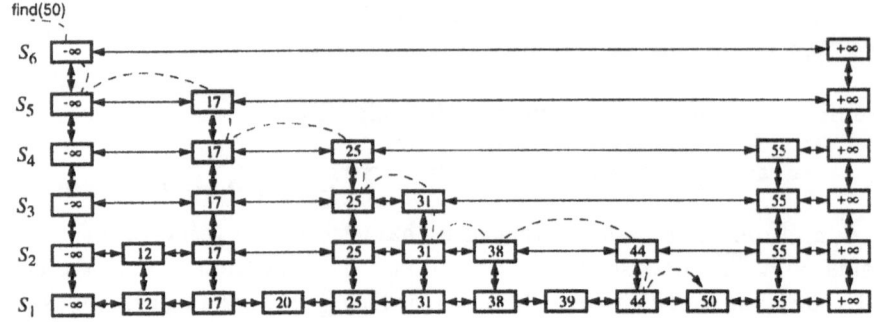

Figure 2.1: Example of a skip list. The dashed lines show the traversal of the structure performed when searching for key 50.

about $n/2$ items, S_2 to have about $n/4$ items, and, in general, S_i to have about $n/2^i$ items. In other words, we expect the height h of S to be about $\log n$.

Using the *position* abstraction used previously by the authors [2] for nodes in sequences and trees, we view a skip list as a two-dimensional collection of positions arranged horizontally into *levels* and vertically into *towers*. Each level corresponds to a sequence S_i and each tower contains positions storing the same item across consecutive sequences. The positions in a skip list can be traversed using the following operations:

> after(p) : the position following p on the same level
> before(p) : the position preceding p on the same level
> below(p) : the position below p in the same tower
> above(p) : the position above p in the same tower

Without going into the details, we note that we can easily implement a skip list by means of a linked structure such that the above traversal methods each take $O(1)$ time, given a skip-list position p.

2.2 SEARCHING

The skip list structure allows for simple dictionary search algorithms. In fact, all of the skip list search algorithms are based on an elegant SkipSearch method that takes a key k and finds the item in a skip list S with the largest key (which is possibly $-\infty$) that is less than or equal to k. Suppose we are given such a key k. We begin the SkipSearch method by setting a position variable p to the top-most, left position in the skip list S. That is, p is set to the position of the special item with key $-\infty$ in S_h. We give a pseudo-code description of the skip-list search algorithm in Code Fragment 1 (see also Figure 2.1).

Algorithm SkipSearch(k):

Input: A search key k

Output: Position p in S_0 such that the item at p has the largest key less than or equal to k

> Let p be the topmost-left position of S (which should have at least 2 levels).
> **while** below$(p) \neq$ **null do**
> > $p \leftarrow$ below(p) {*drop down*}
> > **while** key(after(p)) $\leq k$ **do**
> > > Let $p \leftarrow$ after(p) {*scan forward*}
> > **end while**
> **end while**
> **return** p.

Code Fragment 1: A generic search in a skip list S.

2.3   UPDATE OPERATIONS

Another feature of the skip list data structure is that, besides having an elegant search algorithm, it also provides simple algorithms for dictionary updates.

Insertion

The insertion algorithm for skip lists uses randomization to decide how many references to the new item (k, e) should be added to the skip list. We begin the insertion of a new item (k, e) into a skip list by performing a SkipSearch(k) operation. This gives us the position p of the bottom-level item with the largest key less than or equal to k (note that p may be the position of the special item with key $-\infty$). We then insert (k, e) in this bottom-level list immediately after position p. After inserting the new item at this level we "flip" a coin. That is, we call a method random() that returns a number between 0 and 1, and if that number is less than $1/2$, then we consider the flip to have come up "heads;" otherwise, we consider the flip to have come up "tails." If the flip comes up tails, then we stop here. If the flip comes up heads, on the other hand, then we backtrack to the previous (next higher) level and insert (k, e) in this level at the appropriate position. We again flip a coin; if it comes up heads, we go to the next higher level and repeat. Thus, we continue to insert the new item (k, e) in lists until we finally get a flip that comes up tails. We link together all the references to the new item (k, e) created in this process to create the *tower* for (k, e).

We give the pseudo-code for the above insertion algorithm in Code Fragment 2. The uses an operation insertAfterAbove$(p, q, (k, e))$ that inserts a position storing the item (k, e) after position p (on the same level as p) and above position q, returning the position r of the new item (and setting internal references so that after, before, above, and below methods will work correctly for p, q, and r). We note that this pseudo-code assumes that the insertion process never goes beyond the top level of the skip list. This is not a completely reasonable assumption, for some insertions will most likely continue beyond this level. There are two simple ways of dealing with this occurance, however. The first is to extend the height of the skip list as long as an insertion wishes to go to higher levels. We leave it as a simple exercise to show that it is very unlikely that any insertion will go beyond level $3 \log n$ in this case. The second possibility is to cut off any insertion that tries to go beyond a reasonable notion of what should be the top level of the skip list. For example, one could maintain the top level to be at height $3\lceil \log n \rceil$, and then stop any insertion from going beyond that level. One can show that the expected number of elements that would ever be inserted at this level is actually less than 1.

Algorithm SkipInsert(k, e):
    $p \leftarrow$ SkipSearch(k)
    $q \leftarrow$ insertAfterAbove$(p, null, (k, e))$
    while random$() < 1/2$ **do**
      while above$(p) = null$ **do**
        $p \leftarrow$ before(p) {*scan backward*}
      end while
      $p \leftarrow$ above(p) {*jump up to higher level*}
      $q \leftarrow$ insertAfterAbove$(p, q, (k, e))$
    end while

Code Fragment 2: Insertion in a skip list, assuming random() returns a random number between 0 and 1, and we never insert past the top level.

Removal

Like the search and insertion algorithms, the removal algorithm for a skip list S is quite simple. In fact, it is even easier than the insertion algorithm. Namely, to perform a remove(k) operation, we begin by performing a search for the given key k. If a position p with key k is not found, then we indicate an error condition. Otherwise, if a position p with key k is found (on the bottom level), then we remove all the positions above p, which are easily accessed by using above operations to climb up the tower of this item in S starting at position p (see Figure 2.2).

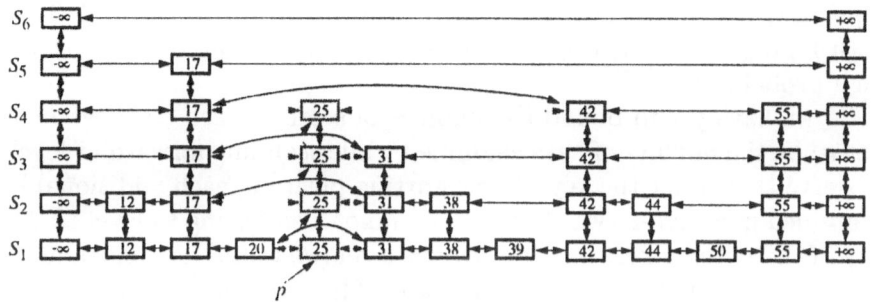

Figure 2.2: Removal of the item with key 25 from a skip list. The positions visited are in the tower for key 25.

2.4   A SIMPLE ANALYSIS OF SKIP LISTS

Our probabilistic analysis of skip lists, which is a simplified version of an analysis of Motwani and Raghavan [4], requires only elementary probability concepts, and does not need any assumptions about input distributions. We begin this analysis by studying the height h of S.

The probability that a given item is stored in a position at level i is equal to the probability of getting i consecutive heads when flipping a coin, that is, this probability is $1/2^i$. Thus, the probability P_i that level i has at least one item is at most

$$P_i \leq \frac{n}{2^i},$$

for the probability that any one of n different events occurs is at most the sum of the probabilities that each occurs.

The probability that the height h of S is larger than i is equal to the probability that level i has at least one item, that is, it is no more than P_i. This means that h is larger than, say, $3 \log n$ with probability at most

$$P_{3 \log n} \leq \frac{n}{2^{3 \log n}} = \frac{n}{n^3} = \frac{1}{n^2}.$$

More generally, given a constant $c > 1$, h is larger than $c \log n$ with probability at most $1/n^{c-1}$. Thus, with high probability, the height h of S is $O(\log n)$.

Consider the running time of a search in skip list S, and recall that such a search involves two nested **while** loops. The inner loop performs a scan forward on a level of S as long as the next key is no greater than the search key k, and the outer loop drops down to the next level and repeats the scan forward iteration. Since the height h of S is $O(\log n)$ with high probability, the number of drop-down steps is $O(\log n)$ with high probability.

So we have yet to bound the number of scan-forward steps we make. Let n_i be the number of keys examined while scanning forward at level i. Observe that, after the key at the starting position, each additional key examined in a scan-forward at level i cannot also belong to level $i + 1$. If any of these items were on the previous level, we would have encountered them in the previous scan-forward step. Thus, the probability that any key is counted in n_i is $1/2$. Therefore, the expected value of n_i is exactly equal to the expected number of times we must flip a fair coin before it comes up heads. This expected value is 2. Hence, the expected amount of time spent scanning forward at any level i is $O(1)$. Since S has $O(\log n)$ levels with high probability, a search in S takes the expected time $O(\log n)$. By a similar analysis, we can show that the expected running time of an insertion or a removal is $O(\log n)$.

Finally, let us turn to the space requirement of a skip list S. As we observed above, the expected number of items at level i is $n/2^i$, which means that the expected total number of items in S is

$$\sum_{i=0}^{h} \frac{n}{2^i} = n \sum_{i=0}^{h} \frac{1}{2^i} < 2n.$$

Hence, the expected space requirement of S is $O(n)$.

3. SORTING

One of the most popular sorting algorithms is the quick-sort algorithm, which uses a *pivot* element to split a sequence and then it recursively sorts the subsequences. One common method for analyzing quick-sort is to assume that the pivot will always divide the sequence almost equally. We feel such an assumption would presuppose knowledge about the input distribution that is typically not available, however. Since the intuitive goal of the partition step of the quick-sort method is to divide the sequence S almost equally, let us introduce randomization into the algorithm and pick as the pivot a *random element* of the input sequence. This variation of quick-sort is called *randomized quick-sort*, and is provided in Code Fragment 3.

Algorithm quickSort(S):

Input: Sequence S of n comparable elements

Output: A sorted copy of S

> if $n = 1$ then
> > return S.
>
> end if
> pick a random integer r in the range $[0, n-1]$
> let x be the element of S at rank r.
> put the elements of S into three sequences:
>
> - S_L, storing the elements in S less than x
> - S_E, storing the elements in S equal to x
> - S_G, storing the elements in S greater than x.
>
> let $S_L' \leftarrow$ quickSort(S_L)
> let $S_G' \leftarrow$ quickSort(S_G)
> return $S_L' + S_E + S_G'$.

Code Fragment 3: Randomized quick-sort algorithm.

There are several analyses showing that the expected running time of randomized quicksort is $O(n \log n)$ (e.g., see [1, 4, 8]), independent of any input distribution assumptions. The analysis we give here simplifies these analyses considerably.

Our analysis uses a simple fact from elementary probability theory: namely, that the expected number of times that a fair coin must be flipped until it shows "heads" k times is $2k$. Consider now a single recursive invocation of randomized quick-sort, and let m denote the size of the input sequence for this invocation. Say that this invocation is "good" if the pivot chosen is such that subsequences L and G have size at least $m/4$ and at most $3m/4$ each. Thus, since the pivot is chosen uniformly at random and there are $m/2$ pivots for which this invocation is good, the probability that an invocation is good is $1/2$.

Consider now the recursion tree T associated with an instance of the quick-sort algorithm. If a node v of T of size m is associated with a "good" recursive call, then the input sizes of the children of v are each at most $3m/4$ (which is the same as $m/(4/3)$). If we take any path in T from the root to an external node, then the length of this path is at most the number of invocations that have to be made (at each node on this path) until achieving $\log_{4/3} n$ good invocations. Applying the probabilistic fact reviewed above, the expected number of invocations we must make until this occurs is at most $2 \log_{4/3} n$. Thus, the expected length of any path from the root to an external node in T is $O(\log n)$. Observing that the time spent at each level of T is $O(n)$, the expected running time of randomized quick-sort is $O(n \log n)$.

4.   SELECTION

The *selection* problem asks that we return the kth smallest element in an unordered sequence S. Again using randomization, we can design a simple algorithm for this problem. We describe in Code Fragment 4 a simple and practical method, called *randomized quick-select*, for solving this problem.

Algorithm quickSelect(S, k):

Input: Sequence S of n comparable elements, and an integer $k \in [1, n]$

Output: The kth smallest element of S

    if $n = 1$ **then**
        return the (first) element of S.
    end if
    pick a random integer r in the range $[0, n - 1]$
    let x be the element of S at rank r.
    put the elements of S into three sequences:

        ■ S_L, storing the elements in S less than x
        ■ S_E, storing the elements in S equal to x
        ■ S_G, storing the elements in S greater than x.

    if $k \leq |S_L|$ **then**
        quickSelect(S_L, k)
    else if $k \leq |S_L| + |S_E|$ **then**
        return x *{each element in S_E is equal to x}*
    else
        quickSelect($S_G, k - |S_L| - |S_E|$)
    end if

Code Fragment 4: Randomized quick-select algorithm.

We note that randomized quick-select runs in $O(n^2)$ *worst-case* time. Nevertheless, it runs in $O(n)$ *expected* time, and is much simpler than the well-known *deterministic* selection algorithm that runs in $O(n)$ worst-case time (e.g., see [1]). As was the case with our quick-sort analysis, our analysis of randomized quick-select is simpler than existing analyses, such as that in [1].

Let $t(n)$ denote the running time of randomized quick-select on a sequence of size n. Since the randomized quick-select algorithm depends on the outcome of random events, its running time, $t(n)$, is a random variable. We are interested in bounding $E(t(n))$, the expected value of $t(n)$. Say that a recursive invocation of randomized quick-select is "good" if it partitions S, so that the size of S_L and S_G is at most $3n/4$. Clearly, a recursive call is good with probability $1/2$. Let $g(n)$ denote the number of consecutive recursive invocations (including the present one) before getting a good invocation. Then

$$t(n) \leq bn \cdot g(n) + t(3n/4),$$

where $b \geq 1$ is a constant (to account for the overhead of each call). We are, of course, focusing in on the case where n is larger than 1, for we can easily characterize in a closed form that $t(1) = b$. Applying the linearity of expectation property to the general case, then, we get

$$E(t(n)) \leq E(bn \cdot g(n) + t(3n/4)) = bn \cdot E(g(n)) + E(t(3n/4)).$$

Since a recursive call is good with probability $1/2$, and whether a recursive call is good or not is independent of its parent call being good, the expected value of $g(n)$ is the same as the expected number of times we must flip a fair coin before it comes up "heads." This implies that $E(g(n)) = 2$. Thus, if we let $T(n)$ be a shorthand notation for $E(t(n))$ (the expected running time of the randomized quick-select algorithm), then we can write the case for $n > 1$ as $T(n) \leq T(3n/4) + 2bn$. Converting this recurrence relation to a closed form, we get that

$$T(n) \leq 2bn \cdot \sum_{i=0}^{\lceil \log_{4/3} n \rceil} (3/4)^i.$$

Thus, the expected running time of quick-select is $O(n)$.

5. CONCLUSION

We have discussed some simplified analyses of well-known algorithms and data structures. In particular, we have presented simplified analyses for skip lists and randomized quick-sort, suitable for a CS2 course, and for randomized quick-select. These simplified analyses, in slightly different form, along with further discussions of simple data structures and algorithms, can be found in the recent book on data structures and algorithms by the authors [2].

References

[1] T.H. Cormen, C.E. Leiserson, and R.L. Rivest, *Introduction to Algorithms*, MIT Press, Cambridge, MA, 1990.

[2] M.T. Goodrich and R. Tamassia, *Data Structures and Algorithms in Java*, John Wiley and Sons, New York, 1998.

[3] P. Kirschenhofer and H. Prodinger, The path length of random skip lists, *Acta Informatica*, 31:775-792, 1994.

[4] R. Motwani and P. Raghavan, *Randomized Algorithms*, Cambridge University Press, New York, NY 1995.

[5] T. Papadakis, J.L. Munro, and P.V. Poblete, Average search and update costs in skip lists, *BIT*, 32:316-332, 1992.

[6] P.V. Poblete, J.L. Munro, and T. Papadakis, The binomial transform and its application to the analysis of skip lists. In *Proceedings of the European Symposium on Algorithms (ESA)*, pages 554-569, 1995.

[7] W. Pugh, Skip lists: a probabilistic alternative to balanced trees, *Commun. ACM*, 33(6):668-676, 1990.

[8] R. Seidel, Backwards analysis of randomized geometric algorithms. In J. Pach, editor, *New Trends in Discrete and Computational Geometry*, volume 10 of *Algorithms and Combinatorics*, pages 37-68, Springer-Verlag, 1993.

Chapter 3

CONCENTRATION OF MEASURE FOR RANDOMIZED ALGORITHMS: TECHNIQUES AND ANALYSIS

Devdatt Dubhashi *
Department of Computer Science and Engineering
Indian Institute of Technology, Delhi
Hauz Khas, New Delhi 110016
India
dubhashi@cs.chalmers.se

Sandeep Sen
Department of Computer Science and Engineering
Indian Institute of Technology, Delhi
Hauz Khas, New Delhi 110016
India
ssen@cse.iitd.ernet.in

1. INTRODUCTION

Randomized algorithms often turn out to be simpler and faster than their deterministic counterparts. For certain problems, like primality testing, there is no matching polynomial-time deterministic algorithm, so the randomized algorithms are the only practical alternative. The above mentioned advantages are sometimes viewed with skepticism because of the inherent uncertainty in the behavior of the randomized algorithm. The uncertainty may be with regards to the correctness (Monte Carlo), or the variation in running time (Las Vegas). In this chapter we will focus on Las Vegas randomized algorithms that always produce the

*Present address: Department of Computing and Mathematical Sciences, Chalmers University of Technology and Göteborg University, Göteborg, Sweden.

S. Rajasekaran et al (eds.), Handbook of Randomized Computing, Volume 1, pp, 35–100.
© 2001 Kluwer Academic Publishers.

correct answer, but manifest some variation in the running times. The most common measure is the expected running time of the algorithm, where the expectation is over the choice of random bits in the algorithm and not the the input distribution. The natural concern is what is an *acceptable* probability of deviating from the expected running time by a certain amount - say a small constant factor. Roughly speaking, the expected running time implies that the probability of exceeding twice the expected time is 1/2. If the hardware itself is subjected to a similar scrutiny, there is empirical evidence that it has a failure probability as high as $\frac{1}{2^k}$ where k is a constant. Therefore it is not unreasonable to aim for a success probability for randomized algorithms that is about an order of magnitude smaller than this.

Over the past decade, the notion of *high probability* bound has gained wide acceptance. An algorithm is said to have a running time $f(n)$ with high probability (whp), if the following holds

$$\Pr(T(n) > c\alpha f(n)) \le \frac{1}{n^\alpha}$$

for some constant c and any α, where $T(n)$ is the running time for an input of size n. This is also referred to as *inverse polynomial* probability of success. It must be noted that there is an implicit trade-off between the success probability and the running time. An inverse-*exponential* probability is even better, namely,

$$\Pr(T(n) > cf(n)) \le \frac{1}{2^n}.$$

However, randomized algorithms that succeed with high probability are considered as reliable as any deterministic algorithm.

1.1   CONCENTRATION OF MEASURE

The simplest and most common method to achieve a higher success probability is to make several independent runs of the algorithm. In this chapter, we investigate and analyze techniques for increasing the success probability of randomized algorithms, without sacrificing efficiency. Sometimes this is possible by analyzing known algorithms more carefully, whereas in other cases, a simple modification leads us to substantially improved bounds.

The performance of a randomized algorithm measured by its running time, work performed or space used is a random variable that can potentially take a wide range of values. However, one observes that in actual fact, the observable behavior is often concentrated in a very sharply defined narrow range. This is one manifestation of the deep and widespread phenomenon of concentration of measure, and the one we are particularly interested in. In this chapter we will discuss a number of general tools of varying power and range of applicability for proving results about concentration of measure and illustrate them with applications to algorithms.

1.2   ORGANIZATION

Recurrences are the most common yet powerful tools for analyzing algorithms. In the context of randomized algorithms, the associated recurrences are far more complex because of some underlying stochastic process that the recurrence models. Certain *ad hoc* methods have been used repeatedly in the past to solve some commonly occurring (probabilistic) recurrences till Karp tried to develop a coherent theory. We start our chapter with a discussion on related techniques, results and applications.

Probabilistic inequalities form the crux of analysis of randomized algorithms. As noted above, the measures like running-time (or working space) of a randomized algorithm are essentially a random variables with (possibly) complicated distributions and our primary interest is to get useful bounds on the tails of the distributions. We discuss some of the more commonly used probabilistic inequalities, including situations where the behavior of an algorithm is related to several such random variables with non-trivial dependencies among them.

Next we illustrate applications of a well-known stochastic process called the branching process and derive concentration bounds in a relatively general framework.

Randomized algorithms have been very influential in the area of computational geometry. In the last two sections, we describe techniques for obtaining high-probability bounds for randomized algorithms in computational geometry with applications to sequential and parallel algorithms.

2.   PROBABILISTIC RECURRENCES

Karp[21] developed an attractive framework for the analysis of randomized algorithms. Suppose we have a randomized algorithm that on input x, performs "work" $a(x)$ and then produces a subproblem of size $H(x)$ which is then solved by recursion. One can analyze the performance of the algorithm by writing down a "recurrence":

$$T(x) = a(x) + T(H(x)). \qquad (3.1)$$

Superficially this looks just the same as the usual analysis of algorithms via recurrence relations. However, the crucial difference is that in contrast with deterministic algorithms, the size of the subproblem produced here, $H(x)$ is a random variable, and so (3.1) is a *probabilistic recurrence* equation.

What does one mean by the solution of such a probabilistic recurrence? The solution $T(x)$ is itself a random variable and we would like as much information about its distribution as possible. While a complete description of the exact distribution is usually neither possible nor really necessary, the "correct" useful analogue to the deterministic solution is a concentration of measure result for $T(x)$. Of course, to do this, one needs some information on the distribution of the subproblem $H(x)$ generated by the algorithm. Karp gives a very easy–to–apply framework that requires only the bare minimum of information on the distribution of $H(x)$, namely (a bound on) the expectation, and yields a concentration result for $T(x)$. Suppose that in (3.1), we have $\mathbf{E}[H(x)] \leq m(x)$ for some function $0 \leq m(x) \leq x$. Consider the "deterministic" version of (3.1) obtained by replacing the random variable $H(x)$ by the deterministic bound $m(x)$:

$$u(x) = a(x) + u(m(x)). \tag{3.2}$$

The solution to this equation is $u(x) = \sum_{i \geq 0} a(m^i(x))$, where $m^0(x) := 0$ and $m^{i+1}(x) = m(m^i(x))$. Karp gives a concentration result around this value $u(x)$:

Theorem 1 (Karp's First Theorem) *Suppose that in (3.1), we have* $\mathbf{E}[H(x)] \leq m(x)$ *for some function* $0 \leq m(x) \leq x$ *and such that* $a(x), m(x),$ *$\frac{m(x)}{x}$ are all non-decreasing. Then*

$$\mathbf{Pr}[T(x) > u(x) + ta(x)] \leq \left(\frac{m(x)}{x}\right)^t.$$

Remark 1 We have stated the result in the simplest memorable form that captures the essence and i essentially correct. However, technically the statement of the theorem above is actually not quite accurate and we have omitted some continuity conditions on the functions involved. These conditions usually hold in all cases where we'd like to apply the theorem. Moreover, as shown in [10], some of these conditions can be discarded at the cost of only slightly weakening the bound. For instance, we can discard the condition that $\frac{m(x)}{x}$ is non-decreasing; in that case, the bound on the right hand side can be essentially replaced by $\left(\max_{0 \leq y \leq x} \frac{m(y)}{y}\right)^t$

Also, in the formulation above, we assumed that the distribution of $H(x)$, the size of the derived subproblem depends only on the input size x. Karp[21] gives a more general formulation where the subproblem is allowed to depend on the actual input instance. Suppose we have a "size" function s on inputs, and on processing an input z, we expend work $a(s(z))$ and get a subproblem $H(z)$ such that $E[s(H(z))] \leq m(s(z))$. The probabilistic recurrence is now

$$T(z) = a(s(z)) + T(H(z)).$$

By considering $T'(x) := \max_{s(z)=x} T(z)$, one can bound this by a recurrence of the earlier form and apply the Theorem to give exactly the same solution. Thus we can apply the Theorem *per se* even in this more general situation.

We illustrate the ease of applicability of this cook–book style recipe by some examples (taken from Karp's paper).

Example 2.1 (Selection) Hoare's classic algorithm for finding the kth smallest element in a n–element set S, proceeds as follows: pick a random element $r \in S$ and by comparing each element in $S \setminus r$ with r, partition $S \setminus r$ into two subsets $L := \{y \in S \mid y < r\}$ and $U := \{y \in S \mid y > r\}$. Then,

- If $|L| \geq k$, recursively find the kth smallest element in L.

- If $|L| = k - 1$, then return r.

- If $|L| < k - 1$, then recursively find the $k - 1 - |L|$th smallest element in U.

The partitioning step requires $n - 1$ comparisons. It can be shown that the expected size of the subproblem, namely the size of L or U is at most $3n/4$, for all k. Thus Karp's Theorem can be applied with $m(x) = 3x/4$. We compute $u(x) \leq 4x$. Thus, if $T(n, k)$ denotes the number of comparisons performed by the algorithm, we have the following concentration result: for all $t \geq 0$,

$$\Pr[T(n, k) > 4n + t(n - 1)] \leq \left(\frac{3}{4}\right)^t.$$

This bound is nearly tight as showed by the following simple argument. Define a *bad* splitter to be one where $\frac{n}{|U|} \geq \log \log n$ or $\frac{n}{|L|} \geq \log \log n$. The probability of this is greater than $\frac{2}{\log \log n}$. The probability of picking $\log \log n$ consecutive bad splitters is $\Omega(\frac{1}{(\log n)^{\log \log \log n}})$.

The work done for $\log\log n$ consecutive bad splitters is

$$n + n\left(1 - \frac{1}{\log\log n}\right) + n\left(1 - \frac{1}{\log\log n}\right)^2 + \ldots n\left(1 - \frac{1}{\log\log n}\right)^{\log\log n}$$

which is $\Omega(n\log\log n)$. Compare this with the previous bound using $t = \log\log n$.

Example 2.2 (Luby's Maximal Independent Set Algorithm)
Luby[23] gives a randomized parallel algorithm for constructing a maximal independent set in a graph. The algorithm works in stages: at each stage, the current independent set is augmented and some edges are deleted form the graph. The algorithm terminates when we arrive at the empty graph. The work performed at each iteration is equal to the number of edges in the current graph. Luby showed that at each stage, the expected number of edges deleted is at least one–eighth of the number of edges in the complete graph. If $T(G)$ is the number of stages the algorithm runs and $T'(G)$ is the total amount of work done, then we get the concentration results:

$$\Pr[T(G) > \log_{8/7} n + t] \leq \left(\frac{7}{8}\right)^t,$$

$$\Pr[T'(G) > (8 + t)n] \leq \left(\frac{7}{8}\right)^t.$$

Example 2.3 (Tree Contraction) Miller and Reif [27] give a randomized *tree contraction* algorithm that starts with a n node tree representing an arithmetic expression and repeatedly applies a randomized contraction operation that provides a new tree representing a modified arithmetic expression. The process eventually reaches a one node tree and terminates. The work performed in the contraction step can be taken to be proportional to the number of nodes in the tree. Miller and Reif show that when applied to a tree with n nodes, the contraction step results in a tree of size at most $4n/5$. However the distribution of the size may depend on the original tree, not just the original size. Define the size function here to be the number of nodes in the tree in order to apply the more general framework. Let $T(z), T'(z)$ denote the number of iterations and the total work respectively when the contraction algorithm is applied to tree z. Then, Karp's Theorem gives the measure concentration results:

$$\Pr[T(z) > \log_{5/4} n + t] \leq (4/5)^t,$$

and

$$\Pr[T'(z) > (5 + t)n] \leq (4/5)^t.$$

Under the weak assumptions on the distribution of the input, Karp's First Theorem is essentially tight. However, if one has additional information on the distribution of the subproblem, say some higher moments, then one can get sharper results which will be explored below in § ref-sec:martingales.

Karp also gives an extension of the framework for the very useful case when the algorithm might generate more than one subproblem. Suppose we have an algorithm that on input x performs work $a(x)$ and then generates a fixed number $k \geq 1$ sub–problems $H_1(x), \ldots, H_k(x)$ each a random variable. This corresponds to the probabilistic recurrence:

$$T(x) = a(x) + T(H_1(x)) + \cdots + T(H_k(x)). \tag{3.3}$$

To obtain a concentration result in this case, Karp uses a different method which requires a certain condition:

Theorem 2 (Karp's Second Theorem) *Suppose that in (3.3), we have that for all possible values $(x_1, \ldots, x_k)$ of the tuple $(H_1(x), \ldots, H_k(x))$, we have*

$$E[T(x)] \geq \sum_i E[T(x_i)]. \tag{3.4}$$

Then, we have the concentration result: for all x and all $t > 0$,

$$\Pr[T(x) > (t+1)E[T(x)]] < e^{-t}.$$

The condition (3.4) says that the expected work in processing *any* sub–problems that can result from the original one can never exceed the expected cost of the processing the original instance. This is a very strong assumption and unfortunately, in many cases of interest, for example in computational geometry, it does not hold. Consequently the theorem is somewhat severely limited in its applicability. A rare case in which the condition is satisfied is for

Example 2.4 (Quicksort) Hoare's Quicksort algorithm is a true classic in Computer Science: to sort a set S of n items, we proceed as in the selection algorithm from above: select a random element $r \in S$ and by comparing it to every other element, partition S as into the sets L of elements less than x and U, the set of elements at least as big as r. Then, recursively, sort L and U. Let $Q(n)$ denote the number of comparisons performed by Quicksort on a set of n elements. Then $Q(n)$ satisfies the probabilistic recurrence:

$$T(n) = n - 1 + Q(H_1(n)) + Q(H_2(n)),$$

where $H_1(n) = |L|$ and $H_2(n) = |U|$. For Quicksort we have "closed-form" solutions for $q_n := E[Q(n)]$ which imply that $q_n \geq q_i + q_{n-i-1} + n - 1$

for any $0 \leq i < n$, which is just the condition needed to apply Karp's Second Theorem. Thus we get the concentration result:

$$\Pr[Q(n) > (t+1)q_n] \leq e^{-t}.$$

Actually one can get a much stronger bound by applying Karp's First Theorem suitably! Charge each comparison made in Quicksort to the non–pivot element, and let $T(\ell)$ denote the number of comparisons charged to a fixed element when Quicksort is applied to a list ℓ. Use the natural size function $s(\ell) := |\ell|$, which gives the number of elements in the list. Then we have the recurrence, $T(\ell) = 1 + T(H(\ell))$, where $s(H(\ell) = |\ell|/2$ since the sublist containing the fixed element (when it's not the pivot) has size uniformly distributed in $[0, |\ell|]$. So applying Karp's First Theorem, we have that for $t \geq 1$,

$$\Pr[T(\ell) > (t+1)\log|\ell|] \leq (1/2)^{t \log |\ell|} = |\ell|^{-t}.$$

Thus any fixed element in a list of n elements is charged at most $(t+1)\log n$ comparisons with probability at least $1 - n^{-t}$. The total number of comparisons is therefore at most $(t+1)n\log n$ with probability at least $1 - n^{t-1}$.

This is an inverse polynomial concentration bound. In a later section we shall get a somewhat stronger and provably optimal bound on the concentration.

It would naturally be of great interest to extend the range of Karp's Second Theorem by eliminating the restrictive hypothesis. For instance, it would be of interest to extend the Theorem under the kind of assumptions in Karp's First Theorem.

3. CHERNOFF–HOEFFDING BOUNDS

Perhaps the most basic and widely used tools for the analysis of randomized algorithms are the *Chernoff-Hoeffding* (CH) bounds on sums of bounded independent random variables. Unlike elementary probabilistic inequalities like Markov and Chebychev, the CH technique make use of the entire moment generating function $M(t) := \mathrm{E}[e^{tX}]$ of a random variable X, and provides much stronger bounds.

$$\begin{aligned}
\Pr[X > s] &= \Pr[e^{tX} > e^{ts}] \\
&\leq \frac{\mathrm{E}[e^{tX}]}{e^{ts}} \quad \text{using Markov's inequality} \quad (3.5)
\end{aligned}$$

By minimizing the right hand side w.r.t. t, we obtain the Chernoff bound for $\Pr[X \geq s]$.

Theorem 3 (Chernoff–Hoeffding Bounds) *Let* $X_1, \ldots, X_n$ *be independent random variables taking values in the unit interval* $[0,1]$ *and let* $X := X_1 + \cdots + X_n$. *Let* $p_i = \mathrm{E}[X_i], i \in [n]$ *and let* $p := (p_1 + \cdots + p_n)/n, q := 1 - p$. *Then*

$$\Pr[X > (p+t)n], \Pr[X < (p-t)n] \leq \exp\left(-nH\left((p+t, q-t) \mid (p, q)\right)\right),$$

where H *is the* **relative entropy**,

$$H\left((p_1, \ldots, p_r), (q_1, \ldots, q_r)\right) := \sum_i p_i \log \frac{p_i}{q_i}.$$

This is not the usual way of stating the CH bounds, nor is it the most useful; we give more useful forms for applications later below. The reasons for stating the bound above are: First, it is the strongest form of the bound from which all the others below can be derived.

Second, it gives the most insight into the bound form which it can be easily remembered. Think of the X_i as identical binary (0/1) variables, like n independent tosses of the same coin. Then X counts the number of heads observed in the n independent coin tosses. From the *a priori* probability, we expect to see np heads. What is the chance that the observed or *a posteriori* number of heads is $(p+t)n$? That is, what is the chance that the *a posteriori* distribution appears as one deriving from a $(p+t, q-t)$ coin when the actual *a priori* coin has a (p, q) distribution? One expects that the probability of this event drops exponentially in the relative entropy "distance" between the *a priori* and *a posteriori* distributions.

Finally, this viewpoint opens the way for many fruitful and far–reaching extensions to the so–called "large deviation" theory. We just give a flavor of this: suppose we have an experiment with r different outcomes with corresponding *a priori* probabilities $p_1, \ldots, p_r$. In coin tossing, $r = 2$ with p_1 and p_2 being the probabilities of heads and tails respectively. If the experiment is tossing a die, $r = 6$ and the probabilities $p_1, \ldots, p_6$ would reflect how the die is loaded (for a fair die, each $p_i = 1/6$). If we now perform n independent and identical trials, what is the probability that the the observed distribution will resemble $(q_1, \ldots, q_r)$? The answer is roughly, that this probability once again falls exponentially in the relative entropy distance. Some forms of the CH bounds that are most useful for applications are summarized in the next theorem; they can be deduced from Theorem 3.

Theorem 4 *Let* $X := X_1 + \cdots + X_n$ *where* $X_i, i \in [n]$ *are independent random variables taking values in the unit interval* $[0,1]$. *Then:*

■

$$\Pr[X > \mathrm{E}[X] + t], \Pr[X < \mathrm{E}[X] - t] < e^{-2t^2/n}.$$

- *For $\epsilon > 0$,*

$$\Pr[X > (1 + \epsilon)\mathrm{E}[X]] < \exp\left(\frac{-\epsilon^2}{3}\mathrm{E}[X]\right),$$

$$\Pr[X < (1 - \epsilon)\mathrm{E}[X]] < \exp\left(\frac{-\epsilon^2}{2}\mathrm{E}[X]\right).$$

4. APPLICATIONS OF CH BOUNDS

Example 4.1 (Skip List) Skip-list is a data structure introduced by Pugh [33] as an alternative to balanced binary search trees for handling dictionary operations on ordered lists. The underlying idea is to substitute complex book-keeping information used for maintaining balance conditions for binary trees by random sampling techniques. It has been shown by Pugh [33] that, given access to random bits, the *expected* search time in a skip-list of n elements is $O(\log n)$ [1] which compares very favourably with balanced binary trees. Moreover, the procedures for insertion and deletion are very simple which makes this data-structure a very attractive alternative to the balanced binary trees.

Since the search time is a stochastic variable (because of the use of randomization), it is of considerable interest to determine the bounds on the tails of its distribution. Often, it is crucial to know the behavior for any individual access rather than a chain of operations since it is more closely related to the real-time response.

Review of Skip-lists. We briefly review the basic data-structure proposed by Pugh. This data-structure is maintained as a hierarchy of sorted linked-lists. The bottom-most level is the entire set of keys S. We denote the linked list at level i from the bottom as L_i and let $|L_i| = N_i$. By definition $L_0 = S$ and $|L_0| = n$. For all $0 \leq i$, $L_i \subset L_{i-1}$ and the topmost level, say level k has constant number of elements. Moreover, correspondences are maintained between common elements of lists L_i and L_{i-1}. For a key with value E, for each level i, we denote by T_i a tuple (l_i, r_i) such that $l_i \leq E \leq r_i$ and $l_i, r_i \in L_i$. We call this tuple *straddling pair* (of E) in level i.

The search begins from the topmost level L_k where T_k can be determined in constant time. If $l_k = E$ or $r_k = E$ then the search is successful else we recursively search among the elements $[l_k, r_k] \cap L_0$. Here $[l_k, r_k]$

denotes the closed interval bound by l_k and r_k. This is done by searching the elements of L_{k-1} which are bounded by l_k and r_k. Since both $l_k, r_k \in L_{k-1}$, the *descendence* from level k to $k-1$ is easily achieved in $O(1)$ time. In general, at any level i we determine the tuple T_i by walking through a portion of the list L_i. If l_i or r_i equals E then we are done else we repeat this procedure by *descending* to level $i-1$.

In other words, we refine the search progressively until we find an element in S equal to E or we terminate when we have determined (l_0, r_0). This procedure can also be viewed as searching in a tree that has variable degree (not necessarily two as in binary tree).

Of course, to be able to analyze this algorithm, one has to specify how the lists L_i are constructed and how they are dynamically maintained under deletions and additions. Very roughly, the idea is to have elements in i-th level point to approximately 2^i nodes ahead (in S) so that the number of levels is approximately $O(\log n)$. The time spent at each level i depends on $[l_{i+1}, r_{i+1}] \cap L_i$ and hence the objective is to keep this small. To achieve these conditions on-line, Pugh [33] uses the following elegant method. The nodes from the bottom-most layer (level 0) are chosen with probability p (for the purpose of our discussion we shall assume $p = 0.5$) to be in the first level. Subsequently at any level i, the nodes of level i are chosen to be in level $i+1$ independently with probability p and at any level we maintain a simple linked list where the elements are in sorted order. If $p = 0.5$, then it is not difficult to verify that for a list of size n, the *expected* number of elements in level i is approximately $n/2^i$ and are spaced about 2^i elements apart. The expected number of levels is clearly $O(\log n)$, (when we have just a trivial length list) and the expected space requirement is $O(n)$.

To insert an element, we first locate its position using the search strategy described previously. Note that a byproduct of the search algorithm are all the T_i's. At level 0, we choose it with probability p to be in level L_1. If it is selected, we insert it in the proper position (which can be trivially done from the knowledge of T_1), update the pointers and repeat this process from the present level. Deletion is very similar and it can be readily verified that deletion and insertion have the same asymptotic run time as the search operation. So we shall focus on this operation.

Analysis. To analyze the run-time of the search procedure, we look at it backwards, i.e., retrace the path from level 0. The search time is clearly the length of the path (number of links) traversed over all the levels. So one can count the number of links one traverses before climbing up a level. In other words the expected search time can be expressed in the following recurrence (from [33])

$$C(k) = (1 - p)(1 + C(k)) + p(1 + C(k - 1))$$

where C(k) is the expected cost for climbing k levels. From the boundary condition C(0) = 0, one readily obtains $C(k) = k/p$. For $k = O(\log n)$, this is $O(\log n)$. The recurrence captures the crux of the method in the following manner. At any node of a given level, we climb up if this node has been chosen to be in the next level or else we add one to the cost of the present level. The probability of this event (climbing up a level) is p which we consider to be a success event. Now the entire search procedure can be viewed in the following alternate manner. We are tossing a coin which turns up heads with probability p - how many times should we toss to come up with $O(\log n)$ heads ? Each head corresponds to the event of climbing up one level in the data structure and the total number of tosses is the cost of the search algorithm. We are done when we have climbed up $O(\log n)$ levels (there is some technicality about the number of levels being $O(\log n)$ but that will be addressed later). The number of heads obtained by tossing a coin N times is given by a Binomial random variable X with parameters N and p. Using Chernoff bounds from Theorem 4, for $N = 15 \log n$ and $p = 0.5$, $\Pr[X \leq 1.5 \log n] \leq 1/n^2$ (using $\epsilon = 9/10$ in equation 1). Using appropriate constants, we can get rapidly decreasing probabilities of the form $\Pr[X \leq c \log n] \leq 1/n^\alpha$ for $c, \alpha > 0$ and α increases with c. These constants can be fine tuned although we shall not bother with such an exercise here.

We thus state the following lemma.

Lemma 1 *The probability that access time for a fixed element in a skip-list data structure of length n exceeds $c \log n$ steps is less than $O(1/n^2)$ for an appropriate constant $c > 1$.*

Proof We compute the probability of obtaining fewer than k (the number of levels in the data-structure) heads when we toss a fair coin ($p = 1/2$) $c \log n$ times for some fixed constant $c > 1$. That is, we compute the probability that our search procedure exceeds $c \log n$ steps. Recall that each head is equivalent to climbing up one level and we are done when we have climbed k levels. To bound the number of levels, it is easy to see that the probability that any element of S appears in level i is at most $1/2^i$, i.e. it has turned up i consecutive heads. So the probability that any fixed element appears in level $3 \log n$ is at most $1/n^3$. The probability that $k > 3 \log n$ is the probability that at least one element of S appears in $L_{3 \log n}$. This is clearly at most n times the probability that any fixed element survives and hence probability of k exceeding $3 \log n$ is less than $1/n^2$.

Given that $k \leq 3 \log n$ we choose a value of c, say c_0 (to be plugged into equation 1 of Chernoff bounds) such that the probability of obtaining fewer than $3 \log n$ heads in $c_0 \log n$ tosses is less than $1/n^2$. The search algorithm for a fixed key exceeds $c_0 \log n$ steps if one of the above events fail; either the number of levels exceeds $3 \log n$ or we get fewer than $3 \log n$ heads from $c_0 \log n$ tosses. This is clearly the summation of the failure probabilities of the individual events which is $O(1/n^2)$. $\square$

Theorem 5 *The probability that the access time for any arbitrary element in skip-list exceeds $O(\log n)$ is less than $1/n^\alpha$ for any fixed $\alpha > 0$.*

Proof: A list of n elements induces $n+1$ intervals. From the previous lemma, the probability P that the search time for a fixed element exceeding $c \log n$ is less than $1/n^2$. Note that all elements in a fixed interval $[l_0, r_0]$ follow the same path in the data-structure. It follows that for any interval the probability of the access time exceeding $O(\log n)$ is n times P. As mentioned before, the constants can be chosen appropriately to achieve this. $\square$

It is possible to obtain even tighter bounds on the space requirement for a skip list of n elements. From Pugh [33] it is known that the expected space is $O(n)$. Moreover it is clear that it does not exceed $O(n \log n)$ with probability $1 - 1/n^2$ (no element survives more than $O(\log n)$ levels with this probability from the previous lemma). One can obtain a much stronger bound by viewing the entire skip list structure as a stochastic experiment each node corresponds to a Bernoulli trial that turns up heads (similar to obtaining the obtaining the query bound). Each element is replicated till the the trial turns up tails. Since there are n elements, the number of nodes corresponds to the number of Bernoulli trials required to obtain n tails. This is a negative binomial distribution and one can use Chernoff bounds (Theorem 4) directly to obtain the following result.

Theorem 6 *For any constant $\alpha > 0$, the probability of the space exceeding $2n + \alpha \cdot n$, is less than $\exp^{\Omega(\alpha^2 n)}$.*

Example 4.2 (Randomized Search Trees and Backward Analysis)

Randomized Search Trees. The class of binary (dynamic) search trees is perhaps the first introduction to non-trivial data-structure in computer science. However, the update operations, although asymptotically very fast are not the easiest to remember. The rules for *rotations* and the *double-rotations* of the AVL trees, the splitting/joining in B-trees and the color-changes of red-black trees are often complex, as well as their correctness proofs. The *Randomized Search Trees* (also known as randomized treaps) provide a practical alternative to the Balanced BST. We still rely on rotations, but no explicit balancing rules are used. Instead we rely on the magical properties of random numbers.

The Randomized Search Tree (RST) is a binary tree that has the keys in an in-order ordering. In addition, each element is assigned a priority (Wlog, the priorities are unique) and the nodes of the tree are heap-ordered based on the priorities. Simultaneously, the key values follow in-order numbering. It is not difficult to see that for a given assignment of priorities, there is exactly one tree. If the priorities are assigned randomly in the range $[1, N]$ for N nodes, the expected height of the tree is *small*. This is the main crux of the following analysis of the performance of the RSTs.

Let us first look at the way search time using a technique known as *backward analysis*. For that we (hypothetically) insert the N elements in a decreasing order of their *priorities* and then count the number of elements that Q can *see* during the course of their insertions. This method (of assigning the random numbers on-line) makes arguments easier and the reader must convince himself that it doesn't affect the final results. Q can *see an element N_i* if there are no previously inserted elements in between.

Observation 1 *The tree constructed by inserting the nodes in order of their priorities (highest priority is the root) is the same as the tree constructed on-line.*

Observation 2 *The number of nodes Q sees is exactly the number of comparisons performed for searching Q. In fact, the order in which it sees corresponds to the search path of Q.*

Theorem 7 *The expected length of search path in RST is $O(H_N)$ where H_N is the N-th harmonic number.*

In the spirit of backward analysis, we pretend that the tree-construction is being reversed, i.e. nodes are being deleted starting from the last node. In the forward direction, we would count the expected number of nodes that Q sees. In the reverse direction, it is the number of times Q's visibility changes (convince yourself that these notions are identical). Let X_i be a Bernoulli rv that is 1 if Q sees N_i (in the forward direction) or conversely Q's visibility changes when N_i is deleted in the reverse sequence. Let X be the length of the search path.

$$X = \sum X_i$$

$$E[X] = E[\sum X_i] = \sum E[X_i]$$

We claim that $E[X_i] = \frac{2}{i}$. Note that the expectation of a Bernoulli variable is the probability that it is 1. We are computing this probability over all permutations of N being equally likely. In other words, if we consider a prefix of length i, all subsets of size i are equally likely. Let us find the probability that $X_i = 1$, *conditioned* on a *fixed* subset $N^i \subset N$. Unconditioning is easy if probability that $X_i = 1$ does not depend on N^i itself. So, given a fixed N^i, all N^{i-1} are equally likely, so the probability that $X_i = 1$ is the probability that one of the (maximum two) neighboring elements was removed in the reverse direction. The probability of that is less than $\frac{2}{i}$ which is independent of any specific N^i. So, the unconditional probability is the same as conditional probability - hence $E[X_i] = \frac{2}{i}$. The theorem follows as $\sum_i E[X_i] = 2\sum_i \frac{1}{i} = O(H_N)$.

The X_i's defined in the previous proof are independent but not identical. So, we can apply Chernoff-Hoeffding bounds to to obtain strong tail-estimates for deviation from the expected bound. From Theorem 4 , it follows that

Theorem 8 *The probability that the search time exceeds* $2\log n$ *comparisons in a randomized treap is less than* $O(1/n)$.

A similar technique can be used for counting the number of rotations required for RST during insertion and deletions. Backward analysis is a very elegant technique for analyzing randomized algorithms, in particular in a paradigm called randomized incremental construction.

Example 4.3 (Random Permutation) A randomized algorithm for generating a random permutation becomes a necessity and we would like to do this efficiently. A natural scheme would be to make a pass through the n elements and try to find a random assignment for each element in the range $[1, n]$. Since the mapping must be 1-1, we have to make sure that no two elements are mapped to the same index, namely there is no *collision*. For this we can use the following algorithm to generate a random permutation Π for elements $\{1, 2 \ldots n\}$.
for each element i do

1. generate a random number in range $[1, n]$, say j. If $\Pi^{-1}(j)$ is less than i repeat the step (there is a collision).

2. Set $\Pi(i)$ to j.

The running time of this algorithm depends on the number of times Step 1 is repeated. A simple analysis shows that for the i-th element, the expected number of repetitions is $\frac{n}{n-i+1}$ for the i-th element yielding a total expected value of $O(n \log n)$. It is known that the time bound is concentrated around its expected value (from the analysis of *coupon collector's problem*). A simple modification makes the expected running time linear. In the first step, generate random numbers in the range $[1, 2n]$ and in the end, compress the indices back to $[1, n]$ (keeping the relative order unchanged). The expected number of repetitions for i-th element is always less than 2. Since the trials are independent, we can apply Chernoff bounds to show that the running time is concentrated around the mean with inverse exponential probability. This technique is also known as *Random Assignment* and is very useful for parallel algorithms, like parallel integer sorting[34].

5.   CH BOUNDS WITH NEGATIVE AND LIMITED DEPENDENCE

In many applications, we are required to give concentration bounds on a sum of variables which fail to be independent. In such cases, we cannot apply the CH bounds directly, and it is of great interest to seek conditions under which they may be salvaged. In this section, we give two general scenarios in which this can be done.

5.1   NEGATIVE DEPENDENCE

One form of dependence in which CH bounds can be intuitively expected to hold is *negative dependence*: when one set of variables is "high", a disjoint set is "low". For, in this kind of systematic dependence, variations of different variables around their means will tend to cancel each other so that their sum is highly concentrated.

We now define two useful formal notions of negative dependence. Random variables $X_1, \ldots, X_n$ satisfy:

(-R) **negative regression** if for every two disjoint subsets I and J of the index set $[n]$ and for every function f which is non–decreasing in each co–ordinate, $\mathbf{E}[f(X_i, i \in I) \mid X_j = x_j]$ is non–increasing in each $x_j, j \in J$.

(-A) **negative association** if for for every two disjoint subsets I and J of the index set $[n]$ and for all functions f and g which are non–decreasing in each co–ordinate,

$$\mathbf{E}[f(X_i, i \in I)g(X_j, j \in J)] \leq \mathbf{E}[f(X_i, i \in I)]\mathbf{E}[g(X_j, j \in J)].$$

The usefulness of these definitions lies in the fact that first they obtain in many natural situations in the analysis of algorithms (some examples follow) and second that

Theorem 9 *The Chernoff–Hoeffding bounds can be applied to sums of variables that satisfy either $(-R)$ or $(-A)$.*

Two of the simplest and most useful examples where these notions of dependence obtain are:

- The *hypergeometric distribution*: the number of red balls in a sample of n balls, drawn *without replacement* from an urn containing N balls, $M \leq N$ of which are red. An intuitive *coupling* argument (that can be esrtablished by more formal arguments) in the context of hypergeometric distribution is of the following kind. Consider a sample of n balls drawn from two populations of N balls each, one of which has m red balls and the other $m' > m$ red balls. It is intuitively clear that the distribution of red balls in the sample drawn from the latter stochastically dominates the distribution drawn from the former.

  Let us sketch how to verify quickly, the seemingly complicated condition $(-R)$. By symmetry, one can assume that the set J represents the first set of trials and the set I the last. Then the conditioning $X_j = x_j, j \in J$ simply represents the sampling experiment with $N - |J|$ balls containing $M - \sum_{j \in J} x_j$ red balls. Since this last number is non–increasing in each $x_j, j \in J$, a simple coupling argument gives the result.

- The *balls and bins experiment*: m (non–identical) balls are thrown independently at random into n (non–identical) bins. Arbitrary probabilities $p_{i,k}$ are assigned to the event that ball k lands in in bin i. Of the various sets of variables of interest in this experiment are the *occupancy numbers* $B_i, i \in [n]$ giving the number of balls in each bin and the *empty bin indicators* $Z_1, \ldots, Z_n$ which are 1 or 0 accordingly as the corresponding bins are empty or not. Both these sets of variables satisfy $(-R)$ as well as $(-A)$. Once again, one can exploit symmetry to verify the $(-R)$ property in the case when all bins and balls are identical. The conditioning $B_j = b_j, j \in J$, simply reduces in this case, to a balls and bins experiment involving fewer balls, namely $m - \sum_{j \in J} b_j$ thrown into the balls indexed by I, and the result follows by an easy coupling. In the general case, i.e. when neither the bins nor the balls are identical, the result is surprisingly non–trivial, see [12].

5.2   LIMITED INDEPENDENCE

Recall that random variables $X_1, \ldots, X_n$ are fully independent if $\Pr[X_i = x_i, i \in [n]] = \prod_i prob[X_i = x_i]$. To reduce the randomness requirements in many randomized algorithms, we are interested in weaker notions of independence. The random variables $X_i, i \in [n]$ are said to be k–wise independent ($k \leq n$) if for each subset $I \subseteq [n]$ of size at most k, $\Pr[X_i = x_i, i \in I] = \prod_{i \in I} prob[X_i = x_i]$.

Suppose we have a set of variables $X_1, \ldots, X_n$ which are k–wise independent, where k is some positive even integer. To derive a concentration result for the sum $X = \sum_i X_i$ around its mean μ, we employ the k–th *moment inequality*:

$$
\begin{aligned}
\Pr[|X - \mu| > t] &= prob[(X - \mu)^k > t^k], \text{since } k \text{ is even} \\
&< \frac{\mathrm{E}[(X - \mu)^k]}{t^k}, \text{by Markov's inequality.} \quad (3.6)
\end{aligned}
$$

To estimate $\mathrm{E}[(X - \mu)^k]$, we observe that by expanding and using linearity of expectation, we only need to compute $\mathrm{E}[\prod_{i \in S}(X_i - \mu_i)]$ for multi–sets S of size k. By the k–wise independence property, this is the same as $\mathrm{E}[\prod_{i \in S}(\hat{X}_i - \mu_i)]$, where $\hat{X}_i, i \in [n]$ are fully independent random variables with the same marginals as $X_i, i \in [n]$. Turning the manipulation on it's head, we now use Chernoff–Hoeffding bounds on $\hat{X} := \sum_i \hat{X}_i$:

$$
\begin{aligned}
\mathrm{E}[(\hat{X} - \mu)^k] &= \int_0^\infty \Pr[(\hat{X} - \mu)^k > t]dt \\
&= \int_0^\infty \Pr[|\hat{X} - \mu| > t^{1/k}]dt \\
&< \int_0^\infty e^{-2t^{2/k}/n}dt, \quad \text{using CH bounds} \\
&= (n/2)^{k/2}\frac{k}{2}\int_0^\infty e^{-y}y^{k/2-1}dy \\
&= (n/2)^{k/2}\frac{k}{2}\Gamma(k/2 - 1) \\
&= (n/2)^{k/2}(k/2)!
\end{aligned}
$$

Now using Stirling's approximation for $n!$ gives the estimate:

$$
\mathrm{E}[(\hat{X} - \mu)^k] \leq 2e^{1/6k}\sqrt{\pi t}\left(\frac{nk}{e}\right)^{k/2},
$$

which in turn, plugged into (3.6) gives the following version of a tail estimate valid under limited i.e. k–wise dependence:

$$
\Pr[|X - \mu| > t] \leq C_k\left(\frac{nk}{t^2}\right)^{k/2},
$$

where $C_k := 2\sqrt{\pi k}e^{1/6k} \leq 1.0004$. This derivation is due to Bellare and Rompel [1].

Using this same basic method with other estimates of C.H. bounds, one can derive other bounds that may be more suitable for the application at hand (for example see [40]).

6. MARTINGALES AND THE METHOD OF BOUNDED DIFFERENCES

Martingales and the so–called Method of Bounded Differences [26] are a very powerful and versatile technique for proving concentration of measure results for arbitrary functions of several variables. In the general set–up, we have a function f and random variables $X_1, \ldots, X_n$ and we would like to establish a concentration result for $f(X_1, \ldots, X_n)$. The power and versatility of the method is highlighted by the fact that

- The random variables $X_1, \ldots, X_n$ need not be independent. This is crucial for applications in the analysis of algorithms where the involved random variables are often the result of complex interactions, and hence often fail to be independent.

- The function f is allowed to be arbitrarily complex and in fact, one does not even require an explicit description of it! Only certain weak conditions are placed on the function, essentially that the function is "well–behaved" or "smooth" in the sense that one can control the effect of individual variables on the function. That is, if we change the values of only a few variables (keeping the remaining fixed), then the resulting change in the value of the function can be bounded well.

We will now give very briefly, the basic definitions and concepts we require. We will give an elementary and simplified account which is sufficient for all our purposes rather than the most general one possible. A sequence of random variables $\mathbf{Y} = Y_0, Y_1, \ldots$ is a martingale with respect to a sequence of random variables $\mathbf{X} = X_0, X_1, \ldots$ if for each $i \geq 0$,

- $Y_i = f_i(X_j, j \leq i)$ for some function f_i.

- $\mathrm{E}[Y_{i+1} \mid X_j, j \leq i] = Y_i$.

Intuitively, the variables $X_1, X_2, \ldots$ are the basic underlying random variables in the stochastic process at hand and their values are being revealed or "exposed" in the indicated sequence. The variables $Y_1, Y_2, \ldots$ are functions of the basic underlying variables which we observe in the indicated sequence as we gather more and more information about the random outcomes of the stochastic system at hand. The first condition merely says that the value of the observation at stage i, Y_i is completely determined by all the random outcomes $X_1, \ldots, X_i$ revealed upto this stage. The second condition says that the value of observations in the future cannot be predicted, on average, better than the currently observed value.

(**Tossing coins or Gambling**) Consider repeated independent trials of tossing a fair coin. The basic underlying random variables are $X_i, i \geq 1$ where X_i is 1 if we get heads on the ith trial and 0 otherwise. Let the observed variables be the partial sums $Y_i := \sum_{j \leq i} X_j$. This representation already shows the first condition. For the second condition, we have,

$$
\begin{aligned}
\mathrm{E}[Y_{i+1} \mid X_j, j \leq i] &= \mathrm{E}[Y_i + X_{i+1} \mid X_j, j \leq i] \\
&= Y_i + \mathrm{E}[X_{i+1} \mid X_j, j \leq i] \\
&= Y_i + \mathrm{E}[X_{i+1}], \quad \text{since the trials are independent,} \\
&= Y_i, \quad \text{since the coin is fair.}
\end{aligned}
$$

Thus the partial sums $Y_i, i \geq 1$ are a martingale with respect to the trial variables $X_i, i \geq 1$. This is the original example from gambling, where the X_i variables give the outcome of the ith gamble and Y_i represents the fortune of the gambler after the ith trial.

The basic inequality is due independently to Azuma and (implicitly) Hoeffding:

Theorem 10 (Azuma's Inequality) *Let* $\mathbf{Y} = Y_0, Y_1, \ldots$ *be a martingale with respect to the variables* $\mathbf{X} = X_0, X_1, \ldots$. *Suppose further that there are non–negative reals* $c_i, i \geq 0$ *such that the following* **bounded differences condition** *holds on the the martingale differences:*

$$
|Y_i - Y_{i-1}| \leq c_i, \quad i \geq 1.
$$

Then

$$
\Pr[Y_n > Y_0 + t], \Pr[Y_n < Y_0 - t] \leq \exp\left(\frac{-2t^2}{\sum_{i \leq i \leq n} c_i^2}\right).
$$

A standard way to define a martingale sequence is via the so–called *Doob process*: let f be an arbitrary function and let $X_1, \ldots, X_n$ be arbitrary random variables. Then the sequence

$$
Y_i := \mathrm{E}[f(X_1, \ldots, X_n) \mid X_1, \ldots, X_i], \quad 0 \leq i \leq n,
$$

is a martingale sequence with respect to $0 =: X_0, X_1, \ldots, X_n$. Applying Azuma's inequality to the Doob martingale gives us the first version of the so–called Method of Bounded Differences, which we shall refer to as the Method of Martingale Differences:

Theorem 11 (Method of Martingale Differences) *Let $X_1, \ldots, X_n$ be arbitrary random variables and let f be an arbitrary function. Suppose there exist non–negative reals $c_i, i \in [n]$ such that*

$$|\mathbf{E}[f \mid X_1, \ldots, X_i] - \mathbf{E}[f \mid X_1, \ldots, X_{i-1}]| \leq c_i. \qquad (3.7)$$

Then

$$\Pr[f > \mathbf{E}[f] + t], \Pr[f < \mathbf{E}[f] - t] \leq \exp\left(\frac{-2t^2}{\sum_{i \leq i \leq n} c_i^2}\right).$$

An averaging argument yields a slightly weaker form of the method which is often more convenient:

Theorem 12 (Method of Bounded Average Differences) *Let $X_1, \ldots, X_n$ be arbitrary random variables and let f be an arbitrary function. Suppose there exist non–negative reals $c_i, i \in [n]$ such that for any two values a, a' that X_i can assume,*

$$\mathbf{E}[f \mid \mathbf{X}_{i-1}, X_i = a] - |\mathbf{E}[f \mid \mathbf{X}_{i-1}, X_i = a']| \leq c_i. \qquad (3.8)$$

Then

$$\Pr[f > \mathbf{E}[f] + t], \Pr[f < \mathbf{E}[f] - t] \leq \exp\left(\frac{-2t^2}{\sum_{i \leq i \leq n} c_i^2}\right).$$

A dramatically simpler *avatara* of the method emerges if we place a very natural condition on f: we will say that f is *Lipschitz* with non–negative constants $c_i, i \in [n]$ if for all $\mathbf{a}, \mathbf{a}'$ that differ only in the ith co–ordinate,

$$f(\mathbf{a}) - f(\mathbf{a}')| \leq c_i. \qquad (3.9)$$

This is in fact the usual definition of the Lipschitz condition when the underlying metric is the weighted Hamming metric:

$$d_H(x, y) := \sum_i c_i [x_i \neq y_i].$$

Under this condition we obtain the best known version of the method:

Theorem 13 (Method of Bounded Differences) *Let $X_1, \ldots, X_n$ be* **independent** *random variables and let f be Lipschitz with non–negative constants $c_i, i \in [n]$. Then,*

$$\Pr[f > \mathbb{E}[f] + t], \Pr[f < \mathbb{E}[f] - t] \leq \exp\left(\frac{-2t^2}{\sum_{i \leq i \leq n} c_i^2}\right).$$

We shall illustrate the three versions of the method with applications below. First however, we make some general remarks on a relative comparison of the three versions:

- The Method of Bounded Differences is the most convenient one to use in applications, for it only involves checking the Lipschitz condition (3.9) on the function f. There are however, two limitations of the method: first the underlying variables are required to be independent, and second, the resulting bound can often be very weak.

- It may be the case in an application that f satisfies the Lipschitz condition with small constants in most cases, but there are some pathological low probability cases in which the constant is much larger. In this case, the Method of Bounded Differences unfairly penalizes the function with the worst case constants rather than the "average case constants". The Method of Bounded Average Differences replaces the worst case constants with a version of "average case constants", according it the name. Since the resulting constants can be much smaller than the constants required in the Method of Bounded Differences, the bound obtained is correspondingly much stronger.

- The Method of Martingale Differences is the most powerful version of the method. While it might not differ perceptibly from the Method of Bounded Average Differences in the bounds obtained, it can be more convenient to apply in certain situations.

6.1   COUPLING

In order to make effective use of the Method of Average Bounded Differences, we need to get a good handle on the bound (3.8), for the difference in the expected values of a function under two different conditioned distributions. A very useful technique for this is the method of *coupling*. Suppose that we can find a joint distribution $\pi(\mathbf{Y}, \mathbf{Y}')$ such that the marginal distribution for Y is the same as the distribution of $\mathbf{X}$ conditioned on $\mathbf{X}_{i-1} = \mathbf{a}_{i-1}, X_i = a_i$ and the marginal distribution for $\mathbf{Y}'$ is the same as the distribution $\mathbf{X}$ conditioned on $\mathbf{X}_{i-1} = \mathbf{a}_{i-1}, X_i = a'_i$. Such a joint distribution is called a coupling of the two original distributions. Then,

$$|\mathrm{E}[f \mid \mathbf{X}_{i-1} = \mathbf{a}_{i-1}, X_i = a_i] - \mathrm{E}[f \mid \mathbf{X}_{i-1} = \mathbf{a}_{i-1}, X_i = a'_i]| =$$
$$|\mathrm{E}_\pi[f(\mathbf{Y})] - \mathrm{E}_\pi[f(\mathbf{Y}')]| = |\mathrm{E}_\pi[f(\mathbf{Y}) - f(\mathbf{Y}')]| \quad (3.10)$$

If the coupling π is well–chosen so that $|f(\mathbf{Y}) - f(\mathbf{Y}')|$ is usually very small, we can get a good bound on the difference (3.8). For example, suppose that

- For any sample point $(\mathbf{y}, \mathbf{y}')$ we have $|f(\mathbf{y}) - f(\mathbf{y}')| \le d$ for some constant $d > 0$; and

- For most sample points $(\mathbf{y}, \mathbf{y}'$, $f(\mathbf{y}) = f(\mathbf{y}')$. That is, $\pi[f(\mathbf{Y}) - f(\mathbf{Y}')] \le p$, for some $p << 1$.

Then, we can conclude using (3.9) that

$$|\mathrm{E}[f \mid \mathbf{X}_{i-1} = \mathbf{a}_{i-1}, X_i = a_i] - \mathrm{E}[f \mid \mathbf{X}_{i-1} = \mathbf{a}_{i-1}, X_i = a'_i]| \le pd.$$

We shall construct suitable couplings to bound the difference in (3.8).

7. APPLICATIONS OF MARTINGALE METHODS

Example 7.1 (Bin Packing) In the *bin packing* problem, we are given items of sizes $a_1, \ldots, a_n$ with $0 \le a_i \le 1$ for each $i \in [n]$, and we are required to "pack" them into a minimum number of bins, each of unit capacity. There are many known heuristics for this problem, which in unlikely to be solvable efficiently because it is NP-complete. One such is *first-fit*: put the next item in the first bin in which it can fit. Consider the average case instance of this problem, where each a_i is drawn independently and uniformly at random from $[0, 1]$. A general theorem from the theory of *subadditive functionals* [43] shows that the optimum number of bins, $B_n := B(a_1, \ldots, a_n)$ satisfies the limit law:

$$\Pr[\lim_n \frac{B_n}{n} = \gamma] = 1,$$

for some constant $\gamma > 0$. In the limit, therefore, the expectation will also be γn. In any case, for any finite n as well, one can easily deduce a sharp concentration bound on $\mathrm{E}[B_n]$ using the simplest form of the method of bounded differences. Just observe that $B(a_1, \ldots, a_n)$ can change by at most 1 if one changes only one item size while keeping the others fixed. In fact this is true also of the number of bins delivered by the first fit heuristic. Thus for either of these random variables, we have the concentration result:

$$\Pr[|B_n - ep[B_n]| > t] \leq \exp\left(-2t^2/n\right),$$

which for $t = \delta E[B_n]$ gives an exponentially decreasing probability.

Example 7.2 (Quicksort) We shall sketch the application of the Method of Martingale Differences to Quicksort. This application is interesting because it is a very natural application of the method and yields a provably optimal tail bound. While conceptually simple, the details required to obtain the full bound are messy, so we shall confine ourselves to indicating the basic method.

Recall that Quicksort can be modeled as a binary tree T, corresponding to the partition around the pivot element performed at each stage. With each node v of the binary tree, we associate the list L_v that needs to be sorted there. At the outset, the root r is associated with $L_r = L$, the input list, and if the the pivot element chosen at node v is X_v, the lists associated with the left and right children of v are the sublists of L_v consisting of, respectively, all elements less than X_v and all elements greater than X_v (for simplicity, we assume that the input list contains all distinct elements). Now, the number of comparisons performed by Quicksort on the input list L, Q_L is a random variable given by some function f of the random choices made for the pivot elements, $X_v, v \in T$:

$$Q_L = f(X_v, v \in T).$$

We shall now expose the variables $X_v, v \in T$ in the natural top–down fashion: level–by–level and left to right within a level, starting with the root. Let us denote this (inorder) ordering of the nodes of T by $<$. Thus, to apply the Method of Martingale Differences, we merely need to estimate for each node $v \in T$,

$$|E[Q_L \mid X_w, w < v] - E[Q_L \mid X_w, w \leq v]|.$$

A moment's reflection shows that this difference is simply

$$|E[Q_{L_v}] - E[Q_{L_v} \mid X_v]|,$$

where L_v is the list associated with v as a result of the previous choices of the partitions given by $X_w, w < v$. That is, the problem reduces to estimating the difference between the expected number of comparisons performed on a given list when the first partition is specified and when it is not. Such an estimate is readily available for Quicksort via the recurrence satisfied by the expected value $q_n := E[Q_n]$, the expected number of comparisons performed on a input list of length n. If the first partition (which by itself requires $n - 1$ comparisons) splits the list into a left part of size $k, 0 \le k < n$ and a right part of size $n - 1 - k$, the expected number of comparisons is $n - 1 + q_k + q_{n-1-k}$ and the estimate is:

$$|q_n - (n - 1 + q_k + q_{n-k-1})| \le n - 1.$$

We shall plug this estimate into the Method of Bounded Differences: thus, if $\ell_v := |L_v|$ is the length of the list associated with node v, then we need to estimate $\sum_v \ell_v^2$. This is potentially problematical, since these lengths are themselves random variables! Suppose, that we restrict attention to levels $k \ge k_1$ for which we can show that $\ell_v \le \alpha n$ for some parameters k_1 and α to be chosen later; then summing over these levels, level by level,

$$
\begin{aligned}
\sum_v \ell_v^2 &= \sum_{k \ge k_1} \sum_{h(v)=k} \ell_v^2 \\
&\le \sum_{k \ge k_1} \sum_{h(v)=k} \alpha n \ell_v \\
&= \sum_{k \ge k_1} \alpha n \sum_{h(v)=k} \ell_v \\
&\le \sum_{k \ge k_1} \alpha n^2.
\end{aligned}
$$

Next we are faced with yet another problem: the number of levels, which itself is again a random variable! Suppose we can show for some $k_2 > k_1$, that the tree has height no more than k_2 with high probability. Then the previously computed sum reduces to $(k_2 - k_1)\alpha n^2$. X Finally all that remains is to choose the parameters carefully: suppose we choose k_1 and α so that the maximum size of the list associated with a node at height at least k_1 exceeds αn with probability at most p_1 and and k_2 so that the overall height of the tree exceeds k_2 with probability at most p_2. This can be done in an elementary way by using the fact that the size of the list at a node at depth $k \ge 0$ is explicitly given by $n \prod_{1 \le i \le k} Z_i$, where each Z_i is uniformly distributed in $[0, 1]$. Then the final result will be:

$$\Pr[Q_n > q_n + t] < p_1 + p_2 + \exp\left(\frac{-2t^2}{(k_2 - k_1)\alpha n^2}\right).$$

We choose the parameters to optimize this sum of three terms. The result whose details are messy (see [25]) is:

Theorem 14 *Let $\epsilon = \epsilon(n)$ satisfy $1/\ln n < \epsilon \leq 1$. Then as $n \to \infty$,*

$$\Pr[|\frac{Q_n}{q_n} - 1| > \epsilon] < n^{-2\epsilon \log \log n}.$$

This bound is slightly better than an inverse polynomial bound and can be shown to be essentially tight.

8.   DISTRIBUTED EDGE COLORING

Vizing's Theorem shows that every graph G can be edge colored sequentially in polynomial time with Δ or $\Delta + 1$ colors, where Δ is the maximum degree of the input graph (see, for instance, [3]). The proof is in fact a polynomial time sequential algorithm for achieving a $\Delta + 1$ coloring.

It is a challenging open problem whether colorings as good as these can be computed fast in a distributed model. Here one has a set of processors connected by an arbitrary network of communication channels. Data is distributed throughout the network amongst the processors. The processors are required to co-operate together in some fashion to compute some function of all the data. For this, they communicate data between each other via the communication channels of the network. In the graph coloring problem, the network of processors is required to compute an edge coloring of itself.

In **many** realistic situations, it is the communication that is expensive rather than internal computation at any single processor. Hence, in the model, communication is at a premium rather than computation. One is therefore required to design algorithms that minimize communication. This poses a *locality* constraint: the algorithm should require each processor to gather only local information i.e. from processors in a small neighborhood around itself.

The difficulty posed by this locality constraint is often compounded by the problem of symmetry-breaking; if every process has the same view of the network– i.e. the same input– it In the absence of such a result one might aim at the more modest goal of computing reasonably good colorings, instead of optimal ones. By a trivial modification of a well-known *vertex* coloring algorithm of Luby it is possible to edge color a graph using $2\Delta - 2$ colors in $O(\log n)$ rounds (where n is the number of processors) [24].

We shall present and analyze two classes of simple localized distributed algorithms that compute near optimal edge colorings. Both algorithms proceed in a sequence of rounds. In each round, a simple randomized heuristic is invoked to color a significant fraction of the edges successfully. The remaining edges are passed over to succeeding rounds. This continues until the number of edges is small enough to employ a brute–force method at the final step. For example, the algorithm of Luby mentioned above can be invoked when the degree of the graph becomes small i.e. when the condition $\Delta \gg \log n$ is no longer satisfied.

8.1   TWO CLASSES OF ALGORITHMS

One of the classes of algorithms involves a standard reduction to bipartite graphs described in [32]: the graph is split into two parts T ("top") and B (bottom). The bipartite graph $G[T, B]$ induced by the edges connecting top and bottom vertices is colored by invoking the Algorithm P described below. The algorithm is then invoked recursively in parallel on $G[T]$ and $G[B]$, the graphs respectively induced by the top and bottom vertices. Both graphs are colored using the same set of colors. Thus it suffices to describe the algorithm used for coloring bipartite graphs.

We describe the action carried out by both algorithms in a single round. For the second class of algorithms, we describe the action only for bipartite graphs; additionally, each vertex knows whether it is top or bottom. At the beginning of each round, there is a palette of fresh new available colors, $[\Delta]$, where Δ is the maximum degree of the graph at the current stage. For simplicity we will assume that the graph is Δ–regular.

Algorithm I(Independent): Each edge *independently* picks a color. This *tentative* color becomes permanent if there are no conflicting edges picking the same tentative color at either endpoint.

Algorithm P(Permutation): There is a two step protocol:

- Each bottom vertex, in parallel, makes a *proposal* independently of other bottom vertices by assigning a random *permutation* of the colors to their incident edges.

- Each top vertex, in parallel, then picks a *winner* out of every set of incident edges that have the same color. Tentative colors of winner edges become final.

- The *losers*– edges who are not winners– are decoloured and passed to the next round.

For the purposes of the high probability analysis below, the exact rule used for selecting the winner edge is unimportant – it can be chosen arbitrarily from any of the edges of the relevant color; we merely require that it should not depend on edges of different colors. This is another illustration of the power of the martingale method.

It is apparent that both algorithms are truly distributed. That is to say, each vertex need only exchange information with the neighbors to execute the algorithm. This and its simplicity make the algorithms amenable for implementations in a distributed environment. Algorithm I is used with some more modifications in a number of edge coloring algorithms [11, 15]. Algorithm P is exactly the algorithm used in [32].

We focus all our attention in the analysis of one round of both algorithms. Let Δ denote the maximum degree of the graph at the beginning of the round and Δ' denote the maximum degree of the leftover graph. One can easily show that both algorithms, $E[\Delta' \mid \Delta] \leq \beta\Delta$, for some constant $\beta < 1$. For algorithm I, $\beta = 1 - e^{-2}$ while for algorithm P, $\beta = 1/e$. The goal is to show that this holds with high probability. This is done in § 8.2 after the relevant tools – the Martingale inequalities – are introduced in the next section.

For completeness, we sketch a calculation of the total number of colors $BC(\Delta)$ used by Algorithm P for the bipartite coloring of a graph with maximum degree Δ: with high probability, it is,

$$
\begin{aligned}
BC(\Delta) &= \Delta + \frac{(1+\epsilon)\Delta}{e} + \frac{(1+\epsilon)^2\Delta^2}{e} + \ldots \\
&\leq \frac{1}{1 - (1+\epsilon)e}\Delta \approx 1.59\Delta \text{ for small enough } \epsilon.
\end{aligned}
$$

To this, one should add $O(\log n)$ colors at the end of the process. As can be seen by analyzing the simple recursion describing the number of colors used by the outer level of the recursion, the overall numbers of colors is the same $1.59\Delta + O(\log n)$, [32].

8.2   HIGH PROBABILITY ANALYSES

Top Vertices. The analysis is particularly easy when v is a top vertex in Algorithm P. For, in this case, the incident edges all receive colors independently of each other. This is exactly the situation of the classical balls and bins experiment: the incident edges are the "balls" that are falling at random independently into the colors that represent the "bins". One can apply the method of bounded differences in the simplest form. Let $T_e, e \in E$, be the random variables taking values in $[\Delta]$ that represent the tentative colors of the edges. Then the number of edges successfully colored around v is a function $f(T_e, e \in N^1(v))$, where $N^1(v)$ denotes the set of edges incident on v.

It is easily seen that this function has the *Lipschitz* property with constant 1: changing only one argument while leaving the others fixed only changes the value of f by at most 1. Note that this is true *regardless oj the rule for choosing winners*, as long as this rule does not depend on edges of different colors. This will also be true of the remaining analyses below and illustrates once again, the power of the martingale methods.

Moreover, the variables $T_e, e \in N^1(v)$ are independent when v is a "top" vertex. Hence, by the method of bounded differences in the simplest form, we get the following sharp concentration result by plugging into Theorem 13:

Theorem 15 *Let v be a top vertex in algorithm P and let f be the number of edges around v that are successfully colored in one round of the algorithm. Then,*

$$\Pr[|f - \mathbb{E}[f]| > t] \leq \exp\left(\frac{-2t^2}{\Delta}\right),$$

For $t := \epsilon\Delta$ ($0 < \epsilon < 1$), this gives an exponentially decreasing probability for deviations around the mean. If $\Delta \gg \log n$ then the probability that the new degree of any vertex deviates far from its expected value is inverse polynomial, i.e. the new max degree is sharply concentrated around its mean.

Other Vertices: The Difficulty.   The analysis for the "bottom" vertices in Algorithm P is more complicated in several respects. It is useful to see why so that one can appreciate the need for using a more sophisticated tool such as the Method of Bounded Average Differences. To start with, one could introduce an indicator random variable X_e for each edge e incident upon a bottom vertex v. These random variable are not independent however. Consider a four cycle with vertices v, a, w, b, where v and w are bottom vertices and a and b are top vertices. Let's refer to the process of selecting the winner (step 2 of the algorithm P) as "the lottery". Suppose that we are given the information that edge va got tentative color red and lost the lottery— i.e. $X_{va} = 0$— and that edge vb got tentative color green. We'll argue intuitively that given this, it is more likely that $X_{vb} = 0$. Since edge va lost the lottery, the probability that edge wa gets tentative color red increases. In turn, this increases the probability that edge wb gets tentative color green, which implies that edge vb is more likely to lose the lottery. So, not only are the X_e's not independent, but the dependency among them is particularly malicious.

One could hope to bound this effect by using the MOBD in it simplest form. This is also ruled out however, for two reasons. The first is that the tentative color choices of the edges around a vertex are not independent. This is because the edges incident on vertex are assigned a permutation of the colors. The second reason applies also to algorithm I where all edges act independently. The new degree of v, a bottom vertex in algorithm P or an arbitrary vertex in algorithm I, is a function $f = f(T_e, e \in N(v))$, where $N(v)$ is the set of edges at distance at most 2 from v. Thus f depends on as many as $\Delta(\Delta - 1) = \Theta(\Delta^2)$ edges. Even if f is Lipshitz with constants $d_i = 2$, this is not enough to get a strong enough bound because $d = \sum_i d_i^2 = \Theta(\Delta^2)$. Applying the method of bounded differences in the simple form, Theorem 13, would give the bound

$$\Pr[|f - \mathbf{E}[f]| > t] \leq 2 \exp\left(-\frac{t^2}{\Theta(\Delta^2)}\right).$$

This bound however is useless for $t = \epsilon \mathbf{E}[f]$ since $\mathbf{E}[f] \approx \Delta/e$.

We will use the Method of Bounded Average Differences, Theorem 12, to get a much better bound. We shall invoke the two crucial features of this more general method. Namely that it does not need the underlying variables to be independent [2], and that, second that as we shall see, it allows us to bound the effect of individual random choices with constants much smaller than those given by the MOBD in simple form.

Let's now move on to the analysis. A similar analysis applies to both cases: when v is a bottom vertex in algorithm P or an arbitrary vertex in algorithm I. Let $N^1(v)$ denote the set of "direct" edges– i.e. the edges incident on v– and let $N^2(v)$ denote the set of "indirect edges" that is, the edges incident on a neighbor of v. Let $N(v) := N^1(v) \bigcup N^2(v)$. The number of edges successfully colored at vertex v is a function $f(T_e, e \in N(v))$. Note that in Algorithm P, even though f seems to depend on edges at distance 3 from v via their effect on edges at distance 2, f can still be regarded as a function of the edges in $N(v)$ only (i.e. f is fixed by giving colors to all edges in $N(v)$ regardless of what happens to other edges) and hence only these edges need be considered in the analysis.

Let us number the variables so that the direct edges are numbered *after* the indirect edges (this will be important for the calculations to follow). We need to compute

$$\lambda_k := |\mathbf{E}[f \mid \mathbf{T}_{k-1}, T_k = c_k] - \mathbf{E}[f \mid \mathbf{T}_{k-1}, T_k = c_k']|. \tag{3.11}$$

We decompose f as a sum to ease the computations later. Introduce the indicator functions $f_e, e \in E$:

$$f_e(\mathbf{c}) := \begin{cases} 1; & \text{if edge } e \text{ is successfully coloured in colouring } \mathbf{c}, \\ 0; & \text{otherwise.} \end{cases}$$

Then $f = \sum_{v \in e} f_e$.

Hence we are reduced, by linearity of expectation, to computing for each $e \in N^1(v)$,

$$|\Pr[f_e = 1 \mid \mathbf{T}_{k-1}, T_k = c_k] - \Pr[f_e = 1 \mid \mathbf{T}_{k-1}, T_k = c'_k]|.$$

For the computations that follows we should keep in mind that in algorithm P bottom vertices assign colors independently of each other. This implies that in either algorithm, the color choices of the edges incident upon a neighbor of v are independent of each other. In Algorithm I, *all* edges have their colors assigned independently.

General Vertex in Algorithm I. To compute a good bound for λ_k in (3.11), we shall construct a suitable coupling of the two different conditioned distributions. The coupling $(\mathbf{Y}, \mathbf{Y}')$ is almost trivial: $\mathbf{Y}$ is distributed as $\mathbf{T}$ conditioned on $\mathbf{T}_{k-1}, T_k = c_k$ and $\mathbf{Y}'$ is identically equal to $\mathbf{Y}$ except that $\mathbf{Y}'_k = c'_k$. It is easily seen that by the independence of all tentative colors, the marginal distributions of $\mathbf{Y}$ and $\mathbf{Y}'$ are exactly the two conditioned distributions $[\mathbf{T} \mid \mathbf{T}_{k-1}, T_k = c_k]$ and $[\mathbf{T} \mid \mathbf{T}_{k-1}, T_k = c'_k]$ respectively.

Now let us compute $|\mathrm{E}[f(\mathbf{Y}) - f(\mathbf{Y}')]|$.

- First, let us consider the case when $e_1, \ldots, e_k \in N^2(v)$, i.e. only the choices of indirect edges are exposed. Let $e_k = (w, z)$, where w is a neighbor of v. Then for a direct edge $e \neq vw$, $f_e(\mathbf{y}) = f_e(\mathbf{y}')$ because in the joint distribution space, $\mathbf{y}$ and $\mathbf{y}'$ agree on all edges incident on e. So we only need to compute $|\mathrm{E}[f_{vw}(\mathbf{Y}) - f_{vw}(\mathbf{Y}')]|$. To bound this simply, we observe first that $f_{vw}(\mathbf{y}) - f_{vw}(\mathbf{y}') \in [-1, 1]$ and second that $f_{vw}(\mathbf{y}) = f_{vw}(\mathbf{y}')$ unless $y_{vw} = c_k \text{ or } c'_k$. Thus we can conclude that

$$\mathrm{E}[f_{vw}(\mathbf{Y}) - f_{vw}(\mathbf{Y}')]| \leq \Pr[Y_e = c_k \vee Y_e = c'_k] \leq \frac{2}{\Delta}.$$

In fact one can do a tighter analysis using the same observations. Let us denote $f_e(\mathbf{y}, y_{w,z} = c_1, y_e = c_2)$ by $f_e(c_1, c_2)$. Note that $f_{vw}(c_k, c_k) = 0$ and similarly $f_{vw}(c'_k, c'_k) = 0$. Hence

$$\begin{aligned}
\mathrm{E}[f_e(\mathbf{Y}) - f_e(\mathbf{Y}') \mid z] =& \\
& (f_{vw}(c_k, c_k) - f_{vw}(c'_k, c_k))\Pr[Y_e = c_k] + \\
& (f_{vw}(c_k, c'_k) - f_{vw}(c'_k, c'_k))\Pr[Y_e = c'_k] \\
=& (f_{vw}(c_k, c'_k) - f_{vw}(c'_k, c_k))\frac{1}{\Delta}
\end{aligned}$$

(Here we used the fact that the distribution of color around v is unaffected by the conditioning around z and that each color is equally likely.) Hence $|\mathrm{E}[f_e(\mathbf{Y}) - f_e(\mathbf{Y}')]| \leq \frac{1}{\Delta}$.

- Now let us consider the case when $e_k \in N^1(v)$, i.e. choices of all indirect edges and of some direct edges have been exposed. In this case, we merely observe that f is Lipshitz with constant 2: $|f(\mathbf{y}) - f(\mathbf{y}')| \leq 2$ whenever $\mathbf{y}$ and $\mathbf{y}'$ differ in only one co-ordinate. Hence we can easily conclude that $|\mathbf{E}[f(\mathbf{Y}) - f(\mathbf{Y}')]| \leq 2$.

Overall,

$$\lambda_k \leq \begin{cases} 1/\Delta; & \text{for an edge } e_k \in N^2(v), \\ 2; & \text{for an edge } e_k \in N^1(v) \end{cases},$$

and we get

$$\sum_k \lambda_k^2 = \sum_{e \in N^2(v)} \frac{1}{\Delta^2} + \sum_{e \in N^1(v)} 4 \leq 4\Delta + 1.$$

We thus arrive at the following sharp concentration result by plugging into Theorem 12:

Theorem 16 *Let v be an arbitrary vertex and let Let f be the number of edges successfully colored around v in one stage of algorithm I. Then,*

$$\Pr[|f - \mathbf{E}[f]| > t] \leq 2 \exp\left(-\frac{t^2}{2\Delta + \frac{1}{2}}\right).$$

Jaikumar Radhakrishnan observed that a similar result can be obtained very simply for Algorithm I by applying Theorem 13: regard f as a function of Δ variables: $\mathbf{T}_w, (v, w) \in E$, where $\mathbf{T}_w := (T_e, w \in e)$ is the vector of tentative colors of edges around w. Since f is Lipshitz with constant 2 with respect to each of these (vector valued) variables, we get the bound:

$$\Pr[|f - \mathbf{E}[f]| > t] \leq 2 \exp\left(-\frac{t^2}{2\Delta}\right).$$

Bottom Vertices in Algorithm P. Once again, to compute a good bound for λ_k in (3.11), we shall construct a suitable coupling [3] of the two different conditioned distributions $\mathbf{T}_{k-1}, T_k = c_k$ and $\mathbf{T}_{k-1}, T_k = c'_k$. Suppose e_k is an edge zy where z is a bottom vertex. The coupling $(\mathbf{Y}, \mathbf{Y}')$ in this case is the following: $\mathbf{Y}$ is distributed as $\mathbf{T}$ conditioned on $\mathbf{T}_{k-1}, T_k = c_k$ and $\mathbf{Y}'$ is identically equal to $\mathbf{Y}$ except on the edges incident on z, where the colors c_k and c'_k are switched. We can think of the distribution as divided into two classes: on the edges incident on a vertex other than z, the two variables $\mathbf{Y}$ and $\mathbf{Y}'$ are identically equal. In particular, when z is not v, they have the same uniform distribution on all permutations of colors on the edges around v. However, on the edges incident on z, the two variables differ on exactly two edges where the colors c_k and c'_k are switched. It is easily seen that by the independence of different vertices, the marginal distributions of $\mathbf{Y}$ and $\mathbf{Y}'$ are exactly the two conditioned distributions $[\mathbf{T} \mid \mathbf{T}_{k-1}, T_k = c_k]$ and $[\mathbf{T} \mid \mathbf{T}_{k-1}, T_k = c'_k]$ respectively.

Now let us compute $|\mathbf{E}[f(\mathbf{Y}) - f(\mathbf{Y}')]|$. Recall that f was decomposed as a sum $\sum_{v \in e} f_e$. Hence by linearity of expectation, we need only bound each $|\mathbf{E}[f_e(\mathbf{Y}) - f_e(\mathbf{Y}')]|$ separately.

- First, let us consider the case when $e_1, \ldots, e_k \in N^2(v)$, i.e. only the choices of indirect edges are exposed. Let $e_k = (w, z)$ for a neighbor w of v. Note that since

$$\mathbf{E}[f(\mathbf{Y}) - f(\mathbf{Y}')] = \mathbf{E}[\mathbf{E}[f(\mathbf{Y}) - f(\mathbf{Y}') \mid \mathbf{Y}_e, \mathbf{Y}'_e, z \in e]],$$

  it suffices to bound $|\mathbf{E}[f(\mathbf{Y}) - f(\mathbf{Y}') \mid \mathbf{Y}_e, \mathbf{Y}'_e, z \in e]|$. Hence, fix some distribution of the colors around z. Recall that $\mathbf{Y}_{w,z} = c_k$ and $\mathbf{Y}'_{w,z} = c'_k$. Suppose $\mathbf{Y}_{z,w'} = c'_k$ for some other neighbor w' of z. Then by our coupling construction, $\mathbf{Y}'_{z,w'} = c_k$ and on the remaining edges $\mathbf{Y}$ and $\mathbf{Y}'$ agree identically. Moreover, by the independence of the other vertices, the distributions of $\mathbf{Y}$ and $\mathbf{Y}'$ on the remaining edges conditioned on the distribution around z is unaffected. let us denote the conditioned joint distribution by $[(\mathbf{Y}, \mathbf{Y}') \mid z]$. We thus need to bound $|\mathbf{E}[f(\mathbf{Y}) - f(\mathbf{Y}') \mid z]|$.

  Then for a direct edge $e \notin vw, vw'$, $f_e(\mathbf{y}) = f_e(\mathbf{y}')$ because in the joint distribution space, $\mathbf{y}$ and $\mathbf{y}'$ agree on all edges incident on e. So we only need to compute $|\mathbf{E}[f_e(\mathbf{Y}) - f_e(\mathbf{Y}')]|$ for $e \in vw, vw'$. To bound this simply, we observe that for either $e = vw$ or $e = vw'$, first, $f_e(\mathbf{y}) - f_e(\mathbf{y}') \in [-1, 1]$ and second that $f_e(\mathbf{y}) = f_e(\mathbf{y}')$ unless $y_e = c_k \text{ or } c'_k$. Thus we can conclude that

$$\mathbf{E}[f_e(\mathbf{Y}) - f_e(\mathbf{Y}')]| \leq \Pr[Y_e = c_k \vee Y_e = c'_k] \leq \frac{2}{\Delta}.$$

  Thus taking the two contributions for vw and vw' together, $|\mathbf{E}[f(\mathbf{Y}) - f(\mathbf{Y}') \mid z]| \leq \frac{4}{\Delta}$.

  In fact one can do a tighter analysis using the same observations. Let us denote $f_e(\mathbf{y}, y_{w,z} = c_1, y_e = c_2)$ by $f_e(c_1, c_2)$. Note that $f_{vw}(c_k, c_k) = 0 = f_{vw}(c'_k, c'_k)$ and similarly $f_{vw'}(c_k, c_k) = 0 = f_{vw'}(c'_k, c'_k)$. Thus, for $e = vw$ or $e = vw'$,

$$\mathbf{E}[f_e(\mathbf{Y}) - f_e(\mathbf{Y}') \mid z] = $$
$$(f_e(c_k, c_k) - f_e(c'_k, c_k))\Pr[Y_e = c_k] +$$

$$(f_e(c_k, c'_k) - f_e(c'_k, y'_e = c'_k))\Pr[Y_e = c'_k]$$
$$= (f_e(c_k, c'_k) - f_e(c'_k, c_k))\frac{1}{\Delta}$$

Hence $|\mathrm{E}[f_e(\mathbf{Y}) - f_e(\mathbf{Y}') \mid z]| \le \frac{1}{\Delta}$. Taking the two contributions for edges vw and vw' together, $|\mathrm{E}[f(\mathbf{Y}) - f(\mathbf{Y}') \mid x]| \le \frac{2}{\Delta}$.

- Now let us consider the case when $e_k \in N^1(v)$, i.e. choices of all indirect edges and of some direct edges have been exposed. In this case, we observe again that $|f(\mathbf{y}) - f(\mathbf{y}')| \le 2$ since $\mathbf{y}$ and $\mathbf{y}'$ differ on exactly two edges. Hence we can easily conclude that $|\mathrm{E}[f(\mathbf{Y}) - f(\mathbf{Y}')]| \le 2$.

Overall,

$$\lambda_k \le \begin{cases} 2/\Delta; & \text{for an edge } e_k \in N^2(v), \\ 2; & \text{for an edge } e_k \in N^1(v) \end{cases},$$

and we get

$$\sum_k \lambda_k^2 = \sum_{e \in N^2(v)} \frac{4}{\Delta^2} + \sum_{e \in N^1(v)} 4 \le 4(\Delta + 1).$$

We thus arrive at the following sharp concentration result by plugging into Theorem 12:

Theorem 17 *Let v be an arbitrary bottom vertex and let f be the number of edges successfully colored around v in one stage of algorithm P. Then,*

$$\Pr[|f - \mathrm{E}[f]| > t] \le 2\exp\left(-\frac{t^2}{2\Delta + 2}\right).$$

Comparing this with the corresponding bound for Algorithm I, we see that the failure probabilities for both algorithms are almost identical. For $t = \epsilon\Delta$, both a probability that decreases exponentially in Δ. As remarked earlier, if $\Delta \gg \log n$, this implies that the new max degree is sharply concentrated around the mean (with failure probability inverse polynomial in n). The constant in the exponent here is better than the one in the analysis in [32].

Extensions. It is fairly clear that the method extends more generally to cover similar scenarios in distributed computing. We sketch such a general setting: One has a distributed randomized algorithm that requires vertices to assign labels to themselves and incident edges. Each vertex acts independently of the others, and furthermore is symmetric with respect to the labels (colors). The function of interest, f depends only on a small local neighborhood around some vertex v, is Lipschitz and satisfies some version of the following *locality* property: the labels on vertices and edges far away from v only effect f if certain events are triggered on nearer vertices and edges; these triggering events correspond to the setting of the nearer vertices and edges to specific values. For example, in edge coloring, the color of an indirect edge only affects f if the incident direct edge has the same color. One can extend the same arguments as above virtually intact for this general setting. This encompasses all the edge coloring algorithms mentioned above as well as the vertex coloring algorithms in [28] and [16].

9.   BRANCHING PROCESSES AND APPLICATIONS

We give a very compact account of the theory of branching process which closely follows the description in Feller [13].

9.1   SUMS OF RANDOM NUMBER OF VARIABLES

Let $\{X_k\}$ be a sequence of i.i.d. random variables with a common distribution function $\Pr[X_k = j] = f_j$ and the generating function $f(s) = \sum_i f_j s^j$. We are often interested in the sums

$$S_N = X_1 + X_2 \dots X_N$$

where N itself is a random variable (independent of X_js). Let $\Pr[N = i] = g_i$ be the distribution of N and $g(s) = \sum_i g_i s^i$ be the generating function. For the distribution $\Pr[S_N = j] = h_j$], we can write using the law of conditional probabilities

$$\Pr[S_N = j] = \sum_{n=0}^{n=\infty} \Pr[N = n] \cdot \Pr[X_1 + X_2 \dots + X_n = j]$$

since N and X_k's are independent. For a fixed n, the distribution of $X_1 + X_2 \dots + X_n$ is given by the n fold convolution of f_i and it is somewhat easier to deal with the generating function given by $f^n(s)$. The generating function of S_N can be written as

$$h(s) = \sum_{j=0}^{\infty} h_j s^j = \sum_{n=0}^{\infty} g_n f^n(s)$$

Lemma 2 *The generating function of the sum $S_N = X_1 + X_2 \dots + X_N$ is given by $g(f(s))$ where g and f are the generating functions of N and X_i respectively.*

9.2   BRANCHING PROCESSES

We can think of branching processes as a tree (possibly infinite), where each node is associated with a distribution function p_k of having k children. Let the mean $\mu = \sum_k k \cdot p_k$. All the nodes whose distance to the root is n form the n-th generation.

Let Z_n denote the size of the n-th generation and P_n denote the generating function of the probability distribution. $Z_0 = 1$ and

$$P_1(s) = P(s) = \sum_{k=0}^{\infty} p_k s^k.$$

We would like to write a recurrence for P_s using the observation that Z_n can be partitioned into *clans* depending on its ancestor in the first generation. Thus Z_n is the sum of Z_1 random variables $Z_n^{(k)}$. Clearly $Z_n^{(k)}$ has the same probability distribution as Z_{n-1} (as the probability distribution at each node are identical and independent). From Lemma 2

$$P_n(s) = P(P_{n-1}(s)) \tag{3.12}$$

Here are some important questions that are relevant to branching process.

- Is the branching process finite ?

- What is the expected size of Z_n ?

- What is the expected size of progeny $Y_n = 1 + Z_1 + Z_2 \ldots Z_n$ and the *total* progeny (if the branching process is finite).

We will state some of the important results known in the context of the above issues.

Theorem 18 *If $\mu \leq 1$, the branching process becomes extinct with probability rapidly converging to 1. If $\mu > 1$, then the probability x_n that the branching process terminates at or before the n-th generation converges to the (unique) root of the equation $x = P(x)$, $x < 1$.*

Proof Sketch: If x_n denotes the probability $\Pr[Z_n = 0]$, then $x_n = P_n(0)$. Clearly $x_1 = p_0$ and from the recursive relation for $P_n(s)$, it follows that

$$x_n = P_n(0) = P(P_{n-1}(0)) = P(x_{n-1})$$

Notice that $x_1 < x_2 < \ldots$ since $x_2 = P(x_1) = P(p_0) > P(0) = x_1$, so there must exist a limit $x \leq 1$ such that $x = P(x)$.

It can be argued that this is the extinction probability.

Remark 2 *By solving the recurrence* $x_n = P(x_{n-1})$, *we can obtain the rate of extinction as a function of n.*

Theorem 19 *The expected size of the n-th generation is* μ^n.

Proof. $E[Z_n] = P'_n(1)$. From equation 3.12 , and chain rule, it follows that

$$P'_n(1) = P'(1) \cdot P'_{n-1}(1) = \mu \cdot E[Z_{n-1}].$$

The result follows by induction. ∎

Remark : The previous theorem gives us a clue about whether the branching process would die out (except for the case $\mu = 1$).

Theorem 20 *The generating function of the total number of descendents is given by the (unique) root of the equation*

$$t = s \cdot P(t)$$

when the branching process is finite. The mean progeny is

$$\frac{1}{1 - \mu}.$$

Proof Sketch: Let $Y_n = 1 + Z_1 + Z_2 \ldots + Z_n$ be the total number of descendents up to and including the n-th generation. The total progeny is Y_n for $n \to \infty$. We will denote the generating function of Y_n by $R_n(s)$. Since $Y_1 = 1 + Z_1$, we have $R_1(s) = s \cdot P(s)$. As in the case of $P_n(s)$, we can write down a recurrence for $R_n(s)$ by adding the root.

$$R_n(s) = s \cdot P(R_{n-1}(s)). \tag{3.13}$$

Notice that for $s < 1$,

$$R_2(s) = sP(R_1(s)) < sP(s) = R_1(s)$$

because of the monotonic nature of the generating functions P and R_n. Inductively, we can argue that $R_n(s) < R_{n-1}(s)$ and so $R_n(s)$ must decrease monotonically to a limit $t(s)$ satisfying equation 3.13, namely,

$$t(s) = s \cdot P(t(s))$$

It can be shown that $t(s)$ is unique and $t(1) = 1$ when the branching process terminates; in fact it is equal to the extinction probability.

By differentiating the previous equation, we obtain, $t'(1) = \frac{1}{1-\mu}$.

Example 9.1 (Min cuts) A *cut* of a given (connected) graph $G = (V, E)$ is set of edges which when removed disconnects the graph. An $s - t$ cut must have the property that the designated vertices s and t should be in separate components. A *mincut* is the minimum number of edges that disconnects a graph and is sometimes referred to as *global* mincut to distinguish is from $s - t$ mincut. The weighted version of the mincut problem is the natural analogue when the edges have non-negative associated weights. A cut can also be represented by a set of vertices S where the cut-edges are the edges connecting S and $V - S$.

It was believed for a long time that the mincut is a harder problem to solve than the $s - t$ mincut - in fact the earlier algorithms for mincuts determined the $s - t$ mincuts for all pairs $s, t \in V$. The $s - t$ mincut can be determined from the $s - t$ maxflow flow algorithms and over the years, there have been improved reductions of the global mincut problem to the $s - t$ flow problem, such that it can now be solved in one computation of $s - t$ flow.

In a remarkable departure from this line of work, first Karger[19], followed by Karger and Stein [20] developed faster algorithms (than maxflow) to compute the mincut with *high probability*. The algorithms produce a cut that is very likely the mincut, i.e., these are Monte Carlo algorithms. Unfortunately, there is yet no known matching verification algorithms. We will describe an algorithm that runs in time $O(n^2 polylog(n)$, $(n = |V|)$ which is nearly best possible for dense graphs[4] This algorithm exploits some properties of branching processes.

The contraction algorithm. The basis of the algorithm is the procedure contraction described below. The fundamental operation $contract(v_1, v_2)$ replaces vertices v_1 and v_2 by a new vertex v and assigns the set of edges incident on v by the union of the edges incident on v_1 and v_2. We do not merge edges from v_1 and v_2 with the same end-point but retain them as multiple edges. Notice that by definition, the edges between v_1 and v_2 disappear.

Procedure Contraction(t)
Input: A multigraph $G = (V, E)$
Output: A t partition of V

Repeat until t vertices remain

       choose an edge (v_1, v_2) at random
       contract(v_1, v_2)

Procedure $Contraction(2)$ produces a cut. Using the observation that, in an n-vertex graph with a mincut value k, the minimum degree of a vertex is k, the following can be shown quite easily.

Lemma 3 *The probability that a specific mincut C survives at the end of Contraction(t) is at least $\frac{t(t-1)}{n(n-1)}$.*

Therefore $Contraction(2)$ produces a mincut with probability $\Omega(\frac{1}{n^2})$.

Lemma 4 *A single iteration of the Procedure Contraction can be carried out in $O(n)$ steps.*

This is done by using an adjacency graph representation (see Karger [19] for details). Therefore using the Procedure Contract to produce mincut is somewhat expensive since we need to repeat it about n^2 times. Instead, we run Procedure $Contraction(\sqrt{n}/2)$ twice independently and repeat it recursively on the contracted graphs. The Algorithm is described below.

Algorithm Fastmincut
Input: A multigraph $G = (V, E)$
Output: A cut C

1. Let $n := |V|$.
2. If $n \le 6$ then compute mincut of G directly else
    2.1 $t := \lceil 1 + n/\sqrt{2} \rceil$.
    2.2 Call $Contraction(t)$ twice (independently) to produce to graphs H_1 and H_2.
    2.3 Let $C_1 = $ **Fastmincut** (H_1) and $C_2 = $ **Fastmincut** (H_2).
    2.4 $C = \min\{C_1, C_2\}$

The running time of algorithm **Fastmincut** satisfies the following recurrence

$$T(n) = 2T\left(\lceil 1 + n/\sqrt{2} \rceil\right) + O(n^2)$$

which yields $T(n) = O(n^2 \log n)$. Perhaps a more interesting question is to ascertain the probability with which **Fastmincut** returns a mincut. The probability that a mincut survives in H_1 after Step 2.2 is

$$\frac{(\lceil 1 + n/\sqrt{2} \rceil)(\lceil 1 + n/\sqrt{2} \rceil - 1)}{n(n-1)} \ge \frac{1}{2}$$

from Lemma 3. The same argument applies to H_2 independently. Therefore we can view the recursive algorithm as a branching process where any node can have zero, one or two children depending on if the mincut survived in zero, one or both children. The distribution function at each node can be approximated by a binomial distribution with two trials, each with success probability greater than $1/2$ (i.e. mean $\mu \geq 1$). Since the algorithm has roughly $2 \log n$ levels of recursion, we can restate the survival probability of the mincut as the complement of the extinction probability at the $2 \log n$ generation.

From Theorem 18, the extinction probability of a branching process with mean $\mu > 1$ converges to the solution of $x = P(x)$ where P is the generating function of the probability distribution. Here, we can approximate $P(s)$ by $\frac{1}{4} + \frac{1}{2}s^2 + \frac{1}{4}s^2$. Solving for x yields $x = 1$ which is an asymptotic solution but does not give us much information about the rate of convergence. For this, we need to solve the recurrence

$$x_n = P(x_{n-1})$$

where x_i is the extinction probability of the i-th generation. Substituting our generating function and simplifying yields

$$x_n = \frac{1}{4}(1 + x_{n-1})^2$$

The solution to this recurrence is $x_n = \Theta(\frac{1}{n})$. So the survival probability of the mincut (after $2 \log n$) levels of recursion is $\Omega(\frac{1}{\log n})$. Repeating the procedure (with independently chosen random bits) $m \log n$ times increases the probability of finding the mincut to $1 - \exp^{-m}$.

Example 9.2 (Fractional Cascading)

The problem of searching for a key in many ordered lists arises very frequently in computational geometry (see [5] for applications). Chazelle and Guibas [4] introduced fractional cascading as a general technique for solving this problem. Their work unified some earlier work in this area and gave a general strategy for improving upon the naive method of doing independent searches for the same key in separate lists. In brief, they devised a data-structure that would enable searching for the same key in n lists in time $O(\log M + n)$ where M is the size of the longest list. If N is the total size of all the lists then this data structure can be built in $O(N)$ preprocessing time and take $O(N)$ space.

Here, we give an alternate implementation of their data-structure that uses randomization. While retaining the salient features of their data-structure, we are able to simplify its construction considerably to an extent that is practical. The motivation of the new technique has been derived from the success of skip-lists (Pugh [33]).

Description of Fractional Cascading. In this section, we give a brief description of the problem setting and the approach taken by

Chazelle and Guibas. Consider a fixed graph $G = (V, E)$ of $|V| = n$ vertices and $|E| = m$ edges. The graph G is undirected and connected and does not contain multiple edges. Each vertex v has a catalog C_v and associated with each edge e of G is a range R_e.

A catalog is an ordered collection of records where each record has an associated value in the set $\Re \cup \{\infty, -\infty\}$. The records are stored in a non-decreasing order of their values and more than one record can have the same value. A catalog is never empty (has at least a ∞ and a $-\infty$). A range is an interval of the form $[x, y], [-\infty, y], [x, \infty], [-\infty, \infty]$. The graph G together with the associated catalogs and ranges is called a catalog graph. This is the basic structure to which fractional cascading is applied.

For notational purposes, if the value k is an end-point of the range $R_{u,v}$, then k appears as the value of some record in both C_u and C_v. Moreover, if two ranges $R_{u,v}$ and $R_{v,w}$ have an end-point in common it appears twice in C_v. The space required to store a catalog graph is $N = \sum_{v \in V} |C_v|$. This includes the space to store the graph itself. A catalog graph G is said to be *locally bounded* by degree d if for each vertex v and each value of $x \in \Re$ the number of edges incident on v whose range includes x is bounded by d.

The input to a query is an arbitrary element k in the universe and a connected subtree $\Pi = (\bar{V}, \bar{E})$ such that $k \in R_e$ for all edges e in the subtree and $\bar{V} \subset V, \bar{E} \subset E$. The output of the query for each vertex $v \in \bar{V}$ is an element y such that $predecessor(y) < k \le y$. We shall refer to such a pair of elements as *straddling pair* of k.

Theorem 21 (Chazelle-Guibas) *Let G be a catalog graph of size N and locally bounded degree d. In $O(N)$ space and $O(dN)$ time it is possible to construct a data-structure which allows multiple look-ups (query) in a subtree of size p in in time $O(pd + \log N)$. If d is fixed this is optimal. In addition, if the underlying catalog graph G is restructured to a constant degree graph, then the search time and the preprocessing time can be improved to $O(p \log d + \log N)$ and $O(N)$ respectively.*

Anatomy of the Data Structure. Our data-structure is very similar to [4] in the sense that we retain their idea of using augmented catalogs A_v for every vertex v such that $C_v \subset A_v$. But we shall use a different method for its construction. An augmented catalog A_v is also a linear list of records whose values form a sorted multiset. Augmented catalogs in neighboring nodes of G will contain a number of records with common values. The corresponding records are linked together to correlate locations in the two catalogs. The objective is that given the location of a record in A_v, we would be able to find its location (the straddling pair) in the augmented catalog of a neighbor of v in constant additional time.

More formally, for each node u and an edge e connecting u and v in G, we maintain a list of 'bridges' from u to v, B_{uv}, which is an ordered subset of records in A_v and lying in the range R_e. The end-points of R_e are the first and last records of B_{uv}. In node v, we maintain for each bridge in B_{uv} a companion bridge in B_{vu}. The value of a record is distinct from the record. Moreover each bridge is associated with a unique edge of G, implying that if a given value in A_u is chosen to be a bridge in both B_{uv} and B_{uw}, then it is duplicated and stored in different records of A_u.

A pair of consecutive bridges associated with the same edge $e = (u, v)$ defines a *gap*. Let a_u and b_u be two consecutive bridges in B_{uv} and a_v (respectively b_v) be the companion bridges in B_{vu}. If **value**$(a_u) <$**value**(b_u), then the gap of b_u includes all elements of A_u positioned strictly between a_u and b_u and all elements of A_v between a_v and b_v (the bridges are not included). By definition, gap of b_v is the same as gap of b_u. One of the key strategy used by [4] is to maintain the invariant that no gap exceeds $6d$ -1 in size.

We now take a closer look at the information maintained with each record. Both C_v and A_v are maintained as linked lists. A record of C_v have the fields *key* and *up-pointer*. The key contains the value and the up-pointer is a pointer to the next record. A_v has several other fields :

(1) key: stores the value k of record r.
(2) C-pointer: holds a pointer $\nu(r)$, the successor of r in C_v.
(3) up-pointer, down-pointer : pointers to successor and predecessor in A_v.

In addition a bridge element also has pointers to its companion bridge and also the label of the edge for which it is a bridge (If r is a bridge in B_{uv} then it stores the label uv).

To answer a multiple look-up query (x, Π) where x is the key value and Π is the subtree, one begins by locating x in the first node of the path Π and then use the following properties:

Lemma 5 (CG) : *If we know the position of value x in A_v, we can compute the position of x in C_v in one step.*

This can be done by using the C-field.

Lemma 6 : *If we know the position of value x in A_v and $e = (v, w)$ is an edge of G such that $x \in R_e$, then we can compute the position of x in A_w in $O(|Gap_e(x)|)$ time. $Gap_e(x)$ is the set of elements in the gap (corresponding to edge e) that x belongs to.*

From the position of x in A_v follow up-pointers until a bridge is found which connects to A_w. Note that Chazelle-Guibas [4] maintained the invariant that all gap sizes are less than $6d$ which yields a search time of $O(\log N + d|\Pi|)$. This invariant was maintained during the construction of all the augmented catalogs which is done incrementally. Their algorithms start with empty catalog and then for each vertex v, the records of C_v are inserted in an increasing order into A_v. Between any two insertions, the gap invariants are restored. Note that a single insertion into A_v can alter the gaps leading to insertion of a new bridge which introduces new records and this could continue as a long chain of events.

We suggest the following modification. Instead of explicitly maintaining gap invariants, we choose a newly inserted element r in A_v to be a bridge with probability p (we shall determine the exact value later). This is repeated for every edge incident on v and whose ranges cover r. If r is chosen to be a bridge for an edge (u, w), it leads to the insertion of a a new record in A_w and (possibly even in A_v if r is already a bridge). These new records are treated exactly the same way as described above (i.e. choose it with certain probability to be a bridge element). For each new record in the augmented catalog, we initialize the following fields:

(i) C-pointer: Can be determined from the predecessor and the successor elements in the augmented catalog. If this element came from C_v, then update the C-pointers for all the elements in A_v between this record and the previous element of C_v.

(ii) Initialize the edge field.

For each new record of C_v, this process is continued until there are no more bridges to insert.

Analysis. Let us first analyze the running time for a multiple look-up query. Given the position of x in A_v we follow the up-pointers until we find a bridge b_v which connects to A_u and then traverse the down-pointers until we locate the straddling pair. Since every element is chosen to be bridge element with probability p, the expected length of a gap is $2/p$. Thus from Lemma 1, the expected search time is $O(\log N + |\Pi|/p)$.

Moreover, if $|\Pi| \geq \log N$, we can show that the search time is $\tilde{O}(\log N + |\Pi|/p)$ using the observation that the search time is a sum of $O(\log N)$ independent random variables with a geometric distribution and parameter p (Sen [41]). There is however an oversight in the above reasoning since the query is a tree of size $|\Pi|$ which can assume various forms. In particular, Π can be any of the S positional trees of maximum degree d. It can be shown that $|S| < 2^{O(d|\Pi|)}$ so that if d is constant then the search time holds with high likelihood. Notice that there are $O(N)$ choices for the root of this tree and $O(N)$ combinatorially distinct search paths given the search tree (corresponding to the $O(N)$ intervals induced by all the key values).

The more interesting aspect of the analysis is to bound the time and space during the data-structure construction. If we look more closely at the way the records are added to the augmented catalogs, the underlying stochastic process can be modeled by a branching process. The root corresponds to an inserted record from C_v and the number of children correspond to the bridges that are created by this record. Each bridge is created with probability p, which corresponds to two children. For each new record inserted from the C_v's the time and the space needed is proportional to the total progeny of this branching process. Each node can have upto 2d children where the number of children is a random variable which takes values 0, 2, 4 ..2d. The probability that this random variable takes value $2k$ is the same as the probability that a binomial random variable with parameters (d, p) takes value k (i.e. there are k bridges created). The mean μ of this distribution is clearly $2pd$ and the generating function $G(s)$ can be worked out as $(q + ps^2)^d$. Here $q = 1 - p$.

From Theorem 18, a branching process is finite if $\mu < 1$. Hence we choose p such that $2pd < 1$, that is $p < 1/2d$. This gives an expected gap length of greater than $4d$. From Theorem 20, if the generating function for the total progeny is denoted by $t(s)$, then $t = sG(t)$. In our case $G(t) = (q + pt^2)^d$. Moreover, the mean total progeny is $\frac{1}{1-\mu}$, which is $\frac{1}{1-2pd}$ in our case. If we choose $p = 1/3d$, this yields a mean progeny of 3. This in turn implies a total expected space bound of $O(\sum_{v \in V} C_v)$ which is $O(N)$.

We can get stronger bounds by estimating the probability of deviating from the mean value. The usual procedure is to use Chernoff bounds but in our case it is complicated by the fact that we cannot get an explicit generating function for $d > 2$ (since it involves solving equations of high degree). Instead we take an indirect approach. The total space and time for data-structure construction is the sum of N independent and identical random variables $Y_i, 1 \leq i \leq N$, each of which is the total progeny of a branching process. If $A = \sum_i X_i$, then from Chernoff Bounds,

$$\Pr[A \geq X] \leq s^{-X} t(s)^N, \text{ for } s \geq 1$$

where $t(s)$ is the generating function for each Y_i. For $X > cN$, for some fixed c, this can be rewritten as

$$\Pr[A \geq cN] \leq \left(\left(\frac{t(s)}{s^c} \right) \right)^N$$

We shall prove that for some $s > 1$, there exists a constant c such that $t(s)/s^c < 1$. Let $F(s, t) = t - s(q + pt^2)^d$. Then

$$F_t(s, t) = 1 - 2tpds(q + pt^2)^{d-1}$$

We have used the notation f_y to denote the partial derivative of f with respect to variable y. Hence $F_t(1,1) = 1 - 2pd$ and for $2pd < 1$, $F_t(1,1) \neq 0$. For completeness, we state the following theorem.

Theorem 22 (Implicit Function Theorem) *Let $f(x,y)$ be continuously differentiable in D. Let (x_o, y_o) be any point in D such that $f_y(x_o, y_o) \neq 0$. Then there exist numbers $\delta > 0$ and $\epsilon > 0$ and a continuously differentiable function $g(x)$ defined for $|x - x_o| \leq \delta$ and $|y - y_o| \leq \epsilon$, then $f(x,y) = f(x_o, y_o) \Leftrightarrow y = g(x)$.*

From *Implicit Function theorem*, it follows that there exists a neighborhood of $(s = 1, t = 1)$, such that

$$F(s,t) = F(1,1) \Leftrightarrow t = t(s).$$

Since $F(1,1) = 0$, i.e., $t(s) = sG(t(s))$ for $t = 1, s = 1$, there is a value $s > 1 + \epsilon$ for which $t(s) < 1 + \delta$ for some $\epsilon, \delta > 0$. By choosing c large enough, $t(s)/s^c$ can be made less than 1 and hence the probability of deviation from mean decreases as $1/2^{O(N)}$.

For each new record in A_v, we need $O(d)$ time to determine if it will be a bridge with respect to any of the d (maximum) neighbors. Moreover inserting a new bridge takes time proportional to gap-size whose expected value is also $O(d)$. To complete the analysis for the time bound for building the data-structure, we have to ensure that the total number of C-pointer updates is also $O(N)$. For any new record inserted into A_v from C_v, the total number of C-pointer updates can be bound by the total number of records (i.e. the space bound). The records of C_v are inserted in an increasing order and so any record in A_v has its C-pointer updated at most once (not including when it is first created). Hence we state the following result

Theorem 23 *Let G be a catalog graph of size N and locally bounded degree d. Our algorithm constructs a data-structure for iterative search in $O(N)$ space and $O(dN)$ time to do iterative search in time $\tilde{O}(\log N + d|\Pi|)$. The bounds for preprocessing time and space hold with probability $1 - 2^{-O(N)}$.*

Remarks

1. The bounds for preprocessing time and search time can be improved to $O(N)$ and $\tilde{O}(\log N + \log d|\Pi|)$ respectively by using the same modifications as Chazelle-Guibas to restructure the catalog graph to a fixed degree graph.

2. The constant associated with the asymptotic bounds for preprocessing time is lower than the deterministic construction.

3. Chazelle-Guibas also arrived at the figure $4d$ for minimum gap size from the observation that otherwise their analysis yields infinite time and space bound for construction. However, they give examples where the actual algorithm halts even when they use gap sizes of less than $4d$. Our analysis captures a more fundamental reason for this phenomenon. Although the mean $\mu < 1$, guarantees that the process is finite, the process dies out with probability x where $x = G(x)$ even when $\mu > 1$. So one can have a gap length of less than $4d$ and still terminate. The motivation for this is clearly a reduction in search time which is inversely proportional to the gap size.

4. To allow insertions/deletions from the catalogs, our procedure for maintaining the augmented catalogs readily dynamize. The arguments for query-time and the space-bound remain identical to the static case. Unfortunately (as in the case of [4]), the bottleneck is maintaining the correspondence between the A_v and C_v. In particular, we are unable to analyze the number of C-pointer updates in the case of inserting or deleting a record from C_v. Using the priority queue of Fries et al. [14] to maintain this correspondence, we get similar bounds. Both the search time and update times are off by an $O(\log \log N)$ factor from the best possible. For the special case of keys chosen randomly, this method leads to optimal dynamic algorithm (Sen [42]).

10.   RANDOM SAMPLING AND POLLING IN COMPUTATIONAL GEOMETRY

Divide-and-conquer is certainly the most commonly used technique for designing parallel algorithms. The idea is analogous to sequential algorithm design where the original problem is sub-divided into smaller subproblems and then the solutions of the subproblems are combined to obtain a solution to the original problem. The smaller subproblems are solved recursively until a sub-problem size becomes smaller than a predetermined threshold. At this stage a direct (usually brute-force) method is used to solve it. For analyzing this procedure it usually suffices to write down the recurrence equation for the time complexity as:

$$T(n) = \begin{cases} \max_i T(n_i) + g(n) & \text{if } n_i > K \\ F(n) & \text{otherwise} \end{cases}$$

where K is a predetermined threshold, n_i is the size of the ith subproblem, F is the complexity of a direct algorithm and $g(n)$ is the cost of dividing the problem and recombining the solutions.

For a number of problems, $\sum_i n_i = n$ and hence it is the size of the largest subproblem (which is of size less than n) that is crucial for determining the running time of the algorithm. The equations for the processor and the space bounds can be written similarly, and the processor complexity is the maximum number of processors used at any step of the algorithm. Since there is a trade-off between the number of processors used and the running time, sometimes it becomes necessary to write a recurrence using two variables namely, the problem size and the number of processors used.

A generalization of the above procedure is to allow for *expected* bounds where it is possible to write down the recurrence equation for expected bound with respect to a specific resource. In the above equation for time bound, we can associate a distribution with the size of the largest subproblem, and the solution would be the expected running time of the algorithm. The exact distribution is often hard to ascertain and in most cases only the expectation is known. For some special forms, one can solve these *probabilistic recurrence relations* satisfactorily as discussed in section 2.

In the context of computational geometry, sorting can be looked upon as a one-dimensional problem. The basic tools that we will develop in this section will enable us to extend some of the techniques to higher-dimensional problems. In some sense, this exercise can be viewed as (although readers are cautioned against over-simplified conclusions) expanding the basic paradigm of quicksort-like algorithms. For the most of this section, the techniques discussed are very general without reference to any particular problem. We shall present some interesting applications in the next section to specific problems where algorithms will be presented more formally.

10.1   RANDOM SAMPLING

The divide-and-conquer mechanism that we will use is based on partitioning the problem using a random subset of the input. Recall that in quicksort, the input was partitioned by splitters that were randomly chosen elements of the input. For the higher-dimensional problems, it is not obvious how these splitters partition the input in the absence of a linear ordering.

Example 1 : For concreteness, consider the problem of constructing the *trapezoidal map* of a given set N of line segments. These segments partition the plane into regions which may have complicated shapes. By passing vertical lines through every end-point and intersection, these regions get partitioned into trapezoids or triangles (degenerate trape-

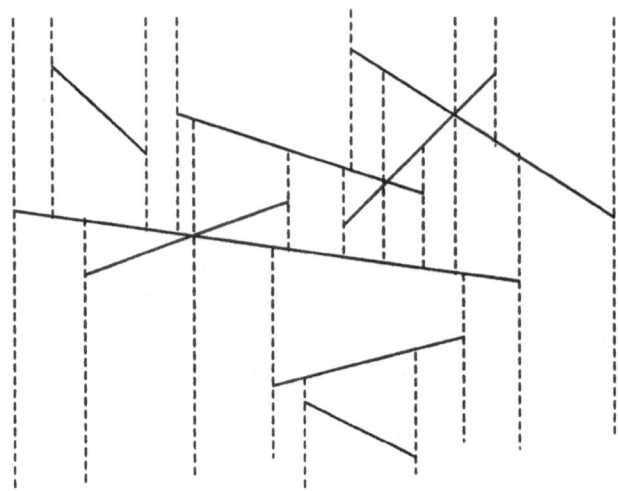

Figure 3.1 Trapezoidal map of a set of line segments

zoids). The vertical lines are not allowed to cross line segments. See Figure 3.1 for an illustration.

We will denote the trapezoidal map of a set S by $\mathcal{T}(S)$ and the set of trapezoids in $\mathcal{T}(S)$ as $H(S)$. By choosing a random subset of line segments $R \subset N$, it is not clear how to partition N as there is no natural ordering between segments (like points in a line). Since it is a two-dimensional problem, a natural solution is to consider the two-dimensional partitions induced by R. For example, consider $\mathcal{T}(R)$. For any trapezoid $\triangle \in H(R)$, consider $\mathcal{T}(N)$ restricted within $\triangle$. The union of $\triangle \cap \mathcal{T}(N)$ for all $\triangle \in \mathcal{T}(R)$ contains all the relevant information about $\mathcal{T}(N)$. Thus we can recursively compute construct $\mathcal{T}(N)$ within each $\triangle \in H(R)$. We will denote the line segments intersecting a trapezoid $\triangle \in H(R)$ by $L(\triangle)$ so that recursively we compute $\mathcal{T}(L(\triangle))$. From our earlier discussion, a bound on $\max\{L(\triangle\}$ will be crucial for the running time of the algorithm. In addition, the quantity $\sum_{\triangle} L(\triangle)$ is also important as a segment $s \in N$ can intersect many $\triangle$s implying that this quantity could exceed n. So it represents the *blow-up* in the overall problem size in a recursive call and would affect the overall efficiency of any recursive algorithm.

Remark 3 *In the case of quicksort, we did not have to worry about it since an element would belong to exactly one interval.*

Example 2: Consider the problem of computing the intersection of a set N of half-spaces in three dimensions. If we adopt the previous approach, we choose a random sample R of half-spaces and construct the intersection $\cap R$. For convenience, we assume it is non-empty and contains the origin in its interior. Unlike the previous case, we do not have regions analogous to the trapezoids. With some thought, it is not difficult to come up with sub-divisions analogous to trapezoids. For example, the pyramids formed by joining the origin to every vertex of the intersection. For technical reasons that will become clear later, we will like to have these regions, which we shall call *ranges* defined by a constant number of input objects or equivalently having constant size. A pyramid defined as above can have a fairly large base if the corresponding face (of $\cap R$ is large). By further subdividing the bases using parallel translates of a fixed plane, we can restrict the pyramid bases to be trapezoids (or triangles), so that these have constant size. The intersection $\cap N$ can be constructed recursively within these pyramids. In this context, $L(\triangle)$ denotes the set of half-spaces whose defining half-planes intersect a pyramid $\triangle$.

The previous examples would have given the reader some intuition about how random sampling gives rise to a natural class of divide-and-conquer algorithms in computational geometry. It is not difficult to come up with a suitable definition of *range* in the context of a given problem. To prove any interesting results about these algorithms, we have to bound the quantities $\max\{L(\triangle)\}$ and $\sum_\triangle L(\triangle)$. The former is crucial to bound the depth of recursion, which determines parallel running time and the latter will determine the efficiency of this approach. We shall prove some useful bounds for these quantities for fairly general situations. From here, we will use $\triangle$ to denote a range in an abstract setting where a range is a subset of the Euclidean space E^d defined appropriately in the context of the problem in E^d. We will require that a range be defined by at most a constant number of input objects, say b which bounds the possible number of ranges by a polynomial in n, namely $O(n^b)$. In the case of trapezoidal maps, the reader can verify that b is 4. We will use $l(\triangle)$ to denote $|L(\triangle)|$ which will be referred to as the *conflict size* of $\triangle$. Note that we are interested in those ranges $\triangle$ where $\triangle$ is a range in $\mathcal{T}(R)$. The following result gives a bound on $l(\triangle)$ for a random sample R.

Lemma 7 ([7, 17]) *For some suitable constant k_1 and large n,*

$$\Pr[\max_{\triangle \in H(R)} l(\triangle) \geq k_1(n/r)\log r] \leq 1/4$$

where the probability is taken over all possible choices of the random sample R such that $|R| = r$.

84

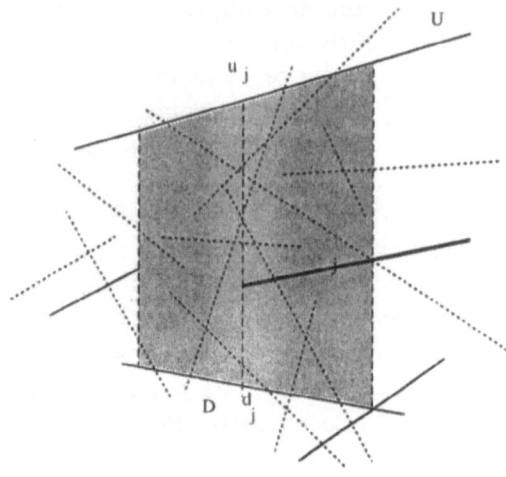

t]

Figure 3.2 Segments intersecting Δ_j

Remark 4 *The lemma uses sampling in a slightly different manner than what we had discussed previously. However, it can be seen easily that $|R|$ is $r + o(r)$ with high probability and the same bound would hold for a random sample of size r exactly. In the subsequent discussions, we will not distinguish between these sampling schemes.*

The next result bounds the expected value of $\sum_\Delta l(\Delta)$.

Lemma 8 ([7, 35]) *For some suitable constant k and large n,*

$$\Pr[\sum_{\Delta \in H(R)} l(\Delta) \geq kn] \leq 1/4$$

where the probability is taken over all possible choices of the random sample R.

Lemma 9 ([7, 31]) *For $|N| = n$ and a random sample R obtained by choosing every element of N independently with probability r/n,*

$$E[\sum_\Delta l^c(\Delta)] = O\left(\left(\frac{n}{r}\right)^c \cdot E[|H(R)|]\right)$$

where $E[|H(R)|]$ is the expected number of ranges in $\mathcal{T}(R)$ and c is a fixed positive integer (independent of n).

Reif and Sen[36] give an alternate proof. The above bound is useful in situations where the algorithm's performance can be expressed as $E[\sum_\Delta l^c(\Delta)]$.

10.2   CONVERTING EXPECTED BOUNDS TO HIGH-PROBABILITY

If the sampling lemma only guarantees a *good* sample with constant probability. The usual method used is that we repeat sampling till we obtain a good sample - the expected number is trials is $O(1)$ in our case. For this we must have a verification procedure $\mathcal{V}$. We can improve the probability of obtaining a good sample further by increasing the number of trials. That is, by choosing $p(n)$ independent sets of samples, we can be assured of a good sample with probability $1 - \frac{1}{2^{p(n)}}$ (at least one of the samples is good). The problem is clearly the extra cost of resampling, namely, we have to invoke the verification procedure $p(n)$ times. To avoid the extra cost, we can think about an *approximate verification* procedure that is much faster and gives us a very good estimate of how good a certain sample is. This is the basic philosophy of *polling*. Instead of testing the goodness of a sample with respect to the entire input (of size n), we actually 'poll' a small fraction of the input, which is typically about $\frac{1}{p(n)}$ - the exact value would depend on the desired success probability and the cost of *calV*. We will discuss some of the technical details in the context of random sampling in computational geometry.

As discussed in the beginning of this section, the probabilistic recurrence relation arising from a recursive randomized algorithm are often hard to solve without the knowledge of the tail distributions. The parallel algorithm's running time is the maximum of the resulting recursive calls unlike the sequential algorithm where it is a sum of the running times of the recursive procedures. The latter becomes easier to bound from the linearity of expectations. No such nice properties are known about the maximum of expected values and it does appear to depend on the tail estimates. This informal discussion is a motivation for the results obtained in this subsection; we do not claim that this is the only way to get around the problem of bounding maximum of expected values.

The naive random sampling gives us a high-probability bound on the maximum size of a subproblem (Lemma 7) and only an expected bound on the sum of subproblems (Lemma 9). We can restate these results in the following manner.

Lemma 10 *For suitable constants k_{total} and k_{max} the following conditions hold with probability at least 1/2 for a sample where each of the n input elements has been selected independently with probability r/n.*

(i) The maximum size of a subproblem is less than $k_{max} \frac{n}{r} \log n$
(ii) The sum of the subproblems is less than $k_{total} \frac{n}{r} \cdot E[|H(R)|]$.

Proof. From Lemma 8 and Markov's inequality we can choose k_{total} such that the probability that (ii) fails is at most 1/3 (i.e., sum of subproblems is thrice the expected value). Choose k_{max} such that failure probability in Lemma 7 is no more than $1/n$. For sufficiently large n, $1/n + 1/3$ is less than $1/2$. Thus the probability that both (i) and (ii) are satisfied is at least $1/2$. ∎

Remark 5 *We can make additional conditions like bounding the higher moments hold with probability at least 1/2 by the above argument. A sample that satisfies a required set of conditions will be called a good sample. In general we may assume that there are some (fixed) number of properties that a good sample must satisfy and a random sample is good with constant probability, say 1/2.*

As a consequence of the previous claim, if we repeat the sampling $p(n)$ times, the probability that the conditions are not satisfied for all samples is less than $2^{-p(n)}$. That is, if we choose independently $p(n) = O(\log n)$ sets of samples, one of them is good with very high likelihood. We assume that $p(n) = \Omega(\log n)$ henceforth. Let us assume that we have a verification procedure $\mathcal{V}$ that verifies if a sample is good using $v(n)$ operations for n input elements. Therefore, to determine if a sample is 'good', we have to run $\mathcal{V}$ $O(\log n)$ times for a total of $v(n) \log n$ operations. Depending on $p(n)$ and $v(n)$, this could make an algorithm inefficient. To make resampling more efficient, we can run $\mathcal{V}$ using only a fraction of the input for each of the samples. For example, we can estimate the size of a subproblem by looking at a fraction of the input.

For exposition, we describe it for the special case of trapezoidal maps - for other problems the same arguments apply with minimal modifications, by replacing trapezoids with appropriate ranges.

We choose $c_0 \cdot \frac{n}{p(n) \cdot f}$ input segments randomly from the n input segments for some constant c_0 and a parameter f. (the actual values will be determined from the required success probability and the efficiency of the verification algorithm). Let X_i^j be the number of segments intersecting trapezoid $\triangle_i$ corresponding to sample R_j, $1 \le j \le p(n)$. A_i^j be the number of segments intersecting $\triangle_i$ out of the $c_0 \cdot \frac{n}{p(n) \cdot f}$ randomly chosen input segments for the same sample. Clearly, A_i^j is a binomial random variable with parameters $c_0 \cdot \frac{n}{p(n) \cdot f}$ (number of trials) X_i^j/n (probability of success).

Assuming that X_i^j is greater than $\bar{c} \cdot p(n) \cdot p(n) \cdot f$, for some constant $\bar{c}$, we can apply Chernoff bounds to estimate X_i^j within a constant multiplicative factor with probability exceeding $1 - \frac{1}{2^{p(n)}}$ (recall $p(n) = \Omega(\log n)$). Since we do it only for $\frac{1}{p(n)f}$ of the input segments, the total number of operations for the $p(n)$ random subsets can be bounded by $O(v(n/(p(n) \cdot f)) \cdot p(n))$. In most cases, we want to ensure that this quantity is $O(v(n))$ and accordingly choose f.

When $X_i^j < \bar{c} \, p(n) \cdot p(n) \cdot f$, the estimates are not as accurate and we must ensure that this does not violate the desired properties of the sample. In the case of line segment intersections and the two properties of Lemma 10 , we must ensure that

$\bar{c} \, p(n) \cdot p(n) \cdot f \le maxsize$, and
$\bar{c} \, p(n) \cdot p(n) \cdot f |H(R)| \le maxsum$.

More formally, for the case $X_i^j < \bar{c} \, p(n) \cdot p(n) \cdot f$ (by invoking Chernoff bounds), for any $\alpha > 0$ (α is a function of c_0), there exists a c_1, independent of n,

$$\Pr[A_i^j \le \alpha c_1 X_i^j / p(n) f)] \le \frac{1}{2^{p(n)}}$$

and

$$\Pr[A_i^j \ge c_2 \alpha c_0 \cdot X_i^j / (p(n) f] < \frac{1}{2^{c_0 p(n)}} < \frac{1}{2^{p(n)}} \text{ for } c_0 > 1).$$

From the last two inequalities, X_i^j is bounded by $L^j = A_i^j p(n) \cdot f / c_0 c_2 \alpha$ from below, and by $U^j = A_i^j p(n) \cdot f / c_1 \alpha$ from above. With appropriate choice of the constants, this condition holds with the desired probability (as defined in section 2.1) for all X_i^j simultaneously. In the context of trapezoidal map, we do the following procedure simultaneously for all the samples R_j and choose the sample R^{j_0} using the following simple test:

Procedure Polling

- *Input:* Samples $R_1 \ldots R_m$ where $m = O(p(n))$.

- *Output:* A good sample R^{j_o}.

- *Notation:* Let $N^j = \sum A_i^j$ and the let actual number of intersections be denoted by T^j and the upper and lower bounds obtained from N^j by U^j and L^j respectively. We will use S to denote $k_{total} n / r E[|H(R)|]$.

> **If $S > U^j$ (clearly good) then accept sample R^j** (since $S \geq U^j \geq T^j$),
>
> **if $S \leq L^j$ then the sample is 'bad'** (since $S \leq L^j \leq T^j$),
>
> **if $L^j \leq S \leq U^j$, (choose the best) then accept the sample R^{j_o} for which N^{j_o} is minimum.** Since both S and T^{j_o} lie in this interval this guarantees that $T^{j_o} \leq c_3 \cdot S$ where $c_3 = U^j / L^j$ which is a constant.

Recall, that from our earlier discussion at least one of the samples would satisfy conditions 1 or 3 with probability $1 - O(\frac{1}{2^{p(n)}})$. Conditions 1 and 3 will fail if all the $p(n)$ are bad or if our estimates are inaccurate and the probability of either of the events is bounded by $\frac{1}{2^{p(n)}}$. We summarize as follows :

Lemma 11 (Polling lemma) *Using procedure* **Polling***, we can obtain a sample that is 'good' with probability $1 - O(\frac{1}{2^{p(n)}})$ where a naive random sample is known to be 'good' with probability 1/2. Given a verification procedure $\mathcal{V}$ that runs in $O(v(n))$ operations for n elements, this procedure uses $O(v(n/p(n))p(n))$ operations.*

For trapezoidal map we can obtain following kinds of bounds

- With probability $1 - \frac{1}{n}$, the maximum size of the subproblem is less than $O(n/r polylog(n))$. Choose $p(n) = 2 \log n$.

- With probability $1 - \frac{1}{2^{n^\delta}}$, $(\delta < 1)$ the maximum subproblem size is less than $O(n^{1+\epsilon}/r)$. Choose $p(n) = n^\delta$.

The sum of the subproblem sizes is $O(n)$ with probabilities $1 - \frac{1}{n}$ and $1 - \frac{1}{2^{n^\delta}}$ respectively.

In the following two sections we illustrate the use of Polling on two algorithms - a sequential iterative algorithm, and a parallel divide-and-conquer algorithm. We will obtain success probability $1 - 1/n$, although stronger bounds can be obtained as shown here.

11. AN OUTPUT SENSITIVE 3-D CONVEX-HULL ALGORITHM

The first known optimal $O(n \log h)$ time output-sensitive algorithm for three dimensional convex hull, where n and h are input and output sizes respectively was a randomized algorithm described in Clarkson and Shor [7]. The expected runing time is over the choice of random bits in the algorithm but did not have any associated tail estimates. Below we show that the use of Polling yields high-probability bounds for practically the same algorithm.

11.1 BRIEF OVERVIEW OF THE ALGORITHM FOR CONVEX HULLS

The problem of constructing the convex hull of points in three dimensions is well known to be equivalent to the problem of finding the *intersection of half-spaces*. Here we give an algorithm for the latter which implies a solution for the former.

Let us denote the input set of half-spaces by S and their intersection by $P(S)$. We construct the intersection $P(R)$ of a random sample R of r half-spaces and filter out the redundant half-spaces i.e. the half-spaces which do not contribute to $P(S)$. Without loss of generality, we can assume that the origin lies inside the intersection. Take an arbitrary (fixed) plane T and partition each face of $P(R)$ into trapezoids using the translates of T that pass through the vertices of the face. We further partition trapezoids into triangles. The convex closure of the origin O with a triangle from the cutting of the faces defines a region which we call the *cones*. These cones will be intersected by bounding planes of a number of half-spaces that were not chosen in the sample. We say that a half-space *intersects* a cone if its bounding plane intersects the cone. We say that a half-space *conflicts* with a cone if its bounding plane intersects the cone. A cone is said to be *critical* if it contains an output point. We delete the half-spaces which do not intersect with any critical cone. The procedure is repeated on the reduced problem.

To prove any interesting result we must determine how quickly the problem size decreases. The random sampling lemmata in the next section show that for a large sample ($> \Omega(h^2)$) the size of the problem decreases very quickly.

11.2   RANDOM SAMPLING PROPERTIES

Let $H(R)$ denote the set of cones induced by a sample R and let $H^*(R)$ denote the set of critical cones. We will denote the set of half-spaces intersecting a cone $\triangle \in H(R)$ by $L(\triangle)$ and its cardinality $|L(\triangle)|$ by $l(\triangle)$. $L(\triangle)$ will also be referred to as the *conflict list* of $\triangle$ and $l(\triangle)$, its *conflict size*. We will use Lemma 8 and Lemma 7 to bound the size of the reduced problem.

A sample is 'good' if it satisfies the properties of Lemma 8 and Lemma 7 simultaneously. Clearly, a sample is 'good' with probability at least $1/16$. Using the method of Polling, we can do the following.

Lemma 12 *We can find a sample R which satisfies both Lemma 8 and Lemma 7 simultaneously with high probability. Moreover this can be done in $O(\log r)$ time and $O(n \log r)$ work with high probability.*

Since $|H^*(R)| \leq h$, a 'good' sample clearly satisfies the following property also

Lemma 13 *For a 'good' sample R,*

$$\sum_{\triangle \in H^*(R)} l(\triangle) = O(nh \log r/r)$$

where $|R| = r$ and $H^*(R)$ is the set of all cones that contain at least one output point.

This will be used repeatedly in the analysis to estimate the non-redundant half-spaces whenever $h \leq r/\log r$.

11.3   ALGORITHM

We give below an algorithm for convex hull and in the following section we analyse each step in details.

Let S be the input set of n half-spaces. The algorithm is iterative. Let n_i (respectively r_i) denote the size of the problem (respectively the sample size) at the ith iteration with $n_1 = n$. A typical iteration of our algorithm is shown in Figure 3.3. Repeat the procedure until $r_i > n^\epsilon$ (this condition guarantees that the sample size is never too big) or $n_i < n^\epsilon$ for some fixed ϵ between 0 and 1. If $n_i < n^\epsilon$ then solve the problem directly else do one more iteration and solve the problem directly.

11.4   OVERVIEW OF ANALYSIS

Let l be the iteration when the sample size exceeds h^2 for the first time. The analysis is actually divided into two phases - work done before

Procedure Rand(i)

1. Use Resampling and Polling to choose a 'good' sample R of size $r_i =$ constant for $i = 1$ and $\max\{r_{i-1}^2, h_{i-1}^*\}$ for $i > 1$ where h_i^* is the maximum output size of a 2d problem (defined later in step 3).

2. Solve the problem for R.

3. Define regions on the basis of the solution obtained in Step 2 — explained in Section 11.1. Let $H(R)$ denote the set of regions induced by R. We say that an input half-space is *redundant* if it does not contribute to the output. Filter out the redundant input as follows.

   (a) For every half space, find out the regions that it intersects.

   (b) Call a region *critical* if it contains an output. Denote the set of critical regions by $H^*(R)$. Find out the set $H^*(R)$ - these are regions that contain only parts of edges/faces of $P(S)$. This can be verified by computing the (two dimensional) polygonal intersection of the faces of $\Delta \in H(R)$ with S. An output sensitive algorithm (like [22, 2]) must be used for this step. The two dimensional polygonal structure of the intersections on the faces of Δ can be used to verify whether the region Δ contains any output vertex. The maximum number of vertices of any polygon defines h_i^* (which is a lower bound on the output size of $P(S)$.

   (c) Delete a half-space if it does not belong to $\cup_{\Delta \in H^*(R)} L(\Delta)$.

4. The input for the next iteration is $\cup_{\Delta \in H^*(R)} L(\Delta)$.

   Size of the reduced problem for the next iteration is $n_{i+1} = |\cup_{\Delta \in H^*(R)} L(\Delta)|$.

   Increment i.

end.

Figure 3.3

the l-th iteration and the work done after iteration l. The work done in Step 1 is bounded by $O(n_i \log r_i)$ with high probability from the polling procedure. Step 3a can be done using an intersection detection for every plane followed by finding all the regions that a half space intersects. These take time $O(n_i \log r_i)$ and $\sum \triangle \in H(R)|L(\triangle)|$ respectively. The second term is $O(n_i)$ from Lemma 8. The work done in Step 3b can be bounded by $O(\sum_{\triangle \in H(R)} |L(\triangle)| \log h_i^*)$ which is $O(n_i \log h_i^*)$ from the property of a good sample (Lemma 12). By our convention, $\log(h_i^*)$ is $O(\log h)$ for $i \leq l$. The remaining steps can be done in $O(n_i)$ time.

The running time of the algorithm is

$$\sum_{i<l} n_i(\log r_i + h_i^*) + \sum_{i \geq l} n_i(\log r_i + h)$$

with high probability. The high probability follows from the use of polling.

The first term can be bounded by $\sum_{i<l} n \log r_i^*$ where $r_i^* \geq (r_{i-1}^*)^2$ and $r_i^* \leq h^2$. Thus this term can be bounded by $O(n \log h)$ with high probability (a geometric sum with leading term $O(n \log h)$).

For the second term, we notice that $r_i \geq h^2$, and so we can simplify it to $\sum_{i \geq l} n_i \log r_i$. Moreover, for $i \geq l$, from Lemma 13

$$n_i \log r_i = O(\frac{n_{i-1}}{r_{i-1}} \log^2 r_i) \leq n_{i-1}/2$$

for $r_i \geq C$ for some constant C. So this term can be bounded by $O(n)$ as n_is decrease geometrically (from Lemma 13).

Thus the entire algorithm terminates in $O(n \log h)$ steps with high probability. The high probability can be improved further to inverse exponential probability by using a stronger tail estimate for polling. The original version of this algorithm described by Clarkson and Shor [7] could only guarantee expected bounds since they only relied on straightforward sampling.

Theorem 24 *Algorithm Rand(i) computes the convex hull of n points in $O(n \log h)$ steps with probability $1 - \frac{1}{2^{n^\epsilon}}$ for some $\epsilon > 0$.*

12. RANDOMIZED DIVIDE-AND-CONQUER

The tools and techniques developed in the previous two sections will be used to design a very general scheme for parallel divide-and-conquer. Random sampling will be used to achieve fairly even partitioning of the problem and the analysis will be done using various properties of random sampling proved earlier. We will illustrate the general methodology using the problem of computing trapezoidal map of line segments that can intersect only in the end-points. The parallel model that will be used is CREW unless otherwise mentioned.

12.1 TRAPEZOIDAL MAP CONSTRUCTION

This problem is the same as defined in previous section except that the segments are non-intersecting except possibly in the end-points. For every end-point, we want to determine the segment lying immediately above and below. This problem is called the *vertical visibility map* and is particularly interesting because of its close connection to triangulation. The algorithm we will describe will construct the visibility map of a given set N of line segments. A very high-level description is as follows:

Algorithm Vertical Visibility

1. Select a 'good' sample R of size $O(n^\epsilon)$ ($\epsilon > 0$ is a constant that will be determined in the analysis). By 'good' sample, we imply that the conditions of Lemma 10 hold. We use the technique of polling to do this step efficiently.

2. Construct $\mathcal{T}(R)$ using a brute-force approach.

3. For each segment of N, determine the trapezoids of $\mathcal{T}(R)$ that it intersects by a procedure we will describe shortly. (The same procedure will be used as the verification algorithm for Polling in step 1).

4. For each trapezoid $\Delta \in \mathcal{T}(R)$, we apply a clean-up phase called *Filtering* to discard some of the segments in $L(\Delta)$. As we will show later, this phase is crucial for bounding the processor complexity.

5. If $l(\Delta) > C$, for some predefined threshold C, then call the algorithm recursively on $L(\Delta)$ else solve the problem directly (We assume that a suitable algorithm already exists - usually a brute-force method suffices).

We now look at the individual steps in some details. The procedures in steps 1-3 are actually quite related. For $\epsilon < 1/2$, the following brute-force method works. For every endpoint, we draw a vertical line and order the segments intersecting this line (sorting in the Y direction of the intersections suffices which can be done in $O(\log n)$ time using n processors for every end-point). From this information, we can determine the segment lying immediate above and below every end-point. From this, we can also compute easily for each segment all the end-points for which it is visible from above (and below). In fact we order the vertical projections of these end-points by sorting. Constructing individual trapezoids can be done by 'walking' around the vertices of a trapezoid using the successor information from the previous computation. So the entire computation can be done in $O(\log n)$ time using $n^\epsilon \cdot n^\epsilon \leq n$ processors.

Partitioning the problem. To determine the segments intersecting a trapezoid, we use a *locus* based approach. This approach involves considering each query as a higher dimensional point and partitioning the underlying space into regions providing the same answer. Thus the query problem is reduced to a point location problem, given sufficient preprocessing time and space. The problem at hand involves preprocessing the trapezoidal subdivisions (induced by the sample) in such a manner that given the end-points of any segment we should be able to list the regions it intersects in $O(\log n)$ time using $\lceil k/\log n \rceil$ processors where k is the number of regions that it intersects. The preprocessing for n segments can be done in $O(\log n)$ time using $O(n^c)$ processors, where c is a fixed constant. Thus any sample of size less than $n^{1/c}$ will suffice. See Reif and Sen [38] for details of this step which we omit from this discussion.

Clean-up for processor bound. Recall that a good sample implies the total sub-problem size is within a constant factor of $n/r \cdot E[|H(R)|]$. In this case (for non-intersecting segments), $H(R)$ is $O(|R|)$, i.e., the total subproblem size will be $\leq kn$ for some constant k. So, after recursion depth i, the total subproblem size can be bounded only by k^i. This algorithm has a recursion depth $O(\log \log n)$ and hence the total sub-problem size could become $\Omega(n\text{polylog}(n))$. This quantity is related to the number of processors; so we wish to bound it by $O(n)$. This is achieved by the following clean-up phase that we call *Filtering*

After partitioning the segments into the trapezoidal regions $\triangle$, we group the segments $L(\triangle)$ into two categories:

(a) $L^1(\triangle)$:Segments (part-segments) that have at least one end-point in the region
(b)$L^2(\triangle)$: Segments that span the region (horizontally)

Notice that number of segments of type (a) is less than $2n$ and $L^2(\triangle)$ in trapezoid $\triangle$, can be completely ordered (with respect to y-coordinate) within $\triangle$. So for the end-points of type $L^1(\triangle)$, a straight-forward binary search suffices to find the nearest segments among $L^2(\triangle)$ in the vertical direction. Consequently, we do not further preprocess the segments in $L^2(\triangle)$, i.e. we can leave them out from further recursive calls. Thus, the total size of the subproblems at any level of the recursive call is no more than $2n$. Processor allocation is achieved by simply allocating processors in a region equal to the number of end-points lying in it. This ensures that the algorithm does not require more than $O(n)$ processors for any step.

Analysis. The algorithm is recursive and we have shown that every step can be done in $\tilde{O}(\log n_i)$ steps for a subproblem of size n_i using n_i processors. The total processor requirement never exceeds n. The reader can verify that the time complexity T_i at depth i satisfies the preconditions of Theorem 26. Hence it follows that

Theorem 25 *Algorithm Vertical Visibility executes in $\tilde{O}(\log n)$ steps using n CREW processors.*

12.2 THE GENERAL STRATEGY

The algorithm developed for visibility maps in the previous section can be used for a number of other problems with some modifications required in the context of the problem. The following gives a bird's eye-view of this approach.

> 1. **Good-sampling** We use the method of Polling to choose a sample satisfying certain properties that holds with only constant probability for a naive sample.
> 2. **Partitioning** For computing the subproblems for a given sample. This step is also used in conjunction with the previous step for verifying if a sample is good. The most common partitioning method used is doing point-location in a parameterized space.
> 3. **Filtering** Keep the total problem size at any stage of the algorithm within the given processor bounds by pruning individual subproblems before recursively solving them. This step is the most problem-dependent and depends on the geometric properties of the problem.
> 4. Solve sub-problems recursively if the size exceeds a certain threshold.

This approach has yielded optimal speed-up algorithms for many problems, the most notable being 3-D convex hull for which no matching deterministic algorithm is known.

We shall now prove the following useful result that gives tail estimates for a class of randomized divide and conquer algorithms :

Theorem 26 *Given a process-tree which has the property that a procedure at depth i from the root takes time T_i such that*

$$\Pr[T_i \geq k c\alpha \log n (\epsilon_0)^i] \leq 2^{-(\epsilon_0)^i c\alpha \log n}$$

then, all the leaf-level procedures are completed in $\tilde{O}(\log n)$ time.

Proof. Setting $t_i = k(\epsilon_0)^i \log n\alpha(c - c_0)$, where c_0 is some constant, we obtain

$$\Pr[T_i \geq k\alpha c(\epsilon_0)^i \log n + t_i] \leq 2^{-(\epsilon_0)^i c\alpha \log n} \leq 2^{-t_i/k}.$$

If T is the total time for this worst case chain of nested calls and $m = 1/(1-\epsilon_0)$, the probability that T exceeds $mk\alpha \log n c_0 + t$ is less than the sum of the probability of events where $\sum_i t_i = t$, $t > 0$, and $\mu = mk\alpha \log nc_o$. We shall compute the probability that $\sum_i t_i = t$.

$\prod_{\sum t_i = t} 2^{-t_i} \leq \sum 2^{-t/k}$ over $t^{O(\log \log n)}$ tuples.

Thus $\Pr[T > km\alpha \log nc_o + t] < 2^{-t/k + O(\log t \log \log n)}$.

Using $t \geq km\alpha(c - c_o) \log n$, for large values of n and $m > 1$, we can rewrite the above expression as

$$\Pr[T > km\alpha c \log n] < 2^{-\alpha(c - c_o) \log n}.$$

For $c > 4c_o$, i.e., $c - c_o > 3/4c$, we have the following required bound,

$$\Pr[T > \alpha log n] \leq 2^{-(3/4)c\alpha \log n} \leq n^{-c_1 \alpha},$$

assuming that k, m, and c are larger than 1. ∎

Remark The proof of this theorem critically uses high probability tail estimates that is possible because of Polling. It is unlikely that any meaningful bound can be obtained only from expected bounds as we are trying to bound the *maximum* of the procedures spawned from any given node.

12.3   FURTHER READING

Random sampling was introduced into parallel computational geometry by Reif and Sen [39] (conference version) independently around the same time as the seminal papers of Clarkson [6, 8] and Haussler and Welzl [17]. All these papers primarily exploited the ε-net property, namely the (almost) even partitioning of the problem using a random subset. The subsequent papers of Clarkson [9] and Clarkson and Shor [7] refined the techniques considerably and developed very general and elegant techniques for designing geometric algorithms, namely, the randomized divide-and-conquer and randomized incremental construction. Mulmuley [29, 30], extended randomized incremental construction for dynamic settings and obtained impressive results for random updates. The textbook of Mulmuley [31] gives an excellent description of these general techniques. One of the most tantalizing open problems in this area, is to obtain concentration measure for the generic randomized incremental construction.

A number of very general properties of random-sampling were proved in Clarkson and Shor [7] that simplified or/and improved existing algorithms. The technique of Polling was first introduced by Reif and Sen [39] (conference version). It is an efficient method to find a sample with certain desirable properties that succeeds with high probability. In particular, the high-probability bounds are critical for divide-and-conquer based parallel algorithms. Even for many sequential algorithms based on Clarkson and Shor's paradigm, Polling can be used to bound the variance of the (otherwise only expected) running times without loss of efficiency as illustrated by the output-sensitive 3-D convex hull algorithm.

For a more detailed treatment of the basic parallel routines, the reader is referred to the textbook of Ja'Ja' [18], and Reif[37].

Notes

1. Note that all logarithms are to base 2 unless otherwise mentioned.

2. Jaikumar Radhakrishnan pointed out that one can redefine variables in the analysis of Algorithm P to make them independent; however, this is unnecessary since Theorem to be applied does not need independence. Moreover, in general, it may not always be possible to make such a redefinition of variables. But the general method will still apply.

3. Edgar Ramos and Jaikumar Radhakrishnan both pointed out an error in an earlier analysis, and the latter also independently developed the analysis given below.

4. A more recent algorithm of Karger improves this to $O(|E|polylog(n)$ using a more sophisticated Monte Carlo algorithm.

References

[1] M. Bellare and J. Rompel. Randomness-efficient oblivious sampling. *Proc. of the 35th Annual Symposium on Foundations of Computer Science*, pages 276 – 287, 1994.

[2] B. Bhattacharya and S. Sen. On a simple pactical output-sensitive randomized planar convex-hull algorithm. *Journal of Algorithms*, 25:177 – 193, 1997.

[3] B. Bollabas. *Graph Theory: An Introductory Course*. Springer Verlag, 1980.

[4] B. Chazelle and L. Guibas. Fractional cascading: I. a data structuring technique. *Algorithmica*, 1:133 – 162, 1986.

[5] B. Chazelle and L. Guibas. Fractional cascading: Ii. applications. *Algorithmica*, 1:163 – 191, 1986.

[6] K. Clarkson. A randomized algorithm for closest point queries. *SIAM Journal on Computing*, 17:830–847, 1988.

[7] Kenneth L Clarkson and Peter W Shor. Applications of random sampling in computational geometry ii. *Discrete Comp. Geom.*, 4:387–421, 1989.

[8] K.L. Clarkson. New applications of random sampling in computational geometry. *Discrete and Computational Geometry*, pages 195 – 222, 1987.

[9] K.L. Clarkson. Applications of random sampling in computational geometry ii. *Proc of the 4th Annual ACM Symp on Computational Geometry*, pages 1 - 11, 1988.

[10] D. Dubhashi and S. Chaudhuri. Probabilistic recurrence relations revisited. *Theoretical Computer Science*, 181:45 - 56, 1997.

[11] D. Dubhashi, D. Grable, and A. Panconesi. Near optimal distributed edge colouring via the nibble method. *Theoretical Computer Science* Special ESA 95 Issue, 203:225 - 251, 1998.

[12] D. Dubhashi and D. Ranjan. Balls and bins: a study in negative dependence. *Random Structures and Algorithm*, 13:99 - 124, 1998.

[13] W. Feller. *An introduction to probability theory and applications, Volume I*. John Wiley, 1968.

[14] O. Fries, K. Mehlhorn, and S. Näher. Dynamization of geometric data structures. In *Proceedings of the ACM Symp. on Computational Geometry*, pages 168 - 176, 1985.

[15] D. Grable and A. Panconesi. Near optimal distributed edge colouring in $o(\log \log n)$ rounds. *Random Structures and Algorithms*, 10:385 - 405, 1997.

[16] D. Grable and A. Panconesi. Brooks and vizing colorings. In *Proc of the SODA*, 1998.

[17] D. Haussler and E. Welzl. ϵ-nets and simplex range queries. *Discrete and Computational Geometry*, 2(2):127 - 152, 1987.

[18] Ja Ja Joseph. *An Introduction to Parallel Algorithms*. Addison-Wesley Publishing Company, 1992.

[19] D. Karger. Global min-cuts in rnc and other ramifications of a simple min-cut algorithm. In *Proc. of SODA*, pages 21 - 30, 1993.

[20] D. Karger and C. Stein. An $\tilde{O}(n^2)$ algorithm for minimum cuts. In *Proc. of the ACM STOC*, pages 757 - 765, 1993.

[21] R. Karp. Probabilistic recurrence relation. *J. ACM*, 41:1136 - 1150, 1994.

[22] D G Kirkpatrick and R Seidel. Output-size sensitive algorithms for finding maximal vectors. *Proc. of ACM Symp. on Computational Geometry*, 1985.

[23] M. Luby. 'a simple parallel algorithm for the maximal independent set problem. *SIAM Journal on Computing*, 15:1036 - 1053, 1986.

[24] M. Luby. Removing randomness in parallel without a processor penalty. *J. Computer and Systems Sciences*, 47:250 - 286, 1993.

[25] C. McDiarmid and R. Hayward. Large deviations for quicksort. *Journal of Algorithms*, 21:476 - 507, 1996.

[26] C.J. McDiarmid. On the method of bounded differences. *Surveys in Combinatorics, London Mathematical Society Lecture Notes Series*, 141, 1989.

[27] G. Miller and J. Reif. Parallel tree contraction and its applications. *Random Structures and Algorithm*, 5:47 – 72, 1989.

[28] M. Molloy and B. Reed. A bound on the strong chromatic index of a graph. *J. Comb. Theory (B)*, 69:103 – 109, 1997.

[29] K. Mulmuley. Randomized multidimensional search trees: Dynamic sampling. *Proc. of the 7th ACMSymp. on Computational Geometry*, pages 121–131, 1991.

[30] K. Mulmuley. Randomized multidimensional search trees: lazy balancing and dynamic shuffling. *Proc. of the 32nd IEEE Foundations of Computer Science*, pages 180–196, 1991.

[31] Ketan Mulmuley. *Computational Geometry : An Introduction through Randomized Algorithms.* Prentice Hall, Englewood Cliffs, NJ, 1994.

[32] A. Panconesi and A. Srinivasan. Randomized distributed edge coloring via an extension of the chernoff–hoeffding bounds. *SIAM J. Computing*, 26:350 – 368, 1997.

[33] W. Pugh. Skip lists: A probabilistic alternative to balanced trees. *Communications of the ACM*, 33:668 – 676, 1990.

[34] S. Rajasekaran and J. Reif. Optimal and sublogarithmic time radomized parallel sorting, algorithms. *SIAM Journal on Computing*, 18:594–607, 1989.

[35] S. Rajasekaran and S. Sen. *Random sampling Techniques and parallel algorithm design.* J.H. Reif editor. Morgan, Kaufman Publishers, 1993.

[36] J. Reif and S. Sen. *Parallel COmputtaional Geometry : An approach using randomization.* J. Sack and J. Urrutia eds. Elsevier, 1999.

[37] J.H. Reif, editor. *Synthesis of Parallel Algorithms.* Morgan, Kaufman Publishers, 1993.

[38] J.H. Reif and S. Sen. Optimal randomized parallel algorithms for computation al geometry. *Algorithmica*, 7(7):91 – 117, 1992.

[39] J.H. Reif and S. Sen. Optimal randomized parallel algorithms for computational geometry. *Algorithmica*, 7:91 – 117, 1992.

[40] J. Schmidt, A. Siegel, and A. Srinivasan. Chernoff-hoeffding bounds for applications with limited independence. *SIAM J. Discrete Math.*, 8:223 – 250, 1995.

100

[41] S. Sen. Some observations on skip lists. *Information Processing Letters*, 39:173 – 176, 1991.

[42] S. Sen. Fractional cascading revisited. *Journal of Algorithms*, 19:161 – 172, 1995.

[43] J.M. Steele. *Probability Theory and Combinatorial Optimization.* CBMS-NSF Regional Conference Series in Applied Mathematics, 69 Society for Industrial and Applied Mathematics, 1997.

Chapter 4

RANDOMIZATION IN GRAPH OPTIMIZATION PROBLEMS: A SURVEY

David R. Karger
MIT Laboratory for Computer Science
Cambridge, MA 02138.
email: **karger@lcs.mit.edu**
URL: **http://theory.lcs.mit.edu/~karger**.

Abstract   Randomization has become a pervasive technique in combinatorial optimization. We survey our thesis and subsequent work, which uses four common randomization techniques to attack numerous optimization problems on undirected graphs.

Keywords: Randomized Algorithm, Maximum Flow, Minimum Cut, Monte Carlo Algorithm, Random Sampling.

1.   INTRODUCTION

Randomization has become a pervasive technique in combinatorial optimization. Randomization has been used to develop algorithms that are faster, simpler, and/or better-performing than previous deterministic algorithms. This article surveys our thesis [Kar94], which presents randomized algorithms for numerous problems on undirected graphs. Our work uses four important randomization techniques:

Random Selection, which lets us easily choose a "typical" element of a set, avoiding rare "bad" elements;

Random Sampling, which provides a quick way to build a small, representative subproblem of a larger problem for quick analysis;

Randomized Rounding, which lets us transform fractional problem solutions into integral ones; and

S. Rajasekaran et al (eds.), Handbook of Randomized Computing, Volume 1, pp, 101–131.
© 2001 *Kluwer Academic Publishers.*

Monte Carlo Simulation, which lets us estimate the probabilities of interesting events.

We apply these techniques to numerous optimization problems on undirected graphs. The graph is one of the most common structures in computer science and optimization, modeling among other things roads, communication and transportation networks, electrical circuits, relationships between individuals, hypertext collections, resource allocations, project plans, database and program dependencies, and parallel architectures. Among the graph problems we address are finding a minimum spanning tree, finding a maximum flow, determining the connectivity (minimum cut) of a graph, network design, graph coloring, and estimating the reliability (disconnection probability) of a network with random edge failures. A great deal of work has been done on all of these problems. Due to space limitations, we are unable to discuss all of this related work. Such a discussion can be found in our thesis.

It is extremely important to note that we are not carrying out "expected case" analysis of algorithms running on random inputs. Rather, we consider *worst case* inputs and use random choices by the algorithm to solve them efficiently.

We begin this article with a survey of many of our applications of the four randomization techniques, sketching the methods, the problems, and the resulting algorithms. Afterwards, we will present a more detailed discussion of a few algorithms and proofs which will hopefully give some flavor of our work.

This survey appeared previously in *Optima* [Kar98c].

1.1   NOTATION

We address only undirected graphs; directed graphs have so far not succumbed to the techniques we apply here. Throughout our discussion, we will consider a graph G with m edges and n vertices. Each graph edge may have a *weight* (reflecting cost or capacity) associated with it. To simplify our presentation here, we often focus on unweighted graphs (that is, graphs with all edge weights equal to one), though many of our algorithms apply equally well to weighted graphs. Unless it has parallel edges (multiple edges with the same endpoints) a graph has $m \leq \binom{n}{2}$.

The notation $\tilde{O}(f)$ denotes $O(f \log^d n)$ for some constant d.

1.2   OVERVIEW OF RESULTS

We show how randomization can be used in several ways on problems of varying degrees of difficulty. For the "easy to solve" minimum spanning tree problem, where a long line of research has resulted in ever closer to linear-time algorithms, random sampling gives the final small increment to a truly linear time algorithm. For harder problems it improves running times by a more significant factor. For example, we improve the time needed to find minimum cuts from $\tilde{O}(mn)$ to $\tilde{O}(m)$, and give the first efficient parallel algorithm for the problem. Addressing some hard $\mathcal{NP}$-complete problems such as network design and graph coloring, where finding an exact solution in polynomial time is thought to be hopeless, we use randomized rounding to give better approximation algorithms than were previously known. Finally, for the problem of determining the reliability of a network (a $\sharp\mathcal{P}$-complete problem which is thought to be "even harder" than $\mathcal{NP}$-complete ones) we use Monte Carlo simulation to give the first efficient approximation algorithm.

1.3   RANDOMIZED ALGORITHMS

Our work deals with randomized algorithms. Our typical model is that the algorithm has a source of "random bits"—variables that are mutually independent and take on values 0 or 1 with probability 1/2 each. Extracting one random bit from the source is assumed to take constant time. If our algorithms use more complex operations, such as flipping biased coins or generating samples from more complex distributions, we take into account the time needed to simulate these operations in our unbiased-bit model. Event probabilities are taken over the sample space of random bit strings produced by the random bit generator. An event occurs *with high probability (w.h.p.)* if on problems of size n it occurs with probability greater than $(1 - \frac{1}{n^k})$ for some constant $k > 1$, and with *low probability* if its complement occurs with high probability.

The random choices that an algorithm makes can affect both its running time and its correctness. An algorithm that has a fixed (deterministic) running time but has a low probability of giving an incorrect answer is called *Monte Carlo (MC)*. If the running time of the algorithm is a random variable but the correct answer is given with certainty, then the algorithm is said to be *Las Vegas (LV)*. Depending on the circumstances, one type of algorithm may be better than the other. However, a Las Vegas algorithm is "stronger" in the following sense.

A Las Vegas algorithm can be made Monte Carlo by having it terminate with an arbitrary wrong answer if it exceeds the time bound $f(n)$. Since the Las Vegas algorithm is unlikely to exceed its time bound, the converted algorithm is unlikely to give the wrong answer. On the other hand, there is no universal method for making a Monte Carlo algorithm into a Las Vegas one, and indeed some of the algorithms we present are Monte Carlo with no Las Vegas version apparent. The fundamental problem is that sometimes it is impossible to check whether an algorithm has given a correct answer. However, the failure probability of a Monte Carlo optimization algorithm can be made arbitrarily small by repeating it several times and taking the best answer; we shall see several examples of this below. In particular, we can reduce the failure probability so far that other unavoidable events (such as a power failure) are more likely than an incorrect answer.

2. A SURVEY OF TECHNIQUES AND RESULTS

In this section, we provide a high level overview of the four randomization techniques and the various algorithms we have been able to develop by using them.

2.1 RANDOM SELECTION

The first and simplest randomization technique we discuss is random selection. The intuition behind this idea is that a single randomly selected individual is probably a "typical" representative of the entire population. Thus, random selection provides a good way to avoid choosing rare "bad" elements. This is the idea behind Quicksort [Hoa62], where the assumption is that the randomly selected pivot will be neither extremely large nor extremely small, and will therefore serve to separate the remaining elements into two roughly equal sized groups.

We apply this idea in a new algorithm for finding minimum cuts in undirected graphs. A *cut* is a partition of the graph vertices into two groups; the *value* of the cut is the number (or total weight) of edges with one endpoint in each group. The minimum cut problem is to identify a cut of minimum value. We distinguish the (global) minimum cut from an *s-t minimum cut* which is required to separate two specific vertices s and t. Finding minimum cuts is of great importance in analyzing network reliability and also plays a role in solving traveling salesman and network design problems.

We present the Recursive Contraction Algorithm (joint work with Clifford Stein [KS96]). The idea behind our algorithm is simple: a randomly selected edge is unlikely to cross a particular minimum cut, so its endpoints are probably on the same side. If we merge two vertices on the same side of this minimum cut, then we will not affect the minimum cut but will reduce the number of graph vertices by one. Therefore, we can find the minimum cut by repeatedly selecting a random edge and merging its endpoints until only two vertices remain and the minimum cut becomes obvious.

An efficient implementation of the above idea leads to a strongly polynomial $\tilde{O}(n^2)$-time algorithm for the minimum cut problem on weighted undirected graphs. In contrast, the best deterministic bound, due to Hao and Orlin [HO94], is $\tilde{O}(mn)$. Our algorithm actually finds *all* minimum cuts with high probability. It extends to enumerating approximately minimum cuts and minimum k-way cuts for constant k, as well as to constructing the *cactus* of a graph (a compact representation of all its minimum cuts). The algorithm is the first with a theoretically good parallelization. A *derandomization* of the algorithm (joint with Rajeev Motwani [KM97]) gave the first proof that there was a fast deterministic parallel algorithm for the minimum cut problem. An implementation experiment shows that the algorithm has reasonable time bounds in practice [CGK+97].

The Contraction Algorithm is Monte Carlo: it gives the correct answer with high probability, but does have a small chance of being wrong. As we are unaware of any algorithm for *verifying* that a cut is minimum, we have been unable to devise a Las Vegas version of the algorithm. This is a case where a willingness to occasionally be wrong seems to provide a significant speedup.

The Contraction Algorithm also gives an important new bound on the number of small cuts a graph may contain; this has important applications in network reliability analysis and graph sampling (see below).

In subsequent work [BK00], we used the Contraction Algorithm to efficiently solve the *graph augmentation problem:* adding the minimum possible capacity to a graph so as to increase its minimum cut to a given value (this work is joint with Andras Benczùr).

2.2   RANDOM SAMPLING

A more general use of randomization than random selection is to generate small representative subproblems. The representative random sample is a central concept of statistics. It is often possible to gather a great deal of information about a large population by examining a small sample randomly drawn from it. This approach has obvious advantages in reducing the investigator's work, both in gathering and in analyzing the data.

Given an optimization problem, it may be possible to generate a small representative subproblem by random sampling (perhaps the most natural sample from a graph is a random subset of its edges). Intuitively, such a subproblem should form a microcosm of the larger problem. Our goal is to examine the subproblem and use it to glean information about the original problem. Since the subproblem is small, we can spend proportionally more time examining it than we would spend examining the original problem. In one approach that we use frequently, an optimal solution to the subproblem may be a nearly optimal solution to the problem as a whole. In some situations, such an approximation might be sufficient. In other situations, it may be easy to refine this good solution into a truly optimal solution.

Floyd and Rivest [FR75] use this approach in a fast and elegant algorithm for finding the median of an ordered set. They select a small random sample of elements from the set and show how inspecting this sample gives a very accurate estimate of the value of the median. It is then easy to find the actual median by examining only those elements close to the estimate. This algorithm, which is very simple to implement, uses fewer comparisons than any other known median-finding algorithm.

The Floyd-Rivest algorithm typifies three components needed in a random-sampling algorithm. The first is a definition of a *randomly sampled subproblem*. The second is an *approximation theorem* that proves that a solution to the subproblem is an approximate solution to the original problem. These two components by themselves will typically yield an obvious approximation algorithm with a speed-accuracy tradeoff. The third component is a *refinement algorithm* that takes the approximate solution and turns it into an actual solution. Combining these three components can yield an algorithm whose running time will be determined by that of the refinement algorithm; intuitively, refinement should be easier than computing a solution from scratch.

In an application of this approach, we present the first (randomized) linear-time algorithm for finding minimum spanning trees in the comparison-based model of computation. This result reflects joint work with Philip Klein and Robert E. Tarjan [KKT95]. A long stream of results reduced the best known deterministic time bound to almost linear [GGST86], but a linear time bound remained elusive. Our fundamental insight is that if we construct a subgraph of a graph by taking a random sample of the graph's edges, then the minimum spanning tree in the subgraph is a "nearly" minimum spanning tree of the entire graph. More precisely, very few graph edges can be used to improve the sample's minimum spanning tree. By examining these few edges, we can refine our approximation into the actual minimum spanning tree at little additional cost.

We also apply sampling to the minimum cut problem and several other related problems involving cuts in graphs, including maximum flows. The maximum flow problem is perhaps the most widely studied of all graph optimization problems, having hundreds of applications. Given vertices s and t and capacitated edges, the goal is to ship the maximum quantity of material from s to t without exceeding the capacities of the edges. The value of a graph's maximum flow is completely determined by the value of the minimum s-t cut in the graph.

We prove a cut sampling theorem that says that when we choose half a graph's edges at random we approximately halve the value of every cut. In particular, we halve the graph's connectivity and the value of all s-t minimum cuts and maximum flows. This theorem gives a random-sampling scheme for approximating minimum cuts and maximum flows: compute the minimum cut and maximum flow in a random sample of the graph edges. Since the sample has fewer edges, the computation is faster. At the same time, our sampling theorems show that this approach gives accurate estimates of the correct values. If we want to get exact solutions rather than approximations, we still can use our samples as starting points to which we can apply inexpensive refinement algorithms. This leads to a simple randomized divide-and-conquer algorithm for finding exact maximum flows.

We put our results on graph sampling into a larger framework by examining sampling from *matroids* [Kar98b]. We generalize our minimum spanning tree algorithm to the problem of finding a minimum cost matroid basis, and extend our cut-sampling and maximum flow results to the problem of matroid basis packing. Our techniques actually give a paradigm that can be applied to any *packing problem* where the goal, given a collection of *feasible* subsets of a universe, is to find a maximum collection of disjoint feasible subsets. For example, in the maximum flow problem, we are attempting to send units of flow from s to t. Each such unit of flow travels along a path from s to t, so the feasible edge-sets are the s-t paths. We apply the sampling paradigm to the problem of packing disjoint bases in a matroid, and get faster algorithms for approximating and exactly finding optimum basis packings.

We have continued to apply random sampling technique in work following our thesis. Our recent results include:

- An $O(m \log^3 n)$-time Monte Carlo algorithm for finding minimum cuts [Kar00],

- A *compression algorithm* that lets us transform any undirected graph into a graph with $\tilde{O}(n)$ edges but roughly the same cut values, speeding up any algorithm that depends only on cut values [BK96],

- As an application, an $O(n^2 \log n)$-time Monte Carlo algorithm for finding any constant factor approximation to an s-t minimum cut [BK96],

- An $\tilde{O}(m\sqrt{n})$-time Las Vegas algorithm for finding any constant factor approximation to an s-t max-flow [Kar98a],

- An $\tilde{O}(n^{2.22})$-time algorithm for finding a maximum flow in a simple (uncapacitated) graph [KL98] (joint work with Matt Levine).

All of these bounds are significantly better than the best general time bound for finding maximum flows in directed graphs ($\tilde{O}(mn)$ for a strongly polynomial bound [GT88], and recently $\tilde{O}(m^{3/2}\log U)$ for a scaling algorithm of Goldberg and Rao [GR97]). This suggests that perhaps a better bound for maximum flow can be achieved, at least on undirected graphs.

2.3 RANDOMIZED ROUNDING

Yet another powerful randomization technique is *randomized rounding*. This approach is used to find approximate solutions to $\mathcal{NP}$-hard *integer programs*. These problems typically ask for an assignment of 0/1 values to variables x_i such that linear constraints of the form $\sum a_i x_i = c$ are satisfied. If we *relax* the integer program, allowing each x_i to take any real value between 0 and 1, we get a linear program that can be solved in polynomial time, giving values p_i such that $\sum a_i p_i = c$. Raghavan and Thompson [RT87] observed that we could treat the resulting values p_i as probabilities. If we randomly set $x_i = 1$ with probability p_i and 0 otherwise, then the *expected value* of $\sum a_i x_i$ is $\sum a_i p_i = c$. Raghavan and Thompson presented techniques for ensuring that the randomly chosen values do in fact yield a sum near the expectation, thus giving approximately correct solutions to the integer program. We can view randomized rounding as a way of sampling randomly from a large space of *answers*, rather than subproblems as before. Linear programming relaxation is used to construct an answer-space in which most of the answers are good ones.

We use our graph sampling theorems to apply randomized rounding to *network design problems*. Such a problem is specified by an input graph G with each edge assigned a cost. The goal is to output a subgraph of G satisfying certain connectivity requirements at minimum cost (measured as the sum of the costs of edges used). These requirements are described by specifying a minimum number of edges that must cross each cut of G. This formulation easily captures many classic problems including perfect matching, minimum cost flow, Steiner tree, and minimum T-join. By applying randomized rounding, we improve the approximation bounds for a large class of network design problems, from $O(\log n)$ (due to Goemans et al [GGP$^+$94]) to $1 + o(1)$ in some cases. Our graph sampling theorems provide the necessary tools for showing that randomized rounding works well in this case.

We also apply randomized rounding to the classic *graph coloring problem*. No linear programs have yet been devised that provide a useful fractional solution, so we use more powerful *semidefinite programming* as our starting point. We show that any 3-colorable graph can be colored in polynomial time with $\tilde{O}(n^{1/4})$ colors, improving on the previous best bound of $\tilde{O}(n^{3/8})$ [Blu94]. We also give presently best results for k-colorable graphs. Along the way, we discover new properties of the *Lovász ϑ-function,* an object that has received a great deal of attention because of its connections to graph coloring, cliques, and independent sets. This work is joint with Rajeev Motwani and Madhu Sudan [KMS98]. We gave a slight improvement with Avrim Blum [BK97].

2.4   MONTE CARLO ESTIMATION

The last randomization technique we consider is Monte Carlo estimation. The technique is applied when we want to estimate the probability p of a given event E over some probability space. Monte Carlo estimation carries out repeated "trials" (samples from the probability space) and measures in what fraction of the trials the event E occurs. This gives a natural estimate of the event probability.

The Monte Carlo approach breaks down when the interesting probability p is very small. To estimate p, we need to carry out enough experiments to see at least a few occurrences of E. But we expect to see a first occurrence only after $1/p$ trials, which may be too many to carry out efficiently. A solution to this problem, explored by Karp, Luby and Madras [KLM89], is to carry out the Monte Carlo simulation in a different, "biased" way that makes the event E more likely to occur, so that we can get by with fewer trials. The trick is to choose the new simulation so that it gives us useful information about the *original* probability space.

We apply this technique to the *network reliability problem*. In this problem, we are interested in estimating the probability that a network is disconnected by random edge failures, so it is perhaps unsurprising that randomization is useful. We are given a graph G whose edges fail randomly and independently with certain specified probabilities. Our goal is to determine the probability that the graph becomes disconnected by edge failures.

Unfortunately, it is known to be $\sharp\mathcal{P}$-hard (even worse than $\mathcal{NP}$-hard) to exactly determine the reliability of a network. But we use Monte Carlo methods to give a *fully polynomial randomized approximation scheme*

(FPRAS) for the network reliability problem [Kar99b]. Given a failure probability p for the edges, our algorithm, in time polynomial in n and $1/\epsilon$, returns a number P that estimates the probability FAIL(p) that the graph becomes disconnected. With high probability, P is in the range $(1 \pm \epsilon)$FAIL(p). The algorithm is Monte Carlo, meaning that the approximation is correct with high probability but that it is not possible to verify its correctness. It generalizes to the case where the edge failure probabilities are different, to computing the probability the graph fails to be k-connected (for any fixed k), and to the more general problem of approximating the *Tutte Polynomial* for a large family of graphs. Our algorithm is easy to implement and appears likely to have satisfactory time bounds in practice [Kar99b, CGK+97, KT97].

A natural way to estimate a network's failure probability is to carry out numerous simulations of the edge failures and check how often the graph is disconnected by them. But as mentioned above, this can take prohibitively many trials if the failure probability is extremely small. However, we use our cut counting and sampling theorems to prove that when P is small, only the small cuts in a graph are significantly likely to fail. We use our cut algorithms to enumerate these small cuts and then use the biased Monte Carlo technique developed by Karp, Luby and Madras [KLM89] to estimate the probability the one of the explicitly enumerated cuts fails.

3.   THE CONTRACTION ALGORITHM

To give some flavor of our results, we begin by describing an algorithm for finding a minimum cut in an undirected graph [KS96]. For simplicity we discuss unweighted graphs, but the algorithm works equally well for graphs with edge weights.

The algorithm is based on the idea of *contracting* edges. Suppose we were somehow able to identify an edge that did not cross the minimum cut. This would tell us that both of its endpoints are on the same side of the cut. We can use this information to simplify the graph by contracting the two endpoints. To contract two vertices v_1 and v_2 we replace them by a vertex v, and let the set of edges incident on v be the union of the sets of edges incident on v_1 and v_2. We do not merge edges from v_1 and v_2 that have the same other endpoint; instead, we allow multiple instances of those edges. However, we remove self loops formed by edges originally connecting v_1 to v_2. Formally, we delete all edges (v_1, v_2), and replace each edge (v_1, w) or (v_2, w) with an edge (v, w). The rest of the graph remains unchanged. We will use $G/(v_1, v_2)$ to denote graph

G with edge (v_1, v_2) contracted (by *contracting an edge*, we will mean contracting the two endpoints of the edge).

Note that a contraction reduces the number of graph vertices by one. We can imagine repeatedly selecting and contracting edges until every vertex has been merged into one of two remaining "metavertices." These metavertices define a cut of the original graph: each side corresponds to the vertices contained in one of the metavertices. It is easy to see that if we never contract an edge that crosses the minimum cut, then the two metavertices we end up with will correspond to the two sides of the minimum cut we are looking for.

So our subgoal is to devise a method for selecting an edge that does not cross the minimum cut. There are some sophisticated deterministic algorithms for doing this [NI92], but unfortunately they are slow. We instead rely on the following observation: almost none of the edges in a graph cross the minimum cut. Thus, if we choose a *random* edge to contract, we probably get a non-min-cut edge! This gives us a very fast edge selection algorithm; the trade-off is that we must be prepared for it occasionally to make mistakes. We describe our algorithm in Figure 4.1. Assume initially that we are given a multigraph $G(V, E)$ with n vertices and m edges. The *Contraction Algorithm*, which is described in Figure 4.1, repeatedly chooses an edge at random and contracts it.

Algorithm Contract(G)

repeat until G has 2 vertices

    choose an edge (v, w) uniformly at random from G

    let $G \leftarrow G/(v, w)$

return the unique cut defined by (the contracted) G

Figure 4.1 The Contraction Algorithm

It is relatively straightforward to implement this algorithm in $O(n^2)$ time.

Lemma 1 *A particular minimum cut in G is returned by the Contraction Algorithm with probability at least* $\binom{n}{2}^{-1}$.

Proof: **Fix attention on some specific minimum cut C with c crossing edges. We will use the term *minimum cut edge* to refer only to edges crossing C. If we never select a minimum cut**

edge during the Contraction Algorithm, then the two vertices we end up with must define the minimum cut.

Observe that after each contraction, the minimum cut value in the new graph must still be at least c. This is because every cut in the contracted graph corresponds to a cut of the same value in the original graph, and thus has value at least c. Furthermore, if we contract an edge (v, w) that does not cross C, then the cut C corresponds to a cut of value c in $G/(v, w)$; this corresponding cut is a minimum cut (of value c) in the contracted graph.

Each time we contract an edge, we reduce the number of vertices in the graph by one. Consider the stage in which the graph has r vertices. Since the contracted graph has a minimum cut of at least c, it must have minimum degree c, and thus at least $rc/2$ edges. However, only c of these edges are in the minimum cut. Thus, a randomly chosen edge is in the minimum cut with probability at most $2/r$. To determine the probability that we *never* contract a minimum cut edge, we simply multiply all of the per-stage probabilities. This shows that the probability that we never contract a minimum cut edge through all $n - 2$ contractions is at least

$$\left(1 - \frac{2}{n}\right)\left(1 - \frac{2}{n-1}\right) \cdots \left(1 - \frac{2}{3}\right) = \left(\frac{n-2}{n}\right)\left(\frac{n-3}{n-1}\right) \cdots \left(\frac{2}{4}\right)\left(\frac{1}{3}\right)$$

$$= \binom{n}{2}^{-1}$$

$$\geq 1/n^2.$$

$\square$

Note that $\binom{n}{2}^{-1} \approx 1/n^2$, so the Contraction Algorithm described above has a relatively small chance of succeeding. But it is large enough to be useful. To improve our chance of success, we may simply repeat the algorithm a large number of times. If we run the Contraction Algorithm $n^2 \ln n$ times, and take the best answer we see, then the probability that we fail to give the right answer is just the probability that *none* of the repetitions of the algorithm yield the right answer, which is at most

$$(1 - 1/n^2)^{n^2 \ln n} \approx 1/n,$$

which means the algorithm works with high probability. This "amplification through repetition" is standard for randomized algorithms: we can get an exponential decrease in the failure probability from a linear slowdown in the running time.

Since the Contraction Algorithm takes $O(n^2)$ time per iteration, we immediately get an algorithm that finds a minimum cut *with high probability* in $O(n^4 \log n)$ time. This is somewhat unsatisfactory, as algorithms based on flow [HO94] can be used to find the minimum cut in $\tilde{O}(mn)$ time.

3.1   THE RECURSIVE CONTRACTION ALGORITHM

By adding another idea we can improve the running time of our minimum cut algorithm to $\tilde{O}(n^2)$. We aim to "share work" among the numerous iterations of the algorithm. Note that the failure probability of the Contraction Algorithm rises as its size decreases. In fact, if we contract G until it has k vertices rather than 2, then the probability that the algorithm does not destroy the minimum cut of G exceeds $(k/n)^2$ (this follows by truncating the product we used to analyze the original Contraction Algorithm). So the real problem with the Contraction Algorithm arises when the graph has gotten small. We might imagine switching over to a deterministic algorithm once the graph is small, and indeed this approach yields improved performance. But we can do even better with another application of the principle that "repetition improves your chances." When the graph gets small, in order to improve our odds of success, we (recursively) carry out *two* executions of the algorithm on what remains.

Let $\texttt{Contract}(G, k)$ denote a subroutine that runs the Contraction Algorithm until G is reduced to k vertices. Consider the *Recursive Contraction Algorithm* in Figure 4.2. As can be seen, we perform two independent trials. In each, we first partially contract the graph, but not so much that the likelihood of the cut surviving is too small. By contracting the graph until it has $n/\sqrt{2}$ vertices, we ensure a 50% probability of not contracting a minimum cut edge, so we expect that on the average one of the two attempts will avoid contracting a minimum cut edge. We then recursively apply the algorithm to each of the two partially contracted graphs. As described, the algorithm returns only a cut value; it can easily be modified to return a cut of the given value. Alternatively, we might want to output every cut encountered, hoping to enumerate all the minimum cuts.

Next we analyze the running time of this algorithm.

Lemma 2 *Algorithm* $\texttt{Recursive-Contract}$ *runs in* $O(n^2 \log n)$ *time.*

Proof: One level of recursion consists of two independent trials of contraction of G to $n/\sqrt{2}$ vertices followed by a re-

---

Algorithm Recursive-Contract(G, n)

input A graph G of size n.

if G has 2 vertices

then

    return the weight of (unique) cut in G

else repeat <u>twice</u>

      $G' \leftarrow$ Contract($G, n/\sqrt{2}$)

      Recursive-Contract($G', n/\sqrt{2}$).

    return the smaller of the two resulting values.

Figure 4.2 The Recursive Contraction Algorithm

cursive call. **Performing a contraction to $n/\sqrt{2}$ vertices can be implemented by Algorithm Contract from the previous section in $O(n^2)$ time. We thus have the following recurrence for the running time:**

$$T(n) = 2\left(n^2 + T\left(n/\sqrt{2}\right)\right). \tag{4.1}$$

This recurrence is solved by

$$T(n) = O(n^2 \log n).$$

☐

We now analyze the probability that the algorithm finds the particular minimum cut we are looking for. We will say that the Recursive Contraction Algorithm *finds* a certain minimum cut if that minimum cut corresponds to one of the leaves in the algorithm's tree of recursive calls. Note that if the algorithm finds any minimum cut then it will output the minimum cut value.

Lemma 3 *The Recursive Contraction Algorithm finds a particular minimum cut with probability $\Omega(1/\log n)$.*

Proof: **We give a recursive argument. The algorithm will find a particular minimum cut if, in one of its two iterations, the following two things happen: (i) the call to Contract($G, n/\sqrt{2}$) preserves the minimum cut and (ii) the recursive call finds the particular minimum cut. The probability that an iteration** succeeds is just the product of the probabilities of events (i) and (ii). The algorithm succeeds if *either* iteration succeeds, and thus fails only if *both* iterations fail. The probability this double failure happens is just the square of the probability that one iteration fails. Thus the success probability is one minus this squared quantity. This yields a recurrence $P(n)$ for a lower bound on the probability of success on a graph of size n:

$$P(2) = 1$$

$$P(n) \geq 1 - \left(1 - \frac{1}{2}P\left(n/\sqrt{2}\right)\right)^2.$$

We solve this recurrence through a change of variables. Write $z_k = 4/P(2^{k/2}) - 1$, so $P(2^{k/2}) = 4/(z_k + 1)$. Plugging this into the above recurrence and solving for z_k yields

$$z_1 = 3$$

$$z_{k+1} = z_k + 1 + 1/z_k.$$

Since clearly $z_k \geq 1$, it follows by induction that

$$k < z_k < 3 + 2k$$

Thus $z_k = \Theta(k)$ and thus that

$$P(n) = 4/(z_{2\log n} + 1) = \Theta(1/\log n).$$

In other words, one trial of the Recursive Contraction Algorithm finds any particular minimum cut with probability $\Omega(1/\log n)$. $\square$

Those familiar with branching processes might see that we are evaluating the probability that the extinction of contracted graphs containing the minimum cut does not occur before depth $2\log n$.

Theorem 4 ([KS96]) *All minimum cuts in an arbitrarily weighted undirected graph with n vertices can be found with high probability in $O(n^2 \log^3 n)$ time.*

Proof: We will see below that there are at most $\binom{n}{2}$ minimum cuts in a graph. Repeating Recursive-Contract $O(\log^2 n)$ times gives an $O(1/n^4)$ chance of missing any particular minimum cut. Thus our chance of missing any one of the at most $\binom{n}{2}$ minimum cuts is upper bounded by $O(\binom{n}{2} \cdot n^{-4}) = O(1/n^2)$. $\square$

3.2   COUNTING CUTS

Besides serving as an algorithm to find minimum cuts, the Contraction Algorithm tells us some interesting things about the number of minimum and, more generally, small cuts in a graph. These results are extremely useful when we consider our next topic, random sampling from graphs.

Definition 5 *An α-minimum cut is a cut whose value is at most α times that of the (global) minimum cut.*

Lemma 6 *There are at most $\binom{n}{2} < n^2$ minimum cuts.*

Proof: We showed that the Contraction Algorithm outputs a given minimum cut with probability at least $\binom{n}{2}^{-1}$. Suppose that there were more than k minimum cuts. Each is output with probability k. Since these output events are disjoint (the algorithm outputs only one cut), the probability that one of them is output is just the sum of their individual probabilities, namely $k/\binom{n}{2}$. This quantity, being a probability, is at most one. So $k \le \binom{n}{2}$. $\square$

Theorem 7 (Cut Counting [KS96]) *In a graph with minimum cut c, there are less than $n^{2\alpha}$ cuts of value at most αc.*

Proof: If we consider a cut of value αc, we can prove (by generalizing the argument we gave for minimum cuts in the obvious way) that the Contraction Algorithm outputs it with probability at least $1/n^{2\alpha}$. The argument then proceeds as in the previous lemma. $\square$

4.   RANDOM SAMPLING

So far we have addressed random selection, which works by finding a "typical" element (eg a non-min-cut edge). We now turn to random sampling, where the goal is to build a small representative model of our input problem. We will describe algorithms for approximating and exactly finding maximum flows and minimum cuts in an undirected graph. For simplicity, we will restrict our discussion to graphs with unit-capacity edges (unweighted graphs) though many of the techniques that we discuss can be applied to weighted graphs as well. Due to space limitations, and because we are focusing on our thesis work rather than later improvements, we present algorithms that only work well when the minimum cut of the graph is large.

In unweighted graphs, the *s-t maximum flow problem* is to find a maximum set, or *packing*, of edge-disjoint *s-t* paths. It is known [FF62] that the value of this flow is equal to the value of the minimum *s-t* cut. In fact, the only known algorithms for finding an *s-t* minimum cut simply identify a cut that is saturated by an *s-t* maximum flow.

In unweighted graphs, a classic algorithm for finding such a maximum flow is the *augmenting path* algorithm (cf. [Tar83, AMO93]). Given a graph and an *s-t* flow of value f, a linear-time depth first search of the so-called *residual graph* will either show how to augment the flow to one of value $f + 1$ or prove that f is the value of the maximum flow. This algorithm can be used to find a maximum flow of value v in $O(mv)$ time by finding v augmenting paths. Of course, since the algorithm's running time depends on the edge count and flow value, we can make it faster by reducing one or both quantities. We show how random sampling can be used to do this.

4.1   A SAMPLING THEOREM

Our algorithms are all based upon the following model of random sampling in graphs. We are given an unweighted graph $G = (V, E)$ with a *sampling probability* p for each edge e, and we construct a random subgraph, or *skeleton,* on the same vertices V by placing each edge e in the skeleton independently with probability p. We denote the skeleton by $G(p)$. Note that if a given cut has k edges crossing it in G, then the expected number of edges crossing that cut in $G(p)$ is pk. In particular, if the *s-t* minimum cut in G has value v, then we might expect that the *s-t* minimum cut in $G(p)$ has value pv.

Unfortunately, samples invariably *deviate* from their expectations. In order to effectively make use of a skeleton, we need to show that these deviations are small. If they are, then the skeleton will tell us things about the original graph that are approximately correct. Let c be the minimum cut of graph G. Our main theorem says that so long as pc (the minimum expected cut value in the skeleton) is sufficiently large, every cut in the skeleton takes on roughly its expected value.

Theorem 8 ([Kar99a]) *Let* $\epsilon = \sqrt{3(d + 2)(\ln n)/pc}$ *(so* $p = \Theta((\ln n)/\epsilon^2 c)$*). If* $\epsilon \leq 1$*, then with probability* $1 - O(1/n^d)$*, every cut in* $G(p)$ *has value between* $1 - \epsilon$ *and* $1 + \epsilon$ *times its expected value.*

This result is somewhat surprising. A graph has exponentially many (2^{n-1}) cuts. Naively, even if each cut is unlikely to deviate far from its expected value, with so many cuts one probably will. We are saved by the cut counting theorem discussed in the previous section. The central limit theorem (as quantified by the Chernoff bound [Che52, MR95b]) says that as the expected value of a sample gets larger, its sample value becomes more and more tightly concentrated about its expectation. In particular, as a cut value grows, its probability of deviating by a given ratio ϵ from its expectation decays exponentially with the cut value. The cut counting theorem says that the number of cuts of a given value increases "only" exponentially with the cut value. The parameters of Theorem 8 are chosen so that the exponential decrease in deviation probability dominates the exponential increase in the number of cuts.

4.2   APPLICATIONS

We now show how the skeleton approach can be applied to minimum cuts and maximum flows. We use the following definitions:

Definition 9 *An α-minimum s-t cut is an s-t cut whose value is at most α times the value of the s-t minimum cut. An α-maximum s-t flow is an s-t flow whose value is at least α times the optimum.*

We have the following immediate extension of Theorem 8:

Theorem 10 *Let G be any graph with minimum cut c and let $p = \Theta((\ln n)/\epsilon^2 c)$ as in Theorem 8. Suppose the s-t minimum cut of G has value v. Then with high probability, the s-t minimum cut in $G(p)$ has value between $(1 - \epsilon)pv$ and $(1 + \epsilon)pv$, and the minimum cut has value between $(1 - \epsilon)pc$ and $(1 + \epsilon)pc$.*

Corollary 11 *Assuming $\epsilon < 1/2$, the s-t min-cut in $G(p)$ corresponds to a $(1 + 4\epsilon)$-minimum s-t cut in G.*

Proof: Assuming that Theorem 10 holds, the minimum cut in G is sampled to a cut of value at most $(1 + \epsilon)c$ in $G(p)$. So $G(p)$ has minimum cut no larger. And (again by the previous theorem) this minimum cut corresponds to a cut of value at most $(1 + \epsilon)c/(1 - \epsilon) < (1 + 4\epsilon)c$ when $\epsilon < 1/2$. $\square$

This means that if we use augmenting paths to find maximum flows in a skeleton, we find them faster than in the original graph for two reasons: the sampled graph has fewer edges, and the value of the maximum flow is smaller. The maximum flow in the skeleton reveals an s-t minimum cut in the skeleton, which corresponds to a near-minimum s-t cut of the original graph. An extension of this idea lets us find near-maximum flows: we randomly partition the graph's edges into many groups (each a skeleton), find maximum flows in each group, and then merge the skeleton flows into a flow in the original graph. Furthermore, once we have an approximately maximum flow, we can turn it into a maximum flow with a small number of augmenting path computations. This leads to an algorithm called DAUG that finds a maximum flow in $O(mv\sqrt{(\log n)/c})$ time, improving on the basic augmenting paths algorithm when c is large.

In the following subsections, we detail the algorithms we just sketched. We lead into DAUG with some more straightforward algorithms.

Approximate *s-t* Minimum Cuts. The most obvious application of Theorem 10 is to approximate *s-t* minimum cuts. We can find an approximate *s-t* minimum cut by finding an *s-t* minimum cut in a skeleton.

Lemma 12 *In a graph with minimum cut c, a $(1 + \epsilon)$-approximation to the s-t minimum cut of value v can be computed in $\tilde{O}(mv/\epsilon^3 c^2)$ time (with a low probability of error).*

Proof: Given ϵ, determine the corresponding $p = \Theta((\log n)/\epsilon^2 c)$ from Theorem 10. Suppose we compute an *s-t* maximum flow in $G(p)$. By Theorem 10, $1/p$ times the value of the computed maximum flow gives a $(1 + \epsilon)$-approximation to the *s-t* min-cut value (with high probability). Furthermore, any flow-saturated (and thus *s-t* minimum) cut in $G(p)$ will be a $(1 + \epsilon)$-minimum *s-t* cut in G.

By the Chernoff bound [Che52, MR95b], the skeleton has $O(pm)$ edges (that is, about its expectation) with high probability. Also, by Theorem 10, the *s-t* minimum cut in the skeleton has value $O(pv)$. Therefore, the standard augmenting path algorithm can find a skeletal *s-t* maximum flow in $O((pm)(pv)) = O(mv \log^2 n/\epsilon^4 c^2)$ time. Our improved augmenting paths algorithm DAUG in Section 4.2 lets us shave a factor of $\Theta(\sqrt{pc/\log n}) = \Theta(1/\epsilon)$ from this running time, yielding the claimed bound. $\square$

Approximate Maximum Flows. A slight variation on the previous algorithm will compute approximate maximum flows.

Lemma 13 *In a graph with minimum cut c and s-t maximum flow v, a $(1 - \epsilon)$-maximum s-t flow can be found in $\tilde{O}(mv/\epsilon c)$ time (with a low probability of error).*

Proof: Given p as determined by ϵ, randomly partition the edges into $1/p$ groups, creating $1/p$ graphs. Each graph looks like (has the distribution of) a p-skeleton, and thus with high probability has an *s-t* minimum cut of value at least $pv(1 - \epsilon)$. It has an *s-t* maximum flow of the same value that can be computed in $O((pm)(pv))$ time as in the previous section (the skeletons are not independent, but even the sum of the probabilities that any one of them violates the sampling theorem is negligible). Adding the $1/p$ flows that result gives a flow of value

$v(1-\epsilon)$. **The running time is** $O((1/p)(pm)(pv)) = O(mv(\log n)/\epsilon^2 c)$. **If we use our improved augmenting path algorithm** DAUG **in Section 4.2, we improve the running time by an additional factor of** $\Theta(1/\epsilon)$, **yielding the claimed bound.** $\square$

A Las Vegas Algorithm. Our max-flow and min-cut approximation algorithms are both Monte Carlo, since they are not *guaranteed* to give the correct output (though the error probability can be made arbitrarily small). However, by combining the two approximation algorithms, we can certify the correctness of our results and obtain a *Las Vegas* algorithm for both problems—one that is guaranteed to find the right answer, but has a small probability of taking a long time to do so. This is a standard example of turning a Monte Carlo (error-prone) algorithm into a Las Vegas (correct but occasionally slow) one by checking the correctness of the output and trying again if it is wrong.

Corollary 14 *In a graph with minimum cut c and s-t maximum flow v, a $(1-\epsilon)$-maximum s-t flow and a $(1+\epsilon)$-minimum s-t cut can be found in $\tilde{O}(mv/\epsilon c)$ time by a Las Vegas algorithm.*

Proof: Run both the approximate min-cut and approximate max-flow algorithms, obtaining (with high probability) a $(1-\epsilon/2)$**-maximum flow of value** v_0 **and a** $(1+\epsilon/2)$**-minimum cut of value** v_1. **We know that** $v_0 \leq v \leq v_1$, **so to verify the correctness of the results all we need do is check that** $(1+\epsilon/2)v_0 \geq (1-\epsilon/2)v_1$, **which happens with high probability. To make the algorithm Las Vegas, we repeat both algorithms until each demonstrates the other's correctness (or switch to a deterministic algorithm if the first randomized attempt fails). We are right on the first try with high probability, so the algorithm runs fast with high probability.** $\square$

Exact Maximum Flows. We now use the above sampling ideas to speed up the familiar augmenting paths algorithm for maximum flows. This section is devoted to proving the following theorem:

Theorem 15 ([Kar99a]) *In a graph with minimum cut value c, a maximum flow of value v can be found in $\tilde{O}(mv/\sqrt{c})$ time by a Las Vegas algorithm.*

We assume for now that $v \geq \log n$. Our approach is a randomized divide-and-conquer algorithm that we analyze by treating each subproblem as a (non-independent) random sample. This technique gives a general approach to solving packing problems with an augmentation algorithm (including packing bases in a matroid [Kar98b]). The flow that we are attempting to find can be seen as a packing of disjoint *s-t* paths. We use the algorithm in Figure 4.3, which we call DAUG (Divide-and-conquer AUGmentation).

---

1. Randomly split the edges of G into two groups (each edge goes to one or the other group with probability $1/2$), yielding graphs G_1 and G_2.

2. Recursively compute *s-t* maximum flows in G_1 and G_2.

3. Add the two flows, yielding an *s-t* flow f in G.

4. Use augmenting paths to increase f to a maximum flow.

Figure 4.3   Algorithm DAUG

Note that we cannot apply sampling in DAUG's cleanup phase (Step 4) because the residual graph we manipulate there is directed, while our sampling theorems apply only to undirected graphs. We have left out a condition for terminating the recursion; when the graph is sufficiently small (say with one edge) we use the basic augmenting path algorithm.

The outcome of Steps 1–3 is a flow. Regardless of its value, Step 4 will transform this flow into a maximum flow. Thus, our algorithm is clearly correct; the only question is how fast it runs. Suppose the *s-t* maximum flow is v. Consider G_1. Since each edge of G is in G_1 with probability $1/2$, we can apply Theorem 10 to deduce that with high probability the *s-t* maximum flow in G_1 is at least $(v/2)(1 - \tilde{O}(\sqrt{1/c}))$ and the global minimum cut is $\Theta(c/2)$. The same holds for G_2 (the two graphs are not independent, but this is irrelevant). It follows that the flow f has value $v(1 - \tilde{O}(1/\sqrt{c})) = v - \tilde{O}(v/\sqrt{c})$. Therefore the number of augmentations that must be performed in G to make f a maximum flow is $\tilde{O}(v/\sqrt{c})$. Each augmentation takes $O(m)$ time on an m-edge graph. Intuitively, this suggests the following recurrence for the running time of the algorithm in terms of m, v, and c:

$$T(m, v, c) = 2T(m/2, v/2, c/2) + \tilde{O}(mv/\sqrt{c}).$$

(where we use the fact that each of the two subproblems expects to contain $m/2$ edges). If we solve this recurrence, it evaluates to $T(m, v, c) = \tilde{O}(mv/\sqrt{c})$.

Unfortunately, this argument does not constitute a proof because the actual running time recurrence is in fact a *probabilistic recurrence*: the number of edges and sizes of cuts in the subproblems are random variables not guaranteed to equal their expectations. In particular, the recursion arguments is likely to be false when $c = o(\log n)$. Actually proving the result requires some additional work [Kar99a].

5. RANDOMIZED ROUNDING

Next we turn to Randomized Rounding. Randomized Rounding is a powerful method for approximately solving integer programming problems. The basic idea is to take the values of some *relaxation* of the problem (eg a linear program) and use them to generate integer values that define a solution to the integer program. There are two elements of a randomized rounding approach: a good relaxation that preserves much of the structure of the original intractable problem but can be solved efficiently, and a rounding strategy that transforms the relaxed solution into an integer one (along with a proof that it works well).

We apply randomized rounding to two $\mathcal{NP}$-complete problems: network design and graph coloring. Both rounding approaches are slightly unusual. In the network design problem, we simultaneously round against exponentially many constraints. For graph coloring, we use *semidefinite programming* instead of the more traditional linear programming to determine a structure-preserving relaxation.

5.1 NETWORK DESIGN

The network design problem is a mirror to the minimum cut problem. The input is a set of vertices and a collection of candidate edges, each of which can be purchased for some specified cost. The goal is to design a network whose cuts are "sufficiently large." For example, one might wish to build (at minimum cost) a network that is k-connected. Alternatively one might want a network with sufficient capacity to route a certain amount of flow v between two vertices s and t (thus, the network must have s-t minimum cut v). Network design also covers many other classic problems, often $\mathcal{NP}$-complete, including perfect matching, minimum cost flow, Steiner tree, and minimum T-join. A minimum cost 1-connected graph is just a minimum spanning tree, but for larger values of k the minimum-cost k-connected graph problem is $\mathcal{NP}$-complete even when all edge costs are 1 or infinity [ET76].

Whenever a network design problem can be formulated in terms of (lower bound) constraints on the capacity or number of edges crossing each cut, one can write it as an integer linear program with a $0/1$ variable for each edge that may be purchased and a constraint for each cut. To make the problem more tractable, we can relax the requirement that variables take $0/1$ values and allow them to take fractional values in the interval $[0, 1]$. This gives rise to a linear programming relaxation that can often be solved in polynomial time. Sometimes the linear programs can be represented compactly and solved with standard methods. At other times, even though the relaxation has exponentially many constraints, it has a good separation oracle (e.g. a minimum cut computation for the k-connected subgraph problem) and can thus be solved with the ellipsoid algorithm.

Solving the relaxation yields a fractional solution. Randomized rounding is used to convert the fractional solution back into an integral one. Given fractional variable values $x_1, \ldots, x_m$, we convert them to integer values $y_1, \ldots, y_m$ by setting $y_i = 1$ with probability x_i and 0 otherwise. Note that $E[y_i] = x_i$. It follows that if $ax = b$ for some constraint vector a and scalar b, then $E[ay] = b$. In other words, y is "expected" to satisfy then same constraint that x did.

The problem, of course, is that random experiment deviate somewhat from their expectation. Raghavan and Thompson [RT87] showed that these deviations are often (provably) small enough that the resulting rounded solution is an approximately optimal solution to the integer program. Unfortunately, their analysis is focused on problems with a small number of constraints, which lets them argue that massive deviations from expectation are unlikely to happen. The network design problem has exponentially many constraints, so even unlikely large deviations are likely to occur in some of them. Fortunately, an analogue to our cut sampling theorem bounds these deviations, with the conclusion that randomized rounding can be applied to "fractional graphs" with much the same approximation guarantees as the original Raghavan-Thompson analysis. Among the results this yields is a $1 + O((\log n)/k)$ approximation algorithm for the minimum k-connected subgraph problem [Kar99a].

5.2   GRAPH COLORING

We also apply randomized rounding to the problem of graph coloring. This problem is $\mathcal{NP}$-complete and has recently been proven extremely hard even to approximate well on graphs with large chromatic number [LY93]. However, there still remains some hope that it might be possible to do reasonably well coloring a graph with small chromatic number. In our thesis, we focus on 3-colorable graphs, and show how to color them with $\tilde{O}(n^{1/4})$ colors. The technique extends to give new performance ratios for graphs with larger chromatic number. This work is joint with Rajeev Motwani and Madhu Sudan [KMS98] and built upon the exciting work of Goemans and Williamson [GW95] on the maximum cut problem. We later improved it in joint work with Avrim Blum [BK97].

To attack graph coloring, we turned to the recently developed technique of *semidefinite programming*. Instead of rounding fractional-valued scalars to integers, we round *vectors*. To illustrate, we describe the relaxation of our graph coloring problem. We aim to assign a unit-length *vector* v_i to each vertex i of our graph such that for any two adjacent vertices i and j, the dot product $v_i \cdot v_j \leq -1/2$. To see that this can be done to any three-colorable graph, consider a "star" of three vectors on the unit circle with 120° angles between them, for example $(1, 0)$, $(-1/2, \sqrt{3}/2)$, and $(-1/2, -\sqrt{3}/2)$. Each has unit length and has dot product $-1/2$ with the other two vectors. Given a 3-colored graph, we can solve the vector problem by assigning the first vector to all red vertices, the second to all green, and the third to all blue vertices. This proves that any 3-colorable graph has a feasible solution to our vector problem, which means that it is a valid relaxation.

Solving the relaxation can be formulated as finding a feasible (vector) solution to the following *semidefinite program* (where E denotes the set of edges in the graph G).

$$
\begin{aligned}
v_i \cdot v_j &\leq -1/2 &&\text{if } (i, j) \in E \\
v_i \cdot v_i &= 1.
\end{aligned}
$$

The fact that such a system of constraints (on any linear combination of dot products) can be solved (to within a negligibly small error) in polynomial time is a difficult result [GLS88] which we can fortunately use as a black box.

Unfortunately, there are many feasible assignments to this semidefinite program—most in a dimension much higher than 2. We cannot constrain the solution to be two dimensional (and still solve the problem in polynomial time) so we must decide how to take a high dimensional relaxed solution and transform it into a coloring. Our method for doing so is quite straightforward: we choose a number of *random* unit vectors as *centers*, and color vertex i with the center closest to v_i. We show that if the number of centers is sufficiently large, no two adjacent vertices are likely to be assigned to the same center—that is, we get a legal coloring. The intuition behind our argument is simple. The vectors for adjacent vertices i and j point "away" from each other thanks to the semidefinite constraints. Thus, if i is "near" a random center, j will be "far" from that center and is thus likely to end up attached to some other center. Some technical arguments involving Gaussian distributions suffice to prove that $\tilde{O}(n^{1/4})$ centers suffice to make the probabilities work out.

6.   MONTE CARLO ESTIMATION

The last randomization technique we consider is Monte Carlo estimation. The technique is applied when we want to estimate the probability p of a given event over some probability space. Monte Carlo estimation carries out repeated "trials" (samples from the probability space) and measures how often the given event occurs. This gives a natural estimate of the event probability.

We use Monte-Carlo estimation to attack the *all-terminal network reliability problem*: given a network on n vertices, each of whose m links is assumed to fail (disappear) independently with some probability, determine the probability that the surviving network is connected. The practical applications of this question to communication networks are obvious, and the problem has therefore been the subject of a great deal of study. A comprehensive survey can be found in [Col87]. As mentioned in Section 2., this problem is $\#\mathcal{P}$-hard to solve exactly, so we give a *fully polynomial randomized approximation scheme (FPRAS)* that gives an answer accurate to within a relative error of ϵ in time polynomial in n and $1/\epsilon$. Although our algorithm is quite general [Kar99b], we restrict discussion here to the case where every edge fails independently with the same probability p. We let FAIL(p) denote the failure probability of G when edges fail with probability p.

The basic approach of our FPRAS is to consider two cases. When FAIL(p) is large, we estimate it in polynomial time by direct Monte Carlo simulation of edge failures. That is, we randomly fail edges and check whether the graph remains connected. Since FAIL(p) is large, a small number of trials gives enough data to estimate it well. When FAIL(p) is small, we show that we can focus on the small cuts in a graph. We enumerate them with our cut algorithms and then use a biased Monte Carlo estimation technique to determine their failure probability.

Observe that a graph becomes disconnected precisely when all of the edges in some cut of the graph fail. If each edge fails with probability p, then the probability that a k-edge cut fails is p^k. Thus, the smaller a cut, the more likely it is to fail. It is therefore natural to focus attention on the small graph cuts. In particular, the probability that the graph becomes disconnected is at least p^c (since this is the probability that a minimum cut fails). At the same time, the probability that any one α-minimum cut fails is $p^{\alpha c}$.

We can now describe our two cases. When FAIL(p) $\geq p^c \geq n^{-3}$, we use direct Monte Carlo simulation to estimate the failure probability. A single experiment consists of flipping coins to see which edges fail and then checking whether the graph is connected. If we carry out roughly

$(\log n)/\epsilon^2 \text{FAIL}(p) = \tilde{O}(n^3/\epsilon^2)$ experiments (a polynomial number), we will see about $(\log n)/\epsilon^2$ failures. This provides enough "evidence" to give a good estimate of the failure probability [Che52, KLM89].

Unfortunately, when $\text{FAIL}(p)$ is small, we need too many simulations to develop a good baseline (note that we do not expect to see a single failure until we perform $1/\text{FAIL}(p)$ experiments; this number can be super-polynomial). We instead turn to an enumeration of the small cuts. When $p^c \leq n^{-3}$, we know that a given α-minimum cut fails with probability $p^{\alpha c} \leq n^{-3\alpha}$. But we argued in our Cut Counting theorem that the number of α-minimum cuts is only $n^{2\alpha}$. It follows that the probability that *any* α-minimum cut fails is less than $n^{-\alpha}$—that is, exponentially decreasing with α. Thus, for a relatively small α, the probability that a greater than α-minimum cut fails is negligible. We can therefore approximate $\text{FAIL}(p)$ by approximating the probability that some less than α-minimum cut fails. We do so by enumerating the α-minimum cuts (using a modification of the Contraction Algorithm [KS96]) and then applying a *DNF counting* algorithm developed by Karp, Luby, and Madras [KLM89]. The algorithm of [KLM89] is also based on Monte-Carlo methods, but uses biased sampling to ensure that we see failures often so that a good estimate of their likelihood can be constructed quickly. The contribution of our work is to show that it is possible to build a small formula that can be fed to the DNF counting algorithm to produce a meaningful answer.

7.   CONCLUSION

Randomization has become an essential tool in the design of optimization algorithms. Randomization leads to algorithms that are faster, simpler, and/or better-performing than their deterministic counterparts. The basic techniques of random selection, random sampling, randomized rounding and Monte Carlo estimation let us draw on our intuitions about common cases and representative samples: whenever we expect that something should "usually" happen or be "typical," randomization may give us a way to turn our suspicion into an algorithm. We have demonstrated this approach on numerous basic optimization problems. But a great deal of work remains to be done.

The most direct open question is how far our particular results can be pushed. The minimum spanning tree and minimum cut problems are essentially "done," one with a linear time algorithm and the other with a linear-times-polylog time algorithm; but our results on s-t minimum cuts and maximum flows seem very incomplete: no lower bounds are evident, and our upper bounds are "odd" (e.g. $\tilde{O}(n^{20/9})$ for flows in simple graphs [KL98]) in ways that suggest that it must be possible to improve them (e.g. to $O(n^2)$). Our approximation algorithms apply well to both capacitated and uncapacitated problems, but our exact algorithms so far apply best to uncapacitated problems. We suspect that more can be done here.

More questionable is whether any of our technology can be applied to *directed* graphs. Absolutely none of the results discussed in this article extend to directed graphs: the Contraction Algorithm fails on them, and as a result we have been unable to prove a sampling theorem, a cut counting theorem (in fact a directed graph can have exponentially many minimum cuts), a sampling theorem, a rounding theorem, or anything about directed reliability. One possible explanation for this is that undirected graphs form natural matroids while directed graphs do not [Kar98b].

Thinking more broadly, a fundamental question about randomization is whether it is truly "necessary." Often, after a randomized algorithm gives insight into a problem, one can devise a deterministic algorithm with some of the same properties. Within theoretical computer science, there is an entire subfield devoted to *derandomization*—the development of techniques that will mechanically convert a randomized algorithm into a deterministic one. For example, the randomized rounding procedure for (polynomial size) linear programs can be made deterministic [Rag88], as can our randomized rounding algorithm for graph coloring [MR95a]. We have also derandomized our Contraction Algorithm [KM97].

Even when it is possible to derandomize an algorithm, it may not be worth doing so. The derandomization can add complexity, either computational (e.g. in the case of the Contraction Algorithm, where the derandomization drastically slows the algorithm) or conceptual (e.g. for randomized rounding, where the intuitive expectation argument is replaced by a more complex numeric calculation).

However, there are still motivations for exploring the derandomization question. Perhaps the strongest is the wish for an algorithm with predictable behavior. In a situation with lives at stake, it would be unsatisfactory to be right *most* of the time, or *usually* fast enough. This problem is particularly acute with our Monte Carlo algorithms, where one cannot even tell whether the answer is correct! An obvious place to begin is the minimum cut problem, where a Monte Carlo algorithm can solve the problem (with high probability) in $\tilde{O}(m)$ time but the best known deterministic running time is $\tilde{O}(mn)$. Another specific question is whether there is a *deterministic* linear-time minimum spanning tree algorithm (which would finally put the problem to rest for good). A more abstract question is the following: we have proven that any graph has a sparse "skeleton" that accurately approximates its cuts; this seems to have assorted uses. Can such a skeleton be constructed deterministically in polynomial time?

Comments and questions on this survey are most welcome.

Acknowledgments

The author would like to thanks the editors of Optima for allowing republication of this survey.

References

[AMO93] Ravindra K. Ahuja, Thomas L. Magnanti, and James B. Orlin. *Network Flows: Theory, Algorithms, and Applications.* Prentice Hall, Englewood Cliffs, New Jersey, 1993.

[BK96] András A. Benczúr and David R. Karger. Approximate s–t min-cuts in $\tilde{O}(n^2)$ time. In *Proceedings of the 28th ACM Symposium on Theory of Computing*, pages 47–55. ACM, ACM Press, May 1996.

[BK97] Avrim Blum and David R. Karger. Improved approximation for graph coloring. *Information Processing Letters*, 61(1):49–53, January 1997.

[BK00] András A. Benczúr and David R. Karger. Augmenting undirected edge connectivity in $\tilde{O}(n^2)$ time. *Journal of Algorithms*, 37:2–36, 2000.

[Blu94] Avrim Blum. New approximation algorithms for graph coloring. *Journal of the ACM*, 41(3):470–516, May 1994.

[CGK+97] Chandra C. Chekuri, Andrew V. Goldberg, David R. Karger, Matthew S. Levine, and Cliff Stein. Experimental study of minimum cut algorithms. In Saks [Sak97], pages 324–333.

[Che52] H. Chernoff. A measure of the asymptotic efficiency for tests of a hypothesis based on the sum of observations. *Annals of Mathematical Statistics*, 23:493–509, 1952.

[Col87] Charles J. Colbourn. *The Combinatorics of Network Reliability*, volume 4 of *The International Series of Monographs on Computer Science*. Oxford University Press, 1987.

[ET76] K. P. Eswaran and Robert E. Tarjan. Augmentation problems. *SIAM Journal on Computing*, 5:653–665, 1976.

[FF62] Lester R. Ford, Jr. and Delbert R. Fulkerson. *Flows in Networks*. Princeton University Press, Princeton, New Jersey, 1962.

[FR75] Robert W. Floyd and Ronald L. Rivest. Expected time bounds for selection. *Communications of the ACM*, 18(3):165–172, 1975.

[GGP+94] Michel X. Goemans, Andrew Goldberg, Serge Plotkin, David Shmoys, Éva Tardos, and David Williamson. Improved approximation algorithms for network design problems. In Daniel D. Sleator, editor, *Proceedings of the 5th Annual ACM-SIAM Symposium on Discrete Algorithms*, pages 223–232. ACM-SIAM, January 1994.

[GGST86] Harold N. Gabow, Zvi Galil, T. H. Spencer, and Robert E. Tarjan. Efficient algorithms for finding minimum spanning tree in undirected and directed graphs. *Combinatorica*, 6:109–122, 1986.

[GLS88] Martin Grötschel, Làszló Lovász, and Alexander Schrijver. *Geometric Algorithms and Combinatorial Optimization*, volume 2 of *Algorithms and Combinatorics*. Springer-Verlag, Berlin, 1988.

[GR97] Andrew Goldberg and Satish Rao. Beyond the flow decomposition barrier. In *Proceedings of the 30th Annual Symposium on the Foundations of Computer Science*, pages 2–11. IEEE, IEEE Computer Society Press, October 1997.

[GT88] Andrew V. Goldberg and Robert E. Tarjan. A new approach to the maximum flow problem. *Journal of the ACM*, 35:921–940, 1988.

[GW95] Michel X. Goemans and David P. Williamson. Improved approximation algorithms for maximum cut and satisfiability problems using semidefinite programming. *Journal of the ACM*, 42(6):1115–1145, May 1995. A preliminary version appeared in Proceedings of the 26th ACM Symposium on Theory of Computing pages 422–431. ACM, ACM Press, May 1994.

[HO94] Jianxiu Hao and James B. Orlin. A faster algorithm for finding the minimum cut in a directed graph. *Journal of Algorithms*, 17(3):424–446, 1994. A preliminary version appeared in Proceedings of the 3rd Annual ACM-SIAM Symposium on Discrete Algorithms.

[Hoa62] C. A. R. Hoare. Quicksort. *Computer Journal*, 5(1):10–15, 1962.

[Kar94] David R. Karger. *Random Sampling in Graph Optimization Problems*. PhD thesis, Stanford University, Stanford, CA

130

94305, 1994. Contact at karger@lcs.mit.edu. Available from http://theory.lcs.mit.edu/~karger.

[Kar98a] David R. Karger. Better random sampling algorithms for flows in undirected graphs. In Howard Karloff, editor, *Proceedings of the 9th Annual ACM-SIAM Symposium on Discrete Algorithms*, pages 490–499. ACM-SIAM, January 1998.

[Kar98b] David R. Karger. Random sampling and greedy sparsification in matroid optimization problems. *Mathematical Programmming B*, 82(1–2):41–81, June 1998. A preliminary version appeared in Proceedings of the 34th Annual Symposium on the Foundations of Computer Science.

[Kar98c] David R. Karger. Random sampling in graph optimization problems: A survey. *Optima*, 58:1–11, 1998.

[Kar99a] David R. Karger. Random sampling in cut, flow, and network design problems. *Mathematics of Operations Research*, 24(2):383–413, May 1999. A preliminary version appeared in Proceedings of the 26th ACM Symposium on Theory of Computing.

[Kar99b] David R. Karger. A randomized fully polynomial approximation scheme for the all terminal network reliability problem. *SIAM Journal on Computing*, 29(2):492–514, 1999. A preliminary version appeared in Proceedings of the 27th ACM Symposium on Theory of Computing.

[Kar00] David R. Karger. Minimum cuts in near-linear time. *Journal of the ACM*, 47(1):46–76, January 2000. A preliminary version appeared in Proceedings of the 28th ACM Symposium on Theory of Computing.

[KKT95] David R. Karger, Philip N. Klein, and Robert E. Tarjan. A randomized linear-time algorithm to find minimum spanning trees. *Journal of the ACM*, 42(2):321–328, March 1995.

[KL98] David R. Karger and Matthew Levine. Finding maximum flows in simple undirected graphs seems faster than bipartite matching. In *Proceedings of the 29th ACM Symposium on Theory of Computing*, pages 69–78, New York, May 23–26 1998. ACM, ACM Press.

[KLM89] Richard M. Karp, Michael Luby, and N. Madras. Monte Carlo approximation algorithms for enumeration problems. *Journal of Algorithms*, 10(3):429–448, September 1989.

[KM97] David R. Karger and Rajeev Motwani. Derandomization through approximation: An $\mathcal{NC}$ algorithm for minimum cuts.

SIAM Journal on Computing, 26(1):255–272, January 1997. A preliminary version appeared in Proceedings of the 25th ACM Symposium on Theory of Computing, pp. 497–506.

[KMS98] David R. Karger, Rajeev Motwani, and Madhu Sudan. Approximate graph coloring by semidefinite programming. *Journal of the ACM*, 45(2):246–265, March 1998.

[KS96] David R. Karger and Clifford Stein. A new approach to the minimum cut problem. *Journal of the ACM*, 43(4):601–640, July 1996. Preliminary portions appeared in SODA 1992 and STOC 1993.

[KT97] David R. Karger and Ray P. Tai. Implementing a fully polynomial time approximation scheme for all terminal network reliability. In Saks [Sak97], pages 334–343.

[LY93] Carsten Lund and Mihalis Yannakakis. On the hardness of approximating minimization problems. In Alok Aggarwal, editor, *Proceedings of the 25th ACM Symposium on Theory of Computing*, pages 286–293. ACM, ACM Press, May 1993.

[MR95a] Sanjeev Mahajan and H. Ramesh. Derandomizing semidefinite programming based approximation algorithms. In *Proceedings of the 36th Annual Symposium on the Foundations of Computer Science*, pages 162–169. IEEE, IEEE Computer Society Press, October 1995.

[MR95b] Rajeev Motwani and Prabhakar Raghavan. *Randomized Algorithms*. Cambridge University Press, New York, NY, 1995.

[NI92] Hiroshi Nagamochi and Toshihide Ibaraki. Linear time algorithms for finding k-edge connected and k-node connected spanning subgraphs. *Algorithmica*, 7:583–596, 1992.

[Rag88] Prabhakar Raghavan. Probabilistic construction of deterministic algorithms: Approximate packing integer programs. *Journal of Computer and System Sciences*, 37(2):130–43, October 1988.

[RT87] Prabhakar Raghavan and C. D. Thompson. Randomized rounding: a technique for provably good algorithms and algorithmic proofs. *Combinatorica*, 7(4):365–374, 1987.

[Sak97] Michael Saks, editor. *Proceedings of the 8th Annual ACM-SIAM Symposium on Discrete Algorithms*. ACM-SIAM, January 1997.

[Tar83] Robert E. Tarjan. *Data Structures and Network Algorithms*, volume 44 of *CBMS-NSF Regional Conference Series in Applied Mathematics*. SIAM, 1983.

Chapter 5

THE DELAY SEQUENCE ARGUMENT

Abhiram Ranade

Department of Computer Science and Engineering
Indian Institute of Technology, Powai, Mumbai 400076
Ranade@cse.iitb.ernet.in

Abstract   The delay sequence argument is a powerful but simple technique for analyzing the running time of a variety of distributed algorithms. We present the general framework in which it may be applied, and give four examples of its use. These examples are drawn from diverse settings: odd-even transposition sort on linear arrays, protocols for tolerating noisy channels in distributed networks, packet routing, and parallelization of backtrack search.

Keywords: delay sequence argument, critical path method, odd-even transposition sort, packet routing, randomized scheduling, randomized resource allocation, fault-tolerant communication, parallel backtrack search.

1.   INTRODUCTION

While investigating a phenomenon, a natural strategy is to look for its immediate cause. For example, the immediate cause for a student failing an examination might be discovered to be that he did not study. While this by itself is not enlightening, we could persevere and ask why did he not study, to which the answer might be that he had to work at a job. Thus an enquiry into successive immediate causes could lead to a good explanation of the phenomenon. Of course, this strategy may not always work. In the proverbial case of the last straw breaking the camel's back, detailed enquiry into why the last straw was put on the camel (it might have flown in with the wind) may not lead to the right explanation of why the camel's back broke. Nevertheless, this strategy is often useful in real life. How it can help in investigating several probabilistic phenomena is the subject of this chapter.

S. Rajasekaran et al (eds.), Handbook of Randomized Computing, Volume 1, pp, 133–150.
© 2001 *Kluwer Academic Publishers.*

In the analysis of probabilistic algorithms and phenomena, this strategy of successively looking for immediate causes has been called the *Delay Sequence Argument*. We will present a general framework in which this argument is applicable, and also illustrate its application to four apparently very diverse problems: Odd-even transposition sort, Distributed computing over noisy networks, Packet routing, and Parallel backtrack search. Odd-even transposition sort is an elementary algorithm for sorting on a linear array of processors[5]. Although the algorithm is deterministic, we have included it because the idea of following back along the chain of immediate causes leads to a very simple proof for the time taken by the algorithm. In our second problem, we are given a network with noisy channels. On this we are required to devise a protocol that simulates the execution of the corresponding noise-less network. Delay sequence arguments provide a very sharp analysis of such protocols. The packet routing problem is also concerned with communication, but without noise. In this we are given a network in which each node holds packets required to be sent to specified destinations while obeying the delay and capacity constraints of the links. This problem is often solved using randomized path selection and randomized scheduling of the packets; the delay sequence argument figures prominently in the analysis of these. In the backtrack search problem we are given a combinatorial optimization problem represented as the root of a tree. The basic operation allowed is to expand a tree node and generate its descendants if any. The goal is to generate the entire tree, the shape of which is revealed only during the execution. The delay sequence argument can be used to analyze certain randomized strategies for parallelizing the tree generation.

The earliest example of the delay sequence argument known to us is by Aleliunas[1] and Upfal[13], both in the context of packet routing. The idea of following back the chain of immediate causes is expressed in this case as "What packet is responsible for the delay suffered by this packet?". This identifies a sequence of immediately responsible packets which "explain" the delay incurred by the first packet. This is the origin of the term *Delay sequence*. The argument has been used by others in packet routing[6, 11], in backtrack search[12], in analysis of noise-tolerant distributed protocols[10], in analysis of asynchronous parallel computation[8, 9]. The earliest use of the basic idea is quite possibly the *Critical Path Method* discussed in Operations Research texts.

Besides the delay sequence argument itself, this chapter will touch upon two powerful tools available to designers of randomized algorithms. The first is *randomized resource allocation:* if a task requires one resource from a pool of resources, the precise one is selected randomly. The second idea is *randomized scheduling:* using randomly chosen priorities to break ties when several tasks contend for a single resource. Both these ideas appear in the papers mentioned relating to packet routing as well as the paper relating to backtrack search.

1.1   OVERVIEW

In Section 2. we will present the basic framework in which the delay sequence argument is applied. In Section 3. we present its use in the odd-even transposition sort. Section 4. discusses distributed computing in the presence of noise. Section 5. discusses packet routing. Section 6. describes parallel backtrack search. Finally Section 7. concludes.

2.   FRAMEWORK

The computational model considered in this paper is a synchronous distributed network of processors: in a single time step each processor can perform $O(1)$ primitive computations and send bounded size messages to immediate neighbours. A *computational problem* is a directed acyclic graph. The nodes represent atomic operations and edges represent precedence constraints between the atomic operations. For example, in the backtrack search problem, each node represents the operation of incremental exploration of the search space for a combinatorial optimization problem. In case of the packet routing problem, each node represents the transmission of a message across some link. We will use the term *computational event* (or just event) to mean the execution of a single node. The central question is *how do we schedule the computational events in a manner consistent with the precedence constraints and resource availability?* We will propose scheduling algorithms and then analyze their performance using the delay sequence argument.

The delay sequence argument may be thought of as a general technique for analyzing such distributed systems. The systems will typically be stochastic; either because they employ randomized algorithms or because their definition includes randomly occurring faults. The technique works by examining execution traces of the system[1]. An execution trace is simply a record of the computational events that happened during execution, including associated random aspects. The idea is to characterize traces in which execution times are high, and show that such traces will occur only rarely.

An important notion we use is that of a *Critical Set* of a (computational) event. Let T_0 denote a fixed integer which we will call the startup

---

[1]Sort of a *post mortem.*.

time for the computation. We will say Y is a critical set for event x iff the occurrence of x at time $t > T_0$ in any execution trace guarantees the occurrence of some $y \in Y$ at time $t-1$ in that execution trace. We could of course declare all events to be in the critical set of every event – this trivially satisfies our definition. However it is more interesting if we can find small critical sets. This is done typically by asking "If x happened at t, why did it not happen earlier? Which event y happening at time $t - 1$ finally enabled x to happen at t?" The critical set of x consists of all such events y.

A delay sequence is simply a sequence of events such that the ith event in the sequence belongs to the critical set of the $i + 1$th event of the sequence. A delay sequence is said to occur in an execution trace if all the events in the delay sequence occur in that trace. Delay sequences are interesting because of the following Lemma.

Lemma 1 *Let T denote the execution time for a certain trace. Then a delay sequence*

$$E = E_{T_0}, E_{T_0+1}, \ldots, E_T$$

occurs in that trace, with event E_t occurring at time t.

Proof: Let E_T be any of the events happening at time T. Let E_{T-1} be any of the events in the critical set of E_T that is known to have happened at time $T - 1$. But we can continue in this manner until we reach E_{T_0}. □

The length of the delay sequence will typically be the same as the execution time of the trace, since typically $T_0 = 1$ as will be seen. Thus, to argue that execution times are small it suffices to argue that long delay sequences are unlikely. This is done by estimating the probability that a particular long delay sequence occurs, multiplied by the number of possible long delay sequences.

To estimate the probability of a delay sequence it is convenient if the events in it are independent. This will not be the case in the last two of our applications. So in this case we will drop some events from the delay sequence to get an abbreviated sequence whose events will all be independent. The probability of the abbreviated sequence is then easily estimated.

The events in the delay sequence will be seen to have substantial structure. For example, consecutive events in the sequence will typically occur either on the same processor, or in processors adjacent in the network. Such structure will constrain the number of possible sequences, and this will be relevant while counting.

We will use the standard notion of "high probability" defined as follows. Suppose N denotes the problem size. Then we will say that $f(N) = O(g(N))$ with high probability if for any given k there exist constants N_0 and c such that $\Pr[f(N) > cg(N)] \le N^{-k}$ whenever $N \ge N_0$.

We will have occasion to use the following well known inequalities:

$$\binom{n}{r} \le \left(\frac{ne}{r}\right)^r \qquad \Pi_i \binom{n_i}{r_i} \le \binom{\sum_i n_i}{\sum_i r_i}$$

3.   ODD-EVEN TRANSPOSITION SORT

Suppose you have a linear array of N processors, each holding a single key. The processors are numbered 1 to N left to right, and it is required to arrange the keys so that the key held in processor i is no larger than the key in processor $i + 1$.

Algorithm. [5] The basic step is called a *comparison-exchange* in which a pair of adjacent processors i and $i + 1$ send their keys to each other. Each processor compares the received key with the key it held originally, and then processor i retains the smaller and processor $i + 1$ the larger. On every odd step of the algorithm in parallel for all i, processors $2i - 1$ and $2i$ perform a comparison exchange step. On every even step, in parallel for all i, processors $2i$ and $2i + 1$ perform a comparison exchange step. Clearly, the smallest key will reach processor 1 in N steps, following which the second smallest will take at most $N - 1$ and so on. The question is, will sorting finish faster than the $O(N^2)$ time suggested by this naive analysis.

Theorem 1 *Odd-even transposition sort on an N processor array completes in N steps.*

The proof uses the Zero-one Lemma[2, 4, 5], as per which we only need consider the case in which each key is either a 0 or a 1. We focus on the movement of the zeros during execution. For the purpose of the analysis, we will number the zeros in the input from left to right. Notice that during execution, zeros do not overtake each other, i.e. the ith zero from the left at the start of the execution continues to have exactly $i - 1$ zeros to its left throughout execution: if the ith and $i + 1$th zero get compared during a comparison-exchange step, we assume that the ith is retained in the smaller numbered processor, and the $i + 1$th in the larger numbered processor.

Let (p, z) denote the event that the zth zero (from the left) arrives into processor p.

Lemma 2 *The critical set for* (p, z) *is* $\{(p+1, z), (p-1, z-1)\}$. *The startup time* $T_0 = 2$.

Proof: Suppose (p, z) happened at time t. We ask why it did not happen earlier. Of course, if $t \leq 2$ then this might be the first time step in which the zth zero participated in a comparison exchange. Otherwise, the reason must be one of the following: (i) the zth zero reached processor $p+1$ only in step $t-1$, or (ii) the zth zero reached processor $p+1$ earlier, but could not move into processor p because $z-1$th zero left processor p only in step $t-1$. In other words, one of the events $(p+1, z)$ and $(p-1, z-1)$ must have happened at time $t-1$. □

$(p+1, z)$ and $(p-1, z-1)$ will respectively be said to cause transit delay and comparison delay for (p, z).

Proof of Theorem 1: Suppose sorting finishes at some step T. A delay sequence $E_{T_0}, \ldots, E_T$ with $T_0 = 2$ must have occurred. We further know that $E_t = (p_t, z_t)$ occurred at time t.

At time 2 we know that there are at least $z_T - z_2$ zeros to the right of zero z_2, and also at least the one that moved right at step 2. Thus $p_2 + 1 + z_T - z_2 \leq N$. Let the number of comparison delays in the delay sequence be c. But this is also the number of distinct zeros occurring in the delay sequence, i.e. $c = z_T - z_2$. Further, a transit delay of an event happens to its right, while a comparison delay to its left. Thus from p_T to p_2 we have $T - 2 - c$ steps to the right, and c steps to the left, i.e. $p_2 - p_T = T - 2 - 2c$. Thus

$$T = p_2 - p_T + 2 + 2c = p_2 - p_T + 2 + 2z_T - 2z_2 \leq N - p_T + 1 + z_T - z_2$$

Noting that $p_T = z_T$ and $z_2 \geq 1$ the result follows. □

4. TOLERATING NOISY CHANNELS

Suppose you have a computation that takes time T on a network in which channels are noise free. How much time is required to complete it if the channels have noise? This problem is considered by Rajagopalan and Schulman[10] who give essentially optimal protocols to simulate execution of noise free networks on noisy networks under the assumption that transmission errors are independent.

Here we will consider a simplified version of their problem involving the simpler *erasure* channels. In an erasure channel, the transmitted message is received correctly with a probability $1 - \pi$, while with probability π a special "error" symbol is received. Note that in an erasure channel the receiver can detect that an error has taken place, while in the more commonly considered model[10] noise may lead to reception of a different but perfectly legal message. Like Rajagopalan and Schulman [10], we also assume that transmission errors are independent.

The computation in a noise free model G is as follows. All processors synchronously execute some τ iterations, in each of which each processor executes a single instruction and sends and receives messages to and from its neighbours. The instruction executed in the ith iteration by processor p is denoted as $I(p,i)$. After executing this instruction processor p sends a message $M(p,i)$ to each of its neighbours. It analogously receives messages sent by its neighbours. The collection of messages received by p is denoted by $R(p,i)$. After receiving $R(p,i)$ the next iteration begins. Note that the data in $R(p,i-1)$ may be needed to execute $I(p,i)$; the results of $I(p,i)$ in turn may be used to construct message $M(p,i)$. We assume for simplicity that the message includes the instruction number i in its body. This computation may be represented by its computational graph C in which the nodes are $I(p,i)$. There are directed edges from $I(p,i)$ to $I(q,i+1)$, where $q = p$ or q is a neighbour of p. Our goal is to execute the computation specified by C in the presence of noise.

Our protocol is as follows. The execution happens in *time steps*, where in each step each processor p executes a single instruction $I(p,i)$ for the largest i possible. Specifically in each time step t, processor p does the following:

t.1 Determine the smallest i for which p has not correctly received $R(p,i)$. i may be viewed as the value of the *program counter* for p, since it says what instruction to execute next.

t.2 Execute the instruction $I(p,i)$.

t.3 Send a message $M(p,i)\|M(p,i-1)$ to all neighbours, where $x\|y$ denotes a single message made by concatenating messages x and y.

t.4 Receive messages sent by neighbours.

Note that in a single time step, different processors might be executing different iterations of the error-free program. Further, an instruction $I(p,i)$ may be executed several times if the program counter of p does not advance because of transmission errors.

The intuition behind the protocol is as follows. First, note that the program counter of neighbouring processors can differ by at most 1; this is because in order for p to have program counter i it must receive messages $M(q,i-1)$ from every neighbour q – but this can happen only if q has program counter at least $i - 1$. Suppose now that we have a configuration in which the smallest program counter does not increase. Let p be a processor that has this smallest program counter i. Then any of its neighbours q will have program counter either i or $i + 1$. Thus the messages q sends will include $M(q, i)$ – which will eventually be correctly received. Thus eventually every message in $R(p, i)$ will be correctly received causing the program counter of p to become $i + 1$.

4.1 EVENTS AND CRITICAL SETS

We will say that an event $(p, i, 1)$ happens at time t if at time t processor p executes instruction $I(p, i)$, and there is no error in receipt of messages at time t at p. We will say that an event $(p, i, 0)$ happens at time t if at time t processor p executes $I(p, i)$ but there is an error while receiving messages from at least one of the neighbours in step t.

Lemma 3 *The critical set for $(p, i, 1)$ or $(p, i, 0)$ consists of event $(p, i, 0)$ and events $(q, i - 1, x)$ where x is either 0 or 1 and q is either p or a neighbour of p. The startup time $T_0 = 1$.*

Proof: Consider any event (p, i, x) happening at time $t > 1$. We will show that some event from the above set must happen at time $t - 1$. At time $t - 1$, either $(p, i - 1, x)$ or (p, i, x) must have happened. The former belongs to the set, so assume that the latter happened. Since the program counters of neighbours differ by at most 1, the program counter of every neighbour q must be either $i - 1$, i or $i + 1$ at time $t - 1$. If the program counter of some q is $i - 1$, then again we have an event from the above mentioned set occurring. If the program counter of all neighbours is i or $i + 1$, then the message transmitted by these neighbours must together contain $R(p, i)$. But then the only reason the program counter of p does not increment at time t could be that one of the messages was not received due to a transmission error. Thus at time $t - 1$ event $(p, i, 0)$ must have happened. $\square$

4.2 ANALYSIS

Lemma 4 *With high probability the time T for finishing the computation is $O(\tau + \log N)$ assuming that the probability π of error for any single transmission is smaller than some constant (that does not depend upon τ and N, but may depend upon the maximum degree Δ of the network).*

Proof: We estimate the probability of occurrence of a delay sequence $E_1, \ldots, E_T$ in which $E_t = (p_t, i_t, x_t)$. Note first that the delay sequence will have only τ events in which $x_t = 1$, because such events are always accompanied by a drop in i_t, and i_t can drop at most τ times. Thus at least $\delta \geq T - \tau$ events have $x_t = 0$.

Given any delay sequence with δ transmission errors, the probability of its occurrence is at most $(\Delta\pi)^\delta$. The number of possible delay sequences can be estimated as follows. We may pick p_1 in N ways, given any p_t we can fix p_{t+1} as either itself or one of its neighbours, i.e. in $\Delta + 1$ ways. Next, we need to identify which transmissions failed. This can be done in at most 2^T ways. The total number of length T delay sequences is thus at most $N(\Delta + 1)^T 2^T$. The probability that some such delay sequence actually occurs is thus at most:

$$(\Delta\pi)^\delta 2^T N(\Delta + 1)^T = \left(2^{\frac{\log N}{\delta}} \Delta (2\Delta + 2)^{\frac{T}{\delta}} \pi\right)^\delta$$

Choosing $T = 2\tau + k \log N$ gives $\delta = \tau + k \log N$ and thus the quantity in the parenthesis is at most $2\Delta(2\Delta + 2)^2\pi$. So if we have $\pi \leq \frac{1}{4\Delta(2\Delta+2)^2}$ we get the probability of the occurrence of some delay sequence as N^{-k}. Thus with high probability, the time to finish is $2\tau + O(\log N)$. $\qquad\square$

5. PACKET ROUTING

Packet routing is a fundamental problem in parallel and distributed computing. Here we consider a formulation of it on *levelled directed networks*[6]. This formulation is interesting because routing problems on many networks can be formulated as routing problems on appropriate levelled directed networks. A levelled directed network with levels $0 \ldots L$ is a directed acyclic graph in which each node v is assigned a number $\text{Level}(v) \in [0, L]$, such that $\text{Level}(v) = 1 + \text{Level}(u)$ for any edge (u, v).

In our packet routing problem, we are given a levelled directed network with a total of N packets distributed amongst its nodes. Each packet has an assigned path in the network along which the packet must be moved, subject to the following restrictions: (i) it takes one time step to cross any edge, (ii) only one packet can cross any edge in one time step, (iii) the decision of which packet to send out along an edge (u, v) must be taken by node u based only on locally available information, i.e. information about paths of packets that are currently residing in u or have already passed through it. The packet routing algorithm may associate small amount of additional information with each packet which can be used while taking routing decisions and which moves with the packets. It is customary to characterize a routing problem in terms of two parameters: d, which denotes the length of the longest packet path, and c, which denotes the maximum number of packet paths assigned through any network edge. Clearly $\max(c, d) = \Omega(c + d)$ is a lower bound on the time needed. The question is how close to this can an algorithm get.

Assuming that each processor has unbounded buffer space for holding messages in transit, we will show that with high probability packet routing can be finished in time $O(c + L + \log N)$, where c denotes the maximum number of packets required to pass through any single edge. If $d = \Omega(L + \log N)$, as is the case in the problems arising in the use of this framework as in [6], our result is clearly optimal to within constant factors. Our time bound actually holds even if the buffer space is bounded[6], but the proof as well as the algorithm are substantially more complex.

Deterministic algorithms are not known to work very well for this problem. In fact, Leighton, Maggs, and Rao[7] show that in general $\Omega(\frac{cd}{\log c})$ time is necessary for a large natural class of deterministic algorithms that includes FIFO (send first the message which arrived first into the node), farthest to go first (send first the message which has the longest distance yet to go) and several others. The best known upper bound for deterministic algorithms is $O(cd)$. This is applicable for almost any algorithm which sends some packet on every link if possible:[2] no packet need wait for more than $c - 1$ steps to let other packets pass at each of the at most d edges along it's path. This bound is substantially worse than $O(c + L + \log N)$ for the typical setting.

5.1 OUR ALGORITHM

Our routing algorithm, following [6, 11], is simple but randomized. For each packet we choose a rank independently and uniformly randomly from the range $[1, \rho]$ where ρ is a number which will be fixed later. Then at each step every node u does the following:

> For every link (u, v) if there are packets waiting in u to be sent on (u, v), select the one with the smallest rank. If there are several with the same rank, then select any one. Send the selected packet (if any) along each link.

5.2 EVENTS AND CRITICAL SETS

By (u, v, p, r) we will denote the event that a packet p gets assigned rank r and gets transmitted on link (u, v). If (u, v, p, r) happens at step

---

[2]Algorithms which may hold back packets even though a link is free are interesting when the buffer space in the receiving nodes is bounded. In fact, in this case it sometimes makes sense to withhold transmission even on occasions when unused buffer space is available in the receiving node[5, 6, 11].

$t > 1$ then it is easily seen that at the previous step either p arrived into u, or if p arrived earlier then some other packet of rank at most r was transmitted on (u, v). Thus for event (u, v, p, r) we will define event (w, u, p, r) as its *transit delay* event, and events (u, v, p', r') with $p' \neq p$ and $r' \leq r$ as its *rank delay* events. Thus the following is obvious.

Lemma 5 *The startup time is $T_0 = 1$. The critical set for any event consists of its transit delay event and its rank delay events.*

Let $E_1, \dots, E_T$ be a delay sequence with $E_t = (u_t, v_t, p_t, r_t)$. We note its following properties:

1. Either $(u_t, v_t) = (u_{t+1}, v_{t+1})$, or $v_t = u_{t+1}$.

2. All (u_t, v_t) are edges on some (directed) path P (not necessarily of any packet).

3. At least $\delta \geq T - L$ events E_{t-1} are rank delays of E_t. This is because we can only have one transit delay per edge in the path P, and there are at most L edges in P.

4. $r_1 \leq r_2 \leq \dots \leq r_T$.

Abbreviated delay sequence.   Consider the events which are rank delays of the preceding events in the delay sequence. Clearly all the packets in such events must be distinct. The abbreviated delay sequence is simply the subsequence of rank delay events in the delay sequence. Clearly this subsequence must have length at least $T - L$, and must satisfy properties 2 and 4 above.

5.3   ANALYSIS

Theorem 2 *With high probability the time T required to finish routing is $O(c + L + \log N)$.*

Proof: We will estimate the probability that some abbreviated delay sequence of length $\delta = T - L$ occurs. This clearly bounds the probability that time T is needed to finish routing.

Number of abbreviated delay sequences.   This is simply the number of ways of choosing (u_i, v_i, p_i, r_i) for $i = 1$ to δ such that properties 2,4 above are satisfied. To ensure this, we use an indirect procedure that first chooses a path in the network, then chooses each edge (u_i, v_i) from the path. Each p_i is then constrained to be one of the packets through the chosen (u_i, v_i). Finally the r_i are chosen so that they are non-decreasing. The path must begin at the origin of some packet. Thus the origin can be chosen in N ways. Each consecutive edge on the path can be chosen in Δ ways, where Δ is the degree of the network. Thus the total number of ways in which the path can be chosen is at most $N\Delta^L \le 2^\delta \Delta^\delta$, assuming $\delta \ge \log N, L$. The edge (u_i, v_i) must be selected such that the edge (u_{i-1}, v_{i-1}) is the same or occurs earlier in the path. Thus all the edges together can be chosen in $\binom{L+\delta-1}{\delta} \le 2^{2\delta}$ ways, assuming $\delta \ge L$. Given that edge (u_i, v_i) is fixed, the packet p_i traversing it can be chosen in at most c ways, for a total of c^δ ways. Finally the ranks of the packets must be chosen in non-decreasing order, i.e. in $\binom{\rho+\delta-1}{\delta} \le \left(\frac{2\rho e}{\delta}\right)^\delta$ ways assuming $\rho \ge \delta$. The number of abbreviated sequences of length δ is thus no more than $\left(2\Delta \cdot 4 \cdot c\frac{2\rho e}{\delta}\right)^\delta$.

Probability of fixed sequence. The events in each sequence concern distinct packets. Any event (u_i, v_i, p_i, r_i) occurs if p_i gets assigned rank r_i. Thus all events together happen with probability at most $\frac{1}{\rho^\delta}$.

The probability of some abbreviated delay sequence occurring is thus at most

$$\left(2\Delta \cdot 4 \cdot c\frac{2\rho e}{\delta}\frac{1}{\rho}\right)^\delta \le \left(\frac{16ec\Delta}{\delta}\right)^\delta$$

So choose $T = 2L + 32ec\Delta + k\log N$. Noting $\delta = T - L$ we have the probability of time T at most N^{-k}. $\qquad\square$

6. BACKTRACK SEARCH

Backtrack search, or complete enumeration of the search space, is often required in solving many combinatorial problems. In this section we consider an algorithm for parallelizing such searches. The model for our problem is due to Karp and Zhang[3], who consider parallelization using a complete network of processors. Here we discuss parallelization using a Butterfly network of processors. This discussion is based on [12].

Karp and Zhang model search problems generically as a rooted tree with a cost function over the leaves. The goal is to find the leaf with the least cost. Each node of the tree represents partial solutions to the search problem at progressively greater level of specification from the root towards the leaves. As an example, tree nodes may represent the state of a printed circuit board, with as many components placed as the distance of the node from the root. In general, at the start of the search we are given only the root of the tree, and also a *node expansion* procedure which given any tree node can generate the children of the node. Note that in general, the tree that results from node expansion need not be balanced, and in fact its shape will not be known *a priori*. Karp and Zhang show that even without knowing the shape, all nodes of an N node tree with height h can be generated in time $O(h + N/P)$ with high probability, using a complete network of P processors with the assumption that it takes unit time to expand a node and a unit time to send a node to other processors. This is optimal within constant factors, because the h nodes on the longest path can be expanded only sequentially, and N/P time is needed even with perfect parallelization.

We consider the problem on a P processor Butterfly. Node expansion takes unit time in our model as well, however, in our model a node can be sent only to nearest neighbours in the Butterfly in a single time step. Further, we assume the tree is binary. For the discussion here, we also assume that the tree is level balanced, i.e. there exists some constant ψ such that every level in the tree has at most $\psi N/h$ nodes. The case of non level balanced trees is considered in [12]; that paper uses the idea of randomly adding dummy nodes in the tree as it expands so that a property similar to level balance is achieved. This aspect is omitted here for simplicity.

A Butterfly network consists of $P = n2^n$ processors indexed as (i, j), for $0 \le j < n$ and $0 \le i < 2^n$. In this discussion the edges will be directed, and will have the form $((i, j), (\text{bitset}(i, j, 0), j + 1 \bmod n))$ and $((i, j), (\text{bitset}(i, j, 1), j + 1 \bmod n))$ where $\text{bitset}(i, j, b)$ denotes the number obtained by setting the jth least significant bit of i to b. We will need the following *unique path* property for Butterfly networks: there is at most one (directed) path from any node u to any node v of length at most n.

A processor (i, j) will be said to belong to level j. The level of a node in the tree will be defined as its distance from the root.

6.1   ALGORITHM

The main question is load balancing: how to distribute the task of expansion amongst the available processors. We use a randomized strategy for this. The randomization is used for deciding where a tree node gets expanded, as well as when. We do not consider details such as how to find the least cost node after the tree is completely expanded, nor how all the processors in the tree determine that the expansion is complete. These are discussed in [12]. Our goal is simply to generate the entire tree.

We start with the root placed on processor $(0, 0)$. The basic execution cycle for each processor is as follows. Each processor has a queue which contains unprocessed nodes. Each node is assigned a rank as discussed later. At each time step, each processor (i, j) picks the smallest rank node in its queue, and expands it. Each expansion can generate upto two new nodes n_0, n_1 – these are not expanded by processor (i, j) but are instead sent to its neighbours. For this, processor (i, j) picks a single random bit b. If n_0 exists then it is sent to neighbour (bitset$(i, j, b), j + 1 \bmod n)$, and if n_1 exists it is sent to neighbour (bitset$(i, j, \bar{b}), j+1 \bmod n)$. Any nodes that are received from its own neighbours are put into the queue, and the cycle then repeats. The following lemma is immediate:

Lemma 6 *If u is expanded on processor (i, j) then* Level$(u) \bmod n = j$ *and the bit chosen in the kth closest ancestor of u is the same as bit $j - k \bmod n$ of i, for $1 \le k \le n$.*

Lemma 7 *If u, v are expanded at processors p_u and p_v, which are connected by a (directed) path π of length less than n, then their n closest ancestors are distinct.*

Proof: Suppose Level$(u) \le$ Level(v), and w is their least common ancestor expanded at p_w. Assume w is within distance n of both u, v. Let P_{wu} denote the path in the Butterfly onto which the path from w to u is placed. Likewise P_{wv}. Now P_{wu} and P_{wv} must diverge at p_w. P_{wu} concatenated with π forms a second path from p_w to p_v. Because of the levelled nature of the Butterfly, the lengths of the paths must differ by multiples of n. Clearly the difference cannot be n or more; hence the lengths must be the same and at most n. This violates the unique path property. □

Ranks, Jobs, and Bands. The rank of each node is made of two parts, a primary rank and a secondary rank. The primary rank of a node at level l is simply $\lfloor l/n \rfloor$. Nodes having the same primary rank k are said to belong to band k. The nodes in any single band induce a forest, and the trees in this forest are called *jobs*. A common secondary rank is assigned to all the nodes in a job. This is simply a random number in the range $[0, \rho - 1]$ chosen at the time the root of the job is created, and then subsequently passed to other nodes in the job as they get created. ρ is fixed large enough so that all jobs get distinct ranks with high probability.[3] The rank of a node u with primary rank $x = \text{band}(u)$ and secondary rank y is defined as $\rho x + y$.

---

[3]It suffices to pick $\rho = \Omega(N^k)$ to ensure distinctness with high probability. In fact, $\rho = n$ actually suffices, and the condition that secondary ranks be distinct is not necessary, as seen in [12]. We use it here to avoid some uninteresting technicalities.

6.2   EVENTS AND CRITICAL SETS

By (u, p, s) we will denote the event that node u is assigned secondary rank s and is expanded on processor p. It is easily seen that the critical set of (u, p, s) consists of (i) event (u', p', s') where (u', u) is an edge in the tree and (p', p) in the Butterfly, and (ii) events (u', p, s') with $\text{band}(u') < \text{band}(u)$, and (iii) events (u', p, s') with $\text{band}(u') = \text{band}(u)$ but $s' < s$. The three kinds of events will respectively be called ancestor delay, band delay (or primary rank delay) and (secondary) rank delay of (u, p, s). Further, the startup time is clearly $T_0 = 1$.

Let $E = E_1, \ldots, E_T$ denote a delay sequence with $E_t = (u_t, p_t, s_t)$. E_t and E_{t-1} satisfy the following relationships:

1. $\text{Level}(u_{t-1}) \leq \text{Level}(u_t)$. This is clearly true if E_{t-1} is an ancestor delay or band delay of E_t. If E_{t-1} is a rank delay, then we know that $\lfloor \text{Level}(u_t)/n \rfloor = \text{band}(u_t) = \text{band}(u_{t-1}) = \lfloor \text{Level}(u_{t-1})/n \rfloor$. But $p_t = p_{t-1}$. Thus $\text{Level}(u_t) \bmod n = \text{Level}(p_t) = \text{Level}(p_{t-1}) = \text{Level}(u_{t-1}) \bmod n$. Thus $\text{Level}(u_t) = \text{Level}(u_{t-1})$.

2. If $\text{Level}(u_t) = \text{Level}(u_{t-1})$, then $p_t = p_{t-1}$. This case only arises if E_{t-1} is a rank delay of E_t.

3. If $p_t \neq p_{t-1}$ then (p_{t-1}, p_t) is an edge in the (directed) Butterfly.

From the above we can conclude that the sequence E is the concatenation of subsequences $\mathcal{E}_0, \ldots, \mathcal{E}_h$ where $E_t \in \mathcal{E}_k$ iff $\text{Level}(u_t) = k$. Further there exists a path $\mathcal{P}_0, \ldots, \mathcal{P}_h$ in the (directed) Butterfly such that $\mathcal{P}_0 = (0, 0)$ and for all $E_t \in \mathcal{E}_k$, $p_t = \mathcal{P}_k$. We will call these properties the $\mathcal{E} - \mathcal{P}$ properties of E.

Let W_b denote the concatenation of subsequences $\mathcal{E}_{bn}, \ldots, \mathcal{E}_{bn+n-1}$. Clearly, W_b contains all nodes belonging to band b. The important property that this sequence satisfies is that it is sorted by the secondary rank. This is because if $E_{t-1}, E_t \in W_b$, then $s_{t-1} = s_t$ if E_{t-1} is an ancestor delay; $s_{t-1} < s_t$ if E_{t-1} is a rank delay, and the case of a band delay does not arise. We will call this the W property of E.

Abbreviated Delay Sequence.   A delay sequence might contain several nodes from a single job. Since these have the same rank, they can only occur as a sequence of ancestor delays. Whenever E_{t-1} is an ancestor delay the non increasing sequence $\text{Level}(u_T), \ldots, \text{Level}(u_1) \in [0, h]$ decreases. Hence the total number of ancestor delays can be at most h. So we drop from the sequence all but one node from every job, leaving behind at least $T - h$ nodes. Next, among the nodes remaining, we retain nodes only from even bands or odd bands – whichever are more.

The total number of nodes that remain in the abbreviated sequence is at least $(T - h)/2$. We note that the abbreviated sequence also satisfies the $\mathcal{E} - \mathcal{P}$ property and the W property.

6.3 ANALYSIS

Theorem 3 *With high probability the time to finish backtrack search is* $O(N/P + h + \log N)$.

Proof: It suffices to upper bound the probability of the occurrence of an abbreviated delay sequence (u_i, p_i, s_i), $i = 1$ to δ satisfying the $\mathcal{E} - \mathcal{P}$ and W properties, for $\delta = (T - h)/2$. We will assume $\delta \geq h$.

Number of abbreviated delay sequences. The path $\mathcal{P}$ must start at $(0, 0)$, and use only h forward edges. Thus it can be fixed in $2^h \leq 2^\delta$ ways. We next fix the sizes of $\mathcal{E}_0, \ldots, \mathcal{E}_h$. These add up to δ and thus can together be fixed in $\binom{h+\delta}{\delta} \leq 2^{2\delta}$ ways. Each u_i such that $(u_i, p_i, s_i) \in \mathcal{E}_k$ can be any node in level k of the tree. Thus it can be fixed in at most $\psi N/h$ ways. All u_i can together be fixed in at most $\Pi_k (\psi N/h)^{|\mathcal{E}_k|} = (\psi N/h)^\delta$ ways. The s_i must be fixed so that they are sorted within each band. Over all the h/n bands, the number of ways in which this can happen is at most $\Pi_b \binom{\rho}{|W_b|} \leq \binom{\rho h/n}{\sum_b |W_b|} = \binom{\rho h/n}{\delta} = \left(\frac{\rho e h}{n\delta}\right)^\delta$. The total number of abbreviated delay sequences is thus at most $\left(2 \cdot 2^2 \cdot \frac{\psi N}{h} \frac{\rho e h}{n\delta}\right)^\delta$

Probability of fixed sequence. Each s_i has $1/\rho$ chance of being chosen as the secondary rank of u_i, and further the bits in the n ancestors of u_i must be chosen correctly so that it lands on p_i. Since each u_i is from a distinct job, their secondary ranks are independently chosen. Further all u_i are either from even bands or all from odd bands. Thus for any $u_i \neq u_j$ either $|\text{band}(u_i) - \text{band}(u_j)| \geq 2$, or $\text{band}(u_i) = \text{band}(u_j)$. The n nearest ancestors of u_i, u_j are clearly distinct in the first case, and distinct because of Lemma 7 in the second case. Thus the probability of the delay sequence is simply $(\rho 2^n)^{-\delta}$.

Hence the probability that some delay sequence occurs is

$$\left(2 \cdot 2^2 \cdot \frac{\psi N}{h} \frac{\rho e h}{n\delta} \frac{1}{2^n \rho}\right)^\delta = \left(\frac{8 e N \psi}{P \delta}\right)^\delta$$

This can be made $O(N^{-k})$ for $\delta = O(N/P + h) + k \log N$, so that $T = O(N/P + h + \log N)$ with high probability. $\qquad \square$

7. CONCLUDING REMARKS

Basically the delay sequence argument allows simple analysis of pipelining in distributed processes. If pipelining were to be ignored, all the four problems considered are amenable to easy analysis. For odd-even sorting, the time without accounting for the pipelining in key movement is $O(N^2)$, while the delay sequence argument implicitly analyzes the pipelining and shows the total time to be just N. For distributed computing in presence of noise, it is easily argued that any single step of the noisefree model can be simulated in $O(\log N)$ steps. This gives a total time of $O(\tau \log N)$. Again, the delay sequence argument estimates the effect of pipelining and gets a better bound. For packet routing, a bound of $O(cL)$ is easily argued, as against the $O(c+L+\log N)$ which we show. Finally, for backtrack search, it is easy to show that each processor is assigned only $O(\log N + \frac{N}{P})$ nodes for expansion with high probability. Thus, we see that the longest path can be completely evaluated in time $O(h(\frac{N}{P} + \log N))$, while the bound we get is $O(h + \frac{N}{P} + \log N)$. We believe that such situations in which pipelining reduces the time occur frequently, and the delay sequence argument might be useful.

Note finally that for simplicity, we have not attempted to get the smallest possible constants in the analyses.

Acknowledgments

The author would like to thank Milind Sohoni, Raghavendra Udupa, Bharat Adsul and Sudeep K. S. for comments on previous versions.

References

[1] R. Aleliunas. Randomized parallel communication. In *Proceedings of the ACM SIGACT-SIGOPS Symposium on Principles of Distributed Computing*, pages 60–72, August 1982.

[2] T. Cormen, C. Leiserson, and R. Rivest. *Introduction to Algorithms*. The MIT Press, 1991.

[3] Richard Karp and Yanjun Zhang. A randomized parallel branch-and-bound procedure. In *Proceedings of the ACM Annual Symposium on Theory of Computing*, pages 290–300, 1988.

[4] D. E. Knuth. *The art of computer programming*, volume 3. Addison-Wesley, 1973.

[5] F. T. Leighton. *Introduction to parallel algorithms and architectures*. Morgan-Kaufman, 1991.

150

[6] Tom Leighton, Bruce Maggs, Abhiram Ranade, and Satish Rao. Routing and Sorting on Fixed-Connection Networks. *Journal of Algorithms*, 16(4):157–205, July 1994.

[7] Tom Leighton, Bruce Maggs, and Satish Rao. Packet routing and job-shop scheduling in O(congestion+dilation) steps. *Combinatorica*, 14(2):167–180, 1994.

[8] M. Luby. On the parallel complexity of symmetric connection networks. Technical Report 214, University of Toronto, CS Dept., 1988.

[9] Charles Martel, Ramesh Subramonian, and Arvin Park. Asynchronous PRAMs are (almost) as good as synchronous PRAMs. In *31st Annual Symposium on Foundations of Computer Science*, volume II, pages 590–599, St. Louis, Missouri, 22–24 October 1990. IEEE.

[10] Sridhar Rajagopalan and Leonard Schulman. A coding theorem for distributed computation. In *Proceedings of the Twenty-Sixth Annual ACM Symposium on the Theory of Computing*, pages 790–799, Montréal, Québec, Canada, 23–25 May 1994.

[11] Abhiram G. Ranade. How to emulate shared memory. *Journal of Computer and System Sciences*, 42(3):307–326, June 1991. An earlier version appeared in the Proceedings of the Symposium on Foundations of Computer Science, 1987.

[12] Abhiram G. Ranade. Optimal speedup for backtrack search on a butterfly network. *Mathematical Systems Theory*, (27):85–101, 1994. Also in the Proceedings of the Third ACM Sympsium on Parallel Architectures and Algorithms, July 1991, pages 40–48.

[13] E. Upfal. Efficient schemes for parallel communication. In *Proceedings of the ACM SIGACT-SIGOPS Symposium on Principles of Distributed Computing*, pages 55–59, August 1982.

Chapter 6

RANDOMIZED ALGORITHMS FOR GEOMETRIC OPTIMIZATION PROBLEMS

Pankaj K. Agarwal *

Center for Geometric Computing, Department of Computer Science, Box 90129, Duke University, Durham, NC 27708-0129, USA.

pankaj@cs.duke.edu

Sandeep Sen

Department of Computer Science and Engineering, IIT Delhi, New Delhi 110016, India.

ssen@cse.iitd.ernet.in

Abstract   This chapter reviews randomization algorithms developed in the last few years to solve a wide range of geometric optimization problems. We review a number of general techniques, including randomized binary search, randomized linear-programming algorithms, and random sampling. Next, we describe several applications of these techniques, including facility location, proximity problems, nearest-neighbor searching, statistical estimators, and Euclidean TSP.

1.   INTRODUCTION

Combinatorial optimization typically deals with problems of maximizing or minimizing a function of one or more variables subject to a large number of inequality constraints. Many problems can be formulated as combinatorial optimization problems, which has made this a very active area of research during the past half century. In several

*Work by the first author was supported by Army Research Office MURI grant DAAH04-96-1-0013, by a Sloan fellowship, by NSF grants EIA–9870724, EIA–997287, and CCR–9732787, and by a grant from the U.S.-Israeli Binational Science Foundation.

S. Rajasekaran et al (eds.), Handbook of Randomized Computing, Volume 1, pp, 151–202.

applications, the underlying optimization problem involves a constant number of variables and a large number of constraints that are induced by a given collection of geometric objects; we refer to such problems as *geometric optimization* problems. In such cases one expects that faster and simpler algorithms can be developed by exploiting the geometric nature of the problem. Much work has been done on geometric optimization problems during the last twenty years. Many new elegant and sophisticated techniques have been developed and successfully applied to a wide range of geometric optimization problems.

Since the seminal paper by Rabin [170], which initiated the study of randomization in developing fast algorithms, randomization has permeated in several areas, including algorithmic number theory, machine learning, distributed computing, and complexity theory. Even though one of the problems studied in Rabin's paper was the closest-pair problem, a central problem in computational geometry, randomization did not become popular in computational geometry until the late 1980s. In the mid 1980s Clarkson was developing his random-sampling technique, which he extended to a surprisingly general framework in his 1988 paper [50, 58]. Around the same time Haussler and Welzl [105] introduced the idea of ε-nets and VC-dimensions. These two techniques brought randomization to the forefront of computational geometry and revolutionized the field. Numerous randomized divide-and-conquer and incremental algorithms, dynamic data structures, and analysis techniques (e.g., backward analysis, Mulmuley's probabilistic games) have been developed in the last decade. Detailed accounts of these developments can be found in the books by Motwani and Raghavan [162] and Mulmuley [163] and in the survey papers [55, 145, 164, 182].

In this chapter we focus on the impact of randomization in geometric optimization. We review a number of randomized techniques that have been successfully applied to geometric optimization problems and examine some of their applications. Like other areas, randomization has led to simpler and improved algorithms for a wide spectrum of geometric optimization problems.

We begin by discussing *randomized binary search*. Several randomized techniques have been developed for searching over the solution space, each step of which discards a fraction of the candidate values with probability at least 1/2. This simple technique leads to fast and simple algorithms for many geometric optimization problems. Next, we review randomized algorithms for linear programming. Seidel [181], Dyer and Frieze [76], and Clarkson [54] proposed randomized algorithms for linear programming whose expected time is linear in any fixed dimension, which are much simpler than their earlier deterministic counterparts.

The dependence on the dimension of the running time of these algorithms is better (though still exponential). Actually, some of these technique are rather general, and are also applicable to a variety of other geometric optimization problems. We discuss some of these techniques in Section 3.. Another significant progress on linear programming was made in the beginning of the 1990s, when new randomized algorithms for linear programming were obtained independently by Kalai [122], and by Matoušek *et al.* [151, 188] (these two algorithms are essentially dual versions of the same technique). The expected number of arithmetic operations performed by these algorithms is subexponential in the input size, and is still linear in any fixed dimension, so they constitute an important step toward the still open goal of obtaining strongly polynomial algorithms for linear programming.[1] This new technique is presented in Section 4.. The algorithm in [151, 188] is actually formulated in a general abstract framework, which fits not only linear programming but also many other problems. Such *LP-type* problems are also reviewed in Section 4..

We also describe a randomized algorithm for set cover, based on the linear-programming algorithm by Clarkson [54]. Since plenty of geometric optimization problems can be formulated as set-cover or hitting-set problems, many of them have benefited from this simple algorithm.

Next, we survey various geometric applications of these techniques and discuss a few additional problem-specific techniques. These applications include problems involving *facility location* (e.g., finding p congruent disks of smallest possible radius whose union covers a given planar point set), *geometric proximity* (e.g., nearest-neighbor searching and computing the diameter of a point set), *statistical estimators and metrology* (e.g., computing the smallest-width annulus that contains a given planar point set), *placement and intersection of polygons and polyhedra* (e.g., finding the largest similar copy of a convex polygon that fits inside a given polygonal environment), and *network design problems* (e.g., traveling salesperson, matching).

Although the common theme of most of the applications reviewed here is that they can be solved efficiently using the general techniques described in the beginning, many of them require a problem-specific, and often fairly sophisticated, approach. For example, the heart of a typical randomized binary search is the design of an efficient algorithm for solving the appropriate problem-specific "decision procedure" (see below for details). We will discuss details of these solutions for some of the problems, but will omit them for most of the applications because of lack of space.

2. RANDOMIZED BINARY SEARCH

In this section we describe a few techniques for performing a randomized binary search. A geometric optimization problem $\mathcal{P} = (\Sigma, \mathcal{W})$ can be described as follows. Σ is a family of "geometric" objects, and $\mathcal{W}$ is a cost function such that, for a finite subset $S \subseteq \Sigma$, $\mathcal{W}(S)$ gives the cost of the optimal solution for S. For example, for the Euclidean 1-center problem, $\mathcal{W}(S)$ is the radius of the smallest disk enclosing S; for the traveling sales person (TSP) problem, $\mathcal{W}(S)$ is the length of the shortest tour of S. Given a subset $S \subseteq \Sigma$, the problem $\mathcal{P}$ ask for computing the value of $\mathcal{W}(S)$.

A subset $A \subseteq S$ is called a *basis* $\mathcal{B}(S)$ of S if A is the smallest subset of S for which $\mathcal{W}(A) = \mathcal{W}(S)$. The size of a basis is constant for many geometric optimization problems. For example, for any given set S of points in the plane, there is a subset $A \subseteq S$ of size at most three so that the smallest disk enclosing S is the same as that of A. Therefore there is a basis of size at most three. For a given S, define

$$\Lambda = \{\mathcal{W}(A) \mid A \subset S \text{ and } |A| \leq |\mathcal{B}(S)|\}.$$

Let $\lambda^* = \mathcal{W}(S)$. By definition, $\lambda^* \in \Lambda$. We can, of course, compute the entire set Λ explicitly and choose λ^*. However, Λ is typically too large to be computed explicitly. We therefore want to search over Λ implicitly. We can associate a decision problem $\mathcal{D}_{\mathcal{P}}$ with the optimization problem $\mathcal{P}$, which, given a value λ, asks whether $\lambda < \lambda^*$, $\lambda = \lambda^*$, or $\lambda > \lambda^*$. Suppose we have an algorithm $\mathcal{A}$ for the decision problem that runs in time $T(n)$. The randomized binary search techniques use $\mathcal{A}$ to compute λ^* in time $O(T(n) \log^c n)$. In some applications even the decision algorithm $\mathcal{A}$ is randomized. We now describe a few techniques for performing an implicit binary search.

2.1 RANDOMIZED HALVING TECHNIQUE

The randomization halving technique is based on the following simple idea, which has been used in a number of problems, including selecting an item from an ordered set [162]. Suppose we know that λ^* lies in an interval $I = [\alpha, \beta]$. Suppose further that we can randomly choose an element $\lambda_0 \in I \cap \Lambda$, where each item is chosen with probability $1/|I \cap \Lambda|$. Then it follows that, by comparing λ^* with a few randomly chosen elements of $I \cap \Lambda$ (i.e., by executing the decision algorithm at these values), we can shrink I to an interval I' that is guaranteed to contain λ^* and that is expected to contain significantly fewer critical values. The difficult part is, of course, choosing a random element from

$I \cap \Lambda$. In many cases, a procedure for computing $|I \cap \Lambda|$ can be converted into a procedure for generating a random element of $I \cap \Lambda$.

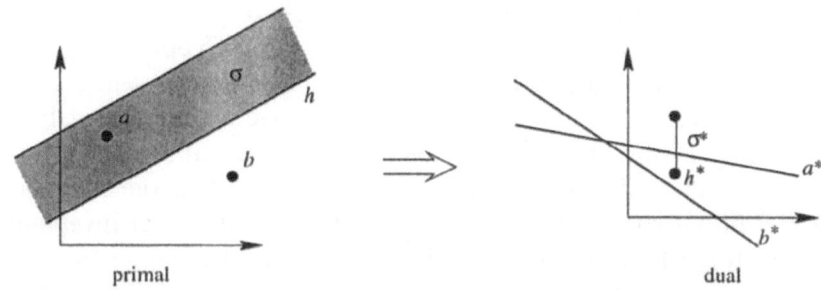

Figure 6.1   The duality transform in two dimensions.

We illustrate this technique by applying it to the so-called *slope-selection* problem. Given a set S of n points in the plane and an integer $k \leq \binom{n}{2}$, determine a line ℓ_k connecting two input points that has the kth smallest slope among all such segments. For $k = \lceil \binom{n}{2}/2 \rceil$, ℓ_k is called the *Theil-Sen estimator* of S. Using the duality transform [78], which maps a point $p = (a, b)$ to the line $p^* : y = -ax + b$ and a line $\ell : y = \alpha x + \beta$ to the point $\ell^* = (\alpha, \beta)$ (see Figure 6.1), we can formulate this problem as follows: We are given a set L of n nonvertical lines in the plane and an integer $1 \leq k \leq \binom{n}{2}$, and we wish to find an intersection point between two lines of L that has the kth smallest x-coordinate. (We assume, for simplicity, *general position* of the lines, so that no three lines are concurrent, and no two intersection points have the same x-coordinate.) We are thus seeking the kth leftmost vertex of the *arrangement* $\mathcal{A}(L)$ of the lines in L;[2] see [13, 78, 186] for more details concerning arrangements. A complicated $O(n \log n)$ deterministic algorithm was developed by Cole *et al.* [59]; a simpler $O(n \log n)$-time deterministic algorithm, based on so-called *cuttings*, was later proposed by Brönnimann and Chazelle [37]. Here we present optimal randomized algorithms.

In this case Λ is the set of x-coordinates of the vertices of $\mathcal{A}(L)$. We describe the decision algorithm and show how it can be used to select a random element of $\Lambda \cap I$, for a given interval I. Given a vertical strip $W = (\alpha, \beta) \times \mathbb{R}$, let $L_\alpha = (\ell_1, \ell_2, \ldots, \ell_n)$ denote the sequence of lines in L sorted by their intercepts with $x = \alpha$, and let $L_\beta = (\ell_{\pi(1)}, \ell_{\pi(2)}, \ldots, \ell_{\pi(n)})$ denote the sequence of these lines sorted by their intercepts with $x = \beta$. An easy observation is that two lines ℓ_i, ℓ_j, with $i < j$, intersect inside W if and only if $\pi(i) > \pi(j)$. In other words, $|I \cap W|$ can be computed, in $O(n \log n)$ time, by counting the number of *inversions* in the permutation π, using the merge-sort procedure as follows [129]. We compute the sequence L_β and then sort it by the intercepts with $x = \alpha$ using the merge sort. We maintain a global counter C to count the number of inversions. Let $A = (a_1, \ldots, a_{n/2})$ be the sequence of the first half of L_β sorted by their intercepts along $x = \alpha$, and let $B = (b_1, \ldots, b_{n/2})$ be the second half of the lines in L_β sorted by their intercepts along $x = \beta$. Suppose that we have recursively computed A and B and that we have also counted the number of inversions (i, j) such that both i, j lie either in A or in B. The merge step merges A and B into a sorted sequence and counts the number of inversions of the form (b_j, a_i). The algorithm keeps track of the last element of B that was added to the overall sorted sequence. For $1 \leq i \leq n/2$, suppose $b_{k(i)}$ was the last element of B inserted in the sequence before adding a_i. Then $b_1, \ldots b_{k(i)} < a_i$, so (b_j, a_i), $1 \leq j \leq k(i)$, is an inversion. We therefore increment the counter C by $k(i)$ after inserting a_i. The overall running time of the algorithm is $O(n \log n)$.

The above algorithm can be extended to generate a multiset of q random vertices of $\mathcal{A}(L) \cap W$ in time $O(n \log n + q)$ as follows. Using the above algorithm, we compute the number ν of vertices of $\mathcal{A}(L)$ lying in W. We then generate a random multiset Q of q integers in the range $[1, \nu]$, and sort it in $O(q)$ time using the bucket-sort algorithm. We run the inversion-counting algorithm once again, but with the following additional twist. Suppose C_i is the value of the counter C after we inserted a_i to the sorted sequence. If Q contains a value $j \in [C_{i-1}+1, C_i]$, we return the intersection point of a_i and $b_{j-C_{i-1}}$. The overall running time of the algorithm is $O(n \log n + q)$. Using this procedure, Matoušek [144] obtained the following simple slope-selection algorithm: Each step of the algorithm maintains a vertical strip $W(a, b) = \{(x, y) \mid a \leq x \leq b\}$ that is guaranteed to contain the kth leftmost vertex; initially $a = -\infty$ and $b = +\infty$. Let m be the number of vertices of $\mathcal{A}(L)$ lying inside W. We repeat the following step until the algorithm terminates.

If $k \leq n$, the kth leftmost vertex of $\mathcal{A}(L)$ can be computed in $O(n \log n)$ by a sweep-line algorithm (through W). Otherwise, set k^* to be the number of vertices lying to the left of the line $x = a$. Let $j = \lfloor (k - k^*) \cdot n/m \rfloor$, $j_a = j - \lfloor 3\sqrt{n} \rfloor$, and $j_b = j + \lfloor 3\sqrt{n} \rfloor$. We choose n random vertices of $\mathcal{A}(L)$ lying inside $W(a, b)$. If the kth leftmost vertex lies in $W(j_a, j_b)$ and the vertical strip $W(j_a, j_b)$ contains at most $cm/\sqrt{n}$ vertices, for some appropriate constant $c > 0$, we set $a = j_a$, $b = j_b$, and repeat this step. Otherwise, we discard the random sample of vertices, and draw a new sample. It can be shown, using Chernoff's bound [162], that the expected running time of the above algorithm is $O(n \log n)$. This technique is quite general and has been used for many other problems.

Shafer and Steiger [184] gave a slightly different $O(n \log n)$ expected-time algorithm for the slope-selection problem. They choose a random subset of $u = O(n \log n)$ vertices of $\mathcal{A}(L)$. Let $a_1, a_2, \ldots, a_u$ be the x-coordinates of these vertices. Using the algorithm by Cole *et al.* [59] for counting the number of inversions approximately, they determine in $O(n \log n)$ time the vertical strip $W(a_i, a_{i+1})$ that contains the kth left-most vertex of $\mathcal{A}(L)$. They prove that, with high probability, $W(a_i, a_{i+1})$ contains only $O(n)$ vertices of $\mathcal{A}(L)$, and therefore the desired vertex can be computed in an additional $O(n \log n)$ time by a sweep-line algorithm. Dillencourt *et al.* [68] proposed yet another randomized slope-selection algorithm.

2.2 CLARKSON-SHOR TECHNIQUE

Clarkson and Shor [58] gave another randomized technique for solving a variety of geometric optimization problems. Although they had originally proposed the algorithm for computing the diameter of a set of points in $\mathbb{R}^3$ (see Section 8.1), their approach works under the following more general setting. Let S be a set of objects, and let $\mu : S \to \mathbb{R}$ be a function such that $\mathcal{W}(S) = \min_{p \in S} \mu(p)$. For example, consider the closest-pair problem: Given a set S of n points in $\mathbb{R}^d$, find the minimum distance between a pair of distinct points of S. We set $\mathcal{W}(S) = \min_{p,q \in S, p \neq q} d(p, q)$. Define $\mu(p) = \min_{q \neq p} d(p, q)$. Then $\mathcal{W}(S) = \min_{p \in S} \mu(p)$. We can then compute $\mathcal{W}(S)$ as shown in Figure 6.2.

```
function procedure OPTIMIZE (S);
    choose a random point p ∈ S;
    ⟨p₁,... ,pₙ⟩:  a random permutation of S.
    curr = ∞
    for i = 1 to n do
      if μ(pᵢ) < curr
         curr = μ(pᵢ)        (⋆)
    return curr
```

Figure 6.2 Clarkson-Shor algorithm.

Note that Step $(\star)$ is executed in the ith iteration only if $\mu(p_i) < \mu(p_j)$ for all $j < i$. Since we choose a random permutation of S,

$$\Pr[\mu(p_i) = \min_{1 \le j \le i} \mu(p_j)] = \frac{1}{i}.$$

Therefore, the expected number of times Step $(\star)$ is executed is $O(\log n)$. The above algorithm is useful when, possibly after some preprocessing, deciding whether $\mu(p_i) < curr$ is easier than computing $\mu(p_i)$. Suppose for a given $curr$, S can be preprocessed in time $P(n)$ so that one can determine in $Q(n)$ time whether $\mu(p) < curr$, then the expected running time of the above algorithm is $O(P(n)\log n + nQ(n))$. For the 3D diameter problem, Clarkson and Shor showed that for a given parameter $curr$, S can be preprocessed in $O(n\log n)$ time into a data structure of size $O(n)$ so that whether $\mu(p) < curr$ can be determined in $O(\log n)$ time, thereby attaining an $O(n\log^2 n)$ time algorithm. In Section 8.1, we will show that the expected running time can further be improved to $O(n\log n)$ using another observation.

Preprocessing S for answering a query of the form "is $\mu(p) < curr$?" is expensive in general. Agarwal and Sharir [11] combined the Clarkson-Shor technique with the random-sampling technique to solve a class of geometric optimization problems that can be formulated as computing a closest pair in some metric.

2.3 CHAN'S ALGORITHM

Chan [42] extended the Clarkson-Shor technique to a more general framework. Suppose $\mathcal{P}$ is decomposable in the sense that we can decompose, in time $D(n)$, a given input S into subsets $S_1, \ldots, S_r$, for some integer $r > 0$, each of size at most αn so that $\mathcal{W}(S) = \min_{1 \le i \le r} \mathcal{W}(S_i)$. Further suppose given a real value $curr$, we can determine whether $\mathcal{W}(A) < curr$ in $D(|A|)$ time. Then we can use the Clarkson-Shor technique as follows. First compute the subproblems $S_1, \ldots, S_r$, and then process these subproblems in a random order. For each subproblem S_i being processed, first determine whether $\mathcal{W}(S_i) < curr$ using the decision algorithm. If the answer is YES, then compute $\mathcal{W}(S_i)$ recursively. The expected number of subproblems for which the optimum value is recursively computed (to reset the current minimum) is only $O(\log r)$. Chan shows, as we will see in the following sections, that in many instances the expected running time of this technique is $O(D(n))$.

3. LINEAR PROGRAMMING

As noted in the introduction, several randomized algorithms have been developed for linear programming in the last decade [1, 54, 76, 151, 181]. These algorithms run in expected linear time for any fixed dimension. In this section we describe the algorithms by Seidel and Clarkson. In the next section we will discuss the algorithm by Matoušek et al. [151]. We are given a set $H = \{h_1, \ldots, h_n\}$ of n halfspaces in $\mathbb{R}^d$, called constraints, and a vector c called the objective function. We wish to compute an $x \in \mathbb{R}^d$ so that cx is minimized over the feasible region $K = \bigcap_{i=1}^n h_i$.

```
function procedure SEIDEL_lp(H, c)                    /* H: n constraints in R^d
    if n ≤ d then
        return Basis (H)                              /* returns v(H)
    else
        choose a random h ∈ H
.       x := SEIDEL_lp(H \ {h})
        if x ∈ h
            return x
            else return SEIDEL_lp ({h ∩ g | g ∈ H \ {h}})            (⋆)
```

Figure 6.3   Seidel's randomized LP algorithm.

Seidel's algorithm is sketched in Figure 6.3. Given a set H of d constraints in general position, the function $\text{Basis}(H)$ computes $x \in \bigcap II$ that minimizes cx. He proved that the probability of executing Step $(\star)$ is at most d/n. Since the recursive call in this step is an instance $(d-1)$-dimensional linear programming, the expected running time of the algorithm is $O(d!n)$.

Dyer and Frieze [76] gave another randomized algorithm based on the random-sampling technique with $O(d^d n)$ expected running time. Agarwal *et al.* [14] later extended this approach to solve a number of other optimization problems.

In the mid 1980s an open question was whether there exists an algorithm whose running time is linear in n but the dependency on d is better than d^d. Clarkson developed two randomized algorithms with faster expected running time. His first algorithm is recursive and is summarized in Figure 6.4.

We call an iteration *successful* if Step $(\star)$ is executed. It can be shown that an iteration is successful with probability at least $1/2$ and that there are at most $d + 1$ successful iterations. Using these observations it can be shown that the expected running time of the algorithm is $O(d^2 n + d^{d/2 + \log \log n + O(1)})$. Chan [40] showed that RECURSIVE_lp can be modified to answer linear-programming queries. That is, preprocess a set H of n constraints, so that for a query objective function c, a point $x \in \bigcap H$ that minimizes cx can be computed efficiently. He sets $r = n/b$ where $b > 9d^2$ is a sufficiently large constant. Consequently, with probability at least $1/2$, $|V_x| \leq \log n$. He preprocesses H and a family of random subsets of H for halfspace range-reporting queries so that V_x can be computed efficiently. He shows how to solve the recursive subproblem for $R \cup V$. For a parameter $n \leq m \leq n^{\lfloor d/2 \rfloor}$, H can be preprocessed in $m \log^{O(1)} n$ time into a data structure of $O(m)$ size so that a linear-programming

```
function procedure RECURSIVE_lp(H, c)              /* H: n constraints in ℝ^d
    if n ≤ 9d² then
        return SIMPLEX_lp (H)                      /* returns v(H)
    else
        V = ∅,  r = d√n
        choose random R ∈ (H over r)
        repeat
            x := RECURSIVE_lp(R ∪ V, c)
            V_x := {h ∈ H | x violates h}
            if |V_x| ≤ 2√n then
                V = V ∪ V_x              (⋆)
        until V = ∅
        return x
```

Figure 6.4 Clarkson's recursive LP algorithm.

query can be answered in $O((n/m^{1/\lfloor d/2 \rfloor}) \log^{2d+1} n)$ expected time. See the original paper for details. The query time was slightly improved by Ramos [173]. In particular, his algorithm can answer a query in time $n^{1-1/\lfloor d/2 \rfloor} 2^{O(\log^* n)}$ using $O(n)$ space.

We now describe the second algorithm by Clarkson. Let H be the set of constraints. We assign a weight $\mu(h) \in \mathbb{Z}$ to each constraint; initially $\mu(h) = 1$ for all $h \in H$. For a subset $A \subseteq H$, let $\mu(A) = \sum_{h \in A} \mu(h)$. The algorithm works in rounds, each of which consists of the following steps. Set $r = 6d^2$. If $|H| \leq 6d^2$, we compute the optimal solution using the simplex algorithm. Otherwise, choose a random sample $R \subset H$ such that $\mu(R) = r$. (We can regard H as a multiset in which each constraint h appears $\mu(h)$ times, and we choose a multiset $R \in \binom{H}{r}$ of r constraints.) We compute the optimal solution x_R for R and the subset $V \subset H \setminus R$ of constraints that x_R *violates* (that is, the subset of constraints that do not contain x_R). If $V = \emptyset$, the algorithm returns x_R. If $\mu(V) \leq 3\mu(H)/d$, we double the weight of each constraint in V; in any case, we repeat the sampling procedure. See Figure 6.5 for a pseudocode of the algorithm.

Let B be the set of d constraints whose boundaries are incident to the optimal solution. A round is called *successful* if $\mu(V) \leq 3\mu(H)/d$. Using the fact that R is a random subset, one can argue that each round is successful with probability at least $1/2$. Every successful round increases $\mu(H)$ by a factor of at most $(1 + 1/3d)$, so the total weight $\mu(H)$ after kd successful rounds is at most $n(1 + 1/3d)^{kd} < ne^{k/3}$. On the other hand, each successful iteration doubles the weight of at least

```
function procedure ITERATIVE_LP (H)                        /* H: n constraints in ℝ^d
    if n ≤ 6d² then
        return SIMPLEX_lp(H)                               /* returns v(H)
    else
        r := 6d²;  μ_h := 1 ∀h ∈ H
        repeat
            choose random R ∈ (H r)
            x_R := SIMPLEX_lp(R)
            V := {h ∈ H | x_R violates h}
            if μ(V) ≤ 3μ(H)/d then
                for all h ∈ V do μ_h := 2μ_h
        until V = ∅
        return x_R
```

Figure 6.5   Clarkson's iterative LP algorithm.

one constraint in B (it is easily verified that V must contain such a constraint), which implies that after kd iterations $\mu(H) \geq \mu(B) \geq 2^k$. Hence, after kd successful rounds, $2^k \leq \mu(H) \leq ne^{k/3}$. This implies that the above algorithm terminates in at most $3d \ln n$ successful rounds. Since each round takes $O(d^d)$ time to compute x_R and $O(dn)$ time to compute V, the expected running time of the algorithm is $O((d^2 n + d^{d+1}) \log n)$. By combining this algorithm with RECURSIVE_lp, i.e., using ITERATIVE_lp inside the loop of RECURSIVE_lp, the expected running time can be improved to $O(d^2 n) + d^{d/2+O(1)} \log n$.

Clarkson showed that ITERATIVE_lp can be extended to integer linear programming by choosing an appropriate value of the sample size and using Lenstra's integer-programming algorithm [137] for the base case. Let b denote the maximum number of bits required to specify the coefficients in a constraint or in the objective function. Then the algorithm performs $O(2^d dn + 8^d d\sqrt{n \log n} \log n) + d^{O(d)} b \log n$ expected number of operations, each involving at most $d^{O(1)} b$ bit numbers. See [123] for an earlier result on integer programming in fixed dimensions.

4.   ABSTRACT LINEAR PROGRAMMING

In this section we present an abstract framework that captures linear programming, as well as many other geometric optimization problems, including computing smallest enclosing balls (or ellipsoids) of finite point sets in $\mathbb{R}^d$, computing largest balls (ellipsoids) inscribed in convex polytopes in $\mathbb{R}^d$, computing the distance between polytopes in $\mathbb{R}^d$, general convex programming, and many other problems. Sharir and Welzl [188] and Matoušek *et al.* [151] presented a randomized algorithm for optimization problems in this framework, whose expected running time is linear in terms of the number of constraints whenever the combinatorial dimension d (whose precise definition, in this abstract framework, will be given below) is fixed. More importantly, the running time is subexponential in d for many of the LP-type problems, including linear programming. This is the first subexponential "combinatorial" bound for linear programming, and is a first step toward the major open problem of obtaining a strongly polynomial algorithm for linear programming. The papers by Gärtner and Welzl [95] and Goldwasser [97] also survey the known results on LP-type problems. A dual version of the algorithm was independently obtained by Kalai [122], but only in the context of linear programming.

4.1   AN ABSTRACT FRAMEWORK

Let us consider optimization problems specified by a pair (H, w), where H is a finite set, and $w : 2^H \to \mathcal{O}$ is a function into a linearly ordered set $(\mathcal{O}, \leq)$; we assume that $\mathcal{O}$ has a minimum value $-\infty$. The elements of H are called *constraints*, and for $G \subseteq H$, $w(G)$ is called the *value of G*. Intuitively, $w(G)$ denotes the smallest value attainable by a certain objective function while satisfying all the constraints of G. The goal is to compute a minimal subset B_H of H with $w(B_H) = w(H)$, assuming the availability of three basic operations, which we specify below.

Such a minimization problem is called *LP-type* if the following two axioms are satisfied:

Axiom 1. (*Monotonicity*) For any F, G with $F \subseteq G \subseteq H$, we have
$$w(F) \leq w(G).$$

Axiom 2. (*Locality*) For any $F \subseteq G \subseteq H$ with $-\infty < w(F) = w(G)$ and any $h \in H$,
$$w(G) < w(G \cup \{h\}) \Rightarrow w(F) < w(F \cup \{h\}).$$

Linear programming is easily shown to be an LP-type problem: Set $w(G)$ to be the vertex of the feasible region that minimizes the objective function and that is coordinate-wise lexicographically smallest (this definition is important to satisfy Axiom 2), and extend the definition of $w(G)$ in an appropriate manner to handle empty or unbounded feasible regions.

A *basis* $B \subseteq H$ is a set of constraints satisfying $-\infty < w(B)$, and $w(B') < w(B)$ for all proper subsets B' of B. For $G \subseteq H$, with $-\infty < w(G)$, a *basis of G* is a minimal subset B of G with $w(B) = w(G)$. (For linear programming, a basis of G is a minimal set of halfspace constraints in G such that the minimal vertex of their intersection is the minimal vertex of G.) A constraint h is *violated by* G if $w(G) < w(G \cup \{h\})$, and it is *extreme in* G if $w(G - \{h\}) < w(G)$. The *combinatorial dimension of* (H, w), denoted as $\dim(H, w)$, is the maximum cardinality of any basis. We call an LP-type problem *basis regular* if for any basis with $|B| = \dim(H, w)$ and for any constraint h, every basis of $B \cup \{h\}$ has exactly $\dim(H, w)$ elements. (Clearly, linear programming is basis-regular, where the dimension of every basis is d.)

We assume that the following primitive operations are available:

> (*Violation test*) h is violated by B: for a constraint h and a basis B, tests whether h is violated by B.
>
> (*Basis computation*) basis(B, h): for a constraint h and a basis B, computes a basis of $B \cup \{h\}$.
>
> (*Initial basis*) initial(H): An initial basis B_0 with exactly $\dim(H, w)$ elements is available.

For linear programming, the first operation can be performed in $O(d)$ time, by substituting the coordinates of the vertex $w(B)$ into the equation of the hyperplane defining h. The second operation can be regarded as a dual version of the pivot step in the simplex algorithm, and can be implemented in $O(d^2)$ time. The third operation is also easy to implement.

We are now in position to describe the algorithm. Using the initial-basis primitive, we compute a basis B_0 and call SUBEX_lp(H, B_0), where SUBEX_lp is the recursive algorithm, given in Figure 6.6, for computing a basis B_H of H.

```
function procedure SUBEX_lp(H, C);          /* H: set of n constraints in R^d;
    if H = C then                               /* C ⊆ H: a basis;
        return C                                /* returns a basis of H.
    else
        choose a random h ∈ H \ C;
        B := SUBEX_lp(H \ {h}, C);
        if h is violated by B then
            return SUBEX_lp(H, basis(B, h))
        else
            return B;
```

Figure 6.6 A randomized algorithm for LP-type problems.

A simple inductive argument shows the expected number of primitive operations performed by the algorithm is $O(2^\delta n)$, where $n = |H|$ and $\delta = \dim(H, w)$ is the combinatorial dimension. However, using a more involved analysis, which can be found in [151], one can show that basis-regular LP-type problems can be solved with an expected number of at most $e^{2\sqrt{\delta \ln((n-\delta)/\sqrt{\delta})} + O(\sqrt{\delta} + \ln n)}$ violation tests and basis computations. This is the subexponential bound that we alluded to. A surprising feature of the LP-type problems is that some of them are instances of nonconvex programming. We will point out a few such instances below. Earlier algorithms for linear programming did not extend to nonconvex programming. Matoušek [146] has given examples of abstract LP-type problems of combinatorial dimension d with $2d$ constraints, for which the above algorithm requires $\Omega(e^{\sqrt{2d}}/\sqrt[4]{d})$ primitive operations.

4.2   LINEAR PROGRAMMING

Returning to linear programming, let H be the given set of n constraints and c the objective function. We can assume that the objective vector is $c = (1, 0, 0, \ldots, 0)$. For a subset $G \subseteq H$, define $w(G)$ to be the lexicographically smallest point (vertex) of the intersection of halfspaces in G. Some care is needed to handle unbounded or empty feasible regions; we omit details concerning this issue.

As noted above, linear programming is a basis-regular LP-type problem, with combinatorial dimension d, and each violation test or basis computation can be implemented in time $O(d)$ or $O(d^2)$, respectively. In summary, we obtain a randomized algorithm for linear programming, which performs $e^{2\sqrt{d \ln(n/\sqrt{d})} + O(\sqrt{d} + \ln n)}$ expected number of arithmetic operations. Using SUBEX_lp instead of the simplex algorithm for solving the small-size problems in the ITERATIVE_lp algorithm (given in Figure 6.5), the expected number of arithmetic operations can be reduced to $O(d^2 n) + e^{O(\sqrt{d \log d})}$.

In view of Matoušek's lower bound on the performance of SUBEX_lp, one should aim to exploit additional properties of linear programming if one wants to obtain a better bound on the performance of the algorithm for linear programming; this is still a major open problem.

4.3   EXTENSIONS

Matoušek [148] has investigated the problem of finding the best solution, for an abstract LP-type problem, that satisfies all but k of the given constraints. He proved that the number of bases that violate at most k constraints in a nondegenerate instance of an LP-type problem is $O((k + 1)^\delta)$, where δ is the combinatorial dimension of the problem, and that they can be computed in time $O(n(k + 1)^\delta)$. In some cases the running time can be improved using appropriate data structures.

For example, given a set H of n halfplanes in the plane and an integer $k \leq n$, the point with the smallest y-coordinate that lies in at least $n - k$ halfplanes can be computed in time $O(n \log k + k^3 \log^2 n)$ [148, 176]. For larger values of k, the running time is $O(n \log n + nk)$ [42]. Using the technique developed in Section 2.3, Chan [42] showed that if the intersection of the halfspaces is nonempty, the running time can be improved to $O(n + k(n/k)^\varepsilon \log n)$ for any $\varepsilon > 0$. Recently, Gärtner and Welzl [96] have proved tail estimates on the running times of some of the LP-type problems. Chazelle and Matoušek [48] gave a deterministic algorithm for solving LP-type problems in time $O(\delta^{O(\delta)} n)$, provided an additional axiom holds together with an additional computational assumption. See [19, 20] for additional extensions of LP-type problems.

5.   GEOMETRIC SET COVER

Let $\Sigma = (X, \mathcal{R})$ be a set system, where X is a set of *objects* and $\mathcal{R}$, a family of subsets of X, is a set of *ranges*. A *set cover* C of Σ is a subset of $\mathcal{R}$ such that $X = \bigcup C$. The *set-cover* problem is to find a set cover of the smallest size. The *hitting-set* problem, which is dual of set cover, asks for computing a subset $H \subseteq X$ that intersects all ranges of $\mathcal{R}$. In a geometric setting, X is a set of geometric objects, e.g., points, lines, hyperplanes, spheres, etc., and $\mathcal{R}$ are geometric ranges. Let $\mathbb{H}$ be the set of all d-dimensional halfspaces. Here are two examples of geometric set systems: (i) $(\mathbb{R}^d, \mathbb{H})$, and (ii) $(\mathbb{H}, \{\{h \in \mathbb{H} \mid p \in h\} \mid p \in \mathbb{R}^d\})$.

It is known that the set-cover problem is NP-hard even in geometric settings. For example, Fowler *et al.* [87] proved that it is NP-hard to decide whether a given set of n points can be covered by k unit squares. The greedy algorithm can be used for computing an $O(\log n)$ approximation. However, one can do slightly better if the VC-dimension of the set system Σ is finite; see [105] for the definition of VC-dimension. Clarkson [52] modified the ITERATIVE_lp algorithm for computing a convex polytope of small complexity that lies between two nested convex polytopes, by reducing it to a geometric set-cover problem. Later Brönnimann and Goodrich [38] showed that Clarkson's algorithm works for any instance of set cover. For simplicity, we describe the algorithm for computing a hitting set. It performs a binary search on the size of the hitting set. At each stage, given an integer k, it either returns a hitting set of size $O(k \log k)$, or concludes that there is no hitting set of size at most k. Figure 6.7 summarizes the algorithm.

We call an iteration *successful* if $\mu(r) \leq \mu(X)/2k$ (see Figure 6.7 for the definition of μ). Following the same argument as for the linear-programming algorithm, Clarkson (and Brönniman and Goodrich) showed

```
function procedure HITTING_SET(X, R);
    t := O(k log k);  successful := 0;
    μ(x) = 1 ∀x ∈ X
    repeat
        choose a random H ∈ (X t);
        if H ∩ r ≠ ∅ for all r ∈ R then
            return H
        choose a r ∈ R s.t.  r ∩ H = ∅
        if μ(r) ≤ μ(X)/2k then
            successful := successful + 1
            for all x ∈ r do μ(x) = 2μ(x)
    until successful := 12k log n
    return NO
```

Figure 6.7 A randomized hitting-set algorithm.

that an iteration is successful with probability at least $1/2$ and that if there exists a hitting set of size at most k, then the algorithm returns a hitting set of size of $O(k \log k)$ within $12k \log n$ successful iterations. If the range space (X, R) admits an ε-net[3] of size $O(1/\varepsilon)$, then the above algorithm can be modified to compute a hitting set of size $O(k)$. The only nontrivial steps in the algorithm are determining whether H intersects all ranges, choosing a range that does not intersect H, and updating the weights of objects. If each iteration can be performed in $T(n) \log n$ time, then the expected running time of the decision algorithm is $O(kT(n) \log n)$. If the set system is maintained implicitly, then $T(n)$ is proportional to $\sum_{r \in R} |r|$. But R could be quite large and in many geometric settings it is defined implicitly. In such cases the main difficulty is to run the above algorithm efficiently without computing R explicitly. Such algorithms have been proposed in some cases [6, 10], and we will mention them below.

6.   FACILITY-LOCATION PROBLEMS

A typical *facility-location* problem is defined as follows: Given a set $D = \{d_1, \ldots, d_n\}$ of n *demand points* in $\mathbb{R}^d$, a parameter p, and a distance function δ, we wish to find a set S of p *supply objects* (points, lines, segments, etc.) so that a given cost function $f(D, S)$ is minimized. A widely studied cost function, known as *k-center*, is the maximum distance between a demand point and its nearest supply object. That is, we minimize, over all possible appropriate sets S of supply objects, the following objective function:

$$c(D, S) = \max_{1 \le i \le n} \min_{s \in S} \delta(d_i, s).$$

Instead of minimizing the above quantity, one can choose other objective functions, such as

$$c'(D, S) = \sum_{i=1}^{n} \min_{s \in S} \delta(d_i, s).$$

In some applications, a weight w_i is assigned to each point $d_i \in D$, and the distance from d_i to a point $x \in \mathbb{R}^d$ is defined as $w_i \delta(d_i, x)$. The book by Drezner [72] describes many other variants of the facility-location problem. For a parameter $\varepsilon > 0$, a solution to the clustering problem is called an *ε-approximate* solution if its cost is at most $(1 + \varepsilon)$ times that of an optimal solution. A useful extension of the facility-location problem, which has been widely studied, is the *capacitated facility-location* problem, in which we have an additional constraint that the size of each cluster should be at most c for some parameter $c \ge n/p$.

The set $S = \{s_1, \ldots, s_p\}$ of supply objects partitions D into p *clusters*, $D_1, \ldots, D_p$, so that s_i is the nearest supply object to all points in D_i. Therefore, a facility-location problem can also be regarded as a clustering problem. These facility-location (or clustering) problems arise in many areas, including operations research, shape analysis [101, 152, 180], data compression and vector quantization [143], information retrieval [62, 63], drug design [86], and data mining [18, 36, 183]. See also a recent survey by Jain *et al.* [118].

If p is considered as part of the input, most facility-location problems are NP-hard, even in the plane or even when only an ε-approximate solution is being sought (provided that ε is a sufficiently small constant) [85, 98, 132, 142, 155, 156]. Although many of these problems can be solved in polynomial time for a fixed value of p, some of them still remain intractable. In this section we review efficient algorithms for a few specific facility-location problems, which can be solved using randomization, especially when p is a small constant.

6.1   EUCLIDEAN *P*-CENTER

Given a set D of n demand points in $\mathbb{R}^d$, we wish to find a set S of p supply *points* so that the maximum Euclidean distance between a demand point and its nearest neighbor in S is minimized. This problem can be solved efficiently, when p is small, using randomized binary search or Megiddo's parametric searching technique [153]. The decision problem in this case is to determine, for a given radius r, whether D can be covered by the union of p balls of radius r. In some applications, S is required to be a subset of D, in which case the problem is referred to as the *discrete p-center* problem.

Euclidean 1-center.. The 1-center problem is to compute the smallest ball enclosing D. The decision procedure for the 1-center problem is thus to determine whether D can be covered by a ball of radius r. The Euclidean 1-center problem is an LP-type problem, with combinatorial dimension $d+1$ [188, 194]. Indeed, the constraints are the given points, and the function w maps each subset G to the radius of the smallest ball containing G. Monotonicity of w is trivial, and locality follows easily from the uniqueness of the smallest enclosing ball of a given set of points. The combinatorial dimension is $d+1$ because at most $d+1$ points are needed to determine the smallest enclosing ball. This problem, however, is not basis regular (the smallest enclosing ball may be determined by any number, between 2 and $d+1$, of points), and a naïve implementation of the basis-changing operation may be quite costly (in d). Nevertheless, Gärtner [94] showed that this operation can be performed in this case using expected $e^{O(\sqrt{d})}$ arithmetic operations. Hence, the expected running time of the algorithm is $O(d^2 n) + e^{O(\sqrt{d \log d})}$.

A natural extension of the 1-center problem is to find a disk of the smallest radius that contains k of the n input points. The best known randomized algorithm runs in $O(n \log n + nk)$ expected time using $O(nk)$ space, or in $O(n \log n + nk \log k)$ expected time using $O(n)$ space [147]. The best known deterministic algorithms are somewhat slower [67, 82, 79]. Matoušek [148] also showed that the smallest disk covering all but k points can be computed in time[4] $O(n \log n + k^3 n^\varepsilon)$. Chan [42] presented a randomized algorithm for computing the discrete 1-center in $\mathbb{R}^3$ whose expected running time is $O(n \log n)$.

There are several other extensions of the smallest-enclosing-ball problem. They include (i) computing the smallest enclosing ellipsoid of a point set [48, 75, 168, 194], (ii) computing the largest ellipsoid (or ball) inscribed inside a convex polytope in $\mathbb{R}^d$ [94], (iii) computing a smallest ball that intersects (or contains) a given set of convex objects in $\mathbb{R}^d$ (see [154], and (iv) computing a smallest area annulus containing a given planar point set. All these problems are known to be LP-type, and thus can be solved using the algorithm described in Section 4.. However, not all of them run in subexponential expected time because they are not all basis regular.

Euclidean 2-center.. In this problem we want to cover a set D of n points in $\mathbb{R}^d$ by two balls of smallest possible common radius. There is a trivial $O(n^{d+1})$-time algorithm for the 2-center problem in $\mathbb{R}^d$, because the clusters D_1 and D_2 in an optimal solution can be separated by a hyperplane [70]. Matoušek [144] gave an algorithm with $O(n^2 \log^2 n)$ expected time by using the randomized halving technique described in Section 2.1. The running time of the decision algorithm was improved by Hershberger [108] to $O(n^2)$, which has been utilized in the best near-quadratic solution, by Jaromczyk and Kowaluk [119], which runs in $O(n^2 \log n)$ time; see also [120].

A major progress on this problem was made by Sharir [185], who gave an $O(n \log^9 n)$-time algorithm, by combining Megiddo's parametric-searching technique with several additional techniques, including a variant of the matrix-searching algorithm of Frederickson and Johnson [90]. Eppstein [81] simplified Sharir's algorithm, using randomization and better data structures, and obtained an improved solution, whose expected running time is $O(n \log^2 n)$. Halperin *et al.* [100] studied the 2-center problem amid obstacles. That is, given a set D of n demand points in $\mathbb{R}^2$ and a set O of pairwise disjoint simple polygons, called *obstacles*, with a total of m vertices, compute two supply points outside O so that the maximum distance between a point its nearest supply point is minimized. Following the same approach as in [81, 185] but using some novel data structures, they presented an $O(mn \log^2(mn) \log n)$ expected time randomized algorithm for this problem. They also presented an ε-approximation algorithm with $O((1/\varepsilon) \log(1/\varepsilon)(m + n \log n) \log(mn))$ expected running time.

Inaba *et al.* [111] have studied the problem of partitioning D into two clusters so that a function on the variance of points within each cluster is minimized. More precisely, let $\bar{x}(S)$ denote the centroid of a point set S. Then the variance of S, denoted as $\mathrm{Var}(S)$, is

$$\mathrm{Var}(S) = \frac{1}{|S|} \sum_{x_i \in S} \|x_i - \bar{x}(S)\|^2.$$

Define $\psi^\alpha(S) = |S|^\alpha \mathrm{Var}(S)$. Inaba *et al.* [111] define the cost of an optimal clustering to be $c(S) = \min \psi^\alpha(S_1) + \psi^\alpha(S_2)$, where the minimum is taken over all partitions S_1, S_2 of S. They consider $\alpha = 1, 2$. For a given $\varepsilon > 0$, they presented an $O(n/\varepsilon^2)$ expected time randomized algorithm for partitioning D into two clusters so that the cost of the clustering is at most $(1+\varepsilon)c(S)$. Recently, their algorithm was derandomized by Matoušek [149]. See [113] for some recent results on 2-clustering in higher dimensions.

Rectilinear p-center.. In this problem the metric is the L_∞-distance, so the decision problem is now to cover the given set D by a set of p axis-parallel cubes, each of length $2r$. The problem is NP-complete if p is part of the input and $d \geq 2$, or if d is part of the input and $p \geq 3$ [87, 155]. Ko *et al.* [132] showed that computing a solution set S with $c(D, S) < 2r^*$, where r^* is the size of an optimal solution, is also NP-complete.

The rectilinear 1-center problem is trivially solved in linear time. See [71, 130, 131, 155] for some earlier results. Sharir and Welzl [189] developed a linear-time algorithm for the rectilinear 3-center problem, by showing that it is an LP-type problem (as is the rectilinear 2-center problem). This is an instance of nonconvex programming that is LP-type. Using the technique described in Section 2.3, Chan [42] proposed an $O(n \log n)$ expected time randomized algorithm for computing a rectilinear 5-center, which is optimal in the worst-case. Chan also presented an $O(n \log n)$ expected time algorithm for computing the smallest square that contains k of a given set of n points in the plane. See [126, 189] for additional related results.

6.2 EUCLIDEAN P-LINE-CENTER

Let D be a set of n points in $\mathbb{R}^d$ and δ be the Euclidean distance function. We wish to compute the smallest real value w^* so that D can be covered by the union of p strips of width w^*. Megiddo and Tamir [157] showed that the problem of determining whether $w^* = 0$ (i.e, D can be covered by p lines) is NP-complete, which not only proves that the p-line-center problem is NP-complete, but also proves that approximating w^* within a constant factor is NP-complete.

Since even approximating w^* is NP-complete, Agarwal and Procopiuc [10] developed an efficient algorithm for the case in which one approximates both w^* and p. In particular, let $w_p^* = w_p^*(D)$ denote the size of the Euclidean p-line center of S. By modifying the hitting-set algorithm described in Section 5., they presented a randomized algorithm that computes $O(p \log p)$ strips of width at most $6w_p^*$ that contain S. The expected running time of their algorithm is $O(np^2 \log^3 n \log(p \log n))$ provided that $p \leq \sqrt{n}$. They also extended their algorithm to higher dimensions in some cases. The main contribution of this result is to improve the expected running time of the algorithm to be near linear as a function of n. In most practical applications, n is quite large and p is a small constant. One can, of course, use the HITTING_SET algorithm to compute $O(p \log p)$ strips of width at most w_p^* in polynomial time, as the problem can be reduced to an instance of the hitting-set problem. If one maintains the underlying set system explicitly, the expected running time of the algorithm is $\Omega(n^3 p)$. Agarwal and Procopiuc showed that by maintaining the set system implicitly, the expected running time can be improved to roughly $n^{4/3}p^{4/3}$, but this algorithm is not practical. A few other approximation algorithms for this problem are given in [103].

6.3   EUCLIDEAN *P*-MEDIAN

Let D be a set of n points in $\mathbb{R}^d$. We wish to compute a set S of p supply points so that the sum of distances from each demand point to its nearest supply point is minimized (i.e., we want to minimize the objective function $c'(D, S)$). This problem can be solved in polynomial time for $d = 1$ (for $d = 1$ and $p = 1$ the solution is the median of the given points, whence the problem derives its name), and it is NP-hard for $d \geq 2$ [156]. The special case of $d = 2, p = 1$ is the classical *Fermant-Weber problem*, which goes back to the seventeenth century. It is known that the solution for the Fermant-Weber problem is unique and algebraic provided that all points of D are not collinear. Several numerical approaches have been proposed to compute an approximate solution. See [44, 195] for the history of the problem and for the known algorithms, and [166] for some heuristics for the p-median problem that work well for a set of random points. Recently, Arora *et al.* [25] described an ε-approximation algorithm for the p-median problem in the plane whose running time is $n^{O(1/\varepsilon)}$. For $d > 2$, the running time of their algorithm is $n^{O((\log n/\varepsilon)^{d-1})}$. Their algorithm is an extension of Arora's approximation algorithm for TSP, which we will describe in Section 11. below. The bound was later improved by Kolliopoulos and Rao [134], who proposed a Monte Carlo ε-approximation algorithm that runs in time $O(2^{1/\varepsilon^d} n \log n \log p)$ with probability at least $1/2$. Ostrovsky and Rabani [165] have proposed a polynomial-time ε-approximation algorithm for comuting a p-median for the case in which p is fixed but d is part of the input. Their algorithm is based on the recent Monte Carlo techniques developed for answer approximate nearest-neighbor queries; see Section 7..

There has also been much work on the p-median problem in arbitrary metric spaces. Lin and Vitter [139] (see also [140]) proposed a randomized algorithm, based on a randomized-rounding scheme, that, for any parameter $\varepsilon > 0$, computes $(1+1/\varepsilon)p$ clusters with a total cost of $2(1+\varepsilon)$ times that of an optimal solution. If we want to return exactly p clusters, Bartal [32] gave an $O(\log n \log \log n)$-approximation algorithm for the p-median problem, which was later improved by Charikar *et al.* [45] to $O(\log p \log \log p)$ using a randomized embedding technique. The best known approximation algorithm is by Charikar and Guha [46] that computes a 3-approximate solution in polynomial time. See also [47].

7.   NEAREST-NEIGHBOR SEARCHING

The *nearest-neighbor query* (*NN query*) problem, also known as the *post-office* problem [129], is defined as follows: Preprocess a set S of points in $\mathbb{R}^d$ into a data structure so that a point in S closest to a query point ξ can be reported quickly. This is one of the most widely studied problems not only in computational geometry but also in several areas of computer science, including pattern recognition [61, 73], data compression [26, 171], information retrieval [84, 178], CAD [158], computational biology [190], image analysis [133, 135], data mining [83, 104], machine learning [60], and geographic information systems [177, 191]. Most applications use so-called *feature vectors* to map a complex object to a point in high dimensions. Examples of feature vectors include color histograms, shape descriptors, Fourier vectors, and text descriptors.

For simplicity, we assume that the distance between points is measured in the Euclidean metric, though a more complicated metric can be used depending on the application. Dobkin and Lipton [69] described a *locus based* method that partitions the space into connected regions such that all points in a region have the same nearest neighbor among the n points. So the problem reduces to point-location for which the method required $n^{2^{O(d)}}$ space and $O(2^d \log n)$ query time. For $d = 2$, one can construct the Voronoi diagram of S and preprocess it for point-location queries in $O(n \log n)$ time using $O(n)$ space so that an NN query can be answered in $O(\log n)$ time [169]. For higher dimensions, Clarkson [51] presented a data structure of size $O(n^{\lceil d/2 \rceil + \varepsilon})$, for any constant $\varepsilon > 0$, that can answer a query in $2^{O(d)} \log n$ time. The data structure can be constructed in $O(n^{\lceil d/2 \rceil + \varepsilon})$ expected time. This paper was one of the earliest applications of random sampling in computational geometry. The query time can be improved to $O(d^5 \log n)$, using a technique of Meiser [159]. A different data structure with $O(n)$ space and $2^{O(d)} n^{1-1/\lfloor d/2 \rfloor}$ query time was proposed by Agarwal and Matoušek [9]. Note that either the query time or the size of the above data structures is exponential in d, so it is impractical even for moderate values of d (say $d \approx 10$). This exponential dependence on dimension is called the *curse of dimensionality*. Several heuristics have been developed, especially in higher dimensions, which use practical data structures such as kd-trees, R-trees, R*-trees, and Hilbert R-trees; see e.g. [92, 83, 104, 109, 135, 133, 177, 191]. Even these algorithms suffer from the curse of dimensionality.

This has led to the development of algorithms for finding approximate nearest neighbors [27, 28, 29, 53, 128, 135] or for special cases, such as when the distribution of query points is known in advance [56, 196]. For a given parameter $\varepsilon > 0$ and a query point ξ, an ε-approximate nearest-neighbor query (ε-*NN query*) asks for returning a point $p \in S$ so that $d(p, \xi) \leq (1 + \varepsilon)d(p', \xi)$ for all $p' \in S$. This relaxation is quite meaningful in the context of the applications mentioned above. Arya *et al.* [29] showed that an ε-NN query can be answered in $O((1/\varepsilon^d) \log n)$ time using $O(n)$ space. Note that the size of their data structure is independent of ε, and that ε can be specified as a part of the query. Although the query bound was later improved by Clarkson [53] and Chan [41] to $O((1/\varepsilon)^{(d-1)/2} \log n)$, ε is fixed for all queries in both the data structures and the size of their data structures depends on ε. The data structure by Arya *et al.* [29] is practical and works well for dimensions up to 20–30, though their empirical results show that an approach [92] based on *kd*-trees works equally well in most of the cases.

Numerous approximation techniques based on *distance-preserving* random projections of points onto lower dimensional subspaces have been proposed, which result in randomized algorithms with query time polynomial in d and $\log n$ [116, 117, 128, 136]. Many of these techniques rely on the following classical result by Johnson and Lindenstrauss [121], which was subsequently improved and simplified in [66, 88, 89]: Any set S of n points in d-dimensional Euclidean space $\mathbb{E}^d$ can be embedded in $O((1/\varepsilon^2) \log n)$ dimensions with at most ε relative error in the pairwise distances of S. Simpler proofs using elementary probabilistic techniques have been proposed [66, 116], which immediately give randomized polynomial-time algorithms for computing such an embedding. In fact, these arguments show that for any point $q \in \mathbb{E}^d$, the distance between the image of q and that of any point $p \in S$ is within ε-relative error of $d(p, q)$, with high probability.

Although distance-preserving hashing had been used earlier for points in $\mathbb{R}^1$ [141] and for searching in higher dimensions [117], Kleinberg [128] was perhaps the first to exploit random projections in the context of ε-NN searching. His algorithm relies on the following observation:

Lemma 7..1 ([128]) *Let $x, y \in \mathbb{R}^d$ be two vectors such that $(1+\gamma)\|x\| \leq \|y\|$ for some $\gamma \leq 1/2$. Then for a random unit vector v chosen uniformly over $\mathbb{S}^{d-1}$,*

$$\Pr[|v \cdot x| \geq |v \cdot y|] \leq \frac{1}{2} - \frac{\gamma}{3}.$$

This lemma implies that if the ratio of the lengths of two vectors is at least $(1 + \gamma)$, then their random projections preserve their ordering in a probabilistic sense. Based on this observation, an ε-NN query can be answered as follows. Choose a sufficiently large set of random unit vectors $V = \{v_1, v_2, \ldots v_L\} \subset \mathbb{S}^{d-1}$ and project the points in S on each of these vectors. For a point $a \in \mathbb{R}^d$, let $a^{(i)}$ denote the projection of a on v_i. For a point $q \in \mathbb{R}^d$, we say that a point a *dominates* b with respect to v_i if $|q^{(i)} - a^{(i)}| \le |q^{(i)} - b^{(i)}|$, and that $a \prec_q b$ if a dominates b with respect to at least $L/2$ vectors in V. To answer an ε-NN query for a point q, we compute the point p in S that is minimal in the $\prec_q$-ordering. Kleinberg showed that S can be stored into a data structure of size $O((n/\varepsilon^2)^{2d} \log^{2d} n)$ so that the minimal point in the $\prec_q$-ordering for a query point q can be computed in time $(d^2/\varepsilon^2) \log^{O(1)} n$. He also proposed another data structure of size $(nd/\varepsilon^2) \log^{O(1)} n$ that answers an ε-NN query in time $O((d/\varepsilon^2) \log d \log n \log(d \log n))$. These algorithms can report k ε-approximate nearest neighbors in additional $O(k)$ time.

Kleinberg's result was subsequently improved by Kushilevitz et al. [136]. Using a clustering argument, they showed that ε-NN searching in $\mathbb{E}^d$ can be reduced to ε'-NN searching on d-dimensional hypercube $\mathbb{Q}^d = \{0, 1\}^d$, for some $\varepsilon' = \varepsilon/O(1)$, under the Hamming metric. (A similar reduction was used by Indyk and Motwani [116].) More precisely, Kushilevitz et al. [136] showed that an ε-NN data structure in $\mathbb{Q}^d$ using $S(n, d, \varepsilon)$ storage and $Q(n, d, \varepsilon)$ query time leads to an ε-NN structure in $\mathbb{E}^d$ using $O(n^2) S(n, \Delta, \varepsilon')$ storage and $(d \log n)/\varepsilon)^{O(1)} + \log n Q(n, \Delta, \varepsilon')$ query time, where $\Delta = O((d/\varepsilon^8) \log^2(d/\varepsilon))$ and $\varepsilon' = \varepsilon/O(1)$. Their data structure for answering an ε-NN query in $\mathbb{Q}^d$ is based on the following observation. Let $H(\cdot, \cdot)$ denote Hamming distance.

Lemma 7..2 ([136]) *Let x, y be two points in $\mathbb{Q}^d$, let $\gamma > 0$ be a constant, let $0 \le \ell \le d$ be an integer, and let q be a point with $H(q, x) \le \ell$ and $H(q, y) > (1 + \gamma)\ell$. Suppose we choose a random vector $r = (r^1, \ldots, r^d) \in \mathbb{Q}^d$ where $\Pr[r_i = 1] = 1/2\ell$, for each $1 \le i \le d$. For two vectors $u, v \in \mathbb{Q}^d$, let*

$$\tau_r(v) = \sum r^i \cdot v^i \quad (\text{mod } 2) \quad \text{and} \quad \Delta(u, v) = \Pr[\tau_r(u) \ne \tau_r(v)].$$

Then there are constants $\delta, \delta_1 > 0$ depending only on γ such that $\Delta(q, x) \le \delta_1$ and $\Delta(q, y) > \delta + \delta_1$.

For a query point $q \in \mathbb{Q}^d$, we can now decide whether there exists a point S within distance ℓ from q, as follows. Choose a set $R = \{r_1, \ldots, r_t\}$ of $t = O((1/\varepsilon^2) \log(n \log d))$ of random vectors as described in the above lemma. Let a be a point in S. Let $\mu = \mu(a)$ be the number of vectors $r \in R$ for which $\tau_r(a) \ne \tau_r(q)$. Using Lemma 7..2 and Chernoff's bound, it can be shown that if $H(q, a) \le \ell$ then $\mu \le (\delta_1 + \delta/3)t$

with probability at least $1 - e^{-2\delta^2 t/9}$, and that if $H(q, a) \geq (1 + \varepsilon)\ell$ then $\mu \geq (\delta_1 + 2\delta/3)$ with probability at least $1 - e^{-2\delta^2 t/9}$. We thus need to determine whether there exists a point $a \in S$ for which $\mu(a) \leq (\delta_1 + \delta/3)t$. Based on this observation, Kushilevitz *et al.* showed that, with probability at least $1 - \rho$, S can be preprocessed in a data structure of size $d(n \log d)^{O(1/\varepsilon^2)}$ so that a query can be answered in time $O((d^2/\varepsilon^2) \log(n \log d/\rho))$. Constructing this structure for every $0 \leq \ell \leq d$ and using the above procedure as the decision procedure in a binary search, we can answer an ε-NN query with probability at least $1 - \rho$ in $O((d^2/\varepsilon^2) \log(n \log d/\rho) \log d)$ using $d^2(n \log d)^{O(1/\varepsilon^2)}$ storage.

A different ε-NN data structure with a slightly worse performance was proposed by Indyk and Motwani [116]. They also proposed another randomized approach for answering ε-NN queries. They develop a data structure called *ring cover trees*, which enables them to reduce the ε-NN searching problem to ε-PLEB (ε-approximate point location in equal balls) at a polylogarithmic overhead on the size and the query time. The latter problem is defined as follows: Given a set S of n points in $\mathbb{R}^d$ and two parameters $\varepsilon, r > 0$, preprocess S into a data structure that for a query point q performs as follows: If there is a point $p \in S$ so that $d(p, q) \leq r$, then it returns YES and a point p' with $d(p, p') \leq (1 + \varepsilon)r$. If $d(p, q) > (1 + \varepsilon)r$ for all $p \in S$, then it returns NO. Otherwise, it either returns NO or a point p with $d(p, q) \leq (1 + \varepsilon)r$. They introduced the notion of *locality-sensitive hashing* to answer ε-PLEB queries. For given parameters $r_1 > r_2 \geq 0$ and $1 > p_1 > p_2 > 0$, a family $\mathcal{H} = \{h : S \to U\}$ of hash functions is called (r_1, r_2, p_1, p_2)-*sensitive* if for every $p, q \in S$: (i) $d(p, q) \leq r_1$ implies that $\mathrm{Pr}_{\mathcal{H}}[h(p) = h(q)] \geq p_1$, and (ii) $d(p, q) > r_2$ implies that $\mathrm{Pr}_{\mathcal{H}}[h(p) = h(q)] \leq p_2$. Using such a family of hash functions, they answer an ε-PLEB query as follows. For simplicity, we will describe the data structure for points in $\mathbb{Q}^d$ under Hamming distance. Set $k = \log_{1/p_2} n$, $\rho = \log p_1 / \log p_2$, and $\ell = n^\rho$. Define a family

$$\mathcal{G} = \{(h_1, h_2, \dots, h_k) : \mathbb{Q}^d \to U^k \mid h_1, h_2, \dots, h_k \in \mathcal{H}\}.$$

Choose a random subset of ℓ functions $g_1, \dots g_\ell \in \mathcal{G}$. Let T be a table of size $|U|^k$. For each $p \in S$ and $i \leq \ell$, we store p in the cell $g_i(p)$ of T, i.e., each point of S is stored in ℓ cells of T. Therefore only $O(n\ell) = O(n^{1+\rho})$ cells of T are nonempty; some of the cells might have multiple entries. We can use the data structure by Fredman *et al.* [91] to store these nonempty entries in a table of size $O(n^{1+\rho})$ so that for a point q, we can access the cell of T corresponding to $g(q)$ in time $O(1)$ after having computed the value of $g(q)$, which in turns requires evaluating k hash functions. For a query point $q \in \mathbb{Q}^d$, the algorithms proceed as follows.

It accesses the cells $g_1(q), \ldots, g_\ell(q)$ of T, and checks whether any of the points stored in these cells is within distance $(1 + \varepsilon)r$. If so, it returns such a point. If more than a total of 2ℓ points (including duplicates) are stored in these cells, the procedure checks at most 2ℓ points. It returns NO, if no point (among at most 2ℓ points) within distance $(1 + \varepsilon)r$ from q was found. The correctness of this procedure follows from the following lemma:

Lemma 7..3 ([116]) *With probability at least $1/2$, for any $p \in S$ so that $d(p, q) \leq r$, $g_j(p) = g_j(q)$ for some $j \leq \ell$, and $\sum_{j=1}^{\ell} |\{p \in S \mid d(p, q) > (1 + \varepsilon)r \wedge g_j(p) = g_j(q)\}| < 2\ell$.*

The size of the data structure is $O(n^{1+\rho})$ and a query requires evaluating ℓ hash functions. Andersson *et al.* [21] showed that for any $r, \varepsilon > 0$, the family of projection functions, $\mathcal{H} = \{h_i : h_i(x_1, \ldots x_d) = x_i \mid 1 \leq i \leq d\}$ is $(r, (1 + \varepsilon)r, 1 - r/d, 1 - (1 + \varepsilon)r/d)$-sensitive. Plugging these values, Indyk and Motwani obtain a data structure of $O(n^{1+1/(1+\varepsilon)})$ size that can answer an ε-PLEB query in $O(dn^{1/(1+\varepsilon)})$ time, with probability at least $1/2$.

All the algorithms described above are Monte Carlo algorithms. Indyk [114] proposed a Las Vegas algorithm that answers an ε-NN query under L_1-metric in $((d/\varepsilon) \log n)^{O(1)}$ expected time using polynomial space. This also yields constant-factor approximation algorithms for ε-NN searching under L_2- and Hamming-metrics. See also [112, 115] for related results. The above data structures have recently been used to answer farthest neighbor queries, to compute the diameter of a point set, and for several other proximity problems [35, 102, 114]. See [35, 31, 39] for recent lower bounds on NN-searching in higher dimensions.

8. PROXIMITY PROBLEMS

8.1 DIAMETER IN R^3

Given a set S of n points in $\mathbb{R}^3$, we wish to compute the *diameter* of S, that is, the maximum distance between any two points of S. The decision procedure here is to determine, for a given radius r, whether the intersection of the balls of radius r centered at the points of S contains S. The intersection of congruent balls in $\mathbb{R}^3$ has linear complexity [99, 107], and it can be computed in $O(n \log n)$ expected time by a randomized incremental algorithm [58]. Checking whether all points of S lie in the intersection can then be performed in additional $O(n \log n)$ time, using straightforward point-location techniques. The technique described in Section 2.2 can therefore compute the diameter in $\mathbb{R}^3$ in $O(n \log^2 n)$ expected time. However, Clarkson and Shor [58] showed that their tech-

nique can be refined so that the diameter can be computed in $O(n \log n)$
expected time. The basic observation is that the size of the problem can
also be reduced in each step. Figure 6.8 gives an outline of the algorithm.

```
function procedure DIAMETER (S);
    choose a random point p ∈ S;
    q = a farthest neighbor of p in S;
    compute I = ∩_{p'∈S} B(p', δ(p,q))
    S₁ = S \ I
    if S₁ = ∅
        then return δ(p,q)
        else return DIAMETER (S₁)
```

Figure 6.8 A randomized algorithm for computing the diameter in 3D.

The correctness of the above algorithm is easy to check. The only non-
trivial step in the above algorithm is computing I and S_1. Since I can
be computed in $O(|S| \log |S|)$ expected time, S_1 can then be computed
in additional $O(|S| \log |S|)$ time, using any optimal planar point-location
algorithm (see, e.g., [179]). Hence, each recursive step of the algorithm
takes $O(|S| \log |S|)$ expected time. Since p is chosen randomly, $|S_1| = i$
with probability $1/n$, which implies that the expected running time of
the overall algorithm is $O(n \log n)$. After several attempts, an $O(n \log n)$
deterministic algorithm was recently obtained by Ramos [172]. Subquadratic algorithms (as a function of n) for computing the diameter in
higher dimensions are known [7], but their running time is exponential
as a function of dimension. Using the recent techniques for nearest-
neighbor searching, Monte Carlo ε-approximation algorithms have been
developed for computing the diameter of a point set in $\mathbb{E}^d$ whose running
time is subquadratic in n and polynomial in d [35, 113].

8.2   DISTANCE BETWEEN POLYTOPES

We wish to compute the Euclidean distance $d(\mathcal{P}_1, \mathcal{P}_2)$ between two
given convex polytopes $\mathcal{P}_1$ and $\mathcal{P}_2$ in $\mathbb{R}^d$. If the polytopes intersect,
then this distance is 0. If they do not intersect, then this distance equals
the maximum distance between two parallel hyperplanes separating the
polytopes; such a pair of hyperplanes is unique, and they are orthog-
onal to the segment connecting two points $a \in \mathcal{P}_1$ and $b \in \mathcal{P}_2$ with
$d(a, b) = d(\mathcal{P}_1, \mathcal{P}_2)$. Gärtner [94] showed that this problem is LP-type,
with combinatorial dimension at most $d+2$ (or $d+1$, if the polytopes do
not intersect). He also showed that the primitive operations can be performed with expected $e^{O(\sqrt{d})}$ arithmetic operations. Hence, the problem

can be solved by the general LP-type algorithm, whose expected number of arithmetic operations is $O(d^2 n) + e^{O(\sqrt{d \log d})}$, where n is the total number of facets in $\mathcal{P}_1$ and $\mathcal{P}_2$.

8.3   SELECTING DISTANCES

Let S be a set of n points in the plane, and let $1 \leq k \leq \binom{n}{2}$ be an integer. We wish to compute the kth smallest distance between a pair of points of S. This can be done using parametric searching. The decision problem is to compute, for a given real r, the sum $\sum_{p \in S} |D_r(p) \cap (S \setminus \{p\})|$, where $D_r(p)$ is the closed disk of radius r centered at p. (This sum is twice the number of pairs of points of S at distance $\leq r$.) Agarwal *et al.* [4] gave a randomized algorithm, with $O(n^{4/3} \log^{4/3} n)$ expected time, for the decision problem, using the random-sampling technique of [58], which yields an $O(n^{4/3} \log^{8/3} n)$ expected-time algorithm for the distance-selection problem. Matoušek [144] showed that the randomized halving technique can be used to solve this problem in time $O(n^{4/3} \log^{5/3} n)$ time.

8.4   SURFACE SIMPLIFICATION

A generic *surface-simplification* problem is defined as follows: Given a polyhedral object P in $\mathbb{R}^3$ and an error parameter $\varepsilon > 0$, compute a polyhedral approximation Π of P with the minimum number of vertices, so that the maximum distance between P and Π is at most ε. There are several ways of defining the maximum distance between P and Π, depending on the application. We will refer to an object that lies within ε distance from P as an *ε-approximation* of P. Surface simplification is a central problem in graphics, geographic information systems, scientific computing, and visualization.

The simplest, but nevertheless an interesting, special case is when P is a convex polytope (containing the origin). In this case we wish to compute another convex polytope Q with the minimum number of vertices so that $(1-\varepsilon)P \subseteq Q \subseteq (1+\varepsilon)P$ (or so that $P \subseteq Q \subseteq (1+\varepsilon)P$). We can thus pose a more general problem: Given two convex polytopes $P_1 \subseteq P_2$ in $\mathbb{R}^3$, compute a convex polytope Q with the minimum number of vertices such that $P_1 \subseteq Q \subseteq P_2$. Das and Joseph [65] had attempted to prove that this problem is NP-hard, but their proof contained an error. It seems a result by Das and Goodrich [64] fixes their proof. Mitchell and Suri [161] have shown that there exists a nested polytope Q with at most $3k_{\text{OPT}}$ vertices, whose vertices are a subset of the vertices of P_2, where k_{OPT} is the minimum number of vertices in a convex polytope lying between P_1 and P_2. The problem can now be formulated as a hitting-set problem, and, using a greedy approach, they presented an $O(n^3)$-time algorithm for computing a nested polytope with $O(k_{OPT} \log n)$ vertices. Clarkson [52] showed that the randomized technique described in Section 5. can compute a nested polytope with $O(k_{OPT} \log k_{OPT})$ vertices in $O(n \log^c n)$ expected time, for some constant $c > 0$. Brönnimann and Goodrich [38] extended Clarkson's algorithm to obtain a polynomial-time deterministic algorithm that constructs a nested polytope with $O(k_{OPT})$ vertices.

A widely studied special case of surface simplification, motivated by applications in geographic information systems and scientific computing, is when P is a polyhedral terrain (i.e., the graph of a continuous piecewise-linear bivariate function). In most of the applications, P is represented as a finite set of n points, sampled from the input surface, and the goal is to compute a polyhedral terrain Q with the minimum number of vertices, such that the vertical distance between any point of P and Q is at most ε. Agarwal and Suri [16] showed that this problem is NP-hard. Agarwal and Desikan [6] have shown that Clarkson's randomized algorithm can be extended to compute a polyhedral terrain of size $O(k_{OPT}^2 \log^2 k_{OPT})$ in expected time $O(n^{2+\delta} + k_{OPT}^3 \log^3 k_{OPT})$. The survey paper by Heckbert and Garland [106] summarizes most of the known results on terrain simplification.

Instead of fixing ε and minimizing the size of the approximating surface, we can fix the size and ask for the best approximation. That is, given a polyhedral surface P and an integer k, compute an approximating surface Q that has at most k vertices, whose distance from P is the smallest possible. Very little is known about this problem, except in the plane. If the vertices of Q are required to be a subset of S, the best known algorithm is by Agarwal and Varadarajan [17]; it is based on the randomized halving technique described in Section 2.1, and its running time is $O(n^{4/3+\varepsilon})$.

9.   STATISTICAL ESTIMATORS AND RELATED PROBLEMS

9.1   LINE FITTING

Fitting a line to a set $S = \{p_1, \ldots, p_n\}$ of n points in the plane is an important problem in statistical estimation. In order to cope with outliers, there has been much interest in defining *robust* line estimators whose slopes do not change much by a few outliers. One such estimator is the Theil-Sen estimator defined in Section 2.1 for which several optimal $O(n \log n)$ algorithms exist. Another commonly used estimator is the *repeated median* (RM) estimator, defined as follows. For each $p_i \in S$, let σ_i be the median of the slopes of the $n - 1$ lines passing through p_i and another point of S, and let σ be the median of $\{\sigma_1, \ldots, \sigma_n\}$. Then the RM estimator of S is the line of slope σ passing through a pair of input points. Using a variant of the randomized halving technique described in Section 2.1 and some sophisticated range-searching data structures, Matoušek *et al.* [150] described an $O(n \log n)$ expected time algorithm for computing the RM estimator. They also described a somewhat simpler randomized algorithm with $O(n \log^2 n)$ expected time.

9.2   PLANE FITTING

Given a set S of n points in $\mathbb{R}^3$, we wish to fit a plane h through S so that the maximum distance between h and the points of S is minimized. This is the same problem as computing the *width* of S — the smallest distance between a pair of parallel supporting planes of S, which is considerably harder than the two-dimensional variant mentioned in Section 6.2. It can be shown that either one of the parallel planes determining the width contains a vertex and the other contains a face of $\text{conv}(S)$, or each of the two parallel planes contains an edge of $\text{conv}(S)$. Houle and Toussaint [110] gave an $O(n^2)$-time algorithm for computing the width in $\mathbb{R}^3$. They show that the first type of pairs of planes can be computed in $O(n \log n)$ time, but there could be $\Theta(n^2)$ *antipodal* pairs of convex hull edges. (A pair of edges e_1, e_2 of $\text{conv}(S)$ is called antipodal if there exist two parallel planes π_1, π_2 supporting them such that S lies between π_1 and π_2.) The problem of computing a closest pair of antipodal edges can be reduced to a number of subproblems, each of which asks for computing a closest pair between two sets L, L' of lines in $\mathbb{R}^3$ (each line containing an edge of the convex hull of S), such that each line in L lies below all the lines of L' [13, 11]. Agarwal and Sharir [11] developed an $O(n^{3/2+\varepsilon})$ expected-time randomized algorithm for this problem, which implies that the width can also be computed in expected time $O(n^{3/2+\varepsilon})$.

9.3   CIRCLE FITTING

Given a set S of n points in the plane, we wish to fit a circle C through S so that the maximum distance between the points of S and C is minimized. This is equivalent to finding an annulus of minimum width that contains S. Ebara *et al.* [77] observed that the center of a minimum-width annulus is a vertex of the closest-point Voronoi diagram of S, a vertex of the farthest-point Voronoi diagram, or an intersection point of a pair of edges of the two diagrams. Based on this observation, they obtained a quadratic-time algorithm. Agarwal and Sharir [11] reduced this problem to computing a bichromatic closest pair in two given sets of lines in $\mathbb{R}^3$, under an appropriate distance function. Using the technique described in Section 2.2, they showed that such a closest pair can be computed in $O(n^{3/2+\varepsilon})$ expected time, which in turn implies that the minimum-width annulus can be computed within that time. If we know the angular ordering of points with respect to the center of the minimum-width annulus, which is the case in some of the applications, Ramos [93] showed that the problem becomes an LP-type problem and can therefore be solved in $O(n)$ expected time. Recently Chan [43] developed an approximation algorithm that using his linear-programming data structure (mentioned in Section 3.) can compute in $O(n + 1/\varepsilon^{16/3} \log n)$ expected time an annulus containing S whose width is at most $(1 + \varepsilon)$ times that of the thinnest annulus.

9.4   CENTER POINTS

Let S be a set of n points in $\mathbb{R}^d$. Let h be a hyperplane, and let S^+, S^- be the subset of points lying in the positive and negative open halfspaces determined by h. For $0 \le \beta \le 1/2$, we say that h β-splits S if $\max\{|S^+|/|S|, |S^-|/|S|\} \le (1 - \beta)$. A point $c \in \mathbb{R}^d$ is a β-center of S if any hyperplane containing c $(1 - \beta)$-splits S. It is a known consequence of Helly's Theorem that a $1/(d+1)$-center always exists. Computing $1/(d+1)$-center is expensive in high dimensions, so approximation algorithms have been proposed. For a given parameter $0 < \delta < 1$, Clarkson *et al.* [57] gave a randomized algorithm that runs in $O((d\log(1/\delta))^{\log d})$ time and computes an $\Omega(1/d^2)$-center with probability at least $1 - \delta$. By combining this approach with linear programming, they developed another algorithm that computes in $O(d/\varepsilon)^{O(d)} \log(1/\delta)$ time a $(1/(d+1) - \varepsilon)$-center with probability at least $1 - \delta$.

9.5 · BUCKETING

Let $S = \{p_1, \ldots, p_n\}$ be a set of n points in $\mathbb{R}^2$ and $1 \le b \le n$ an integer. We want to partition S into b equal-size buckets so that the maximum number of points in a bucket is minimized. We consider two types of buckets. First, we consider the case in which the buckets are strips. That is, we want to find $b+1$ equally spaced parallel lines so that all points of S lie between the extreme lines, the extreme lines contain at least one points of S, and the maximum number of points in a bucket is minimized; see Figure 6.9. We refer to this problem as the *uniform-projection* problem. If the lines have slope θ, we refer to these buckets as the θ-cut of S. For each θ, there is a unique θ-cut of S.

It is convenient to consider the problem in the dual plane (see Section 2.1 for the definition of the duality transform). Let ℓ_i denote the

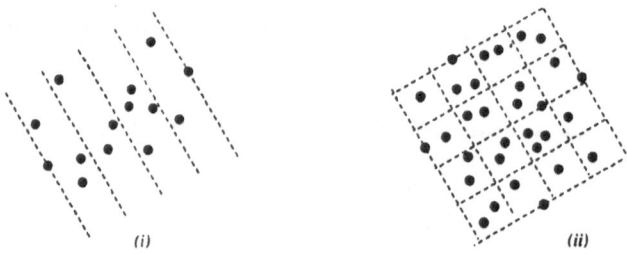

Figure 6.9 (i) Uniform-projection problem; (ii) two-dimensional partitioning problem.

line dual to the point $p_i \in S$, and let $\mathcal{L} = \{\ell_i \mid 1 \leq i \leq n\}$. The dual of a strip σ bounded by two parallel lines ℓ_1 and ℓ_2 is the vertical segment $\sigma^* = \ell_1^* \ell_2^*$; a point p lies in σ if and only if the line σ^* intersects the segment σ^*. Let $\mathcal{A}(\mathcal{L})$ be the arrangement of $\mathcal{L}$. For a fixed x-coordinate θ, let $s(\theta)$ denote the vertical segment connecting the points on the upper and lower envelopes of $\mathcal{A}(L)$. The dual of the θ-cut is the partition of $s(\theta)$ into b equal subsegments. A point $p_i \in S$ lies in the jth bucket of the θ-cut if the dual line ℓ_i intersects the jth subsegment of $s(\theta)$. The goal is therefore to compute θ so that the maximum number of lines intersecting a subsegment of $s(\theta)$ is minimum (over all θ). By sweeping the dual plane with a vertical line from $x = -\infty$ to $x = +\infty$ and maintaining the intersection points of $s(\theta)$ and L during this sweep, the optimal solution can be computed in a straightforward manner. Asano and Tokuyama [30] showed that such a sweep can be performed in $O(n^2)$ time. Agarwal *et al.* [5] developed a Monte Carlo algorithm that computes an optimal solution, with high probability, in subquadratic time in certain cases. They choose a random subset $R \subseteq L$ of lines and compute an optimal uniform projection for this subset. Using this solution, they compute a set of x-intervals that contains an optimal θ-cut with high probability and sweep a vertical line only through these x-intervals. Suppose the maximum number of points in a bucket of an optimal solution is $\frac{n}{b} + \Delta$. By choosing a subset R of appropriate size and by performing the sweep carefully, their algorithm computes an optimal solution in time $O(\min\{bn^{5/3} \log^{7/3} n + (b^2\Delta)n \log^3 n, n^2\})$, with probability at least $1 - 1/n$. In particular, the algorithm can detect $\Delta = 0$ in $O(\min\{bn^{5/3} \log^{7/3} n, n^2\})$ time.

The second type of buckets that one can consider are rectangular buckets. See Figure 6.9 (ii) for an example. A natural extension of the previous approach results in an algorithm that runs in time $O(\min\{b^{1/2}n^{5/3} \log^{7/3} n + (b^{3/2}\Delta)n \log^3 n, n^2\})$, with probability at least $1 - 1/n$, where the optimal value is $(n/b) + \Delta$.

10. PLACEMENT AND INTERSECTION
10.1 INTERSECTION OF POLYHEDRA

Given a set $\mathcal{P} = \{P_1, \ldots, P_m\}$ of m convex polyhedra in $\mathbb{R}^d$, with a total of n facets, is their common intersection $I = \bigcap_{i=1}^{m} P_i$ nonempty? If the answer is yes, return a point in I, say the smallest point v^* in the lexicographical order. We assume that each polytope $P_i \in \mathcal{P}$ is preprocessed so that we can determine in $O(\log n)$ time whether a query point lies inside P_i. Let us call the query procedure **Feasibility**.

Of course, this is an instance of linear programming in $\mathbb{R}^d$ with n constraints, but the goal is to obtain faster algorithms that depend on m more significantly than they depend on n. Note that there exist three polytopes $P_i, P_j, P_k \in \mathcal{P}$ so that v^* is the smallest vertex of $P_i \cap P_j \cap P_k$. Reichling [175] and Eppstein [80] showed that one can compute the smallest vertex of $P_i \cap P_j \cap P_k$ in $O(\log^3 n)$ time using **Feasibility** as a subroutine. Let us call this procedure as **Intersect**. Using **Intersect** and **Feasibility**, we can compute v^* as follows. For each triple $1 \leq i, j, k \leq m$, compute the smallest vertex v_{ijk} of $P_i \cap P_j \cap P_k$ if there exists one and check whether v_{ijk} lies inside all other polytopes of $\mathcal{P}$. Hence, computing v_{ijk} and checking whether $v_{ijk} \in I$ require $O(\log^3 n + m \log n)$ time. We then return the smallest vertex that lies inside I. The total time spent is $O(m^3 \log^3 n + m^4 \log n)$. Eppstein [80] presented a randomized recursive algorithm for computing v^*. At each step his algorithm solves a recursive subproblem for a subset of $\mathcal{P}$ with one less polytope, say P_i, and uses **Feasibility** to check whether the solution returned by the sub problem lies in P_i. At the base case, it invokes **Intersect**. Exploiting randomization and the order in which his algorithm calls recursive subproblems, he proved that the expected number of times his algorithm executes **Intersect** and **Feasibility** procedures are $O(\sqrt{m} \log m)$ and $O(m \log m)$, respectively. Hence, the expected running time of his algorithm is $O(m \log m \log n + \sqrt{m} \log^3 n)$.

10.2 POLYGON PLACEMENT

Let P be a convex m-gon, and let Q be a closed planar polygonal environment with n edges. We wish to compute the largest similar copy of P (under translation, rotation, and scaling) that can be placed inside Q. Using generalized Delaunay triangulation induced by P within Q, Chew and Kedem [49] obtained an $O(m^4 n^2 2^{\alpha(n)} \log n)$-time algorithm. Faster algorithms can be developed using randomization and search parametric searching [3, 187]. The decision problem in this case can be defined as follows: Given a convex polygon B with m edges (a scaled copy of P) and a planar polygonal environment Q with n edges, can B be placed inside Q (allowing translation and rotation)? Each placement of B can be represented as a point in $\mathbb{R}^3$, using two coordinates for translation and one for rotation. Let FP denote the resulting three-dimensional space of all free placements of B inside Q. FP is the union of a collection of cells of an arrangement of $O(mn)$ *contact surfaces* in $\mathbb{R}^3$. Leven and Sharir [138] have shown that the complexity of FP is $O(mn\lambda_6(mn))$, where $\lambda_s(n)$ is the maximum length of a Davenport–Schinzel sequence of order s composed of n symbols [186] (it is almost linear in n for any fixed s). Agarwal *et al.* [3] gave an $O(mn\lambda_6(mn)\log mn)$ expected-time randomized algorithm to compute FP. Plugging these algorithms into the parametric-searching machinery, one can obtain an $O(mn\lambda_6(mn)\log^4 mn)$ expected-time randomized algorithm for computing a largest similar placement of P inside Q.

The *biggest-stick* problem is an interesting special case of the largest-placement problem. In this case, Q is a simple polygon, P is a line segment, and we are interested in finding the longest segment that can be placed inside Q. This problem can be solved using a divide-and-conquer algorithm, developed in [15], and later refined in [2, 11]. It proceeds as follows: Partition Q into two simple polygons Q_1, Q_2 by a diagonal ℓ so that each of Q_1 and Q_2 has at most $2n/3$ vertices. Recursively compute the longest segment that can be placed in each Q_i, and then determine the longest segment that can be placed in Q and that intersects the diagonal ℓ. The decision algorithm for the merge step is to determine whether there exists a placement of a line segment of length w that lies inside Q and crosses ℓ. Agarwal *et al.* [15] have shown that this problem can be reduced to the following: We are given a set S of points and a set Γ of algebraic surfaces in $\mathbb{R}^4$, where each surface is the graph of a trivariate function, and we wish to determine whether every point of S lies below all the surfaces of Γ. They proposed a randomized algorithm with $O(n^{3/2+\varepsilon})$ expected running time for this point-location problem. Using a variant of the Clarkson and Shor approach in conjunction with other ideas, they obtained an $O(n^{3/2+\varepsilon})$ expected-time procedure for the overall merge step (finding the biggest stick that crosses ℓ). The total running time of the algorithm is therefore also $O(n^{3/2+\varepsilon})$.

Another related placement problem is the *penetration-depth* problem: Let P and Q be two polytopes in $\mathbb{R}^3$. The penetration depth of P and Q is the minimum distance by which Q has to be translated in a fixed direction so that P and Q become disjoint. This measures the extent of the intersection of two polytopes. Agarwal *et al.* [8] showed that the algorithm for computing the width can be extended to compute the penetration depth of two convex polytopes in $\mathbb{R}^3$ in expected

$O(K^{1/2+\varepsilon}m^{1/4}n^{1/4} + m^{1+\varepsilon} + n^{1+\varepsilon})$ time, where K is the complexity of the Minkowski sum $A \oplus (-B)$.

11.   NETWORK DESIGN PROBLEMS

In this section, we review a randomized technique that has led to approximation algorithms for several intractable network-design problems in a geometric setting, including Euclidean traveling salesperson, Euclidean Steiner tree, Euclidean k-MST, and Euclidean k-TSP. Until recently, it was not known whether polynomial-time approximation schemes (PTAS) exist for the Euclidean version of these problems even in the planar case (see [167, 24, 34, 197] for the previously best known approximation schemes). For the general problem (including the metric case), no polynomial-time ε-approximation algorithm can be obtained unless $P = NP$ ([24]). Recently it was shown that the some of these problems are MAX SNP-hard even in the Euclidean setting if the dimension is part of the input [192]. In a significant breakthrough, Arora [22, 23] obtained an ε-approximate polynomial-time algorithms for the above problems in any fixed dimension. See also [160]. For simplicity, we describe his technique for Euclidean TSP in $\mathbb{R}^2$.

Let S be a set of n points in the plane, and let $\varepsilon > 0$ be a given parameter. The goal is to compute a tour of S (i.e., a cycle that visits every point of S exactly once) whose length is at most $(1 + \varepsilon)\tau^*$, where τ^* is the length of an optimal tour. Since we are interested in an approximation algorithm, a well-known perturbation argument allows us to assume that the minimum distance between any two points is 8, that the maximum distance between any two points is $O(n/\varepsilon)$, and that the coordinates of points are in the interval $[0, L]$ for $L = O(n/\varepsilon)$ (see e.g. [23]). Let us assume that L is of the form 2^k for some integer k. Let B be the square $[0, 2L] \times [0, 2L]$. Choose two random integers $a, b \in [0, L]$ and translate each point of S by the vector (a, b). We will use S to denote the translated copy of S (i.e., $S = S + (a, b)$). Construct a quad tree Q on B for (the translated copy of) S, i.e., we recursively divide a square into four equal squares, starting from B, until the square contains at most one point; see Figure 6.10.

For two integers m, r, an *m-regular portal set* for Q is a set of points on the edges of each square σ of Q (at all levels) so that each corner of σ contains a portal and there are m other equally spaced points on each edge of σ. A tour π of S is called (m, r)-*light* if it crosses an edge of every square in Q only at portals and if it crosses an edge at most r times. The crux of Arora's algorithm lies in the following lemma.

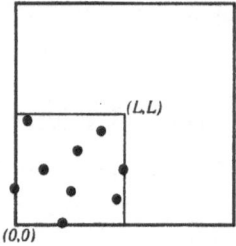 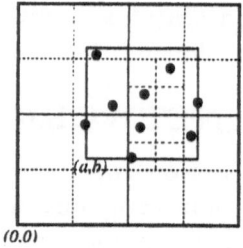

Figure 6.10 (i) Box B and original point set S; (ii) translated point set and the quad tree Q.

Lemma 11..1 *Let S and Q be as above, and let $\varepsilon > 0$ be a parameter. With probability at least 1/2, there is an $((c/\varepsilon) \log n, c/\varepsilon)$-light tour of S of length at most $(1 + \varepsilon)\tau^*$, where c is a constant.*

The proof of this lemma relies on two observations. Let π be a tour of S. First, if π intersects a line segment e of length λ at least three times, then there exists another tour of S of length at most $|\pi| + 3\lambda$ that intersects e at most twice. This argument was used by several heuristics for TSP in the past (see e.g. [33, 125]). Second, if we draw the integer grid inside B, then the number of intersections between π and the grid lines is at most $2|\pi|$. These lemmas and a simple probabilistic argument imply the above lemma.

The above lemma leads to a natural dynamic programming approach for computing an $((c/\varepsilon) \log n, c/\varepsilon)$-light tour π of S. With probability at least 1/2, the length of π is at most $(1 + \varepsilon)\tau^*$. With some care, the dynamic programming can be executed in time $n(\log n)^{O(1/\varepsilon)}$. Since the only randomization step in the above algorithm is choosing the vector (a, b), we can derandomize the algorithm by running the above algorithm for all values of $a, b \in [0, L]$. These algorithms extend to higher dimensions in straightforward manner. Rao and Smith [174] improved Arora's Monte Carlo algorithm by combining his dynamic programming approach with spanners of small overall weight. Their algorithm computes an ε-approximate tour in time $(\sqrt{d}/\varepsilon)^{O(d(\sqrt{d/\varepsilon})^{d-1})} n + O(dn \log n)$ with probability at least 1/2. Note that the running time of these algorithms is doubly exponential in d. In view of the hardness result by Trevisan [192], which shows that the Euclidean TSP problem is MAX SNP-hard in $\mathbb{R}^{\log n}$, this dependence is necessary unless NP has subexponential algorithms.

As mentioned earlier, the same approach yields a polynomial-time ε-approximation algorithm, for any fixed $\varepsilon > 0$, for several other network design problems, include Euclidean Steiner trees and Euclidean

k-MST. This also gives an $n \log^{O(1/\varepsilon)} n$- time ε-approximation algorithm for computing the minimum-weight Euclidean matching of a point set in the plane. This bound was subsequently improved by Varadarajan and Agarwal [193] to $O((n/\varepsilon^3) \log^6 n)$.

12. DISCUSSION

In this chapter we reviewed several randomized techniques and algorithms for a wide range of geometric optimization problems. We mostly focused on summarizing the known techniques and did not state open problems. There are, however, several interesting open problems in this area, e.g., nearest-neighbor searching and clustering algorithms in high dimensions, practical approximation algorithms for network-design problems, strongly polynomial-time algorithms for linear programming, faster algorithms for surface simplification, etc. In the last few years several elegant techniques, e.g., Monte Carlo algorithms using random walks on expander graphs and randomized rounding in conjunction with semidefinite programming, have been developed for nongeometric problems. It would be interesting to explore whether geometric optimization problems could benefit from these techniques. Similarly, the recent work on hardness of approximation algorithms has not been applicable to geometric optimization problems in fixed dimensions.

Although we covered a variety of topics, we did not attempt to cover all applications of randomization in geometric optimization, as it would be an impossible task. Additional applications of randomization in geometric optimization include volume estimation of convex bodies [74], learning geometric concepts, and shape matching. Interested readers can find more material on geometric optimization in [12, 34, 95] and on randomized geometric algorithms in various chapters of this book as well as in [55, 145, 163].

Acknowledgments

The authors thank Ken Clarkson, Sariel Har-Peled, Piotr Indyk, Rohit Khandekar, and Micha Sharir for their invaluable comments and for pointing out a few errors in an earlier version, and Laureen Treacy for proofreading the chapter.

Notes

1. Recall that the polynomial-time algorithms by Khachiyan [127] and by Karmarkar [124] are not strongly polynomial, as the number of arithmetic operations performed by these algorithms depends on the size of the coefficients of the input constraints.

2. The arrangement of L, denoted as $\mathcal{A}(L)$, is the planar subdivision induced by L whose vertices are the intersection points of lines in L, edges are the portions of lines not containing any vertex of $\mathcal{A}(L)$, and whose faces are the maximal connected components of $\mathbb{R}^2 \setminus \bigcup L$.

3. A subset $N \subseteq X$ is called an ε-net of $(X, \mathcal{R})$ if $N \cap r \neq \emptyset$ for every range r of size at least $\varepsilon |X|$.

4. In this chapter, the meaning of complexity bounds that depend on an arbitrary parameter $\varepsilon > 0$, such as the one stated here, is that given any $\varepsilon > 0$, we can fine-tune the algorithm so that its complexity satisfies the stated bound. In these bounds the constant of proportionality usually depends on ε, and tends to infinity when ε tends to zero.

References

[1] I. Adler and R. Shamir, A randomization scheme for speeding up algorithms for linear and convex quadratic programming problems with a high constraints-to-variables ratio, Tech. Report 21-90, Rutgers Univ., New Brunswick, NJ, May 1990.

[2] P. K. Agarwal, B. Aronov, and M. Sharir, Computing envelopes in four dimensions with applications, *SIAM J. Comput.*, 26 (1997), 1714–1732.

[3] P. K. Agarwal, B. Aronov, and M. Sharir, Motion planning for a convex polygon in a polygonal environment, Tech. Report CS-1997-17, Department of Computer Science, Duke University, 1997.

[4] P. K. Agarwal, B. Aronov, M. Sharir, and S. Suri, Selecting distances in the plane, *Algorithmica*, 9 (1993), 495–514.

[5] P. K. Agarwal, B. K. Bhattacharya, and S. Sen, Output-sensitive algorithms for uniform partitions of points, *Proc. 10th Intl. Sympos. Algorithms and Computation*, 1999, pp. 403–414.

[6] P. K. Agarwal and P. K. Desikan, An approximation algorithm for terrain simplification, *Proc. 8th ACM-SIAM Sympos. Discrete Algorithms*, 1997, pp. 139–147.

[7] P. K. Agarwal, H. Edelsbrunner, O. Schwarzkopf, and E. Welzl, Euclidean minimum spanning trees and bichromatic closest pairs, *Discrete Comput. Geom.*, 6 (1991), 407–422.

[8] P. K. Agarwal, L. J. Guibas, S. Har-Peled, A. Rabinovitch, and M. Sharir, Computing the penetration depth of two convex polytopes in 3D, *Proc. 7th Scandanavian Workshop on Algorithmic Theory*, 2000, pp. 328–338.

[9] P. K. Agarwal and J. Matoušek, Ray shooting and parametric search, *SIAM J. Comput.*, 22 (1993), 794–806.

[10] P. K. Agarwal and C. M. Procopiuc, Approximation algorithms for projective clustering, *Proc. 11th ACM-SIAM Sympos. Discrete Algorithms*, 2000, pp. 538–547.

[11] P. K. Agarwal and M. Sharir, Efficient randomized algorithms for some geometric optimization problems, *Discrete Comput. Geom.*, 16 (1996), 317–337.

[12] P. K. Agarwal and M. Sharir, Efficient algorithms for geometric optimization, *ACM Comput. Surv.*, 30 (1998), 412–458.

[13] P. K. Agarwal and M. Sharir, Arrangements and their applications, in: *Handbook of Computational Geometry* (J.-R. Sack and J. Urrutia, eds.), Elsevier Science Publishers, Amsterdam, 2000, pp. 49–119.

[14] P. K. Agarwal, M. Sharir, and S. Toledo, An efficient multidimensional searching technique and its applications, Tech. Report CS-1993-20, Dept. Comp. Sci., Duke University, 1993.

[15] P. K. Agarwal, M. Sharir, and S. Toledo, Applications of parametric searching in geometric optimization, *J. Algorithms*, 17 (1994), 292–318.

[16] P. K. Agarwal and S. Suri, Surface approximation and geometric partitions, *Proc. 5th ACM-SIAM Sympos. Discrete Algorithms*, 1994, pp. 24–33.

[17] P. K. Agarwal and K. R. Varadarajan, Approximating monotone polygonal curves using the uniform metric, *Discrete Comput. Geom.*, 23 (2000), 273–291.

[18] R. Agrawal, A. Ghosh, T. Imielinski, B. Iyer, and A. Swami, An interval classifier for database mining applications, *Proceedings of the 18th Conference on Very Large Databases*, 1992, pp. 560–573.

[19] N. Amenta, Bounded boxes, Hausdorff distance, and a new proof of an interesting Helly theorem, *Proc. 10th Annu. ACM Sympos. Comput. Geom.*, 1994, pp. 340–347.

[20] N. Amenta, Helly-type theorems and generalized linear programming, *Discrete Comput. Geom.*, 12 (1994), 241–261.

[21] A. Andersson, P. B. Miltersen, S. Riis, and M. Thorup, Static dictionaries on AC^0 RAMs: Query time $\Theta(\sqrt{\log n / \log \log n})$ is necessary and sufficient, *Proc. 37th Sympos. on Foundations of Computer Science*, 1996, pp. 441–450.

[22] S. Arora, Polynomial time approximation schemes for Euclidean TSP and other geometric problems, *Proc. 37th Annu. IEEE Sympos. Found. Comput. Sci.*, 1996, pp. 2–11.

[23] S. Arora, Polynomial time approximation schemes for Euclidean traveling salesman and other geometric problems, *J. ACM*, (1998), 753–782.

190

[24] S. Arora and C. Lund, Hardness of approximations, in: *Approximation Algorithms for NP-Hard Problems* (D. S. Hochbaum, ed.), PWS Publishing Company, Boston, MA, 1997, pp. 399–446.

[25] S. Arora, P. Raghavan, and S. Rao, Approximation schemes for Euclidean k-median and related problems, *Proc. 30th Annu. ACM Sympos. Theory Comput.*, 1998, pp. 106–113.

[26] S. Arya and D. M. Mount, Algorithms for fast vector quantization, *Data Compression Conference*, IEEE Press, 1993, pp. 381–390.

[27] S. Arya and D. M. Mount, Approximate nearest neighbor queries in fixed dimensions, *Proc. 4th ACM-SIAM Sympos. Discrete Algorithms*, 1993, pp. 271–280.

[28] S. Arya, D. M. Mount, and O. Narayan, Accounting for boundary effects in nearest-neighbor searching, *Discrete Comput. Geom.*, 16 (1996), 155–176.

[29] S. Arya, D. M. Mount, N. S. Netanyahu, R. Silverman, and A. Wu, An optimal algorithm for approximate nearest neighbor searching in fixed dimensions, *J. ACM*, 45 (1998), 891–923.

[30] T. Asano and T. Tokuyama, Algorithms for projecting points to give the most uniform distribution with application to hashing, *Algorithmica*, 572–590 (1993), 572–590.

[31] O. Barkol and Y. Rabani, Tighter lower bounds for nearest neighbor search and related problems in the cell probe model, *Proc. 32nd Annu. ACM Sympos. Theory Comput.*, 2000, pp. 388–396.

[32] Y. Bartal, Probabilistic approximation of metric spaces and its algorithmic applications, *Proc. 37th Sympos. on Foundations of Computer Science*, 1996, pp. 184–193.

[33] J. Beardwood, J. H. Halton, and J. M. Hammersley, The shortest path through many points, *Math. Proc. Camb. Phil. Soc.*, 55 (1959), 299–327.

[34] M. Bern and D. Eppstein, Approximation algorithms for geometric problems, in: *Approximation Algorithms for NP-Hard Problems* (D. S. Hochbaum, ed.), PWS Publishing Company, Boston, MA, 1997, pp. 296–345.

[35] A. Borodin, R. Ostrovsky, and Y. Rabani, Subquadratic approximation algorithms for clustering problems in high dimensional spaces, *Proc. 31st Annu. ACM Sympos. Theory Comput.*, 1999, pp. 435–444.

[36] T. Brinkhoff and H.-P. Kriegel, The impact of global clusterings on spatial database systems, *Proceedings of the International Conference on Very Large Databases*, 1994, pp. 168–179.

[37] H. Brönnimann and B. Chazelle, Optimal slope selection via cuttings, *Proc. 6th Canad. Conf. Comput. Geom.*, 1994, pp. 99–103.

[38] H. Brönnimann and M. T. Goodrich, Almost optimal set covers in finite VC-dimension, *Discrete Comput. Geom.*, 14 (1995), 263–279.

[39] A. Chakrabarti, B. Chazelle, B. Gum, and A. Lvov, A good neighbor is hard to find, *Proc. 31st Annu. ACM Sympos. Theory Comput.*, 1999, pp. 305–311.

[40] T. M. Chan, Fixed-dimensional linear programming queries made easy, *Proc. 12th Annu. ACM Sympos. Comput. Geom.*, 1996, pp. 284–290.

[41] T. M. Chan, Approximate nearest neighbor queries revisited, *Proc. 13th Annu. ACM Sympos. Comput. Geom.*, 1997, pp. 352–358.

[42] T. M. Chan, Geometric applications of a randomized optimization technique, *Proc. 14th Annu. ACM Sympos. Comput. Geom.*, 1998, pp. 269–278.

[43] T. M. Chan, Approximating the diameter, width, smallest enclosing cylinder, and minimum-width annulus, *Proc. 16th Sympos. Comput. Geom.*, 2000, pp. 300–309.

[44] R. Chandrasekaran and A. Tamir, Algebraic optimization: The Fermat-Weber location problem, *Math. Program.*, 46 (1990), 219–224.

[45] M. Charikar, C. Chekuri, A. Goel, and S. Guha, Rounding via trees: Deterministic approximation algorithms for group Steiner trees and *k*-median, *Proc. 30th Annu. ACM Sympos. Theory Comput.*, 1998, pp. 114–123.

[46] M. Charikar and S. Guha, Improved combinatorial algorithms for facility location and *k*-median problems, *Proc. 40th Sympos. on Foundations of Computer Science*, 1999, pp. 378–388.

[47] M. Charikar, S. Guha, E. Tardos, and D. Shmoys, A constant-factor approximation algorithm for the *k*-median problem, *Proc. 31st Annu. ACM Sympos. Theory Comput.*, 1999, pp. 1–10.

[48] B. Chazelle and J. Matoušek, On linear-time deterministic algorithms for optimization problems in fixed dimension, *J. Algorithms*, 21 (1996), 579–597.

[49] L. P. Chew and K. Kedem, A convex polygon among polygonal obstacles: Placement and high-clearance motion, *Comput. Geom. Theory Appl.*, 3 (1993), 59–89.

[50] K. L. Clarkson, Applications of random sampling in computational geometry, II, *Proc. 4th Annu. ACM Sympos. Comput. Geom.*, 1988, pp. 1–11.

192

[51] K. L. Clarkson, A randomized algorithm for closest-point queries, *SIAM J. Comput.*, 17 (1988), 830–847.

[52] K. L. Clarkson, Algorithms for polytope covering and approximation, *Proc. 3rd Workshop Algorithms Data Struct., Lecture Notes Comput. Sci.*, vol. 709, Springer-Verlag, 1993, pp. 246–252.

[53] K. L. Clarkson, An algorithm for approximate closest-point queries, *Proc. 10th Annu. ACM Sympos. Comput. Geom.*, 1994, pp. 160–164.

[54] K. L. Clarkson, Las Vegas algorithms for linear and integer programming, *J. ACM*, 42 (1995), 488–499.

[55] K. L. Clarkson, Randomized geometric algorithms, in: *Computing in Euclidean Geometry, 2nd edition* (D.-Z. Du and F. Hwang, eds.), vol. 4, World Scientific, Singapore, 1995, pp. 149–194.

[56] K. L. Clarkson, Nearest neighbor queries in metric spaces, *Proc. 29th Annu. ACM Sympos. Theory Comput.*, 1997, pp. 609–617.

[57] K. L. Clarkson, D. Eppstein, G. L. Miller, C. Sturtivant, and S.-H. Teng, Approximating center points with iterative Radon points, *Internat. J. Comput. Geom. Appl.*, 6 (1996), 357–377.

[58] K. L. Clarkson and P. W. Shor, Applications of random sampling in computational geometry, II, *Discrete Comput. Geom.*, 4 (1989), 387–421.

[59] R. Cole, J. Salowe, W. Steiger, and E. Szemerédi, An optimal-time algorithm for slope selection, *SIAM J. Comput.*, 18 (1989), 792–810.

[60] S. Cost and S. Salzberg, A weighted nearest neighbor algorithm for learning with symbolic features, *Machine Learning*, 10 (1993), 57–67.

[61] T. M. Cover and P. E. Hart, Nearest neighbor pattern classification, *IEEE Trans. Inform. Theory*, 13 (1967), 21–27.

[62] D. R. Cutting, D. R. Karger, and J. O. Pedersen, Constant interaction-time scatter/gather browsing of very large document collections, *Proc. 16th Annu. Internat. ACM SIGIR Conf. Research and Develop. in Inform. Retrieval*, 1993, pp. 126–134.

[63] D. R. Cutting, D. R. Karger, J. O. Pedersen, and J. W. Tukey, Scatter/gather: A cluster-based approach to browsing large document collections, *Proc. 16th Annu. Internat. ACM SIGIR Conf. Research and Develop. in Inform. Retrieval*, 1992, pp. 318–329.

[64] G. Das and M. T. Goodrich, On the complexity of optimization problems for 3-dimensional convex polyhedra and decision trees, *Comput. Geom. Theory Appl.*, 8 (1997), 123–137.

[65] G. Das and D. Joseph, The complexity of minimum convex nested polyhedra, *Proc. 2nd Canad. Conf. Comput. Geom.*, 1990, pp. 296–301.

[66] S. Dasgupta and A. Gupta, An elementary proof of the Johnson-Lindenstrauss lemma, Tech. Report TR-99-06, Intl. Comput. Sci. Inst., Berkeley, CA, 1999.

[67] A. Datta, H.-P. Lenhof, C. Schwarz, and M. Smid, Static and dynamic algorithms for k-point clustering problems, *J. Algorithms*, 19 (1995), 474–503.

[68] M. B. Dillencourt, D. M. Mount, and N. S. Netanyahu, A randomized algorithm for slope selection, *Internat. J. Comput. Geom. Appl.*, 2 (1992), 1–27.

[69] D. P. Dobkin and R. J. Lipton, Multidimensional searching problems, *SIAM J. Comput.*, 5 (1976), 181–186.

[70] Z. Drezner, The planar two-center and two-median problem, *Transp. Sci.*, 18 (1984), 351–361.

[71] Z. Drezner, On the rectangular p-center problem, *Naval Res. Logist. Q.*, 34 (1987), 229–234.

[72] Z. Drezner, ed., *Facility Location*, Springer-Verlag, New York, 1995.

[73] R. O. Duda and P. E. Hart, *Pattern Classification and Scene Analysis*, Wiley-Interscience, New York, 1973.

[74] M. Dyer and A. Frieze, Computing the volume of convex bodies: A case where randomness provably helps, in: *Probabilistic Combinatorics and Its Applications* (B. Bollobás, ed.), American Mathematical Society, Providence, RI, 1991, pp. 123–169.

[75] M. E. Dyer, A class of convex programs with applications to computational geometry, *Proc. 8th Annu. ACM Sympos. Comput. Geom.*, 1992, pp. 9–15.

[76] M. E. Dyer and A. M. Frieze, A randomized algorithm for fixed-dimension linear programming, *Math. Program.*, 44 (1989), 203–212.

[77] H. Ebara, N. Fukuyama, H. Nakano, and Y. Nakanishi, Roundness algorithms using the Voronoi diagrams, *Abstracts 1st Canad. Conf. Comput. Geom.*, 1989, p. 41.

[78] H. Edelsbrunner, *Algorithms in Combinatorial Geometry*, Springer-Verlag, Heidelberg, 1987.

[79] A. Efrat, M. Sharir, and A. Ziv, Computing the smallest k-enclosing circle and related problems, *Comput. Geom. Theory Appl.*, 4 (1994), 119–136.

[80] D. Eppstein, Dynamic three-dimensional linear programming, *ORSA J. Comput.*, 4 (1992), 360–368.

[81] D. Eppstein, Faster construction of planar two-centers, *Proc. 8th ACM-SIAM Sympos. Discrete Algorithms*, 1997, pp. 131–138.

[82] D. Eppstein and J. Erickson, Iterated nearest neighbors and finding minimal polytopes, *Discrete Comput. Geom.*, 11 (1994), 321–350.

[83] C. Faloutsos and K.-I. Lin, FastMap: A fast algorithm for indexing, data-mining and visualization of traditional and multimedia databases, *Proc. ACM SIGMOD Conf. on Management of Data*, 1995, pp. 163–173.

[84] C. Faloutsos, M. Ranganathan, and Y. Manolopoulos, Fast subsequence matching in time-series databases, *Proc. ACM SIGMOD Conf. on Management of Data*, 1994, pp. 86–93.

[85] T. Feder and D. H. Greene, Optimal algorithms for approximate clustering, *Proc. 20th Annu. ACM Sympos. Theory Comput.*, 1988, pp. 434–444.

[86] P. Finn, L. E. Kavraki, J.-C. Latombe, R. Motwani, C. Shelton, S. Venkatasubramanian, and A. Yao, RAPID: Randomized pharmacophore identification for drug design, *Proc. 13th Annu. ACM Sympos. Comput. Geom.*, 1997, pp. 324–333.

[87] R. J. Fowler, M. S. Paterson, and S. L. Tanimoto, Optimal packing and covering in the plane are NP-complete, *Inform. Process. Lett.*, 12 (1981), 133–137.

[88] P. Frankl and H. Maehara, The Johnson-Lindenstrauss lemmas and the sphericity of some graphs, *J. Combin. Theory, Ser. B*, 44 (1988), 355–362.

[89] P. Frankl and H. Maehara, Some geometric applications of the beta distribution, *Ann. Inst. Stat. Math*, 42 (1990), 463–474.

[90] G. N. Frederickson and D. B. Johnson, Generalized selection and ranking: Sorted matrices, *SIAM J. Comput.*, 13 (1984), 14–30.

[91] M. L. Fredman, J. Komlos, and E. Szemeredi, Storing a sparse table with $o(1)$ worst case access time, *J. ACM*, 31 (1984), 538–544.

[92] J. H. Friedman, J. L. Bentley, and R. A. Finkel, An algorithm for finding best matches in logarithmic expected time, *ACM Trans. Math. Softw.*, 3 (1977), 209–226.

[93] J. García-Lopez and P. Ramos, Fitting a set of points by a circle, *Proc. 13th Annu. ACM Sympos. Comput. Geom.*, 1997, pp. 139–146.

[94] B. Gärtner, A subexponential algorithm for abstract optimization problems, *SIAM J. Comput.*, 24 (1995), 1018–1035.

[95] B. Gärtner and E. Welzl, Linear programming – randomization and abstract frameworks, *Proc. 13th Sympos. Theoret. Aspects Comput. Sci.*, *Lecture Notes Comput. Sci.*, vol. 1046, Springer-Verlag, 1996, pp. 669–687.

[96] B. Gärtner and E. Welzl, Random sampling in geometric optimization: New insights and applications, *Proc. 16th Sympos. Comput. Geom.*, 2000, pp. 91–99.

[97] M. Goldwasser, A survey of linear programming in randomized subexponential time, *SIGACT News*, 26 (1995), 96–104.

[98] T. Gonzalez, Clustering to minimize the maximum intercluster distance, *Theoret. Comput. Sci.*, 38 (1985), 293–306.

[99] B. Grünbaum, A proof of Vázsonyi's conjecture, *Bull. Research Council Israel, Section A*, 6 (1956), 77–78.

[100] D. Halperin, M. Sharir, and K. Goldberg, The 2-center problem with obstacles, *Proc. 16th Sympos. Comput. Geom.*, 2000, pp. 80–90.

[101] K.-A. Han and S.-H. Myaeng, Image organization and retrieval with automatically constructed feature vectors, *Proc. 16th Annu. Internat. ACM SIGIR Conf. Research and Develop. in Inform. Retrieval*, 1996, pp. 157–165.

[102] S. Har-Peled and P. Indyk, When crossings count: Approximating the minimum spanning tree, *Proc. 16th Annu. ACM Sympos. Comput. Geom.*, 2000. 166–175.

[103] R. Hassin and N. Megiddo, Approximation algorithms for hitting objects by straight lines, *Discrete Appl. Math.*, 30 (1991), 29–42.

[104] T. Hastie and R. Tibshirani, Discriminant adaptive nearest neighbor classification, *IEEE Trans. Pattern Anal. Mach. Intell.*, 18 (1996), 607–616.

[105] D. Haussler and E. Welzl, Epsilon-nets and simplex range queries, *Discrete Comput. Geom.*, 2 (1987), 127–151.

[106] P. S. Heckbert and M. Garland, Fast polygonal approximation of terrains and height fields, Tech. Report CMU-CS-95-181, Carnegie Mellon University, 1995.

[107] A. Heppes, Beweis einer Vermutung von A. Vázsonyi, *Acta Math. Acad. Sci. Hungar.*, 7 (1956), 463–466.

[108] J. Hershberger, A faster algorithm for the two-center decision problem, *Inform. Process. Lett.*, 47 (1993), 23–29.

[109] G. R. Hjaltason and H. Samet, Ranking in spatial databases, *Advances in Spatial Databases — Fourth International Symposium, Lecture Notes Comput. Sci.*, vol 951, 1995, pp. 83–95.

[110] M. E. Houle and G. T. Toussaint, Computing the width of a set, *IEEE Trans. Pattern Anal. Mach. Intell.*, PAMI-10 (1988), 761–765.

[111] M. Inaba, N. Katoh, and H. Imai, Applications of weighted Voronoi diagrams and randomization to variance-based k-clustering, *Proc. 10th Annu. ACM Sympos. Comput. Geom.*, 1994, pp. 332–339.

[112] P. Indyk, On approximate nearest neighbors in non-Euclidean spaces, *Proc. 39th Sympos. on Foundations of Computer Science*, 1998, pp. 148–155.

[113] P. Indyk, A sublinear-time approximation scheme for clustering in metric spaces, *Proc. 40th Sympos. on Foundations of Computer Science*, 1999, pp. 154–159.

[114] P. Indyk, Dimensionality reduction techniques for proximity problems, *Proc. 11th ACM-SIAM Sympos. on Discrete Algorithms*, 2000, pp. 371–378.

[115] P. Indyk and M. Farach-Colton, Approximate nearest neighbor algorithms for Hausdorff metrics via embeddings, *Proc. 40th Sympos. on Foundations of Computer Science*, 1999, pp. 171–180.

[116] P. Indyk and R. Motwani, Approximate nearest neighbors: Towards removing the curse of dimensionality, *Proc. 30th Annu. ACM Sympos. Theory Comput.*, 1998, pp. 604–613.

[117] P. Indyk, R. Motwani, P. Raghavan, and S. Vempala, Locality-preserving hashing in multidimensional spaces, *Proc. 29th Annu. ACM Sympos. Theory Comput.*, 1997, pp. 618–625.

[118] A. K. Jain, M. N. Murty, and P. J. Flynn, Data clustering: A review, *ACM Comput. Surv.*, 31 (1999), 264–323.

[119] J. W. Jaromczyk and M. Kowaluk, An efficient algorithm for the Euclidean two-center problem, *Proc. 10th Annu. ACM Sympos. Comput. Geom.*, 1994, pp. 303–311.

[120] J. W. Jaromczyk and M. Kowaluk, A geometric proof of the combinatorial bounds for the number of optimal solutions to the 2-center Euclidean problem, *Proc. 7th Canad. Conf. Comput. Geom.*, 1995, pp. 19–24.

[121] W. Johnson and J. Lindenstrauss, Extensions of Lipschitz maps into a Hilbert space, *Contemp. Math.*, 26 (1984), 189–206.

[122] G. Kalai, A subexponential randomized simplex algorithm, *Proc. 24th Annu. ACM Sympos. Theory Comput.*, 1992, pp. 475–482.

[123] R. Kannan, Improved algorithms for integer programming and related lattice problems, *Proc. 15th Annu. ACM Sympos. Theory Comput.*, 1983, pp. 193–206.

[124] N. Karmarkar, A new polynomial-time algorithm for linear programming, *Combinatorica*, 4 (1984), 373–395.

[125] R. M. Karp, Probabilistic analysis of partitioning algorithms for the traveling salesman problem in the plane, *Math. Oper. Res.*, 2 (1977), 209–224.

[126] M. J. Katz and F. Nielsen, On piercing sets of objects, *Proc. 12th Annu. ACM Sympos. Comput. Geom.*, 1996, pp. 113–121.

[127] L. G. Khachiyan, Polynomial algorithm in linear programming, *U.S.S.R. Comput. Math. and Math. Phys.*, 20 (1980), 53–72.

[128] J. Kleinberg, Two algorithms for nearest-neighbor search in high dimension, *Proc. 29th Annu. ACM Sympos. Theory Comput.*, 1997, pp. 599–608.

[129] D. E. Knuth, *Sorting and Searching*, Addison-Wesley, Reading, MA, 1973.

[130] M. T. Ko and Y. T. Ching, Linear time algorithms for the weighted tailored 2-partition problem and the weighted rectilinear 2-center problem under L_∞-distance, *Discrete Appl. Math.*, 40 (1992), 397–410.

[131] M. T. Ko and R. C. T. Lee, On weighted rectilinear 2-center and 3-center problems, *Inform. Sci.*, 54 (1991), 169–190.

[132] M. T. Ko, R. C. T. Lee, and J. S. Chang, An optimal approximation algorithm for the rectilinear m-center problem, *Algorithmica*, 5 (1990), 341–352.

[133] V. Koivune and S. Kassam, Nearest neighbor filters for multivariate data, *IEEE Workshop on Nonlinear Signal and Image Processing*, 1995.

[134] S. Kolliopoulos and S. Rao, A nearly linear-time approximation scheme for the Euclidean k-median problem, *Proc. 7th European Sympos. Algorithms*, 1999, pp. 378–389.

[135] F. Korn, N. Sidiropoulos, C. Faloutsos, E. Siegel, and Z. Protopapa, Fast nearest neighbor search in medical image database, *Proceedings of the International Conference on Very Large Databases*, 1996, pp. 215–226.

[136] E. Kushilevitz, R. Ostrovsky, and Y. Rabani, Efficient search for approximate nearest neighbor in high dimensional spaces, *Proc. 30th Annu. ACM Sympos. Theory Comput.*, 1998, pp. 614–623.

[137] H. W. Lenstra, Integer programming with a fixed number of variables, *Math Oper. Research*, 8 (1983), 538–548.

[138] D. Leven and M. Sharir, On the number of critical free contacts of a convex polygonal object moving in two-dimensional polygonal space, *Discrete Comput. Geom.*, 2 (1987), 255–270.

[139] J.-H. Lin and J. S. Vitter, Approximation algorithms for geometric median problems, *Inform. Process. Lett.*, 44 (1992), 245–249.

[140] J.-H. Lin and J. S. Vitter, ϵ-approximations with minimum packing constraint violation, *Proc. 24th Annu. ACM Sympos. Theory Comput.*, 1992, pp. 771–782.

[141] N. Linial and O. Sasson, Non-expansive hashing, *Proc. 28th Annu. ACM Sympos. Theory Comput.*, 1996, pp. 509–517.

[142] W. Maass, On the complexity of nonconvex covering, *SIAM J. Comput.*, 15 (1986), 453–467.

[143] J. Makhoul, S. Roucos, and H. Gish, Vector quantization in speech coding, *Proc. IEEE*, 73 (1985), 1551–1588.

[144] J. Matoušek, Randomized optimal algorithm for slope selection, *Inform. Process. Lett.*, 39 (1991), 183–187.

[145] J. Matoušek, Epsilon-nets and computational geometry, in: *New Trends in Discrete and Computational Geometry* (J. Pach, ed.), Springer-Verlag, 1993, pp. 69–89.

[146] J. Matoušek, Lower bound for a subexponential optimization algorithm, *Random Structures & Algorithms*, 5 (1994), 591–607.

[147] J. Matoušek, On enclosing k points by a circle, *Inform. Process. Lett.*, 53 (1995), 217–221.

[148] J. Matoušek, On geometric optimization with few violated constraints, *Discrete Comput. Geom.*, 14 (1995), 365–384.

[149] J. Matoušek, On approximate geometric k-clustering, Discrete Comput. Geom., 24 (2000), 61–84.

[150] J. Matoušek, D. M. Mount, and N. S. Netanyahu, Efficient randomized algorithms for the repeated median line estimator, *Proc. 4th ACM-SIAM Sympos. Discrete Algorithms*, 1993, pp. 74–82.

[151] J. Matoušek, M. Sharir, and E. Welzl, A subexponential bound for linear programming, *Algorithmica*, 16 (1996), 498–516.

[152] C. Meghini, An image retrieval model based on classical logic, *Proc. 16th Annu. Internat. ACM SIGIR Conf. Research and Develop. in Inform. Retrieval*, 1995, pp. 300–308.

[153] N. Megiddo, Applying parallel computation algorithms in the design of serial algorithms, *J. ACM*, 30 (1983), 852–865.

[154] N. Megiddo, On the ball spanned by balls, *Discrete Comput. Geom.*, 4 (1989), 605–610.

[155] N. Megiddo, On the complexity of some geometric problems in unbounded dimension, *J. Symbolic Comput.*, 10 (1990), 327–334.

[156] N. Megiddo and K. J. Supowit, On the complexity of some common geometric location problems, *SIAM J. Comput.*, 13 (1984), 182–196.

[157] N. Megiddo and A. Tamir, On the complexity of locating linear facilities in the plane, *Oper. Res. Lett.*, 1 (1982), 194–197.

[158] H. Mehrotra and J. E. Gary, Feature-based retrieval of similar shapes, *Proc. 9th IEEE Intl. Conf. on Data Engineering*, 1996, pp. 108–115.

[159] S. Meiser, Point location in arrangements of hyperplanes, *Inform. Comput.*, 106 (1993), 286–303.

[160] J. S. B. Mitchell, A. Blum, P. Chalasani, and S. Vempala, A constant-factor approximation algorithm for the geometric k-MST problem in the plane, *SIAM J. Comput.*, 28 (1998), 771–781.

[161] J. S. B. Mitchell and S. Suri, Separation and approximation of polyhedral objects, *Comput. Geom. Theory Appl.*, 5 (1995), 95–114.

[162] R. Motwani and P. Raghavan, *Randomized Algorithms*, Cambridge University Press, New York, NY, 1995.

[163] K. Mulmuley, *Computational Geometry: An Introduction Through Randomized Algorithms*, Prentice Hall, Englewood Cliffs, NJ, 1993.

[164] K. Mulmuley, Randomized algorithms in computational geometry, in: *Handbook of Computational Geometry* (J.-R. Sack and J. Urrutia, eds.), Elsevier Science Publishers, Amsterdam, 2000, pp. 703–724.

[165] R. Ostrovsky and Y. Rabani, Polynomial time approximation schemes for geometric k-clustering, *Proc. 41st Sympos. on Foundations of Computer Science*, 2000, to appear.

[166] C. H. Papadimitriou, Worst-case and probabilistic analysis of a geometric location problem, *SIAM J. Comput.*, 10 (1981), 542–557.

[167] C. H. Papadimitriou and K. Steiglitz, *Combinatorial Optimization: Algorithms and Complexity*, Prentice Hall, Englewood Cliffs, NJ, 1982.

[168] M. J. Post, Minimum spanning ellipsoids, *Proc. 16th Annu. ACM Sympos. Theory Comput.*, 1984, pp. 108–116.

[169] F. P. Preparata and M. I. Shamos, *Computational Geometry: An Introduction*, Springer-Verlag, New York, 1985.

[170] M. O. Rabin, Probabilistic algorithms, in: *Algorithms and Complexity: New Directions and Recent Results* (J. F. Traub, ed.), Academic Press, New York, NY, 1976, pp. 21–39.

[171] V. Ramasubramanian and K. K. Paliwal, Fast k-dimensional tree algorithms for nearest neighbor search with applications to vector quantization encoding, *IEEE Trans. Signal Processing*, 40 (1992), 518–531.

[172] E. A. Ramos, Deterministic algorithms for 3-D diameter and some 2-D lower envelopes, *Proc. 16th Sympos. Comput. Geom.*, 2000, pp. 290–299.

[173] E. A. Ramos, Linear programming queries revisited, *Proc. 16th Sympos. Comput. Geom.*, 2000, pp. 176–181.

[174] S. Rao and W. D. Smith, Improved approximation schemes for geometric graphs via "spanners" and "banyans", *Proc. 30th Annu. ACM Sympos. Theory Comput.*, 1998. 540–550.

[175] M. Reichling, On the detection of a common intersection of k convex polyhedra, *Computational Geometry and Its Applications*, Lecture Notes Comput. Sci., vol. 333, Springer-Verlag, 1988, pp. 180–186.

[176] T. Roos and P. Widmayer, k-violation linear programming, *Inform. Process. Lett.*, 52 (1994), 109–114.

[177] N. Roussopoulos, S. Kelley, and F. Vincent, Nearest neighbor queries, *Proc. ACM SIGMOD Conf. on Management of Data*, 1995, pp. 71–79.

[178] G. Salton, *Automatic Text Processing*, Addison-Wesley, Reading, MA, 1989.

[179] N. Sarnak and R. E. Tarjan, Planar point location using persistent search trees, *Commun. ACM*, 29 (1986), 669–679.

[180] P. Schroeter and J. Bigün, Hierarchical image segmentation by multi-dimensional clustering and orientation-adaptive boundary refinement, *Pattern Recogn.*, 28 (1995), 695–709.

[181] R. Seidel, Small-dimensional linear programming and convex hulls made easy, *Discrete Comput. Geom.*, 6 (1991), 423–434.

[182] R. Seidel, Backwards analysis of randomized geometric algorithms, in: *New Trends in Discrete and Computational Geometry* (J. Pach, ed.), Springer-Verlag, Heidelberg, Germany, 1993, pp. 37–68.

[183] J. Shafer, R. Agrawal, and M. Mehta, Sprint: A scalable parallel classifier for data mining., *Proceedings of the International Conference on Very Large Databases*, Morgan Kauffman, 1996, pp. 544–555.

[184] L. Shafer and W. Steiger, Randomizing optimal geometric algorithms, *Proc. 5th Canad. Conf. Comput. Geom.*, 1993, pp. 133–138.

[185] M. Sharir, A near-linear algorithm for the planar 2-center problem, *Discrete Comput. Geom.*, 18 (1997), 125–134.

[186] M. Sharir and P. K. Agarwal, *Davenport-Schinzel Sequences and Their Geometric Applications*, Cambridge University Press, New York, 1995.

[187] M. Sharir and S. Toledo, Extremal polygon containment problems, *Comput. Geom. Theory Appl.*, 4 (1994), 99–118.

[188] M. Sharir and E. Welzl, A combinatorial bound for linear programming and related problems, *Proc. 9th Sympos. Theoret. Aspects Comput. Sci., Lecture Notes Comput. Sci.*, vol. 577, Springer-Verlag, 1992, pp. 569 579.

[189] M. Sharir and E. Welzl, Rectilinear and polygonal p-piercing and p-center problems, *Proc. 12th Annu. ACM Sympos. Comput. Geom.*, 1996, pp. 122–132.

[190] B. K. Shoichet, D. L. Bodian, and I. D. Kuntz, Molecular docking using shape descriptors, *J. Computational Chemistry*, 13 (1992), 380–397.

[191] R. F. Sproull, Refinements to nearest-neighbor searching, *Algorithmica*, 6 (1991), 579–589.

[192] L. Trevisan, When Hamming meets Euclid: The approximability of geometric TSP and MST, *Proc. 29th Annu. ACM Sympos. Theory Comput.*, 1997, pp. 21–29.

[193] K. R. Varadarajan and P. K. Agarwal, Approximation algorithms for bipartite and non-bipartite matching in the plane, *Proc. 10th ACM-SIAM Sympos. Discrete Algorithms*, 1999, pp. 805–814.

[194] E. Welzl, Smallest enclosing disks (balls and ellipsoids), in: *New Results and New Trends in Computer Science* (H. Maurer, ed.), *Lecture Notes Comput. Sci.*, vol. 555, Springer-Verlag, 1991, pp. 359–370.

[195] G. Wesolowsky, The Weber problem: History and perspective, *Location Science*, 1 (1993), 5–23.

[196] P. N. Yianilos, Data structures and algorithms for nearest neighbor search in general metric spaces, *Proc. 4th ACM-SIAM Sympos. Discrete Algorithms*, 1993, pp. 311–321.

[197] A. Zelikovsky, Minimum base of weighted k-polymatroid and Steiner tree problem, Report MPI-I-92-121, Max-Planck-Institut Inform., Saarbrücken, Germany, 1992.

Chapter 7

RANDOMIZED GEOMETRY ALGORITHMS FOR COARSE GRAINED PARALLEL COMPUTERS

Xiaotie Deng

Abstract   Randomization has played an important roles in design of parallel and distributed algorithms. The recent emphasis on coarse-grained parallel computing has led to new interesting randomized parallel and distributed algorithms. A specially important property in coarse-grained parallel computing is the possibility of having the same algorithm of the optimal asymptotic performance across various different network architectures. This is achieved by requiring a minimum amount of communication costs. In this article, we discuss some progress in applying randomization to coarse-grained parallel algorithms for some computational geometry problems.

1.   INTRODUCTION

In coarse-grained parallel algorithm design, our main focus is on minimizing communication costs while maintaining the asymptotic optimal speed up in the number of computations per processor. This raises new challenges to the area of design and analysis of parallel algorithms. Traditionally, fine-grained parallel computation models such as the PRAM resolve the issue of communication between processes through a shared memory model and leave the implementation to designers of parallel architectures and communication networks. Therefore, their performance on real machines may not match the speedup predicted for these algorithms and may differ significantly

Valiant's BSP model approaches this problem by introducing slackness in the number of processors, scrambling memory through hashing functions, and routing messages in a pipeline fashion. This allows optimal implementations of PRAM algorithms on a logarithmic fraction of theoretically specified number of processors [20, 21]. Nevertheless, Valiant

S. Rajasekaran et al (eds.), Handbook of Randomized Computing, Volume 1, pp, 203–220.
© 2001 *Kluwer Academic Publishers.*

points out that one may want to design algorithms that utilize local computations and minimize global operations [21]. The LogP model (extending the BSP model) introduced by Culler et al. [5] strongly advocates coarse-grained parallel computing and proposes the LogP model with interesting parallel algorithms of small communication costs. Following this approach, one may further show that, when input size is significantly larger than the number of processors, the same asymptotic optimal performance can be observed for parallel systems with many different communication networks [10].

Since our main interests are to design algorithms that will have roughly the same speedup performance over different parallel architectures, we would like to characterize parallel algorithms with parameters that is independent of parallel systems. In comparison, both the BSP model and the LogP model have parameters that are machine-dependent. Our emphasis here is to expand their excellent ideas in the direction of decoupling the design of parallel programs from physical machines.

We choose the following parameters:

C1. Local computation: the total execution delay if messages were to be delivered instantaneously.

C2. Message traffic: the maximum number of messages generated on or received by one processor.

C3. The number of communication phases: Formally, all the messages generated without receiving any message from other processors belong to Phase 1. For all $i \geq 1$, all the messages computed by one processor, using only messages from other processors of Phase $\leq i$ (at least one from Phase i) and local computation, belong to Phase (i+1).

Obviously, it may not be possible to design algorithms that minimizes all these parameters simultaneously. To achieve this poses new challenges in design and analysis of parallel algorithms. Here, randomization may be a powerful tool to help. In this article, we discuss some computational geometry problems for which this can be done through randomization techniques.

One of the fundamental tools in parallel algorithms is Divide-&-Conquer. A problem is first divided into subtasks and tackled by individual processors. Then solutions for subtasks are merged into the solution for the whole problem. The difficulty is often how to divide the problem so that the merging part can be done efficiently in parallel. Randomization often plays an important role to divide the problem into subtasks. In designing coarse-grained parallel algorithms, the extra effort for minimizing the number of communication phases requires that the task of merging can be divided to individual processors or done by one single processor. In the examples (2D and 3D convex hull as well as Voronoi diagram for line segments) we are going to present, this is done through careful uses of random sampling, the underlying structures of the random sample, as well as probabilistic analysis for the structures in their relations with other points/lines.

Dehne, et al., have discussed coarse-grained parallel algorithms for several computational geometry problems [8]. Most of them require a constant number of communication phases, the only exception is the 2D convex hull algorithm which requires $\log n$ communication phases for input size n. Randomization results in a very simple (yet non trivial) solution [9]. The main idea is to find the convex hull of a small number of sample points. Then we use the sample convex hull to assign no more than $O(\frac{n}{P})$ points to each processor for it to find a convex hull so that every edge of the convex hull is found by one of the processor. Finally, these edges in the convex hull is selected the local convex hull of each processor. Here it is important that randomization ensures that shared points among the processors are those bounded away from the origin by a small number of hyperplanes bounding the sample convex hull, and hence, are of a small number.

This lated evolves into an optimal coarse-grained parallel algorithms for 3D convex hull [7] and for Voronoi diagram [11] by means of graph separator techniques [13] and by means of some insights in for some other computational geometry problems. In Section 2, we will go in depth to present a detailed discussion of the 2D convex hull algorithm. In Section 3, we will sketch the main ideas combining randomization with novel ideas in partitioning for extending this to algorithms for 3D convex hulls and Voronoi diagram for line segments. We conclude our discussion in Section 4.

2.   2D CONVEX HULL

We adopt the notation that a claim holds with high probability, if for all arbitrary constants $c > 0$, it holds with probability at least $1 - \frac{1}{n^c}$. Therefore, a polynomial number of claims hold simultaneously with high probability if each of them holds with high probability.

In this section, we present a randomized algorithm for P processors to find the convex hull of n ($n = \Omega(P^{2+\epsilon})$) points in the plane, almost surely in optimal time $O(\frac{n \log n}{P})$ with $O(\frac{n}{P})$ messages per processor which are sent in $O(1)$ phases. The main idea in the convex hull algorithm by Dehne et al. was a partitioning technique used in [1] for a PRAM model

[8]. Input points were partitioned into P intervals. Convex hull of each interval was found by one processor. Every pair of processors then found the common tangent of their two subconvex hulls, by $\log n$ phases of communications. This type of high levels of communication would also be needed, through shared memories, in several other algorithms (e.g., [1] [12] [14]) for finding convex hulls.

A method to find out convex hull edges with two endpoints belonging to two partitions that are separated far apart is central to our algorithm. Here, input points are roughly equally partitioned. There are some input points shared by several processors. To make the number of shared points small, we apply the random sampling idea for sorting [6] [18], using the Chernoff bound method for probability analysis, similar to those found in [2] [3] [16] [17] [18].

The main idea is as follows: Find an interior point (w.l.o.g., the origin) of input data to partition input points into sector delimited by P rays from the origin in a balanced way. We use the convex hull of a set of sample points to find shared points. There are upto P edges intersecting those P rays. It is important to note that points bounded away from the origin by these edges are enough to find the convex hull edges acrossing several partitions. They are distributed to several different processors to guarantee that the union of convex hulls found by individual processors be the convex hull of all input points. Important to the analysis of the algorithm, the number of points bounded away by a line passing through an edge of the sample convex hull is small with high probability when the sample size is sufficiently large. In fact, if a line passing through two points bounds away a large number of points from the origin, the probability that one of these points be chosen in the sample set will be large. Since there are at most $\binom{n}{2}$ possible line segments passing through two points, the probability none of such line segments (that bound away from the origin a large number of points) is a face of the sample convex hull will be large if the sample size is sufficiently large.

We notice that Goodrich et al. discussed algorithm designs for the external memory model [12] including an adoption of the randomized algorithm of Reif and Sen's [18], and other computational geometry problems problems are discussed in the similar model by Zhu [19].

We introduce the convex hull algorithm with its correctness proof in Subsection 2.1, In Subsection 2.2, we apply Chernoff bounds to complete the probabilistic analysis on time complexity. Our goal is to show that, for an instance of the problem with n points, with high probability, the algorithm runs in $O(\frac{n \log n}{P})$ local computations and $O(\frac{n}{P})$ messages per processor and (absolute) constant phases of communication. For simplicity, in Subsection 2.2, we present an immediate implementation of the above algorithm which requires $P = O(n^{\frac{1}{3}} \log n)$ to achieve the stated time and message bound, as well as the bound for the number of communication phases. In Section 2.3, we relax this restriction to a more efficient implementation, which improves the input/processor ratio to $P = O(n^{\frac{1}{2}} \log n)$.

2.1   THE ALGORITHM

Denote the set of given n points, $p_1, p_2, \cdots, p_n$ in the plane ($p_i = (x_i, y_i)$, $1 \leq i \leq n$) by N. Assume that a known interior point of the convex hull of N is given. This can be obtained by calculating the center of gravity of $\frac{1}{n} \sum_{i=1}^{n} p_i$, which is a vector sum of two coordinates: $\frac{1}{n} \sum_{i=1}^{n} x_i$ and $\frac{1}{n} \sum_{i=1}^{n} y_i$. W.l.o.g., we may assume the origin $(0, 0)$ is the given interior point.

0. Sampling:
Each processor gets n/P input points
and samples each point with an independent probability of
Pk/n, where $k > 0$ will be specified later.
Let $R = \{r_1, r_2, \cdots, r_s\}$ be the whole sample set obtained
collectively.

1. Convex Hull of Samples:
Find the convex hull $CONV(R)$ of the sample set R.
Order the edges of $CONV(R)$ counterclockwisely:
$e_1, e_2, \cdots, e_t, t \leq s$,
where t is the size of $CONV(R)$.

2. Edges hit by Splitters:
Then, pick P of sample points such that
the rays (the splitters) $R_1, R_2, \cdots, R_P$
(for definiteness we adopt the convention that R_1 is toward the positive
direction of x-axis) in counterclockwise order
from $(0, 0)$ through these points
divide, as evenly as possible, the number of sample points in
each of the sectors bounded by R_i and R_{i+1}, $i = 1, 2, \cdots, P$.
Find $e_{R(i)}$, the edge in $CONV(R)$ hit by the ray
R_i. In the degenerate case when R_i passes through a vertex of
$CONV(R)$, we may choose the counterclockwise first edge incident to the
vertex.

3. Shared Points:
Find T_i, the set of all the points separated from $(0, 0)$
by the straight line passing through $e_{R(i)}$, $1 \leq i \leq P$.

4. Partitioning with Splitters:

Find W_i, the set of all the points in the sector bounded between R_i and R_{i+1}.

5. Assigning Points:

Send points in W_i, T_i and T_{i+1},

endpoints of both $e_{R(i)}$ and $e_{R(i+1)}$, and

the origin, to processor P_i.

Denote this set of points by O_i and its convex hull by U_i.

6. Communication About the Convex Hull:

For each local convex hull U_i,

the ray R_j cut through it at no more than one point other than the origin.

If R_j cuts through a vertex on U_i, send it and

the clockwise next vertex to

P_{j-1}, send it and the counterclockwise next vertex to P_j.

Otherwise, R_j cut through an edge on U_i and we send both endpoints of the edge to P_j and P_{j-1}.

7. Update Convex Hull:

Each processor P_i updates

its convex hull (denote it by C_i now)

with points sent over by other processors.

8. Output:

Each processor P_i finds the intersection of

its convex hull C_i with the sector bounded by ray R_i and ray R_{i+1}.

Report edges inside the sector and edges intersecting

ray R_i and ray R_{i+1}. $\square$

Fact 1. The union of local convex hulls U_i's is the convex hull for the set N of given n points.

Proof: We prove any two adjacent points q_1, q_2 in $CONV(N)$ are in one of U_i's. Take the line L which is parallel to $\overline{q_1 q_2}$, tangent to the sample convex hull $CONV(R)$, and separate q_1, q_2 from the origin. We may have two cases:

1. L intersects $CONV(R)$ on one point z.

2. L intersects $CONV(R)$ on edge, let z be the middle point of the edge.

Then z is contained in the sector bounded by R_i and R_{i+1} for some index i. For the simplest case q_1, q_2 are contained in the same W_i, both q_1 and q_2 will be in U_i by Step 4 and 5. Otherwise, L is cut into three piece by R_i and R_{i+1}. One insider the section bounded by R_i and R_{i+1}.

Among the other two pieces, one to the right of R_i which is bounded away from the origin by $e_{R(i)}$, and one to the left of R_{i+1} which is bounded away from the origin by $e_{R(i+1)}$. On the other hand, L bounds q_1 and q_2 away from the origin. Therefore, q_1, q_2 must be both in $W_i \cup T_i \cup T_{i+1}$ and be sent to P_i in Step 5. Thus, they must be on the local convex hull U_i. $\square$

Fact 2. In Step 6, for any $j \neq i$, at most two edges in U_i are sent to P_j.

Proof: Because the origin is in the convex hull U_i and each ray R_j starts at the origin, R_j will hit at most one point other than the origin on the convex hull. Processor P_j receives from P_i at most one edge each which intersects R_{j-1} (R_j, respectively) on U_i. $\square$

The following fact concludes the correctness proof of Algorithm 1.

Fact 3. In Step 8, all the edges on the convex hull $CONV(N)$ are reported in the counterclockwise order along the boundary by processor P_i, $i = 1, 2, \cdots, P$. No other edges are reported.

Proof: First, edges on $CONV(N)$ contained entirely in the sector bounded by R_i and R_{i+1} are reported only by P_i in Step 8 because W_i contains all their endpoints. There are at most two points on $CONV(N)$ intersecting the boundary rays R_i and R_{i+1}: one with R_i and another R_{i+1}. Consider one intersection point with R_{i+1} which is on a convex hull edge (q_1, q_2). By Fact 1, both q_1 and q_2 are contained in some U_j. Moreover (q_1, q_2) is on the convex hull of U_j and R_{i+1} cut through it. By Step 6, the edge (q_1, q_2) will be sent to P_i by P_j. Therefore, it will be reported by P_i. For the same reason, the edge on $CONV(N)$ which intersects R_i is also found by P_i in Step 7. For edges on the convex hull C_i found by P_i not in $CONV(N)$, they must be outside the sector bounded by R_i and R_{i+1} and will be excluded by P_i in Step 8. $\square$

Implementation of the Algorithm: We discuss the implementation of the algorithm, together with a rough estimation of its time complexity. Some detailed probability analysis are left to next subsection after introducing necessary probability lemmas. In calculating $\frac{\sum_{i=1}^{n} p_i}{n}$ to find an interior point, each processor evaluates the partial sum of all its $\frac{n}{P}$ input points. Then, with one round of broadcast of the partial sum, they can find the interior point. The total number of message sent and received by each processor is P, and the computation time is $O(\frac{n}{P})$. (In case of nondegeneracy, we can simply pick up three points.) For Step 0, each processor independently decides whether a point is taken into the sample set R with probability Pk/n. Let s_i be the number of points selected by Processor P_i. Recall that $s = \sum_{i=1}^{n} s_i$. On average, each processor selects k points. Then every processor sends to all other processors its selected points. Thus, each processor would send about k points to $P - 1$ processors, and receive about $(P - 1)k$ points. We need to choose

$$(P - 1)k \leq \frac{n}{P}, \tag{7.1}$$

so that the total number of messages per processor in this phase is dominated by $\frac{n}{P}$. It takes $s \log s$ local computation time for the sorting algorithm to obtain the rays R_1, R_2, $\cdots$, R_P. Then, for Step 1, all the processors have the sample R and they can compute the convex hull $CONV(R)$ individually, which takes $s \log s$ local computation time [15]. Since the expected value of s is Pk, the parallel time would be bounded from above by about $\frac{n}{P} \log \frac{n}{P}$ since $Pk \leq \frac{n}{P}$ as required by Inequality 7.1. Step 2 can also be carried out in the parallel time $P \log t$ which is dominated by $s \log s$. For Step 3, each processor decides which T_i's its input points belong to. Again this takes the parallel time $\frac{n}{P} \log \frac{n}{P}$. Then, all the processors pass their input points in T_i and T_{i+1} to P_i (to be used in Step 5). The number of messages in this phase will be discussed in the next section. Step 4 is done similar to Step 3 and all the points in W_i are passed to processor P_i. P_i constructs U_i after it receives all the points to carry out Step 5. Step 6, 7 and 8 are deterministic steps. They can all be carried out easily, and the time/message bound follows from the above Facts. $\square$

2.2 PROBABILISTIC ANALYSIS AND TIME BOUND

First we introduce the Chernoff bound for binomial distributions (see, e.g., [2] [16]):

Lemma 2..1 *Let $S_{n,p}$ be a binomial random variable with parameters n, p, then, we have*

$$Prob(S_{n,p} \geq (1 + \delta)np) < e^{-\delta^2 np/4}.$$

With the lemma, we discuss sizes of point sets introduced in the algorithm in Section 2 and show the following facts:

Claims:

1. Let s be the sample size.

$$Prob(s \geq \frac{3}{2}Pk) \leq e^{-Pk/16}.$$

2. Let W_i, $i = 1, 2, \cdots, P$, be as defined in Algorithm 1.

$$Prob(|W_i| \geq \frac{9n}{4P}) \leq (e^{-Pk/16} + e^{-k/8}).$$

3. Let T_i, $i = 1, 2, \cdots, P$, be as defined in Algorithm 1.

$$Prob(|T_i| \geq \frac{n}{P}) \leq \binom{n}{2} e^{-k}.$$

4. Let s_i be the size of sample points chosen by processor P_i.

$$Prob(s_i \geq \frac{3}{2}k) \leq e^{-k/16}.$$

5. Denote by T_{ij} the set of input points for P_i which is contained in T_j.

$$Prob(|T_{ij}| \geq \frac{n}{P^2}) \leq n \binom{n}{2} e^{-\frac{k}{P}}.$$

Proof:

P1. Choose $\delta = \frac{1}{2}$ in Lemma 1 since s follows the binomial distribution with $p = \frac{Pk}{n}$,

P2. Denote by l_j the ray from the origin towards the point p_j, $1 \leq j \leq n$. Fix the position of $R_i = l_j$. Consider the set F of $\frac{9n}{4P}$ input points immediately following R_i along the counterclockwise order. Recall that the number of sample points between R_i and R_{i+1} is at most $\frac{s}{P}$ by Step 2. Given the position of R_i, the event that W_i has at least $\frac{9n}{4P}$ points is the same as the event at most s/P points in F are chosen to the sample set R. From Claim 1, $s \leq \frac{3}{2}Pk$ occurs almost surely. Under these conditions, the probability that W_i has at least $\frac{9n}{4P}$ points is no more than the probability that at most $3k/2$ points of F are chosen to the sample set R. This is again a (different) binomial distribution with $\frac{9n}{4P}$ points and probability Pk/n. Applying Lemma 1 with $\delta = \frac{1}{3}$, we have

$$Prob(|W_i| \geq \frac{9n}{4P} \mid R_i = l_j, s \leq 3Pk/2) \leq$$

$$P(S_{\frac{9n}{4P}, \frac{Pk}{n}} \leq 3k/2) \leq e^{-(\frac{1}{3})^2(9k/4)/2}.$$

Apply Claim 1, we have

$$Prob(|W_i| \geq \frac{9n}{4P} \mid R_i = l_j) \leq e^{-Pk/16} + e^{-k/8}, 1 \leq j \leq n.$$

$$Prob(|W_i| \geq \frac{9n}{4P}) = \sum_{j=1}^{n} Prob(|W_i| \geq \frac{9n}{4P} \mid R_i = l_j)Prob(R_i = l_j)$$

$$\leq \sum_{j=1}^{n}(e^{-Pk/16} + e^{-k/8})Prob(R_i = l_j)$$

$$= (e^{-Pk/16} + e^{-k/8})\sum_{j=1}^{n} Prob(R_i = l_j)$$

$$\leq e^{-Pk/16} + e^{-k/8}.$$

P3. Denote the line passing through p_{i_1} and p_{i_2} as $l = l_{i_1 i_2}$. Let n_l be the number points bounded away from the origin by the line l. To evaluate $Prob(|T_i| \geq \frac{n}{P})$, We consider that $Prob(e_{R(i)} = l_{i_1 i_2}, n_l \geq \frac{n}{P})$.

$$Prob(e_{R(i)} = l_{i_1 i_2}, n_l \geq \frac{n}{P}) = Prob(e_{R(i)} = l_{i_1 i_2}|n_l \geq \frac{n}{P})Prob(n_l \geq \frac{n}{P}),$$

where $(n_l \geq \frac{n}{P})$ is a deterministic event and thus $Prob(n_l \geq \frac{n}{P}) = 0, 1$. It then follows that

$$Prob(|T_i| \geq \frac{n}{P}) = \sum_{all \; l} Prob(e_{R(i)} = l_{i_1 i_2}, n_l \geq \frac{n}{P})$$

$$\leq \sum_{all \; l} Prob(e_{R(i)} = l_{i_1 i_2}|n_l \geq \frac{n}{P})$$

The probability $Prob(e_{R(i)} = l_{i_1 i_2}|n_l \geq \frac{n}{P})$ is bounded by the probability i_1 and i_2 are chosen and none of points bounded away from the origin be the line l is chósen. Therefore,

$$Prob(e_{R(i)} = l_{i_1 i_2}|n_l \geq \frac{n}{P}) \leq (1 - \frac{Pk}{n})^{n_l} \leq (1 - \frac{Pk}{n})^{\frac{n}{P}} \leq e^{-k}.$$

Claim 3 follows since there are at most $\binom{n}{2}$ pairs of points.

P4. $Prob(s_i \geq \frac{3}{2}k) \leq e^{-\frac{k}{16}}$. This follows from Lemma 1 since s_i is a variable of the binomial distribution with parameters: $\frac{n}{P}$ points and probability $\frac{kP}{n}$.

P5. Similar to the proof in Claim 3, we show

$$Prob(|T_j| \geq \frac{n}{P^2}) \leq n\binom{n}{2}e^{-\frac{k}{P}}.$$

Claim 5 follows since $|T_{ij}| \leq |T_j|$. $\square$

Proposition 1 *Algorithm 1 can be implemented so that it takes $O(1)$ phases of communication, passes $O(\frac{n}{P})$ messages and uses local time $O(\frac{n \log n}{P})$ in each processor, with high probability to compute the convex hull of n points, for $n = \Omega(P^3 \log P)$.*

Proof: Let $k = Pn^\epsilon$ and we discuss the time bound step by step.

S0. In Step 0, the number of local computation time is $\frac{n}{P}$. Only one phase of communication is done. The probability that no processor selects more than $3k/2$ points is at least $1 - ne^{-k/16}$ by Claim 5. Thus, with this probability, at most $3k(P-1)/2$ messages are sent and received by each processor.

S1. Local time in Step 1 for computing the convex hull of R is $O(kP \log kP)$.

S2. Local time in Step 2 for find $e_{R(i)}$ is $O(kP \log t)$, $t \leq kP$.

S3. For each point p of $\frac{n}{P}$ input points, processor P_i can find out whether p is inside $CONV(R)$ or outside. In the latter case, it finds the two tangents to $CONV(R)$ of p which takes $\log t$ time. These tangents allow P_i to find which T_js, $j = 1, 2, \cdots, P$, this point p belongs to. By Claim 3, each processor will receive $O(n/P)$ points almost surely. However, we also need to bound the number, $\sum_{j=1}^{P} |T_{ij}|$, of points processor P_i sends out to other processors. An important issue here is T_{ij}, $1 \leq j \leq P$, may not be disjoint. Notice that if $\sum_{j=1}^{P} |T_{ij}| \geq \frac{n}{P}$, then for one of j, $|T_{ij}|$, then for at least one of j, $|T_{ij}| \geq \frac{n}{P^2}$. Thus the event that $\sum_{j=1}^{P} |T_{ij}| \geq \frac{n}{P}$ is contained in $\cup_{j=1}^{P} \{|T_{ij}| \geq \frac{n}{P^2}\}$.

$$Prob(\sum_{i=1}^{P} |T_{ij}| \geq \frac{n}{P}) \leq \sum_{i=1}^{P} Prob(|T_{ij}| \geq \frac{n}{P^2})$$

By Claim 5, this sum is greater than n/P with probability no more than $Pe^{-\frac{k}{P}}$, which is exponentially small if we choose $k = Pn^\epsilon$, which forces the condition $n = \Omega(P^{3+\epsilon})$, by Inequality 7.1. This will be relaxed in the next section. Only one phase of messages is sent in this step.

S4. Step 4 is similar to and simpler than Step 3. Each processor will send out at most n/P points since W_i, $1 \leq i \leq P$ are disjoint. Each processor receives $O(n/P)$ points almost surely, by Claim 2. From above the total number of points used for P_i to construct U_i is $O(n/P)$ almost surely.

S5-8. Under the above conditions, Step 5 takes time $O(\frac{n}{P} \log \frac{n}{P})$, and send one phase of $O(\frac{n}{P})$ messages with probability $1 - e^{-\Omega(n^\epsilon)}$. By Fact 2, in Step 6 each processor sends and receives one phase of $O(P)$ messages. Both Step 7 and Step 8 takes time $O(P \log(P + \frac{n}{P}))$.

In summary, let $k = Pn^\epsilon$, the probability that the convex hull algorithm finishes in time $O(\frac{n \log n}{P})$ and in $O(1)$ phases of $O(\frac{n}{P})$ one-word communication is $1 - O(n^2 e^{-cn^\epsilon})$. $\square$

Notice that with the same analysis, we can easily show that, for any $c > 0$, this also holds with probability $1 - n^{-c}$ for $n = \Omega(P^3 \log P)$.

2.3 IMPROVEMENT ON PROBLEM-SIZE/PROCESSOR RATIO

One drawback of the above theorem is that the restriction $P \approx n^{\frac{1}{3}}$ on the number of processors. We can relax this restriction to $P \approx n^{\frac{1}{2}}$. Notice that the restriction $P \approx n^{\frac{1}{3}}$ on the number of processors is imposed by Claim 5 to guarantee that $|T_{ij}| \leq \frac{n}{P^2}$ almost surely. Under this condition, the total number of messages sent out by each processor is no more than $O(\frac{n}{P})$ with high probability. However, in doing so, each processor may have send a same point several times, to different processors. In the following we refine the message passing method for points in T_j's, $1 \leq j \leq P$, to improve the processor bound.

Theorem 2..2 *Algorithm 1 can be implemented in $O(\frac{n}{P} \log n)$ local computation time, with $\frac{n}{P}$ messages per processor, in $O(1)$ phases, with high probability when $n = \Omega(P^2 \log P)$.* $\square$

Proof: In Step 3, processor P_i finds, for each j, among all its input points, which belong to T_j and need to be sent to P_j. One observation is that the indices j's of T_j's a point belongs to are consecutive. W.l.o.g., an input point p belongs to $T_j, T_{j+1}, T_{j+2}, \cdots, T_{j+f(p)}$, with $f(p) \leq P - 2$. This can be found out in time $\log P$ by finding the two tangents of the point p to the sample convex hull $CONV(R)$. $e_{R(j)}$ will be the the clockwise first edge of $CONV(R)$, among those intersecting the P boundary rays, enclosed by these two tangents, and $e_{R(j+f(p))}$ will be the last one. We use the convention that $f(p) = -1$ if p is in the interior of R or the two tangents does not enclose any of the P intersection points of $CONV(R)$ with the P boundary rays. Thus, in $O(\frac{n \log P}{P})$ time, we can find, for each point p, a triple $(p, j, f(p))$, as defined above. By Claim 3, we have, with high probability,

$$\sum_{all\ p}(1+f(p)) = \sum_{j=1}^{P}|T_j| \le n.$$

Then, we can sort all the points p with $f(p) \ge 0$ according to the value $f(p)$: $p_1, p_2, \cdots, p_m$ with $f(p_1) \le f(p_2) \le \cdots \le f(p_m)$ and $m \le n$. We make sure that, in the end of the sorting algorithm, p_i resides at processor P_j with $i \equiv j \bmod (P)$.

If P divides m, then, $\sum_{i=1}^{m/P} f(p_{iP})$ is at least as large as $\sum_{i=0}^{m/P-1} f(p_{(iP+j)})$, $1 \le j < P$. On the other hand, $f(p_{iP}) \le f(p_{(iP+j)})$, $0 \le j \le P-1$. Thus,

$$\sum_{i=1}^{m/P-1} 1 + f(p_{iP}) \le \frac{1}{P} \sum_{i=1}^{m/P-1} \sum_{j=1}^{P} 1 + f(p_{(iP+j)}),$$

which is bounded by

$$\frac{1}{P}\sum_{all\ p}(1+f(p)) = \frac{1}{P}\sum_{j=1}^{P}|T_j| \le \frac{n}{P}.$$

Therefore, if P completely divides m, $\sum_{i=1}^{m/P} f(p_{(iP)}) \le \frac{n}{P} + P$ and $\sum_{i=0}^{m/P} f(p_{(iP+j)}) \le \frac{n}{P} + P$. It immediately follows that, in general, $\sum_{i=0}^{m/P} f(p_{(iP+j)}) \le \frac{n}{P} + 2P$.

Thus, now each processor sends its point p with $(p, j, f(p))$ to processors $P_j, P_{j+1}, \cdots, P_{j+f(p)}$ for all its $\frac{m}{P}$ points. Since sorting can be done with the specified complexity bounds: $O(\frac{n \log n}{P})$ local computations, $O(\frac{n}{P})$ messages sent from or received by each processor, and $O(1)$ communication phases [3] [5] [18], the same bound holds for this step of distributing points in T_j, $1 \le j \le P$.$\square$

3.   EXTENSIONS

Naturally, we would like to know if this method can be extended to 3-dimensional convex hulls, as well as many other problems successfully solved by random sampling methods, to obtain coarse-grained parallel algorithms with a constant number of communication phases.

This has been successfully done for 3D Convex Hull Problem and Voronoi Diagram for Line Segments in the plane [7, 11]. We start with the random sampling set R to obtain a solution for the sample set. We use the sample solution to divide the task into P problems of size $O(n/P)$.

3.1   3D CONVEX HULLS

For the 3D convex hull problem, to every processor, we use $\frac{1}{P}$-th faces of the sample convex hull to collect points enclosed in the cones of the rays from the origin (an interior point) to the vertices of the faces and send to it. In addition, we assign to the processor all the points bounded away from the origin by the faces on the boundary of the region formed by the above faces on the convex hull. A crucial and non-trivial fact is that if two end points of every edge of the convex hull are assigned to one processor. Moreover, the number of point bounded away from the origin by any face of the convex hull for the random sample is small with high probability when the sample size is sufficiently large. The intuition is that, if this is large for a hyperplane, one of the point will be chosen in the sample and the hyperplane will not be a face of the convex hull of the sample set. Since the total number of hyperplanes passing three points is bounded by n^3, with high probability, none of the faces of the sample convex hull will bound away a large number of points.

A major difficulty is that the number of boundary hyperplanes of faces of the sample convex hull that are assigned to two different processors can be as large as $O(\frac{|R|}{P})$. In comparison, this is one for the 2D problem. To control the size of the boundary, we apply the the separator theorem of Lipton and Tarjan [13] for planar graphs, applied to the dual graph of the sample convex hull, to control the size of subproblems. Notice that a 3D convex hull can be viewed as a planar graph: vertices as nodes, edges of the convex hull as edges, and faces of the convex hull as faces (and an arbitrary one as the infinite face). It turns out this works as long as n is approximately $P^{3+\epsilon}$ for any $\epsilon > 0$. Moreover, this can be also implemented as a PRAM algorithm with n processors in $O(\log n)$ time, by choosing appropriate parameters and applying PRAM algorithms to the necessary subproblems.

3.2   VORONOI DIAGRAM FOR LINE SEGMENTS

The general framework can be carried over to Voronoi diagram for line segments on the plane. The new difficulty here is that dividing the problem into subproblems is quite non-trivial since a partition of the plane usually does not constitute a partition of line segments. This is because that a line segment may cut across several faces in a planar partition while a point won't. To divide the problem into P subproblems of size $O(\frac{n}{P})$ each, we use Voronoi vertices of the sample sets. One observes that a planar partition is a partition of Voronoi vertices and the number of Voronoi diagram is proportional to the number of line segments. We make sure that any set of three line segments defining a Voronoi vertex are distributed to at least one subproblem. Several fundamental geometry problems are useful in this algorithm: the shortest path problem and the ray shooting problem to polygons with circular edges. Similarly, we can prove that the union of Voronoi vertices for subproblems contains all the Voronoi vertices for the original problem. The next step is to identify the Voronoi vertices for the original problem from a total of $O(n)$ candidates obtained in the subproblems. Similar to the 3D problem, this also leads to a conceptually simple optimal CRCW PRAM algorithm [11]. In comparison, the first optimal parallel algorithm for Voronoi Diagram of line segments by S. Rajasekaran and S. Ramaswami [17] was developed from the pooling idea of J. Reif, and S. Sen [18].

4.   REMARKS AND DISCUSSION

In summary, the requirement of minimizing communication phases in design and analysis of coarse-grained parallel algorithm may need new insight and techniques in dividing tasks. Our examples show that randomization can play an important role here when combined with new ideas in design and analysis.

An immediate open problem here is whether we can improve the ratio of sizes of input data and processors. Next, can we de-randomize the above algorithms, or is it possible to have a deterministic algorithm for 3D convex hulls, *optimal* in our criteria? Our conjecture is that, when C1 and C2 are satisfied, any deterministic algorithm would need $\Omega(\log P)$ phases of communication. More generally, can we generalize this to higher dimensional convex hull problems? A similar approach would require a higher dimensional version of the graph separator theorem for convex hulls for higher dimensions, which might also be useful in other problems.

Beyond computational geometry problems, randomization has started to be applied to coarse-grained parallel algorithms for graph problems [10, 4]. Research into this direction is still not fully explored though some limited progress has been made. In the emergence of coarse-grained parallel and distributed computer systems [20, 5], this is a very promising research direction.

Acknowledgments

I would like to acknowledge the support of a grant from the Rearch Grants Council of the Hong Kong Special Administration Region, China (Project No. CityU 1049/98E) and a grant of City University of Hong Kong (Project No. 7000763).

References

[1] A. Aggarwal, B. Chazelle, L. Guibas, C. Ó'Dúnlaing and C. Yap. "Parallel computational geometry", *Algorithmica*, **3**, pp. 293-327, 1988.

[2] N. Alon, and N. Meggido. Parallel Linear Programming in Fixed Dimension Almost Surely in Constant Time. *The Thirty First Symposium on Foundations of Computer Science*, pp.574-582, 1990.

[3] G.E. Blelloch, C.E. Leiserson, B.M. Maggs, C.G. Plaxton, S. J. Smith, and M. Zagha. "A Comparison of Sorting Algorithms for the Connection Machine CM-2," *The third Annual ACM Symposium on Parallel Algorithms and Architecture*, pp.3-16, 1991.

[4] E. Caceres, F. Dehne, A. Ferreira, P. Flocchini, I. Rieping, A. Roncato, N. Santoro, and S. W. Song, "Efficient parallel graph algorithms for coarse grained multicomputers and BSP," *Proc. 24th International Colloquium on Automata, Languages and Programming (ICALP'97)*, Bologna, Italy, 1997, Springer Verlag Lecture Notes in Computer Science, Vol. 1256, pp. 390-400.

[5] D. Culler, R. Karp, D. Patterson, A.Sahay, K.E. Schauser, E.Santos, R. Subramonian, and T. von Eicken. "LogP: Towards a Realistic Model of Parallel Computation," *Proceedings of the 4th ACM SIGPLAN Symposium on Principles and Practice of Parallel Programming*, pp.235-261, 1993.

[6] K.L. Clarkson, and P. Shor. Applications of Random Sampling in Computational Geometry, II. *Discrete and Computational Geometry 4*, pp. 387-421, 1989.

[7] F. Dehne, X. Deng, P. Dymond A. Fabri, and A. Khokhar. "A Randomized Parallel 3D Convex Hull Algorithm For Coarse Grained Multicomputers," *Proceedings of the Sixth ACM Symposium on Parallel Architectures and Algorithms*, Santa Barbara, pp. 27-33, 1995. Journal version in *Theory of Computing Systems*, Vol. 30, 1997, pp. 547-558.

[8] F. Dehne, A. Fabri, and A. Rau-Chaplin. Scalable Parallel Geometric Algorithms for Coarse Grained Multicomputers. *The 9th Annual ACM Symposium on Computational Geometry*, pp.298-307, 1993. Journal version in *International Journal on Computational Geometry*, Vol. 6, No. 3, pp. 379-400, 1996.

[9] X. Deng, "A Convex Hull Algorithm on Multiprocessor," *Lecture Notes in Computer Science 834, Proceedings of the 5th International Symposium on Algorithms and Computations*, Beijing, pp. 634–642, 1994.

[10] X. Deng and P. Dymond. " Efficient Routing and Message Bounds for Implementing Parallel Algorithms," *International Parallel Processing Symposium*, pp.556-562, 1995.

[11] X. Deng and B. Zhu. " A Randomized Algorithm for Voronoi Diagram of Line Segments on Coarse Grained Multiprocessors," Proceedings of the 10th IEEE International Parallel Processing Symposium, Honolulu, Hawaii, April 1996, pp. 192-198. Journal version at: Algorithmica Vol 24, No 3/4, pp. 270–286, 1999, a special issue on coarse grained parallel algorithms.

[12] M.T. Goodrich. "Finding the convex hull of a sorted point set in parallel," *Inform. Process. Lett.*, **26**, pp.176-179, 1987

[13] R. J. Lipton, and R. E. Tarjan, "A Separator Theorem for Planar Graphs," *SIAM J. Appl. Math.*, Vol 36, pp.177–189, 1979.

[14] R. Miller, and Q.F. Stout. "Efficient parallel convex hull algorithms," *IEEE Trans. Comput.*, C-37, 12, pp.1605-1618, 1988.

[15] F. P. Preparata, and M.I. Shamos, *Computational Geometry, An Introduction*, Springer-Verlag, New York, 1985.

[16] P. Raghavan, *Lecture Notes on Randomized Algorithms*, IRM Research Report, 1990, RC15340(#68237)1/9/90.

[17] S. Rajasekaran and S. Ramaswami, Optimal Parallel Randomized Algorithms for the Voronoi Diagram of Line Segments in the Plane and Related Problems, Proc. *10th Annual ACM Computational Geometry Conference*, June 1994, pp. 57-66.

[18] J. Reif, and S. Sen. Polling: A New Randomized Sampling Technique for Computational Geometry, *The Twenty First Annual ACM Symposium on Theory of Computing*, pp.394-404, 1989.

[19] Binghai Zhu. "Further Computational Geometry in Secondary Memory," *Proceedings of 5th International Symposium*, ISSAC'94, Beijing, P. R. China, August, pp.514-522, 1994.

[20] L. G. Valiant, "A Bridging Model for Parallel Computation", *CACM* , Vol. 33, No. 8, pp. 103-111, August, 1990.

[21] L. G. Valiant, "General Purpose Parallel Architectures", *Handbook of Theoretical Computer Science*, Edited by J. van Leeuwen, the MIT Press/Elsevier, pp. 943-972, 1990.

Chapter 8

A RANDOMIZED APPROACH TO ROBOT PATH PLANNING BASED ON LAZY EVALUATION

Robert Bohlin

Department of Mathematics
Chalmers University of Technology
SE-412 96 Göteborg, Sweden

Lydia E. Kavraki

Department of Computer Science
Rice University
Houston, TX 77005, USA

Abstract   Path planning addresses the problem of finding collision-free paths for moving objects – robots – among obstacles. Randomized techniques have shown great performance in high-dimensional configuration spaces and are now the methods of choice for complex problems. In this paper we describe the Probabilistic Roadmap Method (PRM), variations of PRM, and other closely related algorithms. PRM is a simple and widely used planner applicable to virtually any kind of robot. The underlying idea is to build a roadmap in the configuration space. The nodes in the roadmap correspond to feasible configurations selected at random, and the edges correspond to feasible path segments. In the query phase, the initial and goal configurations are connected to the roadmap, and then the planner searches for a shortest path.

The contribution of this paper is a new scheme for the lazy evaluation of the feasibility of the roadmap. We apply the scheme to PRM and obtain a planner called Lazy PRM. The overall theme of the algorithm is to increase speed by only exploring the part of the roadmap that is necessary for the current query. Lazy PRM is tailored to efficiently answer single planning queries, but can also be used for multiple queries. Experimental results provided in this paper show that our lazy method is very efficient in practice.

Keywords: Collision avoidance, motion planning, path planning, probabilistic roadmaps, randomized path planning, robotics.

221

S. Rajasekaran et al (eds.), Handbook of Randomized Computing, Volume 1, pp, 221–253.
© 2001 *Kluwer Academic Publishers.*

1.   INTRODUCTION AND MOTIVATION

Path planning for robots is a broad and intensively studied problem. The general problem is to find collision-free paths for moving objects among a set of obstacles. The moving objects are called robots and they operate in an environment called the workspace. A robot can be a single rigid body or a collection of rigid bodies connected by joints.

Path Planning Applications.   Planners are used in a wide variety of applications. Industrial robots weld, paint, and assemble products and their actions could be programmed automatically to a large extent. Many Automated Guided Vehicles (AGVs) plan their paths in real-time while they transport assembly parts between different stations in a workshop. AGVs use sensors to navigate and to detect unknown moving obstacles (like humans). Planners are integrated with Computer-Aided Design (CAD) software to support the design engineer in deciding whether or not it is possible to assemble or maintain a product.

More applications can be found in computer graphics where planners generate motions for animated humans, vehicles and other objects [26]. The task may also involve coordination of several robots. One important issue here is to generate motions that look natural and smooth. In medical surgery, planners are used to generate paths for operations so that the damage in tissues involved is minimized. Neurosurgeons, for example, use the planner in [40] to compute how to generate radiation beams that destroy brain tumors without damaging surrounding tissues. Last but not least, planning has interesting applications in biomedicine. For example, a path planner can be used to compute if a drug molecule (the robot) can find a path into a large protein molecule, dock to its active site, and in this way prevent an undesired reaction in the active site [15, 28].

Need for Efficient Path Planners.   There are several reasons for using path planners in the above applications. In many industrial problems and in computer graphics animation, the main reason is to reduce programming time. Manual calculation of paths can be very tedious. For example, a car body may have thousands of spots to weld and there the potential time savings from the use of a planner are huge. The quality of the produced paths is also an important issue. Often, execution time is also critical, and planners can be used to optimize paths in that respect.

A planner may also allow the computation of very complex motions. An industrial example may be the coordination of fixture motion and robot motion in an arc welding application. In computer graphics, one may want to have a large number of virtual characters interacting, e.g., a crowd of tourists can be made to follow a guide. A car-like robot has curvature constraints on the paths it can follow due to its limited turning radius. Constraints of a different type arise when a dextrous robot grasps and manipulates an object, or when cooperating robots simultaneously manipulate an object. Then closed kinematic chains appear and these must always be maintained, otherwise the object will fall down. Another difficulty that appears in real scenarios is dealing with uncertainties. We may have uncertainties in the control of the robot, in sensing, or incomplete knowledge of the environment (e.g., the Pathfinder rover that was sent to explore the surface of Mars [34]).

Our Work and Outline of the Paper. Different path planning algorithms have been developed for addressing the issues above. We concentrate on the most basic version of the path planning problem, that of moving a robot in a static environment. Efficient solutions of that problem translate to improvements in the solution of problems with more constraints such as the ones mentioned above. The high computational complexity of the basic path planning problem (see Section 2) dictates the use of randomized techniques for its solution.

In the first part of this paper, we revisit some general randomized path planning techniques for robots with many degrees of freedom (dof). The class of algorithms we are mostly interested in are the Probabilistic Roadmap Methods (PRMs). These algorithms are simple, general and can be used for a broad range of problems. In the second part of the paper, we present a new planner based on the PRM framework. The planner tries to minimize the number of collision checks done in the PRM framework and hence increase speed. We call the new planner Lazy PRM. We demonstrate that Lazy PRM is particularly effective in standard industrial applications. The latter are high-dimensional problems characterized by complex geometry and relatively uncluttered spaces.

This paper is organized as follows. Section 2 discusses the complexity of the basic path planning problem, motivating the need for randomized techniques for its solution. Section 3 establishes the terminology that will be used in the rest of the paper. Section 4 gives a description of the basic PRM, a number of variations of the algorithm, and other closely related algorithms. In Section 5 we describe Lazy PRM. We draw our discussion from [6, 7, 8]. Related ideas about lazy evaluation have been developed concurrently and independently in [35]. An analysis of the planner is given in Section 6 while experimental results are given in Section 7. We conclude in Section 8 with a discussion of the capabilities and limitations of Lazy PRM. Some of our comments apply to randomized approaches to path planning in general.

2.   COMPLEXITY ISSUES

In the previous section we gave some examples of path planning applications with different characteristics, and some problems that arise in practice. Unfortunately, there is no planning technique that deals with all of those conditions and constraints. For some special cases there exist good planners, but in general there is still a way to go until planners can be practically useful.

An algorithm is called *complete* if it always will find a solution or determine that none exists. Complete algorithms for the basic path planning problem exist, but they can not be used in practice due to their high complexity. So far, the fastest complete algorithm for arbitrary robots is given in [12] and it is exponential in the number of dof of the robot. The algorithm is too slow to be useful in practice, and it is mostly used in theoretical analysis as an upper bound on the complexity of the path planning problem. Some versions of the path planning problems have been proved PSPACE-hard [38]. Various other complexity results are described in [15, 27].

An alternative to complete planners are *probabilistically complete* planners. If a collision-free path exists, a probabilistically complete planner finds a solution with a probability approaching 1 given enough time for the computation [15]. Trading completeness for speed, randomized techniques have been successfully applied to generate path planners that are now the methods of choice for complex planning tasks. In this paper, we will explore in depth one such technique, the PRM framework, which has delivered excellent experimental results in the last five years.

3.   NOTATION

We need a way to describe the position of the robot in a convenient way. For instance, if the robot is a single rigid object in a three-dimensional workspace, it has six dof and we can specify its position and orientation by six parameters; three coordinates for the translation, and three angles for the orientation. Similarly, the pose of an articulated robot arm (with a fixed base) is completely specified by the joint values.

More generally, a *configuration* is a set of independent parameters such that the position of every point of the robot can be determined relative to a fixed frame in the workspace. The set of all configurations is called the *configuration space* and is denoted by C. The dimension of C is equal to the number of dof of the robot. The obstacles in the workspace can be mapped onto a subset of $\mathcal{C}$ – the configuration space obstacles – by the following definition: a configuration belongs to the configuration space obstacles if the robot intersects any obstacle in the workspace. The open subset of collision-free configurations is denoted by $\mathcal{F}$.

A path for the robot is simply a continuous curve in $\mathcal{C}$. We will also refer to a path as a sequence of points in $\mathcal{C}$, in which case the path is the piecewise linear curve obtained by linearly interpolating subsequent points.

With this notation, we can rephrase the basic path planning problem as follows: given an initial configuration q_{init} and a goal configuration q_{goal} in $\mathcal{F}$, find a continuous curve in $\mathcal{F}$ connecting these points, or determine that none exists [27]. This formulation of the problem is usually favorable, since it is stated in terms of navigating a point rather than objects in the workspace. In that sense, the planning problem becomes independent of the geometry and kinematics of the robot, and may appear simpler. However, as soon as the robot is allowed to rotate or has revolute joints, the kinematics is non-linear, and seemingly simple obstacles in the workspace are mapped onto very complex obstacles in the configuration space. If the dimension of $\mathcal{C}$ is low, say four or less, then it is possible to obtain an approximate representation of the configuration space obstacles by discretization. But if the dimension is higher there is no convenient way to represent $\mathcal{C}$.

So, how can we plan in a space where the obstacles cannot be represented? We do have an implicit representation of the obstacles. Given a configuration it is generally easy to use the forward kinematics function to calculate the position of the robot in the workspace. Then we can test for intersection with the workspace obstacles to decide whether the configuration is feasible or not. Thus, we can point-wise determine if we are in $\mathcal{F}$. As a consequence, we cannot completely verify that a path, i.e. a curve in $\mathcal{C}$, is entirely collision-free, but we consider a path being feasible if it is collision-free to a certain resolution.

4. RANDOM ROADMAPS FOR FAST PATH PLANNING

The idea of using randomization to solve the path planning problem is not new. The Randomized Path Planner (RPP) [5] was one of the first algorithms that used randomization. The algorithm was a breakthrough that made it possible to solve difficult problems in high-dimensional configuration spaces. At the same time, RPP possessed completeness properties. The planner uses a potential field in the configuration space and calculates the potential without an explicit representation of the configuration space obstacles. The potential field is composed of one attractive field guiding the search towards the goal, and one repulsive field preventing from collisions with the obstacles. Starting from q_{init}, the algorithm follows the steepest descent direction towards q_{goal}. However, the potential is not perfect (calculating an ideal potential is probably as difficult as the path planning problem in itself) and generally contains local minima. To escape local minima the RPP combines descent motions and random walks.

One of the most general and widely used methods for path planning is the Probabilistic Roadmap Method (PRM). The algorithm is easy to implement and has been successful in many applications, also in high-dimensional configuration spaces. PRM is practically independent of the geometry and the kinematics of the robot, and can be used with virtually any kind of robot. In this section we will start with a detailed description of a basic PRM and continue with some variations of PRM.

4.1 THE PROBABILISTIC ROADMAP METHOD (PRM)

The idea behind the Probabilistic Roadmap Method (PRM), described in [24, 25, 37], is to represent and capture the connectivity of $\mathcal{F}$ by a random network – a *roadmap*. The nodes in the roadmap correspond to randomly selected configurations, and the edges correspond to path segments between the nodes. In a preprocessing step, or a *learning phase*, a large number of points are distributed uniformly at random in $\mathcal{C}$, and those found to be in $\mathcal{F}$ are retained as nodes in the roadmap. A local planner is then used to find paths between each pair of nodes that are sufficiently close together. If the planner succeeds in finding a path between two nodes, they are connected by an edge in the roadmap. In the *query phase*, the user specified start and goal configurations are connected to the roadmap by the local planner. Then the roadmap is searched for a shortest path between the given points, see Figure 8.1.

Even though a powerful local planner will require few nodes to obtain a well connected roadmap, most implemented PRMs show that it is computationally more efficient to distribute nodes densely and use a relatively weak, but fast, local planner, see [25, 37]. The local planner may for instance only check the straight line between two nodes. Other local planners are discussed and evaluated in [1].

Often the learning phase of PRM has a *node enhancement* step in order to increase the connectivity of the roadmap by adding more nodes in difficult regions of $\mathcal{F}$. Different techniques are used to identify these

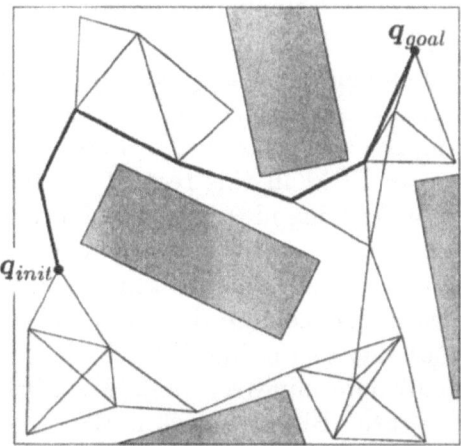

Figure 8.1 A random roadmap created by basic PRM in a simple two-dimensional configuration space with rectangular obstacles (grey). The shortest path is the thick line.

regions; one way is to distribute new points close to a number of *seeds* randomly selected among the existing nodes. In [24], the probability that a node is selected is proportional to $\frac{1}{1+b}$, where b is the number of edges connected to the node. An alternative selection can be based on a node's ratio of failed attempts by the local planner to find paths to other nodes [25]. Other techniques to increase the connectivity of the roadmap are described in [2] and [17].

PRM has shown to work well in practice in high-dimensional config-uration spaces, see [25]. Indeed, it is useful for multiple queries, since once an adequate roadmap has been created, queries can be answered very quickly.

4.2   VARIATIONS OF PRM

The idea of using randomization, and the simplicity and generality of randomized algorithms, have inspired further development of PRM and other closely related methods. A few algorithms similar to PRM do not divide the planning process into a learning phase and a query phase. Given an initial and a goal configuration, the planner in [36] inserts randomly distributed nodes in $\mathcal{F}$, one at a time, and connects them to the different components of the roadmap by a local planner. New nodes are inserted until the initial and goal configurations can be found in the same connected component of the roadmap. See also [13] and [21] for related algorithms. The latter paper gives an adaptive scheme for adjusting the power of the local planner.

Although the node enhancement step was developed to increase the connectivity of the roadmap, basic PRM still has weaknesses in finding paths through narrow passages in $\mathcal{F}$. Several recent approaches are intended to improve PRM in this respect by using different sampling strategies. The underlying idea is to distribute nodes close to the boundary of $\mathcal{F}$. The planner in [18] initially allows the robot to penetrate the obstacles to a certain extent. Small neighborhoods around the configurations just in collision are then re-sampled in order to place nodes close to the boundary of $\mathcal{F}$. The Obstacle Based PRM (OBPRM) in [2] and [3], repeatedly determines a configuration in collision to be the origin of a number of rays. Binary search is then used along each ray to find points on the boundary of $\mathcal{F}$, where roadmap nodes are placed. In [9], another idea is presented. The planner identifies the boundary of $\mathcal{F}$ by distributing points in pairs. Each pair is generated by first picking one point uniformly at random in $\mathcal{C}$, and then picking another point close to the first one. One of the points is added to the roadmap only if it is in $\mathcal{F}$ and the other point is not. Yet another technique to increase the number of nodes in narrow passages of $\mathcal{F}$ is presented in [41]. Points are picked uniformly at random in $\mathcal{C}$ and then retracted onto the medial axis of $\mathcal{F}$. Retraction to the medial axis of the workspace is done in [10].

Randomized approaches related to PRM are described in [19] and [30]. These build two trees rooted at the initial and goal configurations respectively. As soon as the trees intersect, a feasible path can be extracted. What differs these two methods is the way of expanding the trees. In [19], the trees are expanded by generating new nodes randomly in the vicinity of the two trees, and connecting them to the trees by a local planner. The planner in [30] iteratively generates a configuration, an attractor, uniformly at random in $\mathcal{C}$. Then, for both trees, the node closest to the attractor is selected and a local planner searches for a path of a certain maximum length towards the attractor. A new node is placed at the end of both paths. A new attractor is selected until the two trees intersect.

The algorithm in [29] is a method to keep the number of nodes in the roadmap to a minimum. Candidate nodes are generated uniformly at random, one at a time. A node is inserted to the roadmap only if it can be connected to at least two components of the roadmap, or if it does not see any other node. In the former case the components are merged, and in the latter case a new component is created. Variations of PRM have also been used for manipulation planning and for robots with closed kinematic chains, see [16, 31, 35].

5. DESIGNING AN EFFICIENT RANDOMIZED PLANNER USING A LAZY EVALUATION SCHEME

5.1 MOTIVATION

In many applications, the configuration space changes frequently. For example, as soon as the robot changes tools, grasps or deforms an object, or when a new obstacle enters the workspace, the feasible part $\mathcal{F}$ is affected. A planner useful in practice must be able to plan in new configuration spaces instantly, so long preprocessing must be avoided. Ideally, the time required for planning should relate to the difficulty of the planning task, i.e., a simple path in an uncluttered environment should be found quickly, while a more complicated path may require more time.

In a similar way, the planning time should relate to the desired quality of the solution path. The quality of a path is difficult to quantify (see further discussion in Section 5.4), but in general we prefer short paths in $\mathcal{C}$, with respect to some metric.

We would also like the planner to learn to some extent, i.e., to use information from previous queries in order to speed up subsequent queries. For example, if the algorithm finds a path through a narrow passage in $\mathcal{F}$, it should be able to use that information when searching for a new path back through the passage.

The general theme for roadmap algorithms is to construct a network of paths verified to be collision-free by a local planner. Unfortunately, it is difficult to find a global strategy that can use these local planners efficiently in order to avoid traps and dead ends. In environments with complex geometry and expensive collision checks, this often means that too much time is spent on planning local paths that will not appear in the final path. So, even though PRM fulfills most of the above requirements, it is too slow in many applications.

Our solution is to avoid using local planners as much as possible, and instead keep a global view through the entire planning process. In this section we present Lazy PRM – a path planning algorithm tailored for single queries in high-dimensional, relatively uncluttered configuration spaces. We address the problem of finding simple paths quickly in industrial environments with complex geometry. In these environments collision checking is computationally expensive, so to make the planner fast, the main theme is to minimize the number of collision checks.

5.2   OVERALL SCHEME OF LAZY PRM

This section describes a new algorithm for single query path planning. The algorithm is similar to the basic PRM in [25] in the sense that the aim is to find the shortest path in a roadmap generated by randomly distributed configurations. In contrast with existing PRMs, we do not build a roadmap of feasible paths, but rather a roadmap of paths *assumed* to be feasible. The idea is to lazily evaluate the feasibility of the roadmap as planning queries are processed.

In other words, let q_{init}, q_{goal}, and a number of uniformly distributed configurations form nodes in a roadmap. We connect by edges each pair of nodes being sufficiently close together. Lazy PRM finds a shortest feasible path in the roadmap by repeatedly searching for a shortest path, and then checking whether it is collision-free or not. Each time a collision occurs, the corresponding node or edge is removed from the roadmap, and then Lazy PRM searches for a new shortest path.

This procedure can terminate in either of two ways. If there exist feasible paths in the roadmap between q_{init} and q_{goal}, we will find a shortest one among them. Otherwise, if there is no feasible path, we will eventually find q_{init} and q_{goal} in two disjoint components of the roadmap. In the latter case, we can either report failure, or, if we still have time, add more nodes to the roadmap in a similar way to the node enhancement in [24, 25], and start searching again. A high-level description of the algorithm is given in Figure 8.2.

The point by using this scheme for lazy evaluation is that we only explore the part of the roadmap that is needed for the current query. The scheme is simple, general and can be applied also to other roadmap planners in order to increase performance. The strength is to either find a collision-free path or to conclude that none exists in the roadmap by using a small number of collision checks. It is always an advantage to use lazy evaluation since we can never do more work, in terms of collision checking, than basic PRM would do.

The rest of this section explains the different steps of the algorithm in more detail, Section 6 gives a proof of its probabilistic completeness, and Section 7 shows some experimental results.

5.3   BUILDING THE INITIAL ROADMAP

The first step in the algorithm is to build a roadmap $\mathcal{G}$ in $\mathcal{C}$. There are two parameters that determine the size of $\mathcal{G}$; the number of nodes, N_{init}, and the expected number of neighbors, M_{neighb}, connected to each node.

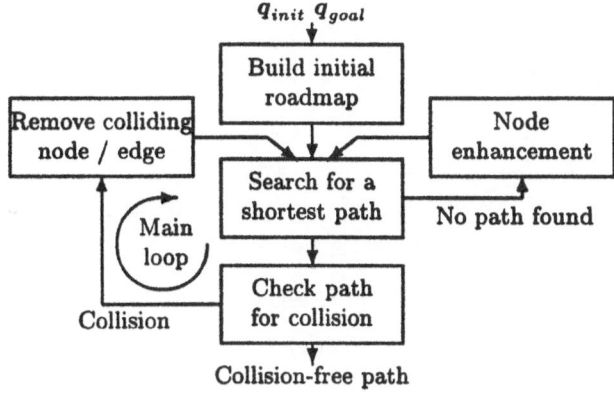

Figure 8.2 High-level description of Lazy PRM.

Initial Distribution of Nodes. Initially, we distribute N_{init} points uniformly at random in $\mathcal{C}$. These points, together with $q_{init} \in \mathcal{F}$ and $q_{goal} \in \mathcal{F}$, form nodes in $\mathcal{G}$. An important issue is the choice of N_{init}. The initial density of nodes, determined by N_{init}, is strongly correlated to the probability of finding a short path, if one exists. The correlation is hard to quantify, but the following example may give an illustration. Assume there exist only two ways to get to the goal configuration; either a short path through a rather narrow corridor, or a somewhat longer path through a wide corridor. If $\mathcal{G}$ is sufficiently dense, the algorithm will find a short path through the narrow passage, see Figure 8.3(b). If $\mathcal{G}$ is sparse, the algorithm will find a longer path through the wide passage, see Figure 8.3(a). In the worst case, if the roadmap is too sparse, there will be no feasible path at all in the roadmap, and the algorithm has to go to the enhancement step to generate more nodes. On the other hand, if N_{init} is too large, we will distribute more nodes than necessary, see Figure 8.3(c). Although we may obtain better paths, this will lead to somewhat longer planning times.

However, the idea behind the algorithm is that only a small fraction of the nodes in the roadmap will be necessary to check for collision. This makes the algorithm relatively insensitive to high density of nodes, so we can choose N_{init} relatively large. (In our experiments we start with $N_{init} = 10000$ nodes and check on average 322 nodes in the most difficult planning task, see Task $D \rightarrow E$ in Table 8.1(a), Section 7.) The number of nodes required to find a path is further explored in Section 6.

Selecting Neighbors. To build the roadmap we connect each node in $\mathcal{G}$ by edges to a set of neighbor nodes. An edge represents the straight

232

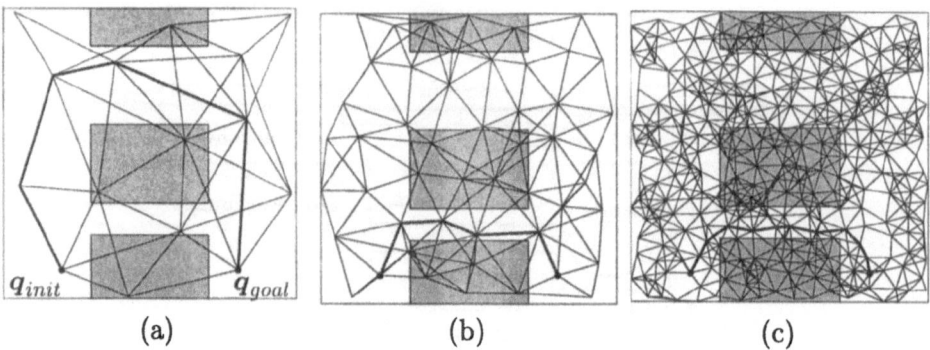

(a)   (b)   (c)

Figure 8.3 Example of a two-dimensional configuration space with rectangular obstacles (grey). The thick lines show the shortest feasible paths between q_{init} and q_{goal} in three different roadmaps. The roadmap in (a) is too sparse and no short feasible path exists. The roadmap in (c) is very dense, and the shortest feasible path will take longer time to find than the shortest feasible path in (b).

line path in C between two nodes. Neither the nodes nor the edges are being checked for collision in the initial step, but we want, of course, to have edges which are likely to be feasible. Since it would require far too much memory to connect all pairs of nodes, and it is unlikely that the straight line path between two nodes far apart is feasible, it is natural to only consider nodes which are sufficiently close together.

In order to select appropriate neighbors, we need a metric $\rho_{coll} : C \times C \to [0, \infty)$ such that the distance between two configurations under this metric reflects the difficulty of connecting them by a collision-free straight line path. Then we connect each pair of nodes (q, q') such that $\rho_{coll}(q, q') \leq R_{neighb}$. For any fixed radius R_{neighb}, the number of neighbors of a node is a random variable, so depending on the initial number of nodes N_{init}, we choose R_{neighb} such that the expected number of neighbors equals the parameter M_{neighb} introduced in the beginning of Section 5.3.

In many cases it is harder to make feasible connections in certain directions than in others. Consider for instance an articulated robot arm; then it is more likely that a collision occurs when the base joint is moving one unit, than if a joint close to the end-effector is moving one unit. With this in mind, we let ρ_{coll} be a weighted Euclidean metric,

$$
\begin{aligned}
\rho_{coll}(x, y) &= \Big(\sum_{i=1}^{d} w_i^2 (x_i - y_i)^2 \Big)^{1/2} \\
&= ((x - y)^T W (x - y))^{1/2}, \qquad (8.1)
\end{aligned}
$$

where d is the dimension of C, $\{w_i\}_{i=1}^d$ are positive weights, $W = diag(w_1^2, \ldots, w_d^2)$, and x^T is the transpose of x. The weights are chosen in proportion to the maximum possible distance (Euclidean distance in the workspace) traveled by any point on the robot, when moving one unit in C along the corresponding axis. This metric is easy to use and has been shown to work well in our experiments presented in Section 7.

5.4   SEARCHING FOR A SHORTEST PATH

The second step in the algorithm is to find a shortest path in G between q_{init} and q_{goal}, or determine that none exists. We use the A^* algorithm [33], and a metric $\rho_{path} : C \times C \to [0, \infty)$ to measure the length of a path and the remaining distance to q_{goal}.

If the search procedure succeeds in finding a path, we need to check it for collision. Otherwise, if no path exists in the roadmap, we either report failure, or go to the node enhancement step to add more nodes to the roadmap and start searching again depending on the overall time allowed to solve the problem.

Choosing an Appropriate Metric for A^*. The tool available to give preference to certain paths and reject others is the metric ρ_{path}. Thus, by defining this metric we decide which paths are assumed to be of high quality and which paths are assumed to be of poor quality.

In this paper we focus on articulated robots and use the Euclidean configuration space $I_1 \times \cdots \times I_d$, where I_i is the range of joint i and d is the number of dof. Thus, we do not identify angles equal modulo 2π as being equal, although they define the same position in the workspace. This is because a real robot in general has supply wires, etc., which otherwise would be entangled. The metric ρ_{path} is a weighted Euclidean metric, similar to (8.1), where the weights are equal to $\frac{1}{v_i}$, $i = 1, ..., d$, where v_i is the maximum angular velocity of joint i. This tends to give preference to paths with short execution time, which in many applications is the most interesting response variable.

In the general case, however, there are a large number of other response variables to consider. Some of them are measurable such as energy consumption, dynamic forces on joints, etc. Others are more subjective; for example, the motion should look natural and smooth from the user's point of view. Under any Euclidean metric, the straight line path in C between two configurations is the shortest, but considering all of these response variables, the straight line path is not necessarily optimal. Thus, the choice of a configuration space parameterization and an appropriate metric is an intricate task in itself.

5.5 CHECKING PATHS FOR COLLISION

When the A^* algorithm has found a shortest path in the roadmap between q_{init} and q_{goal}, we need to check the nodes and edges along the path for collision. In most applications it is straightforward to perform a collision check for a given configuration, i.e. determine whether a point is in $\mathcal{F}$ or not [11, 32, 39]. It is considerably more expensive to check whether a path segment is entirely in $\mathcal{F}$ or not. To keep the planner as simple and general as possible, we only use a collision checker for points in $\mathcal{C}$; path segments, i.e. edges in the roadmap, are discretized and checked with a certain resolution. However, in [6, 7] we describe a variation of the algorithm which makes use of a function giving the minimum distance between the robot and the obstacles.

The overall purpose of the Search, Check, and Remove steps of our algorithm (the main loop in Figure 8.2), is roughly to identify and remove colliding nodes and edges from the roadmap until the shortest path between q_{init} and q_{goal} is feasible. Accordingly, when checking a path for collision, we are not primarily interested in verifying whether an individual node or edge is in $\mathcal{F}$ or not, but rather to remove colliding nodes and edges as efficiently as possible. Since a removal of a node implies all its connected edges to be removed, it seems reasonable to check the feasibility of the nodes along the path before checking the edges.

Checking Nodes. Starting respectively with the first and the last node on the examined path and working toward the center, we alternately check the nodes along the path. As soon as a collision is found, we remove the corresponding node and its connected edges from the roadmap, and search for a new shortest path.

The reason for checking the nodes in this order is that the probability of having the shortest feasible path via a particular node is higher if the node is close to either q_{init} or q_{goal}. Consider, for instance, the nodes connected to q_{init}; a shortest feasible path (if one exists) must pass through at least one of them. Since, in a cluttered space, we cannot give preference to certain directions, the probability of having the shortest feasible path via a particular neighbor of q_{init} is at least $1/b$, where b is the number of neighbors of q_{init}. Nodes connected to q_{goal} have a similar probability, whereas nodes further away from both q_{init} and q_{goal} have a much lower probability of being in the shortest feasible path. Therefore, we check the nodes along a path starting from the end-nodes and working toward the center.

Checking Edges. If all nodes along the path are in $\mathcal{F}$, we start checking the edges in a similar fashion; working from the outside in. However, to minimize the risk of doing unnecessary collision checks, we first check all edges along the path with a coarse resolution, and then do stepwise refinements until the specified resolution is reached. As with the nodes, if a collision is found, we remove the corresponding edge, and search for a new shortest path. If no collision is found along the path, the algorithm terminates and returns the collision-free path. Figure 8.4 gives an illustration. To make the overall algorithm efficient, we record which nodes have been checked for collision, and to which resolution each edge has been checked, in order to avoid checking any point in $\mathcal{C}$ more than once.

The total number of collision checks depends on the resolution with which the edges along the path are checked. Again, since ρ_{coll} reflects the probability of collision, we determine the resolution with respect to this metric. The resolution is quantified by a step-size δ, but we prefer not to let the user specify the step-size by a certain number, because the resolution should depend on the scale of $\mathcal{C}$ and the weights defining the metric. A better way is to introduce a parameter M_{coll}, specifying the number of collision checks required to check the longest possible straight line path in $\mathcal{C}$. In other words, assuming that $\mathcal{C}$ is a d-dimensional rectangle and q and q' are two opposite corners, the step-size is related to the length of the diagonal of $\mathcal{C}$ according to

$$\delta = \frac{\rho_{coll}(q, q')}{M_{coll}}.$$

5.6   NODE ENHANCEMENT

If the search procedure in Figure 8.2 fails, no feasible path between q_{init} and q_{goal} exists in the roadmap, and more nodes are necessary in order to find one. In the node enhancement step, we generate N_{enh} new nodes, add them to $\mathcal{G}$, and select neighbors in the same way as when $\mathcal{G}$ was initially built.

We may not only distribute the new nodes uniformly, but rather use the information available in the roadmap (or what is left of the roadmap), in order to distribute new nodes in difficult regions of $\mathcal{C}$. In a method similar to the node enhancement in [24, 25], we randomly select a number of points in $\mathcal{G}$, called *seeds*, and then distribute a new point close to each of them. Our experience is that it is better to select many seeds and distribute one new node around each of them, instead of selecting few seeds and distribute several nodes around each of them; the latter method is more dependent on the selection of seeds.

Although the seeds may help us identify difficult regions of $\mathcal{C}$, we still want to maintain a smooth distribution all over $\mathcal{C}$, because the knowledge

236

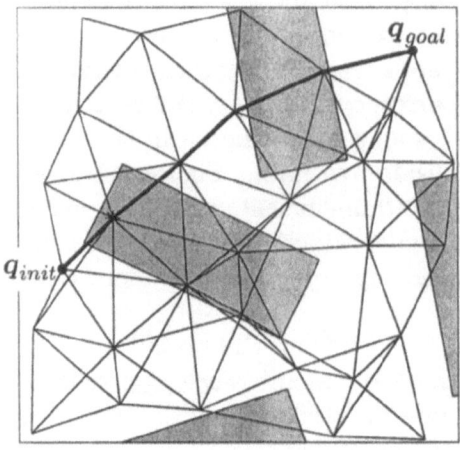

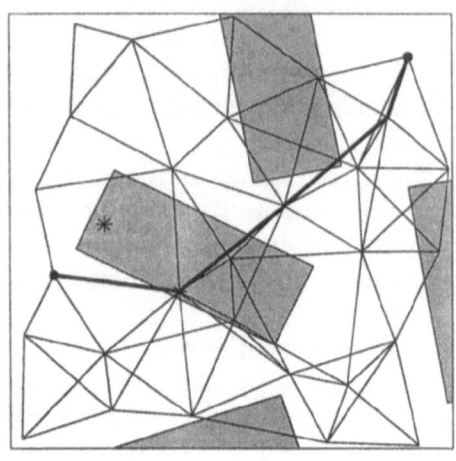

(a): Lazy PRM searches for a shortest path and checks the nodes. A collision is detected (∗) and corresponding node is removed from the roadmap.

(b): Then Lazy PRM searches for a new shortest path, detects a new collision (∗) and removes corresponding node.

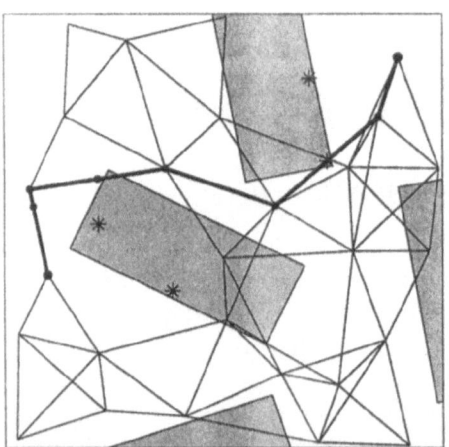

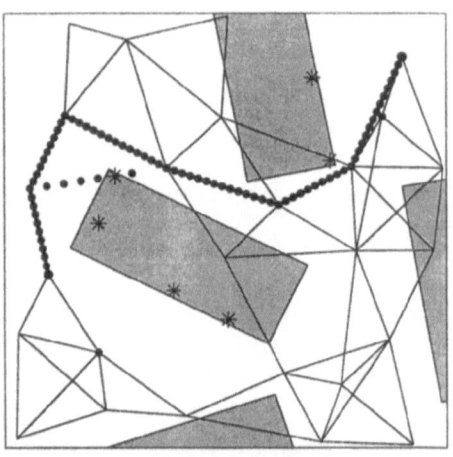

(c): After a few iterations, a sequence of feasible nodes is found. When checking the edges with a coarse resolution a collision is found (∗). The edge is removed from the roadmap, and the planner searches for a new shortest path.

(d): Eventually, the planner finds a path whose nodes are collision-free, and whose edges are collision-free to a specified resolution.

Figure 8.4 Example of a planning query in a two-dimensional configuration space with rectangular obstacles (grey). *All* collision checks performed are marked with ∗ . (collision) or • (collision-free).

about $\mathcal{C}$ is limited and we do not want to rely too much on the selection of seeds. To ensure probabilistic completeness (see Section 6), we also distribute new nodes uniformly at random in each step. In our algorithm, we let half of the enhancement nodes be uniformly distributed, and the rest distributed around seeds.

Selecting Seeds.   The set of edges which have been removed from the roadmap and have at least one end-point in $\mathcal{F}$ will certainly intersect the boundary of $\mathcal{F}$. Using the mid-points of these edges as seeds may help us distribute points close to the boundary of $\mathcal{F}$, thus increase the probability of finding paths through narrow passages in $\mathcal{F}$.

However, if the enhancement step is executed several times, this may cause problems with clustering of nodes. Assume that we add a new node q. This node will give rise to a number of edges which in the next enhancement step may increase the probability of adding even more nodes close to q. Thus, the distribution of new enhancement nodes depends on the preceding enhancement steps, and may eventually cause undesired clusters of nodes. To avoid this phenomenon, we only use edges whose end-nodes are generated uniformly at random when selecting seeds.

Distributing New Nodes.   When distributing a new point q around a seed η, we use the multivariate normal distribution. This distribution is smooth, easy to use, and allows us to control the distribution of q in terms of the metric ρ_{coll}. Hence, we can stretch the distribution in directions where the probabilities of making feasible connections are higher.

Introducing two parameters $\alpha \in (0,1)$ and $\lambda > 0$, we can choose the distribution such that

$$\rho_{coll}(\boldsymbol{q}, \boldsymbol{\eta}) \leq \lambda R_{neighb} \tag{8.2}$$

is an event with probability $1 - \alpha$, see Figure 8.5. R_{neighb} is the maximum length of an edge defined in Section 5.3. To achieve this property, we define a covariance matrix Σ as follows:

$$\Sigma = \frac{\lambda^2 R_{neighb}^2}{\chi_d^2(\alpha)} W^{-1}. \tag{8.3}$$

Here W is the same as in (8.1) and $\chi_d^2(\alpha)$ is the upper α percentile of a χ^2-distribution with d dof. Then we let the new point $\boldsymbol{q} \sim N_d(\boldsymbol{\eta}, \Sigma)$, i.e., q is multivariate normally distributed with d dof, mean $\boldsymbol{\eta}$, and covariance matrix Σ. Since Σ is diagonal, this simply means that each component $q_i, i = 1, \ldots, d$, of q is normally distributed with mean η_i and variance $\Sigma_{i,i}$.

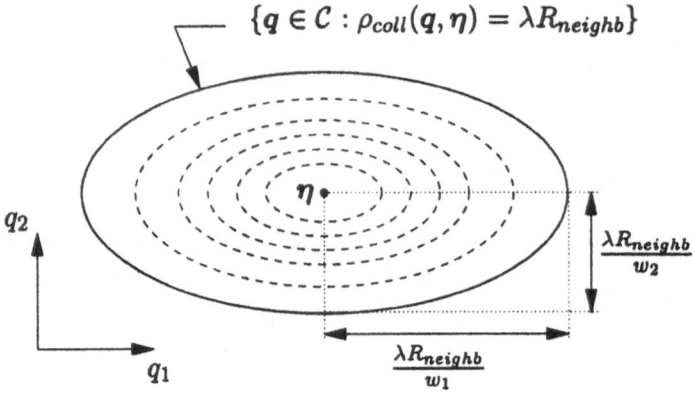

Figure 8.5 Example of a seed η in a two-dimensional configuration space. If a new point q is distributed according to $N_d(\eta, \Sigma)$, with Σ as in (8.3), then q is distributed within the confidence ellipse (solid line) with probability $1 - \alpha$. The dashed ellipses are contours of the distribution function. w_1 and w_2 are the weights defined in (8.1).

To show (8.2), we use that $(q - \eta)^T \Sigma^{-1} (q - \eta)$ is χ^2-distributed with d dof [22]. Thus, the event

$$(q - \eta)^T \Sigma^{-1} (q - \eta) \le \chi_d^2(\alpha)$$

has probability $1 - \alpha$. Using (8.1) and (8.3) gives the confidence ellipsoid in (8.2).

We see in (8.3) that Σ depends on the the ratio $\lambda^2/\chi_d^2(\alpha)$. Since both λ^2, $\lambda > 0$, and $\chi_d^2(\alpha)$, $\alpha \in (0, 1)$, are continuous functions whose ranges are $(0, \infty)$, one of the two parameters α and λ is redundant, so we can without loss of generality choose $\alpha = 0.05$. Then, the parameter λ controls the size of the 95% confidence ellipsoid relative to R_{neighb} as shown in Figure 8.5. In our experiments we found that $\lambda = 1$ is a suitable choice.

Another possibility of distributing the new point q, is to let it be uniformly distributed in a rectangular box centered at η. If we let the sides of the box be of equal length under ρ_{coll}, we stretch the box in a similar way as the ellipsoids above. In our path planning algorithm, however, the normal distribution has a major advantage compared to the uniform distribution; the contours of the distribution function are ellipsoids around η (see Figure 8.5). Hence, under the metric ρ_{coll}, which reflects the difficulty of making connections, the distribution is symmetric around η. In contrast, the uniform distribution favors the directions of the corners of the box, and nodes are more frequently distributed there than in other directions.

5.7 MULTIPLE QUERIES

When the planner has found a collision-free path, it terminates and returns the path. The information about which nodes and edges have been checked for collision is stored in the roadmap. As long as the configuration space remains the same, we use the same roadmap when processing subsequent queries. Thus, we benefit from the information obtained in each planning query. The new initial and goal configurations are simply added to the roadmap, and the same algorithm, except for the initial generation of nodes, is run again.

As several queries are processed, more and more of the roadmap will be explored, and the planner will eventually find paths via nodes and edges which have already been checked for collision. This makes the planner efficient for multiple queries.

Even in the long run, many nodes and edges may never be explored since they are located in odd regions of $\mathcal{C}$. Thus, given a fixed size of the roadmap, the number of collision checks performed by Lazy PRM will never exceed the number of collision checks performed by the basic PRM described in Section 4.1. Accordingly, there is no reason to entirely evaluate the roadmap unless we explicitly want it. The lazy evaluation scheme will find the shortest feasible path in the roadmap by using less collision checks.

6. PROBABILISTIC COMPLETENESS

In this section we give a proof of probabilistic completeness of Lazy PRM. Before stating the theorem, which also can be found in [7], we need some notation. Let $\gamma : [0, L] \to \mathcal{F}$ be a curve (also called path) parameterized by arc length and with continuous tangent. A *tube* τ of radius r around $\gamma(s)$ is the set of points at distance r from γ measured perpendicular to the tangent $\gamma'(s)$. Similarly, the corresponding *solid tube* is the set of points at distance $\leq r$ from γ. For simplicity, we usually omit the word solid.

A *regular tube* is a tube that does not intersect itself. If γ is enclosed by a regular tube of radius r, this particularly implies that its curvature, $\kappa(s) = |\gamma''(s)|$, is bounded from above by $1/r$. Otherwise the tube would be folded. The following lemma, proved in [14], states a useful property of regular tubes.

Lemma 1 *The volume enclosed by a regular tube around a curve in a d-dimensional Euclidean space is the product of the length of the curve and the $(d-1)$-dimensional area of a cross-section.*

In other words, if B_r^d is the ball of radius r in a d-dimensional space, and μ_d the Lebesgue measure, we can express the volume of a regular tube τ of radius r around γ as

$$\mu_d(\tau) = L\mu_{d-1}(B_r^{d-1}) = Lr^{d-1}\mu_{d-1}(B_1^{d-1}), \tag{8.4}$$

where L is the length of γ.

Assuming there exists a path between q_{init} and q_{goal}, enclosed by a regular tube in $\mathcal{F}$, the following theorem gives an upper bound on the probability of failure to find a path between q_{init} and q_{goal}. The assumption of an enclosing tube in $\mathcal{F}$ is relevant since $\mathcal{F}$ is an open subset of $\mathcal{C}$. Moreover, the theorem says that the probability of failure decreases exponentially in the *total* number of uniformly distributed nodes N. Since N increases in each enhancement step (Figure 8.2 and Section 5.6), the probability of failure vanishes as time tends to infinity. This is equivalent to the definition of *probabilistic completeness*, see [20]. Thus, Lazy PRM is a probabilistically complete path planner. Since the configuration space is at least two-dimensional (otherwise path planning is trivial), we assume that $d \geq 2$. Recalling the parameter R_{neighb} from Section 5.3 and the matrix W defined in (8.1) with norm $\|W\|$, we formulate the theorem as follows.

Theorem 1 *Let N be the total number of nodes generated uniformly at random in $\mathcal{C}$. If there exists a path γ between q_{init} and q_{goal}, enclosed by a regular tube τ of radius $R \leq \frac{1}{2\|W\|^{1/2}}R_{neighb}$ entirely in $\mathcal{F}$, then Lazy PRM will fail to find a path with probability at most*

$$\frac{Ld}{R}e^{-\beta N},$$

where $\beta = \frac{R^d \mu_{d-1}(B_1^{d-1})}{3d\, \mu_d(\mathcal{C})}$ and L is the length of γ.

Proof. Let $u = R/d$, $r = R(1 - 1/d)$, and $k = \lfloor L/u \rfloor$. The idea of the proof is to take a tube of radius r, divide it into $k - 1$ cells of length u, and calculate the probability of having at least one node in each cell. We will show that any two points in adjacent cells can be connected by a straight line, and that one node in each cell is enough for the planner to succeed. Assume first that $k \geq 2$. The case $k < 2$ is trivial and will be considered at the end of the proof.

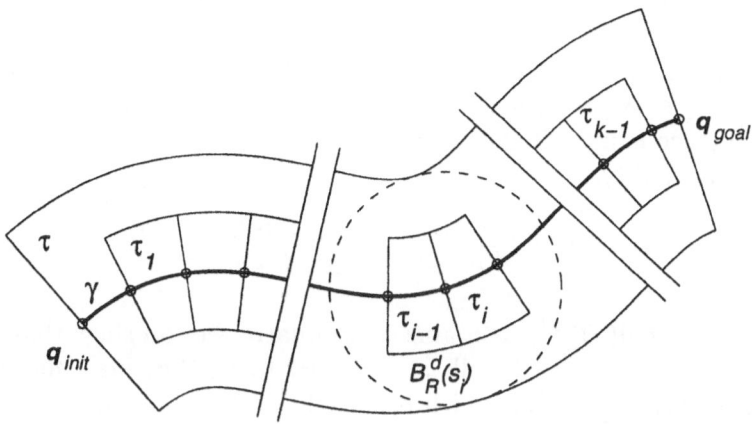

Figure 8.6   Illustration to the proof of Theorem 1.

Let $s_i = iu$, $i = 1, \ldots, k$, and let τ_i be the tube segment around $\gamma(s)$ for $s \in [s_i, s_{i+1})$, $i = 1, \ldots, k-1$, see Figure 8.6. The tube segments $\{\tau_i\}_{i=1}^{k-1}$ are pairwise disjoint and, by (8.4),

$$\frac{\mu_d(\tau_i)}{\mu_d(\mathcal{C})} = ur^{d-1} \frac{\mu_{d-1}(B_1^{d-1})}{\mu_d(\mathcal{C})}.$$

Now, for $d \geq 2$, $(1 - 1/d)^{d-1}$ is a decreasing function whose limit is e^{-1}, and since

$$ur^{d-1} = \frac{R^d}{d} \left(1 - \frac{1}{d}\right)^{d-1} \geq \frac{R^d}{d} e^{-1} \geq \frac{R^d}{3d},$$

we get that

$$\frac{\mu_d(\tau_i)}{\mu_d(\mathcal{C})} \geq \frac{R^d \mu_{d-1}(B_1^{d-1})}{3d \, \mu_d(\mathcal{C})} = \beta. \tag{8.5}$$

The N points generated by the algorithm are uniformly and independently distributed in $\mathcal{C}$. Thus, the probability that τ_i is empty equals $\left(1 - \frac{\mu_d(\tau_i)}{\mu_d(\mathcal{C})}\right)^N$, which, by (8.5), can be estimated:

$$\left(1 - \frac{\mu_d(\tau_i)}{\mu_d(\mathcal{C})}\right)^N \leq (1 - \beta)^N. \tag{8.6}$$

Let $B_R^d(s)$ be a ball of radius R centered at $\gamma(s)$, i.e., $B_R^d(s)$ has the same radius as τ. Unless $B_R^d(s)$ is close to the end-points of γ, it will be covered by τ, see Figure 8.6. If it is close to the end-points, however, it might intersect the circular discs at the ends of the tube. Nevertheless, the intersection between $B_R^d(s)$ and τ is still convex, a property we will need later.

Now, let $q_{i-1} \in \tau_{i-1}$ and $q_i \in \tau_i$. By the definition of a tube there exists an $\sigma_i \in [s_i, s_{i+1})$ such that $|q_i - \gamma(\sigma_i)| \leq r$. Since γ is parameterized by arc length, it follows that $|\gamma(s) - \gamma(t)| \leq |s - t|$, and, by the triangle inequality,

$$
\begin{aligned}
|q_i - \gamma(s_i)| &\leq |q_i - \gamma(\sigma_i)| + |\gamma(\sigma_i) - \gamma(s_i)| \\
&\leq r + u = R.
\end{aligned}
$$

Hence, the ball $B_R^d(s_i)$ contains τ_i. Similarly, we can show that it also contains τ_{i-1}. Since both cells are covered by τ, they are contained in the convex set $B_R^d(s_i) \cap \tau$ which is entirely in $\mathcal{F}$. Thus, q_{i-1} and q_i are at most $2R$ apart and the straight line between them lies entirely in $\mathcal{F}$. From (8.1) we get that

$$
\begin{aligned}
\rho_{coll}(q_{i-1}, q_i) &\leq |q_{i-1} - q_i| \|W\|^{1/2} \\
&\leq 2R \|W\|^{1/2} \\
&\leq R_{neighb},
\end{aligned}
$$

i.e., any node in τ_{i-1} is in the neighborhood of any node in τ_i and will therefore be interconnected by Lazy PRM. Moreover, since $q_{init} \in B_R^d(s_1)$ and $q_{goal} \in B_R^d(s_k)$, they will be connected to any node in τ_1 and τ_{k-1} respectively. Consequently, it is enough to have at least one node in each of the cells $\tau_1, \ldots, \tau_{k-1}$, in order for Lazy PRM to find a collision-free path between q_{init} or q_{goal}.

The probability of failure for our algorithm, $P_{failure}$, can now be estimated:

$$
\begin{aligned}
P_{failure} &\leq P(\text{some } \tau_i \text{ is empty}) \\
&\leq \sum_{i=1}^{k-1} P(\tau_i \text{ is empty}) \\
&\leq (k-1)(1-\beta)^N,
\end{aligned}
$$

where we used Boole's inequality and (8.6) in the second and third step respectively. Using that $k - 1 \leq Ld/R$ and $(1 - \beta)^N \leq e^{-\beta N}$ gives the desired estimation.

What remains is the case $k < 2$, i.e., $L < 2u = 2R/d \leq 2R$. Then both q_{init} and q_{goal} are contained in the convex set $B_R^d(L/2) \cap \tau$ which is entirely in $\mathcal{F}$. This guarantees that Lazy PRM will find the straight line path between q_{init} and q_{goal}, so the probability of failure is zero. □

Note that a related theorem regarding basic PRM can be found in [4] and [23]. Both theorems give a bound on the failure probability expressed in terms of, among other variables, the density of nodes. An important difference is that Lazy PRM has to reach a certain density of nodes in C, while basic PRM has to reach approximately the same density in $\mathcal{F}$. This seems like a weakness of our method, but looking at how the nodes in basic PRM are generated, we see that this is not the case. In order to reach the desired density in $\mathcal{F}$, basic PRM has to distribute nodes uniformly all over C and exclude those in collision. Consequently, for both algorithms to reach the same density, the number of nodes checked for collision in the learning phase of basic PRM has to be the same as the number of uniformly distributed nodes in Lazy PRM. So whether the density is specified in $\mathcal{F}$ or in C does not matter. The difference of practical significance is that Lazy PRM avoids checking *all* of the nodes for collision.

7.   EXPERIMENTAL RESULTS

In this section we present performance tests of Lazy PRM when applied to a six dof robot in a realistic industrial environment. The planner has been implemented in C++ as a plug-in module to RobotStudio[1] – a simulation and off-line programming software running under Windows NT. The collision checks are handled internally in RobotStudio. The experiments have been run on a PC with a 400 MHz Pentium II processor and 512 MB RAM. In all tests we let $N_{init} = 10000$, $M_{neighb} = 60$, $M_{coll} = 200$, and $N_{enh} = 500$.

7.1   PATH PLANNING TASKS

The test example is a part of a real manufacturing process in which an ABB 4400 robot is tending press breaking. Metal sheets are formed by the hydraulic press shown in Figure 8.7. In this particular example, plane sheets of metal are picked at a pallet, bent once at the hydraulic press, and then placed at another pallet.

The process is divided into several steps, and our aim is to automatically plan the unconstrained paths of the robot. We let A to E denote five different configurations shown in Figures 8.7 and 8.8. These are used as either initial or goal configurations in four planning tasks, denoted for

---

[1]RobotStudio is developed by ABB Robotics, Göteborg, Sweden.

Figure 8.7 The work cell used in the experiments. The robot is in its home configuration denoted by A.

example $A \to B$, where A is the initial configuration and B is the goal configuration.

The scenario is as follows. Starting from the home configuration A, the robot picks a sheet of metal from the pallet at B (task $A \to B$) and puts the sheet-metal at the press C (task $B \to C$). After the breaking, the robot grasps the sheet-metal at D, places the sheet-metal at the pallet E (task $D \to E$), and then returns to the home configuration A (task $E \to A$).

Note that during this series of steps, the configuration space changes several times. As soon as we grasp or place a sheet of metal, the collision-free part, $\mathcal{F}$, is changing. Accordingly, we have four different configuration spaces in which to plan, and we have to build one roadmap in each of them.

The results include the number of collision checks, the number of enhancement steps, and the planning time. The minimum, average, and maximum values, based on 20 consecutive runs for each task, are shown in Table 8.1(a). The average number of collision checks performed on nodes and edges respectively are presented, as well as the average number of collision checks performed on the collision-free paths that the planner returned. Since paths are checked for collision with a certain resolution

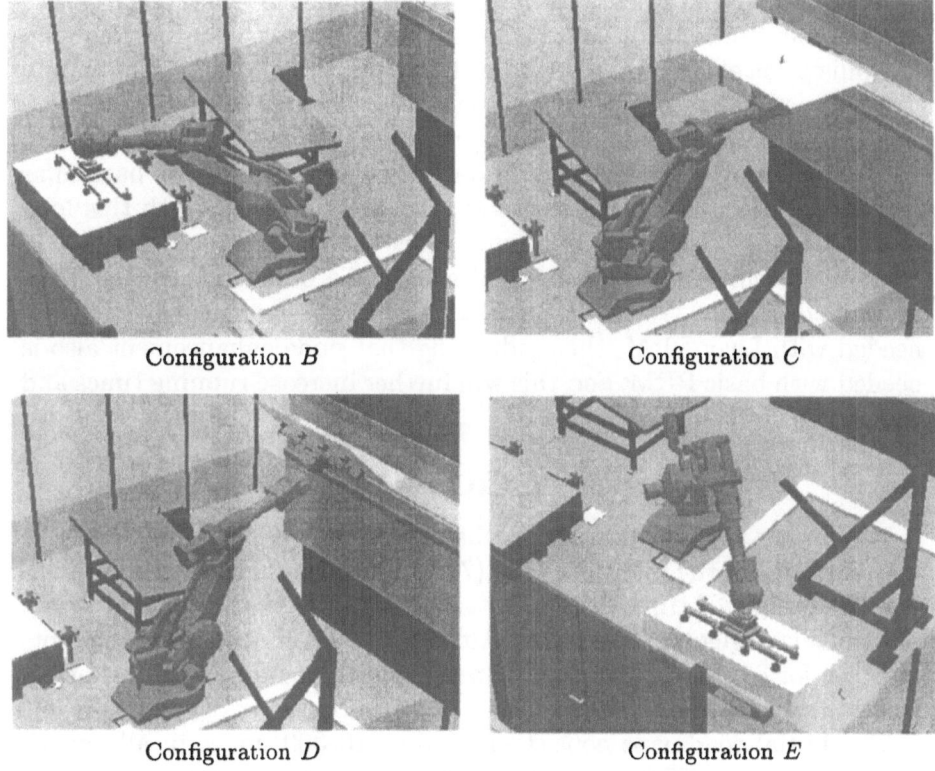

Configuration B Configuration C

Configuration D Configuration E

Figure 8.8 Configurations B to E used in the experiments.

(see Section 5.5), the latter figures correspond to the lengths of the collision-free paths.

The running times in Table 8.1(a) are divided into three parts. First, graph building, which includes distance calculations between nodes in $\mathcal{C}$ and node and edge adding, second, graph searching, and finally collision checking.

In the last column of Table 8.1, the average values of the recorded data are summed up. Thus, the last column indicates the average number of collision checks and average planning times for the entire press breaking operation.

In Table 8.1(b), we have included some results corresponding to the learning phase without node enhancement of basic PRM. For each task, we generated a roadmap in exactly the same way as Lazy PRM generates roadmaps in the initial step. Then we checked *all* nodes for collision, deleted the colliding ones, and then checked *all* of the remaining edges as described in Section 5.5. In other words, we checked the *entire* roadmap for collision as efficiently as possible. Due to the long running times, only one full roadmap was explored for each task. The result gives an indication of how large fraction of the roadmap that really has to be explored, and the amount of work saved by our lazy approach, in this particular example. Note this is a conservative estimate since even with this long preprocessing, there is no guarantee that the remaining roadmap will contain a feasible path. Table 8.1(b) shows whether a collision-free path was found or not. We see in 8.1(a) that several enhancement steps are needed with Lazy PRM, thus indicating that node enhancement also is needed with basic PRM, and this will further increase running times and the number of collision checks.

7.2   INTERPRETATION OF RESULTS

We clearly see in Table 8.1(a) that collision checking represents the vast majority of the planning time (79%), but also that the graph building takes a lot of time (19%). Interestingly, the time spent on graph searching is negligible, about 2%. Although we carefully select the points to check for collision by frequently searching the roadmap for a shortest path, the total time spent on that is very short.

The initial roadmaps consist of $N_{init} = 10,000$ nodes in all experiments. We see in Table 8.1(b) that the number of collision checks required to explore one entire roadmap is of order 500,000. Table 8.1(a) shows, on the other hand, that Lazy PRM in average solves the different planning tasks in 92 to 693 collision checks. Thus, Lazy PRM only explores a small fraction, less than 0.1%, of the roadmap. This is the strength of the algorithm; to either find a collision-free path or to conclude that none exists in the roadmap by using a small number of collision checks.

We also see in Table 8.1(a) that a large percentage, 27%, of the total number of collision checks are actually performed on the collision-free solution paths, and are therefore inevitable. This large percentage can be explained by two reasons. First, the algorithm finds a sequence of collision-free nodes before edges are being checked. This prevents from planning local paths in dead ends and in regions from where no way out exists. Secondly, we check the edges along the path starting from both ends with increasing resolution and stop as soon as a collision occurs. The colliding edge is removed from the roadmap, and a new shortest path is found. Thus, we avoid using a local planner and instead keep a global view throughout the planning process. As a consequence, very few edges – often only the edges along the final path – are checked with the finest resolution. This also makes the algorithm relatively insensitive to the resolution with which the paths are checked.

Since all of the nodes in the initial roadmap are uniformly distributed, the number of collision-free nodes found by basic PRM will give a good estimation of the relative size of $\mathcal{F}$. We see in Table 8.1(b) that for the tasks $A \to B$ and $E \to A$ approximately 40% of C is collision-free. For the other tasks approximately 30% of C is collision-free. As expected, the free part of C is reduced when the robot grasps a sheet of metal.

Furthermore, from the planner's point of view, the robot's tool includes both the gripper and possibly also a sheet of metal attached to it. If the tool is large and irregularly shaped, then its orientation becomes more important, whereas if the tool is small (e.g. the gripper only), the wrist motions of the robot, which basically determine the orientation, become less important. In this kind of environment, the planning problem is significantly easier if the tool is small. This explains why the tasks $A \to B$ and $E \to A$ are successfully planned without any node enhancement, and reveals the strength of our method in adapting to the difficulty of the problem.

8. DISCUSSION

The aim of Lazy PRM is essentially to minimize the number of collision checks while searching for the shortest feasible path in a roadmap in the context of a PRM planner. This is done on the expense of frequent graph search. For a complex robot working in a complex workspace, like our six dof example, collision checking is an expensive operation, and careful selection of the points being checked for collision reduces the planning time considerably.

However, if the robot and the obstacles have a very simple geometry, then collision checking is very fast. Frequent graph searching may, instead of speeding up the planning, become a bottleneck. Trading some collision checking for less graph searching may increase performance of Lazy PRM. So, instead of re-planning the entire path every time a collision is found, we can try to remove several nodes from the roadmap in each iteration of the main loop, see Figure 8.2. A simple way would be to always check *all* nodes along a path before searching for a new path.

Another modification of Lazy PRM is necessary when the configuration space is very cluttered. This is, for instance, the case with the ten dof robot in [24], where more than 99% of the configuration space is infeasible. If we run our algorithm, we would need a large number of nodes in the initial roadmap, and then remove from the roadmap ap-

Table 8.1 Performance data for Lazy PRM based on 20 consecutive runs for each task. Table 8.1(b) shows data for basic PRM based on one run for each task. The initial number of nodes, N_{init}, is 10000 in all tests.

			Task			
		$A{\to}B$	$B{\to}C$	$D{\to}E$	$E{\to}A$	*Total*
Lazy PRM						
Number of collision checks						
-for nodes	ave	9	209	322	18	558(37%)
-for edges	ave	83	387	371	124	965(63%)
	min	74	169	151	81	
-total	ave	92	596	693	142	1523
	max	131	1114	1301	299	
-for returned path	ave	78	136	117	82	413(27%)
Number of enh. steps						
	min	0	0	0	0	
	ave	0	1.3	1.7	0	
	max	0	3	4	0	
Running time (sec.)						
-graph building	ave	6.6	7.9	8.5	6.6	29.6(19%)
-graph searching	ave	0	0.9	3.0	0	3.9(2%)
-coll. checking	ave	6.1	45.7	62.4	11.6	125.8(79%)
	min	11.2	19.1	21.4	13.0	
-total	ave	12.7	54.5	73.9	18.2	159.3
	max	16.2	102.7	133.5	31.2	

Table 8.1(a).

	$A{\to}B$	$B{\to}C$	$D{\to}E$	$E{\to}A$
PRM				
Number of collision checks				
-for nodes	10000	10000	10000	10000
of which in $\mathcal{F}$	4085	2980	3041	4121
-for edges	763063	419613	446782	787507
-total	773063	429613	456782	797507
Running time (sec.)				
-total	56625	32088	35115	56234
Found feasible path	yes	yes	no	yes

Table 8.1(b).

proximately 99% of the nodes being checked, which would take a lot of time. Fortunately, we can easily modify Lazy PRM to check all nodes *before* we insert them into the roadmap. This would certainly cause unnecessary nodes to be checked for collision, but, on the other hand, we would save many inserting and removing operations in the roadmap. After that, we still have the efficient way of exploring the edges along paths. In this way, the lazy approach can be employed with most of the existing sampling schemes and variations of PRM discussed in Section 4.2.

Our primary interest in this project has been path planning in industrial environments, and the experimental results show that Lazy PRM works well in practice. By using either or both of the two modifications of the algorithm suggested above, we can tune the amount of graph search according to the application and the time required to perform a collision check, so that Lazy PRM becomes efficient for an even wider range of problems.

As far as future work is concerned let us note the following. Lazy PRM has essentially one parameter that is critical for the performance – N_{init}, the initial number of nodes. As indicated in Theorem 1, N_{init} is strongly correlated to the probability of finding a feasible path without using the node enhancement step. The optimal choice depends on the dimension of C, the workspace, the planning task, and the desired quality of the solution path. Our future work includes an investigation of the dependence between N_{init} and the planning time in different environments, as well as different distributions of the nodes.

Randomized techniques, like Lazy PRM, often give very fast planning. In Table 8.1(a), however, we can see that the maximum planning time is approximately twice as long as the average planning time. New improved enhancement techniques, in order to make the algorithms more robust in the sense that the worst case performance is improved, will also be a topic of our future research.

9.   SUMMARY

The advantage of considering the path planning problem in the configuration space is that even the most complicated robot is transformed into a single point. On the other hand, simple obstacles in the workspace generally become very complicated in the configuration space. Thus, we trade complex robots and simple obstacles for simple robots and complex obstacles. The new problem is essentially to explore an unknown space. But without a priori knowledge, or assumptions of the properties of the space, it is hard to construct powerful heuristic algorithms that can solve the problem. This is where randomized techniques have shown to be very useful, particularly in high-dimensional configuration spaces. Randomized planners (like PRM and Lazy PRM) are generally easy to implement and very efficient in exploring unknown environments, which make them popular and applicable to a wide variety of problems.

In this paper we further develop randomized planning techniques in the direction of achieving general and practically useful single query planners. We address standard industrial applications characterized by complex geometry and high-dimensional, relatively uncluttered configuration spaces. Our algorithm – called Lazy PRM – is based upon a general scheme for lazy evaluation of the feasibility of the roadmap. The scheme is simple and general and can be applied to any graph that needs to be explored. In addition to Lazy PRM, most other existing variations of PRM, and other related algorithms, can benefit from this scheme and significantly increase performance.

Acknowledgments

The authors would like to thank Bo Johansson and ABB Robotics for supporting the project and for providing suitable software. The core of this work was performed during the visit of Robert Bohlin to the Physical Computing Group at the Computer Science Department of Rice University. Robert Bohlin was supported by NUTEK, the Swedish National Board for Industrial and Technical Development, project P10499. Work on this paper by Lydia Kavraki was supported by NSF CAREER Award IRI-970228, NSF CISE SA1728-21122N and a Sloan Fellowship.

References

[1] N.M. Amato, O.B. Bayazit, L.K. Dale, C. Jones, and D. Vallejo. Choosing good distance metrics and local planners for probabilistic roadmap methods. In *Proc. IEEE Int. Conf. on Rob. & Aut.*, 1998.

[2] N.M. Amato, O.B. Bayazit, L.K. Dale, C. Jones, and D. Vallejo. OBPRM: An obstacle-based PRM for 3D workspaces. In P. K. Agarwal, L. E. Kavraki, and M. Mason, editors, *Robotics: The Algorithmic Perspective*, pages 630–637. AK Peters, 1998.

[3] N.M. Amato and Y. Wu. A randomized roadmap method for path and manipulation planning. In *Proc. IEEE Int. Conf. on Rob. & Aut.*, pages 113–120, 1996.

[4] J. Barraquand, L. E. Kavraki, J. C. Latombe, T.-Y. Li, R. Motwani, and P. Raghavan. A random sampling scheme for path planning. *Int. J. of Robotics Research*, 16(6):759–775, 1997.

[5] J. Barraquand and J.C. Latombe. Robot motion planning: A distributed representation approach. *Int. J. of Rob. Research*, 10:628–

649, 1991.

[6] R. Bohlin. *Motion Planning for Industrial Robots*. Licentiate thesis, Chalmers University of Technology, 1999.

[7] R. Bohlin and L.E. Kavraki. A lazy probabilistic roadmap planner for single query path planning. Submitted to Int. J. on Robotics Research.

[8] R. Bohlin and L.E. Kavraki. Path planning using lazy PRM. In *Proc. IEEE Int. Conf. on Rob. & Aut.*, 2000.

[9] V. Boor, M.H. Overmars, and F. van der Stappen. The Gaussian sampling strategy for probabilistic roadmap planners. In *Proc. IEEE Int. Conf. on Rob. & Aut.*, pages 1018–1023, 1999.

[10] L. Kavraki C. Holleman. A framework for using the workspace medial axis in PRM planners. In *Proc. IEEE Int. Conf. on Rob. & Aut.*, 2000.

[11] S. Cameron. Enhancing GJK: Computing minimum distance and penetration distanses between convex polyhedra. In *Proc. IEEE Int. Conf. on Rob. & Aut.*, pages 3112–3117, 1997.

[12] J.F. Canny. *The Complexity of Robot Motion Planning*. MIT Press, Cambridge, MA, 1988.

[13] B. Glavina. Solving findpath by combination of goal-directed and randomized search. In *Proc. IEEE Int. Conf. on Rob. & Aut.*, pages 1718–1723, 1990.

[14] A. Gray. *Tubes*. Addison-Wesley, Redwood City, CA, 1990.

[15] K. Gupta and A. P. del Pobil. *Practical Motion Planning in Robotics*. John Wiley, West Sussex, England, 1998.

[16] L. Han and N.M Amato. Kinematics-based probabilistic roadmap method for closed chain systems. In *Wokshop on the Algorithmic Foundations of Robotics*, 2000.

[17] T. Horsch, F. Schwarz, and H. Tolle. Motion planning for many degrees of freedom - random reflections at C-space obstacles. In *Proc. IEEE Int. Conf. on Rob. & Aut.*, pages 3318–3323, 1994.

[18] D. Hsu, L.E. Kavraki, J.C. Latombe, R. Motwani, and S. Sorkin. On finding narrow passages with probabilistic roadmap planners. In P. Agarwal, L. Kavraki, and M. Mason, editors, *Robotics: The Algorithmic Perspective*, pages 141–154. A K Peters, 1998.

[19] D. Hsu, J. C. Latombe, and R. Motwani. Path planning in expansive configuration spaces. In *Proc. IEEE Int. Conf. on Rob. & Aut.*, pages 2719–2726, 1997.

[20] Y.K. Hwang and N. Ahuja. Gross motion planning - a survey. *ACM Comp. Surveys*, 24(3):219–291, 1992.

[21] P. Isto. A two-level search algorithm for motion planning. In *Proc. IEEE Int. Conf. on Rob. & Aut.*, pages 2025–2031, 1997.

[22] R.A. Johnson and D.W. Wichern. *Applied Multivariate Statistical Analysis*. Prentice Hall, New Jersey, 1998.

[23] L.E. Kavraki, M.N. Kolountzakis, and J.C. Latombe. Analysis of probabilistic roadmaps for path planning. In *Proc. IEEE Int. Conf. on Rob. & Aut.*, pages 3020–3025, 1996.

[24] L.E. Kavraki and J.C. Latombe. Randomized preprocessing of configuration space for fast path planning. In *Proc. IEEE Int. Conf. on Rob. & Aut.*, pages 2138–2145, 1994.

[25] L.E. Kavraki, P. Švestka, J.C. Latombe, and M. Overmars. Probabilistic roadmaps for fast path planning in high dimensional configuration spaces. *IEEE Tr. on Rob. & Aut.*, 12:566–580, 1996.

[26] J.J Kuffner and J.C. Latombe. Fast synthetic vision, memory, and learning models for virtual humans. In *IEEE Computer Animation*, pages 118 –127, 1999.

[27] J.C. Latombe. *Robot Motion Planning*. Kluwer, Boston, MA, 1991.

[28] J.C. Latombe. Motion planning: A journey of robots, molecules, digital actors, and other artifacts. *Int. J. of Rob. Research*, 18(11):1119–1128, 1999.

[29] J.P. Laumond and T. Siméon. Notes on visibility roadmaps and path planning. In *Wokshop on the Algorithmic Foundations of Robotics*, 2000.

[30] S.M. LaValle and J.J. Kuffner. Randomized kinodynamic planning. In *Proc. IEEE Int. Conf. on Rob. & Aut.*, pages 473–479, 1999.

[31] S.M. LaValle, J.H. Yakey, and L.E. Kavraki. A proababilistic roadmap approach for systems with closed kinematic chains. In *Proc. IEEE Int. Conf. on Rob. & Aut.*, pages 1671–1676, 1999.

[32] M.C. Lin and J.F. Canny. A fast algorithm for incremental distance computation. In *Proc. IEEE Int. Conf. on Rob. & Aut.*, pages 1008–1014, 1991.

[33] G.F. Luger and W.A. Stubblefield. *Artificial intelligence and the design of expert systems*. Benjamin/Cummings, Redwood City, CA, 1989.

[34] A.H. Mishkin, J.C. Morrison, T.T. Nguyen, B.K. Cooper H.W. Stone, and B.H. Wilcox. Experiences with operations and

autonomy of the mars pathfinder microrover. In *IEEE Aerospace Conference*, pages 337 –351, 1998.

[35] C.L. Nielsen and L.E. Kavraki. A two level fuzzy PRM for manipulation planning. Technical Report TR2000-365, Rice University, 2000.

[36] M. Overmars. A random approach to motion planning. Technical Report RUU-CS-92-32, Utrecht University, the Netherlands, 1992.

[37] M. Overmars and P. Švestka. A probabilistic learning approach to motion planning. In K.Y. Goldberg, D. Halperin, J.C. Latombe, and R.H. Wilson, editors, *Algorithmic Foundations of Robotics*, pages 19–37. A K Peters, 1995.

[38] J. Reif. Complexity of the mover's problem and generalizations. In *Proc. 20th IEEE Symp. on Found. of Comp. Sci.*, pages 421–427, 1979.

[39] F. Thomas and C. Torras. Interference detection between non-convex polyhedra revisited with a practical aim. In *Proc. IEEE Int. Conf. on Rob. & Aut.*, 1994.

[40] R.Z. Tombropoulos, J.R. Adler, and J.C. Latombe. CARABEAMER: A treatment planner for a robotic radiosurgical system with general kinematics. *Medical Image Analysis*, 3(3):237–264, 1999.

[41] S.A. Wilmarth, N.M. Amato, and P.F. Stiller. MAPRM: A probabilistic roadmap planner with sampling on the medial axis of the free space. In *Proc. IEEE Int. Conf. on Rob. & Aut.*, pages 1024–1031, 1999.

Chapter 9

THE POWER OF TWO RANDOM CHOICES: A SURVEY OF TECHNIQUES AND RESULTS

Michael Mitzenmacher

Computer Science, Harvard University

*Cambridge, MA 02138,*michaelm@eecs.harvard.edu.

Andréa W. Richa

Department of Computer Science and Engineering

Arizona State University, Tempe

AZ 85287-5406, aricha@asu.edu.

Ramesh Sitaraman

Department of Computer Science, University of Massachusetts

*Amherst, MA 01003, and Akamai Technologies Inc.*ramesh@cs.umass.edu.

1. INTRODUCTION

To motivate this survey, we begin with a simple problem that demonstrates a powerful fundamental idea. Suppose that n balls are thrown into n bins, with each ball choosing a bin independently and uniformly at random. Then the *maximum load*, or the largest number of balls in any bin, is approximately $\log n / \log \log n$ with high probability.[1] Now suppose instead that the balls are placed sequentially, and each ball is placed in the least loaded of $d \geq 2$ bins chosen independently and uniformly at random. Azar, Broder, Karlin, and Upfal showed that in this case, the maximum load is $\log \log n / \log d + \Theta(1)$ with high probability [ABKU99].

The important implication of this result is that even a small amount of choice can lead to drastically different results in load balancing. Indeed, having just two random choices (i.e., $d = 2$) yields a large reduction in the maximum load over having one choice, while each additional choice

S. Rajasekaran et al (eds.), Handbook of Randomized Computing, Volume 1, pp. 255–312.

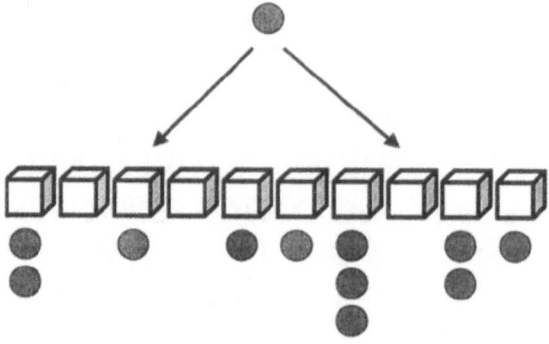

Figure 9.1 Two choices, the balls-and-bins model.

beyond two decreases the maximum load by just a constant factor. Over the past several years, there has been a great deal of research investigating this phenomenon. The picture that has emerged from this research is that the power of two choices is not simply an artifact of the simple balls-and-bins model, but a general and robust phenomenon applicable to a wide variety of situations. Indeed, this *two-choice paradigm* continues to be applied and refined, and new results appear frequently.

1.1   APPLICATIONS OF THE TWO-CHOICE PARADIGM

The two-choice paradigm and balls-and-bins models have several interesting applications. We outline a few here and we point out more applications in the succeeding sections.

Hashing.   Although the balls-and-bins models we discuss may appear simplistic, they have many interesting applications to hashing. In particular, the two-choice paradigm can be used to reduce the maximum time required to search a hash table. The standard hash table implementation [Knu73] uses a single hash function to map keys to entries in a table. If there is a collision, i.e., if two or more keys map to the same table entry, then all the colliding keys are stored in a linked list called a *chain*. Thus, each table entry is the head of a chain and the maximum time to search for a key in the hash table is proportional to the length of the longest chain in the table. If the hash function is *perfectly random* — i.e., if each key is mapped to an entry of the table independently and uniformly at random, and n keys are sequentially inserted into a table with n entries — then the length of the longest chain is $\Theta(\log n / \log \log n)$

with high probability. This bound follows from the analogous bound on the maximum load in the classical balls-and-bins problem where each ball chooses a single bin independently and uniformly at random.

Now suppose that we use *two* perfectly random hash functions. When inserting a key, we apply both hash functions to determine the two possible table entries where the key can be inserted. Then, of the two possible entries, we add the key to the shorter of the two chains. To search for an element, we have to search through the chains at the two entries given by both hash functions. If n keys are sequentially inserted into the table, the length of the longest chain is $\Theta(\log \log n)$ with high probability, implying that the maximum time needed to search the hash table is $\Theta(\log \log n)$ with high probability. This bound also follows from the analogous bound for the balls-and-bins problem where each ball chooses two bins at random.

In general, using multiple hash functions can be advantageous when the key parameter is the maximum number of keys located on a chain at a table entry. Such situations also arise naturally in practice when one hopes to fit each chain in a single cache line, as described for example in [BM00].

The two-choice paradigm applied to hashing has several advantages over other proposed hashing techniques (e.g., [BK90, DKM+88, FKS84]) in that it uses only two hash functions, it is easy to parallelize, and it is on-line (i.e., it does not involve re-hashing of data). Furthermore, it is not necessary to have perfectly random hash functions: similar results hold by choosing our hash functions randomly from smaller families of hash functions; see [KLM96].

Shared memory emulations on DMMs. One of the earliest applications of the two-choice paradigm is in the study of algorithms to emulate shared memory machines (as, for example, PRAMs) on distributed memory machines (DMMs) [CMS95, KLM96, MSS96]. In such emulations, the processors and the memory cells of the shared memory machine are distributed to the processors and memory modules of the DMM using appropriately chosen (universal) hash functions. Typically, the goal of the emulation algorithm is to minimize slowdown, or delay, of the emulation, which is the time needed by the DMM to emulate one step of the shared memory machine. Several of the balls-and-bins ideas and analysis are relevant in this context since minimizing slowdown involves orchestrating the communication between the processors (the balls) and the memory modules (the bins) so as to avoid memory contention, caused by several processors attempting to access the same memory module.

Load balancing with limited information. Another area where the two-choice paradigm has proven useful is the problem of dynamically assigning tasks to servers (e.g., disk servers or network servers). For simplicity, suppose that all the servers and all the tasks are identical, and that any task can be assigned to any server. Furthermore, suppose that the tasks arrive sequentially and need to be assigned to a server. Clearly, in the interest of response time of the tasks, we would like to keep the maximum load (where here load refers to the number of tasks) of any server as small as possible. Ideally, when a task arrives requesting a server, we would like to assign it to the least loaded server. However, complete information about the loads of all the servers may be expensive to obtain. For instance, querying a server for its load may involve sending a message and waiting for a response, processing an interrupt at the server, etc. An alternative approach that requires no information about the server loads is to simply allocate each task to a random server. If there are n tasks and n servers, using the balls-and-bins analogy, some server is assigned $\Theta(\log n / \log \log n)$ tasks with high probability. If instead each task obtains limited information by querying the load of two servers chosen independently and uniformly at random, and allocates itself to the least loaded of these two servers, then the maximum load on the n servers is only $\Theta(\log \log n)$ with high probability.

Low-congestion circuit routing. Many of the early applications of the two-choice approach have a distinct load balancing flavor. Cole et al. [CMM+98] show that the two-choice paradigm can be applied effectively in a different context, namely, that of routing virtual circuits in interconnection networks with low congestion. They show how to incorporate the two-choice approach to a well-studied paradigm due to Valiant for routing virtual circuits to achieve significantly lower congestion, as we discuss in Section 3..

1.2   A BRIEF HISTORY

We now provide a brief history of research on the two-choice paradigm. In the sections that follow, we discuss the results in more detail.

The earliest work we know that applies the two-choice paradigm to load balancing is that of Eager, Lazowska, and Zahorjan [ELZ86a]. The authors provide empirical evidence that a load balancing system based on allowing tasks to migrate to the least loaded of randomly selected processors improves performance. They also derive analytical results based on an appropriate Markov model. Their approach is related to fluid limit models. Recent work by Vvedenskaya, Dobrushin, and Karpelevich [VDK96] and Mitzenmacher [Mit96b, Mit96a] has led to an enduring technique for analysis of these load balancing systems based on fluid limit models, as described in Section 4..

The first rigorous analytical demonstration of the power of two choices is due to Karp, Luby, and Meyer auf der Heide [KLM92, KLM96], who considered the possibility of using two hash functions in the context of PRAM emulation by DMMs. Subsequent work on shared memory emulations on DMMs [CMS95, MSS96] has given rise to a powerful technique for analysis called the witness tree method. (See Section 3. for more details on this technique.)

The balls-and-bins problem has proven to be a fertile ground for investigating the power of two choices. The *classical balls-and-bins problem*, where each ball is thrown into a bin chosen independently and uniformly at random, has been studied for several decades [JK77]. Azar, Broder, Karlin, and Upfal [ABKU99] first considered the *sequential multiple-choice* balls-and-bins problem, where each ball chooses $d \geq 2$ bins independently and uniformly at random, and the balls are thrown sequentially into the least loaded of its d bin choices. This seminal paper introduced an important and intuitive technique for analyzing algorithms that use the two-choice paradigm, known as the layered induction method. In Section 2., we present this technique in more detail. Adler et al. [ACMR95] introduced the *parallel multiple-choice balls-and-bins problem* where each ball chooses $d \geq 2$ random bins independently and uniformly at random, but the balls must be assigned to bins in parallel by performing a limited number of rounds of communication. Since this survey focuses on results that show the power of having multiple choices in balls-and-bins problems, whenever we refer to a balls-and-bins problem in the remainder of this chapter, we will implicitly be referring to a *multiple-choice* balls-and-bins problem (i.e., a problem where each ball is assigned $d \geq 2$ random bin choices), unless we clearly state otherwise.

A balls-and-bins problem (such as those described above) where balls are inserted but never deleted from the system is referred to as a *static* problem. In a *dynamic problem*, balls can also be deleted from the system. Azar et al. [ABKU99] introduced a simple dynamic model for the sequential balls-and-bins problem, in which at each step a random ball is deleted and a new ball is inserted into the system; each time a ball is inserted, it is placed in the least loaded of two bins chosen independently and uniformly at random. An alternative to *random deletions* is *adversarial deletions* where an oblivious adversary decides on the sequence of insertions and deletions of the balls. (In this context, an *oblivious adversary* is one that specifies the sequence of insertions and deletions of the balls in advance, without knowledge of the random bin choices of the balls.) Only recently, this more powerful dynamic model has been analyzed [CMM+98, CFM+98]. Dynamic models have also been explored in connection with the parallel balls-and-bins problem. For instance, Adler et al. [ABS98] consider the situation where the balls are queued in first-in, first-out (FIFO) order at each bin and the first ball in each queue is deleted at each time step.

Finally, a *uniform balls-and-bins problem* (again, such as the ones described above) is a d-choice balls-and-bins problem where the d random choices assigned to a ball are independent and uniform. Vöcking [Vöc99] was the first to show how nonuniform ball placement strategies can help reduce the maximum bin load.

1.3 THE THREE MAJOR TECHNIQUES

One interesting aspect of the work in this rich area is that several different techniques have proven useful. The main techniques used to analyze balls-and-bins problems are layered induction, witness trees, and fluid limits via differential equations. Our survey is organized by successively discussing these three techniques, so that our focus is at least as much on the techniques as on the results obtained by using them. In fact, we demonstrate all of these approaches with examples. In presenting our survey in this manner, we hope to provide the reader with the necessary tools to pursue further research in this area.

In Section 2., we discuss the *layered induction* technique pioneered by Azar, Broder, Karlin, and Upfal [ABKU99]. In this approach, we bound the maximum load by bounding the number of bins with k or more balls via induction on k. The layered induction approach provides nearly tight results and a straightforward attack for handling balls-and-bins problems. It has proven effective for much more than the original problem studied in [ABKU99]. For example, as we explain, the layered induction approach can be used in a dynamic setting where an adversary inserts and deletes balls from the system over time.

In Section 3., we discuss an alternative technique for handling these problems called the *witness tree* method. The key idea of this approach is to show that if a "bad event" occurs — in our case, if some bin is heavily loaded — one can extract from the history of the process a suitable "tree of events" called the witness tree. The probability of the bad event can then be bounded by the probability of occurrence of a witness tree. Generally, witness tree arguments involve the most complexity, and they have proven to be the most challenging in terms of obtaining tight results. This complexity, however, yields strength: witness tree arguments tend to provide the strongest results, especially for dynamic settings that include both deletions and re-insertions of items. Furthermore, the more sophisticated uses of the two-choice paradigm in the design of communication protocols are more easily amenable to witness tree analyses.

In Section 4., we discuss a final technique, which studies algorithms that use the two-choice paradigm via *fluid limit models*. If one pictures the size (in this case, the number of bins) of the system growing to infinity, the resulting system can then be described by an appropriate family of differential equations. The fluid limit approach is more standard in the queueing theory literature, where it is used more widely, and it has proven especially useful for studying variations of the balls-and-bins problems that map naturally to queueing problems. A major weakness of the fluid limit approach is that even minor dependencies in the system can make the approach untenable. Also, its theoretical basis is at times incomplete; one must often return to a more concrete probabilistic argument to obtain the desired results. In many ways, however, it is the simplest and most flexible of the three methods. Moreover, when a problem can be put in this framework, the differential equations generally yield extremely accurate numerical results.

2.   THE LAYERED INDUCTION APPROACH

In this section, we address the results in the balls-and-bins literature that follow the layered induction approach introduced by Azar, Broder, Karlin, and Upfal in [ABKU99]. In this approach, we inductively bound the number of bins that contain at least j balls conditioned on the number of bins that contain at least $j - 1$ balls. Azar et al. show that, in the sequential d-choice balls-and-bins problem, the maximum load of a bin is $\log \log n / \log d + \Theta(1)$ with high probability. They also show that this bound is optimal (up to an additive constant term) among all the uniform multiple-choice placement strategies.

The layered induction approach in fact proved to be also useful in dynamic scenarios. For example, Azar et al. analyze the situation where at each step a random ball is deleted and a new ball is inserted in the system using layered induction [ABKU99]. Recently some progress was made in analyzing the behavior of the balls-and-bins problem under more realistic deletion scenarios. In [CFM+98], Cole et al. consider two natural situations: the particular case where balls that have been in the system for the longest time are deleted, and the more general case where an adversary specifies the sequence of insertions and deletions in advance. They show how to use layered induction arguments to provide simple proofs of upper bounds on the maximum bin load for these two scenarios. One of the main contributions of Cole et al. in [CFM+98] was to demonstrate how the layered induction techniques can yield interesting results in realistic deletion scenarios.

This section is organized as follows: in Section 2.1, we show the layered induction approach introduced by Azar et al. for the sequential balls-and-bins problem, and show how this approach can be modified to handle some extensions of the sequential problem; Section 2.2 shows how the layered induction approach can be adapted to actually prove lower bound results.

2.1 THE APPROACH

In this section, we describe the main results in the balls-and-bins literature that use layered induction for placing an upper bound on the maximum bin load. Our main goal is to make the reader understand the basic layered induction techniques in detail, so we start by presenting the simple layered induction argument for the sequential balls-and-bins problem due to Azar et al. [ABKU99]. Then we present other balls-and-bins results obtained by layered induction, and show how to modify the original argument of Azar et al. to hold for these results. The proof and notation we present here are very close to the original paper by Azar et al. We have made some minor changes in the interest of clarity.

Theorem 1 *Suppose that n balls are sequentially placed into n bins. Each ball is placed in the least full bin at the time of the placement, among d bins, $d \geq 2$, chosen independently and uniformly at random. Then after all the balls are placed, with high probability the number of balls in the fullest bin is at most $\log \log n / \log d + O(1)$.*

Azar et al. also show in [ABKU99] that the maximum bin load is at least $\log \log n / \log d - O(1)$ (see Section 2.2), proving that the maximum bin load for this problem is in fact equal to $\log \log n / \log d + \Theta(1)$.

Before presenting the proof, which is somewhat technical, we briefly sketch an intuitive analysis. For any given i, instead of trying to determine the number of bins with load *exactly* i, it is easier to study the number of bins with load *at least* i. The argument proceeds via what is, for the most part, a straightforward induction. Let the *height* of a ball be one more than the number of balls already in the bin in which the ball is placed. That is, if we think of balls as being stacked in the bin by order of arrival, the height of a ball is its position in the stack. Suppose we know that the number of bins with load at least i, over the entire course of the process, is bounded above by β_i. We wish to find a β_{i+1} such that, with high probability, the number of bins with load at least $i+1$ is bounded above by β_{i+1} over the course of the entire process with high probability. We find an appropriate β_{i+1} by bounding the number of balls of height at least $i + 1$, which gives a bound for the number of bins with at least $i + 1$ balls.

A ball has height at least $i+1$ only if, for each of the d times it chooses a random bin, it chooses one with load at least i. Conditioned on the value of β_i, the probability that each choice finds a bin of load at least i is $\frac{\beta_i}{n}$. Therefore the probability that a ball thrown any time during the process joins a bin already containing i or more balls is at most $\left(\frac{\beta_i}{n}\right)^d$. For $d \geq 2$, we can conclude that the sequence β_i/n drops at least quadratically at each step in the following manner. The number of balls with height $i+1$ or more is stochastically dominated by a Bernoulli random variable, corresponding to the number of heads with n (the number of balls) flips, with the probability of a head being $\left(\frac{\beta_i}{n}\right)^d$ (the probability of a ball being placed in a bin with i or more balls). We can find an appropriate β_{i+1} using standard bounds on Bernoulli trials, yielding $\beta_{i+1} \leq cn \left(\frac{\beta_i}{n}\right)^d$, for some constant c. The fraction $\frac{\beta_i}{n}$ therefore drops at least quadratically at each step, so that after only $j = O(\log \log n)$ steps the fraction drops below $1/n$, and we may conclude that $\beta_j < 1$. The proof is technically challenging primarily because one must handle the conditioning appropriately.

We shall use the following notation: the state at time t refers to the state of the system immediately after the tth ball is placed. $B(n, p)$ is a Bernoulli random variable with parameters n and p. The variable $h(t)$ denotes the height of the tth ball, and $\nu_i(t)$ and $\mu_i(t)$ refer to the number of bins with load at least i and the number of balls with height at least i at time t, respectively. We use ν_i and μ_i for $\nu_i(n)$ and $\mu_i(n)$ when the meaning is clear.

In preparation for the detailed proof, we make note of two elementary lemmas. The first statement can be proven by standard coupling methods:

Lemma 2 *Let $X_1, X_2, \ldots, X_n$ be a sequence of random variables in an arbitrary domain, and let $Y_1, Y_2, \ldots, Y_n$ be a sequence of binary random variables, with the property that $Y_i = Y_i(X_1, \ldots, X_{i-1})$. If*

$$\mathbf{Pr}(Y_i = 1 \mid X_1, \ldots, X_{i-1}) \leq p,$$

then

$$\mathbf{Pr}(\sum_{i=1}^{n} Y_i \geq k) \leq \mathbf{Pr}(B(n, p) \geq k);$$

and similarly, if

$$\mathbf{Pr}(Y_i = 1 \mid X_1, \ldots, X_{i-1}) \geq p,$$

then

$$\mathbf{Pr}(\sum_{i=1}^{n} Y_i \leq k) \leq \mathbf{Pr}(B(n,p) \leq k).$$

∎

The second lemma presents some useful Chernoff-type bounds; proofs may be found in [HR90].

Lemma 3 *If X_i $(1 \leq i \leq n)$ are independent binary random variables, $\mathbf{Pr}[X_i = 1] = p$, then the following hold:*

$$\text{For } t \geq np, \quad \mathbf{Pr}\left(\sum_{i=1}^{n} X_i \geq t\right) \leq \left(\frac{np}{t}\right)^t e^{t-np}. \tag{9.1}$$

$$\text{For } t \leq np, \quad \mathbf{Pr}\left(\sum_{i=1}^{n} X_i \leq t\right) \leq \left(\frac{np}{t}\right)^t e^{t-np}. \tag{9.2}$$

In particular, we have

$$\mathbf{Pr}\left(\sum_{i=1}^{n} X_i \geq enp\right) \leq e^{-np}, \quad \text{and} \tag{9.3}$$

$$\mathbf{Pr}\left(\sum_{i=1}^{n} X_i \leq np/e\right) \leq e^{(\frac{2}{e}-1)np}. \tag{9.4}$$

∎

Proof:[Proof of Theorem 1:] Following the earlier sketch, we shall construct values β_i so that $\nu_i(n) \leq \beta_i$, for all i, with high probability. Let $\beta_6 = \frac{n}{2e}$, and $\beta_{i+1} = \frac{e\beta_i^d}{n^{d-1}}$, for $6 \leq i < i^*$, where i^* is to be determined. We let $\mathcal{E}_i$ be the event that $\nu_i(n) \leq \beta_i$. Note that $\mathcal{E}_6$ holds with certainty. We now show that, with high probability, if $\mathcal{E}_i$ holds then $\mathcal{E}_{i+1}$ holds, for $6 \leq i \leq i^* - 1$.

Fix a value of i in the given range. Let Y_t be a binary random variable such that

$$Y_t = 1 \text{ iff } h(t) \geq i+1 \text{ and } \nu_i(t-1) \leq \beta_i.$$

That is, Y_t is 1 if the height of the tth ball is at least $i+1$ and at time $t-1$ there are fewer than β_i bins with load at least i.

Let ω_j represent the bins selected by the jth ball. Then

$$\mathbf{Pr}(Y_t = 1 \mid \omega_1, \ldots, \omega_{t-1}) \leq \frac{\beta_i^d}{n^d} \stackrel{\text{def}}{=} p_i.$$

Thus, from Lemma 2, we may conclude that

$$\mathbf{Pr}(\textstyle\sum_{i=1}^n Y_t \geq k) \leq \mathbf{Pr}(B(n, p_i) \geq k).$$

Conditioned on $\mathcal{E}_i$, we have $\sum Y_t = \mu_{i+1}$. Thus

$$
\begin{aligned}
\mathbf{Pr}(\nu_{i+1} \geq k \mid \mathcal{E}_i) &\leq \mathbf{Pr}(\mu_{i+1} \geq k \mid \mathcal{E}_i) \\
&= \mathbf{Pr}(\textstyle\sum Y_t \geq k \mid \mathcal{E}_i) \\
&\leq \frac{\mathbf{Pr}(\sum Y_t \geq k)}{\mathbf{Pr}(\mathcal{E}_i)} \\
&\leq \frac{\mathbf{Pr}(B(n, p_i) \geq k)}{\mathbf{Pr}(\mathcal{E}_i)}
\end{aligned}
$$

We bound the tail of the binomial distribution using Equation (9.3). Letting $k = \beta_{i+1}$ in the above, we have that

$$\mathbf{Pr}(\nu_{i+1} \geq \beta_{i+1} \mid \mathcal{E}_i) \leq \frac{\mathbf{Pr}(B(n, p_i) \geq e n p_i)}{\mathbf{Pr}(\mathcal{E}_i)} \leq \frac{1}{e^{p_i n} \mathbf{Pr}(\mathcal{E}_i)},$$

or that

$$\mathbf{Pr}(\neg \mathcal{E}_{i+1} \mid \mathcal{E}_i) \leq \frac{1}{n^2 \mathbf{Pr}(\mathcal{E}_i)}$$

whenever $p_i n \geq 2 \log n$.

Hence, whenever $p_i n \geq 2 \log n$, we have that if $\mathcal{E}_i$ holds with high probability, then so does $\mathcal{E}_{i+1}$. To conclude we need to handle the case where $p_i n \leq 2 \log n$ separately: we shall show that if this is the case, then with high probability there are no balls of height at least $i + 2$. Let i^* be the smallest value of i such that $\frac{\beta_i^d}{n^d} \leq \frac{2 \log n}{n}$. It is easy to check inductively that $\beta_{i+6} \leq n/2^{d^i}$, and hence that $i^* \leq \frac{\log \log n}{\log d} + O(1)$.

We have

$$\mathbf{Pr}(\nu_{i^*+1} \geq 6 \log n \mid \mathcal{E}_{i^*}) \leq \frac{\mathbf{Pr}(B(n, 2\log n/n) \geq 6 \log n)}{\mathbf{Pr}(\mathcal{E}_{i^*})} \leq \frac{1}{n^2 \mathbf{Pr}(\mathcal{E}_{i^*})},$$

where the second inequality again follows from Equation (9.3). Also,

$$
\begin{aligned}
\mathbf{Pr}(\mu_{i^*+2} \geq 1 \mid \mu_{i^*+1} \leq 6 \log n) &\leq \frac{\mathbf{Pr}(B(n, (6 \log n/n)^d) \geq 1)}{\mathbf{Pr}(\mu_{i^*+1} \leq 6 \log n)} \\
&\leq \frac{n(6 \log n/n)^d}{\mathbf{Pr}(\mu_{i^*+1} \leq 6 \log n)},
\end{aligned}
$$

where the second inequality comes from applying the crude union bound. We remove the conditioning using the fact that

$$\mathbf{Pr}(\neg\mathcal{E}_{i+1}) \le \mathbf{Pr}(\neg\mathcal{E}_{i+1} \mid \mathcal{E}_i)\mathbf{Pr}(\mathcal{E}_i) + \mathbf{Pr}(\neg\mathcal{E}_i),$$

to obtain that

$$\mathbf{Pr}(\mu_{i^*+2} \ge 1) \le \frac{(6 \log n)^d}{n^{d-1}} + \frac{i^* + 1}{n^2} = O\left(\frac{1}{n}\right),$$

which implies that with high probability the maximum bin load is less than $i^* + 2 = \log \log n / \log d + O(1)$. ∎

We now present some of the extensions of the sequential balls-and-bins problem which were analyzed using layered induction. For each of these extensions, we give a brief sketch on how to modify the argument in the proof of Theorem 1 to hold for the new balls-and-bins problem. We refer to the respective papers for the complete proofs.

We start by considering the extensions of the sequential problem which appear in [ABKU99]. Azar et al. consider the case when the number of balls may not be equal to the number of bins in the system. Let m denote the number of balls to be sequentially inserted into the n bins, where each ball makes d bin choices independently and uniformly at random, and is placed in the least filled of the d bins. Azar et al. show that the maximum bin load is now $(\log \log n / \log d)(1 + o(1)) + \Theta(m/n)$ with high probability. The major changes in the proof of Theorem 1 in order to hold for this case are in the definition of the values β_i and in the choice of the base case for our inductive process (in Theorem 1, we chose the base case to be $i = 6$). Here we let $\beta_x = n^2/(2em)$, for some convenient choice of base case x, and we require that $\mathbf{Pr}(\nu_x \ge \frac{n^2}{2em})$ holds with high probability. Then we define the variable β_{i+x} so as to be less than or equal to $\frac{n}{2^{d^i}}$, for all i, thus obtaining (using the same analysis as in the proof of Theorem 1) that

$$\mathbf{Pr}(\mu \ge x + \log \log n / \log d + 2) = o(1).$$

The main challenge needed to complete the proof is to show that x can be chosen to be $O(m/n) + o(\log \log n / \log d)$. Note that when $m \gg n$, the bound on the maximum bin load is asymptotically optimal. The heavily loaded case where $m \gg n$ was also recently studied in more detail in [BCSV00].

Azar et al. also consider a dynamic extension of the sequential problem in [ABKU99], as described in the following theorem:

Theorem 4 *Consider the infinite process where at each step, a ball is chosen independently and uniformly at random to be removed from the* *system, and a new ball is inserted in the system. Each new ball inserted in the system chooses $d \geq 2$ possible destination bins independently and uniformly at random, and is placed in the least full of these bins. This process may start at any arbitrary state, provided we have at most n balls in the system. For any fixed $T > n^3$, the fullest bin at time T contains, with high probability, fewer than $\log \log n / \log d + O(1)$ balls.*

The analysis of the case $d = 1$ for the infinite stochastic process defined in Theorem 4 is simple, since the location of a ball does not depend on the locations of any other balls in the system. Thus for $d = 1$, in the stationary distribution, with high probability the fullest bin has $\Theta(\log n / \log \log n)$ balls. The analysis of the case $d \geq 2$ is significantly harder, since the locations of the current n balls might depend on the locations of balls that are no longer in the system. By the definition of the process, the number of balls of height i cannot change by more than 1 in a time step. Hence the variable $\mu_{\geq i}(t)$ can be viewed as a random walk on the integers ℓ, $0 \leq \ell \leq n$. The proof of Theorem 4 is based on bounding the maximum values taken by the variables $\mu_{\geq i}(t)$ by studying the underlying process.

Only recently, Cole et al. [CFM+98] showed how to use layered induction to address the more realistic deletion scenarios in Theorem 5 below.

Theorem 5 *Consider the polynomial time process where in the first n steps, a new ball is inserted into the system, and where at each subsequent time step, either a ball is removed or a new ball is inserted in the system, provided that the number of balls present in the system never exceeds n. Each new ball inserted in the system chooses $d \geq 2$ possible destination bins independently and uniformly at random, and is placed in the least full of these bins. Suppose that an adversary specifies the full sequence of insertions and deletions of balls in advance, without knowledge of the random choices of the new balls that will be inserted in the system (i.e., suppose we have an oblivious adversary). If this process runs for at most n^c time steps, where c is any positive constant, then the maximum load of a bin during the process is at most $\log \log n / \log d + O(1)$, with high probability.*

Cole et al. show that the original argument of Azar et al. for the sequential balls-and-bins problem can in fact be made to hold in this dynamic scenario: the key difference between this result and that of [ABKU99] is that Azar et al. find a dominating distribution of heights on one set of n balls, whereas Cole et al. use a distribution that applies to every set of n balls present in the system as it evolves. As it happens, the bounds and the proof are essentially the same; the most significant changes lie in the end game, where we must bound the number of bins containing more than $\log \log n / \log d$ balls.

Cole et al. also consider a situation where items that have been in the system for the longest time are deleted, again using a variant of the layered induction argument in [ABKU99]. In this case initially $2n$ balls are inserted, and then repeatedly the oldest n balls are deleted and n new balls are inserted. This argument makes use of a two-dimensional family of random variables, similar in spirit to the work of [Mit00] (which we address in Section 4.). The bounds are the same as in Theorem 5, and hence the results are actually already implied by this theorem. However, the approach used in the proof for this specialized case may provide interesting results when applied to other problems, not only in the balls-and-bins domain. See [CFM+98] for the complete proofs.

Bounds on the recovery time. Suppose we start with a situtation where n balls are allocated to n bins in some *arbitrary* fashion. Now, consider the infinite process of Theorem 4 where at each time step a ball chosen independently and uniformly at random is deleted, and a new ball is inserted into the least loaded of d bins chosen independently and uniformly at random. How many time steps does it take for the system to approach steady-state (i.e., typical) behavior? More specifically, how many time steps does it take for the maximum load to be $\log \log n / \log d + O(1)$ with high probabilty? This quantity that is related to the mixing time of the underlying Markov process is called the *recovery time*. The recovery time quantifies the transient behavior of the system and is a useful measure of how quickly the system recovers from an arbitrarily bad configuration. It turns out that the bound of n^3 time steps for the recovery time in Theorem 4 is not tight. Czumaj and Stemann [CS97] provide a tight bound using a standard probabilistic tool called the coupling method [Lin92]. Specifically, they show that after $(1 + o(1))n \ln n$ steps the maximum load is $\log \log n / \log d + O(1)$ with high probability. The proof of this result was later simplified significantly by Czumaj [Czu98] via the use of the path coupling method[2] of Bubley and Dyer [BD97]. Czumaj also considers a variation of the infinite process in Theorem 4 where deletions are performed differently. In the new process instead of deleting a random ball, each deletion is performed by choosing a non-empty bin independently and uniformly at random and deleting a ball from the chosen bin. He shows that even though the new deletion process does not significantly affect steady-state behavior, the recovery time of the new process is at least $\Omega(n^2)$ and at most $O(n^2 \ln n)$, i.e., the recovery time of the new process is significantly larger.

2.2   HOW TO USE LAYERED INDUCTION TO PROVE LOWER BOUNDS

In this section, we illustrate how to use layered induction to prove lower bounds. We show how we can adapt the argument in the proof of Theorem 1 to provide a lower bound on the maximum number of balls in a bin for the sequential balls-and-bins problem. More specifically, a corresponding lower bound of $\log \log n / \log d - O(1)$ is presented, based on the following idea: first we bound the number of bins with load at least 1 after the $(n/2)$th ball in inserted, then we bound the number of bins of height 2 after the $(3n/4)$th ball, etc. This lower bound, combined with the results in Theorem 1, demonstrates that the maximum bin load for the sequential d-choice balls-and-bins problem is in fact $\log \log n / \log d + \Theta(1)$ with high probability. The proof is taken from [ABKU99].

Before proving this result, we note that Azar et al. actually proved that the greedy strategy is stochastically optimal among all possible multiple-choice uniform placement strategies [ABKU99]. (Recall that a d-choice uniform placement strategy is a placement strategy where all d random bin choices assigned to a ball are independent and uniform). Equivalently, the probability that the maximum height exceeds any value z for any uniform placement strategy based on d choices is smallest when the bin with the least number of balls is chosen. Hence their result is the best possible, for uniform placement strategies. (It is worth noting that Vöcking [Vöc99] uses a placement strategy that is not uniform to beat this lower bound, as we discuss in Section 3..) They show this by establishing a one-to-one correspondence between the possible results under the proposed greedy strategy and any other fixed strategy. This one-to-one correspondence matches results so that the maximum load for each possible result pair is smaller using Azar et al.'s greedy placement strategy. This is an example of a simple stochastic comparison; for more on this area, see [Sto83, SS94].

Theorem 6 *Suppose that n balls are sequentially placed into n bins. Each ball is placed in the least full bin at the time of the placement, among d bins, $d \geq 2$, chosen independently and uniformly at random. Then after all the balls are placed the number of balls in the fullest bin is at least $\log \log n / \log d - O(1)$ with high probability.*

Proof: Let $\mathcal{F}_i$ be the event that $\nu_{\geq i}(t) \geq \gamma_i$, where the variables γ_i are such that $\gamma_{i+1} < \gamma_i / 2$ (the variables γ_i will be revealed shortly). In fact, each $\gamma_i < n/2^{2^i}$. We want to upper bound

$$\mathbf{Pr}(\neg \mathcal{F}_{i+1} \mid \mathcal{F}_i).$$

Our goal is to show that, given $\mathcal{F}_i$, $\mathcal{F}_{i+1}$ holds with high probability.

We fix $i > 0$ and define the binary random variables Z_t for t in the range $R = [(1 - 1/2^i)n, (1 - 1/2^{i+1})n)$ so that

$$Z_t = 1 \text{ iff } h(t) = i + 1 \text{ or } \nu_{\geq i+1}(t - 1) \geq \gamma_{i+1}.$$

That is, the value Z_t is 1 if and only if the height of the tth ball equals $i+1$ or there are already γ_{i+1} bins with load at least $i+1$ at time $t-1$. Note that, as i increases, we consider the values of Z_t over shorter but further out time intervals. The intuition here is that in order to show that there are at least so many bins with load $i+1$ at time $(1-1/2^{i+1})n$, we start counting balls with that height from time $(1 - 1/2^i)n$; we wait until that point in time in order to ensure that there are sufficiently many bins with load i to make counting balls with height $i + 1$ worthwhile. We can get away with decreasing the amount of time we count balls as i increases, since the values γ_i decrease so fast.

Our definition of Z_t implies that as long as $\nu_{\geq i+1}(t - 1) \leq \gamma_{i+1}$, then $Z_t = 1$ precisely when all d choices have load at least i, and at least one of the d choices for the tth ball has load exactly i. Let ω_j represent the choices available to the jth ball. Then

$$\mathbf{Pr}(Z_t = 1 \mid \omega_1, \dots, \omega_{t-1}) \geq \frac{\gamma_i^d}{n^d} - \frac{\gamma_{i+1}^d}{n^d} \geq \frac{1}{2}\frac{\gamma_i^d}{n^d} \stackrel{\text{def}}{=} p_i.$$

Hence

$$\mathbf{Pr}\left(\sum_{t \in R} Z_t \leq k \mid \mathcal{F}_i\right) \leq \mathbf{Pr}(B(n/2^{i+1}, p_i) \leq k).$$

By choosing

$$\gamma_0 = n;$$
$$\gamma_{i+1} = \frac{\gamma_i^d}{2^{i+3}n^{d-1}} = \frac{n}{2^{i+3}}\left(\frac{\gamma_i}{n}\right)^d = \frac{1}{2}\frac{n}{2^{i+1}}p_i,$$

we may conclude that

$$\mathbf{Pr}(B(n/2^{i+1}, p_i) \leq \gamma_{i+1}) = o(1/n^2)$$

as long as $p_i n/2^{i+1} \geq 17\ln n$ by using a tail bound such as [AS92]

$$\mathbf{Pr}(B(N, p) < Np/2) < e^{-Np/8}.$$

Let i^* be the largest integer for which the tail bound holds. Clearly $i^* = \ln\ln n/\ln d - O(1) = \log\log n/\log d - O(1)$.

Now by the definition of Z_t, the event $\{\sum_{t \in R} Z_t \geq \gamma_{i+1}\}$ implies $\mathcal{F}_{i+1}$. Hence

$$\mathbf{Pr}(\neg \mathcal{F}_{i+1} \mid \mathcal{F}_i) \leq \mathbf{Pr}(\sum_{t \in R} Z_t < \gamma_{i+1} \mid \mathcal{F}_i) = o(1/n^2).$$

Thus for sufficiently large n

$$\begin{aligned}
\mathbf{Pr}(\mathcal{F}_{i^*}) &= \mathbf{Pr}(\mathcal{F}_{i^*} \mid \mathcal{F}_{i^*-1}) \cdot \mathbf{Pr}(\mathcal{F}_{i^*-1} \mid \mathcal{F}_{i^*-2}) \cdot \ldots \cdot \mathbf{Pr}(\mathcal{F}_1 \mid \mathcal{F}_0) \cdot \mathcal{F}_0 \\
&\geq (1 - 1/n^2)^{i^*} = 1 - o(1/n).
\end{aligned}$$

■

3. THE WITNESS TREE METHOD

Another powerful technique for analyzing balls-and-bins problems is the witness tree method. Suppose that we would like to bound the probability of the occurrence of some "bad event", such as the probability of the occurrence of a "heavily-loaded" bin. The key idea is to show that the occurrence of the bad event implies the occurrence of a "tree of events" called the witness tree. Thus, the probability that the bad event occurs is at most the probability that some witness tree occurs. The latter probability can in turn be bounded by enumerating all possible witness trees and summing their individual probabilities of occurrence.[3]

One of the earliest uses of the witness tree method occurs in the study of algorithms to emulate shared memory machines (as for example, PRAMs) on distributed memory machines (DMMs) [CMS95, MSS96]. Besides shared memory emulations, witness trees were independently discovered and used in the context of the parallel balls-and-bins problem [ACMR95].

3.1 THE SEQUENTIAL BALLS-AND-BINS PROBLEM

We start by providing a simple analysis of a variant of the sequential balls-and-bins problem using the witness tree technique. The proof provided here is adapted from [CFM+98], but all the essential ideas in the proof were used earlier in the analysis of randomized circuit-switching algorithms [CMM+98].

The problem that we wish to study can be described formally as a random process $Q_d(\vec{v}, \vec{w})$, where $\vec{v} = (v_1, v_2, \cdots)$, and $\vec{w} = (v_1, v_2, \cdots)$ are (infinite) vectors that specify the identity of the balls to be deleted and inserted respectively. The process begins with n insertions, where n is the total number of bins, followed by an alternating sequence of deletions and insertions specified by $\vec{v}$ and $\vec{w}$ respectively.[4] We assign each ball a unique ID number, and without loss of generality we assume the first n balls have ID numbers 1 through n. At time $n + j$, the ball with ID number v_j is deleted and then the ball with ID number w_j is inserted. If ball w_j has never been inserted before, then it is placed in the least loaded of d bins chosen independently and uniformly at random. If the ball has been inserted before, it is placed in the least loaded (at time $n + j$, after the deletion of ball v_j) of the d bins chosen when it was first inserted; that is, the bin choices of a ball are fixed when it is first inserted in the system. We assume that $\vec{v}$ and $\vec{w}$ are consistent, so there is only one ball with a given ID number in the system at a time. Note also that $\vec{v}$ and $\vec{w}$ must be chosen by the adversary before the process begins, without reference to the random choices made during the course of the process. For simplicity, we now consider only the special case $d = 2$.

Theorem 7 *At any time t, with probability at least $1 - 1/n^{\Omega(\log \log n)}$, the maximum load of a bin achieved by process $Q_2(\vec{v}, \vec{w})$ is $4 \log \log n$.*

Proof: We prove the theorem in two parts. First, we show that if there is a bin r at time t with 4ℓ balls, where $\ell = \log \log n$, then there exists a degree ℓ pruned witness tree. Next, we show that with high probability, no degree ℓ pruned witness tree exists.

Constructing a witness tree. A witness tree is a *labeled* tree in which each node represents a bin and each edge (r_i, r_j) represents a ball whose two bin choices are r_i and r_j. Suppose that some bin r has load 4ℓ at time t. We construct the witness tree as follows. The root of the tree corresponds to bin r. Let $b_1, \dots, b_{4\ell}$ be the balls in r at time t. Let r_i be the other bin choice associated with ball b_i (one of the choices is bin r). The root r has 4ℓ children, one corresponding to each bin r_i. Let $t_i < t$ be the last time b_i was (re-)inserted into the system. Without loss of generality, assume that $t_1 < t_2 < \dots < t_{4\ell}$. Note that the height of ball b_i when it was inserted at time t_i is at least i (since balls $b_1, \dots, b_{i-1}$ were already in bin r at time t_i). Therefore, the load of bin r_i, the other choice of b_i, is at least $i - 1$ at time t_i. We use this fact to recursively grow a tree rooted at each r_i.

The witness tree we have described is irregular. However, it contains as a subgraph an ℓ-ary tree of height ℓ such that

- The root in level 0 has ℓ children that are internal nodes.

- Each internal node on levels 1 to $\ell - 2$ has two children that are internal nodes and $\ell - 2$ children that are leaves.

- Each internal node on level $\ell - 1$ has ℓ children that are leaves.

For convenience we refer to this subtree as the actual witness tree henceforth.

Constructing a pruned witness tree. If the nodes of the witness tree are guaranteed to represent distinct bins, proving our probabilistic bound is a relatively easy matter. However, this is not the case; a bin may reappear several times in a witness tree, leading to dependencies that are difficult to resolve. This makes it necessary to *prune* the tree so that each node in the tree represents a distinct bin. Consequently, the balls represented by the edges of the pruned witness tree are also distinct. In this regard, note that a ball appears at most once in a pruned witness tree, even if it was (re-)inserted multiple times in the sequence.

We visit the nodes of the witness tree iteratively in *breadth-first* search order starting at the root. As we proceed, we remove (i.e., prune) some nodes of the tree and the subtrees rooted at these nodes – what remains is the pruned witness tree. We start by visiting the root. In each iteration, we visit the next node v in breadth-first order that has not been pruned. Let $B(v)$ denote the set of nodes visited *before* v.

- If v represents a bin that is different from the bins represented by nodes in $B(v)$, we do nothing.

- Otherwise, prune all nodes in the subtree rooted at v. Then, we mark the edge from v to its parent as a *pruning edge*.

Note that the pruning edges are not part of the pruned witness tree. The procedure continues until either no more nodes remain to be visited or there are ℓ pruning edges. In the latter case, we apply a final pruning by removing all nodes that are yet to be visited. (Note that this final pruning produces no new pruning edges.) The tree that results from this pruning process is the pruned witness tree. After the pruning is complete, we make a second pass through the tree and construct a set C of *pruning balls*. Initially, C is set to $\emptyset$. We visit the pruning edges in BFS order and for each pruning edge (u, v) we add the ball corresponding to (u, v) to C, if this ball is distinct from all balls currently in C and if $|C| \leq \lceil p/2 \rceil$, where p is the total number of pruning edges.

Lemma 8 *The pruned witness tree constructed above has the following properties.*

1. *All nodes in the pruned witness represent distinct bins.*

2. *All edges in the pruned witness tree represent distinct balls. (Note that pruning edges are not included in the pruned witness tree.)*

3. *The pruning balls in C are distinct from each other, and from the balls represented in the pruned witness tree.*

4. *There are $\lceil p/2 \rceil$ pruning balls in C, where p is the number of pruning edges.*

Proof: The first three properties follow from the construction. We prove the fourth property as follows. Let b be a ball represented by some pruning edge, and let v and w be its bin choices. Since v and w can appear at most once as nodes in the pruned witness tree, ball b can be represented by at most two pruning edges. Thus, there are $\lceil p/2 \rceil$ distinct pruning balls in C. ∎

Enumerating pruned witness trees. We bound the probability that a pruned witness tree exists by bounding both the number of possible pruned witness trees and the probability that each such tree could arise. First, we choose the shape of the pruned witness tree. Then, we traverse the tree in breadth-first order and bound the number of choices for the bins for each tree node and the balls for each tree edge; we also bound the associated probability that these choices came to pass. Finally, we consider the number of choices for pruning balls in C and the corresponding probability that they arose. Multiplying these quantities together yields the final bound – it is important to note here that we can multiply terms together only because all the balls in the pruned witness tree and the pruning balls in C are all distinct.

Ways of choosing the shape of the pruned witness tree. Assume that there are p pruning edges in the pruned tree. The number of ways of selecting the p pruning edges is at most

$$\binom{\ell^2 2^\ell}{p} \le \ell^{2p} 2^{\ell p},$$

since there are at most $\ell^2 2^\ell$ nodes in the pruned witness tree.

Ways of choosing balls and bins for the nodes and edges of the pruned witness tree. The enumeration proceeds by considering the nodes in BFS order. The number of ways of choosing the bin associated with the root is n. Assume that you are considering the ith internal node v_i of the pruned witness tree whose bin has already been chosen to be r_i. Let v_i have δ_i children. We evaluate the number of ways of choosing a distinct bin for each of the δ_i children of v_i and choosing a distinct ball for each of the δ_i edges incident on v_i and weight it by multiplying by the appropriate probability. We call this product E_i.

There are at most $\binom{n}{\delta_i}$ ways of choosing distinct bins for each of the δ_i children of v_i. Also, since there are at most n balls in the system at any point in time, the number of ways to choose distinct balls for the δ_i edges incident on v_i is also at most $\binom{n}{\delta_i}$. (Note that the n balls in the system may be different for each v_i; however, there are still at most

$\binom{n}{\delta_i}$ possibilities for the ball choices for any vertex.) There are $\delta_i!$ ways of pairing the balls and the bins, and the probability that a chosen ball chooses bin r_i and a specific one of δ_i bins chosen above is $2/n^2$. Thus,

$$E_i \leq \binom{n}{\delta_i}\binom{n}{\delta_i}\delta_i!\left(\frac{2}{n^2}\right)^{\delta_i} \leq (2e)^{\delta_i}/\delta_i!. \tag{9.5}$$

Let m be the number of internal nodes v_i in the pruned witness tree such that $\delta_i = \ell$. Using the bound in Equation 9.5 for only these m nodes, the number of ways of choosing the bins and balls for the nodes and edges respectively of the pruned witness tree weighted by the probability that these choices occurred is at most $n \cdot ((2e)^\ell/\ell!)^m$.

Ways of choosing the pruning balls in C. Using Lemma 8, we know that there are $\lceil p/2 \rceil$ distinct pruning balls in C. The number of ways of choosing the balls in C is at most $n^{\lceil p/2 \rceil}$, since at any time step there are at most n balls in the system to choose from. Note that a pruning ball has both its bin choices in the pruned witness tree. Therefore, the probability that a given ball is a pruning ball is at most

$$\binom{\ell^2 2^\ell}{2}\frac{2}{n^2} \leq \ell^4 2^{2\ell}/n^2.$$

Thus the number of choices for the $\lceil p/2 \rceil$ pruning balls in C weighted by the probability that these pruning balls occurred is at most

$$n^{\lceil p/2 \rceil}(\ell^4 2^{2\ell}/n^2)^{\lceil p/2 \rceil} \leq (\ell^4 2^{2\ell}/n)^{\lceil p/2 \rceil}.$$

Putting it all together. The probability at time t that there exists a pruned witness tree with p pruning edges, and m internal nodes with $\ell = \log\log n$ children each, is at most

$$\ell^{2p}2^{\ell p} \cdot n \cdot ((2e)^\ell/\ell!)^m \cdot (\ell^4 2^{2\ell}/n)^{\lceil p/2 \rceil} \leq n \cdot ((2e)^\ell/\ell!)^m \cdot (\ell^8 2^{4\ell}/n)^{\lceil p/2 \rceil}$$
$$\leq n \cdot (2e^2/\log\log n)^{m\log\log n} \cdot (\log\log^8 n \log^4 n/n)^{\lceil p/2 \rceil} \tag{9.6}$$

Observe that either the number the pruning edges, p, equals ℓ or the number of internal nodes with ℓ children, m, is at least $2^{\ell-2} = \log n/4$. Thus, in either case, the bound in Equation 9.6 is $1/n^{\Omega(\log\log n)}$. Furthermore, since there are at most $\ell^2 2^\ell$ values for p, the total probability of a pruned witness tree is at most $\ell^2 2^\ell \cdot 1/n^{\Omega(\log\log n)}$ which is $1/n^{\Omega(\log\log n)}$. This completes the proof of the theorem. ∎

A similar approach can be used to show that the maximum load of $Q_d(\vec{v}, \vec{w})$ is $O(\log\log n/\log d)$, with high probability, for arbitrary values of d. The witness tree method can be used to analyze several complex problems that are not easily amenable to layered induction or fluid limit models. The analysis presented above of the sequential balls-and-bins problem with adversarial insertions and deletions is a good example of such a problem. However, due to their enumerative nature, it is difficult (though often possible) to obtain the best constants using witness tree arguments. For instance, the layered induction technique can be used to provide a tighter high-probability bound of $\log\log n/\log d + O(1)$ on the maximum load of $Q_d(\vec{v}, \vec{w})$, even though the analysis holds only when deletions are performed by removing a random ball currently in the system [ABKU99].

Extensions. The basic sequential balls-and-bins problem and the multiple-choice approach has been extended in several natural ways. We review two such extensions that are insightful and perhaps counter-intuitive.

Czumaj and Stemann [CS97] consider the multiple-choice approach with a small twist. Suppose you throw n balls into n bins sequentially, and each ball chooses d bins independently, uniformly and at random. Now, suppose that when a ball is thrown into one of its d chosen bins you are allowed to reallocate the balls in the d chosen bins so that the loads in these bins are as evenly balanced as possible, i.e., their loads differ at most by 1 after the rebalancing. Does this rebalancing decrease the maximum load? If so, by how much? Czumaj and Stemann show that even though the maximum load of the rebalancing algorithm is smaller than that of the original multiple-choice algorithm, the difference is no more than an additive constant! In particular, the maximum load of the rebalancing algorithm is also $\log\log n/\log d + \Omega(1)$, with high probability. Thus, rebalancing produces no significant additional benefit.

Vöcking [Vöc99] considers a variation of the multiple-choice method where each ball makes d independent but *nonuniform* choices. In particular, the bins are divided into d groups with n/d bins each, and each ball makes its ith choice uniformly from the bins in the ith group, for $1 \le i \le d$. As before, the ball is placed in a bin with the smallest number of balls. (If there are several bins with the smallest number of balls, we choose one of them randomly.) Does this make any difference to the minimum load? One can show that the maximum load is still $\Theta(\log\log n/\log d)$, with high probability, using a witness tree argument that is similar to the proof of Theorem 7.

Now, Vöcking considers an additional twist. Suppose the balls choose bins independently in the nonuniform manner described above, and in addition, we introduce the following tie-breaking rule called "always-go-left". The always-go-left rule states that a ball must be placed in the bin with the minimum load of its d choices, and if there are several bins with the smallest load it must be placed in the *leftmost* of these bins. Now, what happens to the maximum load? At first glance, it may appear that the tie-break rule should not make a big difference, and it should if anything increase the load. But, surprisingly, the combination of the nonuniform choices and always-go-left rule actually *decreases* the maximum load to $\frac{\ln \ln n}{d \cdot \ln \phi_d} + O(1)$ with high probability, where here ϕ_d corresponds to the exponent of growth for a generalized Fibonacci sequence. (For reference, $\phi_2 = 1.61 < \phi_3 = 1.83 < \phi_4 = 1.92 \ldots < 2$.) It should be pointed out that if the balls make independent uniform choices, any tie-breaking rule including always-go-left, does not make a difference; i.e., the maximum load is still $\Theta(\log \log n / \log d)$ with high probability [ABKU99].

In view of these results, it is natural to ask if there is a method of choosing the d bins and a rule for allocating each ball to one of its chosen bins that provides an even smaller maximum load. Vöcking shows that no significant decrease in the maximum load is possible. In particular, he shows that if each ball chooses its d bins according to an *arbitrary* distribution on $[n]^d$ and the ball is placed in one of its chosen bins using an *arbitrary* rule, the maximum load is $\frac{\ln \ln n}{d \cdot \ln \phi_d} - O(1)$, with high probability.

3.2 THE PARALLEL BALLS-AND-BINS PROBLEM

In this section, we illustrate how to use the witness tree approach to analyze collision protocols for the parallel balls-and-bins problem. The parallel version of the balls-and-bins problem was first studied by Adler et al. [ACMR95]. Unlike the sequential case where balls are thrown into bins one after another, we consider the situation when n balls choose each d bins independently and uniformly at random, in parallel. The balls choose their final destinations by performing ρ rounds of communication. Each round consists of two stages. In the first stage each ball can send messages, in parallel, to any of its d chosen bins. In the second stage, each bin can respond by sending messages, in parallel, to any of the balls from which it received a message.

A natural class of protocols for the parallel balls-and-bins problem is the class of *collision protocols*. Collision protocols have been used widely for contention resolution in message routing [KLM96, GMR94, MPR98, CMS95, MSS96]. Such protocols were first used for the parallel balls-and-bins problem in [ACMR95]. The algorithm we present here is due to Stemann [Ste96] and can be described as follows. We set a threshold τ such that each bin can accept no more than a total of τ balls during the entire process — i.e., τ is the maximum load of any bin. The collision protocol proceeds as follows. (For simplicity, we study the case when $d = 2$.)

- In parallel each ball picks two bins independently and uniformly at random.

- While there is a ball that has not been allocated, do the following.

  - In parallel, each unallocated ball sends a request to its two chosen bins.

  - In parallel, each bin that would have load at most τ if it accepted all balls requesting the bin in that round sends an acknowledgment to all the requesting balls. (A bin that would achieve a load greater than τ does nothing.)

  - Each ball that receives an acknowledgment is allocated to the respective bin (ties are broken by randomly selecting one of the bins that sent an acknowledgment).

We illustrate how we can analyze this simple protocol using the witness tree method.

Theorem 9 *For any $1 \leq \rho \leq \log \log n$, the collision protocol described above with threshold $\tau = O(\sqrt[\rho]{\frac{\log n}{\log \log n}})$ finishes after ρ rounds with probability at least $1 - \frac{1}{n^{\Omega(1)}}$.*

Proof Sketch: As in the proof of Theorem 7, we start by building a witness tree. Suppose that there are unallocated balls at the end of round ρ. This implies that some bin r received more than τ requests in the ρth round. The root of the tree corresponds to bin r. Let $b_1, \ldots, b_{\tau+1}$ be the balls that sent a request to r in round ρ. (We assume that the balls b_i are ordered in the ascending order of their IDs.) For each $1 \leq i \leq \tau + 1$, both of the bin choices of b_i received at least $\tau + 1$ requests in round $\rho - 1$. Let r_i be the other bin choice associated with ball b_i (one of the choices is bin r). The root r has $\tau + 1$ children, one corresponding to each bin r_i. Now, we use that fact that each bin r_i had $\tau + 1$ requests in round $\rho - 1$ to recursively grow a depth-$(\rho - 1)$ tree rooted at each r_i. Thus, we have constructed a complete $\tau + 1$-ary tree of depth ρ as our witness tree.

The next step is to enumerate all possible witness trees and prove that the probability that some witness tree occurs is at most $1/n^{\Omega(1)}$. It is instructive to first consider the situation where all the nodes in the witness tree represent distinct bins. In this situation, the enumeration proceeds as follows. Let m be the number of nodes in the witness tree.

- The number of ways of choosing a distinct bin for each node of the tree is at most $n \cdot (n-1) \cdots (n-m+1) = n^{\underline{m}} \leq n^m$.

- The number of ways of choosing distinct balls for each of the $\tau + 1$ edges of an internal node of the tree is $\binom{n}{\tau+1} \leq n^{\tau+1}/(\tau+1)!$. Note that once the balls are chosen, they are paired up in the ascending order (from left to right) of their IDs, i.e., there is only one way of pairing them up with the bins. Since at least $\frac{m-1}{\tau+1}$ nodes of the tree are internal nodes, the total number of ways of labeling each edge of the tree with balls is $n^{m-1}/((\tau+1)!)^{\frac{m-1}{\tau+1}}$.

- Once the entire tree is labeled with balls and bins, the probability that the labeled tree occurs is at most $\left(\frac{2}{n^2}\right)^{m-1}$, since there are $m-1$ edges and since each edge corresponds to the event that a particular ball chooses two particular bins.

Putting it all together, the probability of occurrence of a witness tree (provided each node represents a distinct bin) is at most

$$n^m \cdot \frac{n^{m-1}}{((\tau+1)!)^{\frac{m-1}{\tau+1}}} \cdot \left(\frac{2}{n^2}\right)^{m-1} \leq \frac{n \cdot 2^{m-1}}{((\tau+1)!)^{\frac{m-1}{\tau+1}}}.$$

Observing that $m = \frac{(\tau+1)^{\rho+1}-1}{\tau}$, and setting $\tau = c\sqrt[\rho]{\frac{\log n}{\log \log n}}$, for a suitably large constant c, the above bound is at most $1/n^{\Omega(1)}$.

Unfortunately, the above simplified analysis does not always hold since the bins in the witness tree may not be distinct. We resolve this problem in a manner similar to the proof of Theorem 7. We prune the witness tree so that the resulting tree contains only bins that are distinct (and, hence, the balls are distinct also). If there are are too "few" pruning edges, then there exists a "large" subtree where the bins are distinct. In this case, we perform an analysis similar to the one outlined above to derive the bound. Otherwise, if there are a "large" number of pruning edges, both the random choices of the balls corresponding to the pruning edges fall within the set of bins in the witness tree. Since the size of the witness tree is small compared to n, i.e., the bins in the witness tree are a small subset of all the bins, it is unlikely that there are a "large" number of pruning edges. Thus, the probability that a witness tree exists is small in either case. ∎

A consequence of Theorem 9 is that the collision protocol achieves a maximum load of $O\left(\sqrt{\frac{\log n}{\log \log n}}\right)$ in two rounds, and a maximum load of a constant in $O(\log \log n)$ rounds, with high probability.

The basic parallel balls-and-bins problem can be extended in various natural ways. We now look at some of these extensions.

Weighted Balls. Berenbrink et al. [BMS97] generalize the parallel balls-and-bins problem to the situation where m balls are thrown into n bins in parallel, and each ball has a *weight* associated with it. The load of any bin in the weighted version of the problem is the sum of the weights of the balls allocated to that bin. They show that a more sophisticated collision protocol achieves a maximum load of $O\left(\frac{m \cdot w_{avg}}{n} + w_{max}\right)$ in

$$O\left(\frac{\log\log n}{\log\left(\frac{m \cdot w_{avg}}{n \cdot w_{max}} + 1\right)}\right)$$ rounds, with high probability, where w_{avg} and w_{max} are the average and maximum weight of a ball respectively. Note that by the pigeonhole principle, some bin receives a load of at least $\frac{m \cdot w_{avg}}{n}$, and the bin with the maximum-weighted ball has load at least w_{max}. Thus, the protocol of Berenbrink et al. achieves the optimal maximum load to within constant factors, with high probability.

Dynamic Arrivals. Adler et al. [ABS98] consider a natural generalization of the parallel balls-and-bins problem. In their model, m balls arrive in *each round* and must be allocated in parallel to n bins. (This model should be distinguished from the dynamic but sequential arrival of balls considered in Sections 4.2 and 3.1.) Each bin has a first-in, first-out (FIFO) queue where the balls wait to be "served". In each round, each bin *serves* and *removes* the ball at the head of its queue. The goal is to allocate balls in a manner that minimizes the number of rounds that a ball spends waiting to be served by a bin. Adler at al. study a natural protocol for allocating the balls. Each arriving ball chooses two bins independently and randomly, and adds itself to the queues of *both* bins. When a ball is served and removed by one of its queues, the ball is also deleted from the other queue.

Theorem 10 *For the protocol outlined above, any given ball waits at most $O(\log\log n)$ rounds before being served, with high probability, provided $m \leq \frac{n}{6e}$.*

Proof Sketch: The proof of the result uses the classical witness tree method except that we view the nodes of the tree as representing balls (instead of bins). Suppose that a ball b arrives at time T and waits more than τ rounds before being served. A depth-τ witness tree can be constructed as follows. The root of the witness tree is associated with ball b. Consider the two bins r' and r'' chosen by ball b. Since b is not served by either bin at time $T + \tau$, there must exist two balls b' and b'' that are served by r' and r'' respectively at time $T + \tau$. We make balls b' and b'' the two children of ball b in the witness tree. Since each queue uses the FIFO protocol, b' and b'' arrived at time T or earlier. Hence balls b' and b'' waited more than $\tau - 1$ rounds before being served. Thus, we can recursively grow depth-$(\tau - 1)$ trees rooted at b' and b''.

We now enumerate and bound the probability of occurrence of a depth-τ witness tree, for $\tau = O(\log \log n)$. As always, one has to deal with the fact that a ball can appear multiple times in the tree. But, perhaps the greater technical difficulty is that, unlike our other examples of witness tree proofs, we have no a priori bound on the number of balls present in the system at a given time. Therefore, we need to explicitly characterize the balls that can appear in the witness tree. The reader is referred to [ABS98] for further details.   ■

Load Balancing in Parallel Environments.   Much of the recent interest in the balls-and-bins problem is due to its applicability to scheduling tasks in a parallel or distributed environment. The examples we have seen so far apply to the so-called *client-server* paradigm where the clients generate tasks (i.e., balls), sequentially or in parallel, and the tasks must be allocated to servers (i.e., bins) so as to balance the load on each server.

In this section, we explore a somewhat different paradigm that is relevant to load balancing in a parallel computer. In a parallel computer, unlike the client-server model, the processors play a *dual role* in that they both generate and execute tasks. We seek distributed algorithms that ensure that the maximum load of any processor remains "small", i.e., we would like the tasks to be as evenly distributed among the processors as possible. In addition, we would like to avoid excessive communication between processors and would like to execute the tasks in the processors where they are generated as much as possible. This additional locality constraint is an important distinguishing feature that is absent in the client-server model.

The load balancing problem can be classified according to the nature of the tasks themselves. The problem is more complex if the tasks have explicit dependencies; for instance, the tasks may represent a multi-threaded computation modeled as a directed acyclic graph [BL94, ABP98], or the tasks may be generated by a backtrack search or branch-and-bound algorithm [KZ93]. The situation where the tasks are independent is somewhat simpler and several models for generating and consuming independent tasks are considered in the literature [RSAU91, BFM98, BFS99]. In the *random load model* each processor in each step generates a task with a fixed probability λ and consumes a task with a fixed probability μ, where $0 \le \lambda < \mu \le 1$. Whereas in the *adversarial load model*, the load of each processor at each time can be modified arbitrarily by an adversary, provided the net change in load is at most a given parameter δ.

The literature in this area can also be classified by the nature of the proposed algorithm. In a *work sharing* algorithm, a "heavily-loaded" processor seeks to donate some of its excess tasks to a suitable processor [BFM98, BFS99]. A key issue in work sharing is how a heavily-loaded processor finds one or more "lightly-loaded" processors to which to donate its excess tasks. The matching of the heavily-loaded processors to lightly-loaded processors must be performed efficiently and in a distributed manner. In a *work stealing* algorithm, a "lightly-loaded" processor "steals" tasks from a suitable processor (e.g., [ELZ86b, FMM91, FM87, HZJ94, Mit98, BL94, ABP98]). A particular example of this approach is idle-initiated work stealing where a processor that becomes idle seeks to obtain tasks from nonidle processors.

Randomized algorithms have proven to be a critical tool in this matching process since the earliest investigations in this area. More recently, there has been an effort to use collision algorithms and related ideas to perform this matching [BFM98, BFS99]. Berenbrink et al. [BFS99] show how to maintain a load of $O(\log \log n)$ on an n-processor parallel machine, with high probability, in the random load model. Collision protocols are used to construct a tree rooted at each heavily-loaded processor, and each such processor communicates down its tree to search for an "unattached" lightly-loaded processor. Note that one can easily achieve the same bound on the load by migrating each task as soon as it is generated using a variant of the algorithm described in Section 3.2. However, such an algorithm would entail a large amount of communication. The primary contribution of Berenbrink et al. is that their work sharing algorithm ensures that processors send tasks only if they are heavily-loaded, reducing the total communication performed by a factor of $\Theta(\log \log n)$ with high probability.

3.3 A LOWER BOUND USING WITNESS TREES

We have seen how to use witness trees for proving upper bounds on the maximum load of a bin. However, witness trees are useful in proving lower bounds as well. In the upper bound proofs of Theorem 7 and 9 we observed that if there is a "heavily-loaded" bin there exists a witness tree whose node-degree and height are "large". The key idea in deriving lower bounds using witness trees is that, in some circumstances, the converse is also true: if a witness tree with "large" degree and height occurs then some bin is expected to receive a "large" number of balls. We illustrate this technique by proving a lower bound on the maximum load for the parallel balls-and-bins problem.

The collision protocol that we outlined in Section 3.2 is *nonadaptive* in that the possible destinations for the balls are chosen *before* any communication takes place. Furthermore, the protocol is *symmetric* in that all balls and bins perform the same algorithm and the bins are chosen independently and at random. A natural question to ask is if there exists a nonadaptive and symmetric algorithm that achieves a smaller expected maximum load than the collision protocol outlined in Section 3.2. Adler et al. [ACMR95] show that the expected maximum load of a bin is $\Omega\left(\sqrt[\rho]{\frac{\log n}{\log\log n}}\right)$ for any protocol that performs ρ rounds of communication, provided that ρ is a constant and the protocol belongs to a natural *subclass* of nonadaptive and symmetric protocols. (Most known nonadaptive and symmetric protocols belong to this subclass. We describe the restrictions that define this subclass in Theorem 11 below.) This lower bound result was later extended to all values of ρ by Berenbrink et al. [BMS97], which we state below.

Theorem 11 *The expected maximum load of a bin is* $\Omega\left(\sqrt[\rho]{\frac{\log n}{\log\log n}}\right)$ *for any protocol that performs ρ rounds of communication, $1 \le \rho \le \log\log n$, provided that the protocol satisfies the following conditions:*

- *the protocol is nonadaptive and symmetric,*

- *removing a set of balls before the protocol begins cannot increase the expected maximum load achieved by the protocol, and*

- *if a ball cannot distinguish between its two bin choices after ρ rounds of communication, the protocol allocates the ball randomly with probability 1/2 to either of its choices.*

Proof Sketch: Let a (T, r)-ary tree be a tree of depth r such that the root has T children, and every other internal node has $T - 1$ children. The first step in our proof is to show that with constant probability there exists a $(\tau+1, \rho+1)$-ary witness tree W such that all the nodes of the tree represent distinct bins, for some $\tau = \Theta\left(\sqrt[\rho]{\frac{\log n}{\log\log n}}\right)$. It is easy to show that the expected number of such witness trees is $\Omega(n)$ using an enumeration very similar to the one in the proof of Theorem 9. (The primary difference is that we now seek a lower bound on the expectation.) The number of ways of labeling a $(\tau+1, \rho+1)$-ary tree with distinct balls and bins is at least $n^{\underline{m}} \cdot n^{\underline{m-1}} \ge (n-m)^{2m-1}$, where m is the number of nodes in the tree. The probability of occurrence of each labeled tree is $\left(\frac{2}{n^2}\right)^{m-1}$. Thus the expected number of $(\tau+1, \rho+1)$-ary witness trees is (using linearity of expectation) at least

$$(n - m)^{2m-1} \cdot \left(\frac{2}{n^2}\right)^{m-1} = \Omega(n),$$

since $m = (\tau + 1)\frac{\tau^{\rho+1}-1}{\tau-1} = o(n)$. The fact that the expected number of witness trees is $\Omega(n)$ does not immediately imply that there exists such a witness tree with constant probability. We need to show that the variance of the number of witness trees is "small". The lower and upper bounds on the expectation and variance respectively, in conjunction with Chebyshev's inequality, yields the fact that a $(\tau + 1, \rho + 1)$-ary witness tree exists with constant probability. We refer to Czumaj et al [CMS95] and Stemann [Ste96] for the details of the existence proof.

Let W be the $(\tau + 1, \rho + 1)$-ary witness tree with distinct bins that we have shown exists with constant probability. The initial random choices made by the balls can be represented by an *access graph* with n nodes and n edges. Each of the n nodes represent a distinct bin. Each of the n edges represents the pair of bin choices made by a distinct ball. Note that the witness tree W is a subgraph of the access graph. In addition to assuming that the protocol is nonadaptive and symmetric, we now use our second additional assumption that *removing a set of balls, i.e., deleting some edges of the access graph, cannot increase the expected maximum load achieved by the protocol.* More precisely, given two access graphs G and G' such that the edges of G are a superset of the edges of G', the expected[5] maximum load achieved by the protocol on G' is at most the expected maximum load achieved on G. This assumption is intuitively reasonable and holds for most natural algorithms considered in the literature, including the collision protocol outlined in Section 3.2. We utilize this assumption to remove all edges, i.e., balls, in the access graph that do not belong to W.

Henceforth, we consider only tree W and any lower bound we derive on the expected maximum load of the protocol on W is a lower bound on the expected maximum load for the original access graph. Let b be the ball that corresponds to an edge incident on root r of W. We argue that ball b has probability $\frac{1}{2}$ of being allocated to r.

The key to the argument is quantifying the amount of knowledge b can acquire by performing ρ rounds of communication. Initially, ball b knows nothing about the choices made by any other ball. After the first round, assuming in the worst case that the two bins chosen by b convey all their information to b, b knows about the choices of all the balls that chose one of its bins, i.e., b knows about the edges in the neighborhood $N(b)$, where $N(b)$ is the set of all edges of W incident to an endpoint of b. (We define $N(S)$, where S is a set of edges, to be $\cup_{b\in S}N(b)$.)

Inductively, after $\rho > 1$ rounds, ball b knows about the ρ-neighborhood $N_\rho(b)$ which is recursively defined to equal $N(N_{\rho-1}(b))$.

Let ball b correspond to edge (r, r'), where r is the root of W and r' is its child. Removing edge (r, r') from the neighborhood set $N_\rho(b)$ splits it into two connected components, $N_{\rho,r}(b)$ that contains r and $N_{\rho,r'}(b)$ that contains r'. Since W is a $(\tau + 1, \rho + 1)$-ary tree, both $N_{\rho,r}(b)$ and $N_{\rho,r'}(b)$ are both complete τ-ary trees of depth ρ, i.e., they are *identical*. Since the neighborhood set containing *either* bin is identical, there is no reason for ball b to choose one bin over another. We now use our third additional assumption that in this case the *protocol must choose a bin randomly with probability* $\frac{1}{2}$. Thus ball b chooses root r with probability $\frac{1}{2}$. Since this holds for each of the $\tau + 1$ children of r, the expected load of r is at least $\frac{\tau+1}{2} = \Omega\left(\sqrt[\rho]{\frac{\log n}{\log \log n}}\right)$.

Since a $(\tau + 1, \rho + 1)$-ary tree W occurs with constant probability, it follows that the expected maximum load is $\Omega(\sqrt[\rho]{\frac{\log n}{\log \log n}})$ for any non-adaptive symmetric protocol that obeys the two additional assumptions stated in the theorem.   ∎

3.4   RANDOMIZED PROTOCOLS FOR CIRCUIT ROUTING

Much of the recent interest in the balls-and-bins problem derives from the straightforward analogy of scheduling tasks (i.e., balls) on multiple servers or processors (i.e., bins). Can the multiple-choice approach be used effectively in problem domains other than task scheduling? In this section, we show how we can use a variant of the multiple-choice method to perform low-congestion circuit routing in multistage interconnection networks. The results presented in this section are based on Cole et al. [CMM+98] and represent some of the more sophisticated applications of witness tree arguments. It is worth pointing out that these results do not appear to be amenable to either layered induction or fluid limit models.

Several modern high-speed multimedia switches and ATMs utilize (virtual) circuit-switching to route communication requests [RCG94, TY97]. In a circuit-switched network, requests arrive at the input nodes of the network and require a path to some output node of the network. A *circuit-routing algorithm* allocates a path through the network for each request. When the request is completed, the path allocated to it is freed up. The goal is to devise routing algorithms that minimize congestion, where *congestion* is the maximum number of paths that must be simultaneously supported by a link of the network.

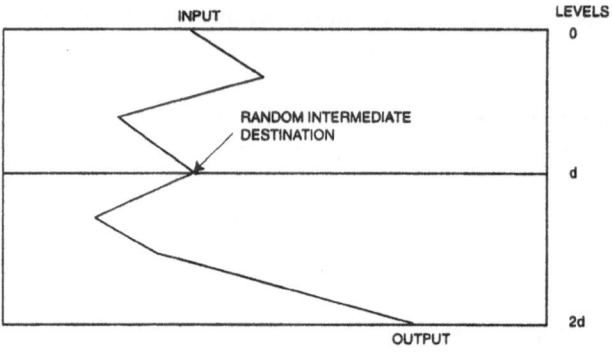

Figure 9.2 Valiant's paradigm

A canonical circuit routing problem often studied in the literature is the permutation routing problem. In a *permutation routing problem* at most one request originates at each input of the network and at most one request is destined for each output of the network. Furthermore, we distinguish between two kinds of permutation routing problems: static and dynamic. In a *static problem*, all the requests that constitute a permutation routing problem are present at time 0, before the routing begins. The routing algorithm constructs paths for all of the requests in a "batch" mode. All of the requests in a batch complete before routing of the next batch of requests begins. In contrast, in a *dynamic problem*, requests arrive and leave over time, according to a sequence constructed by an oblivious adversary. The routing algorithm routes a path for each arriving request in an on-line fashion with no knowledge of future request arrivals. We assume that at any time, the requests being routed form a partial permutation; that is, each input and output node correspond to at most one routed request.

The results in this section apply to variants of a popular type of multi-stage interconnection network called the *butterfly network* (See [Lei92] For a description of its structure.). An n-input butterfly B_n has $n(\log n + 1)$ nodes arranged in $\log n + 1$ *levels* of n nodes each.

Furthermore, there is a unique path of length $\log n$ in B_n from each input node to each output node. A *two-fold butterfly* BB_n consists of two copies of B_n placed one after the other such that each output node in the first copy is identified with the corresponding input node of the second copy. The inputs of BB_n are at level 0 while the outputs are at level $2d$, where $d = \log n$. (See Figure 9.2).

An early example of the use of randomized algorithms for communication problems is the work of Valiant [Val82, VB81]. Valiant showed that any permutation routing problem can be transformed into two random problems by first routing a path for each request to a random intermediate destination, chosen independently and uniformly, and then on to its true destination (See Figure 9.2). This routing technique known as *Valiant's paradigm* is analogous to the classical balls-and-bins problem where each request (i.e., ball) chooses a random intermediate destination (i.e., bin). It follows from this analogy that the congestion achieved by Valiant's paradigm corresponds to the maximum load of the classical balls-and-bins problem, and is $\Theta(\log n / \log \log n)$, with high probability.

Valiant's Paradigm with Two Random Choices. A question that begs asking is if the two-choice method can be incorporated into Valiant's paradigm to significantly reduce the congestion, just as two random choices can be used to significantly reduce the maximum load of a bin in the balls-and-bins problem. The simplest way to incorporate the two-choice approach into Valiant's algorithm is to let each request choose *two* random intermediate destinations (instead of one), and choose the path that has the smaller congestion. (The congestion of a path is the maximum congestion of all its edges.) But this simple approach fails to decrease the congestion significantly. The problem lies in the fact that even though a request can choose any of the n intermediate nodes in level d to be on its path, it has very few nodes that it can choose in levels that are close to its input or output. Therefore, it is quite likely that there exists a set of $m = \Theta(\log n / \log \log n)$ requests such that all the paths chosen by these requests intersect at some node at level $\log m$ of BB_n. Thus, any allocation of requests to paths causes congestion of at least $m = \Omega(\log n / \log \log n)$.

The key to applying the two-choice approach to circuit routing is to avoid creating hot spots of congestion near the inputs and outputs of the network. To achieve this we select two random paths for each request as follows. The nodes on levels $0, \ldots, d/2 - 1$ and $d + d/2 + 1, \ldots, 2d$ are flipped randomly, where $d = \log n$. In particular, each input and output node maps the *first path* p of a request to its straight edge and its *second path* p' to its cross edge with probability $\frac{1}{2}$, and with probability $\frac{1}{2}$ the order is reversed. Similarly, each node on levels $1, \ldots, d/2 - 1$ and $d + d/2 + 1, \ldots, 2d - 1$ with probability $\frac{1}{2}$ connects its input straight edge with its output straight edge and its input cross edge with its output cross edge, and with probability $\frac{1}{2}$ the connections are reversed. (See Figure 9.3.) Note that these random choices completely determine the two paths p and p' of each request, because there is exactly one path connecting a node on level $d/2$ with a node on level $d + d/2$ in the BB_n network. For a path p, the other path p' connecting the same input and

288

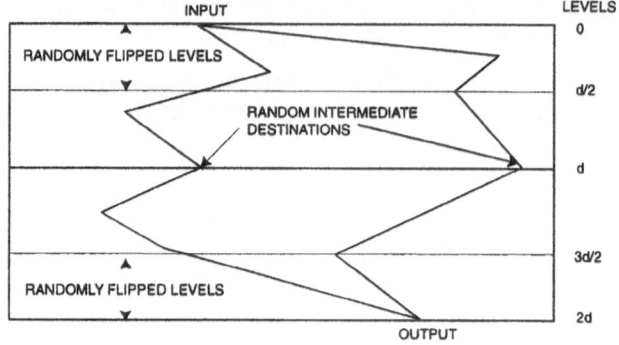

Figure 9.3 Valiant's paradigm with two random destinations

output nodes is called the *buddy* of p. The random switching ensures that any edge on the levels $1, \ldots, d/2$ and $d + d/2 + 1, \ldots, 2d$ is traversed by at most one of the randomly-generated paths.[6] However, each edge that originates at a node in an interior level, i.e., levels $d/2 + 1, \ldots, d + d/2$, is potentially traversed by several of these paths. Note that the random paths chosen for distinct requests are no longer independent, because the paths of the two requests may share one or more randomly-flipped switches. This is a key technical difficulty that complicates the witness tree analyses in this section.

Permutation Routing. Recall that in a dynamic permutation routing problem, the requests arrive and leave according to a sequence constructed by an oblivious adversary. When a request r arrives, our routing algorithm chooses two random paths as described in Section 9.2, evaluates the congestion of the two paths, and allocates r to the path with smaller congestion.

Theorem 12 *The routing algorithm described above finds paths for every request of a dynamic permutation routing problem in network BB_n such that the congestion at any given time t is $O(\log \log n)$, with high probability.*

Proof Sketch: The overall structure of the proof is similar to that of Theorem 7, though the details are much more complicated. First, we fix the settings of the randomly-flipped switches. This determines two choices of paths for each request. Assume that there is an edge e with congestion larger than $4c$ at some time t, where $c = \lceil \log \log n \rceil$. Let p denote the last path mapped to edge e on or before time t. When p was mapped to e there were already $4c$ other paths present at this edge.

Let $p_1, \ldots, p_{4c}$ denote these paths such that p_i was mapped to e at time step t_i with $t_i < t_{i+1}$. The root of the tree is the request corresponding to p and the requests corresponding to $p_1, \ldots, p_{4c}$ are its children. Now we consider the buddies $p'_1, \ldots, p'_{4c}$ of these paths. Path p'_i traverses an edge with congestion at least $i-1$ at time step t_i, because the congestion of p_i is not larger than the congestion of p'_i at time i, and when p_i was mapped to e there were already $i-1$ other paths present at this edge. As a consequence, we can construct a tree by applying the argument above recursively to $p'_2, \ldots, p'_{4c}$. The tree constructed in this fashion is the witness tree.

The technical challenge is performing the next step of enumerating and bounding the probability that a witness tree occurs. Recollect that the paths are not chosen independently for each request, since paths belonging to distinct requests may share one or more randomly-flipped switches in the first or last $d/2$ levels. Therefore, when the witness tree is pruned, besides ensuring that the requests in the pruned tree are distinct, it is also necessary to ensure that the paths of the requests in the pruned tree share only a "limited" number of randomly-flipped switches, i.e., the paths in the pruned tree represent "almost" independent random choices. The bound follows by enumerating the paths in the witness trees and their respective probabilities of occurrence. ■

The results in this section show that the two-choice method can be used to significantly reduce the congestion of Valiant's algorithm for dynamic permutation routing. The two-choice approach can also be adapted to route any static permutation routing problem on BB_n with congestion $O(\log \log n / \log \log \log n)$, with high probability [CMM+98].

4.   A DIFFERENTIAL EQUATIONS APPROACH: FLUID LIMITS

We now describe a third technique that has proven useful for analyzing randomized load balancing schemes based on the two-choice paradigm. This technique was developed in parallel in the computer science and queueing theory communities. The approach relies on determining the behavior of the system as its size grows to infinity. For example, in the balls-and-bins model, the size of the system naturally corresponds to the number of bins, so we consider what happens as the number of bins grows towards infinity. Often we can describe the behavior naturally as a differential equation or a family of differential equations.

Once we have a family of differential equations that describe the process, we can try to solve it to obtain a closed form solution. Then, in many cases, we can relate this solution to the behavior of a system of finite size using a concentration result. That is, we may show that the behavior of a system of finite size is close to the behavior given by the solutions of the differential equations, with high probability.

Sometimes we cannot obtain a closed form from the differential equations. In such cases, however, we can often solve the differential equations using numerical methods.

4.1   BALLS AND BINS

Empty bins.   To introduce the approach, let us consider the sequential multiple-choice balls-and-bins problem, where $m = cn$ balls are thrown into n bins with each ball having two bin choices. We first ask what fraction of the bins remain empty. This question was first considered by Hajek [Haj88].

The problem can be solved by developing a Markov chain with a simple state that describes the balls-and-bins process. We first establish a concept of time. Let $Y(T)$ be the number of nonempty bins after T balls have been thrown. Then $\{Y(i)\}$, $0 \le i \le m$, is a Markov chain, since the choices for each ball are independent of the state of the system. Moreover

$$E[Y(T+1) - Y(T)] = 1 - \left(\frac{Y(T)}{n}\right)^d, \tag{9.7}$$

since the probability that a ball finds all nonempty bins among its d choices is $(Y(T)/n)^d$.

The notation becomes somewhat more convenient if we scale by a factor of n. If t is the time at which exactly nt balls have been thrown, and $y(t)$ is the fraction of nonempty bins, then equation (9.7) becomes

$$\frac{E[y(t+1/n) - y(t)]}{1/n} = 1 - (y(t))^d. \tag{9.8}$$

We claim the random process described by equation (9.8) is well approximated by the trajectory of the differential equation

$$\frac{dy}{dt} = 1 - y^d. \tag{9.9}$$

This equation has been obtained from equation (9.8) by replacing the left hand side with the appropriate limiting value as n grows to infinity, dy/dt. That is, we think of each ball as being thrown during a small interval of time dt of duration $1/n$. In equation (9.8) we replace the expected change in y over this interval by dy, with the intuition that the behavior of the system tends to follow its expectation over each step.

This claim, that the system behaves close to what we might expect simply by looking at the expected change at each step, follows from a theorem similar to the law of large numbers for special types of Markov chains. The important feature is that the differential equation obtained is independent of the number of bins n; such Markov chains are called *density dependent*, as their behavior depends essentially on the density (in this case y) of objects in the state rather than the total number of objects. Here, the objects are the nonempty bins. For such a system there are bounds similar to Azuma's inequality for martingales [MR95]. Indeed, the derivation of the theorem is based on a suitable martingale. For example, for the balls-and-bins problem above, if y^* is the solution of the differential equation, then

$$\mathbf{Pr}\left(\sup_{0 \le t \le T} |y(t) - y^*(t)| \ge \epsilon\right) \le C_1 e^{-nC_2(\epsilon)}$$

for constants C_1 and C_2 that may depend on T. This approach is also often referred to as the *fluid limit* approach, since the discrete process is replaced by an often simpler continuous process reminiscent of the behavior of physical fluid models.

These theorems apparently first appeared in the work of Kurtz [EK86, Kur70, Kur81], and were eventually applied to algorithmic problems related to random graphs [Haj88, KS81, KVV90] as well as to queueing problems [CH91]. Recently these techniques have resurged in the random graph community, initiated by the work of Wormald [Wor95]. The text by Shwartz and Weiss on large deviations provides a solid introduction into the entire area of large deviations, including Kurtz's work [SW95]. There are by now many examples of works that use large deviation bounds and differential equations for a variety of problems, including but in no way limited to [AM97, AH90, AFP98, KMPS95, LMS+97].

Given this framework, it is easy to find the limiting fraction of empty bins after $m = cn$ balls have been thrown, by solving the differential equation $\frac{dy}{dt} = 1 - y^d$ with the initial condition $y(0) = 0$ at time c. This can be done by using the nonrigorous high school trick of writing the equation as $\frac{dy}{1-y^d} = dt$ and integrating both sides.

Theorem 13 *Let c and d be fixed constants. Suppose cn balls are sequentially thrown into n bins, each ball being placed in the least full of d bins chosen independently and uniformly at random. Let Y_{cn} be the number of nonempty bins when the process terminates. Then $\lim_{n \to \infty} \mathbb{E}[\frac{Y_{cn}}{n}] =$*

y_c, where $y_c < 1$ satisfies

$$c = \sum_{i=0}^{\infty} \frac{y_c^{id+1}}{(id+1)}.$$

Using this we may solve for y_c. (Closed form expressions exist for some values of d, but there does not seem to be a general way to write y_c as a function of c.)

We may actually use Kurtz's Theorem to obtain a concentration result.

Theorem 14 *In the notation of Theorem 13, $\left|\frac{Y_{cn}}{n} - y_c\right|$ is* $O\left(\sqrt{\frac{\log n}{n}}\right)$ *with high probability, where the constant depends on c.*

One can also obtain entirely similar bounds for Y_{cn} using more straightforward martingale arguments; however, the martingale approach does not immediately lead us to the value to which Y_{cn}/n converges. This is a standard limitation of the martingale approach: in contrast, the fluid limit model allows us to find the right limiting value.

Nonempty bins. The previous analysis can be extended to find the fraction of bins with load at least (or exactly) k for any constant k as $n \to \infty$. To establish the appropriate Markov chain, let $s_i(t)$ be the fraction of bins with load *at least* i at time t, where again at time t exactly nt balls have been thrown. Note that this chain deals with the tails of the loads, rather than the loads themselves. This proves more convenient, as we found in Section 2.. The differential equations regarding the growth of the s_i (for $i \geq 1$) are easily determined [Mit96b, Mit99b]:

$$\begin{cases} \dfrac{ds_i}{dt} &= (s_{i-1}^d - s_i^d) \quad \text{for } i \geq 1; \\ s_0 &= 1. \end{cases} \tag{9.10}$$

The differential equations are easily interpreted as denoting that an increase in the number of bins with at least i balls occurs when the d choices of a ball about to be placed must all be bins with load at least $i-1$, but not all bins with load at least i.

The case of n balls and n bins corresponds to time 1. Interestingly, the differential equations reveal the double exponential decrease in the tails in this situation quite naturally:

$$
\begin{aligned}
s_i(1) &= \int_{t=0}^{1} \left[(s_{i-1}(t))^d - (s_i(t))^d \right] dt \\
&\leq \int_{t=0}^{1} (s_{i-1}(t))^d \, dt \\
&\leq \int_{t=0}^{1} (s_{i-1}(1))^d \, dt \\
&= (s_{i-1}(1))^d.
\end{aligned}
$$

Hence

$$
s_i(1) \leq (s_1(1))^{d^{i-1}}.
$$

The above argument shows that the tails $s_i(1)$ in the limiting case given by the differential equations decrease doubly exponentially; of course, since the results for finite n are tightly concentrated around these s_i, the implication is that the behavior also occurs in finite systems. The astute reader will notice that the steps taken to bound the integral expression for $s_i(1)$ above entirely mimics the original proof of Azar, Broder, Karlin, and Upfal. That is, we bound $s_i(1)$ based only on $s_{i-1}(1)$. Indeed, the differential equations provide an appealing natural intuition for their proof, and a similar approach can be used to mimic their lower bound argument as well. Although this intuition implies a maximum load of $\log \log n / \log d + \Theta(1)$ in the case of n balls and n bins, the general theory for density dependent chains does not seem to get one there immediately. An immediate problem is that Kurtz's theorem, as generally stated, requires a fixed number of dimensions. That is, we can only work with s_i for $i \leq K$ for some fixed constant K. Hence considering loads of up to $O(\log \log n)$ requires more than a simple application of the theorem. A further problem is that as the tail gets small, the probabilistic bounds are not strong enough. Hence one apparently needs an explicit argument, such as that given by Azar et al., to achieve such an explicit bound.

While the fluid limit approach does not obviate the need for detailed probabilistic arguments, it provides a remarkably useful tool. In particular, when applicable it generally provides natural intuition and remarkable accuracy in predicting the behavior of even moderate sized systems. (See, e.g., [Mit96b].) Moreover, it offers tremendous flexibility. Many variations on the balls-and-bins problem can be placed into the differential equations framework. We provide a short introduction to some of the more interesting ones.

4.2   QUEUEING THEORY MODELS

We consider the fluid limit model of a natural queueing system that generalizes the static multiple-choice balls-and-bins problem. Suppose we think of the bins as FIFO (First-In, First-Out) servers and the balls as tasks that enter and leave after being processed. In this case, we assume that tasks arrive as a Poisson process of rate λn proportional to the number of servers n, with $\lambda < 1$. We also assume that tasks require an amount of service distributed exponentially with mean 1. This model, a generalization of the natural M/M/1 queueing model to many servers, has been widely studied in the case where incoming customers are placed at the server with the shortest queue. (See, for example, the essential early work of Weber [Web78], Whitt [Whi86], and Winston [Win77], as well as the more recent work by Adan and others [AWZ90, Ada94].) Of course, such a load balancing scheme requires some means of centralization. In a completely decentralized environment, attempting to determine the shortest queue might be expensive, in terms of time or other overhead. Rather than just assigning tasks randomly, we might consider having each task examine a small number of servers and go to the least loaded of the queues examined. This idea appeared in early work by Eager, Lazowska, and Zahorjan [ELZ86a].

We consider the case where each task chooses $d \geq 2$ servers at random. The fluid limit analysis for this setting and its surprising implications were found independently by Vvedenskaya, Dobrushin, and Karpelevich [VDK96] and Mitzenmacher [Mit96b, Mit96a].

As previously, we let $s_i(t)$ be the fraction of queues with load *at least* i at time t. The differential equations describing the fluid limit process are easily established.

$$
\begin{cases}
\dfrac{ds_i}{dt} & = \quad \lambda(s_{i-1}^d - s_i^d) - (s_i - s_{i+1}) \ \text{ for } \ i \geq 1; \\
s_0 & = \quad 1.
\end{cases}
\tag{9.11}
$$

Let us explain the reasoning behind the system in (9.11). Consider a system with n queues, and determine the expected change in the number of servers with at least i customers over a small period of time of length dt. The probability a customer arrives during this period is $\lambda n \, dt$, and the probability an arriving customer joins a queue of size $i-1$ is $s_{i-1}^d - s_i^d$. (This is the probability that all d servers chosen by the new customer are of size at least $i-1$ but not all are of size at least i.) Thus the expected change in the number of queues with at least i customers due to arrivals is exactly $\lambda n(s_{i-1}^d - s_i^d)dt$, and the expected change in the fraction of queues with at least i customers due to arrivals is therefore $\lambda(s_{i-1}^d - s_i^d)dt$. Similarly, as the number of queues with i customers is

$n(s_i - s_{i+1})$, the probability that a customer leaves a server of size i in this period is $n(s_i - s_{i+1})dt$. Thus the expected change in s_i due to departures is $-(s_i - s_{i+1})dt$. Putting it all together, and replacing the expected change by ds_i, we obtain the system (9.11).

To determine the long range behavior of the system above requires looking for a *fixed point*. A fixed point (also called an *equilibrium point* or a *critical point*) is a point where for all i, $\frac{ds_i}{dt} = 0$. Intuitively, if the system reaches its fixed point, it will stay there.

Lemma 15 *The system (9.11) with $d \geq 2$ and $\lambda < 1$ has a unique fixed point with $\sum_{i=1}^{\infty} s_i < \infty$ given by*

$$s_i = \lambda^{\frac{d^i-1}{d-1}}.$$

Proof: It is easy to check that the proposed fixed point satisfies $\frac{ds_i}{dt} = 0$ for all $i \geq 1$. Conversely, from the assumption $\frac{ds_i}{dt} = 0$ for all i we can derive that $s_1 = \lambda$ by summing the equations (9.11) over all $i \geq 1$. (Note that we use $\sum_{i=1}^{\infty} s_i < \infty$ here to ensure that the sum converges absolutely. The condition corresponds to the natural condition that expected number of tasks in the system is finite at the fixed point. That $s_1 = \lambda$ at the fixed point also follows intuitively from the fact that at the fixed point, the rate at which customers enter and leave the system must be equal.) The result then follows from (9.11) by induction. ∎

Intuitively, we would expect a well-behaved system to converge to its fixed point. In fact one can show that the trajectory of the fluid limit process given by the system (9.11) indeed converges to its fixed point [Mit96a, VDK96]; in fact, it does so exponentially quickly [Mit96a]. That is, the L_1 distance to the fixed point decreases like $c_1 e^{-c_2 t}$ for some constants c_1 and c_2. These results imply that in the limit as n gets large, the equilibrium distribution of a system with n queues is tightly concentrated around the fixed point [VDK96]. In fact, this behavior is readily seen even when the number of servers n is around 100.

Looking at the fixed point, we see that the tails decrease doubly exponentially in d. Hence we expect to see much shorter queues when $d \geq 2$ as opposed to when a server is chosen uniformly at random, i.e. $d = 1$. In fact, both the maximum queue length and the expected time in the system decrease exponentially! More formally, we know that the expected time in the system for an M/M/1 queue in equilibrium is $\frac{1}{1-\lambda}$, and hence this is the expected time in the setting with n servers when tasks choose servers uniformly at random. If we let $T_d(\lambda)$ be the expected time for a task in the system corresponding to the fluid limit model (i.e. as n grows to infinity) for $d \geq 2$, then

$$\lim_{\lambda \to 1^-} \frac{T_d(\lambda)}{\log \frac{1}{1-\lambda}} = \frac{1}{\log d}.$$

That is, as the system is saturated, the average time a task spends in the system when it queues at the shortest of $d \geq 2$ choices grows like the logarithm of the average time when just one choice is made. The result is remarkably similar in flavor to the original result by Azar et. al. [ABKU99]; a more compelling, simple explanation of this connection would certainly be interesting.

It is worth emphasizing, however, the importance of this result in the queueing model. In the case of the static balls-and-bins problem, the difference between one choice and two choices is relatively small, even when the number of balls and bins is large: with one million balls and one million bins, when the balls are distributed randomly, the maximum load is generally at most 12, while using two choices the maximum load drops to 4. Because the average time spent waiting before being served depends on the load λ, even with a small number of servers, having two choices can have a great effect under high load. For example, with one hundred servers at an arrival rate of $\lambda = 0.99$ per server, randomly choosing a server leads to an average time in the system of 100 time units; choosing the shortest of two reduces this to under 6!

Simple variations. The flexibility of the fluid limit approach allows many variations of this basic scheme to be examined. For example, suppose there are two classes of tasks entering the system. High priority tasks choose the shortest of two servers, while low priority tasks choose a random server; the servers, however, are still FIFO. In this case the corresponding fluid limit model is governed by the following set of differential equations:

$$\begin{cases} \dfrac{ds_i}{dt} &= \lambda p(s_{i-1}^2 - s_i^2) + \lambda(1-p)(s_{i-1} - s_i) - (s_i - s_{i+1}) \quad \text{for } i \geq 1; \\ s_0 &= 1. \end{cases} \quad (9.12)$$

The fixed point is given by $s_1 = \lambda$, $s_i = \lambda s_{i-1}(1 - p + p s_{i-1})$. There does not appear to be a convenient closed form for the fixed point for general values of p. Note that at the fixed point, it is easy to determine the distribution of the queue length customers of each priority join.

Surprisingly, the effect of increasing the fraction of customers with two choices has a nonlinear effect on the average time a customer spends in the system that is dramatic at high loads. Figure 9.4, which was

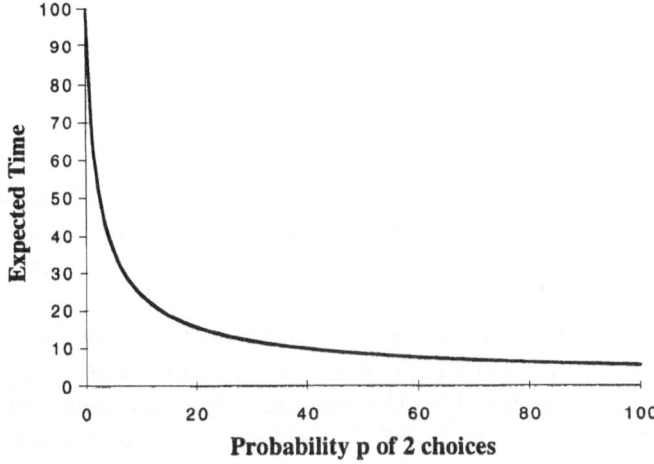

Figure 9.4 Expected time in the system versus probability (p) of that a customer chooses two locations ($\lambda = 0.99$).

obtained by numerically solving for the fixed point, demonstrates this phenomenon at $\lambda = 0.99$; most of the gain occurs when only 20% of the customers have two choices. Simulation results verify this behavior. The intuition for this effect is that the average queue length is not linear in the load; at high loads small increases in the load can dramatically increase average queue lengths. Hence, even giving a small fraction of the incoming tasks additional information greatly reduces the average queue length. This example demonstrates how the fluid limit approach can be used to gain significant insights into the original problem by studying variations in a simple, natural way.

As another example, we consider the variation considered by Vöcking described earlier in Section 3.1 [Vöc99]. We simplify by focusing on the case where $d = 2$. In this case, there are two sets of servers, with half of the bins on the left and half on the right. An incoming task chooses two servers, one uniformly at random from the left and one from the right, and queues at the server with fewer customers. Ties are broken always in favor of the bins on the left. Let $y_i(t)$ be the fraction of the n servers that have load at least i and are in the group on the left. Similarly, let $z_i(t)$ be the fraction of the n bins that have load at least i and are in the group on the right. Note $y_i(t), z_i(t) \leq 1/2$, and $y_0(t) = z_0(t) = 1/2$ for all time. Also, if we choose a random bin on the left, the probability that it has load at least i is $\frac{y_i}{1/2} = 2y_i$. The differential equations describing the limiting process are thus

$$\frac{dy_i}{dt} = 4\lambda \left(y_{i-1} - y_i\right) z_{i-1} - \left(y_i - y_{i+1}\right) ; \tag{9.13}$$

$$\frac{dz_i}{dt} = 4\lambda \left(z_{i-1} - z_i\right) y_i - \left(z_i - z_{i+1}\right). \tag{9.14}$$

That is, for y_i to increase, our choice on the left must have load $i - 1$, and the choice on the right must have load at least $i - 1$. For z_i to increase, our choice on the right must have load $i - 1$, but now the choice on the right must have load at least i. For y_i to decrease, a task must leave a server on the left with load i, and similarly z_i decreases when a task leaves a server on the right with load i. This system appears substantially more complex than the standard dynamic two-choice model; in fact, there is as yet no proof that there is a unique fixed point, although experiments suggest that this is the case. Indeed, even if the fixed point is unique, it does not appear to have a simple closed form. We shall assume that the fixed point for this system is unique from here on. An argument in [MV98] demonstrates that such a fixed point must be strictly better than the fixed point for uniform selection of two servers, in the following sense. If u_i represents the fraction of servers with load at least i at the fixed point for the system given by (9.13) and (9.14), then $u_i \leq \lambda^{2^i - 1}$ for all i.

Using the fluid limit, it is simple to consider the following natural variation: suppose we split the left and the right sides unevenly. That is, suppose the left contains $\alpha \cdot n$ bins, and the right contains $(1 - \alpha) \cdot n$ bins. Then $y_0 = \alpha$, $z_0 = 1 - \alpha$ for all time, and by the same reasoning as for equations (9.13) and (9.14),

$$\frac{dy_i}{dt} = \frac{1}{\alpha(1 - \alpha)} \left(y_{i-1} - y_i\right) z_{i-1} - \left(y_i - y_{i+1}\right). ; \tag{9.15}$$

$$\frac{dz_i}{dt} = \frac{1}{\alpha(1 - \alpha)} \left(z_{i-1} - z_i\right) y_i - \left(z_i - z_{i+1}\right). \tag{9.16}$$

Interestingly, an even split is not best! As shown in Figure 9.4, for $\lambda = 0.9$, the tails fall slightly faster using a somewhat uneven split. In general the right value of α depends on λ; as λ increases to 1, the best value of α approaches $1/2$. Increasing the fraction of processors on the left appears to mitigate the tendency of processors on the left to be more heavily loaded than those on the right.

Other variations that can be easily handled include constant service times; other service distributions can also be dealt with, although with more difficulty [Mit99a, VS97]. Threshold-based schemes, where a second choice is made only if a first choice has high load, are easily examined

i	s_i	u_i $\alpha = 0.5$	u_i $\alpha = 0.53$
1	0.9000	0.9000	0.9000
2	0.7290	0.7287	0.7280
3	0.4783	0.4760	0.4740
4	0.2059	0.1998	0.1973
5	0.0382	0.0325	0.0315
6	0.0013	0.0006	0.0005

Figure 9.5   The tails of the distribution at $\lambda = 0.9$.

[Mit99a, VS97]. Closed models where customers recirculate require only minor changes [Mit96b]. Similar simple load stealing models, such as those those developed in [ELZ86b], can also be attacked using the fluid limit approach, as in [Mit98]. Tackling these variations with differential equations allows insight into how the changes affect the problem and yields a simple methodology for generating accurate numerical results quickly.

We now consider two variations on the basic queueing model that appear both particularly interesting and open for further research.

Dealing with stale information.   Thus far, we have assumed that the load information obtained by a task when deciding at which server to queue is completely accurate. This may not always be the case. For example, there may be an *update delay*, if load information may be updated infrequently. Alternatively, if there is some delay in transferring a task to its queue choice, the load information it obtained will necessarily not reflect the load when it actually joins the queue. If these delays are of the same order as the processing time for a job, it can have dramatic effects. This model was considered by Mirchandaney, Towsley, and Stankovic in [MTS89]. Further work using fluid limit models appeared in [Mit00], which we follow. Additional simulation studies and novel models appear in [Dah99].

For example, let us again consider a system of n FIFO servers and Poisson arrivals at rate λn of tasks with exponentially distributed service requirements. Suppose that queue length information is available on a global bulletin board, but it is updated only periodically, say every T units of time. We might choose to ignore the bulletin board and simply have each task a choose a random server. We might allow a task to peek at the bulletin board at a few random locations, and proceed to the server with the shortest posted queue from these random choices. Or a task might look at the entire board and proceed to the server with the shortest posted queue. How does the update delay effect the system behavior in these situations?

In this case, an appropriate limiting system utilizes a two-dimensional family of variables to represent the state space. We let $P_{i,j}(t)$ be the fraction of queues at time t that have true load j but have load i posted on the bulletin board. We let $q_i(t)$ be the rate of arrivals at a queue of size i at time t; note that, for time-independent strategies (that is, strategies that are independent of the time t), the rates $q_i(t)$ depend only on the load information at the bulletin board and the server selection strategy used by the tasks. In this case, if we denote the time that the last phase began by T_t, then $q_i(t) = q_i(T_t)$, and the rates q_i change only when the bulletin board is updated.

We first consider the behavior of the system during a phase, or at all times $t \neq kT$ for integers $k \geq 0$. Consider a server showing i customers on the bulletin board, but having j customers: we say such a server is in state (i, j). Let $i, j > 1$. What is the rate at which a server leaves state (i, j)? A server leaves this state when a customer departs, which happens at rate $\mu = 1$, or a customer arrives, which happens at rate $q_i(t)$. Similarly, we may ask the rate at which customers enter such a state. This can happen if a customer arrives at a server with load i posted on the bulletin board but having $j - 1$ customers, or a customer departs from a server with load i posted on the bulletin board but having $j + 1$ customers. This description naturally leads us to model the behavior of the system by the following set of differential equations:

$$\frac{dP_{i,0}(t)}{dt} = P_{i,1}(t) - P_{i,0}(t)q_i(t) \, ; \tag{9.17}$$

$$\frac{dP_{i,j}(t)}{dt} = (P_{i,j-1}(t)q_i(t) + P_{i,j+1}(t)) - (P_{i,j}(t)q_i(t) + P_{i,j}(t)) \, , \tag{9.18}$$

These equations simply measure the rate at which servers enter and leave each state. (Note that the case $j = 0$ is a special case.)

At times t where the board is updated, a discontinuity occurs. At such t, necessarily $P_{i,j}(t) = 0$ for all $i \neq j$, as the load of all servers is correctly portrayed by the bulletin board. If we let $P_{i,j}(t^-) = \lim_{z \to t^-} P_{i,j}(z)$, so that $P_{i,j}(t^-)$ represents the state just before an update, then

$$P_{i,i}(t) = \sum_j P_{j,i}(t^-).$$

Experiments with these equations suggest that instead of converging to a fixed point, because of the discontinuity at update times, the system converges to a *fixed cycle*. That is, there is a state such that if the limiting system begins a phase in that state, then it ends the phase in the same state, and hence repeats the same cycle for every subsequent phase. Currently, however, there is no known proof of conditions that guarantee this cyclic behavior.

The age of the load information can have dramatic effect on the performance of the system. We provide a representative example that demonstrates the issues that arise. Figure 9.6 presents simulation results for $n = 100$ server at $\lambda = 0.9$. (The case of one random choice was not simulated; since each server in this case acts as an $M/M/1$ queue, the equilibrium distribution is fully known.) The results from numerically solving the fluid limit model are not included simply because they would be difficult to distinguish from the simulation results; the simulation results are within 1-2% of the results obtained from the fluid limit model, except in the case of choosing the shortest queue, where the simulations are within 8-17% of the fluid limit model. (Modeling the shortest queue system requires an additional approximation that causes some inaccuracy; see [Mit00] for details.) Simulations were performed for 50,000 time steps, with data collected only after the first 5,000 steps to allow the dependence on the initial state to not affect the results. The values of T simulated were $T = 0, 0.1, 0.5, 1.0, 2.0, 3.0, 4.0, 5.0, 10.0, 15.0, 20.0, 25.0, 50.0$. The results presented are the average of three separate simulations.

An interesting, and at first glance counter-intuitive, behavior that immediately manifests is that going to the apparently shortest queue can be a terribly bad strategy. The intuition explaining this phenomenon becomes clear when we recall that the board contains out of date information about the loads. The problem is that all of the incoming tasks seek out the small number of queues with small load, so that a rush of tasks all head to a processor until the board later appropriately updates its load. The tasks essentially exhibit a herd behavior, moving together in the same fashion, due to the unfortunate feedback given by the system.

Another way to describe this intuition is to consider what happens at a market when it is announced that "Aisle 7 is now open." Very often Aisle 7 quickly becomes the longest queue. This herd behavior has been noticed in real systems that use old information in load balancing; for example, in a discussion of the TranSend system, Fox et al. note that initially they found "rapid oscillations in queue lengths" because their system updated load information periodically [FGC+97][Section 4.5].

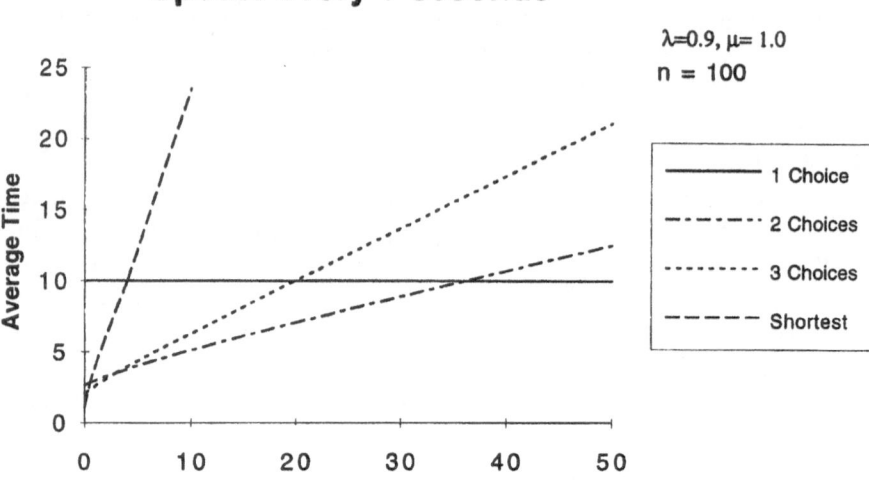

Figure 9.6 Comparing different strategies at $\lambda = 0.90$, 100 queues.

Even given this intuition, it is still rather surprising that even for reasonably small delays, choosing the shortest of two randomly selected processors is a better global strategy than having all tasks choose the shortest from three! The reasoning remains the same: choosing the shortest from three processors skews the distribution towards a smaller set of processors, and when the updates are not quick enough to reflect this fact, poor balance ensues. Of course, as the delay between updates reaches infinity, even two choices performs worse than simple random selection!

Of course, whenever there is more information, better performance can result. For example, suppose that whenever a task is placed on a server the board is appropriately marked. In this case the periodic updates simply provide information about how many tasks have finished. In this setting, going to the apparently shortest queue again becomes a worthwhile strategy. This solution was the one adopted by Fox et al. [FGC+97]; note, however, that it requires that tasks be able to update the board.

Although these initial results provide some insight into the problems that arise in using stale information about queue lengths, the general problem of how to cope with incomplete or inaccurate load information and still achieve good load balancing performance appears to be an area with a great deal of research potential.

Networks of Server Banks.   Once the case of a bank of servers has been handled, one might ask about the case of networks of such systems. In particular, we review the well-understood case of Jackson networks. A standard Jackson network consists of J servers labeled $1, \ldots, J$. Each server has an infinite buffer and services tasks at rate μ_j. For each server j there is an associated Poisson arrival process of rate λ_j. A task repeatedly queues for service, according to the rule that when it finishes at server j, it moves to server k with probability p_{jk} and leaves the system with probability $1 - \sum_k p_{jk}$. (For convenience we assume that the p_{jk} are such that the expected number of servers visited is always finite.)

The state of a Jackson network can be given by a vector representing the queue lengths, $\vec{n}(t) = (n_1(t), \ldots, n_J(t))$, where $n_i(t)$ is the length of the queue at the ith server at time t. Under this formulation the state $\vec{n}(t)$ is a Markov process. The stationary distribution may be described as follows. We consider the *effective arrival rate* ρ_j at each server j. This is a combination of the external arrival rate plus the arrival rate from nodes within the network. These effective arrival rates ρ_j satisfy

$$\rho_j = \lambda_j + \sum_k \rho_k p_{kj}.$$

The tails of the queue lengths for the system of queues satisfy

$$\mathbf{Pr}(n_j \geq r_j \text{ for all } j) = \prod_j \left(\frac{\rho_j}{\mu_j} \right)^{r_j}.$$

The standard interpretation for this result is that in equilibrium each server looks like an independent M/M/1 queue with the appropriate arrival and service rates.

Suppose instead of a standard Jackson network of single servers we have a similar network consisting of server banks, with each bank having a large number of servers n. Of course the arrival rates are scaled so that the arrival rate at each server bank is $n\lambda_j$, and the effective arrival rate at each server bank is $n\rho_j$. A job seeking service at a bank of servers selects a server from that bank to queue at by taking the shortest of $d \geq 2$ random choices. (When $d = 1$, the system is easily seen to be equivalent to a standard Jackson network.) Given our previous results on server systems and the results for standard Jackson networks one might hope that in the limit as n grows to infinity, the load n_{ij} at the

ith server of the jth bank would have the distribution:

$$\mathbf{Pr}(n_{ij} \geq r_j) = \left(\frac{\rho_j}{\mu_j}\right)^{\frac{d^{r_j}-1}{d-1}}.$$

Moreover, we would hope that the Jackson-like network continues to have the property that in equilibrium each bank of servers appears to be an independent bank of servers with the appropriate arrival and service rates.

Indeed, Martin and Suhov prove this to be the case in [MS99]. They call the resulting system a *Fast Jackson network*, since the queue lengths decrease doubly exponentially, and hence the expected time in a system is naturally faster than in a standard Jackson network. Further work by Martin examines the stochastic processes that arise from such networks in greater detail [Marb, Mara].

Given the results of Vöcking [Vöc99], it seems clear that a Jackson-like network of servers using the tie-breaking scheme leads to *Faster Jackson Networks*. A full proof of this fact, however, will require extending the work of [MS99] to this situation. Another direction to take is to try to extend these results to broader classes of networks, or use these results to bound the performance of networks that may not exhibit such pleasant properties. Given the wealth of results showing that two choices are substantially better than one in both static and dynamic situations, it seems that extending this idea in more directions in network scenarios would be worthwhile. Although the results for butterfly-like networks presented in Section 3.4 suggest that the analysis of the two-choice paradigm can be significantly more complex, these results suggest that the fluid limit approach still has potential in this area.

Notes

1. We use *with high probability* to mean with probability at least $1 - O(1/n^{\alpha})$ for some constant α; generally this will be 1. A precise analysis shows that the expected maximum load is $\Gamma^{-1}(n) - 3/2 + o(1)$ [Gon81].

2. The path coupling technique when applicable, is easier to use than standard coupling; see [Jer98] for a good survey of this technique.

3. The witness tree method is similar in spirit to the delay sequence method used to bound the message latencies of routing algorithms [Upf84, Ale82, LMRR94, MS92].

4. The fact that insertions and deletions alternate is not crucial except to ensure that the total number of balls in the system at any given time is at most n.

5. The expectation is taken over different runs of the protocol on a *given* access graph, i.e., the bins chosen by the balls are fixed.

6. The idea of using randomly-flipped switches to control congestion was first used by Ranade [Ran87] for packet routing and was later adapted to circuit-switching algorithms by Maggs and Sitaraman [MS92].

References

[ABKU99]   Y. Azar, A. Z. Broder, A. R. Karlin, and E. Upfal. Balanced allocations. *SIAM Journal on Computing*, 29:180–200, 1999. A preliminary version of this paper appeared in *Proceedings of the Twenty-Sixth Annual ACM Symposium on the Theory of Computing*, 1994.

[ABP98]   N. S. Arora, R. D. Blumofe, and C. G. Plaxton. Thread scheduling for multiprogrammed multiprocessors. In *Proceedings of the Tenth ACM Symposium on Parallel Algorithms and Architectures*, pages 119–129, June 1998.

[ABS98]   M. Adler, P. Berenbrink, and K. Schröder. Analyzing an infinite parallel job allocation process. *Lecture Notes in Computer Science (Proceedings of ESA 1998)*, 1461:417–428, 1998.

[ACMR95]   M. Adler, S. Chakrabarti, M. Mitzenmacher, and L. Rasmussen. Parallel randomized load balancing. In *Proceedings of the Twenty-Seventh Annual ACM Symposium on the Theory of Computing*, pages 238–247, May 1995.

[Ada94]   I. J. B. F. Adan. *A compensation approach for queueing problems*. CWI (Centrum voor Wiskunde en Informatica), 1994.

[AFP98]   J. Aronson, A. Frieze, and B. Pittel. Maximum matchings in sparse random graphs: Karp-Sipser revisited. *Random Structures and Algorithms*, 12(2):111–177, March 1998.

[AH90]   M. Alanyali and B. Hajek. Analysis of simple algorithms for dynamic load balancing. *Mathematics of Operations Research*, 22(4):840–871, 1990.

[Ale82]   R. Aleliunas. Randomized parallel communication. In *Proceedings of the ACM SIGACT-SIGOPS Symposium on Principles of Distributed Computing*, pages 60–72, August 1982.

[AM97]   D. Achlioptas and M. Molloy. The analysis of a list-coloring algorithm on a random graph. In *Proceedings of the Thirty-Eighth Annual Symposium on Foundations of Computer Science*, pages 204–212, 1997.

[AS92]   N. Alon and J. H. Spencer. *The Probabilistic Method*. John Wiley and Sons, 1992.

[AWZ90]   I. J. B. F. Adan, J. Wessels, and W. H. M. Zijm. Analysis of the symmetric shortest queue problem. *Stochastic Models*, 6:691–713, 1990.

306

[BCSV00] P. Berenbrink, A. Czumaj, A. Steger, and B. Vöcking. Balanced allocations: The heavily loaded case. In *Proceedings of the Thirty-Second Annual ACM Symposium on Theory of Computing*, pages 745–754, 2000.

[BD97] R. Bubley and M. Dyer. Path coupling: a technique for proving rapid mixing in Markov chains. In *Proceedings of the Thirty-Eighth Annual Symposium on Foundations of Computer Science*, pages 223–231, 1997.

[BFM98] P. Berenbrink, T. Friedetzky, and E. W. Mayr. Parallel continuous randomized load balancing. In *Proceedings of the Tenth ACM Symposium on Parallel Algorithms and Architectures*, pages 192–201, 1998.

[BFS99] P. Berenbrink, T. Friedetzky, and A. Steger. Randomized and adversarial load balancing. In *Proceedings of the Eleventh ACM Symposium on Parallel Algorithms and Architectures*, pages 175–184, 1999.

[BK90] A. Z. Broder and A. Karlin. Multi-level adaptive hashing. In *Proceedings of the First Annual ACM–SIAM Symposium on Discrete Algorithms*, pages 43–53, 1990.

[BL94] R. D. Blumofe and C. E. Leiserson. Scheduling multithreaded computations by work stealing. In *Proceedings of the Thirty-Fifth Annual Symposium on Foundations of Computer Science*, pages 356–368, November 1994.

[BM00] A. Broder and M. Mitzenmacher. Using multiple hash functions to improve ip lookups. Technical Report TR–03–00, Department of Computer Science, Harvard University, Cambridge, MA, 2000.

[BMS97] P. Berenbrink, F. Meyer auf der Heide, and K. Schröder. Allocating weighted jobs in parallel. In *Proceedings of the Ninth ACM Symposium on Parallel Algorithms and Architectures*, pages 302–310, June 1997.

[CFM+98] R. Cole, A. Frieze, B.M. Maggs, M. Mitzenmacher, A. W. Richa, R. K. Sitaraman, and E. Upfal. On balls and bins with deletions. In *Second International Workshop on Randomization and Approximation Techniques in Computer Science (RANDOM)*, number 1518 in Lecture Notes in Computer Science, pages 145–158. Springer–Verlag, October 1998.

[CH91] J. P. Crametz and P. J. Hunt. A limit result respecting graph structure for a fully connected loss network with alternative routing. *The Annals of Applied Probability*, 1(3):436–444, 1991.

[CMM+98] R. Cole, B. M. Maggs, F. Meyer auf der Heide, M. Mitzenmacher, A. W. Richa, K. Schröder, R. K. Sitaraman, and B. Vöcking. Randomized protocols for low-congestion circuit routing in multistage interconnection networks. In *Proceedings of the Thirtieth Annual ACM Symposium on Theory of Computing*, pages 378–388, May 1998.

[CMS95] A. Czumaj, F. Meyer auf der Heide, and V. Stemann. Shared memory simulations with triple-logarithmic delay. *Lecture Notes in Computer Science*, 979:46–59, 1995.

[CS97] A. Czumaj and V. Stemann. Randomized allocation processes. In *Proceedings of the Thirty-Eighth Annual Symposium on Foundations of Computer Science*, pages 194–203, October 1997.

[Czu98] A. Czumaj. Recovery time of dynamic allocation processes. In *10th Annual ACM Symposium on Parallel Algorithms and Architectures*, pages 202–211, 1998.

[Dah99] M. Dahlin. Interpreting stale load information. In *Proceedings of the Nineteenth Annual IEEE International Conference on Distributed Computing Systems*, 1999. To appear in *IEEE Transactions on Parallel and Distributed Systems*.

[DKM+88] M. Dietzfelbinger, A. Karlin, K. Mehlhorn, F. Meyer auf der Heide, H. Rohnert, and R. Tarjan. Dynamic perfect hashing — upper and lower bounds. In *Proceedings of the Twenty-Ninth Annual Symposium on Foundations of Computer Science*, 1988.

[EK86] S. N. Ethier and T. G. Kurtz. *Markov Processes: Characterization and convergence*. John Wiley and Sons, 1986.

[ELZ86a] D. L. Eager, E. D Lazowska, and J. Zahorjan. Adaptive load sharing in homogeneous distributed systems. *IEEE Transactions on Software Engineering*, 12:662–675, 1986.

[ELZ86b] D. L. Eager, E. D. Lazowska, and J. Zahorjan. A comparison of receiver-intiated adaptive load sharing. *Performance evaluation Review*, 16:53–68, March 1986.

[FGC+97] A. Fox, S. D. Gribble, Y. Chawathe, E. A. Brewer, and P. Gauthier. Cluster-based scalable network services. In *Proceedings of the Sixteenth ACM Symposium on Operating Systems Principles*, October 1997.

[FKS84]   M. L. Fredman, J. Komlós, and E. Szemerédi. Storing a sparse table with $o(1)$ worst-case access time. *Journal of the ACM*, 31:538–544, 1984.

[FM87]   R. Finkel and U. Manber. DIB – A distributed implementation of backtracking. *ACM Transactions on Programming Languages and Systems*, 9(2):235–256, April 1987.

[FMM91]   R. Feldmann, P. Mysliwietz, and B. Monien. A fully distributed chess program. In *Advances in Computer Chess 6*, pages 1–27, Chichester, UK, 1991. Ellis Horwood.

[GMR94]   L. A. Goldberg, Y. Matias, and S. Rao. An optical simulation of shared memory. In *Proceedings of the Sixth ACM Symposium on Parallel Algorithms and Architectures*, pages 257–267, June 1994.

[Gon81]   G. H. Gonnet. Expected length of the longest probe sequence in hash code searching. *Journal of the ACM*, 28(2):289–304, 1981.

[Haj88]   B. Hajek. Asymptotic analysis of an assignment problem arising in a distributed communications protocol. In *Proceedings of the 27th Conference on Decision and Control*, pages 1455–1459, 1988.

[HR90]   T. Hagerup and C. Rüb. A guided tour of Chernoff bounds. *Information Processing Letters*, 33:305–308, February 1990.

[HZJ94]   M. Halbherr, Y. Zhou, and C. F. Joerg. MIMD-style parallel programming based on continuation-passing thread. Memo CSG Memo-335, Massachussets Institute of Technology,Laboratory for Computer Science, Computation Structures Group, March 1994. Also appears in *Proceedings of Second International Workshop on Massive Parallelism: Hardware, Software and Applications*, October 1994.

[Jer98]   M. Jerrum. Mathematical foundations of the markov chain monte carlo method. In *Probabilistic Methods for Algorithmic Discrete Mathematics, Algorithms and Combinatorics*, pages 116–165. Springer–Verlag, 1998.

[JK77]   N. Johnson and S. Kotz. *Urn Models and Their Application.* John Wiley and Sons, 1977.

[KLM92]   R. M. Karp, M. Luby, and F. Meyer auf der Heide. Efficient PRAM simulation on a distributed memory machine. In *Proceedings of the Twenty-Fourth Annual ACM Symposium on the Theory of Computing*, pages 318–326, May 1992.

[KLM96]   R. M. Karp, M. Luby, and F. Meyer auf der Heide. Efficient PRAM simulation on a distributed memory machine. *Algorithmica*, 16:245–281, 1996.

[KMPS95] A. Kamath, R. Motwani, K. Palem, and P. Spirakis. Tail bounds for occupancy and the satisfiability threshold conjecture. *Random Structures and Algorithms*, 7(1):59–80, August 1995.

[Knu73]   D. E. Knuth. *The Art of Computer Programming I–III*. Addison–Wesley, Reading, MA, second edition, 1973.

[KS81]   R. M. Karp and M. Sipser. Maximum matchings in sparse random graphs. In *Proceedings of the Twenty-Second Annual Symposium on Foundations of Computer Science*, pages 364–375, 1981.

[Kur70]   T. G. Kurtz. Solutions of ordinary differential equations as limits of pure jump Markov processes. *Journal of Applied Probability*, 7:49–58, 1970.

[Kur81]   T. G. Kurtz. *Approximation of Population Processes*. SIAM, 1981.

[KVV90]   R. M. Karp, U. V. Vazirani, and V. V. Vazirani. An optimal algorithm for on-line bipartite matching. In *Proceedings of the Twenty-Second Annual ACM Symposium on Theory of Computing*, pages 352–358, 1990.

[KZ93]   R. M. Karp and Y. Zhang. Randomized parallel algorithms for backtrack search and branch-and-bound computation. *Journal of the ACM*, 40(3):765–789, July 1993.

[Lei92]   F. T. Leighton. *Introduction to Parallel Algorithms and Architectures: Arrays • Trees • Hypercubes*. Morgan Kaufmann, San Mateo, CA, 1992.

[Lin92]   T. Lindvall. *Lectures on the Coupling Method*. John Wiley and Sons, 1992.

[LMRR94] F. T. Leighton, Bruce M. Maggs, Abhiram G. Ranade, and Satish B. Rao. Randomized routing and sorting on fixed-connection networks. *Journal of Algorithms*, 17(1):157–205, July 1994.

[LMS+97] M. Luby, M. Mitzenmacher, M. A. Shokrollahi, D. Spielman, and V. Stemann. Practical loss-resilient codes. In *Proceedings of the Twenty-Ninth Annual ACM Symposium on Theory of Computing*, pages 150–159, 1997.

[Mara]   J. B. Martin. Point processes in fast Jackson networks. Submitted for publication.

[Marb]   J. B. Martin. Stochastic bounds for fast Jackson networks. Submitted for publication.

[Mit96a]   M. Mitzenmacher. Load balancing and density dependent jump Markov processes. In *Proceedings of the Thirty-Seventh Annual Symposium on Foundations of Computer Science*, pages 213–222, 1996.

[Mit96b]   M. D. Mitzenmacher. *The power of two choices in randomized load balancing*. PhD thesis, University of California at Berkeley, Department of Computer Science, Berkeley, CA, 1996.

[Mit98]   M. Mitzenmacher. Analyses of load stealing models using differential equations. In *Proceedings of the Tenth ACM Symposium on Parallel Algorithms and Architectures*, pages 212–221, 1998.

[Mit99a]   M. Mitzenmacher. On the analysis of randomized load balancing schemes. *Theory of Computing Systems*, 32:361–386, 1999.

[Mit99b]   M. Mitzenmacher. Studying balanced allocations with differential *Combinatorics, Probability, and Computing*, 8:473–482, 1999.

[Mit00]   M. Mitzenmacher. How useful is old information? *IEEE Transactions on Parallel and Distributed Systems*, 11(1):6–20, 2000.

[MPR98]   P. D. Mackenzie, C. G. Plaxton, and R. Rajaraman. On contention resolution protocols and associated probabilistic phenomena. *Journal of the ACM*, 45(2):324–378, March 1998.

[MR95]   R. Motwani and P. Raghavan. *Randomized Algorithms*. Cambridge University Press, 1995.

[MS92]   Bruce M. Maggs and Ramesh K. Sitaraman. Simple algorithms for routing on butterfly networks with bounded queues (extended abstract). In *Proceedings of the Twenty-Fourth Annual ACM Symposium on the Theory of Computing*, pages 150–161, May 1992. To appear in SIAM Journal on Computing.

[MS99]   J. B. Martin and Y. M Suhov. Fast Jackson networks. *Annals of Applied Probability*, 9:854–870, 1999.

[MSS96]   F. Meyer auf der Heide, C. Scheideler, and V. Stemann. Exploiting storage redundancy to speed up randomized shared memory. *Theoretical Computer Science*, 162, 1996.

[MTS89]   R. Mirchandaney, D. Towsley, and J. A. Stankovic. Analysis of the effects of delays on load sharing. *IEEE Transactions on Computers*, 38:1513–1525, 1989.

[MV98]   M. Mitzenmacher and B. Vöcking. The asymptotics of selecting the shortest of two, improved. In *Proceedings of the 37th Annual Allerton Conference on Communication, Control, and Computing*, pages 326–327, 1998. Full version available as Harvard Computer Science TR-08-99.

[Ran87]   A. G. Ranade. Constrained randomization for parallel communication. Technical Report YALEU/DCS/TR-511, Department of Computer Science, Yale University, New Haven, CT, 1987.

[RCG94]   R. Rooholamini, V. Cherkassky, and M. Garver. Finding the right ATM switch for the market. *Computer*, 27(4):16–28, April 1994.

[RSAU91] L. Rudolph, M. Slivkin-Allalouf, and E. Upfal. A simple load balancing scheme for task allocation in parallel machines. In *Proceedings of the Third Annual ACM Symposium on Parallel Algorithms and Architectures*, pages 237–245, July 1991.

[SS94]   M. Shaked and J. Shantikumar. *Stochastic Orders and Their Applications*. Academic Press, Inc., 1994.

[Ste96]   V. Stemann. Parallel balanced allocations. In *Proceedings of the Eighth ACM Symposium on Parallel Algorithms and Architectures*, pages 261–269, June 1996.

[Sto83]   D. Stoyan. *Comparison Methods for Queues and Other Stochastic Models*. John Wiley and Sons, 1983.

[SW95]   A. Shwartz and A. Weiss. *Large Deviations for Performance Analysis*. Chapman & Hall, 1995.

[TY97]   J. Turner and N. Yamanaka. Architectural choices in large scale ATM switches. Technical Report WUCS–97–21, Department of Computer Science, Washington University, St. Louis, MO, 1997.

[Upf84]   E. Upfal. Efficient schemes for parallel communication. *Journal of the ACM*, 31(3):507–517, July 1984.

[Val82]   L. G. Valiant. A scheme for fast parallel communication. *SIAM Journal on Computing*, 11(2):350–361, May 1982.

[VB81]   L. G. Valiant and G. J. Brebner. Universal schemes for parallel communication. In *Conference Proceedings of the*

312

Thirteenth Annual ACM Symposium on Theory of Computing, pages 263–277, May 1981.

[VDK96]   N.D. Vvedenskaya, R.L. Dobrushin, , and F.I. Karpelevich. Queueing system with selection of the shortest of two queues: an asymptotic approach. *Problems of Information Transmission*, 32:15–27, 1996.

[Vöc99]   B. Vöcking. How asymmetry helps load balancing. In *Proceedings of the Fortieth Annual Symposium on Foundations of Computer Science*, pages 131–140, 1999.

[VS97]   N.D. Vvedenskaya and Y.M. Suhov. Dobrushin's mean-field approximation for a queue with dynamic routing. *Markov Processes and Related Fields*, 3(4):493–526, 1997.

[Web78]   R. Weber. On the optimal assignment of customers to parallel servers. *Journal of Applied Probabiblities*, 15:406–413, 1978.

[Whi86]   W. Whitt. Deciding which queue to join: Some counterexamples. *Operations Research*, 34:55–62, 1986.

[Win77]   W. Winston. Optimality of the shortest line discipline. *Journal of Applied Probabilities*, 14:181–189, 1977.

[Wor95]   N. C. Wormald. Differential equations for random processes and random graphs. *Annals of Applied Probabilities*, 5:1217–1235, 1995.

Chapter 10

RANDOMIZED TECHNIQUES FOR MODELLING FAULTS AND ACHIEVING ROBUST COMPUTING

Sotiris E. Nikoletseas
Computer Technology Institute and
Patras University, Greece
nikole@cti.gr

Paul G. Spirakis
Computer Technology Institute and
Patras University, Greece
spirakis@cti.gr

Abstract   Computing in the presence of faults is a challenging problem. Modelling of faults in communication networks leads to graph models where some edges (or vertices) may be missing with a certain probability. Faults in computations can be modeled by stochastic patterns of unavailability of some processing power. In this chapter we discuss various structural and simulation results of the area of robust computations, which make heavy use of randomization techniques.

Introduction

Graphs are simple models of communication networks. Each edge (that is a communication line) may have a certain probability of failing. In the simplest case it is assumed that each communication line transmits information both ways and that the lines fail independently and with a common probability. Thus one is lead to various random graph models. For example, the $G_{n,p}$ model $(0 < p < 1)$ consists of all graphs of n vertices in which the edges are chosen independently and with probability p. Another useful model for structured networks is the G_n^r, i.e. the class of all regular graphs of n vertices and degree r $(0 < r < n)$. We can turn this into a probability space by giving

S. Rajasekaran et al (eds.), Handbook of Randomized Computing, Volume 1, pp, 313–399.

all the elements of the class the same probability. If, in addition, we remove each edge of a random element of G_n^r with probability $1 - p$, then we have the model of random regular graphs with edge faults, denoted by $G_{n,p}^r$. For an excellent summary of work on these models see the book "Random Graphs" of B. Bollobás ([10]).

Perhaps the most celebrated theorem in Random Graph Theory is that of the Giant Component: Erdös and Renyi ([26]) first showed that $G_{n,p}$ has a giant (i.e. $\Theta(n)$-sized) connected component when, roughly, $p > \frac{1}{n}$. In fact, many of the results in Random Graphs were inspired by the work of Uncle Paul Erdös.

Advances in computer technology made parallel and distributed machines a reality. It is a challenge to design such systems, or protocols for such systems, that compute correctly and efficiently in the presence of faults. The difficulty associated with combining fault-tolerance with efficiency is that the two have conflicting means, as observed by Kanellakis and Shvartsman in [44]: fault-tolerance is achieved by introducing redundancy, while efficiency is achieved by removing redundancy. It is interesting that randomized techniques can help to achieve both means (or to sharply express their tradeoff).

In this chapter, we take a partial view of the models of random faults and randomized robustness techniques. In the first half, we examine structural properties of random graphs and random regular graphs with edge faults, such as connectivity, expansion, dominance and trade-offs between density and robustness. In the second half, we survey recent work on robust parallel and distributed computing and present an interesting case study. The writing is biased by the work of the authors and aims at motivating further work in this fascinating field where randomization techniques are essential for both modeling and analysis.

1. BASIC PROBABILISTIC TECHNIQUES

1.1 INTRODUCTION

The common, unifying concept of the various techniques comprising the Probabilistic Method is the non-constructive proof of the existence of combinatorial structures satisfying certain desired properties.

The basic probabilistic technique, which we call the method of positive probability, can be described in the following way: in order to prove the existence of a combinatorial structure satisfying certain desired properties, we first construct (by an abstract random experiment) an appropriate probability space, whose sample points correspond to the combinatorial structures under examination. Then we show that the desired property is satisfied in this space with positive (non-zero) probability. This implies, by the axiomatic definition of the concept of probability, the existence of at least one point in the space (that is, a corresponding combinatorial structure) with the desired property.

In this section, we will present some basic probabilistic techniques. In each case, we will also provide a characteristic example from recent research. More specifically:

1. we will present the use of Markov's inequality in proving non-existence results. As an example, we will show that in dense random graphs there do not exist dominating sets of size less than $\ln n$, almost certainly.

2. we will then employ the famous second moment method in order to prove that dominating sets of size $\lceil \log n \rceil$ exist in dense random graphs, with high probability.

3. we will present Chernoff bounds, a strong probability-theoretic tool for bounding the large deviation probability of a sum of independent random variables from its mean, widely used in the analysis of randomized algorithms. As an example, we provide an efficient randomized algorithm for finding near-optimal dominating sets in dense random graphs.

4. we also present the notion of random configurations and use it in order to prove various connectivity properties (and also, by a BFS approach, to prove the existence of a giant component) in random regular graphs with edge faults.

1.2   FIRST MOMENT ARGUMENTS FOR NON-EXISTENCE PROOFS

We will use the following important inequality:

Theorem 1 (Markov's inequality) *Let X any non-negative random variable. For every $t > 0$,*

$$\Pr\{X \geq t\} \leq \frac{E(X)}{t}$$

□

As an immediate consequence of the above theorem we get that:

Corollary 1 *For any non-negative, integer-valued random variable $X(n)$,*

$$\lim_{n \to \infty} E[X(n)] = 0 \quad \Rightarrow \quad \lim_{n \to \infty} \Pr\{X(n) = 0\} = 1$$

□

Thus, in order to show the almost certain non-existence of desired combinatorial structures, it suffices to prove that the mean of a random variable X counting their number tends to 0.

1.3   LOWER BOUNDS FOR DOMINATING SETS IN DENSE RANDOM GRAPHS

The dominating set problem is the following: given a graph $G(V, E)$ and a positive integer $k < |V|$, is there a subset $V' \subseteq V$ (called a dominating set) with $|V'| \leq k$ such that for all u in $V - V'$ there is a v in V' for which $\{u, v\} \in E$? The problem is NP-Complete (see [34]), even for planar graphs with maximum degree 3. In [60], Nikoletseas and Spirakis have studied the "average case" complexity of the problem, following the line of research of [25]. More specifically, they have examined the existence and efficient finding of small dominating sets in dense random graphs. They show, for the model $G_{n,p}$ with $p = 1/2$, that:

1. The probability of existence of dominating sets of size less than $\ln n$ tends to zero as n tends to infinity.

2. Dominating sets of size $\lceil \log n \rceil$ exist almost surely.

3. They provide two algorithms which construct small dominating sets in $G_{n,1/2}$ and run in $O(n \log n)$ time (on the average and also with high probability). These algorithms almost surely construct a dominating set of size at most $(1 + \epsilon) \ln n$, for any fixed $\epsilon > 0$.

These results extend easily to the case $G_{n,p}$ with p fixed to any constant < 1.

Recently, Gkantsidis, Hatzis, Nikoletseas and Spirakis ([39]) have studied the same problem in the (technically much more difficult) case of sparse random graphs, providing both lower bounds and algorithms.

In this section we prove the following result.

Theorem 2 *For any $k < \ln n$ the probability of existence of a dominating set of size k in $G_{n,1/2}$ tends to 0 as n tends to infinity.*

Proof: Let us consider a $G_{n,1/2}$ graph and let S be any fixed set of k vertices of this graph. Because of the stochastic independence characterizing the $G_{n,p}$ model, the mean of the random indicator variable

$$X_S = \begin{cases} 1 & \text{if } S \text{ is dominating} \\ 0 & \text{otherwise} \end{cases}$$

is equal to

$$E(X_S) = \left(1 - \frac{1}{2^k}\right)^{n-k}$$

since the probability that a certain vertice out of S connects to at least one vertice in it is $\left(1 - \frac{1}{2^k}\right)$ and this must happen for all $n - k$ vertices out of S. The random variable

$$X = \sum_{S, |S|=k} X_S$$

counts the number of dominating sets of size k and, because of linearity of expectation, its mean is

$$E(X) = \sum_{S,|S|=k} E(X_S) = \binom{n}{k}\left(1 - \frac{1}{2^k}\right)^{n-k}$$

But $\binom{n}{k} \le n^k$ and $1 - \frac{1}{2^k} \le e^{-\frac{1}{2^k}}$ so that

$$E(X) \le e^{\frac{k}{2^k}}\left(e^{k\ln n - \frac{n}{2^k}}\right)$$

We remark that

$$k = \frac{\ln n}{c} \quad \text{where} \quad c > 1 \Rightarrow F = k\ln n - \frac{n}{2^k} \to -\infty$$

that is the exponent F tends to infinity when $k < \ln n$ and finally

$$k < \ln n \Rightarrow E(X) \to 0$$

In other words, the expected value of the number of the dominating sets of size less than $\ln n$ tends to zero, thus there do not exist such dominating sets, almost certainly.   □

1.4   EXISTENCE RESULTS WITH THE SECOND MOMENT METHOD

The second moment method is a powerful and quite general technique widely used in proving the almost certain existence of combinatorial structures satisfying desired properties. For a complete presentation of the second moment method we suggest the excellent books of J. Spencer ([69]) and N. Alon, J. Spencer ([1]). These references also contain characteristic applications of this method, such as Paul Turán's proof on the prime factors of a number, results on distinct sums of sets of positive integers, concentration results on the clique number of dense random graphs and also thresholds for the existence of balanced subgraphs in $G_{n,p}$ random graphs.

From the probability-theoretic point of view, the second moment method is based on Chebyshev's inequality for bounding the deviations (in terms of the variance) of the values of a random variable from its mean.

Theorem 3 (Chebyshev's inequality) *For any random variable X with mean μ and variance σ^2, it is*

$$\Pr\{|X - \mu| \ge \lambda\sigma\} \le \frac{1}{\lambda^2}$$

where $\lambda > 0$ any constant.   □

Chebyshev's inequality is much stronger compared to Markov's inequality and best possible when only the mean and the variance of the random variable are known. This fact is the probabilistic-theoretic reason for why second moment arguments are much stronger than first moment ones (though technically more difficult to apply, since we must calculate the variance, whose linearity needs stochastic independence). Still, we note that there are techniques and inequalities (e.g. the Local Lemma, Janson's inequality, the Martingale technique) achieving (in the case when, roughly speaking, the stochastic dependencies are small) exponentially small bounds (compared to the only polynomial ones provided by Chebyshev's inequality) on this large deviation probability, and thus are much stronger (though less general) than the second moment method.

We now give the technical theorem the second moment method is based on:

Theorem 4 (Second Moment Method) *For any random variable X if*

$$E(X) \to \infty \quad and \quad Var(X) = o(E^2(X))$$

then

$$\Pr\{X = 0\} \to 0$$

(the limits taken with respect to an independent variable n, that is $X = X(n)$).
□

Thus, in order to deliver existence proofs using the second moment method, we must not only show that the expected value of the random variable X counting the number of the combinatorial structures under examination tends to infinity, but also that its variance grows asymptotically slower than the square of its mean.

The random variable X can be usually expressed as a sum of random indicator variables, which are in most cases not independent. We employ the notion of the covariance of two random variables, in order to capture these stochastic dependencies:

Definition 1 *The covariance of two random variables X, Y is given by*

$$Cov(X, Y) = E(XY) - E(X) E(Y)$$

From the above definition we remark that the covariance of two variables provides a measure of their stochastic dependencies, taking value zero in the case of independent variables.

We can now express the variance of the sum X of the random indicator variables mentioned earlier in terms of the covariances of the (ordered) pairs of these variables.

Theorem 5 *Let* $X = X_1 + \cdots + X_n$, *where the variances* $Var(X_i)$ *of the random indicator variables* $X_i, 1 \le i \le n$ *exist and are finite. Then*

$$Var(X) = \sum_{1 \le i,j \le n} Cov(X_i, X_j) \tag{10.1}$$

     □

As an immediate consequence, we get the following theorem, providing an upper bound for the variance in terms of the expected value.

Theorem 6 *If* $X = \sum_i X_i$, *where* $X_i, 1 \le i \le n$, *are random indicator variables, then*

$$\begin{aligned} Var(X) &= \sum_{i=1}^{n} Var(X_i) + \sum_{1 \le i \ne j \le n} Cov(X_i, X_j) \\ &\le E(X) + \sum_{1 \le i \ne j \le n} Cov(X_i, X_j) \end{aligned} \tag{10.2}$$

     □

We provide here two techniques (presented by N. Alon and J. Spencer in [1, 69]) in order to further bound the variance from above):

- using quantities Δ or Δ^*

- using quantity $f(S, T)$

These techniques are essentially the same and only suggest different ways to carry out calculations. Still, the first one might be considered as more general, since it deals with quantities Δ, Δ^* that are also used in other probabilistic techniques (e.g. Janson inequality).

Let us now define the following important quantity Δ: Let A_i, A_j the events corresponding to the random indicator variables X_i, X_j and the relation $i \sim j$, holding when $i \ne j$ and the events A_i, A_j are stochastically dependent. Then

$$\Delta = \sum_{i \sim j} \Pr\{A_i \wedge A_j\}$$

Since

$$Cov(X_i, X_j) \le E(X_i X_j) = \Pr\{A_i \wedge A_j\}$$

we get, from inequality (10.2), that

$$Var(X) \le E(X) + \Delta$$

and finally the basic theorem of the second moment method becomes:

Theorem 7

$$E(X) \to \infty \quad \text{and} \quad \Delta = o(E^2(X)) \quad \Rightarrow \quad \Pr\{X = 0\} \to 0$$

$\square$

We also note the following useful variation of the above theorem: we call the events A_i, A_j symmetric, if

$$\Pr\{X_j | X_i = 1\} = \Pr\{X_i | X_j = 1\}$$

Note that in most cases the above condition is satisfied, especially when studying graph-theoretic functions where the events A_i, A_j concern properties of subgraphs where "symmetry" appears naturally, in the sense that order does not matter and we can, without loss of generality, assume any of the two events as the event which we condition on.

In the case of symmetric events, let

$$\Delta^* = \sum_{j \sim i} \Pr\{A_j | A_i\}$$

Then, by using the definition of conditional probability, we get

$$\Delta = \Delta^* E(X)$$

and finally

Theorem 8

$$E(X) \to \infty \quad \text{and} \quad \Delta^* = o(E(X)) \quad \Rightarrow \quad \Pr\{X = 0\} \to 0$$

$\square$

We now provide the second alternative way to carry out technical calculations in order to bound the variance from above (see also J. Spencer's presentation in [69]).

Let X be expressed as a sum of random indicator variables $X_1, X_2, \ldots, X_m$, each having mean μ. According to equation (10.1), we get

$$Var(X) = \sum_{S,T} Cov(X_S, X_T)$$

where S, T are the sets corresponding to i, j, respectively. Clearly

$$Cov(X_S, X_T) = \mu^2 f(S, T)$$

where

$$f(S, T) = \frac{E(X_T | X_S = 1)}{E(X_T)} - 1$$

is a quantity measuring the dependency of X_T from X_S (as an example of this, remark that it takes value 0 if X_S, X_T are independent).

Since there are m^2 ways to select ordered pairs from the m variables, we get

$$Var(X) = m^2 \mu^2 E_{S,T} \left[f(S,T) \right]$$

(where $E_{S,T} \left[f(S,T) \right]$ is the expected value of $f(S,T)$ when set S is fixed and set T is randomly chosen among all possible sets) and because $E(X) = m\mu$ we get

$$\frac{Var(X)}{E^2(X)} = E_{S,T} \left[f(S,T) \right]$$

Hence, the basic theorem of this variation of the second moment method becomes:

Theorem 9 *Let X a random variable expressed as a sum of indicator variables. We have*

$$E(X) \to \infty \quad and \quad E_{S,T} \left[f(S,T) \right] = o(1) \quad \Rightarrow \quad \Pr\{X = 0\} \to 0$$

$\square$

We note that $E_{S,T} \left[f(S,T) \right]$ depends on the size of intersection $S \cap T$ and is actually a sum of the products of the probabilities of each intersection size times the corresponding $f(S,T)$.

In the next section we present, as an example of the use of the second moment method (and its $f(S,T)$ variation), a result of Nikoletseas and Spirakis ([60]), proving the almost certain existence of dominating sets of size $\lceil \log n \rceil$ in $G_{n,\frac{1}{2}}$ dense random graphs.

1.5   LOGARITHMIC DOMINATING SETS IN DENSE RANDOM GRAPHS

We first remark (see also section 1.3) that the expectation of the number $X^{(k)}$ of dominating sets of size k is:

$$E\left(X^{(k)} \right) = \binom{n}{k} \left(1 - \frac{1}{2^k} \right)^{n-k}$$

Next we easily get the following theorem:

Theorem 10

$$E\left(X^{(\log n)} \right) \to +\infty$$

$\square$

Thus, it is enough to show that $Var\left(X^{(k)}\right) = o\left(E^2\left(X^{(k)}\right)\right)$ when $k = \log n$. By fixing a set S of size $\log n$ and by randomly choosing sets T of the same size, we get for $k = \log n$:

$$\frac{Var\left(X^{(k)}\right)}{E^2\left(X^{(k)}\right)} = \sum_{i=0}^{k} a(i) \, f_i'(S,T) \tag{10.3}$$

where:

$$a(i) = \Pr\{|S \cap T| = i\}$$

and

$$f_i(S,T) = f(S,T) = \frac{E(X_T|X_S = 1)}{E(X_T)} - 1$$

when $|S \cap T| = i$ and X_T, X_S are indicator variables for the sets T, S with respect to being dominating sets.

We will now bound appropriately from above the terms of the products in the sum. Clearly:

$$\alpha(i) = \frac{\binom{k}{i}\binom{n-k}{k-i}}{\binom{n}{k}}$$

Lemma 1 *There exists a constant γ and a positive integer n_0 : $\forall n \geq n_0$ and $\forall i = 1, \ldots, \log n$:*

$$f_i(S,T) \leq \gamma$$

Proof: Since X_T is an indicator variable, we get $E(X_T|X_S = 1) \leq 1$. Thus

$$f_i(S,T) \leq \frac{1}{E(X_T)} - 1$$

But

$$E(X_T) = \left(1 - \frac{1}{2^{\log n}}\right)^{n - \log n} \geq \left(1 - \frac{1}{n}\right)^n$$

But $\forall n \geq 2 : \left(1 - \frac{1}{n}\right)^n \geq 1/4$, thus $E(X_T) \geq 1/4$, hence $f_i(S,T) \leq 3$ and choose $\gamma = 3$. $\qquad \square$

Lemma 2 *$\alpha(i)$ is monotone decreasing on i for $i = 1, \ldots, \log n$.*

Proof: Simply remark that

$$\Lambda = \frac{\alpha(i+1)}{\alpha(i)} = \frac{k-i}{i+1} \frac{k-i}{n-2k+i+1} \leq \frac{k^2}{n-2k+1}$$

for $i \geq 0$. For $k = \log n$, we get

$$\Lambda \leq \frac{(\log n)^2}{n - 2\log n + 1} < 1 \quad \text{for all} \ \ n > 37$$

We note here, since we are investigating asymptotic results for large graphs $(n \to \infty)$, that small finite graphs of a few nodes are not of interest and can be in any case processed by brute force. $\qquad \square$

Lemma 3 *For $i = 2, 3, \ldots, \lceil \log n \rceil$ we have*

$$\alpha(i) \leq \frac{4}{n}$$

Proof: From previous lemma we have $\alpha(i) \leq \alpha(2)$. But

$$\alpha(2) = \frac{\binom{k}{2}\binom{n-k}{k-2}}{\binom{n}{k}} \leq k^4 \; \frac{1}{n-k+1} \; \frac{1}{n-k+2}$$

i.e. for $k = \log n$ and sufficiently large n

$$\alpha(2) \leq \frac{(\log n)^4}{(n - \log n + 1)^2} \leq \frac{4}{n}$$

$\qquad \square$

We also bound the first term of sequence $a(i)$ from above:

Lemma 4

$$\alpha(1) \leq \frac{(\log n)^2}{n - \log n + 1}$$

Proof:

$$\alpha(1) = \frac{\binom{k}{1}\binom{n-k}{k-1}}{\binom{n}{k}} \leq \frac{k^2}{n-k+1}$$

$\qquad \square$

We know get:

Theorem 11 *As $n \to +\infty$*

$$\frac{Var\left(X^{(\log n)}\right)}{E^2\left(X^{(\log n)}\right)} \to 0$$

Proof: By equation (10.3) and previous lemmas we get

$$\frac{Var\left(X^{(\log n)}\right)}{E^2\left(X^{(\log n)}\right)} \leq \gamma\alpha(1) + \gamma \sum_{i=2}^{\log n} \alpha(i)$$

$$\leq \gamma \frac{(\log n)^2}{n - \log n + 1} + \gamma \frac{4\log n}{n}$$

$$\leq \frac{18 \, (\log n)^2}{n} \to 0$$

□

Thus we finally get:

Theorem 12 *There exists a dominating set of size $\lceil \log n \rceil$ in $G_{n,1/2}$ with probability at least $1 - 18 \frac{(\log n)^2}{n}$ i.e. almost surely.* □

1.6   CHERNOFF BOUNDS

Chernoff bounds ([15]) are strong probability-theoretic inequalities bounding from above the large deviation probability of a sum of independent random variables from its mean, thus proving sharp concentration of the values of such sums around their expected value. These bounds are widely used in the analysis of randomized algorithms, where the random variables of such sums are indicator variables corresponding to Poisson (or Bernoulli, in the case of same probability) trials representing the repetitive execution of independent experiments with two possible outcomes (success, failure). These outcomes correspond to the realization or not of certain events during the repetitive random choices which the algorithm's evolution is based on.

For a detailed presentation of Chernoff bounds see the excellent book of R. Motwani and P. Raghavan on Randomized Algorithms ([58]) and also the detailed survey in [40]. We here provide the following general form of the Chernoff bound for the tail probability of exceeding the mean.

Theorem 13 *Let $X_1, \ldots, X_n$ independent random variables corresponding to Poisson trials with success probabilities $\Pr\{X_i\} = p_i$, $1 \leq i \leq n$. The probability that their sum $X = \sum_{i=1}^{n} X_i$ exceeds its expected value $\mu = \sum_{i=1}^{n} p_i$ is bounded from above according to the following inequality*

$$\Pr\{X \geq (1 + \beta)\mu\} \leq \left(\frac{e^{\beta}}{(1 + \beta)^{1+\beta}} \right)^{\mu}$$

where $\beta > 0$ a constant. □

We also present the following (less general but easier to use and very popular) form, when $\beta \in (0, 1)$:

Theorem 14 *For any $\beta \in [0, 1]$ we have*

$$\Pr\{X \geq (1 + \beta)\mu\} \leq e^{-\beta^2 \mu/3}$$

□

Next we provide a similar theorem for the tail probability of values below the mean.

Theorem 15 *Let $X_1, \ldots, X_n$ independent variables corresponding to Poisson trials with success probabilities $\Pr\{X_i\} = p_i$, $1 \leq i \leq n$. The probability that their sum $X = \sum_{i=1}^{n} X_i$ takes values below its mean $\mu = \sum_{i=1}^{n} p_i$ is bounded from above according to the following inequality*

$$\Pr\{X \leq (1 - \beta)\mu\} \leq e^{-\beta^2 \mu/2}$$

where $\beta \in [0, 1]$ a constant. □

Combining the above theorems, we easily get the following bound for the tail probabilities of large deviations from the mean.

Theorem 16 (Chernoff bound) *Let $X_1, X_2, \ldots, X_n$ be independent random variables corresponding to Poisson trials with success probabilities $\Pr\{X_i\} = p_i$, $1 \leq i \leq n$. The probability that their sum $X = \sum_{i=1}^{n} X_i$ is concentrated around its mean $\mu = \sum_{i=1}^{n} p_i$ is at least*

$$\Pr\{X \in (1 \pm \beta)\mu\} \geq 1 - 2e^{-\beta^2 \mu/3}$$

where $\beta \in [0, 1]$ a constant. □

In the next section, we give an example of the use of Chernoff bounds, presenting the analysis of a fast algorithm for finding near-optimal dominating sets in dense random graphs, with high probability and also on the average.

1.7   A GREEDY ALGORITHM FOR NEAR-OPTIMAL DOMINATING SETS IN DENSE RANDOM GRAPHS

The greedy approach each time chooses a vertex to be put in a dominating set under construction and then it deletes all its neighbors (and the corresponding edges) from the graph, since these vertices are already covered by the set. Note that, because of deleting these inspected edges, the remaining graph is each time still random, since its edges are not exposed. The method achieves dominating sets of size $(1 + \epsilon) \log n$ almost surely and runs in $O(n \log n)$ expected time.

In the following technical description of the algorithm, D is the dominating set under construction, while V_i, E_i are each time the remaining sets of vertices, edges respectively. We also each time note by N_i the neighboring vertices of selected vertex u_i in the graph $G(V_i, E_i)$ and we use set V' to temporarily store the remaining vertex set $V_i - N_i$.

ALGORITHM "GREEDY"

Input: A random graph $G(V, E)$ of $G_{n,1/2}$ and a constant $\epsilon > 0$
(1) $i \leftarrow 0$; $V_i \leftarrow V$; $D \leftarrow \emptyset$
(2) **until** $|V_i| \leq \epsilon \log n$ **do**
  begin
    select a vertex $u_i \in V_i$
    $V' \leftarrow V_i - N_i$
    $D \leftarrow D \cup \{u_i\}$
    $i \leftarrow i + 1$
    $V_i \leftarrow V'$
  end
(3) $D \leftarrow D \cup V_i$
(4) **return** D

The basic idea of the Chernoff bound analysis of the algorithm above is the following: while the number of vertices $|V_i|$ in the remaining graph is still at least logarithmic, Chernoff bounds prove that with high probability the selected vertex each time connects with about half of the remaining, and thus the number of remaining vertices is divided by two each time. Hence this greedy process performs well for a logarithmic number of repetitions, after which the number $|V_i|$ of remaining vertices falls below $\epsilon \log n$ and Chernoff bounds do not provide satisfactory probability anymore. We then finish, by explicitly including the remaining vertices in the dominating set under construction D.

We now provide technical lemmas for the analysis of the algorithm. By using Theorem 15, we get for the $|V_i| - 1$ Bernoulli trials with success probability $\frac{1}{2}$ corresponding to set N_i that:

Lemma 5 *Let $\beta \in [0, 1]$ a constant. For any, arbitrarily small, constant $\epsilon > 0$ and given $|V_i| \geq \epsilon \log n$ we have*

$$\Pr\left\{|N_i| \geq (1 - \beta)\frac{|V_i| - 1}{2}\right\} \geq 1 - \alpha n^{-\frac{\beta^2}{4}\epsilon}$$

where $\alpha = 1.28$.   □

We now define the event the above lemma is conditioned on.

Definition 2 *Let $\varepsilon_i = $ "during the i-th execution of the loop of algorithm GREEDY, $|N_i| \geq (1 - \beta)\frac{|V_i| - 1}{2}$ "*

Definition 3 *Let $\varepsilon = \varepsilon_1 \cap \varepsilon_2 \cap \cdots \cap \varepsilon_t$ until $|V_t| < \epsilon \log n$*

We next provide the following theorem bounding the number of repetitions of the loop of the algorithm, and thus the number of vertices put in the dominating set during phase (2).

Lemma 6 *Conditioning on event* ε, *the number* t *of repetitions of the loop of the algorithm is at most* $(1 + \epsilon') \log n$, *where* $\epsilon' > 0$ *a constant.*

Proof: Clearly $|V_{i+1}| \leq (1 - \gamma)|V_i|$ after i-th repetition, where $\gamma = \frac{1-\beta}{2}$, and cumulatively $|V_t| \leq (1 - \gamma)^t n$. Since $(1 - \gamma)^t n \leq \epsilon \log n$ we get $t \geq \frac{\log n}{\log(\frac{1}{1-\gamma})} - \Theta(\log \log n)$. We note that $\frac{1}{1-\gamma} = \frac{2}{1+\beta}$. By choosing, for any $\epsilon' > 0$,

$$\beta = 2^{\frac{\epsilon'}{1+\epsilon'}} - 1$$

we finally get $t \leq (1 + \epsilon') \log n$.

□

We now prove high probability for the event ε.

Lemma 7 *Let any constants* $\beta \in [0, 1], \epsilon > 0$ *and* $\alpha = 1.28$. *We have*

$$\Pr\{\varepsilon\} \geq 1 - \alpha n^{-\frac{\beta^2}{8}\epsilon}$$

Proof: By lemmas 5, 6 we get

$$\Pr\{\overline{\varepsilon}\} \leq \sum_{j=1}^{t} \Pr\{\overline{\varepsilon_j}\} \leq \alpha t n^{-\frac{\beta^2}{4}\epsilon}$$

hence

$$\Pr\{\overline{\varepsilon}\} \leq \alpha(1 + \epsilon')(\log n) \, n^{-\frac{\beta^2}{4}\epsilon} \leq \alpha n^{-\frac{\beta^2}{8}\epsilon}$$

for any constants $\beta \in [0, 1], \epsilon > 0, \epsilon' > 0$.

□

Combining the above results we get the following final theorem:

Theorem 17 *Algorithm "GREEDY" constructs a dominating set of size* $(1 + \epsilon' + \epsilon) \log n$ *in time* $O((1 + \epsilon')n \log n)$ *with probability at least* $1 - n^{-\frac{\beta^2}{8}\epsilon}$.
□

We remark that algorithm "GREEDY" constructs a near-optimal dominating set, in small polynomial time, both with high probability and on the average.

However, we remark that the probability of success of algorithm "GREEDY", though tending to 1 with n, is not polynomially big, since in $1 - n^{-\frac{\beta^2}{8}\epsilon}$ constant $\frac{\beta^2}{8}\epsilon$ is just positive.

In [60], Nikoletseas and Spirakis provide also a better algorithm for the same problem (algorithm "REPEATED TRIALS"), with success probability $1 - n^{-\alpha}$, where $\alpha > 1$ a constant. This algorithm is based on the fact that any fixed vertex set of size $(1 + \epsilon) \log n$ ($\epsilon > 0$ constant) is a dominating set with high probability.

Although this abundance of near-optimal dominating sets implies that by choosing such a set we may succeed in finding it to be dominating with high probability, the repetition of this experiment provides trials that are not independent since the edges between previously and currently tried sets have been exposed already.

The main idea of this algorithm is to exploit the abundance of dominating sets of size $(1 + \epsilon) \log n$ by repetitively trying new vertex sets in such a way that the independence of the trials is ensured, while "backward" dominance is guaranteed by construction.

1.8 RANDOM REGULAR GRAPHS WITH EDGE FAULTS AND RANDOM CONFIGURATIONS

We note that it is not as easy (from the technical point of view) as in the $G_{n,p}$ case to argue about random regular graphs, because of the stochastic dependencies on the existence of the edges due to regularity. The notion of *configurations*, which we present in the next subsection, was introduced by B. Bollobás ([11, 10]) to translate statements for random regular graphs to statements for the corresponding configurations which avoid the edge dependencies due to regularity and thus are much easier to deal with.

To model edge faults we use the notion of type$-H$ random graphs. Let H be an undirected graph. A *random graph of type$-H$* is obtained by selecting edges of H independently and with probability p. Let $f = 1 - p$ be called the *link failure probability*.

We study here a simple but fundamental type of H: H being a *random* member of the class of all regular graphs of n nodes and degree r, leading to *random regular graphs with edge faults*, a new class which we call $G_{n,p}^r$. In the model $G_{n,p}^r$, the random selection of a regular graph is capturing the *random allocation of a subnetwork* to a computation (e.g. by the operating system) and the edge failure probability $f = 1 - p$ is capturing the link faults that may happen.

An explicit definition of the new probability space $G_{n,p}^r$ follows.

Definition 4 (the $G_{n,p}^r$ probability space) *Let G_n^r be the probability space of all random regular graphs with n vertices where the degree of each vertex is r and each such graph is selected equiprobably. The probability space $G_{n,p}^r$ of random regular graphs with edge faults is constructed by the following two subsequent random experiments: first, we choose a random regular graph from the space G_n^r and, second, we randomly and independently delete each edge from this graph with probability $f = 1 - p$.*

A preliminary presentation ([62]) of this new model has already inspired relevant research: Nikoletseas et al investigate in [62, 63] important connectivity properties of $G_{n,p}^r$ graphs by estimating the ranges of r, p for which, with high probability, $G_{n,p}^r$ graphs a) are highly connected b) become disconnected and c) admit a giant connected component of small diameter. Also, recently Andreas Goerdt ([35]) continued the work presented in [62] and estimated the threshold probability for the existence of a linear sized component in the faulty version of almost all random regular graphs.

In order to deal with this new model $G_{n,p}^r$, we extend the notion of configurations and the translation lemma between configurations and random regular graphs provided by B. Bollobás ([11, 10]), by introducing here the concept of *random configurations* to account for edge faults, and by also providing an *extended translation lemma* between random configurations and random regular graphs with edge faults.

1.8.1   Bollobás translation lemma.   The notion of *configurations* was introduced by B. Bollobás ([11], [10]) to translate statements for random regular graphs to statements for the corresponding configurations which avoid the edge dependencies due to regularity and thus are much easier to deal with:

Definition 5 (configurations, Bollobás, [11]) *Let G be a graph with degrees $d_i, i = 1, \ldots, n$. Let $w = \cup_{j=1}^n w_j$ be a fixed set of $2m = \sum_{j=1}^n d_j$ labeled vertices where $|w_j| = d_j$. A configuration F is a partition of w into m pairs of vertices, called edges of F.*

Given a configuration F, let $\theta(F)$ be the (multi)graph with vertex set V in which (i,j) is an edge if and only if F has a pair (edge) with one element in w_i and the other in w_j.

Bollobás notes that every regular graph $G \in G_n^r$ is of the form $G = \theta(F)$ for exactly $(r!)^n$ configurations. However not every configuration F with $d_j = r$ for all j corresponds to a $G = \theta(F) \in G_n^r$ since F may have an edge entirely in some w_j or parallel edges joining w_i and w_j.

Let ϕ be the set of all configurations F and let G_n^r be the set of all regular graphs. Given a property (set) $Q \subseteq G_n^r$ let $Q^* \subseteq \phi$ such that $Q^* \cap \theta^{-1}(G_n^r) = \theta^{-1}(Q)$.

In order to identify the set $\{\theta^{-1}(G) : G \in G_n^r\}$, Bollobás considers possible cycles of length one (self-loops) and two (loops) among pairs w_i, w_j of a configuration. Let $X_1(F), X_2(F)$ be the numbers of self-loops, loops in a configuration F, respectively.

Clearly, $\theta^{-1}(G_n^r) = \{F : X_1 = X_2 = 0\}$. Thus, if we turn θ into a probability space by giving all configurations the same probability, we can estimate the probability of a property Q, from the corresponding set of configurations Q^* such that $Q^* \cap \theta^{-1}(G_n^r) = \theta^{-1}(Q)$, by using the trivial fact that:

$$\Pr\{Q\} = \Pr\{Q^* | X_1 = X_2 = 0\} \geq 1 - \frac{1 - \Pr\{Q^*\}}{\Pr\{X_1 = X_2 = 0\}} \qquad (10.4)$$

Bollobás proves, in an ingenious way, by examining joint factorial moments, that the distributions of X_1, X_2 are asymptotically independent Poisson distributions, with mean $E(X_i) = \frac{\lambda^i}{2i}, i = 1, 2$ where

$$\lambda = \frac{1}{m} \sum_{i=1}^{n} \binom{d_i}{2} = r - 1$$

thus

$$\Pr\{X_1 = X_2 = 0\} \sim e^{-\frac{\lambda}{2} - \frac{\lambda^2}{4}}$$

So, Bollobás shows the following important lemma:

Lemma 8 (Bollobás, [10]) *If $r \geq 2$ is fixed and Q^* holds for a.e. configuration, then Q holds for a.e. r-regular graph.* □

The main importance of the above lemma is that when we study random regular graphs whose degree does not grow too fast with n, instead of considering the set of all random regular graphs, we can study the (much more easier to deal with) set of configurations.

1.8.2   An extended translation lemma.   In order to deal with edge failures, we present here the following extension of the notion of configurations:

Definition 6 (random configurations, Nikoletseas et al, [62]) *Let G a graph with degrees $d_i, i = 1, \ldots, n$. Let $w = \cup_{j=1}^{n} w_j$ be a fixed set of $2m = \sum_{j=1}^{n} d_j$ labeled "vertices" where $|w_j| = d_j$. Let F be any configuration of the set ϕ. For each edge of F, remove it with probability $1 - p$, independently. Let $\hat{\phi}$ be the new set of objects and $\hat{F}$ the outcome of the experiment. $\hat{F}$ is called a random configuration.*

Now let $\bar{Q}$ be a property for $G_{n,p}^r$ graphs and $\hat{Q}$ the corresponding property for random configurations.

By introducing probability p in every edge, direct translation of the proof of Lemma 8 leads (since in both $\bar{Q}$ and $\hat{Q}$ each edge has the same probability and independence to be deleted, thus the modified spaces follow the properties of Q and Q^*) to the following extension to random configurations.

Lemma 9 (extended translation lemma) *Let $r \geq 2$ fixed and $\bar{Q}$ be a property for $G_{n,p}^r$ graphs. If $\hat{Q}$ holds for a.e. random configuration, then the corresponding property $\bar{Q}$ holds for a.e. graph in $G_{n,p}^r$.* □

Important Remark: We note that, *for monotone increasing properties*, results proved using the above extended translation lemma for $r \geq 2$ fixed, *obviously extend to greater (unbounded) r*, since (because of the monotonicity) increasing r can only strengthen the probability of such properties. In this work, we use this remark for the monotone properties a) of graph connectivity and b) of the existence of a giant connected component in a graph.

In the next two sections, we present the work of Nikoletseas et al ([62, 63]) concentrating on the investigation of connectivity properties of $G_{n,p}^r$. We show that:

1. $G_{n,p}^r$ is r−connected, except for $O(1)$ vertices, with high probability, for all failure probabilities $f \leq n^{-\epsilon}$ ($\epsilon > 0$ fixed).

2. $G_{n,p}^r$ is disconnected, almost certainly, for constant f and any $r \leq \frac{1}{2}\sqrt{\log n}$, but is highly connected, with high probability, when $r \geq \alpha \log n$, where $\alpha > 0$ a constant.

3. Even when $G_{n,p}^r$ becomes disconnected, it still has a giant connected component of small diameter, even when $r = O(1)$.

1.9 CONNECTIVITY PROPERTIES IN RANDOM REGULAR GRAPHS WITH EDGE FAULTS

Let G_n^r be the class of all regular graphs on n vertices of degree r. It is known (see e.g. [10]) that G_n^r is almost certainly r−connected for $r \geq 3$. In the next section we show the impressive effect of constant faults in connectedness, since we prove that the corresponding graphs with faults become, in the case of small degrees, disconnected, with high probability.

1.9.1 $G_{n,p}^r$ becomes disconnected in the case of constant edge failure probability and small degrees . Using the notion of random configurations presented in the previous section, we can now prove the following theorem:

Theorem 18 *When $2 \leq r \leq \frac{\sqrt{\log n}}{2}$ and $p = \Theta(1)$ then $G_{n,p}^r$ has at least one isolated node with probability at least $1 - n^{-k}, k \geq 2$.*

Proof: Consider the random configuration $\hat{F}$ corresponding to $G_{n,p}^r$. In the configuration F (out of which $\hat{F}$ was produced) there exist $n/(r+1)$ "vertices" that do not share an "edge". The probability that each particular one of them (say v) is isolated in $\hat{F}$ is:

$$\Pr\left\{v \text{ isolated in } \hat{F}\right\} = (1-p)^r$$

With respect to the above event, the $n/(r+1)$ vertices define $n/(r+1)$ *independent* experiments. Let $\hat{E}$ the event "there exists at least one isolated vertex in $\hat{F}$ out of the $\frac{n}{r+1}$ vertices". Thus we have:

$$\Pr\left\{\hat{E}\right\} = 1 - (1 - (1-p)^r)^{\frac{n}{r+1}} \geq 1 - \exp\left(-\frac{n}{r+1}(1-p)^r\right)$$

But for $r \leq \frac{\sqrt{\log n}}{2}$ and by setting $f = 1 - p$, we get:

$$\Pr\left\{\hat{E}\right\} \geq 1 - \exp\left(-\frac{n^{1 - \frac{\log(1/f)}{2\sqrt{\log n}}}}{0.5\sqrt{\log n} + 1}\right) \geq 1 - n^{-c}$$

for any constant $c > 0$. Thus, by the remark above and lemma 6, if $\bar{E}$ is the event that $G_{n,p}^r$ has an isolated point, then:

$$\Pr\left\{\bar{E}\right\} \geq 1 - \frac{1 - \Pr\{\hat{E}\}}{\Pr\{\hat{X}_1 = \hat{X}_2 = 0\}} \geq 1 - \frac{n^{-c}}{\Pr\{X_1 = X_2 = 0\}}$$

But $\Pr\{X_1 = X_2 = 0\} \sim e^{-\frac{\lambda}{2} - \frac{\lambda^2}{4}}$. Since $\lambda = r - 1$, we get $e^{-\frac{\lambda}{2} - \frac{\lambda^2}{4}} \sim n^{-\frac{1}{16}}$. Thus, $\Pr\{\bar{E}\} \geq 1 - n^{-k}$, for $k \geq c - \frac{1}{16}$. The proof is completed by choosing $c \geq 2 + \frac{1}{16}$. $\quad\square$

In the next section we study constant link failure probability f, which represents a worst case for the connectivity preservation. Still, we are able to show that logarithmic degrees suffice to guarantee that $G_{n,p}^r$ remains almost certainly highly connected, despite these constant edge failures.

1.9.2 Large degree random regular graphs are robust even for constant failure probability.

Let $r \geq \alpha \log n$, for α being a suitably chosen constant, and let the probability of edge failures f be constant (w.l.o.g. let $f = 1/2$).

Then, for any particular vertex v, whose degree after the failures is denoted by $deg(v)$, we have:

$$\Pr\left\{(1 - \beta)rp \leq deg(v) \leq (1 + \beta)rp\right\} \geq 1 - 2e^{-\frac{\beta^2}{2}rp} \geq 1 - 2n^{-\gamma}$$

where $\gamma \geq 2$ a constant, by Chernoff bounds, and by choosing $\alpha \geq 2\gamma/(\beta^2 p)$. Clearly then:

$$\Pr\left\{\text{for all } v, (1 - \beta)rp \leq deg(v) \leq (1 + \beta)rp\right\} \geq 1 - 2n^{-(\gamma - 1)}$$

We note that, interestingly, this property is shared by the graphs $G_{n,p'}$ for $p' \geq (\alpha p \log n)/n$, by the same technical argument. Indeed, remark that in $G_{n,p}^r$ graphs, the neighbors of any vertex are determined by first selecting r edges at random among all $n - 1$ possible, and by then sampling them randomly and independently with the same probability p. In $G_{n,p'}$ graphs, all $n - 1$ possible adjacent edges of a vertex are sampled independently and with the same probability p'. In either case, we end up, with high probability, with almost the *same number* of edges adjacent to any vertex, which are *randomly selected*. We note that it is true that the spaces $G_{n,p}^r$ and $G_{n,p'}$ are different, but the difference includes only the graphs where although p' is small, the degree of a vertex may exceed r, and this event is of negligible probability compared to the rest, from Chernoff bounds. In fact, by choosing (for an $\beta \in (0,1)$) $np' = \frac{rp}{1+\beta}$, the model $G_{n,p'}$ is weaker that the $G_{n,p}^r$. That is, in this case the "randomly truncated" random regular graph $G_{n,p}^r$ and $G_{n,p'}$ produce almost *the same graph outcomes and have the same properties*, with extremely high probability. The minimum necessary probability p_c so that $G_{n,p}$ graphs are a.c. k-connected when $p \geq p_c$, is given in [10] (p. 152, lemma 5). By calculating the maximum value of k such that

$$p' = \frac{\alpha \log n}{2(1+\beta)n} \geq p_c = \frac{\log n + (k-1) \log \log n - \omega(n)}{n}$$

where $\omega(n) \to \infty$, $\omega(n) \leq \log \log \log n$, we get the following theorem.

Theorem 19 *Let G be an instance of $G_{n,p}^r$ where $p = \Theta(1)$ and $r \geq \alpha \log n$, where $\alpha > 0$ an appropriate constant. Then G is almost certainly k-connected, where*

$$k = O\left(\frac{\log n}{\log \log n}\right)$$

$\square$

We remark here that the constant link failure probability dramatically alters the connectivity structure of the regular graph. While G_n^r is a.c. r−connected for any $r \geq 3$, we get that $G_{n,p}^r$ is *disconnected* w.h.p. for all $r \leq \frac{\sqrt{\log n}}{2}$ and that large degrees $r \geq \alpha \log n$ are needed for $G_{n,p}^r$ to be a.c. connected. The gap $\left(\frac{1}{2}\sqrt{\log n}, \alpha \log n\right)$ is due to the translation theorem from random configurations to $G_{n,p}^r$ (we recall that the extended translation lemma is valid for $r \geq 2$ fixed). In order to handle this gap, we must avoid the translation lemma. In fact, first consider a graph $G \in G_n^r$. By Turán's celebrated theorem ([72]), it contains an independent set S with cardinality $|S| \geq \frac{n}{r+1}$. Now delete edges from the graph with probability p, and let X_v denote the event that in the new graph G_p obtained in this way a vertex v is isolated. Then, for each pair v, w of vertices of S the events X_v and X_w are independent. Now simple calculations and Chebyshev's inequality reveal that w.h.p. the graph G_p contains an isolated vertex. Then we can repeat the calculations that we did in the proof of Theorem 1 (exactly the same argument but not in the configuration graph, so we avoid the translation lemma in this case) and get that $G_{n,p}^r$ is disconnected a.c. for any $r = o(\log n)$ and constant f.

1.9.3 $G_{n,p}^r$ is r−connected for $f = O(n^{-\epsilon})$ except for $O(1)$ vertices.

We now consider the (more practical) case in which $f = 1 - p = o(1)$ and we prove that the nice properties of random regular graphs are preserved despite link failures.

The technical lemmas related to random configurations are still valid. By a similar result in [10] (p. 174, Theorem 32) on G_n^r and by the fact that a $G_{n,p}^r$ graph can not have more edges than the corresponding G_n^r graph, the following also holds:

Lemma 10 *There is a constant $c = c(r)$ such that $G_{n,p}^r$ graphs do not have a subgraph with $k < c \log n$ vertices and at least $k + 1$ edges; this holds with probability at least $1 - e^{-n}$.* □

Consider now in $G_{n,p}^r$ a partition of $V : V = A \cup S \cup B$, where $s_1 = |S| \geq (r-2)\alpha_0, \alpha_0 \geq 3$. Let $\alpha_1 = 4(r-2)\alpha_0$ and consider all partitions in which $|A| = \alpha \geq \alpha_1$. Consider the event: $\varepsilon_1(\alpha) = $ "There is a partition $V = A \cup S \cup B$ as above with no $A - B$ edges, i.e. with S a separating set and $|A| = \alpha$".

Theorem 20 *Almost every $G_{n,p}^r$ graph is such that if a smaller than cardinality r set separates it, then only $O(1)$ vertices break away and the rest of the graph (of $O(n)$ size) is r-connected.*

Proof: It suffices to show that $\sum_\alpha Pr\{\varepsilon_1(\alpha)\} \to 0$. By Lemma 10, $G_{n,p}^r$ is such that w.h.p. every set of $k < c(r) \log n$ vertices spans at most k edges.

We have $\binom{n}{s_1}$ choices for S and $\binom{n-s_1}{\alpha}$ choices for A. By using random configurations we see that a configuration edge incident with an element of $\hat{A}$ joins this element to $\hat{B}$ with probability at least $(\frac{n-s_1-\alpha}{n})p$.

Let $S(\alpha) = Pr\{\varepsilon_1(\alpha)\}$. By pairing off elements of $\hat{A}$ one by one and by conditioning on the events of Lemma 10 we get:

$$S(\alpha) \leq \binom{n}{s_1}\binom{n-s_1}{\alpha}\left(1 - \frac{n-s_1-\alpha}{n}p\right)^{\frac{\alpha r}{2}}$$

Now, we use $f = 1 - p$ to get :

$$S(\alpha) \leq S'(\alpha) = n^{s_1}\binom{n-s_1}{\alpha}\left(\frac{s_1+\alpha}{n}(1-f)+f\right)^{\frac{\alpha r}{2}}$$

We will study only the worst, with respect to connectivity issues, case where $r = O(\sqrt{\log n})$. For bigger r, things are even better.

Clearly, the hardest case for connectivity is when f is large. So, let us assume that $f = \Omega\left(\frac{\sqrt{\log n}}{n}\right)$. Then $S'(\alpha) \leq n^{s_1+\alpha}((1+\gamma)f)^{\alpha r}$ for some $\gamma > 0$. We remark that if $f \leq \frac{1}{1+\gamma}n^{-\frac{2s_1+\alpha}{\alpha r}}$ then $S'(\alpha) \leq n^{-s_1}$.

Let $\alpha_2 = \lfloor\frac{n-\alpha_1}{2}\rfloor$. Clearly

$$\sum_\alpha Pr\{\varepsilon_1(\alpha)\} \leq \sum_{\alpha=\alpha_1}^{\alpha_2} S'(\alpha) \tag{10.5}$$

Since we assumed $f = \Omega\left(\frac{\sqrt{\log n}}{n}\right)$, we have that

$$\frac{\sqrt{\log n}}{n} < f \leq \frac{1}{1+\gamma}\, n^{-\frac{2s_1+\alpha}{\alpha r}}$$

Note that: $\frac{S'(\alpha+1)}{S'(\alpha)} \sim nf^r$ i.e. $\frac{S'(\alpha+1)}{S'(\alpha)} \leq n(\frac{1}{1+\gamma})^r\, n^{-\frac{2s_1+\alpha}{\alpha}} < \frac{1}{\sqrt{n}}$, since $4s_1 \geq \alpha$. Thus, from inequality (10.5) above, we easily get

$$\sum_{\alpha=\alpha_1}^{\alpha_2} S'(\alpha) \to 0$$

as $n \to +\infty$. $\qquad\square$

From the above theorem we get:

Theorem 21 *Let* $r \geq 3$ *and* $f = 1 - p = O(n^{-\epsilon})$ *for* $\epsilon \geq \frac{3}{2r}$. *Then* $G_{n,p}^r$ *is* $r-$*connected, except for* $O(1)$ *vertices, with probability tending to 1 as n tends to* $+\infty$. $\qquad\square$

1.10 GIANT COMPONENTS IN RANDOM REGULAR GRAPHS WITH EDGE FAULTS

Since $G_{n,p}^r$ is a.c. disconnected for $r = o(\log n)$ and $1 - p = f = \Theta(1)$, we would like to know whether at least a large part of the network represented by $G_{n,p}^r$ is still connected, i.e. whether the biggest connected component of $G_{n,p}^r$ is large. Perhaps the most celebrated theorem in Random Graph Theory is that of the Giant Component: $G_{n,p}$ has a giant (i.e. $\Theta(n)$-sized) connected component when, roughly, $p > \frac{1}{n}$. This was first remarked by Erdös and Renyi ([26]). We will here show that:

Theorem 22 *When* $f < 1 - \frac{32}{r}$ *then* $G_{n,p}^r$ *admits a giant (i.e.* $\Theta(n)$*-sized) connected component with probability at least* $1 - O\left(\frac{\log^2 n}{n^{\alpha/3}}\right)$, *where* $\alpha > 0$ *a constant that can be selected.*

Proof: We will prove the Theorem for r fixed by using the extended translation lemma. The result then obviously extends to greater (unbounded) r by the monotonicity of the giant connected component property.

We will consider the random configuration $\hat{F}$ corresponding to $G_{n,p}^r$. We first select each existing edge of $\hat{F}$ with probability 1/2, independently. This partitions the edges of $\hat{F}$ into two sets of edges (E_1 and E_2).

Since $f < 1 - \frac{32}{r}$ we have $p > \frac{32}{r}$ i.e. $pr > 32$. In the graph G_n^r we have a total number of edges $\frac{rn}{2}$ and after the experiment of *independently* deleting each edge with probability f and the experiment of the partition, the remaining number of edges in E_1, E_2 satisfies the Bernoulli $B\left(\frac{nr}{2}, \frac{p}{2}\right)$ i.e. by Chernoff bounds we get that $|E_1|, |E_2| \geq (1-\beta)\frac{nrp}{4}$, for any $\beta \in (0, 1)$ w.h.p. and since $pr > 32$ we get $|E_1|, |E_2| \geq (1-\beta)8n$ with probability $\geq 1 - \exp\left(-\frac{\beta^2}{2}8n\right)$. We now condition, for E_1, on the event $A_1 = ``|E_1| \geq (1-\beta)8n''$.

Let $\hat{F}(E_i)$ the random configuration corresponding to the edge set E_i, $i = 1, 2$. The following technical claim will be needed here:

Claim 1 $\hat{F}(E_1)$ *contains a "path" (i.e. a sequence of sets w_i, of r vertices, each connected to the next by an edge) of length at least $\alpha \log n$ ($\alpha > 0$ an appropriate constant that can be selected) with probability at least $1 - n^{-\lambda}, \lambda \geq 2$.*

Proof: Consider now the set E_1 partitioned into g groups $g_1, g_2, \ldots$ of $\alpha \log n$ ($\alpha > 0$ an appropriate constant) edges in each group. For each group g_i let q_i be the probability that it will form a path in $\hat{F}(E_1)$ if thrown alone in the $\binom{rn}{2}$ possible "places". Let a random indicator variable $X_i = 1$ with probability q_i and $X_i = 0$ else. Clearly the random variables X_i are *not independent*. The number of ways for g_i to form a path is calculated as follows:

(a1) Choose the supervertices of the path in $\binom{n}{\alpha \log n}$ ways

(a2) Order them in the path in $(\alpha \log n)!$ ways

(a3) In this order $(r^2)^{\alpha \log n - 1}$ paths are defined, because any two specific supervertices can be joined by an edge in r^2 ways.

But there are

$$\binom{\binom{rn}{2}}{\alpha \log n}$$

places for the edges of g_i. Thus

$$q_i = \frac{\binom{n}{\alpha \log n}(\alpha \log n)! \; r^{2\alpha \log n - 2}}{\binom{\binom{rn}{2}}{\alpha \log n}}$$

The denominator is at most $(rn)^{2\alpha \log n} = r^{2\alpha \log n} n^{2\alpha \log n}$, thus

$$q_i \geq \frac{\frac{n!}{(\alpha \log n)!(n - \alpha \log n)!}(\alpha \log n)!}{r^2 \; n^{2\alpha \log n}}$$

i.e.

$$q_i \geq \frac{n(n-1)\cdots(n - \alpha \log n + 1)}{r^2 n^{2\alpha \log n}}$$

and since $r \leq n - 1$ we get

$$q_i \geq \frac{(n-2)\cdots(n - \alpha \log n + 1)}{n^{2\alpha \log n}} \geq \frac{\left(\frac{n}{2}\right)^{\alpha \log n - 2}}{n^{2\alpha \log n}} \geq \frac{4}{n^{\alpha \log n + 2 + \alpha}}$$

But there are $N \geq \binom{4n}{\alpha \log n}$ groups g_i i.e.

$$N \geq \frac{(4n)!}{(\alpha \log n)!(4n - \alpha \log n)!}$$

i.e. (crudely, and for sufficiently large n)

$$N \geq \frac{(3.9n)^{\alpha \log n}}{(\alpha \log n)!} \geq (3n)^{\alpha \log n}$$

Notice that forall g_i

$$N \cdot q_i \geq 4(1.5)^{\alpha \log n}$$

i.e.

$$N \cdot q_i \geq 4 \left(\frac{3}{2}\right)^{\alpha \log n} = 4n^{\alpha \log(\frac{3}{2})}$$

As remarked in [24], "a frequently occurring scenario underlying the analysis of many randomized algorithms and processes involves random variables that are, intuitively, dependent in the following *negative* way: if one subset of the variables is "high", then a disjoint subset of the variables is "low". One specific paradigm involving negative dependency is the classical ball and bins experiment".

Definition 7 (Negative Association, [24]) *Let* $X := (X_1, \ldots, X_n)$ *be a vector of random variables. The random variables* X_i *are negatively associated if for every two disjoint index sets* $I, J \subseteq [n]$,

$$E[f(X_i, i \in I)g(X_j, j \in J)] \leq E[f(X_i, i \in I)]E[g(X_j, j \in J)]$$

for all functions $f : R^{|I|} \to R$ *and* $g : R^{|J|} \to R$ *that are both non-decreasing or both non-increasing.*

Now we notice that in our case the random indicator variables X_i and X_j (showing whether a path is formed or not in group g_i, g_j) are negatively associated for all i, j, because if $X_j = 1$ (high) then X_i is lower since some places are already occupied. That is

$$E(X_i|X_j) \leq E(X_i)$$

i.e.

$$E(X_iX_j) \leq E(X_i)E(X_j)$$

(negative covariance).

We are interested in the sum $X = \sum_i X_i$ i.e. in the number of paths of length $\alpha \log n$ in $\hat{F}(E_1)$. From [24] we have Theorem 23. For its proof we will use the following Lemma.

Lemma 11 (Dubhashi and Ranjan, [24]) *Let $X_1, \ldots, X_n$ random variables satisfying the negative association condition. Then for any non-decreasing functions $f_i, i \in [n]$,*

$$E\left(\prod_{i \in [n]} f_i(X_i)\right) \leq \prod_{i \in [n]} E(f_i(X_i))$$

Proof: Take the non-decreasing functions $f(X_i, i < n) := \prod_{i<n} f_i(x_i)$ and $g(x_n) := f_n(x_n)$ to deduce that

$$E\left(\prod_{i \in [n]} f_i(X_i)\right) \leq E\left(\prod_{i<n} f_i(X_i) E(f_n(X_n))\right)$$

and now use induction.
(end of proof of Lemma 11).

Theorem 23 (Dubhashi and Ranjan, [24]) *The Chernoff-Hoeffding bounds are applicable to sums of random variables that satisfy the negative association condition, as in definition 7.*

Proof: Let $X_1, \ldots, X_n$ be negatively associated (and bounded) variables. To show that the Chernoff-Hoeffding bounds apply to the sum $X = X_1 + \cdots + X_n$, we use the standard proof of the Chernoff-Hoeffding bound, see for example [57]. The only change needed is in a crucial step, where one uses the fact for *independent* variables, $E(e^{tX}) = E(\prod_i e^{tX_i}) = \prod_i E(e^{tX_i})$. For negatively associated variables, we have, for $t > 0$, $E(e^{tX}) = E(\prod_i e^{tX_i}) \leq \prod_i E(e^{tX_i})$, by Lemma 11 applied with each $f_i(x) := e^{tx}$. The rest of the proof is unchanged, and gives the upper tail bound. For the lower tail, we apply the same argument to the variables $b_i - X_i$, where b_i is an upper bound on the variable X_i. Note that if the X_i variables are negatively associated, then so are the variables $b_i - X_i$. (In fact, because of the monotonicity of the exponential function, the negative covariance relation among the variables suffices to prove the theorem).
(end of proof of Theorem 23)

Thus, the Chernoff-Hoeffding bounds are applicable in this case and we have that, for any $\gamma \in (0, 1)$,

$$\Pr\{X \geq (1-\gamma)Nq_i\} \geq 1 - e^{-\frac{\gamma^2}{2}Nq_i}$$

i.e. the number of paths of length $\alpha \log n$ in $\hat{F}(E_1)$ is at least $4n^{\alpha \log(\frac{3}{2})}$ with probability at least

$$1 - e^{-\frac{\gamma^2}{2}4n^{\alpha \log(\frac{3}{2})}}$$

As an easy conclusion, $\hat{F}(E_1)$ contains at least one such path of length $\alpha \log n$ with probability at least

$$\left(1 - e^{-\frac{\gamma^2}{2} 4n^{\alpha \log\left(\frac{3}{2}\right)}}\right)\left(1 - e^{-\frac{\beta^2}{2} 8n}\right) \geq 1 - n^{-\lambda}$$

for any $\lambda \geq 1$.

(end of proof of Claim 1)

Let us now condition the rest of the proof of the theorem on the existence of such a path $\Pi = (P_1, P_2, \ldots, P_{\alpha \log n})$, where each "supervertex" P_i (of r vertices) is connected to P_{i+1} by an edge of E_1. We now define a Breadth-First-Search Expansion Process in $\hat{F}(E_2)$ as follows:

BFS Expansion Process
1. Let $B_0 = \{P_1, P_2, \ldots, P_{\alpha \log n}\}$; $i = 0$.
2. Let $i \leftarrow i + 1$.
3. Let B_i be the set of supervertices not in S_{i-1} that are connected to at least one supervertex in S_{i-1} by an edge.
4. If $B_i \neq \emptyset$ then go to 2.

For each set of supervertices B_i, define:

Definition 8 *Let* $S_i = \cup_{j \leq i} B_j$.

Definition 9 *Let* Active(B_i) *be the set of those supervertices in* B_i *each of which has only at most* $r/2$ *edges connected to* S_i.

Definition 10 *Let* Out(B_i) *be the set of edges of* B_i *connecting to vertices which do not belong to* S_i.

Definition 11 *Let* R_i *be the event* $|Active(B_i)| \geq \frac{|B_i|}{2}$.

Note that the edges we use in the expansion process are independent of the edges we used to get the existence of the path Π. Let $i \leq \log n - o(\log n)$. Choose $\alpha \geq \frac{6 \cdot 32}{rp}$. Assume inductively that

$$\Pr\{R_i\} \geq \left(1 - \frac{1}{n^{\alpha/3}}\right)^i$$

For $i = 0$, $\Pr\{R_i\} = 1$ (basis). By the induction hypothesis and conditioning on R_i, we get that at least $\frac{r}{2}\frac{|Bi|}{2}$ edges of S_i go out of S_i. Each of them will exist with probability p. Thus, by the Bernoulli $B\left(\frac{r}{2}\frac{|B_i|}{2}, p\right)$, and because $rp > 32$, the event $\epsilon_1(i) = $ "at least $(1 - \beta_1)8|B_i|$ edges exist and go out of S_i departing from S_i" has (by Chernoff bounds) probability at least $1 - \exp\left(-\frac{\beta_1^2}{2}8|B_i|\right)$ and since $|B_i| \geq \alpha \log n$ we get that

$$\Pr\{\epsilon_1(i)\} \geq 1 - e^{-\frac{\beta_1^2}{2}8\alpha \log n} \geq 1 - n^{-\beta_1^2 4\alpha}$$

for any $\beta_1 \in (0, 1)$.

Now, condition also on $\epsilon_1(i)$. Then let us see how many new supervertices these edges hit. We can safely assume that during the process, $|S_i|$ (i.e. the number of supervertices in the component in the i-th stage of the BFS) is less than $\frac{n}{8}$ (else we have a giant component already). But then consider the edges out of S_i in the order $e_1, e_2, \ldots, e_x$ ($x \geq (1 - \beta_1)8|B_i|$). Let the set $\{e_1, e_2, \ldots, e_\kappa\}$ have hit m_1 new supervertices. We may again assume $m_1 < \frac{n}{8}$ (since else we are done). Consider $e_{\kappa+1}$. The probability that it hits another *new* supervertex is at least $\frac{n - \frac{n}{8} - \frac{n}{8}}{\frac{7n}{8}} \geq \frac{6}{7}$ and this holds for all $\kappa \leq x$. Consider the Bernoulli $B\left(x, \frac{6}{7}\right)$. By Chernoff bounds, the number of edges that each hits a supervertex separate from those hit before is $(1 - \beta_2)x\frac{6}{7}$ thus at least $(1 - \beta_2)(1 - \beta_1)8|B_i|\frac{6}{7}$ i.e. at least $(1 - \beta_3)6|B_i|$ with probability at least

$$1 - e^{-\frac{\beta_2^2}{2}6|B_i|} \geq 1 - n^{-3\alpha\beta_2^2}$$

since $|B_i| \geq \alpha \log n$.

Note that by appropriately choosing β_1, β_2 we get that the number of edges out of S_i that each hits a new supervertex separate from before, is at least $6|B_i|$ with probability at least $1 - n^{-3\alpha\frac{\beta_2^2}{2}}$. But each of these new supervertices (they belong to B_{i+1}) is *active by construction*, because each has only one edge into S_i, thus $r - 1$ edges free to expand. So, given that R_i holds, the conditional probability $\Pr\{R_{i+1}|R_i\}$ is at least

$$\Pr\{R_{i+1}|R_i\} \geq \Pr\{\epsilon_1(i)\}\left(1 - n^{-3\alpha\beta_2^2}\right) \geq 1 - n^{-\frac{\alpha}{3}}$$

Thus

$$\Pr\{R_{i+1}\} \geq \Pr\{R_i\}\Pr\{R_{i+1}|R_i\} \geq \left(1 - n^{-\frac{\alpha}{3}}\right)^{i+1}$$

(end of proof of induction)

So, in each stage we get an expansion of the number of connected new supervertices by 6, provided $i \leq \gamma \log n - o(\log n)$ ($\gamma \geq 1$). Thus, at the end we get a giant component of size $\Theta(n)$, *provided all events R_i hold*.

The probability of the event "$\exists i : \bar{R}_i$" is the probability of failure of the process. This is at most

$$\Pr\left\{\exists i : \bar{R}_i\right\} \leq \sum_{i=0}^{\log n} \Pr\{\bar{R}_i\}$$

But

$$\Pr\{R_i\} \geq \left(1 - \frac{1}{n^{\frac{\alpha}{3}}}\right)^i \geq 1 - \frac{i}{n^{\frac{\alpha}{3}}}$$

thus

$$\Pr\{\bar{R}_i\} \leq \frac{i}{n^{\frac{\alpha}{3}}} \leq \frac{\log n}{n^{\frac{\alpha}{3}}}$$

and finally

$$\sum_{i=0}^{\log n} \Pr\{\bar{R}_i\} \leq \frac{\log^2 n}{n^{\frac{\alpha}{3}}}$$

This ends the proof of Theorem 22.

Recently, Andreas Goerdt ([35]) continued the work presented in a preliminary version of this paper ([62]) and showed the following results: if the degree r is fixed then $p = \frac{1}{r-1}$ is a threshold probability for the existence of a linear sized component in the faulty version of almost all random regular graphs. In fact, he further shows that if each edge of an *arbitrary* graph G with maximum degree bounded above by r is present with probability $p = \frac{\lambda}{r-1}$, when $\lambda < 1$, then the faulty version of G has only components whose size is at most logarithmic in the number of nodes, with high probability. His result implies some kind of optimality of random regular graphs with edge faults.

2. TRADE-OFFS BETWEEN DENSITY AND ROBUSTNESS IN RANDOM INTERCONNECTION GRAPHS

In this section, we discuss the work of Ph. Flajolet, K. Hatzis, S. Nikoletseas and P. Spirakis ([27]) applying combinatorial and symbolic-analytic techniques in order to characterize the interplay between two parameters of a random graph: its density (number of edges) and its robustness to link failures, where robustness means multiple connectivity by short paths. A triple (G, s, t), where G is a graph and s, t are designated vertices, is called $\ell - robust$ if s and t are connected via at least two edge-disjoint paths of length at most ℓ. Flajolet et al determine in [27] the expected number of ways to get from s to t via two edge-disjoint paths in the random graph model $G_{n,p}$ and also derive bounds on related threshold functions.

2.1   INTRODUCTION

Given a triple (G, s, t), where G is a $G_{n,p}$ random graph and s, t are two of its nodes, a natural notion of robustness is to require at least two edge disjoint paths of short length (say, exactly ℓ or at most ℓ) between s and t, so that connectivity by short paths survives, even in the event of a link failure. We next give the following formal definition of $\ell - robustness$.

Definition 12 (ℓ-robustness) *A random graph G of the model $G_{n,p}$ is $\ell -$ robust for two vertices s, t when there are, with high probability, two edge-disjoint paths of length at most ℓ between s, t in G.*

In this section, we investigate the expected number of such paths between two vertices of the random graph, as well as bounds for the threshold probability $p_{n,\ell}$ (as functions of ℓ and n) for the existence of such paths in the $G_{n,p}$ random graph G.

It turns out that even the enumeration of paths among the vertices 1 and n that avoid all edges of the graph $(1, 2 \ldots, n)$ but pass through its vertices, is a non-trivial task. In fact, such an enumeration is a special case of enumerating permutations $(\sigma_1, \sigma_2, \ldots, \sigma_n)$ of $(1, 2, \ldots, n)$ where certain gaps $\sigma_{i+1} - \sigma_i$ are forbidden. In our case, $\sigma_{i+1} - \sigma_i$ must not be in the set $\{-1, 1\}$.

In this section, we provide a precise estimate of the expected number of unordered pairs of paths in a random graph that connect a common source to a common destination, and have no edge in common, though they may share some nodes. Thus, for any given set of values of n, p, ℓ (where ℓ represents the length) we estimate precisely the mean number of avoiding pairs in graphs of a given size. This leads to a tight bound for the probability of non-existence of such paths between any fixed pair of nodes and also to a bound for the probability of the existence of many pairs of edge disjoint paths of length ℓ.

In order to achieve this, Flajolet et al devise a finite state mechanism that describes classes of permutations with free places and exceptions. The finite-state description allows for a direct construction of a multivariate generating function. The generating function is then subjected to an integral transform that implements an inclusion-exclusion argument and an explicit enumeration result; see Theorems 24 and 25. This enables them to quantify the trade-off between ℓ-robustness (as defined above) and the density of the graph (i.e. the number of its edges). The originality of our approach consists in introducing in this range of problems methods of analytic combinatorics and recent research in automatic analysis (based on symbolic computation). For context, see [28], [29], [19]. Additional threshold estimates regarding properties of multiple source-destination pairs are discussed in the last two sections of the paper.

From earlier known results ([10], [62]) and the results in this section, a picture of robustness under the $G_{n,p}$ model emerges. (As usually in random graph theory, various regimes for $p = p(n)$ are considered). Start with an initially totally disconnected graph, corresponding to $p = 0$. As p increases, the graph becomes connected near the connectivity threshold $p_c(n) \simeq (\log n)/n$. All the following results hold for any integer ℓ: $6 \leq \ell < n$. Any fixed s, t pair (or equivalently a random s, t pair, given the invariance properties of $G_{n,p}$) is *"likely" to be ℓ-robust* when p crosses the value

$$p_{n,\ell}(n) = 2^{\frac{1}{2\ell}} \, n^{-1+\frac{1}{\ell}}$$

(see Theorem 26 and Equation 10.11). (Here "likely" means that the expected number of edge-disjoint pairs is at least 1 when $n \to \infty$). As long as $p \leq p_L(n, \ell)$, where

$$p_L(n, \ell) = n^{-1+\frac{1}{\ell}} \left(\log \frac{n^2}{\log n} \right)^{\frac{1}{\ell}}$$

we know, w.h.p., the *existence* of s, t pairs that are *not* connected by short (of length at most ℓ) paths; see Theorem 27. (The function $p_L(n)$ is in fact a threshold for diameter). However, one only needs a tiny bit more edges, namely $p \geq p_U$:

$$p_U(n, \ell) = n^{-1+\frac{1}{\ell}} \, 2 \left(\log \left(n^2 \log n \right) \right)^{\frac{1}{\ell}}$$

to ensure that *almost all* s, t-pairs are ℓ-robust; see Theorem 28.

2.2   THE ENUMERATION OF "AVOIDING" CONFIGURATIONS

The problem at hand is that of estimating the expected number of "avoiding pairs" of length ℓ between a random source and a random destination in a $G_{n,p}$ random graph G. (An avoiding pair of length ℓ means an unordered pair of paths, each of length ℓ, that connect a common source to a common destination, and have no edge in common though they may share some nodes). This problem first necessitates the solution of enumeration problems that involve two major steps:

— Enumerate simple paths called "avoiding permutations" of length ℓ that can be viewed as hamiltonian paths on the set of nodes $[1, \ldots, \ell + 1]$, connecting the source 1 and the destination $\ell + 1$, and having no edge of type $[i, i + 1]$ or $[i, i - 1]$.

— Enumerate so-called "avoiding paths", that are simple paths allowed to contain outer nodes taken from outside the segment $[1, \ldots, \ell + 1]$. This situation is closer to the random graph problem since it allows nodes taken from the pool of vertices available in the graph $G \in G_{n,p}$.

The first problem is of independent combinatorial interest as it is equivalent to counting special permutations with restrictions on adjacent values. It also serves as a way to introduce the methods needed for the complete random graph problem. Both problems rely on the inclusion-exclusion principle that is familiar from combinatorial analysis and counting by generating functions (GF's). Applications of these results to the $G_{n,p}$ model are treated in the next section.

2.2.1 Symbolic Enumeration Methods. Flajolet et al ([27]) use a symbolic approach to combinatorial enumeration, according to which many general set-theoretic constructions have direct translations over generating functions. A specification language for elementary combinatorial objects is defined for this purpose. The problem of enumerating a class of combinatorial structures then simply reduces to finding a proper specification, a sort of a formal grammar, for the class in terms of basic constructions. (This general method has been carefully explained in the fundamental work of F. Chyzak, Ph. Flajolet and B. Salvy ([5])).

In this framework, classes of combinatorial structures are defined either iteratively or recursively in terms of simpler classes by means of a collection of elementary combinatorial constructions. The approach followed resembles the description of formal languages by means of context-free grammars, as well as the construction of structured data types in classical programming languages.

The approach developed here is direct, more "symbolic", as it relies on a specification language for combinatorial structures. It is based on so-called admissible constructions that have the important feature of admitting direct translations into generating functions. We specifically examine constructions whose natural translation is in terms of ordinary generating functions.

The ordinary generating function (OGF) of a sequence $\{A_n\}$ is, we recall, $A(z) = \sum_{n=0}^{\infty} A_n \cdot z^n$.

Definition 13 (Admissible Constructions) *Assume that Φ is a binary construction that associates to two classes of combinatorial structures B and C a new class $A = \Phi(B, C)$ in a finite way (each A_n depends on finitely many of the B_n and C_n). The Φ is admissible iff the counting sequence $\{A_n\}$ of A is a function of the counting sequences $\{B_n\}$ and $\{C_n\}$ of B and C only: $\{A_n\} = \Xi[\{B_n\}, \{C_n\}]$. In that case, there exists a well defined operator Ψ relating the corresponding ordinary generating functions: $A(z) = \Psi[B(z), C(z)]$.*

In this work, we will basically use three important constructions: union, product and sequence, which we describe below.

Definition 14 (Union Construction) *The disjoint union A of two classes B, C is the union (in the standard set-theoretic sense) of two disjoint copies, B^o and*

C^o, of B and C. *Formally, we introduce two distinct "markers" ϵ_1 and ϵ_2, each of size zero, and define the (disjoint) union $A = B + C$ of B, C by $B + C = (\{\epsilon_1\} \times B) \cup (\{\epsilon_2\} \times C)$. The ordinary generating function is clearly $A(z) = B(z) + C(z)$. We represent the disjoint union construction by* Union.

Definition 15 (Product Construction) *If construction A is the cartesian product of classes B and C ($A = B \times C$), then, considering all possibilities, the counting sequences corresponding to A, B, C are related by the convolution relation: $A_n = \sum_{k=0}^{n} B_k \cdot C_{n-k}$ and the ordinary generating function is clearly $A(z) = B(z) \cdot C(z)$. We represent the product construction by* Prod.

Definition 16 (Sequence Construction) *If C is a class of combinatorial structures then the sequence class $\mathcal{G}\{C\}$ is defined as the infinite sum $\mathcal{G}\{C\} = \{\epsilon\} + C + (C \times C) + \cdots$ with ϵ being a "null structure", meaning a structure of size 0. The null structure plays a role similar to that of the empty word in formal language theory and the sequence construction is analogous to the Kleene star operation (C^*). The ordinary generating function is clearly given by $A(z) = 1 + B(z) + B^2(z) + \cdots = \frac{1}{1-B(z)}$ where the geometric sum converges in the sense of formal power series since $[z^0]B(z) = 0$. We represent the sequence construction by* Sequence.

2.2.2 Avoiding Permutations. An avoiding permutation of length ℓ is a sequence $\tau = [\tau_1, \tau_2, \ldots, \tau_\ell, \tau_{\ell+1}]$ that is a permutation of $[1, \ldots, \ell + 1]$ and that satisfies the following conditions: $\tau_1 = 1$, $\tau_{\ell+1} = \ell+1$, and $\tau_{i+1} - \tau_i \neq \pm 1$ for all i such that $1 \leq i \leq \ell$. Clearly, such a permutation encodes a path from 1 to $\ell + 1$ that has no edge in common with the graph $(1, 2, \ldots, \ell + 1)$. The parameter $\ell + 1$ is referred to as the size. There is no avoiding permutation for sizes $2, 3, 4, 5$. Surprisingly, the first nontrivial configurations occur at size 6, where the 2 possibilities are $[1, 4, 2, 5, 3, 6]$ and $[1, 3, 5, 2, 4, 6]$, while for size 7, there are 10 possibilities

$[1, 3, 6, 4, 2, 5, 7], [1, 3, 5, 2, 6, 4, 7], [1, 4, 6, 2, 5, 3, 7], [1, 4, 2, 6, 3, 5, 7], [1, 5, 3, 6, 4, 2, 7],$
$[1, 5, 3, 6, 2, 4, 7], [1, 5, 2, 4, 6, 3, 7], [1, 4, 6, 3, 5, 2, 7], [1, 6, 3, 5, 2, 4, 7], [1, 6, 4, 2, 5, 3, 7].$

The goal in this subsection is to determine the number Q_n of avoiding permutations of size n (that is, of length $n - 1$). We prove:

Theorem 24 *Avoiding permutations have ordinary generating function*

$$Q(z) := \sum_n Q_n z^n = \frac{\left(z - 1 + e^{\frac{1+z}{z(z-1)}} \, Ei\left(1, \frac{1+z}{z(z-1)}\right)\right) z}{(z-1)(1+z)} =$$

$$\frac{z\left(z + z\,hypergeom\left([1,1],[\,],-\frac{z(z-1)}{1+z}\right)+1\right)}{(1+z)^2}$$

where *Ei* is the exponential integral and *hypergeom* represents the hypergeometric series. Equivalently, Q_n is expressible as a double binomial sum:

$$Q_{n+2} = (-1)^{n-1} + \sum_{k_2=0}^{n}\sum_{k_1=0}^{n-k_2}(-1)^{k_1+k_2}(n-k_1-k_2)!\binom{n-k_1-k_2}{k_1}\binom{n+1-k_1}{k_2}$$

Proof: By the inclusion-exclusion principle, we need to determine the number $F_{n,j}$ of permutations $[\tau_1 = 1, \tau_2, \ldots, \tau_{n-1}, \tau_n = n]$, with at least j "exceptions", among which j distinguished, that are successions of values of the form $\tau_j - \tau_{j-1} = \pm 1$. The number of permutations with no exception is then:

$$Q_n = \sum_{j=0}^{n-1}(-1)^j F_{n,j} \tag{10.6}$$

A permutation with exceptions includes a subcollection of "exceptional" edges that belong to the graph with edges $(1,2),(2,3),\ldots,(n-1,n)$. If we scan from left to right and group such exceptions by blocks, we get a *template*; a template thus represents a possible pattern of exceptional edges.

A template can be defined directly as made of blocks that are either: (i) isolated points; (ii) contiguous unit intervals oriented left to right (LR); (iii) contiguous unit intervals oriented right to left (RL). There is the additional constraint that the first and last blocks cannot be of type RL. For instance, for $n = 13$, the template $[[1,2,3], [4], [5,6], [7], [8], [11,10,9], [12,13]]$ will correspond to any permutation that has successions of values (in the cycle traversal) $1, 2$; $2, 3$; $5, 6$; $11, 10$; $10, 9$; $12, 13$ as distinguished exceptions to the basic constraint of avoiding permutations.

We next provide the combinatorial specification for avoiding permutations.

Let $\{a, b\}$ be a binary alphabet. We now describe the grammar of templates.

The collection of strings beginning and ending with a letter a is described by the following rule:

sp0 := S = Prod(Sequence(Prod(a, Sequence(b))), a)

(It suffices to decompose according to each occurrence of the letter a). Now, the thre types of blocks in a template are described by the following rules:

sp1 := Prod(*begin_blockP, Z, end_blockP*)
sp2 := Prod(*begin_blockLR, Z,* Sequence(Prod(*mu_length, Z*), $1 \leq card$), *end_blockLR*)
sp3 := Prod(*begin_blockRL,* Sequence(Prod(*mu_length, Z*), $1 \leq card$), *Z, end_blockRL*)

Clearly, *sp2* and *sp3* are combinatorially isomorphic. For reasons related to the application of the inclusion-exclusion argument, we keep track of the size (number of nodes $l = 1$ of the basic interval graph, denoted by *card*) as well as of the length of blocks and of their *LR* or *RL* character, denoted by *mu_length*. Then, the grammar of templates is completed by substituting into $sp0$

$$a = \text{Union}(sp1, sp2) \quad \text{and} \quad b = sp3$$

Let $F_{n,k,j}$ be the number of templates with size n, k blocks and j exceptional edges. Then, counting the number of ways of linking blocks together, yields:

$$F_{n,j} = \sum_k F_{n,k,j}\,\phi(k) \tag{10.7}$$

where $\phi(k)$ is the Gamma integral:

$$\phi(k) \equiv (k - 2)! = \int_0^\infty e^{-u}\, u^k\, \frac{du}{u^2}$$

for $k \geq 2$, and $\phi(1) = 1$ (since any such linking is determined by an arbitrary permutation of the $k - 2$ intermediate blocks). Observe that the extension of ϕ by linearity to an arbitrary series h(u) in u is given by

$$\phi(\text{h}(u)) = \int_0^\infty \frac{e^{-u}\left(\text{h}(u) - (u - u^2)\left(\frac{\partial}{\partial u}\text{h}(u)\right)_{u=0}\right)}{u^2}\, du$$

That is to say, we just replace in expansions $u \to u^2$ and apply the Euler integral

$$\int_0^\infty e^{-u} u^k\, du = k!$$

Thus, with F$(z, u, v) = \sum F_{n,k,l}\, z^n u^k v^l$, the OGF Q$(z) = \sum Q_n z^n$ satisfies

$$Q(z) = \phi(\text{F}(z, u, -1))$$

Thus, from (10.6) and (10.7) above, everything boils down to obtaining the $F_{n,k,j}$.

Template enumeration. The approach to determining the sequence $F_{n,k,j}$ consists in introducing the trivariate GF, which immediately results from the above combinatorial specification.

$$F(z, u, v) = \sum_{n,k,\ell} F_{n,k,\ell}\, z^n u^k v^l$$

There, z records size, u records the total number of blocks (needed for subsequent permutation enumerations since blocks should be chained to each other), and v records the total length of *LR* or *RL* blocks (the number of distinguished exceptions needed for inclusion-exclusion).

We now carefully employ the generating functions for the union, product and sequence constructions in the grammar rules of the combinatorial specification for avoiding permutations defined above.

The set of words made of a's and b's that start and end with an a is described symbolically by

$$W = \frac{1}{1 - \frac{a}{1-b}} \cdot a \tag{10.8}$$

This is because $(1 - f)^{-1} = 1 + f + f^2 + f^3 + \cdots$ generates symbolically all sequences of objects of type f. Thus, W represents a sequence of objects of type $\frac{a}{1-b}$ that start with an a. On the other hand, $\frac{a}{1-b}$ represents a sequence of objects of type b that end with an a. Take now the three types of blocks: isolated, LR, and RL. The GF's are, respectively, z, $LR(z) = z^2/(1 - z)$, $RL(z) = z^2/(1 - z)$. This is because isolated points are always of size 1, while LR and RL objects must be of size at least 2 (we have thus to multiply with z^2). Since the first and the last blocks can only be isolated points or LR blocks, the univariate GF for blocks is obtained by substituting a by $z + LR$ (isolated point or LR block) and b by RL in W. Thus we get the following trivariate GF:

$$F(z, u, v) = \left(1 - \frac{u\left(z + \frac{z^2 v}{1 - vz}\right)}{1 - \frac{uz^2 v}{1 - vz}} \right)^{-1} \cdot u\left(z + \frac{vz^2}{1 - vz}\right) =$$

$$= -\frac{uz(-1 + vz + uz^2 v)}{1 - 2vz - uz + v^2 z^2 + v^2 z^3 u}$$

Path counting. For the inclusion-exclusion argument, it is easy to observe that the desired sum $\sum_{n,k,\ell} F_{n,k,\ell} \, z^n u^k (-1)^\ell$ corresponds to the specialization $F(z, u, -1)$. This yields:

$$F(z, u, -1) = -\frac{uz(-1 - z - uz^2)}{1 + 2z - uz + z^2 + z^3 u} \tag{10.9}$$

Application of the ϕ-transformation (that counts the number of ways to connect the blocks) requires the modified form and so we get:

$$F(z, u, -1) = \frac{zu^2(uz^2 + 2z - uz + 1)}{(1 + z)(uz^2 + z - uz + 1)}$$

The corresponding ordinary generating function is

$$Q(z) = \int_0^\infty \frac{z(uz^2 + 2z - uz + 1)}{(1 + z)(uz^2 + z - uz + 1)} \cdot e^{-u} \, du$$

The quantity $Q(z)$ can be expressed in terms of the exponential integral

$$\int_0^\infty e^{-u} u^k \, du = k!$$

in the following closed form

$$Q(z) := \frac{\left(z - 1 + e^{\frac{1+z}{z(z-1)}} \operatorname{Ei}\left(1, \frac{1+z}{z(z-1)}\right)\right) z}{(z-1)(1+z)}$$

Since one deals with ordinary generating functions, this is to be taken as a formal (asymptotic) series. Note also that the exponential integral (Ei) involves the divergent series of factorials $\sum_{n=0}^\infty n!(-y)^{(-n-1)}$ which is also a hypergeometric series. This gives rise to a general conversion procedure from exponential integrals to hypergeometric forms. Hence, another closed form for the OGF of the Q_n is

$$Q(z) := \frac{z\left(z + z \text{ hypergeom}\left([1,1],[\,],-\frac{z(z-1)}{1+z}\right) + 1\right)}{(1+z)^2}$$

Thus, with $F(z, u, v) = \sum F_{n,k,l} z^n u^k v^l$ and for OGF $Q(z) = \sum Q_n z^n$, by recalling that $Q(z) = \phi(F(z, u, -1))$ we get the expression for $Q(z)$ as stated in the Theorem. The expression can then be expanded using the binomial theorem, and double combinatorial sums result for coefficients. $\square$

Though they have no direct bearing on the graph problem at hand, we mention two interesting consequences of this theorem.

Corollary 2 *The quantities Q_n satisfy the recurrence*

$$(n + 1)Q_n + Q_{n+1} - 2nQ_{n+2} + 4Q_{n+3} + (n + 3)Q_{n+4} - Q_{n+5} = 0,$$

where $Q(0) = 0$, $Q(1) = 1$, $Q(2) = Q(3) = Q(4) = Q(5) = 0$ and the asymptotic estimate

$$\frac{Q_n}{(n-2)!} = e^{-2}\left(1 + O\left(\frac{1}{n}\right)\right)$$

Proof: To get the recurrence relation, we use the following holonomic descriptions (introduced by Zeilberger), that is sequences that satisfy linear recurrences with polynomial coefficients:

$$\left(z^4 + z^5 + 4z^3 - 1 - z + 4z^2\right) Y(z) + \left(-2z^4 + z^2 + z^6\right)\left(\frac{\partial}{\partial z} Y(z)\right) -$$

$$-z^4 - 2z^3 \_C_0 - 2z^3 - z^4 \_C_0 + \_C_0 z^2 - z^5 + z - z^2 = 0$$

where $Y(z) = Q(z)$. By putting $C_0 = 1$, we get:

$$(z^4 + z^5 + 4z^3 - 1 - z + 4z^2)Y(z) + (-2z^4 + z^2 + z^6)\left(\frac{\partial}{\partial z}Y(z)\right) - 2z^4 - 4z^3 - z^5 + z = 0$$

We can now get (by elementary properties of the z-transform) the following transformation to a linear recurrence:

$$u(0) = 0, \ u(1) = 1, \ u(2) = 0, \ u(3) = 0, \ u(4) = 0, \ u(5) = 0,$$

$$(n+1)u(n) + u(n+1) - 2n\,u(n+2) + 4u(n+3) + (n+3)u(n+4) - u(n+5) = 0$$

We note that this provides an algorithm that uses a linear number of arithmetic operations to determine the quantities Q_n. By using the following principle based on the generating function method:

$$coeff_{z^n}\mathrm{hypergeom}\left([1,1],[\,],z + dz^2 + \mathrm{O}\left(z^3\right)\right) = n!\,e^d(1 + o(1))$$

provided that the argument of the hypergeometric is a function that is analytic at the origin, we have proved

$$Q(z) = \frac{z\left(z + z\,\mathrm{hypergeom}\left([1,1],[\,],-\frac{z(z-1)}{1+z}\right) + 1\right)}{(1+z)^2}$$

and since

$$-\frac{z(z-1)}{1+z} = z - 2z^2 + 2z^3 - 2z^4 + 2z^5 + \mathrm{O}(z^6)$$

we have proved that the asymptotic proportion of legal permutations is *exactly* equal to e^{-2}. $\square$

   The recurrence above implies the non-obvious fact that the number of avoiding permutations Q_n are computable in linear time. The asymptotic estimate extends properties known for permutations with excluded patterns (e.g., derangements have asymptotic density e^{-1}). Consequently, a nonzero proportion (about 13.53%) of all permutations that start with 1 and end with n are avoiding.

2.2.3   Avoiding Paths. We consider now the problem of counting the number $Q_{n,j}$ of avoiding paths of type (n, j), where n is the size (the number of nodes) and j is the number of "outer nodes". Such avoiding paths are defined by the fact that they satisfy the basic constraints of avoiding permutations regarding the base line $(1, 2, \ldots, n)$, but contain in addition j outer nodes taken to be indistinguishable (unlabelled) and conventionally represented by the symbol '$\star$'. For instance, for types $(n, j) = (3, 1), (4, 1), (4, 2)$, the listings are respectively

$$\{[1, \star, 3]\} \qquad \{[1, 3, \star, 4], [1, \star, 2, 4]\} \qquad \{[1, \star, \star, 4]\}$$

Theorem 25 *The number of avoiding paths is expressible as*

$$Q_{n+2,j} = \sum_{k_2=0}^{n-j} \sum_{k_1=0}^{n-j-k_2} (-1)^{k_1+k_2} (n - k_1 - k_2)! \binom{n - j - k_1 - k_2}{k_1}$$
$$\binom{n - j + 1 - k_1}{k_2} \binom{n - k_1 - k_2}{j}^2$$

(where $j \geq 0$)

Proof: We first define templates on which an inclusion-exclusion argument is applied. The specifications are a simple modification of the templates associated to avoiding permutations.

Let $\{a, b, x\}$ be a ternary alphabet. We now define the grammar of templates.

The collection of strings beginning with a and containing only one x that occurs at the end is described by the rule:

$$sp0 := S = Prod(Sequence(Prod(a, Sequence(b))), x)$$

(It suffices to decompose according to each occurrence of the letter a). We first need so-called "outer points" that are taken from outer space:

$$Outerpoints := Sequence(Prod(Z, mu\_outerpoint))$$

We also need "inner points":

$$Innerpoints := Sequence(Prod(Z, mu\_innerpoint))$$

Size is defined as the cumulative number of points in the pair of paths that underlies an avoiding path in the sense above: it is thus equal to the length of the avoiding path plus the number of $\star$ symbols corresponding to the outer nodes. We thus introduce a special notation for nodes of the integer line that are shared by the two paths:

$$Z2 := Prod(Z, Z)$$

Now, the three types of blocks are described by the following rules:

$sp1 := Prod(mu\_block, Z2, Outerpoints, Innerpoints)$
$sp2 := Prod(mu\_block, Z2, Sequence(Prod(mu\_length, Z2), card \geq 1),$
$\qquad\qquad\qquad\qquad\qquad\qquad\qquad Outerpoints, Innerpoints)$
$sp3 := Prod(Sequence(Prod(mu\_length, Z2), card \geq 1), Z2, Outerpoints,$
$\qquad\qquad\qquad\qquad\qquad\qquad\qquad Innerpoints, mu\_block)$

(Clearly, $sp2$ and $sp3$ are combinatorially isomorphic). The blocks that can occur at the end are of type x and can only be of type $sp1$ or $sp2$ but without outer points nor inner points.

$$sp1x := \mathrm{Prod}(mu\_block, Z2)$$
$$sp2x := \mathrm{Prod}(mu\_block, Z2, \mathrm{Sequence}(\mathrm{Prod}(mu\_length, Z2), 1 \le card))$$

The above grammar is completed (to give S) by substituting into $sp0$

$$
\begin{aligned}
a &= \mathrm{Union}(sp1, sp2) \\
b &= sp3 \quad \text{and} \\
x &= \mathrm{Union}(sp1x, sp2x)
\end{aligned}
$$

The 5-variate GF immediately results from the above specification:

$$F(z, u, v, w_1, w_2) :=$$

$$\frac{-u\, z^2(-1 + z\, w_2 + z\, w_1 - z^2\, w_1\, w_2 + v\, z^2 - v\, z^3\, w_2 - v\, z^3\, w_1 + v\, z^4\, w_1\, w_2 + u\, z^4\, v)}{1 - z(w_2 + w_1) + z^2(w_1\, w_2 - u) + v\, z^2(-2 + 2z\, w_1 - 2\, z^2\, w_1\, w_2 + v\, z^2(1 - z\, w_2 - z\, w_1 + z^2\, w_1\, w_2 + z^2\, u))}$$

where u, v, w_1, w_2 represent the blocks, the length, the outer nodes and the inner nodes, respectively.

For inclusion-exclusion, we set $v = -1$. Application of the ϕ-transformation (that counts the number of ways to connect the blocks) requires the modified form

$$F(z, u, -1, w_1, w_2) :=$$

$$\frac{u^2\, z^2(1 + 2z^2 - z^3\, w_1 + u\, z^4 + z^4\, w_1\, w_2 - z^3\, w_2 - u\, z^2 - z\, w_2 - z\, w_1 + z^2\, w_1\, w_2)}{(1 + z^2)(z^4\, w_1\, w_2 + u\, z^4 - z^3\, w_2 - z^3\, w_1 + z^2\, w_1\, w_2 + z^2 - u\, z^2 - z\, w_2 - z\, w_1 + 1)}$$

The ordinary generating function is here

$$Q(z) := \int_0^\infty \frac{z^2(1 + 2z^2 - z^3\, w_1 + u\, z^4 + z^4\, w_1\, w_2 - z^3\, w_2 - u\, z^2 - z\, w_2 - z\, w_1 + z^2\, w_1\, w_2)\, e^{-u}}{(1 + z^2)(z^4\, w_1\, w_2 + u\, z^4 - z^3\, w_2 - z^3\, w_1 + z^2\, w_1\, w_2 + z^2 - u\, z^2 - z\, w_2 - z\, w_1 + 1)}\, du$$

And this can be expressed in terms of the exponential integral as follows:

$$Q(z) := z^2 \cdot e^{-\frac{(1+z^2)(w_2 + w_1)}{z(z-1)(1+z)}}.$$

$$
\frac{\mathrm{Ei}\left(1, \frac{z^4\, w_1\, w_2 - z^3\, w_2 - z^3\, w_1 + z^2\, w_1\, w_2 + z^2 - z\, w_2 - z\, w_1 + 1}{z^2(z^2-1)}\right) e^{\frac{(1+z^2)(z^2\, w_1\, w_2 + 1)}{z^2(1-z^2)}}}{(z^2 - 1)(1 + z^2)} +
$$

$$
+ z^2 \cdot e^{-\frac{(1+z^2)(w_2+w_1)}{z(z-1)(1+z)}} \cdot \frac{z^2\, e^{\frac{(1+z^2)(w_2+w_1)}{z(1-z^2)}} - e^{\frac{(1+z^2)(w_2+w_1)}{z(1-z^2)}}}{(z^2 - 1)(1 + z^2)}
$$

Again, there is an "explicit form" of the OGF of the problem

$$Q(z) := z^2 \cdot$$

$$
\frac{z^3(z\, w_1\, w_2 - w_2 - w_1) + z^2\, \mathrm{hypergeom}\left([1, 1], [\,], -\frac{z^2\,(z-1)(1+z)}{(1+z^2)(z\, w_2-1)(z\, w_1-1)}\right)}{(1 + z^2)^2(z\, w_2 - 1)(z\, w_1 - 1)} +
$$

$$
+ z^2 \cdot \frac{z^2(1 + w_1\, w_2) - z(w_2 + w_1) + 1}{(1 + z^2)^2(z\, w_2 - 1)(z\, w_1 - 1)}
$$

and also

$$Q(z) := \frac{z^4 \text{ hypergeom}\left([1,1],[\,],-\frac{z^2(z-1)(1+z)}{(1+z^2)(z\,w_2-1)(z\,w_1-1)}\right)}{(1+z^2)^2(z\,w_2-1)(z\,w_1-1)} + \frac{z^2}{1+z^2}$$

The coefficient $c(n,j,k)$ of $z^n\,w_1{}^j\,w_2{}^k$ is obtained by straight expansion and avoiding paths are then enumerated by $C(n,j) = c(n,j,j)$. The corresponding formulae of the Theorem statement are obtained directly by symbolic expansions.

The computations are rather intensive and, for instance, the 4-variable GF that "lifts" $F(z,u,-1)$ is found to be

$$\frac{u\,z^2\left(1 - z\,w_2 - z\,w_1 + z^2\,w_1\,w_2 + z^2 - z^3\,w_2 - z^3\,w_1 + z^4\,w_1\,w_2 + u\,z^4\right)}{(1+z^2)\left(z^4\,w_1\,w_2 + u\,z^4 - z^3\,w_2 - z^3\,w_1 + z^2\,w_1\,w_2 + z^2 - u\,z^2 - z\,w_2 - z\,w_1 + 1\right)} \tag{10.10}$$

It is to be noted that computations have been performed with the help of the computer algebra packages Combstruct and Gfun that are dedicated to automating computations in combinatorial analysis and have been developed in the Maple system for symbolic computation. $\square$

2.3 AVERAGE-CASE ANALYSIS FOR THE RANDOM GRAPH MODEL

We show now how to estimate the robustness to link failures in a random graph that obeys the $G_{n,p}$ model. An *avoiding pair* of length ℓ in a graph is an *unordered* pair of paths, each of length ℓ, with a common source and a common destination, that may share some nodes, but are edge disjoint. We have:

Theorem 26 *The mean number of avoiding pairs of length ℓ between a random source and a random destination in a random graph obeying the $G_{n,p}$ model is*

$$N_\ell(n,p) := \frac{p^{2\ell}}{2n(n-1)} \sum_{j=0}^{\ell} Q_{\ell+1,j} \binom{n}{l+1+j}(l+1+j)!$$

where the coefficients $Q_{n,j}$ are given by Theorem 25.

Proof: The coefficient $1/2$ corresponds to the fact that one takes unordered pairs of paths; the coefficient $1/(n(n-1))$ averages over all possible sources and destinations; the factor $p^{2\ell}$ provides the edge weighting corresponding to $G_{n,p}$; the arrangement numbers account for the number of ways to embed an avoiding path into a graph by choosing certain nodes and assigning them in some order to an avoiding path; the coefficients $Q_{\ell+1,j}$ provide the basic counting of avoiding paths that build up avoiding pairs. $\square$

Note: Since the $G_{n,p}$ model implies isotropy, the quantity $N_\ell(n,p)$ is also the mean number of avoiding pairs between *any fixed* source and destination s,t.

Robustness.. A short table of initial values of $N_\ell(n,p)$ follows:

$$N_2 = \tfrac{1}{2}(n-2)(n-3)p^4 \qquad N_3 = \tfrac{1}{2}(n-2)(n-3)^2(n-4)p^6$$
$$N_4 = \tfrac{1}{2}(n-1)(n-2)(n-3)(n-4)(n-5)^2 p^8$$
$$N_5 = \tfrac{1}{2}(n-2)(n-3)(n-4)(n-5)^2(n^3-11n^2+25n+32)p^{10}$$

From developments in the previous section, the formulæ are computable in low polynomial time (as a function of ℓ). They make it possible to estimate the mean number of avoiding pairs in graphs of a given size for all reasonable values of n, p, ℓ. Take for instance a graph with $n = 10^5$ nodes and an edge probability $p = 5 \cdot 10^{-5}$. This corresponds to a mean node degree that is extremely close to 5, so that, on average, each node has 5 neighbors. Then the mean values are

$$N_2 = 3.1 \cdot 10^{-6},\ N_3 = 7.8 \cdot 10^{-5},\ N_4 = 1.9 \cdot 10^{-5},\ N_5 = 4.8 \cdot 10^{-4},\ N_6 = 1.2 \cdot 10^{-2}$$
$$N_7 = 0.30,\ N_8 = 7.6,\ N_9 = 190,\ N_{10} = 4763,\ N_{11} = 119062,\ N_{12} = 2.9 \cdot 10^7$$

Thus, in this example, one expects to have short and multiple connections between source and destination provided paths of length 8 are allowed. This numerical example also shows that there are rather sharp transitions. The formula of Theorem 26, that entails the following rough approximation

$$N_\ell(n,p) \approx \frac{1}{2}\, n^{2\ell-2} p^{2\ell} \tag{10.11}$$

precisely accounts for such a sharpness phenomenon.

In the introduction, we have defined ℓ-robustness as multiple connectivity by edge-disjoint paths of length *at most* ℓ. In fact, Equation 10.10 leads to explicit expressions for generalized avoiding pairs of type (ℓ_1, ℓ_2) that are made of two paths, of lengths ℓ_1, ℓ_2. It can then be seen that the bottleneck for existence of pairs (ℓ_1, ℓ_2) with $\ell_1, \ell_2 \le \ell$ is in fact the case (ℓ, ℓ). Thus, since $N_\ell(n,p) \to 0$ when $\frac{p}{p_r(n,\ell) \to 0,}$ the function

$$p_r(n, \ell) = 2^{\frac{1}{2\ell}} n^{-1+\frac{1}{\ell}}$$

is a cut-off point for ℓ-robustness and an $(\le \ell, \le \ell)$-avoiding pair is expected or not depending on whether p/p_r tends to 0 or to ∞.

Corollary 3 *Any fixed pair in a $G_{n,p}$ graph is almost certainly not $\ell - robust$ if $p/p_r(n) \to 0$.*

Proof: When $\frac{p}{p_r(n,\ell)} \to 0$, then the expected number $N_\ell(n,p)$ of the desired pairs of paths tends to 0 and so does the probability of existence of at least one such pair of paths (since this probability, by Markov Inequality, is bounded from above by the expectation). Thus, with probability tending to 1, there is no pair of edge disjoint paths between the two vertices and these two vertices are, almost certainly, not $\ell - robust$. $\square$

2.4 THRESHOLDS IN THE RANDOM GRAPH MODEL

In this section we provide bounds for the probability (and thus the threshold, if it exists) of the existence, between any fixed pair of vertices, of two edge-disjoint paths of length at most ℓ, by proving the following:

- We give an estimation of the value $p_L \equiv p_L(n, \ell)$ such that $G_{n,p}$ graphs with $p \leq p_L$ do not satisfy the desired property of the existence, between any fixed pair of nodes, of two edge-disjoint paths between some pair of vertices, with probability tending to 1 as n goes to infinity.

- We present a value $p_U \equiv p_U(n, \ell)$ such that almost every $G_{n,p}$ graph with $p \geq p_U$ has almost all its source-destination pairs of vertices connected by at least two edge-disjoint paths of length at most ℓ.

Theorem 27 *Define*

$$P_L(n, \ell) = \left(\log \frac{n^2}{\log n} \right)^{\frac{1}{\ell}} n^{-1+\frac{1}{\ell}}$$

Then, for $p \leq P_L(n, \ell)$, almost surely, there exists a pair of vertices in the $G_{n,p}$ graph that does not have the $\ell - robustness$ property.

Proof: By using the threshold function for diameter. $\square$

Theorem 28 *Define*

$$P_U(n, \ell) = 2 \left(\log \left(n^2 \log n \right) \right)^{\frac{1}{\ell}} n^{-1+\frac{1}{\ell}}$$

Then, for $p \geq P_U(n, \ell)$, almost surely, almost all pairs of vertices of a $G_{n,p}$ graph have the $\ell - robustness$ property.

Proof: Consider two independent distributions G_{n,p_1} and G_{n,p_2} on the same set of vertices. Let $E_i (i = 1, 2)$ be the events "G_{n,p_i} has diameter ℓ".

Consider the graph $\tilde{G}$ obtained when we superimpose an instance $G' \in G_{n,p_1}$ and an instance $G'' \in G_{n,p_2}$ and OR them (i.e., $\tilde{G}$ has an edge joining u, v iff at least one of G', G'' has). Clearly $\tilde{G} \in G_{n,p}$ with

$$p = p_1(1 - p_2) + p_2(1 - p_1) + p_1 p_2 = p_1 + p_2 - p_1 p_2$$

In fact, if u, v are joined in G' by a path P_1 and in G'' by a path P_2, then these two paths both exist in $\tilde{G}$. For p around the threshold for diameter ℓ of $G_{n,p}$ and $\ell = o(n)$, the number of pairs u, v of $\tilde{G}$ for which the paths of G', G'' overlap is $o(n^2)$, thus the vast majority of pairs of vertices ($n^2 - o(n^2)$ of them) in $\tilde{G}$ is connected via two edge disjoint paths of length $\leq \ell$.

This gives approximately a $p_U \leq p_1 + p_2 - p_1 p_2$ and if $p_1 = p_2 = p_0^{(\ell)}$ (p_0 a threshold for diameter ℓ or $\ell + 1$) then

$$p_U \leq 2p_0^{(\ell)} - \left(p_0^{(\ell)}\right)^2 \leq 2(2\log n - \log c)^{\frac{1}{\ell}} n^{\frac{1}{\ell}-1}$$

from [10], where c can be adjusted so that the diameter is almost surely ℓ (see [10], Corollary 12, p. 237). □

3. RANDOM WALKS AND EXPANDER PROPERTIES IN RANDOM REGULAR GRAPHS WITH EDGE FAULTS

3.1 INTRODUCTION

The study of random walks has many important applications to the design and analysis of randomized algorithms. The main reason is that random walks can be used to simulate the evolution of randomized algorithms, where certain typical questions about the walk (such as the expected number of steps to go from a given vertex to another or the expected number of steps to visit all vertices) are directly related to the expected time complexity of the simulated algorithm.

In order to facilitate the systematic study of random walks, we also employ an abstract representation of them, by using the powerful tool of Markov chains. Random walks and corresponding Markov chains are widely used in Computer Science. There is recently great interest in Markov Chains that are "rapidly mixing", that is they get close (in terms of the variance distance) to the limit distribution after a polylogarithmic number of steps, as a function of the total number of states of the chain (see e.g. [58, 67]). It is known that if G is an expander then the Markov chain associated with P is rapidly mixing. One important application of the rapid mixing property is in the almost uniform generation and approximate counting problems (see [67]).

We also note the close relation of certain important graph properties (such as connectedness, regularity, expansion) to the eigenvalues of the adjacency matrix of the graph. It is also well known that random regular graphs of the model G_n^d are almost surely efficient certifiable expanders. In this section, we investigate whether their partially destroyed due to edge faults counterparts (of the model $G_{n,p}^d$, see also section 1.8) have still expander properties that can be efficiently verified, by studying the second largest eigenvalue of their adjacency matrix. Recall (see also section 1.10) that $G_{n,p}^d$ has a giant component of small diameter, even when in the (worst) case of constant edge failures and constant degrees $d = O(1)$. Here we wish to determine the minimum necessary value of p for which the giant component of $G_{n,p}^d$ remains a certifiable expander with high probability. In the next section we show that the second eigenvalue of the adjacency matrix of the giant component of $G_{n,p}^d$ is concentrated in an interval of small width around its mean, and that its mean is $O((dp)^{1/2})$, provided that $dp > 33$ (in fact $dp > 31^{\frac{1}{1-\epsilon}}$, where $\epsilon > 0$ a very small constant. By appropriate tuning of parameters, we get $\epsilon = 0.0166$ and $dp > 32.86$). Thus, the giant component of a random member of $G_{n,p}^d$ remains, with high probability, a certifiable efficient expander, despite the link faults, provided that $dp > 33$. Recently, A. Goerdt ([35]) has continued the work of Nikoletseas and Spirakis ([64]), using their model of random configurations and strengthening their results.

3.2 DEFINITIONS AND STATE OF THE ART

Let $G(V, E)$ be a graph of n vertices. We allow self-loops and multiple edges. Let A be its adjacency matrix. Its entries, $a_{i,j}$, are positive integers and count the number of edges from i to j (self-loops are counted twice). A is a $n \times n$ symmetric matrix. The graph G is d–regular if $\sum_j a_{i,j} = d$, for $1 \leq i \leq n$.

Let $\lambda_1 \geq \lambda_2 \geq \cdots \geq \lambda_n$ the eigenvalues of the graph $G(V, E)$. For any d–regular graph, $\lambda_1 = d$. The subdominant eigenvalues λ_2 and λ_n provide important information about expansion properties of G. For d–regular graphs, one wants the size of $\rho = max\{d^{-1}|\lambda_2|, d^{-1}|\lambda_n|\}$ to be as small as possible.

For the graph G it is sometimes useful to consider the matrix P with entries $p_{i,j} = \frac{1}{d_i} a_{i,j}$ where d_i is the degree of vertex i. The matrix P can be viewed as the transition matrix of the Markov chain of n states, associated with a random walk on the graph G.

If G is not connected, then the Markov Chain is separated into a number of aperiodic and irreducible chains, one per connected component.

There is considerable recent interest in Markov chains that are "rapidly mixing" i.e. they get close (in terms of the variance distance) to the limit distribution after a polylogarithmic number of steps, as a function of the total number of states of the chain (see e.g. [67]). It is known that if G is an expander then the Markov chain associated with P is rapidly mixing.

Definition 17 *A graph $G(V, E)$ is a c–expander if for every $X \subseteq V, |X| \leq n/2$ implies that $|N(X) - X| \geq c|X|$ where $N(X) = \{y | (x, y) \in E$ and $x \in X\}$. The constant c is called the expansion constant.*

Expanders are the building blocks of optimal networks and algorithms for many purposes such as sorting, routing, superconcentrators etc (see [65]). For d–regular graphs it is true that $c \geq \alpha(1 - \lambda_2/d)$ where $\alpha > 0$ is a constant.

Random d−regular graphs were shown to be almost surely very powerful expanders. Broder and Shamir ([14]) have shown that for most d−regular graphs, $|\lambda| \leq cd^{3/4}$ where $\lambda = max\{|\lambda_2|, |\lambda_n|\}$ and $c > 0$ a constant. Slightly better estimates were produced by ([30]). The above result leads to a quick randomized algorithm for generating certified efficient expanders:

Repeat
Construct a random member G of G_n^d ;
Evaluate its λ_2 ;
until a G is found with $\lambda_2 \leq cd^{3/4}$

By the result of ([14]), the expected number of repetitions is bounded.

Let now G be a random member of $G_{n,p}^d$. Let now λ_i be the eigenvalues of the adjacency matrix of the giant component of G. How does p affect λ_2? In fact, if the giant component of $G_{n,p}^d$ remains a certifiable efficient expander, despite edge faults, then this would lead to an efficient construction of robust fat-trees and other universal networks (see e.g. [55]). Here we prove that for almost any member of $G_{n,p}^d$, $|\lambda| \leq c(dp)^{1/2}$ for a constant $c > 0$. In fact $c^2 \leq \sqrt{33}$, provided $dp > 33$. Thus the efficient certifiable construction of robust expanders despite faults is indeed possible, when p is such that $dp > 33$.

3.3   THE MODEL AND THE GENERAL FRAMEWORK

The work of Nikoletseas and Spirakis ([64]) closely follows, but judiciously modifies and extends, the framework of Broder and Shamir ([14]).

We construct an undirected random $2d$−regular graph G_n^{2d} by choosing independently d permutations uniformly randomly among all possible permutations of the n vertices. For each of the chosen permutations π and for each vertex i ($1 \leq i \leq n$) we add to the graph the edge $(i, \pi(i))$ with probability p, independently. Then an outcome of $G_{n,p}^{2d}$ is produced where each vertex i has a degree $d_i \leq 2d$.

We recall (see also sections 1.9 and 1.10) the following results:

Fact 1 $G_{n,p}^d$ is d−connected except for $O(1)$ vertices, with high probability, for all failure probabilities $f = 1 - p \leq n^{-\epsilon}$ ($\epsilon > 0$ fixed).

Fact 2 When $G_{n,p}^d$ is disconnected, it still has a giant (i.e. $\Theta(n)$−sized) connected component of $\Theta(\log n)$ diameter for any $f < 1 - \frac{32}{d}$, with high probability.

These facts are proved for the (different) d−regular graph model of ([10]) by first showing the results in the configuration space with edge faults. A configuration H is a collection of n labeled groups (supervertices) each group having d labeled (small) vertices and a partition of the set of nd vertices into pairs called edges. We then remove each edge independently with probability p.

One can translate the proofs of Facts 1, 2 to the permutation model by appropriately replacing k−th group of n edges of the partition by k−th permutation. Furthermore, all the theorems proved in this paper for the permutation space hold also for the configuration space with the same mapping working in the opposite direction.

Let A be the incidence matrix of the giant component, C, of G (an outcome of $G_{n,p}^{2d}$) and P its Markov Chain matrix (irreducible, aperiodic) as before. Let C have $n' = \Theta(n)$ vertices. In the case where the graph G is connected, we simply take as C the graph G itself and $n' = n$.

Let the eigenvalues of P be $\rho_1, \ldots, \rho_{n'}$ where we consider them sorted: $\rho_1 = 1 \geq \rho_2 \geq \cdots \geq \rho_{n'}$. Let $\rho = max\{\rho_2, |\rho_{n'}|\}$. P has real eigenvalues, because the Markov Chain it defines is reversible (see [67]).

Lemma 12 *The Markov Chain defined by P (i.e. a random walk on the giant component C of G) is (time) reversible.*

Proof : It is enough to show that for all $i, j \in \{1, \ldots, n'\}$, $p_{ij}\pi_i = p_{ji}\pi_j$ where $\{1, \ldots, n'\}$ are the vertices of C, π_i is the (steady state) probability of state i and $p_{i,j} = \frac{\alpha_{ij}}{d_i}$ is the probability of walking in one step from state i to state j. It is well known (see [73]) that $\pi_i = \frac{d_i}{2e}$ where e is the number of edges of C. Thus $p_{ij}\pi_i = \frac{\alpha_{ij}}{2e} = \frac{\alpha_{ji}}{2e} = p_{ji}\pi_j$ (because C is undirected and A is symmetric). ☐

Thus, the eigenvalues of P are also those of a similar symmetric matrix (see [67]) and so are all real.

Since

$$trace\left(P^{2k}\right) = \sum_i \rho_i^{2k}$$

and since P has real eigenvalues we get:

$$\rho^{2k} \leq trace\left(P^{2k}\right) - 1$$

for $k \geq 1$. Thus, by taking expectations:

$$E(\rho) \leq \left(E\left(\rho^{2k}\right)\right)^{1/2k} \leq \left[E\left(trace(P^{2k})\right) - 1\right]^{1/2k} \qquad (10.12)$$

by Jensen's inequality.

Let $P_{ii}^{(2k)}$ be, for the graph C, the probability that a particle starting at vertex i and moving by following the random walk defined by P, is again at vertex i after $2k$ steps. Note that the probability $P_{ii}^{(2k)}$ is a random variable, since it depends on the (randomly produced) graph. Clearly $E(P_{ii}^{(2k)})$ does not depend on i when C is taken over $G_{n,p}^{2d}$, due to uniformity. Thus

$$E\left(trace\left(P^{(2k)}\right)\right) = n'E\left(P_{11}^{(2k)}\right) = \Theta(n)\,E\left(P_{11}^{(2k)}\right) \qquad (10.13)$$

where the expectation is over all graphs $G \in G_{n,p}^{2d}$, and for their giant components C.

We view G as produced in two steps: First we choose a random member $G_{reg} \in G_n^{2d}$. Then we construct G by selecting each edge of G_{reg} with probability p. Fix a G_{reg}.

In order to study P_{11}, first we formally represent the particle motion on the C of G produced from the G_{reg} by a sequence of moves $S = \sigma_1\sigma_2\cdots\sigma_{2k}$. Each S is a word of length $2k$ in the free monoid M generated by the alphabet $A = \{f_1, f_1^{-1}, \ldots, f_d, f_d^{-1}\}$. Here the mapping f_i is "what remains" from $\pi(i)$ (out of which G_{reg} was constructed) after applying edge faults to edges of G_{reg} to get G. In fact f_i are "partial permutations" since some edges have been removed. Second, we assign to each f_i a partial permutation by choosing, for each vertex v in G, equiprobably one of its (remaining) neighbors. The sequence S then determines, for each vertex x, a trajectory starting at x: $T = (x, x_1, x_2, \ldots, x_{2k})$ where $x_1 = f_1(x), \ldots, x_i = f_i(x_{i-1})$.
In the free group F generated by the letters $\{f_1, \ldots, f_d\}$ S is just an element. Let 1_F (e.g. equal to $f_1 f_1^{-1}$) be the identity element of F.

Definition 18 S *is an identity sequence if* $S \equiv 1_F$.

Definition 19 *Given* S, *let* $R(S)$ *be the unique sequence of minimum length equivalent to* S *obtained by removing from* S *all identity subsequences.* $R(S)$ *is called the reduction of* S.

Definition 20 *Given* S, *if* $S = R(S)$ *then* S *is irreducible. If* S^2 *is irreducible, we call* S *strongly irreducible.*

Then for G:

Lemma 13 $P_{11}^{(2k)}$ *is the probability that the particle starting at vertex 1 either returns to 1 after* $2k$ *steps because* S *is an identity sequence or it does so because* S *reduces to some non-empty irreducible sequence* S' *having a trajectory* $T(S')$ *with* $x'_{2s} = 1$, *where* $2s$ *is the length of* S'). $\qquad\square$

We note that the above events are exclusive (the second event implies that S is not an identity sequence), thus $P_{11}^{(2k)}$ is the sum of their corresponding probabilities. We now bound from above these probabilities.

3.4 ESTIMATION OF $E(P_{11}^{(2K)})$ AND OF THE MEAN OF THE SECOND LARGEST EIGENVALUE

In the sequel, we consider vertex 1 belonging to the giant component C of G.

Lemma 14 *Let S be a random word of length $2k$ in M. Then, assuming $k = \Theta(\log n)$, $\Pr\{S$ is an identity sequence$\} \leq \left(\frac{31}{(dp)^{1-\epsilon}}\right)^{k/2}$.*

Proof: A balanced string of parentheses is equivalent to an identity sequence (just assign σ to every left parenthesis and σ^{-1} to the corresponding right and vice-versa). The number of balanced strings of parentheses is the Catalan number: $\frac{1}{2k+1}\binom{2k+1}{k}$. Let d_i be the degree of vertex x_i in G, for each vertex x_i in the trajectory $T(S)$. Consider first G_{reg} fixed and let G be an outcome of the experiment of sampling edges out of G_{reg}.

Let $X_v(k)$ be the number of paths of length k starting from vertex v in the giant component C of G. The fraction, y, of the number of all identity sequences of length $2k$ over the number of all paths of length $2k$ in G, will be bounded by

$$y \leq \frac{1}{2k+1}\binom{2k+1}{k}\frac{X_1(k)}{X_1(2k)}$$

Let $c_1(k)$ be the number of all cycles of length k starting (and returning) to vertex 1 of C. Then, clearly

$$X_1(2k) \geq c_1(k)\,X_1(k)$$

Thus

$$y \leq \frac{1}{2k+1}\binom{2k+1}{k}\frac{1}{c_1(k)}$$

In order to estimate a lower bound on $c_1(k)$, we must consider the degrees of the vertices of C. Let $d_1, d_2, \ldots, d_k$ the degrees in a cycle including vertex 1. From the experiment in G_{reg} we have $E(d_i) = 2dp$ for each vertex separately, because the edges are sampled in independently (of course there are dependencies among neighbors). For a particular vertex i, then

$$\Pr\{d_i > 2dp(1-\beta)\} \geq 1 - \exp\left(-\frac{\beta^2}{3}2dp\right)$$

by Chernoff bounds in the Bernoulli of $2d$ experiments with success probability p. Fix a $\beta \in (0,1)$ and let $\gamma = 1 - \exp\left(-\frac{\beta^2}{3}2dp\right)$ $(0 < \gamma < 1)$. We shall call a "success" the outcome "$d_i > 2dp(1-\beta)$" for the corresponding vertex i. The probability of success is γ.

Fix a cycle in C of length k and let $k = 2\alpha \log n$ ($\alpha > 0$ a constant to be determined). Consider on the cycle the $\frac{k}{2} = \alpha \log n$ vertices which are not neighbors. Then $c_1(k)$ is at least the number of cycles in the subgraph of C where all vertices have degree at least 2 except for the vertices i where a success happened, and they have degree at least $2dp(1 - \beta)$. The number of such successes in a single cycle is at least $g(k) \geq \frac{k}{2}\gamma(1 - \beta')$ (where $\beta' \in (0, 1)$) with probability at least $1 - \exp\left(-\frac{\beta'^2}{3}\gamma\frac{k}{2}\right)$ by Chernoff bounds.

By choosing appropriate values for the constants β, β', α, we get high probability for having a number of successes which is at least $\frac{k}{2}$ minus a very small constant.

Lemma 15 *Let $\mathcal{E}$ be the event of "at least $\frac{k}{2}(1 - \epsilon)$ successes", where $\epsilon = 1 - \gamma(1 - \beta') > 0$ a constant. By choosing appropriate values for the constants β, β', α, we can make ϵ as small as desired and get*

$$\Pr\{\mathcal{E}\} \geq 1 - \frac{1}{n}$$

(As an example, when $\beta = 0.5, \beta' = 0.01, \alpha \geq 30000$, then $\epsilon = 0.0166$).

(end of proof of lemma 15) □

Thus, conditioning on event $\mathcal{E}$, we get, per cycle, at least $g(k) = \frac{k}{2}(1 - \epsilon)$ vertices of degree $d_i \geq dp$ with high probability. So, $c_1(k) \geq [2dp(1 - \beta)]^{g(k)}$ gives $c_1(k) \geq (dp)^{\frac{k}{2}(1-\epsilon)}$. We then have, by using also $\binom{n}{k} \leq \left(\frac{ne}{k}\right)^k$, that

$$\Pr\{S \equiv 1_F | \mathcal{E}\} \leq \frac{1}{2k+1}\binom{2k+1}{k}\frac{1}{(dp)^{k/2(1-\epsilon)}} \leq \frac{5.5^k}{(dp)^{k/2(1-\epsilon)}}$$

Thus

$$\Pr[S \equiv 1_F | \mathcal{E}] \leq \left(\frac{5.5^2}{(dp)^{1-\epsilon}}\right)^{k/2}$$

When $(dp)^{1-\epsilon} > 31$ (in fact, $\frac{31}{(dp)^{1-\epsilon}} = 1 - \epsilon'$, we get that

$$\Pr\{S \equiv 1_F | \mathcal{E}\} \leq n^{-x}$$

where $x > 0$ a constant. In fact $x = \alpha \log\left(\frac{1}{1-\epsilon'}\right)$.

We get the high probability result by removing the conditioning. Clearly,

$$\Pr\{S \equiv 1_F\} = \Pr\{S \equiv 1_F | \mathcal{E}\}\Pr\{\mathcal{E}\} + \Pr\{S \equiv 1_F | \bar{\mathcal{E}}\}\Pr\{\bar{\mathcal{E}}\} \leq \Pr\{S \equiv 1_F | \mathcal{E}\}\left(1 - \frac{1}{n}\right)$$

for each G_{reg}, and by summing over all the G_{reg} (their total probability is 1) we get

$$\Pr[S \equiv 1_F] \leq \left(\frac{31}{(dp)^{1-\epsilon}}\right)^{k/2}$$

(end of proof of lemma 14)   □

Let S be a fixed sequence and let its trajectory be $T(S) = (1, x_1, \ldots, x_s)$. Clearly $x_1, \ldots, x_s$ are random variables depending on G and the assignment made to $f_1, \ldots, f_d$. We can view $x_1, \ldots, x_s$ constructed as follows: $\sigma_1(1)$ is chosen uniformly randomly in $\{1, \ldots, n\}$. (This does not depend on d or p since, if vertex 1 is connected to some vertex this could be any vertex in $\{1, \ldots, n\}$ equiprobably. Indeed, the regularity constraint on G_{reg} fixes for each vertex only the number of its neighbors. These neighbors are randomly selected and they clearly remain random after the faults, because faults sample edges randomly). For $i \geq 2$, if $\sigma_i(x_{i-1})$ is already fixed in S then $x_i \leftarrow \sigma_i(x_{i-1})$ else x_i is randomly chosen, etc. We call the former situation a forced choice, else a free choice. Because of the fact that these probabilities do not depend on either d or p, we can easily extend the corresponding lemmas proved in [14]:

Lemma 16 (extension of [14], Broder and Shamir) *Given any sequence S of length s, the probability that its $T(S)$ induces a subgraph of more than one loop is $O\left(\frac{s^4}{n^2}\right)$, provided $dp \geq 2$.*   □

By the same argument we get:

Lemma 17 (extension of [14], Broder and Shamir) *For any irreducible sequence S of length s, the probability that its trajectory starting at 1 returns to 1 by traversing a single loop once is $\frac{1}{n} + O\left(\frac{s}{n^2}\right)$.*   □

Next we prove the following upper bound on the probability of reductions of a given size.

Lemma 18 (extension of [14], Broder and Shamir) *The probability that induced loops are traversed at least twice (given that $R(S)$ is an identity) cannot be more than $\frac{k}{n} + O\left(\frac{k^2}{n^2}\right)$.*   □

We also get that:

Lemma 19 *Let S be a sequence of length $2k$ chosen uniformly and randomly. The probability that S:*

1. *is not an identity and*

2. *the trajectory of S, $T(S)$ induces a subgraph of exactly one loop and*

3. *$T(S)$ returns to 1 by traversing the loop at least twice is*

$$O\left(2\,\frac{k}{n}\left(\frac{31}{(dp)^{1-\epsilon}}\right)^{k/2}\right)$$

By adding up we get the following basic theorem:

Theorem 29 *If* $k = \alpha \log n$, $\alpha \geq 30000$ *then*

$$E\left(P_{11}^{(2k)}\right) \leq \left(\frac{31}{(dp)^{1-\epsilon}}\right)^{k/2} + \frac{1}{n} + O\left(\frac{k^4}{n^2}\right) + O\left(2\,\frac{k}{n}\left(\frac{31}{(dp)^{1-\epsilon}}\right)^{k/2}\right)$$

Now, from relations 10.12 and 10.13 in previous section, we get

$$E(\rho) \leq \left[n\frac{31}{(dp)^{1-\epsilon}}\right)^{k/2} + \frac{1}{n} + O\left(\frac{k^4}{n^2}\right) + O\left(2\,\frac{k}{n}\left(\frac{31}{(dp)^{1-\epsilon}}\right)^{k/2}\right)\right]^{1/(2k)}$$

$$\leq \left[n\left(\frac{31}{(dp)^{1-\epsilon}}\right)^{k/2}(1+o(1))\right]^{1/(2k)}$$

and by using a suitable k we have:

Corollary 4 *Let P be the Markov chain probability matrix of a random member of $G_{n,p}^{2d}$. Then the second largest eigenvalue of the adjacency matrix of the giant component satisfies*

$$E(\rho) \leq \left(\frac{31}{(dp)^{1-\epsilon}}\right)^{1/2}(1+o(1))$$

Corollary 5 *The second largest eigenvalue, λ, of the adjacency matrix of $G_{n,p}^{2d}$ satisfies:*

$$E(|\lambda|) \leq (31)^{1/2}(dp)^{(1/2)(1-\epsilon)}(1+o(1))$$

provided that $(dp)^{1-\epsilon} > 31$

Note that in the proof of Lemma 14 we required paths of length $2k$, $k \geq 30000 \log n$. Since $k < n$, our result is asymptotic and holds for $n > k \geq 30000 \log n$ i.e. $\frac{n}{\log n} \geq 30000$.

By using a martingale argument (for the technique of martingale sequences see [1]) we can also show sharp concentration of the eigenvalue around its mean, thus proving that the eigenvalue is bounded away from 1.

4.   ROBUST DECENTRALIZED COMPUTATIONS THROUGH RANDOMIZATION[1]

4.1   INTRODUCTION

The issue of fault tolerance in the framework of Parallel and Distributed Computations, tries to capture phenomena where some nodes (or communication links) of a target parallel machine corrupt (become unavailable or stall) during the execution of a parallel algorithm. The issue has become very intriguing in last years, due to the demand for execution of parallel algorithms over arbitrary sets of machines that work as a whole. In addition, the necessity of exploiting off-the-self computational power has lead to the consideration of arbitrary environments of Parallel and Distributed Computing that may vary with time, according to the availability of their building blocks. Thus, it would be very interesting to devise techniques that use an environment which is prone to failures, for the emulation of a similar environment which is guaranteed to be fault-free during the execution of a parallel algorithm. Fault tolerance in the concept of parallel and distributed computations can be provided at various levels of such a decentralized computing environment:

- At machine level, where the underlying environment is actually a fixed network of processing elements that tries to overcome the corruption of a specific node (or edge). Such works are usually based on two major techniques: the first is the technique of embedding of an ideal parallel machine into a fixed, fault-tolerant underlying network of processing elements which can be fault-tolerant with only constant slowdown (e.g. in [54] it is showed that an N-node butterfly or shuffle-exchange network can emulate a fault free network of the same type and size, with only constant slowdown). The second technique for providing fault tolerance at machine level, is the technique of redundant computations. Albeit it seems that redundancy in parallel computations is rather a waste of computational power, this technique is proved as powerful as randomization in some cases, especially when we have to deal with static faults (see in subsection 4.1.1 the categorization of the fault occurrences).

- At cost model level, where the abstract machine model that is considered by the programmer, is itself fault-tolerant. In particular, the cost model tries to exploit the underlying realistic machine in such a way that the overall execution of the input algorithm will not be affected by the corruption of arbitrary processing elements at runtime. The major difference between this and the previous category of fault tolerance is that, in the latter category, the augmented cost models may become the middle-tier technologies between the actual (arbitrary) environments that provide parallel computation capabilities and the programming environments that need to consider general-purpose parallel architectures in order to implement parallel algorithms transferable from parallel machine to parallel machine. Most works of this kind focus on the PRAM cost model, which used to be the most popular model of Parallel Computations until the early nineties (e.g. [45, 46, 48, 49]). Recently some new works on fault tolerant versions of more realistic cost models than PRAM (such as BSP) have come up, which seem to provide general solutions for fault tolerant versions of these realistic cost models (see for example [52, 53]).

- At programming environment level, in which the programming environment itself takes over the responsibility to provide fault-tolerant primitives to the designers of parallel algorithms (e.g. synchronization operations, end-to-end safe communication, robustness of the storage scheme, agreement protocols among the live processors, etc.). Some of these works are [32, 22, 21].

4.1.1   Categorization of faults.   The major distinction of fault-tolerance problems is based on the kind of the fault occurrences they consider. In the area of Parallel Computing, the prevailing model of faults is the fail-stop model, introduced by Kanellakis and Shvartsman ([46]), according to which whenever a processor dies it is excluded from the remaining simulation process. In [45] the processing elements are allowed to restart at arbitrary times (this is the so-called restartable fail-stop model). Of course, in that case, serious problems with the coordination of work might arise, which are usually dealt with by definite synchronization operations, or timestamping techniques.

Similarly, in the area of Distributed Computing we may have to come up with a large range of faults, from crash faults (which are equivalent to the fail-stop model), to omission faults, or even malicious[1] faults, where the faulty processors join forces to affect the rest of the simulation process). In most cases the crash faults case is considered, while the malicious faults case has to do mainly with issues such as virus attacks in parallel or distributed settings, secure storage in distributed environments, etc.

Additionally the faults may be classified as static (i.e. known at the beginning of the simulation process), or dynamic (i.e. they may occur at arbitrary points of the simulation process). Both these cases are quite interesting. More specifically, the static case reflects the adaptation of a certain cost model or parallel computing environment, over an unknown (but fixed from that point on) working environment. On the other hand, the problem of tolerating dynamic processor faults in pragmatic settings, can be seen as a {Safe Storage & Checkpoint & (dynamic) Scheduling} problem. More specifically, the issue of the fault tolerance in such environments can be modeled as a problem of scheduling sequential sets of independent jobs, on a dynamically changing working environment. Hence, our task is to reassign the amount of pending work because of new processor (or communication links) corruptions on a dynamically decreasing set of live processors, in such a way, that as long as there is at least a fraction of the initial processors still live, the total amount of work will be successfully executed. The challenge lies in proposing an efficient strategy that will achieve an almost work-preserving, robust execution of the input algorithm, and will also assure a balanced split of the work load among the operational processors. This strategy will also have to exploit a robust storage scheme that will tolerate arbitrary processor failures, and provide a periodic Checkpoint procedure to commit the work of the simulation process at runtime. Typical examples of this approach are [53, 21].

4.1.2   Fault tolerance on PRAM machines.   There have been many works in the area of fault tolerance on the PRAM cost model, especially in early nineties ([49, 48, 45, 46, 18]). In these works the fail-stop model is adopted. The reason was that the PRAM model considers that the input algorithms are executed in a lock-step fashion, and the processing elements are totally synchronized by a global clock. Thus, there is no chance of dealyed action of a processing element, that might completely mislead the whole computation. On the other hand, the malicious (or byzantine) faults necessitate the coordination of work among the live processors via the robust implementation of agreement protocols, which are unnecessary in the PRAM case, due to the complete synchrony of the processing elements, and the shared-memory feature.

In [48] a general strategy for simulating arbitrary CRCW PRAM steps[2] over a setting that allows dynamic processor faults is provided, by solving a core problem for this cost model, the `Certified Write All` problem. In [49] the same problem is dealt with over the restartable fail-stop model, by using a combination of tentative computations (i.e. computations that are most likely to be correct) and definite computations (guaranteed computations against any sequence of fault occurrences). This approach achieves constant amortized slowdown per CRCW PRAM step, for many reasonable fault distributions. This is done by having the processors tentatively simulate the fault-free execution of the input algorithm, while a definite auditing procedure monitors the simulation process at specific points. In [46, 45] some strategies are provided for dealing with the `Write All` problem in the fail-stop, non-restartable and restartable cases, where the faults may occur dynamically during the execution of an input CRCW PRAM algorithm.

The reader is referred to the monograph of Paris Kanellakis and Alex Shvartsman ([44]) for an overview of the most important simulation strategies that deal with the issue of fault-tolerance on PRAM machines, and an excellent classification of the instances of fault tolerance on PRAM machines in the fail-stop model.

4.1.3   Fault tolerance on arbitrary machines.   In the case of arbitrary computing environments that consist of processing elements communicating via an underlying network infrastructure, there are several crucial parameters other than the computational power provided by the machine, that affect the performance of fault-tolerant strategies. Such parameters are the latency of the communication infrastructure, the bandwidth per processor or the synchronization cost, that are not accounted for in the PRAM cost model. Thus, proprietary solutions for providing fault tolerance in such settings should be provided, or more appropriate models should be chosen, that better estimate the actual overhead of a simulation strategy.

In [22] the case of dynamic processor faults is considered over an arbitrary message-passing underlying computing environment of synchronous machines. In this setting, an optimal strategy is provided for executing a set of independent tasks. In this work it is stated that the core of any simulation strategy over a synchronous message-passing environment, is a CHECKPOINT routine of the remaining live processors, along with a BALANCED ALLOCATION strategy.

In [32], another primitive operation of distributed computing is considered, namely the BYZANTINE AGREEMENT. In this work the BYZANTINE AGREEMENT is provided against crash failures, within optimal message complexity. Then this primitive operation is used for the provision of a family of early stopping agreement protocols with improved message complexity, and a new solution to the CHECKPOINTING procedure that was provided in [22].

Another work that deals with matters of synchronization over computing environments of limited asynchrony is [33], where a general strategy for simulating a completely synchronous network by a network of limited asynchrony. Despite the fact that this work considers a totally reliable underlying network, it is interesting that it uses the notion of tentative computations and definite auditing (or checkpointing) procedures for the safe progress of the simulation process, which is a strategy that was exploited in many works of fault tolerance (e.g., [49, 53]).

Although such proprietary approaches achieve great efficiency (and in some cases optimality) in the general case of an arbitrary synchronous message-passing computing environment, they cannot exploit the feature of bulk synchrony provided by some new cost models (e.g. BSP, QSM) that seem to prevail in the area of Parallel Computation the last few years. It should also be noted that bulk synchrony (or limited asynchrony) is inherent in the parallel algorithms themselves in many cases, and this gives raise to the provision of bulk synchronous, fault-tolerant environments, that focus their power on other features than the continuous synchronization and agreement protocols, such as Adaptive Load Balancing and Robust Storage Schemes (e.g. see [53]).

4.1.4   Fault tolerance on BSP machines. The BSP cost model focuses mainly on the {computation & communication & bulk synchrony} cost during the execution of an algorithm, rather than on the continuous synchronization procedure of a completely synchronous setting, as in the cases of [32, 22]. Thus new strategies are necessitated that exploit this special characteristic of bulk synchrony and will provide fault tolerance on BSP machines.

In [52], the issue of fault tolerance over BSP machines has been addressed. Simulations for two different cases were considered. In the static case, the faulty or unavailable processors are already known at the beginning of the computation and no processor changes its status afterwards. On the other hand, in the (semi) dynamic case, each processor may fail or become unavailable with a fixed probability during the computation and remains so until the end of the computation; however, some critical periods during the computation where no processor was allowed to fail, could not be avoided. In this work some Monte Carlo constructions based on embedding of the Virtual BSP machine on the operational subset of the Real BSP machine (for the static case) and of work redundancy (for the semi-dynamic case) assure efficient executions of BSP algorithms over fault-prone BSP machines.

In [53], fault tolerant BSP computations under "fully dynamic processor faults" are considered. Namely, the faults may happen online at any point of the computation. To tackle the problem, the issue of the fault tolerance on BSP is modeled as an independent job scheduling problem, on a dynamically changing working environment. The goal is to choose an efficient strategy that will achieve an almost work preserving, robust execution of the BSP algorithm, and will also assure a balanced split of the work load among the operational processors. A modular and efficient simulation scheme is proposed, which, compared to an optimal offline adversarial computation under the same sequence of fault occurrences, achieves an $\mathcal{O}((\log n \cdot \log\log n)^2)$-factor times the optimal work. This scheme combines an ADAPTIVE LOAD BALANCING scheme with a MIXED STORAGE scheme (based on Rabin's Information Dispersal Algorithm, [66]) and a CHECKPOINTING procedure (that exploits a BSPAGREEMENT Protocol for periodic synchronization among the processing elements).

4.1.5   Fault tolerance on NETWORKS OF WORKSTATIONS. The NETWORKS OF WORKSTATIONS ([21, 4]) platform tries to satisfy the demand for construction of parallel systems using off-the-self workstations that actually deliver and even surpass the power and reliability of many large scale machines. It actually represents an inherently asynchronous (or bulk-synchronous) environment for the execution of parallel algorithms. It seems that a cost model such as QSM or BSP would be very easily applicable to this parallel setting, because it consists of processing elements communicating via a specific communication infrastructure, and operate in an asynchronous or bulk-synchronous mode. In the work of [21] the NETWORK OF WORKSTATIONS is modeled as a completely asynchronous multiprocessor, shared-memory system, augmented with a fault-tolerant mechanism that treats even the slow workstations as failed ones. The architecture of this system is centralized, in the sense that there are specialized processors that perform specific operations (i.e. there is one task manager that schedules the pending work, some processors that deposit the necessary information, and some workers that actually execute the tasks that are assigned to them). Nevertheless, as stated in this work, an optimal fault-tolerant strategy should minimize job migration and should be integrated into the parallel system itself. This work actually tries to exploit the techniques that have appeared in the literature for providing fault tolerance over PRAM, combined with robust storage schemes based on information dispersal techniques (see [66]). Of course fault tolerant strategies on more relevant cost models will be much more realistic.

In the sequel, we present as a case study a simulation strategy for handling processor failures on the BSP cost model, which is a bulk-synchronous model of parallel computations. This strategy has to face fully dynamic processor faults and a more complex approach is adopted, that combines a balancing scheme, a storage scheme and a checkpointing procedure.

4.2 CASE STUDY: ROBUST BSP COMPUTATIONS UNDER DYNAMIC PROCESSOR FAULTS

In this section we present as a case study the work of Kontogiannis, Pantziou, Spirakis and Yung for providing BSP simulations over faulty BSP computing environments ([53]).

4.2.1 The BSP cost model.
The Bulk Synchronous Model (BSP) was introduced by L. Valiant ([73]) as a bridging model that tries to close the gap between the domains of decentralized (i.e. parallel or distributed, according to the terminology of [44]) architectures and parallel algorithms.

The applicability of the BSP cost model lies in the fact that, apart from the cost of the parallelism that is accounted by the traditional PRAM cost model, it also considers the communication and synchronization limitations that are imposed by the realistic decentralized architectures. Yet, it does not limit the interoperability of the model among different decentralized computing environments, by abstracting away from the designers detailed architectural features such as the topology of the processing elements, or the synchronization procedures. Thus, the objective of this model is to allow the design of parallel algorithms that can be efficiently executed on a variety of decentralized architectures, at a predictable cost, w.r.t. some architectural parameters that reflect the capabilities of the underlying decentalized machine.

A BSP algorithm $\mathcal{A}$ consists of a sequence of supersteps that are separated by Bulk Synchronization operations (SYNC in short). Each superstep consists of a Local Computation phase and a Communication Phase (LC-phase and Comm-Phase respectively). During an LC-phase each processor performs a sequence of operations on data held in its local memory at the beginning of the superstep, while Comm-phase takes over the transmission of the outgoing messages of each processing element to their destinations, through the underlying communication infrastructure. At the end of the superstep a SYNC operation indicates the end of the current superstep. A BSP machine consists of the following components:

- A collection of n identical processor/memory elements which are distinguished by their unique identification numbers.

- A communication infrastructure that takes over the point-to-point communication process. This infrastructure is characterized by the bandwidth parameter g and the latency parameter L (which are explained in the next paragraph).

- A barrier synchronization mechanism among the n processing elements.

The two major parameters of the decentralized architecture are the bandwidth g, i.e. the (per-processor) ratio of the total throughput of the whole system in terms of local computation operations, to the throughput of the underlying communication network in terms of words of information delivered, and the latency L, which is the minimum time interval between two consecutive SYNC operations.

Thus, the running time of a single superstep on the BSP cost model is characterized by the parameters n, g, and L, and is given by the following formula:

$$T_{\text{superstep}} = \max\{L, T_{\text{LC}} + T_{\text{Comm}}\}, \qquad (10.14)$$

where T_{LC} is the maximum (among the processing elements) cost for local computations, and T_{Comm} is the maximum time needed for transmitting all the outgoing messages to their target processors. If we consider that during the Comm-phase each processor sends and receives at most h $1-$word messages (i.e. an $h-$relation has to be implemented during the Comm-phase) then $T_{\text{Comm}} = g \cdot h$, according to W.F. McColl ([56]). In that case, we have:

$$T_{\text{superstep}} = \max\{L, T_{\text{LC}} + g \cdot h\} \qquad (10.15)$$

Remark: In some cases the underlying machine charges the implementation of an $h-$relation as $g \cdot \max\{h, h_0\}$ for some h_0 that depends on the machine.

372

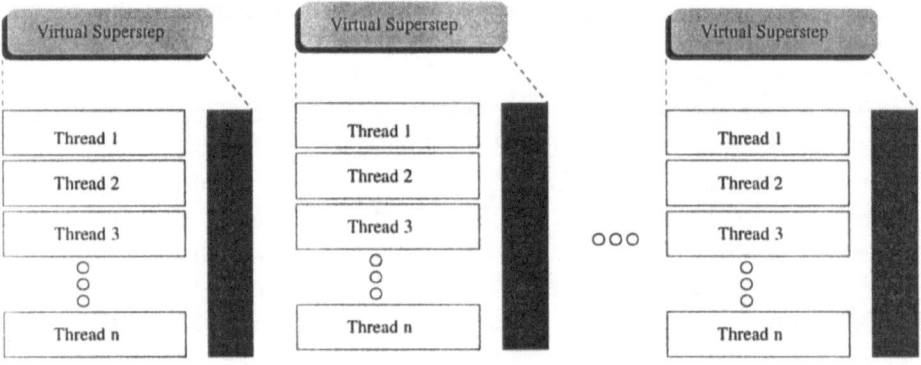

Figure 10.1 Representation of a BSP algorithm as a multithreaded decentralized computation.

This is due to a fixed communication initialization cost, which is irrelevant to the size of the h−relation to be implemented by the network infrastructure.

Finally, the following fact will be used in our time estimations in the sequel (for a justification of this fact, the reader is referred to [36, 6]):

Fact 3 *There exists a BSP algorithm that broadcasts a k−word message to N processors, that requires time at most $\mathcal{O}\left(\log N \cdot \max\{L, \frac{gk}{\log N}\}\right)$. Moreover, if $L \leq \frac{g \cdot k}{\log N}$ then the algorithm needs time $\mathcal{O}(g \cdot k)$.*

4.2.2   The Dynamic Simulation Model. Suppose that we are given a BSP algorithm $\mathcal{A}$, which is designed to run over a hypothetical, fault free n−processor BSP machine (the Virtual Machine, $\mathcal{VM}$). Because of its BSP nature, $\mathcal{A}$ assigns atomic work loads and imposes some demands for communication (e.g. the implementation of an h−relation) among the virtual processing elements, in each Virtual Superstep on $\mathcal{VM}$. Let's call the atomic work load of a specific virtual processor Q_i, along with the portion of communication on behalf of Q_i for the current Virtual Superstep, thread T_i. In that case, the execution of each Virtual Superstep (consisting of an LC-phase and a Comm-phase) may be considered as a balanced, multithreaded computation, while the execution of $\mathcal{A}$ may be considered as a sequence of multithreaded computations, distinguished by periodic barrier synchronization (SYNC) operations among the processing elements (see figure 10.1).

Suppose also that we are given an n−processor BSP machine which is prone to processor failures (the Real Machine, $\mathcal{RM}$)[3]. Then the task of $\mathcal{RM}$ will be to assure the progress of the simulation process until the end of the execution of the input algorithm $\mathcal{A}$, but also keep the total work evenly balanced among the remaining live processors and assure recovery from any sequence of faults at any time.

Let the (uniquely) assigned work load of a real processing element P_i in $\mathcal{RM}$ comprise its PRIMARYJOB. This means that the major task of P_i in each virtual superstep will be to simulate the execution of the threads which correspond to the virtual processors residing in its *PrimaryJobQueue*. Any additional work load that may be taken over by P_i (at its own initiative) during the simulation process, will comprise a SECONDARYJOB. This load is not necessarily uniquely assigned to the specific processor.

Initially we assume that there is a "1-1" correspondence between the processing elements of $\mathcal{RM}$ and $\mathcal{VM}$. This means that any thread of virtual processor Q_i is initially considered to be executed by the real processor P_i (which is its physical destination), unless P_i dies, in which case Q_i will have to migrate to another real processor that will have to execute its forthcoming threads. The necessary information for executing a specific thread T_m, is stored in a LOCALSTORAGE scheme comprised by the real processors. This information is stored using as a key the Virtual Identification number ($\mathcal{V}id$) of the corresponding virtual processor. This storage scheme is based on a BSP implementation of the well known Information Dispersal Algorithm (IDA, [66]) and assures the robustness against small bursts of processor failures.

Additionally, a GLOBALSTORAGE scheme has been adopted, for the periodic creation of some robust replicas of specific instances of $\mathcal{RM}$'s status during the simulation process (we call these securely stored instances *SafeStates*). The *SafeStates* play the role of reference points to which our simulation process will backtrack, in case of locally unrecoverable situations.

We also assume that processors may share a seed from a strong pseudorandom number generator (i.e., we have "coordinated randomness" among the operational nodes of $\mathcal{RM}$).

Definition 1 *A thread is called assigned if it is included in a live real processor's* PrimaryJobQueue, *otherwise it is called pending. Additionally, if a real processor has already completed the simulation of a specific thread (i.e., it has simulated its LC-phase and its Comm-phase and has safely stored the new status of the corresponding virtual processor), then this thread is considered to be completed for the current Virtual Superstep. Finally, a thread that corresponds to a virtual processor Q_i which is assigned to a real processor P_j with $i \neq j$ (i.e., P_j is not the physical destination of Q_i), is called migrated.*

For the reader's easier understanding of the measured quantities, we adopt the following terminology:

A_k: the number of currently live processors at the beginning of virtual superstep VS_k (estimation).

$C_i(\mathcal{P}_i/\mathcal{A}_i)$:   the number of completed (pending/assigned) threads at the end of the SECONDARYJOB J_i of the current Virtual Superstep.

h: the size of the current virtual communication phase (i.e., the size of the $h-$relation among the n threads).

F_k: the number of faults that occur during Virtual Superstep VS_k.

$[n]$:   the set $\{1, \ldots, n\}$.

For the representation of the *CurrentStatus* of the Virtual BSP Machine ($\mathcal{VM}$), we use the following data structures, which are robustly stored among the live processors of $\mathcal{RM}$, using a simple combination of a tentative storage scheme (the LOCALSTORAGE) and a definite one (the GLOBALSTORAGE), (the reader is referred to the full version of this work, for the description of the mixed storage scheme adopted):

VLM_i: the contents of the local memory used by the virtual processing element Q_i. The size of the VLMs is an input parameter for our simulation strategy. A reasonable assumption might be to consider a polylogarithmic size (e.g. $\mathcal{O}(\log^2 n)$) for these local memories.

LoU_i: a list of still undelivered outgoing messages, on behalf of the threads hosted by the live processor P_i.

PJ_i: the *PrimaryJobQueue* of P_i, containing the (uniquely assigned) threads that it executes during each PRIMARYJOB.

N_u: the size of a local neighborhood of real processors after u BACKTRACK operations have occurred, with $N_0 = \log n$ and $N_u = 2 \cdot N_{u-1} = 2^u \cdot \log n$.

4.2.3 Assumptions. Regarding the model of faults, we consider that an oblivious adversary determines a sequence of processor failures that are randomly distributed all over the whole simulation of a specific input algorithm $\mathcal{A}$, and these faults are unrecoverable (i.e., we adopt the non-restartable fail-stop model, where the dynamic faults are distributed uniformly during the whole simulation process).

This fault model tries to capture the behavior of a decentralized setting, such as a NETWORK OF WORKSTATIONS, or an arbitrary distributed computing environment, that behaves like a dynamically changing parallel machine through a virtual parallel interface, (e.g. MPI or PVM). This dynamic computing environment initially allocates processors to the simulation process according to the demands of the input algorithm, and then reclaims resources (according to the demands of the whole parallel setting) because of processor unavailability or temporary machine stalls.

It is also assumed that the amount of processor failures during the simulation process is upper-bounded by a fraction a, which is an input parameter to our simulation strategy. Clearly, this bound immediately affects the cost for creating *SafeStates* during the simulation process, to which the strategy will backtrack, when some unexpected behavior of the parallel setting occurs, without jeopardizing a total corruption of the whole simulation process.

From the above two restrictions (i.e. the random distribution of the faults during the simulation process and the existence of an overall upper-bound on the fraction of faults), it becomes apparent that if we divide the simulation time into sufficiently large (e.g. polylog(n)) time intervals, there is an upper bound $r = r(a)$ on the fraction of faults that may occur in each of these intervals[4] and it is easily shown that the concentration around this bound is very sharp (this is easily understood by a simple application of the well known Coupon Collector's Problem). This implies that if a virtual superstep VS_k needs polylog(n) time to be executed, then $F_k \leq rn$.

For the communication part, we suppose that the BSP algorithm to be executed imposes regular $h-$relation implementations (for some h, which is a parameter of the input algorithm), which means that the corresponding communication graph among the n threads is an $h-$regular digraph. In case that this is not true, some dummy messages could be used so as to have a regular communication digraph.

4.2.4 The Robust-$\mathcal{BSP}$ simulation strategy.

Our simulation strategy for the dynamic case considers a BSP algorithm $\mathcal{A}$ of polynomial number of virtual supersteps, and divides it into epochs of K_0 virtual supersteps[5]. It then tries to robustly simulate the execution of the next epoch of $\mathcal{A}$ and create a new *SafeState* at the end of it, so as to have the work done up to this point committed.

In the sequel we deal with the simulation of a single epoch of virtual supersteps. The presentation begins with the simulation of a single virtual superstep, and then we describe the major techniques that we use for the mixed storage scheme, the dynamic load balancing of the work load and a tentative agreement protocol among the live processors of $\mathcal{RM}$. All these techniques are used by the Robust-$\mathcal{BSP}$ simulation strategy.

Simulation of a single virtual superstep

The purpose of the Robust-$\mathcal{BSP}$ strategy (see figure 10.2) is to initially let the live processors of $\mathcal{RM}$ try to execute the current virtual superstep VS_k as if there were no more processor failures and the work load were evenly distributed among the live processors of $\mathcal{RM}$ (this is done during the routine PRIMARY JOB). In case that VS_k is not yet completed, a number of SECONDARYJOBS[6] follows, so that the work of VS_k be finished and the pending work load be evenly redistributed among the remaining live processors of $\mathcal{RM}$.

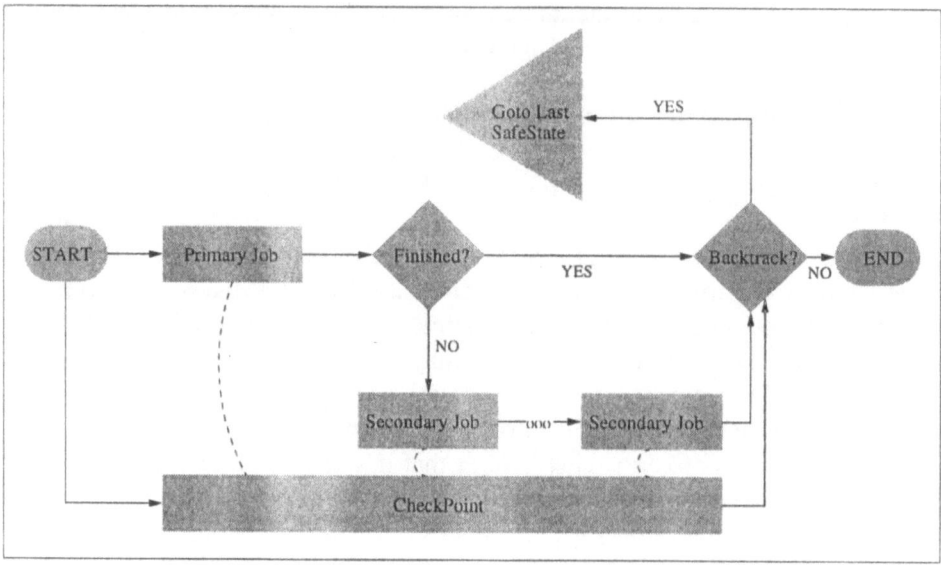

Figure 10.2 The simulation of a single virtual superstep.

As it will be demonstrated in the description of the PRIMARY JOB routine, the *PrimaryJobQueues* of the remaining live processors at the end of the previous Virtual Superstep (VS_{k-1}) comprise a partition of the set of virtual processors into disjoint and almost evenly balanced sets of threads to be executed. Thus, provided that no more processor failures occur during the PRIMARY JOB of VS_k and the work load is already evenly balanced among the live processors of $\mathcal{RM}$, this part would be sufficient for the completion of the work of VS_k. Notice that no work replication is imposed at this part of our simulation strategy.

To assure the robustness of the simulation process against processor failures, a mixed storage scheme has been adopted (see figure 10.3). This storage scheme consists of two basic schemes: a tentative LOCAL STORAGE scheme[7], according to which the *CurrentStatus* of $\mathcal{VM}$ is stored during each superstep in some properly constructed neighborhoods of real processors[8], and a definite GLOBAL STORAGE scheme for creating a *SafeState* at the end of each epoch: that is a robustly stored replica of a specific instance of $\mathcal{VM}$'s *CurrentStatus* so as to be able to reconstruct it at any time, in case that a neighborhood of processors of the LOCAL STORAGE scheme has corrupted. The LOCAL STORAGE procedure is based on an implementation of the well known Information Dispersal Algorithm (IDA, [66]), while the GLOBAL STORAGE procedure creates the necessary replicas of LOCALSTORAGES so that up to arbitrary $a \cdot n$ pro-

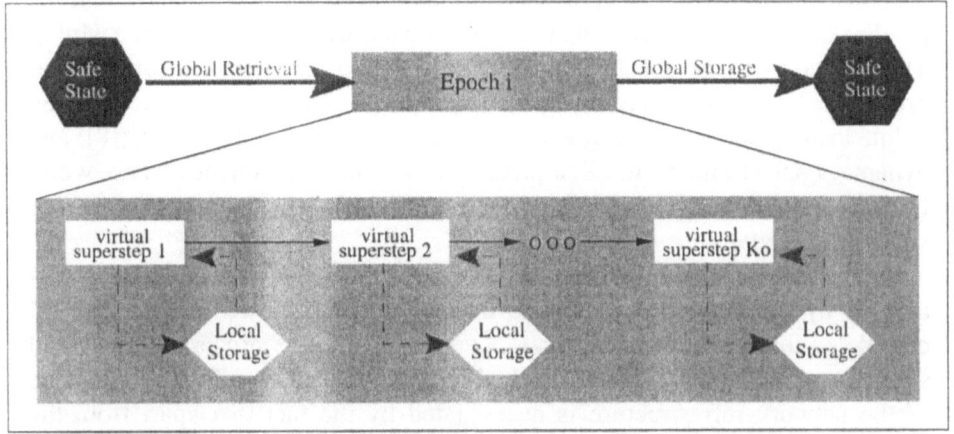

Figure 10.3   The MIXED STORAGE scheme

cessor failures can be overcome. The following lemma is taken from the work of Kontogiannis et al. and gives the time costs for this MIXED STORAGE scheme:

Lemma 20 *Supposing that each live processor is responsible for at most λ threads, the time costs for the procedures of our storage scheme are:*

$$
\begin{align}
T_{\text{LocalStorage}} &= 2\lambda|VLM|(N + \max\{L, g\}) & (10.16)\\
T_{\text{GlobalStorage}} &= \mathcal{O}(\lambda an|VLM|) & (10.17)\\
T_{\text{LocalRetrieval}} &= \lambda(N_u|VLM| + 2|VLM| \cdot \max\{L, g\}) & (10.18)\\
T_{\text{GlobalRetrieval}} &= \mathcal{O}(\lambda an|VLM|) & (10.19)
\end{align}
$$

Remark: A possible new processor failure (death) that will not let the corruption information be disseminated to all the live processors, or a failure of the GLOBAL STORAGE may cause no problem at all. This is because the problematic situation will be discovered during the next superstep, and a new BACKTRACK procedure will fix this abnormality. Additionally, each time a new BACKTRACK occurs, the probability of having a new corruption decreases significantly because of the redistribution of faults in the new, double-sized neighborhoods of processors, and thus there is no chance of an infinite loop in the same epoch.

In our simulation strategy an ADAPTIVE LOAD BALANCING scheme (ALB in short) is inherent. It tries to evenly distribute the work load of $\mathcal{VM}$ among the currently live processors of $\mathcal{RM}$. The purpose of ALB is to make the

PRIMARY JOB routine work as close to optimal as possible, in case that the parallel setting is at a stable state (i.e. when no more processors die for a while). This is a consequence of the "almost even work loads among the operational processors" property, achieved by our balancing scheme. This is actually an online load balancing technique based on ideas such as work stealing ([8]) and dynamic contest among the live processors for the assignment of new work. In section 4.2.5, ALB is shown to have a very good performance w.r.t. our simulation model.

The "FINISHED?" condition of the Robust-$\mathcal{BSP}$ strategy can be implemented by the network infrastructure that will have to add the number of completed threads in the *PrimaryJobQueues* of the live processors, and will signify the end of VS_k if n is reached by this sum. This added intelligence of the network infrastructure is necessitated by the fact that apart from the periodic synchronization of the real processors (that is provided by the underlying BSP machine, $\mathcal{RM}$) a periodic synchronization operation among the virtual processors is also necessary, that will signify the end of the virtual supersteps in the ideal machine, $(\mathcal{VM})$. Another approach could be the use of a tentative estimation for this boolean condition, by having the processors apply a BSP AGREEMENT protocol to decide whether the simulation of VS_k has been completed or not. Of course, a solution like this would require an assured Load Balancing property (i.e. very tight upper and lower bounds on the *PrimaryJobQueues*) so as for the failure probability to be extremely small. On the other hand, some checkpointing procedures such as those proposed in [32, 22] could be adapted to provide this virtual SYNC operation that would assure the integrity of $\mathcal{VM}$. Recall that a possible failure at this point would not be catastrophic, since it would be discovered anyway during the next virtual superstep.

As for the BACKTRACK? condition at the end of the currecnt virtual superstep, if a new corruption of the LOCAL STORAGE has been discovered by a processor, the dissemination of this exceptional information will occur the next time a BSP AGREEMENT protocol will be executed by the remaining live processors of $\mathcal{RM}$. This is done at the beginning of the next job (either a PRIMARY JOB or a SECONDARY JOB). At the end of the current superstep, a BACKTRACK routine is performed by all the live processors to the last *SafeState*, in case that a locally unrecoverable error has occurred, otherwise the simulation process proceeds with a new virtual superstep, or with the creation of a new *SafeState* and the work of the current epoch is then completed.

Primary Job

Suppose that the current virtual superstep that has to be executed is VS_k. As mentioned above, the task of PRIMARY JOB is to let all the live processors behave as if everything were fine and no new processor failures will occur during VS_k. In this phase, each live processor P_i executes its own (unique)

> [1] BSPAGREEMENT: Find an estimation of the currently
>   live processors, A_k.
> [2] **IF** $|PJ_i| > \frac{c \cdot n}{A_k}$ **THEN** Discard some randomly chosen
>   of the assigned threads, until $|PJ_i| \leq \frac{c_1 \cdot n}{A_k}$.
> [3] **LC-phase:** Execute all the LC-phases of the threads
>   in PJ_i.
> [4] **Comm-phase:** Send all the outgoing messages to
>   their physical destinations.
> [5] Register all the undelivered messages to the List
>   of Undelivered messages, LoU_i.
> [6] SDOU: Try to send as many as possible of the
>   undelivered messages, to their destinations.
> [7] LOCALSTORAGE: Store the VLMs of all the completed
>   threads into the proper neighborhoods.

Figure 10.4 The Primary Job procedure.

chunk of threads, held in PJ_i. Before starting the execution of the threads held in its *PrimaryJobQueue*, each live processor P_i upper-bounds the work load that it will execute, according to an estimation A_k of the currently live processors, so that no processor be unevenly overloaded. This estimation is based on an application of the tentative BSP AGREEMENT protocol.

Consequently, each live processor executes its own portion of work, sends the outgoing messages to their physical destinations, and tries (employing the SDOU routine, see section 4.2.6) to forward the undelivered messages which are heading for migrated threads. Finally, it stores the VLM of each completed thread in the proper neighborhood of the LOCAL STORAGE scheme, according to the corresponding virtual processor's (unique) Vid.

According to the analysis of this routine in [53], the overall cost of each PRIMARY JOB is upper-bounded by

$$T_{\text{PrimJob}} = \mathcal{O}\left(\mu_k \cdot \max\{h \cdot \log n, \log^4 n, T_{\text{LC}} + T_{\text{Comm}}\}\right), \quad (10.20)$$

where $\mu_k \equiv \frac{c_1 n}{A_k}$ indicates the maximum load of each *PrimaryJobQueue* after the discarding of the excessive loads from the overloaded processors.

Secondary Job

The goal of each SECONDARY JOB J_i is to complete the outstanding work of VS_k and redistribute it among the live processors as evenly as possible. The new work assignment is done as follows: Each of the A_k live processing elements chooses $\mathbf{b} \equiv g(n) \cdot \frac{n}{A_k}$ threads to cover (that is, to execute, if not yet completed) either at random, or derived by previous unsuccessful communication attempts with pending threads. For the purposes of the following analysis we consider all these choices independent and random, although it is our strong belief that the "biased" choices would make the processing elements focus exactly on the remaining unsatisfied threads.

Lemma 21 *The number of individual processors that cover a specific (pending) thread during J_i, is $c_3 \cdot \log n$, with probability $1 - n^{-c_4}$, $\forall c_3 > 0$, and c_4 depending on c_3.*

Proof: The total number of random choices for covering the n threads, is $M \equiv n \cdot g(n)$. If we have $g(n) = \ln n + c_2$ then it is assured by the well known `Coupon Collector's Problem` (CCP) that with overwhelming probability $(e^{-e^{-c_2}}$, see section 3.6 of [58]) all the n threads will be covered by a live processor at least once. Here we want to prove a less strict property, i.e. that all the threads will be covered by at least $\mathcal{O}(\log n)$ processors with high probability (which implies that the failure probability will be polynomially small).

Let $X_{i,j}$ be the indicator variable of the ith trial hitting thread T_j. Then $Y_j = \sum_{i=1}^{M} X_{i,j}$ is the random variable indicating the total number of "hits" for the thread T_j. Obviously, $E[Y_j] = g(n)$, and by applying Chernoff Bounds on these random Bernoulli trials, it is easily shown that

$$\forall j \in [n], \Pr[Y_j < c_3 \cdot \log n] < \exp\left(-\frac{g(n)}{2} - c_3 \cdot \log n + \frac{c_3^2 \cdot \log n}{2g(n)}\right)$$

We want to show that

$$\Pr[\exists j \in [n], Y_j < c_3 \cdot \log n] = \mathcal{O}(n^{-c_4}), \tag{10.21}$$

for some positive constant c_4 that depends on the number of independent trials.

$$\Pr[\exists j \in [n], Y_j < c_3 \cdot \log n] \leq \sum_{j \in [n]} \Pr[Y_j < c_3 \cdot \log n] \leq n \cdot e^{-K},$$

where $K \equiv \frac{g(n)}{2} + c_3 \cdot \log n - \frac{c_3^2 \cdot \log n}{2g(n)}$. We now have that

$$n \cdot e^{-K} = n^{-c_4} \Rightarrow c_4 = \frac{K}{\ln n} - 1 \Rightarrow$$

$$c_4 = \frac{g(n)}{2 \ln n} + \frac{c_3 \cdot \log n}{\ln n} - \frac{c_3^2 \cdot \log n}{2g(n) \cdot \ln n} - 1$$

Clearly, for $g(n) = \Theta(\log n)$ we have shown that the joint probability of each thread being "hit" by less than $c_3 \cdot \log n$ trials is polynomially small. But we would like to know the number of distinct processing elements that cover a specific thread. As it is shown by the `Sparse Occupancy Problem` (see [53]), each processor hits a specific thread at most once when the number of trials $g(n)$ is significantly smaller than n, with probability that tends to unity. So

---

[1] BSPAGREEMENT: Check for new LOCALSTORAGE
      corruptions.

[2] Each live processor P_i creates a bucket B_i
      of b randomly chosen threads.

[3] INFORMATIONGATHERING: for all the threads in B_i,
      get the corresponding VLMs from the
      appropriate neighborhoods.

[4] **LC-phase:** Execute the $LC - phases$ of the threads
      in B_i.

[5] **Comm-phase:** Send all the outgoing messages to
      their physical destinations.

[6] Register all the undelivered messages to the List
      of Undelivered messages, LoU_i.

[7] SDoU: Try to send as many of the undelivered
      messages as possible, to their destinations.

[8] LOCALSTORAGE: Store the VLMs of all the completed
      threads into the proper neighborhoods.

Figure 10.5   The Secondary Job procedure.

we can guarantee that each thread is covered by at least $c_3 \cdot \log n$ processing elements with probability $1 - n^{-c_4}$. $\square$

Conditioning now on the number of distinct processors that cover a random thread during a SECONDARY JOB and then applying the Markov Inequality, we can show that:

Theorem 1 *The probability of having at least c_5 pending threads at the end of a* SECONDARY JOB *is given by*

$$\Pr[\mathcal{P}_i \geq c_5] \leq \frac{n^{-\min[c_3 \cdot \log(\frac{1-a}{r}), c_4] + 1}}{c_5} \qquad (10.22)$$

The time analysis of each SECONDARY JOB is quite similar to the analysis of a PRIMARY JOB and thus its cost is given by

$$T_{\text{SecJob}} = \mathcal{O}\left(b \cdot \max\{h \cdot \log n, \log^4 n, T_{\text{LC}} + T_{\text{Comm}}\}\right) \qquad (10.23)$$

As for the number of SECONDARY JOBS that are necessary for the whole simulation process to work, this is determined by the following statement of [53]:

Lemma 22 *The failure probability of a Virtual Superstep VS_k to complete its work is $n^{-\mathcal{O}(\log\log n)}$, considering that $y = \log\log n$ SECONDARYJOBS are executed, if necessary.*

Having a subpolynomially small failure probability for each of the virtual supersteps that comprise an epoch, it is now apparent that an epoch can contain $K_0 = \Theta(n)$ virtual supersteps, and still have subpolynomially small failure probability.

Checkpointing and Backtracking

The CHECKPOINT procedure is actually a virtual process that is done by any processor during the simulation of the input algorithm. More specifically, CHECKPOINT signifies the failure of some LOCAL STORAGE or INFORMATIONGATHERING procedure call, discovered by a live processor, either because of an update failure of the LOCAL STORAGE routine, or because of an unsuccessful attempt to retrieve a specific *VLM*. The discovery of a problematic situation is disseminated to the rest of the live processors, the next time that a BSP AGREEMENT protocol is executed, as an exception code to the specific live processor's value. Notice that a new processor failure may not cause any trouble at all, because the problematic situation will be rediscovered by $\mathcal{O}(\log n)$ live processors that will try to cover the corresponding pending (because of the new death) thread during the next SECONDARY JOB.

When such an interruption to the flow of the simulation process is done, the BACKTRACK operation simply makes the live processors set their program counters to the last *SafeState*, set a new value $N_u = 2 \cdot N_{u-1}$ (u is the number of BACKTRACKS having occurred up to now), consider a new, random hash function for the size-N_u equi-partition of the real processing elements into $\frac{n}{N_u}$ neighborhoods of real processors (see the description of the LOCAL STORAGE scheme in the full paper), and continue with the simulation process after having retrieved the *CurrentStatus* from the last *SafeState*.

The choice of a new (pseudo-) random hash function for the construction of the new neighborhoods of processors is done in order to redistribute the up to this point occurred faults evenly among the new neighborhoods, and protect the simulation from a malicious behavior of an adversary that would try to focus its power on a specific neighborhood. On the other hand, the size of each neighborhood is doubled in order to bound the total probability of having too many corruption, as indicated by the following technical lemma[9]:

Lemma 23 *The probability of having a new neighborhood corruption after* $\log \log n$ BACKTRACK *operations, is*

$$\mathcal{O}(n^{-\log n} \cdot \log^{-3} n).$$

Proof: Consider that we have n processors and we choose a random equi-partition of them into n/N_u neighborhoods of size N_u (u indicates the number of BACKTRACK occurrences, up to now). Suppose also that D processors have already died (and have caused the u BACKTRACKS) and $A = n - D$ remain live. We say that an equi-partition corrupts, if there exists a neighborhood that has at least $\frac{N_u}{2} + 1$ dead processors. The following claim will shed some light to our discussion:

Claim 2 *The probability* $\Phi(D, N)$ *that a randomly chosen size$-N$ equi-partition corrupts because of D processor failures, is given by*

$$\Phi(D, N) = \begin{cases} 0, \text{ if } D < N/2 + 1, \\ \frac{n \cdot 2^{N+\frac{1}{2}}}{\sqrt{\pi N} \cdot (N+2)} \cdot \exp[-(N/2+1)(H_n - H_D)], \text{ otherwise,} \end{cases}$$

where H_n, H_D are the corresponding harmonic numbers.

Proof: For the proof of this claim the reader is referred to [53]. □ The corruption probability of a size-N_u equi-partition ($N_u = 2^u \cdot \log n$) is given by $\Phi(D, N_u)$, which is always at most equal to $\Phi(an, N_u)$. Some calculation will help to see that this is a very good bound for the failure probability, which actually implies that at most $\log\log n$ backtrack operations may occur during the simulation of the input algorithm, with very high probability (wvhp). More specifically, considering that $H_n - H_{a \cdot n} = \ln(1/a) + \Theta(1) = \ln(1/a) + \omega_1$, we have

$$\Phi(D, N_u) \leq \Phi(an, N_u) \leq \frac{2^{\omega_2}}{\sqrt{\pi N_u} \cdot (N_u + 2)} \tag{10.24}$$

where,

$$\omega_2 \equiv N_u + 1/2 + \log n - N_u \cdot \left(\frac{\log(1/a)}{2} + \frac{\omega_1}{2\ln 2} \right) -$$
$$- \left(\log(1/a) + \frac{\omega_1}{\ln 2} \right) = \mathcal{O}\left(-2^u \cdot \log n \right)$$

□

The Performance of the Robust-$\mathcal{BSP}$ simulation strategy

In this subsection we estimate the amortized cost of a virtual superstep executed by the Robust-$\mathcal{BSP}$ simulation strategy, and give a bound on the competitive ratio of our strategy, against an optimal offline strategy, that always lets the live processors execute a fully balanced work load[10].

Theorem 2 *The amortized cost for the simulation of a single virtual superstep is given by*

$$T_{VS_k} = \mathcal{O}\left((\log n \cdot \log\log n)^2 \cdot T_{OPT} + \text{polylog(n)} \right) \tag{10.25}$$

with probability at least $\mathcal{O}(1 - n^{-\log n} \cdot \log^{-3} n)$.

Proof: According to the up to now analysis of the Robust-$\mathcal{BSP}$ simulation strategy, the parallel time of a single epoch of K_0 virtual supersteps is upper-bounded by the following equation:

$$
\begin{aligned}
T_{\text{epoch}} \;=\; & (\mathcal{B}+1)K_0\left(T_{\text{PrimJob}} + \log\log n \cdot T_{\text{SecJob}}\right) + \\
& + \mathcal{B}\cdot(\mu_{\text{k}} + \mathbf{b}\cdot\log\log n)\cdot \\
& \cdot(T_{\text{GlobalStorage}} + T_{\text{GlobalRetrieval}})
\end{aligned}
\tag{10.26}
$$

where at most $\mathcal{B}$ BACKTRACKS occur during this epoch. But we know that with subpolynomially small failure probability, the number of BACKTRACK operations in the whole simulation process is no more than $\log\log n$. We also know that in each superstep, a live processor may request for at most $\mu_{\text{k}} + b\cdot\log\log n$ threads.

Notice also that if we suppose that $T_{\text{LC}} + T_{\text{Comm}} = \mathcal{O}(g\cdot h) \geq \log^4 n$, then $T_{\text{PrimJob}} = \mu_{\text{k}}\cdot\max\{h\cdot\log n, (T_{\text{LC}} + T_{\text{Comm}})\} = \mathcal{O}(\frac{\log n}{g}\cdot T_{\text{OPT}})$ and $T_{\text{SecJob}} = \mathbf{b}\cdot\max\{h\cdot\log n, (T_{\text{LC}} + T_{\text{Comm}})\} = \mathcal{O}(\frac{\log^2 n}{g}\cdot T_{\text{OPT}})$. In that case, the amortized cost of a single virtual superstep is given by

$$
\begin{aligned}
T_{\text{VS}_{\text{k}}} \;=\; & (\mathcal{B}+1)\cdot T_{\text{PrimJob}} + \frac{an\cdot\log\log n}{K_0}\cdot T_{\text{SecJob}} + \\
& + \frac{\mathcal{B}\cdot(\mu_{\text{k}} + \mathbf{b}\cdot\log\log n)}{K_0}\cdot(T_{\text{GlobalStorage}} + T_{\text{GlobalRetrieval}}) \\
\;=\; & \mathcal{O}((\log n\cdot\log\log n)^2)\cdot T_{\text{OPT}} + \text{polylog(n)}
\end{aligned}
\tag{10.27}
$$

$\square$

Remark: Since the probability of having more than $\log\log n$ BACKTRACK occurrences is $\mathcal{O}(n^{-\log n}\cdot\log^{-3} n)$, the expected (amortized) cost of each virtual superstep converges to the above value, since the subpolynomially small failure probability dominates over the cost of some extra BACKTRACK occurrences.

In the following sections we present some fundamental randomized techniques that were exploited by the Robust-$\mathcal{BSP}$ strategy. More specifically, we present the ADAPTIVE LOAD BALANCING scheme (section 4.2.5), the SDoU procedure that implements the indirect communication among the migrated threads (section 4.2.6), and the BSP AGREEMENT protocol exploited by the live processors to achieve a common estimation on the number of live processors at the beginning of a new virtual superstep (section 4.2.7).

4.2.5   The Adaptive Load Balancing strategy. A major result of this work which is also of independent interest, is the proposed algorithm for balancing the work of the n virtual processors among the currently live processors of $\mathcal{RM}$. In fact, this is a randomized, adaptive load balancing technique, since the sequence of the fault occurrences is not known a priori to the simulation process and the live processors contest for the pending threads and get some of them according to a probability depending on their current loads.

Starting from a balanced situation (at the last *SafeState*), we show in this subsection how we keep the work of the live processors balanced (in a tentative fashion) for a whole epoch of K_0 virtual supersteps[11]. Let $\mathcal{Z}_0(i, k)$ denote the size of P_i's queue after discarding the excessive work (in case of overloaded real processors), and $\mathcal{Z}(i, k)$ denote its size at the end of VS_k. Our ADAPTIVE LOAD BALANCING scheme (ALB in short), which is inherent in the Robust-$\mathcal{BSP}$ simulation strategy, is the following:

[1] At the beginning of each virtual superstep VS_k, each overloaded real processor P_i cuts off the excessive work (i.e. keeps at most $\frac{c_1 n}{A_k}$ threads at random in PJ_i, according to the estimation A_k of the number of live processors in $\mathcal{RM}$, and makes the remaining threads pending). This is the Discarding step at the beginning of each PRIMARY JOB.

[2] All the pending threads of $\mathcal{VM}$ (due to either new deaths, or discardings) are rescheduled to still live processors of $\mathcal{RM}$ as follows:

(a) Live processors contest for $b \cdot \log\log n$ randomly chosen threads to cover, during the $\log\log n$ SECONDARY JOBS.

(β) Each mailbox that takes over a thread T_λ during a round of SDoU, assigns it (if it is pending) to one of the live processors that contests for it, with probability

$$\Pr[P_{\lambda_i} \text{ gets } T_\lambda] = \frac{\frac{1}{\mathcal{Z}_0(\lambda_i, k)}}{\sum_{j \in C} \frac{1}{\mathcal{Z}_0(j, k)}}$$

where $C = \{P_{\lambda_1}, P_{\lambda_2}, \ldots\}$ is the set of processors contesting for T_λ.

Theorem 3 *Let $\xi > 1$ be a constant. Then the load of each live processor P_i in $\mathcal{RM}$ at the end of VS_k is $\mathcal{Z}(i, k) \leq \frac{c_1 n}{A_k} \cdot \Theta(\log\log n)$, with probability at least*

$$\left(1 - n^{-\xi}\right)^{(k+1)} \geq 1 - (k+1)n^{-\xi}$$

Proof: Let $\xi > 0$ be a constant. We shall prove the good behavior of ALB using induction on the epochs of the input algorithm $\mathcal{A}$.

Initial Step. At the beginning of our simulation process, each virtual processor is assigned to its **physical destination**, and thus, each live processor in $\mathcal{RM}$ has exactly one thread in its *PrimaryJobQueue*.

Inductive Hypothesis. Suppose that at the beginning of epoch e_i, we have

$$\forall j \in [n], \; \mathcal{Z}(i, k-1) \leq \frac{c_1 n}{A_{k-1}} \cdot \omega(n) \qquad (10.28)$$

where $\omega(n)$ is a constant times $\log\log n$.

Inductive Step. Recall that a *PrimaryJobQueue* PJ_j is overloaded iff

$$\frac{c_1 n}{A_k} < \mathcal{Z}(i, k-1) \leq \frac{c_1 n}{A_{k-1}} \cdot \omega(n) \qquad (10.29)$$

The amount $\mathcal{P}$ of pending threads during VS_k will have to be rescheduled among the remaining operational processors of $\mathcal{RM}$. In $\mathcal{P}$ we measure only assignments of threads to processors that remain live until their completions, because a thread assigned to a processor that dies before achieving its completion has already been included in $\mathcal{P}$, either as a pending thread because of a new death, or as a discarded thread at the beginning of VS_k.

Let f_k denote the set of newly dead processors during VS_k, and D the number of discarded threads at the beginning of VS_k. Then $F_k = |f_k| \leq rn$. Clearly, $\mathcal{P} = D + \sum_{i \in f_k} \mathcal{Z}_0(i, k) \leq a \cdot n$. Consider now a specific thread T_λ and suppose that the operational processors $P_{\lambda_1}, P_{\lambda_2}, \ldots, P_{\lambda_m}$ contest for it during a SECONDARYJOB. We already know that T_λ is covered by less than $c_3 \cdot \log n$ initially operational processors of $\mathcal{RM}$ with probability n^{-c_4}, for some positive constant c_3 and c_4 depending on c_3. So we can go on with our proof conditioning on the event "$m \geq \log n$". By the rescheduling rule we have:

$$\forall i \in [\log n], \; \Pr[P_{\lambda_i} \text{ gets } T_\lambda] \leq \frac{\frac{1}{\mathcal{Z}_0(\lambda_i, k)}}{\sum_{j=1}^{\log n} \frac{1}{\mathcal{Z}_0(\lambda_j, k)}} \qquad (10.30)$$

It is now obvious that $\mathcal{Z}_0(\lambda_j, k) \leq \frac{c_1 n}{A_k}, \forall j \in [\log n]$. So, $\sum_{j=1}^{\log n} \frac{1}{\mathcal{Z}_0(\lambda_j, k)} \geq \frac{A_k \cdot \log n}{c_1 n}$ and thus we have:

$$\forall \, i \in [\log n], \; \Pr[P_{\lambda_i} \text{ gets } T_\lambda] \leq \frac{c_1 n}{\mathcal{Z}_0(\lambda_i, k) \cdot \log n \cdot A_k} \qquad (10.31)$$

Now, each live processor P_i will randomly (with replacement) choose $b \cdot \log\log n = \frac{c_1 n}{A_k} \cdot \log n \log\log n$ threads to cover during the SECONDARYJOBS of VS_k. Each random choice has a probability $\frac{\mathcal{P}}{n}$ of hitting a pending thread and is independent of the other choices. Thus the number of pending threads that P_i will contest for, is the number of successes from $b \cdot \log\log n$ Bernoulli trials $B(q, \frac{\mathcal{P}}{n})$, where $q \equiv \frac{c_1 n}{A_k} \cdot \log n \cdot \log\log n$. For each such thread, the probability of P_i prevailing over the other contestants (in a specific round of

SDoU), has already been shown to be

$$\Pr[P_i \text{ gets } T_\lambda] \leq \frac{c_1 n}{A_k \cdot \log n} \cdot \frac{1}{\mathcal{Z}_0(\lambda_i, k)} \equiv \varphi$$

Let Ξ_i be the number of pending threads that P_i contests for. Then

$$\forall \varepsilon \in (0, 1), \Xi_i \in \left[(1 - \varepsilon) \cdot q \cdot \frac{\mathcal{P}}{n}, (1 + \varepsilon) \cdot q \cdot \frac{\mathcal{P}}{n} \right]$$

with probability at least $1 - \exp\left(-\frac{\varepsilon^2}{2} \cdot q \cdot \frac{\mathcal{P}}{n}\right)$, by a simple application of Chernoff Bounds on the Bernoulli trials. But recall that $\mathcal{P} \leq a \cdot n$, and so

$$\forall \varepsilon \in (0, 1), \Xi_i \in [(1 - \varepsilon) \cdot q \cdot a, (1 + \varepsilon) \cdot q \cdot a]$$

with probability at least

$$1 - \exp\left(-\frac{\varepsilon^2}{2} \cdot \frac{ac_1}{1 - a} \cdot \log n \cdot \log\log n\right) = 1 - n^{-\frac{\varepsilon^2 \cdot c_1 \cdot a \cdot \log e}{2(1-a)} \cdot \log\log n}$$

Now, given Ξ_i, due to our rescheduling rule, P_i will actually add at most $\Xi_i \cdot \varphi$ new threads in its own *PrimaryJobQueue* with probability at least $1 - n^{-c_9}$ as it is easily shown by a new application of Chernoff Bounds, considering the worst case for P_i, $\mathcal{Z}_0(i, k) = 1$, for which φ is maximized). So, with total probability at most $(1 - n^{-c_9}) \cdot \left(1 - n^{-\frac{\varepsilon^2 \cdot c_1 \cdot a \cdot \log e}{2(1-a)} \cdot \log\log n}\right)$ each operational processor P_i gets at most $\Xi_i \cdot \varphi$ threads, i.e.,

$$\Delta\mathcal{Z}(i, k) \leq (1 + \varepsilon) \cdot a \cdot q \cdot \varphi \leq \frac{\beta}{\mathcal{Z}_0(i, k)}$$

with $\beta = (1 + \varepsilon)a \left(\frac{c_1}{1-a}\right)^2 \cdot \log\log n$. Hence,

$$\mathcal{Z}(i, k) = \mathcal{Z}_0(i, k) + \Delta\mathcal{Z}(i, k) = \mathcal{Z}_0(i, k) + \frac{\beta}{\mathcal{Z}_0(i, k)}$$

which is maximized at $\max\{\mathcal{Z}(i, k)\} = \max\{1 + \beta, 1 + \gamma \cdot \log\log n\} \cdot \frac{c_1 n}{A_k}$ and thus it satisfies the inductive hypothesis with probability of success at least

$$\left(1 - n^{-\xi}\right)^k \cdot (1 - n^{-c_9}) \cdot \left(1 - n^{-\frac{\varepsilon^2 \cdot c_1 \cdot a \cdot \log e}{2(1-a)} \cdot \log\log n}\right) \geq \left(1 - n^{-\xi}\right)^{k+1}$$

with $\xi \leq \min\left\{c_9, \frac{\varepsilon^2 \cdot c_1 \cdot a \cdot \log e}{2(1-a)} \cdot \log\log n\right\}$, and $k \leq K_0$.

Remark: Notice that $\left(1 - n^{-\xi}\right)^{k+1} \geq 1 - (k + 1) \cdot n^{-\xi}$ which implies that our robust system can tolerate computations of polynomial length.

□

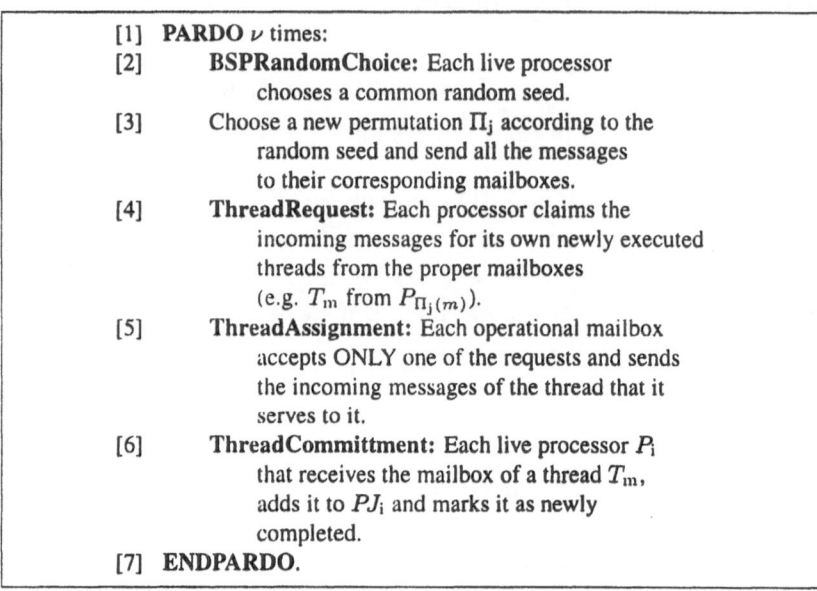

Figure 10.6 The Secure Delivery of messages to migrated threads.

4.2.6 Secure Delivery of Undelivered Messages.

In the SDoU procedure the primary objective is the safe delivery of messages heading for migrated threads, that have been kept in the LoUs of the live processors up to this point. Additionally, in the case of a SECONDARY JOB having preceded, this is also the rescheduling procedure, for the pending threads.

The basic idea of this procedure (see figure 10.6) is the use of multiple mailboxes for each thread, until all threads have been served by at least one live mailbox. That is, using ν random permutations of $[n]$, $(\Pi_1, \Pi_2, \ldots, \Pi_\nu)$, we proceed in ν rounds of indirect communication attempts, where in round j, each live processor P_i tries to successfully send any message for T_m, held in LoU_i, to the mailbox $P_{\Pi_j(m)}$. Consequently (in the same round), each mailbox sends the incoming messages of the corresponding thread to a (possibly one chosen among many) requester of T_m, according to the following rescheduling rule: In the case of multiple contesting processors for a specific thread, the mailbox sends it to one of the contestants (and thus assigns the corresponding virtual processor to it), according to the contestants' current work load.

Lemma 24 *If the number ν of required rounds of implicit communication attempts is $\frac{c_6+1}{\log(1/a)} \cdot \log n$, then the failure probability is $\Pr[SDoU\ fails] \leq n^{-c_6}$, $\forall c_6 > 0$.*

Proof: Let Π_0 be the identical permutation of $[n]$ and let $\Pi_1, \Pi_2, \ldots, \Pi_\nu$ be the required permutations so that each of the n threads be served by at least one live processor (to be its mailbox). For the following proof we shall need some definitions from the Theory of Negative Dependence of Random Variables (see [24, 23]):

Definition 2 *Let n be a positive integer.*

1. *The random variables $J_1, \ldots, J_n$ are said to have the permutation distribution on $[n]$, if they take values in $[n]$, and for any permutation $\sigma : [n] \to [n]$,*

$$\Pr[J_1 = \sigma(1), \ldots, J_n = \sigma(n)] = \frac{1}{n!}.$$

2. *Let $x_1, \ldots, x_n$ arbitrary reals. The random variables $X_1, \ldots, X_n$ are said to have a permutation distribution on $(x_1, \ldots, x_n)$, if there is a set of random variables $J_1, \ldots, J_n$ with the permutation distribution on $[n]$ and $X_i = x_{J_i}, \forall i \in [n]$.*

Definition 3 *The random variables $\mathbf{X} = (X_1, \ldots, X_n)$ are negatively associated if for every index set*

$$I \subseteq [n], \mathrm{Cov}[f(X_i, i \in I), g(X_j, j \in [n] - I)] \leq 0,$$

for all non-decreasing functions $f : \mathbb{R}^{|I|} \to \mathbb{R}$ and $g : \mathbb{R}^{|[n]-I|} \to \mathbb{R}$.

Consider now the following $\nu \cdot n$ indicator variables:

$$\forall i \in [\nu], j \in [n], X_{i,j} = \begin{cases} 1, & \text{if } P_{\Pi_i(j)} \text{ is live,} \\ 0, & \text{otherwise.} \end{cases}$$

For any fixed i, the vector $\mathbf{X}_i = (X_{i,1}, X_{i,2}, \ldots, X_{i,n})$ have the permutation distribution on

$$Status = \{Status(1), Status(2), \ldots, Status(n)\}.$$

Thus, for any fixed $i \in [\nu]$, the vector $\mathbf{X}_i$ follows the negative association property. Additionally, the vector $\mathbf{X} = (X_1, X_2, \ldots, X_\nu)$ consists of mutually independent components (since they correspond to randomly and independently chosen permutations), and thus $\mathbf{X}$ also has the negative association property.

Consider now the functions $Z_j = \sum_{i=1}^{\nu} X_{i,j}$. Since these variables are actually non-decreasing functions on disjoint sets of negatively associated r.v.'s, the vector $\mathbf{Z} = (Z_1, \ldots, Z_n)$ follows also the negative association property. Now we can proceed with the failure probability of the SDoU routine to successfully serve all the assigned or pending threads:

$$\Pr[\text{SDoU fails}] = \Pr[\exists j \in [n] : Z_j = 0] \leq \sum_{j=1}^{n} \Pr[Z_j = 0]$$

$$\Pr[Z_j = 0] = \Pr[X_{i,j} \leq 0, \forall i \in [\nu]] = \Pr[\bar{X}_{i,j} \geq 1, \forall i \in [\nu]]$$

Since $(X_{1,j}, \ldots, X_{\nu,j})$ are negatively associated, the same goes for the vector of complementary random variables $(\bar{X}_{1,j}, \ldots, \bar{X}_{\nu,j})$. Thus we have:

$$
\left.
\begin{aligned}
\Pr[\bar{X}_{i,j} \geq 1, \forall i \in [\nu]] &\leq \prod_{i \in [\nu]} \Pr[\bar{X}_{i,j} \geq 1] \\
\forall i \in [\nu], j \in [n], \Pr[\bar{X}_{i,j} \geq 1] &= \Pr[X_{i,j} \leq 0] \leq a
\end{aligned}
\right\} \Rightarrow \Pr[Z_j = 0] \leq a^\nu
$$

$$(10.32)$$

Now the total failure probability of SDoU is given by

$$
\Pr[\text{SDoU fails}] \leq n \cdot a^\nu = n^{-c_6}
$$

where

$$
n \cdot a^\nu = n^{-c_6} \Rightarrow \nu = \frac{c_6 + 1}{\log(1/a)} \cdot \log n
$$

$$(10.33)$$

$\square$

As indicated in [53], the time cost for such an indirect message propagation is given by

$$
T_{\text{SDoU}} = \mathcal{O}\left(h \cdot \log n \cdot \max\{\mathbf{b}, \mu_k\}\right)
$$

$$(10.34)$$

4.2.7   A tentative agreement protocol. The purpose of an agreement protocol in a decentralized, fault-prone setting, is to have all the live processors agree in a specific value, or participate in the evaluation of a function whose outcome will be available to all the participants.

According to the BSP AGREEMENT protocol, each live processor achieves an InitialValue and subsequently, an agreement rule (e.g. majority, median, e.t.c.) given as an input of this protocol is employed, in order to have all the live processors agree in a unique value. For example, consider that we have to make a unique estimation of the number of currently live processors, common for all the live processors. At the beginning of the protocol, the InitialValue is calculated by each live processor (e.g. a RANDOM SAMPLING is applied so as each live processor to have an estimation of the number of live processors). Subsequently, some rounds of agreement are performed, in order for all the live processors to end up with the same value. In each round, each live processor estimates a new value, according to the agreement rule, that depends on the nature of the problem to be solved. For example, a "median" agreement rule for the choice of a unique random seed would destroy the randomness of common seed. On the other hand, if the problem is the estimation of the number of the currently live processors, A_k, then a "weighted median" among sampled values seems more reasonable than a "majority" rule that would have no sense in this case. As an example, we have used this technique for the estimation of the live processors of $\mathcal{RM}$ at the beginning of a new PRIMARY JOB, before discarding the excessive load of each live processor, according to this estimation.

---

[1] **InitialValues:** Each processor estimates
     independently an initial value (e.g. for estimating the
     number of currently live processors, P_i sends
     $k_1 \log n$ messages to randomly chosen destinations).
[2] **PARDO** x times:
[2.1]   **Agreement:** Each processor sends $\log n$
         messages to random target processors.
[2.2]   Each live processor calculates the new estimation as
         $NewEstimation = AgreementRule(samples)$

Figure 10.7 The BSP Agreement Protocol.

Lemma 25 *The requested number of rounds of the* BSP AGREEMENT *protocol until all the processors have agreed (with polynomially small deviation probability) in a estimation of A_k, is $\mathcal{O}(\frac{\log n}{\log N})$, where N is the size of the samples of the* RANDOM SAMPILNG *applications.*

Proof: We proceed in our analysis by following the BSP AGREEMENT routine step by step (see figure 10.7):

STEP 1. Each live processor P_j samples N real processors from $\mathcal{RM}$. Consider one indicator variable $X_{i,j}$ for the ith sample of P_j, which is

$$X_{i,j} = \begin{cases} 1, & \text{if the } i^{\text{th}} \text{ sample of } P_j \text{ corresponds to a live processor} \\ 0, & \text{otherwise.} \end{cases}$$

The RANDOM SAMPLING is actually the implementation of 2 $\mathcal{O}(N)$-relations, one during which each live processor polls N processors chosen u.a.r. from $\mathcal{RM}$ (with replacement), and one for getting the answers by the live polled processors. If at the beginning of VS_k the number of live processors is A_k and the RANDOM SAMPLING is performed at the beginning of VS_k, then the probability of each sample hitting a live processor will be upper-bounded by $p_a \equiv \frac{A_k}{n}$. Thus, $\Pr[X_{i,j} = 1] \leq p_a$, and $\Pr[X_{i,j} = 0] \geq 1 - p_a$. And since the RANDOM SAMPLING is a relatively fast routine p_a may be considered to be a very good approximation of the actual ratio of live processors in $\mathcal{RM}$ at the end of the RANDOM SAMPLING. Thus, these indicator variables are clearly i.i.d. (random samples with replacement from the same sample space) with success probability very close to p_a, which implies that

$$E[X_{i,j}] = p_a, \text{ and } Var[X_{i,j}] = p_a \cdot (1 - p_a), \forall i \in [N], j \in [n].$$

So P_j's first estimation of the ratio of live processors in $\mathcal{RM}$ is given by the following relation:

$$R_j^{(0)} = \frac{1}{N} \cdot \sum_{i=1}^{N} X_{i,j}.$$

Obviously, the expected value and the variance of these new random variables are

$$E[R_j^{(0)}] \;=\; p_a, \text{ and}$$

$$Var[R_j^{(0)}] \;=\; \frac{1}{N^2} \cdot \left(\sum_{i=1}^{N} Var[X_{i,j}] - 2 \cdot \sum_{i>k} Cov(X_{i,j}, X_{k,j}) \right) = \frac{p_a \cdot (1 - p_a)}{N}$$

STEP 2. Each live processor P_j randomly selects processors from $\mathcal{RM}$, until it gets N live answers (this implies that each live processor will have to send $\frac{c}{1-a} \cdot N$ polling messages to randomly chosen targets, and will accept the first N answers). The new estimation of P_j is

$$R_j^{(1)} = \frac{1}{N} \cdot \sum_{i=1}^{N} R_{\lambda_i}^{(0)}$$

(P_j has accepted the answers of $P_{\lambda_1}, P_{\lambda_2}, \ldots, P_{\lambda_N}$). For this new estimation we have

$$E[R_j^{(1)}] = p_a, \text{ and } Var[R_j^{(1)}] = \frac{1}{N^2} \cdot \left(\sum_{i=1}^{N} Var[R_{\lambda_i}^{(0)}] - 2 \cdot \sum_{i>k} Cov(R_{\lambda_i}^{(0)}, R_{\lambda_k}^{(0)}) \right)$$

Since $R_{\lambda_1}^{(0)}, P_{\lambda_2}^{(0)}, \ldots, R_{\lambda_N}^{(0)}$ are variables randomly chosen (with replacement) from the same sample space, they are independent with each other, and this implies that $Cov(R_{\lambda_i}^{(0)}, R_{\lambda_k}^{(0)}) = 0, \forall j, k \in [N]$. So, we have that

$$Var[R_j^{(1)}] \;=\; \frac{1}{N^2} \cdot \sum_{i=1}^{N} Var[R_{\lambda_i}^{(0)}] = \frac{p_a \cdot (1 - p_a)}{N^2} \qquad (10.35)$$

STEP 3:. Repeat STEP 2 x times. After these x repetitions we shall have:

$$E[R_i^{(x+1)}] \;=\; p_a$$
$$Var[R_j^{(x+1)}] \;=\; \frac{p_a \cdot (1 - p_a)}{N^{x+2}} \qquad (10.36)$$

Each processor P_j that is live estimates the number of live processors in $\mathcal{RM}$ at the beginning of the Virtual Superstep VS_k as $A_{j,k} = n \cdot R_j^{(x+1)}$, where

$$E[A_{j,k}] = n \cdot p_a, \text{ and } Var[A_{j,k}] = n^2 \cdot Var[R_j^{(x+1)}] = \frac{n^2 p_a(1 - p_a)}{N^{x+2}} \le \frac{n^2}{4N^{x+2}}$$

In order to have small deviations among the live processors' estimations, the following must hold:

$$\lim_{n \to \infty} \frac{n^2}{4N^{x+2}} = 0 \Rightarrow x = \mathcal{O}\left(\frac{\log n}{\log N} \right), \qquad (10.37)$$

considering that $N = \log n$. For example, if $x = \frac{c \log n}{\log N} - 2$ then $Var[A_{j,k}] = \frac{n^{-c+2}}{4}$, and by applying Chebysev's Inequality we have:

$$\Pr[|A_{j,k} - p_a \cdot n| \geq 1] \leq \frac{n^{-2c+4}}{16} \qquad (10.38)$$

$\square$

For the RANDOM SAMPLING that estimates the Initial Values of the live processors, an $\mathcal{O}(\log n)$-relation is necessary for the polling of a size-$(\log n)$ sample (each live processor throws independently and randomly $\mathcal{O}(\log n)$ polling messages which are sent to distinct targets with overwhelming probability (see the Sparse Occupancy Problem in full paper), and an $\mathcal{O}(\log n)$-relation for the responses of the sampled processors (since the choices of samples are random, the expected number of pollings for each processor is $\mathcal{O}(\log n)$ and the deviation is very sharp around the expected number).

Thus the cost for RANDOM SAMPLING is

$$T_{\text{RandomSampling}} = \mathcal{O}(\log n) \cdot \max\{L, g\} \qquad (10.39)$$

For each of the BSP AGREEMENT steps, which is actually a new RANDOM SAMPLING among the processing elements of $\mathcal{RM}$, $\mathcal{O}(\log n)$-relations are required for each round, and the agreement rule is a "weighted median" operation among the samples of each live processor.

Remark: Similar uses of such an agreement protocol among live processors that cannot possibly know eahc other might be useful for the probabilistic estimation of other functions, such as boolean conditions. For example, the FINISHED? condition of the Robust-$\mathcal{BSP}$ simulation strategy might be able to be estimated with small failure probability by such an agreement protocol.

Acknowledgments

The authors wish to thank Dr. Spyros C. Kontogiannis (email: kontog@cti.gr) who has also co-authored Section 4 of this Chapter. This work was partially supported by the EU IST FET-OPEN Project ALCOM-FT. The authors wish to thank the Computer Technology Institute (CTI) for generous support.

Notes

1. In some works the malicious faults are also called byzantine faults.

2. The existence of smulations of arbitrary PRIORITY CRCW PRAM steps (which is the strongest PRAM variant) by at most $\mathcal{O}(\log n)$ EREW PRAM steps, implies corresponding results for the weakest variant of this cost model as well.

3. Notice that we do not exploit any slackness in our setting, which might indicate an "optimal" use of a subset of $\mathcal{RM}$ for the improvement of our results. This is because we want to have a realistic measure of the performance of our simulation process.

4. For example, if T_s is the total time of the simulation process, then $r(a) = \mathcal{O}\left(\frac{\text{polylog}(n)}{T_s} \cdot a\right)$.

5. The size of each epoch can be as large as $K_0 = \Theta(n)$, as it will be demonstrated later on.

6. As it will be shown in the sequel, a double logarithmic number of SECONDARY JOBS is enough for our strategy to work.

7. The calls of LOCAL STORAGE are not shown in the Robust-$\mathcal{BSP}$ strategy, because they are performed at the end of each PRIMARY JOB and each SECONDARY JOB.

8. The size of these neighborhoods of real processors depends on the fault occurrences up to the specific point of the simulation process.

9. This also implies that the size of the neighborhoods will be at most $\mathcal{O}(\log^2 n)$ with subpolynomially small failure probability.

10. Recall that the optimal work load would be $\frac{n}{A_k}$ threads per live processor, and thus $T_{\text{OPT}} = \frac{n}{A_k} \cdot (T_{\text{LC}} + T_{\text{Comm}})$.

11. Recall that K_0 can be as large as $\Theta(n)$.

References

[1] N. Alon and J. Spencer, "The Probabilistic Method", John Wiley and Sons, 1992.

[2] D. Angluin and L. Valiant, "Fast Probabilistic Algorithms for Hamiltonian Circuits and Matchings", J. of Comput. Systems Sci., vol. 18, pp. 155–193, 1979.

[3] Y. Aumann, M. Bender and L. Zhang, "Efficient execution of non-deterministic parallel programs on asynchronous systems", Proc. of the 8th ACM Symposium on Parallel Algorithms and Architectures (SPAA), pp. 270–276, 1996.

[4] Y. Aumann, Z. Kedem, K. Palem and M. Rabin, "Highly efficient asynchronous execution of large-grained parallel programs", Proc. of the 34th Annual Symposium on Foundations of Computer Science (FOCS), 1993.

[5] D. Bauer, F. Boesch, C. Suffel and R. Tindell, "Connectivity extremal problems and the design of reliable probabilistic networks", The Theory of Applications of Graphs, John Wiley and Sons, 1981.

[6] A. Baumker, W. Dittrich and F. Meyer auf der Heide, "Truly efficient parallel algorithms: c-optimal multisearch for an extension of the BSP model", Proc. of the 3rd Ann. European Symposium on Algorithms (ESA), Springer Verlag LNCS 979, pp. 17–30, 1995.

[7] P. Berenbrink, F. Meyer auf der Heide and V. Stemann, "Fault-tolerant shared memory simulations", Proc. of the 13th Ann. Symp. on Theoretical Aspects of Computer Science (STACS), Springer Verlag, pp. 181–192, 1996.

[8] R. Blumofe and C. Leiserson, "Scheduling multithreaded computations by work stealing", Proc. of the 35th Annual Symposium on Foundations of Computer Science (FOCS), pp. 356–368, 1994.

[9] F. T. Boesch, F. Harary and J. A. Cabell, "Graphs and models of communication networks vulnerability: connectivity and persistence", Networks, vol. 11, pp. 57–63, 1981.

[10] B. Bollobás, "Random Graphs", Academic Press, 1985.

[11] B. Bollobás, "A probabilistic proof of an asymptotic formula for the number of labeled regular graphs", Europ. J. Combinatorics, vol.1, pp.311–316, 1980.

[12] B. Bollobás, "The distribution of the maximum degree of a random graph", Discrete Math. 32, pp. 201–203, 1980.

[13] B. Bollobás and A. G. Thomason , "Random Graphs of small order", Random Graphs, Annals of Discr. Math., pp.47–97, 1985.

[14] A. Broder and E. Shamir, "On the second eigenvalue of random regular graphs", Proc. 19st ACM Symp. on Theory of Computing (STOC), pp. 286–294, 1987.

[15] H. Chernoff, "A measure of the asymptotic efficiency for tests of a hypothesis based on the sums of observations", Annals of Mathematical Statistics, 23, pp. 493–509, 1952.

[16] B. Chlebus, A. Gambin and P. Indyk, "Shared-Memory Simulations on a Faulty DMM", Proc. of Intl. Colloquium on Automata, Languages and Programming (ICALP), 1996.

[17] B. Chlebus, A. Gambin and P. Indyk, "PRAM computations resilient to memory faults", Proc. of the 2nd Ann. European Symp. on Algorithms (ESA), Springer Verlag LNCS 855, pp. 401–412, 1994.

[18] B. Chlebus, L. Gasieniec and A. Pelc, "Fast Deterministic Simulation of Computations on Faulty Parallel Machines", Proc. of the 3rd Ann. European Symp. on Algorithms, Springer Verlag LNCS 979, pp. 89–101, 1995.

[19] F. Chyzak, Ph. Flajolet and B. Salvy, "Studies in automatic Combinatorics, Volume II", INRIA 1998. Available at URL http://pauillac.inria.fr/algo/libraries/autocomb/autocomb.html

[20] T. Cormen, C. Leiserson and R. Rivest, "Introduction to Algorithms", MIT Press, 1990.

[21] P. Dasgupta, Z. Kedem and M. Rabin, "Parallel processing on networks of workstations: a fault tolerant, high performance approach", Proc. of the the 15th International Conference on Distributed Systems, 1995.

[22] R. De Prisco, A. Mayer and M. Yung, "Time-Optimal Message-Efficient Work Performance in the Presence of Faults", Proc. of the ACM Symposium of Distributed Computing, 1994.

[23] D. Dubhashi, V. Priebe and D. Ranjan, "Negative Dependence through the FKG Inequality", BRICS Report Series, RS-96-27, ISSN 0909-0878.

[24] D. Dubhashi and D. Ranjan, "Balls and Bins: A study in Negative Dependence", BRICS Report Series, RS-96-25, ISSN 0909-0878.

[25] M.Dyer and A.Frieze, "The Solution of Some Random NP-hard Problems in Polynomial Expected Time", Journal of Algorithms, vol. 10, pp. 451–489, 1989.

[26] P. Erdös and A. Renyi, "On the evolution of random graphs", Magyar Tud. Akad. Math. Kut. Int. Kozl. 5, pp. 17–61, 1960.

[27] P. Flajolet, K. Hatzis, S. Nikoletseas and P. Spirakis, "Trade-offs between Density and Robustness in Random Interconnection Graphs", to appear in the Proceedings of the IFIP International Conference on Theoretical Computer Science (TCS2000), Sendai, Japan, 2000.

[28] Ph. Flajolet and R. Sedgewick, "An introduction to the Analysis of Algorithms", Addison Wesley, 1996.

[29] Ph. Flajolet and R. Sedgewick, "Analytic Combinatorics", book in preparation, expected in 1999. (Individual chapters are available as INRIA Research Reports 1888, 2026, 2376, 2956, 3162).

[30] J. Friedman, J. Kahn and E. Szemeredi, "On the second eigenvalue of random regular graphs", Proc. 21st ACM Symp. on Theory of Computing (STOC), pp. 286–294, 1989.

[31] A. Frieze, "Probabilistic Analysis of Graph Algorithms", Feb. 1989.

[32] Z. Galil, A. Mayer, M. Yung, "Resolving Message Complexity of Byzantine Agreement and beyond", Proc. of the 36th IEEE Annual Symposium on Foundations of Computer Science (FOCS), 1995.

[33] J. Garofalakis, S. Rajsbaum, P. Spirakis and B. Tampakas, "Tentative and Definite Distributed Computations: An Optimistic Approach to Network Synchronization", Journal of Theoretical Computer Science (TCS), Volume 128 (Issues 1-2), pp. 63–74, 1994.

[34] M.Garey and D.Johnson, "Computers and Intractability: A Guide to the Theory of NP-Completeness", W.H.Freeman and Co., NY, 1979.

[35] A. Goerdt, "Random Regular Graphs with Edge Faults: Expansion through Cores", Proc. of ISAAC 1998.

[36] A. Gerbessiotis and C. Siniolakis, "Communication efficient data structures on the BSP model with Applications", Oxford Univ. Technical Report, PRG-TR-13-96, May 1996.

[37] M. Gereb-Graus and T. Tsantilas, "Efficient optical communication in parallel computers", Proc. of the 4th Ann. ACM Symp. on Parallel Algorithms and Architectures (SPAA), pp. 41–48, 1992.

[38] A. Gerbessiotis and L. Valiant, "Direct Bulk-Synchronous Parallel Algorithms", Journal of Parallel and Distributed Computing, 22, pp. 251–267, 1994.

[39] C. Gkantsidis, K. Hatzis, S. Nikoletseas and P. Spirakis, "Dominating Sets in Sparse Random Graphs", Computer Technology Institute, Technical Report, November 1998.

[40] T. Hagerup and C. Rüb, "A guided tour of Chernoff bounds", Information Processing Letters, 33, pp. 305–308, 1989/1990.

[41] J. Hromkovic, R. Klasing, E. Stoehr and H. Wagener, "Gossiping in Vertex-Disjoint Paths Mode in d-dimensional Grids and Planar Graphs", Proc. 1st ESA, pp. 200-211, LNCS vol. 726, 1993.

[42] S.Janson, D. Knuth, T. Łuczak and B. Pittel, "The Birth of the Giant Component", Random Structures and Algorithms, vol. 4, pp. 232–355, 1993.

[43] A. Kamath, R. Motwani,K. Palem and P. Spirakis, "Tail bound for occupancy and the satisfiability threshold conjecture", Proc. of the 35th IEEE Annual Symposium on Foundations of Computer Science (FOCS), pp. 592–603, 1994.

[44] P. Kanellakis and A. Shvartsman, "Fault-Tolerant Parallel Computation", Kluwer Academic Publishers, ISBN 0-7923-992-6, 1997.

[45] P. Kanellakis and A. Shvartsman, "Efficient parallel algorithms on restartable fail-stop processors", In Proceedings of the 10th Ann. ACM Symposium on Principles of Distributed Computing (PODC), pp. 23–36, 1991.

[46] P. Kanellakis and A. Shvartsman, "Efficient parallel algorithms can be made robust", Distributed Computing, 5, pp. 201-217, 1992.

[47] R. Karp and M. Sipser, "Maximum matching in sparse random graphs", 22nd FOCS, pp. 364–375, 1981.

[48] Z. Kedem, K. Palem, and P. Spirakis, "Efficient Robust Parallel Computations", Proc. 22nd ACM Symp. on Theory of Computing (STOC), pp. 138–148, 1990.

[49] Z. Kedem, K. Palem, A. Raghunathan and P. Spirakis, "Combining Tentative and Definite Executions for Very Fast Dependable Parallel Computing", Proc. 23nd ACM Symp. on Theory of Computing (STOC), 1991.

[50] Z. Kedem, K. Palem, P. Spirakis and M. Yung, "Faulty Random Graphs: reliable efficient-on-the-average network computing", Computer Technology Institute (Patras, Greece), Technical Report, 1993.

[51] D. E. Knuth, "The Art of Computer Programming", vol. 1, 2nd edition, Addison Wesley, 1973.

[52] S. Kontogiannis, G. Pantziou and P. Spirakis, "Efficient Computations on Fault-Prone BSP Machines", Proc. of the 9th ACM Symposium on Parallel Algorithms and Architectures, 1997. (Full version at http://www.ceid.upatras.gr/~kontog/bsp/spaa97/).

[53] S. Kontogiannis, G. Pantziou, P. Spirakis and M. Yung, "Dynamic -Fault-Prone BSP: a paradigm for robust computations in changing environments", Full version at
http://www.ceid.upatras.gr/~kontog/bsp/spaa98/

[54] F. T. Leighton, B. M. Maggs and R. K. Sitaraman, "On the Fault Tolerance of some popular Bounded-Degree networks", SIAM J. on Computing, Vol. 27 (5), 1998.

[55] C. Leiserson, "Fat-trees: Universal networks for hardware-efficient supercomputing", IEEE Transactions on Computers, C-34 (10):892–900, 1985.

[56] W.F. McColl, "Scaleable Parallel Computing: A grand unified theory and its practical development", Proc. of the IFIP World Congress, 1, pp. 539–546, Hamburg, August, 1994.

[57] R. Motwani, "Expanding graphs and the average-case analysis of algorithms for matching and related problems", 21st STOC, pp. 550–561, 1989.

[58] R. Motwani and P. Raghavan, "Randomized Algorithms", Cambridge University Press, 1995.

[59] S. Nikoletseas, G. Pantziou, P. Psycharis and P. Spirakis, "On the reliability of fat-trees", 3rd International European Conference on Parallel Processing (Euro-Par), pp. 208 – 217, Passau, Germany, 1997.

[60] S. Nikoletseas and P. Spirakis, "Near-Optimal Dominating Sets in Dense Random Graphs in Polynomial Expected Time", 19th International Workshop on Graph-Theoretic Concepts in Computer Science (WG), 1993.

[61] S. Nikoletseas, J. Reif, P. Spirakis and M. Yung, "Stochastic Graphs Have Short Memory: Fully Dynamic Connectivity in Poly-Log Expected Time", 22nd ICALP, pp. 159 – 170, 1995.

[62] S. Nikoletseas, K. Palem, P. Spirakis and M. Yung, "Short Vertex Disjoint Paths and Multiconnectivity in Random Graphs: Reliable Network Computing", 21st International Colloquium on Automata, Languages and Programming (ICALP), Jerusalem, pp. 508 – 515, 1994.

[63] S. Nikoletseas, K. Palem, P. Spirakis and M. Yung, "Connectivity Properties in Random Regular Graphs with Edge Faults", in the Special Issue on Randomized Computing of the International Journal of Foundations of Computer Science (IJFCS), 2000.

[64] S. Nikoletseas and P. Spirakis, "Expander Properties in Random Regular Graphs with Edge Faults", 12th Annual Symposium on Theoretical Aspects of Computer Science (STACS), München, pp. 421 – 432, 1995.

[65] N. Pippenger, "Telephone switching networks", The Mathematics of Networks, AMS, Providence, 1982.

[66] M. Rabin, "Efficient dispersal of information for security, load balancing and fault tolerance", Journal of the Association for Computing Machinery, Vol. 36, No. 2, pp. 335-348, April 1989.

[67] A. Sinclair, "Algorithms for random generation and Counting", ed. Birkhauser, 1992.

[68] M. Sipser and D. Spielman, "Expander codes", Proc. of 35th Ann. Symp. on Foundations of Computer Science, pp. 566–576, 1994.

[69] J. Spencer, "Ten Lectures on the Probabilistic Method", SIAM, 1987.

[70] P. Spira and A. Pan, "On finding and updating spanning trees and shortest paths", SIAM J. Comput., 4, pp. 375–380, 1975.

[71] R. Tarjan, "Data structures and Network Algorithms", 1985.

[72] P. Turán, "On an extremal problem in graph theory", *Matematicko Fizicki Lapok*, 48, pp. 436–452, 1941.

[73] L. Valiant, "A Bridging Model for Parallel Computation", CACM, vol. 33, no. 8, pp. 103–111, 1990.

[74] L. Valiant. "General purpose parallel architectures", Handbook of Theoretical Computer Science, J. van Leeuwen ed., North Holland, 1990.

[75] N. Wormald, "The asymptotic distribution of short cycles in random regular graphs", Journal of Combinatorial Theory (B) 31, pp. 168 – 182, 1981.

Chapter 11

RANDOMIZED COMMUNICATION IN RADIO NETWORKS

Bogdan S. Chlebus

Instytut Informatyki
Uniwersytet Warszawski
Banacha 2
Warszawa 02-097
Poland
chlebus@mimuw.edu.pl

Abstract   A communication network is called a radio network if its nodes exchange messages in the following restricted way. First, a send operation performed by a node delivers copies of the same message to all directly reachable nodes. Secondly, a node can successfully receive an incoming message only if exactly one of its neighbors sent a message in that step. It is this semantics of how ports at nodes send and receive messages that defines the networks rather than the fact that only radio waves are used as a medium of communication; but if that is the case then just a single frequency is used. We discuss algorithmic aspects of exchanging information in such networks, concentrating on distributed randomized protocols. Specific problems and solutions depend a lot on the topology of the underlying reachability graph and how much the nodes know about it. In single-hop networks each pair of nodes can communicate directly. This kind of networks is also known as the multiple access channel. Popular broadcasting protocols used on such channels are Aloha and the exponential backoff. Multi-hop networks may have arbitrary topology and packets need to be routed hopping through a sequence of adjacent nodes. Distributed protocols run by such networks are usually robust enough not to expect the nodes to know their neighbors. These ad-hoc networks and protocols model the situation when nodes are mobile and do not rely on a fixed infrastructure.

S. Rajasekaran et al (eds.), Handbook of Randomized Computing, Volume 1, pp, 401–456.
© 2001 *Kluwer Academic Publishers.*

1. INTRODUCTION

The advent of new technologies in both computers and telecommunication has caused a proliferation of computer networks. The ubiquitous cellular phones and portable computers have made wireless communication and mobile computation part of our daily experience.

Computer/communication networks are collections of information-processing nodes which communicate among themselves. Nodes are often independent units and the purpose of communication is to carry out distributed computation by sharing distributed resources. Such networks need communication protocols that are versatile enough to handle arbitrary patterns of communication and bursty traffic.

Various taxonomies of communication networks are possible. One of them is based on scale, that is, the size and distance among nodes; most popular categories are *local area networks* (LANs), like all the computers on a campus, and *wide area networks* (WANs), like all the hosts of academic institutions in a country. Another classification may be based on the technology used for transmission; nodes may be connected by copper wire or optical fiber or a network may be wireless. Yet another distinction is by the criterion if all the nodes are stationary or rather some are mobile.

Another classification is into the following two types: *broadcast networks* and *point-to-point networks*. The former are a number of nodes which share a communication channel; a packet sent by any node is received by all the nodes. A point-to-point network is a collection of nodes of which certain pairs are connected by transmission links; a packet sent from a node and destined at some specific node may need to hop through a sequence of nodes, along the links that connect them, until eventually it arrives at the destination. Not unexpectedly, LANs are often broadcast networks and WANs usually point-to-point.

Some of these taxonomies have little relevance to the principles governing design of algorithms for communication protocols, on either the network or media-access levels. For instance, the fact that a packet comes over a copper wire link and has covered a large distance is less relevant than what happens if two packets arrive simultaneously at a node, because it may be the case that both of them cannot be successfully received, and hence need to be sent again.

We consider mainly synchronous networks, this means the following. Each node has access to a clock. A clock cycle is called a *step* or a *slot*. The clocks are assumed to start at the same moment, with possibly different initial clock values. All the clocks tick simultaneously at the same rate. This defines the *locally synchronous* model. If, moreover, all the clocks show the same number, in other words, have access to a *global clock*, then the model is *globally synchronous*. We assume this stronger model, unless stated otherwise.

A *message* is a finite string of bits. A *packet* is a message supplied with additional information to facilitate its traversal through a network. A packet is assumed to have such a size that it can be transmitted between two nodes in one step.

A useful (partial) specification of a communication network is given by a directed graph, in which the nodes correspond to processing and communicating units, and the arcs denote ability to send messages directly. If there is an arc from v_1 to v_2 then node v_1 is a *neighbor* of node v_2, while v_2 is said to be *reachable* from v_1. A node can send packets directly only to these nodes that are reachable from it. If communication in a network is over physical links then an arc corresponds to such a connection, and if nodes communicate by electromagnetic waves then all the nodes in the range of a node are reachable from it.

Additionally, we need to make clear how nodes handle multiple messages concurrently. In general, a node of a network may have the capability to send different messages to any subset of adjacent nodes in a step. Similarly, a node may be either able to successfully receive all the incoming messages in a step, or it may accommodate just a certain subset of them, the size of which may depend on the capacity of its buffer, or in an extreme case, if many packets come to a node in a step then none of them is successfully received.

The area of our considerations is radio networks. What distinguishes them from other communication networks is the way nodes send and receive messages in a step.

> In a radio network: If a node performs a send operation then it sends out just one message, and copies of this message are delivered by the next step to all the reachable nodes.

The mode of receiving information by the nodes of a radio network is restricted, a node can successfully receive a message only if exactly one was sent to it in a step. If many neighbors send messages simultaneously to a node then all of them are delivered to the recipient but they are received as garbled. If a message is received by node v in its correct form then we say that v can *hear* the message.

> In a radio network: A node can hear a message only if it was sent by its neighbor, and it was the only neighbor that performed a send operation in that step.

It is convenient to assume that if a node does not hear a message then instead it can hear some *noise*, which is distinct from any meaningful message. If no message has been sent to a node then it hears the *back-ground noise*. If a node receives more than one message then we say that a *collision* or a *conflict* occurred at the node. If a collision happens then the node hears the *interference noise*.

Radio networks are categorized into four groups by the following two independent taxonomies. If the nodes of a network can distinguish the background noise from the interference noise then the network is said to be *with collision detection*, otherwise it is *without collision detection*. The next categorization is with respect to the topology of the underlying graph. If this graph is complete bi-directional, that is, for each pair of nodes v_1 and v_2 there is an arc from v_1 to v_2, then the network is said to be *single-hop*, otherwise it is *multi-hop*.

Single-hop radio networks with collision detection are also known as *multiple access channels* or *broadcast channels*, and are a special case of broadcast networks. A node is usually called a *station* in the context of single-hop radio networks. Since a broadcast performed by a single station makes the message heard by all the stations, such a broadcast is called *successful*. Multi-hop radio networks are also called *packet* or *point-to-point* ones.

Mobile users/nodes can form networks without any fixed infrastructure. Then the nodes may have a limited knowledge about the current topology of the network. We assume that the size n of the network is known, unless stated otherwise. Moreover, the nodes are assumed to have been assigned unique identifying numbers (IDs), the range of IDs being $[1..N]$, where $N = \Theta(n)$. Each node knows its ID. Distributed communication protocols are often robust enough to handle a situation when nodes do not know the IDs of their adjacent nodes. If we discuss such protocols then the underlying network is said to be either *ad-hoc* or *unknown* or *of unknown topology*. We assume that multi-hop radio networks are ad-hoc, when considering distributed communication protocols for them, unless stated otherwise.

A model related to radio networks is that of optical communication (see [44, 47, 48]). To explain the relationship, notice that a radio network can be interpreted as operating under just one wave frequency used by all the nodes. In an *optical communication parallel computer* (OCPC) each node v is assigned its own frequency $F(v)$. Node v can simultaneously receive any message placed on $F(v)$ (possibly garbled) and send a single message to the channel of any other node. Node v can *hear* a message delivered by $F(v)$ only if it was just a single message placed on it in a step. We can see that OCPC is similar to a single-hop radio network in two respects: first, it can be interpreted as a communication network with the topology of a complete bi-directional graph, and secondly, if

many messages are send to a node then it cannot hear any of them. The main difference between the models of OCPC and single-hop radio-networks is that the operation of sending in an OCPC network delivers the message to just a single recipient node rather than to all the stations. A node v_1 sending to channel $F(v_2)$ cannot detect the collisions at $F(v_2)$ but at $F(v_1)$. If a message is successfully delivered in a single-hop radio network then the sender also can hear the message, which serves as an acknowledgement. This mechanism can be implemented indirectly in the OCPC: let each packet carry both a message and the identification of its sender, then let after each 'sending step' another 'acknowledgement step' follow, when all the nodes that have successfully received messages send confirmation messages to the senders. Our discussion of optical communication protocols will be limited to situations in which there is some connection with radio-network communication.

2.   SINGLE-HOP NETWORKS

Most of the work done about randomized communication in radio networks has concerned multiple access channels. In this section we first give a historical background, then discuss various methodologies used to investigate problems in this area.

2.1   HISTORY AND EXISTING SYSTEMS

The first communication network with the semantics of a (single-hop) radio network seems to be the system Alohanet [2] developed in the beginning of the 1970's at the University of Hawaii. The system used packet radio communication between the central computer and its terminals on the campuses on remote islands. That is how 'broadcast channel' was born. This historical development is a justification to classify as 'radio communication' all modes of communication in which conflicting packets cannot be heard, even if the electromagnetic waves are not used as a medium to propagate signals.

The work on the Alohanet system has started already at the end of the 1960's. As Abramson [1] put it, the key idea was: "not the use of radio communications for computers, but the use of a broadcast communication architecture for the radio channel". The developers of Alohanet realized that a protocol was needed to handle bursty traffic, with most of the stations being idle for most of the time, hence techniques like time-division multiplexing were not considered suitable. The protocol invented for Alohanet is called Aloha. Its most relevant feature is that control is distributed among the attached stations, and arbitration of access is statistical. This is done in Aloha in an exceptionally simple way: the stations needing to broadcast backlogged packets keep tossing independent coins to decide if to perform a broadcast in a step or rather to pause. Actually Alohanet used the so called unslotted Aloha protocol, with the stations not assumed to be synchronized and making broadcast attempts at arbitrary times (see [15]).

The use of satellite and other channels implementing the multiple access channel followed in subsequent years. Among them was Ethernet [77] developed in the mid 1970's. It is a communication system facilitating broadcast of data packets among locally distributed computers. The medium of communication is usually a coaxial cable. In Ethernet again the arbitration of conflicts is distributed and randomized, the conflict-resolution protocol is the so called *truncated binary exponential backoff*. It operates as follows: a new packet is transmitted immediately, the time of the next attempt, after k unsuccessful transmissions, is selected with a uniform distribution on $\{1, 2, 3, \ldots, 2^{\min(10,k)}\}$, until $k = 16$ when the packet is discarded. Results of measurements of the performance of existing Ethernet installations can be found in [101]. This paper reports results of an experiment that was carried out on 120 stations; very low latency and few collisions were experienced under normal loads, the channel utilization approached 98% under artificially generated very heavy load. It is because of this remarkable performance that Ethernet is so popular. The stability of the protocol is guaranteed by a possibility to discard packets, when the channel gets really jammed, so it is not a 'clean' protocol from a theoretical point of view.

2.2   FORMAL MODELS

Consider a single-hop radio network with some of the stations storing packets that need to be broadcast on the channel. The stations compete for access to the channel, and if randomized and decentralized control is used then conflicts for access are inevitable. Resolving such conflicts is the problem we want to be able to handle. Distributed protocols used to provide such control and arbitrate among competing stations are called *conflict resolution protocols*, we will simply call them *protocols*.

A station is said to be *active* if it is ready to transmit a packet. We assume that there is some external mechanism to generate messages and assign them to stations with the purpose of broadcasting them on the channel. The process of assigning new messages may be either performed in advance or done on-line. In the latter case, if a station does not hold a message in the beginning of step i and it is assigned a message in step i then we say that the station was *activated in step i*. In the case when new messages are allocated to stations on-line some confusion with ter-

minology may occur. Namely, if new messages are assigned dynamically to stations then we say that they *arrive* to stations, which is consistent with the term 'arrival rate'. On the other hand, the nodes of a radio network communicate among themselves by sending packets containing messages, the incoming packets also are acquired by nodes, but in such a situation we say that they are *received*. We hope that 'received packets' will not be confused with 'arriving messages'. A message/packet that has been attempted to be broadcast in vain is called a *backlogged* one. A message which in step $t - 1$ became the one that a station will attempt to broadcast starting from step t is a *new* message at step t.

The problem of conflict resolution may be investigated either in a static or a dynamic scenario. In a former one, all the messages are created and allocated to the stations prior to the beginning of a protocol. Additionally, a station is allocated either just a single message or no messages at all. In a dynamic scenario, new messages keep being generated all the time and assigned to stations, which then try to broadcast them on the channel. Independently, we may have either an infinite or a finite number of users, depending on whether the set of stations is potentially unbounded or finite, respectively. Having an infinite collection of stations may seem to be strange at first, but the motivation is to treat messages as running protocols themselves, independently from the other messages. We distinguish the following four settings.

In a *static finitely-many users model* there are a total of n stations, some k stations among them are allocated messages prior to the start of a protocol. Each of the numbers k and n may either be known or not to the stations.

In a *static infinitely-many users model* there are countably infinitely many stations. Each of them performs a Bernoulli trial with some probability p of success prior to the start of a protocol. A success results in a message placed at the station, whereas a failure means that there is no message located there and never will. The probability p is referred to as *the probability to hold a message*.

A *dynamic finitely-many users model* is specified, first, by the number n of stations, and secondly, by the mode of arrivals of messages. Usually the arrival process is Bernoulli, that is, in each step at each station a message is generated with some constant probability p, independently of the other stations and steps. Each station has a buffer of potentially unbounded capacity, to store backlogged messages. It operates as a queue, that is, on the first-come-first-served basis. Otherwise, it it operated on the first-come-last-served basis, then some protocols in which each message is eventually dispatched successfully with a queue would deteriorate in such a way that the probability that a message is eventually sent is less than one. The current size of the queue of the i-th station is usually denoted as q_i.

In a *dynamic infinitely-many users model* the number of stations is again countably infinite. In each step a number of new messages are generated. Each of the messages is allocated to a new station. A station is identified by the time it obtained a message to transmit. The number of messages generated in a step has a fixed distribution, exactly k new messages are generated in a step with some probability λ_k. If it is not stated otherwise then the distribution is Poisson with parameter λ, that is, $\lambda_k = e^{-\lambda}\lambda^k/k!$ (throughout this article e is the base of the natural logarithm).

Variations on the theme of models are possible. For instance, a static finitely-many users model may be defined by stipulating that some N messages have arrived, with the arrival times being real numbers in the interval $(0,1)$ generated by a Poisson process, where simultaneously there are N stations, each station learns its corresponding message at its arrival time (see [90]).

For the static finitely-many users model two tasks are of interest. First is to make some station successfully transmit its packet, which is the *static selection problem*. Second is to make all the k stations with packets transmit successfully, which is the *static all-broadcast problem*. For the static infinitely-many users model, the task is to make each message broadcast successfully with probability one, simultaneously achieving the maximum transmission rate.

For the dynamic case, the basic question is whether a given protocol is stable, in the sense that it is practically operational, without the channel eventually getting clogged. We associate a Markov chain with each of the dynamic models. In the finitely-many users case, a state description includes the sequence of numbers of packets waiting at the stations. In the infinitely-many users model, a state includes the total number of backlogged packets. Stability of protocols is usually expressed in terms of properties of the underlying Markov chain.

One might expect differences in performance of similar protocols if they are analyzed in both the infinite and finite-numbers-of-users models. Observe that with infinitely many stations a newly generated message may be immediately broadcast on the channel, what contributes to burstiness of the traffic and hence is destabilizing. With finitely many stations, a message is first put into a buffer of the station to wait till its predecessors in the buffer have been handled, what stabilizes the traffic.

2.3   CONFLICT RESOLUTION

First let us discuss a certain basic feature of protocols, which is related to synchrony. A protocol is *slotted* if attempts to broadcast are always performed at the beginnings of slots. We will consider only slotted protocols, mainly to simplify the analysis. They need at least a local synchronization, which is an additional requirement on a network, so we may ask a question if it can be harnessed to improve performance of protocols. This indeed is the case, the following is a heuristic explanation. Suppose that we have a certain unslotted protocol $\mathcal{UP}$. Consider the slotted protocol $\mathcal{SP}$ obtained from $\mathcal{UP}$ by using the following more stringent regime to send packets: a packet to be broadcast during a slot is forced to be broadcast exactly at the first coming border between two slots. Let a slot last some t time units, and suppose that it takes the full slot size to perform the operation of transmitting a packet. A collision of packets sent by $\mathcal{SP}$ happens if they were sent during the same t time units constituting a slot. On the other hand, two packets processed by $\mathcal{UP}$ collide if they were sent in such a way that the time period from the start of the transmission of the earlier one till the completion of transmission of the later one comprise less than $2t$ consecutive time units. Hence one may expect that the slotted protocol $\mathcal{SP}$ has twice as high a transmission rate as $\mathcal{UP}$ (see [63]).

A protocol may grant transmission rights to a number of stations at the beginning of a step, we say then that the protocol *enables* these stations. Each enabled station transmits a message, provided it is active. Protocols may refer to stations in various ways, depending on the formal model in case. In a static model, each station knows its unique identification number (ID), and protocols enable stations by referring to these IDs. In a dynamic model, stations are not identifiable by unique IDs, and protocols refer to them by their activation time. In a static infinitely-many users model a protocol is assumed to know the probability to hold a message by a station. In a dynamic case, the knowledge of the rate of arrival of messages is usually not needed by protocols. This is because if the arrival rate is smaller than the maximum stable transmission rate of a protocol then the actual transmission rate automatically adjusts to the arrival rate.

Packets carry messages which bring some useful information to the stations. Our philosophy is to work with 'clean' protocols in which this information does not affect the future schedule of broadcasts. This also simplifies implementation of protocols. In particular, we do not want to consider any reservation mechanism, the reasons for this are as follows. A static reservation system, like time division multiplexing, would assign slots to stations on a fixed basis, what would be a waste of time if the traffic is bursty. A dynamic reservation system, say when a packet carries a message about reservation of some of the following slots, may make single stations dominate the channel for long periods of time.

The more restricted a protocol is, in terms of the information retrieved from the channel that is uses in scheduling future broadcasts, the more viable and versatile it is. The categorization of protocols is usually made based on the kind of feedback they receive and use. Consider first the single-hop radio network without collision detection. Then two kinds of feedback in a step are possible to be received by every station:

> `meaningful message`: a successful broadcast;
>
> `silent channel`: either no attempt of broadcast has been made or a conflict among multiple attempts occurred.

If a station is not making an attempt to broadcast in a step and it hears a silent channel then it gains no information. This is because it is our assumption that packets have the same size and it takes a full slot to broadcast a packet, hence a busy slot following a silent one does not mean that processing of a packet will be continued and the channel will be busy. A station attempting a broadcast can learn if it was successful by listening to the channel: it can either hear its own message (a success) or silence (a collision).

Protocols that schedule broadcasts of the current packet based only on the history of the previous attempts to broadcast it are called *acknowledgement based*. The explanation for this term is that each station relies only on the acknowledgement of a successful broadcast of its own packet. Such protocols may operate in both the channel with collision detection and without it, since a station performing a broadcast can always detect a collision. An acknowledgement based protocol refers to the following restricted history, which is a sequence $\langle t_1, t_2, t_3, \ldots, t_k \rangle$, with t_i being the number of step of the i-th unsuccessful attempt to broadcast the current packet, among k unsuccessful attempts made so far.

In a channel with collision detection every station may have the following feedback:

> `meaningful message`: a successful broadcast;
>
> `silent channel`: no attempt of broadcast has been made;
>
> `interference noise`: a collision caused by many simultaneous attempts to broadcast.

Because of these three possibilities such channels are often called *ternary feedback*. For the channel with collision detection, it may be useful if every station listens to the channel all the time and remembers the events that happened. Usually the channel history is stored as a sequence $\langle u_1, u_2, u_3, \ldots \rangle$, where value $u_i = 0$ means a silent channel with no attempt to broadcast during the i-th step, value $u_i = 1$ corresponds to a successful broadcast during step i by some station, and finally value $u_i = *$ means a collision in step i. Additionally, a station might remember its own attempts to broadcast. Protocols that rely on such a history are called *full sensing*.

We may also consider scenarios when stations obtain more information from the channel. If each station gets to know the number of stations involved in a collision in the current step then the channel and protocols are said to be *with collision-size detection* or *with control*.

Let us consider a dynamic scenario. A protocol is said to be *stable* if the associated Markov chain is ergodic, that is, irreducible, positive recurrent and aperiodic (see [38]). Usually natural protocols are both irreducible and aperiodic, so only positive recurrence is an issue. The Markov chain of a stable protocol converges with time to a stationary distribution of states, each probability of arriving at a state being the inverse of its expected return time. In particular, the expected time to wait for the backlog of packets to disappear is finite, as is the time to wait for a given message in the system to be sent successfully.

A related issue is that of buffer size. One can construct an example of a Markov chain with the states $\{0, 1, 2, \ldots\}$ which is ergodic and such that the average state number for the stationary distribution is infinite. Interpreting the size of buffer as the state number we can see that a stable protocol may have an infinite expected buffer size, with respect to the stationary distribution. The time spent by a message waiting depends on the size of the buffer at the step when the message was assigned to the station. Both this dependence and the stationary distribution determine the expected waiting time of a newly generated message. If the expected buffer size is infinite then such is the expected waiting time. Hence if a protocol has a finite average waiting time then this is a stronger property than stability, we call it *strong stability*.

If a protocol is not stable then the associated Markov chain can be either null recurrent or transient. The former means that with probability one the backlog will become empty eventually but the expected time for this to happen is infinite. The latter means that with probability greater than zero the system will eventually be jammed, in the sense that the backlog will never be empty.

It is possible that each message is eventually delivered but the protocol is not stable. For instance, consider the example of a finite number of users and a stable protocol $\mathcal{SP}$. Convert $\mathcal{SP}$ to $\mathcal{UP}$ by the following modification: if a station has just a single message then it stops running

$\mathcal{SP}$ temporarily and does not attempt to broadcast the message until a new message is generated, then it resumes $\mathcal{SP}$. This protocol is unstable in the sense that the state of empty backlog never happens after a finite number of steps, but the expected time to broadcast each message is finite. Needless to say, protocols are designed so that such a strange behavior does not happen. Clearly the problem with protocol $\mathcal{UP}$ is that the associated Markov chain is not irreducible.

A protocol that is not stable may still be considered good if one of the following properties hold:

(a) The expected delay between two consecutive messages sent is finite.

(b) Each message is sent eventually with probability one.

(c) The number of messages sent successfully is infinite with probability one.

We say that a protocol is *weakly stable* if it has property (c) above, which we consider to be the weakest satisfactory behavior of a protocol.

A protocol is *first-come first-served* (FCFS) if for any two messages m_1 and m_2, such that m_1 is generated at step t_1 and successfully broadcast at step t'_1, while m_2 is generated at step t_2 and broadcast at step t'_2, we have that if $t_1 \leq t_2$ then $t'_1 \leq t'_2$. For FCFS protocols, property (a) implies weak stability, and property (b) is equivalent to it.

In a dynamic model, during each step i some $A(i)$ new messages arrive. If the limit $\lambda = \lim_{t \to \infty} \frac{1}{t} \sum_{i \leq t} A(i)$ exists then λ is called the (average) *arrival rate*; throughout this chapter the letter λ will denote the arrival rate, unless stated otherwise. In the infinitely-many users model, if new messages are generated with the Poisson distribution with parameter λ, or similarly in the finitely-many users model of n stations, if each station generates a new message with probability λ/n independently over the steps and the stations, then this λ is the average arrival rate. Similarly, suppose some protocol $\mathcal{P}$ governing the operation of a channel is used, and at step i some $Z(i)$ packets are successfully broadcast over the channel, where $Z(i)$ is a random variable equal to either zero or one. If the limit $\sigma = \sigma_{\mathcal{P}} = \lim_{t \to \infty} \frac{1}{t} \sum_{i \leq t} Z(i)$ exists then σ is called the (average) *transmission rate* or *departure rate* of protocol $\mathcal{P}$. The ratio σ/λ is a measure of efficiency of the protocol and is called its *throughput*. Clearly a throughput is at most one. It is necessary for a protocol to attain a throughput of exactly one if it is to perform well in terms of stability, because otherwise the backlog increases and the channel is eventually clogged. On the other hand, it is possible that both the arrival and departure rates are one and the protocol is not stable ([58]): consider a single station, which always broadcasts a message, provided it has one, and such that during each step either two new messages arrive or none, each such event happening with probability 1/2 independently over the steps. This system behaves like a symmetric random walk on a line, hence it is null recurrent.

The throughput is a relative departure rate, that is, the transmission rate related to the arrival rate. For a static model, there is no notion of time regarding the generation of new messages, and hence no arrival rate. However, for the static infinitely-many users model, the transmission rate has the same intuitive meaning as the throughput in the dynamic case.

The throughput of a protocol is an incidental measure of the utilization of a channel, because it may happen that the same protocol has either the maximum unit throughput or a throughput smaller than one, depending on the magnitude of the arrival rate λ. It would be good to have the 'ultimate performance' measure of a channel, indicating the limits of its performance in absolute terms. Such a measure will be called capacity. We assume the infinitely-many users model, otherwise this measure becomes parametrized by the number of stations.

In a static model, where each station holds a message with some probability p, the performance of a protocol $\mathcal{Q}$ is measured by its departure rate. Notice that it is easy to attain a transmission rate equal to p by trying all the stations one by one; then the transmission rate approaches one if p approaches one. This indicates that the challenge for a protocol is when the probability to hold a message is small rather than large, and is a justification of the following definition of capacity. Let $\mathcal{Q}$ be a protocol for a static model. Let $\sigma_{\mathcal{Q}}(p)$ be the departure rate of the protocol $\mathcal{Q}$, with p being the probability to hold a message; recall that we assume that a protocol knows the probability p. We define the *tenacity* of protocol $\mathcal{Q}$, denoted $\mathrm{TN}(\mathcal{Q})$, to be the limit $\lim_{p \to 0} \sigma_{\mathcal{Q}}(p)$ if it exists. The *capacity* of a channel is defined to be the supremum of the tenacities $\mathrm{TN}(\mathcal{Q})$ over all protocols $\mathcal{Q}$.

The capacity of a channel in a dynamic case can be defined similarly. Let $\mathcal{R}$ be a protocol. Suppose new messages are generated with a Poisson distribution with the arrival rate λ. Let $\sigma_{\mathcal{R}}(\lambda)$ be the transmission rate of protocol $\mathcal{R}$ for this model of arrivals. Define the *tenacity* of protocol $\mathcal{R}$, denoted again as $\mathrm{TN}(\mathcal{R})$, to be the supremum of the arrival rates λ for which the throughput of $\mathcal{R}(\lambda)$ is also λ. The capacity of a channel is now defined to be the supremum of tenacities $\mathrm{TN}(\mathcal{R})$, over all the protocols $\mathcal{R}$. We might say briefly that, in the dynamic case, the capacity of a channel is its maximum stable transmission rate. The following is an alternative definition of capacity: it is the supremum of the throughputs

$\sigma_{\mathcal{R}}(1)$ attained for the Poisson arrivals with the unit rate, over all the protocols $\mathcal{R}$.

There is a relationship between Bernoulli distributions of messages for the static model and Poisson distributions of new arrivals for the dynamic model. The Bernoulli distribution of the number of successes in n trials, each with the probability p of success, converges to the Poisson distribution with parameter λ if $n \to \infty$ and $p \cdot n = \lambda$ (see [38]). Hence when the probability p to hold a message converges to zero, we may model the distribution of messages in the static model, in the limit, by a dynamic model with a Poisson distribution.

Sometimes it is useful to interpret the messages waiting at stations as customers waiting in queues, to apply the methods of queuing theory (see [65]). Suppose $L(t)$ is the number of customers at step t. The limit $L = \lim_{t \to \infty} L(t)/t$ is the *average number of customers*. If a message m arrives at step t_1 and is successfully broadcast at step t_2 then $t_2 - t_1$ is its *waiting time*. Let W_t be the average waiting time of the messages that have been sent through the channel by step t. If the limit $W = \lim_{t \to \infty} W_t$ exists then it is called the *average waiting time* of a customer. If any two among the parameters W, L and λ are finite then the third one is finite too, and they satisfy the Little's formula $L = \lambda \cdot W$ (see [102]). Hence if a protocol behaves nicely then we do not need to study separately L and W for a given arrival rate, estimates of one of them immediately yield estimates for the other one.

Many protocols, especially acknowledgement-based ones, operate by making each active station broadcast a packet at step t with some probability f_t, which depends on the history of feedback from the channel, independently of the other steps and stations. An especially simple scheme of such an algorithm is determined by a fixed sequence of probabilities $\langle h_0, h_1, h_2, h_3, \ldots \rangle$ as follows. When a new packet needs to be handled, in each step a coin is tossed with probability h_0 of obtaining heads, until the first heads come up, then the packet is broadcast immediately; usually it is assumed that $h_0 = 1$. In case of a failure, a new sequence of coin tosses is performed, a toss per step, each with the probability h_1 of obtaining heads, a broadcast is performed when heads come up. This is continued through a sequence of failed broadcast attempts, using a coin with probability h_i after the i-th failed attempt, until eventually the packet is successfully broadcast. So the protocol backoffs from the probability h_{i-1} after the i-th failure and tries the next probability h_i, that is why such protocols are called *backoff*. We may also call them *randomly oblivious*, because the sequence of probabilities $\langle h_i \rangle$ is fixed in advance. The number of failed attempts of station i to broadcast the current packet is maintained in the *backoff counter*, denoted b_i. The probabilities $\langle h_i \rangle$ are called *backoff probabilities*. The function assigning the probability h_i to number i is referred to as *backoff rate*. Backoff protocols are interpreted naturally as Markov chains, by defining the states to be either sequences of the backoff counters b_i of the stations with backlogged packets, in the case of infinitely-many users model, or vectors $\langle b_1, \ldots, b_n, q_1, \ldots, q_n \rangle$ of the backoff counters b_i together with the queue sizes q_i in the case of finitely-many users model of n stations.

Certain full-sensing algorithms operate in a similar fashion as randomly oblivious ones and use coin tosses to decide whether to attempt a broadcast, but they modify the probability f_t of heads coming up in step t on-line, by a function of the history of the ternary feedback of the channel up to step t; we call such algorithms *randomly adaptive*. Such randomized algorithms for the channel with collision-size detection have this probability $f_t = f_t(k)$ depending also on the number k of stations colliding in step $t - 1$.

A protocol called *Aloha* is the simplest randomly oblivious one, it operates as follows. A new packet is broadcast immediately. A backlogged packet is transmitted repeatedly in the following steps until success, each attempt made independently with some constant probability f. *Aloha with control* is a class of protocols in which each station with a message transmits with the probability equal to about $1/k$, where k is the current number of stations with messages. Such protocols need to be further specified, since the strongest feedback from a channel that we consider provides merely the number of colliding packets, if any. A backoff protocol with $h_i = \Theta(a^{-i})$, for $a > 1$, is called an *exponential backoff*, the number a is called its *factor*, in particular if the factor is $a = 2$ then the resulting protocol is *the binary exponential backoff*. A backoff protocol with $h_i = \Theta(i^{-b})$, for a constant $b > 0$, is called *polynomial*, which in turn is *sublinear* if $b < 1$, *linear* if $b = 1$, and finally *superlinear* if $b > 1$.

An alternative approach to define backoff protocols is by specifying an increasing sequence $\langle s_0 = 1, s_1, s_2, \ldots \rangle$ of integers. A station with a new message acquired at step t makes a sequence of attempts $\langle t_0, t_1, t_2, \ldots \rangle$ to broadcast it, until the first success. First $t_0 = t + s_0$, then t_1 is a random element of the interval $[t + s_0 + 1, t + s_1]$, and in general the time step t_i, for $i > 0$, is randomly selected from the interval $[t + s_{i-1} + 1, t + s_i]$. This is similar to how the protocol of Ethernet operates, which uses the sequence $\langle 2^i \rangle$, for $i = 0, 1, 2, \ldots$. Consider both such a protocol determined by the sequence $\langle s_0 = 1, s_1, s_2, \ldots \rangle$, and also the one determined by the probabilities $\langle h_i \rangle$, where $h_0 = 1$ and $h_i = 1/(s_i - s_{i-1})$, for $i > 0$. We expect that they are equivalent, as far as properties like stability are concerned. Backoff protocols defined by a sequence of time intervals may be called *unmodified*, because historically the first backoff protocol was that used by Ethernet, which is determined by intervals. Accordingly, a backoff protocol defined by a sequence $\langle h_i \rangle$ of probabilities is called *modified*.

3.   CLASSIFYING MARKOV CHAINS

To answer questions about stability of protocols, we need criteria to answer questions about either positive recurrence or null recurrence or transience of a given Markov chain.

Let $\langle X_i \rangle$ be a Markov chain with the set of positive integers as the set of states. Let $\langle P_{ij} \rangle$ be the transition matrix, that is, $\mathsf{P}[X_{k+1} = j \mid X_k = i] = P_{ij}$. The chain is recurrent if for each i the event $X_n = i$ occurs infinitely often with probability one. The chain is transient if it diverges to infinity with probability one. In what follows we will consider only irreducible and aperiodic Markov chains, they will also have a countable state space.

Sufficient conditions for the ergodicity and recurrence of (irreducible and aperiodic) Markov chains were given by Foster [40]. Various extensions have been found later, a very simple one is by Pakes [86]. Let $\langle X_n \rangle$ be a Markov chain, with states being the nonnegative integers. Let the *drift at state i* be the quantity

$$\mathcal{D}(i) = \mathsf{E}[X_n - X_{n-1} \mid X_{n-1} = i] \ .$$

The Pakes criteria, for an irreducible and aperiodic Markov chain, are as follows:

1. If the drift is finite for all states, and bounded above by a negative constant for all but finitely many states, then the chain is ergodic.

2. If the drift is bounded above by zero, for all but finitely many states, then the chain is recurrent.

Various criteria to classify Markov chains have been formulated in terms of a positive real function F defined on the set of states. Such functions are called either *test* or *Lyapunov* or *potential* ones. For instance, the test functions used in [58], to investigate backoff protocols for the finitely-many users model, are of the form

$$F(\langle q_1, \ldots, q_n, b_1, \ldots, b_n \rangle) = H_1(n) \cdot \sum_{i=1}^{n} q_i + \sum_{i=1}^{n} H_2(b_i) + H_3(n) \ ,$$

for certain functions H_1, H_2, and H_3, where numbers q_i are queue sizes and numbers b_i are backoff counters.

Another example of a useful criterion is the one which states that, for a given Markov chain and test function $F \geq 1$, the following are equivalent:

1. The chain is ergodic with the stationary probability π and a finite expected value of F with respect to π.

2. There is a finite set of states C such that the expected values of the sums

$$\sum_{t=1}^{T(C)} F(s_t) \tag{11.1}$$

   are finite for all $s \in C$, if the chain starts at s and proceeds through the states $s_1, \ldots, s_{T(C)}$, where $T(C)$ is the time needed for the chain to return to C after the start.

It was used in [58], see [78] for a proof.

Given a test function F, the drift can be generalized to *F-drift*, which at state i is the quantity

$$\mathcal{D}_F(i) = \mathsf{E}\left[F(X_n) - F(X_{n-1}) \mid X_{n-1} = i\right].$$

The Foster's criterion generalized with such a drift is as follows:

A Markov chain is ergodic if and only if there exists a positive real function F and a finite set of states C such that:

(a) the drifts $\mathcal{D}_F(i)$ are uniformly bounded above by a negative constant, for each state i outside C, and

(b) the expectancies $\mathsf{E}\left[F(X_{n+1}) \mid X_n = i\right]$ are all finite for each state i in C.

This criterion gives some flexibility by allowing the drift to be calculated with respect to a test function, but still it requires that its value decreases in just one step. This requirement may be relaxed by allowing the decrease to happen in a number of steps, which is bounded by another function. Given two functions F an G, where G is integer valued, define drift by the formula:

$$\mathcal{D}_{F,G}(i) = \mathsf{E}\left[F(X_{n+G(X_n)}) - F(X_n) \mid X_n = i\right].$$

The following is a further generalization of the criterion:

A Markov chain is ergodic if and only if there are two positive functions F and G, where G is integer-valued, a positive constant ϵ, and a finite set of states A such that the following holds:

1. The inequalities $\mathcal{D}_{F,G}(i) \leq -\epsilon\, G(i)$ hold for all states i not in set A.

2. The quantity $\mathsf{E}\left[F(X_{n+G(X_n)}) \mid X_n = i\right]$ is finite for all states i in set A.

This criterion was used in [6] to prove stability of the binary exponential backoff for certain arrival rates in the finitely-many users model. For proofs of these generalizations of the Foster's criterion see [37].

Kaplan [61] showed that if a Markov chain has the following two properties, then it is *not* egodic:

1. The drift is positive, for all but finitely many states.

2. There are constants $B \geq 0$ and $0 \leq c \leq 1$ such that the following inequality

$$z^i - \sum_j P_{ij} z^j \geq -B(1-z)$$

holds for $z \in [c, 1]$ and all but finitely many states i.

Property 2 above holds if the downward transitions are uniformly bounded, that is, $P_{ij} = 0$ whenever $j < i - k$, for some constant $k > 0$. For instance, let $X_n = X_{n-1} + a_n - b_n$, for $n \geq 1$, be the number of packets in the system, with a_n arrivals and b_n packets successfully dispatched in step n. If b_n is either 0 or 1, like in the multiple access channel, then the downward transitions are uniformly bounded. The system is unstable if

$$\mathsf{E}\left[a_n \mid X_{n-1} = i\right] > \mathsf{E}\left[b_n \mid X_{n-1} = i\right],$$

for all but finitely many i.

These criteria have later been extended in [99, 100, 104, 105, 106].

A criterion to prove that a Markov chain is transient was given by Rosenkrantz and Towsley [96]. It is as follows:

Suppose the Markov chain $\langle X_i \rangle$ is irreducible and aperiodic, and that a nonconstant positive test function F exists such that the inequalities

$$\mathsf{E}\left[F(X_{i+1}) \mid X_i = k\right] \leq F(k)$$

hold for every state k. Then the Markov chain $\langle X_i \rangle$ is transient.

A proof can be based on the martingale convergence theorem. Namely observe that by the assumption in the criterion the sequence $F(X_i)$ is a nonnegative supermartingale with respect to $\langle X_i \rangle$. Suppose, to the contrary, that the chain $\langle X_i \rangle$ is recurrent. Take any two states j_0 and j_1

such that $F(j_0) \neq F(j_1)$. It follows from the recurrence that the events $X_i = j_0$ and $X_i = j_1$ occur infinitely often with probability one, and hence that $F(X_i) = F(j_0)$ and $F(X_i) = F(j_1)$ infinitely often with probability one. It follows that the limit of $F(X_i)$ does not exist, what contradicts the martingale convergence theorem (see [32]).

A systematic exposition of related topics can be found in books by Fayolle, Malyshev and Menshikov [37] and Meyn and Tweedie [78].

4. INFINITELY-MANY USERS MODELS

In this section we consider single-hop networks, and concentrate on the setting in which each newly generated message receives a dedicated station, which then disappears after the message has been handled successfully. This is a natural model to study the issue of channel capacity. We present upper bounds, which reflect the inherent limitation on a randomized conflict resolution, and lower bounds, provided by efficient algorithms.

4.1 UPPER BOUNDS ON CAPACITY

We assume that the channel gives a ternary feedback, unless stated otherwise. The first upper bound on the capacity of the multiple access channel, for the infinitely-many users model with Poisson arrivals, was shown to be at most 0.774 by Pippinger [90] by an information theoretic argument. Improvements of this estimate were given in [29, 79, 80, 110]. The best known upper bound appears to be 0.568, it was proved by Tsybakov and Likhanov [110]. Some specialized bounds are known, if the class of protocols is restricted. Panwar, Towsley and Wolf [87] showed that 0.5 is an upper bound for the class of FCFS algorithms. Goldberg, Jerrum, Kannan, and Paterson [46] showed that every backoff protocol is transient if the arrival rate is at least 0.42, and for every acknowledgement-based protocol the same holds if the arrival rate is greater than 0.531.

We sketch a reasoning given by Molle [80] to estimate the capacity. Consider the static infinitely-many users model with the probability p to hold a message. The trick is to analyze a conceptual algorithm which resorts to a helpful genie. The algorithm is in phases, a new one starts when the algorithm enables some new N stations. If a collision happens then the genie reveals the two smallest stations with packets involved in the collision, which are then enabled in the two following steps. We use the following notation: S_N is the probability of success, I_N of an idle slot, $C_N = 1 - S_N - I_N$ of a collision. The transmission rate conditioned on N is

$$\rho_N = \frac{S_N + 2C_N}{I_N + S_N + 3C_N}. \tag{11.2}$$

An optimal number $N(p)$ of points to be enabled simultaneously at the beginning of a new phase is a nonincreasing function of p. If p is sufficiently large then enabling just a single point is optimal, and the departure rate is p. Comparing this with the throughput conditioned on N we obtain that this happens for $p > 1/\sqrt{2}$. Similarly, enabling two points, that is $N = 2$, is optimal for the probability p between $1/\sqrt{2}$ and the solution to the equation $\rho_2 = \rho_3$, which is $p \approx 0.568$. Suppose we go to larger N and resort to help from the genie. We may model Poisson arrivals by assuming $p \to 0$ while N is chosen such that $p \cdot N$ converges to a constant value λ. Then the transmission rate given by equation (11.2) takes the form

$$\frac{2 - (2 + \lambda)e^{-\lambda}}{3 - 2(1 + \lambda)e^{-\lambda}},$$

which is maximized at $\lambda \approx 2.89$, yielding the departure rate approximately equal to 0.6731. This is an upper bound on the capacity of the multiple access channel.

4.2   ALGORITHMS

Conflict resolution protocols process the feedback from the channel and schedule the future broadcasts accordingly. We consider three possible kinds of feedback: that giving the size of the set of stations in conflict, the ternary feedback, and finally that of a single-hop network without collision detection.

4.2.1   Protocols with control. Let us start the discussion of protocols from those that can rely on the most informative feedback: if a collision occurs then every station receives the number of colliding packets sent to the channel in the step. This follows historical developments, because the study of properties of conflict-resolution protocols was started with investigating Aloha with control.

Dynamic model. Aloha with control is strongly stable for the arrival rate smaller than $1/e \approx 0.368$ (see [36, 39, 65]). To show that this is an ergodic process, one can proceed as follows. Let $n(t)$ be the number of active stations at step t. Suppose first that each active station knows the number $n(t)$ at the beginning of step t and enables itself with the probability $f(n(t))$. The probability that exactly one active station will enable itself is

$$n(t) \cdot f(n(t)) \cdot (1 - f(n(t))^{n(t)-1}.$$

This is maximized for $f(n(t)) = n(t)^{-1}$ and the probability of a successful broadcast becomes

$$\left(1 - \frac{1}{n(t)}\right)^{n(t)-1} \approx \frac{1}{e} \ .$$

The sequence $\langle n(i) \rangle$ is a Markov chain. If Y_i is the number of arrived messages and Z_i the number of successfully broadcast packets at step i then the chain is ergodic for $\lambda = \mathsf{E}\,Y_i < \mathsf{E}\,Z_i = 1/e$, by the Pakes' criterion (see Section 3.).

Now consider the real situation when the feedback at step t does not provide the number of active stations $n(t)$ but merely the number of the colliding packets at step t. We consider the following specific protocol. The slots are partitioned conceptually into dynamic contiguous blocks. In the first slot of a block, that is, just after all the stations trying to broadcast in the previous block have succeeded, all the currently active stations broadcast, then the block continues until all these stations successfully dispatch their packets. The protocol requires all the stations to listen to the channel all the time to be aware of the traffic and to be able to decide when a new block begins.

It also follows directly by the Kaplan criterion (see Section 3.) that this version of Aloha with control is not stable for the arrival rate greater than $1/e \approx 0.368$, notice that the downward transitions are uniformly bounded. One can show that the protocol is actually transient.

Consider a more general scenario when the process of arrival of new packets has distribution $\langle \lambda_i \rangle$, that is, λ_i is the probability that exactly i new packets arrive in a step. We consider such algorithms in which each among the n stations active at a step enables itself with some probability $f(n)$, independently across steps and stations. Then $a(n) = (1 - f(n))^n$ is the probability of a silent channel, and $b(n) = n \cdot f(n) \cdot (1 - f(n))^{n-1}$ is the probability of a successful broadcast. Fayolle, Gelenbe and Labetoulle [36] showed that if the limit

$$d = \lim_{n \to \infty} \lambda_1 \cdot a(n) + \lambda_0 \cdot b(n)$$

exists then the following holds: for $\lambda < d$ the protocol is stable and for $\lambda > d$ it is unstable. Consider such protocols for which both limits $\lim_{n \to \infty} f(n)$ and $\lim_{n \to \infty} n \cdot f(n)$ are finite, as well as d, and also suppose that $\sum_{i \to \infty} i \cdot \lambda_i = \lambda > \lambda_1$. As a corollary, we obtain that for such a protocol to be stable it is necessary that both of the following conditions:

1. $\lim_{n \to \infty} f(n) = 0$;

2. $\lim_{n \to \infty} n f(n) > 0$;

hold. Namely, if the first condition does not hold then $d = 0$. On the other hand, if the second one does not hold then $d = \lambda_1 < \lambda$. In both cases the protocol is unstable.

Static model. Pippenger [90] showed that the capacity of a channel with the size-detection mechanism equals one under the static scenario. The model considered in [90] is that of finitely many messages with arrival times in the interval $(0, 1)$ generated by a Poisson process, but the approach translates in a natural way into the static infinitely-many users model with Bernoulli arrivals, which we consider in this exposition. The algorithm developed in [90] is not constructive, the proof is by the probabilistic method ([10]). The algorithm operates in stages, each broadcasts the messages of the stations in a segment of $\Theta(N)$ stations. Consider such a consecutive segment X, and let N be the exact number of messages in the stations in X, which can be found in one step by the size detection. Let $B = 2\lceil N(\log N)^{1/2} \rceil$. The segment X is conceptually partitioned into B subsegments $X_1, \ldots, X_B$, of approximately equal sizes. Next the algorithm proceeds through two phases: during the first one the number N_i of messages in each segment B_i is found, during the next one the actual broadcasting is performed. The first phase is to gather information, although incidental successful transmissions may happen, and the fact that this can be done in time $\mathcal{O}(N/(\log N)^{1/2})$ is the crux of the proof. The second phase takes time $N + \mathcal{O}(N/(\log N)^{1/2})$. The design of phase one is based on the following combinatorial fact: for a natural number L there is a $K \times L$ binary matrix F, where $K = \mathcal{O}(L/\log L)$ is a natural number, such that each sequence $G = \langle G_1, \ldots, G_L \rangle$ with the property $G_1 + \ldots + G_L \leq L$ is uniquely determined by the vector FG of length K. The existence of such a matrix is proved by random coding and extends a result of Erdős and Rényi [34]. The first phase begins with determining the numbers of messages $N^{(1)} = N_1 + \ldots + N_{B/2}$ in $X^{(1)} = X_1 \cup \ldots \cup X_{B/2}$ and $N^{(2)} = N_{B/2+1} + \ldots + N_B$ in $X^{(2)} = X_{B/2+1} \cup \ldots \cup X_B$. If some of them are at most one, what may happen with exponentially small probability, then the respective $X^{(i)}$ are handled either directly or recursively. Otherwise the numbers N_i are found without causing any successful broadcasts. To this end the combinatorial fact mentioned above is used with $L = B/2$, so that $K = \mathcal{O}(N/(\log N)^{1/2})$. In the second phase the algorithm first performs successful broadcasts by enabling the stations in X_i for $N_i = 1$, and handles the remaining segments X_i recursively. It takes $N + \mathcal{O}(N/(\log N)^{1/2})$ steps to broadcast N messages, so an arbitrarily high transmission rate can be attained by taking N sufficiently large.

4.2.2   Full sensing protocols.   Next we consider protocols which rely on the ternary feedback. First we present the randomly adaptive algorithms, then the tree and splitting ones.

The first randomly adaptive algorithm was proposed by Hajek and van Loon [57]. Each station maintains parameter f and each active station broadcasts a packet with probability f. This parameter f is not changed after a success, and is multiplied by a constant after each step without a successful broadcast, that is why it may be called a *multiplicative* protocol. These constants are as follows: 1.518 when the channel is silent, and 0.559 in the case of a collision. This protocol was proved by Hajek [56] to be stable for $\lambda < e^{-1} \approx 0.368$.

Kelly [62] proposed randomly adaptive *additive* protocols, in which at each step t each station maintains parameter $A(t)$ interpreted as the estimated current number of active stations. An active station performs a broadcast at step t with probability $1/A(t)$. This parameter $A(t)$ is updated after a step by adding a suitable constant among some fixed three ones, depending on the respective feedback from the channel. Rivest [94] proposed a specific additive adaptive protocol in which parameter $A(t)$ is updated as follows: if during step t there is no collision then $A(t+1) = \max\{1, A(t) - 1 + \frac{1}{e}\}$ otherwise $A(t+1) = A(t) + \frac{1}{e-2} + \frac{1}{e}$. Tsitsiklis [108] proved that this protocol is stable for the arrival rate $\lambda < e^{-1} \approx 0.368$.

Another paradigm of full-sensing algorithms was proposed independently by Capetanakis [17], Hayes [59] and Tsybakov and Mikhailov [112]. The algorithms they developed are usually called *tree algorithms*. They operate as follows. Let A be a positive integer interpreted as the arity of a tree. The algorithm run by a station is either in a waiting phase or a conflict resolution phase, switching between them accordingly. A *conflict resolution phase* starts after a conflict occurs. A *waiting phase* starts after all the packets of the stations involved in a conflict resolution have been transmitted successfully. If a new packet is generated then it is assigned to a station, which broadcasts it immediately if being in a waiting phase, otherwise it waits for the beginning of a waiting phase. Conflict resolution is performed as follows. First each among the involved stations selects randomly and independently an integer in the segment $[1..A]$, we say that they *branch out*. Then A stages follow, the stations that selected number i participate in the i-th stage. They begin with broadcasting their packets. If a conflict occurs then it is resolved recursively, which may be interpreted as going down one level in a tree. A stage terminates when all the participating stations have succeeded. Parameter A is often taken equal to 2. If the channel is quiet during all the branchings 1 through $A - 1$, then stage A need not be performed, it will certainly result in a collision, hence the stations might branch out immediately. Tree algorithms are stable for sufficiently small arrival rates, namely their variants were shown to be stable for $\lambda < 3/8$ in [112] and for $\lambda < 0.430$ in [17]. Notice that tree algorithms are not FCFS.

A modification of the tree-algorithm paradigm, called *splitting algorithm*, was proposed independently by Gallager and by Tsybakov and Mikhailov (see [42]). This algorithm is FCFS. The idea is to abandon the underlying tree structure and refer to the stations directly by their activation times. The stations join the conflict resolution in the order of activation. This may be interpreted as if they were stored in a stack. Suppose that at step k the protocol has just successfully transmitted all the messages that arrived before some time $T(k)$. All the messages that arrived in the time period $[T(k), T(k) + a(k)]$ are broadcast in slot k. If there is a collision then this time interval is partitioned into subintervals, namely into *the left interval* $[T(k), T(k) + b(k)]$ and *the right interval* $[T(k) + b(k), T(k) + a(k)]$, where $0 < b(k) < a(k)$. The stations from the left interval broadcast first. If a collision happens again then the left interval is subdivided into two subintervals. It is now when the departure from the tree-algorithm paradigm happens. Namely, only these two subintervals obtained from the left interval are remembered, the stations from the right interval are put back on the stack, and are not distinguished from the other stations still there. Otherwise, if there is no collision in the left interval, then the right interval is considered in a similar fashion. The function $a(k)$ is of the form $\min[c, k - T(k)]$, where the constant c is taken to be $c = 2.6$ to maximize the stable transmission rate. The simplest function $b(k)$ to be considered is the one that halves the interval $[T(k), T(k) + a(k)]$, this determines the *binary* version of the splitting algorithm. Binary splitting algorithm was shown to be strongly stable for $\lambda < 0.4871$.

This binary algorithm was later streamlined by Mosely and Humblet, and Tsybakov and Mikhailov, who showed that by partitioning the interval in an optimal way increases the tenacity to 0.4878 (see [82, 113, 114]). Vvedenskaya and Pinsker [115] developed a protocol with tenacity still better by 3.6×10^{-7}. More on the analysis of parameters of the tree and splitting algorithms can be found in [35, 52, 76]. In particular, they are strongly stable for sufficiently small arrival rates.

One might wonder if the best full-sensing protocols are FCFS. This hypothesis was shown to be doubtful by Vvedenskaya and Pinsker [115] whose protocol is not FCFS. The issue of optimality among subclasses of full-sensing algorithms was also studied by Molle [81] and then by Panwar, Towsley and Wolf [87]. In particular, they considered *nested*

FCFS algorithms which enable stations in the following conservative way:

1. If no set of stations is known in a step to contain more than one message and the set of stations C enabled in the next step contains station s then set C contains also all the stations preceding s whose status, with respect to holding a message, has not been clarified yet (this is the FCFS property).

2. If a set B of stations is known in a step to contain more than one message then in the next step a subset of B is enabled (this is the property of being nested).

A method to construct an optimal nested FCFS algorithm for a given probability to hold a message (Bernoulli arrivals for static model) has been developed in [87]. It is based on the Markovian decision theory, in particular on the value iteration algorithm (see [60]) and the Odoni bound (see [84]). Panwar, Towsley and Wolf [87] developed a scheme of *mixing algorithms*, which are not FCFS, and showed that they have higher transmission rates than optimal nested FCFS ones, for certain small probabilities to hold messages. To obtain such an algorithm, run an optimal nested FCFS algorithm augmented with the following trick. Suppose that we have identified a set $\{s_1, s_2, s_3\}$ of three stations, of which we know only that at least two of them hold packets. Then instead of enabling either $\{s_1\}$ or $\{s_1, s_2\}$, as a FCFS algorithm would do, enable $\{s_1, s_4\}$, where station s_4 is the next station after s_3, and which may or may not hold a message. This approach works because, for sufficiently small probability to hold a message, station s_4 will likely turn out not to hold a message, so with a single try we will eliminate two stations.

4.2.3   **Acknowledgement-based protocols.** Now we consider protocols which rely on the weakest feedback, that of a network without collision detection. They can be run on any single-hop network, they simply ignore the information from the channel, if any, obtained while staying idle. Such protocols usually erase their history after a successful broadcast, so they are acknowledgement-based. Because of that, we will not distinguish between possibly more general arbitrary protocols for the channel without collision detection and acknowledgement-based protocols.

The oldest protocols, namely Aloha and the binary exponential back-off, are in this class. The infinitely-many users model was initially considered with a hope to show in a clean setting that these two protocols were stable, at least for sufficiently small arrival rates. First, it was realized that the expected packet delay in Aloha is infinite (see [3, 65]). Then it was shown that Aloha is unstable for any arrival rate (see [36]). This follows directly from the Kaplan's criterion (see Section 3.). Namely, first notice that the downward transitions are uniformly bounded. Secondly, let $0 < f < 1$ be the probability that a station with a backlogged packet performs a broadcast in a step, and let i be the number of such stations. Then the drift is at least

$$2 \cdot (1 - e^{-\lambda} - \lambda e^{-\lambda}) - (e^{-\lambda} \cdot i \cdot f \cdot (1 - f)^{i-1}).$$

So it is positive and bounded away from zero by a constant for sufficiently large i.

Rosenkrantz and Towsley [96] proved that Aloha is transient, by their criterion based on the martingale convergence theorem (see Section 3.). This result was later strengthened by Kelly [62] who showed that with probability one the channel transmits a finite number of messages and then becomes jammed forever; moreover one can show that the expected time for this to happen is finite. Hence Aloha is not weakly stable for any arrival rate. The reasoning in [62] was as follows. Let $N(t)$ be the number of stations with backlogged packets at step t, and $Z(t)$ be the number of broadcasts at step t. Let $p(n)$ be the probability that the event $Z(t) \leq 1$ happens before the backlog increases from n. It satisfies the equation

$$\mathsf{P}\left[Z(t) \leq 1 \mid N(t) = n\right] = p(n) \cdot (1 - \mathsf{P}\left[N(t+1) = n, Z(t) > 1 \mid N(t) = n\right]).$$

This implies, by straightforward calculations, that

$$p(n) \sim \frac{nf(1 - f)^{n-1}}{1 - e^{-\lambda}},$$

with $n \to \infty$. Let $r(i)$, for $i = 1, 2, 3 \ldots$, be the times at which the backlog reaches record values, that is, $r(1) = 1$, and

$$r(i + 1) = \min\{t > r(i) : N(t) > N(r(i))\}.$$

Then $\sum_{i \geq 1} p(N(r(i))) < \infty$ and by the Borel-Cantelli lemma we obtain that with probability one the channel will jam itself forever after a finite number of steps.

Kelly and MacPhee [62, 63] generalized this approach to cover acknowledgement based protocols, a sketch of their analysis follows. Let us fix an acknowledgement protocol. For the purpose of argument let us assume that the channel is externally jammed from the very beginning so that no packet is successfully broadcast. Let $g(k)$ be the probability that a station broadcasts a packet in the k-th step since it was generated, conditioned on the event that the previous attempts to broadcast this packet failed. Let $s(t) = \sum_{k=1}^{t} g(k)$. The number of transmissions made in step t is Poisson with parameter $\lambda \cdot s(t)$. Hence the probability that less than two broadcasts are made in a slot is

$$u(t) = (1 + \lambda s(t)) \exp(-\lambda s(t)) .$$

The expected number of such slots is

$$U(\lambda) = \sum_{t \geq 1} u(t) .$$

The assumption about the external jamming can be removed by considering arbitrarily small additional Poisson traffic. Function $U(\lambda)$ is nonincreasing and it may be equal to infinity for sufficiently small λ. It categorizes the arrival rates as follows: if $U(\lambda) < \infty$ then with probability one the channel has only finitely many successful transmissions, and if $U(\lambda) = \infty$ then the protocol is weakly stable.

Define the critical arrival rate as follows:

$$\lambda_c = \inf\{\lambda > 0 : U(\lambda) < \infty\} .$$

Hence if $\lambda > \lambda_c$ then the number of packets successfully transmitted is finite with probability one, and if $\lambda < \lambda_c$ then the protocol is weakly stable. In particular, if $s(t) = o(\ln t)$ then $\lambda_c = 0$, and if $\ln t = o(s(t))$ then $\lambda_c = \infty$. Consider specific protocols as examples. For instance, we have $s(t) \sim f \cdot t$ for the Aloha scheme, hence $\lambda_c = 0$. The estimation is $s(t) \sim \log_a t$ for the exponential backoff with factor a, so we have $\lambda_c = \ln a$; in particular $a = 2$ and $\lambda_c \approx 0.693$ for the binary exponential backoff. Finally, any backoff protocol, with the backoff rate asymptotically smaller than that of an exponential backoff, like a polynomial one, is not weakly stable.

Aldous [7] showed that the binary exponential backoff is not stable for any arrival rate. More precisely, he showed that this protocol is transient and has zero throughput. Combining the results from [7, 62, 63] we can see that it is possible for a backoff protocol to be weakly stable, for sufficiently small traffic, and simultaneously non-recurrent and of zero throughput. This also shows that for the infinitely-many users model the exponential backoff protocols may be considered superior to the polynomial backoffs, in the sense that the former can be weakly stable if the traffic is sufficiently small, while the latter never are. This is in contrast with the finitely-many users model, see Subsection 5.2, where arguments for the opposite preferences are given.

We present a sketch of the approach from [7]. The interpretation used is in terms of pebbles and boxes, with messages called pebbles. At each step new messages (pebbles) are placed in box number zero, and a pebble in box i is either moved to box $i+1$ with probability 2^{-i} or remains in i, independently for different pebbles. This can be interpreted as a Markov chain, with the numbers of pebbles in the boxes giving the state. It has the stationary distribution which places a number of pebbles in box i with the Poisson distribution with parameter $\lambda 2^i$. Start the chain with this distribution. Let $Y_i(t)$ be the number of pebbles in box i. This process $Y(t) = \langle Y_1(t), Y_2(t), \ldots \rangle$ models an externally jammed channel. There is another process $X(t)$, which uses a coloring scheme. A new pebble is red, and if exactly one red pebble is moved in a step then it is recolored white. The number of white pebbles in box i at step t is denoted by $X_i(t)$, this defines the process $X(t) = \langle X_1(t), X_2(t), \ldots \rangle$, which captures the real behavior of the channel, because the white pebbles correspond to successful transmissions. For the externally jammed channel the expected number of packets with i unsuccessful transmissions is $\mathsf{E}\, Y_i = \lambda 2^i$. The real channel is represented by $X(t)$. One of its properties is that if $X_i(t) \geq \lambda 2^i$ then a positive chance exists than the backlog of packets increases. To argue more precisely, let us introduce the notation

$$f(x) = \sum_{i \geq 1} x_i \cdot 2^{-i}$$

for a sequence of nonnegative integers $x = \langle x_0, x_1, x_2, \ldots \rangle$. A key technical observation is that the probability of the event that some pebble is recolored in the step $t+1$, conditioned on the state being represented by x, is at most $2\exp(-f(x)/2)$, which is proved by the Chernoff bound (see [72, 83]). Then it is shown that $f(X(t))$ converges to infinity with the probability one, as t goes to infinity, what yields the result by a probability estimate.

Kelly and MacPhee [63] asked the question if there exists an acknowledgement based protocol which is recurrent in the infinitely-many users model with Poisson arrivals. This was answered in the affirmative by Goldberg, MacKenzie, Paterson, and Srinivasan [50], who developed a protocol with a constant expected delay, and hence strongly stable, for arrival rates smaller than $1/e$. We give an overview of the protocol. There is a conceptual tree of infinitely many levels, and infinitely many nodes at each level. With each node v there is associated a *trial set*, denoted by $Trial(v)$, of time slots. The first and last elements of $Trial(v)$ are denoted as $L(v)$ and $R(v)$, respectively. The number of elements in $Trial(v)$ is referred to as the *size* of v. The trial sets make a partition of the positive integers (are disjoint and cover all these integers) with these properties:

1. If either u is a proper descendant of v, or u and v are on the same level with u to the left of v, then $R(u) < L(v)$.

2. The sizes of nodes at the same level i are all equal to br^i, for some constant parameters b and r.

The arity of the tree is some positive integer k, which is a parameter satisfying $k > r$. The protocol operates as follows. When a message m arrives at time step t then it is assigned to the leaf v which is the leftmost one among those with the property $L(v) > t$. As soon as m is assigned to a node v then it picks a time t uniformly at random from $Trial(v)$, and is attempted to be sent onto the channel at time step t. If a failure occurs then message m is moved immediately to the parent of v. This defines the way message m keeps moving up the tree along the ancestor links until eventually it is broadcast successfully.

Notice that the protocol described above resembles the unmodified exponential backoff protocols in that it uses a partition of time steps into disjoint subsets $S_{i,t}$ of sizes growing exponentially in i. Messages that arrive at time step t jump from one subset $S_{i,t}$ to the next one $S_{i+1,t}$ after having failed to use the channel at a random time step in $S_{i,t}$. The efficiency of the protocol stems from the fact that groups of many messages arriving together are spread among a number of leaves and hence handled independently; this is because the trial sets of leaves are not contiguous segments of time steps but are intertwined with the trial sets of the nodes from higher levels.

5. FINITELY-MANY USERS MODELS

In this section we continue our discussion of the single-hop networks, concentrating on the case of finitely many stations, which may be considered as more realistic. Also it allows to study natural problems for the static arrivals which have no counterparts in the infinitely-many users model, namely when the packets are distributed arbitrarily among the stations. These problems are considered in the next subsection. We assume the ternary feedback, unless stated otherwise.

5.1 STATIC CASE

Let n be the total number of stations, and suppose some k of them have messages. The static selection problem is to broadcast just *any* single message successfully. The all-broadcast problem is to make *all* the k stations with messages to broadcast successfully. Notice first that if the value of k is known to the participating stations then they may run controlled Aloha, that is, each would broadcast with probability $1/k$ in a step, independently, and a successful broadcast will happen after the expected time $\mathcal{O}(1)$. Hence we assume in what follows that the number k is not known to the stations.

Willard [116] developed a randomized protocol for the static selection problem. It is in two versions. The first one covers the scenario when number n is known, it has the expected time $\lg \lg n + \mathcal{O}(1)$. The second version assumes that n is unknown and terminates in the expected time $\lg \lg k + \mathcal{O}(\lg \lg \lg k)$.

We present a brief overview of the approach in [116]. Suppose first that number n is known. The algorithm is in two phases. The underlying idea of the first phase is to perform a binary search on a space of size $\log n$ to estimate the number k of stations with messages. This is done by adaptively adjusting the individual probabilities to broadcast. More precisely, let K be an integer, and let $\text{TEST}(K)$ be a procedure in which each stations performs a broadcast in a step with the probability $1/K$; it returns either silence or success or collision. Let $\lg()$ be the binary logarithm. The binary search of the first phase is as follows:

1. Set $L = 0$ and $U = \lceil \lg n \rceil + 1$;

2. while $L \neq U$ do:

    (a) set $i = \lceil (L + U)/2 \rceil$ and $K = 2^i$;
    (b) set $t = \text{TEST}(K)$;
    (c) if $t = $ success then terminate the whole protocol;
    (d) if $t = $ silence then $U = i$;
    (e) if $t = $ collision then $L = i$.

Then the second phase is performed, in which we plug the latest K as the estimate of the number of stations into controlled Aloha and hope for the best. The first phase is never performed for more than $\lceil \lg \lg n \rceil$ steps. A crucial thing in the analysis is to show that the expected number of steps during the second phase is $\mathcal{O}(1)$.

Consider now the scenario when the number of stations n is unknown. The binary search cannot be applied, but it can be replaced by the search algorithm of Bentley-Yao [14] yielding a second version of the algorithm, which has the expected time $\lg \lg k + \mathcal{O}(\lg \lg \lg k)$.

Willard [116] also proved a matching lower bound to show the optimality of the first version of the protocol for a known number n. However the considered protocols were assumed to be *fair*: all the stations that toss coins to decide if to broadcast use coins with the same probability of heads to come up. The following was shown: there is an absolute constant c such that any fair protocol has the expected time at least $\lfloor \lg \lg n \rfloor - c$, for some k.

The algorithm of Willard operates on a single-hop radio network with collision detection. Kushilevitz and Mansour [68] showed a lower bound $\Omega(\log n)$ for the static selection problem if collision detection is not available. This yields an exponential gap between the two models.

Next we consider a related problem of finding maximum: suppose some k among n stations hold keys, we need to find the maximum among them with respect to some ordering among keys. A combination of the binary search with respect to the ordering of keys and the selection algorithm of Willard gives an algorithm with the expected time $\mathcal{O}(\log \log k \cdot \log k)$. Martel and Vayda [73, 74] showed that this can be streamlined to $\mathcal{O}(\log k)$. To this end a binary search is applied, which proceeds by creating a sequence of subsets $\langle A_i \rangle$ of diminishing sizes. The next subset A_{i+1} of a current subset A_i is created by partitioning A_i according to the key of a random element in A_i. Such an element is selected by running a modified selection procedure. The idea is to start the first phase from the estimate of the size of the previous subset. This results in an amortization of the binary search among the consecutive successful broadcasts.

5.1.1 Deterministic solutions.
First we consider the selection problem. Notice that the tree algorithm, adapted to a finite number of stations with identification numbers, solves the selection problem among n stations in $\lceil \lg n \rceil$ steps. Martel and Vayda [74] showed a lower bound of at least $\lg(n - k) - c$ steps on any deterministic algorithm, for a certain constant c. Both this lower bound and the randomized protocols of Willard demonstrate that randomization allows to develop selection protocols with asymptotically better expected performance than the worst-case performance of any deterministic protocol.

Next we consider the all-broadcast problem. For unknown k and known n, the problem can be also solved by the tree algorithm, with the worst-case time bound $\mathcal{O}(k + k \log(n/k))$. Komlós and Greenberg [66] developed a nonadaptive deterministic algorithm to solve the all-broadcast problem in time $\mathcal{O}(k + k \log(n/k))$, where both numbers n and k are known. The proof is nonconstructive, by the probabilistic method. Greenberg and Winograd [53] proved a lower bound of $k + \lg(n/k)$ steps on the time to solve the all-broadcast problem deterministically, which they generalized to a lower bound $\Omega(k(\log n)/(\log k))$.

5.1.2 Optical communication.

A channel even weaker than a single-hop radio network without collision detection can be considered, we call it *optical channel*. It operates as follows.

> In an optical channel:
>
> A station that performs a broadcast onto the channel, and is the only station broadcasting in a step, receives the message success as feedback.
>
> If at least two stations broadcast simultaneously in a step then each of them can hear only the background noise.
>
> A station that does not attempt to broadcast in a step can hear only the background noise.

The property that none of the idle stations receive the messages sent successfully to the channel by other stations is what distinguishes the optical channel from the single-hop radio network without collision detection. Notice that the stations that actually perform a broadcast in a step can recognize a collision when it happens.

A problem similar to the all-broadcast problem for the multiple access channel, when each of some k among the n stations needs to broadcast its message on the optical channel, and each of the stations with messages does know the number k, has been called the *control tower problem* in [70]. Geréb-Graus and Tsantilas [44] developed a randomized protocol solving this problem in time $\mathcal{O}(k + \lg n \lg k)$ with the probability polynomially close to one. MacKenzie, Plaxton and Rajaraman [70] proved the following lower bound: if a randomized algorithm solving the control tower problem operates in time $T(n, k)$ with probability at least $1 - n^{-3/4}$ then $T(n, k) = \Omega(\log k \log n)$. The control tower problem can be solved in the expected time $\mathcal{O}(k)$.

The OCPC model of optical communication can be interpreted as consisting of some n stations, each with a dedicated optical channel, the owner of a channel receives messages broadcast on it successfully. A basic communication problem for the OCPC model is to *realize an h-relation*: each node is a source of at most h packets, and also a destination of at most h packets, we need to deliver the packets successfully. Goldberg, Jerrum, Leighton and Rao [47] developed a randomized routing protocol realizing h-relations in the expected time $\mathcal{O}(h + \log \log n)$. A routing algorithm is *direct* if nodes may send packets only to their destination nodes. The algorithm presented in [47] is not direct. A direct randomized protocol to realize h relations was developed in [44], it operates in time $\Theta(h + \lg n \lg h)$ with high probability; actually only $h \geq \log n$ was considered and the obtained time was $\Theta(h + \log n \log \log n)$. A lower bound on direct algorithms was shown in [47]: a direct randomized algorithm that can realize any 2-relation with the success probability of at least $1/2$ needs time $\Omega(\log n)$ on some 2-relation. An $\Omega(h + \sqrt{\log \log n})$ lower bound, for routing h-relations in the OCPC model, was shown by Goldberg, Jerrum and MacKenzie [48], with no restriction on algorithms, in particular it covers indirect algorithms.

5.2   DYNAMIC ARRIVALS

Let the number of stations be denoted by n. The average arrival rate at station i is λ_i, and $\lambda = \sum_{1 \leq i \leq n} \lambda_i$ is the total arrival rate.

5.2.1   Aloha.   If arrivals are diversified among the stations, then we need the probability of performing a broadcast in Aloha also diversified, that is, station i performs a broadcast in a step with some probability f_i, provided its queue is nonempty, independently of the other steps and stations. Tsybakov and Mikhailov [111] considered arbitrary arrival processes, not necessary Bernoulli, and showed that Aloha was stable for certain configurations of parameters. For more on the topic see [97, 105, 106]. These results are in contrast with the situation for the infinitely-many users model where Aloha is not even weakly stable (see Subsection 4.2.3). However this is not that surprising if we realize that with the number of station n fixed and known to all the stations, using $f_i = \Theta(1/n)$ yields a process that is quite similar to controlled Aloha, hence it might be expected to be stable, at least for certain arrival rates.

Consider the Markov chain $\mathcal{M}_o$ which has the sequences of sizes of the buffers $\langle q_1, \ldots, q_n \rangle$ as its states. This chain is not state homogeneous, in the sense that the probability of moving from state a to b by vector $c = b - a$ does not depend only on c. The approach in [111] was by way of considering another state homogeneous Markov chain, which was suggested by the observation that Aloha has its worst time when all the buffers are nonempty. First let us present intuitions of this approach, inspired by [111]. For the simplicity of calculations, let us consider the special case when the arrivals are Bernoulli and distributed evenly among the stations, that is $\lambda_i = \lambda/n$, and also where the probabilities $f_i = f/n$ are all equal, for $i < n$. Define the protocol *Aloha with jamming* to run like Aloha but if a station has an empty buffer then let it immediately generate itself a dummy message and run Aloha until the packet with this message is successfully broadcast. It is intuitively clear that if the Aloha with jamming is stable so is the original Aloha. We compute the drift in order to apply the Pakes' criterion. It is equal to the average arrival rate minus the average departure rate. The departure rate is

$$n \cdot \frac{f}{n} \cdot \left(1 - \frac{f}{n}\right)^{n-1} e^{-\lambda} \approx f e^{-f} e^{-\lambda} \,.$$

However the arrival rate is not just λ but also the average arrival rate of the dummy packets. This additional rate can be estimated directly by finding the stationary distribution of the Markov chain, which requires solving a linear recurrence with constant coefficients. After combining all this we obtain a bound on the arrival rate which guarantees ergodicity, details are omitted.

The approach presented in [111] was direct, by way of generating functions and systems of equations to determine probabilities, rather than by applying any ergodicity criteria. The authors considered Markov chains $\mathcal{M}_k$, for $k = 1, \ldots, n$, with the state space $0, 1, 2, \ldots$, where state i corresponds to the total number of packets in the buffer of the k-th station, and the probability p_{ij} of going from i to j is defined as the probability of changing the buffer of the k-th station by $j - i$ packets in chain $\mathcal{M}_o$, conditioned on all the buffers not being empty. It was shown in [111] that chain $\mathcal{M}_k$ is ergodic if the inequality $\lambda_k < \gamma_k$ holds, where

$$\gamma_k = f_k \prod_{i \neq k} (1 - f_i) \,.$$

Then it was proved that if all the inequalities $\lambda_k < \gamma_k$ hold, for $1 \leq k \leq n$, then Aloha is stable. As a corollary, it follows that if $\lambda < e^{-1}$ then Aloha is stable, for probabilities $f_i = e\lambda_i(1 + e)^{-1}$.

5.2.2 Backoff protocols.

Recall that neither polynomial nor exponential backoff protocols are stable in the infinitely-many users model (see Subsection 4.2.3).

Håstad, Leighton and Rogoff [58] showed that the binary exponential backoff is not stable if the arrival rates at stations are equal and their sum λ exceeds $0.567 + (1/(4n - 2))$. The method used was by considering the following test function:

$$F(\langle q_1, \ldots, q_n, b_1, \ldots, b_n \rangle) = (2n - 1) \sum_{i=1}^{n} q_i + \sum_{i=1}^{n} 2^{b_i} - n,$$

where $\langle b_i \rangle$ are backoff counters, and $\langle q_i \rangle$ are queue sizes. It was shown in [58] that this function is expected to grow by at least a fixed positive amount, from which the instability follows. This result was strengthened asymptotically in [58] to show that if the arrival rate is bounded away from $1/2$ by any constant $c > 1/2$ then the system is not stable for the number of stations sufficiently large, depending on the constant c.

It was also shown in [58] that any linear or sublinear polynomial back-off protocol is unstable for any arrival rate and sufficiently large number of stations. The proof is by investigating the behavior of the following test function

$$F(\langle q_1, \ldots, q_n, b_1, \ldots, b_n \rangle) = n^{3/2} \sum_{i=1}^{n} q_i - \sum_{i=1}^{n} (b_i + 1)^{a+1} \, ,$$

where $h_i = i^{-a}$ are the backoff probabilities.

On the other hand, Goodman, Greenberg, Madras and March [51] showed that, for Poisson arrivals, the binary exponential backoff is stable when $\lambda < n^{-a \log n}$, for a certain constant $a > 0$. They also studied the special case of two stations in detail, with a specific bound on λ. Al-Ammal, Goldberg and MacKenzie [6] improved this result by showing that the binary exponential backoff is stable for λ at most $\mathcal{O}(n^{-\delta})$, for $\delta > 0.75$. This was done by applying the generalized Foster's criterion which employs two functions (see Section 3.). Unfortunately the lower bound on λ decreases to zero with $n \to \infty$.

Surprisingly enough, Håstad, Leighton and Rogoff [58] showed that any superlinear polynomial backoff algorithm is strongly stable for any arrival rate smaller than one. The proof was by investigating the properties of the test function

$$F(\langle q_1, \ldots, q_n, b_1, \ldots, b_n \rangle) = \sum_{i=1}^{n} q_i + \sum_{i=1}^{n} (b_i + 1)^{a+\frac{1}{2}} - n \, ,$$

where the backoff rate has the form $(x + 1)^{-a}$, for $a > 1$, as a function of x. Then sums of the form (11.1) (given in Section 3.) of the values of F over some states of the chain were investigated, and the respective criterion was applied.

This shows that exponential backoff protocols are not necessarily better than polynomial ones, but rather vice versa, what is in contrast with the infinitely-many users model, see Subsection 4.2.3. An essential property of backoff protocols in the finitely-many users model is that if a station broadcasts successfully then it keeps sending messages until eventually a collision occurs, so a success allows the station to grab the channel for a possibly substantial amount of time and hence to load off a lot of messages from its queue. This effect needs to be tuned to the rate of backing off, that is why linear backoff is too slow, quadratic just perfect and exponential rather too fast. The effect of holding the channel over a significant period of time cannot happen in the infinitely-many users model because a station dies after a success.

Goldberg and MacKenzie [49] considered a system of clients and servers, each server as a multiple access channel. Clients generate requests to servers with some Bernoulli distributions. The client-server request rate is the maximum, over all the pairs of a client and a server, of all the request rates associated with either the client or the server. It was shown in [49] that any superlinear polynomial backoff protocol is strongly stable if the request rate is smaller than one. This result can be applied to routing in optical networks, like the OCPC.

5.2.3   Short-delay protocols.

The proof of a strong stability of polynomial backoff protocols given in [58] yields exponential upper bounds on the waiting time in the number of stations. It was also shown in [58] that for any superlinear polynomial backoff protocol the expected waiting time is superlinear, more precisely, if the backoff rate is $f(x) = (1 + x)^{-a}$ then the lower bound is $\Omega(n^{\frac{a+1}{a}})$.

Raghavan and Upfal [91] developed an acknowledgement-based protocol which is strongly stable for sufficiently small arrival rates and which has $\mathcal{O}(\log n)$ expected delay. This was the first strongly stable protocol with a provably sublinear waiting time of messages. An acknowledgement-based protocol, with a constant expected delay for arrival rates smaller than $1/e$, was developed in [50], it is an adaptation of the infinitely-many users version as discussed in Subsection 4.2.3. Moreover, the stations need not be fully synchronized, provided that each station survives for polynomially many steps every time it restarts.

The following fairly general lower bound was showed in [91]: for each acknowledgement based protocol, there is a number $0 < a < 1$ such that if the arrival rate λ satisfies $\lambda > a$ then the expected delay is $\Omega(n)$. A sketch of the proof is as follows. Suppose that the arrivals at stations are all equal to λ/n. Let X_t be the current state of the system at time t, which is the total number of backlogged messages. Suppose also that there is a stationary distribution, otherwise the system is not positive recurrent and the bound holds. In the remaining part of the proof the probabilities are calculated with respect to this distribution. Let us consider the case $P[X_t < n/2] > 1/2$, otherwise the bound clearly holds. Suppose that if $X_t < n/2$ then the probability of a collision at step $t + n/3$ is at least q. The probability of a collision in a given step is at least $q/2$, and the system cannot be stable if the arrival rate λ is greater than $1 - q/2$. So it is sufficient to show that the probability q is at least a constant. There are at least $n/2$ senders with no packets with probability at least $1/2$, we call them *empty*. The protocol determines the probability distribution $\langle p_i \rangle$ such that when a message arrives to an empty sender then it transmits this message at step $t + i$ with probability p_i. The number $\sum_{i=0}^{n/3} p_i$ is at least a constant β, since otherwise the delay is at least $n(1 - \beta)$. Thus the probability that any empty sender transmits at time $t + n/3$ is at least $\beta\lambda/n$, and the probability q of a collision is at least a constant.

6.   MULTI-HOP NETWORKS

In the case of general multi-hop radio networks, various forms of dissemination of information can be studied. The simplest one is that of *broadcasting*. In this context it does not mean an attempt to send a single message to the recipients of a station, as in the multiple access channel case, but rather the task of delivering a message originally stored by some *source node* to all the other nodes in the network. The source does not need an acknowledgement of the completion of broadcasting, otherwise the problem becomes *broadcasting with acknowledgement*. More complex communication problems concern many concurrent instances of either point-to-point or broadcast communication tasks. We refer to all of them as *multiple communication*. In particular, the problem when all the nodes need to perform broadcasting simultaneously is called *gossiping*: initially each node knows its message, and all these messages need to be acquired by every node.

We always assume that individual messages are of such size that each of them can be sent between adjacent nodes in one step. In the context of multiple communication, a node may find it useful to transfer many messages simultaneously. It depends on the capacity of links if this is feasible. If any number of individual source messages can be lumped together into a bigger one that still can be transitted in one step then we say that *messages can be combined*. If each of such messages has to be sent separately then we refer to the model as that of *separate messages*. When gossiping is considered, then the model assumed is that of combined messages, unless stated otherwise.

All distributed protocols discussed in this section are designed for ad-hoc networks, with nodes not assumed to know their adjacent ones. The model is without collision detection, and the size of the network is known by the nodes, all this holds, unless stated otherwise.

Randomized algorithms are usually categorized as either *Monte Carlo* or *Las Vegas* (see [83]). The former perform their required tasks successfully with some positive probability only, the latter terminate after having completed their task. If an execution of a randomized algorithm results in performing the required task then we say that it is *correct*.

We use the following notation in this section: n is the number of nodes, D is the diameter of the network, Δ_{in} is the maximum in-degree of a node, Δ_{out} is the maximum out-degree of a node, and $\Delta = \max\{\Delta_{in}, \Delta_{out}\}$ is the maximum degree.

6.1 COMMUNICATION ALGORITHMS

6.1.1 **Randomized broadcasting.** An algorithm which performs broadcasting in multi-hop radio networks without collision detection was developed by Bar-Yehuda, Goldreich and Itai [12]. It operates in time $\mathcal{O}((D + \log n/\epsilon) \log n)$, and succeeds with probability $1 - \epsilon$. The number $\epsilon > 0$ is a parameter, which is part of the code.

A brief exposition of the algorithm is as follows. There are two procedures Decay and Broadcast. An essential property of procedure Decay is that if node x has some d neighbors, and they all know the message and simultaneously perform Decay(k), where $k \geq 2 \lg d$, then node x will hear the message by time step k with the probability greater than $1/2$.

```
procedure Decay(k);
        repeat at most k times :
                transmit the message ;
                toss a coin ;
        until tails come up .
```

The broadcasting algorithm itself begins with the source node transmitting the message to the nodes reachable from it directly, then all the nodes perform procedure Broadcast, which consists of procedure Decay(k) iterated about $\log(n/\epsilon)$ times, where number k is about $\log \Delta$. A precise description of Broadcast is as follows:

```
procedure Broadcast ;
        set k = 2⌈lg Δ⌉ ;
        wait until receiving the message ;
        repeat 2⌈lg(n/ε)⌉ times :
                wait until the time step is a multiple of k ;
                call Decay(k) ;
        end repeat .
```

The expected number of transmissions performed by the algorithm is $\mathcal{O}(n \cdot \log(n/\epsilon))$.

Observe that the information reaching a node might have traversed the network via paths of different lengths, so the broadcasting algorithm cannot be used directly to find distances to the source node. A Monte Carlo algorithm to construct the Breadth-First Search (BFS) tree was developed in [12], which also finds the distances of each node to the root. The algorithm is again based on procedure Decay, which now is slowed down to increase the probability to detect distances correctly from the times of arrivals. The algorithm refers to local variable `distance`, to store the distance to the source node, and the current time step given by `time`. Each node runs the following procedure:

procedure BFS :
    set $k = 2\lceil \lg \Delta \rceil$;
    wait until receiving the message ;
    set distance $= \lfloor \text{time}/(k\lceil \lg(n/\epsilon)\rceil)\rfloor$;
    repeat $\lceil \lg(n/\epsilon)\rceil$ times :
        wait until the time step is a multiple of $k\lceil \lg(n/\epsilon)\rceil$;
        call Decay(k) ;
    end repeat .

The algorithm operates in time $\mathcal{O}(D \log \Delta \log(n/\epsilon))$ with the probability $1 - \epsilon$, and is correct with the same probability.

6.1.2   Deterministic broadcasting.

Deterministic broadcasting algorithms in multi-hop radio networks that have been presented in the literature are usually *oblivious*, in the sense that a node performs transmissions at steps known in advance, once it receives the message. A precise definition of such algorithms is as follows. Let a set of nodes be called a *transmission*. A sequence of transmissions $\langle T_1, T_2, \ldots \rangle$ is called a *schedule*. Given a schedule, a broadcasting algorithm can be obtained as follows: a node v broadcasts the message in step i if v has received the message before step i and v is in T_i. This is a natural class of algorithms, which do not make the nodes send anything prior to receiving the source message. All deterministic broadcasting algorithms that we discuss are of this form, and the lower bounds we mention are for the algorithms in this class.

Schedules are usually obtained by taking families of sets of nodes $\mathcal{F}_1, \mathcal{F}_2, \ldots, \mathcal{F}_k$ and making the transmissions to be elements of these families. One way to arrange this is by *interleaving*, in which we cycle through the families, and consecutive selections from a family are also done in some cyclic order. More precisely, in the i-th step a set in $\mathcal{F}_\ell$ is selected, where $i \equiv \ell \bmod k$. Similarly, in the j-th selection form $\mathcal{F}_\ell$ the element is taken whose index is congruent to j modulo the size of $\mathcal{F}_\ell$. Families are needed which have useful combinatorial properties, to guarantee correctness of broadcasting, and be as small as possible, to provide efficiency. This paradigm was first applied by Chlebus, Gąsieniec, Gibbons, Pelc and Rytter [22], where the notion of a selective family was introduced. A family $\mathcal{F}$ of subsets of $[1..n]$ is k-*selective* if, for any subset $A \subseteq [1..n]$ such that $|A| \leq k$, there is $B \in \mathcal{F}$ such that $|A \cap B| = 1$. The family of singletons is k-selective, for any number $k \leq n$, its size is n. The algorithm whose transmissions cycle through it is called ROUNDROBIN, it performs broadcasting in time $\mathcal{O}(n^2)$. Smaller selective families can be defined as follows. Let i be a positive integer, $i \leq \lg n$. Let $C \subseteq [1..\lceil \lg n \rceil]$ be a set of size i. Let D be a string of 0-s and 1-s of length i. Define

$S(C, D)$ to consist of these nodes whose IDs in binary have the bit on the j-th position in C equal to the j-th element in D, for $1 \leq j \leq i$. Let family $\mathcal{R}(i)$ consist of all such sets $S(C, D)$. One can show that $\mathcal{R}(i)$ is 2^i-selective. A deterministic broadcasting algorithm obtained by interleaving ROUNDROBIN with a certain specific $\mathcal{R}(j)$, where j depends on n, was shown in [22] to operate in time $\mathcal{O}(n^{11/6})$. The algorithm obtained by interleaving all the families $\mathcal{R}(i)$ such that $|\mathcal{R}(i)| = \mathcal{O}(n)$ was shown in [23] to work in time $\mathcal{O}(n^{2-\lambda+\epsilon})$, for any ϵ, where $\lambda \cong 0.22709$ is defined by the equation $\lambda + H(\lambda) = 1$, where H is the binary entropy function. It was shown [25, 27, 89] that for each $k \leq n$ there is a k-selective family of size $\mathcal{O}(k \log n)$. Various selective families, their combinatorial properties and applications to broadcasting were studied in [22, 23, 25, 27, 30, 89]. The fastest currently known deterministic broadcasting algorithm was developed by Chrobak, Gąsieniec and Rytter [25], it works in time $\mathcal{O}(n \log^2 n)$ and is based on a selective family, with a certain additional property, shown to exist by the probabilistic method. A constructive deterministic broadcasting algorithm operating in time $\mathcal{O}(n^{3/2})$ was given by Chlebus, Gąsieniec, Östlin and Robson [23], it is currently the fastest constructive algorithm known.

Brusci and Del Pinto [16] proved a lower bound $\Omega(D \log n)$, for D up to $\Theta(n)$, for deterministic broadcasting in multi-hop radio networks. Clementi, Monti and Silvestri [27] showed a similar lower bound $\Omega(n \log D)$, for D up to $\Theta(n)$. An impossibility of acknowledgement, while broadcasting with no collision detection in general graphs, was showed in [22].

6.1.3   Multiple communication.   Bar-Yehuda, Israeli and Itai [13] considered multiple instances of point-to-point communication and broadcast. They developed randomized Las Vegas algorithm to achieve this in the model of separate messages. Networks are assumed to be symmetric and nodes know their neighborhood. Their algorithm begins with preprocessing. First a leader is elected, by resorting to the algorithm described in Subsection 6.2, then a BFS tree is found, rooted at the leader. The expected time of preprocessing is $\mathcal{O}((n + D \log n) \log \Delta)$. The protocol itself consists of two phases: during the first Collection phase all the messages are collected at the root, then the Distribution phase follows, during which the messages are sent from the root towards their destinations. The idea is to send messages to the root of the BFS tree and then to the target nodes along the edges of the tree, pipelining messages along the branches of the BFS tree. A deterministic local acknowledgement is used, which immediately confirms a successful receipt of a message along an edge. A probabilistic analysis is done by interpreting the system as a line of queues, each with Bernoulli service and arrival times (see [21] for an alternative approach to similar problems). A set of k point-to-point transmissions require the expected time $\mathcal{O}((k + D)\log\Delta)$. Similarly, k broadcasts require the expected time $\mathcal{O}((k + D)\log\Delta\log n)$.

The broadcasting algorithm presented in Subsection 6.1.1 is Monte Carlo, but it may be converted to be Las Vegas, as was shown in [13]. The method is as follows. Since the size n is known, choose $\epsilon = 1/n$ and build a BFS tree. After the tree is expected to have been completed, collect the information along its edges from all the nodes in the tree by the procedure Collection discussed in the preceding paragraph. This is done within a certain time interval, and if not successful, then the whole algorithm is repeated from the very beginning.

A method to obtain a gossiping algorithm from one performing broadcasting was developed in [25]. If the broadcasting algorithm works in time $\mathcal{O}(B(n))$ then the resulting gossiping algorithm operates in time $\mathcal{O}(\sqrt{B(n)}n\log n)$. Applying this method with the broadcasting algorithm from [25] with $B(n) = n\log^2 n$ yields a gossiping algorithm working in time $\mathcal{O}(n^{3/2}\log^2 n)$. Similarly, the constructive algorithm from [23] with $B(n) = n^{3/2}$ gives a deterministic constructive gossiping algorithm working in time $\mathcal{O}(n^{7/4}\log n)$. A randomized Las-Vegas gossiping algorithm, working in the expected time $\mathcal{O}(n\log^4 n)$ was given in [24].

6.2   SIMULATIONS

Bar-Yehuda, Goldreich and Itai [11] developed a simulation of the multiple access channel with ternary feedback on the multi-hop radio network without collision detection. The simulation algorithm uses the procedures Decay and Broadcast presented in Subsection 6.1.1. The slowdown of the simulation is $B_\epsilon = \mathcal{O}((D + \log(n/\epsilon))\log\Delta)$, that is, B_ϵ is the time it takes to perform a single broadcast of the channel on the packet network, with the probability at least $1 - \epsilon$, which is also the probability of correctness of the protocol. Additionally, an implementation of conflict-detection mechanism was developed for a multi-hop radio network without conflict detection, in time $\mathcal{O}(\log(1/\epsilon)\log\Delta)$ for one step, and with the probability $1 - \epsilon$. It was also shown that any broadcast-channel algorithm operating in t steps can be emulated in time $\mathcal{O}(t(\log t\log\Delta + B_\epsilon))$ on a packet-network, with the probability $1 - \epsilon$. The main application is leader election in time $\mathcal{O}((D + \log(n/\epsilon))\log\Delta\log\log n)$, obtained by a combination of the simulation and Willard's algorithms, optimality of this algorithm is an open problem.

Alon, Bar-Noy, Linial and Peleg [9] developed step-by-step simulations of message-passing networks on packet radio networks. Two models were considered. In the *general model* a node may send in one round arbitrary distinct messages, and in the *uniform model* only copies of one message.

Consider first the general model. The results are as follows. There are graphs that require slowdown $\Omega(\Delta_m^2)$ for any simulation algorithm. Schedules of $\mathcal{O}(\Delta_{in}\Delta_{out})$ rounds can be found by a polynomial (centralized) algorithm. A randomized Las Vegas algorithm was developed with slowdown $\mathcal{O}(\Delta_{in}\Delta_{out}\log(n/\epsilon))$ with probability $1 - \epsilon$.

Next consider the uniform model. It was shown that there are graphs that require the slowdown of $\Omega(\Delta_m \log \Delta_m)$, for any algorithm. There is a polynomial time (centralized) constructible schedule with slowdown $\mathcal{O}(\Delta_{in}\log n)$. Randomized Las Vegas algorithm was developed with slowdown $\mathcal{O}(\Delta_{in}\log(n/\epsilon))$ with probability $1 - \epsilon$.

6.3 OPTIMALITY

There are two relevant issues concerning optimality of randomized communication. The first one concerns the question whether randomness can improve the performance of protocols, as compared with purely deterministic ones. The second task is to find an optimal randomized communication algorithm in the class of randomized ones.

The problem of broadcasting in ad-hoc radio networks is an example of a problem for which randomness can cause an exponential speedup in its complexity. A lower bound $\Omega(n)$ for deterministic algorithms performing broadcasting was shown in [12]. It holds even if the networks are restricted to be of diameter equal to 3 and of especially simple form: such a network of n nodes consists of three layers, the first layer contains only the source, the second layer contains the remaining nodes except one *sink*, which is the only node in the third layer. All the nodes from the second layer are reachable from the source, but only a proper subset of the nodes in the second layer are neighbors of the sink. Notice that the algorithm presented in Subsection 6.1.1 performs broadcasting in the expected time $\mathcal{O}(\log n)$ on such a network, what gives an exponential gap.

To assess optimality of the broadcasting algorithm (of Section 6.1.1), in the class of randomized algorithms, we may compare its expected performance with the following two lower bounds. The first one is $\Omega(\log^2 n)$, it holds for a family of graphs of diameter two, it was proved by Alon, Bar-Noy, Linial and Peleg [8]. The other lower bound is $\Omega(D \log(n/D))$, it was proved by Kushilevitz and Mansour [68]. It follows that the expected complexity $\mathcal{O}(\log^2 n + D \cdot \log n)$ of the broadcasting algorithm is optimal for networks of small diameter. Namely, the upper bound becomes $\mathcal{O}(\log^2 n)$ for networks of diameter $D = \mathcal{O}(\log n)$, which is optimal by the first lower bound, and if the diameter D satisfies $D = \mathcal{O}(n^a)$, for a constant $a > 0$, then the upper bound becomes $\mathcal{O}(D \log n)$, which is again optimal by the second lower bound.

The lower bound $\Omega(\log^2 n)$ is actually in the following strong form: there is a family of graphs, with n nodes and of diameter two, for which any algorithm requires time $\Omega(\log^2 n)$. It follows that the bound holds for randomized algorithms and the graph may be even assumed to be known. The proof is by the probabilistic method. The lower bound $\Omega(D \log(n/D))$ is of a weaker form: for any randomized broadcasting algorithm, and for each numbers n and suitable D, there exists a network of n nodes and of diameter D, such that the expected broadcasting time of the algorithm is $\Omega(D \log(n/D))$. The bound is valid for a class of algorithms which are arbitrary functions determining the probability of broadcasting of a node in a step, depending on both its ID and its history. The proof is based on the following lemma: for a single-hop radio network without collision detection, there is a subset of stations such that if exactly they are active then the expected time till the first successful broadcast is $\Omega(\log n)$. The main lower bound is obtained by constructing a layered network, in which the nodes are partitioned into layers in such a way that two consecutive layers induce a complete binary graph, and there are no other edges except for those connecting any two nodes in consecutive layers. The number D of layers is the depth, and the lemma is applied to each layer of size n/D.

7. RELATED WORK

In this section we present additional relevant results concerning communication algorithms in radio networks.

Gąsieniec, Pelc and Peleg [43] compared the locally and globally synchronous models of single-hop radio networks in the context of the *wakeup* problem, in which the time when each station joins the protocol is controlled by an adversary, and the goal of an algorithm is to perform a successfull broadcast as soon as possible. If the stations have access to a global clock then wakeup can be realized in time $\mathcal{O}(\log n \log(1/\epsilon))$ with the probability at least $1 - \epsilon$, by iterating procedure **Decay**. If the local clocks are not synchronized, then processors may flip coins with the probability $1/n$ of heads to come up, to decide if to transmit in a step, this yields an algorithm operating in time $\mathcal{O}(n \log(1/\epsilon))$ with the probability at least $1 - \epsilon$. Deterministic algorithms in such setting require time $\Omega(n)$, and deterministic schedules are known to exist achieving time $\mathcal{O}(n \log^2 n)$.

Dynamic aspects of radio networks include not only the mobility of nodes, but also the possibility of changing the ranges of nodes by modifying their power consumption. To study such aspects, the notion of a radio network needs to be modified, for instance by assuming that the nodes are in some metric space and the reachability relation is determined by ranges that the nodes can cover.

A multi-hop network is called *power-controlled* if hosts are able to change their transmission ranges, and in this way change also the topology of the underlying reachability graph. Adler and Scheideler [5] introduced a general formal model of power-controlled networks, in which each pair of vertices is assigned the lowest transmission power that allows to maintain a direct connection between the vertices. They developed efficient strategies for permutation routing in such networks. They considered also a special case where the nodes are located in a square region in the plane, distributed randomly, and the power consumption to maintain a connection is proportional to the distance. They showed that permutation routing on such n-node networks can be performed in $\mathcal{O}(\sqrt{n})$ steps, which was then proved to be optimal by a matching lower bound $\Omega(\sqrt{n})$.

Diks, Kranakis, Krizanc and Pelc [31] developed deterministic algorithms and lower bounds for broadcasting in the case of nodes located on a line, the nodes know their coordinates. Ravishankar and Singh [92, 93] studied broadcasting and gossiping when stations are randomly placed on a line. Kranakis, Krizanc and Pelc [67] studied deterministic fault-tolerant broadcasting in radio networks with the stations located either on a line or on na two-dimensional grid, some of them faulty.

Kushilevitz and Mansour [69] considered communication in noisy single-hop networks. The specific problem they studied is that of computing a threshold function of the bits stored in the nodes, one bit per node. The protocols are assumed to be oblivious, so the bits cannot be encoded by silence/transmission sequences. Every bit transmitted is flipped randomly and independently over the receiving stations, due to noise, the probability that a wrong bit is received is below $1/2$, but this probability is not assumed to be known by the stations. Let n be the number of stations. A protocol is developed which, for a given natural number $k \leq n$ and $\epsilon > 0$, decides if the number of bits equal to 1 is at least k with the probability at least $1 - \epsilon$ by performing $\mathcal{O}(n)$ transmissions.

The following are some results on centralized off-line algorithms computing broadcast schedules, the networks are known then, and are inputs to such sequential algorithms. Chlamtac and Weinstein [20] developed a deterministic algorithm which finds in polynomial time a schedule giving broadcasting time $\mathcal{O}(D \cdot \log^2)$. Gaber and Mansour [41] showed that for any packet radio network there is a schedule of broadcasting giving time $\mathcal{O}(D + \log^5 n)$, and that it can be found by a deterministic polynomial algorithm. Chlamtac and Kutten [18] and Sen and Hun [98] showed the NP-completeness of finding optimal broadcasting schedules in packet radio networks, in particular it was shown in [98] that the problem remains NP-complete even if the nodes are points in the plane and the reachability is determined by distances. Kirousis, Kranakis, Krizanc and Pelc [64] studied the problem of assigning transmission ranges to nodes located on a line so as to minimize the total power consumption and preserve strong connectedness. This was later extended by Clementi, Ferreira, Penna, Perennes and Silvestri [26]. Chlamtac and Kutten [19] showed the NP-completeness of minimizing the average time per node during broadcasting when nodes are blocked from hearing messages other than those resulting from a broadcast algorithm.

8.   DISCUSSION

Most of the material in this chapter concerns single-hop radio networks. This is not surprising as they have been investigated since the beginning of the 1970's, while the research concerning distributed randomized communication in ad-hoc data radio networks is much newer and goes back to the middle of the 1980's.

While the multiple access channel has been investigated over the years, first the infinitely-many users model was dominant but later the research shifted to the finitely-many users model. The original popularity of the infinitely-many users model seems to have stemmed from the belief that it was a clean model to prove a good behavior of nice protocols in a situation when the number of stations may be arbitrarily large. Later on, after it has been shown that such protocols like Aloha and both the polynomial and exponential backoffs are not stable in this model, the finitely-many users model became more popular. Does the existing body of research provide evidence that results for the infinitely-many users model are less meaningful in general because the model 'distorts' the true behavior of protocols? We believe that this is not the case, the following are some of the reasons. First, the stability of superlinear polynomial backoff protocols in the finitely-many users model simply confirms the stabilizing effect, for this class of protocols, of a scenario when messages can be queued, as opposed to a more challenging situation when messages run protocols on their own. Secondly, the existence of constant-delay protocols for both of these models shows that such ultimate good behavior is possible in both models.

The multiple access channel is widely used in existing distributed systems. This popularity has created a common belief that certain specific protocols used, like the one governing conflict resolution in Ethernet, and exponential backoff protocols in general, are the best to provide simple and efficient communication in local area networks. This is in contrast with recent theoretical work, let us quote [58]: "Based on our analysis, it would appear that the most popular and well-studied protocols are precisely the wrong protocols." This statement is apparently meant to be restricted to backoff protocols. Only recently protocols with a constant expected message delay have been found, which are not backoff but still are acknowledgement-based, their development is a major progress in our understanding how to operate the multiple access channel.

To sum this up, issues related to communication in single-hop radio networks seem to be quite well understood, due to the intensive research effort over the past thirty years. Multi-hop radio networks have started to be investigated much later, and are a fertile ground for future research.

8.1   FUTURE WORK

There are many technical open problems in radio networks, and possible ramifications of known results, which concern both deterministic and randomized communication modes. Some of such problems have been mentioned while discussing specific topics. We conclude our overview with additional open problems which seem to be especially interesting.

Consider first the single-hop networks. The best known upper bound on capacity of the multiple-access channel with ternary feedback is slightly above 1/2, and the lower bound of the best algorithm known is a little below 1/2. It is a natural question to ask how is the optimum capacity related to 1/2, in particular if it is simply equal to 1/2. Next, no stable backoff protocol for the infinitely-many users model is known. What we know is that the binary exponential backoff is unstable for any arrival rate and that unstability always happens for sufficiently large arrival rate. It is an open question if there is a backoff protocol that would be stable for some positive arrival rate in this setting, this question was already posed by MacPhee [71].

Our knowledge about the optimal complexity of broadcasting in multi-hop radio networks is not complete. The best lower bound for deterministic algorithms is $\Omega(n \log n)$, whereas the fastest deterministic algorithm known works in time $\mathcal{O}(n \log^2 n)$. This creates a gap of a multiplicative factor $\log n$. The upper bound is not constructive, we only know an existence proof by the probabilistic method. The best constructive design yields an algorithm of performance $\mathcal{O}(n^{3/2})$, so here the gap is much bigger. Next the class of randomized broadcasting algorithms. The lower bound $\Omega(D \log(n/D))$ shows that an algorithm of the expected time performance $\mathcal{O}((D + \log n/\epsilon) \cdot \log n)$ is optimal provided $D = \mathcal{O}(n^a)$, for a constant $a > 0$. If the diameter D is close to the size n of the network then the question of optimality is open. And do we know if randomization helps in broadcasting? For networks of small diameter D, in particular if D is polylog in n, then indeed there is an exponential gap in performance between deterministic and randomized algorithms. However if D is $\Theta(n)$, then the upper bound $\mathcal{O}((D + \log n/\epsilon) \cdot \log n)$ becomes $\mathcal{O}(n \log n)$. As mentioned above, there is a gap between the best upper and lower bounds for the deterministic algorithms, and we can see that the best upper bound for randomized algorithms in arbitrary networks is within this gap, actually it is on its boundary determined by the function $n \log n$. Hence solving the optimality question for deterministic algorithms would also answer the question if randomization helps in broadcasting when the diameter is close to the size of a network.

We still know little about multiple communication in the model allowing combined messages. The gap between the best known deterministic and randomized algorithms for gossiping is asymptotically bigger than for broadcasting. This may be not necessarily due not to the fact that gossiping is much more inherently time demanding, but rather because the known deterministic gossiping algorithms are much further from optimal. What are the optimal time complexities of both deterministic and randomized distributed gossiping in ad-hoc radio networks are interesting open questions.

Most of the work done on communication in radio networks assumes the globally synchronous model. Investigations of the locally synchronous one include the work on the wakeup problem [43], also some of the constant-delay protocols developed in [50] were shown to be robust enough to allow stations to stop and then restart. Sometimes protocols for the stronger model can be redesigned to be able to rely on the local synchrony only without a loss in performance, this, for instance, holds for broadcasting, as noticed by Peleg [89]. On the other hand, analysis of stability issues of conflict-resolution protocols for the multiple-access channel typically relies on the global synchrony. Clarifying the extent to which communication protocols need global synchronization is an interesting topic of research.

8.2   SURVEY INFORMATION

General overviews of networking technologies can be found in books by Bertsekas and Gallager [15] and Tanenbaum [107].

Issues of wireless and mobile communication are discussed in handbooks edited by Gibson [45], Imielinski and Korth [54], Stojmenovic [103], and in a book by Pahlavan and Levesque [85].

For more on the multiple-access channel see books by Bertsekas and Gallager [15], and by Rom and Sidi [95]. A special issue on random-access communications [75] contains survey articles by Gallager [42] and Tsybakov [109], which cover the investigations on the multiple access channel up to the early 1980's.

Ephremides and Hajek [33] gave an overview of connections between the information theory and communication in networks, the paper contains a section on the multiple access channel. A book by Cover and Thomas [28] includes a discussion of the multiple access channel from the point of view of the information theory.

A recent survey by Pelc [88] covers broadcasting in multi-hop radio networks in detail.

Acknowledgments

I thank Leslie Goldberg and Andrzej Pelc, whose comments on a preliminary version helped to clarify some points. This work was partly done while visiting the University of Liverpool.

References

[1] N. Abramson, Development of the ALOHANET, *IEEE Trans. on Information Theory* 31 (1985) 119–123.

[2] N. Abramson, The Aloha system, in [4] .

[3] N. Abramson, The throughput of packet broadcasting channels, *IEEE Trans. on Communications* 25 (1977) 117–128.

[4] N. Abramson, and F. Kuo, (Eds.), "Computer-Communication Networks," Prentice-Hall, 1973.

[5] M. Adler, and C. Scheideler, Efficient communication strategies for ad-hoc wireless networks, in *Proc. 10th ACM Symp. Parallel Algorithms and Architectures,* Puerto Vallerta, Mexico, 1998, pp. 259–268.

[6] H. Al-Ammal, L.A. Goldberg, and P. MacKenzie, Binary Exponential Backoff is stable for high arrival rates, in *Proc. 17th Symp.*

on *Theoretical Aspects of Computer Science*, Lille, France, 2000, Springer LNCS 1770, pp. 169–180.

[7] D.J. Aldous, Ultimate unstability of exponential back-off protocol for acknowledgement-based transmission control of random access communication channels, *IEEE Trans. on Information Theory* 32 (1987) 219–223.

[8] N. Alon, A. Bar-Noy, N. Linial, and D. Peleg, A lower bound for radio broadcast, *J. Computer and System Sciences* 43 (1991) 290–298.

[9] N. Alon, A. Bar-Noy, N. Linial, and D. Peleg, Single round simulation of radio networks, *J. Algorithms* 13 (1992) 188–210.

[10] N. Alon, J.H. Spencer, and P. Erdős, "The Probabilistic Method," John Wiley, 1992.

[11] R. Bar-Yehuda, O. Goldreich, and A. Itai, Efficient emulation of single-hop radio network with collision detection on multi-hop radio network with no collision detection, *Distributed Computing* 5 (1991) 67–72.

[12] R. Bar-Yehuda, O. Goldreich, and A. Itai, On the time complexity of broadcast in radio networks: An exponential gap between determinism and randomization, *J. Computer and System Sciences* 45 (1992) 104–126.

[13] R. Bar-Yehuda, A. Israeli, and A. Itai, Multiple communication in multi-hop radio networks, *SIAM J. on Computing* 22 (1993) 875–887.

[14] J. Bentley, and A. Yao, An almost optimal algorithm for unbounded search, *Information Processing Letters* 5 (1976) 82–87.

[15] D. Bertsekas, and R.G. Gallager, "Data Networks," Prentice Hall, 1991.

[16] D. Brusci, and M. Del Pinto, Lower bounds for the broadcast problem in mobile radio networks, *Distributed Computing* 10 (1997) 129–135.

[17] J.I. Capetanakis, Tree algorithms for packet broadcast channels, *IEEE Trans. on Information Theory* 25 (1979) 505–515.

[18] I. Chlamtac, and S. Kutten, On broadcasting in radio networks - problem analysis and protocol design, *IEEE Trans. on Communications* 33 (1985) 1240–1246.

[19] I. Chlamtac, and S. Kutten, Tree based broadcasting in multihop radio networks, *IEEE Trans. on Computers* 36 (1987) 1209–1223.

[20] I. Chlamtac, and O. Weinstein, The wave expansion approach to broadcasting in multihop radio networks, in *Proc. 6th Joint Conference of the IEEE Computer and Communication Societies (INFOCOM)*, 1987, pp. 1116-1124.

[21] B.S. Chlebus, K. Diks, and A. Pelc, Sorting on a mesh-connected computer with delaying links, *SIAM J. on Discrete Mathematics* 7 (1994) 119-132.

[22] B.S. Chlebus, L. Gąsieniec, A.M. Gibbons, A. Pelc, and W. Rytter, Deterministic broadcasting in unknown radio networks, in *Proc. 11th ACM-SIAM Symp. on Discrete Algorithms*, San Francisco, California, 2000, pp. 861-870.

[23] B.S. Chlebus, L. Gąsieniec, A. Östlin, and J.M. Robson, Deterministic radio broadcasting, in *Proc. 27th Int. Colloquium on Automata, Languages and Programming*, Geneva, Switzerland, 2000, Springer LNCS 1853, pp. 717-728.

[24] M. Chrobak, L. Gąsieniec, and W. Rytter, A randomized algorithm for gossiping in radio networks, 2000, a manuscript.

[25] M. Chrobak, L. Gąsieniec, and W. Rytter, Fast broadcasting and gossiping in radio networks, in *Proc. 41st IEEE Symp. on Foundations of Computer Science*, Redondo Beach, California, 2000, to appear.

[26] A.E.F. Clementi, A. Ferreira, P. Penna, S. Perennes, and R. Silvestri, The minimum range assignment problem on linear radio networks, in *Proc. 8th European Symposium on Algorithms*, Saarbrücken, Germany, 2000, Springer LNCS 1879, pp. 143-154.

[27] A.E.F. Clementi, A. Monti, and R. Silvestri, Selective families, superimposed codes, and broadcasting in unknown radio networks, in *Proc. 12th ACM-SIAM Symp. on Discrete Algorithms*, Washington, DC, 2001, to appear.

[28] T.M. Cover, and J.A. Thomas, "Elements of Information Theory," John Wiley, 1991.

[29] R. Cruz, and B. Hajek, A new upper bound on the throughput of a multi-access broadcast channel, *IEEE Trans. on Information Theory* 28 (1982) 402-405.

[30] G. De Marco, and A. Pelc, Faster broadcasting in unknown radio networks, *Information Processing Letters*, to appear.

[31] K. Diks, E. Kranakis, D. Krizanc, and A. Pelc, The impact of knowledge on broadcasting time in radio networks, in *Proc. 7th European Symposium on Algorithms*, Prague, Czech Republic, 1999, Springer LNCS 1643, pp. 41-52.

[32] J.L. Doob, "Stochastic Processes," John Wiley, 1953.

[33] A. Ephremides, and B. Hajek, Information theory and communication networks: an unconsummated union, *IEEE Trans. on Information Theory* 44 (1998) 2416–2432.

[34] P. Erdős, and A. Rényi, On two problems in information theory, *Publ. Hung. Acad. Sci.* 8 (1963) 241–254.

[35] G. Fayolle, P. Flajolet, M. Hofri, and P. Jacquet, Analysis of a stack algorithm for random multiple-access communication, *IEEE Trans. on Information Theory* 31 (1985) 244–254.

[36] G. Fayolle, E. Gelenbe, and J. Labetoulle, Stability and optimal control of the packet switching broadcast channel, *J. ACM* 24 (1977) 375–386.

[37] G. Fayolle, V.A. Malyshev, and M.V. Menshikov, "Topics in the Constructive Theory of Countable Markov Chains," Cambridge University Press, 1995.

[38] W. Feller, "An Introduction to Probability Theory and Its Applications," vol. I, John Wiley, 1961.

[39] M.J. Ferguson, On the control, stability, and waiting time in a slotted ALOHA random-access system, *IEEE Trans. on Communications* 11 (1975) 1306–1311.

[40] F.G. Foster, On stochastic matrices associated with certain queuing processes, *Ann. Math. Stat.* 26 (1953) 355–360.

[41] I. Gaber, and Y. Mansour, Broadcast in radio networks, in *Proc. 6th ACM-SIAM Symp. on Discrete Algorithms*, 1995, pp. 577–585.

[42] R.G. Gallager, A perspective on multiaccess channels, *IEEE Trans. on Information Theory* 31 (1985) 124–142.

[43] L. Gąsieniec, A. Pelc, and D. Peleg, The wakeup problem in synchronous broadcast system, in *Proc. 19th ACM Symp. on Principles of Distributed Computing*, Portland, Oregon, 2000, pp. 113–122.

[44] M. Geréb-Graus, and T. Tsantilas, Efficient optical communication in parallel computers, in *Proc. 4th ACM Symp. on Parallel Algorithms and Architectures*, 1992, pp. 41–4.

[45] J.D. Gibson (Ed.), "The Mobile Communications Handbook," CRC Press, 1996.

[46] L.A. Goldberg, M. Jerrum, S. Kannan, and M. Paterson, A bound on the capacity of backoff and acknowledgement-based protocols, in *Proc. 27th Int. Colloquium on Automata, Languages and Programming*, Geneva, Switzerland, 2000, Springer LNCS 1853, pp. 705–716.

452

[47] L.A. Goldberg, M. Jerrum, T. Leighton, and S. Rao, Doubly logarithmic communication algorithms for optical communication parallel computers, *SIAM J. on Computing* 26 (1997) 1100–1119.

[48] L.A. Goldberg, M. Jerrum, and P. MacKenzie, An $\Omega(\sqrt{\log\log n})$ lower bound for routing in optical networks, *SIAM J. on Computing* 27 (1998) 1083–1098.

[49] L.A. Goldberg, and P. MacKenzie, Analysis of practical backoff protocols for contention resolution with multiple servers, *J. Computer and System Sciences* 58 (1999) 232–258.

[50] L.A. Goldberg, P. MacKenzie, M. Paterson, and A. Srinivasan, Contention resolution with constant expected delay, *J. ACM*, to appear.

[51] J. Goodman, A.G. Greenberg, N. Madras, and P. March, Stability of binary exponential backoff, *J. ACM* 35 (1988) 579–602.

[52] A.G. Greenberg, P. Flajolet, and R. E. Ladner, Estimating the multiplicities of conflicts to speed their resolution in multiple access channels, *J. ACM* 34 (1987) 289–325.

[53] A.G. Greenberg, and S. Winograd, A lower bound on the time needed in the worst case to resolve conflicts deterministically in multiple access channels, *J. ACM* 32 (1985) 589–596.

[54] T. Imielinski, and H. Korth, (Eds.), "Mobile Computing," Kluwer Academic Publishers, 1996.

[55] M. Habib, C. McDiarmid, J. Ramirez-Alfonsin, and B. Reed, (Eds.), "Probabilistic Methods for Algorithmic Discrete Mathematics," Springer-Verlag, 1998.

[56] B. Hajek, Hitting-time and occupation-time bounds implied by drift analysis with applications, *Adv. Appl. Prob.* 14 (1982) 502–525.

[57] B. Hajek, and T. van Loon, Decentralized dynamic control of a multiaccess broadcast channel, *IEEE Trans. on Automatic Control* 27 (1982) 559–569.

[58] J. Håstad, T. Leighton, and B. Rogoff, Analysis of backoff protocols for multiple access channels, *SIAM J. on Computing* 25 (1996) 740–774.

[59] J. Hayes, An adaptive technique for local distributions, *IEEE Trans. on Communications* 26 (1978) 1178–1186.

[60] R. Howard, "Dynamic Programming and Markov Processes," MIT, Cambridge, Massachusetts, 1960.

[61] M. Kaplan, A sufficient condition of nonergodicity of a Markov chain, *IEEE Trans. on Information Theory* 25 (1979) 470–471.

[62] F.P. Kelly, Stochastic models of computer communication systems, *J. R. Statist. Soc.* B 47 (1985) 379–395.

[63] F.P. Kelly, and I.M. MacPhee, The number of packets transmitted by collision detect random access schemes, *The Annals of Probability* 15 (1987) 1557–1568.

[64] L. Kirousis, E. Kranakis, D. Krizanc, and A. Pelc, Power consumption in packet radio networks, in *Proc. 14th Symposium on Theoretical Aspects of Computer Science*, Lubeck, Germany, 1997, Springer LNCS 1200, pp. 363-374.

[65] L. Kleinrock, "Queuing Systems, II: Computer Applications," John Wiley, 1976.

[66] J. Komlós, and A.G. Greenberg, An asymptotically nonadaptive algorithm for conflict resolution in multiple-access channels, *IEEE Trans. on Information Theory* 31 (1985) 303–306.

[67] E. Kranakis, D. Krizanc, and A. Pelc, Fault-tolerant broadcasting in radio networks, in *Proc. 6th European Symposium on Algorithms*, Venice, Italy, 1998, Springer LNCS 1461, pp. 283–294.

[68] E. Kushilevitz, and Y. Mansour, An $\Omega(D\log(N/D))$ lower bound for broadcast in radio networks, *SIAM J. on Computing* 27 (1998) 702–712.

[69] E. Kushilevitz, and Y. Mansour, Computation in noisy radio networks, in *Proc. 9th ACM-SIAM Symp. on Discrete Algorithms*, 1998, pp. 236–243.

[70] P.D. MacKenzie, C.G. Plaxton, and R. Rajaraman, On contention resolution protocols and associated probabilistic phenomena, *J. ACM* 45 (1998) 324–378.

[71] I.M. MacPhee, On optimal strategies in stochastic decision processes, D. Phil. Thesis, University of Cambridge, 1987.

[72] C. McDiarmid, "Concentration," a chapter in [55], pp. 195–248.

[73] C.U. Martel, Maximum finding on a multiple access broadcast network, *Information Processing Letters* 52 (1994) 7–13.

[74] C.U. Martel, and T.P. Vayda, The complexity of selection resolution, conflict resolution, and maximum finding on multiple access channels, in *Proc. 3rd Int. Workshop on Parallel Computation and VLSI Theory*, 1988, pp. 401–410.

[75] J.L. Massey, (Ed.), Special issue on random-access communications, *IEEE Trans. on Information Theory* 31 (1985).

[76] P. Mathys, and P. Flajolet, Q-ary collision resolution algorithms in random-access systems with free or blocked channel access, *IEEE Trans. on Information Theory* 31 (1985) 217–243.

[77] R. Metcalfe, and D. Boggs, Ethernet: Distributed packet switching for local computer networks, *Comm. ACM* 19 (1976) 395–404.

[78] S.P. Meyn, and R.L. Tweedie, "Markov Chains and Stochastic Stability," Springer-Verlag, 1993.

[79] V.A. Mikhailov, and T.S. Tsybakov, Upper bound for the capacity of a random multiple access system, *Probl. Information Transmission* 17 (1981) 63–67.

[80] M.L. Molle, On the capacity of infinite population multiple access protocols, *IEEE Trans. on Information Theory* 28 (1982) 396–401.

[81] M.L. Molle, Unifications and extension of the multiple access communication problem, Computer Science Department, University of California, Los Angeles, Rep. No. CSD-810730, 1981.

[82] J. Mosely, and P.A. Humblet, A class of efficient contention resolution algorithms for multiple access channels, *IEEE Trans. on Communications* 33 (1985) 145–151.

[83] R. Motwani, and P. Raghavan, "Randomized Algorithms," Cambridge University Press, 1995.

[84] A.R. Odoni, On finding the maximal gain for Markov decision process, *Operations Research* 17 (1969).

[85] K. Pahlavan, and A. Levesque, "Wireless Information Networks," Wiley-Interscience, 1995.

[86] A.G. Pakes, Some conditions on ergodicity and recurrence of Markov chains, *Operations Research* 17 (1969) 1058–1061.

[87] S.S. Panwar, D. Towsley, and J.K. Wolf, On the throughput of degenerate intersection and first-come first-served collision resolution algorithms, *IEEE Trans. on Information Theory* 31 (1985) 274–279.

[88] A. Pelc, Broadcasting in radio networks, a chapter in [103].

[89] D. Peleg, Deterministic radio broadcast with no topological knowledge, 2000, a manuscript.

[90] N. Pippinger, Bounds on the performance of protocols for a multiple-access broadcast channel, *IEEE Trans. on Information Theory* 27 (1981) 145–151.

[91] P. Raghavan, and E. Upfal, Stochastic contention resolution with short delays, *SIAM J. on Computing* 28 (1998) 709–719.

[92] K. Ravishankar, and S. Singh, Asymptotically optimal gossiping in radio networks, *Discr. Appl. Math.* 61 (1995) 61–82.

[93] K. Ravishankar, and S. Singh, Broadcasting on [0, *L*], *Discrete Mathematics* 53 (1994) 299–319.

[94] R.L. Rivest, Network control by Bayesian broadcast, *IEEE Trans. on Information Theory* 33 (1987) 323–328.

[95] R. Rom, and M. Sidi, "Multiple access protocols: performance and analysis," Springer-Verlag, 1990.

[96] W.A. Rosenkrantz, and D. Towsley, On the instability of the slotted Aloha multiaccess algorithm, *IEEE Trans. on Information Theory* 28 (1983) 994–996.

[97] T. Saadawi, and A. Ephremides, Analysis, stability and optimization of slotted ALOHA with a finite number of buffered users, *IEEE Trans. on Automatic Control* 26 (1981) 680–689.

[98] A. Sen, and M.L. Huson, A new model for scheduling packet radio networks, in *Proc. 15th Joint Conference of the IEEE Computer and Communication Societies (INFOCOM)*, 1996, pp. 1116-1124.

[99] L.I. Sennott, Conditions for the non-ergodicity of Markov chains with application to a communication system, *J. Appl. Prob.* 24 (1987) 339–346.

[100] L.I. Sennott, P.A. Humblet, and R.L. Tweedie, Mean drifts and the non-ergodicity of Markov chains, *Operations Research* 21 (1983) 783–789.

[101] J.F. Shoch, and J.A. Hupp, Measured performance of an Ethernet local network, *Comm. ACM* 23 (1980) 711–720.

[102] S. Stidham, Jr., A last word on $L = \lambda W$, *Operations Research* 17 (1974) 417–421.

[103] I. Stojmenovic, (Ed.), "Handbook of Wireless Networks and Mobile Computing," John Wiley, 2001, to appear.

[104] W. Szpankowski, Some sufficient conditions for non-ergodicity of Markov chains, *J. Appl. Prob.* 22 (1985) 138–147.

[105] W. Szpankowski, Stability conditions for multidimensional queueing systems with computer applications, *Operations Research* 36 (1988) 944–957.

[106] W. Szpankowski, and V. Rego, Some theorems on instability with applications to multiaccess protocols, *Operations Research* 36 (1988) 958–966.

[107] A.S. Tanenbaum, "Computer Networks," Prentice Hall, 1996.

456

[108] J.N. Tsitsiklis, Analysis of a multiaccess control scheme, *IEEE Trans. on Automatic Control* 32 (1987) 1017–1020.

[109] B.S. Tsybakov, Survey of USSR contributions to random multiple-access communications, *IEEE Trans. on Information Theory* 31 (1985) 143–165.

[110] B.S. Tsybakov, and N.B. Likhanov, Upper bound on the capacity of a random multiple-access system, *Problemy Peredachi Informatsii* 23 (1987) 64–78.

[111] B.S. Tsybakov, and V.A. Mikhailov, Ergodicity of a slotted Aloha system, *Probl. Information Transmission* 16 (1980) 301–312.

[112] B.S. Tsybakov, and V.A. Mikhailov, Free synchronous packet access in a broadcast channel with feedback, *Probl. Information Transmission* 14 (1978) 259–280.

[113] B.S. Tsybakov, and V.A. Mikhailov, Random multiple access of packets: Part and try algorithm, *Probl. Information Transmission* 16 (1980) 65–79.

[114] S. Verdu, Computation of the efficiency of the Mosely-Humblet contention resolution algorithm: a simple method, *Proc. of the IEEE* 74 (1986) 613–614.

[115] N.D. Vvedenskaya, and M.S. Pinsker, Non-optimality of the part-and-try algorithm, in *Abstracts of the Int. Workshop on Convolutional Codes and Multiuser Communication*, Sochi, USSR, 1983, pp. 141-144.

[116] D.E. Willard, Log-logarithmic selection resolution protocols in a multiple access channel, *SIAM J. on Computing* 15 (1986) 468–477.

Chapter 12

A GUIDE TO CONCENTRATION BOUNDS

Josep Díaz, Jordi Petit and Maria Serna
Departament de Llenguatges i Sistemes Informàtics
Universitat Politècnica Catalunya
Jordi Girona Salgado 1-3
08034-Barcelona.
diaz,jpetit,mjserna@lsi.upc.es

Abstract   In this chapter we survey some of the concentration bounds that are used in the field of combinatorics and theoretical computer science. We have tried to emphasize the intuition behind the use of the different bounds. Throughout the text, the reader can follow the application of the different bounds to the same examples, comparing each time with the previous results.

1.   INTRODUCTION

Worst case analysis of an algorithm gives information on hard instances that may not be representative of its real behavior on the data of interest. In such situations, the analysis of its *average behavior* may give more useful information. Given a probability distribution on the set of inputs and a performance measure, we wish to analyze the expected value of this performance, seen as a random variable on the probability space defined by the inputs of a given size. Relevant performance measures may be running time, memory size, solution quality, number of executions of a type of instructions, etc. For instance, in a sorting algorithm, the uniform distribution on the symmetric group S_n is a natural distribution when the input to be sorted has n keys. As one of the basic operations used by many sorting algorithms is to compare values, the number of comparisons performed along the algorithm's execution is a possible measure to analyze. For the Quicksort algorithm, it is a classi-

S. Rajasekaran et al (eds.), Handbook of Randomized Computing, Volume 2, pp, 457–507.
© 2001 *Kluwer Academic Publishers.*

cal result that the the expected number of comparisons is $O(n \log n)$ [5]. However, even the expected behavior may not give enough information: the dispersion of the random variable must be taken into account. A way to solve this problem is to show that the value of the random variable is concentrated around the expected value, so that the probability of getting an instance such that the performance deviates very much from the expected value, is very small. In the case of Quicksort, a recent result states that the probability that the real number of comparisons performed by the algorithm deviates more than ϵ times their expectation is $O(n^{-\epsilon \ln \ln n})$ [30].

Another important technique in computer science and discrete mathematics is the *probabilistic method*. The basic idea behind the probabilistic method is to show that a certain property holds for some combinatorial structure. To do so, one constructs an appropriate probability space, and shows that the property has non-zero probability on it. A lot of research has been devoted to the probabilistic method, see for example [31] for a survey and [1] for a reference book. Although the goal of every application of the probabilistic method is to show that a particular desirable event occurs with positive probability, frequently it is needed to show first that the probability of some other related events are very small. For example, in most of the applications of the *Lovász local lemma*, in order to show that the probability of the union of a set of events is less than one, there is a first step proving that each event has very small probability [1, Chapter 5]. Concentration bounds are a useful tool to prove that the probability of a sequence of events becomes very small.

We have mentioned two areas where showing concentration around the expected value is important. In general, *concentration bounds* are used to prove that the probability that a random variable deviates significatively from its expectation is sufficiently small.

Theorems like the *law of large numbers*, the *central limit theorem* or the *de Moivre–Laplace limit theorem* are some of the most basic theorems in probability theory (see e.g. [10, 14]). However, it turns out that these classical results are not the most adequate tools to study the problems associated with the analysis of algorithms and discrete mathematics. A first reason is that, in probability, classical limit theorems provide asymptotic results that only apply in the infinite limit, but for most of the applications in discrete mathematics and algorithmics we are interested in large but finite cases. A second reason is that we are interested in explicit rates of convergence for the concentration bounds, while the above mentioned theorems do not deal with rates of convergence.

The survey is organized as follows. In Section 2, we present basic notation and results that the reader should know. In Section 3, we present several basic concentration bounds for a random variable. In Section 4, we derive results for the sum of independent random variables. In Section 5, we extend the concentration results to the case of functions of independent random variables. In Section 6, we present the martingale concept and use it to derive bounds for the case of dependent random variables that form a martingale. Finally, in Section 7, we give a quick glance to the topic of isoperimetric inequalities.

Some of the basic results shown here are included as parts of textbooks, such as Chapter 7 and Appendix A of [1] or Chapter 4 of [32]. There are also two encyclopedic and excellent surveys [28, 29]; however the goal of this survey is to give a "basic introduction" to the topic of concentration bounds to the researcher in the field of theoretical computer science, without strong mathematical background.

2.   BASIC TERMINOLOGY AND RESULTS

Let us review some basic definitions and facts from probability theory; for a more complete presentation, refer to any standard reference [10, 14]. For any $n \in \mathbb{N}$, let $[n]$ denote the set $\{1, 2, \dots, n\}$. For any set A, let $\mathcal{P}(A)$ denote the *power set* of A. For any Boolean expression B, we will denote by (B) the indicator function of B, which equals 1 if B and 0 if $\neg B$.

A *sample space* Ω is the set of possible outcomes of an experiment. Any element $\omega \in \Omega$ is a *basic event* and any subset $A \subseteq \Omega$ is an *event*. A *σ-field* $\mathcal{A}$ is a nonempty set of subsets of Ω such that:

1. if $A \in \mathcal{A}$ then $\overline{A} \in \mathcal{A}$,

2. the countable union of elements in $\mathcal{A}$ also is in $\mathcal{A}$, and

3. $\emptyset \in \mathcal{A}$.

A *probability measure* $\mathbf{Pr} : \mathcal{A} \to \mathbb{R}$ is a function that satisfies the following conditions:

1. $0 \leq \mathbf{Pr}\,[E] \leq 1$, for any $E \in \mathcal{A}$

2. $\mathbf{Pr}\,[\Omega] = 1$, and

3. $\mathbf{Pr}\,[\cup_{n \geq 1} E_n] = \sum_{n \geq 1} \mathbf{Pr}\,[E_n]$ for any sequence $(E_n)_{n \geq 1}$ of events in $\mathcal{A}$ such that $E_n \cap E_m = \emptyset$ for $n \neq m$.

A *probability space* is a triple $(\Omega, \mathcal{A}, \mathbf{Pr})$ where Ω is a sample space, $\mathcal{A}$ is a σ-field and $\mathbf{Pr}$ is a probability measure. Observe that $\mathcal{P}(\Omega)$ is a σ-field.

Boole's inequalities are useful to bound the probability that, given a set of events, at least one of them happens:

Proposition 12.1 (Boole's inequalities). *Let $E_1, \ldots, E_n$ be a collection of events; then*

$$\mathbf{Pr}\left[\bigcup_{i=1}^{n} E_i\right] \leq \sum_{i=1}^{n} \mathbf{Pr}\left[E_i\right] \quad and \quad \mathbf{Pr}\left[\bigcap_{i=1}^{n} E_i\right] \geq 1 - \sum_{i=1}^{n} \mathbf{Pr}\left[\overline{E}_i\right].$$

Assume that a probability space $(\Omega, \mathcal{A}, \mathbf{Pr})$ is fixed. A collection of events $A_1, \ldots, A_n$ is *independent* if $\mathbf{Pr}\left[\cap_{i=1}^{n} A_i\right] = \prod_{i=1}^{n} \mathbf{Pr}\left[A_i\right]$. Given two events A and B such that $\mathbf{Pr}\left[B\right] \neq 0$, the *conditional probability* of event A given event B is defined as

$$\mathbf{Pr}\left[A|B\right] = \frac{\mathbf{Pr}\left[A \cap B\right]}{\mathbf{Pr}\left[B\right]}.$$

A *random variable* $X : \Omega \to \mathbb{R}$ is a real-valued function such that, for any $x \in \mathbb{R}$, the set $\{\omega \in \Omega \mid X(\omega) < x\} \in \mathcal{A}$. We shall give basic terminology only for discrete random variables, that is, when the set $X(\Omega)$ is discrete. Recall that the set $\{X = x\} = \{\omega \in \Omega \mid X(\omega) = x\}$ is an event and $\mathbf{Pr}\left[X = x\right] = \sum_{X(\omega)=x} \mathbf{Pr}\left[\omega\right]$. Given a random variable X, the function $f_X(x) = \mathbf{Pr}\left[X = x\right]$ is called the *density function* of X (some authors call it *probability mass function* to differentiate it from the continuous case).

A sequence of random variables $X_1, \ldots, X_n$ is *independent* if, for all vectors $(x_1, \ldots, x_n)$, the collection of events $(\{X_i = x_i\})_{i \in [n]}$ is independent.

Given two random variables X and Y on $(\Omega, \mathcal{A}, \mathbf{Pr})$ their *joint distribution* is defined by $f_{X,Y}(x, y) = \mathbf{Pr}\left[\{X = x\} \cap \{Y = y\}\right]$; and the *conditional distribution function* of Y given X is defined by $f_{Y|X}(x|y) = \mathbf{Pr}\left[Y = y|X = x\right]$ for any $x \in \mathbb{R}$ such that $\mathbf{Pr}\left[X = x\right] > 0$. From the definitions, we have

$$f_{Y|X}(y|x) = \frac{\mathbf{Pr}\left[\{X = x\} \cap \{Y = y\}\right]}{\mathbf{Pr}\left[X = x\right]} = \frac{f_{X,Y}(x, y)}{f_X(x)}. \tag{12.1}$$

Given a random variable X, its *expectation*, *expected* or *mean* value is defined as $\mathbf{E}\left[X\right] = \sum_{x \in X(\Omega)} x \, \mathbf{Pr}\left[X = x\right]$, its *variance* is defined as $\mathbf{Var}\left[X\right] = \mathbf{E}\left[(X - \mathbf{E}\left[X\right])^2\right]$ and its *standard deviation* is $\sigma_X = \sqrt{\mathbf{Var}\left[X\right]}$. If k is a positive integer, the k-th *moment* of X is $\mathbf{E}\left[X^k\right]$. The *median* of X is defined to be the set of all $x \in X(\Omega)$ such that $\mathbf{Pr}\left[X \leq x\right] \geq 1/2$ and $\mathbf{Pr}\left[X \geq x\right] \geq 1/2$. In our examples, the set of median values will have an unique element that will be refereed to by $M[X]$.

The following proposition, whose proof is left to the reader, presents some basic properties of expectations and variances.

Proposition 12.2. *For any random variables X and Y and any $c \in \mathbb{R}$,*

$$\mathbf{E}[X + Y] = \mathbf{E}[X] + \mathbf{E}[Y],$$
$$\mathbf{E}[cX] = c\,\mathbf{E}[X],$$
$$\mathbf{Var}[X] = \mathbf{E}[X^2] - \mathbf{E}[X]^2,$$
$$\mathbf{Var}[cX] = c^2\,\mathbf{Var}[X].$$

Moreover, if X and Y are independent,

$$\mathbf{E}[XY] = \mathbf{E}[X]\,\mathbf{E}[Y],$$
$$\mathbf{Var}[X + Y] = \mathbf{Var}[X] + \mathbf{Var}[Y].$$

The *conditional expectation* of a random variable Y given the random variable X is defined for every $x \in X(\Omega)$ such that $\mathbf{Pr}[X = x] > 0$ as:

$$\mathbf{E}[Y|X = x] = \sum_{y \in Y(\Omega)} y f_{Y|X}(y|x) = \sum_{y \in Y(\Omega)} y\,\mathbf{Pr}[Y = y|X = x].$$

Therefore, $\mathbf{E}[Y|X]$ as function of X is a random variable.

Some useful properties of the conditional expectation are given in the following proposition.

Proposition 12.3. *For any random variables X and Y, and any function f for which the corresponding expectations exist,*

$$\mathbf{E}[\mathbf{E}[Y|X]] = \mathbf{E}[Y],$$
$$\mathbf{E}[X\,Y] = \mathbf{E}[X\,\mathbf{E}[Y|X]],$$
$$\mathbf{E}[f(X,Y)|X = x] = \mathbf{E}[f(x,Y)|X = x]$$
$$\mathbf{E}[Y|X = x] = \mathbf{E}[Y], \qquad \text{if } X \text{ and } Y \text{ are independent.}$$

Proof. To prove the first equality, consider any basic event $\omega \in \Omega$ such that $X(\omega) = x$ and $Y(\omega) = y$. Then, $\mathbf{E}[Y|X = x] = \sum_{y \in Y(\Omega)} y f_{Y|X}(y|x)$, so that

$$
\begin{aligned}
\mathbf{E}\left[\mathbf{E}\left[Y|X\right]\right] &= \sum_{x \in X(\Omega)} \mathbf{E}\left[Y|X = x\right] f_X(x) \\
&= \sum_{x \in X(\Omega)} \sum_{y \in Y(\Omega)} y\, f_{Y|X}(y|x) f_X(x) \\
&= \sum_{x \in X(\Omega)} \sum_{y \in Y(\Omega)} y\, f_{X,Y}(x,y) \\
&= \sum_{y \in Y(\Omega)} y\, f_Y(y) = \mathbf{E}\left[Y\right].
\end{aligned}
$$

The proofs of the remaining inequalities are left as exercises.　　❧

Later, we shall need the notion of conditional expectation of a multi-variate random variable Y defined on a product space Ω^n given the values of the variables in a random vector $\mathbf{X} = (X_1, \ldots, X_n)$. In such a case, the definition of conditional expectation is extended, in the obvious way, to $\mathbf{E}\left[Y|X_1 = x_1, \ldots, X_i = x_i\right]$. Observe that $\mathbf{E}\left[Y|X_1 = x_1, \ldots, X_i = x_i\right]$ is a function on $\mathbb{R}^{n-i}$. To simplify notation, given a n-dimensional vector $\mathbf{X} = (X_1, \ldots, X_n)$, $\mathbf{X}_i$ will denote the vector formed by the first i components. The equality $\mathbf{X}_i = \mathbf{x}_i$ represents $X_1 = x_1, \ldots, X_i = x_i$, and the inequality $\mathbf{X}_i > \mathbf{x}_i$ represents $X_1 > x_1, \ldots, X_i > x_i$. The following proposition extends Proposition 12.3.

Proposition 12.4. *For any random variable Y defined on Ω^n, any random vector $\mathbf{X} = (X_1, \ldots, X_n)$ and any function $f : \mathbb{R}^n \to \mathbb{R}$ for which the corresponding expectations exist,*

$$
\begin{aligned}
\mathbf{E}\left[\mathbf{E}\left[Y|\mathbf{X}\right]\right] &= \mathbf{E}\left[Y\right], \\
\mathbf{E}\left[\mathbf{E}\left[Y|\mathbf{X}_i\right]|\mathbf{X}_j\right] &= \mathbf{E}\left[X|X_1, \ldots, X_j\right], \qquad 1 \le j < i \le n, \\
\mathbf{E}\left[Yf(\mathbf{X})|\mathbf{X} = \mathbf{x}\right] &= f(\mathbf{x})\mathbf{E}\left[Y|\mathbf{X} = \mathbf{x}\right]
\end{aligned}
$$

Let us now recall the definition and properties of some probability distributions. A random variable X has the *Bernoulli distribution* with parameter $p \in [0,1]$ if $\mathbf{Pr}\left[X = 1\right] = p$ and $\mathbf{Pr}\left[X = 0\right] = 1 - p = q$. This distribution has expectation p and variance pq. When $X_1, \ldots, X_n$ is a sequence of independent Bernoulli random variables, the random variable $X = \sum_{i=1}^{n} X_i$ has the *binomial distribution* with parameters n and p and it is denoted as $\mathrm{BIN}(n,p)$. The binomial distribution has expectation np and variance npq.

A continuous random variable with the *normal distribution* with mean μ and standard deviation σ has density function

$$
f(x) = \frac{1}{\sigma\sqrt{2\pi}} e^{-\frac{1}{2}(x-\mu)^2/\sigma^2}
$$

and it is denoted as $N(\mu, \sigma^2)$. As a particular case, $\Phi(x)$ will denote the distribution function of $N(0,1)$, that is,

$$\Phi(x) = \mathbf{Pr}\left[N(0,1) \le x\right] = \int_{-\infty}^{x} \frac{1}{\sqrt{2\pi}} \exp(-\tfrac{1}{2}u^2)\mathrm{d}u.$$

A sequence of events $(E_i)_{i \ge 1}$ occurs *with high probability* (whp) if $\mathbf{Pr}\left[E_n\right] \to 1$ as $n \to \infty$. Some authors use the term whp when the condition $\mathbf{Pr}\left[E_n\right] \ge 1 - n^{-c}$ holds for some $c > 1$, and the term *with overwhelming probability* when $\mathbf{Pr}\left[E_n\right] \ge 1 - 2^{-cn}$ for some $c > 0$ [18].

Let us finish this review with three classical limit theorems of probability theory (see for example [10, 14]). Recall that, given a sequence of random variables $X_1, X_2, \ldots$ and a random variable X, X_n *converges in distribution* to X if, for all x, $\mathbf{Pr}\left[X_n < x\right] \to \mathbf{Pr}\left[X < x\right]$ as $n \to \infty$.

Theorem 12.1 (Law of large numbers). *Let $X_1, X_2, \ldots$ be a sequence of independent and identically distributed random variables with finite expectation μ. For each $n \ge 1$, set $S_n = \sum_{i=1}^{n} X_i$. Then, S_n/n converges in distribution to μ.*

Theorem 12.2 (Central limit theorem). *Let $X_1, X_2, \ldots$ be a sequence of independent and identically distributed random variables with finite expectation μ and finite variance σ^2. For each $n \ge 1$, set $S_n = \sum_{i=1}^{n} X_i$. Then, $(S_n - n\mu)/\sigma\sqrt{n}$ converges in distribution to $N(0,1)$.*

Theorem 12.3 (de Moivre–Laplace limit theorem). *Let $X_1, X_2, \ldots$ be a sequence of independent and identically distributed Bernoulli random variables with parameter p. For each $n \ge 1$, set $S_n = \sum_{i=1}^{n} X_i$. Then, for any fixed constants $c_1 < c_2$, as $n \to \infty$,*

$$\mathbf{Pr}\left[np + c_1\sqrt{npq} \le S_n \le np + c_2\sqrt{npq}\right] \to \Phi(c_2) - \Phi(c_1),$$

where $q = 1 - p$ and Φ denotes $N(0,1)$.

3.   BASIC TAIL BOUNDS FOR A RANDOM VARIABLE

Let X be a random variable and let x be a real value, then $\mathbf{Pr}\left[X > x\right]$ and $\mathbf{Pr}\left[X < x\right]$ are called the upper and lower tails, respectively. Figure 12.1 illustrates this definition. Concentration bounds give upper bounds for the tails of the random variable $X - \mathbf{E}\left[X\right]$. We start by presenting the most basic tail bounds.

Theorem 12.4. *Let $t > 0$ and let X be a random variable, defined on a probability space Ω. Then, for any integer $k > 0$,*

$$\mathbf{Pr}[|X| > t] \le \frac{\mathbf{E}\left[|X|^k\right]}{t^k}.$$

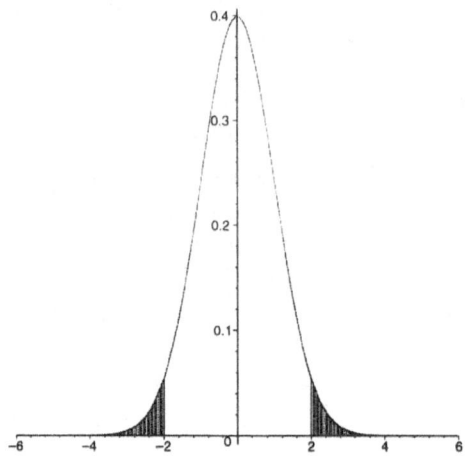

Figure 12.1 Tails of a random variable.

Proof.

$$
\mathbf{E}\left[|X|^{k}\right] = \sum_{x \in X(\Omega)} |x|^{k} \mathbf{Pr}[X = x]
$$

$$
= \sum_{|x| \leq t} |x|^{k} \mathbf{Pr}[X = x] + \sum_{|x| > t} |x|^{k} \mathbf{Pr}[X = x]
$$

$$
\geq \sum_{|x| > t} |x|^{k} \mathbf{Pr}[X = x]
$$

$$
\geq t^{k} \sum_{|x| > t} \mathbf{Pr}[X = x]
$$

$$
= t^{k} \mathbf{Pr}[|X| > t].
$$

The first classical bound on the deviation from the expected value follows from the previous theorem. It states a well known bound for the upper tail and is usually referred as the *first moment method* or *Markov inequality*.

Theorem 12.5 (Markov inequality). *Let* X *be a positive random variable. Then, for any* $t > 0$,

$$
\mathbf{Pr}[X > t\mathbf{E}[X]] \leq 1/t.
$$

Proof. Take $k = 1$ and substitute t by $t\mathbf{E}[X]$ in Theorem 12.4.

Markov inequality does not provide information on the other tail. Notice that this inequality cannot be used with the random variable $-X$; even if we would consider the random variable $Y = -X + \max_{\omega \in \Omega} X(\omega)$ (which is a positive random variable), Markov inequality gives no information on the lower tail. On the other hand, Markov inequality becomes vacuous when $t \leq 1$. Markov inequality is mainly an intermediate tool to get better bounds.

The following example is taken from [6].

Example 12.1 (Points into boxes). Let us consider a collection of n points in the unit square $([0,1]^2)$. Divide the unit square in $n/\log^2 n$ squares boxes, each of side $\log n/\sqrt{n}$. For the sake of simplicity, assume $\sqrt{n}/\log n$ is an integer. Given $\epsilon \in (0,1)$, we say that a the collection of points is ϵ-*nice* if each box contains at least $(1-\epsilon)\log^2 n$ points and at most $(1+\epsilon)\log^2 n$ points. We would like to prove that, with high probability, a collection of n points independently and uniformly distributed in the unit square is ϵ-nice.

Let X be a random variable counting the number of points of a fixed box. Then X follows a binomial distribution $\mathrm{BIN}(n,p)$ with $p = \log^2 n/n$. Therefore, $\mu = \mathbf{E}[X] = np = \log^2 n$. According to Markov inequality,

$$\mathbf{Pr}[X > (1+\epsilon)\mu] \leq \frac{1}{1+\epsilon},$$

which is independent of n. With that bound we cannot even say that a single fixed box contains less than $(1+\epsilon)\mu$ points with high probability!

Let us derive another classical bound, attributed to Chebyshev.

Theorem 12.6 (Chebyshov inequality). *Let X be a random variable with expected value μ and standard deviation σ. Then, for all $t > 0$,*

$$\mathbf{Pr}[|X - \mu| > t\sigma] \leq 1/t^2.$$

Proof. In Theorem 12.4, substitute t by $t\sigma$, X by $X - \mathbf{E}[X]$, k by 2 and use the definition of variance.

The use of Chebyshov inequality indicates that with probability greater than $1 - 1/t^2$, a random variable X will fall within t times the standard deviation around μ. For example, 75% of the times a random value falls in the interval $[\mu - 2\sigma, \mu + 2\sigma]$.

466

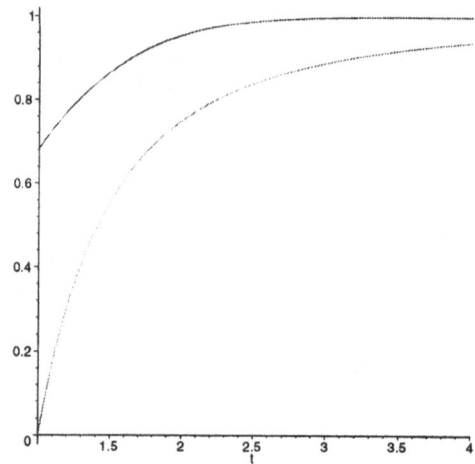

Figure 12.2 Comparing Chevyshov (thin line) and de Moivre–Laplace (thick line) bounds for Example 12.2

⏎ **Example 12.2 (Coin tosses).** Let us study the concentration of the random variable that counts the number of heads after tossing n times a fair coin. Let $X_1, \ldots, X_n$ be the random variables produced by the outcome of each toss, where $X_i =$ (the i-th toss is a head). We wish to study the concentration of $S_n = \sum_{i=1}^{n} X_i$. Observe that $\mathbf{E}[S_n] = \frac{1}{2}n$ and $\mathbf{Var}[S_n] = \frac{1}{4}n$.

If we apply Chebyshov inequality, we get

$$\mathbf{Pr}[|S_n - \mathbf{E}[S_n]| > t\sigma] < 1/t^2, \qquad \forall t > 0.$$

If we apply the classical de Moivre–Laplace Theorem, then

$$\mathbf{Pr}[|S_n - \mu| \leq t\sigma] \to \Phi(t) - \Phi(-t), \qquad \forall t > 0.$$

Comparing the two bounds (see Figure 12.2), we can observe that, asymptotically, we get a better concentration bound for S_n around its expectation using the de Moivre–Laplace theorem than with Chebyshov inequality. But while the application of Chebyshov inequality is valid for all n, the de Moivre–Laplace Theorem only applies in the limit. ⏎

⏎ **Example 12.1 (cont).** Recall that X counts the number of points in a fixed box. As X is $\text{BIN}(n,p)$, $\mu = \mathbf{E}[X] = \log^2 n$ and $\sigma^2 = \sigma_X^2 = npq = \log^2 n(1 - \log^2 n/n)$. Using Chebyshov inequality, we have

$$\mathbf{Pr}[|X - \mu| > \epsilon\mu] \leq \frac{\sigma^2}{\epsilon^2\mu^2} = \frac{n - \log^2 n}{\epsilon^2 n \log^2 n},$$

which tends to 0 as $n \to \infty$. Therefore, we have that, with high probability, a fixed box has at least $(1-\epsilon)\log^2 n$ points and at most $(1+\epsilon)\log^2 n$ points.

Unfortunately, we cannot yet affirm that this happens for all boxes: If we use Boole's inequality, we have that the probability that some box has less than $(1-\epsilon)\log^2 n$ points or more than $(1+\epsilon)\log^2 n$ points is less than

$$\frac{n}{\log^2 n} \cdot \frac{n - \log^2 n}{\epsilon^2 n \log^2 n} = \frac{n - \log^2 n}{\epsilon^2 \log^4 n}$$

which diverges when n goes to infinity.   ✏

Observe that Chebyshov inequality justifies the fact that if the variance of a random variable goes to zero then, its real value is close to its expectation with high probability. Unfortunately, this happens rarely, and usually, variances are difficult to compute.

4.   SUMS OF BOUNDED INDEPENDENT RANDOM VARIABLES

Our purpose in this section is to obtain concentration inequalities for the sum of bounded independent random variables. Assuming that $X_1, \ldots, X_n$ are independent identical distributed Bernoulli random variables, with probability of success p, an application of Chebyshov inequality gives

$$\mathbf{Pr}\left[\sqrt{\frac{n}{p(1-p)}} \left| \frac{1}{n} \sum_{i=1}^{n} X_i - p \right| > t \right] \le \frac{1}{t^2}.$$

By the central limit theorem, we know that $\sum_{i=1}^{n} X_i$ approaches the normal distribution; so one may expect that

$$\mathbf{Pr}\left[\sqrt{\frac{n}{p(1-p)}} \left(\frac{1}{n} \sum_{i=1}^{n} X_i - p \right) > t \right] \longrightarrow 1 - \Phi(t) \le \frac{1}{\sqrt{2\pi}} \frac{e^{-t^2/2}}{t},$$

showing that the bound given by using Chebyshov inequality is far off from what one should see asymptotically, according to the central limit theorem.

The next method is an extension of Markov inequality. In this case, the bounds of the two distribution tails will be a negative exponential function to the number of trials. The bound for independent Bernoulli random variables is due to Chernoff [4], although it seems that the idea and the proof technique was already present in the work of Bernstein [28,

Page 161]. Chernoff's bounds have become standard material in most books and surveys on probabilistic techniques for combinatorics and computer science (see [1, 32, 29, 31, 12] among others). In particularly, a good survey on the relations between the most used versions of Chernoff's bounds is presented in [15].

Lemma 12.1. *Let $X_1, \ldots, X_n$ be a sequence of independent random variables corresponding to Bernoulli experiments, each variable with probability p_i of success and probability $q_i = 1 - p_i$ of failure. Let X be the random variable defined by $X = \sum_{i=1}^{n} X_i$. Then, for any $c > 0$ and any $t > 0$,*

$$\mathbf{Pr}[X > t] \le e^{-ct} \prod_{i=1}^{n} (p_i e^c + q_i) \tag{12.2}$$

and

$$\mathbf{Pr}[X < t] \le e^{ct} \prod_{i=1}^{n} (p_i e^{-c} + q_i). \tag{12.3}$$

Proof. The key of the proof is to consider the random variable e^{cX} instead of X. Notice that for $c > 0$, e^{cX} is a positive monotone increasing function of X and therefore $\mathbf{Pr}[X > t] = \mathbf{Pr}[e^{cX} > e^{ct}]$. Using Markov inequality,

$$\mathbf{Pr}[X > t] = \mathbf{Pr}[e^{cX} > e^{ct}] \le \frac{\mathbf{E}\left[e^{cX}\right]}{e^{ct}}$$

but

$$\mathbf{E}\left[e^{cX}\right] = \mathbf{E}\left[\exp\left(c \sum_{i=1}^{n} X_i\right)\right] = \mathbf{E}\left[\prod_{i=1}^{n} e^{cX_i}\right].$$

As $X_1, \ldots, X_n$ are independent, then $e^{cX_1}, \ldots, e^{cX_n}$ are also independent, and so we have $\mathbf{E}\left[\prod_{i=1}^{n} e^{cX_i}\right] = \prod_{i=1}^{n} \mathbf{E}\left[e^{cX_i}\right]$. This implies that

$$\mathbf{Pr}[X > t] \le e^{-ct} \prod_{i=1}^{n} \mathbf{E}\left[e^{cX_i}\right].$$

For each $i \in [n]$, with probability p_i we have $e^{cX_i} = e^c$ and with probability q_i we have $e^{cX_i} = 1$. The statement of the lemma follows from the definition of the expected value.

To get the bound on the lower tail, we take into account the equality $\mathbf{Pr}[X < t] = \mathbf{Pr}[-X > -t]$, so we can work as before starting from $\mathbf{Pr}\left[e^{-cX} > e^{-ct}\right]$ to get the bound for the lower tail. ✍

The following result gives some other bounds for the concentration around the expected value of a sum of Bernoulli variables.

Theorem 12.7 (General Chernoff's bounds). *Let $X_1, \ldots, X_n$ be independent random variables corresponding to Bernoulli experiments, each variable with probability p_i of success and probability $q_i = 1 - p_i$ of failure. Let $X = \sum_{i=1}^{n} X_i$ be a random variable with expected value $\mu = \sum_{i=1}^{n} p_i$. Then, for any $d > 0$,*

$$\Pr[X > (1+d)\mu] \leq \left(\frac{e^d}{(1+d)^{1+d}} \right)^{\mu} . \tag{12.4}$$

Moreover, in the case $0 < d < 1$,

$$\Pr[X < (1-d)\mu] \leq \left(\frac{e^{-d}}{(1-d)^{1-d}} \right)^{\mu} . \tag{12.5}$$

Proof. Let us start with the upper tail. Observe that for every $x > 0$ we have that $\ln(1+x) < x$ and that $1 + x \leq e^x$.

Substitute $t = (1+d)\mu$ and $c = \ln(1+d)$ in Equation (12.2) in Lemma 12.1. Then, we have

$$\Pr[X > (1+d)\mu] \leq e^{-\ln(1+d)(1+d)\mu} \prod_{i=1}^{n} \mathbf{E}\left[e^{\ln(1+d)X_i} \right]$$

$$= (1+d)^{-(1+d)\mu} \prod_{i=1}^{n} \mathbf{E}\left[(1+d)^{X_i} \right] .$$

As $\mathbf{E}\left[(1+d)^{X_i} \right] = p_i(1+d) + (1-p_i) = 1 + d\, p_i \leq e^{d\, p_i}$ we get

$$\Pr[X > (1+d)\mu] \leq (1+d)^{-(1+d)\mu} \prod_{i=1}^{n} e^{dp_i}$$

$$\leq \left(\frac{e^d}{(1+d)^{1+d}} \right)^{\mu} .$$

For $d < 1$ the value $t = (1-d)\mu$ is positive, so, taking $c = -\ln(1-d)$ and using the same argument, we get the bound on the lower tail. ∎

The previous theorem is the basis to develop customized Chernoff's bounds, which usually are weaker but easier to apply.

Corollary 12.1. *Let $X_1, \ldots, X_n$ be independent random variables corresponding to Bernoulli experiments, each variable with probability p_i of* *success and probability* $q_i = 1 - p_i$ *of failure. Let* $X = \sum_{i=1}^n X_i$ *be a random variable with expectation* $\mu = \sum_{i=1}^n p_i$*. Then, for any* $d \in (0,1)$*,*

$$\mathbf{Pr}[X < (1-d)\mu] \le e^{-d^2\mu/2} \quad \textit{and} \quad \mathbf{Pr}[X > (1+d)\mu] \le e^{-d^2\mu/3}.$$

Therefore,

$$\mathbf{Pr}[|X - \mu| > d\mu] \le 2e^{-d^2\mu/3} \tag{12.6}$$

Proof. Using Taylor's expansion around 0 we have

$$(1-d)\ln(1-d) = -d + \tfrac{1}{2}d^2 + O(d^3),$$

which implies $(1-d)^{1-d} > e^{d^2/2-d}$. Substituting in the denominator of (12.5) we get

$$\mathbf{Pr}[X < (1-d)\mu] \le \left(\frac{e^{-d}}{e^{d^2/2-d}}\right)^{\mu} \le e^{-\frac{d^2\mu}{2}}.$$

We claim that for $0 \le d \le 1$ we have $d - (1+d)\ln(1+d) + d^2/3 \le 0$. It is true for $d = 0$ and $d = 1$. To prove it for $d \in (0,1)$, let $f(d) = d - (1+d)\ln(1+d) + d^2/3$. Then, $f'(d) = \frac{2}{3}d - \ln(1+d)$. But as f is continuous, $f(0) = 0$, $f(1) = -0.529$ and the roots of f' are $d = 0$ and $d = 1.144$, f must be decreasing in the real interval $(0,1)$, which together with the fact that $f(0) = 0$ implies $f(d) \le 0$ for $0 \le d \le 1$.

As a consequence, we have $(1+d)\ln(1+d) \ge d + d^2/3$ and therefore, taking exponentials, $(1+d)^{(1+d)} \ge e^{d+d^2/3}$. Substituting in equation (12.4) in Theorem 12.7,

$$\mathbf{Pr}[X > (1+d)\mu] \le \left(\frac{e^{d}}{e^{d+\frac{d^2}{3}}}\right)^{\mu} \le e^{-\frac{d^2\mu}{3}}.$$

Equation (12.6) is obtained using Boole's inequality. ✌

To finish Example 12.1, we will use Corollary 12.1.

✏ **Example 12.1 (cont).** Recall that X counts the number of points in a fixed box and that $\mu = \mathbf{E}[X] = \log^2 n$. Using Chernoff's bounds, we have

$$\mathbf{Pr}\left[|X - \mu| > \epsilon\mu\right] \le 2\exp(-\epsilon^2\mu/3) = 2n^{-\epsilon^2 \log n/3}$$

which tends to 0 as $n \to \infty$. Therefore, we have that, with high probability, a fixed box has at least $(1-\epsilon)\log^2 n$ points and at most $(1+\epsilon)\log^2 n$ points.

Using Boole's inequality, we have now that the probability that some box has less than $(1 - \epsilon) \log^2 n$ points or more than $(1 + \epsilon) \log^2 n$ points is less than

$$\frac{n}{\log^2 n} 2n^{-\epsilon^2 \log n/3} \leq 2n^{1-\epsilon^2 \log n/3}$$

which tends to 0. As a consequence, for any fixed ϵ, with high probability, a random configuration of points in the unit square is ϵ-nice.

We leave to the reader the generalization of this result when the assumption on the divisibility of n by $\log^2 n$ is removed. This can be done by slightly modifying the sides of the boxes using the floor or the ceil function to completely fit the unit square.   ⊜

The next example is taken from [40].

⊜ **Example 12.3 (Mating lemma).**   Let L be a sequence of n individuals, where each one can be male (M) or female (F) with probability $\frac{1}{2}$ and independently of the sex of the other members. Let us construct the subsequence S of L in the following way: an individual x of L does not belong to S when it is male and its immediate predecessor is female. We wish to prove that, with overwhelming probability, $|S| < \frac{15}{16}n$.

Let $x \in L$ be an element in an even position and let y be its predecessor in L. Then,

$$\mathbf{Pr}[x \notin S] = \mathbf{Pr}[x = M \wedge y = F] = \mathbf{Pr}[x = M]\mathbf{Pr}[y = F] = \tfrac{1}{4}.$$

Therefore, there are $\frac{1}{2}n$ candidates to be excluded from S with probability $\frac{1}{4}$. Let X be the random variable that counts the number of excluded elements. Then, $\mu = \mathbf{E}[X] \geq \frac{1}{8}n$, and using Corollary 12.1 with $d = \frac{1}{2}$, we get

$$\mathbf{Pr}\left[X > \tfrac{1}{16}n\right] = \mathbf{Pr}\left[X > (1 - \tfrac{1}{2})\tfrac{1}{8}n\right] \leq \mathbf{Pr}\left[X > (1 - d)\mu\right]$$
$$\leq e^{-d^2\mu/2} \leq e^{-n/64}.$$

This means

$$\mathbf{Pr}[|S| > \tfrac{15}{16}n] > 1 - e^{-n/64},$$

which implies that, with overwhelming probability, $|S| < \frac{15}{16}n$ (see Page 463 for the definition of wop).   ⊜

The following result gives further variations on basic Chernoff's bounds when variables are restricted to be identically distributed, that is, to the binomial distribution. Equations (12.8) and (12.9) are due to Hoeffding [17].

Corollary 12.2. *Let X be a random variable following a binomial distribution $BIN(n,p)$ with expectation $\mu = np$. Then,*

$$\mathbf{Pr}[|X - \mu| > t] \ \ \leq 2e^{-\frac{t^2}{3\mu}}, \qquad\qquad\qquad 0 < t \leq \mu, \quad (12.7)$$

$$\mathbf{Pr}[|X - \mu| > tn] \leq 2e^{-2nt^2}, \qquad\qquad\qquad t \geq 0, \quad (12.8)$$

$$\mathbf{Pr}[X > an] \ \ \leq \left(\left(\frac{p}{a}\right)^a \left(\frac{1-p}{1-a}\right)^{1-a}\right)^n, \qquad p \leq a < 1. \quad (12.9)$$

Proof. Equation (12.7) is obtained by taking $t = d\mu$ in (12.6). To prove (12.8), let $q = 1 - p$ and $\delta = n(p+t)$. Observe that from (12.2) we get

$$\mathbf{Pr}[X > \delta] \leq e^{-c\delta} \prod_{i=1}^{n}(pe^c + q) = e^{-c\delta}(pe^c + q)^n.$$

As we wish to make the bound as small as possible, we want to make

$$\frac{d}{dc}\left(e^{c\delta}(pe^c + q)^n\right) = 0$$

but as $q = 1 - p$ we obtain

$$\delta e^{-c\delta}(pe^c + q)^n + ne^{-c\delta}pe^c(pe^c + q)^{n-1} = 0 \implies (\delta p + np)e^c + q\delta = 0$$

$$\implies e^c = \frac{-\delta q}{\delta p + np}.$$

Substituting δ by $n(p+t)$, we get

$$e^c = \frac{n(p+t)(p-1)}{n(p+t)p - np} = \frac{(p+t)(p-1)}{p(p+t-1)},$$

which holds for all values $0 < t < q$. Therefore,

$$\mathbf{Pr}[|X - \mu| > tn] \leq e^{-n(p+t)\ln\frac{(p+t)(1-p)}{p(p+t-1)}}\left(1 - p + pe^{\frac{(p+t)(1-p)}{p(p+t-1)}}\right)^n$$

$$= e^{-n(p+t)\ln\frac{(p+t)(1-p)}{p(p+t-1)}}\left(1 - p + \frac{(p+t)(1-p)}{p+t-1)}\right)^n$$

$$= \left(\frac{(p+t)(1-p)}{p(1-p-t)}\right)^{-n(p+t)}$$

$$= \exp\left(-(p+t)\ln\frac{p+t}{p} - (1-p+t)\ln\frac{1-p+t}{1-p}\right).$$

Set

$$f(t) = -(p+t)\ln\frac{p+t}{p} - (1-p+t)\ln\frac{1-p+t}{1-p}.$$

Deriving we get,

$$f'(t) = \ln\frac{p+t}{p} - \ln\frac{1-p-t}{1-p} \quad \text{and} \quad f''(t) = \frac{1}{(p+t)(1-p-t)}.$$

Observe that $f(0) = f'(0) = 0$ and that $xy \leq \frac{1}{4}$ for any non-negative reals summing up to 1, using Taylor's expansion we get $f(t) \geq 2t^2$, from which Equation (12.8) follows.

To prove (12.9), let us consider a sequence $X_1, \ldots, X_n$ of Bernoulli variables with success probability p. Then, $X = \sum_{i=1}^{n} X_i$. For any given $a \in [p, 1)$, set $d = a/p - 1$ so that $an = (1 + d)\mu$. It follows from Lemma 12.1 that for any $c > 0$,

$$\Pr[X > an] \leq e^{-acn} \prod_{i=1}^{n} \mathbf{E}\left[e^{cX_i}\right].$$

As $\mathbf{E}\left[e^{cX_i}\right] = (pe^c + (1-p))$,

$$\Pr[X > an] \leq e^{-acn}(pe^c + (1-p))^n.$$

Let $c = \ln\left(\frac{a(1-p)}{p(1-a)}\right)$; taking logarithms in the previous equation, we get

$$\Pr[X > an] \leq \left(\frac{p(1-a)}{a(1-p)}\right)^{an} \left(\frac{a(1-p)}{1-a} + (1-p)\right)^n,$$

and (12.9) is obtained by simplifying the previous equation. 🌶

The next example is taken from [22].

☞ **Example 12.4 (MaxSat).** The well known **NP**-hard problem named MAXSAT is defined as follows: Given a set of n literals and a DNF Boolean formula ϕ with m clauses, we wish to find an assignment that satisfies the maximal number of clauses. Let T^* be an optimal assignment satisfying the maximal number of clauses. For every literal x_i in ϕ, define a random assignment $T(x_i)$ where

$$\Pr[T(x_i) = 0] = \Pr[T(x_i) = 1] = \frac{1}{2}.$$

We wish to find the value ϵ such that we can state that, with high probability, T coincides with T^* in at least $(1 - \epsilon)\frac{1}{2}n$ assignments.

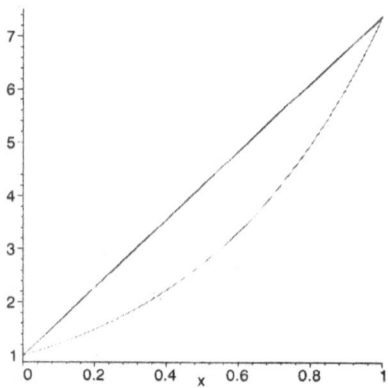

Figure 12.3 A convex function

Let Z be a random variable counting the number of literals x_i such that $T^*(x_i) = T(x_i)$. Define $Z_i = (T^*(x_i) = T(x_i))$. Then, $Z = \sum_{i=1}^{n} Z_i$ and $\mathbf{E}[Z] = \sum_{i=1}^{n} \mathbf{E}[Z_i] = \frac{1}{2}n$. By Corollary 12.1,

$$\mathbf{Pr}\left[Z < (1-\epsilon)\tfrac{1}{2}n\right] \le e^{-\epsilon^2 n/4}.$$

Thus, in order to guarantee that the previous probability goes to zero, we need $\epsilon > \sqrt{b(n)/n}$ where $b(n) \to \infty$ and $b(n)/n \to 0$ as $n \to \infty$.

The inequalities of Corollary 12.2 were extended by Hoeffding to the sum of bounded independent random variables. Let us introduce first a bound on $\mathbf{E}\left[e^{tX}\right]$.

Lemma 12.2. *Let X be a random variable such that $a \le X \le b$ with $\mathbf{E}[X] = 0$. Then, for any $t > 0$,*

$$\mathbf{E}\left[e^{tX}\right] \le exp\left(\frac{(b-a)^2}{8}t^2\right).$$

Proof. By the convexity of the exponential function (see Figure 12.3),

$$e^{tX} \le \frac{x-a}{b-a}e^{tb} + \frac{x-b}{a-b}e^{ta}$$

for $a \le x \le b$. Using linearity of expectation and $\mathbf{E}[X] = 0$, we have

$$\mathbf{E}\left[e^{tX}\right] \le \frac{-a}{b-a}e^{tb} + \frac{b}{b-a}e^{ta}.$$

Taking $p = -a/(b-a)$ and $1 - p = b/(b-a)$, we obtain

$$\mathbf{E}\left[e^{tX}\right] \leq pe^{tb} + (1-p)e^{ta}$$

$$\leq \left(pe^{t(b-a)} + (1-p)\right)e^{ta}$$

$$\leq \left(pe^{t(b-a)} + (1-p)\right)e^{-pt(b-a)}$$

$$\leq e^{-pt(b-a)+\log\left(1-p+pe^{t(b-a)}\right)}.$$

To upper bound the exponent, define $u = t(b-a)$, so that the exponent is the function

$$g(u) = -pu + \log\left(1 - p + pe^{u}\right).$$

Therefore,

$$g'(u) = -p + \frac{p}{p + (1-p)e^{-u}} \quad \text{and} \quad g''(u) = \frac{p(1-p)e^{-u}}{(p + (1-p)e^{-u})^2} \leq \tfrac{1}{4}.$$

Using Taylor's expansions, for some $\theta \in [0, u]$, we have

$$g(u) = g(0) + ug'(0) + \tfrac{1}{2}u^2 g''(\theta) \leq \tfrac{1}{8}u^2 = \tfrac{1}{8}t^2(b-a)^2,$$

which proves the claimed result. ✄

In the case that not all the variables have the same ranges, the following result due to Hoeffding [17] might be useful.

Theorem 12.8 (Hoeffding's inequalities). *Let $X_1, \ldots, X_n$ be independent random variables such that, for all $i \in [n]$, there are values a_i and b_i such that $a_i \leq X_i \leq b_i$. Let $X = \sum_{i=1}^{n} X_i$ and $\mu = \mathbf{E}[X]$. Then, for any $t \geq 0$,*

$$\mathbf{Pr}\left[X - \mu > t\right] \leq exp\left(-\frac{2t^2}{\sum_{i=1}^{n}(b_i - a_i)^2}\right) \quad and$$

$$\mathbf{Pr}\left[X - \mu < -t\right] \leq exp\left(-\frac{2t^2}{\sum_{i=1}^{n}(b_i - a_i)^2}\right).$$

Therefore,

$$\mathbf{Pr}\left[|X - \mu| > t\right] \leq 2exp\left(-\frac{2t^2}{\sum_{i=1}^{n}(b_i - a_i)^2}\right)$$

Proof. By Markov inequality, and Lemma 12.2, we have that for any $\lambda > 0$,

$$\mathbf{Pr}[X - \mu > t] \leq e^{-\lambda t}\mathbf{E}\left[e^{\lambda(X-\mu)}\right]$$

$$\leq e^{-\lambda t}\prod_{i=1}^{n}\mathbf{E}\left[e^{\lambda(X_i - \mathbf{E}[X_i])}\right].$$

As the random variables are independent, we get

$$\mathbf{Pr}[X - \mu > t] \le e^{-\lambda t} \prod_{i=1}^{n} e^{\lambda^2 (b_i - a_i)^2 / 8}$$

$$= e^{-\lambda t} e^{\lambda^2 \sum_{i=1}^{n} (b_i - a_i)^2 / 8}$$

taking $\lambda = 4t / \sum_{i=1}^{n} (b_i - a_i)$, we get

$$\mathbf{Pr}[X - \mu > t] \le e^{-2t^2 / \sum_{i=1}^{n} (b_i - a_i)^2}.$$

The same proof applies for the other tail, and the last inequality is obtained using Boole's inequality. &

➲ **Example 12.5 (Sum of random numbers).** Let $X_1, \ldots, X_n$ be independent random variables, all of them uniformly distributed in the interval $[0, 1]$. Let $X = \sum_{i=1}^{n} X_i$. As for all $i \in [n]$, $\mathbf{E}[X_i] = \frac{1}{2}$, then by linearity of the expectation, $\mathbf{E}[X] = \frac{1}{2}n$. We can apply Hoeffding's inequalities to bound the probability that the sum of n random numbers deviates from its expected value taking $a_i = 0$ and $b_i = 1$ for all $i \in [n]$:

$$\mathbf{Pr}[|X - \mathbf{E}[X]| > t] \le 2e^{-2t^2 / \sum_{i=1}^{n} (b_i - a_i)^2} \le 2e^{-2t^2 / n}.$$

Observe that this probability goes to zero when $t = b(n)\sqrt{n}$ where $b(n) \to \infty$ as $n \to \infty$.

For a numeric setting, take $n = 200$. Then, the probability that $90 \le X \le 110$ is greater than $2/e \approx 0.735$. ✏

The following longer example is a classical application of Chernoff's bounds taken from [26].

➲ **Example 12.6 (Maximal Independent Set).** Recall that given a graph, a set of vertices is a an *independent set* if there are no edges among the elements of the set. A *maximal independent set* is an independent set that is not contained in any other independent set. It is well known that the problem of finding a maximal independent set is solvable in polynomial time (belongs to the class **P**). However, it is not know to be solvable in polylogarithmic parallel time using a polynomial number of process (in the class **NC**).

We present a simple randomized PRAM algorithm, due to Luby, to solve the problem in polylogarithmic time using a polynomial number of processors [26]. The example is self contained, and the reader does not need to know about parallel algorithms (the interested reader can find more about parallel algorithms in [19]).

function *Luby's-Algorithm* (G)
    begin
        $I := \emptyset$
        $G' := G$
        while $V' \neq \emptyset$ **do**
            $S := \emptyset$
            for all $v \in V'$ **do in parallel**
                with probability $\frac{1}{2d(v)}$ **do** $S := S \cup \{v\}$
            end for
            for all $v, u \in V'$ **do in parallel**
                if $d(v) \leq d(u)$ **then**
                    $S := S - \{v\}$
                else
                    $S := S - \{u\}$
                    $I := S \cup I$
                end if
            end for
            $G' :=$ induced subgraph of G on $V' - (S \cup N(S))$
        end while
        return I
    end

Figure 12.4   Luby's algorithm for maximal independent set

Luby's algorithm to find a maximal independent set I is given in Figure 12.4. The basic strategy is iterative, choosing at each iteration, with a small probability, a set S of vertices that are candidates to be in I.

For every vertex pair $v_1, v_2 \in S$ such that $(v_1, v_2) \in E$, discard the vertex with lower degree. Recall that for any $v \in V$, its *degree* is defined as $d(v) = |N(v)|$, where $N(v)$ denotes the set of neighbors of v in G.

Call a vertex $v \in V$ *good* when $\sum_{u \in N(v)} \frac{1}{2d(u)} \geq 1/6$. Let us first prove that in the given graph, if a vertex v is good, then $\mathbf{Pr}\left[v \in N(I)\right] \geq \frac{1}{36}$, where $N(I)$ is the set of vertices in V connected to some vertex inside of I.

We break the proof into two parts. Given v good, first we prove it for the case that $u \in N(I)$ has $d(u) \leq 2$ and after for the remaining vertices in $N(v)$ with degree greater than 2.

For any $v \in V$, let $L(v) = \{u \in N(v) \mid d(u) \geq d(v)\}$, then

$$\mathbf{Pr}\left[v \notin I | v \in S\right] \leq \mathbf{Pr}\left[\exists u \in (L(v) \cap S) | v \in S\right] \leq \sum_{u \in L(v)} \mathbf{Pr}\left[u \in S | v \in S\right]$$

$$= \sum_{u \in L(v)} \mathbf{Pr}\left[u \in S\right] \leq \sum_{u \in L(v)} \frac{1}{2d(v)} \leq \frac{d(v)}{2d(v)} = \tfrac{1}{2}.$$

Therefore,

$$\mathbf{Pr}\left[v \in I\right] \geq \mathbf{Pr}\left[v \in I \mid v \in S\right]\mathbf{Pr}\left[v \in S\right] \geq \frac{1}{4d(v)},$$

which when there is any $u \in N(v)$ with $d(u) \leq 2$, gives

$$\mathbf{Pr}\left[v \in N(v)\right] \geq \mathbf{Pr}\left[u \in I\right] \geq \frac{1}{4d(u)} \geq \tfrac{1}{8} > \tfrac{1}{36}.$$

For every $u \in N(v)$ with $d(u) \geq 3$, we have that $\frac{1}{2d(u)} \leq \tfrac{1}{6}$. But as by hypothesis v is good, then $\sum_{u \in N(v)} \frac{1}{2d(u)} \geq \tfrac{1}{6}$ which implies that there exists $M(v) \subset N(v)$ such that

$$\tfrac{1}{6} \leq \sum_{u \in M(v)} \frac{1}{2d(u)} \leq \tfrac{1}{3} \tag{12.10}$$

Using inclusion-exclusion we have,

$$\mathbf{Pr}\left[v \in N(I)\right] \geq \mathbf{Pr}\left[\exists u \in (M(v) \cap I)\right]$$

$$\geq \sum_{u \in N(v)} \mathbf{Pr}\left[u \in I\right] - \sum_{u \neq w \in M(v)} \mathbf{Pr}\left[u \in I \cap w \in I\right]$$

but as the variables are pairwise independent, on this last term, we get

$$\mathbf{Pr}\left[v \in N(I)\right] \geq \sum_{w \in M(v)} \frac{1}{4d(u)} - \sum_{u \neq w \in M(v)} \mathbf{Pr}\left[u \in S\right]\mathbf{Pr}\left[w \in S\right],$$

using equation (12.10),

$$\mathbf{Pr}\left[v \in N(I)\right] \geq \tfrac{1}{36}.$$

Now we can prove that if X is the random variable counting the number of edges deleted at each iteration of the algorithm, then

$$\mathbf{E}\left[X\right] \geq \tfrac{1}{72}|E|.$$

Let $B \subseteq E$ be the set of good edges. For every $e \in E$ define the indicator variable $X_e = 1$ if and only if e is deleted. As $X = \sum_e X_e$, and using linearity of the expected value together with the previous lemma and the easy to prove fact that at least half of the edges in a graph are good, we get

$$\mathbf{E}\left[X\right] = \sum_{e \in E} \mathbf{E}\left[X_e\right] \geq \sum_{e \in B} \mathbf{E}\left[X_e\right]$$

$$= \sum_{e=(u,v) \in B} \mathbf{Pr}\left[v \in N(I)\right] \geq \tfrac{1}{36}|B| \geq \tfrac{1}{72}|E|.$$

It remains to prove that Luby's algorithm will find a maximal independent set, within $O(\log n)$ iterations, with high probability.

Each iteration is a success if at least $\frac{|E'|}{144}$ edges are removed, where E' is the set of edges at the beginning of the iteration. Let X be the variable counting the number of edges deleted. The by Markov,

$$\mathbf{Pr}\left[\text{success}\right] = \mathbf{Pr}\left[X > \tfrac{1}{144}\right] \leq \tfrac{1}{143},$$

which is small but does not tend to 0, so we can not say that happens whp.

Consider $\log n$ iterations and use Chernoff's,

$$\mathbf{Pr}\left[\text{success} > \tfrac{1}{2}\log n\tfrac{1}{144}\right] \leq e^{-\log n/288}$$

that indeed tends to 0 as n grows.   ✐

5.   THE INDEPENDENT BOUNDED DIFFERENCES METHOD

In this section we present bounds on the tails of a function defined on a product space. The main result, due to McDiarmid [28], requires that the function satisfies the so-called *Lipschitz* or *independent bounded differences* condition. Whereas the canonical way to prove this result is through the use of martingales, we present here a direct proof.

Definition 12.1 (Independent bounded differences condition).

A function $f : \mathbb{R}^n \to \mathbb{R}$ satisfies the independent bounded differences condition *or is said to be* Lipschitz, *if there exists a vector* $\mathbf{c} = (c_1, \ldots, c_n)$ *such that*

$$|f(\mathbf{x}) - f(\mathbf{x}')| \leq c_i$$

for all vectors $\mathbf{x}$ and $\mathbf{x}'$ that differ only at the i-th coordinate ($i \in [n]$).

☞ **Example 12.2 (cont).** Consider again the example of counting the number of heads when tossing n times a fair coin. The corresponding probability space is $\Omega = \{0,1\}^n$. Let $\mathbf{X} = (X_1, \ldots, X_n)$ be a random vector where, for each $i \in [n]$, the probability that the i-th toss is a head is $\mathbf{Pr}[X_i = 1] = \frac{1}{2}$. Define $f \colon \Omega \to \mathbb{R}$ assigning to any $\mathbf{x} = (x_1, \ldots, x_n)$ the value $f(\mathbf{x}) = \sum_{i=1}^{n} x_i$, so that f is a random variable counting the number of heads. Then, f is Lipschitz, as if $\mathbf{x}$ and $\mathbf{x}'$ only differ at the i-th coordinate, this means that the number of heads in $\mathbf{x}$ and $\mathbf{x}'$ differs only by one, i.e. $|f(\mathbf{x}) - f(\mathbf{x}')| \leq 1$. ▱

Lemma 12.3. *Let $X_1, \ldots, X_n$ be a collection of independent random variables. Assume $f \colon \mathbb{R}^n \to \mathbb{R}$ is a function that satisfies the independent bounded differences condition with $\mathbf{c} = (c_1, \ldots, c_n)$. Define the random variable Y as*

$$Y = f(X_1, \ldots, X_n) - \mathbf{E}[f(X_1, \ldots, X_n)]$$

and, for any $i \in [n]$, define the random variable Y_i as

$$Y_i = \mathbf{E}[f(X_1, \ldots, X_n)|\mathbf{X}_i] - \mathbf{E}[f(X_1, \ldots, X_n)|\mathbf{X}_{i-1}].$$

Then, $|Y_i| \leq c_i$.

Proof. Let $\mathbf{X} = (X_1, \ldots, X_n)$. According with the definition of conditional expectation, and taking into account the independence,

$$|Y_i| = |\mathbf{E}[f(\mathbf{X})|\mathbf{X}_i] - \mathbf{E}[f(\mathbf{X})|\mathbf{X}_{i-1}]|$$

$$= \left| \sum_{x_{i+1}, \ldots, x_n} f(\mathbf{X}_i, x_{i+1}, \ldots, x_n) \prod_{j=i+1}^{n} \mathbf{Pr}[X_j = x_j] \right.$$

$$\left. - \sum_{x_i, \ldots, x_n} f(\mathbf{X}_{i-1}, x_i, \ldots, x_n) \prod_{j=i}^{n} \mathbf{Pr}[X_j = x_j] \right|$$

$$= \left| \sum_{x_{i+1}, \ldots, x_n} \left(\sum_{y_i \neq x_i} f(\mathbf{X}_i, x_{i+1}, \ldots, x_n) \right. \right.$$

$$\left. \left. - f(\mathbf{X}_{i-1}, x_i, \ldots, x_n) \right) \prod_{j=i}^{n} \mathbf{Pr}[X_j = x_j] \right|$$

$$\leq \sum_{x_i, \ldots, x_n} c_i \prod_{j=i}^{n} \mathbf{Pr}[X_j = x_j] = c_i,$$

which proves the result.   ✄

The next result is known as the *method of independent bounded differences* [28].

Theorem 12.9 (McDiarmid's bound). *Given a product space defined by $\Omega = \prod_{i=1}^{n} \Omega_i$, let $X_1, \ldots, X_n$ be independent random variables on $\Omega_1, \ldots, \Omega_n$. Assume $f : \mathbb{R}^n \to \mathbb{R}$ satisfies the independent bounded differences condition with $\mathbf{c} = (c_1, \ldots, c_n)$. Then, for all $t > 0$,*

$$\mathbf{Pr}\left[f(X_1, \ldots, X_n) - \mathbf{E}\left[f(X_1, \ldots, X_n)\right] > t\right] < exp\left(-\frac{t^2}{\sum_{i=1}^{n} c_i^2}\right).$$

Proof. Define the random variable $Y = f(X_1, \ldots, X_n) - \mathbf{E}\left[f(X_1, \ldots, X_n)\right]$ and, for any $i \in [n]$, define the random variables $Y_i = \mathbf{E}\left[f(X_1, \ldots, X_n)|X_i\right]$. Then, $Y = \sum_{i=1}^{n} Y_i$ and furthermore $\mathbf{E}\left[Y_i\right] = 0$. By Lemma 12.3, we have $-c_i \leq Y_i \leq c_i$. By Lemma 12.2 we have

$$\mathbf{E}\left[e^{\lambda Y_i}\right] \leq e^{\lambda^2 (2c_i)^2/8} \leq e^{\lambda^2 c_i^2/2}.$$

Therefore, $\mathbf{Pr}[Y > t] \leq \mathbf{E}\left[e^{\lambda Y}\right]/e^{\lambda t}$. To bound this expression, observe that setting $S_i = \sum_{j=1}^{i} Y_j$ we have $Y = S_n$ and $\lambda S_i = \lambda S_{i-1} + \lambda Y_i$ for any $i \in [n]$. Thus, using the fact that, given $\mathbf{X}_i$, the random variable S_i is totally determined, we get

$$\mathbf{Pr}[Y > t] \leq \mathbf{E}\left[e^{\lambda S_n}\right]/e^{\lambda t}$$
$$\leq \mathbf{E}\left[e^{\lambda S_{n-1}} e^{\lambda Y_n}\right]/e^{\lambda t}$$
$$\leq \mathbf{E}\left[e^{\lambda S_{n-1}} \mathbf{E}\left[e^{\lambda Y_n}|X_1, \ldots, X_n\right]\right]/e^{\lambda t}$$
$$\vdots$$
$$\leq \prod_{i=1}^{n} \mathbf{E}\left[e^{\lambda Y_i}|\mathbf{X}_i\right]/e^{\lambda t}$$
$$\leq e^{\lambda^2 \sum_{i=1}^{n} c_i^2/2}/e^{\lambda t}.$$

Taking $\lambda = 1/t$ we have the minimum in the denominator, and thus,

$$\mathbf{Pr}[Y - \mathbf{E}\left[Y\right] \geq t] < e^{-t^2/\sum_{i=1}^{n} c_i^2},$$

which proves the theorem.   ✄

Observe that the inequality in the above theorem is only one-sided, but applying it to the function $-f$ we get

$$\mathbf{Pr}\left[f(X_1, \ldots, X_n) - \mathbf{E}\left[f(X_1, \ldots, X_n)\right] < -t\right] < e^{-t^2/\sum_{i=1}^{n} c_i^2},$$

and therefore,

$$\mathbf{Pr}\left[|f(X_1,\ldots,X_n) - \mathbf{E}\left[f(X_1,\ldots,X_n)\right]| < t\right] < 2e^{-t^2/\sum_{i=1}^{n} c_i^2}.$$

In the rest of this section, we present some application examples of the above bounds.

☞ **Example 12.2 (cont).** Recall that in the example of counting the number of heads in n tosses of a fair coin, f is Lipschitz with $\mathbf{c} = 1$. For any $t > 0$, using McDiarmid's bound we get

$$\mathbf{Pr}\left[f(\mathbf{X}) - n/2 > t\right] \le e^{-t^2/n}.$$

and using Chernoff's bounds (with $t = d\mu$ in Corollary 12.1) we get

$$\mathbf{Pr}\left[f(\mathbf{X}) - n/2 > t\right] \le e^{-\frac{2}{3}t^2/n}.$$

In this case, both bounds are equivalent up to the constants.   ▭

The next example is taken from [8]. See also [1, Theorem 5.1] for a related example.

☞ **Example 12.7 (Balls and bins).** Consider the following experiment: throw independently and uniformly m balls into n bins. We wish to count the number of empty bins.

In this case, $\Omega = [n]^m$. Let Z be a random variable counting the number of empty bins. Let $X_1,\ldots,X_m$ be a sequence of independent random variables, where each X_i takes values in $[n]$ and indicates the bin where the i-th ball falls. It is well known that

$$\mu = \mathbf{E}\left[Z\right] = n\left(1 - \frac{1}{n}\right)^m.$$

Notice that Z is Lipschitz, because if the i-th ball is moved from one bin to another, keeping all other balls in the same position, the number of empty bins can increase or decrease at most by 1. So, applying McDiarmid's bound, we get that, for any $t > 0$,

$$\mathbf{Pr}\left[|Z - \mu| > t\right] \le 2e^{-t^2/m}. \tag{12.11}$$

A concentration bound tending to zero as $m \to \infty$ is obtained when t grows faster than $\sqrt{m}$.   ▭

The next example is taken from [21]

☞ **Example 12.8 (Bin packing).** The BINPACKING problem is another classical **NP**-hard problem: Given n objects with sizes $x_1, \ldots, x_n$ (such that $0 < x_i < 1$ for all $i \in [n]$) and an unlimited collection of bins, each of size 1, what is the minimal number of bins required to pack the objects? In the randomized version of this problem, we suppose that the sizes of the objects are generated at random (uniformly and independently). Given $X_1, X_2, \ldots$, we wish to study the random variable $Y_n = f(X_1, \ldots, X_n)$ that represents the minimal number of bins required to pack the first n items. It has been shown that Y_n grows linearly in n, that is, there exists a constant β such that

$$\frac{\mathbf{E}\,[Y_n]}{n} \to \beta, \qquad \text{as } n \to \infty.$$

Observe that f is Lipschitz with $\mathbf{c} = 1$. From Theorem 12.9, it follows that

$$\mathbf{Pr}\,[|Y_n - \mathbf{E}\,[Y_n]| > t] < 2e^{-t^2/n}.$$

As a consequence, with high probability $|Y_n - \mathbf{E}\,[Y_n]| > b(n)\sqrt{n}$ where $b(n)$ is any function so that $b(n) \to \infty$.   ☜

Let us consider another example the *longest increasing subsequence* problem for a random permutation [36]

☞ **Example 12.9 (Longest increasing subsequence).** A permutation π of $[n]$ induces a sequence with a reordering of the elements $\{1, \ldots, n\}$, an *increasing subsequence* in π is a subsequence in which each element is smaller than the next. Let L_π denote the length of the longest increasing subsequence in π.

We are interested in the random variable L_π when the permutation π is random with uniform distribution. It is known that, asymptotically, $\mathbf{E}\,[L_\pi] = 2\sqrt{n}$ [25, 39]. Using the Hoeffding-Azuma inequality, Frieze [13] gave a concentration bound for L_π within $(2\sqrt{n})^{2/3}$ of its expected value. His proof is non trivial. However, changing a bit the setting, we could easily apply the independent bounded differences method. Consider n uniformly and independently chosen real random numbers in $[0, 1]$, $X_1, \ldots, X_n$. As they are uniform random real numbers, with probability 1, all of them would be different, therefore the order on the reals induces a permutation π of $[n]$. Let L_n be the random variable measuring the length of the longest increasing sequence on these reals. Notice that by changing only one number, the length of the longest increasing sequence may be increased or decreased by one, therefore using Theorem 12.9, for any $t \geq 0$

$$\mathbf{Pr}\,[|L_n - \mathbf{E}\,[L_n]| > t] \leq e^{-2t^2/n}.$$

Better concentration bounds will be given later.   ✑

6.   TAIL BOUNDS
FOR DEPENDENT VARIABLES

Chernoff bounds and the independent bounded differences method require the independence of the the random variables of interest. In this section, we introduce a powerful tool, *martingale* theory, and use it to derive concentration results, for the case in which the random variables under consideration are not necessarily independent.

The study of martingales is an old topic in probability theory and provides one of the most useful tools to establish limit theorems, because the convergence of martingales is guaranteed. We will restrict ourselves to finite martingale sequences, as we are not interested in convergence results. We shall introduce only the basic definitions and results needed for our purposes. For a more complete introduction to martingales see [14].

The name martingale comes from the old game of martingale: A gambler has a large amount of money, he puts 1 euro in the first bet. While he loses the bet, he doubles the amount in the next bet. The play continues until the gambler wins a bet. Observe that if the gambler arrives to the n-th play, then he bets 2^n euros on the $n + 1$-th bet. So, if he wins at the n-th round, there will be a profit of $2^n - \sum_{i=0}^{n-1} 2^i$, that is just 1 euro!

If X_i denotes the amount that the gambler could win or lose at the i-th hand, as at any play round i, the gambler wins X_i with probability $1/2$ and loses X_i with the same probability, then $\mathbf{E}[X_n | X_1, \dots, X_{n-1}] = X_{n-1}$, for any n. The book [14] contains an amusing history about G. Casanova and the game of martingales in 18th century Venice.

Let us start by introducing the concept of martingale.

Definition 12.2. *A sequence of random variables $Y_0, \dots, Y_n$ is a* martingale *with respect to the sequence $X_1, \dots, X_n$, if for all $i \in [n]$,*

$$\mathbf{E}[Y_i | X_1, \dots, X_{i-1}] = Y_{i-1}.$$

The canonical example of martingale is the sum of bounded independent random variables with expected value 0.

✑ **Example 12.10 (Independent sums).** Let $X_1, \dots, X_n$ be a sequence of bounded independent variables with expected value 0. For instance $X_i = 1$ if the toss of a fair coin results in a head and $X_i = -1$ otherwise. For any i, $0 \le i \le n$, define $S_i = \sum_{k=1}^{i} X_k$. Notice that in this example the sum is the function that defines S_n. To prove that

$S_0, \ldots, S_n$ is a martingale with respect to $X_1, \ldots, X_n$, we have to show that for any i, $\mathbf{E}[S_i | X_1, \ldots, X_{i-1}] = S_{i-1}$.

$$\mathbf{E}[S_i | X_1, \ldots, X_{i-1}] = \mathbf{E}[S_{i-1} + X_i | X_1, \ldots, X_{i-1}]$$
$$= \mathbf{E}[S_{i-1} | X_1, \ldots, X_{i-1}] + \mathbf{E}[X_i | X_1, \ldots, X_{i-1}].$$

As X_i is independent from $X_1, \ldots, X_{i-1}$, we get

$$\mathbf{E}[S_i | X_1, \ldots, X_{i-1}] = \mathbf{E}[S_{i-1} | X_1, \ldots, X_{i-1}] + \mathbf{E}[X_i] = S_{i-1} + 0.$$

Often, it will be the case that we have a martingale with respect to itself.

Definition 12.3. *A sequence of random variables $Y_0, Y_1, \ldots, Y_n$ is a martingale if for all $i \in [n]$,*

$$\mathbf{E}[Y_i | Y_1, \ldots, Y_{i-1}] = Y_{i-1}.$$

☞ **Example 12.10 (cont).**   It is easy to show that the process $S_0, \ldots, S_n$ defined as the prefix sums of a sequence of bounded independent variables with mean 0, is a martingale with respect to itself. In fact the events $\mathbf{X}_i = \mathbf{x}_i$ and $\mathbf{S}_i = \mathbf{s}_i$ are in a one to one correspondence.

Let us end our presentation of martingales with two properties:

Proposition 12.5. *Let $Y_0, \ldots, Y_n$ be a martingale with respect to the sequence $X_1, \ldots, X_n$. Then, for all $i \in [n]$*

1. $\mathbf{E}[Y_{i+j} | X_1, \ldots, X_i] = Y_i$, for any j such that $i + j \in [n]$, and

2. $\mathbf{E}[Y_i] = \mathbf{E}[Y_0]$.

Proof. By Proposition 12.4

$$\mathbf{E}[Y_{i+j} | X_1, \ldots, X_{i+j}] = \mathbf{E}[\mathbf{E}[Y_{i+j} | X_1, \ldots, X_{i+j}] | X_1, \ldots, X_i]$$

by definition of martingale we get

$$\mathbf{E}[Y_{i+j} | X_1, \ldots, X_{i+j}] = \mathbf{E}[Y_{i+j-1} | X_1, \ldots, X_i].$$

An iteration of the above argument gives the first claim.

Recall the $\mathbf{E}[Y_i] = \mathbf{E}[\mathbf{E}[Y_i | X_1, \ldots, X_j]]$ for any $j < i$, therefore, using the first part, we get $\mathbf{E}[Y_i] = \mathbf{E}[\mathbf{E}[Y_i | X_1]] = \mathbf{E}[\mathbf{E}[Y_1 | X_1]] = \mathbf{E}[Y_0]$. ⚥

Another useful form of a martingale is presenting the process as a sequence of differences.

Definition 12.4. *A process* $Y_1, \dots, Y_n$ *is a* martingale differences sequence *with respect to* $X_1, \dots, X_n$ *if for every* $i \in [n]$*, the random variable* Y_i *verifies* $\mathbf{E}[Y_i | X_1, \dots, X_{i-1}] = 0$.

The name of differences comes from the fact that any martingale sequence $Y_0, \dots, Y_n$ with respect to a sequence $X_1, \dots, X_n$ provides a natural way to define a sequence of differences, $Z_1, Z_2, \dots, Z_n$, by setting $Z_i = Y_i - Y_{i-1}$ for any $i \in [n]$. Indeed, the sequence $Z_1, \dots, Z_n$ is a martingale differences:

$$\mathbf{E}[Z_i | X_1, \dots, X_{i-1}] = \mathbf{E}[Y_i - Y_{i-1} | X_1, \dots, X_{i-1}]$$

by linearity of expectation, the definition of martingale and Proposition 12.4

$$\mathbf{E}[Z_i | X_1, \dots, X_{i-1}] = \mathbf{E}[Y_i | X_1, \dots, X_{i-1}] - \mathbf{E}[Y_{i-1} | X_1, \dots, X_{i-1}]$$
$$= Y_{i-1} - Y_{i-1} = 0.$$

We will refer to such construction as the *martingale differences sequence*.

Conversely, from any given random variable X_0 and a martingale differences sequence $Y_1, \dots, Y_n$, we obtain a martingale by setting $X_i = X_0 + \sum_{k=1}^{i} Y_k$. Furthermore, observe that any sequence of independent random variables with mean expected value zero is a martingale differences sequence, because $\mathbf{E}[X_{i+1} | X_i, \dots, X_1] = \mathbf{E}[X_{i+1}] = 0$.

6.1 EXPOSURE MARTINGALE

Our usual setting will be a random structure (a graph, a tree, some balls and bins experiment, etc.) on which we have defined a measure. Often the structure can be described by an incremental process, in such a way that at each step some portion of the structure is revealed. The conditional expectation given the so far exposed parts allows the definition of a random process, that is usually referred to as the *Doob's process*. The formal setting to introduce this process is when studying a random variable on a product space.

Definition 12.5 (Doob's process). *Let* $X_1, \dots, X_n$ *be a sequence of random variables, where, for any* $i \in [n]$*,* X_i *is defined on the probability space* $(\Omega_i, \mathcal{A}_i, \mathbf{Pr}_i)$*. Let* $f : \mathbb{R}^n \to \mathbb{R}$ *be a function. The process* $Y_0, \dots, Y_n$ *defined as*

$$Y_i = \mathbf{E}[f(\mathbf{X}) | X_1, \dots, X_i], \qquad for \ 0 \le i \le n \qquad (12.12)$$

is called the Doob's process.

Observe that for any i, $0 \le i \le n$, Y_i is a random variable on the product space $\Omega = \prod_{i=1}^{n} \Omega_i$.

Using Proposition 12.4 we have that

$$\mathbf{E}\left[Y_i|\mathbf{X}_{i-1}\right] = \mathbf{E}\left[\mathbf{E}\left[f(\mathbf{X})|\mathbf{X}_i\right]|\mathbf{X}_{i-1}\right] = \mathbf{E}\left[f(\mathbf{X})|\mathbf{X}_{i-1}\right] = Y_{i-1}.$$

Therefore the Doob's process $Y_0, \ldots, Y_n$ is indeed a martingale with respect to $X_1, \ldots, X_n$. We will refer to such construction as a *Doob's martingale*. Each Y_i measures the expected value of $f(\mathbf{X})$ after discovering the first i portions of the structure.

Thus, Doob's construction is useful to define a martingale from any function of a combinatorial object, provided that the random construction of the object can be performed incrementally.

☞ **Example 12.11 (Three tosses).** Consider the experiment of tossing three times a fair coin. Assume that we are interested in knowing the number of heads after the three tosses.

To analyze this random variable, we will define a Doob's martingale. To do so we first need to describe the sample space as a product space. In this particular case, the task is easy, as we can describe the outcome of the experiment using for each $i \in [3]$ the indicator variables X_i of the event: a head in the i-th toss. For instance, the event $\mathbf{X} = (0, 1, 0)$ indicates tail, head, tail.

Let f be the sum function; then $f(X_1, X_2, X_3)$ counts the number of heads. The Doob's process Y_0, Y_1, Y_2, Y_3 corresponds to follow the experiment stopping after each toss, to compute the expected number of heads, given that we already known the outcome of this and the previous tosses. The Doob's martingale sequence is the following:

$$Y_0(\omega) = \mathbf{E}\left[f(\mathbf{X})\right] = 3/2,$$

$$Y_1(\omega) = \mathbf{E}\left[f(\mathbf{X})|X_1\right](\omega) = \begin{cases} 2 & X_1(\omega) = 1 \\ 1 & X_1(\omega) = 0 \end{cases}$$

$$Y_2(\omega) = \mathbf{E}\left[f(\mathbf{X})|X_1, X_2\right](\omega) = \begin{cases} 5/2 & X_1(\omega) = 1, X_2(\omega) = 1 \\ 3/2 & X_1(\omega) = 1, X_2(\omega) = 0 \\ 3/2 & X_1(\omega) = 0, X_2(\omega) = 1 \\ 1/2 & X_1(\omega) = 0, X_2(\omega) = 0 \end{cases}$$

$$Y_3(\omega) = \mathbf{E}\left[Y_3|X_1, X_2, X_3\right](\omega) = f(X_1(\omega), X_2(\omega), X_3(\omega)).$$

Notice that the last random variable in the sequence corresponds to the function in whose analysis we are interested, and the first one (Y_0) to its expected value.   ☜

☞ **Example 12.7 (cont).** The random variable X that counts the number of empty bins, after throwing independently and uniformly at random m balls into n bins, can be described incrementally by the process of throwing the balls one after the other. Throwing a ball can be represented by a random variable X_i uniformly distributed on $[n]$ where $X_i = b_i$ denotes the event: "the i-th ball falls into bin b_i", for any $i \in [n]$. From this representation we can consider the Doob's martingale $Y_0, \ldots, Y_n$ where

$$Y_i = \mathbf{E}\left[X|X_1, \ldots X_i\right],$$

that will allow us to study concentration of the random variable X, by analyzing the incremental process.   ☜

Random graphs provide a natural setting to define exposure martingales. In fact there are two usual ways to grow up a graph, by discovering edges or vertices.

☞ **Example 12.12 (Edge exposure).** Let $G = (V, E)$ be a random graph in the $G_{n,p}$ model, with vertex set $V = [n]$. Recall that in the $G_{n,p}$ a graph with n vertices is obtained by the independent selection of edges. Each edge appears with probability p.

Taking $m = \binom{n}{2}$, and labeling the edges with numbers in $[m]$, any graph G with n vertices can be represented by an m-tuple $(a_1, a_2, \ldots, a_m)$ where $a_j = 1$ if and only if the edge j appears in the graph. The sequence of variables $X_1, \ldots, X_m$, where for each $j \in [m]$, X_j is the indicator variable of the event: "the edge j appears in the graph", is the basis to define the associated Doob's martingale for any graph theoretical function f. In this case each random variable X_j, is a Bernoulli variable with probability of success p.

The *Doob's edge exposure martingale* for a given graph function f, is defined as usual by $Y_j = E[f(G)|\mathbf{X}_j]$, $0 \leq j \leq m$. Observe that the random variables Y_j's measure the expected value of $f(G)$, when the existence of the first i edges of the graph is known.   ☜

☞ **Example 12.13 (Vertex exposure).** Let $G = (V, E)$ be a random graph in the $G_{n,p}$ model. We now want to consider a way of generating the graph by exposing its vertices. Starting from 1, add a new vertex i and the edges that connect i with the previous $i - 1$ vertices. When adding vertex i, the set of possible end-points of its edges is $[i - 1]$, this can be represented by a string of $i - 1$ bits. For each $i \in [n]$ define the random variable X_i uniformly distributed on the set $B_i = \{0, 1\}^{i-1}$ of numbers represented with $i - 1$ bits. So, any graph G with n vertices will be specified by a sequence of bit strings, one for each addition of a vertex.

Given any graph theoretical function f, we can consider the *Doob's vertex exposure martingale* by $Y_i = E[f(G)|\mathbf{X}_i]$. In this case the random variable Y_i, gives the expected value of $f(G)$, when the first i vertices and their corresponding internal edges are known. It is not difficult to see, that with an appropriate ordering of the set of possible edges, the vertex exposure martingale is a subsequence of the edge exposure martingale.

⬳

The following construction was used in [7] to show that the *monotone circuit value problem* is in Average **NC**.

⬳ **Example 12.14 (Circuit's representation depth).** Consider the following process to generate an unlabeled binary random circuit. Gate 1 holds a 0 and gate 2 holds a 1, these are the input gates. At step i $(i > 2)$, select two nodes from $[i - 1]$ as inputs to gate i. The depth of a gate g is the length of the longest directed path from g to the inputs. We are interested in analyzing the depth of the circuit, that is the maximum of the depths of its gates. This measure corresponds to the time needed to evaluate the circuit in presence of parallelism.

The incremental process suggests the use of a vertex exposure filter and the definition of the corresponding martingale.   ⬳

6.2   THE BOUNDED DIFFERENCES CONDITION

For the case of independent random variables, we asked for the Lipschitz condition. In the case of dependent random variables, that form a martingale, in order to get concentration results, we need some sort of guarantee of bounded increments. We will only need that the process does not make big jumps in one step.

Definition 12.6 (Bounded differences condition). *A martingale sequence* $Y_0, \ldots, Y_n$ *is said to satisfy the* bounded differences condition *if for any* $i \in [n]$ *there exists a vector* $\mathbf{c} = (c_1, \ldots, c_n)$ *such that*

$$|Y_i - Y_{i-1}| \leq c_i.$$

The next theorem says that when the bounded differences condition holds, the process itself does not go far away from its starting point. The theorem is known as Hoeffding-Azuma's inequality and has appeared in different forms in papers by Hoeffding [17], Azuma [2], Steiger [35], Freedman [11] and others.

Theorem 12.10 (Hoeffding-Azuma's inequality.). *Let $Y_0, \dots, Y_n$ be a martingale with respect to $X_1, \dots, X_n$ that satisfies the bounded differences condition with* $\mathbf{c} = (c_1, \dots, c_n)$. *Then, for any $t > 0$,*

$$\mathbf{Pr}\left[|Y_n - Y_0| > t\right] \leq 2\exp\left(-\frac{t^2}{2\sum_{i=1}^n c_i^2}\right).$$

Proof. Let us prove the upper bound $\mathbf{Pr}\left[Y_n > Y_0 + t\right] \leq 2\exp(-\frac{t^2}{2c})$. Without loss of generality we can assume $Y_0 = 0$. Define the martingale differences $D_i = Y_i - Y_{i-1}$, so $Y_n = Y_{n-1} + D_n$. Using Markov inequality we get that, for all $\lambda > 0$, $\mathbf{Pr}\left[Y_n > t\right] \leq \mathbf{E}\left[e^{\lambda Y_n}\right]/e^{\lambda t}$. We wish to find the value of λ that minimizes this ratio.

$$\mathbf{E}\left[e^{\lambda Y_n}\right] = \mathbf{E}\left[e^{\lambda(Y_{n-1}+D_n)}\right] = \mathbf{E}\left[\mathbf{E}\left[e^{\lambda(Y_{n-1}+D_n)}|X_0, \dots, X_{n-1}\right]\right]$$

but as Y_{n-1} is a function of $X_0, \dots, X_{n-1}$ using Proposition 12.4 we get

$$\mathbf{E}\left[e^{\lambda Y_n}\right] = \mathbf{E}\left[e^{\lambda Y_{n-1}}\mathbf{E}\left[e^{\lambda D_n}|X_0, \dots X_{n-1}\right]\right]. \tag{12.13}$$

Now, $\mathbf{E}\left[\mathbf{E}\left[D_n|X_0, \dots X_{n-1}\right]\right] = 0$ and $|\mathbf{E}\left[D_n|X_0, \dots X_{n-1}\right]| \leq c_n$. Applying Lemma 12.2 to $\mathbf{E}\left[D_n|X_0, \dots X_{n-1}\right]$ we get,

$$\mathbf{E}\left[e^{\lambda D_n}|X_0, \dots X_{n-1}\right] \leq e^{\lambda^2 c_n^2/2}.$$

Substituting this expression and iterating we get

$$\mathbf{E}\left[e^{\lambda Y_n}\right] \leq ne^{\lambda^2 c/2},$$

where $c = \sum_{i=1}^n c_i^2$. But $\frac{e^{\lambda^2 c/2}}{e^{\lambda t}}$ attains the minimum at $\lambda = t/c$ and the proof of the upper tail follows.

In order to get the proof of the lower tail bound, just consider the symmetrical martingale $-Y_0, \dots, -Y_n$. ☙

The main point of the Hoeffding-Azuma's inequality with respect to the bounds in the previous section are the following: First, we don't need to make any assumption on the independence of the random variables. Second, we stress that the c_i in the bounded differences condition may depend on i.

A theorem that is often refereed as Azuma's inequality is the particular case of Theorem 12.10, when the martingale is defined with respect to itself. See for example [29, Theorem 3.10] or [32, Theoremm 4.16]. Observe that in the case of the Doob's Martingale, we have $Y_0 = \mathbf{E}\left[f(\mathbf{X})\right]$, so Theorem 12.10, gives a concentration around the mean result.

An useful corollary to Hoeffding-Azuma's inequality arises when there is a value c for which $c_i = c$ for any $i \in [n]$ [21]:

Corollary 12.3. *Let $Y_0, \ldots, Y_n$ be a martingale satisfying the bounded differences condition such that for a vector $\mathbf{c} = (c, \ldots, c)$. Then for any $t > 0$*

$$\mathbf{Pr}\left[|Y_n - Y_0| \geq ct\sqrt{n}\right] \leq 2\exp\left(-\tfrac{1}{2}t^2\right).$$

☞ **Example 12.2 (cont).**   We can use the previous inequality to study the concentration of the random variable that counts the number of heads after tossing a fair coin n times. Let $Y_0, \ldots, Y_n$, be the corresponding Doob's martingale with respect to $X_1, \ldots, X_n$, the indicator variables that record the outcome of each toss. Changing the value of X_i can affect the difference $Y_i - Y_{i-1}$ by at most by 1. Therefore $|Y_i - Y_{i-1}| \leq 1$ and we can apply the previous corollary with $c = 1$ and $Y_0 = n/2$, to get

$$\mathbf{Pr}[|Y_n - \mu| \geq t\sqrt{n}] \leq 2e^{-\frac{t^2}{2}}. \tag{12.14}$$

To compare with Chernoff and the bound obtained from the independent bounded condition, take $t = \frac{d\sqrt{n}}{2}$ so, we get,

$$\mathbf{Pr}\left[|S_n - \mathbf{E}\left[S_n\right]| \geq dn/2\right] \leq \begin{cases} 2\exp(-\tfrac{1}{8}nd^2) & \text{Hoeffding-Azuma's} \\ & \text{inequality} \\ 2\exp(-\tfrac{2}{3}nd^2) & \text{Chernoff inequality} \\ 2\exp(-\tfrac{1}{4}nd^2) & \text{independent bounded} \\ & \text{differences} \end{cases}$$

The three bounds give the same behavior, but with different constants in the exponent. The better bound in this case is obtained with Chernoff inequality.   ☜

In general, for binomial distribution, the appropriate use of Chernoff bounds will yield better concentration bounds than Hoeffding-Azuma's inequality. On the other hand, Hoeffding-Azuma's does not require independence.

As in most situations we will have a set of random variables and a function of them, we state the bounded differences condition in such terms.

Definition 12.7. *A sequence of random variable $X_1, \ldots, X_n$ and a function $f : \mathbb{R}^n \to \mathbb{R}$ satisfy the* bounded differences condition *if, for any $i \in [n]$, there is a vector $\mathbf{c} = (c_1, \ldots, c_n)$ such that*

$$|\mathbf{E}\left[f(\mathbf{X})|\mathbf{X}_i\right] - \mathbf{E}\left[f(\mathbf{X})|\mathbf{X}_{i-1}\right]| < c_i.$$

This leads the Hoeffding-Azuma's inequality given in Theorem 12.10, for the case of the Doob's martingale.

Corollary 12.4. *Let $X_1, \ldots, X_n$ be a sequence of random variables and let $f : \mathbb{R}^n \to \mathbb{R}$ be a function that satisfies the bounded differences condition for $\mathbf{c} = (c_1, \ldots, c_n)$. Then, for any $t > 0$,*

$$\mathbf{Pr}\left[|f(\mathbf{X}) - \mathbf{E}\left[f(\mathbf{X})\right]| > t\right] \leq 2exp\left(-\frac{t^2}{2c}\right)\right]$$

where $c = \sum_{i=1}^n c_i^2$.

☛ **Example 12.7 (cont).** The Doob's martingale for the number of empty bins, after throwing m balls into n bins, verifies the bounded differences condition with all $c_i = 1$, because by throwing an additional ball we may only fill an empty bin. It is well known that $\mathbf{E}\left[X\right] = ne^{-m/n}$, therefore,

$$\mathbf{Pr}\left[|X - \mathbf{E}\left[X\right]| > t\right] \leq 2\exp\left(\frac{-t^2}{2m}\right).$$

Letting t growing to ∞ a bit faster than $\sqrt{m}$ we get a sharp concentration result. ☚

Doob's construction for functions that satisfy the bounded differences condition give directly concentration results. However, it does not give any information about the value of the expected value. The following example is from [33].

☛ **Example 12.15 (Chromatic number).** Given a graph G its chromatic number $\chi(G)$ is the minimum number of colors that are needed to color the vertices of G, in such a way that no two adjacent vertices have the same color.

Let us consider the $G_{n,p}$ model, according to Example 12.13, for any graph G, we can define a Doob's vertex exposure martingale for the function $\chi(G)$. Observe that a single additional vertex can increase only by one the number of colors needed to color the graph exposed so far. Therefore, the Doob's vertex exposure martingale for chromatic number satisfies the bounded differences condition, with $c_i = 1$ for any $i \in [n]$.

Therefore taking $t = d\sqrt{n}$, we can apply Corollary 12.4 to say:

$$\mathbf{Pr}\left[|\chi(G) - \mathbf{E}\left[\chi(G)\right]| > d\sqrt{n}\right] \le 2e^{-d^2/2}$$

Letting d go arbitrary slowly to ∞ we get a sharp concentration result. However there is no no hint on the value of $\mathbf{E}\left[\chi(G)\right]$. ✐

Sometimes, it is necessary to give a clever definition of function to guarantee the bounded differences condition. The following example is from [1]

☞ **Example 12.16 (Clique number).** Given a graph $G = (V, E)$ its clique number $w(G)$ is the size of the maximum clique in G. Notice that by the addition of a single edge or vertex, we can join many vertices, increasing in more than a constant the size of the biggest clique. Therefore, the vertex or edge exposure martinagale for the clique number will not show concentration.

Let $f(G)$ be the maximal size of a family of edge disjoint cliques of size k in G. Let us consider the $G \in G_{n,p}$ model, according to Example 12.12 define the Doob's edge exposure martingale for f. Adding a single edge can only add at most one clique to a family of edge disjoint cliques. Therefore Corollary 12.4 gives that the clique number is concentrated around its expected value.

When $G \in G_{n,1/2}$ it can be shown, that

$$\mathbf{E}\left[f(G)\right] \ge \frac{n^2}{2k^4}(1 + o(1)),$$

and combining this fact with the concentration result we get,

$$\mathbf{Pr}\left[w(G) < k\right] < \exp\left(-(d + o(1))\frac{n^2}{\ln^8 n}\right)$$

for a positive constant d. This is the basic step used by Bollobás [3] to show that $\chi(G)$ is almost surely $n/2\log n$ where G is a random graph with n nodes and $p = 1/2$. ✐

☞ **Example 12.14 (cont).** In the greedy process of constructing a representation of a circuit, adding a new gate may only increase the depth of the circuit by one, in the worst case. Therefore we can obtain tight concentration results. Again, Hoeffding-Azuma inequality does not give any information on the expected value. It is known that the expected depth of the representation of a circuit is a polylogarithmic function of the number of gates [7, 38]. ✐

It is possible to consider an intermediate notion of bounded condition, that falls between the independent bounded differences condition, and the bounded differences condition.

Definition 12.8. *A sequence of random variables $X_1, \ldots, X_n$ and a function $f : \mathbb{R}^n \to \mathbb{R}$ satisfy the the* average bounded differences *condition if, for any $i \in [n]$, there exists a vector $\mathbf{c} = (c_1, \ldots, c_n)$ such that for any $a_i, b_i \in X_i(\Omega_i)$*

$$|\mathbf{E}\left[f(\mathbf{X})|\mathbf{X}_{i-1}, X_i = a_i\right] - \mathbf{E}\left[f(\mathbf{X})|\mathbf{X}_{i-1}, X_i = b_i\right]| < c_i.$$

The average bounded differences condition implies the bounded differences condition. Therefore, as a corollary of Theorem 12.10 we get,

Corollary 12.5. *Let $X_1, \ldots, X_n$ be a sequence of random variables and let $f : \mathbb{R}^n \to \mathbb{R}$ be a function that satisfy the average bounded differences condition for $\mathbf{c} = (c_1, \ldots, c_n)$, then for any $t > 0$,*

$$\mathbf{Pr}\left[|f(\mathbf{X}) - \mathbf{E}\left[f(\mathbf{X})\right]| > t\right] \leq 2exp\left(-\frac{t^2}{2c}\right),$$

where $c = \sum_{i=1}^n c_i^2$

Proof. Let us show that the average bounded differences condition implies the bounded differences condition, with the same constants. Assume that for any $i \in [n]$ suitable c_i and any $a_i, b_i \in X_i(\Omega_i)$ we have

$$|\mathbf{E}\left[f(\mathbf{X})|\mathbf{X}_{i-1}, X_i = a_i\right] - \mathbf{E}\left[f(\mathbf{X})|\mathbf{X}_{i-1}, X_i = b_i\right]| \leq c_i$$

Then we can express

$$\mathbf{E}\left[f(\mathbf{X})|\mathbf{X}_{i-1}\right] = \sum_{a \in X_i(\Omega_i)} \mathbf{E}\left[f(\mathbf{X})|\mathbf{X}_{i-1}, X_i = a\right]\mathbf{Pr}\left[X_i = a|X_{i-1}\right]$$

$$\mathbf{E}\left[f(\mathbf{X})|\mathbf{X}_i\right] = \sum_{a \in X_i(\Omega_i)} \mathbf{E}\left[f(\mathbf{X})|\mathbf{X}_i\right]\mathbf{Pr}\left[X_i = a|X_{i-1}\right]$$

Therefore

$$|\mathbf{E}\left[f(\mathbf{X})|\mathbf{X}_{i-1}\right] - \mathbf{E}\left[f(\mathbf{X})|\mathbf{X}_i\right]|$$

$$= \left|\sum_a (\mathbf{E}\left[f(\mathbf{X})|\mathbf{X}_{i-1}, X_i = a\right] - \mathbf{E}\left[f(\mathbf{X})|\mathbf{X}_i\right])\mathbf{Pr}\left[X_i = a|X_{i-1}\right]\right|$$

$$\leq \sum_a |\mathbf{E}\left[f(\mathbf{X})|\mathbf{X}_{i-1}, X_i = a\right] - \mathbf{E}\left[f(\mathbf{X})|\mathbf{X}_i\right]|\mathbf{Pr}\left[X_i = a|X_{i-1}\right]$$

$$\leq \sum_a c_i\mathbf{Pr}\left[X_i = a|X_{i-1}\right] = c_i,$$

which proves the statement.

The previous corollary is refereed to as *the method of bounded average differences*. Notice that the real improvement is not in the application of Hoeffding-Azuma's inequality but in the fact the value c in the average bounded differences method is usually better that the corresponding value in the bounded differences method.

➥ **Example 12.7 (cont).** To apply the average bounded differences condition, to the function that counts the number of empty bins, we will follow [8]. In order to use the method of bounded average differences, we need to compute, for any $i \in [m]$,

$$c_i = |\mathbf{E}\left[Z|\mathbf{X}_{i-1} = \mathbf{b}_{i-1}, X_i = b_i\right] - \mathbf{E}\left[Z|\mathbf{X}_{i-1} = \mathbf{b}_{i-1}, X_i = b_i'\right],$$

for fixed $b_1, \dots, b_{i-1}, b_i, b_i' \in [n]$ and $b_i \neq b_i'$. The variable Z can be decomposed as the sum of indicator random variables, that is for $i \in [n]$ set Z_i to be 1 when the bin i is empty and 0 when it is not.

Let $S = \{b_1, \dots, b_{i-1}\}$. As $Z = \sum_{j \in [n]} Z_j$, by linearity of the expectation, the computation of c_i boils down to compute for each $j \in [n]$,

$$c_{ij} = \left|\mathbf{E}\left[Z_j|\mathbf{X}_{i-1} = \mathbf{b}_{i-1}, X_i = b_i\right] - \mathbf{E}\left[Z_j|\mathbf{X}_{i-1} = \mathbf{b}_{i-1}, X_i = b_i'\right]\right|.$$

so $c_i = \sum_j c_{ij}$. Let us consider some cases:
Case 1: For $j \in S$ then

$$\mathbf{E}\left[Z_j|\mathbf{X}_{i-1} = \mathbf{b}_{i-1}, X_i = b_i\right] = \mathbf{E}\left[Z_j|\mathbf{X}_{i-1} = \mathbf{b}_{i-1}, X_i = b_i'\right],$$

so $c_{ij} = 0$.
Case 2: For $j \notin S \cup \{b_i', b_i\}$

$$\mathbf{E}\left[Z_j|\mathbf{X}_{i-1} = \mathbf{b}_{i-1}, X_i = b_i\right] \mathbf{E}\left[Z_j|\mathbf{X}_{i-1} = \mathbf{b}_{i-1}, X_i = b_i'\right] = (1 - 1/n)^{m-i}.$$

Therefore, again $c_{ij} = 0$.
Case 3: For $j = b_i \notin S$, and $b_i' \notin S$ then

$$\mathbf{E}\left[Z_j|\mathbf{X}_{i-1} = \mathbf{b}_{i-1}, X_i = b_i\right] = 0$$

and

$$\mathbf{E}\left[Z_j|\mathbf{X}_{i-1} = \mathbf{b}_{i-1}, X_i = b_i'\right] = (1 - 1/n)^{m-i},$$

and in this case $c_{ij} = (1 - 1/n)^{m-i}$.

Putting all together, we get $c_i = \sum_{j,i} c_{ij} \le (1 - 1/n)^{m-i}$, so that

$$c = \sum_{i \in [m]} c_i^2 \le \frac{1 - (1/n)^{2m}}{1 - (1 - 1/n)^2} = \frac{n^2 - (n(1/n)^m)^2}{2n - 1} = \frac{n^2 - \mu^2}{2n - 1}.$$

Thus taking $\mu = \mathbf{E}[Z] = ne^{-m/n}$, from Corollary 12.4 we get,

$$\mathbf{Pr}[|Z - \mu| \geq t] = \mathbf{Pr}[|Z - \mu| \geq t] \leq 2\exp\left(-\frac{t^2(2n-1)}{2(n^2 - \mu^2)}\right).$$

Recall that using the method of bounded differences we got

$$\mathbf{Pr}[|Z - \mu| > t] \leq 2\exp\left(\frac{-t^2}{2m}\right).$$

As indicated in [8], if we take $t = d\mu$ we get the same bounds that in [20].

7. ISOPERIMETRIC INEQUALITIES

Isoperimetric inequalities have become an important topic in graph theory. Moreover, some of the techniques used to show isoperimetric inequalities provide a nice example for the application of the independent bounded differences method. We start this section by presenting a brief introduction to isoperimetry. After we move to present Talagrand's isoperimetric inequality and show some of its applicability. For the interested reader, we recommend [23, 24] and [28, Sec. 7].

In graph theoretical terms, given a graph, an isoperimetric inequality of a subgraph is an inequality relating the size of a subgraph with the size of its boundary. To formalize and generalize the above intuition, let (Ω, d) be a metric space with distance d, together with a measure ν. Recall that for any subset $A \subseteq \Omega$ the distance of a point $\mathbf{x}$ to A is defined as

$$d(\mathbf{x}, A) = \inf\{d(\mathbf{x}, \mathbf{y}) \mid \mathbf{y} \in A)\},$$

and the distance between two subsets A and B of Ω is

$$d(A, B) = \inf\{d(\mathbf{x}, B) \mid \mathbf{x} \in A\}.$$

For a given $A \subseteq \Omega$ and any real $t > 0$, define the *t-neighborhood* A_t of A as

$$A_t = \{\mathbf{x} \in \Omega \mid d(\mathbf{x}, A) \leq t\}.$$

Such a subset A_t is also known in the literature as *t-fattening* [28] or *t-boundary* [23]. Notice that $A \subseteq A_t$. An *isoperimetric inequality* is a lower bound of the size $\nu(A_t)$ depending on $\nu(A)$ and t. This relationship is usually presented as an upper bound on the expression $\nu(A)(\nu(\Omega) - \nu(A_t))$.

Considering again graphs, let $G = (V, E)$ be a finite graph; for any $u, v \in V$, let $d(u, v)$ be the shortest distance between u and v in G, and for any $A \subseteq V$, let the measure $\nu(A)$ be the number of vertices in A. For any A, the set A_t consists of the vertices that can be joined to a vertex in A by a path of length at most t. An isoperimetric inequality would be an inequality of the form $|A_t| \geq f(|A|, t)$, where we would like to have the value of the function $f(|A|, t)$ as close to the actual value of $|A_t|$ as possible. Recall that the *boundary* ∂ of a subgraph are the set of vertices connected by an edge to some vertex in A. Therefore, in this particular example, $\partial(A) = A_1 \setminus A$.

Let us begin by considering the case of isoperimetric inequalities in a product space, where the distance is the Hamming distance and the probability of a set gives its measure. Formally, let $\Omega_1, \ldots, \Omega_n$ be probability spaces, and let $\Omega = \prod_{i=1}^{n} \Omega_i$ be the corresponding product space. The Hamming distance between any two points $\mathbf{x} = (x_1, \ldots, x_n)$ and $\mathbf{y} = (y_1, \ldots, y_n)$ in Ω, denoted by $d_H(\mathbf{x}, \mathbf{y})$, is the number of indices i such that $x_i \neq y_i$. Given $\mathbf{x}$ and $\mathbf{y}$ as before, for each $i \in [n]$, define the function $\delta_i(\mathbf{x}, \mathbf{y}) = 1$ if $x_i \neq y_i$ and $\delta_i(\mathbf{x}, \mathbf{y}) = 0$ otherwise. Thus we can express the Hamming distance as

$$d_H(\mathbf{x}, \mathbf{y}) = \sum_{i=1}^{n} \delta_i(\mathbf{x}, \mathbf{y}).$$

The t-neighborhood of a set A under the Hamming distance will be denoted by A_t^H. For any $A \in \Omega$ consider the measure $\nu(A) = \mathbf{Pr}\,[\mathbf{X} \in A]$. In this case, the independent bounded differences inequality can be used to show that when the probability that a point belongs to a set A is not to small, then with probability tending to 1, any random point is close to A. The following isoperimetric inequality is from McDiamird [29].

Theorem 12.11 (McDiamird). *Let* $\mathbf{X} = (X_1, \ldots, X_n)$ *be a sequence of random variables, each* X_i *defined on* Ω_i. *Then for any* $t > 0$,

$$\mathbf{Pr}\,[\mathbf{X} \in A]\,(1 - \mathbf{Pr}\,[\mathbf{X} \in A_t^H]) \leq e^{-t^2/4n}.$$

Proof. Notice that the function $d_H(\mathbf{X}, A)$ is a random variable defined on Ω. As usual let $\mu = \mathbf{E}\,[d_H(\mathbf{X}, A)]$. By Theorem 12.9, for $t > 0$, we have

$$\mathbf{Pr}\,[d_H(\mathbf{X}, A) - \mu < -t] \leq e^{-2t^2/n}.$$

Notice that $d_H(\mathbf{X}, A) = 0$ only holds for those points that belong to A. Taking $t = \mu$ we have

$$\mathbf{Pr}\,[\mathbf{X} \in A] = \mathbf{Pr}\,[d_H(\mathbf{X}, A) - \mu \leq -\mu] \leq e^{-2\mu^2/n},$$

498

making

$$t_0 = \left(\tfrac{1}{2} n \ln \left(\frac{1}{\mathbf{Pr}\,[\mathbf{X} \in A]} \right) \right)^2,$$

we have $\mu \leq t_0$. Therefore, for $t > t_0$ we have

$$\mathbf{Pr}\,[d_H(\mathbf{X}, A) > t] \leq e^{-2(t-t_0)^2/n}.$$

As for $t \geq 2a$, $(t-a)^2 \geq t^2/4$ for $t \geq 2a$, we have that for $t \geq 2t_0$,

$$\mathbf{Pr}\,[d_H(\mathbf{X}, A) > t] \leq e^{-2(t-t_0)^2/n}.$$

When $0 \leq t \leq 2t_0$, $\mathbf{Pr}\,[d_H(\mathbf{X}, A) > t] \geq \mathbf{Pr}\,[\mathbf{X} \in A]$. Thus

$$\min\{\mathbf{Pr}\,[d_H(\mathbf{X}, A) > t]\,, \mathbf{Pr}\,[\mathbf{X} \in A]\} \leq e^{-t^2/2n},$$

which is the isoperimetric inequality in the statement of the theorem. ✄

In plain words Theorem 12.11 says that for any subset $A \subseteq \Omega$ such that $\mathbf{Pr}\,[\mathbf{X} \in A]$ is not too small, say at least $1/2$, with high probability, a random point is close to A under the Hamming distance, that is

$$\mathbf{Pr}\,\left[\mathbf{X} \in A_t^H\right] \geq 1 - 2e^{-t^2/4n}. \tag{12.15}$$

☞ **Example 12.17 (Isoperimetry on the Hypercube).** Let us consider the particular example of the n-dimensional discrete hypercube Q_n with vertex set $\{0,1\}^n$ and edges between pairs of vertices whose Hamming distance is 1. Consider the probability space $\Omega = \prod_{i=1}^n \Omega_i$ where $\Omega_i = \{0,1\}$ for all $i \in [n]$. Let $\mathbf{X} = (X_1, \ldots, X_n)$ be a random vertex of Q_n such that $\mathbf{Pr}\,[X_i = 0] = \mathbf{Pr}\,[X_i = 1] = \tfrac{1}{2}$ for all $i \in [n]$.

Then, for any set A of Ω, $\mathbf{Pr}\,[X \in A] = |A|/2^n$. Thus, if we choose $|A| = 2^{n-1}$ so $\mathbf{Pr}\,[X \in A] \geq 1/2$, by equation (12.15), we get the following isoperimetric inequality for the hypercube: For any $t \geq 0$,

$$|A_t| \geq 2^n(1 - 2e^{-t^2/4n}). \tag{12.16}$$

Historically, L. Harper [16], was the first to prove a tight isoperimetric inequality and he did it for the n-hypercube Q_n. His technique was based in the following: Given a hypercube Q_n and a real $t \geq 0$, define the *Hamming ball* of radius t centered at a vertex $\mathbf{x}$ as the set of vertices in Q_n at distance less or equal than t. Harper showed that for all two subsets $A_1, A_2 \subset Q_n$, there exist two Hamming balls B_1 and B_2 centered at vertex $\mathbf{0}$ and $\mathbf{1}$ respectively, such that $|B_1| = |A_1|$, $|B_2| = |A_2|$ and $d_H(B_1, B_2) \geq d_H(A_1, A_2)$. His final result is the following,

Theorem 12.12 (Harper). *For any given* $0 \leq r \leq n$, *let* $A \subseteq Q_n$ *be such that* $|A| \geq \sum_{k=0}^{r} \binom{n}{k}$. *Then, for any* $t > 0$, *we have* $|A_t^H| \geq \sum_{k=0}^{r+t} \binom{n}{k}$

Harper's result is deterministic, with $\nu(A) = |A|$. If we consider the probabilistic setting in which $\mathbf{Pr}\,[\mathbf{X} \in A] = 1/2$, a straightforward application of Chernoff's bounds (in the form of equation (12.8)), gives the following isoperimetric inequality for Q_n and any $t > 0$,

$$|A_t^H| \geq 2^n \left(1 - 2e^{-2t^2/n}\right). \tag{12.17}$$

Notice that this last equation is an improvement on (12.16).   ✐

Let us consider now another definition of distance, for which we are going to derive an isoperimetric inequality due to Talagrand [36, 37]. We shall survey the main features of this inequality and present some of its applications. In order to keep the exposition at a basic level, we will skip the proof of the inequality. We refer the interested reader to [34, Chap. 6], [41, Chap. 6] or [29, Sec. 4] for details of the proofs. Most of our presentation follows the line in the last reference.

Recall that the L_2 norm of a vector $\mathbf{a} = (a_1, \ldots, a_n) \geq \mathbf{0}$, is defined as

$$\|\mathbf{a}\| = \left(\sum_i^n a_i^2\right)^{1/2}.$$

As usual, let $\Omega = \prod_{i=1}^{n} \Omega_i$ be a probability product space, let $\mathbf{X}$ be a sequence of random variables $(X_1, \ldots, X_n)$ in the corresponding spaces Ω_i, and let $\mathbf{a} = (a_1, \ldots, a_n)$ be a point in $[0, \infty)^n$, such that not all its components are 0. For any two points in Ω, $\mathbf{x} = (x_1, \ldots, x_n)$ and any $\mathbf{y} = (y_1, \ldots, y_n)$, define the distance $d_{\mathbf{a}}$ as:

$$d_{\mathbf{a}}(\mathbf{x}, \mathbf{y}) = \sum_{i=1}^{n} a_i \, \delta_i(\mathbf{x}, \mathbf{y}).$$

Notice, that if we consider the specific vector $\mathbf{a} = \mathbf{1} = (1, 1, \ldots, 1)$ then,

$$d_{\mathbf{a}}(\mathbf{x}, \mathbf{y}) = \sum_{i=1}^{n} \delta_i(\mathbf{x}, \mathbf{y}) = d_H(\mathbf{x}, \mathbf{y}).$$

Given $A \subseteq \Omega$ define $d_{\mathbf{a}}(\mathbf{x}, A)$ as usual. The *Talagrand's convex distance* $d_T(\mathbf{x}, A)$ from a point $\mathbf{x}$ to a subset A, is the supremum over all possible

n-dimensional vectors $\mathbf{a}$ with $||\mathbf{a}|| = 1$ of $d_{\mathbf{a}}(\mathbf{x}, A)$, that is,

$$d_T(\mathbf{x}, A) = \sup_{||\mathbf{a}||=1} \left\{ \inf_{y \in A} \left\{ \sum_{i=1}^{n} a_i \, \delta_i(\mathbf{x}, \mathbf{y}) \right\} \right\}.$$

In other words, the Talagrand's distance is the best convex distance over *all* choices of non-negative unit vectors.

Given $A \subseteq \Omega$ and $t > 0$, the *Talagrand's t-neighborhood of A*, A_t^T, is defined in the usual way:

$$A_t^T = \{\mathbf{x} \in \Omega : d_T(\mathbf{x}, A) \le t\}.$$

With these definitions, we are ready to state Talagrand's inequality [36]:

Theorem 12.13 (Talagrand's isoperimetric inequality). *Let Ω be a product space, let $\mathbf{X} = (X_1, \dots, X_n)$ be a sequence of random variables, with each X_i defined on Ω_i. Then, for any $A \subseteq \Omega$ and any $t \ge 0$,*

$$\mathbf{Pr}[\mathbf{X} \in A]\left(1 - \mathbf{Pr}\left[\mathbf{X} \in A_t^T\right]\right) \le e^{-t^2/4}.$$

In plain words, Talagrand's inequality states that under the Talagrand's convex distance, if $A \subseteq \Omega$ is large enough, then A_t^T covers a large portion of Ω.

Let us give some intuition for the inequality in the previous Theorem with respect to the one obtained in Theorem 12.11. If we consider the vector $\mathbf{a}' = (1/\sqrt{n}, \dots, 1/\sqrt{n})$, then $||\mathbf{a}'|| = 1$, and for any $\mathbf{x} \in \Omega$

$$d_T(\mathbf{x}, A) = \sup_{||\mathbf{a}'||=1} \inf_{y \in A} \left\{ \sum_{i=1}^{n} a_i \delta_i(\mathbf{x}, \mathbf{y}) \right\}$$

$$\ge \inf_{y \in A} \sum_{i=1}^{n} \frac{1}{\sqrt{n}} \delta_i(\mathbf{x}, \mathbf{y})$$

$$= \inf_{y \in A} d_H(\mathbf{y}, A) = \frac{1}{\sqrt{n}} d_H(\mathbf{x}, A),$$

therefore, for any $t > 0$, we get

$$A_{t/\sqrt{n}}^T \subseteq A_t^H. \tag{12.18}$$

➡ **Example 12.17 (cont).** Let us return to the hypercube. Consider a random vertex $\mathbf{X} \in \Omega = Q_n$. Given a set $A \subseteq \Omega$ with $\mathbf{Pr}[\mathbf{X} \in A] \ge 1/2$, and using Talagrand's inequality together with (12.18), we have

$$\mathbf{Pr}\left[\mathbf{X} \in A_t^H\right] \ge \mathbf{Pr}\left[\mathbf{X} \in A_{t/\sqrt{n}}^T\right] \ge 1 - 2e^{t^2/4n}.$$

Therefore, we obtain the isoperimetric inequality of equation (12.17). ✏

Once again, we stress than the real strength of Talagrand's inequality over the inequality in Theorem 12.11 lies in the fact that Talagrand's refers simultaneously to all possible unit vectors $\mathbf{a}$.

Talagrand's isoperimetric inequality has proved to be a useful tool to simplify concentration bounds on several kind of problems. However, its application is far from obvious. The basic setting to use Talagrand's is the following: given a product space Ω, and given function $f : \mathbb{R}^n \to \mathbb{R}$, and a collection of random variables $\mathbf{X} = (X_1, \ldots, X_n)$, we wish to study the concentration of $f(\mathbf{X})$ around the median of $f(\mathbf{X})$. Notice that the set of points with measure above the median has probability at least $1/2$.

To apply the method, we need first to find a bound on the Talagrand's distance between any two $\mathbf{x}, \mathbf{y}$ points in Ω. Then, considering the set A of points such that $f(\mathbf{x}) \leq \mathbf{M}[f(\mathbf{X})]$ and choosing a $t > 0$, we can apply Theorem 12.13, to show that the number of $\mathbf{x} \notin A_t^T$ is very small, so most of the points in Ω are in A_t^T, which implies that whp $f(\mathbf{x})$ and $\mathbf{M}[f(\mathbf{X})]$ are within distance t.

Let us return to the example of the longest increasing subsequence in a random permutation.

✏ **Example 12.9 (cont).**   We have seen, that the problem is equivalent to the problem of generate independently and uniformly n random reals in $[0, 1]$ and compute the length of the longest increasing subsequence. Let $\Omega = [0, 1]^n$, and let $\mathbf{X} \in \Omega$, be a random real sequence $\mathbf{X} = (X_1, \ldots, X_n)$. We wish to bound the concentration of the random variable $L(\mathbf{X})$ measuring the longest increasing subsequence.

Let $\mathbf{x} = (x_1, \ldots, x_n)$ a sequence of reals and let $K_{\mathbf{x}} \subseteq [n]$ be the set of subindex corresponding to the longest increasing subsequence, so $L(\mathbf{x}) = |K_{\mathbf{x}}|$. For any other sequence $\mathbf{y} \in [0, 1]^n$, we have

$$L(\mathbf{y}) \geq L(\mathbf{x}) - |\{i \in K_{\mathbf{x}} \mid x_i \neq y_i\}|.$$

To bound the Talagrand distance between any two points $\mathbf{x}$ and $\mathbf{y}$, let us consider the vector $\mathbf{a}' = (a_1', \ldots, a_n')$ where for every $i \in [n]$, $a_i' = 1/\sqrt{L(\mathbf{x})}$ if $i \in K_{\mathbf{x}}$ and $a_i' = 0$ otherwise. Then $\|\mathbf{a}'\| = 1$. Moreover, by definition of $\mathbf{a}'$, $|\{i \in K_{\mathbf{x}} \mid x_i \neq y_i\}| = \sqrt{L(\mathbf{x})}d_{\mathbf{a}'}(\mathbf{x}, \mathbf{y})$ and we get

$$L(\mathbf{y}) \geq L(\mathbf{x}) - \sqrt{L(\mathbf{x})}d_{\mathbf{a}'}(\mathbf{x}, \mathbf{y}). \tag{12.19}$$

For instance, take

$$\mathbf{x} = (0.7, 0.8, 0.3, 0.1, 0.5, 0.6, 0.2, 0.7, 0.4) \quad \text{and}$$
$$\mathbf{y} = (0.3, 0.7, 0.4, 0.9, 0.5, 0.2, 0.6, 0.8, 0.1).$$

We have $K_{\mathbf{x}} = \{3, 5, 6, 8\}$, $L(\mathbf{x}) = 4$ and $\mathbf{a}' = (0, 0, \frac{1}{2}, 0, \frac{1}{2}, \frac{1}{2}, 0, \frac{1}{2}, 0)$. Furthermore, $|\{i \in K_{\mathbf{x}} \mid X_i \neq Y_i\}| = 3$ and $5 = L(\mathbf{y}) \geq 4 - 2$.

If we choose $A = \{\mathbf{x} \mid L(\mathbf{x}) \leq m\}$, where $m = \mathbf{M}[L(\mathbf{X})]$, by definition of median and for any random variable $\mathbf{X}$ on Ω,

$$\mathbf{Pr}[\mathbf{X} \in A] = \mathbf{Pr}[L(\mathbf{X}) \leq m] \geq \frac{1}{2}$$

and

$$\mathbf{Pr}[\mathbf{X} \notin A] = \mathbf{Pr}[L(\mathbf{X}) \geq m] < \frac{1}{2}.$$

For any $t \in [n]$ define the usual way A_t^T. Let $\mathbf{y} \in A$, that means $L(\mathbf{y}) \leq m$, and let $\mathbf{x} \notin A_t^T$. From Equation (12.19), taking the infimum over all possible $\mathbf{a} \in A$ and replacing $L(\mathbf{y})$ by m, we get $L(\mathbf{x}) \leq m + \sqrt{L(\mathbf{x})} \, d_{\mathbf{a}}(\mathbf{x}, A) \leq m + \sqrt{L(\mathbf{x})} \, d_T(\mathbf{x}, A)$. So,

$$d_T(\mathbf{x}, A) \geq \frac{L(\mathbf{x}) - m}{\sqrt{L(\mathbf{x})}} \geq \frac{t}{\sqrt{t + m}},$$

where this last inequality follows from the fact that the function $\frac{L(\mathbf{x}) - m}{\sqrt{L(\mathbf{x})}}$ is increasing for values of $L(\mathbf{x}) > m$. Thus,

$$\begin{aligned}
\mathbf{Pr}[\mathbf{X} \notin A_t^T] &= \mathbf{Pr}[L(\mathbf{X}) \geq m + t] \\
&\leq \mathbf{Pr}\left[d_T(\mathbf{X}, A) \geq \frac{t}{\sqrt{t + m}}\right] \\
&\leq e^{-\frac{t^2}{4(m+t)}}.
\end{aligned} \tag{12.20}$$

Using Talagrand's inequality, for any $t \in [m]$,

$$\begin{aligned}
\mathbf{Pr}[L(\mathbf{X}) \leq m]\,\mathbf{Pr}[L(\mathbf{X}) \geq m + t] &= \mathbf{Pr}[\mathbf{X} \in A]\,\mathbf{Pr}[\mathbf{X} \notin A_t^T] \\
&\leq \mathbf{Pr}[\mathbf{X} \in A]\,\mathbf{Pr}\left[d_T(\mathbf{x}, A) \geq \frac{t}{\sqrt{t + m}}\right] \\
&\leq 2e^{-\frac{t^2}{4(m+t)}}
\end{aligned}$$

and, as by definition of the set A, we have $\mathbf{Pr}[L(\mathbf{x}) \leq m] \geq 1/2$, then for any $t \geq 0$

$$\mathbf{Pr}[L(\mathbf{X}) \geq m + t] \leq 2e^{-\frac{t^2}{4(m+t)}}, \tag{12.21}$$

and repeating the argument for $m - t$ in the place of m, we get that for any $t \in [m - 1]$,

$$\mathbf{Pr}\left[L(\mathbf{X}) \geq m - t\right] \leq 2e^{-\frac{t^2}{4m}}, \tag{12.22}$$

which gives a good concentration result. ✏

We have used the median, to ensure that the set A is "large enough". However, it turns out, that under the condition of Talagrand's inequality, the same inequalities imply that the expectation μ and the median m are very close: $|\mu - m| \leq O(\sqrt{m})$. The following theorem is from [29] and generalizes this relationship. The proof uses only classical probability theory and it is left to the reader.

Theorem 12.14 (McDiarmid). *Let the random variable $f(\mathbf{X})$ have mean μ and median m, and let $a, b > 0$ be constants. Then for any $t > 0$,*

$$\mathbf{Pr}\left[f(\mathbf{X}) - m > t\right] \leq ae^{-t^2/b} \implies |\mu - m| \leq O(a\sqrt{b})$$
$$\mathbf{Pr}\left[f(\mathbf{X}) - m \geq t\right] \leq ae^{-t^2/b(m+t)} \implies |\mu - m| \leq O(\sqrt{m}).$$

There have been several studies on families of problems, where similar concentration theorems based on Talagrand's isoperimetric inequality, can be used. For instance, the above proof is taken from a more general theorem for functions f satisfying certain conditions, very similar to equation 12.19. See [29, Theorem 4.3]. Furthermore, for problems that could be defined in terms of hereditary properties of index sets, there exists a framework to apply a concentration theorem based in Talagrand's inequalities. Examples of problems in this class are increasing weighted sequences, balls and bins, counting extensions and representation of a graph, and graph edge coloring. See [9] and [34, Chapter 6]. Geometrical problems are another class of problems that can be treated with a unified approach using this method. Among them, the Euclidean traveling salesman problem, the Steiner tree problem and minimum spanning tree problem have received much attention. Again, see [9] and [34, Chap. 6]. A particularly handy but weaker form of Talagrand's inequality, was given in [31].

Theorem 12.15 (Molloy). *Let $f(\mathbf{X})$ be a random variable determined by n independent experiments, satisfying the following two conditions:*

1. *Changing the outcome of any one experiment can affect f by at most a small constant c, and*

2. *for any s, if $f(\mathbf{X}) \geq s$, by changing the outcomes of all the exper-*
iments, except s of them, the inequality $f(\mathbf{X}') \geq s$, still holds.

Then, for any $0 < t \leq m$,

$$\mathbf{Pr}\left[|f(\mathbf{X}) - m| > t\right] \leq e^{-\frac{t^2}{16c^2m}},$$

where m is the median of f.

Notice, that to prove the second condition, it is sufficient to show that the outcomes of the selected s trials provide a lower bound for f. Let us apply Molloy's version of Talagrand's, to the the longest increasing subsequence example.

☞ **Example 12.9 (cont).** Let $\mathbf{X} = (X_1, \ldots, X_n)$ be a random sequence of reals, let $s = L(\mathbf{X})$ and $m = \mathbf{M}[L(\mathbf{X})]$. Notice that changing the value of a single X_i, it may affect $L(\mathbf{X})$ by at most one. Moreover, the $\{X_i\}$ such that $i \in K_{\mathbf{X}}$ are by themselves alone witnesses to the fact that $s = L(\mathbf{X})$. Therefore, we can apply Therefore, Theorem 12.15 to get

$$\mathbf{Pr}\left[|L(\mathbf{X}) - m| > t\right] \leq 2e^{-t^2/16m}.$$

That shows concentration around the mean value m.   ✏

Coming back to the balls and bins, experiment, Molloy's version of Talagrand's inequality can be applied to show concentration of the number of non-empty bins.

☞ **Example 12.7 (cont).** Let us consider now a symmetrical version of the balls and bins experiment. Let Y be the random variable that counts the number of non empty bins. Notice that changing the outcome of one throw only can change the function by at most 1. If we have that the number of non empty bins in a configuration is at least s, there must exits s balls that fall in s different bins, by fixing these positions, changing the outcomes of the other balls will maintain s as lower bound. From Theorem 12.15 we get

$$\mathbf{Pr}\left[|Y - \mathbf{M}[Y]| > t\right] \leq 2\exp\left(-\frac{t^2}{16\mathbf{M}[Y]}\right).$$

That shows concentration around the median. Using Theorem 12.14 and taking into account that $Z + Y = n$, the above result may be used to show concentration around the mean for the number of empty bins.   ✏

A final remark is that in the statement of Talagrand's isoperimetric inequality, the variables $X_1, \ldots, X_n$ must be independent. Recent work by Marton extends Talagrand's result to enable some sort of limited dependence [27].

References

[1] N. Alon, J. H. Spencer, and P. Erdős. *The probabilistic method.* Wiley-Interscience, New York, 1992.

[2] K. Azuma. Weighted sums of certain dependent random variables. *Tokuku Math. Journal*, 19:357–367, 1967.

[3] B. Bollobás. The chromatic number of random graphs. *Combinatorica*, 8:49–55, 1988.

[4] H. Chernoff. A measure of asymptotic efficiency for tests of a hypotheses based on the sum of observations. *Annals of Mathematics Statistics*, 23:493–507, 1952.

[5] T. H. Cormen, Ch. Leiserson, and R. Rivest. *Introduction to Algorithms.* The MIT Press, Cambridge, Mass., 1989.

[6] J. Díaz, M. D. Penrose, J. Petit, and M. J. Serna. Linear orderings of random geometric graphs. In P. Widmayer, G. Neyer, and S. Eidenbenz, editors, *Graph Theoretic Concepts in Computer Science*, volume 1665 of *Lecture Notes in Computer Science*, pages 291–302. Springer-Verlag, 1999.

[7] J. Díaz, M. J. Serna, P. Spirakis, J. Torán, and T. Tsukiji. On the expected depth of boolean circuits. Technical Report LSI 94-7-R, Universitat Politècnica de Catalunya, 1994.

[8] D. Dubhashi. Simple proofs of occupancy tail bounds. *Random Structures and Algorithms*, 11:119–123, 1997.

[9] D. Dubhashi. Talagrand's inequality in hereditary settings. Technical report, Dept. CS, Indian Institute of Technology, 1998.

[10] W. Feller. *An Introduction to Probability Theory and its Applications. Vol I.* J. Wiley, New York, 1957.

[11] D. A. Freedman. On tail probabilities for martingales. *Annals Probability*, 3:100–118, 1975.

[12] A Frieze and B. Reed. Probabilistic analysis of algorithms. In H. Habib, C. McDiarmid, J. Ramirez, and B. Reed, editors, *Probabilistic Methods for Algorithmic Discrete Mathematics*, pages 36–92. Springer-Verlag, 1998.

506

[13] Alan Frieze. On the length of the longest monotone subsequence in a random permutation. *Annals Applied Probability*, 1:301–305, 1991.

[14] G. Grimmett and D. R. Stirzaker. *Probability and Random Processes. (2nd. edition)*. Oxford University Press, 1992.

[15] T. Hagerup and C. Rub. A guided tour of Chernoff bounds. *Information Processing Letters*, 33:305–308, 1990.

[16] L. H. Harper. Optimal numberings and isoperimetric problems on graphs. *Journal of Combinatorial Theory*, 1(3):385–393, 1966.

[17] W. Hoeffding. Probability inequalities for sums of bounded random variables. *Journal American Statistics Association*, 58:13–30, 1963.

[18] M. Jerrum and G. Sorkin. Simulated Annealing for Graph Bisection. In *34th Symposium on Foundations of Computer Science*, pages 94–103, 1993.

[19] J. JáJá. *An introduction to parallel algorithms*. Addison-Wesley, Reading, Mass., 1992.

[20] A. Kamath, R. Motwani, K. Palem, and P. Spirakis. Tail bounds for occupancy problems and the satisfiability conjecture. *Random Structures and Algorithms*, 7:59–80, 1995.

[21] R. M. Karp. Class notes for CS-292F. 1989.

[22] E. Koutsupias and X. Papadimitriu. On the greedy algorithm for satisfiability. *Information Processing Letters*, 43:53–55, 1992.

[23] I. Leader. Discrete isoperimetric inequalities. In B. Bollobas, editor, *Probabilistic Combinatorics and its Applications*, volume 44 of *Proceedings of Symposia in Applied Mathematics*, pages 57–80. American Mathematical Society, Rhode Island, 1991.

[24] M. Ledoux. Isoperimetry and gaussian analysis. In P. Groeneboom R. Dobrushin and M. Ledoux, editors, *Lectures in Probability Theory. Ecole d'Ete de Probabilites de Saint-Flour XXIV*, volume 1648 of *Lectures Notes in Mathematics*, pages 165–294. Springer-Verlag, Berlin, 1994.

[25] B. F. Logan and L. A. Shepp. A variational problem for Young tableaux. *Advances in Mathematics*, 26:206–222, 1977.

[26] M. Luby. A simple parallel algorithm for the maximal independent set. *SIAM Journal on Computing*, 15:1036–1056, 1986.

[27] K. Marton. On a measure concentration inequality of Talagrand for dependent random variables. Technical report, Math. Institute of the Hungarian Academy of Sciences, 1996.

[28] C. McDiarmid. On the method of bounded differences. In J. Siemons, editor, *Surveys in Combinatorics, 1989*, pages 669–188. London Mathematical Society Lectures Notes Series, 1989.

[29] C. McDiarmid. Concentration. In H. Habib, C. McDiarmid, J. Ramirez, and B. Reed, editors, *Probabilistic Methods for Algorithmic Discrete Mathematics*, pages 195–248. Springer-Verlag, 1998.

[30] C. McDiarmid and R. B. Hayward. Large deviations for quicksort. *Journal of Algorithms*, 21:476–507, 1996.

[31] M. Molloy. The probabilistic method. In H. Habib, C. McDiarmid, J. Ramirez, and B. Reed, editors, *Probabilistic Methods for Algorithmic Discrete Mathematics*, pages 1–35. Springer-Verlag, 1998.

[32] R. Motwani and P. Raghavan. *Randomized Algorithms*. Cambridge University Press, 1995.

[33] E. Shamir and J. Spencer. Sharp concentration of the chromatic number in random graphs. *Combinatorica*, 7:121–130, 1987.

[34] J. M. Steele. Probabilistic algorithm for the directed traveling salesman problem. *Mathematics of Operation Research*, 11:343–350, 1986.

[35] W. Steiger. Some Kolmogoroff-type inequalities for bounded random variables. *Biometrica*, 54:641–648, 1967.

[36] M. Talagrand. Concentration of measure and isoperimetric inequalities in product spaces. *Publications Mathématiques de l'I.H.E.S.*, 81:73–1205, 1995.

[37] M. Talagrand. A new look at independence. *Annals of Probability*, 24:1–34, 1996.

[38] T. Tsukiji and F. Xhafa. On the depth of randomly generated circuits. In J. Díaz and M. Serna, editors, *4th European Symposium on Algorithms*, volume 1136 of *Lecture Notes in Computer Science*, pages 208–219. Springer-Verlag, Berlin, 1996.

[39] A. M. Vershik and C. V. Kerov. Asymptotics of the Plancherel measure of the symmetric group and a limiting form from Young Tableau. *Dokl. Akad. Nauk. U.S.S.R.*, 233:1024–1027, 1977.

[40] U. Vishkin. Randomized speed-ups in parallel computation. In *15th. ACM Symposium on Theory of Computing*, pages 230–239. ACM Society, 1984.

[41] J. E. Yukich. *Probability Theory of classical Euclidian Optimization Problems*. Springer–Verlag, Heidelberg, 1998.

Chapter 13

BOUNDED ERROR PROBABILISTIC FINITE STATE AUTOMATA

Anne Condon

The Department of Computer Science

University of British Columbia

Vancouver, B.C. Canada

condon@cs.ubc.ca

1. INTRODUCTION

What power does randomness confer to computing devices? In this article, we focus on this question in what is perhaps its simplest form, namely when the computing device is a finite state automaton.

Some of the oldest studies of probabilistic computations, dating as far back as the 40's, implicitly concern probabilistic finite state devices. A beautiful theory of probabilistic finite state automata was developed starting in the early 60's [Rabin, 1963, Paz, 1971]. This work primarily concerned automata with 1-way heads on the input tape, where an input w is considered to be accepted if the probability of reaching the accept state from the initial configuration is greater than some threshold, say $1/2$. The class of languages thus accepted is known as the stochastic languages.

A pfa for a stochastic language may err by rejecting inputs in the language with probability that approaches $1/2$ as the input size increases. It is natural to consider the language-recognition power of pfa's whose probability of error is bounded away from $1/2$. The theory of so-called *bounded error* (or isolated threshold) pfa's was initiated by Rabin [Rabin, 1963], but much progress in understanding these pfa's has been made more recently, and so we focus on bounded error pfa's in this survey. The study of bounded error pfa's is motivated by questions such as the following: if one wants to recognize patterns with small degree of error,

S. Rajasekaran et al (eds.), Handbook of Randomized Computing, Volume 2, pp, 509–531.

© 2001 *Kluwer Academic Publishers.*

can it be done with fewer states then when recognizing a pattern exactly? What kinds of patterns can be recognized when a small probability of error is allowed, but that cannot be recognized exactly?

The purpose of this survey is to describe a coherent set of results on bounded error pfa's. In Section 2. we describe pfa's and define associated language classes. In Section 3. we present a result of Freivalds that pfa's with a 2-way input head can accept nonregular languages. In particular, the language $\{a^n b^n \mid n \geq 0\}$ is accepted with bounded error by a 2pfa, although the 2pfa has worst case expected running time that is exponential in the length of its input. We also describe work by Ravikumar [Ravikumar, 1992] that builds on Freivalds work to show two classes of languages that are recognized with bounded error by 2pfa's.

Freivald's work raises the question: if the expected running time of a pfa is limited to polynomial time, can nonregular languages still be recognized? In Section 4. we show that the answer is no. Rabin [Rabin, 1963] showed pfa's with a 1-way input head accept exactly the regular languages. Dwork and Stockmeyer [Dwork and Stockmeyer, 1992] and independently Kaneps and Freivalds [Kaneps and Freivalds, 1990] showed that 2pfa's that run in subexponential expected time only accept the regular languages. The techniques used to prove this result include fundamental results on Markov chains that are interesting in their own right [Dwork and Stockmeyer, 1990, Greenberg and Weiss, 1986, Leighton and Rivest, 1983]. We conclude this section with a brief summary of results on succinctness of 2pfa's in terms of number of states needed to recognize languages compared with deterministic or nondeterministic finite automata.

Section 5. concerns undecidability results for pfa's. Freivalds showed that the problem of determining whether a bounded error 2pfa with known acceptance threshold accepts the empty language is undecidable. Freivalds' work was generalized to prove a related undecidability result for 1pfa's [Condon and Lipton, 1989]. At the end of this section, we describe how these undecidability results were applied to show that problems in probabilistic planning and Markov decision processes are undecidable [Madani et al., 1999].

2.   PFA MODEL

We assume that the reader is familiar with the definitions of deterministic and nondeterministic finite state automata (dfa's and nfa's). We denote the class of languages accepted by such automata, with either 1-way or 2-way input heads, as **Regular**.

A 2-way head probabilistic finite state automaton (pfa or 2pfa), like a 2-way dfa or nfa, has a finite set of states including an initial state and zero or more accept states, a finite input alphabet, and a transition function. Also associated with a pfa are a left endmarker, #, and a right endmarker, $, which are used to mark the left and right ends, respectively, of the input and are not in the input alphabet. The transition function maps (state, symbol) pairs to a (real-valued) probability distribution over (state, head direction) pairs, where the symbol may be from the input alphabet or an endmarker and the head direction determines whether the head moves one position left (L) on the input tape, one position right (R), or remains in the same position (N). Transitions are defined so that the head never moves left from the left endmarker or right from the right endmarker. For example, the pfa of Figure 1 has input alphabet $\{a\}$, left endmarker #, and right endmarker $. On input symbol a, state 2 from state 1 is reached with probability 1/5 and state 1 is reached with probability 4/5; in both cases the head moves right.

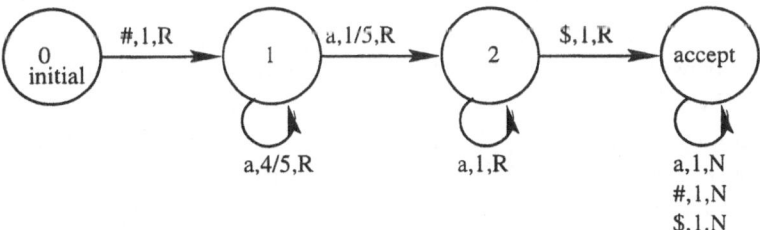

Figure 13.1 Example pfa with initial state 0. An edge labeled (σ, p, X) denotes a transition with probability p in which σ is under the tape head. When $X = R$, the head moves right and when $X = N$ the head does not move; since on no transition does the head move left, this is a 1pfa. The pfa has an additional reject state which is not shown, and all transitions not shown are to this reject state. For example, from state 0 on either $ or a, the reject state is reached with probability 1.

Throughout, we will use the following conventions. We assume that a pfa has a unique accepting state which is halting, that is, all transitions from the accept state lead back to this state without moving the head. Similarly, there is a unique halting rejecting state. A 1pfa is a pfa with no transitions that cause the tape head to move left. Whenever we refer to an input w of a pfa, we mean that the pfa has $\#w\$$ on its tape.

We say that input w is *accepted* by the pfa if the probability that the accept state of the pfa is eventually reached, starting with the head in the initial state on the left endmarker, is greater than 1/2. The language accepted by pfa A is denoted by $L(A)$. It is not hard to see that the pfa of Figure 1 accepts the language $\{a^k \mid k \geq 4\}$. The class

512

of languages accepted by pfa's is known as the stochastic languages, and we denote it by **Stochastic**. Replacing the acceptance threshold by any real number in the range $(0, 1)$ in the definition of string acceptance does not change the class **Stochastic** [Paz, 1971]. Rabin [Rabin, 1963] showed that **Stochastic** contains nonregular languages. For examples of other nonregular stochastic languages, including unary languages, see [Dwork and Stockmeyer, 1990, Paz, 1971, Salomaa and Soittola, 1978]. The class Stochastic was initially defined for 1pfa's only, but Kaneps [Kaneps, 1989], building on work of Turakainen [Turakainen, 1969], showed that 2pfa's accept exactly those languages accepted by 1pfa's. Macarie [Macarie, 1998] showed that the question of whether a given string x is accepted with probability $> 1/2$ by a given 1pfa P which has rational transition probabilities can be decided in deterministic logarithmic space.

Consider a pfa A with the property that for all inputs in $L(A)$, the probability of reaching the accept state is in fact greater than $1/2 + \gamma$ for some constant $\gamma > 0$ (independent of the input length), while the probability that the accept state is reached on inputs not in $L(A)$ remains at most $1/2$. We say that such a pfa accepts its language with bounded error, and we refer to $1/2 + \gamma$ as the *acceptance threshold* of the pfa. We denote the languages accepted by bounded error 1pfa's by 1PFA and the languages accepted by bounded error 2pfa's by 2PFA. The class of languages accepted by pfa's which are restricted to run in polynomial expected time is denoted by 2PFA-polytime.

Finally, we say that a pfa is *coin-flipping* if all transition probabilities are in the set $\{0, 1/2, 1\}$. We note that for any pfa with rational transition probabilities there is an equivalent coin-flipping pfa, in the sense that acceptance probabilities on all strings are the same. Figure 2 shows a coin-flipping pfa that is equivalent to that of Figure 1. All of the 2pfa constructions described in Sections 3. and 5. are coin-flipping.

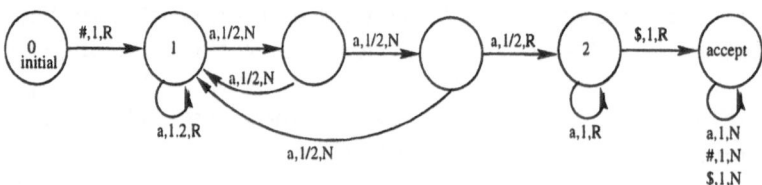

Figure 13.2 Coin-flipping pfa that is equivalent to that in Figure 1.

3.   2PFA CONTAINS NON-REGULAR LANGUAGES

Before describing results on the limitations of pfa's, it is useful to first see a pfa that does something surprising. We describe here a coin-flipping 2pfa of Freivalds [Freivalds, 1981] that accepts the language $\{a^n b^n \mid n > 0\}$ with bounded error.

Freivalds' 2pfa does the following. First, (deterministically) check that the input is of the form $a^n b^m$ for some $n > 0, m > 0$ and that $n = m \bmod (k + 1)$ for some constant k and if not, reject. Here, k controls the error probability of the 2pfa, as we discuss later.

Then, while scanning the input repeatedly, do the following. On each scan, flip a fair coin for each a and b; call these a-flips and b-flips, respectively. A scan is said to be a success for the a's if all a-flips are heads but at least one b-flip is a tail, and is a success for the b's if all b-flips are heads but at least one a-flip is a tail. If there are L successes for the a's before any success for the b's or vice versa, then reject, else accept. Here again L is a constant that controls the error probability, as discussed later.

If $n = m$ then on each scan the probability of success for the a's is the same as the probability of success for the b's. However, if $n > m+k$, then the probability of success for the b's is at least 2^k times the probability of success for the a's. A similar statement holds if $m > n + k$.

Thus, if $n = m$, the chance of L successes for the a's before any success for the b's is at most $(1/2)^L$ (and vice versa), and so the 2pfa accepts a string in the language with probability at least $1 - (1/2)^{L-1}$. For example, to obtain a machine with acceptance threshold $3/4$, it is sufficient to choose $L = 3$.

But if $n > m + k$, then in a single trial,

$$\frac{\text{Prob[success for the } a\text{'s]}}{\text{Prob[success for the } a\text{'s or } b\text{'s]}} \leq \frac{1}{2^k + 1}.$$

Therefore, the probability that the first L successful trials are all successes for the b's, and thus that the 2pfa rejects, is at least $(1 - 1/(2^k + 1))^L$. By choosing k sufficiently large (depending on L), this constant can be made as close to 1 as desired. For example, if $L = 3$ and $k = 2$ then $(1 - 1/(2^k + 1))^L = .512$ and so A rejects inputs not in L with probability at least $1/2$.

B. Ravikumar [Ravikumar, 1992] extended Freivald's construction to show that all bounded semilinear languages and all languages recognizable by deterministic blind counter machines are contained in 2PFA. These language include non-context free languages such as $\{a^n b^n c^n \mid n \geq 0\}$.

4.   LIMITS ON THE POWER OF TIME-RESTRICTED PFA'S

The expected running time of Freivalds' 2pfa is exponential in the input size, since the probability that a single scan is a success for the a's or b's is exponentially small in the length of the input. If one is interested in efficient finite state pattern recognizers, it is necessary to restrict attention to pfa's that run in polynomial expected time. As it turns out, such pfa's accept only the regular languages. The earliest result along these lines is due to Rabin [Rabin, 1963] and is for 1pfa's. We describe this first, and then return to the class 2PFA-polytime.

4.1   1PFA = REGULAR

Rabin's proof uses the following fundamental property of the regular languages. Define two strings x and x' to be Myhill-Nerode-equivalent with respect to a given language L if for all strings y, xy is in L if and only if $x'y$ is in L. Then L is regular if and only if the number of Myhill-Nerode equivalence classes with respect to L is finite (see the text by Hopcroft and Ullman [Hopcroft and Ullman, 1979]). The proof of one direction of the Myhill-Nerode theorem is easy: if A is a dfa for L, define x and x' to be A-equivalent if x and x' lead to the same state of A from the initial state. A-equivalence implies Myhill-Nerode equivalence, and hence the number of Myhill-Nerode equivalence classes is at most the number of states of A.

We'd like to use the behavior of a 1pfa P on two strings x and x' to define some notion of P-equivalence between x and x', such that if x and x' are P-equivalent, then they are Myhill-Nerode-equivalent. A natural approach is to base a notion of P-equivalence on the probability distribution of states reached when the head moves to the right off x or x', when starting in the initial configuration. For simplicity, assume without loss of generality that the 1pfa P does not enter the accept or reject state unless the head reaches the right end-marker. However P may loop forever without reaching either the accept or reject state, perhaps by cycling among several states without ever moving its head. Let the non-halting states of 1pfa P be numbered $1, \ldots, c$ and let $p(x)$ be a c-vector whose ith entry is the probability of being in state i when the tape head moves off x. Intuitively, if the vectors $p(x)$ and $p(x')$ are close then for any string y, the probability that xy is accepted is close to the probability that $x'y$ is accepted.

More precisely, suppose that $||p(x) - p(x')|| \leq \epsilon$, where $\epsilon > 0$ and $||x||$ denotes the max of the absolute values of the entries of the vector x. Let $q(y)$ be the column c-vector with ith entry equal to the probability that the accept state of P is reached from a starting configuration in which P is in state i and the head of P is on the first (leftmost) symbol of $y\$$. Then, the inner product $p(x)q(y)$ is the probability that P accepts string xy and $p(x')q(y)$ is the probability that P accepts $x'y$. Therefore,

$$|\text{Prob}[P \text{ accepts } xy] - \text{Prob}[P \text{ accepts } x'y]|$$
$$\leq ||p(x) - p(x')q(y)|| \leq c\epsilon. \tag{1.1}$$

Let $\epsilon > 0$ be sufficiently small so that the acceptance threshold of P is $> 1/2 + c\epsilon$. Then if P accepts xy, $\text{Prob}[P \text{ accepts } xy] > 1/2 + c\epsilon$. Combining this with inequality (1.1), we have that $\text{Prob}[P \text{ accepts } x'y] > 1/2$. Hence, P must also accept $x'y$.

Now, partition the space of all vectors in $[0, 1]^c$ into a constant number of cells of dimension at most ϵ, so that for any pair of vectors v, w in the same cell, $||v - w|| \leq \epsilon$. Define x and x' to be P-equivalent if $p(x)$ and $p(x')$ are both in the same cell. The argument of the previous paragraph shows that if x and x' are P-equivalent (with ϵ chosen appropriately depending on the acceptance threshold of P), then for any string y, xy is in L if and only if $x'y$ is in L. Thus, the number of equivalence classes of the language accepted by P is bounded by the number of cells in the partition of $[0, 1]^c$, and thus is finite. By the Myhill-Nerode theorem, the language accepted by P is therefore regular.

4.2   2PFA-POLYTIME = REGULAR

Dwork and Stockmeyer [Dwork and Stockmeyer, 1992] and independently Kaneps and Freivalds [Kaneps and Freivalds, 1990] generalized Rabin's result to show that only the regular languages are recognized by 2pfa's that are restricted to have expected running time that is subexponential. In particular, 2PFA-polytime = Regular. We describe their techniques in this section. These build on earlier work of Greenberg and Weiss [Greenberg and Weiss, 1986], who showed that exponential time is needed to recognize $\{a^n b^n\}$.

Let P be a 2pfa that halts in polynomial expected time; in particular P halts with probability 1, and assume again for simplicity that the accept or reject states are only entered when the head is on the right endmarker. As in Rabin's proof, a key idea is to model the behavior of P on x in such a way that if the models for x, x' are close (according to some measure), then xy is accepted if and only if $x'y$ is. In the case of a 1pfa, the behavior on x is modeled by the vector $p(x)$, but this is inadequate for 2pfa's, since, during a computation of a 2pfa on input xy, the tape head may repeatedly cross left from y back onto x. Thus, to fully capture the behavior of P on x, in addition to including the probabilities in $p(x)$, we need to include, for each pair of nonhalting states q and q', the probability that, starting from q with the head on the right end of $\#x$, P is in state q' when its head moves off the right end of $\#x$ for the first time. Note that these probabilities are independent of whatever string is to the right of x.

It is convenient to represent these probabilities as weights on edges of a graph, which we denote by $M[x]$. $M[x]$ has $2c + 1$ nodes and directed weighted edges. Of these nodes, c are of the form $(q, 1), 1 \leq q \leq c$, representing the configuration in which P's head is on the right end of x in state q. Also, c are of the form $(q', 2)$, representing the configuration in which the state of P is q' and the head is on the first symbol of the string that is to the right of $\#x$. The weight of the edge $(q, 1) \rightarrow (q', 2)$ is the probability that, starting from configuration $(q, 1)$, the first configuration reached, among those configurations that are represented as nodes of $M[x]$ of the form $(*, 2)$, is $(q', 2)$. The remaining node in the graph is Initial, denoting the initial configuration of P on $\#x$. The weight of the edge Initial $\rightarrow (q', 2)$ denotes the probability that the first configuration reached from Initial, among those configurations that are represented as nodes of $M[x]$ of the form $(*, 2)$, is $(q', 2)$. Similarly, we can model the behavior of P on $y\$$ as a graph $M[y]$ with $2c + 2$ nodes. Of these nodes, c are of the form $(q, 1)$, representing the configuration in which the head is on the symbol just to the left of $y\$$, c are of the form $(q, 2)$, representing the configuration in which the head is on the leftmost symbol of $y\$$, and the weight of edge $(q', 2) \rightarrow (q, 1)$ is the probability that, starting from $(q', 2)$, the first configuration reached, among those represented as nodes of $M[y]$ that are not of the form $(*, 2)$, is $(q, 1)$. The two additional nodes are Accept and Reject, where edge $(q', 2) \rightarrow$ Accept (Reject) has weight equal to the probability of reaching the accept (reject) state from configuration $(q', 2)$ before reaching any configuration of type $(*, 1)$. Also, there is an edge of weight 1 from Accept to itself and from Reject to itself.

Let $M[x, y]$ denote the graph with $2c + 3$ nodes obtained from the "union" of the graphs $M[x]$ and $M[y]$. That is, the set of nodes (resp. edges) of $M[x, y]$ is the union of the set of nodes (resp. edges) of $M[x]$ and $M[y]$. Note that the sum of the weights of edges from any node of $M[x, y]$ is 1 since P halts with probability 1. The probability of eventually reaching Accept from Initial in $M[x, y]$, while transitioning between states according to the edge probabilities, equals the probability that the accept state of P is reached from the initial configuration on input xy.

Now, consider a sequence of random variables $X_0, X_1, \ldots$ over the nodes of $M[x, y]$, where $X_0 = $ Initial and for $i \geq 1$, $\text{Prob}[X_i = N \mid X_{i-1} = N']$ is the weight of edge $N \rightarrow N'$. This sequence is a (discrete, time-inhomogeneous) Markov chain; we also use $M[x, y]$ to refer to this Markov chain. Associated with a Markov chain M is a square transition probability matrix $[p_{ij}]$ with dimension equal to the number of states of M. Entry p_{ij} is the probability of reaching state j from state i in one step. To define equivalence between strings x and x', we need to know how small perturbations in the transition probabilities of a Markov chain affect the probability of eventually reaching Accept from Initial.

Let's look at two examples, to get some intuition on this. In our first example, let M have states 1, 2, and 3, with 1 being the initial state and states 2 and 3 being absorbing, that is, the probability of reaching 2 from 2 is 1 and similarly for 3. Let $a(M)$ and $a(M')$ denote the probability that state 2 is eventually reached, starting from state 1. Let p_2, p_3 be the probabilities of going from state 1 to state 2 and from state 1 to state 3, respectively. The probability of looping at state 1 is thus $1 - p_2 - p_3$ and $a(M)$ is $p_2/(p_2 + p_3)$. Let M' be the same as M, except that p_i is replaced by p_i' everywhere. Suppose that $p_2 = p_3$, $p_2' = 2p_2$, and $p_3' = p_3$. Then $a(M) = 1/2$ and $a(M') = 2/3$. By choosing p_2 small enough, the quantity $|p_2' - p_2| = 2p_2 - p_2 = p_2$ can be made arbitrarily small, yet the difference $a(M) - a(M')$ remains equal to $3/2 - 1/2$.

This example suggests that the *ratio*, rather than the difference, of corresponding transition probabilities needs to be close to 1 in order for two Markov chains M and M' to have similar absorbtion probabilities at their (common) absorbing states.

Let $\beta \geq 1$. We say that two real numbers p and p' are β-close if either (i) $p = p' = 0$, or (ii) $p > 0, p' > 0$, and $\beta^{-1} \leq p/p' \leq \beta$. Let M and M' be two Markov chains with associated transition probability matrices $[p_{ij}]$ and $[p_{ij}']$. We say that M and M' are β-close if for all i, j, p_{ij} and p_{ij}' are β-close.

Let M and M' be β-close Markov chains over the same state space and suppose that $a(M)$ and $a(M')$ are the probabilities of reaching a (common) absorbing state. We would like an upper bound on the ratio $a(M)/a(M')$, as a function of β and m, where m is the number of states of M and M'. The following proposition provides such a bound. It was first proved by Greenberg and Weiss [Greenberg and Weiss, 1986] and follows from the Markov Chain Tree Theorem of Leighton and Rivest [Leighton and Rivest, 1983].

Proposition 1 *Let M and M' be β-close. Then $a(M)$ and $a(M')$ are β^{2m}-close.*

For example, suppose that M has m states, of which $m - 1$ and m are absorbing, and 1 is the initial state. Let $a(M)$ be the probability of eventually reaching $m - 1$ from 1. Let the probability of going from state i to $i + 1, 1 \leq i \leq m - 2$, be p, where $0 < p < 1/2$. Let the probability of going from state i to state m, $1 \leq i \leq m - 2$, be $1 - p$. Thus, $a(M) = p^{m-2}$. If M' is the same as M except that p is replaced everywhere by $2p$, then $a(M')$ is $(2p)^{m-2}$. In this example, M and M' have m states and are β-close for $\beta = 2$, and the ratio $a(M)/a(M')$ is β^m.

Proposition 1 does not seem too promising: for M and M' to be close, the difference between logs of (non-zero) transition probabilities needs to be small. But since non-zero transition probabilities lie in the range (0,1], the logs of these probabilities lie in the range $(-\infty, 1]$. Thus, it is not possible to partition the space of Markov chains into a finite number of classes such that all of those in the same class are β-close, where β is a small constant.

One ray of hope is that if x has length at most n, then the non-zero transition probabilities of $M[x]$ cannot be less than 2^{-cn-1}. (Recall that $M[x]$ is that part of the graph $M[x, y]$ with $2c + 1$ states determined by x.) To see this, note that, for example, if configuration $(q', 1)$ is reachable from configuration $(q, 2)$ on some execution of the pfa P on input $\#x$, then there is an execution for which $(q', 1)$ is reachable from $(q, 2)$ in at most $cn - 1$ steps. Since P is a coin-flipping pfa, the probability of this execution is at least 2^{-cn-1}.

If $|x| \leq n$, then the non-zero transition probabilities of $M[x]$ lie in the range $[2^{-cn-1}, 1]$ and so the logs (to base 2) of these probabilities lie in the range $[-cn - 1, 0]$. As Theorem 1 below makes precise, it is possible to partition the graphs $M[x]$, and thus the strings x of length $\leq n$, into a number of equivalence classes that is bounded by a polynomial function of n, such that if x and x' are in the same equivalence class then xy is accepted by P if and only if $x'y$ is accepted by P. There is some hope that this might be a useful step towards the goal of showing that P accepts only regular languages, if it can be combined with a quantitative version of the Myhill-Nerode theorem. Roughly, the latter would provide a lower bound on the number of words which must be distinguished by any recognizer of the strings of length at most n in a regular language.

Theorem 1 *Let P be a 2pfa that accepts its language with bounded error. There is a partition of the strings of length at most n into a number of classes that is bounded by a polynomial in n, such that if x and x' are in the same class then for all y, xy is accepted by P if and only if $x'y$ is.*

Since $M[x, y]$ models a 2pfa that always halt with probability 1, the following sketch of the proof of Theorem 1 applies only to such 2pfa's, but the techniques can easily be extended to apply to 2pfa's that do not necessarily halt with probability 1. Partition the interval $[-cn-1, 1]$ into subintervals of length at most ϵ, for some constant $\epsilon < 0$. Define x and x' to be ϵ-close if for each pair of corresponding edge weights p and p' in $M[x]$ and $M[x']$, either $p = p' = 0$ or $\log_2 p$ and $\log_2 p'$ are both in the same subinterval. Note that ϵ-closeness is an equivalence relation, and the number of equivalence classes is at most $(\lceil (cn+1)/\epsilon \rceil + 1)^{(2c+1)^2}$.

Let $a(xy)$ and $a'(xy)$ be the probability of reaching state Accept from Initial in $M[x, y]$ and $M[x', y]$, respectively. From the fact that x and x' are ϵ-close, it follows that $M[x, y]$ and $M[x', y]$ are 2^ϵ-close. By Proposition 1, it follows that $a(xy)$ and $a(x'y)$ are $2^{2(2c+3)\epsilon}$-close. Therefore,

$$\frac{a(xy)}{a(x'y)} \geq 2^{-2(2c+3)\epsilon}.$$

If $x'y$ is accepted then $a(x'y) > 1/2 + \delta$, for some constant δ. Hence,

$$a(xy) \geq (1/2 + \delta)2^{-2(2c+3)\epsilon}.$$

By choosing ϵ sufficiently small (depending on the number c of states of P and on δ), the right hand side of the above inequality can be made greater than $1/2$, and so xy must be accepted by P.

Corollary 1 *The language $\{w\&w \mid w$ is in $\{a, b\}^*\}$ is not in 2PFA.*

To see this, suppose that P is a 2pfa that accepts this language with bounded error. Let $p(n)$ be the polynomial of Theorem 1. Let n be such that $2^n > p(n)$. Consider the 2^n distinct strings of length n. All of these must be in different classes (see statement of Theorem 1), which is impossible, contradiction.

However, we cannot hope to strengthen it to show that *no* nonregular language is in 2PFA, since Freivalds has shown that $\{a^n b^n \mid n \geq 0\}$ is in 2PFA. In order to prove something about 2PFA-polytime, we need to strengthen Theorem 1 in the case that P halts in polynomial expected time. The following result does this.

Theorem 2 *Let P be a 2pfa that accepts its language with bounded error. Let the expected running time of P be bounded by a polynomial $t(n)$. Then there is a partition of the strings of length at most n into a number of classes that is bounded by a polynomial in $\log n$, such that if x and x' are in the same class then for all y such that $|xy| \leq n, |xy'| \leq n$, xy is accepted by P if and only if $x'y$ is.*

Roughly, the proof of Theorem 2 is as follows. Let (s, s') be a transition in $M[x, y]$ with probability p where $p \leq t(n)^{-2}$. Since P halts in expected time $t(n)$, it is unlikely that $M[x, y]$ ever transitions from s to s' before halting. Let $M'[x]$ be the graph obtained by changing transition probabilities of $M[x]$ that are $< t(n)^{-2}$ to 0. Since the edge weights of $M'[x]$ are in the range $[t(n)^{-2}, 1]$, it is possible to partition the strings x of length at most n into a number of classes that is polylogarithmic in n, such that if x and x' are in the same class then for all y with $|xy| \leq n$ and $|x'y| \leq n$, xy is accepted by P if and only if $x'y$ is.

We can use Theorem 2 to show that 2PFA-polytime accepts only regular languages if we can show that for some function $g(n)$ that grows faster than any polylogarithmic function of n, for all nonregular languages L, for infinitely many n, there is a set of strings of size at least $g(n)$, each of length at most n, such that each pair of strings x, x' in the set is n-$dissimilar$ with respect to L. By this, we mean there is some y for which $|xy| \leq n, |xy'| \leq n$, and xy is in L if and only if $x'y$ is not in L.

Towards this end, we define a more quantitative measure of the "non-regularity" of a language L than the notion of inequivalence defined by Myhill and Nerode. Let $N_L(n)$ be the maximum k such that there exist k distinct words which are pairwise n-dissimilar with respect to L. It is not hard to show that $N_L(n) = O(1)$ if and only if L is regular. The following theorem states that if L is non-regular, then $N_L(n)$ is bounded below by a linear function of n for infinitely many n. For related results, see the work of Shallit et al. [Glaister and Shallit, 1998, Pomerance et al., 1997, Shallit and Breitbart, 1996].

Theorem 3 *If L is non-regular then for infinitely many n, $N_L(n) \geq n/2 + 1$.*

The proof of Theorem 3 is based on an old result of Moore [Moore, 1956]. We say that a dfa D recognizes the initial n-fragment of L if $L(M) \cap \Sigma_{\leq n} = L \cap \Sigma_{\leq n}$. Let $\phi_L(n)$ be the size of a minimal 1dfa that recognizes the initial n-fragment of L. Kaneps and Freivalds [Kaneps and Freivalds, 1990] showed that $N_L(n) = \phi_L(n)$.

Karp [Karp, 1967] showed that if L is non-regular, then for infinitely many n, $\phi_L(n) > n/2 + 1$. To see this, it is sufficient to show that, for any positive integer r, there is $n > r$ such that $\phi_L(n) > n/2+1$. Given r, let n $(n > r)$ be the unique integer such that $\phi_L(r) = \phi_L(n-1) < \phi_L(n)$.

Such an n exists, since $\phi_L()$ is monotone and unbounded. Let minimal finite-state machines M and N be chosen so that M recognizes the initial $(n-1)$-fragment of L and N recognizes the initial n-fragment of L. Note that M and N cannot be identical, since M does not have enough states needed to recognize the initial n-fragment of L, and so they disagree on a string z of length n. Moore [Moore, 1956] proved the following result, which provides an upper bound on the length of a string on which M and N disagree as a function of the number of their states.

Proposition 2 *Let M and N be dfa's with j and k states, respectively, such that $L(M) \neq L(N)$. Then there is a string z of length at most $j + k - 2$ such that z is in $L(M)$ if and only if z is not in $L(N)$.*

It follows that there is a string z length at most $\phi_L(n-1) + \phi_L(n) - 2$ such that z is in $L(M)$ if and only if z is not in $L(N)$. Hence, $\phi_L(n-1) + \phi_L(n) - 2 \geq n$. Since we also have that $\phi_L(n-1) \leq \phi_L(n) - 1$, we obtain

$$(\phi_L(n) - 1) + \phi_L(n) > n + 1, \text{ or } \phi_L(n) > n/2 + 1,$$

as required.

The fact that 2PFA-polytime = Regular follows immediately from Theorems 2 and 3.

4.3   RELATED RESULTS

Let 2PFA(rat) be the class of languages recognized with bounded error by 2pfa's that have rational transition probabilities (or equivalently, are coin-flipping 2pfa's). Wang [Wang, 1992] showed that 2PFA(rat) is contained in the class of deterministic, context-sensitive languages; we don't include details of his proof here but it is based on the techniques of the Markov Chain Tree Theorem.

Finally, a note on succinctness of polynomial time bounded 2pfa's. Dwork and Stockmeyer [Dwork and Stockmeyer, 1990] extended the above techniques to address the question of whether bounded error 2pfa's that run in polynomial expected time can recognize languages with fewer states than dfa's or nfa's. Roughly, they showed that if P is a bounded error 2pfa with c states that runs in polynomial expected time, then there is a 1dfa for $L(P)$ that has $c^{O(c^2)}$ states, where the constant hidden in the big-O notation depends on the acceptance threshold of P and the degree of the polynomial bound on P's running time. They also showed that the c^2 in the exponent cannot be replaced by a function of c which grows more slowly than $\sqrt{c/\log c}$, even if 1dfa is replaced by 2nfa.

5.   UNDECIDABILITY RESULTS FOR BOUNDED ERROR PFA'S AND THEIR CONSEQUENCES

We now turn to the following question: given a pfa P, does it accept a regular language? In particular, does P accept the empty language?

In this section, we describe some results on this question, known as the emptiness problem for finite state automata. Later in this section, we describe applications of these results to show undecidability of probablistic planning problems and problems on partially observable and unobservable Markov decision processes.

For dfa's or nfa's there is an efficient algorithm for the emptiness problem, since a dfa or nfa accepts the empty language if and only if there is no path from the initial state to an accept state in the transition graph for the automaton.

In the case of a 1pfa P that accepts its language with bounded error and the acceptance threshold is known, then the question is decidable: ¿From Section 4.1, we know that there is a 1nfa D that accepts the same language as P and has a number of states at most exponential in the size of P, where the exponential bound can be calculated explicitly from the acceptance threshold γ and the number of states c of P. Denote this bound by $B(\gamma, c)$. Thus, a naive procedure to solve the emptiness problem in this case is to enumerate all nfa's D of size at most $B(\gamma, c)$; for each, determine whether $L(D) = L(P)$ and, if so, whether D accepts the empty language. Note that if $L(D) \neq L(P)$ then Moore's result (Proposition 2 above) provides a bound on the length of a witness to this fact. Thus, by enumerating all potential witnesses and checking whether each is accepted by P and by D, it is possible to determine whether $L(D) = L(P)$.

5.1 UNDECIDABILITY OF THE EMPTINESS PROBLEM FOR 2PFA'S

Since 2pfa's accept nonregular languages, it is perhaps not surprising that the emptiness problem for 2pfa's is undecidable. This result was proved by Freivalds [Freivalds, 1981].

Theorem 4 *The following problem is undecidable. Given a natural number $k \geq 2$ and a 2pfa P recognizing a language L with acceptance threshold $1/2 + 1/k$, determine whether the language L is empty.*

The proof of Theorem 4 is via a reduction from the halting problem for Turing machines on empty input. A first try at a reduction would be to use the description of a Turing machine M to produce a 2pfa $A = A(M)$ that can recognize (on its input tape) halting computations of M on the empty string. This 2pfa may err, as long as the error is bounded away from $1/2$. Assuming that M has a single, 1-way infinite tape, a standard representation of a computation of M is a string of the form $C_0 \& C_1 \& \ldots \& C_t$, where each C_i describes a configuration of M. If at the ith move of M on empty input the contents of M's tape is uv, the head is placed on the rightmost symbol of u, and the state of M is s, then C_i is the string usv.

Note that C_{i+1} differs from C_i in a constant number of places around the tape head, and a dfa can check that these differences between C_{i+1} and C_i correspond to a valid transition of M. A dfa can also check that C_0 represents the initial configuration of M on the empty string and that C_t is a halting configuration. The remaining task of the 2pfa is easily reducible to repeatedly recognizing strings of the form $w\&w$. Unfortunately, from Corollary 1, a 2pfa cannot recognize strings of the form $w\&w$. Unless, of course, w is a string over a unary alphabet.

Are Turing machines with a unary worktape alphabet equivalent in power to general Turing machines? It turns out that if the Turing machine has two worktapes, the answer is yes [Hopcroft and Ullman, 1979]. Specifically, the emptiness problem for the following model, called the 2-counter machine, is undecidable. A 2-counter machine is a Turing machine with two one-way infinite read-only worktapes (in addition to a read-only input tape). The leftmost symbol on both worktapes is \$ and all remaining symbols are blank. In effect, the 2-counter machine can use the worktapes as counters, simulating an increment or decrement by moving the head right or left, respectively, and testing for zero by checking if \$ is under the tape head.

Given a description of a 2-counter machine M, which is an instance of the halting problem for 2-counter machines, the reduction constructs a 2pfa $P = P(M)$ with the following property. P accepts with probability $\geq 1 - \epsilon$ any string that represents a halting computation of M on the empty input, and rejects with probability $1 - \epsilon$ any string that represents a non-halting computation of M on the empty input, where ϵ may be any constant in the range $(0, 1)$. A computation of M is represented as a string of the form

$$u_1 c^{k_1} d^{m_1} \& u_2 c^{k_2} d^{m_2} \& \dots \& u_t c^{k_t} d^{m_t},$$

where each u_i represents the state of M, and k_i, m_i denote the values of the 2 counters of M at the ith step of the computation. A dfa can check that the states u_i are in accordance with the transitions of the 2-counter machine, that $u_1 c^{k_1} d^{m_1}$ is the initial configuration, and that u_t is a halting state. The remaining task is to check that the counters are updated correctly. Since a counter's value changes by at most 1 on each transition, this task is easily reduced to that of recognizing the following language:

$$A^* = \{0^{n_1} 1^{n_1} 0^{n_2} 1^{n_2} \dots 0^{n_k} 1^{n_k}, \mid k \geq 0 \text{ and each } n_i \text{ is a positive integer}\}.$$

The proof that this language can be recognized by a 2pfa with bounded error is a natural extension of Freivald's algorithm for $\{a^n b^n\}$ and can be found in [Freivalds, 1981].

5.2 UNDECIDABILITY OF THE EMPTINESS PROBLEM FOR 1PFA'S

The emptiness problem for 1pfa's is as follows: given a 1pfa A and a constant $\epsilon, 0 < \epsilon < 1/2$, is there some input that the machine accepts with probability at least $1 - \epsilon$?

Theorem 5 *The emptiness problem for unbounded error 1pfa's is undecidable.*

This was proved by Paz [Paz, 1971]. The proof presented here is different than that of Paz, and is due to Lipton [Condon and Lipton, 1989, Lipton, 1989], and actually shows the following stronger result.

Theorem 6 *The following promise problem is undecidable: Given a constant $\epsilon, 0 < \epsilon < 1$ and a 1pfa P that either accepts some string with probability at least $1 - \epsilon$ or accepts all strings with probability at most ϵ, decide which is the case.*

Recall that the key to Freivalds' proof was a bounded error 2pfa for the language

$$A^* = \{0^{n_1} 1^{n_1} 0^{n_2} 1^{n_2} \ldots 0^{n_k} 1^{n_k}, \mid k \geq 0 \text{ and } n_i \text{ is a positive integer}\}.$$

At the heart of the proof is a simple game that is played on a string of the form $0^i 1^j$ and is inspired by Freivald's work. The game works as follows. If $i \neq j \mod k$ where k is some constant, then the outcome is "bad". Otherwise, four tests are performed: (i) flip a coin for each 0 and 1; (ii) flip a coin for each 0 and 1; (iii) flip two coins for each 0; and (iv) flip two coins for each 1. Say that sums win if all coins in either (i) or (ii) are heads and say that doubles win if all coins in either (iii) or (iv) are heads. If sums win, but not doubles, the outcome is good, and if doubles win but not sums, the outcome is bad. Otherwise the outcome is inconclusive. A key property of this game is that if $i = j$, then the outcomes good and bad are equally likely, but if i is not equal to j then the outcome bad is $f(k)$ times more likely than the outcome good, where $f(k)$ goes to infinity as k goes to infinity.

Lipton devised a 1pfa, call it R, that plays the above game independently on successive strings of the form $0^i 1^j$ and has the following property. R has an associated constant parameter $C > 0$ which can be chosen to be arbitrarily large. R has three outcomes: accept, reject, or inconclusive. Also, if x is in A^*, then $\text{Prob}[R \text{ accepts } x] = \text{Prob}[R \text{ rejects } x]$ and if x is not in A^*, then $\text{Prob}[R \text{ rejects } x] > C\text{Prob}[R \text{ accepts } x]$.

We can now describe the reduction from the halting problem for 2-counter machines to the emptiness problem for 1pfa's. Let M be a 2-counter machine. The reduction produces a 1pfa $P = P(M)$ which works as follows. If the input to P is of the form $E_1 \& E_2 \& E_3 \& \ldots E_t$, then, working from left to right, P does a probabilistic check that successive E_i's represent halting computations of M on the empty input. This check has three outcomes: good, bad, or inconclusive. If E_i is indeed a halting computation of M, then the probability that the outcome of P on E_i is good equals the probability that the outcome of P on E_i is bad. But if E_i is not a valid halting computation of M, then the probability that the outcome of P on E_i is bad is C times as great as the probability that the outcome of P on E_i is good. This check can easily be implemented using the 1pfa R described above, similar to Freivalds' proof. If k bad outcomes occur before any good outcome, then P rejects the input; otherwise P accepts the input.

Let $0 < \epsilon < 1$. If C and k are chosen appropriately, then the 1pfa P has the following properties: If M halts on the empty input, then P accepts some input with probability at least $1 - \epsilon$, namely the input that is the concatenation of sufficiently many halting computations of M on the empty input. However, if M does not halt on the empty input, then P accepts any input with probability at most ϵ.

5.3   MARKOV DECISION PROCESSES

In this section we show how undecidability results for 1pfa's lead to undecidability results for central problems on unobservable Markov decision processes (MDP's). Such problems are are abstractions of general planning problems; finding good heuristics for solving planning problems is an active research area in Artificial Intelligence [Boutilier et al., 1999, Littman, 1997, Papadimitriou and Tsitsiklis, 1987].

The following description of a Markov decision process is largely from Derman [Derman, 1970]. Consider a finite state system with state space S that is observed at times 1,2, ... Let $Y_0, Y_1, Y_2, \ldots$ be the sequence of observed states. After each observation of the system, one of a possible number of actions are taken. Let $A_0, A_1, \ldots, A_t, \ldots$ denote an infinite sequence of actions. A rule, or policy, or strategy, is a prescription for taking actions at each point in time. The most general type of strategy for taking an action at time t may be a function of the entire "history" of the system up to time t and may also involve randomization. Thus, a strategy is a set of functions

$$D_a(H_{t-1}, Y_t),$$

where H_t denotes the history up to time $t-1$, that is, the sequence $Y_0, A_0, Y_1, A_1, \ldots Y_t, A_t$, a is a possible action when the system is in state $Y_t, t = 0, 1, \ldots, 0 \le D_a \le 1$, and the sum of the D_a is 1. The interpretation is: if H_{t-1} denotes the history up to time $t-1$ and Y_t denotes the observed state at time t, then the probability of taking action a at time t is $D_a(H_{t-1}, Y_t)$.

The system has an associated transition function that maps (state, action) pairs to a probability distribution over (state, head direction) pairs. If $Y_{t-1} = i$ and action $A_t = a$, then the probability that $Y_t = j$ is determined by the transition function. Given a distribution over the initial observed states of the system and a strategy R, the sequence $\{Y_t, A_t, t = 0, 1, \ldots\}$ is a stochastic process called a Markov decision process.

A certain reward structure is imposed on a Markov decision process: whenever the system is in state i and action a is taken, we assume that a known reward w_{ia} is incurred. In order to consider the accumulated rewards of a process over time, we need the following notation. Let W_t be the random variable which is equal to w_{ia} if $Y_t = i$, $A_t = a$. Let $E_R[W_t]$ be the expected reward of W_t when the strategy is R. That is,

$$E_R[W_t] = \sum_{i,a} \text{Prob}[Y_t = i, A_t = a \mid \text{strategy is} R] w_{ia}$$

The following quantities are among those most commonly studied for Markov decision processes.

- *Finite Horizon Reward:* Let M be a MDP. Let

$$\text{Reward}(M, T) = \max_R \{ \sum_{t=0}^{T} E_R[W_t] \}$$

denote the maximum total expected reward of M that can be accumulated over the first T steps, where the maximum is taken over all strategies R.

- *Total (Infinite Horizon) Reward:* The quantity

$$\text{Total-Reward}(M, T) = \lim_{T \to \infty} \text{Reward}(M, T),$$

if it exists, is called the total reward of the MDP M. Note that the limit may be infinity.

- *Discounted Reward*: Let $0 \leq \alpha < 1$; α is referred to as the discount factor. Let

$$D(M, R, \alpha) = \sum_{t=0}^{\infty} \alpha^t E_R[W_t]$$

be the expected discounted reward of M with strategy R. Let

$$\text{Discounted-Reward}(M, \alpha) = \sup_R D(M, R, \alpha)$$

denote the supremum, over all strategies R, of the expected discounted reward of M with strategy R.

A variant of the MDP model is the unobservable MDP, or UMDP, model. In this model, the strategy is restricted so that the choice of action may depend on t and also on the history of actions up to time $t-1$, but may not depend on the history of states. Thus, a strategy for a UMDP is an infinite sequence of actions. Both UMDP's and MDP's are a special case of POMDP's or Partially Observable Markov Decision Processes, in which the strategy may depend on partial information about the history of states, in addition to the history of actions.

The Total Reward Problem for UMDP's is: given a UMDP M, decide whether Total-Reward$(M) \geq 1/2$ (where in the definition of Total Reward, Reward(M, T) is maximized over all strategies that are restricted as for UMDP's). Similarly, the Discounted Reward Problem for UMDP's is: given a UMDP M and a discount factor α, decide whether Discounted-Reward$(M) \geq 1/2$.

UMDP's and 1pfa's are closely related, as we now explain. Let P be a 1pfa and without loss of generality assume that P always moves its head to the right at every step. Furthermore, assume that P always rejects if the initial symbol under the input head is not # or if # is read when the head of P has moved right off the leftmost #, and that P enters an absorbing state (accept or reject) when $ is reached. Let Σ be the alphabet of P. Now let M be a UMDP with the same state space as P with the following properties. The number of actions of M associated with each state is $|\Sigma| + 2$ (the "+2" includes the endmarkers), and the transition function of M is identical to that of P. Any transition that causes M to enter the accept state from a state other than the accept state itself has a reward of 1 and all other transitions have a reward of 0. Think of the accept state as a goal state; then the total reward of M on a given strategy is the probability of eventually reaching the goal state on that strategy. Note that there is a 1-1 correspondence between inputs #$x$$ of P and strategies of M that start with #$x$$. Let R_x be any strategy corresponding to input string x. Then the probability that P accepts x equals the total reward of M on strategy R_x.

Theorem 7 *The Total Reward problem for UMDP's is undecidable.*

Theorem 7 follows immediately from the following stronger result.

Theorem 8 *The following promise problem for UMDP's is undecidable: Given a constant $\epsilon, 0 < \epsilon < 1$ and a UMDP M that either has total-reward at least $1 - \epsilon$ or at most ϵ, decide which is the case.*

This result follows directly from the relationship between 1pfa's and UMDP's described above, together with Theorem 6. Let $\epsilon, 0 < \epsilon < 1$ be given and let P be a 1pfa that either accepts some string with probability at least $1 - \epsilon$ or accepts all strings with probability at most ϵ. Let M be the corresponding Markov decision process as described above. Then, P accepts some string with probability at least $1 - \epsilon$ if and only if the unobservable total reward of M is at least $1 - \epsilon$, and P accepts all strings with probability at most ϵ if and only if the unobservable total reward of M is at most ϵ.

Theorem 9 *The Discounted Reward Problem for UMDP's is undecidable.*

The proof is a simple modification of the argument for the Total Reward problem. Let $1/2 < d < 1$, and consider the discounted reward of a UMDP obtained in the previous reduction. Effectively, the discount d acts as a penalty at every step, analogous to entering the reject state with probability $1 - d$ at each step. A natural idea, then, is to try to *balance out this penalty by adding a bonus* at each transition of the UMDP. Specifically, let M' be obtained from M by decreasing the probability of each transition of M from all states by a factor of $(1 - d)$ and then increasing the probability of entering the accept (goal) state by d.

Fortuitously, this modification has the desired effect: if the UMDP M has a strategy whereby the goal state is reached with probability greater than $1/2$, then the discounted reward of M' is greater than $1/2$. Furthermore, if on all strategies the total reward of M is less than $1/2$, then similarly on all strategies the discounted reward of M' is less than $1/2$. A key property of M that ensures this is that, on an input w, P may only enter the accept state upon reaching the right endmarker symbol, \$, for the first time.

Acknowledgments

This research was supported by National Sciences and Engineering Research Council (NSERC) of Canada.

References

[Boutilier et al., 1999] Boutilier, C., Dean, T., and Hanks, S. (1999). Decision theoretic planning: Structural assumptions and computational leverage. *Journal of Artificial Intelligence Research*, 11:1–94.

[Condon and Lipton, 1989] Condon, A. and Lipton, R. (1989). On the complexity of space bounded interactive proofs. In *Proceedings of the 30th Annual IEEE Symposium on the Foundations of Computer Science*, pages 462–467.

[Derman, 1970] Derman, C. (1970). *Finite State Markovian Decision Processes*, volume 67 of *Mathematics in Science and Engineering (Edited by Bellman)*. Academic Press.

[Dwork and Stockmeyer, 1990] Dwork, C. and Stockmeyer, L. (1990). A time-complexity gap for two-way probabilistic finite state automata. *SIAM J. Comput.*, 19:1011–1023.

[Dwork and Stockmeyer, 1992] Dwork, C. and Stockmeyer, L. (1992). Finite state verifiers i: the power of interaction. *J. ACM*, 39(4):800–828.

[Freivalds, 1981] Freivalds, R. (1981). Probabilistic two-way machines. In *Proc. of the International Symposium on Mathematical Foundations of Computer Science Springer-Verlag Lecture Notes in Computer Science, 188*, pages 33–45.

[Glaister and Shallit, 1998] Glaister, I. and Shallit, J. (1998). Automaticity iii: Polynomial automaticity and context-free languages. *Comput. complex.*, 7:371–387.

[Greenberg and Weiss, 1986] Greenberg, A. G. and Weiss, A. (1986). A lower bound for probabilistic algorithms for finite state machines. *J. Comput. Syst. Sci.*, 33:88–105.

[Hopcroft and Ullman, 1979] Hopcroft, J. E. and Ullman, J. D. (1979). *Introduction to Automata Theory, Languages, and Computation*. Addison Wesley.

[Kaneps, 1989] Kaneps, J. (1989). Stochasticity of the languages acceptable by two-way finite probabilistic automata. *Diskretnaya Matematika*, 1:63–77.

[Kaneps and Freivalds, 1990] Kaneps, J. and Freivalds, R. (1990). Minimal nontrivial space complexity of probabilistic one-way turing machines. In *Proc. of the Conference on Mathematical Foundations of Computer Science, Springer Verlag Lecture Notes in Computer Science, 452*, pages 355–361.

[Karp, 1967] Karp, R. M. (1967). Some bounds on the storage requirements of sequential machines and turing machines. *J. ACM*, 14(3):478–489.

[Leighton and Rivest, 1983] Leighton, F. T. and Rivest, R. L. (1983). The markov chain tree theorem. Technical Report MIT/LCS/TM-249, Laboratory for Computer Science, Massachusetts Institute of Technology, Cambridge, MA (Also in IEEE Transactions on Information Theory, IT-37(6), (1986) 733-742).

[Lipton, 1989] Lipton, R. J. (1989). Recursively enumerable languages have finite state interactive proofs. Technical Report CS-TR-213-89, Department of Computer Science, Princeton University, Princeton, N.J.

[Littman, 1997] Littman, M. L. (1997). Probabilistic propositional planning: Representations and complexity. In *Proc. Fourteenth National Conference on AI, AAAI Press*, pages 748–754.

[Macarie, 1998] Macarie, I. (1998). Space-efficient deterministic simulation of probabilistic automata. *SIAM J. Comput.*, 27(2):448–465.

[Madani et al., 1999] Madani, O., Condon, A., and Hanks, S. (1999). On the undecidability of probabilistic planning and infinite-horizon partially observable markov decision process problems. In *Proc. Sixteenth Annual Conference on Artificial Intelligence*.

[Moore, 1956] Moore, E. F. (1956). *Gedanken-experiments on sequential machines*, pages 129–153. Annals of Math. Studies No. 34. Princeton University Press, Princeton, N. J.

[Papadimitriou and Tsitsiklis, 1987] Papadimitriou, C. H. and Tsitsiklis, J. N. (1987). The complexity of markov decision processes. *Mathematics of Operations Research*, 12(3):441–450.

[Paz, 1971] Paz, A. (1971). *Introduction to Probabilistic Automata*. Academic Press.

[Pomerance et al., 1997] Pomerance, C., Robson, J. M., and Shallit, J. (1997). Automaticity ii: Descriptional complexity in the unary case. *Theoretical Computer Science*, 180(1-2):181–201.

[Rabin, 1963] Rabin, M. O. (1963). Probabilistic automata. *Information and Control*, 6:230–245.

[Ravikumar, 1992] Ravikumar, B. (1992). Some observations on 2-way probabilistic finite automata. In *Lecture Notes in Computer Science*, volume 652, pages 392–403. Springer-Verlag.

[Salomaa and Soittola, 1978] Salomaa, A. and Soittola, M. (1978). *Automata-theoretic aspects of formal power series*. Texts and Monographs in Computer Science. Springer-Verlag, New York.

[Shallit and Breitbart, 1996] Shallit, J. and Breitbart, Y. (1996). Automaticity i: Properties of a measure of descriptional complexity. *J. Comput. Systems Sci.*, 53:10–25.

[Turakainen, 1969] Turakainen, P. (1969). Generalised automata and stochastic languages. In *Proc. American Math. Soc.*, volume 21(2), pages 303–309.

[Wang, 1992] Wang, J. (1992). A note on two-way probabilistic automata. *Information Processing Letters*, 43:321–326.

Chapter 14

COMMUNICATION PROTOCOLS - AN EXEMPLARY STUDY OF THE POWER OF RANDOMNESS

Juraj Hromkovič

Lehrstuhl für Informatik I, RWTH Aachen,

Ahornstraße 55, 52074 Aachen, Germany

jh@cs.rwth-aachen.de

Abstract   There are not many computing models for which the power of randomized computation is as well understood as for two-party communication protocols and their communication complexity. Moreover, communication complexity theory contains several simple and transparent examples illustrating some paradigms of the design of randomized algorithms. They are real jewels for teaching an introductory course on randomization in computation.

The aim of this chapter is to present the contributions of the study of randomized communication complexity on the level of intuitive ideas as well as on the level of formal mathematical proofs. In this way, we hope to contribute to the transfer of technical investigation to teaching folklore in the area of randomized computation.

1.   INTRODUCTION

The communication complexity of two-party protocols was introduced by Yao [Ya79] in 1978-79. (Note that communication complexity was implicitly considered by Abelson [Ab78], too.) The initial goal was to develop a method for proving lower bounds on the complexity of distributed and parallel computations, with a special emphasis on VLSI computations [Hr97, KN97, Th79, Th80, Sa81, Lei80, Lo89, Hr88a, ĎG93].

---

*The work on this article was supported by INTAS-96-753 project "Boolean Functions: Complexity and Applications".

S. Rajasekaran et al (eds.), Handbook of Randomized Computing, Volume 2, pp. 533–596.
© 2001 *Kluwer Academic Publishers.*

In the 20 years of its existence the study of communication complexity contributed to computer science much more than one had expected at the beginning in the early eighties. Due to the well developed mathematical machinery for determining the communication complexity of concrete problems (see, for instance [AUY83, DHS96, Hr97, KN97, Le90, NW95, PS82]), communication complexity has established itself as a subarea of complexity theory. The contributions of the study of communication complexity can be approximately partitioned into the following three streams.

(i) *Lower bounds*

To use communication complexity for proving lower bounds on the complexity of specific problems was the primary reason to introduce it. Analogously to the applications of Kolmogorov complexity in proving lower bounds on different complexity measures of sequential computations, communication complexity has been developed as a method for the study of the complexity of concrete computing tasks in parallel information processing. Because of the success in developing methods for proving good lower bounds on the communication complexity of concrete problems, and in proving strong relations between communication complexity and many fundamental complexity measures, the approach based on communication complexity becomes one of the most powerful methods for proving lower bounds, even in the area of sequential computations. The following, not exhaustive list presents some fundamental complexity measures for which the approach based on communication complexity was used to prove nontrivial lower bounds (for some more involved surveys see [Hr97, KN97]).

- VLSI circuits (also multilective and three-dimensional)
  - trade-offs of area and time complexity
  - area complexity
- Boolean circuits
  - depth of general Boolean circuits and monotone ones
  - combinational complexity of planar Boolean circuits (also multilective)
  - area complexity of Boolean circuits
  - length of Boolean formulae
- Interconnection networks
  - trade-offs between size and topology
- Finite automata

      – size of finite automata
- Turing machines
    – time complexity
    – space complexity
    – trade-offs between time and space
- Branching programs and decision trees
    – depth
    – size of restricted versions

In several applications the communication complexity caused a breakthrough in the long efforts in proving lower bounds. To illustrate the progress included in the above list we mention lower bounds on the depth of monotone Boolean circuits [KW88, RW90], the first superpolynomial lower bounds on the size of linear depth oblivious branching programs [AM86] and recently even on the size of linear depth general branching programs [Aj99, BST98], and the first superlinear lower bound on the combinational complexity of planar Boolean circuits [Hr91, Tu89].

(ii) *Comparison of the power of different modes of computation*
   One of the most fundamental research problems of current theoretical computer science is the investigation of the computational power of nondeterministic and randomized computations, especially in comparison with the deterministic ones. The fundamental questions about polynomial-time computations like P versus NP, P versus ZPP, or P versus RP are long stated open problems. For communication complexity the research has been successful and several of the relations between different computation modes were fixed. This has essentially contributed to the understanding of the nature of randomness and nondeterminism. Some of the main results are the following ones.

   1. *Determinism and nondeterminism*
      There is an exponential gap between deterministic communication complexity and nondeterministic one [PS82, PS84].

   2. *Las Vegas randomization*
      There is a polynomial relation between Las Vegas and determinism for communication complexity [MS82, DHRS97], and there is an exponential gap between Las Vegas and nondeterminism.

   3. *Monte Carlo randomization*
      There are exponential gaps between one-sided-error Monte

Carlo randomization and determinism [Fr77, JPS84], and also between one-sided-error Monte Carlo randomization and two-sided-error Monte Carlo randomization [JPS84]. Monte Carlo and nondeterminism are incomparable [BFS86a, KS87, Raz90] in the following strong sense. For some function f the non-deterministic communication complexity of f is exponential in the Monte Carlo communication complexity of f, and for some function g the situation is vice versa.

4. *Number of random bits*
$O(\log_2 n)$ random bits are sufficient to reach the full power of randomized (bounded error) protocols, where n is the input size [New91].

5. *Number of advice bits*
In contrast to the randomized case, the number of advice bits of nondeterministic communication protocols cannot be bounded by any polylogarithmic function. Surprisingly, there exist functions f_n with high threshold on the amount of non-determinism in the sense that their nondeterministic communication complexity with restricted number of advice bits by some threshold T_n is equal to the deterministic communication complexity of f_n, but taking more than T_n advice bits one jumps to the full power of nondeterminism [HS96].

(iii) *Cryptography*
The study of communication protocols contributed to the development of new cryptographic ideas and concepts. The most interesting one is the concept of privacy [Ya82] that considers a new dimension of requirements on security. The protection against adversaries is not sufficient, one even wants protection against the counterpart in communication. The general scenario is that two parties want to compute a function depending on the data of both parties. The parties have to compute the result in such a way that no party learns the data of the other party in the communication. Surprisingly, many practically interesting functions can be evaluated in this way.

This article focuses on the randomized communication protocols and so on the part (ii) of the communication complexity theory. But the goal is to show some applications of the results about randomized communication complexity for the investigation of randomization in other models of computation, too.

An important advantage of the communication complexity theory is that several of its crucial results have nice proofs based on simple, transparent ideas and so it can be used for an exemplary illustration of the basic concepts and paradigms of the design of randomized algorithms.

This chapter is organized as follows. Section 2 contains the definitions of two-party (communication) protocols and their communication complexity. The randomized protocols are introduced and examples of randomized protocols, that transparently illustrate some basic paradigms of the design of randomized algorithms (foiling an adversary, abundance of witnesses, fingerprinting), are given. In Section 3 we give a short presentation of methods for proving lower bounds on the communication complexity of specific functions. We need them for the separation of different computation modes in the subsequent sections. Section 4 is devoted to the Las Vegas randomization. We show there that determinism and Las Vegas are polynomially related for communication protocols. An exponential gap between nondeterminism and Las Vegas is presented there, too. Section 5 is devoted to the one-sided-error Monte Carlo randomization and to the two-sided-error one. We relate the computational power of these error-bounded randomization modes to the power of other modes of computation there. In Section 6 we consider the possibility to bound the number of random bits of protocols without any essential increase of their communication complexity. The last Section 7 shows that some of the results on randomized protocols can be exported to other computing models. Namely, we transfer the polynomial relation between Las Vegas and determinism to the size of finite automata and to the size of ordered binary decision diagrams.

2. DEFINITIONS AND BASIC CONCEPTS

Informally, a **two-party (communication) protocol** consists of two computers C_I and C_{II} and a communication link between them. A protocol computes a finite function $f : U \times V \to Z$ in the following way. At the beginning C_I obtains an input $\alpha \in U$ and C_{II} obtains an input $\beta \in V$. Then C_I and C_{II} communicate according to the rules of the protocol by exchanging binary messages until one of them knows $f(\alpha, \beta)$. The **communication complexity of the computation on an input** (α, β) is the sum of the lengths of messages exchanged. The communication complexity of the protocol is the maximum of the complexities over all inputs from $U \times V$. The **communication complexity of f**, **cc(f)**, is the complexity of the best protocol for f.

Typically, one considers the situation that $U = \{0,1\}^n$, $V = \{0,1\}^n$ and $Z = \{0,1\}$, i.e. f is a Boolean function of $2n$ variables. Since this restriction is sufficient for our purposes we give the formal definition for protocols computing Boolean functions only. In what follows we describe the computation of a protocol on an specific input by a string $c_1\$c_2\$\ldots\$c_k\$c_{k+1}$, where $c_i \in \{0,1\}^+$ for $i = 1, \ldots, k$ are the messages exchanged (c_1 is the first message sent from C_I to C_{II}, c_2 is the second message submitted from C_{II} to C_I, etc.) and c_{k+1} is the result of the computation. The part $c_1\$c_2\$\ldots\$c_k$ is called the **history of communication**.

Definition 1 *Let $f : \{0,1\}^n \times \{0,1\}^n \to \{0,1\}$ be a Boolean function over a set $X = \{x_1, x_2, \ldots, x_{2n}\}$ of Boolean variables. A **protocol** P over X (or over $\{0,1\}^n \times \{0,1\}^n$) is a function from $\{0,1\}^n \times \{0,1,\$\}^*$ to $\{0,1\}^+ \cup \{\text{accept}, \text{reject}\}$ such that*

(i) *P has the **prefix-freeness property**:*
  For each $c \in \{0,1,\$\}^*$ and any two different $\alpha, \beta \in \{0,1\}^n$, $P(\alpha, c)$ is no proper prefix of $P(\beta, c)$.
  {The next message $P(\alpha, c)$ is computed either by C_I or by C_{II} in the dependence of the input α of C_I (C_{II}) and the history c of communication. The prefix-freeness property assures that the messages exchanged between the two computers are self-delimiting, and no extra "end of transmission" symbol is required. To visualize the end of messages in the history of communication c, we write the special symbol $\$$ at the end of every message.}

(ii) *If $\Phi(\alpha, c) \in \{\text{accept}, \text{reject}\}$ for an $\alpha \in \{0,1\}^m$, and $c \in (\{0,1\}^+\$)^{2p}$ for some $p \in I\!N$ [for $c \in (\{0,1\}^+\$)^{2p+1}$], then for all $q \in I\!N$, $\gamma \in \{0,1\}^m$, $d \in (\{0,1\}^+\$)^{2q+1}$ [$d \in (\{0,1\}^+\$)^{2q}$], $P(\gamma, d) \notin \{\text{accept}, \text{reject}\}$ for every communication history $d \in \{0,1,\$\}^*$.*
  {This property assures that the output value is always computed by the same computer independently of the input assignment.}

(iii) *For every $c \in \{0,1,\$\}^*$, if $P(\alpha, c) \in \{\text{accept}, \text{reject}\}$ for some $\alpha \in \{0,1\}^n$, then $P(\beta, c) \in \{\text{accept}, \text{reject}\}$ for every $\beta \in \{0,1\}^n$.*
  {This property assures that if the computer C_I (C_{II}) computes the output for an input, then the other computer C_{II} (C_I) knows that C_I (C_{II}) knows the result, and so it does not wait for further communication.}

*A **computation of P** on an input $(\alpha, \beta) \in \{0,1\}^n \times \{0,1\}^n$ is a string $c = c_1\$c_2\$\ldots\$c_k\$c_{k+1}$, where*

(1) *$k \geq 0$, $c_1, \ldots, c_k \in \{0,1\}^+$, $c_{k+1} \in \{\text{accept}, \text{reject}\}$, and*

(2) *for every integer l, $0 \leq l \leq k$,*

  (2.1) *if l is even, then $c_{l+1} = P(\alpha, c_1\$c_2\$\ldots\$c_l\$)$, and*
    {The message c_{l+1} is sent by C_I, and C_I computes c_{l+1} in

dependence of its input part α and of the whole current communication history $c_1\$c_2\$\ldots\$c_l\$$.}

(2.2) *if l is odd, then* $c_{l+1} = P(\beta, c_1\$c_2\$\ldots\$c_l\$)$.
{*The message c_{l+1} is sent from C_{II} to C_I, and C_{II} computes C_{l+1} in the dependence of its input β and the communication history $c_1\$c_2\$\ldots\$c_l\$$.*}

For every computation $c = c_1\$c_2\$\ldots\$c_k\$c_{k+1}$,

$$\mathbf{Com}(c) = c_1\$c_2\$\ldots\$c_k\$$$

is the **communication** *(or* **communication history***) of c.*

We say that **P computes f** *if, for every $(\alpha, \beta) \in \{0,1\}^n \times \{0,1\}^n$, the computation of P on (α, β) is finite and ends with "accept" if and only if $f(\alpha, \beta) = 1$. In what follows we also say that a computation is* **accepting** *(***rejecting***) if it ends with* accept*(*reject*).*

The **length of a computation** *c is the total length of all messages in c (ignoring \$'s and the final* accept*/*reject*). The* **communication complexity of the protocol P, $\mathrm{cc}(P)$,** *is the maximum of all computation lengths over all inputs $(\alpha, \beta) \in \{0,1\}^n \times \{0,1\}^n$.*

The **communication complexity of f** *is*

$$\mathbf{cc}(f) = \min\{\mathrm{cc}(P)\,|\,P \text{ computes } f\}.$$

□

We use the notation **$P(\alpha, \beta)$** for the output $\in \{\text{accept}, \text{reject}\}$ of the computation of the protocol P on the input (α, β). Note that "accept" is used to denote the result "1", and "reject" is used to denote the result "0". We use the notations "accept" and "reject" instead of "1" and "0" in order to distinguish between the communication bits and the results. Besides the reason above it will be convenient to be able to speak about accepting and rejecting computations in what follows.

Observe that if one wants to formally define a protocol for a finite function $f : U \times V \to Z$ as a protocol $P = (C_I, C_{II})$ over $U \times V$, then it is convenient to consider two functions C_I and C_{II} instead of one function P as in Definition 1. In this case C_I is a mapping from $U \times \{0, 1, \$\}$ to $\{0,1\}^+ \cup Z$ and C_{II} is a mapping from $V \times \{0, 1, \$\}^*$ to $\{0,1\}^+ \cup Z$. All the properties (i), (ii), and (iii) must be then expressed in this notation.

To illustrate the above definition we consider the following simple examples. Let

$$\mathrm{Eq}_n(x_1, x_2, \ldots, x_n, y_1, y_2, \ldots, y_n) = \bigwedge_{i=1}^{n} x_i \equiv y_i$$

be the Boolean function of $2n$ variables from $\{0,1\}^n \times \{0,1\}^n$ to $\{0,1\}$ that takes the value 1 iff the first half of the input is equal to the second half of the input. A protocol P computing Eq_n can simply work as follows. C_I sends its whole input $\alpha \in \{0,1\}^n$ to C_{II} and C_{II} compares whether α is identical with its input $\beta \in \{0,1\}^n$. Formally,

$$P(\alpha, \lambda) = \alpha, \text{ and}^1$$

$$P(\beta, \alpha\$) = \begin{cases} \text{accept} & \text{if } \alpha \equiv \beta. \\ \text{reject} & \text{if } \alpha \not\equiv \beta. \end{cases}$$

We observe that the above described strategy (C_I sends its whole input to C_{II}) works for every function and so

$$\text{cc}(f) \le n$$

for every Boolean function $f : \{0,1\}^n \times \{0,1\}^n \to \{0,1\}$ of $2n$ variables.

Now, consider the symmetric function $s_{2n} : \{0,1\}^{2n} \to \{0,1\}$ that takes the value 1 if and only if the number of 1's in the input is equal to the number of 0's in the input. A simple way to compute s_{2n} by a protocol follows. C_I sends the binary representation of the number $\#_1(\alpha)$ of 1's in its input $\alpha \in \{0,1\}^n$. C_{II} checks whether $\#_1(\alpha) + \#_1(\beta) = n$, where $\beta \in \{0,1\}^n$ is the input of C_{II}. Obviously, the communication complexity of this protocol is $\lceil \log_2(n+1) \rceil$.

Note that we shall later show that the above protocols for Eq_n and s_{2n} are optimal. We observe that these protocols are very simple because the whole communication consists of the submission of one message. The protocols whose computations contains at most one message are called **one-way protocols** in what follows. For every Boolean function f

$$\textbf{cc}_1(f) = \min\{\text{cc}(P) \,|\, P \text{ is a one-way protocol computing } f\}$$

is the **one-way communication complexity of f**.

In Section 3 we shall see that there can be an exponential gap between $\text{cc}_1(f)$ and $\text{cc}(f)$. The following function $f_{\text{ind}(n)}(x_1, \ldots, x_n, y_1, \ldots, y_n)$ for $n = 2^k$, $k \in \mathbb{N} - \{0\}$, is an example of a computing problem where one can profit from the exchange of more than one message between C_I and C_{II}. Let, for every binary string $\alpha = \alpha_0\alpha_1 \ldots \alpha_k$,

$$\textbf{Number}(\alpha) = \sum_{i=0}^{k} \alpha_{k-i} \cdot 2^i$$

---

[1] λ denotes the empty word.

be the number with the binary representation α. The function $f_{\text{ind}(n)}$ takes the value 1 iff

$$x_{\text{Number}(y_{\text{Number}(x_1 x_2 \ldots x_{\lceil \log_2 n \rceil})+1} \cdots y_{(\text{Number}(x_1 x_2 \ldots x_{\lceil \log_2 n \rceil})+\lceil \log_2 n \rceil) \bmod n})+1} = 1.$$

Informally, the first $\log_2 n$ values of the variables $x_1, x_2, \ldots, x_{\log_2 n}$ determine a position (an index) $a = \text{Number}(x_1, \ldots, x_{\log_2 n}) + 1$. y_a and the following $\log_2 n - 1$ values of y variables determine an index b, and one requires $x_b = 1$ for the result 1. We describe a protocol P that computes $f_{\text{ind}(n)}$. C_I sends $x_1 x_2 \ldots x_{\lceil \log_2 n \rceil}$ to C_{II}. After that C_{II} sends the message $y_a y_{(a+1) \bmod n} \cdots y_{(a+\log_2 n-1) \bmod n}$ to C_I, where $a = \text{Number}(x_1 \ldots x_{\lceil \log_2 n \rceil}) + 1$. Now, C_I accepts iff

$$x_{\text{Number}(y_a y_{(a+1) \bmod n} \cdots y_{(a+\lceil \log_2 n \rceil-1) \bmod n})+1} = 1.$$

The communication complexity of this protocol is $2 \cdot \lceil \log_2 n \rceil$.

In what follows we use, for all nonnegative integers l, k, $l \geq \lceil \log_2 k \rceil$, $l \geq 1$, $\mathbf{BIN}_l(k)$ to denote the binary representation of k by a binary string of length l. This means that if $l > \lceil \log_2 k \rceil$, $l - \lceil \log_2 k \rceil$ leading 0's are added to the representation.

One can introduce nondeterminism for protocols in the usual way. Because of this we prefer to give an informal description of nondeterministic protocols rather than an exact formal definition.

Let $f : U \times V \rightarrow \{0, 1\}$ be a finite function. A **nondeterministic protocol** P computing on inputs from $U \times V$ consists of two nondeterministic computers C_I and C_{II}. At the beginning C_I obtains an input $\alpha \in U$, and C_{II} obtains an input $\beta \in V$. As in the deterministic case, the computation consists of a number of communication rounds, where in one round one computer sends a message to the other one. The computation finishes when one of the computers decides to accept or to reject the input (α, β). In contrast to the deterministic case, C_I can be viewed as a relation on

$$(U \times \{0, 1, \$\}^*) \times (\{0, 1\}^+ \cup \{\text{accept,reject}\})$$

and C_{II} can be viewed as a relation on

$$(V \times \{0, 1, \$\}^*) \times (\{0, 1\}^+ \cup \{\text{accept,reject}\}).$$

This means that, for every argument (α, c), C_I (C_{II}) nondeterministically chooses a message from a finite set of possible messages determined by the argument (α, c).

We say that P **computes** f if for every $(\alpha, \beta) \in U \times V$,

   (i) if $f(\alpha, \beta) = 1$, then there exists an accepting computation of P on the input (α, β), and

(ii) if $f(\alpha, \beta) = 0$, then all computations of P on (α, β) are rejecting ones.

Again, we require that the prefix-freeness property and the property that exactly one computer takes the final decision for all inputs are satisfied.

The **nondeterministic communication complexity of P**, denoted by **ncc(P)**, is the maximum of the lengths of all accepting computations of P. The **nondeterministic communication complexity of f** is

$$\text{ncc}(f) = \min\{\text{ncc}(P) \mid P \text{ is a nondeterministic protocol computing } f\}.$$

To show the power of nondeterminism in communication consider, for every positive integer n, the function $\text{Ineq}_n : \{0, 1\}^n \times \{0, 1\}^n \to \{0, 1\}$ defined by

$$\text{Ineq}_n(x_1, x_2, \ldots, x_n, y_1, y_2, \ldots, y_n) = \bigvee_{i=1}^{n}(x_i \oplus y_i),$$

where $\oplus$ denotes the exclusive or (plus mod 2).

A nondeterministic protocol P that accepts all inputs $(\alpha, \beta) \in \{0, 1\}^n \times \{0, 1\}^n$ with $\alpha \neq \beta$ can work as follows. For every input $\alpha = \alpha_1\alpha_2\ldots\alpha_n$, C_I nondeterministically chooses a number $i \in \{1, \ldots, n\}$ and sends the message $\alpha_i \text{BIN}_{\lceil \log_2 n\rceil}(i)$ to C_{II}, where $\text{BIN}_{\lceil \log_2 n\rceil}(i)$ is the binary representation of i of the length[2] $\lceil \log_2 n\rceil$. Now, for every input $\beta = \beta_1 \ldots \beta_n$ of C_{II}, C_{II} accepts if and only if $\alpha_i \neq \beta_i$. In the case $\alpha_i = \beta_i$, C_{II} rejects the input. Obviously, for all $\alpha, \beta \in \{0, 1\}^n$ with $\alpha \neq \beta$, there exists a j such that $\alpha_j \neq \beta_j$ and so there exists an accepting computation of P on (α, β). On the other hand if $\alpha = \beta$, then all n different computations of P on (α, β) are rejecting. Thus, P computes Ineq_n within the communication complexity $\lceil \log_2 n\rceil + 1$, i.e.,

$$\text{ncc}(\text{Ineq}_n) \leq \lceil \log_2 n\rceil + 1.$$

Similarly as in the deterministic case, one can consider **one-way nondeterministic protocols** whose computations contain at most one message. Let **ncc$_1$(f)** denote the **one-way nondeterministic communication complexity of f**. In contrast to the deterministic case, we show that there is no difference in the power of nondeterministic protocols and one-way nondeterministic protocols.

---

[2]This means, that additional 0's are taken to achieve the required length, if necessary.

Theorem 1 *For every finite function f,*

$$\mathrm{ncc}(f) = \mathrm{ncc}_1(f).$$

Proof. Let f be a function from $U \times V$ to $\{0, 1\}$, and let $D = (C_I, C_{II})$ be a nondeterministic protocol computing f. We construct a one-way nondeterministic protocol D_1 that computes f within the communication complexity $\mathrm{ncc}(D)$. In what follows we say that a computation $C = c_1 \$ c_2 \$ \ldots \$ c_k \$ c_{k+1}$ is **consistent** for C_I with an input $\alpha \in U$ (or from the point of view of C_I and α) if C_I with the input α has the possibility to send the messages $c_1, c_3, c_5, \ldots$ when receiving the messages $c_2, c_4, c_6, \ldots$ from C_{II}. Similarly, one can define the consistency of C according to C_{II} with an input $\beta \in V$. Obviously, if C is a consistent computation for C_I on an input α and for C_{II} on an input β, then C is a possible computation of (C_I, C_{II}) on the input (α, β).

Let $C_1, C_2, \ldots, C_k$ be all accepting computations of D over all inputs from $U \times V$. Clearly, $k \leq 2^{\mathrm{ncc}(D)}$. Let D_1 consist of the computers C_I^1 and C_{II}^1. For every $\alpha \in U$, C_I^1 nondeterministically chooses an $i \in \{1, \ldots, k\}$, and it sends the message $\mathrm{BIN}_{\mathrm{ncc}(D)}(i)$ to C_{II}^1 if C_i is a possible accepting computation of D from the point of view of C_I and α. In other words, C_I^1 can send the binary representation of i if and only if there exists an $\gamma \in V$ such that C_i is the accepting computation of D on (α, γ). For every input β of C_{II}^1, when C_{II}^1 receives the binary representation of a number i, then C_{II}^1 accepts if C_i is a consistent accepting computation from the C_{II} and β point of view.

So, D_1 has exactly the same number of accepting computations as D and $\mathrm{ncc}_1(D_1) = \mathrm{ncc}(D)$. $\qquad\square$

There are two distinct ways to introduce randomized protocols. One possibility is to take a nondeterministic protocol and to consider a probability distribution for every possible nondeterministic guess. Such randomized protocols are called **private** because each of the computers takes its random bits from a separate source; i.e., C_I (C_{II}) does not know the random bits of C_{II} (C_I). If one of the computers wants to know the random bits influencing the choice of the message submitted by the other computer, then these bits must be communicated. Another possibility to define randomized protocols is to say that a randomized protocol is a probability distribution over a set of deterministic protocols. Such randomized protocols are called **public** randomized protocols because this model corresponds to the situation when both computers have the same source of random bits (i.e. everybody sees the random bits of the other one for free). This second approach represents the well-known paradigm *"foiling an adversary"* of the design of randomized algorithms. So, every efficient public randomized protocol can be viewed as a successful application of this paradigm.

Clearly, public randomized protocols are at least as powerful as private ones. Newman [New91] has proved that the relations between the communication complexities of public randomized protocols and private ones are linear for every bounded-error model.[3] We shall present this result in Section 6. Because of this and in order to simplify the matters we formally define the public versions of randomized protocols only.

In the following definitions we use also the notation $P(\alpha, \beta) = 1$ (0) instead of $P(\alpha, \beta) = $ accept (reject).

Definition 2 *Let U and V be finite sets. A **randomized protocol over $U \times V$** is a pair $R = (Prob, S)$, where*

(i) $S = \{D_1, D_2, \ldots, D_k\}$ is a set of (deterministic) protocols over $U \times V$, and

(ii) Prob is a probability distribution over the elements of S.

*For $i = 1, 2, \ldots, k$, **$Prob(D_i)$** is the probability that the protocol D_i is randomly chosen to work on a given input.*

*For an input $(\alpha, \beta) \in U \times V$, the **probability that R computes an output z** is*

$$Prob(R(\alpha, \beta) = z) = \sum_{\substack{D \in \{D_1, \ldots, D_k\} \\ D(\alpha, \beta) = z}} Prob(D).$$

*The **communication complexity of R** is*

$$\mathbf{cc}(R) = \max\{cc(D) \mid D \in S\}.$$

*A randomized protocol $R = (Prob, S)$ is called **one-way** if all elements of S are one-way protocols.*

*For every randomized protocol $R = (Prob, \{D_1, D_2, \ldots, D_k\})$ with a uniform probability distribution Prob, the **degree of randomness** of P is $\lceil \log_2 k \rceil$. Since one can unambiguously identify every D_i with a binary string of length $\lceil \log_2 k \rceil$, we call $\lceil \log_2 k \rceil$ the **number of random bits of R**, too.*

In what follows we consider only randomized protocols computing Boolean functions. In contrast to the previous protocol models we also allow a "**neutral**" output "?", whose meaning is that the protocol was

---

[3]So, there is no difference between these two models from the asymptotic point of view.

not able to compute any final answer in the given computation (random attempt). We consider this possibility for Las Vegas protocols that never err but may produce the answer "?" with a bounded probability.

Note that in what follows we use also the notation $R(\alpha, \beta) = 1$ ($R(\alpha, \beta) = 0$) instead of $R(\alpha, \beta) = $ "accept" ($R(\alpha, \beta) = $ "reject"). This is convenient because we obtain the possibility to speak about the probability of the event $R(\alpha, \beta) = f(\alpha, \beta)$ in this way.

Definition 3 *Let* $f : U \times V \to \{0, 1\}$ *be a finite function. We say, that a randomized protocol* $R = (Prob, S)$ *is a* **(public) Las Vegas protocol** *for* f *if*

(i) for every $(\alpha, \beta) \in U \times V$ *with* $f(\alpha, \beta) = 1$,
$Prob(R(\alpha, \beta) = 1) \geq \frac{1}{2}$ *and* $Prob(R(\alpha, \beta) = 0) = 0$, *and*

(ii) for every $(\alpha, \beta) \in U \times V$ *with* $f(\alpha, \beta) = 0$,
$Prob(R(\alpha, \beta) = 0) \geq \frac{1}{2}$ *and* $Prob(R(\alpha, \beta) = 1) = 0$.

The **Las Vegas communication complexity of** f *is*

$$\mathbf{lvcc}(f) = \min\{cc(R) \,|\, R \text{ is a Las Vegas protocol for } f\}.$$

The **one-way Las Vegas communication complexity of** f *is*

$$\mathbf{lvcc_1}(f) = \min\{cc(R) \,|\, R \text{ is a one-way Las Vegas protocol for } f\}.$$

We present a simple example of a one-way Las Vegas protocol. More involved ideas for the design of Las Vegas protocols can be found in Section 4. Consider the function $\mathrm{Index}_n : \{0, 1\}^n \times \{1, 2, \ldots, n\} \to \{0, 1\}$ defined as follows[4]

$$\mathrm{Index}_n((x_1, x_2, \ldots, x_n), j) = x_j.$$

A Las Vegas one-way protocol D for Index_n can be described as the pair $(Prob, \{D_1, D_2\})$, where

(i) $Prob(D_1) = Prob(D_2) = \frac{1}{2}$,

(ii) $D_1 = (D_{I,1}, D_{II,1})$, where $D_{I,1}$ sends the first half $\alpha_1 \ldots \alpha_{\lceil \frac{n}{2} \rceil}$ of its input $\alpha_1 \ldots \alpha_n$ to $D_{II,1}$. $D_{II,1}$ outputs α_j if the input j of $D_{II,1}$ belongs to $\{1, 2, \ldots, \lceil \frac{n}{2} \rceil\}$. If $j > \lceil \frac{n}{2} \rceil$, then $D_{II,1}$ outputs "?",

(iii) $D_2 = (D_{I,2}, D_{II,2})$ where $D_{I,2}$ sends the second half $\alpha_{\lceil \frac{n}{2} \rceil + 1} \ldots \alpha_n$ of its input $\alpha_1 \ldots \alpha_n$ to $D_{II,2}$. $D_{II,2}$ outputs α_j if the input j of

---

[4]Observe, that Index_n can be also viewed as a Boolean function if one represents the numbers $1, 2, \ldots, n$ by binary strings

$D_{II,2}$ belongs to $\{\lceil \frac{n}{2} \rceil + 1, \ldots, n\}$. If $j \leq \lceil \frac{n}{2} \rceil$, then $D_{II,2}$ outputs "?".

Another possibility to describe D is as follows.

Las Vegas one-way protocol $D = (D_I, D_{II})$ for Index$_n$.

Input: (α, j), $\alpha = \alpha_1 \ldots \alpha_n \in \{0,1\}^n$, $j \in \{1, \ldots, n\}$.
  { D_I gets the input α, and D_{II} gets the input j.}

Step 1: D_I chooses a random bit $r \in \{0,1\}$.
  { Note, that D_{II} knows r, too.}
  If $r = 0$, then D_I sends the message $\alpha_1 \alpha_2 \ldots \alpha_{\lceil \frac{n}{2} \rceil}$.
  If $r = 1$, then D_I sends the message $\alpha_{\lceil \frac{n}{2} \rceil + 1} \alpha_{\lceil \frac{n}{2} \rceil + 2} \ldots \alpha_n$.

Step 2: If $r = 0$ and $j \in \{1, 2, \ldots, \lceil \frac{n}{2} \rceil\}$, then D_{II} outputs α_j.
  If $r = 1$ and $j > \lceil \frac{n}{2} \rceil$, then D_{II} outputs α_j.
  Else, D_{II} outputs "?".

In what follows we shall prefer the second form of the description of randomized protocols. Clearly, D never errs, and the probability of giving the output "?" is $\frac{1}{2}$ for every input $(\alpha, j) \in \{0,1\}^n \times \{1, \ldots, n\}$. The communication complexity of D is $\lceil \frac{n}{2} \rceil$.

Note, that the constant $\frac{1}{2}$ bounding the probability of the output "?" in the definition of (two-way) Las Vegas protocols is not essential from the asymptotic point of view. Instead of giving the output "?" a Las Vegas protocol may start a new communication from the beginning with new random bits. If it outputs "?" only if it reaches "?" in k independent computation attempts, then the probability to obtain the output "?" decreases from $\frac{1}{2}$ to $\frac{1}{2^k}$, but the communication complexity increases only by a factor of k in comparison with the original protocol.

Definition 4 *Let $f : U \times V \to \{0,1\}$ be a finite function. We say that a randomized protocol $R = (Prob, S)$ is a* **(public) one-sided-error Monte Carlo protocol for f** *if*

(i) *for every $(\alpha, \beta) \in U \times V$ with $f(\alpha, \beta) = 1$,*
   $Prob(R(\alpha, \beta) = 1) \geq \frac{1}{2}$, *and*

(ii) *for every $(\alpha, \beta) \in U \times V$ with $f(\alpha, \beta) = 0$,*
   $Prob(R(\alpha, \beta) = 0) = 1$.

We say that a randomized protocol R is a **(public) two-sided-error Monte Carlo protocol for f** *if, for every $(\alpha, \beta) \in U \times V$,*

$$Prob(R(\alpha, \beta) = f(\alpha, \beta)) > \frac{2}{3}.$$

The **one-sided-error Monte Carlo communication complexity of** f *is*

$$\mathbf{1mccc}(f) = \min\{\text{cc}(R) \quad | \quad R \text{ is a one-sided-error}$$
$$\text{Monte Carlo protocol for } f\}.$$

The **two-sided-error Monte Carlo communication complexity of** f *is*

$$\mathbf{2mccc}(f) = \min\{\text{cc}(R) \quad | \quad R \text{ is a two-sided-error}$$
$$\text{Monte Carlo protocol for } f\}.$$

Because of the condition (ii) of one-sided-error Monte Carlo protocols it is clear that private one-sided-error Monte Carlo protocols are a restricted version of nondeterministic ones. For the public randomized protocols defined here we obtain

$$\text{ncc}(f) \leq 1\text{mccc}(f) + \text{ the number of random bits}$$

for every finite function f.

Similarly as in the case of Las Vegas, the constant $\frac{1}{2}$ in the inequality $Prob(R(\alpha, \beta) = 1) \geq \frac{1}{2}$ is not essential for one-sided-error Monte Carlo from the asymptotic point of view and so one-sided-error Monte Carlo protocols can be viewed as a restricted version of two-sided-error Monte Carlo protocols, too.

To show the power of Monte Carlo randomization we present a one-sided-error Monte Carlo one-way protocol for Ineq_n with the communication complexity in $O(\log_2 n)$.

Example 1 (based on [Fr77]) The idea of a randomized protocol is based on the **abundance of witnesses** method for the design of randomized algorithms. Let f be a Boolean function. A **witness for** $f(\gamma) = a$ is any binary string δ, such that using δ there is an efficient way to prove (verify) that $f(\gamma) = a$. For instance, any factor (nontrivial divisor) y of a number x is a witness of the claim "x *is composite*". Obviously, to check whether $x \bmod y = 0$ is much easier than to prove that x is a composite without any additional information. In general, one considers witnesses only if they essentially decrease the complexity of computing the result. For many functions the difficulty with finding a witness deterministically is that the witness lies in a search space that is too large to be searched exhaustively. However, by establishing that the space contains a large number of witnesses, it often suffices to choose an element at random from the space. The randomly chosen item is likely to be a witness. If this probability is not high enough, an independent random choice of several items reduces the probability that no witness is found.

The framework of this approach is very simple. One has for every input γ a set $\mathrm{CandW}(\gamma)$ that contains all items candidating to be a witness for the input γ. Often $\mathrm{CandW}(\gamma)$ is the same set for all inputs of the same size as γ. Let $\mathrm{Witness}(\gamma)$ contain all witnesses for γ that are in $\mathrm{CandW}(\gamma)$. The aim is to reach a situation where the cardinality of $\mathrm{Witness}(\gamma)$ is proportional to the cardinality of $\mathrm{CandW}(\gamma)$.

To design a randomized protocol for Ineq_n, we say that a **prime p is a witness for $\alpha \neq \beta$, $\alpha, \beta \in \{0,1\}^n$**, if

$$\mathrm{Number}(\alpha) \bmod p \neq \mathrm{Number}(\beta) \bmod p.$$

For every input $(\alpha, \beta) \in \{0,1\}^n \times \{0,1\}^n$, $\mathrm{CandW}(\alpha, \beta)$ is the set of all primes from $\{2, 3, \ldots, n^2\}$. Due to the Prime Number Theorem we know that $|\mathrm{CandW}(\alpha, \beta)|$ is approximately $\frac{n^2}{\ln n^2}$. Now, we estimate the lower bound on $|\mathrm{Witness}(\alpha, \beta)|$. Let $\alpha \neq \beta$. If, for a prime p,

$$\mathrm{Number}(\alpha) \bmod p = \mathrm{Number}(\beta) \bmod p, \tag{1}$$

then p divides $h = \mathrm{Number}(\alpha) - \mathrm{Number}(\beta)$. Since $h < 2^n$, h has fewer than n different prime divisors.[5] This means that at most $n - 1$ primes from $\mathrm{CandW}(\alpha, \beta)$ have the property (1). Thus,

$$|\mathrm{Witness}(\alpha, \gamma)| \geq |\mathrm{CandW}(\alpha, \beta)| - n + 1. \tag{2}$$

Now, we use (2) to design our randomized protocol.

One-sided-error Monte Carlo protocol $R = (R_I, R_{II})$ for Ineq_n.

Input: $(\alpha, \beta) \in \{0,1\}^n \times \{0,1\}^n$

Step 1: R_I chooses uniformly a prime $p \in \{2, 3, \ldots, n^2\}$ at random.
{Note, that R_{II} knows this choice of R_I.}

Step 2: R_I computes $s = \mathrm{Number}(\alpha) \bmod p$ and sends the binary representation of s to R_{II}.
{Note, that the length of the message is $\lceil \log_2 n^2 \rceil \leq 2 \cdot \lceil \log_2 n \rceil$.}

Step 3: R_{II} computes $q = \mathrm{Number}(\beta) \bmod p$.
If $q \neq s$, then R_{II} outputs 1 ("accept").
If $q = s$, then R_{II} outputs 0 ("reject").

---

[5] Observe, that $n! > 2^n$.

We show that R is a one-sided-error Monte Carlo protocol for Ineq_n. If $\alpha = \beta$, for an input $(\alpha, \beta) \in \{0,1\}^n \times \{0,1\}^n$, then $\text{Number}(\alpha) \bmod p = \text{Number}(\beta) \bmod p$ for every prime p. So,

$$Prob(R(\alpha, \beta) = \text{``reject''}) = 1.$$

Let $\alpha \neq \beta$, i.e. $\text{Ineq}_n(\alpha, \beta) = 1$. Due to the inequality (2), the probability that R chooses a prime with the property (1) is at most

$$\frac{|\text{CandW}(\alpha, \beta)| - |\text{Witness}(\alpha, \gamma)|}{|\text{CandW}(\alpha, \beta)|} \leq \frac{n-1}{|\text{CandW}(\alpha, \beta)|}.$$

Since $|\text{CandW}(\alpha, \beta)| \geq \frac{n}{2 \ln n^2}$ already for small n's, the probability that R rejects (α, β) is at most

$$\frac{n-1}{|\text{CandW}(\alpha, \beta)|} \leq \frac{n-1}{n^2/2 \ln n^2} \leq \frac{2 \ln n^2}{n}.$$

Thus,

$$Prob(R(\alpha, \beta) = \text{``accept''}) \geq 1 - \frac{2 \ln n^2}{n},$$

that even tends to 1 for sufficiently large n's. Thus, we have proved

$$1\text{mccc}(\text{Ineq}_n) = O(\log_2 n).$$

Note, that the above protocol R is also an illustration of the paradigm of **fingerprinting** that is typical for hashing. Fingerprinting means that one represents an object of a large representation size by a short **fingerprint** using a random mapping. In Example 1 we represented a binary string $\alpha \in \{0,1\}^n$ by the fingerprint $\text{Number}(\alpha) \bmod p$ for a randomly chosen prime $p \in \{1, \ldots, n^2\}$. The length of this fingerprint is $\lceil \log_2 n^2 \rceil$ which is essentially smaller than n. This approach works very well if the probability that the fingerprint of α is equal to the fingerprint of β is small for different α and β. Thus, fingerprinting may provide an efficient method for recognizing identities as we did in Example 1.

Exercise 1 Change the protocol R of Example 1 by choosing a prime from $\{2, 3, \ldots, n^d\}$ for some constant d. Which influence does this have on the error probability of R and on the communication complexity of R?

3. LOWER BOUND TECHNIQUES

In this section we do not deal with randomized protocols. The aim here is to present methods for proving lower bounds on the communication complexity of deterministic and nondeterministic protocols. These methods will be used in the subsequent sections to show the power of randomization in the comparison with deterministic and nondeterministic communications.

The most transparent way to explain the methods for proving lower bounds on communication complexity is to consider the representation of functions to be computed in the form of so-called communication matrices.

Definition 5 *Let $U = \{\alpha_1, \ldots, \alpha_k\}$, $V = \{\beta_1, \ldots, \beta_m\}$ be two sets, and let $f : U \times V \to \{0,1\}$. The* **communication matrix** *of f is the $|U| \times |V|$ Boolean matrix $M_f = [a_{ij}]_{i=1,\ldots,k, j=1,\ldots,m}$ defined by*

$$a_{ij} = a_{\alpha_i \beta_j} = f(\alpha_i, \beta_j).$$

For every $\gamma \in U$, the row corresponding to the input part γ is called the **row of γ** *and it is denoted by* $\mathbf{row}_\gamma = (a_{\gamma\beta_1}, a_{\gamma\beta_2}, \ldots, a_{\gamma\beta_m})$. *Similarly, the column corresponding to an input part $\delta \in V$ is called the* **column of δ**, *and it is denoted by* $\mathbf{column}_\delta = (a_{\alpha_1\delta}, a_{\alpha_2\delta}, \ldots, a_{\alpha_k\delta})$.

For all non-empty index sets $\{i_1, \ldots, i_l\} \subseteq \{1, \ldots, k\}$, $\{j_1, \ldots, j_r\} \subseteq \{1, \ldots, m\}$,

$$M_f(\{\alpha_{i_1}, \alpha_{i_2}, \ldots, \alpha_{i_l}\}, \{\beta_{j_1}, \beta_{j_2}, \ldots, \beta_{j_r}\})$$

denotes the $l \times r$ submatrix of M_f obtained by the intersection of the rows of M_f that correspond to inputs $\alpha_{i_1}, \alpha_{i_2}, \ldots, \alpha_{i_l}$ and the columns corresponding to the inputs $\beta_{j_1}, \beta_{j_2}, \ldots, \beta_{j_r}$.

Observe that one needs an ordering on the elements of U and V in order to unambiguously determine M_f for a function $f : U \times V \to \{0,1\}$. Since we usually work with Boolean functions one may always consider the lexicographic order on the Boolean vectors (binary words). So, for the Boolean function $f_{\mathrm{pr}} : \{0,1\}^3 \times \{0,1\}^2 \to \{0,1\}$ defined by

$$f_{\mathrm{pr}}(u, v) = 1 \text{ iff } v \text{ is a prefix of } u$$

(i.e., $v_1 \equiv u_1, v_2 \equiv u_2$ if $u = u_1 u_2 u_3$, $v = v_1 v_2$), $M_{f_{\mathrm{pr}}}$ is the following matrix

	00	01	10	11
000	1	0	0	0
001	1	0	0	0
010	0	1	0	0
011	0	1	0	0
100	0	0	1	0
101	0	0	1	0
110	0	0	0	1
111	0	0	0	1

All lower bound techniques we consider are based on some algebraic or combinatorial properties of communication matrices. Let P be a protocol computing a function $f : U \times V \to \{0,1\}$. As already mentioned above, any row of M_f (an element of U) corresponds to an input of C_I and any input of C_{II} (an element of V) corresponds to a column of M_f. So, the work of P is the game where C_I knows M_f and the row of M_f corresponding to its input and C_{II} knows M_f and the column corresponding to its input. The goal is to determine the value on the intersection between this row and this column. Note that C_I and C_{II} do not necessarily need to determine the exact position of the intersection of their row and column, because if the actual row (seen by C_I) contains only 0's, then C_I knows already the output value. The communication between C_I and C_{II} may be viewed as taking smaller and smaller submatrices of M_f until one obtains a monochromatic submatrix of M_f. To support this intuition consider that C_I sends the first message $c_1 \in \{0,1\}^+$ to C_{II}. Because C_{II} knows the communication protocol, it knows the set $S(c_1)$ of rows for which C_I sends the message c_1. So, C_{II} knows that the output value lies in the intersection of its column and the rows of $S(c_1)$. So, the first communication of this game reduces M_f to a submatrix $M_f(c_1)$ of M_f consisting of the rows of $S(c_1)$. Next, C_{II} sends a message c_2 to C_I. Since C_I knows the protocol and the messages c_1 and c_2, it can determine for which columns C_{II} sends c_2 if receiving c_1. So, the matrix $M_f(c_1\$c_2)$ remaining in the game becomes smaller again. Following this game one may suggest that every computation of P determines a set of inputs from $U \times V$, that corresponds to a monochromatic submatrix of M_f. So, all computations of any protocol P for f partition M_f into monochromatic submatrices. We confirm this intuition by using the following fundamental observation.

Observation 1 *Let $f : U \times V \to \{0,1\}$ be a function, and let P be a protocol. For all $\alpha_1, \alpha_2 \in U$, $\beta_1, \beta_2 \in V$, if P has the same computation C on the inputs (α_1, β_1) and (α_2, β_2), then P has this computation C on inputs (α_1, β_2) and (α_2, β_1), too.*

Particularly it means, that if P accept [rejects] (α_1, β_1) and (α_2, β_2), then P accepts [rejects] (α_1, β_2) and (α_2, β_1), too.

Proof. Let $C = c_1\$c_2\$ \ldots \$c_k\$c_{k+1}$, $c_i \in \{0,1\}^+$ for $i = 1, \ldots, k$, $c_{k+1} \in \{\text{accept}, \text{reject}\}$, be the computation of P on the inputs (α_1, β_1) and (α_2, β_2). Assume $\alpha_1 \neq \alpha_2$ and $\beta_1 \neq \beta_2$ (see Figure 1) because in the opposite case the claim of Observation 1 is obvious. Let us show that C is the computation of P on (α_1, β_2), too. The proof for (α_2, β_1) is analogous.

552

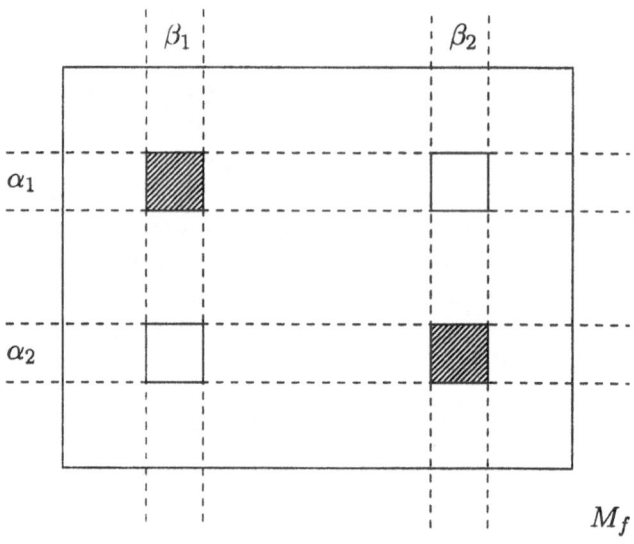

Figure 1

At the beginning C_I has the same argument α_1 for both inputs (α_1, β_1) and (α_1, β_2) and so C_I sends c_1 to C_{II} in both cases. Since C is the computation of P on (α_2, β_2), C_{II} sends the message c_2 to C_I for the arguments β_2 and $c_1\$$. So, C_{II} sends c_2 to C_I also if P works on (α_1, β_2) because the arguments of C_{II} are β_2 and $c_1\$$, too. Again, since C is the computation of P on (α_1, β_1), C_I sends c_3 to C_{II} if C_I has arguments α_1 and $c_1\$c_2\$$, i.e., also if P works on the input (α_1, β_2), etc. So, we see that C is the computation of P on (α_1, β_2), too. $\qquad\square$

Observation 1 provides two powerful techniques for proving lower bounds on communication complexity. Figure 1 provides a transparent explanation of the following claim.

Corollary 1 *Let* $f : U \times V \to \{0, 1\}$ *be a function that is computed by a protocol* P. *Let, for every computation* C *of* P, **Set**(C) *be the set of all inputs from* $U \times V$ *that are proceeded by* C. *Then,* $\mathrm{Set}(C)$ *corresponds to a monochromatic submatrix* $\boldsymbol{M_f}(\boldsymbol{C})$ *of* M_f *and* $M_f(C)$ *is a 1-monochromatic [0-monochromatic] submatrix of* M_f *iff* C *is an accepting [rejecting] computation.*

Figure 2 depicts a 1-monochromatic submatrix that is the intersection of rows 001, 010, 100, 111 and the columns 000, 011, 101, and 110 of the communication matrix M_f for the function $f : \{0, 1\}^3 \times \{0, 1\}^3 \to \{0, 1\}$ defined by

$$f(x_1, x_2, x_3, y_1, y_2, y_3) = x_1 \oplus x_2 \oplus x_3 \oplus y_1 \oplus y_2 \oplus y_3.$$

	000	001	010	011	100	101	110	111
000	0	1	1	0	1	0	0	1
001	1	0	0	1	0	1	1	0
010	1	0	0	1	0	1	1	0
011	0	1	1	0	1	0	0	1
100	1	0	0	1	0	1	1	0
101	0	1	1	0	1	0	0	1
110	0	1	1	0	1	0	0	1
111	1	0	0	1	0	1	1	0

Figure 2

The computation corresponding to this 1-monochromatic submatrix is involved in the protocol P with $P(x_1x_2x_3) = x_1 \oplus x_2 \oplus x_3$ and $P(y_1y_2y_3, a\$) = $ accept [reject] iff $y_1 \oplus y_2 \oplus y_3 \oplus a = 1$ [0].

Corollary 1 immediately implies that every protocol P computing a function f with d different computations unambiguously determines a partition of M_f into d pair-wise disjoint monochromatic submatrices.[6] So, the number of computations of any protocol computing a function f is at least the cardinality of the minimal cover of M_f by pair-wise disjoint monochromatic submatrices. Let us formalize this idea in what follows.

Definition 6 *Let M be a Boolean matrix and let $S = \{M_1, \ldots, M_k\}$ be a set of monochromatic submatrices of M. We say that S is a **cover** of M if, for every element a_{ij} of M. there exists an $m \in \{1, \ldots, k\}$ such that a_{ij} is an element of M_m. We say that S is an **exact cover** of M if S is a cover of M and $M_r \cap M_s = \emptyset$ for every $r \neq s$, $r, s \in \{1, \ldots, k\}$.*

*The **tiling complexity** of M is*

$$\mathrm{Tiling}(M) = \min\{|S| \mid S \text{ is an exact cover of } M\}.$$

Theorem 2 *For every finite function $f : U \times V \to \{0, 1\}$,*

$$\mathrm{cc}(f) \geq \lceil \log_2 \lceil \mathrm{Tiling}(M_f) \rceil \rceil - 1.$$

Proof. Above we have proved that every protocol P for f determines an exact cover S of M_f, where $|S|$ is the number of different computations

---

[6]Note, that a protocol computing f must have exactly one computation for every entry of the matrix M_f.

of P. Thus, $\text{Tiling}(M_f)$ is a lower bound on the number of computations of any protocol computing f. Because of the prefix-freeness property of the protocols, every protocol for f must have a computation C of length at least $\log_2\lceil\text{Tiling}(M_f)\rceil$. So, $|\text{Com}(C)| \geq \log_2\lceil\text{Tiling}(M_f)\rceil - 1$ and $\text{cc}(f) \geq \log_2\lceil\text{Tiling}(M_f)\rceil - 1$. $\qquad\square$

One can observe that the matrix M_f depicted at Figure 2 can be exactly covered by two 1-monochromatic submatrices

$$M_f(\{001, 010, 100, 111\}, \{000, 011, 101, 110\}),$$

$$M_f(\{000, 011, 101, 110\}, \{001, 010, 100, 111\}),$$

and by two 0-monochromatic submatrices

$$M_f(\{000, 011, 101, 110\}, \{000, 011, 101, 110\}),$$

and

$$M_f(\{001, 010, 100, 111\}, \{001, 010, 100, 111\}).$$

Thus, the communication protocol sending the bit $x_1 \oplus x_2 \oplus x_3$ for an input $x_1 x_2 x_3$ is an optimal one and $\text{cc}(f) = 1$ for the function f whose matrix M_f is depicted in Figure 2.

A simple transparent example of the application of the tiling method is proving the optimal lower bound on $\text{cc}(\text{Eq}_n)$. Since M_{Eq_n} is the diagonal $2^n \times 2^n$ matrix, one needs 2^n 1-monochromatic submatrices of M_{Eq_n} to cover the diagonal elements (1's) and 2^n 0-monochromatic submatrices to cover all 0's of M_{Eq_n}. So, $\text{Tiling}(M_{\text{Eq}_n}) \geq 2^{n+1}$, and $\text{cc}(\text{Eq}_n) \geq n$.

Note, that in general, to find an optimal exact cover for a given Boolean matrix is a nontrivial combinatorial task and so in some cases the search for $\text{Tiling}(M_f)$ is no simplification of the search for the communication complexity of an optimal protocol f.

The next method may be very practical in some cases because it is a constructive method. To get a lower bound on $\text{cc}(f)$ it is sufficient to find a set of input instances with some special property. This method is also based on Observation 1. We search for such a set of inputs that any two inputs from this set cannot be covered by the same monochromatic submatrix. So, the cardinality of every exact cover must be at least the cardinality of this set. In this relation this method can be viewed as a constructive method for proving lower bounds on $\text{Tiling}(M_f)$, too.

Definition 7 *Let $f : U \times V \to \{0, 1\}$ be a function. For every $\delta \in \{0, 1\}$, we say that a set $\mathcal{A} \subseteq U \times V$ is a δ-**fooling set** for f if, for all $(\alpha_1, \beta_1), (\alpha_2, \beta_2) \in \mathcal{A}$,*

(i) $f(\alpha_1, \beta_1) = f(\alpha_2, \beta_2) = \delta$, and

(ii) $f(\alpha_1, \beta_2) \neq \delta$ *or* $f(\alpha_2, \beta_1) \neq \delta$.

We define

Fool$(f) = \max\{|\mathcal{A}| \mid \mathcal{A} \text{ is } \delta\text{-fooling set for } f \text{ for some } \delta \in \{0, 1\}\}.$

Looking at Figure 2 we see that $\mathcal{A} = \{(001, 011), (011, 100)\}$ is a 1-fooling set because $f(001, 011) = f(011, 100) = 1$ and $f(001, 100) = f(011, 011) = 0$. So, there does not exist any 1-monochromatic submatrix of M_f that would cover both elements of $\mathcal{A}$.

Theorem 3 *For every finite function* $f : U \times V \to \{0, 1\}$,

$$cc(f) \geq \lceil \log_2(\text{Fool}(f)) \rceil.$$

Proof. Let $\mathcal{A}$ be a δ-fooling set for f. Following Figure 1 or our consideration above, no pair of inputs $(\alpha_1, \beta_1), (\alpha_2, \beta_2)$ of $\mathcal{A}$ can be covered by one δ-monochromatic submatrix. So, only to cover the elements of $\mathcal{A}$ in M_f, one needs at least $|\mathcal{A}|$ δ-monochromatic submatrices. This is the same as to say that, for $\delta = 1$ [0], every protocol for f must have at least $|\mathcal{A}|$ different accepting [rejecting] computations. But $|\mathcal{A}|$ different accepting [rejecting] computations mean $|\mathcal{A}|$ different communications. As already observed, every protocol with $|\mathcal{A}|$ different communications must have a communication of length at least $\lceil \log_2 |\mathcal{A}| \rceil$. Thus

$$cc(f) \geq \lceil \log_2 |\mathcal{A}| \rceil$$

for every δ-fooling set $\mathcal{A}$ for f. $\qquad\square$

A nice example is again M_{Eq_n} for every positive integer n. $\mathcal{A} = \{(\alpha, \alpha) \mid \alpha \in \{0, 1\}^n\}$ is a 1-fooling set for Eq_n because $\text{Eq}_n(\alpha, \beta) = 0$ for all $\alpha, \beta \in \{0, 1\}^n$, $\alpha \neq \beta$, and so $cc(\text{Eq}_n) \geq \log_2(2^n) = n$. Analogously, $\mathcal{A}$ is a 0-fooling set for Ineq_n because $\text{Ineq}_n(\alpha, \beta) = 1$ for all $\alpha, \beta \in \{0, 1\}^n$, $\alpha \neq \beta$.

Another example is the proof of the optimality of the protocol presented in Section 2 for the symmetric function s_{2n} that takes the value 1 iff exactly n of its $2n$ arguments are 1's. Consider, for $i = 0, 1, \ldots, n$, $\alpha_i = 1^i 0^{n-i}$ and $\beta_i = 0^i 1^{n-i}$. We claim that the set $\mathcal{A}(s_{2n}) = \{(\alpha_i, \beta_i) \mid i = 0, 1, \ldots, n\}$ is a 1-fooling set for s_{2n} because $s_{2n}(\alpha_i, \beta_i) = 1$ for $i = 0, 1, \ldots, n$ and $s_{2n}(\alpha_i, \beta_j) = 0$ for $i \neq j$. Thus, $cc(s_{2n}) = \lceil \log_2(n + 1) \rceil$.

Note that the method based on the fooling sets is nothing else than the classical crossing sequence argument used many times for many different models in complexity theory.

The third method for proving lower bounds on communication complexity is based on the rank of communication matrices. The advantage of this method is that the rank of a matrix can be computed efficiently.[7]

Let, for any field F with identity elements 0 and 1 and any Boolean matrix M, $\mathbf{rank}_F(M)$ denote the rank[8] of the matrix M over the field F. $\mathbf{rank}(M) = \mathrm{rank}_{GF(2)}(M)$ denotes the rank of M over the Galois field $GF(2)$, that contains only the elements 0 and 1. We define, for every Boolean matrix M,

$$\mathbf{Rank}(M) = \max\{\mathrm{rank}_F(M) \mid F \text{ is a field with identity} $$
$$\text{elements 0 and 1}\}.$$

Observation 2 *For every Boolean matrix M,*

$$\mathrm{Rank}(M) = \mathrm{rank}_Q(M),$$

where Q is the set of all rational numbers.

There are two different arguments for the claim

$$cc(f) \geq \lceil \log_2(\mathrm{Rank}(M_f)) \rceil.$$

Below, we present one of them in an informal way, and the second one is used to give the formal proof of the next theorem. Let P be a protocol computing a finite function f. Then one can split M_f into two parts $M_f(0)$ and $M_f(1)$ according to the first bit communicated in the following way. $M_f(0)$ consists of rows[9] for which C_I send a message beginning with 0, and $M_f(1)$ consists of rows for which C_I sends a message with the prefix 1. Obviously, at least one of the matrices $M_f(0)$ and $M_f(1)$ must have the rank at least $\lceil \mathrm{Rank}(M_f)/2 \rceil$. Let $M_f(c_1)$ be this submatrix for some $c_1 \in \{0, 1\}$. Then the next bit communicated again splits $M_f(c_1)$ into two submatrices. If the second bit is submitted by C_I, then the partition is done by a row-split again[10], if the second bit is sent by C_{II}, then the partition of $M_f(c_1)$ is done by some column-split. In this way one obtains a matrix $M_f(c_1 c_2)$ of the rank at least $\lceil \mathrm{Rank}(M_f)/4 \rceil$. Continuing in this way one always finds a submatrix

---

[7]On the other hand, to find a minimal exact cover or a maximal fooling set may be hard optimization problems.

[8]Remember, that the rank is the number of linearly independent rows (columns) of M over F.

[9]Better to say, the inputs for which C_I sends a message $0c$, $c \in \{0,1\}^*$, determine the rows of $M_f(0)$.

[10]Remember, that because of the prefix-freeness property it is unambiguously determined which of C_I and C_{II} sends the second bit for every computation beginning with the bit c_1.

$M_f(c_1 c_2 \ldots c_i)$ of M_f whose rank is at least $\lceil \text{Rank}(M_f)/2^i \rceil$. Obviously, the communication procedure can stop only if the resulting matrix either consists of monochromatic rows (if C_I decides about the acceptance) or of monochromatic columns (if C_{II} decides about the acceptance).[11] Thus, the above procedure may stop only if the rank of $M_f(c_1 c_2 \ldots c_i)$ is 1. But this means that i must be at least $\log_2(\text{Rank}(M_f))$, and so there exists a communication of P consisting of at least $\log_2(\text{Rank}(M_f))$ bits.

Theorem 4 *For every finite function* $f : U \times V \to \{0,1\}$,

$$cc(f) \geq \lceil \log_2(\text{Rank}(M_f)) \rceil.$$

Proof. Let P be a protocol computing f, and let F be any field with identity elements 0 and 1. It is sufficient to show that the number of accepting computations of P is at least $\text{rank}_F(M_f)$.

Let $C_1, C_2, \ldots, C_k$ be all accepting computations of P. We already know, that, for every $i \in \{1, 2, \ldots, k\}$, C_i determines the 1-monochromatic submatrix $M_f(C_i)$ of M_f. Let $\bar{M}_f(C_i) = [a_{rs}]$ be the $|U| \times |V|$ Boolean matrix defined by

$$a_{rs} = 1 \text{ iff } a_{rs} \text{ is an element of } M_f(C_i)$$

for $i = 1, \ldots, n$. Obviously

$$M_f = \sum_{i=1}^{k} \bar{M}_f(C_i).$$

By the properties of the rank over any field we have

$$\text{rank}_F(M_f) \leq \sum_{i=1}^{k} \text{rank}_F(\bar{M}_f(C_i)).$$

Thus, $k \geq \text{rank}_F(M_f)$ and to get k accepting computations P must contain a communication of at least $\log_2 k$ bits. $\square$

Again, M_{Eq_n} and M_{Ineq_n} are simple examples showing that the rank method can provide optimal lower bounds on the complexity of some functions.

Since the comparison of the lower bound proof methods introduced above is not the matter of this work, we only survey the known results on this topic. All formal proofs can be found in Section 2.2.2 of [Hr97].

---

[11] Remember, that it is impossible that one of the computers produces the answer from {accept, reject} for some inputs of the submatrix corresponding to the actual communication history and continues to communicate for other inputs of this submatrix.

(i) $cc(f)$ and $\text{Tiling}(f)$ are polynomially related, because

$$\lceil \log_2(\text{Tiling}(M_f)) \rceil \leq cc(f) \leq (\lceil \log_2(\text{Tiling}(M_f)) \rceil + 1)^2$$

for every f.

(ii) $\text{Tiling}(M_f) \leq 2^{\text{rank}(M_f)}$ and there exists a Boolean function f with an exponential gap between $\text{Tiling}(M_f)$ and $\text{rank}(M_f)$, i.e. the rank method over the Galois field $GF(2)$ may be much weaker than the tiling method.

(iii) $\text{Fool}(f) \leq \text{Tiling}(M_f)$ for every finite function f and there exists a Boolean function g with an exponential gap between $\text{Fool}(g)$ and $\text{Tiling}(M_g)$, i.e. the fooling set method can be much weaker than the tiling method.

(iv) For every finite function f and every field F with the identity elements 0 and 1,

$$\text{Fool}(f) \leq (\text{rank}_F(M_f) + 2)^2,$$

i.e., the fooling set method cannot be essentially better than the rank method.

There exists a Boolean function g with an exponential gap between $\text{Fool}(g)$ and $\text{Rank}(M_g)$, i.e., the rank method may be essentially better than the fooling set method.

It is an open problem whether an exponential gap between $cc(f)$ and $\log_2(\text{Rank}(M_f))$ is possible. A more involved discussion about this problem can be found in [KN97].

Observe that the suitability of the presented methods in concrete situations does not depend only on the relations between these methods and their relations to communication complexity. To find an optimal exact cover or to find a maximal fooling set for a given matrix may be hard tasks. Since we usually do not search for the communication complexity of a specific finite function f only, but for lower bounds on $cc(f_n)$ for an infinite sequence of Boolean functions $\{f_n\}_{n=1}^{\infty}$, $f_n : \{0,1\}^n \times \{0,1\}^n \to \{0,1\}$, it may be also hard to establish the ranks of all infinitely many matrices M_{f_n}. In what follows we always choose the method that reasonably works for the considered function.

Now, we look for lower bounds on one-way communication complexity. Here, the situation is quite simple.

Definition 8 *Let* $f : U \times V \to \{0,1\}$ *be a finite function. A set* $\mathcal{A} \subseteq U$ *is called a* **one-way fooling set** *for* f, *if, for every* $\alpha, \beta \in \mathcal{A}$, $\alpha \neq \beta$, *there exists a $\gamma \in V$ such that*

$$f(\alpha, \gamma) \neq f(\beta, \gamma).$$

We define

One-Way-Fool(f) $= \max\{|\mathcal{A}| \,|\, \mathcal{A} \text{ is a one-way fooling set for } f\}$.

*For every Boolean matrix M, we define **Row(M)** to be the number of different rows of M.*

Theorem 5 *For every finite function $f : U \times V \to \{0, 1\}$,*

$$2^{\mathrm{cc}_1(f)} \geq \text{One-Way-Fool}(f) = \text{Row}(M_f).$$

Proof. One can easily observe that two rows α and β of M_f are different if and only if there is a column γ such that $f(\alpha, \gamma) \neq f(\beta, \gamma)$. So, One-Way-Fool($f$) = Row($M_f$) because every one-way fooling set $\mathcal{A}$ for f determines $|\mathcal{A}|$ different rows of M_f and, vice versa, any k different rows of M_f determine a one-way fooling set of the cardinality k.

Let P be an one-way protocol for f. Let $P(\alpha) = P(\beta)$ for some $\alpha, \beta \in U$, $\alpha \neq \beta$ (i.e. C_I sends the same message to C_{II} for both α and β). For every $\gamma \in V$, P must compute the same output for both inputs (α, γ) and (β, γ) because $P(\gamma, P(\alpha)\$) = P(\gamma, P(\beta)\$)$. So, $P(\alpha) = P(\beta)$ can be true only if the rows α and β are equal. Thus, the number of different messages (communications) of P must be at least Row(M_f). $\square$

We can immediately observe that there may exist an exponential gap between $\mathrm{cc}_1(f)$ and $\mathrm{cc}(f)$ because there are Boolean matrices with an exponential gap between their Row(M) and Rank(M). For instance, for any positive integer n, one can construct a $2^n \times n$ Boolean matrix M consisting of 2^n different rows, i.e. containing all n-dimensional Boolean vectors as the rows. Obviously, Row(M) = 2^n and Rank(M) = n.

To see the difference for a function from $\{0, 1\}^n \times \{0, 1\}^n$ to $\{0, 1\}$ consider the function $f_{\mathrm{ind}(n)}$ introduced in Section 2.

Lemma 1 *For every positive integer $n = 2^k$, $k \in \mathbb{N} - \{0\}$,*

$$\mathrm{cc}_1(f_{\mathrm{ind}(n)}) = n \text{ and } \mathrm{cc}(f_{\mathrm{ind}(n)}) \leq 2 \cdot \log_2 n.$$

Proof. A protocol computing $f_{\mathrm{ind}(n)}$ within the communication complexity $2 \cdot \log_2 n$ was presented in Section 2. The fact $\mathrm{cc}_1(f_{\mathrm{ind}(n)}) \leq n$ is obvious.

It remains to show $\mathrm{cc}_1(f_{\mathrm{ind}(n)}) \geq n$. To do this observe that the $2^n \times 2^n$ matrix $M_{f_{\mathrm{ind}(n)}}$ has 2^n different rows. Let $\alpha = \alpha_1 \ldots \alpha_n \in \{0, 1\}^n$,

$\beta = \beta_1 \ldots \beta_n \in \{0,1\}^n$. $\alpha \neq \beta$ be two input parts determining two rows of $M_{f_{\mathrm{ind}(n)}}$. Since $\alpha \neq \beta$. there exists an integer $j \in \{1, \ldots, n\}$ such that $\alpha_j \neq \beta_j$. Without loss of generality assume $\alpha_j = 1$ and $\beta_j = 0$. Let j be the smallest integer with the above property. We distinguish two possibilities according to whether $j \geq \log_2 n$ or $j \leq \log_2 n$.

If $j > \log_2 n$, $\alpha_1 \ldots \alpha_{\log_2 n} = \beta_1 \ldots \beta_{\log_2 n}$, and so

$$a = \mathrm{Number}(\alpha_1 \ldots \alpha_{\log_2 n}) + 1 = \mathrm{Number}(\beta_1 \ldots \beta_{\log_2 n}) + 1.$$

We choose an $\gamma = \gamma_1 \ldots \gamma_n$ such that

$$\mathrm{Number}(\gamma_a \gamma_{(a+1) \bmod n} \cdots \gamma_{(a+\lceil \log_2 n \rceil) \bmod n}) + 1 = j.$$

Obviously $f(\alpha, \gamma) = 1$ and $f(\beta, \gamma) = 0$ and so the rows corresponding to α and β are different.

The proof for the case $j \leq \log_2 n$ is left to the reader. $\qquad\square$

Finally, we deal with proving lower bounds on nondeterministic communication complexity. First of all we observe that Observation 1 holds for nondeterministic protocols, too. This means, that if C is an accepting computation on inputs (α_1, β_1) and (α_2, β_2), then C is a possible accepting computation on the inputs (α_1, β_2) and (α_2, β_1). So, we have again the situation that every accepting computation corresponds to a 1-monochromatic submatrix. The situation differs from the deterministic case in that the nondeterministic protocol has possibly several computations on one input and so the monochromatic matrices determined by computations may overlap. Thus, any nondeterministic protocol determines a cover of 1's of the communication matrix by possibly non-disjoint 1-monochromatic submatrices.[12]

Definition 9 *Let M be a Boolean matrix, and let $S = \{M_1, \ldots, M_k\}$ be a set of 1-monochromatic submatrices of M. We say that S is a 1-cover of M if every 1 of M is contained in at least one of the 1-submatrices of S. We define*

$$\mathbf{Cover}(M) = \min\{|S| \mid S \text{ is a 1-cover of } M\}.$$

Theorem 6 *For every finite function $f : U \times V \to \{0,1\}$,*

$$\mathrm{ncc}(f) = \lceil \log_2(\mathrm{Cover}(M_f)) \rceil.$$

Proof. We already argued that $\mathrm{ncc}(f) \geq \lceil \log_2(\mathrm{Cover}(M_f)) \rceil$ because every nondeterministic protocol with k accepting computations determines a 1-cover of M_f of size k.

---

[12]Observe that a nondeterministic protocol does not necessarily determine any cover of 0's of M_f.

Thus, it remains to prove $\text{ncc}(f) \leq \lceil \log_2(\text{Cover}(M_f)) \rceil$. Let $S = \{M_1, M_2, \ldots, M_k\}$ be a 1-cover of M_f, and let $k = \text{Cover}(M_f)$. We construct a nondeterministic one-way protocol $D = (C_I, C_{II})$ for f as follows. For every input $\alpha \in U$, C_I looks for all submatrices of S with nonempty intersections with row$_\alpha$. C_I chooses one of these 1-submatrices M_i and sends the message $\text{BIN}_{\lceil \log_2 k \rceil}(i)$. If C_{II} receives $\text{BIN}_{\lceil \log_2 k \rceil}(i)$ and column$_\beta$ of the input $\beta \in V$ of C_{II} intersect M_i, then C_{II} accepts. Otherwise, C_{II} rejects. □

Observe, that the proof of Theorem 6 provides also an alternative proof of the fact $\text{ncc}(f) = \text{ncc}_1(f)$ for every f. Since the cardinality of any 1-fooling set for f is a lower bound[13] on $\text{Cover}(M_f)$, we obtain the following result.

Theorem 7 *Let* $f : U \times V \to \{0, 1\}$ *be a finite function. For every 1-fooling set* $\mathcal{A}$ *for* f.

$$\text{ncc}(f) \geq \lceil \log_2 |\mathcal{A}| \rceil.$$

We see that the function Eq_n is hard for nondeterministic protocols because one needs 2^n 1-monochromatic submatrices to cover the 2^n diagonal elements of the $2^n \times 2^n$ diagonal matrix M_{Eq_n}.

For Ineq_n the situation essentially differs. The ones of an $2^n \times 2^n$ 0-diagonal matrix can be covered by $2n$ 1-monochromatic submatrices $M_1, M_2, \ldots, M_n, M_1', M_2', \ldots, M_n'$, where M_i (M_i') is the intersection of all rows whose i-th bit is 1 (0) with rows whose i-th bit is 0 (1). On the other hand to cover 0's of M_{Ineq_n} one needs 2^n 0-monochromatic matrices. So, this implies the following exponential gap between deterministic communication complexity and nondeterministic communication complexity.

Theorem 8 *For every positive integer* n

(i) $\text{ncc}(\text{Ineq}_n) \leq \lceil 2 \log_2 n \rceil$, *and*

(ii) $\text{cc}(\text{Ineq}_n) = n$.

4.   LAS VEGAS PROTOCOLS

In this section we relate Las Vegas and determinism for communication complexity. The results show that Las Vegas can be more efficient than determinism, but the difference between the power of these two

---

[13]Remember that no pair of elements of a 1-fooling set can be covered by one 1-monochromatic submatrix.

models of computation is not very large. One conjectures that this kind of relation should hold for several fundamental computing models and their complexity measures, including the time complexity of Turing machines. Unfortunately, one fails in the effort to establish such relations for most complexity measures investigated. In this section we show that there is an at most quadratic gap between $cc(f)$ and $lvcc(f)$ and a linear relation between one-way communication complexity and Las Vegas one-way communication complexity. Examples of concrete functions show that Las Vegas can do something better than determinism in the framework of relations established above. In Section 7 we show how to apply these results to show close relations between Las Vegas and determinism for finite automata and some restricted models of branching programs.

First of all we show that there is a polynomial relation between Las Vegas communication complexity and deterministic communication complexity. To get this result one proves a more powerful result claiming that if both a function f and its complement $\overline{f}$ are easy for a nondeterministic protocol, then f is easy for a deterministic protocol, too. In what follows, for every function $f : U \times V \to \{0, 1\}$, the **complement of f** is the function $\overline{f}$ defined by

$$\overline{f}(\alpha, \beta) = \Gamma(f(\alpha, \beta))$$

for all $(\alpha, \beta) \in U \times V$. Γ denotes the unary Boolean operation called **negation**.

Theorem 9 ([AUY83]) *For every finite function $f : U \times V \to \{0, 1\}$,*

$$cc(f) \le ncc(f) \cdot (ncc(\overline{f}) + 2).$$

Proof. Let $r_1 = ncc(f)$ and $r_0 = ncc(\overline{f})$. This directly implies that the 1's of M_f can be covered by at most 2^{r_1} 1-monochromatic submatrices, and that the 0's of M_f can be covered by at most 2^{r_0} 0-monochromatic submatrices. We shall prove that $cc(f) \le r_1 \cdot (r_0 + 2)$.

The proof relies on the following property of monochromatic submatrices. Let $M(R, S)$ be a 0-monochromatic submatrix, and let $M(S', R')$ be a 1-monochromatic submatrix of M_f. Then either $R \cap R' = \emptyset$ or $S \cap S' = \emptyset$. The proof for this claim is straightforward because if $R \cap R' \ne \emptyset$ and $S \cap S' \ne \emptyset$, then there exists a pair $(\alpha, \beta) \in (R \cap R') \times (S \cap S')$, i.e. a pair that belongs to both $M(R, S)$ and $M(R', S')$. However, this is impossible because (α, β) in $M(R, S)$ implies $f(\alpha, \beta) = 0$, whereas (α, β) in $M(R', S')$ implies $f(\alpha, \beta) = 1$.

Before starting the proper proof we fix some useful notation. Let $\overline{C} = \{C_1, \ldots, C_m\}$, $m \le 2^{r_0}$, be a set of 0-monochromatic submatrices of M_f such that $\overline{C}$ covers all 0's of M_f. Let $\overline{H} = \{H_1, \ldots, H_l\}$, $l \leq 2^{r_1}$, be a set of 1-monochromatic submatrices of M_f covering all 1's in M_f. Let A_i denote the submatrix of M_f formed by those rows of M_f that are contained in C_i (i.e., if $C_i = M(R, S)$, then $A_i = M(R, V)$). Let B_i denote the submatrix of M_f formed by the columns of M_f that meet C_i (i.e., if $C_i = M(R, S)$, then $B_i = M(U, S)$). Let $\mathbf{int(A_i)}$ and $\mathbf{int(B_i)}$ respectively denote the number of 1-monochromatic submatrices from $\overline{H}$ that have a non-empty intersection with A_i and B_i respectively. Since the intersection of A_i and B_i is exactly the 0-monochromatic submatrix C_i, the claim stated above implies that no matrix $H_j \in \overline{H}$ has non-empty intersections with both A_i and B_i. Thus,

$$\text{int}(A_i) + \text{int}(B_i) \leq l = |\overline{H}|$$

for every $i \in \{1, 2, \ldots, m\}$. Set

$$\overline{C}_1 = \{C_k \in \overline{C} \mid \text{int}(A_k) \leq \lceil l/2 \rceil\}, \text{ and}$$

$$\overline{C}_2 = \{C_s \in \overline{C} \mid \text{int}(B_s) \leq l/2\} = \overline{C} - \overline{C}_1.$$

Now, we describe the first two rounds of a deterministic protocol $D = \langle D_I, D_{II} \rangle$ for f. For every input $(\alpha, \beta) \in U \times V$, D works as follows.

Round 1. D_I looks on the row of M_f that corresponds to α in order to see whether it intersects any of the 0-monochromatic submatrices in $\overline{C}_1$. If so, it sends the message "1BIN$_{r_0}(j)$", where j is the smallest index such that $C_j \in \overline{C}_1$ and C_j intersects the row of α. If row$_\alpha$ does not intersect any submatrix of $\overline{C}_1$, then D_I sends the message "0".

Round 2. If D_{II} receives "0", it looks whether the column corresponding to its input β intersects any of the 0-monochromatic submatrices in $\overline{C}_2$. If so, it sends the message "1BIN$_{r_0}(k)$", where k is the smallest index such that $C_k \in \overline{C}_2$ and C_k intersects the column of β.
Otherwise, D_{II} sends the message "0". If D_{II} receives "1BIN(j)", then it sends "1" to D_I.

Now, let us distinguish and discuss three possible situations after the first two rounds of D.

Case 1. The current communication history is 0\$0\$, i.e. both computers failed to find an appropriate 0-monochromatic submatrix. Since $\overline{C}_1 \cup \overline{C}_2 = \overline{C}$ and the set $\overline{C}$ covers all zeros in M_f, we get

$f(\alpha, \beta) = 1$. So, both D_I and D_{II} know that the output has to be "accept".

Case 2. The current communication history is $1\text{BIN}_{r_0}(j)\$1\$$. In this case both D_I and D_{II} know, that the input (α, β) belongs to A_j. Since $C_j \in \overline{C}_1$, $\text{int}(A_j) \leq \lceil l/2 \rceil \leq 2^{r_1-1}$, i.e., all ones in A_j can be covered by at most $2^{r_1-1} = 2^{r_1}/2$ 1-monochromatic submatrices of A_k. (Note that these 1-monochromatic submatrices are all intersections of A_k with 1-monochromatic submatrices $H_1, H_2, \ldots, H_l$ of M_f.)

Case 3. The current communication history is $0\$\text{BIN}_{r_0}(k)\$$. In this case both D_I and D_{II} know that the input (α, β) lies in B_k. Since $C_k \in \overline{C}_2$, they know that $\text{int}(B_k) \leq l/2 \leq 2^{r_1-1}$, i.e., all ones in B_k can be covered by at most 2^{r_1-1} 1-monochromatic submatrices.

Thus, after this first two rounds either both computers know the output value $f(\alpha, \beta)$ or both D_I and D_{II} know that (α, β) lies in a matrix $M_1 \in \{A_k, B_m\}$ whose ones can be covered by at most 2^{r_1-1} 1-monochromatic submatrices of M_1, and whose zeros can be covered by at most 2^{r_0} 0-monochromatic submatrices. Following the same communication strategy as described above for M_f in the next two rounds for the matrix M_1, D_I and D_{II} either learn the result $f(\alpha, \beta)$ or they agree that (α, β) is in a submatrix M_2 such that

(i) all ones of M_2 can be covered by at most 2^{r_1-2} 1-monochromatic submatrices of M_2, and

(ii) all zeros of M_2 can be covered by at most 2^{r_0} 0-monochromatic submatrices of M_2.

Continuing in this way, D_I and D_{II} learn $f(\alpha, \beta)$ after at most r_1 rounds. Since every information exchange in rounds $2i$ and $2i + 1$ has the length $2 + r_0$, $\text{cc}(D) \leq r_1 \cdot (2 + r_0)$. $\qquad \square$

Obviously, $\text{lvcc}(f) = \text{lvcc}(\overline{f})$ for every function f. Since $\text{ncc}(f)$ is a lower bound on the private Las Vegas communication complexity, there is at most an quadratic gap between determinism and private Las Vegas communication protocols. Since the relation between the private randomized communication complexity and the public one is linear if the randomized communication complexity is at least $\Omega(\log_2 n)$ for Boolean functions of n variables[14], we can express the consequences of Theorem 9 in terms of public Las Vegas communication complexity as follows.

---

[14]See Section 6 for the exact formulation of the relation and its proof.

Theorem 10 *For every Boolean function* $f : \{0,1\}^n \times \{0,1\}^n \to \{0,1\}$,

$$\mathrm{cc}(f) = O\left((\mathrm{lvcc}(f))^2 + \log_2 n\right).$$

Our next aim is to show, that the at most quadratic gap between determinism and Las Vegas presented in Theorem 10 can be achieved for a concrete Boolean function. Let, for every positive integer n, $n = m^2$,

$$\mathrm{ExAll}_{2n}(x_{1,1}, x_{1,2}, \ldots, x_{1,m}, \ldots, x_{m,1}, x_{m,2}, \ldots, x_{m,m},$$
$$y_{1,1}, y_{1,2}, \ldots, y_{1,m}, \ldots, y_{m,1}, y_{m,2}, \ldots, y_{m,m})$$

be a Boolean function from $\{0,1\}^n \times \{0,1\}^n$ to $\{0,1\}$ defined by

$$\mathrm{ExAll}_{2n}(\alpha_{1,1}, \ldots, \alpha_{m,m}, \beta_{1,1}, \ldots, \beta_{m,m}) = 1$$

iff $\exists j \in \{1, \ldots, m\}$ such that $\alpha_{j,1}, \ldots, \alpha_{j,m} = \beta_{j,1}, \ldots, \beta_{j,m}$.

Theorem 11 **([MS82])** *For every positive integers* n, m, $n = m^2$,

(i) $\mathrm{cc}(\mathrm{ExAll}_{2n}) = n = m^2$, *and*

(ii) $\mathrm{lvcc}(\mathrm{ExAll}_{2n}) \le 2m(\lceil \log_2 m \rceil^2 + 1)$ *for sufficiently large* m.

Proof.

(i) The fact $\mathrm{cc}(\mathrm{ExAll}_{2n}) \le n$ is obvious. To prove the lower bound it is sufficient to prove $\mathrm{cc}(\overline{\mathrm{ExAll}_{2n}}) \ge n$. Note that

$$\overline{\mathrm{ExAll}_{2n}}(\alpha_{1,1}, \ldots, \alpha_{m,m}, \beta_{1,1}, \ldots, \beta_{m,m}) = 1$$

iff, for every $i \in \{1, \ldots, m\}$, $\alpha_{i,1}, \ldots, \alpha_{i,m} \ne \beta_{i,1}, \ldots, \beta_{i,m}$. To prove the lower bound we use the rank method. More precisely, we show that the rows of the $2^n \times 2^n$ diagonal matrix are linear combinations of the row of $M_{\overline{\mathrm{ExAll}_{2n}}}$ and so $\mathrm{rank}(M_{\overline{\mathrm{ExAll}_{2n}}}) = 2^n$. Consider the function $q_{2n} : \{0,1\}^n \times \{0,1\}^n \to \{0,1\}$ defined for all $\delta_1, \ldots, \delta_m, \gamma_1, \ldots, \gamma_m \in \{0,1\}^m$ as follows:

$$q_{2n}(\delta_1 \delta_2 \ldots \delta_m, \gamma_1 \gamma_2 \ldots \gamma_m) =$$

$$\sum_{\alpha_1 \ne \delta_1} \sum_{\alpha_2 \ne \delta_2} \cdots \sum_{\alpha_m \ne \delta_m} \overline{\mathrm{ExAll}_{2n}}(\alpha_1 \alpha_2 \ldots \alpha_m, \gamma_1 \gamma_2 \ldots \gamma_m) \bmod 2,$$

where $\alpha_i \in \{0,1\}^m$ for $i = 1, 2, \ldots, m$. Our aim is to show that

$$q_{2n}(\alpha, \beta) = \mathrm{Eq}_n(\alpha, \beta)$$

for every positive integer n and all $(\alpha, \beta) \in \{0,1\}^n \times \{0,1\}^n$. Let us prove this. First, consider inputs $(\delta, \delta) = (\delta_1 \delta_2 \ldots \delta_m, \delta_1 \delta_2 \ldots \delta_m)$ for any $\delta_i \in \{0,1\}^m$ for $i = 1, \ldots, m$.

$$q_{2n}(\delta, \delta) = q_{2n}(\delta_1 \delta_2 \ldots \delta_m, \delta_1 \delta_2 \ldots \delta_m) = (2^m - 1)^m \bmod 2 = 1$$

because there are exactly $2^m - 1$ words from $\{0,1\}^m$ different from δ_i and $\overline{\mathrm{ExAll}_{2n}}(w, \delta_1, \ldots, \delta_m) = 1$ for all

$$w \in \{w_1 w_2 \ldots w_m \mid w_i \in \{0,1\}^m - \{\delta_i\} \text{ for } i = 1, \ldots, m\}.$$

Now, consider inputs $(\delta, \gamma) = (\delta_1 \delta_2 \ldots \delta_m, \gamma_1 \gamma_2 \ldots \gamma_m)$, $\delta_i, \gamma_i \in \{0,1\}^m$ for $i = 1, \ldots, m$, with $\delta \neq \gamma$. Let $S(\delta, \gamma) \subseteq \{1, \ldots, m\}$ such that, for all $j \in S(\delta, \gamma)$, $\delta_j \neq \gamma_j$, and for every $k \in \{1, \ldots, n\} - S(\delta, \gamma)$, $\delta_k = \gamma_k$. Let $|S(\delta, \gamma)| = r$. We know that $r \geq 1$. Then

$$q_{2n}(\delta, \gamma) = (2^m - 2)^r \cdot (2^m - 1)^{m-r} \bmod 2 = 0.$$

We proved that the communication matrix $M_{q_{2n}}$ is the diagonal matrix of size $2^n \times 2^n$, and so $\mathrm{rank}(M_{q_{2n}}) = 2^n$. Following the definition of q_{2n} we see that every row of $M_{q_{2n}}$ is a linear combination of rows of $M_{\overline{\mathrm{ExAll}_{2n}}}$. Thus,

$$\mathrm{rank}(M_{\overline{\mathrm{ExAll}_{2n}}}) \geq \mathrm{rank}(M_{q_{2n}}) = 2^n.$$

(ii) First, we describe a randomized protocol $D = (D_I, D_{II})$ for ExAll_{2n} and then we analyze its communication complexity. Let $\mathrm{Prim}_m = \{p \in \mathbb{N} \mid p \leq m \text{ and } p \text{ is a prime}\}$.

Protocol D

Input: $(\alpha, \beta) = (\alpha_1 \alpha_2 \ldots \alpha_m, \beta_1 \beta_2 \ldots \beta_m)$, $\alpha_i, \beta_i \in \{0,1\}^m$ for $i = 1, 2, \ldots, m$.

Step 1. $d = 2 \cdot \lceil \log_2 m \rceil$ prime numbers $s_1, s_2, \ldots, s_d$ are uniformly chosen from Prim_m at random.
{ So, both D_I and D_{II} know $s_1, s_2, \ldots, s_d$.}

Step 2. The first computer D_I computes $\mathrm{Number}(\alpha_i) \bmod s_j$ for all $i \in \{1, \ldots, m\}$ and all $j \in \{1, \ldots, d\}$ and sends the binary representation of all $m \cdot d$ results to D_{II}.

Step 3. If, for every $i \in \{1, \ldots, m\}$, there exists j_i such that

$$\mathrm{Number}(\alpha_i) \bmod s_{j_i} \neq \mathrm{Number}(\beta_i) \bmod s_{j_i},$$

then D_{II} outputs "reject".
Else, let $k \in \{1, \ldots, m\}$ be the smallest integer such that

$$\mathrm{Number}(\alpha_k) \bmod s_j = \mathrm{Number}(\beta_k) \bmod s_j$$

for all $j \in \{1, 2, \ldots, d\}$. Then D_{II} sends the binary representation of k to D_I.

Step 4. When D_I receives the binary representation of an positive integer k, then D_I send α_k to D_{II}.

Step 5. Receiving α_k, D_{II} compares α_k with β_k.
If $\alpha_k = \beta_k$, then D_{II} outputs "accept".
If $\alpha_k \neq \beta_k$, then D_{II} outputs "?".

We again use the fact presented in Section 2 that, for all $\gamma, \delta \in \{0,1\}^m$ and sufficiently large m's, $\gamma \neq \delta$ implies

$$\left| \{p \in \mathrm{Prim}_m \,|\, \mathrm{Number}(\gamma) \bmod p \neq \mathrm{Number}(\delta) \bmod p\} \right| \geq \frac{|\mathrm{Prim}_m|}{2}.$$

Consider an input (α, β) such that $\mathrm{ExAll}_{2n}(\alpha, \beta) = 0$. This means $\alpha = \alpha_1 \alpha_2 \ldots \alpha_m$, $\beta = \beta_1 \beta_2 \ldots \beta_m$ and $\alpha_i \neq \beta_i$ for $i = 1, \ldots, m$. For every $i \in \{1, \ldots, m\}$, the probability that

$$\mathrm{Number}(\alpha_i) \bmod s_j = \mathrm{Number}(\beta_i) \bmod s_j$$

for all $j \in \{1, \ldots, d\}$ is 2^{-d}. So, the probability that D_{II} recognizes $\alpha_i \neq \beta_i$ in Step 3 is $1 - 2^{-d}$. This implies that the probability to recognize $\alpha_i \neq \beta_i$, for all $i \in \{1, \ldots, m\}$ and so to reject (α, β) is at least $(1 - \frac{1}{2^d})^m = (1 - \frac{1}{m^2})^m > \frac{1}{2}$ for sufficiently large m. Obviously, if D_{II} does not reject (α, β) in Step 4, then D_{II} must output "?" in Step 5.

Now, consider $\mathrm{ExAll}_{2n}(\alpha, \beta) = 1$, i.e. there exists a $k \in \{1, \ldots, n\}$ such that $\alpha_k = \beta_k$. If there are several such numbers, let k be the smallest one with this property. Obviously,

$$\mathrm{Number}(\alpha_k) \bmod s_j = \mathrm{Number}(\beta_k) \bmod s_j$$

for all $j \in \{1, 2, \ldots, m\}$ and so D_{II} cannot reject (α, β) in Step 3. If D_{II} sends the binary representation of k to D_I, then D_I sends α_k to D_{II} and D_{II} accepts the input. This scenario fails to work only if D_{II} sends $\mathrm{BIN}_{\lceil \log_2 m \rceil}(j)$ to D_I for some $j < k$ despite of the fact that $\alpha_j \neq \beta_j$. In that case D_{II} outputs "?". But this can happen only with probability at most

$$\frac{1}{2^d} \cdot \left(1 - \frac{1}{2^d}\right)^{j-1} < \frac{1}{2^d} \cdot \left(1 - \frac{1}{m^2}\right)^m < \frac{1}{2^d} = \frac{1}{m^2}$$

for every $j \in \{1, \ldots, m\}$. So,

$$Prob\,(D(\alpha, \beta) = \text{"?"}) \leq \sum_{l=1}^{k-1} \left(\frac{1}{2^d}\left(1 - \frac{1}{2^d}\right)^{l-1}\right) < \sum_{l=1}^{m} \frac{1}{m^2} = \frac{1}{m}.$$

Thus,

$$Prob\,(D(\alpha,\beta) = \text{"accept"}) > 1 - \frac{1}{m}.$$

We conclude that D is a Las Vegas protocol for ExAll_{2n}.

It remains to calculate the communication complexity of D. In Step 2 D_I sends $m \cdot d$ binary representations of length $\lceil \log_2 m \rceil$ and so the length of the first message is always exactly $m \cdot 2 \cdot \lceil \log_2 m \rceil^2$. If D_{II} sends a message to D_I, then this message has the length $\lceil \log_2 m \rceil$. Then in Step 4 D_I answers with a message of length m. So, the longest communication takes

$$2 \cdot m \cdot \lceil \log_2 m \rceil^2 + \lceil \log_2 m \rceil + m \le 2m(\lceil \log_2 m \rceil^2 + 1)$$

bits.

$\square$

Observe that Theorem 8 and Theorem 9 together imply that there is an exponential gap between nondeterminism and Las Vegas for communication complexity.

The last question we want to consider in this section is whether the established relation between $cc(f)$ and $lvcc(f)$ is valid also for one-way communication protocols. Surprisingly, there is even a linear relation between $cc_1(f)$ and $lvcc_1(f)$, i.e. determinism is almost as powerful as Las Vegas for one-way protocols. We shall show nice applications of this tight relation between $cc_1(f)$ and $lvcc_1(f)$ for proving polynomial relations between determinism and Las Vegas for other computing models in Section 7.

Theorem 12 *For every function $f : U \times V \to \{0,1\}$ with finite sets U and V,*

$$lvcc_1(f) \ge cc_1(f)/2.$$

Proof. First, we give an informal idea of the proof. Let $f : U \times V \to \{0,1\}$ be a finite function. We represent f by its communication matrix $M_f = [a_{u,v}]_{u \in U, v \in V}$ with $a_{u,v} = f(u,v)$. Remember that the number of different messages of an optimal one-way protocol P computing f is exactly the same as the number $\mathbf{Row}(M_f)$ of different rows of M_f, i.e. $cc_1(f) = \lceil \log_2(\text{Row}(M_f) \rceil$.

Any public one-way Las Vegas protocol P' may be considered as a collection of deterministic one-way protocols $P_1, P_2, \ldots$ with probabilities $p_1, p_2, \ldots$. For any input α, P_i may produce the results $0, 1$ or "?" (i.e., *"don't know"*). Since P' is a Las Vegas protocol, no protocol P_i ever errs and for every $(u,v) \in U \times V$, the protocols $P_1, P_2, \ldots$ produce the output "?" with probability at most $\frac{1}{2}$. To any protocol P_i $(i = 1, 2, \ldots)$, one can assign its $0/1/$"?" communication matrix $M(P_i) = [b^i_{uv}]_{u \in U, v \in V}$, where $b^i_{uv} = a_{uv}$ if P_i does not give output "?" and $b^i_{uv} = $ "?" otherwise.

Our goal is to find one protocol P_i such that $M(P_i)$ has at least $\sqrt{\text{Row}(M_f)}$ different rows. In order to reduce the number of different rows of these deterministic protocols we will have to replace certain entries of M_f by a "?" in a clever way. Obviously, replacing certain entries of M_f by "?" will help reducing the number of different rows far more than the replacement of other entries by "?". For the identity matrix the diagonal entries play this "helper role". For instance, we can reduce the number of different rows to two by setting the upper left and the lower right quarter to "?". Observe that this radical reduction in the number of different rows is obtained after replacing only one half of the entries by "?". On the other hand, any significant reduction in the number of different rows has to involve the diagonal entries and any such entry has to stay untouched with probability at least one half. An obvious averaging argument shows that one deterministic protocol exists with at least $N/2$ different rows (if we consider the $N \times N$ identity matrix).

In the above example the diagonal entries form a fooling set and any Las Vegas communication protocol has to have at least $\frac{|F|}{2}$ messages for a fooling set F. However, we cannot expect to find large fooling sets in general. In particular, the $n \times \log_2 n$ communication matrix M^* whose ith row contains the binary representation of i possesses only fooling sets of logarithmic size. But it can be shown in this case that any Las Vegas one-way protocol has to have $\sqrt{n}$ messages.

Our proof will introduce a new notion of fooling sets. Set $M_f = M$ and assume that M has r pairwise different rows and c pairwise different columns. Our new notion of fooling sets is based on a real-valued weight assignment

$$\text{weight} : \{1, \ldots, r\} \times \{1, \ldots, c\} \to \mathbb{R}$$

for M. Let $I = \{1, \ldots, r\}$. We define the function weight recursively, processing M column after column in a clever way.

Case 1: If column 1 is monochromatic for all rows in I, then set, for $i \in I$,

$$\text{weight}(i, 1) = 0$$

and $I = I - \min\{i \in I\}$.

Case 2: If column 1 of M is not monochromatic for the rows in I, then there is a d, with $0 < d < 1$, such that $d \cdot |I|$ rows have a 0 in column 1 (and $(1 - d) \cdot |I|$ rows have a 1 in column 1). We set

$$\text{weight}(i, 1) = \begin{cases} \log_2(\frac{1}{d}) & \text{if } M[i, 1] = 0 \\ \log_2(\frac{1}{1-d}) & \text{otherwise,} \end{cases}$$

and define

$$I_0 = \{i \in I \mid M[i, 1] = 0\} \text{ and } I_1 = \{i \in I \mid M[i, 1] = 1\}.$$

The procedure recursively continues with I_0 and I_1. Observe that the procedure stops if the row sets are singletons, since then all columns will be monochromatic.

We begin our analysis with the following technical fact. In what follows $0 \cdot \log 0$ is defined to be 0.

Fact 1 *For any $x, y \geq 0$ and $d \in (0, 1)$,*

$$x \cdot \log_2 \frac{x}{d} + y \cdot \log_2 \frac{y}{1 - d} \geq (x + y) \cdot \log_2(x + y).$$

Proof. The cases where x and y are 0 are trivial. Write $p = x/(x+y)$ and $q = y(x + y)$. Then $p + q = 1$. The fundamental fact that the *informational divergence*

$$\sum_{i=1}^{n} p_i \log_2 \frac{p_i}{q_i}$$

for any two probability distributions $(p_1, \ldots, p_n)$ and $(q_1, \ldots, q_n)$ is always nonnegative [CK86] tells us that

$$p \cdot \log_2(p/d) + q \cdot \log_2(q/(1 - d)) \geq 0$$

for every $d \in (0, 1)$. Adding the obvious equality (remember that $p+q = 1$)

$$p \cdot \log_2(x + y) + q \cdot \log_2(x + y) = \log_2(x + y)$$

and canceling yields

$$p \cdot \log_2(x/d) + q \cdot \log_2(y/(1 - d)) \geq \log_2(x + y).$$

Multiplying by $x + y$ yields the claimed inequality. $\qquad \square$

Proof of Theorem 12 continued. For a subset $R \subseteq \{1, \ldots, r\}$ set

$$\text{differ}(R) = \{j \mid \exists i_1, i_2 \in R : M[i_1, j] \neq M[i_2, j]\}.$$

Now, we are ready to analyze the properties of our weight assignment.

Lemma 2   *(i)* *For each* $(i,j) \in \{1,\ldots,r\} \times \{1,\ldots,c\}$,

$$\text{weight}(i,j) \geq 0.$$

(ii) *For each* $i \in \{1,\ldots,r\}$,

$$\sum_{j=1}^{c} \text{weight}(i,j) = \log_2 r.$$

(iii) *For any* $R \subseteq \{1,\ldots,r\}$,

$$\sum_{j \in \text{differ}(R)} \sum_{i \in R} \text{weight}(i,j) \geq |R| \cdot \log_2 |R|.$$

Proof. Part **(a)** is immediate by construction. We verify part **(b)** by induction on r. The basis for $r = 1$ is trivial. For the inductive step we can assume without loss of generality that column 1 is not monochromatic. Let I_0 (resp. I_1) be the set of those rows with a zero (resp. one) in column 1 and assume that $|I_0| = c \cdot r$.

We apply the induction hypothesis to the rows in I_0 and I_1. For a row $i \in I_0$ we obtain

$$\sum_{j=2}^{c} \text{weight}(i,j) = \log_2(c \cdot r).$$

But $\text{weight}(i,1) = \log_2(\frac{1}{c})$ and

$$\sum_{j=1}^{c} \text{weight}(i,j) = \log_2\left(\frac{1}{c}\right) + \log_2(c \cdot r) = \log_2 r.$$

Part **(b)** follows with a symmetric argument for the rows in I_1.

We apply induction on the size of R to verify part **(c)**. The basis for $|R| = 1$ is again trivial. We assume for the inductive step that column 1 is not monochromatic for the rows in R. Hence R splits into the subsets R_0 resp. R_1 of those rows in R with value zero (resp. one) in column 1. Since we can apply the induction hypothesis to R_0 and R_1, we obtain

$$\sum_{j \in \text{differ}(R)} \sum_{i \in R} \text{weight}(i,j)$$

$$= \sum_{i \in R} \text{weight}(i,1) + \sum_{j \in \text{differ}(R), j \neq 1} \sum_{i \in R} \text{weight}(i,j)$$

$$\geq |R_0| \cdot \log_2\left(\frac{1}{c}\right) + |R_1| \cdot \log_2\left(\frac{1}{1-c}\right)$$

$$+ |R_0| \cdot \log_2(|R_0|) + |R_1| \cdot \log_2(|R_1|).$$

Thus part **(c)** follows from Fact 1. □

Proof of Theorem 12 continued. Assume we have a one-way Las Vegas protocol P' for a Boolean function f represented by the matrix M_f with r pairwise different rows. Let the function weight be defined for M_f with the above three properties. Then there is a deterministic one-way protocol $P \in \{P_1, P_2, \ldots\}$ such that

(*) the sum of all weights of entries of $M(P)$ with value "?" is at most one half of the sum of all weights (i.e., at most $\frac{1}{2} \cdot \sum_{i=1}^{r} \sum_{j=1}^{c}$ weight$(i,j) = \frac{r}{2} \cdot \log_2 r$ according to property (b) of the weight assignment).

This follows, since for every input the output of P' is equal to "?" with probability at most one half. The deterministic protocol P partitions the set of all rows of M_f into classes $R_1, \ldots, R_k$ of identical rows (after replacing certain entries by "?"). By property Lemma 2 (c) of the function weight we obtain for any class R_s

$$\sum_{j \in \text{differ}(R_s)} \sum_{i \in R_s} \text{weight}(i, j) \geq |R_s| \cdot \log_2 |R_s|.$$

Let $M(R_s)$ be the restriction of M_f to the rows in R_s. The quantity on the left hand side is a lower bound for the weight of all entries of $M(R_s)$ with value "?". Since the function $\sum_{s=1}^{k} x_s \log_2 x_s$ with $x = \sum_{s=1}^{k} x_s$ is minimized for $x_1 = \ldots = x_k = \frac{x}{k}$,

$$\sum_{s=1}^{k} x_s \log_2 x_s \geq k \cdot \left(\frac{x}{k} \cdot \log \frac{x}{k} \right) = x \log_2 x - x \log_2 k.$$

Hence the sum of weights of entries of $M(P)$ with value "?" is at least

$$\sum_{s=1}^{k} |R_s| \log_2 |R_s| \geq r \log_2 r - r \cdot \log_2 k.$$

¿From the above inequality and from (*), it follows that

$$\frac{r \log_2 r}{2} \geq \sum_{s=1}^{k} |R_s| \log_2 |R_s| \geq r \log_2 r - r \cdot \log_2 k$$

and hence that $k \geq \sqrt{r}$. In other words, $M(P)$ has at least $\sqrt{r}$ different rows, so that the deterministic protocol P has to consist of at least $\sqrt{r}$ messages. □

In the proof of Theorem 12 we have assumed that the probability of the output "?" is bounded by $\frac{1}{2}$ for every input. Since for Las Vegas computing models the size of the upper bound on failure is not essential, one can use any upper bound $\varepsilon < 1$ of computing "?" in the definition of Las Vegas protocols. One can easily observe that the exchange of the upper bound $\frac{1}{2}$ on failure for an arbitrary upper bound ε, $0 < \varepsilon < 1$, would result in the fact that the matrix $M(P)$ has at least $r^{1-\varepsilon}$ different rows. Thus, in this general case the claim of Theorem 12 would be $\text{lvcc}_1(f) \geq (1 - \varepsilon) \cdot \text{cc}_1(f)$.

5. MONTE CARLO PROTOCOLS

In the previous section we showed that Las Vegas communication complexity is closely related to deterministic communication complexity. This is not true for Monte Carlo randomization. In this section we show that

(i) already one-sided-error Monte Carlo protocols may be more powerful than deterministic ones, but the gap between one-sided-error Monte Carlo communication complexity and (deterministic) communication complexity can be at most exponential,

(ii) two-sided-error Monte Carlo protocols can be more efficient than nondeterministic ones, and

(iii) nondeterminism can be more powerful than Monte Carlo randomization.

We present the results in the sequence as written above. Observe that the power of one-sided-error Monte Carlo randomization lies between determinism and nondeterminism. Our first result shows that an exponential gap between determinism and one-sided-error Monte Carlo is possible for communication complexity.

Theorem 13 ([JPS84])

(i) For every positive integer n, $\text{cc}(\text{Ineq}_n) = n$, and

(ii) $\text{1mccc}(\text{Ineq}_n) \leq 2\lceil \log_2 n \rceil$ for sufficiently large integer n.

Proof.

(i) We showed already that all three lower bound techniques (Tiling, Rank, Fooling sets) provide $\text{cc}(\text{Ineq}_n) \geq n$.

(ii) In Example 1 we proved $\text{1mccc}(\text{Ineq}_n) \leq 2 \cdot \lceil \log_2 n \rceil$ for sufficiently large n.

$\square$

A natural question is whether there is a possibility to bound the difference between $cc(f)$ and $1mccc(f)$ for all functions f (i.e. whether a larger gap than the exponential gap presented in Theorem 13 is possible). The following theorem shows that the gap cannot be larger than exponential because nondeterminism can be simulated by determinism with an exponential blow-up of communication complexity. Since one-sided-error Monte Carlo protocols can be viewed as restricted nondeterministic protocols, we are done.

Theorem 14 ([PS82]) *For every finite function* $f : U \times V \to \{0,1\}$,

$$cc_1(f) \leq 2^{ncc(f)}.$$

Proof. We know that $ncc(f)$ is exactly the cardinality of an optimal cover of 1's of M_f by 1-monochromatic submatrices of M_f. Let $\{M_1, M_2, \ldots, M_{ncc(f)}\}$ be such an optimal cover. One can construct a (deterministic) one-way protocol $P = (C_I, C_{II})$ that computes f as follows.

Input: A pair $(\alpha, \beta) \in U \times V$.
  { α is the input part of C_I, and β is the input part of C_{II}.}

Step 1. For every input $\alpha \in U$, C_I sends the message $d_1 d_2 \ldots d_{ncc(f)} \in \{0,1\}^{ncc(f)}$ to C_{II}, where, for all $i \in \{1, \ldots, ncc(f)\}$, $d_i = 1$ if row$_\alpha$ has a nonempty intersection with the 1-monochromatic submatrix M_i.
  { In this way C_I tells C_{II} the names of all 1-monochromatic submatrices that candidate to contain the element (α, β) from the C_I point of view.}

Step 2. After receiving $d_1 d_2 \ldots d_{ncc(f)}$, C_{II} accepts, if there exists an $i \in \{1, \ldots, ncc(f)\}$ such that $d_i = 1$ and column$_\beta$ has a nonempty intersection with the 1-monochromatic submatrix M_i. Otherwise, C_{II} rejects.

Obviously, P computes f and its communication complexity is exactly $ncc(f)$. □

The next result continues to demonstrate the power of Monte Carlo randomization by showing that two-sided-error Monte Carlo protocols may be much more efficient than their nondeterministic counterparts.

Theorem 15 ([JPS84]) *For every positive integer* n,

  (i) $ncc(Eq_n) = n$, *and*

(ii) $2\mathrm{mccc}(\mathrm{Eq}_n) \leq 2\lceil \log_2 n \rceil$ *for sufficiently large n.*

Proof.

(i) The fact $\mathrm{ncc}(\mathrm{Eq}_n) \leq n$ is obvious and the lower bound $\mathrm{ncc}(\mathrm{Eq}_n) \geq n$ was already observed as an application of the 1-fooling set technique and the cover technique in Section 3.

(ii) Consider the one-sided-error Monte Carlo protocol R of Example 1 for Ineq_n. If one modifies R to $\overline{R}$ in such a way that $\overline{R}$ accepts iff R rejects, then $\overline{R}$ computes $\mathrm{Eq}_n = \overline{\mathrm{Ineq}_n}$. This is because, for every input $(\alpha, \beta) \in \{0, 1\}^n \times \{0, 1\}^n$,

(1) if $\alpha = \beta$ (i.e. $\mathrm{Eq}_n(\alpha, \beta) = 1$), then

$$Prob(\overline{R}(\alpha, \beta) = \text{"accept"}) = 1,$$

(2) if $\alpha \neq \beta$ (i.e. $\mathrm{Eq}_n(\alpha, \beta) = 0$), then

$$Prob(\overline{R}(\alpha, \beta) = \text{"reject"}) \geq 1 - \frac{2 \ln n^2}{n}.$$

Thus, for sufficient large n, $\overline{R}$ is a two-sided-error Monte Carlo protocol for Eq_n. Since R works within $2 \cdot \lceil \log_2 n \rceil$ communication complexity, $\overline{R}$ works with $2 \cdot \lceil \log_2 n \rceil$ communication complexity, too.

$\square$

Observe that Theorem 15 implies that there is an exponential gap between two-sided-error Monte Carlo randomization and one-sided-error Monte Carlo randomization for communication complexity.

The following result together with Theorem 15 shows that bounded-error Monte Carlo randomization and nondeterminism are incomparable. For some problems one computing mode can be essentially better than the other one, and for another problem the situation may be vice versa. To see it, consider the following Boolean function

$$\mathrm{Disj}_n(x_1, \ldots, x_n, y_1, \ldots, y_n) : \{0, 1\}^n \times \{0, 1\}^n \to \{0, 1\}$$

defined by

$$\mathrm{Disj}_n(\alpha_1, \alpha_2, \ldots, \alpha_n, \beta_1, \beta_2, \ldots, \beta_n) = \Gamma\left(\min\left\{\sum_{i=1}^{n} \alpha_i \beta_i, 1\right\}\right).$$

Thus, $\mathrm{Disj}_n(\alpha_1, \alpha_2, \ldots, \alpha_n, \beta_1, \beta_2, \ldots, \beta_n) = 1$ iff $\sum_{i=1}^{n} \alpha_i \beta_i = 0$. In other words, $\overline{\mathrm{Disj}_n}(\alpha_1, \alpha_2, \ldots, \alpha_n, \beta_1, \beta_2, \ldots, \beta_n) = 1$ iff there exists a $j \in \{1, 2, \ldots, n\}$ such that $\alpha_j = \beta_j = 1$.

Theorem 16 ([BFS86a, KS87, Raz90]) *For every positive integer n*

   (i) $\mathrm{ncc}(\overline{Disj}_n) \leq \lceil \log_2 n \rceil$, *and*

   (ii) $2\mathrm{mccc}(\overline{Disj}_n) = \Omega(\sqrt{n})$.

Proof.

(i) A nondeterministic protocol $D = (D_I, D_{II})$ for $\overline{Disj}_n$ can work as follows.

    Input. A pair $(\alpha, \beta) = (\alpha_1 \ldots \alpha_n, \beta_1 \ldots \beta_n) \in \{0,1\}^n \times \{0,1\}^n$.

    Step 1. For every $\alpha = \alpha_1 \ldots \alpha_n$, D nondeterministically chooses a $j \in \{1, \ldots, n\}$ such that $\alpha_j = 1$ and sends the message $\mathrm{BIN}_{\lceil \log_2 n \rceil}(j)$ to D_{II}.

    Step 2. When D_{II} receives $\mathrm{BIN}_{\lceil \log_2 n \rceil}(j)$, and $\beta_j = 1$, then D_{II} accepts.
Otherwise, D_{II} rejects.

    Obviously, D computes $\overline{Disj}_n$ and its communication complexity is $\lceil \log_2 n \rceil$.

(ii) The proof of the lower bound $2\mathrm{mccc}(\overline{Disj}_n) = \Omega(\sqrt{n})$ is too technical and cannot be done by any straightforward application of the lower bound techniques presented in Section 3. Because of this we omit its presentation here. The original proof of this lower bound can be found in [KS87], and a simplified version of this proof is presented in [Raz90].

□

    There are several further papers devoted to randomized protocols (see, for instance, [Abl96, CG85, HR88, Hr97, KA86, KN97, MW95, NW91, Ya83]). For all we mention here only that Ablayev [Abl96] showed a nice combinatorial characterization of Boolean functions for which Monte Carlo one-way protocols cannot be much better than the one-way deterministic protocols.

6.   A BOUND ON THE NUMBER OF RANDOM BITS

    The number of random bits used is an important characteristic of every randomized algorithm. To consider the number of random bits as a complexity measure of randomized computation, especially in some tradeoffs with other computational complexity measures, is one of the central research tasks on randomization. There are at least two important general reasons for this.

(i) The random bits are not for free and the cost of producing a sequence of random bits grows essentially with the length of the sequence.

(ii) If the number of random bits used in a randomized computation is not too large, then there is a possibility of an efficient derandomization (i.e. one can create an efficient deterministic algorithm doing the same job as the original randomized one).

So, a typical question for a randomized computing model is whether one can take an upper bound on the number of random bits without paying for this with an essential decrease of the computational power (efficiency). The upper bound one searches for is usually a function of the input size.

Besides this two general reasons above we have one more special reason to investigate the number of random bits for communication protocols. As already discussed in Section 2, we have the two fundamental possibilities to randomize protocols, namely either by using one common random source (public randomized protocols) or by using two independent random sources (private randomized protocols). The difference between these two randomization approaches is that to simulate public randomized protocols by private ones, it may happen that the private random bits need to be communicated. Thus, any upper bound on the number of random bits is also an upper bound on the possible difference between the communication complexity of public randomized protocols and private randomized protocols.

In this section we present the result of Newman [New91] who showed that $O(\log_2 n)$ bits are enough to exploit the full power of bounded error (Las Vegas, one-sided-error Monte Carlo, two-sided-error Monte Carlo) protocols.

In what follows, for any finite function f, we denote by **priv-lvcc(f)**, **priv-1mccc(f)**, and **priv-2mccc(f)** resp. the private counterparts of the public randomized communication complexities $\mathrm{lvcc}(f)$, $1\mathrm{mccc}(f)$, and $2\mathrm{mccc}(f)$ respectively.

Theorem 17 *Let D be a public x-randomized protocol for a Boolean function $f : \{0,1\}^n \times \{0,1\}^n \to \{0,1\}$, $n \in I\!N$, where $x \in \{Las Vegas, one-sided-error Monte Carlo, two-sided-error Monte Carlo\}$. Then, there exists an equivalent public x-randomized protocol D' for f that uses at most $O(\log_2 n)$ random bits and*

$$\mathrm{cc}(D') = O(\mathrm{cc}(D)).$$

Proof. Because the error probability of randomized protocols can be decreased to an arbitrary constant by a few repetitions of the work of randomized protocols, we may assume that there is a protocol $\tilde{D}$ that computes f with error probability at most $\frac{1}{6}$ and $cc(\tilde{D}) = O(cc(D))$. The idea of the proof is to use some kind of derandomization of $\tilde{D}$ in order to construct D' that computes f with error probability at most $\frac{1}{3}$ and $O(\log_2 n)$ random bits.

Let Π be the set of all random sequences used by $\tilde{D}$, and let $Prob$ be the probability distribution over Π. So, we can write $\tilde{D} = (Prob, \{D_r | r \in \Pi\})$, where D_r is the protocol corresponding to r. Let $Z(\alpha, \beta, r)$ be a random variable that gets the value 1 if $\tilde{D}$ gives the wrong answer ("?" in the case of Las Vegas protocols) on the input $(\alpha, \beta) \in U \times V$, if choosing the random sequence r (i.e. the (deterministic) protocol D_r corresponding to r gives a wrong answer). Otherwise (if $D_r(\alpha, \beta) = f(\alpha, \beta)$), $Z(\alpha, \beta, r) = 0$. Because $\tilde{D}$ computes f with error at most $\frac{1}{6}$, we have

$$E_{r \in \Pi}[Z(\alpha, \beta, r)] = \sum_{r \in \Pi} Prob(r) \cdot Z(\alpha, \beta, r) \leq \frac{1}{6} \tag{3}$$

for all $(\alpha, \beta) \in U \times V$.

Our aim is to show that there exist $t = 36n$ deterministic protocols $D'_1, D'_2, \ldots, D'_t$ in $\{D_r \mid r \in \Pi\}$ such that $D' = (Pr, \{D'_1, D'_2, \ldots, D'_t\})$ with $Pr(D_i) = Pr(D_j) = \frac{1}{t}$, for all $i, j \in \{1, 2, \ldots, t\}$, computes f with error at most $\frac{1}{3}$. To prove the existence of $D'_1, D'_2, \ldots, D'_t$ we use the probabilistic method.

Let $D_{r_1}, D_{r_2}, \ldots, D_{r_t}$ be arbitrary t (deterministic) protocols from $\{D_r \mid r \in \Pi\}$. Consider the following randomized protocol $D_{r_1 r_2 \ldots r_t} = (Pr, \{D_{r_1}, D_{r_2}, \ldots, D_{r_t}\})$ with the uniform probability distribution Pr. We use

$$E_{Pr}[Z(x, y, r_i)] = \sum_{i=1}^{t} Pr(D_{r_i}) \cdot Z(x, y, r_i) = \sum_{i=1}^{t} \frac{1}{t} \cdot Z(x, y, r_i)$$

to denote the expected error of $D_{r_1 r_2 \ldots r_t}$. By the Chernoff inequality and the bound (3), we obtain

$$\Pr_{r_1, \ldots, r_t}\left[\left(\frac{1}{t} \cdot \sum_{i=1}^{t} Z(\alpha, \beta, r_i) - \frac{1}{6}\right) > \frac{1}{6}\right] \leq 2e^{-2(\frac{1}{6})^2 t} < 2^{-2n} \tag{4}$$

for every $(\alpha, \beta) \in U \times V$. The inequality (4) implies that, for a random choice of $r_1, \ldots, r_t$, the probability that $E_{Pr}[Z(x, y, r_i)] > \frac{1}{6} + \frac{1}{6} = \frac{1}{3}$ for some input (α, β) is smaller than $2^{2n} \cdot 2^{-2n} = 1$. This implies that there exists a choice of $r_1, \ldots, r_t$ where, for every $(\alpha, \beta) \in U \times V$, the error $E_{Pr}[Z(x, y, r_i)]$ of the protocol $D_{r_1 r_2 \ldots r_t}$ is at most $\frac{1}{3}$.

Since the protocol D' is a uniform probability distribution over t deterministic protocols, the number of random bits of D' is

$$\lceil \log_2 t \rceil = \lceil \log_2 (36n) \rceil \leq 5 \cdot \lceil \log_2 n \rceil.$$

The communication complexity of D' is at most the communication complexity of $\widetilde{D}$, and so[15] $cc(D') = O(cc(D))$. $\square$

A direct consequence of Theorem 17 is the linear relation between private randomized communication complexities and the corresponding public randomized communication complexities, provided that the communication complexities are at least of order $\log_2 n$. This result is formulated in the following theorem.

Theorem 18 *For every $x \in \{lv, 1mc, 2mc\}$, and for every Boolean function $f : \{0,1\}^n \times \{0,1\}^n \to \{0,1\}$, $n \in \mathbb{N} - \{0\}$,*

$$priv\text{-}xcc(f) = O(xcc(f) + \log_2 n).$$

7.   SOME APPLICATIONS

The goal of this section is to present two examples that show how the results about the power of randomized protocols can be transfered to other computing models in order to answer questions about the power of randomized computation in these models. If one wants to apply communication complexity for the investigation of the complexity of concrete problems in other computing models, then usually it is not sufficient to consider the simple partition of the input into some prefix and some suffix. Note that we have always (implicitly) considered a partition of the input by the definition of the finite function $f : U \times V \to \{0,1\}$. The typical case $f : \{0,1\}^n \times \{0,1\}^n \to \{0,1\}$ means that the input is partitioned into the first half and the second half. But one can also consider f as a Boolean function from $\{0,1\}^{2n}$ to $\{0,1\}$ of $2n$ variables and to take a partition, where C_I gets the odd bits of the input and C_{II} gets the even bits of the input. Another possibility is that C_I takes the first and the third quarter of the input and C_{II} gets the rest. Under such a partition the function Eq_n becomes easy because it can be computed with 1 communication bit. In order to prove lower bounds on the complexity of a specific Boolean function f in some circuit models one usually needs to prove that the communication complexity of f is large

---

[15]Note that for Las Vegas and one-sided-error Monte Carlo one can write $cc(D') \leq 3 \cdot cc(D)$.

for every balanced (or almost balanced) partition of the input, where **balanced** means that both C_I and C_{II} gets the same number of input bits. Obviously, to prove such a lower bound over the set of all balanced partitions is harder than to prove a lower bound on the communication complexity of a function according to the standard partition considered up till now. Observe, that the symmetric function s_{2n} has the property that its communication complexity is the same for every balanced partition of the input. In general, different applications may require to consider different sets of partitions.

In this section we use the result of Theorem 12 (establishing the linear relation between determinism and Las Vegas) in order to show polynomial relations between determinism and Las Vegas for the sizes of finite automata and for the sizes of ordered binary decision diagrams (OBDDs, also called one-time-only oblivious branching programs). We shall see that these two different applications require to consider distinct sets of input partitions. We start with showing that the gap between Las Vegas finite automata and (deterministic) finite automata can be at most quadratic.

Here, we consider the standard models of one-way deterministic finite automata (**DFA**) and one-way nondeterministic finite automata (**NFA**). In what follows $L(A)$ denotes the regular language accepted be the automaton A. We consider the **one-way Las Vegas finite automata (LVFA)** as introduced in [ĎHRS97]. A **randomized finite automaton** can be viewed as a NFA A, where

(i) The set of states of A is partitioned into three disjoint groups: **accepting states**, **rejecting states**, and **neutral states**.

(ii) For every state q of A and every symbol a of the input alphabet, there is a probability distribution over the set of edges leaving q and labeled by a. The probability of a computation of A is the product of the transition probabilities along the path of the computation.

We say that a randomized finite automaton A is a LVFA recognizing a language $L = L(A)$ if

(i) $w \in L$ implies that A working on w reaches an accepting state with probability at least $\frac{1}{2}$ ($Prob(A(w) = "accept") \geq \frac{1}{2}$), and the probability that A finishes the work in a rejecting state is zero ($Prob(A(w) = "reject") = 0$), and

(ii) $w \notin L$ implies that A working on w reaches a rejecting state with probability at least $\frac{1}{2}$ ($Prob(A(w) = "reject") \geq \frac{1}{2}$), and the probability that A finishes the work in an accepting state is zero ($Prob(A(w) = "accept") = 0$).

If A is a LVFA recognizing a language $L(A)$, then by declaring the neutral states to be rejecting states one obtains a NFA recognizing the same language $L(A)$. So, LVFA recognize regular languages only. Obviously, LVFAs recognize all regular languages because a DFA is a special case of a LVFA.

For every regular language L we define $\mathbf{s(L)}$, $\mathbf{ns(L)}$, and $\mathbf{lvs(L)}$ respectively as the size of a minimal deterministic, nondeterministic, and Las Vegas finite automaton for L. The complexity measures $s(L)$, $ns(L)$, and $lvs(L)$ are of principal interest in formal language theory because finite automata are considered as finite representations of potentially infinite objects — regular languages. It is a well-known fact that $s(L) \leq 2^{ns(L)}$ for every regular language L, and that there exists, for every positive integer k, a regular language H_k such that $s(H_k) = 2^{ns(H_k)} = 2^k$ [MF71]. Our aim is to show

$$lvs(L) \geq \sqrt{s(L)}$$

for every regular language L, and to exhibit a language with a quadratic gap between the size of the minimal DFA and a minimal LVFA.

The idea of the communication complexity approach is to find a strong connection between the number of messages of one-way protocols and the number of states of finite automata in such a way that Theorem 12 can be applied. Such a connection between one-way protocols and finite automata has been observed already in [Hr86] in the form

$$cc_1(h_n(L)) \leq \lceil \log_2(s(L)) \rceil$$

for every regular language $L \subseteq \{0,1\}^*$ and every $n \in I\!N$, where $\mathbf{h_n(L)}$ is a function from $\Sigma^{\lfloor n/2 \rfloor} \times \Sigma^{\lceil n/2 \rceil} \to \{0,1\}$ defined by $h_n(L)(\alpha,\beta) = 1$ iff $\alpha\beta \in L$ for all $(\alpha,\beta) \in \Sigma^{\lfloor n/2 \rfloor} \times \Sigma^{\lceil n/2 \rceil}$. More precisely, the number of different messages of the best one-way protocol is a lower bound on the number of states of finite automata. The argument for this claim is very simple. A one-way protocol $D_n = (D_I, D_{II})$ computes $h_n(L)$ by the following strategy. For every $\alpha \in \Sigma^{\lfloor n/2 \rfloor}$, D_I sends the code of the state q_α reached by the minimal DFA A for L when processing on α to D_{II}. For every $\beta \in \Sigma^{\lceil n/2 \rceil}$, D_{II} accepts iff A reaches an accepting state when working on β from the state q_α. Obviously this relation holds in the nondeterministic and randomized cases, too. Unfortunately, the difference between these two complexity measures may be arbitrarily large because communication complexity is defined in terms of a non-uniform computing model, whereas automata are a uniform computing model. Consider, for instance unary languages L where we always have $cc_1(h_n(L)) \leq 1$, but for any constant c we have a unary regular language L' with $s(L') > c$.

To overcome this difficulty we introduce one-way uniform protocols. The idea is to allow an arbitrary partition of any input into a prefix and a suffix.

Definition 10 *Let Σ be an alphabet and let $L \subseteq \Sigma^*$. A* **one-way uniform protocol over** Σ *is a pair $D = \langle C_I, C_{II} \rangle$, where:*

(i) $C_I : \Sigma^* \to \{0,1\}^*$ is a function with the prefix-freeness property ($C_I(\alpha)$ is no proper prefix of $C_I(\beta)$ for any $\alpha, \beta \in \Sigma^*$), and

(ii) $C_{II} : \Sigma^* \times \{0,1\}^* \to \{\text{accept}, \text{reject}\}$ is a function.

We say that $D = \langle C_I, C_{II} \rangle$ **accepts** L, $L(D) = L$, *if for all $\alpha, \beta \in \Sigma^*$:*

$$C_{II}(\beta, C_I(\alpha)) = \text{accept} \iff \alpha\beta \in L.$$

The **message complexity of the protocol D** *is*

$$\mathbf{mc}(D) = |\{C_I(\alpha) \mid \alpha \in \Sigma^*\}|$$

and we define the **message complexity of L** *as*

$$\mathbf{mc}(L) = \min\{\mathrm{mc}(D) \mid D \text{ is a one-way uniform protocol accepting } L\}.$$

Here we consider special infinite communication matrices, that describe a language L in a similar way as M_f describes a finite function f.

Definition 11 *Let Σ be an alphabet and let $L \subseteq \Sigma^*$. We define the infinite Boolean matrix $M_L = (a_{uv})_{u,v \in \Sigma^*}$ so that*

$$a_{uv} = 1 \iff uv \in L.$$

Let $\mathbf{row}_L = \mathrm{Row}(M_L)$ *be the number of different rows of M_L.*

Observe, that in contrast to communication matrices of finite functions, M_L contains exactly $|w| + 1$ elements for every input $w \in \Sigma^*$ (namely $a_{\alpha\beta}$ for all $\alpha, \beta \in \Sigma^*$ such that $w = \alpha\beta$, i.e. one element for every possible partition of w into a prefix and a suffix). While a finite communication matrix always unambiguously determines a finite function (assuming the names of rows and columns are fixed), there exist infinite Boolean matrices that do not represent any language. This is the consequence of our effort to take one matrix for the representation of a language under all possible (allowed) partitions, while to represent a finite function f one takes a finite matrix M_f that represents f under one fixed input partition.

Now, we formulate the crucial observation claiming that the message complexity of a regular language L is the same complexity measure as the size of the minimal finite automaton for L.

Lemma 3 ([ĎHRS97]) *For every regular language L over an alphabet Σ,*

$$s(L) = \mathrm{mc}(L) = \mathrm{row}_L.$$

Proof. The equality $s(L) = \mathrm{row}_L$ is just a reformulation of the Myhill-Nerode theorem, because row_L is exactly the index of the right invariant relation on Σ^* according to L.

Since $s(L)$ is finite for every regular language L, and according to Theorem 5 the number of different rows of a communication matrix is equal to the number of different messages used by the best one-way protocol computing this matrix, we obtain $\mathrm{mc}(L) = \mathrm{row}_L$. □

Now, we have all notations and technicalities needed to successfully apply Theorem 12 for proving a close relation between Las Vegas and determinism for finite automata.

Theorem 19 (([ĎHRS97])) *For every regular language L,*

$$\mathrm{lvs}(L) \geq \sqrt{s(L)}.$$

Proof. Let L be a regular language over an alphabet Σ. Since row_L is finite, one can easily find a finite submatrix $M = M_L(U, V)$, $U, V \subseteq \Sigma^*$, of M_L with $\mathrm{row}_L = s(L) = \mathrm{Row}(M)$ different rows. Let $f : U \times V \to \{0, 1\}$ be the finite function of two arguments that corresponds to M. Then the optimal one-way protocol for f uses exactly row_L different messages. Let A be a LVFA for L with $s(A)$ states. In the obvious way described above, this automata induces a one-way Las Vegas protocol for the function f that uses $s(A)$ different messages. In the proof of Theorem 12 it was shown that in this situation we must have $s(A) \geq \sqrt{\mathrm{Row}(M)} = \sqrt{\mathrm{row}_L}$. By Lemma 3 the result follows. □

The language $L_k = \{w \in \{0, 1\}^* \mid w = u1v \text{ and } |v| = k - 1\}$ is a well known example of a language producing an exponential gap between $s(L)$ and $\mathrm{ns}(L)$ [MF71]. We use L_k to show that Theorem 19 cannot be essentially improved.

Theorem 20 (([ĎHRS97])) *For every positive integer k,*

(i) $s(L_k) = 2^k$,

(ii) $\mathrm{lvs}(L_k) \leq 4 \cdot 2^{k/2+1} = O(\sqrt{s(L_k)})$,

(iii) $\mathrm{ns}(L_k) = k + 1$.

Proof. (i) and (ii) are well-known facts [MF71]. Note that to prove (i) one can apply communication complexity, too. Consider $h_{2k}(L_k)$: $\{0,1\}^k \times \{0,1\}^k \to \{0,1\}$. The set $\mathcal{A}_k = \{0,1\}^k$ is a one-way fooling set for $h_{2k}(L_k)$ because for all $\alpha, \beta \in \{0,1\}^k$, $\alpha \neq \beta$, there exists a $\gamma \in \{0\}^*$ such that $h_{2k}(L_k)(\alpha, \gamma) \neq h_{2k}(L_k)(\beta, \gamma)$ (If $\alpha \neq \beta$, $\alpha = \alpha_1 \ldots \alpha_k$, $\beta = \beta_1 \ldots \beta_k$, then there exists a $j \in \{1, \ldots, k\}$ such that $\alpha_j \neq \beta_j$. Without loss of generality assume $1 = \alpha_j \neq \beta_j = 0$. Then we choose $\gamma = 0^{j-1}$. Obviously $\alpha\gamma \in L_k$ and $\beta\gamma \notin L_k$.)

To show (ii) we consider the following strategy of a LVFA A. The computation of A starts by randomly guessing whether the important bit (the k-th bit from the end) is on an even or odd bit position. Since one can easily construct a DFA of $2^{k/2+1}$ states accepting $L_k^{\mathrm{odd}} = \{w \in \{0,1\}^* \mid w = u1v, |v| = k-1,$ and $|u|$ is odd$\}$ and a DFA of $2^{k/2+1}$ states accepting $L_k^{\mathrm{even}} = \{w \in \{0,1\}^* \mid w = u1v, |v| = k - 1,$ and $|u|$ is even$\}$, we obtain a LVFA for L_k of $1 + 2 \cdot 2^{k/2+1}$ states. $\qquad\square$

An exponential relation between the sizes of minimal DFAs and minimal one-sided-error Monte Carlo finite automata was originally established without using communication complexity (for more information about the study of distinct versions of Monte Carlo finite automata see [RS59, Ra63, Fr82, Po79, AF86, Abl88]). But one can simply get it by using our results about the communication complexity of Ineq_n. So, the corresponding finite language is

$$\mathrm{In}_n = \{\alpha_1\alpha_2 \ldots \alpha_n\beta_1\beta_2 \ldots \beta_n \mid \alpha_i, \beta_i \in \{0,1\} \text{ for } i = 1, \ldots, n,$$
$$\text{and there exists } j \in \{1, \ldots, n\}$$
$$\text{such that } \alpha_j \neq \beta_j\}$$

(Obviously, $h_{2n}(\mathrm{In}_n)$ is exactly Ineq_n). Since $\mathrm{cc}_1(\mathrm{Ineq}_n) = n$, $\mathrm{s}(\mathrm{In}_n) = 2^n$. To construct a one-sided-error Monte Carlo finite automaton A for In_n it is sufficient to use the uniform probability distribution from the initial state in order to choose one of the DFAs A_p, where $p \in \{2, \ldots, n^2\}$ is a prime. Every A_p computes, for every input $\alpha_1\alpha_2 \ldots \alpha_n\beta_1\beta_2 \ldots \beta_n$,

$$\mathrm{Number}(\alpha_1 \ldots \alpha_n) \bmod p \text{ and } \mathrm{Number}(\beta_1 \ldots \beta_n) \bmod p$$

and accepts the input iff these two numbers are different. The calculation in Example 1 proves that A is a one-sided-error Monte Carlo finite automaton for In_n. Since A consists of approximately $\frac{n^2}{2\ln n}$ DFAs and every DFA A_p can be realized in size $O(n^3)$, A is of polynomial size.

Finishing the application of communication complexity for finite automata we formulate one interesting open problem. Public randomized protocols can be seen as one random choice over deterministic protocols, i.e., one starts with a random choice and continues to work in a completely deterministic way. Observe, that the randomized finite automata designed above have also this property, i.e., one takes a random choice over DFAs. The question is whether every randomized finite automaton can be converted into such a "normalized" finite automaton that does not contain any random decision in a cycle. Note that this question is equivalent to the question whether a constant (independent on the input size) number of random bits is enough for every randomized finite automaton. Let us formulate this question more precisely for LVFAs.

Open Problem Does there exist a polynomial p such that, for every LVFA A, there exists an equivalent[16] LVFA B with the following properties?

(i) $s(B) = p(s(A))$, and

(ii) the number of random decisions of B is bounded by a constant for every input (i.e. B does not contain any random choice in a cycle of its transition diagram).

For more information about the computational power of different versions of Las Vegas finite automata (two-way FA, two-dimensional FA) we refer to [HS99, ĎHI00].

In what follows we apply Theorem 12 to get a polynomial relationship between Las Vegas and determinism for ordered binary decision diagrams (OBDDs). The difference to the previous application is that now we also need to consider input partitions that cannot be described by cutting the input into a prefix and a suffix. The similarity to the case of finite automata is in the very strong relation between (deterministic) one-way communication complexity and the size of OBDDs.

OBDDs [Br85, Br86] are highly restricted branching programs [PŽ83]. A **branching program** A over a set $X = \{x_1, \ldots, x_n\}$ of Boolean variables is a directed acyclic graph with one source and two sinks labeled by Boolean constants 0 and 1. The non-sink nodes are labeled by Boolean variables from X and have exactly two outgoing edges labeled by 0 and 1. The computation of a branching program A on an input $a = \alpha_1 \alpha_2 \ldots \alpha_n \in \{0, 1\}^n$ starts at the source. At an inner node labeled by x_i the outgoing edge with the label α_i is chosen. The label of the sink that is reached defines $f_A(\alpha)$ for the Boolean function f_A computed by A. An **OBDD** is a branching program that satisfies the following restrictions:

---

[16]Equivalent means $L(A) = L(B)$.

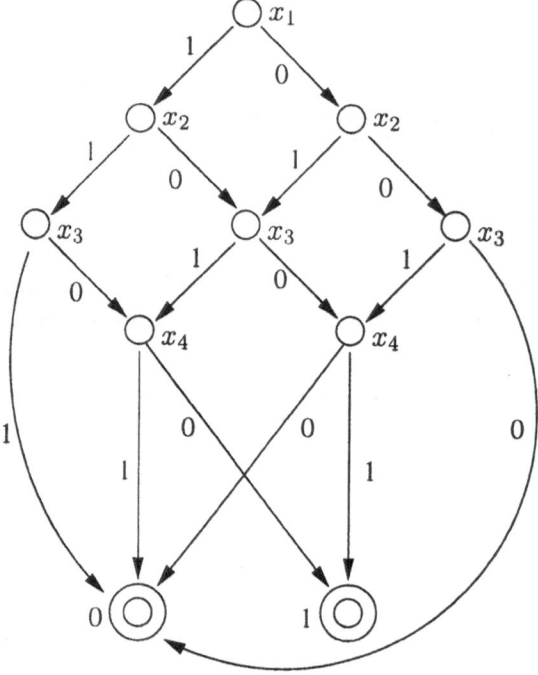

Figure 3

(i) for every input variable x and for every directed path P of A, P contains at most one node labeled by x {every input bit is read at most once in any computation of the branching program}, and

(ii) there exists an ordering $X_A = x_{i_1}, x_{i_2}, \ldots, x_{i_n}$ of the input variables $x_1, x_2, \ldots, x_n$ such that the sequence of the labels of nodes on every directed path of A is a subsequence of X_A.

If one considers only the restriction (i), then we speak about one-time-only branching programs. The restriction (ii) assures that the computing model is oblivious. Because of this people use sometimes the term one-time-only oblivious branching programs instead of OBDDs.

The most important complexity measure for OBDDs is their **size**, namely the number of their nodes. For any Boolean function f, we denote by **size(f)** the size of the best OBDD for f.

Recently, (see, e.g. [AK96, Abl97, AK98, AK98a, AKM98, KM98, AT98, ĎHJSS00, HS98, HSa00, Sa98, Sa99, We00]) randomized branching programs and OBDDs have been investigated. We consider Las Vegas OBDDs (LV-OBDDs). The extension of OBDDs to LV-OBDDs is straightforward. One adds a new sink labeled by "?" (*"don't know"*). For any non-sink node v one allows several edges $(v, u_1), \ldots, (v, u_k)$, $k \geq 1$, labeled by the same Boolean constant. To each edge (v, u_i) the probability $Prob(v, u_i)$ is assigned. The only requirements are $0 < Prob(v, u_i) \leq 1$ for every i, and $\sum_{i=1}^{k} Prob(v, u_i) = 1$ (i.e. $Prob$ is a probability distribution over the branching from the node v). The meaning is that during the computation of a LV-OBDD at each inner node labeled by x_i one of the edges labeled by α_i is chosen with probability given by $Prob$. Thus, the probability of a path $s_1, s_2, \ldots, s_n, s_{n+1}$ is $\prod_{i=1}^{n} Prob(s_i, s_{i+1})$.

We say that a LV-OBDD A computes a Boolean function f if for every input α the probability of reaching the sink labeled by $f(\alpha)$ is at least $\frac{1}{2}$ and the probability to reach the sink labeled by $\overline{f(\alpha)}$ is 0. Let **LV-size(f)** denote the size of the best Las Vegas OBDD for f.

To establish the relationship between Las Vegas and determinism for OBDDs we present a few simple facts. In what follows we need to consider one-way protocols for arbitrary input partitions. A partition of a set $X = \{x_1, \ldots, x_n\}$, $n \in \mathbb{N}$, of input variables is any pair $\Pi = (\Pi_I, \Pi_{II})$ where $\Pi_I \cup \Pi_{II} = X$ and $\Pi_I \cap \Pi_{II} = \emptyset$. A one-way protocol according to a partition $\Pi = (\Pi_I, \Pi_{II})$ is the usual one-way protocol; the only difference is that the computer C_I obtains values of variables from Π_I and C_{II} obtains values of variables from Π_{II}. Let **mc(D)** denote the number of different messages of a deterministic one-way protocol D. Let the **one-way communication complexity of f according to Π, cc$_1(f, \Pi)$**, be the one-way communication complexity of the best protocol computing f according to Π, and let

$$\mathbf{mc}(f, \Pi) = \min\{\mathrm{mc}(D) \mid D \text{ is a one-way protocol computing } f \text{ according to } \Pi\}.$$

Analogously, consider **lvmc(f, Π)** for Las Vegas one-way protocols.

For formal definitions and details of the study of branching programs see [PŽ83, Sa99, We87, We00]. An example of an OBDD is given in Figure 3. This OBDD has the variable ordering x_1, x_2, x_3, x_4, and it computes the symmetric function s_4.

Let $f : \{0,1\}^n \to \{0,1\}$ be a Boolean function over $X = \{x_1, \ldots, x_n\}$, and let $X_{i_1 i_2 \ldots i_n} = x_{i_1}, x_{i_2}, \ldots, x_{i_n}$ be a variable ordering of X. For every $j = 1, \ldots, n-1$, we consider the partition

$$\Pi_j(i_1 i_2 \ldots i_n) = (\{i_1, \ldots, i_j\}, \{i_{j+1}, \ldots, i_n\}).$$

Now, we give two simple observations relating one-way communication complexity and the size of OBDDs. Let

$$\text{size}(f, X_{i_1 i_2 \ldots i_n}) = \{\text{size}(A) \quad | \quad A \text{ is an OBDD that computes } f$$
$$\text{with the ordering } X_{i_1 i_2 \ldots i_n}\},$$

and let **lv-size**$(f, X_{i_1 \ldots i_n})$ have the same meaning for Las Vegas OBDDs.

Observation 3 (([HS98])) *Let* f *be a Boolean function over* $X = \{x_1, \ldots, x_n\}$, *and let* $X_{i_1 i_2 \ldots i_n}$ *be a variable ordering. For every* $j \in \{1, \ldots, n - 1\}$,

$$\text{size}(f, X_{i_1 i_2 \ldots i_n}) \geq \text{mc}(f, \Pi_j(i_1 i_2 \ldots i_j)), \text{ and}$$

$$\text{lv-size}(f, X_{i_1 i_2 \ldots i_n}) \geq \text{lvmc}(f, \Pi_j(i_1 i_2 \ldots i_j)).$$

Proof. The proof is similar as for the finite automata. Let A be a minimal OBDD that computes f for every $j \in \{1, \ldots, n - 1\}$. We construct a one-way protocol $D_j = (C_I^j, C_{II}^j)$ that acts as follows. For every input $\alpha_1 \alpha_2 \ldots \alpha_n$, C_I^j sends the name of the vertex v that is reached by A when working on $\alpha_{i_1} \alpha_{i_2} \ldots \alpha_{i_j}$ and whose label is in $\{x_{i_{j+1}}, \ldots, x_{i_n}\}$ (i.e. v is the last vertex that can be reached be reading the values of the variables from $\{x_{i_1}, \ldots, x_{i_j}\}$ only). C_{II}^j accepts iff A starting from v with the input values $\alpha_{i_j+1} \ldots \alpha_{i_n}$ finishes the computation in the sink labeled by 1. Clearly, the number of different messages of D_j is bounded by size(A).

Obviously, this strategy works for any computation mode and so for the Las Vegas randomization, too. □

Observation 4 (([HS98])) *Let* f *be a Boolean function over* $X = \{x_1, \ldots, x_n\}$. *For every variable ordering* $X_{i_1 i_2 \ldots i_n}$ *of* X,

$$\text{size}(f, X_{i_1 i_2 \ldots i_n}) \leq \sum_{j=1}^{n-1} \text{mc}(f, \Pi_j(i_1 i_2 \ldots i_n)) + 3.$$

Proof. Let, for $j = 1, \ldots, n - 1$, D_j be a one-way protocol that computes f according to $\Pi_j(i_1 i_2 \ldots i_n)$ within $\text{mc}(D_j) = \text{mc}(f, \Pi_j(i_1 \ldots i_n))$. For every $j \in \{1, \ldots, n - 1\}$, we set Set_j to be the set of all messages of D_j. Now, we construct an OBDD for f. The set of nodes of A consists of a root labeled by x_{i_1}, two sinks, and all nodes (j, c), where $j \in \{1, \ldots, n - 1\}$ and $c \in \text{Set}_j$. For $j = 1, \ldots, n - 1$, all nodes from $\{(j, c) \mid c \in \text{Set}_j\}$ are labeled by $x_{i_{j+1}}$. Now, starting from the root one can add the edges in a unique straightforward way (see [HS98, SW93] for more details). □

Now, we use the above observations to relate OBDDs and Las Vegas OBDDs. Observe that we have obtained a situation similar to that for finite automata. The size of (deterministic) OBDDs is strongly related to the message complexity of (deterministic) one-way protocols, and Las Vegas one-way protocols provide a lower bound on the size of Las Vegas OBDDs.

Theorem 21 ([HS98, KM99]) *For any Boolean function* $f : \{0,1\}^n \to \{0,1\}$, $n \in \mathbb{N} - \{0\}$,

$$\text{lv-size}(f) \geq \frac{\sqrt{\text{size}(f)}}{\sqrt{n}}.$$

Proof. Let $X = \{x_1, x_2, \ldots, x_n\}$ be the set of input variables of f. Let A be an optimal LV-OBDD for f, i.e. $\text{size}(A) = \text{lv-size}(f)$. Let $X_A = X_{i_1 i_2 \ldots i_n}$ for a permutation $(i_1, \ldots, i_n)$ of $(1, \ldots, n)$. Following Observation 3 we have

$$
\begin{aligned}
\text{lv-size}(f) &= \text{lv-size}(f, X_{i_1 i_2 \ldots i_n}) = \text{size}(A) \\
&\geq \max\{\text{lvmc}(f, \Pi_j(i_1 \ldots i_n)) \mid j = 1, \ldots, n-1\}. \quad (5)
\end{aligned}
$$

Following Theorem 12 we have

$$\text{lvmc}(f, \Pi_j(i_1 \ldots i_n)) \geq \sqrt{\text{mc}(f, \Pi_j(i_1 \ldots i_n))} \qquad (6)$$

for every $j \in \{1, \ldots, n\}$. (5) and (6) together imply

$$\text{lv-size}(f) \geq \sqrt{\max\{\text{mc}(f, \Pi_j(i_1 \ldots i_n)) \mid j = 1, \ldots, n-1\}}. \qquad (7)$$

Observation 4 implies

$$\text{size}(f, X_{i_1 \ldots i_n}) \leq n \cdot \max\{\text{mc}(f, \Pi_j(i_1 \ldots i_n)) \mid j = 1, \ldots, n-1\}. \qquad (8)$$

So, inserting (8) into (7) we obtain

$$\text{lv-size}(f) \geq \frac{\sqrt{\text{size}(f, X_{i_1 \ldots i_n})}}{\sqrt{n}} \geq \frac{\sqrt{\text{size}(f)}}{\sqrt{n}}.$$

□

If a function f depends on all its variables, then $\text{size}(f)$ is at least the number of its variables. So, we obtain the following consequence of Theorem 21.

590

Corollary 2 *Let f be a Boolean function that depends on all its n variables, $n \in I\!N - \{0\}$. Then*

$$(\text{lv-size}(f))^3 \geq \text{size}(f) \geq \text{lv-size}(f).$$

We finish this section by briefly listing some further applications of communication complexity for the study of the power of randomization of branching programs. These results are in contrast with the above presented relations because they show that randomization (even Las Vegas) may be substantially more powerful than determinism for some models of branching programs. Sauerhoff [Sa99] showed an exponential gap between Las Vegas and determinism for one-time-only branching programs that are a non-oblivious generalization of OBDDs. A superpolynomial gap $n^{\log n}$ between Las Vegas and determinism was established in [HSa00] for a generalization of OBDDs, where every variable can be read at most constant many times by using constant many different variable orderings. One of the most impressive examples of the power of randomization was proved in [ĎHJSS00]. For the characteristic function of linear codes, there is a small two-sided-error OBDD (even co-RP OBDD) but any nondeterministic one-time-only branching program computing this function is of exponential size. This means that bounded error (oblivious) randomization may be much more powerful than nondeterminism and non-obliviousness together. Several further results of the power of randomized branching programs can be found in [AK96, Abl97, AK98, AK98a, AKM98, AT98, KM98, Sa99].

Acknowledgements

I would like to thank Hans-Joachim Böckenhauer and Dirk Bongartz for their help during the whole time I worked on this chapter and to Farid Ablayev for some comments on the previous version of this work.

References

[Ab78] Abelson, H.: Lower bounds on information transfer in distributed computations. *Proc. 19th IEEE FOCS*, IEEE 1978, pp. 151–158.

[Abl88] Ablayev, F.M.: The complexity properties of probabilistic automata with isolated cut point. *Theoretical Computer Science* 57 (1988), pp. 87–95.

[Abl93] Ablayev, F.M.: Lower bounds for one-way probabilistic communication complexity. *Proc. 20th ICALP'93, Lecture Notes in Computer Science* 700, Springer-Verlag 1993, pp. 241–252.

[Abl96] Ablayev, F.M.: Lower bounds for one-way probabilistic communication complexity and their application to space complexity. *Theoretical Computer Science* 157 (1996), pp. 139–159.

[Abl97] Ablayev, F.M.: Randomization and nondeterminism are incomparable for polynomial ordered binary decision diagrams. *Proc. 24th ICALP, Lecture Notes in Computer Science* 1256, Springer-Verlag 1997, pp. 195–202.

[AF86] Ablayev, F.M., Freivalds, R.: Why sometimes probabilistic algorithms can be more effective. *Proc. MFCS'86, Lecture Notes in Computer Science* 233, Springer-Verlag 1986, pp. 1–14.

[AK96] Ablayev, F.M., Karpinski, M.: On the power of randomized branching programs. *Proc. 23rd ICALP, Lecture Notes in Computer Science* 1099, Springer-Verlag 1996, pp. 348–356.

[AK98] Ablayev, F.M., Karpinski, M.: A lower bound for integer multiplication on randomized read-once branching programs. *Technical Report TR98-011, Electronic Colloquium on Computational Complexity* 1998.

[AK98a] Ablayev, F.M., Karpinski, M.: On the power of randomized ordered branching programs. *Technical Report TR98-004, Electronic Colloquium on Computational Complexity* 1998.

[AKM98] Ablayev, F.M., Karpinski, M., Mubarakzjanov, R.: On BPP versus NP ∪ coNP for ordered read-once branching programs. *Proc. International Workshop on Randomized Algorithms*, Brno 1998, pp. 25–34.

[AT98] Agrawal, M., Thierauf, T.: The satisfiability problem for probabilistic ordered branching programs. *Proc. 13th IEEE Conference on Computational Complexity*, IEEE 1998, pp. 81–90.

[AUY83] Aho, A.V., Ullman, J.D., Yannakakis, M.: On notions of informations transfer in VLSI circuits. *Proc. 15th ACM STOC*, ACM 1983, pp. 133–139.

[Aj99] Ajtai, M.: A non-linear time lower bound for Boolean branching programs. *Proc. 40th IEEE FOCS*, IEEE 1999, pp. 60–70.

[AM86] Alon, N., Maas, W.: Meanders, Ramsey theory and lower bounds for branching programs. *Proc. 27th IEEE FOCS*, IEEE 1986, pp. 410–417.

[AM88] Alon, N., Maas, W.: Meanders and their applications in lower bound arguments. *Journal of Computer and System Sciences* 37 (1988), pp. 118–129.

[BFS86a] Babai, L., Frankl, P., Simon, J.: Complexity classes in communication complexity theory. *Proc. 27th IEEE FOCS*, IEEE 1986, pp. 337–347.

[BGS75] Baker, T., Gill, J. and Solovay, R.: Relativization of the P=?NP question. *SIAM Journal of Computing* 4 (1975), pp. 431–442.

[BST98] Beame, P.W., Saks, M., Thathachar, J.S.: Time-space tradeoffs for branching programs. *Proc. 39th IEEE FOCS*, IEEE 1998, pp. 254–263.

[BC94] Bovet, D.P. and Crescenzi, P.: *Introduction to the Theory of Complexity*. Prentice Hall 1994.

[Br85] Bryant, R.E.: Symbolic manipulation of Boolean functions using a graphical representation. *22nd ACM/IEEE Design Automation Conference*, 1985, pp. 688–694.

[Br86] Bryant, R.E.: Graph-based algorithms for Boolean function manipulation. *IEEE Transactions on Computers* 35 (1986), pp. 677–691.

[CG85] Chor, B., Goldreich, O.: Unbiased bits from sources of weak randomness and probabilistic communication complexity. *Proc. 26th IEEE FOCS*, IEEE 1985, pp. 429–442.

[CK86] Csiszar, I. and Körner, J.: *Information theory: coding theorems for discrete memoryless systems*. Academic Press 1986.

[DHS96] Dietzfelbinger, M., Hromkovič, J., Schnitger, G.: A comparison of two lower bounds methods for communication complexity. *Theoretical Computer Science* 168 (1996), pp. 39–51.

[DKR94] Dietzfelbinger, M., Kutylowski, M. and Reischuk, R.: Exact lower bounds for computing Boolean functions on CREW PRAMs. *Journal on Computer and System Sciences* 48 (1994), pp. 231–254.

[ĎG93] Ďuriš, P., Galil, Z.: On the power of multiple reads in chip. *Information and Computation* 104 (1993), pp. 277–287.

[ĎHI00] Ďuriš, P., Hromkovič, J., Inoue, K.: A separation of determinism, Las Vegas and nondeterminism for picture recognition. *Proc. 15th IEEE Conference on Computational Complexity*, IEEE 2000, pp. 214–228.

[ĎHJSS00] Ďuriš, P., Hromkovič, J., Jukna, S., Sauerhoff, M., Schnitger, G.: On multipartition communication complexity. Unpublished manuscript, 2000.

[ĎHRS97] Ďuriš, P., Hromkovič, J., Rolim, J.D.P., Schnitger, G.: Las Vegas versus determinism for one-way communication complexity, finite automata and polynomial-time computations. *Proc. STACS'97, Lecture Notes in Computer Science* 1200, Springer-Verlag 1997, pp.

117–128. (also *Technical Report TR97-029, Electronic Colloquium on Computational Complexity* 1997)

[Fr77] Freivalds, R.: Probabilistic machines can use less running time. *Information Processing 1977, IFIP*, North Holland 1977, pp. 839–842.

[Fr82] Freivalds, R.: On the growing number of states in determinization of finite probabilistic automata. *Automaika a Vichislitelnaya Technika* 3 (1982), pp. 39–42 (in Russian).

[HR88] Halstenberg, B., Reischuk, R.: On different modes of communication. *Proc. 20th ACM STOC*, ACM 1988, pp. 162–172.

[Hr86] Hromkovič, J.: Relation between Chomsky hierarchy and communication complexity hierarchy. *Acta Math. Univ. Com.* 48-49 (1986), pp. 311–317.

[Hr88a] Hromkovič, J.: Some complexity aspects of VLSI computations. Part 1. A framework for the study of information transfer in VLSI circuits. *Computers and Artificial Intelligence* 7 (1988), pp. 229–252.

[Hr91] Hromkovič, J.: Nonlinear lower bounds on the number of processors of circuits with sublinear separators. *Information and Computation* 95 (1991), pp. 117–128.

[Hr97] Hromkovič, J.: *Communication Complexity and Parallel Computing*. Springer-Verlag 1997.

[Hr99] Hromkovič, J.: Communication complexity and lower bounds on multilective computations. *Theoretical Informatics & Applications* 33(2), (1999), pp. 193-211.

[HS96] Hromkovič, J., Schnitger, G.: Nondeterministic communication with a limited number of advice bits. *Proc. STOC'96, ACM* 1996, pp. 551–560.

[HS98] Hromkovič, J., Schnitger, G.: On the power of Las Vegas computation. *Information and Computation*, to appear.

[HS99] Hromkovič, J., Schnitger, G.: On the power of Las Vegas II: Two-way finite automata. *Proc. ICALP'99, Lecture Notes in Computer Science* 1644, Springer-Verlag 1999, pp. 433–443. (also *Theoretical Computer Science*, to appear)

[HSa00] Hromkovič, J., Sauerhoff, M.: Tradeoff between nondeterminism and complexity for communication protocols and branching programs. *Proc. STACS '00, Lecture Notes in Computer Science* 1770, Springer-Verlag 2000, pp. 145–156.

[JPS84] Ja'Ja, J., Prassanna Kamar, V.K., Simon, J.: Information transfer under different sets of protocols. *SIAM Journal on Computing* 13 (1984), pp. 840–849.

594

[KS87] Kalyanasundaram, B., Schnitger, G.: The probabilistic communication complexity of set intersection. *Proc. 2nd Annual Conference on Structure in Complexity Theory*, 1987, pp. 41–47.

[KW88] Karchmer, M., Wigderson, A.: Monotone circuits for connectivity require superlogarithmic depth. *Proc. 20th ACM STOC*, ACM 1988, pp. 539–550. (also *SIAM Journal on Discrete Mathematics* 3 (1990), 718–727).

[KM98] Karpinski, M., Mubarakzjanov, R.: Some separation problems on randomized OBDDs. Manuscript, 1998.

[KM99] Karpinski, M., Mubarakzjanov, R.: A note on Las Vegas OBDDs. *Technical Report TR99-009, Electronic Colloquium on Computational Complexity* 1999.

[KA86] King, F.P., Abhasel, G.: Communication complexity of computing the Hamming distance. *SIAM Journal on Computing* 15 (1986), pp. 932–946.

[KN97] Kushilevitz, E., Nisan, N.: *Communication Complexity*, Cambridge University Press 1997.

[Lei80] Leiserson, C.E.: Area efficient graph algorithms (for VLSI). *Proc. 21st IEEE FOCS*, IEEE 1980, pp. 270–281.

[Le90] Lengauer, Th.: VLSI Theory. In: *Handbook of Theoretical Computer Science, Vol. A, Algorithms and Complexity*, Elsevier 1990, pp. 835–868.

[Lo89] Lovász, L.: Communication Complexity: A Survey. *Technical Report CS-TR-204-89*, Princeton University, 1989 (also *Paths, Flows and VLSI Layout* (Korte, Lovász, Promel, and Schrijver, eds.), Springer-Verlag 1990, pp. 235–266).

[MS82] Mehlhorn, K., Schmidt, E.: Las Vegas is better than determinism in VLSI and distributed computing. *Proc. 14th ACM STOC*, ACM 1982, pp. 330–337.

[MW95] Meinel, Ch., Waack, S.: Lower bounds for the majority communication complexity of various graph accessibility problems. *Proc. 20th MFCS'95, Lecture Notes in Computer Science* 969, Springer-Verlag. 1995, pp.299–308.

[MF71] Meyer, A. R. and Fischer, M. J.: Economies of description by automata, grammars and formal systems. *Proc. 12th IEEE Symposium on Switching and Automata Theory*, 1971, pp. 188–191.

[New91] Newman, I.: Private versus common random bits in communication complexity. *Information Processing Letters* 39 (1991), pp. 67–71.

[NW91] Nisan, N., Wigderson, A.: Bounds in communication complexity revised. *32nd ACM STOC*, ACM 1991, pp. 419–429 (also in *SIAM Journal on Computing* 22 (1993), pp. 211–219).

[NW95] Nisan, N., Wigderson, A.: On ranks vs. communication complexity. *Combinatorica* 15 (1995), 557–565.

[PS82] Papadimitriou, Ch., Sipser, M.: Communication complexity. *Proc. 14th ACM STOC*, San Francisco, ACM 1982, pp. 196–200.

[PS84] Papadimitriou, Ch. and Sipser, M.: Communication complexity. *Journal of Computer and System Sciences* 28 (1984), pp. 260–269.

[Po79] Pokrovskaya, I.: Some bounds on the number of states of probabilistic automata recognizing regular languages. *Problemy Kibernetiki* 36 (1979), pp. 209–224 (in Russian).

[PŽ83] Pudlák, P. and Žák, S.: Space complexity of computation. Techn. Report, Prague 1983.

[Ra63] Rabin, M.O.: Probabilistic automata. *Information and Control* 6 (1963), pp. 230–245.

[RS59] Rabin, M.O.. Scott, D.: Finite automata and their decision problems. *IBM J. Research & Development* 3 (1959), pp. 114–125.

[Raz90] Razborov, A.A.: On the distributed complexity of disjointness. *Proc. 17th ICALP* , Springer-Verlag 1990, pp. 249–253. (also *Theoretical Computer Science* 106 (1992), 385–390.)

[RW90] Raz, R., Wigderson, A.: Monotone circuits for matching require linear depth. *Proc. 22nd ACM STOC*, ACM 1990, pp. 287–292. (also *Journal of Association for Computing Machinery* 39 (1992), pp. 736–744.)

[Sa98] Sauerhoff, M.: Lower bounds for randomized read-k-times branching programs, *Proc. 15th Symposium on Theoretical Aspects in Computer Science, Lecture Notes in Computer Science* 1373, Springer-Verlag 1998, pp. 105–115.

[Sa99] Sauerhoff, M.: *Complexity Theoretical Results for Randomized Branching Programs.* PhD thesis, University of Dortmund, Shaker 1999.

[Sa81] Savage, J.E.: Area-time tradeoffs for matrix multiplication and related problems in VLSI models. *Journal of Computer and System Sciences* 20 (1981), pp. 230–242.

[Sa70] Savitch, W. J.: Relationships between nondeterministic and deterministic tape complexities. *Journal of Computer and System Sciences* 4 (1970), pp. 177–192.

596

[SW93] Sieling, D., Wegener, I.: NC-algorithms for operations on binary decision diagrams. *Parallel Processing Letters* 3 (1), (1993), pp. 3–12.

[Th79] Thompson, C.D.: Area-time complexity for VLSI. *Proc. 11th ACM STOC*, ACM 1979, pp. 81–88.

[Th80] Thompson, C.D.: A complexity theory for VLSI. Doctoral dissertation. CMU-CS-80-140, Computer Science Department, Carnagie-Mellon University, Pittsburgh, August 1980, 131 p.

[Tu89] Turán, G.: Lower bounds for synchronous circuits and planar circuits. *Information Processing Letters* 130 (1989), pp. 37–40.

[We87] Wegener, I.: *The Complexity of Boolean Functions.* Wiley-Teubner Series in Computer Science, John Wiley and Sons Ltd., and Teubner, B.G., Stuttgart 1987.

[We00] Wegener, I.: *Branching Programs and Binary Decision Diagrams - Theory and Applications.* Monographs on Discrete and Applied Mathematics, SIAM 2000.

[Ya79] Yao, A.C.: Some complexity questions related to distributive computing. *Proc. 11th ACM STOC*, ACM 1979, pp. 209–213.

[Ya81] Yao, A.C.: The entropie limitations on VLSI computations. *Proc. 13th ACM STOC*, ACM 1981, pp. 308–311.

[Ya82] Yao, A.C.: Protocols for secure computations. *Proc. 23rd IEEE FOCS*, IEEE 1982, pp. 160–164.

[Ya83] Yao, A.C.: Lower bounds by probabilistic arguments. *Proc. 25th ACM STOC*, ACM 1983, pp. 420–428.

Chapter 15

PROPERTY TESTING

Dana Ron
Dept. of EE – Systems
Tel Aviv University
Ramat Aviv, ISRAEL
danar@eng.tau.ac.il

1.   INTRODUCTION

In broad terms, property testing is the study of the following class of problems:

> Given the ability to perform (local) queries concerning a particular object (e.g., a function, or a graph), the task is to determine whether the object has a predetermined (global) property (e.g., linearity or bipartiteness), or is far from having the property. The task should be performed by inspecting only a small (possibly randomly selected) part of the whole object, where a small probability of failure is allowed. [1]

In order to define a property testing problem, we need to specify the type of queries that the testing algorithm can perform, and a distance measure between objects. The latter is required in order to define what it means that the object is *far* from having the property. We assume that the algorithm is given a *distance* parameter ϵ. The algorithm should accept with probability at least 2/3 every object that has the property, and should reject with probability at least 2/3 every object that has distance more than ϵ (according to the selected distance measure) from any object having the property. [2]

When the object in question is a function $f : X \to Y$ (for finite X and Y), then the natural form of queries is: "What is the value of $f(x)$?" for

---

[1] A more general definition in which the algorithm cannot necessarily perform queries but rather is given "samples" from the object distributed according to some fixed (possibly unknown) distribution, will be discussed subsequently.

[2] The choice of success probability 2/3 is of course arbitrary, and any constant strictly greater than 1/2 can be used. In order to obtain success probability of $1 - \delta$ for any $\delta < 1/3$, the algorithm should be executed $\Theta(\log(1/\delta))$ times, and the majority output taken.

S. Rajasekaran et al (eds.), Handbook of Randomized Computing, Volume 2, pp. 597–649.

any choice of $x \in X$. A natural distance measure between two functions is the fraction of domain elements on which the functions differ. Thus, for example, when testing linearity of functions, the testing algorithm can obtain the value of the tested function f on x's of its choice. If f is a linear function then the algorithm should accept it with probability at least 2/3. However, if the value of f must be modified on more than an ϵ fraction of the domain elements so that it becomes linear, then the algorithm should reject it with probability 2/3. (If f is close to being linear, then the algorithm can either accept or reject.)

When studying graph properties, the form of queries, and in some cases the distance measure, depend on the graph representation. For instance, if graphs are represented by their adjacency matrix, then the queries are of the form: "Is there an edge between vertex v and vertex u"? The distance measure between graphs in this case is the fraction of adjacency-matrix entries on which the two graphs differ. Thus, for example, when testing bipartiteness[3] of graphs represented by their adjacency matrix, then the algorithm is only allowed to accept graphs $G = (V, E)$ for which at most $\epsilon |V|^2$ edges should be removed so that they become bipartite.

In this tutorial we mainly focus on testing graph properties, though we shall briefly survey other results as well. For sake of the presentation, some of the following motivational discussion refers to graphs. However, much of it is relevant to testing other types of objects.

1.1 MOTIVATION

The task of testing a certain graph property is a relaxation of the task of deciding *exactly* whether a graph has the property. Namely, an exact decision procedure is required to accept every graph that has the property and reject every graph that *does not have* the property. A testing algorithm is still required to accept every graph that has the property, but is only required to reject every graph that *is far from having* the property. While relaxing the task, we expect the algorithm to observe only a small part of the graph and to run significantly faster than any exact decision procedure. Specifically, we aim at spending time that is *sub-linear in* or even *independent of* the size of the graph.

In fact, as we shall elaborate later, many graph properties have very fast property testing algorithms whose query complexities do not depend at all on the size of the graph. This should be put in contrast to known

---

[3] A graph is bipartite if its set of vertices can be partitioned into two disjoint subsets such that there are not edges within the subsets.

lower bounds on the complexity of exactly deciding graph properties. Rivest and Vuillemin [RV76] showed that any deterministic procedure for deciding any non-trivial monotone N-vertex graph property must examine $\Omega(N^2)$ entries in the adjacency matrix representing the graph, thus resolving the Aanderaa–Rosenberg Conjecture [Ros73]. The query complexity of *randomized* decision procedures was conjectured by Yao to be also $\Omega(N^2)$. Progress towards proving this conjecture was made by Yao [Yao87], King [Kin91] and Hajnal [Haj91] culminating in an $\Omega(N^{4/3})$ lower bound.

1.1.1 A Tradeoff Between Accuracy and Efficiency.

It follows from the above discussion that Property Testing trades *accuracy* for *efficiency*, where accuracy is measured in terms of the distance parameter ϵ. Since we expect the running time of the algorithm to increase as a function of $1/\epsilon$, the algorithm is more accurate as ϵ decreases, but its running time increases.

This paradigm may be useful is several scenarios.

1. A fast property tester can be used to speed up a slow exact decision procedure as follows. Before running the decision procedure, run the tester. If the tester rejects, then we know with high confidence that the property does not hold and it is unnecessary to run the (slower) decision procedure. In fact, it is often the case that when the testing algorithm rejects, it provides a *witness* that the graph does not have the property. On the other hand, if the tester accepts, then an exact decision procedure will determine whether the property is close to holding or actually holds. We thus save time in applications where *typical* graphs are either good (have the property), or very bad (far from having the property).

2. Furthermore, if it is *guaranteed* that graphs are either good or very bad then we may not even need the exact algorithm at all.

3. There are circumstances in which knowing that a property nearly holds is good enough and consequently exact decision is unnecessary.

4. In some cases (e.g., connectivity) there are algorithms for "fixing" the graph (that is, modifying it so that it will have the property). If the graph is accepted then we know with high confidence that the number of required modification is not too large, and assuming there is a cost associated with each edge modification, the total cost is not too large.

5. The graph may be too large to fully scan, so one *must* make a decision without observing the whole graph.

6. It may be NP-hard to answer the question exactly, and so (even if scanning the graph is feasible), some form of approximation is inevitable

1.1.2 Relation to Other Notions of Approximation.

In the study of approximation algorithms, and in particular approximation algorithms for graph optimization problems, the following is the dominant approach. For each instance (graph) there is a set of feasible solutions (e.g., subsets of vertices that form cliques). With each feasible solution there is an associated cost or utility (e.g., the size of the clique). The goal is to approximate the value of the minimum cost or maximum utility of a feasible solution. In some cases, the goal is to actually *find* a solution whose cost or utility is close to optimal.

Property testing is related to an alternative notion of approximation, namely that of *dual* approximation [HS87, HS88]. Instead of approximating the maximum utility of a feasible solution, dual approximation tries to approximate the distance (in terms of edge-modifications) to having a feasible solution with a certain utility. In the example of the clique for instance, the goal is to approximate the number of edges that must be added in order to obtain a clique with a certain size.

The preferred notion of approximation is naturally dependent on the context in which it is applied. We note that in some cases the two notions coincide. This is true for example in the case of Max-Cut. In the more standard approach, the goal is to approximate the size of a maximum cut (that is, the maximum number of edges crossing any two-way partition). In the dual approach, the goal is to approximate the number of edges that should be added in order to obtain some cut with a given size. In this case, an algorithm for the latter problem can be used to solve the former problem.

1.1.3 Property Testing, Program Testing and PCP.

Property testing of functions was first explicitly defined by Rubinfeld and Sudan [RS96] in the context of *program testing*. The goal of a program testing algorithm is to test whether a given program computes a specified function. Here one may choose to test that the program satisfies a certain property (which the function holds) before checking that it computes the specified function itself. This paradigm has been followed both in the theory of program testing [BLR93, RS96, Rub99], and in practice where often programmers first test their programs by verifying that the programs satisfy properties that are known to be satisfied by the function they compute.

Property testing also emerges naturally in the context of probabilistically checkable proofs (PCP). In this context the property being tested is whether the function is a codeword of a specific code. This paradigm, explicitly introduced in [BFLS91], has shifted from testing codes defined by low-degree polynomials [BFL91, BFLS91, FGL+96, AS98, ALM+98] to testing Hadamard codes [ALM+98, BGLR93, BS94, BCH+95, Kiw96, Tre98], and to testing the "long code" [BGS98, Hås96, Hås97, Tre98].

We note that in both the above contexts, the properties tested were algebraic.

1.1.4   Property Testing and Learning.

One of the initial motivations for the study of property testing is its relation to Computational Learning Theory. Let the objects we are interested in be functions (in particular, boolean functions), and consider the following variant of our initial description of property testing: Instead of allowing the algorithm to query the tested function f on inputs of its choice, it is provided with a *labeled sample* $\{(x^1, f(x^m)), \ldots, (x^m, f(x^m))\}$, where the x^i's are distributed according to some fixed but unknown distribution D over the domain X. In this case, distance between functions is measured with respect to the distribution D. That is, the distance between functions g and h is $\Pr_{x \sim D}[g(x) \neq h(x)]$.

The above definition of property testing is inspired by the Probably Approximately Correct (PAC) learning model proposed by Valiant [Val84]. In the PAC model, a learning algorithm is given a random sample labeled by an unknown function f as defined above, and is required to output (with high probability) a *hypothesis* h that approximates f well. That is, $\Pr_{x \sim D}[h(x) \neq f(x)] \leq \epsilon$. In the standard PAC model, it is assumed that f belongs to a known class of functions F, and it is either required that h belong to F as well, or to some specified hypothesis class $H \supseteq F$.[4]

Thus, property testing as defined in this subsection can be viewed as a relaxation of PAC learning – instead of requiring a good approximation of the function f, we only ask whether such a good approximation *exists* in a given class (the class of functions having the tested property). Our original definition of property testing which allows queries and defines distance with respect to the uniform distribution can be seen as a relaxation of a variant of the PAC model: Learning with queries under the uniform distribution.

---

[4]In the *agnostic* model [KSS94], nothing is assumed about f, and so the distance between h and f is required to be not much larger than the distance between f and the closest function in H.

Given the above view, it is reasonable to expect that for some classes of functions (properties), testing can be done much more efficiently than learning (in terms of sample/query complexity and/or running time). This is true for example in the case of linear functions [BLR93], multivariate polynomials [RS96], and monotone functions [GGL+00]. If we have fast testing algorithms that require relatively small samples, we may be able to use such algorithms in the context of learning. Namely, we can use them to choose between alternative hypothesis representations without actually incurring the expense of running the corresponding learning algorithms. For example, suppose that we are considering running C4.5 (a fast algorithm) to find a decision tree hypothesis (a relatively weak representation). But we may also want to consider running backpropagation (a slow algorithm) to find a multilayer neural network (a relatively powerful representation, requiring more data, but with perhaps greater accuracy). Ideally, we would like a fast, low-data test that informs us whether this investment would be worthwhile.

1.2 TESTING GRAPH PROPERTIES

The study of *testing graph properties* was initiated by Goldreich Goldwasser and Ron [GGR98]. As noted previously, the precise definition of testing graph properties is dependent on the representation of graphs. There are two standard representations of graphs, *adjacency matrices* and *incidence lists*, and we discuss the related testing models below. We restrict our attention to *undirected* graphs.

- *Adjacency-Matrix Model.* Goldreich *et. al.* [GGR98] consider the adjacency-matrix representation of graphs, where the testing algorithm is allowed to probe into the matrix. That is, the algorithm can query whether there is an edge between any two vertices of its choice. In this representation the distance between graphs is the *fraction of entries* in the adjacency matrix on which the two graphs differ. By this definition, for a given distance parameter ϵ, the algorithm should reject every graph that requires more than $\frac{\epsilon}{2} \cdot |V|^2$ edge modifications in order to acquire the tested property (the factor of $\frac{1}{2}$ is because each edge is represented twice in the matrix). This representation is most appropriate for dense graphs, and the results for testing in this model are most meaningful for such graphs.

- *Incidence-Lists Models.* Goldreich and Ron [GR97] consider the incidence-lists representation of graphs. In the model they consider, graphs are represented by lists of *length d*, where d is a bound on the degree of the graph. Here the testing algorithm can query, for every vertex v and index $i \in \{1, \ldots, d\}$, which vertex is the i'th neighbor of

v. If no such neighbor exists then the answer is '0'. Analogously to the adjacency-matrix model, the distance between graphs is defined to be the fraction of entries on which the graphs differ according to this representation. Since the total number of incidence-list entries is $d \cdot |V|$, a graph should be rejected if the number of edge modifications required in order to obtain the property is greater than $\frac{\epsilon}{2} \cdot d|V|$. (Once again, the factor of $\frac{1}{2}$ is because each edge (u, v) is represented both as an entry $[u, i]$ and as an entry $[v, j]$).

A variant of the above model allows the incidence lists to be of varying lengths [PR99a]. In such a case, the distance between graphs is defined with respect to the total number of edges in the graph (or an upper bound on this number). This model is suitable for testing graphs that are not dense but for which there is large variance in the degrees of the graph vertices. Furthermore, some problems are more interesting in this model, in the sense that removing the degree bound makes them less restricted. For example, testing whether a graph has a diameter of at most a bounded size, is less interesting in the bounded degree model, since a bound d on the degree implies a lower bound on the diameter of a graph. Intuitively, testing in this model is at least as hard as testing in the bounded-degree model described above, and in fact in some cases it is strictly harder.

We note that for both the adjacency-matrix model and the bounded-degree incidence-lists model, the representations can be viewed as *functional* representations of graphs. Namely, in the first model it is a function from all $|V|^2$ vertex-pairs to $\{0, 1\}$, and in the second case it is a function from all $|V| \cdot d$ pairs of vertex and index, to the set of vertices. Furthermore, the definition of distance between graphs in these models is determined by the representation: it is the symmetric difference between the functions representing the graphs, divided by the size of the domain of the functions. The unbounded-degree incidence-lists model is not a functional representation, and the notion of distance is divorced from the representation.

1.2.1   Techniques. The applicable techniques depend on the choice of representation. A central technique that is used for the adjacency-matrix representation is *random sampling*. Specifically, the algorithm randomly selects a small set U of vertices from the graph G, finds the edges interconnecting the vertices, and determines whether the property holds (or "almost holds") for the small subgraph induced by U. If so, then the algorithm accepts. If not, the algorithm rejects. It is typically straightforward to show that any graph having the property is accepted (always, or with high probability). The crux of the proof is in showing that a graph that is far from having the property is rejected with high probability.

The following general analysis technique is often used for proving the above claim. The sample is viewed as consisting of two disjoint subsamples. The first sample is viewed as implicitly inducing certain *constraints* on all other graph vertices. These constraints are such that if the graph is far from having the property then many vertices (or pairs of vertices) do not obey the constraints. The function of the second part of the sample is to show evidence to the unsatisfied constraints.

The incidence-lists representation requires a different set of techniques, more applicable to sparse graphs. Specifically, because the graphs have few edges, a small random sample of vertices typically has no edges internal to the sample, that is, it is just the empty graph. Thus, algorithms for this setting use additional techniques besides pure random sampling. In particular, some algorithms apply various forms of exhaustive local search [GR97, PR99a] (such as performing a breadth-first-search until a particular number of vertices are observed). Other algorithms use random walks starting from randomly selected vertices [GR99].

Organization

The rest of the paper consists of three sections. The first (and main) two sections deal with testing graph properties, and the last with other properties. In particular, the first section is dedicated to the adjacency-matrix model. We provide a summary of results in this model, and present in detail the algorithm for testing bipartiteness and its analysis. We also sketch the ideas for testing whether a graph has a clique of a given size, and give some further details for a few other results. The second section is dedicated to testing in the incidence-lists model. Here too we give a summary of the results in this model, and provide more details for two properties: k-connectivity and bipartiteness. The last section gives a summary of other property-testing results that do not deal with graphs.

2. TESTING GRAPH PROPERTIES IN THE ADJACENCY MATRIX MODEL

2.1 DEFINITIONS

We consider undirected, simple graphs (no multiple edges or self-loops). For a graph G, we denote by $V(G)$ its vertex set and by $E(G)$ its edge set (whenever it is clear from the context, we shall simply use

V and E). The size of V(G) is denoted by N. We assume, without loss of generality, that $V(G) = \{1, ..., N\}$. Graphs are represented by their (symmetric) adjacency matrix. Thus, graphs are associated with the (symmetric) boolean function f_G corresponding to this matrix. That is, $f_G(u, v) = 1$ if $(u, v) \in E(G)$, and $f_G(u, v) = 0$ otherwise. For two (not necessarily disjoint) sets of vertices, X_1 and X_2, we let $E(X_1, X_2) \overset{\text{def}}{=} \{(u, v) \in E(G) : u \in X_1, v \in X_2\}$.

The distance between two N-vertex graphs G_1 and G_2 is defined as the number of unordered pairs $(u, v) \in [N]^2$ such that $f_{G_1}(u, v) \neq f_{G_2}(u, v)$, divided by the total number of pairs,[5] N^2.

Definition 2.1.1 *For any graph property $\mathcal{P}$, and $0 \leq \epsilon \leq 1$, we say that a graph G is ϵ-far from (having) property $\mathcal{P}$, if it has distance greater than ϵ from every graph that has the property. Otherwise it is ϵ-close.*

Definition 2.1.2 *A property testing algorithm for property $\mathcal{P}$ working in the adjacency matrix model is given a distance parameter ϵ and can perform queries concerning the existence of edges between any pair of vertices of its choice. If the tested graph has the property, the algorithm should accept with probability at least 2/3, and if it is ϵ-far from having the property then the algorithm should reject with probability at least 2/3.*

2.2 SUMMARY OF RESULTS

The following graph properties were studied in [GGR98] and were shown to have testing algorithms with query complexity poly($1/\epsilon$) and time complexity at most exp(poly($1/\epsilon$)). In what follows, N denotes the number of graph vertices.

- **Bipartiteness.** The algorithm has query complexity and running time $\tilde{O}(\epsilon^{-3})$.[6] Recently, Alon and Krivelevich [AK99] improved the analysis of the algorithm and obtained a bound of $\tilde{O}(\epsilon^{-2})$ on the query complexity and running time.

- **k-colorability, $k \geq 3$.** The algorithm has query complexity $\tilde{O}\left(k^4/\epsilon^6\right)$ and running time $\exp\left(\tilde{O}\left(k^2/\epsilon^3\right)\right)$. Recently, Alon and Krivelevich [AK99] improved the analysis of the algorithm and obtained

---

[5]In [GGR98] (and in the introduction) the distance was defined as the number of such *ordered* pairs (entries in the matrix) divided by N^2. While this seems more appropriate as N^2 is the number of ordered pairs, it implies that each undirected edge in the symmetric difference between the graphs is counted twice, causing slight cumbersomeness in the analysis of the algorithms. Thus, we have chosen here a less natural definition that makes the analysis later simpler.

[6]The $\tilde{O}(\cdot)$ notation, which is used for sake of succinctness, "hides" logarithmic factors (which in all our algorithms are at most quadratic).

a bound of $\tilde{O}\left(k^2/\epsilon^4\right)$ on the query complexity, and $\exp\left(\tilde{O}\left(k/\epsilon^2\right)\right)$ on the running time.

- ρ-Clique. The property is having a clique of size $\rho \cdot N$, where $0 < \rho < 1$ is a constant. The query complexity of the algorithm is $\tilde{O}\left(\rho^2/\epsilon^6\right)$ and the running time is $\exp\left(\tilde{O}\left(\rho/\epsilon^2\right)\right)$.

- ρ-Cut. The property is having a 2-way cut with ρN^2 crossing edges. The query complexity of the algorithm is $\tilde{O}\left(\epsilon^{-7}\right)$ and the running time is $\exp\left(\tilde{O}\left(\epsilon^{-3}\right)\right)$. The algorithm generalizes to k-way cuts, at a multiplicative cost of $O(\log^2(k))$ in the query complexity and in the exponent of the running time. The algorithm can also be modified to test ρ-Bisection. This property is similar to ρ-Cut except that the partition is to equal size subsets. The query complexity is $\tilde{O}\left(\epsilon^{-8}\right)$ and the running time is $\exp\left(\tilde{O}\left(\epsilon^{-3}\right)\right)$.

NOTES

1. For all the above properties (except bipartiteness) it is very unlikely that there is a testing algorithm having running time $\text{poly}(1/\epsilon)$. If such an algorithm exists, by setting $\epsilon = 1/N$ one would be able to obtain an exact (randomized) decision procedure that runs in polynomial time, and this would imply that $\mathcal{NP} \subseteq \mathcal{BPP}$.

2. The bipartiteness and k-colorability algorithms have one sided error: they always accept graphs that have the property. Furthermore, whenever a graph is rejected, the algorithm supplies *evidence* that it does not have the property. Evidence is in the form of a small subgraph that is not bipartite/k-colorable. All other algorithms have two-sided error and this can be shown to be unavoidable within $o(N)$ query-complexity.

A testing algorithm for k-Colorability whose complexity is independent of N was already implicit in work of Alon *et. al.* [ADL$^+$94]. They build on a constructive version of the Regularity Lemma of Szemerédi [Sze78] which they prove, and the complexity of the resulting testing algorithm is a tower of $\text{poly}(1/\epsilon)$ exponents.

Constructing (Good) Partitions. For all the above properties, in case the graph has the desired property, the testing algorithm outputs some auxiliary information which allows to construct, in $\text{poly}(1/\epsilon) \cdot N$ time, a partition that approximately obeys the property. For example, for ρ-Clique, the algorithm will find a subset of vertices of size ρN, such that at most ϵN^2 edges need to be added so that it becomes a clique. In the case of ρ-Cut, the algorithm will construct a partition with at least $(\rho - \epsilon)N^2$ crossing edges. The basic idea is that the partition of the sample that caused the algorithm to accept is used to partition the whole graph. This idea was later also used by Frieze and Kannan [FK99] to obtain polynomial time approximation schemes for various problems, where they apply a relaxed constructive version of Szemerédi's Regularity Lemma [Sze78].

General Graph Partition Properties.   All the above properties are special cases of a class of graph partition properties. Each property in the class is parameterized by an integer k and by $k + k^2$ pairs of lower and upper bounds in the interval $[0, 1]$. A graph has the property if its vertices can be partitioned into k subsets having relative sizes within the designated upper and lower bounds, and such that the edge densities between the parts are within the required bounds as well. A more precise definition is given in Subsection 2.5.

Not surprisingly, generality has a price, and the testing algorithm for the above class of properties, presented in [GGR98], has query complexity $\left(\tilde{O}(k^2)/\epsilon\right)^{2k+8}$ and running time $\exp\left(\tilde{O}(k^2)/\epsilon\right)^{k+1}$.

First Order Graph Properties.   Alon, Fischer, Krivelevich and Szegedy [AFKS99], study the class of *first order* graph properties. These are properties that can be formulated by first order expressions about graphs. That is, expressions that contain quantifiers over vertices, boolean connectives, equality of vertices, and adjacency relations. They show that all first order graph expressions containing at most one quantifier, as well as all first order graph expression of the type "$\exists\forall$" (i.e., $\exists x_1, \ldots, x_t \, \forall y_1, \ldots, y_s \, A(x_1, \ldots, x_t, y_1, \ldots, y_s)$ where A is a quantifier-free first order graph expression and t and s are constants), can be tested with query complexity and running time independent of N. Once again generality has a (steep) price: the dependence on the distance parameter ϵ is either a tower of $\text{poly}(1/\epsilon)$ exponents or a tower of towers of $\text{poly}(1/\epsilon)$ exponents. They also prove that there exist first order graph expressions of the type "$\forall\exists$" that cannot be tested with query complexity and running time independent of N. In particular this is true of a natural property based on graph isomorphism, where the required number of queries is $\Omega(\sqrt{N})$.[7]

We give some more details in Subsection 2.6

Properties of Directed Graphs. The adjacency-matrix model can be naturally extended to deal with directed graphs. The algorithm may perform queries of the form: "is there an edge *from* vertex u *to* vertex u", and distance between graphs is defined as the number of *ordered* pairs that are an edge in one graph and not in the other, divided by N^2.

Some properties of undirected graphs have analogies in directed graphs. Furthermore, in some cases the testing algorithms for undirected graphs can be extended to directed graphs. This is true for example in the case of ρ-Cut (see [GGR98, Sec. 10.1]). However, in other cases, the testing problems are quite different. This is true for example for the cycle-freeness property. In order for an undirected graph to be cycle-free it must be very sparse. Namely, it may contain at most $N - 1$ edges. Hence, in order to test cycle-freeness of undirected graphs (in the adjacency-matrix model), all that is required is to roughly estimate the number of edges in the graph by sampling. This is not true of directed graphs. Namely, a directed graph may be very dense but still acyclic. Hence, for this property a non-trivial analysis is needed. In [BR00] it is shown that the "natural" algorithm that takes a sample of vertices and accepts or rejects based on the acyclicity of the induced subgraphs, is a testing algorithm. The required sample size is $O(\log(1/\epsilon)/\epsilon)$. The special case in which there is a directed edge (in some direction) between *every* pair of vertices, was treated in [EKK+98] when dealing with total orders. They present an algorithm having complexity $\text{poly}(1/\epsilon)$.

2.3 TESTING BIPARTITENESS

Recall that a graph $G = (V, E)$ is *bipartite* if its set of vertices V can be partitioned into two (disjoint) subsets V_1 and V_2 so that there are no edges between vertices that belong to the same subset. If there is no such partition, then the graph is not bipartite. Deciding whether a graph is bipartite or not can be done in time linear in the number of graph edges by performing a Breadth First Search (BFS) starting from an arbitrary vertex.

What does it mean that a graph is ϵ-far from bipartite? For any two-way partition (V_1, V_2) of V, we say that an edge $(u, v) \in E$ is a *violating*

---

[7]Previous hardness results in [GGR98] only showed the *existence* of such hard-to-test properties (in $\mathcal{NP}$).

edge with respect to (V_1, V_2), if either $u, v \in V_1$ or $u, v \in V_2$. We say that (V_1, V_2) is ϵ-*bad* if the number of violating edges with respect to (V_1, V_2), is greater than ϵN^2, otherwise it is ϵ-good. Given the above definition, G is ϵ-far from bipartite if and only if *every* partition (V_1, V_2) of V is ϵ-bad.

Suppose we fix a *particular* partition (V_1, V_2). If we now take a sample of size $\Theta(1/\epsilon)$ of vertices (or pairs of vertices) and obtain the edges between them, then with high probability we shall see evidence to the badness of the partition in the form of a violating edge. Since in case G is ϵ-far from bipartite, all its partitions are ϵ-bad, the naive use of the above observation is to take a sufficiently large sample of vertices such that with high probability for *every* partition there exists an edge between vertices in the sample that violates the partition. Such a sample induces a subgraph that is necessarily not bipartite, and so the algorithm could simply run a BFS on the sample to detect this. Unfortunately, the number of all two-way partitions of V is exponential in N. Therefore, a straightforward analysis of the above algorithm (which applies a probability union bound) would require that the size of the sample be logarithmic in the number of partitions, that is, linear in N.

Nonetheless, as we shall see below, the naive algorithm that simply takes a uniformly selected sample of vertices and checks whether the induced subgraph is bipartite does work, but requires a slightly more refined analysis.

Test-Bipartite

1. Uniformly and independently select $m = \Theta\left(\frac{\log(1/\epsilon)}{\epsilon^2}\right)$ vertices.

2. For every pair of vertices v and u selected, query whether there is an edge between v and u, thus obtaining the induced subgraph.

3. Perform a (BFS) to determine whether the subgraph induced by the sample is bipartite. If it is bipartite output **accept** otherwise output **reject**.

Theorem 1 *The algorithm* Test-Bipartite *is a testing algorithm for bipartiteness. In particular, if the graph is bipartite, then it is always accepted, and if it is ϵ-far from bipartite, then it is rejected with probability at least 2/3. Furthermore, whenever the algorithm rejects a graph, it outputs a certificate to the non-bipartiteness of the graph in form of a non-bipartite subgraph having $O\left(\frac{\log(1/\epsilon)}{\epsilon^2}\right)$ vertices.*

Proof: The first claim in the theorem, concerning bipartite graphs is obvious: if G is bipartite then *every* subgraph of G is bipartite, and so the graph is accepted for any choice of the sample. The heart of the proof is hence in the second claim, that is, showing that if a graph is ϵ-far from bipartite then with probability at least 2/3 over the choice of the sampled vertices, the resulting induced subgraph is not bipartite. We thus focus on the second part of the theorem, and assume from now on that G is ϵ-far from bipartite. Recall that this implies that for every two-way partition of V there are at least ϵN^2 violating edges. The rough outline of the proof is the following:

1. We shall view the sample of vertices as consisting of two parts, which we refer to as U and S. The set U consists of the first $m_1 = \Theta((\log(1/\epsilon))/\epsilon)$ vertices selected, and the set S of the latter $m_2 = \Theta((\log(1/\epsilon))/\epsilon^2)$ vertices. (Since the vertices are selected independently, repetitions may occur.)

2. We show that with probability at least 5/6 over the choice of U, it can be used to implicitly induce a relatively small number of partitions of the whole graph, that are in a way *consistent with* U (this notion will be clarified later).

3. We then show that with probability at least 5/6 over the choice of the second part of the sample, S, it will contain violating edges with respect to each of the partitions implicitly induced by U.

4. Putting the above two items together we conclude that with probability at least 2/3 over the choice of both parts of the sample, the induced subgraph is not bipartite.

We start with some definitions.

Definition 2.3.1 *A vertex v is* influential *if its degree in G is at least $\frac{\epsilon}{4}N$. Otherwise it is not influential.*

If a vertex v is not influential, then by definition, for every partition (V_1, V_2) of V, the number of edges incident to v that are violating with respect to (V_1, V_2) is at most $(\epsilon/4)N$. Furthermore, since there are at most N non-influential vertices, the total number of edges incident to non-influential vertices that violate some partition, is at most $(\epsilon/4)N^2$. Intuitively, the means that we shall "not rely" on non-influential vertices for giving us evidence to the fact that G is far from bipartite.

Definition 2.3.2 *For any vertex v and set of vertices U, we say that U* covers v *if v has at least one neighbor in U.*

We are now ready for our first lemma concerning the first part of the sample U.

Lemma 2.3.1 *With probability at least 5/6 over the choice of the vertices in U, all but at most $(\epsilon/4)N$ influential vertices are covered by U.*

Proof: Consider any fixed influential vertex v. Recall that the vertices in U are selected uniformally and independently. Hence, the probability that U does not cover v, that is, that U contains none of the at least $(\epsilon/4)N$ neighbors of v is at most

$$\left(1 - \frac{\epsilon}{4}\right)^{m_1} < \exp\left(\frac{\epsilon}{4} \cdot m_1\right)$$

If we set $m_1 = \frac{4}{\epsilon} \cdot \ln(24/\epsilon)$ then the above probability is at most $(\epsilon/24)$. Since there are at most N influential vertices, this implies that the expected number of influential vertices that are not covered by U is at most $(\epsilon/24)N$. By Markov's inequality, the probability that there are more than $(\epsilon/4)N$ such influential vertices, is at most $1/6$, as required. ∎

We assume from now on that U in fact covers all but at most $(\epsilon/4)N$ of the influential vertices. Let C be the set of vertices in V that are covered by U and let R be the remaining vertices. (The sets C and R also contain the vertices of U, but the size of U should be thought of as negligible compared to $|C \cup R| = N$.) By our assumption, R contains at most $(\epsilon/4)N$ influential vertices, and possibly all non-influential vertices. Consider a fixed partition (U_1, U_2) of U. Then (U_1, U_2) can be used to induce a partition (C_1, C_2) of C as follows: every vertex in C that has a neighbor in U_1, belongs to C_2, and all other vertices in C (that necessarily have a neighbor in U_2), belong to C_1. Let (R_1, R_2) be an arbitrary partition of R (with the only restriction that $U_1 \cap R \subseteq R_1$ and $U_2 \cap R \subseteq R_2$), and consider the partition $(C_1 \cup R_1, C_2 \cup R_1)$ of V.

Since all partitions of V are ϵ-bad, this is in particular true of the partition $(C_1 \cup R_1, C_2 \cup R_1)$. Where do the at least ϵN^2 edges reside? Since R contains at most $(\epsilon/4)N$ influential vertices, each incident to at most N edges, and at most N non-influential vertices, each incident to at most $(\epsilon/4)N$ edges, the total number of edges incident to vertices in R is at most $(\epsilon/4)N \cdot N + N \cdot (\epsilon/4)N = (\epsilon/2)N^2$. Thus, the number of violating edges with respect to $(C_1 \cup R_1, C_2 \cup R_1)$ that are incident to vertices in R is at most $(\epsilon/2)N^2$, and this is true for *every* possible partition (R_1, R_2). This implies that (no matter how R is partitioned) there must be at least $(\epsilon/2)N^2$ violating edges that are incident only to vertices within C_1 or within C_2. As we show in the next lemma, if we now take an additional (sufficiently large) sample (the sample S), then with high probability it will contain a pair of vertices connected by a

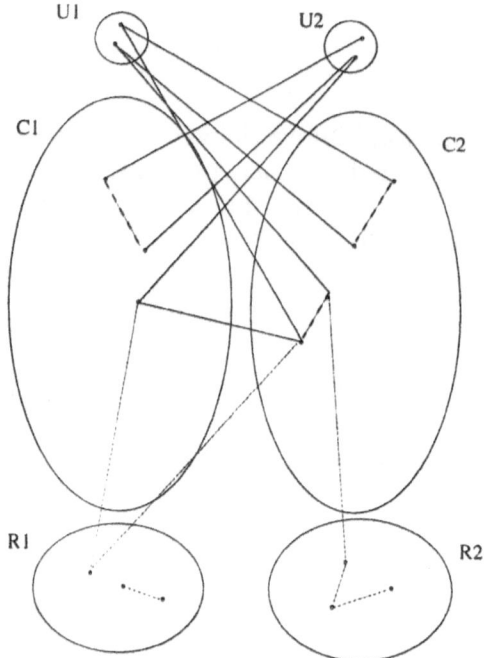

Figure 15.1 An illustration of the partition induced by (U_1, U_2). The vertices in C_2 each have at least one neighbor in U_1, and the vertices in C_1 at least one neighbor in U_2. The vertices in R have no neighbor in either U_1 or U_2, and are partitioned arbitrarily. The edges incident to R (of which there are at most $(\epsilon/2)N^2$, both violating and not violating), are dotted, and the violating edges residing in either C_1 or C_2 (of which there are at least $(\epsilon/2)N^2$), are dashed.

violating edge (with respect to (C_1, C_2)). This is then shown to imply that for every partition (S_1, S_2) of S, there is some edge between the sample vertices that is violating with respect to $(U_1 \cup S_1, U_2 \cup S_2)$. For an illustration of the partition and the violating edges, see Figure 15.1.

Lemma 2.3.2 *Let* $G = (V, E)$ *be a graph that is ϵ-far from bipartite, U a subset of V that covers all but at most $(\epsilon/4)N$ of the influential vertices in G, and (U_1, U_2) a fixed partition of U. Let S be a uniformly and independently selected sample of $m_2 = \Theta(|U|/\epsilon)$ vertices. Then, with probability at least $1 - 2^{-|U|}/6$ over the choice of S, for every partition (S_1, S_2) of S, there is some edge between vertices in $U \cup S$ that is violating with respect to $(U_1 \cup S_1, U_2 \cup S_2)$.*

Proof: It will be convenient to view S as $m_2/2$ pairs of vertices. By the discussion preceding the lemma, for every such pair (v, w), the probability that (v, w) constitutes a violating edge with respect to (C_1, C_2)

is at least $\epsilon/2$. The probability that among the $m_2/2$ pairs, there is no violating edge, is at most $(1 - (\epsilon/2))^{m_2/2}$ which for $m_2 = (16|U|/\epsilon)$ is less than $2^{-|U|}/6$. To complete the claim we need to show that if S contains such a pair, then it is not possible to partition S into (S_1, S_2) so that $(U_1 \cup S_1, U_2 \cup S_2)$ has no violating edges.

Consider an edge (v, w) such that $v, w \in S$ that violates (C_1, C_2), and without loss of generality, assume $v, w \in C_2$. If we put both vertices either in S_1 or in S_2 then (v, w) is violating with respect to (S_1, S_2). However, since v and w belong to C_2, by our definition of the partition (C_1, C_2), v has some neighbor $u \in U_1$ and w has some neighbor $u \in U_1$. Therefore, if we put $v \in S_1$ and $w \in S_2$, then the edge (u, v) will be violating, and if we put $w \in S_1$ and $v \in S_2$, then the edge (u', w) will be violating. The lemma follows. ∎

Combining Lemma 2.3.2 with the fact that there are $2^{|U|}$ partitions of U, it follows that with probability at least 5/6 over the choice of S, for *every* partition (U_1, U_2) of U, and for every partition (S_1, S_2) of S, the sample contains edges that violate $(U_1 \cup S_1, U_2 \cup S_2)$. In other words, the sample $U \cup S$ cannot be partitioned without violations. Combining this with Lemma 2.3.1, the theorem follows. ∎ (Theorem 1)

The Query and Time Complexities. As described, the algorithm Test-Bipartite has query and time complexities that are quadratic in the size of the sample. That is, $\Theta(\log^2(1/\epsilon)/\epsilon^4)$. However, given the analysis, we can slightly improve on this bound. Assume the algorithm actually partition the sample into two parts U and S of sizes $m_1 = \Theta(\log(1/\epsilon)/\epsilon)$ and $m_2 = \Theta(\log(1/\epsilon)/\epsilon^2)$, respectively. It views S as consisting of $m_2/2$ pairs of vertices, and queries whether an edge exists only between all pairs of vertices in $U \times S$, and between the $m_2/2$ pairs in S. It then checks whether the resulting subgraph is bipartite. Then by the analysis this suffices to obtain the desired success probability while decreasing the complexities to $\Theta(\log^2(1/\epsilon)/\epsilon^3)$.

2.4   TESTING ρ-CLIQUE

In this subsection we give the basic underlying ideas of a testing algorithm for the ρ-Clique property. A graph is said to have the property if it contains a clique of size $\rho \cdot N$, for a given constant $0 \leq \rho \leq 1$. Hence, by definition, a graph is ϵ-far from having a ρ clique, if for *every* subset X of the vertices of size ρN, the number of pairs of vertices from X that do not have an edge between them is more than ϵN^2.

While it can be shown that the "natural" algorithm, which simply takes a sample S of vertices and accepts if an only if S contains a clique of size slightly smaller than $\rho|S|$, will work, there is no direct proof for its correctness. Rather, its correctness follows from that of another somewhat "unnatural" algorithm, which we discuss below. In the case of the natural algorithm, it is very easy to show that a graph having a clique of size ρN will be accepted with high probability, and the difficulty is in proving that a graph which is ϵ-far from having the property is rejected with probability at least $2/3$. As we shall see, for the unnatural algorithm it is relatively easy to show that a graph which is ϵ-far from having the property is rejected with probability at least $2/3$, but we need to work harder to show that a graph having the property is accepted with probability at least $2/3$.

We start by showing how, given a graph that *has a clique* of size ρN it is possible to find a subset of vertices having size ρN that is an *approximate* clique. Namely, the number of pairs of vertices in the set that do not have an edge between them is at most ϵN^2.

An Oracle Aided Procedure. Consider first the following mental experiment. Let C be a clique of size ρN in G, and suppose we have access to an oracle that provides us with the number of neighbors that any given vertex $v \in V$ has in C. We would like to use this information in order to construct a clique C' of size ρN (which may differ from C). Let $d_C(v)$ denote the number of neighbors that vertex v has in C. By definition of a clique, for every $v \in C$, $d_C(v) = \rho N - 1$. However, there may be other vertices, outside of C, for which the same is true. Consider all vertices v for which $d_C(v) \geq \rho N - 1$, and let their set be denoted by $T(C)$. Assume we order these vertices according to the number of neighbors they have in $T(C)$ (i.e., according to the degree they have in the subgraph induced by $T(C)$), and let C' be the first ρN vertices according to this order (breaking ties arbitrarily). Then we claim that C' is a clique.

To see this, observe that by definition of $T(C)$, each vertex in C neighbors every vertex in $T(C)$ (except itself). Thus, each vertex in C has degree $|T(C)| - 1$ in the subgraph induced by $T(C)$, which is the maximum possible. Since $|C| = \rho N$, every vertex in C' must have degree $|T(C)| - 1$ as well (because the vertices in C are all candidates for the set C' whose size is ρN as well). In other words, every vertex in C' neighbors every (other) vertex in $T(C)$, and in particular it neighbors every other vertex in C', making C' a clique.

Suppose next that instead of having an oracle as described above, we were provided with a uniformly chosen set U' in C of sufficient size (i.e., of size $\Theta(\rho^2 \cdot \log(1/\epsilon))/\epsilon^2)$). Let $T(U')$ be the set of vertices that neighbor every vertex in U'. Then, with high probability over the choice of U', almost every vertex in T(U') neighbors almost all vertices in C (where "almost" is all but a fraction of ϵ^2). Similarly to the above, we could order the vertices in T(U') according to their degree in the graph induced by T(U'), and take the first ρN vertices, whose set we denote by F(T(U')). By extending the argument given above for the oracle-aided procedure, it can be shown that F(T(U') is close to being a clique.

The Actual Approximate-Clique Finding Algorithm. Since a uniformly chosen sample in C is not provided to the algorithm, it instead "guesses" such a set. More precisely, it uniformly selects a set U from all graph vertices, and it considers all its subsets U' of size $\frac{\rho}{2}|U|$. Since with high probability $|U \cap C| \geq \frac{\rho}{2}|U|$, there exists a subset U' contained in C. Since U is selected uniformly, the set U' is uniformly distributed in in C. Hence, with high probability, for this U', almost all vertices in T(U') (the set of vertices that neighbor every vertex in U'), neighbor almost every vertex in C. If we now order the vertices in T(U') according to their degree in the subgraph they induce and select the first ρN vertices according to this order, then the resulting set F(T(U')) is close to being a clique (with high probability over the choice of U). For an illustration, see Figure 15.2.

Thus, the algorithm considers all subsets U' of U having size $\frac{\rho}{2}|U|$. For each subset U' it finds the vertices in T(U') and among them those in F(T(U')). It selects the set F(T(U')) that is closest to being a clique. By the above discussion, if the graph has a clique of size ρN, then with high probability the selected set will in fact be close to being a clique. On the other hand, if the graph is ϵ-far from having a ρ clique, then, by definition, none of these sets will be close to a clique.

The Testing Algorithm. The testing algorithm is based on the above approximate-clique finding algorithm. Similarly to that algorithm, it uniformly selects a set U and considers each of its subsets U' having size $\frac{\rho}{2}|U|$. The algorithm then: Samples from T(U'); Approximates the number of neighbors that each sampled vertex has in T(U'); Orders the sampled vertices according to this approximate degree; Takes the first ρ fraction according to this order; And checks how close they are to being a clique.

More precisely, the algorithm performs the following steps:

1. Uniformly and independently select three samples, U, S, and W, where the first two are of size $\Theta(\rho \cdot \log(1/\epsilon))/\epsilon^2))$, and the third of size $\Theta(1/\epsilon^3)$.

2. For each subset U' of U having size $\frac{\rho}{2}|U|$, do:

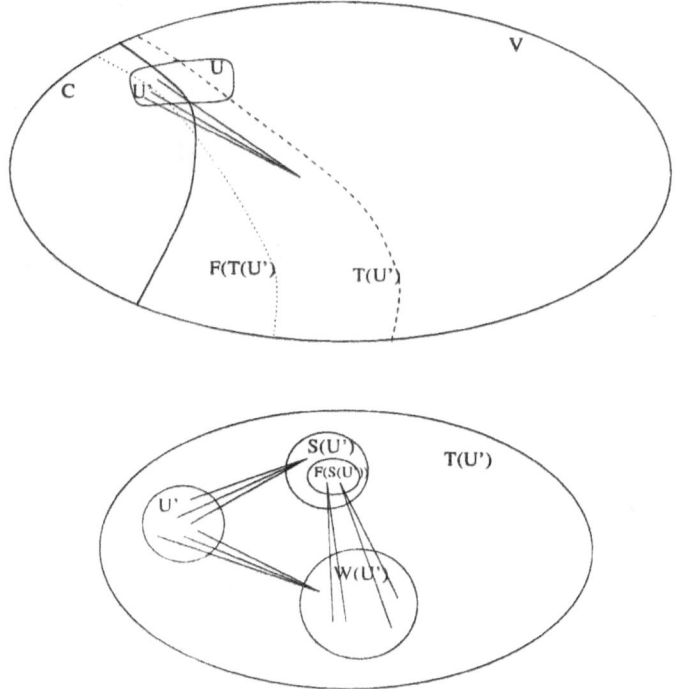

Figure 15.2 An illustration of the approximate-clique finding algorithm (top figure), and of the testing algorithm (bottom figure). Top figure: the clique C is separated from the rest of the graph vertices by a bold line. The uniformally selected sample U intersects with C and the intersection is denoted by U' (more precisely, U' is a subset of this intersection having size $(\rho/2)|U|$ but for sake of brevity we assume it coincides with the intersection). The set T(U'), denoted by a dashed separating line, is the set of all vertices that neighbor every vertex in U' (it is a superset of C). Finally, F(T(U')) are the first ρN vertices with highest degree in the subgraph induced by T(U'). Bottom figure: U', and T(U') are as in the top figure. S(U') and W(U) are the subsets of vertices in the samples S and W, respectively, that intersect T(U'). (S and W are not illustrated). Finally, F(S(U')) are the first $\rho|S|$ vertices in S(U') with the largest number of neighbors in W(U').

(a) Let $S(U') \subseteq S$ be the set of vertices in S that neighbor all vertices in U'. Note that since S is uniformly distributed in V, this set is uniformly distributed in T(U').

(b) Let $W(U') \subseteq W$ be the set of vertices in W that neighbor all vertices in U'. The vertices in W(U') are also uniformly distributed in T(U'), and they will serve to approximated the degree of the vertices in S(U').

(c) For each vertex $v \in S$, let $\hat{d}(v)$ be the number of neighbors v has in W(U').

(d) Order the vertices in $S(U')$ according to $\hat{d}(\cdot)$. Let $F(S(U'))$ be the first $\rho|S|$ vertices according to this order. If $S(U') < \rho|S|$, then $F(S(U')) = S(U')$.

(e) View the sample S as consisting of pairs $\{s_{2i-1}, s_{2i}\}_{i=1}^{m}$ (where the pairing is predetermined), and let $mis(U')$ be the number of pairs $(s_{2i-1}, s_{2i}) \in F(S(U')) \times F(S(U'))$ that *do not* have an edge between them.

If for any one of the subsets U', the set $F(S(U'))$ is of size at least $(\rho - \epsilon/80)|S|$, and $mis(U') \leq \frac{2\epsilon}{3}m$, then **accept**, otherwise it **reject**.

For an illustration, see Figure 15.2.

If the graph is ϵ-far from having a ρ-clique then for *every* U and $U' \subset U$, the set $F(T(U'))$ is far from being a clique. On the other hand, if the graph has a ρ-clique, then as we claimed above, with high probability over the choice of U, for *some* subset U', the set $F(T(U'))$ is is close to being a clique. A probabilistic argument shows that with high probability over the choices of S and W, in the former case, for every U' there are many missing edges between pairs of vertices in $F(S(U'))$, while in the latter case, for some U', there are few missing edges. The above constitutes a rough sketch of the following theorem.

Theorem 2 *The algorithm described above is a testing algorithm for the ρ-Clique property: If the graph has a clique of size ρN, then it is accepted with probability at least $2/3$, and if it is ϵ-far from having such a clique, then it is rejected with probability at least $2/3$.*

2.5 A GENERAL CLASS OF PARTITION PROPERTIES

The following framework of a general partition problem captures any graph property which requires the existence of partitions satisfying certain fixed density constraints. These constraints may refer both to the number of vertices in each component of the partition and to the number of edges between each pair of components.

Let $\Phi \stackrel{\text{def}}{=} \left\{ \rho_j^{\text{LB}}, \rho_j^{\text{UB}} \right\}_{j=1}^{k} \cup \left\{ \varrho_{j,j'}^{\text{LB}}, \varrho_{j,j'}^{\text{UB}} \right\}_{j,j'=1}^{k}$ be a set of non-negative parameters so that $\rho_j^{\text{LB}} \leq \rho_j^{\text{UB}}$ $(\forall j)$ and $\varrho_{j,j'}^{\text{LB}} \leq \varrho_{j,j'}^{\text{UB}}$ $(\forall j, j')$. (LB stands for Lower Bound, and UB stands for Upper Bound.) Let $\mathcal{GP}_\Phi$ be the class of graphs which have a k-way partition $(V_1, \ldots, V_k)$ with the following conditions being satisfied.

$$\forall j \quad \rho_j^{\text{LB}} \cdot N \leq |V_j| \leq \rho_j^{\text{UB}} \cdot N \qquad (15.1)$$

and

$$\forall j, j' \qquad \varrho_{j,j'}^{\text{LB}} \cdot N^2 \leq |\text{E}(\text{V}_j, \text{V}_{j'})| \leq \varrho_{j,j'}^{\text{UB}} \cdot N^2 \qquad (15.2)$$

where recall that $\text{E}(\text{V}_j, \text{V}_{j'})$ is the set of edges between vertices in V_j and vertices in $\text{V}_{j'}$. That is, Equation (15.1) places lower and upper bounds on the relative sizes of the various components of the partition; whereas Equation (15.2) imposes lower and upper bounds on the density of edges among the various pairs of components.

Below we show how the various properties mentioned in Subsection 2.2 are defined as special cases.

1. **Bipartiteness.** We set $k = 2$ (as we are interested in a two-way partition), $\rho_1^{\text{LB}} = \rho_2^{\text{LB}} = 0$, and $\rho_1^{\text{UB}} = \rho_2^{\text{UB}} = 1$ (as there are no restrictions on the sizes of the partition subsets), $\varrho_{1,1}^{\text{LB}} = \varrho_{2,2}^{\text{LB}} = \varrho_{1,1}^{\text{UB}} = \varrho_{2,2}^{\text{UB}} = 0$, enforcing the main constraint that there be no edges within the partition subsets, and finally $\varrho_{1,2}^{\text{LB}} = 0$, $\varrho_{1,2}^{\text{UB}} = 1$, since the number of edges between the two subsets is not restricted.

2. **k-colorability, $k \geq 3$.** This is a generalization of bipartiteness, and the important density bounds are $\varrho_{j,j}^{\text{UB}} = 0$ for every $1 \leq j \leq k$. All lower bounds are 0 and all other upper bounds are 1.

3. **ρ-Clique.** Here $k = 1$, $\rho_1^{\text{LB}} = \rho_1^{\text{UB}} = \rho$, enforcing the restriction that one subset should be of size ρN, and $\varrho_1^{\text{LB}} = \frac{1}{2}(\rho^2 - \rho/N)$ enforcing the restriction that the subset be a clique.

4. **ρ-Cut.** Here $k = 2$, and $\varrho_{1,2}^{\text{LB}} = \rho$. For the case of ρ-bisection we add the constraints that $\rho_j^{\text{LB}} = \rho_j^{\text{UB}} = 1/2$ for $j \in \{1, 2\}$.

The testing algorithm for the class of partition properties and its analysis are somewhat complex and we refer the reader interested in further details to [GGR98].

2.6 FIRST ORDER GRAPH PROPERTIES

Let $A(x_1, \ldots, x_t, y_1, \ldots, y_s)$ be a quantifier free graph expression. That is, it contains equality of vertices, adjacency relations between vertices, and boolean connectives. We say that an expression is of the type "$\exists\forall$" if it has the form $\exists x_1, \ldots, x_t \forall y_1, \ldots, y_s A(x_1, \ldots, x_t, y_1, \ldots, y_s)$. An expression of the type "$\forall\exists$" is defined analogously. Alon, Fischer, Krivelevich and Szegedy [AFKS99] show that it is possible to test any graph property that can be described as an "$\exists\forall$" statement with query complexity and running time independent of the size of the graph. The bound on the number of queries is a tower of towers of $\text{poly}(1/\epsilon, t + s)$ exponents. They also show that there exists a (natural) property concerning isomorphism between (sub)graphs, which can be expressed as a

"∀∃" expression, for which the required query complexity (for constant ϵ) is $\Omega(\sqrt{N})$.

To prove the first (main) result, Alon *et. al.* introduce the following notion of *indistinguishability* between graph properties, which may be useful in general.

Definition 2.6.1 *Two graph properties P and P′ are said to be* indistinguishable *if for every $\epsilon > 0$ there exists an integer N_ϵ for which the following holds. For every graph G with $N > N_\epsilon$ vertices such that G has property P, there exists an N-vertex graph G′ having property P′ such that the distance between G and G′ is at most ϵ. Similarly, for every graph H′ with $N > N_\epsilon$ vertices such that H′ has property P′, there exists an N-vertex graph H having property P such that the distance between H′ and H is at most ϵ.*

For example, the property of being k-colorable is indistinguishable for the first order property that there exist k vertices $v_1, \ldots, v_k$ such that every other vertex is adjacent to exactly one of $v_1, \ldots, v_k$, but no two vertices that have an edge between them are adjacent to the same v_i. To verify this, consider first the case in which the first order expression holds for G. Let G′ be the graph in which all edges between the v_i's and the rest of the graph are removed from G. The distance between G and G′ is less than $\frac{k}{N}$ which is less than any constant ϵ provided N is sufficiently large. Then we can color each v_i and the vertices it is adjacent to in G with the i'th color, and obtain a k-coloring in G′. In the other direction, assume H is k-colorable. Then to obtain H′ from H we can pick k vertices $v_1, \ldots, v_k$, each colored by a different color, and add edges between each v_i and the other vertices having the same color. The first order expression holds for H′ and the distance between H and H′ is smaller than ϵ.

Given the above definition they then show:

Lemma 2.6.1 *If P and P′ are indistinguishable graph properties, then P is testable using a number of queries independent of N if and only if P′ is testable using a number of queries independent of N. Specifically, if the number of queries used in testing one property is $Q(\epsilon)$, then the other property can be tested using at most $3Q(\epsilon/2)$ queries.*

Next they define the following generalization of the notions of colorability and subgraph-freeness, and show that every "∃∀" first order property is indistinguishable from such a generalized property.

Definition 2.6.2 *Let $\mathcal{F}$ be a family of graphs (where repetitions are allowed), such that each graph in the family comes with a (not necessarily*

*proper) coloring by at most c colors. A coloring of a graph G by c colors
(which, again, need not be proper) is called an $\mathcal{F}$-coloring, if no member
of $\mathcal{F}$ appears as an induced subgraph of G with an identical coloring. A
graph G is called $\mathcal{F}$-colorable if there exists an $\mathcal{F}$-coloring of G.*

The notion of a (proper) k-coloring discussed above, is a special case of
this definition. Another, more complex property for example, is having
a coloring with 2 colors without any monochromatic triangle.

Lemma 2.6.2 *For every first order property P of the form:*

$$\exists x_1, \ldots, x_t \, \forall y_1, \ldots, y_s \, A(x_1, \ldots, x_t, y_1, \ldots, y_s)$$

*there exists a family $\mathcal{F}$ of $(2^{t+\binom{t}{2}} + 1)$-colored graphs, each with at most
$\max\{2, t+1, s\}$ vertices such that the property P is indistinguishable
from the property of being $\mathcal{F}$-colorable.*

Finally they show:

Theorem 3 *For every constant c and every family $\mathcal{F}$ of c-colorable
graphs, the property of being $\mathcal{F}$-colorable can be tested with query complexity and running time independent of the size of the graph.*

In order to prove Theorem 3 they prove their central technical lemma
which is a variant of Szemerédi's Regularity Lemma [Sze78].

3. TESTING GRAPH PROPERTIES IN THE INCIDENCE-LISTS MODELS

3.1 DEFINITIONS

As in the adjacency-matrix model, we use the notation G, V, E, and
N to denote the graph, its vertex set, its edge set, and the number of
graph vertices, respectively.

The Bounded-Degree Incidence-Lists Model. In this model, a
graph G whose vertices have degree at most d is represented by a two-
dimensional array of size $N \times d$ (which can be viewed as N *lists*), where
for each vertex v and integer $i \in \{1, \ldots, d\}$ the value of the corresponding
entry is the ith neighbor of v. If v has less than d neighbors then this
value may be 0 (where $0 \notin V$). Distance between graphs is defined as
the number of pairs of vertices that have an edge between them in one
graph but not in the other, divided by $d \cdot N$.[8] The notion of ϵ-far (ϵ-

---

[8]Once again, for sake of simplicity, this definition slightly differs from that discussed in the
introduction and in [GR97]. There, distance is defined as the fraction of entries in the $N \times d$

close) to having a property is defined analogously to the definition in the adjacency matrix model (Definition 2.1.1).

The Unbounded-Degree Incidence-Lists Model.   In this model a graph G is represented by N incidence lists of possibly varying lengths, where the length of each list (i.e., the degree of the vertex the list corresponds to) is provided at the head of the list. We denote by $d(v)$ the degree of vertex v. Distance between graphs is defined with respect to a given upper bound M on the number of graph edges. Namely, the distance between graph G_1 and G_2 with respect to the bound M is the number of (unordered) pairs of vertices that are an edge in one graph but not in the other, divided by M. When using this notion of distance to define the distance of a graph G to having a property, we assume G has at most M edges, but do not necessarily assume that the closest graph having the property has at most this many edges. Thus the unbounded-degree model is more general than the bounded degree model except in this technical aspect (as in the bounded-degree model the closest graph having the property must have at most dN edges and degree bound d as well).

In both models the testing algorithm can query, for each vertex v and every i (where $1 \leq i \leq d$ in the first model and $1 \leq i \leq d(v)$ in the second), what vertex is the i'th neighbor of v. The acceptance and rejection requirements of a testing algorithm are as in the adjacency-matrix model (Definition 2.1.2), with the appropriate notions of distance.

3.2   SUMMARY OF RESULTS

As noted previously in the introduction, intuitively, testing in the bounded-degree model is easier (not harder) than testing in the unbounded-degree model. In fact, the bound on the degree of every vertex can make testing strictly easier (as we see below in the case of cycle-freeness) and there is no known property for which it is actually harder. However, as mentioned in Subsection 3.1, this bound also has a price: the definition of distance between a graph and the property (class of graphs having the property) is more restricted. This sometimes raises the following difficulty when testing in the bounded-degree

---

matrix representation on which the two graphs differ. According to that definition, each (undirected) edge (v, u) in the symmetric difference between the graphs is counted twice - once as an entry $[v, i]$ and once as an entry $[u, j]$, while here we count each edge only once. (Distance as defined here is also insensitive to the ordering of the edges incident to a vertex, though this issue looses its relevance once we talk about the distance to having the property.)

model. Suppose the property is such that if a graph does not have the property then edges should be added so that it have the property (e.g. connectivity). Assume we would like to show that if a graph is accepted with probability greater than 1/3 then it is close to having the property (which is equivalent to showing that if it is far from having the property then it is rejected with probability at least 2/3). Then we would like to show that by adding a relatively small number of edges, the graph can be made to have the property. The difficulty is that in doing so we must maintain the degree bound of the graph, and this may sometimes be possible only by *removing* other edges. Hence, this technical difficulty sometimes makes the analysis simpler in the unbounded degree model (as is the case for connectivity).

The following is a list of results concerning testing properties in the incidence-lists model.

3.2.1 Bounded-Degree Model.

- Connectivity. The algorithm has query complexity and running time $\tilde{O}(\epsilon^{-1})$ [GR97].

- k-Edge-Connectivity. The algorithm has query complexity and running time $\tilde{O}(k^3 \cdot \epsilon^{-3+\frac{2}{k}})$ [GR97]. For $k = 2, 3$ there are improved algorithms whose running-times are $\tilde{O}(\epsilon^{-1})$ and $\tilde{O}(\epsilon^{-2})$, respectively.

- 2 and 3 -Vertex-Connectivity. The algorithms have query complexity and running time $\tilde{O}(\epsilon^{-2})$ and $\tilde{O}(\epsilon^{-3})$, respectively [GR97].

- Eulerian. The algorithm has query complexity and running time $\tilde{O}(\epsilon^{-1})$ [GR97].

- Cycle-Freeness. The algorithm has query complexity and running time $O\left(\frac{1}{\epsilon^3} + \frac{d}{\epsilon^2}\right)$ [GR97].

  As we shortly discuss in Subsection 3.5, in the case of acyclicity of *directed* graphs, there is a lower bound of $\Omega(N^{1/3})$ (for constant ϵ and d) on the query complexity of any testing algorithm for the property [BR00].

- Bipartiteness. There is a lower bound of $\Omega(\sqrt{N})$ on the query complexity of testing this property (for constant ϵ and d) [GR97]. There is an algorithm having query complexity and running time $\sqrt{N} \cdot \text{poly}(\log N/\epsilon)$ [GR99].

- Expansion. There is a lower bound of $\Omega(\sqrt{N})$ on the query complexity of testing whether a graph has a given constant expansion (for constant ϵ and d) [GR97].

3.2.2 Unbounded-Degree Model.

- Connectivity, k-edge-connectivity, 2 and 3-vertex-connectivity, Eulerian. The algorithms for these properties that work in the bounded-degree model can be adapted to work in the unbounded-degree model at a multiplicative cost of $O(\epsilon^{-1})$ in the query complexity and running time (while actually simplifying the analysis).

- Cycle-Freeness. Any algorithm for testing this property (in the unbounded-degree model) has query complexity $\Omega(\sqrt{N})$ (for constant ϵ and d). This can easily be verified by observing that the following two graphs (families of graphs) cannot be distinguished using $o(\sqrt{N})$ queries: One graph is the empty graph, which is cycle free, and the other graph contains only a single clique of size $\sqrt{N}$, and is very far from being cycle-free [PR99a].

- Diameter. There is a family of algorithms that test whether the diameter of a graph is bounded by a given parameter D, or is ϵ-far from any graph with diameter at most $\beta(D)$ [PR99a]. The function $\beta(D)$ ranges between $D + 4$ and $4D + 2$, depending on the algorithm. All algorithms have query complexity and running time $\tilde{O}(\epsilon^{-3})$, but differ in the ranges ϵ values for which they are applicable.

3.3 TESTING K-EDGE-CONNECTIVITY

In this subsection we describe and analyze the algorithm for testing connectivity (i.e., $k = 1$). We then sketch the ideas required for testing k-Edge-Connectivity for $k \geq 2$. We present the algorithms in the unbounded-degree model. Let the tested graph G have at most M edges (where $M \geq N - 1$ or otherwise the algorithm could immediately reject), and in what follows, whenever we say ϵ-far (or ϵ-close), we mean with respect to the bound M.

3.3.1 Testing Connectivity.
The algorithm is based on the following simple observation concerning the connected components (i.e., the maximal connected subgraphs) of a graph.

Lemma 3.3.1 *If* G *is ϵ-far from being connected than it has more than ϵM connected components.*

Proof: Assume contrary to the claim that G has at most ϵM connected components, and denote these components by $C_1, \ldots, C_k$, $k \leq \epsilon M$. Then by adding an edge between some vertex in C_i and some vertex in C_{i+1}, for every $1 \leq i < k$, the graph can be made connected. As the total number of edges added is $k - 1 < \epsilon M$, we obtain a contradiction to the premise of the lemma that G is ϵ-far from being connected. ∎

The following notation will be used throughout this subsection: $\tilde{\epsilon} \stackrel{\text{def}}{=} \frac{M}{N} \cdot \epsilon$. Thus, $\tilde{\epsilon}$ is ϵ times the average degree of G. As an immediate corollary of Lemma 3.3.1 we get:

Corollary 3.3.2 *If G is ϵ-far from being connected, then G has at least $\frac{\epsilon M}{2}$ connected components each containing less than $2/\tilde{\epsilon}$ vertices.*

Proof: By Lemma 3.3.1, G has more than ϵM connected components. The number of connected components containing at least $2/\tilde{\epsilon}$ vertices is at most $\frac{N}{2/\tilde{\epsilon}} = \frac{\epsilon M}{2}$. So the remaining ones are at least $\epsilon M - \frac{\epsilon M}{2} = \frac{\epsilon M}{2}$ in number, and each contains less than $2/\tilde{\epsilon}$ vertices. ∎

An implicit implication of Lemma 3.3.1 is that for $\epsilon \geq \frac{N}{M}$, *every* graph is ϵ-close to being connected (as otherwise the lemma would imply the existence of an N-vertex graph with more than N connected components). Thus we may assume that $\epsilon < \frac{N}{M}$. By using the fact that each connected component contains at least one vertex we conclude that if G is ϵ-far from being connected then the probability that a uniformly selected vertex belongs to a connected component which contains less than $2/\tilde{\epsilon}$ vertices, is at least $\frac{\epsilon M/2}{N} = \tilde{\epsilon}/2$. Therefore, if G is ϵ-far from being connected and we uniformly select $m = 4/\tilde{\epsilon}$ vertices, then the probability that no selected vertex belongs to a component of size less than $2/\tilde{\epsilon}$ is bounded above by

$$\left(1 - \frac{\tilde{\epsilon}}{2}\right)^m < \exp\left(-\frac{\tilde{\epsilon}}{2} \cdot m\right) = \exp(-2) < \frac{1}{3}$$

Once we select such a vertex, we may detect that it belongs to a small connected component by performing a BFS until no new vertices are reached. This gives rise to the following testing algorithm, where we assume that $N \geq 2/\tilde{\epsilon}$ (since otherwise a connected graph having less than $2/\tilde{\epsilon}$ vertices would be rejected in Step (2)). If $N < 2/\tilde{\epsilon}$, we can determine if the graph is connected by simply inspecting the whole graph (which takes time $O(M) = O(\epsilon^{-1})$).

Connectivity Testing Algorithm

1. Uniformly and independently select $m = 4/\tilde{\epsilon}$ vertices in G;

2. For each vertex v selected perform a Breadth First Search (BFS) starting from v until $2/\tilde{\epsilon}$ vertices have been reached or no more new vertices can be reached (a small connected component has been found);

3. If any of the above searches finds a small connected component, then output REJECT, otherwise output ACCEPT.

Since a connected graph consists of a single component, the algorithm never rejects a connected graph. By the discussion preceding the algorithm and Corollary 3.3.2, if a graph is ϵ-far from connected then it is rejected with probability at least $2/3$. Since the number of edges traversed in each BFS is at most the number of vertices visited, squared,[9] the query complexity and running time of the algorithm are $m \cdot (2/\tilde{\epsilon})^2 = O(\tilde{\epsilon}^{-3})$. We note that the choice to perform a BFS is quite arbitrary, and that any other linear-time (in the number of edges) search method (e.g., DFS) will do.

The complexity of the Connectivity Tester can be improved by applying Corollary 3.3.2 more carefully. Above, when analyzing the probability that the algorithm selects a vertex in a small component, we considered the extreme case in which the component consists of a single vertex. On the other hand, when analyzing the complexity of scanning the component, we considered the extreme case in which the component consists of $\Theta(1/\tilde{\epsilon})$ vertices. Instead, suppose that all components in the conclusion of Corollary 3.3.2 were of the same size, denoted s. Then the probability that a vertex in such a component is selected is at least $s \cdot \frac{\epsilon M/2}{N} = \frac{s \cdot \tilde{\epsilon}}{2}$. This implies that it suffices to set $m = \Theta(1/(s\tilde{\epsilon}))$ in Step (1) of the algorithm above, and that in Step (2) it suffices to stop the search after $s + 1$ vertices (at a cost of $O(s^2)$). Thus, the overall complexity would be $O(s/\tilde{\epsilon})$, provided that such s exists and is given to the algorithm.

Since the latter assumption does not hold, we use a relaxed generalization of the above idea: That is, suppose that G has at least $L \stackrel{\text{def}}{=} \frac{\tilde{\epsilon}}{2} \cdot M$ connected components each of size less than $2/\tilde{\epsilon}$. Then, there exists an $i \leq \ell \stackrel{\text{def}}{=} \log(2/\tilde{\epsilon})$ so that G has at least $\frac{L}{\ell}$ connected components of size ranging between 2^{i-1} and $2^i - 1$ (see the proof of Lemma 4 for details). We do not know this i, but we may try them all. This suggests the following improved algorithm.

Connectivity Testing Algorithm – Improved Version:

---

[9]Here is where the algorithm in the bounded-degree model saves, as in that model the complexity of each BFS is only $O(d/(\epsilon d)) = O(\epsilon^{-1})$.

1. For $i = 1$ to $\log(2/(\tilde{\epsilon}))$ do:

   (a) Uniformly and independently select $m_i = \frac{8 \cdot \log(2/(\tilde{\epsilon}))}{2^i \cdot \tilde{\epsilon}}$ vertices in G;

   (b) For each vertex v selected, perform a BFS starting from v until 2^i vertices have been reached or no new vertices can be reached.

2. If any of the above searches finds a small connected component then output REJECT, otherwise output ACCEPT.

Once again, if the graph is connected, then it is always accepted. It thus remains to show:

Theorem 4 *If G is ϵ-far from being connected, then the improved connectivity testing algorithm will reject it with probability at least $\frac{2}{3}$. The query complexity and running time of the algorithm are $O\left(\frac{\log^2(1/(\tilde{\epsilon}))}{\tilde{\epsilon}^2}\right)$.*

Proof: Let B_i be the set of connected components in G which contain at most $2^i - 1$ vertices and at least 2^{i-1} vertices. Let $\ell \overset{\text{def}}{=} \lceil \log(2/\tilde{\epsilon}) \rceil$. By Corollary 3.3.2 we know that $\sum_{i=1}^{\ell} |B_i| \geq \frac{\epsilon \cdot M}{2} = \frac{\tilde{\epsilon}}{2} \cdot N$. Hence, there exists an index $j \in \{1, 2, ..., \ell\}$ so that $|B_j| \geq \frac{\tilde{\epsilon} \cdot N}{2 \cdot \ell}$. Thus, the number of vertices residing in components belonging to B_j is at least $2^{j-1} \cdot |B_j|$. It follows that the probability that a uniformly selected vertex resides in one of these components is at least

$$\frac{2^{j-1} \cdot |B_j|}{N} \geq \frac{\tilde{\epsilon} \cdot 2^j}{4 \cdot \ell} = \frac{2}{m_j}$$

(where m_j is as defined in Step (1a) of the improved connectivity testing algorithm). Thus, with probability at least $1 - (1 - \frac{2}{m_j})^{m_j} > 1 - e^{-2} > \frac{2}{3}$, a vertex v belonging to a component in B_j is selected in iteration j of Step (2), and the BFS starting from v will discover a small connected component (of size smaller than 2^j), leading to the rejection of G. The query complexity and running-time of the algorithm are bounded by $\sum_{i=1}^{\ell} m_i \cdot 2^{2i} = O\left(\frac{\log(1/(\tilde{\epsilon}))}{\tilde{\epsilon}^2}\right)$. ∎

3.3.2 Testing k-Connectivity for $k > 1$. A subset of vertices $S \subseteq V$ is said to be k–edge-connected if there are k edge-disjoint paths between each pair of vertices in S. A graph G $= (V, E)$ is k-(edge-)connected if V is k-edge-connected. The structure of the testing algorithm for k-Connectivity where $k > 1$ is similar to the structure of the Connectivity Tester (i.e., the case $k = 1$): The algorithm uniformly select a set of vertices and for each vertex selected, it tests if it belongs to a small component of the graph that is separated from the rest of the graph by an edge-cut of size less than k. Similarly to the $k = 1$ case, it can be shown that if a graph is ϵ-far from being k–connected then it has many such components. This can be shown by defining an auxiliary graph [DW98] whose nodes are components of the graph and that is based on the *cactus* structure of [DKL76]. In addition, there are efficient procedures for recognizing such a component given a vertex that resides in it. In what follows we sketch these procedures, for the different values of k. For simplicity we assume the tested graph is $k - 1$-connected (where this assumption can be removed).

2-Connectivity and 3-Connectivity. In the case of 2-connectivity, the procedure is given a vertex v, and is required to output found in case v belongs to a 2-connected subset of vertices C of size at most $n < N$, that is separated from the rest of the graph by a cut of size 1. (If the graph is 2-connected, the procedure should output not-found). The procedure performs the following steps.

1. Starting from v, perform a Depth First Search (DFS) until $n + 1$ vertices have been reached. Let T be the directed tree defined by the search, and let E(T) be its tree edges.

2. Starting once again from v, perform another search (using either DFS or BFS) until n vertices are reached or no new vertices can be reached. This search is restricted as follows: If (u, v) is an edge in T, where u is the parent of v, then (u, v) cannot be used to get from u to v in the second search (but can be used to get from v to u). Let S_2 be the set of vertices reached.

3. If there is a single edge with one end-point in S_2 and the other outside of S_2 (i.e. $(S_2, V \setminus S_2)$ is a cut of size 1), then output found, otherwise output not-found.

The running time of the procedure is $O(n^2)$, and clearly, if the graph is 2-connected, then it outputs not-found. Thus assume v belongs to a 2-connected subset of vertices C of size at most n, that is separated from the rest of the graph by a cut of size 1. Since the first DFS terminates after seeing $n + 1$ vertices, it must visit at least one vertex outside of C. This is possible only by traversing the single edge (u, v) from $u \in C$ to $v \notin C$. Thus, (u, v) must be a edge in the DFS tree T (with u being the parent). This ensures that the second search will never exit C. In other words, $S_2 \subseteq C$. Using the fact that C is 2-connected, it can be shown that the second search will reach *every* vertex in C (that is, $S_2 = C$), and hence will detect the cut.

The procedure for 3-connectivity is given a vertex v that belongs to a subset of vertices C that is 3-connected. If the cut $(C, V \setminus C)$ has size 2 and $|C| \leq n$ $(n < N)$, then the algorithm should output found. The procedure first performs a DFS until $n+1$ vertices are discovered. Next, for each edge e in this DFS-tree (which contains n edges), it "omits" e from the graph and invokes the 2-connectivity procedure on the residual graph. The procedure has running time $O(n^3)$, and its correctness is argued as follows.

Clearly the initial DFS must cross an edge of the cut $(C, V \setminus C)$ and so its DFS-tree has at least one cut edge. Let the graph resulting from omitting this cut edge from G be denoted G'. In G' the cut $(C, V \setminus C)$ contains a single edge in the resulting graph, denoted G'. While the removal of this edge might decrease the connectivity of C (which was 3 in G), it is at least 2–connected in G'. By the correctness of the procedure for 2-connectivity, we are done.

k-Connectivity, $k \geq 2$. The following applies to any $k \geq 2$, but for $k = 2, 3$ we have described more efficient procedures above. The procedure for detecting whether a vertex v belongs to a k-connected subset C of size at most n such that the cut $(C, V \setminus C)$ has size $k - 1$, is based on Karger's Contraction Algorithm [Kar93] which is a randomized algorithm for finding a minimum cut in a graph.

Given a vertex v and a size bound n, the following randomized search process is performed $\Theta(n^{2-\frac{2}{k}})$ times, or until a cut $(S, V \setminus S)$ of size less than k is found:

> Random search process: Starting from the singleton set $\{v\}$, the procedure maintains the set, denoted S, of vertices it has visited. In each step, as long as $|S| < n$ and the cut $(S, V \setminus S)$ has size at least k, the procedure selects at random (as specified below) an edge to traverse among the cut edges in $(S, V \setminus S)$ and adds the new vertex reached to S. In case the cut $(S, V \setminus S)$ has size less than k, output found. If $|S| = n$ then the current search is completed. Otherwise, proceed to the next step (i.e., select a new random edge from the cut $(S, V \setminus S)$).

In case none of the $\Theta(n^{2-\frac{2}{k}})$ invocations of the above process has detected a small cut, output not-found.

The random selection of edges to traverse is done as follows. We think of uniformly and independently assigning each edge in the graph a cost in [0, 1]. Then, at each step of the procedure, we select the edge with lowest cost in the current cut $(S, V \setminus S)$. This is implemented as follows: Whenever a new vertex is added to S, its incident edges that were not yet assigned costs are each assigned a random cost uniformly in [0, 1]. Thus, whenever we need to select an edge from the current cut $(S, V \setminus S)$, all edges in the cut have costs, and we select the edge with lowest cost (just as in the mental experiment in which all graph edges are assigned uniform costs at the beginning).

The correctness of the procedure follows from a probabilistic analysis that bounds the probability that an edge in the cut $(C, V \setminus C)$ is selected before S = C. For further details see [GR97].

3.4 TESTING BIPARTITENESS

As defined in Subsection 2.3, a graph is said to be *bipartite* if its set of vertices can be partitioned into two disjoint sets having no violating edges. An equivalent characterization of bipartite graphs, which we use in this subsection, is that they contain no odd-length cycles. In what follows we sketch both the lower bound on testing bipartiteness in the bounded-degree incidence-lists model, and the (almost matching) upper bound.

3.4.1 The Lower Bound. The following theorem is proved in [GR97].

Theorem 5 *Testing Bipartiteness (in the bounded-degree incidence-lists model) with distance parameter 0.01 requires $\frac{1}{4} \cdot \sqrt{N}$ queries.*

The proof describes two families of degree-3 N-vertex graphs that are hard to distinguish by any algorithm that makes less than $\sqrt{N}/4$ queries: A typical member of one family is 0.01-far from being bipartite, whereas all members of the second family are bipartite graphs. Specifically, let us fix any testing algorithm that makes less than $\sqrt{N}/4$ queries, and consider its decision when given a graph uniformly selected in one of these families. The indistinguishability claim implies that on the average, such an algorithm will accept the random input graph, with about the same probability regardless of the family it was selected from. But this contradicts the requirement from a testing algorithm, since it should accept every member of the second family with probability at least 2/3 while for almost all members of the second family it is allowed acceptance probability smaller than 1/3.

The two families are defined as follows. Let N be an even integer.[10]

1. The first family, denoted $\mathcal{G}_1^N$, consists of all degree-3 graphs that are composed of the union of a Hamiltonian cycle and a perfect matching. That is, there are N edges connecting the vertices in a cycle, and the other $N/2$ edges are a perfect matching.

2. The second family, denoted $\mathcal{G}_2^N$, is the same as the first *except* that the perfect matchings allowed are restricted as follows: the distance on the cycle between every two vertices that are connected by a perfect matching edge must be odd.

In both cases we assume that the edges incident to any vertex are labeled in the following fixed manner: Each cycle edge is labeled 1 in one endpoint and 2 in the other. This labeling forms an orientation of the cycle. The matching edges are labeled 3.

Clearly, all graphs in $\mathcal{G}_2^N$ are bipartite as all cycles in the graph are of even length. Consider a graph uniformly selected in $\mathcal{G}_1^N$. We would like to show that with high probability it is far from bipartite. Such a graph can be selected by selecting a random permutation of the vertices $1, \ldots, N$ on the cycle, and then uniformly selected a matching between the vertices. For any fixed permutation of the vertices on the cycle, consider any fixed two-way partition of the vertices. If there are many violating cycle edges with respect to this partition (give the ordering of the vertices on the cycle), then we are done. Thus, assume there aren't many. Then it can be shown that with high probability over the choice of the matching edges, many of these edges will violate the partition. By bounding the number of such partitions, and using a probability union bound, the claim follows.

The remainder of the proof is focused on showing that a testing algorithm that performs less than $\frac{1}{4}\sqrt{N}$ queries is not able to distinguish (with sufficient probability) between a graph chosen randomly from $\mathcal{G}_2^N$ (which is always bipartite) and a graph chosen randomly from $\mathcal{G}_1^N$ (which with high probability is far from bipartite). This is done by defining two processes, one for each class of graphs. Each process answers the queries of the testing algorithm while randomly constructing a graph in the respective class (where the distribution on the resulting graphs can be shown to be uniform in the class). The crux of the proof is that for any testing algorithm, the two distributions on the query-answer sequences induced by the two processes, are statistically indistinguishable as long

---

[10] For odd N, every graph (in both families) contains one degree-0 vertex, and the rest of the vertices are connected as in the even case.

as the sequence is of length less than $\alpha\sqrt{N}$. This essentially follows from the fact that in sequences of such length, if we consider the subgraph induced by the query-answer sequence, then it does not contain a cycle (either even or odd).

3.4.2   The Algorithm. Since the algorithm can make queries of the form: "who is the i'th neighbor of vertex v", it can perform *walks* on G. Namely, starting from any vertex s, it can obtain the sequence of vertices lying on any path $i_1, i_2, \ldots, i_t$ (where each i_j is an edge label) that originates from s by querying: who is the i_1^{th} neighbor of s, who is the i_2^{th} neighbor of the vertex returned, and so on. In particular, the algorithm described below performs *random* walks on G: At each step, if the degree of the current vertex v is $d' \le d$, then the walk *remains* at v with probability $1 - \frac{d'}{2d} \ge \frac{1}{2}$, and for each $u \in \Gamma(v)$, the walk *traverses* to u with probability $\frac{1}{2d}$. Thus, the stationary distribution over the vertices is uniform.

For every walk (or, more generally, any sequence of steps), there corresponds a *path* in the graph. The path is determined by those steps in which an edge is traversed (while ignoring all steps in which the walks stays in the same vertex). Such a path is not necessarily simple, but does not contain self loops. Note that when referring to the length of a walk, we mean the total number of steps taken, including steps in which the walk remains at the current vertex, while the length of the corresponding path does not include these steps.

Test-Bipartite (Incidence-Lists model)

- Repeat $T = \Theta\left(\frac{1}{\epsilon}\right)$ times:

    1. Uniformly select s in V.
    2. (a) Let $K = \text{poly}((\log N)/\epsilon) \cdot \sqrt{N}$, and $L = \text{poly}((\log N)/\epsilon)$;
       (b) Perform K random walks starting from s, each of length L;
       (c) If some vertex v is reached (from s) both on a prefix of a random walk corresponding to an even-length path and on a prefix of a walk corresponding to an odd-length path then reject.

- In case the algorithm did not reject in any one of the above iterations, it accepts.

Theorem 6 *The algorithm* Test-Bipartite *is a testing algorithm for bipartiteness. In particular, if the graph is bipartite it is always accepted,* *and if it is ϵ-far from bipartite it is rejected with probability at least 2/3. Furthermore, whenever the algorithm rejects a graph it outputs a certificate to the non-bipartiteness of the graph in form of an odd-length cycle of length* $\text{poly}(\epsilon^{-1} \log N)$.

Clearly, a bipartite graph is always accepted as it contains no odd-length cycles. Hence the heart of the proof is showing that if a graph is ϵ-far from bipartite it is rejected with probability at least 2/3. This is shown by proving the contrapositive statement: If a graph is accepted with probability greater than 1/3, then it is ϵ-close to bipartite. Namely, by removing at most $\epsilon \cdot dN$ edges, it can be made bipartite. The proof of this statement is somewhat complex, and here we only provide the underlying ideas.

The Rapidly–Mixing Case. To gain intuition, consider first the following "ideal" case: From each starting vertex s in G, and for every $v \in V$, the probability that a random walk of length $L = \text{poly}((\log N)/\epsilon)$ ends at v is at least $\frac{1}{2N}$ and at most $\frac{2}{N}$ – i.e., approximately the probability assigned by the stationary distribution. (Note that this ideal case occurs when G is an expander). Let us fix a particular starting vertex s. For each vertex v, let p_v^0 be the probability that a random walk (of length L) starting from s, ends at v and corresponds to an even-length path. Define p_v^1 analogously for odd-length paths. Then, by our assumption on G, for every v, $p_v^0 + p_v^1 \geq \frac{1}{2N}$.

We consider two cases regarding the sum $\sum_{v \in V} p_v^0 \cdot p_v^1$ — In case the sum is (relatively) "small", we show that there exists a partition (V_0, V_1) of V that is ϵ-good, and so G is ϵ-close to being bipartite. Otherwise (i.e., when the sum is not "small"), we show that the probability that the algorithm finds an odd cycle when performing the random walks starting from s, is constant. This implies that in case G is accepted with probability greater than $\frac{1}{3}$, then G is ϵ-close to being bipartite. In what follows we give some intuition concerning the two cases.

Consider first the case in which $\sum_{v \in V} p_v^0 \cdot p_v^1$ is smaller than $c \cdot \frac{\epsilon}{N}$ for some suitable constant $c < 1$. Let the partition (V_0, V_1) be defined as follows: $V_0 = \{v : p_v^0 \geq p_v^1\}$ and $V_1 = \{v : p_v^1 > p_v^0\}$. Consider a particular vertex $v \in V_0$. By definition of V_0 and our rapid-mixing assumption, $p_v^0 \geq \frac{1}{4N}$. Assume v has neighbors in V_0. Then for each such neighbor u, $p_u^0 \geq \frac{1}{4N}$ as well. However, since there is a probability of $\frac{1}{2d}$ of taking a transition from u to v in walks on G, we can infer that each neighbor u contributes $\Omega(\frac{1}{2d} \cdot \frac{1}{4N})$ to the probability p_v^1. (This inference is not completely straightforward since both p_u^0 and p_v^1 correspond to walks of length exactly L, but this slight difficulty can be overcome.) Thus, if there are many (more than ϵdN) violating edges with respect to (V_0, V_1), then the sum $\sum_{v \in V} p_v^0 \cdot p_v^1$ is large (greater than $\epsilon dN \cdot \frac{1}{4N} \cdot \frac{1}{8dn} \geq c \cdot \frac{\epsilon}{N}$), contradicting our case hypothesis.

We now turn to the second case $(\sum_{v \in V} p_v^0 \cdot p_v^1 \geq c \cdot \frac{\epsilon}{N})$. For every fixed pair $i, j \in \{1, \dots, K\}$, (recall that $K = \Omega(\sqrt{N})$ is the number of walks taken from s), consider the 0/1 random variable that is 1 if and only if both the ith and the jth walk end at the same vertex v but correspond to paths with different parity. Then the expected value of each random variable is $\sum_{v \in V} 2 \cdot p_v^0 \cdot p_v^1$. Since there are $K^2 = \Omega(N)$ such variables, the expected value of their sum is greater than 1. These random variables are not pairwise independent, nonetheless we can obtain a constant bound on the probability that the sum is 0 using Chebyshev's inequality (cf., [AS92, Sec. 4.3]).

The General Case. Unfortunately, we may not assume in general that for every (or even some) starting vertex, all (or even almost all) vertices are reached with probability $\Theta(1/N)$. Instead, for each vertex s, we may consider the set of vertices that are reached from s with relatively high probability on walks of length $L = \text{poly}((\log N)/\epsilon)$. As was done above, we could try and partition these vertices according to the probability that they are reached on random walks corresponding to even-length and odd-length paths, respectively. The difficulty that arises is how to combine the different partitions induces by the different starting vertices, and how to argue that there are few violating edges between vertices partitioned according to one starting vertex and vertices partitioned according to another.

To overcome this difficulty, we proceed in a slightly different manner. Let us call a vertex s *good*, if the probability that the algorithm finds an odd cycle when performing random walks starting from s, is at most 0.1. Then, assuming G is accepted with probability greater than $\frac{1}{3}$, all but at most $\frac{\epsilon}{16}$ of the vertices are *good*. We define a partition in stages as follows. In the first stage we pick any *good* vertex s. What we can show is that not only is there a set of vertices S that are reached from s with high probability and can be partitioned without many violations (due to the goodness of s), but also that there is a small cut between S and the rest of the graph. Thus, no matter how we partition the rest of the vertices, there cannot be many violating edges between S and $V \setminus S$. We therefore partition S (as above), and continue with the rest of the vertices in G.

In the next stage, and those that follow, we consider the subgraph H induces by the yet "unpartitioned" vertices. If $|H| < \frac{1}{4}N$ then we can partition H arbitrarily and stop since the total number of edges adjacent to vertices in H is less than $\frac{\epsilon}{4} \cdot dN$. If $|H| \geq \frac{\epsilon}{4} N$ then it can be shown that any *good* vertex s in H that has a certain additional property (which at least half of the vertices in H have), determines a set S (whose vertices are reached with high probability from s) with the following properties: S can be partitioned without having many violating edges among vertices in S; and there is a small cut between S and the rest of H. Thus, each such set S accounts for the violating edges between pairs of vertices that both belong to S as well as edges between pairs of vertices such that one vertex belongs to S and one to $V(H) \setminus S$. Adding it all together, the total number of violating edges with respect to the final partition is at most $\epsilon \cdot dN$.

THE SET S. To prove the existence of such sets S, consider first the initial stage in the partition process (i.e., here H = G). Recall that in this stage we are looking for a subset of vertices $S \subseteq V$, all reached with relatively high probability from some good vertex s, that are separated from the rest of G by relatively few edges. ¿From the previous discussion we know that if for all (or almost all) vertices v in G, a random walk of length $\text{poly}((\log N)/\epsilon)$ starting from s ends at v with probability $\Theta(1/N)$ then we can define a good partition of all of G and be done. Thus assume we are not in this case. Namely, there is a significant fraction of vertices that are reached from s with probability that differs significantly from $1/N$. In other words, the distribution on the ending vertices (when starting from s) is far from stationary. What can be shown (using techniques of Mihail [Mih89]) is that this implies the existence of a small cut between some set of vertices S that are each reached from s with probability that is roughly $1/\sqrt{|S| \cdot N}$ and the rest of G. Furthermore, it can be shown that S has an additional property that combined with the fact that s is good implies that it can be partitioned without having many violating edges.

In the next stages of the partition process, we would have liked to apply the same techniques to determine small cuts (with other desired properties) in subgraphs H of G. If we could at each stage "cut-away" the subgraph H from the rest of G and perform walks only inside H then we would have proceeded as in the first stage. However, these subgraphs H are only determined by the analysis while the algorithm, oblivious to the analysis, always performs random walks on all of G. Therefore we would like to have a way to map walks in G to walks in H so that probabilities of events occurring in imaginary walks on H can be related to events occurring in the real walks on G. This is done by defining a special Markov chain given H and relating walks on the Markov chain to walks on G. For further details see [GR99].

3.5   DIRECTED GRAPHS

As noted at the end of Subsection 2.2, for some properties of undirected graphs that have analogous properties in directed graphs, algorithms that work on undirected graphs can be transformed to work on directed graphs. An example in the incidence-lists model is connectivity. A directed graph is (strongly) connected if there is a directed path in the graph from any vertex to any other vertex. The algorithm of [GR97] presented in Subsection 3.3 can be extended to the directed case *if the algorithm can also perform queries about the incoming edges to each vertex* [PR99b]. Otherwise, (the algorithm can only perform queries about outgoing edges), a simple lower bound of $\sqrt{N}$ on the number of queries can be obtained. In the case of testing acyclicity of directed graphs the situation is more complex. Here the cycle-freeness test for undirected graphs [GR97] can not be transformed to work on directed graphs, as it is (partly) based on the observation that that an undirected graph contains no cycles only if it has at most $N-1$ edges. Furthermore, in the case of directed graphs, even if there is access to both incoming and outgoing edges, every algorithm for testing acyclicity must use $\Omega(N^{\frac{1}{3}})$ queries [BR00].

4.   TESTING OTHER PROPERTIES

In this section we provide a brief summary of results on testing properties of objects other than graphs. In particular, most results concern functions.

Definition 4.0.1 *For a given function* $f : X \to Y$, *and a property* $\mathcal{P}$ *(defined over functions with domain* X *and range* Y*), we say that* f *is* ϵ-far *from having property* $\mathcal{P}$, *if for every* $g : X \to Y$, $\mathrm{Pr}_{x \sim U}[f(x) \neq g(x)] > \epsilon$, *(where* U *denotes the uniform distribution), otherwise it is* ϵ-close *to* $\mathcal{P}$.

4.1   TESTING ALGEBRAIC PROPERTIES

4.1.1   Testing Linearity.   In this subsection we present a test due to Blum, Luby and Rubinfeld [BLR93], with a slightly modified analysis due to Sudan [Sud99].

Definition 4.1.1 (linearity of a function) *Let* F *be a finite field.   A function* $f : F^m \to F$ *is called* linear *(or more precisely,* multi-linear*) if there exist constants* $a_1, \ldots, a_m \in F$ *s.t. for all* $x = (x_1, \ldots, x_m) \in F^m$ *it holds that* $f(x) = \sum_{i=1}^{m} a_i x_i$.

It is not hard to verify the following fact, which provides an alternative definition of linearity.

Fact 4.1.1 (alternative definition of linearity) *A function* $f : F^m \to F$ *is* linear *if and only if for every* $x, y \in F^m$ $f(x) + f(y) = f(x + y)$.

The following test uniformly selects pairs of elements in the field, and checks whether linearity (according to the second definition) is violated.

Linearity Test

1. Uniformly and independently select $m = \Theta(\epsilon^{-1})$ pairs of elements $x, y \in F$.

2. For every pair of elements selected, check whether $f(x) + f(y) = f(x + y)$.

3. If for any of the selected pairs linearity is violated (that is $f(x) + f(y) \neq f(x + y)$), then reject, otherwise, accept.

By Fact 4.1.1, if f is linear then it is always accepted. It thus remains to prove:

Theorem 7 *If f is ϵ-far from linear then with probability at least 2/3, Linearity Test rejects it.*

Here we shall give a simple proof of the theorem for the case in which f is "not too far" from linear. Namely, that its distance from some linear function is bounded away from $\frac{1}{2}$ (that is, the distance is at most $\frac{1}{2} - \gamma$ for some constant γ).

Proof: We say that a pair of elements x, y are a *violating* pair, if $f(x) + f(y) \neq f(x + y)$. Let δ denote the (exact) distance of f from linearity (so that in particular, $\delta > \epsilon$). We shall show that the probability that a single uniformly selected pair of elements is a violating pair, is at least $3\delta(1 - 2\delta)$. For δ bounded away from $\frac{1}{2}$, this probability is $\Omega(\delta)$. Since the test selects $\Theta(1/\epsilon) = \Omega(1/\delta)$ pairs, the probability that no violating pair is selected is at most $1/3$ (for the appropriate choice of constant in the $\Omega(\cdot)$ notation).

Let g be a linear function at distance δ from f. Let $G \stackrel{\text{def}}{=} \{x : f(x) = g(x)\}$ be the set of *good* elements in F on which f and g agree. For any pair x, y, if among the three elements, x, y, and $(x + y)$ two of them belong to G while the third doesn't, then x, y are a violating pair. Hence,

$$\Pr[x, y \text{ are a violating pair }] \geq$$
$$\Pr[x \notin G, y \in G, (x + y) \in G] +$$

$$\Pr[x \in G, \, y \notin G, \, (x+y) \in G] \, +$$
$$\Pr[x \in G, \, y \in G, \, (x+y) \notin G] \qquad (15.3)$$

Consider the first probability in the above sum (the treatment of other two is analogous, as the important property of any triplet is that every two of the elements are pairwise independent).

$$\Pr[x \notin G, \, y \in G, \, (x+y) \in G] =$$
$$\Pr[x \notin G] \cdot \Pr[y \in G, \, (x+y) \in G \mid x \notin G] =$$
$$\delta \cdot (1 - \Pr[y \notin G \text{ or } (x+y) \notin G \mid x \notin G]) \qquad (15.4)$$

By using a probability union bound, and the fact that both y and $(x+y)$ are uniformly distributed,

$$1 - \Pr[y \notin G \text{ or } (x+y) \notin G \mid x \notin G] \geq 1 - 2\Pr[y \notin G \mid x \notin G] \quad (15.5)$$

Since x and y are chosen independently,

$$\Pr[y \notin G \mid x \notin G] = \Pr[y \notin G] = \delta \qquad (15.6)$$

and so by combining Equations (15.3)– (15.6), the probability of selecting a violating pair is at least $3 \cdot \delta \cdot (1 - 2\delta)$. ■

4.1.2   Testing (Low-Degree) Polynomials.

We present a test for univariate polynomials that is based on a basic property of polynomials, where we follow the presentation of Sudan [Sudan-PhD]. There are also tests for multivariate polynomials but their analysis is more complex (see for example [GLR+91, RS96, RS97, AS97]).

Definition 4.1.2 *Let* F *be a finite field. A function* $f : F \to F$ *is a (univariate)* polynomial of degree d, *if there exist coefficients* $c_0, \ldots, c_d \in$ F, *such that* $f(x) = \sum_{i=0}^{d} c_i \cdot x^i$.

Recall that given any $d+1$ pairs $\{(x_i, y_i)\}_{i=0}^{d}$, where $x_i, y_i \in F$, there exists a *unique* degree d polynomial h such that $h(x_i) = y_i$ for every $i \in \{0, \ldots, d\}$, and h can be found by interpolation.

Low-Degree Test

1. Repeat the following $m = 2/\epsilon$ times:

    (a) Uniformly and independently select $d + 2$ distinct points $x_0, \ldots, x_{d+1} \in F$.

(b) Check (by interpolating) whether there exists a degree d polynomial q such that $q(x_i) = f(x_i)$ for every $i \in \{0, \ldots, d+1\}$.

2. If in any of the iterations evidence was found that f is not a degree d polynomial, then reject, otherwise, accept.

Clearly, if f is a degree d polynomial, then it is always accepted. It thus remains to prove:

Theorem 8 *If f is ϵ-far from any degree d polynomial then with probability at least $2/3$, Low-Degree Test rejects it.*

Proof: Let δ be the distance between f and a closest degree d polynomial (so that $\delta > \epsilon$). We show that in each iteration of the algorithm, the probability that the check in Step (1b) fails, is at least δ. Since $m = 2/\epsilon > 2/\delta$ iterations are performed, the probability that all checks succeed is $(1 - \delta)^{2/\epsilon} < \exp(-2) < 1/3$.

Let g be a degree d polynomial closest to f (so that the distance between f and g is δ). We fix $z_0, \ldots, z_d$, and let h be the unique degree d polynomial such that $h(z_i) = f(z_i)$ for every $i \in \{0, \ldots, d\}$. By definition of δ, we have that the probability over a uniformly chosen point z_{d+1} in F that $h(z_{d+1}) = f(z_{d+1})$, is at most $1 - \delta$. Now, the probability over $x_0, \ldots, x_{d+1}$ that there exists a degree d polynomial q s.t. $\forall i \in \{0, \ldots, d\}$, $q(x_i) = f(x_i)$, is upper bounded by the maximum over $x_0, \ldots, x_d$ of the probability over x_{d+1} that a degree d poly that agrees with f on $x_0, \ldots, x_d$, agrees on x_{d+1}, which is at most $1 - \delta$. and so the probability that such a polynomial q does *not* exist is at least δ as claimed. ∎

4.1.3   Testing Other Algebraic Properties.

Functional Equations.   Rubinfeld [Rub99] studies properties of functions $f : X \to Y$ that can be characterized by *(quantified) functional equations* of the form: $\forall x, y \in X$, $F[f(x), f(y), f(x + y), f(x - y)] = 0$. For example, linearity falls into this framework since it can be characterized by $\forall x, y \in X$, $f(x + y) - f(x) - f(y) = 0$.

Such a characterization is said to be *robust* [RS96] if whenever the functional equation holds for f *for most* x, y, there exists a function g that has the property and is *close* to f. This implies that the property can be tested by verifying that the functional equation holds on a sample of uniformly selected pairs x, y.

Rubinfeld shows several sufficient conditions for robustness. In particular, functional equations of the (additive) form $\forall x, y$, $f(x + y) = G[f(x), f(y)]$ are robust, where many trigonometric functions can be characterized by such equations. d'Alembert's equation: $\forall x, y$, $f(x + $

$y) + f(x - y) = 2f(x)f(y)$ is also robust, and so are several variants of it. Rubinfeld also provides necessary conditions for robustness (that have a combinatorial form).

Group Operations. Let G be a finite set. Let ∘ be an operation on pairs of elements in G, so that for every $x, y \in$ G, $x \circ y \in$ G. The operation ∘ is *associative*, if for every $x, y, z \in$ G, $x \circ (y \circ z) = (x \circ y) \circ z$. An *identity* element with respect to ∘, is an element $e \in$ G such that for every $x \in$ G, $x \circ e = x$. An *inverse* of an element $x \in$ G, is an element $x' \in$ G such that $x \circ x' = e$.

The operation ∘ is a *group* operation if it is associative, has a unique identity element with respect to ∘, and every element has an inverse under ∘. Ergun et. al. [EKK+98] describe an algorithm having complexity $\tilde{O}(|G|/\epsilon)$ for testing whether ∘ is a group operation, given access to the value of $x \circ y$ on pairs of elements of its choice under the assumption that the operation ∘ is cancelative.[11] This assumption can be removed at a further multiplicative cost of $O(\sqrt{G})$.

4.2 TESTING REGULAR LANGUAGES

In this subsection we sketch the result of Alon, Krivelevich, Newman, and Szegedy [AKNS99], showing that for every regular language L $\subseteq$ $\{0, 1\}^*$, there exists a testing algorithm for L. Namely, the algorithm accepts every word $w \in$ L, and rejects with probability at least 2/3 every word w that differs on more than $\epsilon \cdot |w|$ bits from any $w' \in$ L. The running time of the algorithm is $\tilde{O}(\epsilon^{-1})$, that is, independent of the length n of w. (The running time is dependent on the size of the (smallest) finite automaton accepting M, but this size is a fixed constant with respect to n). Recently, Newman [New00] extended this result and gave an algorithm having query complexity poly($1/\epsilon$) for testing whether a word w is accepted by a given constant-width branching program. We note that Alon *et. al.* also show that a very simple context free language (of all strings of the form vv^Ru, where v^R denotes the reversal of v), cannot be tested using $o(\sqrt{N})$ queries.

We start by recalling some definitions.

Definition 4.2.1 *A deterministic finite automaton (DFA)* M *over the alphabet* $\{0, 1\}$ *is defined by a set of* states Q $= \{q_0, \ldots, q_{m-1}\}$, *a subset* F $\subseteq$ Q *of accepting states, and a transition function* $\delta :$ Q $\times \{0, 1\} \mapsto$ Q. *The state* q_0 *is called the* initial *state. The transition function* δ *is extended to be defined on* $\{0, 1\}^*$ *in the following recursive manner: For*

---

[11]An operation ∘ is *cancelative* if $a \circ c = b \circ c$ implies $a = b$, and $a \circ b = a \circ c$ implies $b = c$

every $q \in Q$, $u \in \{0,1\}^*$, *and* $\sigma \in \{0,1\}$, $\delta(q, u\sigma) = \delta(\delta(q,u), \sigma)$, *where* $\delta(q, \lambda) = q$ *(λ denoting the empty string).*
We say that M accepts *a word* $w \in \{0,1\}^*$, *if* $\delta(q_0, w) \in$ F *(otherwise it* rejects *it).*

We shall use the definition of regular languages that is based on DFA.

Definition 4.2.2 *A language* L $\subseteq \{0,1\}^*$ *is said to be* regular *if there exists a DFA* M *such that* M *accepts all words* $w \in$ L, *and no other words.*

We thus assume that a regular language L is given by providing the (smallest) DFA M that accepts it. Given a DFA M, it induces a directed graph $G(M) = (V, E)$ in a straightforward manner: $V = Q$, and $E = \{(q_i, q_j) : \exists \sigma \in \{0,1\}, \delta(q_i, \sigma) = q_j\}$. We shall refer to the vertices of $G(M)$ as *states*.

Given a word $w \in \{0,1\}^n$, we first assume that the language L contains words of length n (or otherwise w can be directly rejected). Let u be a sub-word of w that starts at position i. That is $w = u'uu''$ where $|u'| = i$. We say that u is *feasible with respect to the DFA* M *starting from position* i if there exists a state q such that q can be reached in $G(M)$ from q_0 in exactly $i - 1$ steps and there is some path in $G(M)$ from $q' = \delta(q, w)$ to an accepting state. When the index i is clear from the context we just say that u feasible. It is possible to verify whether u is feasible, in time that depends only on the size of M. Clearly, if a word w contains a sub-word u that is *not* feasible, then $w \notin$ L. The algorithm tries to find evidence to w not belonging to L in the form of infeasible sub-words.

Following Alon *et. al.* [AKNS99], we describe a special case of regular languages and show that for these languages, every word that is ϵ-far from belonging to the language, contains many short infeasible sub-words. Hence, an algorithm that simply samples such sub-words and checks whether they are feasible, will, with high probability, detect that a word w is ϵ-far from belonging to the language. The analysis of Alon *et. al.* reduces the general case to this special case. (For further details on the general case see [AKNS99].)

We make the following assumptions concerning the DFA M that accepts the language L. First, M contains a single accepting state, denoted q_{acc}. Second, the set of states Q of M can be partitioned into to subsets C and D such that:

1. The subset C contains both q_0 and q_{acc}.

2. The subgraph of $G(M)$ induced by C is strongly connected. We denote this subgraph by $G(C)$.

3. There are no edges in G(M) going from states in D to states in C (though there may be edges going in the other direction).

We further assume for simplicity that the greatest common divisor (GCD) of cycle lengths in G(C) is 1. This implies that there exists a constant $r = r(G(C))$. referred to as the *reachability* constant of G(C), such that for every two states x, y in G(M), and for every $n \geq r$, there exists a directed path from x to y in G(C) of length n. The size of r is at most quadratic in $|C|$. (If the GCD if cycle lengths is not 1 then a slightly different notion of reachability constant is required).

Lemma 4.2.1 *Let* M *be constrained as described above, and let* w *be a word of length* n *that is* ϵ-*far from the language* L *accepted by* M *(where* M *accepts some words of length* n*). Then the number of infeasible subwords of* w *having length at most* $\frac{4r}{\epsilon}$ *is at least* $\frac{\epsilon n}{4r}$*.*

Proof: We shall construct a sequence of disjoint minimal-length infeasible sub-words of w. That is, each sub-word is infeasible, but each of its prefixes (and in particular its longest prefix) is feasible. Let the starting position of the j'th sub-word, u_j, be s_j, then we shall select the sub-words so that for very j, $s_j \geq r + 1$, and $s_j + |u_j| \leq n - r$.

We constructs these sub-words in a *greedy* manner. The first subword, u_1, starts at position $s_1 = r + 1$, and is the shortest sub-word of w, starting at position s_1, that is infeasible. The next sub-word, u_2, starts at position $s_2 = s_1 + |u_1|$, and is the shortest infeasible sub-word that starts at position s_2. In general, the j'th sub-word starts at position $s_j = s_{j-1} + |u_{j-1}|$ and is the shortest infeasible sub-word that starts at this position. The procedure terminates when position $n - r$ is reached, and the last sub-word is "cut-off" at this position. Hence, the last subword may actually be feasible. Note that for every position $i \leq n - r$, the empty sub-word is always feasible, and so each infeasible sub-word has length at least 1. Let the number of sub-words obtained be h. Then,

$$|w| = n = 2 \cdot r + \sum_{j=1}^{h} |u_j| \tag{15.7}$$

For each $1 \leq j \leq h$, let $\bar{u}_j$ be the prefix of u_j of length $|u_j| - 1$ (so that $\bar{u}_j$ may be empty) Recall that by definition of u_j, $\bar{u}_j$ is feasible. For every $1 \leq j \leq h$ we fix a state $q_{i_j} \in C$ so that q_{i_j} is reachable from q_0 in $s_j - 1$ steps, and so that $\delta(q_{i_j}, \bar{u}_j) \in C$. Note that such a state in C must exist because $\bar{u}_j$ is feasible, and by our assumption that there are no edges going from D to C. Not also that because u_j is infeasible, $\delta(q_{i_j}, u_j) \in D$ (so that the last bit in u_j "forces" a transition to D from which there is no way to reach the accepting state in C).

For a given word $w' \in \{0,1\}^n$, we denote by $\text{dist}(w, w')$ the number of bits on which w and w' differ. We shall show that there exists a word $w^* \in L$, having length n, such that $\text{dist}(w, w^*) \leq (2 \cdot h \cdot r)/n$. This will give us a lower bound on h. The construction is done inductively, where in the j'th step we obtain a word w_j of length $s_j - 1$ that is feasible from position 1. Based on our assumptions on M this in particular means that the sequence of states traversed given w_j are all in C. The basic idea is to modify w so that each bit (at the end of an infeasible sub-word u_j) that causes a transition from a state in C to a state in D, is replaced by a bit that causes a transition to another state in C (from which the accepting state can be reached). In order to "glue" these modifications together, a little more work is needed. Details follow.

The initial word, w^0, is some word of length r that is feasible from position 1. In general, we construct w^j based on w^{j-1} in the following manner. Let $\tilde{w}^{j-1}$ be the prefix of length $s_j - 1 - r$ of w^{j-1}, and let $p_j = \delta(q_0, \tilde{w}^{j-1})$. Since there exists some path of length r from p_j to q_{i_j} (where q_{i_j} was defined above), by modifying (some of) the last r bits of w^{j-1} we can obtain a word z_j such that $\delta(q_0, z_j) = q_{i_j}$. If $j < h$ we let w^j be the concatenation of z_j, $\bar{u}_j$, and some bit b_j such that $\delta(q_0, w^{j-1}\bar{u}_j b_j) \in C$ (since G(C) is strongly connected, such a bit must exist). In case $j = h$ we let $w^* = w^h$ be the concatenation of z_h, $\bar{u}_h$, and some word v of length r so that $\delta(q_0, w^{h-1}\bar{u}_j v) = q_{\text{acc}}$, implying that $w^* \in L$. By this construction,

$$\text{dist}(w, w^*) \leq \frac{1}{n}\left((h-1) \cdot r + 2r\right) = \frac{hr + r}{n} \leq \frac{2hr}{n}$$

By our assumption on w, $\text{dist}(w, w^*) \geq \epsilon$, and hence $h \geq \frac{\epsilon n}{2r}$.

Since all the infeasible sub-words, $u_1, \ldots, u_{h-1}$, are disjoint, the number of infeasible sub-words having length greater than $4r/\epsilon$ is less than $n/(4r/\epsilon) = (\epsilon n)/(4r)$. Since the total number of infeasible sub-words is at least $\frac{\epsilon n}{2r} - 1$, the lemma follows. ∎

4.3 TESTING MONOTONICITY

A function $f : \{0,1\}^n \to \{0,1\}$ is said to be monotone if $f(x) \leq f(y)$ for every $x \prec y$, where $\prec$ denotes the natural partial order among strings (i.e., $x_1 \cdots x_n \prec y_1 \cdots y_n$ if $x_i \leq y_i$ for every i and $x_i < y_i$ for some i).

The algorithm for testing monotonicity presented in [GGL+00] whose query complexity and running time are linear in n and $1/\epsilon$ performs a simple local test: It verifies whether monotonicity is maintained for randomly chosen pairs of strings that differ exactly on a single bit. More precisely,

Monotonicity Testing Algorithm

On input n, ϵ and oracle access to $f : \{0,1\}^n \to \{0,1\}$, repeat the following steps up to n/ϵ times

1. Uniformly select $x = x_1 \cdots x_n \in \{0,1\}^n$ and $i \in \{1, \ldots, n\}$.

2. Obtain the values of $f(x)$ and $f(y)$, where y results from x by flipping the ith bit (that is, $y = x_1 \cdots x_{i-1} \bar{x}_i x_{i+1} \cdots x_n$).

3. If $x, y, f(x), f(y)$ demonstrate that f is not monotone then reject.

   That is, if either $(x \prec y) \wedge (f(x) > f(y))$ or $(y \prec x) \wedge (f(y) > f(x))$ then reject.

If all iterations are completed without rejecting then accept.

Thus the algorithm has a similar structure to the linearity testing algorithm. In the analysis of the algorithm, the probability of observing a local violation of monotonicity is related to the global measure relevant to testing – the minimum distance of the function to any monotone function. For further details see [GGL$^+$00].

The definition of monotonicity can be extended in a straightforward manner to monotonicity of functions $f : \Sigma^n \to \{0,1\}$ where there is a total order over Σ. The algorithm can be modified so as to yield a testing algorithm having query complexity and running time $O\left(\frac{n \cdot \log |\Sigma|}{\epsilon}\right)$ [GGL$^+$00]. The notion of monotonicity can be further extended to functions mapping to totally ordered ranges and the testing algorithm adapted in a corresponding manner. The dependence of the query complexity and running time of the modified algorithm are logarithmic in $|\Xi|$ [DGL$^+$99]. The *spot-checker for sorting* presented in [EKK$^+$98, Sec. 2.1] implies a tester for monotonicity with respect to. functions from any fully ordered domain to any fully ordered range, having query and time complexities that are logarithmic in the size of the domain. This corresponds to the special case $n = 1$ (with general Σ and Ξ).

Another extension of testing monotonicity of boolean functions is testing *unateness* of functions. A function $f : \{0,1\}^n \to \{0,1\}$ is said to be unate if for every $i \in \{1, \ldots, n\}$ exactly one of the following holds: whenever the ith bit is flipped from 0 to 1 then the value of f does not decrease; *or* whenever the ith bit is flipped from 1 to 0 then the value of f does not decrease. Thus, unateness is a more general notion than monotonicity. The algorithm for testing monotonicity of boolean functions over $\{0,1\}^n$ can be extended to test whether a function is unate or far from any unate function at an additional cost of a (multiplicative) factor of $\sqrt{n}$ [GGL$^+$00].

4.4 TESTING USING RANDOM EXAMPLES

As noted in Subsection 1.1.4, one of the initial motivations for the study of property testing is its relation to Computational Learning Theory. In particular, while in some learning models queries are allowed, it is usually preferable that the learning algorithm be only provided with random examples. In analogy, here we shall assume that the testing algorithm is given a labeled sample $\{(x^1, f(x^1)), \ldots, (x^m, f(x^m))\}$, where the x^i's are distributed according to some fixed (possibly unknown) distribution D over the domain X. Distance between functions is defined as the weight, according to D of the symmetric distance between the functions. Namely, $\text{dist}(f, g) = \Pr_{x \sim D}[f(x) \neq g(x)]$.

In particular we consider the case in which the underlying distribution D is uniform (and so the distance measure between functions is as in the case where queries are allowed). In some cases, such as testing monotonicity, allowing only random examples makes the problem essentially intractable [GGL$^+$00]. In other cases, while testing with queries is more efficient, there are still efficient testing algorithms that use only random examples.

This is in particular true of testing for decision trees over $[0, 1]^d$ (for constant d) [KR98]. That is, the property is belonging to the class of decisions trees over $[0, 1]^d$ having at most s nodes. This class of functions is defined as follows. Given an input $\vec{x} = (x_1, \ldots, x_d)$, the (binary) decision at each node of the tree is whether $x_i \geq a$ for some $i \in \{1, \ldots, d\}$ and $a \in [0, 1]$. The labels of the leaves of the decision tree are in $\{0, 1\}$. We define the *size* of such a tree to be the number of leaves, and we let DT_s^d denote the class of decision trees of size at most s over $[0, 1]^d$. Thus, every tree in DT_s^d determines a partition of the domain $[0, 1]^d$ into at most s axis aligned rectangles, each of dimension d (the leaves of the tree), where all points belonging to the same rectangle have the same label.

The testing algorithm for decision trees decides whether to accept or reject a function f by pairing "nearby" points in the sample, and checking that such pairs have common labels. More precisely, it will consider a certain collection of d-dimensional grids that partition the domain into *cells*. For each grid the algorithm computes the fraction of pairs of points that fall into the same grid cell and have the same label. If for some grid this fraction is above a certain threshold then it accepts, otherwise it rejects. The heart of the analysis is a combinatorial argument, which shows that there exists a (not too large) set of (relatively coarse) d-dimensional grids $G_1, \ldots, G_k$ for which the following holds: for every function $f \in \text{DT}_s^d$, there exists a grid G_i such that a "significant" fraction of the cells in G_i "fit inside" the leaves of f — that is, there are not too many cells of G_i that intersect a decision boundary of f.

The following theorem is proved in [KR98]. Note that it uses a more relaxed notion of testing: First the algorithm needs to distinguish between functions in DT_s^d and functions that are far from any function in $\mathrm{DT}_{s'}^d$, where $s' > s$ (though for constant d, s' is not much larger). Second, it does not work for every distance parameter ϵ, but only for values bounded away from $1/2$ (and so can be seen as analogous to *weak learning*).

Theorem 9 *For any size* s, *dimension* d *and constant* $C \geq 1$, *let* $s' = s'(s, d, C) \overset{\text{def}}{=} 2^{d+1}(2s)^{1+1/C}$. *Then there exists an algorithm that uses uniformly distributed examples, and with probability at least* $2/3$ *accepts functions* $f \in \mathrm{DT}_s^d$ *and rejects functions that are* $\left(\frac{1}{2} - \frac{1}{2^{d+5}(Cd)^d}\right)$-*far from any decision tree in* $\mathrm{DT}_{s'}^d$. *The algorithm uses* $\tilde{O}\left((2Cd)^{2.5d} \cdot s^{\frac{1}{2}(1+1/C)}\right)$ *examples, and its running time is at most* $(2\log(2s))^d$ *times the number of examples used.*

A version of the algorithm that performs queries, has query complexity and running time $O\left((2Cd)^{2d+1} \cdot \log(s)^{d+1}\right)$.

References

[ADL+94] N. Alon, R. A. Duke, H. Lefmann, V. Rodl, and R. Yuster. The algorithmic aspects of the regularity lemma. *Journal of Algorithms*, 16:80–109, 1994.

[AFKS99] N. Alon, E. Fischer, M. Krivelevich, and M Szegedy. Efficient testing of large graphs. In *Proceedings of the Fortieth Annual Symposium on Foundations of Computer Science*, pages 645–655, 1999.

[AK99] N. Alon and M. Krivelevich. Testing k-colorability. Manuscript, 1999.

[AKNS99] N. Alon, M. Krivelevich, I. Newman, and M Szegedy. Regular languages are testable with a constant number of queries. In *Proceedings of the Fortieth Annual Symposium on Foundations of Computer Science*, pages 656–666, 1999.

[ALM+98] S. Arora, C. Lund, R. Motwani, M. Sudan, and M. Szegedy. Proof verification and intractability of approximation problems. *JACM*, 45(3):501–555, 1998.

[AS92] N. Alon and J. H. Spencer. *The Probabilistic Method*. John Wiley & Sons, Inc., 1992.

[AS97]   S. Arora and S. Sudan. Improved low degree testing and
           its applications. In *Proceedings of the Thirty-First Annual
           ACM Symposium on the Theory of Computing*, pages 485–
           495, 1997.

[AS98]   S. Arora and S. Safra. Probabilistic checkable proofs: A new
           characterization of NP. *JACM*, 45(1):70–122, 1998.

[BCH+95]   M. Bellare, D. Coppersmith, J. Håstad, M. Kiwi, and M. Su-
           dan. Linearity testing in characteristic two. In *Proceedings
           of the Thirty-Sixth Annual Symposium on Foundations of
           Computer Science*, pages 432–441, 1995.

[BFL91]   L. Babai, L. Fortnow, and C. Lund. Non-deterministic ex-
           ponential time has two-prover interactive protocols. *Compu-
           tational Complexity*, 1(1):3–40, 1991.

[BFLS91]   L. Babai, L. Fortnow, L. Levin, and M. Szegedy. Checking
           computations in polylogarithmic time. In *Proceedings of the
           Twenty-Third Annual ACM Symposium on Theory of Com-
           puting*, pages 21–31, 1991.

[BGLR93]   M. Bellare, S. Goldwasser, C. Lund, and A. Russell. Effi-
           cient probabilistically checkable proofs and applications to
           approximation. In *Proceedings of the Twenty-Sixth Annual
           ACM Symposium on the Theory of Computing*, pages 294–
           304, 1993.

[BGS98]   M. Bellare, O. Goldreich, and M. Sudan. Free bits, PCPs and
           non-approximability – towards tight results. *SIAM Journal
           on Computing*, 27(3):804–915, 1998.

[BLR93]   M. Blum, M. Luby, and R. Rubinfeld. Self-testing/correcting
           with applications to numerical problems. *JACM*, 47:549–595,
           1993.

[BR00]   M. Bender and D. Ron. Testing acyclicity of directed graphs
           in sublinear time. In *Proceedings of ICALP*, pages 809–820,
           2000.

[BS94]   M. Bellare and M. Sudan. Improved non-approximability re-
           sults. In *Proceedings of the Twenty-Sixth Annual ACM Sym-
           posium on the Theory of Computing*, pages 184–193, 1994.

[DGL+99]   Y. Dodis, O. Goldreich, E. Lehman, S. Raskhodnikova,
           D. Ron, and A. Samorodnitsky. Improved testing algorithms
           for monotonocity. In *Proceedings of RANDOM*, pages 97–
           108, 1999.

[DKL76]   E. A. Dinic, A. V. Karazanov, and M. V. Lomonosov. On
           the structure of the system of minimum edge cuts in a graph.

Studies in Discrete Optimizations, pages 290–306, 1976. In Russian.

[DW98]   Y. Dinitz and J. Westbrook.   Maintaining the classes of 4-edge-connectivity in a graph on-line. *Algorithmica*, 20(3):242–276, 1998.

[EKK+98] F. Ergun, S. Kannan, S. R. Kumar, R. Rubinfeld, and M. Viswanathan. Spot-checkers. In *Proceedings of the Thirty-Second Annual ACM Symposium on the Theory of Computing*, pages 259–268, 1998.

[FGL+96] U. Feige, S. Goldwasser, L. Lovász, S. Safra, and M. Szegedy. Approximating clique is almost NP-complete. *JACM*, 43(2):268–292, 1996.

[FK99]   A. Frieze and R. Kanan. Quick approximation to matrices and applications. *Combinatorica*, 19(2):175–220, 1999.

[GGL+00] O. Goldreich, S. Goldwasser, E. Lehman, D. Ron, and A. Samordinsky. Testing monotonicity. *Combinatorica*, 20(3):301–337, 2000.

[GGR98] O. Goldreich, S. Goldwasser, and D. Ron. Property testing and its connection to learning and approximation. *JACM*, 45(4):653–750, 1998.

[GLR+91] P. Gemmell, R. Lipton, R. Rubinfeld, M. Sudan, and A. Wigderson. Self-testing/correcting for polynomials and for approximate functions. In *Proceedings of the Twenty-Third Annual ACM Symposium on Theory of Computing*, pages 32–42, 1991.

[GR97]   O. Goldreich and D. Ron. Property testing in bounded degree graphs. In *Proceedings of the Thirty-First Annual ACM Symposium on the Theory of Computing*, pages 406–415, 1997. To appear in *Algorithmica*. A long version is available from http://www.eng.tau.ac.il/~danar/papers.html.

[GR99]   O. Goldreich and D. Ron. A sublinear bipartite tester for bounded degree graphs. *Combinatorica*, 19(3):335–373, 1999.

[Haj91] P. Hajnal. An $\Omega(n^{4/3})$ lower bound on the randomized complexity of graph properties. *Combinatorica*, 11(2):131–144, 1991.

[Hås96] J. Håstad. Testing of the long code and hardness for clique. In *Proceedings of the Thirtieth Annual ACM Symposium on the Theory of Computing*, pages 11–19, 1996.

648

[Hås97]   J. Håstad. Getting optimal in-approximability results. In *Proceedings of the Thirty-First Annual ACM Symposium on the Theory of Computing*, pages 1–10, 1997.

[HS87]   D. S. Hochbaum and D. B. Shmoys. Using dual approximation algorithms for scheduling problems: Theoretical and practical results. *JACM*, 34(1):144–162, January 1987.

[HS88]   D. S. Hochbaum and D. B. Shmoys. A polynomial approximation scheme for machine scheduling on uniform processors: Using the dual approximation approach. *SIAM Journal on Computing*, 17(3):539–551, 1988.

[Kar93]   D. Karger. Global min-cuts in $\mathcal{RNC}$ and other ramifications of a simple mincut algorithm. In *Proceedings of the Fourth Annual ACM-SIAM Symposium on Discrete Algorithms*, pages 21–30, 1993.

[Kin91]   V. King. An $\Omega(n^{5/4})$ lower bound on the randomized complexity of graph properties. *Combinatorica*, 11(1):23–32, 1991.

[Kiw96]   M. Kiwi. *Probabilistically Checkable Proofs and the Testing of Hadamard-like Codes*. PhD thesis, MIT, 1996.

[KR98]   M. Kearns and D. Ron. Testing problems with sub-learning sample complexity. In *Proceedings of the Eleventh Annual ACM Conference on Computational Learning Theory*, pages 268–277, 1998.

[KSS94]   M. J. Kearns, R. E. Schapire, and L. M. Sellie. Toward efficient agnostic learning. *Machine Learning*, 17(2-3):115–141, 1994.

[Mih89]   M. Mihail. Conductance and convergence of Markov chains - A combinatorial treatment of expanders. In *Proceedings 30th Annual Conference on Foundations of Computer Science*, pages 526–531, 1989.

[New00]   I. Newman. Testing of functions that have small width branching programs. In *Proceedings of the Forty-First Annual Symposium on Foundations of Computer Science*, 2000.

[PR99a]   M. Parnas and D. Ron. Testing the diameter of graphs. In *Proceedings of RANDOM*, pages 85–96, 1999.

[PR99b]   M. Parnas and D. Ron. Testing the diameter of graphs. Available from http://www.eng.tau.ac.il/~danar, 1999.

[Ros73]   A. L. Rosenberg. On the time required to recognize properties of graphs: A problem. *SIGACT News*, 5:15–16, 1973.

[RS96]   R. Rubinfeld and M. Sudan. Robust characterization of polynomials with applications to program testing. *SIAM Journal on Computing*, 25(2):252–271, 1996.

[RS97]   R. Raz and S. Safra. A sub-constant error-probability low-degree test, and a sub-constant error-probability PCP characterization of NP. In *Proceedings of the Thirty-First Annual ACM Symposium on the Theory of Computing*, pages 475–484, 1997.

[Rub99] R. Rubinfeld. Robust functional equations and their applications to program testing. *SIAM Journal on Computing*, 28(6):1972–1997, 1999.

[RV76]   R. L. Rivest and J. Vuillemin. On recognizing graph properties from adjacency matrices. *Theoretical Computer Science*, 3:371–384, 1976.

[Sud99] M. Sudan. Private communications, 1999.

[Sze78] E. Szemeredi. Regular partitions of graphs. In *Proc. Colloque Inter. CNRS*, pages 399–401, 1978.

[Tre98] L. Trevisan. Recycling queries in pcps and in linearity tests. In *Proceedings of the Thirty-Second Annual ACM Symposium on the Theory of Computing*, pages 299–308, 1998.

[Val84] L. G. Valiant. A theory of the learnable. *CACM*, 27(11):1134–1142, November 1984.

[Yao87] A. C. C. Yao. Lower bounds to randomized algorithms for graph properties. In *Proceedings of the Twenty-Eighth Annual Symposium on Foundations of Computer Science*, pages 393–400, 1987.

Chapter 16

THE RANDOM PROJECTION METHOD

Santosh Vempala
Department of Mathematics, MIT
Cambridge MA 02139
Email: **vempala@math.mit.edu**
Keywords: Projection, subspace, robust learning, nearest neigbors.

1.   INTRODUCTION

Random projection refers to the technique of projecting a set of points (or a distribution) from a high-dimensional space to a randomly chosen low-dimensional subspace. In recent years this technique has been used in algorithms for problems from a variety of areas, including combinatorial optimization, information retrieval and machine learning. In this chapter I will describe the technique and its basic properties, and survey its application to the design of efficient algorithms.

2.   HOW?

Let $u = (u_1, \ldots, u_n)^T$ be a vector in n-dimensional Euclidean space. Let k, a positive integer less than n, be the *target* dimension, i.e. our goal is to project u to a k-dimensional Euclidean space. To do this we first need to select a target subspace. A k-dimensional subspace can be represented as an $n \times k$ matrix whose columns are orthogonal unit vectors (such a matrix is called an *orthonormal* matrix). The columns of the matrix are a basis of the k-dimensional space. To carry out the projection, we choose a random orthonormal matrix R and then multiply R and u and scale the resulting vector. The projected vector will simply be

$$v = \sqrt{\frac{n}{k}} R^T u \qquad (16.1)$$

651

S. Rajasekaran et al (eds.), Handbook of Randomized Computing, Volume 2, pp, 651–671.
© 2001 *Kluwer Academic Publishers.*

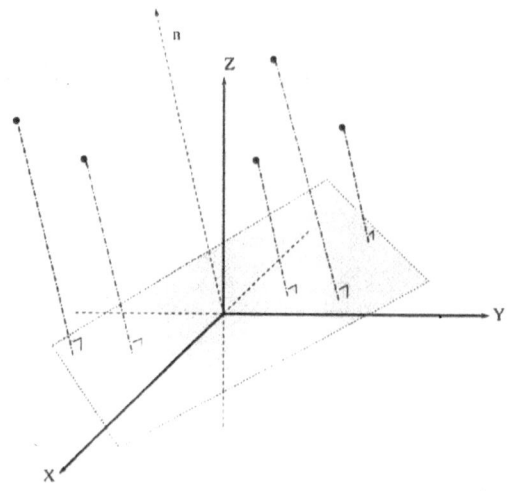

Figure 16.1 Random projection

3. WHY?

Random projection, while reducing dimensionality, approximately preserves pairwise distances with high probability. The following lemma due to Johnson and Lindenstrauss makes this precise.

Lemma 1 *[21] For any ϵ such that $\frac{1}{2} > \epsilon > 0$, and any set of points S in R^n, with $|S| = m$, upon random projection to R^k where $k \geq \frac{9 \ln m}{\epsilon^2 - \frac{2}{3}\epsilon^3} + 1$ the following property holds: with high probability, for every pair $u, v \in S$,*

$$(1 - \epsilon)|u - v|^2 \leq |f(u) - f(v)|^2 \leq (1 + \epsilon)|u - v|^2.$$

Frankl and Maehara [16] show that the above property is true when the projection matrix R is a random orthonormal matrix. A similar statement holds when each entry of the random matrix is chosen independently from the standard normal distribution, $N(0, 1)$. In fact, it is significantly easier to prove, as noted in [3, 20].

Lemma 2 *Let each entry of an $n \times k$ matrix R be chosen independently from $N(0, 1)$. Let $v = \frac{1}{\sqrt{k}} R^T u$ for $u \in R^n$. Then*

1. $E(|v|^2) = |u|^2$.

2. $Pr(||v|^2 - |u|^2| > \epsilon|u|^2) \leq 2e^{-\frac{\epsilon^2 k}{4}}$.

Proof.   Let $X = \frac{k}{|u|^2}|v|^2$. A straightforward calculation yields that $E(X) = k$ and thus $E(|v|^2) = |u|^2$. To obtain the concentration bounds, observe that

$$X = \sum_{j=1}^{k} \frac{(R_j^T \cdot u)^2}{|u|^2}$$

where R_j denotes the jth column of R. Each term $R_j^T \cdot u/|u|$ has the standard normal distribution (since each component of R_j does). The sum of their squares has the *chi-squared* distribution with k degrees of freedom [13], and its moment-generating function is

$$f(\lambda) = E(e^{\lambda X}) = \frac{1}{(1 - 2\lambda)^{\frac{k}{2}}}$$

when $\lambda \leq \frac{1}{2}$. Applying Markov's inequality, we get that for this range of λ,

$$Pr(X \geq (1 + \epsilon)k) = Pr(e^{\lambda X} \geq e^{(1+\epsilon)k\lambda}) \leq \frac{e^{-(1+\epsilon)k\lambda}}{(1 - 2\lambda)^{\frac{k}{2}}} \leq e^{-(\epsilon\lambda - \lambda^2)k}$$

Using $\lambda = \frac{\epsilon}{2}$ we have that

$$Pr(X \geq (1 + \epsilon)k) \leq e^{-\frac{\epsilon^2 k}{4}}$$

Similarly,

$$Pr(X \leq (1 - \epsilon)k) \leq e^{-\frac{\epsilon^2 k}{4}}$$

Combining the two proves the second part of the lemma. $\square$

4.   WHEN?

Random projection is useful in many settings. A natural setting is when the input data is in a high-dimensional space, and it is possible to preserve the essential properties of the data (for the particular problem at hand) while reducing the dimensionality. Another, perhaps less intuitive scenario, is when projection to a lower dimensional space actually *highlights* the essential properties. Our first example, the celebrated algorithm of Goemans and Williamson for the *maxcut* problem, is of the latter type.

4.1   MAXIMUM CUTS

Given an undirected graph $G = (V, E)$, the maximum cut problem (maxcut for short) is to find a bipartition of the vertices that maximizes the number of edges going across the partition. The problem appears in Karp's original list of NP-complete combinatorial problems and has a very simple $\frac{1}{2}$ approximation — choose a random cut. In fact, such a cut will contain, in expectation, half the edges of the graph.

For two decades improving on this approximation factor remained an open problem. In 1993, Goemans and Williamson [19] used semidefinite programming along with random projection to obtain a 0.87856 approximation, currently the best known.

To understand their algorithm, let us first consider the following integer program for maxcut.

$$\text{max} \quad \frac{1}{4} \sum_{ij \in E} (x_i - x_j)^2 \tag{16.2}$$

$$x_i \quad \in \quad \{-1, 1\} \quad \forall i \in V \tag{16.3}$$

The program assigns a label x_i that is either 1 or -1 to each vertex i. This defines a cut. The objective function value of such a cut is number of edges that have endpoints with different labels. If $x_i = x_j$ for an edge (i, j) then this contributes zero to the value of the cut. If $x_i \neq x_j$ then the edge contributes 1 to the value of the cut. Thus the optimum solution to the integer program is the maxcut of the graph.

Now consider the following relaxation of the integer program, obtained by relaxing the constraint (16.3) that forces each x_i to be 1 or -1 to one that allows each x_i to be some vector in n-dimensional Euclidean space, $\boldsymbol{R}^n$.

$$\text{max} \quad \frac{1}{4} \sum_{ij \in E} |x_i - x_j|^2 \tag{16.4}$$

$$|x_i|^2 \quad = \quad 1 \quad \forall i \in V \tag{16.5}$$

$$x_i \quad \in \quad \boldsymbol{R}^n \quad \forall i \in V \tag{16.6}$$

This program can be solved in polynomial-time. To see this, consider the change of variables $x_{ij} = x_i \cdot x_j$, i.e. the inner product of x_i and x_j. Then each term in the objective function can be rewritten as $\frac{1}{4}|x_i - x_j|^2 = \frac{1}{4}(x_{ii} + x_{jj} - 2x_{ij}) = \frac{1}{2}(1 - x_{ij})$. The constraint that each x_i is a vector in $\boldsymbol{R}^n$ is equivalent to the constraint that the $n \times n$ matrix X with x_{ij}'s as its entries is positive semidefinite. Thus the relaxation is equivalent to the following semi-definite program.

$$\text{max} \quad \frac{1}{2} \sum_{ij \in E} 1 - x_{ij} \tag{16.7}$$

$$x_{ii} \quad = \quad 1 \quad \forall i \in V \tag{16.8}$$

$$X = (x_{ij}) \quad \geq \quad 0 \tag{16.9}$$

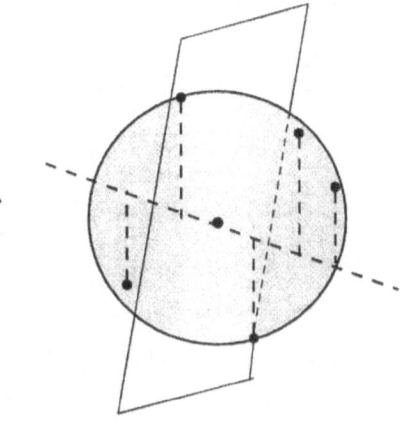

Figure 16.2   Cutting with a random hyperplane

Since semi-definite programs can be solved in polynomial-time [18, 2], we can solve the above relaxation and obtain a set of unit vectors x_i, one for each vertex i. How do we go from this to a cut of the graph, and hopefully to one with a lot of edges?

One way to get a cut is to simply project the vectors to a random one-dimensional line passing through the origin and define the two shores of the cut as the vectors that fall on the positive half of the line and the vectors that fall on the negative half of the line. This is equivalent to picking a random $n - 1$-dimensional hyperplane passing through the origin, which automatically divides the vectors into the two shores of a cut, one in each half-space.

Why is the solution obtained any good? In other words, why will it have a good number of edges compared to the optimum solution? The answer is given by the next two lemmas.

Lemma 3 *Let x_i and x_j be two unit vectors at an angle θ to each other. On projection to a random line r (a unit vector chosen uniformly at random), the probability that $r \cdot x_i$ has a different sign from $r \cdot x_j$ is exactly $\frac{\theta}{\pi}$.*

Thus the expected number of edges in the cut is

$$\sum_{ij \in E} \frac{\theta_{ij}}{\pi}.$$

And hence the approximation factor is

$$\frac{\sum_{ij\in E}\frac{\theta_{ij}}{\pi}}{\sum_{ij\in E}\frac{1}{2}(1-x_{ij})} = \frac{2}{\pi}\frac{\sum_{ij\in E}\theta_{ij}}{\sum_{ij\in E}(1-\cos(\theta))}$$

$$\geq \max\frac{2\theta}{\pi(1-\cos(\theta))}$$

Lemma 4

$$\max\frac{2\theta}{\pi(1-\cos(\theta))} > 0.87856$$

Theorem 5 *The maximum cut of a graph can be approximated to within a factor of* 0.87856 *of the optimum.*

In this algorithm, we used random projection to a line, i.e. a one-dimensional subspace. Several related problems have been tackled using this approach, notably max-k-cut [14], minimum bandwidth and other vertex ordering problems [12, 7]. Random projection to d-dimensional subspaces for $d = 2$ and higher has also turned out to be useful — for VLSI layout problems [32].

4.2   APPROXIMATE NEAREST NEIGHBORS

Given a distance function on pairs of points, a fixed set of points V, and a new point q, the nearest neighbor problem is to find the point in V that is nearest to q. The problem can be solved in polynomial-time simply by computing the distance of q to each point in V. The challenge is to find an algorithm that can solve the problem significantly faster. Not surprisingly, this usually involves some pre-processing (storage) of the fixed point set. The quality of an algorithm is then measured using two parameters: the storage space, and the query-processing time.

The choice of distance function plays a crucial role. We will focus on the case of points in d-dimensional Euclidean space. Ideally, one would like an algorithm that takes time polynomial in d and $\log n$, where n is the total number of points in V; and space polynomial in n and d. However, the best-known algorithms for finding the *exact* nearest neighbors in this setting are either much slower (time is exponential d or $\log n$, e.g. calculate the distance to every point) or use much more space (exponential in d, e.g. compute the Voronoi diagram of V [11, 27]).

By relaxing the problem to one of finding *approximate* nearest neighbors the time bounds improve dramatically. The ϵ-approximate nearest neighbor (or ϵ-near neighbor) problem is to find a point whose distance from q is within $(1 + \epsilon)$ times the distance from q to the true nearest point.

We now describe a solution to this problem proposed by [20]. The ratio of the largest distance in V to the smallest, denoted by R will be assumed to be at most n. This assumption makes it easier to explain the main ideas but is not essential (for the general case, see [20]). The first step is to reduce it to the following subproblem: *Given a query point q and a distance r, find a point in V at a distance of less than r if one exists; else report that none exists.* Suppose this problem can be solved efficiently, i.e. suppose we can set up a data structure that is space-efficient and has small query time. Then we can solve the nearest neighbor problem by searching for the smallest r at which there is a point in V. This can be done via a binary search, incurring an overhead of $O(\log R)$.

A similar idea works for ϵ-near neighbors. Suppose we have a data structure to solve the following point location problem (PL): Given $\epsilon > 0$, a query point q, and a distance r,

- If the nearest neighbor of q is within distance r, then it returns some point within distance $r(1 + \epsilon)$ of q.

- If the nearest neighbor of q is at a distance greater than $r(1 + \epsilon)$, then it reports that there are no neighbors within distance r.

Lemma 6 *There exists a data structure to solve the point location problem that takes up $nO(\frac{1}{\epsilon})^d$ space and answers queries in $O(d)$ time.*

Proof.   For each point v in the point set, cover the ball of radius r around v with cubes of side length $\frac{r\epsilon}{\sqrt{d}}$. Given a query point, check if it lies in one of the cubes around a point. If it does, then report the point; else report that there is no neighbor within a distance of r. The diameter of each cube is $r\epsilon$, so the reported point is within a distance of $r(1 + \epsilon)$.

To bound the total number of cubes, let $Vol(B_d(r))$ denote the volume of a d-dimensional ball of radius r. Then the number of cubes per point is at most

$$\frac{Vol(B_d(r(1 + \epsilon)))}{(\frac{r\epsilon}{\sqrt{d}})^d}$$

This follows from the observation that all the cubes around a point are contained in a ball of radius $r(1 + \epsilon)$. Using the well-known formula

$$Vol(B_d(r)) = \frac{2\pi^{\frac{d}{2}} r^d}{d\Gamma(d/2)}$$

where $\Gamma(x)$ is the gamma function (for a positive integer x, $\Gamma(x) = (x-1)!$), and Stirling's aproximation, we have that

$$\frac{Vol(B_d(r(1+\epsilon)))}{(\frac{r\epsilon}{\sqrt{d}})^d} = \frac{2\pi^{\frac{d}{2}}r^d(1+\epsilon)^d}{d\Gamma(d/2)}(\frac{\sqrt{d}}{r\epsilon})^d$$

$$= \frac{2\pi^{\frac{d}{2}}}{d\Gamma(d/2)}(\frac{(1+\epsilon)\sqrt{d}}{\epsilon})^d$$

$$\leq \frac{1}{\epsilon^d}\left((1+\epsilon)\sqrt{\frac{2\pi de}{d}}\right)^d$$

$$\leq O(\frac{1}{\epsilon})^d$$

Thus the total space requirement is $nO(\frac{1}{\epsilon})^d$. □

We will set up such a data structure for each r in the set $\{r_0, r_0(1+\epsilon), r_0(1+\epsilon)^2, \ldots, r_0(1+\epsilon)^t\}$, where r_0 is the smallest distance in V, and the last value of r exceeds R, the largest distance in V. To find an ϵ-near neighbor for a query point q, we search for the smallest r such that there is a point within r of q. There are $O(\log R)$ data structures, so the space is $O(\log R)$ times the space taken up by a single one. The time to answer a query is $O(d\log\log R)$.

The query time is nearly the best possible ($O(d)$ is a trivial lower bound). The best exact algorithm with similar query time takes up $n^{O(d)}$ space. So the approximate data structure reduces the space requirement considerably, but it is still exponential in the dimension. This is where random projection comes in. We first project V to a randomly chosen subspace of dimension $O(\frac{\log n}{\epsilon^2})$. With high probability, all pairs of distances will be approximately preserved. Build data structures for answering queries in the lower-dimensional space. To answer a query, project it to the random subspace and report a near neighbor of the projected point.

Theorem 7 *Approximate ϵ-near neighbors can be found with high probability in $O(d\log n + \log n\log\log R)$ time using $O(n^{O(\frac{\log\frac{1}{\epsilon}}{\epsilon^2})}\log R)$ space.*

Proof. The distances between pairs of points in the given set (with n points) are approximately preserved upon projection to a random subspace of dimension $O(\frac{\log n}{\epsilon^2})$. That is the distance after projection is at least $(1-\epsilon)$ and at most $(1+\epsilon)$ of the distance before projection. Given a new query point q we project it to the same subspace and compute its ϵ-near neighbors. From Lemma 6, computing near neighbors in the lower dimensional space takes a data structure of space complexity

$O((\frac{1}{\epsilon})^{O(\frac{\log n}{\epsilon^2})} \log R) = O(n^{O(\frac{\log \frac{1}{\epsilon}}{\epsilon^2})} \log R)$. The time to answer a query is $O(\frac{d \log n}{\epsilon^2})$ to project the query point plus $O(\log n \log \log R)$ to search the data structure. $\qquad\qquad\qquad\qquad\qquad\qquad\qquad\qquad\qquad\qquad\qquad$ □

The space requirement can be improved to $O(n^{O(\frac{1}{\epsilon^2})})$ by using random projection in conjunction with Hamming spaces [22]. The best known algorithms with polynomial space usage have query times that are nearly linear in n (see [20] for an $O(n^{\frac{1}{1+\epsilon}})$ bound).

4.3   FAST LOW-RANK APPROXIMATION

Our next application is towards a fast algorithm for *low-rank approximation*. Given an $n \times m$ matrix A, a rank k approximation of A is a matrix B of the same dimensions and rank at most k that minimizes (some measure of) the difference with A. In general low-rank approximations of data matrices serve two purposes: they reduce space requirements (always), and they provide a more transparent representation (often) [5, 26]. Standard methods to compute low-rank approximations via the *singular value decomposition (SVD)* take $O(\min\{nm^2, mn^2\})$ time in the worst case. Using random projection, this can be improved to $O(nm \log n)$ with a slight loss in accuracy.

SVD background.   Let A be an $n \times m$ matrix of rank r whose singular values (the eigenvalues of AA^T) are $\sigma_1 \geq \sigma_2 \geq \ldots \geq \sigma_r$ (not necessarily distinct). The singular value decomposition of A expresses A as the product of three matrices $A = UDV^T$, where $D = \text{diag}(\sigma_1, \ldots, \sigma_r)$ is an $r \times r$ matrix, $U = (u_1, \ldots, u_r)$ is an $n \times r$ matrix whose columns are orthonormal, and $V = (v_1, \ldots, v_r)$ is an $m \times r$ matrix which is also column-orthonormal.

A rank-k approximation to A is obtained by omitting all but the k largest singular values in the above decomposition. In a real application k should be small enough to enable fast retrieval, and large enough to adequately capture the structure of A (in practice, k is much smaller than m or n).

Let $D_k = diag(\sigma_1, \ldots, \sigma_k)$, $U_k = (u_1, \ldots, u_k)$ and $V_k = (v_1, \ldots, v_k)$. Then

$$A_k = U_k D_k V_k^T$$

is a matrix of rank k, which is our approximation of A. In other words, the column vectors of A (documents) are projected to the k-dimensional space spanned by the column vectors of U_k, i.e. the top k singular vectors of A. How good is this approximation? The following well-known theorem tells us that for a very natural measure of approximation, it is the *best* possible.

Theorem 8 *(Eckart and Young, see [17].) Among all $n \times m$ matrices C of rank at most k, A_k is the one that minimizes $\|A - C\|_F^2 = \sum_{i,j}(A_{i,j} - C_{i,j})^2$.*

The algorithm. Random projection preserves distance approximately, while reducing the dimensionality. This naturally suggests the following *two-phase* approach:

1. Apply a random projection to the initial corpus to l dimensions, for some small $l > k$, to obtain, with high probability, a much smaller representation, which is still very close (in terms of distances and angles) to the original matrix.

2. Find the top $O(k)$ singular vectors of the projected matrix and use these to find a low-rank approximation of the original matrix (because of the random projection, the number of singular values that need to be kept may have to be increased a little).

Let B be the $n \times l$ matrix obtained after projection. The SVD's of A and B can be expressed as

$$A = \sum_{i=1}^{r} \sigma_i u_i v_i^T, \quad B = \sum_{i=1}^{t} \lambda_i a_i b_i^T$$

where the u_i's and a_i's are left singular vectors and the v_i's and b_i's are right singular vectors of A and B respectively. Our first observation is that the singular values are approximately preserved.

Lemma 9 *Let ϵ be an arbitrary positive constant. If $l \geq C\frac{\log n}{\epsilon^2}$ for a sufficiently large constant C then, for $p = 1, \ldots, t$*

$$\lambda_p^2 \geq \frac{1}{k}[(1 - \epsilon)\sum_{i=1}^{k} \sigma_i^2 - \sum_{j=1}^{p-1} \lambda_j^2].$$

Proof. The pth eigenvalue of B can be written as

$$\lambda_p^2 = max_{|v|=1} v^T[B - \sum_{j=1}^{p-1} a_j b_j^T]^T[B - \sum_{j=1}^{p-1} a_j b_j^T]v$$

Consider the above expression for $v_1, \ldots, v_k$, the first k eigenvectors of A. For the ith eigenvector v_i it can be reduced to

$$v_i^T (B^T B - \sum_{j=1}^{p-1} \lambda_j^2 b_j b_j^T) v_i$$

$$v_i^T B^T B v_i - \sum_{j=1}^{p-1} \lambda_j^2 (b_j \cdot v_i)^2$$

$$\sigma_i^2 |u_i^T R|^2 - \sum_{j=1}^{p-1} \lambda_j^2 (b_j \cdot v_i)^2$$

$$\geq (1-\epsilon)\sigma_i^2 - \sum_{j=1}^{p-1} \lambda_j^2 (b_j \cdot v_i)^2$$

Summing this up for $i = 1, \ldots, k$,

$$\sum_{i=1}^{k} v_i^T B^T B v_i \geq (1-\epsilon) \sum_{i=1}^{k} \sigma_i^2 - \sum_{j=1}^{p-1} \lambda_j^2 \sum_{i=1}^{k} (b_j \cdot v_i)^2$$

Since the v_i's are orthogonal and the b_j's are unit vectors,

$$\geq (1-\epsilon) \sum_{i=1}^{k} \sigma_i^2 - \sum_{j=1}^{p-1} \lambda_j^2$$

Hence

$$\lambda_p^2 \geq max_v, v_i^T B^T B v_i \geq \frac{1}{k}[(1-\epsilon) \sum_{i=1}^{k} \sigma_i^2 - \sum_{j=1}^{p-1} \lambda_j^2]$$

$\square$

As a corollary we have:

Corollary 10

$$\sum_{p=1}^{2k} \lambda_p^2 \geq (1-\epsilon)\|A_k\|_F^2.$$

Our rank $2k$ approximation to the original matrix is

$$B_{2k} = A \sum_{i=1}^{2k} b_i b_i^T.$$

We can now state the main result of this section:

Theorem 11 *If $l > C\frac{\log n}{\epsilon^2}$ for a large enough constant C, then*

$$||A - B_{2k}||_F^2 \leq ||A - A_k||_F^2 + 2\epsilon||A||_F^2$$

The measure $||A - A_k||_F$ tells us how much of the original matrix is recovered by directly computing the best rank-k approximation. So, in other words, the theorem says that the matrix obtained via random projection followed by a rank-$2k$ approximation is almost as good as the matrix obtained by the best rank-k approximation.

Proof of Theorem 11.   We have,

$$A = \sum_{i=1}^{n} \sigma_i u_i v_i^T, \quad A_k = \sum_{i=1}^{k} \sigma_i u_i v_i^T$$

And also,

$$B = \sum_{i=1}^{l} \lambda_i a_i b_i^T, \quad B_{2k} = A\sum_{i=1}^{2k} b_i b_i^T$$

Since $b_1, \ldots, b_n$ are an orthonormal set of vectors,

$$||A - B_{2k}||_F^2 = \sum_{i=1}^{n} |(A - B_{2k})b_i|^2$$

But, for $i = 1, \ldots, 2k$,

$$(A - B_{2k})b_i = Ab_i - Ab_i = 0$$

And for $i = 2k+1, \ldots, n$

$$(A - B_{2k})b_i = Ab_i$$

Hence,

$$||A - B_{2k}||_F^2 = \sum_{i=2k+1}^{n} |Ab_i|^2 = \sum_{i=1}^{n} |Ab_i|^2 - \sum_{i=1}^{2k} |Ab_i|^2$$

$$= ||A||_F^2 - \sum_{i=1}^{2k} |Ab_i|^2$$

On the other hand,

$$||A - A_k||_F^2 = \sum_{i=k+1}^{n} \sigma_i^2 = ||A||_F^2 - ||A_k||_F^2$$

Hence

$$||A - B_{2k}||_F^2 - ||A - A_k||_F^2 = ||A_k||_F^2 - \sum_{i=1}^{2k} |Ab_i|^2$$

That is,

$$||A - B_{2k}||_F^2 = ||A - A_k||_F^2 + (||A_k||_F^2 - \sum_{i=1}^{2k} |Ab_i|^2)$$

Next, we will show that

$$(1 + \epsilon) \sum_{i=1}^{2k} |Ab_i|^2 \geq \sum_{i=1}^{2k} \lambda_i^2$$

We have

$$\sum_{i=1}^{2k} \lambda_i^2 = \sum_{i=1}^{2k} |Bb_i|^2$$

$$= \sum_{i=1}^{2k} |\sqrt{\frac{n}{l}} R^T (Ab_i)|^2$$

Now from lemma 1 we have that for l large enough, i.e., $l = \Omega(\frac{\log n}{\epsilon^2})$, with high probability,

$$(1 - \epsilon)|Ab_i|^2 \leq \frac{n}{l}|R^T(Ab_i)|^2 \leq (1 + \epsilon)|Ab_i|^2$$

for each i. Therefore with high probability,

$$\sum_{i=1}^{2k} \lambda_i^2 \leq (1 + \epsilon) \sum_{i=1}^{2k} |Ab_i|^2$$

From corollary 10

$$\sum_{i=1}^{2k} \lambda_i^2 \geq (1 - \epsilon)||A_k||^2$$

$$\sum_{i=1}^{2k} |Ab_i|^2 \geq \frac{1}{(1 + \epsilon)} \sum_{i=1}^{2k} \lambda_i^2 \geq \frac{1 - \epsilon}{1 + \epsilon}||A_k||_F^2 \geq (1 - 2\epsilon)||A_k||_F^2$$

Substituting this above, we have,

$$||A - B_{2k}||_F^2 \leq ||A - A_k||_F^2 + 2\epsilon||A_k||_F^2$$

Hence

$$||A - B_{2k}||_F^2 \leq ||A - A_k||_F^2 + 2\epsilon||A||_F^2$$

$\square$

What are the computational savings achieved by the two-step method? Let A be an $n \times m$ matrix. The time to compute the SVD is $O(mnc)$ if A is sparse with c non-zero entries per column. The time needed to compute the random projection to l dimensions is $O(mcl)$. After the projection, the time to compute the SVD is $O(ml^2)$. So the total time is $O(ml(l + c))$. To obtain an ϵ approximation we need l to be $O(\frac{\log n}{\epsilon^2})$. Thus the running time of the two-step method is asymptotically superior: $O(m \log n(c + \log n))$ compared to $O(mnc)$.

In [15], Frieze, Kannan and Vempala propose an alternative way to speed up low-rank approximations. They compute an approximate Singular Value Decomposition from a randomly chosen submatrix of A. For any given k, ϵ, δ, their Monte-Carlo algorithm finds *the description of* a matrix D of rank at most k so that

$$||A - D||_F \leq ||A - A_k||_F + \epsilon||A||_F$$

holds with probability at least $1 - \delta$. The algorithm takes time polynomial in $k, 1/\epsilon, \log(1/\delta)$ only, i.e. *independent of m, n*. In [10] it is shown how to find such a low-rank approximation explicitly constructed in $O(mk^2)$ time.

4.4   ROBUST CONCEPTS

Our last application is drawn from learning theory, and is motivated by the following question: How does the brain effectively learn concepts from a relatively small number of examples, when each example consists of a huge amount of information?

Let us consider the problem in Valiant's *probably approximately correct* (PAC) framework. Here the learner is presented with examples drawn from an unknown distribution. The examples are labelled with one of two labels: positive or negative, denoting whether the example belongs to the concept or not. The problem is to learn a hypothesis (e.g. a half-space), from a fixed family of hypotheses (e.g. the set of half-spaces in R^n), that is likely to correctly label new examples drawn from the same distribution.

A basic question in the PAC model is the number of examples needed to learn using a hypothesis from a particular family. The VC-dimension, of a hypothesis class, introduced by Vapnik and Chervonenkis [31] gives a tight bound on the number of examples. For concepts and examples involving n attributes (dimensions), the VC-dimension is typically n or greater. This is not consistent with observed reality: humans typically take a small ("constant") number of examples to learn a wide variety of concepts.

There are two approaches to explaining this phenomenon. The first, due to Valiant, is *attribute-efficient learning* [30, 23, 24]. Here it is assumed that the target concept is simple in a specific manner: it is a function of only a small subset of the set of attributes, called the *relevant* attributes, while the rest are *irrelevant*. From this assumption one can typically argue that the VC-dimension of the resulting concept class is a function of only the number of relevant attributes and the logarithm of the total number of attributes. Unfortunately, although the model is theoretically appealing, it is not known how anything more complex than a disjunction of variables can be learned in polytime (without membership queries) [1].

A second approach, called *robust concept learning* [3], is based on the simple idea illustrated in the following example. Imagine a child learning the concept of an "elephant". We point the child to pictures of elephants or to real elephants a few times and say "elephant", and perhaps to a few examples of other animals and say their names (i.e. "*not elephant*"). From then on, the child will almost surely correctly label only elephants as elephants. On the other hand, imagine a child learning the concept of "African elephant" (as opposed to the Indian elephant) just from examples. It will probably take many more examples, and perhaps even be necessary to explicitly point out the bigger ears of the African elephant.

The crucial difference in the two concepts above is not in the number of attributes, or even in the number of relevant attributes of the examples presented, but in the *robustness* of the concepts. There is a clearer demarcation between elephants and non-elephants than there is between African elephants and Indian elephants. This is the basis of the robust concept model, and will be formulated precisely below. An alternative perspective of robustness is that it is a measure of how much the attributes of an example can be altered without affecting the concept.

The robustness of a concept is measured by a parameter ℓ (defined below). The main feature of robust concepts is that the number of examples and the time required to learn a robust concept can be bounded as a function of ℓ and do not depend on the total number of attributes. In the rest of this section we will formalize the model and prove the main theorem for the concept class of half-spaces in n dimensions.

---

[1] Further, it is NP-hard to learn a disjunction of k variables as a disjunction of fewer than $k \log n$ variables.

The model. To describe the model, we adopt the terminology used in the literature. We assume that attributes are real valued; an *example* is a point in R^n; a *concept* is a subset of R^n. An example that belongs to a concept is labelled *positive* for the concept, and an example that lies outside the concept is labelled a *negative* example.

Given a set of labelled examples drawn from an unknown distribution $\mathcal{D}$ in R^n, and labelled according to an unknown *target* concept the learning task is to find a hypothesis with low error. A *hypothesis* is a polynomial-time computable function. The error of a hypothesis h with respect to the target concept is the probability that h disagrees with the target function on a random example drawn from $\mathcal{D}$. Thus, if h has error ϵ, then the probability for a random x that $h(x)$ disagrees with the target concept is at most ϵ. So, given an error parameter ϵ and a confidence parameter δ, with probability at least $1 - \delta$, the algorithm has to find a concept that has error at most ϵ on $\mathcal{D}$. Then the algorithm is said to PAC-learn the concept class [29].

The basic insight is the idea of robustness (implicit in earlier work, see [4] for related work). Intuitively, a concept is "robust" if it is immune to attribute noise. That is, modifying the attributes of an example by some bounded amount does not change its label. Another interpretation is that points with different labels are far apart. This is formalized below:

Definition 1 *For any real number $\ell > 0$, a concept C in conjunction with a distribution $\mathcal{D}$, is said to be ℓ-robust, if*

$$\Pr_{\mathcal{D}}[x | \exists y : label(x) \neq label(y), |x - y| \leq \ell] = 0$$

The norm $|x - y|$ is the distance between x and y. Here we use the Euclidean norm. The probability is over all points x with the property that there is some point y with a different label within a distance ℓ. In other words, a concept is ℓ-robust if there is zero probability of points being within ℓ of the boundary of the concept. The definition could be weakened by requiring only that the above probability should be negligible (e.g. $1/2^n$).

When $\mathcal{D}$ is over a discrete subset of R^n, then this has a simple interpretation. A ball of radius ℓ around any point x of non-zero probability lies entirely on one side of the concept, i.e., every point in the ball has the same label as x.

In what follows, we present tools and algorithms for learning robust concepts. It is important to note that "robustness" refers only to the target concept. That is, it is not required of all concepts in the class. Thus the algorithms can also be used when the robustness parameter is not known a priori — we can iteratively reduce the robustness parameter and still learn in complexity bounds related to the (a priori unknown) robustness of the target concept.

Learning robust half-spaces. The generic algorithm for learning robust concepts is based on two high-level ideas:

1. Since the target concept is robust, *random projection* to a much lower dimensional subspace will "preserve" the concept.

2. In the lower-dimensional space, the number of examples and time required to learn concepts are relatively low.

We now focus on the problem of learning a half-space in R^n (a linear threshold function). This is one of the oldest problems studied in learning theory. The problem can be solved in polytime by using a (polytime) algorithm for linear programming on a sample of $O(n)$ examples. Typically, however, it is solved by using simple greedy methods. A commonly-used greedy algorithm is the *Perceptron Algorithm* [28, 1], which has the following guarantee [25]. Given a collection of data points in R^n, each labeled as *positive* or *negative*, the algorithm will find a vector w such that $w \cdot x > 0$ for all positive points x and $w \cdot x < 0$ for all negative points x, if such a vector exists[2]. The running time of the algorithm depends on a separation parameter (described below). However, in order for the hypothesis to be reliable, we need to use a sample of $O(n)$ points.

Let $\mathcal{H}_n$ be the class of homogenous half-spaces in R^n (i.e. half-spaces passing through the origin). Let $(h, \mathcal{D})$ be a concept-distribution pair such that the half-space $h \in \mathcal{H}_n$ is ℓ-robust with respect to the distribution $\mathcal{D}$ over R^n. We restrict $\mathcal{D}$ to be over the unit sphere (i.e. all the examples are at unit distance from the origin). The latter condition is not really a restriction since examples can be scaled to have length 1 without changing their labels. The parameters k and f in the algorithm below will be specified later.

Half-space Algorithm:

1. Choose an $n \times k$ random matrix R for projection.

2. Obtain $f(\ell)$ examples from $\mathcal{D}$ and project them to R^k using R.

3. Run the Perceptron Algorithm to find a half-space $w \cdot x = 0$ in R^k: Set $w = 0$. Then perform the following operation until all examples are correctly classified:
   Pick an arbitrary misclassified example x and let $w \leftarrow w + label(x)x$.

---

[2] A non-zero threshold can be achieved by adding an extra dimension to the space

4. Output R and w.

A future example x is labelled positive if $w \cdot (R^T x) \geq 0$ and negative otherwise. This is of course the same as checking if $(wR^T) \cdot x > 0$, i.e. a half-space in the original n-dimensional space.

To prove that this agrees with h on most of $\mathcal{D}$, we first show that $R^T h$ agrees with h on most of $\mathcal{D}$.

Lemma 12 *Assume $\ell < 1$. For any $\alpha > 0$ and any integer $k \geq \frac{8\alpha}{\ell^2}$,*

$$Pr_{x \in \mathcal{D}}[label_h(x) \neq label_{R^T h}(R^T x)] \leq e^{-\alpha}$$

Proof. The target half-space h can be assumed to be of unit length, i.e., $|h| = 1$. The theorem follows by applying lemma 1 to the vector $h - x$ for an arbitrary x drawn from $\mathcal{D}$. $\square$

Next we bound the number of examples required for a hypothesis found by the algorithm to be reliable.

Theorem 13 *Let w be a half-space in $\mathbf{R}^k$ that correctly classifies a $(1 - \epsilon/4)$ fraction of a projected sample S of $|S| = \Omega(\frac{k \log(1/\delta)}{\epsilon})$ points drawn from $\mathcal{D}$. Then with probability at least $1 - \delta$, w correctly classifies a $1 - \epsilon$ fraction of $\mathcal{D}$.*

Proof (outline). To prove the theorem, we basically mimic the proof of the fundamental VC theorem. The only difference is that in that theorem, it is assumed that there is a hypothesis consistent with the entire sample. Here we can only assume that there is a hypothesis that correctly classifies $1 - \epsilon/4$ fraction of the sample. The proof proceeds by defining two events A and B. A is the event that there is a bad consistent hypothesis, i.e., a hypothesis that has error less than $\epsilon/4$ on the sample and error greater than ϵ on the distribution. B is the event that in a set of $2m$ examples, there is a concept that has error less than $\epsilon/4$ on the first m but has error greater than ϵ on the rest. Then we show that the probability of B is low and this implies that the probability of A is low. In bounding the probability of B, a crucial observation is that there are only $\binom{2m}{k}$ distinct ways to label $2m$ points using half-spaces in $\mathbf{R}^k$. $\square$

A choice of α greater than $4 \log \frac{1}{\epsilon \ell} \log \frac{1}{\delta}$ in Lemma 12 guarantees that with high probability ($> 1 - \delta$), there is a half-space that correctly classifies *all of* of a sample of $|S|$ points, and further, is $\ell/2$ robust with respect to the sample S. Given the existence of such a robust hypothesis in the lower-dimensional space, a consistent half-space will be quickly found by the Perceptron Algorithm. A classic theorem (see [25]) describes the convergence properties of the standard Perceptron Algorithm.

Theorem 14 *[25] Suppose the data set S can be correctly classified by some unit vector w^*. Then, the Perceptron Algorithm converges in at most $1/\sigma^2$ iterations, where $\sigma = min_{x \in S}|w^* \cdot \hat{x}|$.*

Using lemma 12, and theorems 13, 14, we have the main result of this section.

Theorem 15 *An ℓ-robust half-space in R^n can be PAC learned using $O(\frac{\log \frac{1}{\ell}}{\ell^2}\frac{1}{\epsilon}\log \frac{1}{\delta})$ examples in $O(n \cdot poly(\frac{1}{\ell}))$ time.*

The Perceptron algorithm is known to be tolerant to various types of classification noise [8, 6]. It is a straightforward consequence that these properties continue to hold for the above algorithm. In [3], similar results are described for learning other robust concept classes.

5. CONCLUSION

Random projection is a general, easy-to-use technique with a surprising variety of applications. Usually its application is a time or space saving exercise. However, there are cases where it is a crucial step in solving a problem. One example of this is the algorithm of [33] for learning an intersection of half-spaces. There, random projection is used to identify an "irrelevant" subspace. Another application that we have not discussed in this chapter is learning a mixture of Gaussians [9].

References

[1] S. Agmon, "The relaxation method for linear inequalities," Canadian Journal of Mathematics, 6(3):382–392, 1954.

[2] F. Alizadeh, Interior point methods in semidefinite programming with applications to combinatorial optimization. Siam J. Optimization, 5(1), 13-51, 1995.

[3] R. I. Arriaga and S. Vempala, "An algorithmic theory of learning: Robust concepts and random projection," Proc. of FOCS 1999.

[4] P. Bartlett and J. Shawe-Taylor, "Generalization Performance of Support Vector Machines and Other Pattern Classifiers," In *Advances in Kernel Methods - Support Vector Learning* (Ed.s B. Schvlkopf, C. Burges, and A. J. Smola) MIT Press, 1998.

[5] M. W. Berry, S. T. Dumais, and G. W. O'Brien. Using linear algebra for intelligent information retrieval. SIAM Review, 37(4), 1995, 573-595, 1995.

[6] A. Blum, A. Frieze, R. Kannan, and S. Vempala, "A Polynomial-time Algorithm for Learning Noisy Linear Threshold Functions," Algorithmica, 22(1), 35-52, 1998.

[7] A. Blum, G. Konjevod, R. Ravi and S. Vempala, "Semi-definite relaxations for minimum bandwidth and other vertex-ordering problems," Proc. STOC 1998.

[8] T. Bylander, "Learning linear threshold functions in the presence of classification noise," Proc. 7th Workshop on Computational Learning Theory, 1994.

[9] S. Dasgupta, "Learning a mixture of Gaussians," Proc. of FOCS 1999.

[10] P. Drineas, A. Frieze, R. Kannan, S. Vempala, and V. Vinay. Clustering in large graphs and matrices. To appear, *Proc. of the ACM-SIAM SODA*, 1999.

[11] H. Edelsbrunner. Algorithms in Combinatorial Geometry. Springer.

[12] U. Feige, "Approximating the bandwidth via volume-respecting embeddings," Proc. of STOC 1998.

[13] W. Feller, *An Introduction to Probability Theory and Its Applications*, John Wiley and Sons, Inc., 1957.

[14] A. Frieze and M. R. Jerrum. Improved approximation algorithms for MAX k-cut and MAX BISECTION. Algorithmica 18, 61-77, 1997.

[15] A. Frieze, R. Kannan and S. Vempala, "Fast Monte-Carlo Algorithms for finding low-rank approximations," Proc. 1998 FOCS, pp. 370-378, 1998.

[16] P. Frankl and H. Maehara. The Johnson-Lindenstrauss Lemma and the Sphericity of some graphs, J. Comb. Theory B **44** (1988), 355-362.

[17] G. Golub and C. Reinsch. Handbook for matrix computation II, Linear Algebra. Springer-Verlag, New York, 1971.

[18] M. Grötschel, L. Lovász, A. Schrijver, *Geometric Algorithms and Combinatorial Optimization*, Springer, 1988.

[19] M. Goemans and D. Williamson, "Improved Approximation Algorithms for Maximum Cut and Satisfiability Problems Using Semidefinite Programming," J. ACM, 42, 1115–1145, 1995.

[20] P. Indyk and R. Motwani, "Approximate nearest neighbors: Towards removing the curse of dimensionality," Proc. of ACM STOC 1998.

[21] W. B. Johnson and J. Lindenstrauss, "Extensions of Lipshitz mapping into Hilbert space," Contemporary Mathematics, 26, 189-206, 1984.

[22] E. Kushilevitz, R. Ostrovsky and Y. Rabani, "Efficient search for approximate nearest neighbors in high-dimensional spaces," Proc. of STOC 1998.

[23] N. Littlestone, "Learning Quickly When Irrelevant Attributes Abound: A New Linear-threshold Algorithm," Machine Learning, 2:285-318, 1987.

[24] N. Littlestone, "Redundant noisy attributes, attribute errors, and linear threshold learning using winnow," Proc. 4th Workshop on Computational Learning Theory, 1991.

[25] M. Minsky and S. Papert. *Perceptrons: An Introduction to Computational Geometry.* The MIT Press, 1969.

[26] C. Papadimitriou, P. Raghavan, H. Tamaki and S. Vempala, "Latent Semantic Indexing: A Probabilistic Analysis," Proc. of the 17th ACM Symposium on the Principles of Database Systems, Seattle, 1998.

[27] F. Preparata and M. Shamos. Computational Geometry: An Introduction. Springer.

[28] F. Rosenblatt. *Principles of Neurodynamics.* Spartan Books, 1962.

[29] L. G. Valiant, "A Theory of the Learnable," Communications of the ACM, 27(11), 1134-1142.

[30] L. G. Valiant, "A Neuroidal Architecture for Cognitive Computation," Proc. of ICALP 98.

[31] V. N. Vapnik and A. Ya. Chervonenkis, "On the uniform convergence of relative frequencies of events to their probabilities," Theory of Probability and its applications, XVI(2):264–280, 1971.

[32] S. Vempala, "Random Projection: A New Approach to VLSI Layout," Proc. of the 39th IEEE Foundations of Computer Science, 1998.

[33] S. Vempala, "A random sampling based algorithm for learning the intersection of half-spaces," Proc. of FOCS 1997.

Chapter 17

ERROR ESTIMATIONS FOR INDIRECT MEASUREMENTS: RANDOMIZED VS. DETERMINISTIC ALGORITHMS FOR "BLACK-BOX" PROGRAMS

Vladik Kreinovich
and Raúl Trejo

Abstract   In many real-life situations, it is very difficult or even impossible to directly measure the quantity y in which we are interested: e.g., we cannot directly measure a distance to a distant galaxy or the amount of oil in a given well. Since we cannot measure such quantities *directly*, we can measure them *indirectly*: by first measuring some relating quantities $x_1, \ldots, x_n$, and then by using the known relation between x_i and y to reconstruct the value of the desired quantity y.

In practice, it is often very important to estimate the error of the resulting indirect measurement. In this paper, we describe and compare different deterministic and randomized algorithms for solving this problem in the situation when a program for transforming the estimates $\tilde{x}_1, \ldots, \tilde{x}_n$ for x_i into an estimate for y is only available as a *black box* (with no source code at hand).

We consider this problem in two settings: *statistical*, when measurements errors $\Delta x_i = \tilde{x}_i - x_i$ are independent Gaussian random variables with 0 average and known standard deviations σ_i, and *interval*, when the only known information about Δx_i is that $\Delta x_i \in [-\Delta_i, \Delta_i]$ for a known bound Δ_i. In statistical setting, we describe the optimal error estimation algorithm; in interval setting, we describe a new algorithm which may be not optimal but which is better than the previously known ones.

S. Rajasekaran et al (eds.), Handbook of Randomized Computing, Volume 2, pp, 673–729.
© 2001 *Kluwer Academic Publishers.*

1. ERROR ESTIMATION FOR INDIRECT MEASUREMENTS – FORMULATION OF THE PROBLEM

What Are Indirect Measurements. In many real-life situations, it is difficult or even impossible to directly measure the quantity y in which we are interested. For example, it is, at present, practically impossible to directly measure a distance to a distant quasar, or the amount of oil in a given area. Since we cannot measure such quantities *directly*, we have to measure them *indirectly*: Namely, we measure some other quantities $x_1, \ldots, x_n$ which are related to y by a known dependence $y = f(x_1, \ldots, x_n)$, and then apply the known function f to the results $\tilde{x}_1, \ldots, \tilde{x}_n$ of measuring x_i.

For example, to determine the amount of oil y in a given area, we measure the results x_i of sending ultrasound signals between the two parallel wells, and then estimate y by solving the appropriate system of partial differential equations (in this example, $f(x_1, \ldots, x_n)$ is a program for solving this system).

In general, such a two-stage procedure (measurement followed by compu- tations) is called an *indirect measurement*, and the value $\tilde{y} = f(\tilde{x}_1, \ldots, \tilde{x}_n)$ resulting from this two-stage procedure is called the *result* of indirect measure- ment.

Toy Example. To make the exposition clearer, we will illustrate these notions on the following toy example: Suppose that we are interested in the voltage V, but we have no voltmeter at hand. One possibility of measuring V indirectly follows from Ohm's law: we can measure the current I and the resistance R, and compute V as $I \cdot R$. In this case, x_1 is the current, $x_2 = R$, and $f(x_1, x_2) = x_1 \cdot x_2$.

If the measured value $\tilde{x}_1$ of the current is 1.0, and the measured value of the resistance is $\tilde{x}_2 = 2.0$, then the result of the corresponding indirect measurement is $\tilde{y} = 1.0 \cdot 2.0 = 2.0$.

Error Estimation for Indirect Measurements: A Real-Life Problem. Measurements are never 100% accurate; hence, the result $\tilde{x}_i$ of each direct measurement is, in general, somewhat different from the actual value of the measured quantity. As a result of these measurement errors $\Delta x_i = \tilde{x}_i - x_i$, the result $\tilde{y} = f(\tilde{x}_1, \ldots, \tilde{x}_n)$ of applying f to the measurement result will be, in general, different from the actual value $y = f(x_1, \ldots, x_n)$ of the desired quantity.

> For example, in our toy problem, the actual value of the current may be $x_1 = 0.9 \neq \tilde{x}_1 = 1.0$, and the actual value of the resistance is $x_2 = 2.05 \neq \tilde{x}_2 = 2.0$. In this case, the actual value of the voltage is $y = x_1 \cdot x_2 = 0.9 \cdot 2.05 = 1.845 \neq 2.0$.

Since the result $\tilde{y}$ of indirect measurement is, in general, different from the actual value y, it is desirable to know the characteristics of the error $\Delta y = \tilde{y} - y$ of indirect measurement. How can we estimate these characteristics?

Possible Information Available for Estimating the Error of Indirect Measurements.   First, we know the function $f(x_1, \ldots, x_n)$. This function may be given as an analytical expression, or, more frequently, as an algorithm. It may be a program written in a high-level programming language (i.e., a *source code*), which can be translated into an executable file ready for computations, or it may be only an executable file, with no source code provided.

Second, we know the results $\tilde{x}_1, \ldots, \tilde{x}_n$ of direct measurements.

Finally, we need some information about the errors of the direct measurements. The errors Δx_i come from the imperfection of the corresponding measuring instruments. For an instrument to be called *measuring*, its manufacturer must supply some (well-defined) information about the measurement errors. Ideally, this information must include the probability distribution of different measurement errors.

The knowledge of these probabilities is desirable but not always required and not always possible. In many practical cases, we only know the upper bounds Δ_i for the possible measurement errors, i.e., we only know that $|\Delta x_i| \leq \Delta_i$. In such cases, after each direct measurement, the only information that have about the actual value x_i of the measured quantity is that this value belongs to the *interval* $[\tilde{x}_i - \Delta_i, \tilde{x}_i + \Delta_i]$.

> For example, in our toy case, the manufacturer of the measuring instruments may guarantee that the measurement error Δx_1 of measuring current cannot exceed $\Delta_1 = 0.1$, and the measurement error Δx_2 of measuring resistance cannot exceed $\Delta_2 = 0.05$. If no other information about the measurement accuracy is given, then, after we got the measurement results $\tilde{x}_1 = 1.0$ and $\tilde{x}_2 = 2.0$, the only information we have about the actual value of the current x_1 is that $x_1 \in [1.0 - 0.1, 1.0 + 0.1] = [0.9, 1.1]$. Similarly, the only information we have about the actual value of the resistance x_2 is that $x_2 \in [2.0 - 0.05, 2.0 + 0.05] = [1.95, 2.05]$. (The actual values $x_1 = 0.9$ and $x_2 = 1.05$, of course, belong to these intervals; if they did not, this would mean that the manufacturer's bounds are incorrect.)

In the situations when we only know the upper bounds on the measurement errors, the problem of estimating the error of indirect measurement is called the problem of *interval computations*; for details and examples of practical applications, see, e.g., [31, 65]. The setting when we only know intervals will be one of the settings considered in this paper.

Another setting which we will consider is a setting described in standard engineering textbooks on measurement (see, e.g., [25, 56]; see also [13, 28]). In this setting, the measurement error Δx_i of each direct measurement is normally distributed with 0 average and known standard deviation σ_i, and measurement errors of different direct measurements are independent random variables. These assumptions can be justified by two related reasons:

- The first reason is more *theoretical*. Usually, the manufacturers of the measuring instruments have made their best effort to eliminate the major sources of measurement error. The resulting measurement error comes from a variety of small independent error sources, and thus, can be described as a sum of a large number of small independent random variables. According to the central limit theorem, such a sum, under reasonable conditions, converges to normal distribution. Thus, if there are sufficiently many small random components, the resulting error distribution is indeed close to normal.

- The second reason for choosing Gaussian distribution is *empirical*: for the majority of measuring instruments, the measurement error is indeed normally distributed (see, e.g., [52, 54]).

In This Paper, We Will Only Consider Situations When the Measurements Are Reasonably Accurate. In this paper, we will only consider situations in which the direct measurements are accurate enough, so that the resulting measurement errors Δx_i are small, and terms which are quadratic (or of higher order) in Δx_i can be safely neglected, and so, the dependence of the desired value $y = f(x_1, \ldots, x_n) = f(\widetilde{x}_1 - \Delta x_1, \ldots, \widetilde{x}_n - \Delta x_n)$ on Δx_i can be safely assumed to be linear.

> In our toy example, $f(x_1, x_2) = x_1 \cdot x_2$, so $y = f(\widetilde{x}_1 - \Delta x_1, \widetilde{x}_2 - \Delta x_2) = \widetilde{x}_1 \cdot \widetilde{x}_2 - \widetilde{x}_2 \cdot \Delta x_1 - \widetilde{x}_1 \cdot \Delta x_2 + \Delta x_1 \cdot \Delta x_2$. In our case, $\widetilde{x}_1 = 1.0$, $\widetilde{x}_2 = 2.0$, so $y = 2.0 - 2\Delta x_1 - \Delta x_2 + \Delta x_1 \cdot \Delta x_2$. The only non-linear term in this expansion is the quadratic term $\Delta x_1 \cdot \Delta x_2$.

> Here, $\Delta x_1 = \widetilde{x}_1 - x_1 = 1.0 - 0.9 = 0.1$, $\Delta x_2 = 2.0 - 2.05 = -0.05$, and $\Delta y = 2.0 - 1.845 = 0.155$. If we ignore the quadratic term, we get approximate values $y_{\text{approx}} = 2.0 - 2\Delta x_1 - \Delta x_2 = 1.85$ and $\Delta y_{\text{approx}} = \widetilde{y} - y_{\text{approx}} = 0.15$. The error of this linear approximation is $0.005 \ll 0.15$.

Comments.

- To avoid possible confusion, we must emphasize the following. We are *not* talking about functions f which are linear for *all* possible values of input data. In this paper, we are considering data processing functions f

which can be approximated by linear ones in the close vicinity of every measurement result $\vec{\tilde{x}} = (\tilde{x}_1, \ldots, \tilde{x}_n)$. These linear approximations, however, are *different* for different measurement results.

- There are practical situations when the accuracy of the direct measurements is not high enough, and hence, quadratic terms cannot be safely neglected (see, e.g., [31] and references therein). In this case, the problem of error estimation for indirect measurements becomes computationally difficult (NP-hard) even when the function $f(x_1, \ldots, x_n)$ is quadratic [43, 62]. However, in most real-life situations, the possibility to ignore quadratic terms is a reasonable assumption, because, e.g., for an error of 1% its square is a negligible 0.01%.

With the above restriction in place, we can easily deduce the explicit expression for the error Δy of indirect measurement.

Indirect Measurement Error: Derivation and the Resulting Formula. Due to the accuracy requirement, we can simplify the expression for $\Delta y = \tilde{y} - y = f(\tilde{x}_1, \ldots, \tilde{x}_n) - f(x_1, \ldots, x_n)$ if we expand the function f in Taylor series around the point $(\tilde{x}_1, \ldots, \tilde{x}_n)$ and restrict ourselves only to linear terms in this expansion. As a result, we get the expression

$$\Delta y = c_1 \cdot \Delta x_1 + \ldots + c_n \cdot \Delta x_n, \tag{1}$$

where by c_i, we denoted the value of the partial derivative $\partial f / \partial x_i$ at the point $(\tilde{x}_1, \ldots, \tilde{x}_n)$:

$$c_i = \frac{\partial f}{\partial x_i}\bigg|_{(\tilde{x}_1, \ldots, \tilde{x}_n)}. \tag{2}$$

Probability Distribution of the Indirect Measurement Error: Derivation and the Resulting Formula. In the statistical setting, the desired measurement error Δy is a linear combination of independent Gaussian variables Δx_i, and hence, Δy is also normally distributed, with 0 average and the standard deviation

$$\sigma = \sqrt{c_1^2 \cdot \sigma_1^2 + \ldots + c_n^2 \cdot \sigma_n^2}. \tag{3}$$

Comment. A similar formula holds if we *do not* assume that Δx_i are normally distributed: it is sufficient to assume that they are independent variables with 0 average and known standard deviations σ_i.

Interval of Possible Values of the Indirect Measurement Error: Derivation and the Resulting Formula. In the interval setting, we do not know the probability of different errors Δx_i; instead, we only know that $|\Delta x_i| \leq \Delta_i$. In this case, the sum (1) attains its largest possible value if each term $c_i \cdot \Delta x_i$ in this sum attains the largest possible value:

- If $c_i \geq 0$, then this term is a monotonically non-decreasing function of Δx_i, so it attains its largest value at the largest possible value $\Delta x_i = \Delta_i$; the corresponding largest value of this term is $c_i \cdot \Delta_i$.

- If $c_i < 0$, then this term is a decreasing function of Δx_i, so it attains its largest value at the smallest possible value $\Delta x_i = -\Delta_i$; the corresponding largest value of this term is $-c_i \cdot \Delta_i = |c_i| \cdot \Delta_i$.

In both cases, the largest possible value of this term is $|c_i| \cdot \Delta_i$, so, the largest possible value of the sum Δy is

$$\Delta = |c_1| \cdot \Delta_1 + \ldots + |c_n| \cdot \Delta_n. \tag{4}$$

Similarly, the smallest possible value of Δy is $-\Delta$.

Hence, the interval of possible values of Δy is $[-\Delta, \Delta]$, with Δ defined by the formula (4).

Comment. In our toy problem, it is easy to compute the actual interval of possible values of $y = x_1 \cdot x_2$ when $x_1 \in [0.9, 1.1]$ and $x_2 \in [1.95, 2.05]$: Indeed, the function $f(x_1, x_2) = x_1 \cdot x_2$ is monotonically increasing as a function of each of its variables (for $x_1 > 0$ and $x_2 > 0$). Thus, the largest possible value of $y = f(x_1, x_2)$ is attained when both input variables take their largest possible values, i.e., when $x_1 = 1.1$ and $x_2 = 2.05$, and is equal to $1.1 \cdot 2.05 = 2.255$. Similarly, the smallest possible value of $y = f(x_1, x_2)$ is attained when both input variables take their smallest possible values, i.e., when $x_1 = 0.9$ and $x_2 = 1.95$; this smallest value of y is equal to $0.9 \cdot 1.95 = 1.755$. So, the interval of possible values of y is equal to $[1.755, 2.255]$. Hence, the interval of possible values for $\Delta y = \tilde{y} - y = 2 - y$ is $[-0.255, 0.245]$.

On the other hand, applying formula (4), we get $\Delta = 2.0 \cdot 0.1 + 1.0 \cdot 0.05 = 0.25$ and the interval $[-0.25, 0.25]$. (We can see that this is indeed a good approximation to the actual interval.)

Error Estimation for Indirect Measurement: a Precise Computational Formulation of the Problem. As a result of the above analysis, we get the following explicit formulation of the problem: given a function $f(x_1, \ldots, x_n)$, n numbers $\tilde{x}_1, \ldots, \tilde{x}_n$, and n positive numbers $\sigma_1, \ldots, \sigma_n$ (or $\Delta_1, \ldots, \Delta_n$), compute the corresponding expression (3) or (4).

Let us describe how this problem is solved now.

Textbook Case: The Function f is Given by Its Analytical Expression. If the function f is given by its analytical expression, then we can simply explicitly differentiate it, and get an explicit expression for (3) and (4). This is the case which is typically analyzed in textbooks on measurement theory (see, e.g., [25, 56]).

A More Complicated Case: Analytical Differentiation. In many practical cases, we do not have an explicit analytical expression, we only have an *algorithm* for computing the function $f(x_1, \ldots, x_n)$, an algorithm which is too complicated to be expressed as an analytical expression.

When this algorithm is presented in one of the standard programming languages such as Fortran or C, we can apply one of the existing analytical differentiation tools (see, e.g., [8, 27]), and automatically produce a program which computes the partial derivatives c_i. These tools analyze the code and produce the differentiation code as they go.

In Many Practical Applications, We Must Treat the Function $f(x_1, \ldots, x_n)$ As a Black Box. In many other real-life applications, an algorithm for computing $f(x_1, \ldots, x_n)$ may be written in a language for which an automatic differentiation tool is not available, or a program is only available as an executable file, with no source code at hand. In such situations, when we have no easy way to analyze the code, the only thing we can do is to take this program as a *black box*: i.e., to apply it to different inputs and use the results of this application to compute the desired value σ.

In this paper, we will analyze such black-box situations, and describe the optimal algorithm for computing σ, and a new algorithm for computing Δ. Before we describe these algorithms, we must do two things: formulate the black-box approach in mathematical terms, and describe known black-box-oriented algorithms.

Comment. Algorithms designed for the black-box situation can be also used when we do have a code of a program f, and we can, in principle, use automatic differentiation tools; we can always ignore this knowledge and treat the program as a black box. In some practical situations, this turned out to be useful, because both the automatic differentiation algorithms and the black-box-oriented algorithms come with a substantial computational overhead, and the overhead for black-box-oriented algorithms is sometimes smaller.

Error Estimation for Indirect Measurements: Towards a Mathematical Reformulation of the Black-Box Approach. The general idea of a black-box approach is that, in addition to computing the result $\tilde{y} = f(\tilde{x}_1, \ldots, \tilde{x}_n)$ of the indirect measurement, we also apply the function f to other inputs which are different from $(\tilde{x}_1, \ldots, \tilde{x}_n)$.

Our ultimate goal is to analyze how errors in x_i induce errors in the value of the function f. We have already assumed that the errors in x_i are small, so the vector $(x_1, \ldots, x_n)$ of the actual values is close to the measured values

$(\tilde{x}_1, \ldots, \tilde{x}_n)$ (so close that the squares of measurement errors $\Delta x_i = \tilde{x}_i - x_i$ can be neglected). From the viewpoint of this goal, we are only interested in the values of the function f in the small vicinity of the measurement point, i.e., for the values $x_i = \tilde{x}_i + \delta_i$ for some small δ_i (so small that their squares are negligible). Since quadratic terms are negligible, we can expand f in Taylor series and only keep linear terms in this expansion. As a result, when we apply the function f to the values $x_i + \delta_i$, we get the result

$$f(\tilde{x}_1 + \delta_1, \ldots, \tilde{x}_n + \delta_n) = \tilde{y} + c_1 \cdot \delta_1 + \ldots + c_n \cdot \delta_n. \tag{5}$$

So, in effect, for sufficiently small vectors $\vec{\delta}$ (i.e., for vectors for which $\left\|\vec{\delta}\right\| \leq \delta_0$ for some real number $\delta_0 > 0$), we get the *dot (scalar)* product $\vec{c} \cdot \vec{\delta}$ between the input vector $\vec{\delta} = (\delta_1, \ldots, \delta_n)$ and the (unknown) vector $\vec{c} = (c_1, \ldots, c_n)$.

Thus, for an arbitrary *small* vector $\vec{\delta}$, we can find its dot product $\vec{c} \cdot \vec{\delta}$ with the (unknown) gradient vector $\vec{c}$ by a single call to the black-box computing f: namely, as $f\left(\tilde{\vec{x}} + \vec{\delta}\right) - \tilde{y}$. We can use a slightly more complex idea to compute the product $\vec{c} \cdot \vec{\delta}$ for an *arbitrary* (not necessarily small) vector $\vec{\delta}$ as follows:

- First, we compute the length $\delta = \left\|\vec{\delta}\right\|$ of the given vector $\vec{\delta}$. Then:

- If $\delta \leq \delta_0$, then we compute $\vec{c} \cdot \vec{\delta}$ as $f\left(\tilde{\vec{x}} + \vec{\delta}\right) - \tilde{y}$.

- If $\delta > \delta_0$, then we compute $\vec{c} \cdot \vec{\delta}$ as follows:
  - first, we compute the normalization coefficient $K_{\text{norm}} = \delta/\delta_0$;
  - then, we compute the auxiliary vector $\vec{\beta} = \vec{\delta}/K_{\text{norm}}$ with components $\beta_i = \delta_i/K_{\text{norm}}$; due to normalization, for this auxiliary vector, $\left\|\vec{\beta}\right\| = \delta_0$;
  - third, we compute $\vec{c} \cdot \vec{\beta} = f\left(\tilde{\vec{x}} + \vec{\beta}\right) - \tilde{y}$;
  - finally, we compute $\vec{c} \cdot \vec{\delta}$ as $K_{\text{norm}} \cdot \left(\vec{c} \cdot \vec{\beta}\right)$.

Let us explicitly repeat the resulting mathematical reformulation.

Error Estimation for Indirect Measurements: A Mathematical Reformulation of the Black-Box Approach. We know the values $\tilde{y}, \sigma_1, \ldots, \sigma_n$ (or $\tilde{y}$ and $\Delta_1, \ldots, \Delta_n$); we know that there is a vector $\vec{c} = (c_1, \ldots, c_n)$ but we do not know the values of its components. For any given vector $\vec{\delta} = (\delta_1, \ldots, \delta_n)$, we can compute the dot product $\vec{c} \cdot \vec{\delta} = c_1 \cdot \delta_1 + \ldots + c_n \cdot \delta_n$. Based on the results of these computations, we must compute, correspondingly, the value (3) or (4).

A Straightforward Method of Solving This Problem: Numerical Differentiation. The most straightforward algorithm for solving this problem is to compute the derivatives c_i one-by-one, and then use the corresponding formula (3) or (4) to compute the desired σ. To compute the i-th partial derivative, we change the i-th input x_i to $\tilde{x}_i + h_i$ for some h_i, and leave other inputs unchanged, i.e., we take $\delta_i = h_i$ for this i and $\delta_j = 0$ for all $j \neq i$. Then, we estimate c_i as

$$c_i = \frac{1}{h_i} \cdot (f(\tilde{x}_1, \ldots, \tilde{x}_{i-1}, \tilde{x}_i + h_i, \tilde{x}_{i+1}, \ldots, \tilde{x}_n) - \tilde{y}).$$

This algorithm is called *numerical differentiation*.

We want the change h_i to be small (so that quadratic terms can be neglected); we already know that changes of the order σ_i are small. So, it is natural to take $h_i = \sigma_i$ (or, correspondingly, $h_i = \Delta_i$). In other words, to compute c_i, we use the following values: $\delta_1 = \ldots = \delta_{i-1} = 0$, $\delta_i = \sigma_i$ (or $\delta_i = \Delta_i$), $\delta_{i+1} = \ldots = \delta_n = 0$.

This choice of δ_i has an additional advantage: it decreases the computation time. Namely, no matter which values h_i we use, we need the same number of calls to f (i.e., n calls), but by using the above values h_i, we decrease the total number of additional computations. Indeed, e.g., in interval setting:

- For *general* h_i, we need three arithmetic operations to compute each value c_i: one addition to compute $\tilde{x}_i + h_i$, one subtraction to compute the difference $f(\tilde{x}_1, \ldots, \tilde{x}_{i-1}, \tilde{x}_i + h_i, \tilde{x}_{i+1}, \ldots, \tilde{x}_n) - \tilde{y}$, and one division to divide this difference by h_i. So, to compute all n partial derivatives, we need $3n$ arithmetic operations. Then, we need n takings of absolute value, n multiplications and $n - 1$ additions to compute the sum (4), which brings the total to $3n + n + n + (n - 1) = 6n - 1$ arithmetic operations.

- For the above values of h_i, the difference $f(\tilde{x}_1, \ldots, \tilde{x}_{i-1}, \tilde{x}_i + h_i, \tilde{x}_{i+1}, \ldots, \tilde{x}_n) - \tilde{y}$ is equal to $c_i \cdot h_i = c_i \cdot \Delta_i$, so there is no need to first divide it by h_i and then multiply it by Δ_i: we can directly estimate $|c_i| \cdot \Delta_i$ as $|f(\tilde{x}_1, \ldots, \tilde{x}_{i-1}, \tilde{x}_i + h_i, \tilde{x}_{i+1}, \ldots, \tilde{x}_n) - \tilde{y}|$. Thus, to compute Δ, we now need n additions $\tilde{x}_i + h_i$, n subtractions $f(\tilde{x}_1, \ldots, \tilde{x}_{i-1}, \tilde{x}_i + h_i, \tilde{x}_{i+1}, \ldots, \tilde{x}_n) - \tilde{y}$, n absolute values to compute the terms $|c_i| \cdot \Delta_i$, and $n - 1$ additions to add all these terms, to the total of $4n - 1 (< 6n - 1)$ arithmetic operations.

For statistical setting, we get a similar decrease in computation time.

Comment. The above numerical differentiation algorithm is the simplest possible. There are other algorithms that give better results for non-linear functions (see, e.g., [26]); e.g., we can estimate c_i as

$$f_{,i} = \frac{1}{h_i} \cdot \left(f\left(\tilde{x}_1, \ldots, \tilde{x}_{i-1}, \tilde{x}_i + \frac{h_i}{2}, \tilde{x}_{i+1}, \ldots, \tilde{x}_n\right) - \right.$$

$$\left. f\left(\tilde{x}_1, \ldots, \tilde{x}_{i-1}, \tilde{x}_i - \frac{h_i}{2}, \tilde{x}_{i+1}, \ldots, \tilde{x}_n\right)\right).$$

Such algorithms take longer to compute (e.g., the above algorithm requires $2n$ calls to f instead of n). Spending this extra time make sense if terms quadratic in h_i cannot be neglected. However, since we assumed that we can neglect these quadratic terms, even our simplest numerical differentiation algorithm leads to exact value of c_i and therefore, there is no need to spend time on more sophisticated differentiation techniques.

A natural next question is: can we use an even simpler numerical differentiation algorithm which would allow us to compute the quantity (3) or (4) with fewer than n calls to f? As we will soon see, if we restrict ourselves to *deterministic* algorithms, the answer is: No, n is the smallest we can get.

Sometimes, Numerical Differentiation Takes Too Long. If a function $f(x_1, \ldots, x_n)$ is simple and fast-to-compute (e.g., if it is given by an explicit analytical expression), then we do not need the black-box-oriented algorithms at all. We only need these algorithms when the program f is itself time-consuming (e.g., computing f may involve solving an inverse problem). In this case, applying the function f is the most time-consuming part of this algorithm. So, the total time T that it takes us to compute σ is (approximately) equal to the running time T_f for the program f multiplied by the number of times N_f that we call the program f.

For numerical differentiation, $N_f = n$ (we call f n times to compute n partial derivatives). Hence, if the program f takes a long time to compute, and n is huge, then the resulting time T may be too long. For example, if we are determining some parameters of an oil well from the geophysical measurements, we may get n in the thousands, and T_f in minutes. In this case, $T = T_f \cdot n$ may take several weeks. This may be OK for a single measurement, but too long if we want more on-line results.

If We Do Not Have Enough Time for Numerical Differentiation, We May Use Randomized Algorithms. The only way to save on computation time is to limit the number of times we call f, i.e., to limit N_f.

If we limit this number of calls to $N_f < n$, we cannot get an answer which is always correct: Indeed, if applied $N_f < n$ different vectors $\vec{\delta} = (\delta_1, \ldots, \delta_n)$ and computed the values $\vec{\delta} \cdot \vec{c}$, then we get $\leq n - 1$ equations for n variables $c_1, \ldots, c_n$; therefore, this system does not have a single solution; it has at least a 1-D line of possible solutions $\vec{c}$, and for points of this line which go to ∞, both expressions (3) and (4) tend to ∞ as well. Thus, we can have arbitrarily large values of σ and Δ which are consistent with the results of black-box testing, and we cannot, therefore, deduce the exact value of σ or Δ from these testing results.

Since we cannot use *deterministic* algorithms, i.e., algorithms which always guarantee an answer, we can try *randomized* algorithms, which provide us with an answer with a certain probability. In real-life applications, the fact that these algorithms have a (small) probability of erring is OK, because the measuring instruments themselves have a small probability of failing anyway.

Randomized algorithms can indeed speed up computations: e.g., in statistical setting, a straightforward simulation (Monte-Carlo type) saves time drastically:

Monte-Carlo Simulation for Statistical Setting.   In this algorithm, we use a computer-based random number generator to simulate the normally distributed error. A standard normal random number generator usually produces a normal distribution with 0 average and standard deviation 1. So, to simulate a distribution with a standard deviation σ_i, we multiply the result α_i of the standard random number generator by σ_i. In other words, we take $\delta_i = \sigma_i \cdot \alpha_i$.

As a result of N Monte-Carlo simulations, we get N values $c^{(1)} = \vec{c} \cdot \vec{\delta}^{(1)}, \ldots, c^{(N)} = \vec{c} \cdot \vec{\delta}^{(N)}$ which are normally distributed with the desired standard deviation σ. So, we can determine σ by using the standard statistical estimate

$$\sigma = \sqrt{\frac{1}{N-1} \cdot \sum_{k=1}^{N} \left(c^{(k)}\right)^2}.$$

The relative error of this estimate depends only on N (as $\approx 1/\sqrt{N}$), and not on the number of variables n. Therefore, the number of steps N_f needed to achieve a given accuracy does not depend on the number of variables at all.

The error of the above algorithm is asymptotically normally distributed, with a standard deviation $\sigma_e \sim \sigma/\sqrt{2N}$. Thus, if we use a "two sigma" bound, we conclude that with probability 95%, this algorithm leads to an estimate for σ which differs from the actual value of σ by $\leq 2\sigma_e = 2\sigma/\sqrt{2N}$.

This is an error with which we estimate the error of indirect measurement; we do not need too much accuracy in this estimation, because, e.g., in real life, we say that an error is $\pm 10\%$ or $\pm 20\%$, but *not* that the error is, say, $\pm 11.8\%$. Therefore, in estimating the error of indirect measurements, it is sufficient to estimate the characteristics of this error with a relative accuracy of, say, 20%.

For the above "two sigma" estimate, this means that we need to select the smallest N for which $2\sigma_e = 2\sigma/\sqrt{2N} \leq 0.2 \cdot \sigma$, i.e., to select $N_f = N = 50$.

In many practical situations, it is sufficient to have a standard deviation of 20% (i.e., to have a "two sigma" guarantee of 40%). In this case, we need only $N = 13$ calls to f.

On the other hand, if we want to guarantee 20% accuracy in 99.9% cases, which correspond to "three sigma", we must use N for which $3\sigma_e = 3 \cdot \sigma/\sqrt{2N} \leq 0.2 \cdot \sigma$, i.e., we must select $N_f = N = 113$, etc.

For $n \approx 10^3$, all these values of N_f are much smaller than $N_f = n$ required for numerical differentiation.

So, if we have to choose between the (deterministic) numerical differentiation and the randomized Monte-Carlo algorithm, we must select:

- a deterministic algorithm when the number of variables n satisfies the inequality $n \leq N_0$ (where $N_0 \approx 50$), and

- a randomized method if $n \geq N_0$.

These two algorithms are the most widely used. A natural question is: are they optimal? Our answer is: no, for many different values of n, it is possible to find an algorithm which is better than both.

What We Are Planning To Do. In this paper, we will describe the *best* randomized algorithm for estimating σ (which is better than the Monte-Carlo simulation), and a new algorithm for estimating Δ.

Our main ideas and the preliminary (restricted) versions of our results were first announced in [33, 37, 40, 41, 42, 44, 45].

2. ERROR ESTIMATION FOR INDIRECT MEASUREMENT: STATISTICAL SETTING

2.1 ERROR ESTIMATION FOR INDIRECT MEASUREMENT REFORMULATED AS A TOMOGRAPHY PROBLEM

When $\sigma_1 = \ldots = \sigma_n = 1$, The Error Estimation Problem Has a Natural Geometric (Tomographic) Interpretation. When $\sigma_1 = \ldots = \sigma_n = 1$, then the formula (3) becomes a formula for the length of a vector $\vec{c} = (c_1, \ldots, c_n)$. Therefore, the above error estimation problem for indirect measurements can be reformulated in the following geometric terms: We have an (unknown) vector $\vec{c}$. We want to know its length $c = \|\vec{c}\|$. In order to compute this length, we can pick an arbitrary vector $\vec{\delta}$ and compute a scalar product $\vec{c} \cdot \vec{\delta}$. The question is: how to estimate the desired length by picking the smallest possible number of vectors. Alternatively, if the number N of picked vectors is fixed, the question is: How to select these vectors $\vec{\delta}^{(1)}, \ldots, \vec{\delta}^{(N)}$ in such a way that from the scalar products $c^{(1)}, \ldots, c^{(N)}$, we will able to get the best possible estimate $\tilde{c}$ for the desired length c.

From the mathematical viewpoint, knowing the scalar product $\vec{c} \cdot \vec{\delta}$ of the unknown vector $\vec{c}$ with a known vector $\vec{\delta}$ is equivalent to knowing the *projection*

$\pi_{\vec{\delta}}(\vec{c}) = (\vec{c} \cdot \vec{\delta}) / \|\vec{\delta}\|$. Therefore, the above problem means that we need to reconstruct the length of a vector from its projections to different directions.

In general, problems in which we want to reconstruct a certain property of an unknown object based on the characteristics of its projections are called *tomography problems*. Therefore, the above problem can be viewed as a (simple) case of tomography problems.

The General Case Can Be Naturally Reduced to the (Geometrically Interpretable) Case When $\sigma_1 = \ldots = \sigma_n = 1$. Let us show that the general case of the error estimation problem for indirect measurements can be naturally reduced to the case of $\sigma_i = 1$. Indeed, in the general case, we must find the expression (2) for unknown values $c_1, \ldots, c_n$. Instead of considering the values c_i as unknowns, we can introduce new unknowns: $c'_i = c_i \cdot \sigma_i$. Then, the desired expression (3) is simply the length $\sigma = \left\| \vec{c'} \right\|$ of the new unknown vector $\vec{c'} = (c'_1, \ldots, c'_n)$.

To complete the reduction to the case of $\sigma_i = 1$, we need to reformulate the expression for the observed quantity $\sum c_i \cdot \delta_i$ in terms of the new variables $c'_i = c_i \cdot \sigma_i$. Such a reformulation is straightforward: $\sum c_i \cdot \delta_i = \sum c'_i \cdot \delta'_i$, where $\delta'_i = \delta_i / \sigma_i$. If we know δ'_i, then we can reconstruct δ_i as $\delta_i = \delta'_i \cdot \sigma_i$.

Therefore, if we know how to solve the geometrizable problem (with $\sigma_i = 1$), we can then solve the problem with arbitrary σ_i as follows:

- first, we generate a vector $\vec{\delta'}$ which is appropriate for the geometrized problem, and

- then use the values $\delta_i = \delta'_i \cdot \sigma_i$ to solve the given problem.

This reduction is very natural, because in both above algorithms – numerical differentiation and Monte-Carlo – the values δ_i were actually of the form $\delta_i = \sigma_i \cdot \alpha_i$ for some values α_i which did not depend on σ_i.

Precise Formulation of the Corresponding Geometric Problem. In view of the above reduction, it is sufficient to solve the geometrizable particular case of the error estimation problem. Let us formulate this problem in precise mathematical terms.

Definition 1. *Let positive integers n and N be given. By an algorithm which solves the n-variable error estimation problem in N calls (or simply an (n, N)-algorithm, for short), we mean a tuple $\mathcal{U} = (p_1, \ldots, p_N, \tilde{e})$, where:*

- *for every k from 1 to N, p_k is a function which assigns, to every element from $R^{n \times (k-1)} \times R^{k-1}$, a probability measure on R^n, and*

- *$\tilde{e}$ is a function which assigns to every element of $R^{n \times N} \times R^N$ a probability measure on R.*

Definition 2. *Let* $n, N > 0$. *By a* (n, N)-computational trajectory, *we mean a sequence* $\vec{\delta}^{(1)}, c^{(1)}, \vec{\delta}^{(2)}, c^{(2)}, \ldots, \vec{\delta}^{(N)}, c^{(N)}, \tilde{c}$, *where* $\tilde{c}$ *is a real number, and for each k from 1 to N*, $\vec{\delta}^{(k)} \in R^n$ *is an n-dimensional vector and* $c^{(k)}$ *is a real number.*

Definition 3. *Let* $\mathcal{U} = (p_1, \ldots, p_N, \tilde{e})$ *be an* (n, N)-algorithm. *Then, for each n-dimensional vector* $\vec{c} \in R^n$, *the algorithm* $\mathcal{U}$ *generates the following probability distribution on the set of all possible* (n, N)-computational trajectories:

- *for* $k = 1$, *a random vector* $\vec{\delta}^{(1)}$ *is generated according to the probability distribution* p_1, *and we compute* $c^{(1)} = \vec{c} \cdot \vec{\delta}^{(1)}$;

- *next, for* $k = 2, 3, \ldots, N$, *a random vector* $\vec{\delta}^{(k)}$ *is generated according to the probability distribution* $p_k \left(\vec{\delta}^{(1)}, \ldots, \vec{\delta}^{(k-1)}, c^{(1)}, \ldots, c^{(k-1)} \right)$; *then, we compute* $c^{(k)} = \vec{c} \cdot \vec{\delta}^{(k)}$;

- *finally, a random value* $\tilde{c}$ *is generated according to the probability measure* $\tilde{e} \left(\vec{\delta}^{(1)}, \ldots, \vec{\delta}^{(N)}, c^{(1)}, \ldots, c^{(N)} \right)$.

Let $\Phi(\tilde{c}, c)$ *be a function which, for each c, attains its minimum for* $\tilde{c} = c$ *and is convex in c.*

- *This function will be called the* inaccuracy *of approximating c by* $\tilde{c}$.

- *For each vector* $\vec{c} \in R^n$ *of length* $c = \|\vec{c}\| \neq 0$, *the inaccuracy* $I(\mathcal{U}, \vec{c})$ *of an algorithm* $\mathcal{U}$ *on this vector is defined as* $I(\mathcal{U}, \vec{c}) = E(\Phi(\tilde{c}, c))$.

- *By an* inaccuracy $I(\mathcal{U})$ *of an algorithm* $\mathcal{U}$, *we mean the largest possible value of its inaccuracy on any vector* $\vec{c} \neq 0$.

Example. For example, we can take

$$\Phi(\tilde{c}, c) = \left(\frac{\tilde{c} - c}{c} \right)^2.$$

Comment. In the above definitions, we define a randomized algorithm for solving our problem. In particular, if all probability distributions are degenerate (i.e., concentrate on a single vector with probability 1), we get a deterministic algorithm.

Definition 4. *Let* $n, N > 0$. *We say that an* (n, N)-algorithm *is* optimal *if it has the smallest possible inaccuracy.*

Now, we are ready to formulate and to prove the main result of this section:

2.2   THE MAIN RESULT FOR STATISTICAL SETTING: OPTIMAL ERROR ESTIMATION ALGORITHM FOR INDIRECT MEASUREMENTS

In this subsection, we will explicitly describe, for each n and N and for each inaccuracy criterion, the optimal (n, N)-algorithm. For readers' convenience, all the proofs are located in a special (last) section.

Theorem 1. *When $N \geq n$, numerical differentiation is the optimal (n, N)-algorithm.*

When $N < n$, the following (n, N)-algorithm $\mathcal{U} = (p_1, \ldots, p_k, \widetilde{e})$ is optimal:

- p_1 *is a uniform distribution on a unit sphere in R^n;*

- *for each $k > 1$, $p_k\left(\vec{\delta}^{(1)}, \ldots, \vec{\delta}^{(k-1)}, c^{(1)}, \ldots, c^{(k-1)}\right)$ is a uniform distribution on the set of all unit vectors of R^n which are orthogonal to $(k-1)$ vectors $\vec{\delta}^{(1)}, \ldots, \vec{\delta}^{(k-1)}$ (this set is an $(n-k)$-dimensional sphere);*

- *finally, the probability distribution $\widetilde{e}\left(\vec{\delta}^{(1)}, \ldots, \vec{\delta}^{(N)}, c^{(1)}, \ldots, c^{(N)}\right)$ is concentrated, with probability 1, on a real number*

$$\widetilde{c} = g\left(\sum_{k=1}^{N} \left(c^{(k)}\right)^2\right), \tag{6}$$

where $g : R \to R$ is a function of one variable which depends on the criterion Φ.

Comments.

- So, if we allow at least n calls to a black-box program f, then the deterministic numerical differentiation algorithm is optimal for error estimation; when we allow fewer than n calls, we have to use a randomized algorithm.

- As we will see from the proof, a similar result holds if instead of estimating the length c of the vector $\vec{c}$, we estimate an arbitrary *function* of this length. For example, if we want to estimate $\sigma^2 = c^2$, then it is natural to choose a criterion

$$\Phi(\widetilde{c}, c) = \left(\frac{(\widetilde{c})^2 - c^2}{c^2}\right)^2.$$

For this criterion, the optimal estimate comes from

$$g(z) = \sqrt{\frac{n+2}{N+2}} \cdot z :$$

688

that $g(z) = k \cdot z$ follows from the fact that the problem is invariant under scaling $z \to \lambda \cdot z$ for an arbitrary $\lambda > 0$, and the specific optimal choice of k comes from the explicit computations.

- A similar theorem holds if instead of *arbitrary* (n, N)-algorithms, we will only consider algorithm which produce *un-biased* estimates for c or for an appropriate function of c. For example, if we are estimating $c^2 = \sigma^2$, and we want to choose an unbiased estimate which is optimal under the above criterion, then we get the estimate

$$(\widetilde{c})^2 = \frac{n}{N} \cdot \sum_{k=1}^{N} \left(c^{(k)} \right)^2 .$$

Here, $g(z) = (n/N) \cdot z$.

- When we choose the desired accuracy of an error estimation algorithm, then we select the algorithm which requires the smallest number of calls to the black-box program $f(x_1, \ldots, x_n)$. For the three algorithms that we have so far described – numerical differentiation, Monte-Carlo simulation, and the new (optimal) algorithm – a typical dependence of the corresponding number of calls on the number of variables n is depicted in Figure 1.1.

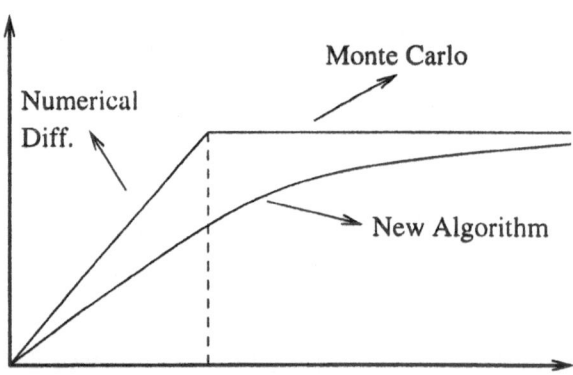

Figure 17.1 Comparison of different algorithms

2.3 HOW TO ACTUALLY PROGRAM THIS OPTIMAL ALGORITHM·

The mathematical description of the above optimal algorithm may look somewhat complex, but in reality, this algorithm is rather easy to program. In this programming, we must know the value δ_0 such that for vectors $\vec{\delta} =$

$(\delta_1, \ldots, \delta_n)$ with $\left\| \vec{\delta} \right\| \leq \delta_0$, we can neglect quadratic terms in the expansion of the (black-box) function $f\left(\vec{\tilde{x}} + \vec{\delta} \right)$ in δ_i.

First Auxiliary Algorithm: Gram-Schmidt Orthogonalization. In the first auxiliary algorithm, we use the well-known Gram-Schmidt orthogonalization procedure: If we have N vectors $\vec{\delta}^{(1)}, \ldots, \vec{\delta}^{(N)}$, then the following formulas produce orthogonal unit vectors $\vec{e}^{(1)}, \vec{e}^{(2)}, \ldots$ which span the same linear space as the vectors $\vec{\delta}^{(k)}$:

$$\vec{e}^{(1)} = \frac{\vec{\delta}^{(1)}}{\left\| \vec{\delta}^{(1)} \right\|}, \tag{7}$$

and then for each $k > 1$,

$$\vec{e}^{(k)} = \frac{\vec{\varphi}^{(k)}}{\left\| \vec{\varphi}^{(k)} \right\|},$$

where

$$\vec{\varphi}^{(k)} = \vec{\delta}^{(k)} - \sum_{i=1}^{k-1} \left(\vec{\delta}^{(k)} \cdot \vec{e}^{(i)} \right) \cdot \vec{e}^{(i)}. \tag{8}$$

Second Auxiliary Algorithm: Gaussian Random Number Generator. We must use a random number generator for generating a normally distributed random variable with 0 average and unit standard deviation, so that consequent calls to this program produce independent and equally distributed random variables (some programming languages have explicit commands for such a generator, others have known algorithms for it).

The Algorithm Itself.

- First, we apply the above random number generator n times, and get n components of the first vector $\vec{a}^{(1)}$. Then, we apply this same random number generator n more times, and get n components of the second vector $\vec{a}^{(2)}$, etc.

- After we have generated N such vectors, we apply Gram-Schmidt orthogonalization to generate the vectors $\vec{e}^{(1)}, \ldots, \vec{e}^{(N)}$.

- Then, for each k from 1 to N, we compute $c^{(k)} = \vec{c} \cdot \vec{e}^{(k)}$ by applying the function f to the values $\vec{\tilde{x}} + \delta_0 \cdot \vec{e}^{(k)}$, and then computing $c^{(k)}$ as

$$c^{(k)} = \frac{1}{\delta_0} \cdot \left(f\left(\vec{\tilde{x}} + \delta_0 \cdot \vec{e}^{(k)} \right) - \tilde{y} \right).$$

(This computation is slightly easier than in the general case, because we know that $\vec{\delta} = \vec{e}^{(k)}$ is a unit vector, so its length $\delta = \left\| \vec{e}^{(k)} \right\|$ equals 1.)

- Finally, we compute the estimate $\tilde{c}$ by using the formula (6).

3. ERROR ESTIMATION FOR INDIRECT MEASUREMENT: INTERVAL SETTING

3.1 MONTE-CARLO SIMULATION FOR INTERVAL SETTING: A KNOWN ALGORITHM

Main Idea Behind the Known Randomized Algorithm. In statistical setting, Monte-Carlo simulation is a known error estimation algorithm. It is somewhat less well known that a similar simulation algorithm can be designed for interval setting. Let us explain how such an algorithm can be designed. In this explanation, we will try to show that the resulting algorithm is not an unnatural mathematical trick, it does appear naturally. (A reader who is only interested in the algorithm itself, not in its motivations, is welcome to go directly to the next subsection.)

In Monte-Carlo algorithm for statistical setting, we start with n independent variables $\alpha_1, \ldots, \alpha_n$ which are normally distributed with 0 average and standard deviation 1. The variable α_i simulates the error Δx_i when the standard deviation σ_i of Δx_i equals 1. To simulate the error Δx_i for a general σ_i, we use $\delta_i = \sigma_i \cdot \alpha_i$. The simulation algorithm is based on the fact that for all possible sequences of real numbers $c_1, \ldots, c_n$, and $\sigma_1 \geq 0, \ldots, \sigma_n \geq 0$, the distribution of the linear combination

$$c = c_1 \cdot \delta_1 + \ldots + c_n \cdot \delta_n = c_1 \cdot \sigma_1 \cdot \alpha_1 + \ldots + c_n \cdot \sigma_n \cdot \alpha_n$$

is the same as for $\sigma \cdot \alpha$, where σ is determined by the formula (3).

To design a similar algorithm for the interval setting, it is therefore desirable to find a new probability distribution which will play the same role for this setting as a standard normal distribution plays for the statistical setting. In other words, we would like to be able to start with n independent variables $\alpha_1, \ldots, \alpha_n$ which are all distributed according to this new distribution. The variable α_i simulates the error Δx_i when the bound Δ_i corresponding to Δx_i equals 1. To simulate the error Δx_i for a general Δ_i, we use $\delta_i = \Delta_i \cdot \alpha_i$.

For this idea to work, we must select the distribution in such a way that if $\alpha_1, \ldots, \alpha_n$ are n independent random variables distributed according to this distribution, then for all possible sequences of real numbers $c_1, \ldots, c_n$, and $\Delta_1 \geq 0, \ldots, \Delta_n \geq 0$, the distribution of the linear combination

$$c = c_1 \cdot \delta_1 + \ldots + c_n \cdot \delta_n = c_1 \cdot \Delta_1 \cdot \alpha_1 + \ldots + c_n \cdot \Delta_n \cdot \alpha_n$$

is the same as for $\Delta \cdot \alpha$, where Δ is determined by the formula (4). Let us find distributions which satisfy this condition.

A natural way to describe a probability distribution is to describe its probability density function $\rho(z)$. However, this may not be the best way to represent the (unknown) desired probability distribution, because the probability density for the sum of several independent random variables is described by a convolution, and hence, in terms of $\rho(z)$, the above condition becomes a complex integral equation. We therefore need to look for alternative descriptions of a probability distribution which will enable us to handle the sum of independent random variables easier.

Independence of two random variables x and x' means, in particular, that the mathematical expectation $E[x \cdot x']$ of their product is equal to the product of their mathematical expectations: $E[x \cdot x'] = E[x] \cdot E[x']$. Independence of x and x' also means that for any two functions $f(z)$ and $f'(z)$, the variables $f(x)$ and $f'(x')$ are also independent and therefore,

$$E[f(x) \cdot f'(x')] = E[f(x)] \cdot E[f'(x')].$$

Therefore, to use the condition that c is a sum of independent random variables $x_n = c_n \cdot \Delta_n \cdot \alpha_n$, we can use the fact that for an exponential function $f(z) = \exp(a \cdot z)$, we have $f(c) = f(x_1 + \ldots + x_n) = f(x_1) \cdot \ldots \cdot f(x_n)$, and thus, $E[f(c)] = E[f(x_1)] \cdot \ldots \cdot E[f(x_n)]$, i.e.,

$$E[\exp(a \cdot c)] = E[\exp(a \cdot c_1 \cdot \Delta_1 \cdot \alpha_1)] \cdot \ldots \cdot E[\exp(a \cdot c_n \cdot \Delta_n \cdot \alpha_n)].$$

The requirement that c has the same distribution as $\Delta \cdot \alpha$ implies that

$$E[\exp(a \cdot \Delta \cdot \alpha)] = E[\exp(a \cdot c_1 \cdot \Delta_1 \cdot \alpha_1)] \cdot \ldots \cdot E[\exp(a \cdot c_n \cdot \Delta_n \cdot \alpha_n)].$$

When the real part $\mathrm{Re}(a)$ of the coefficient a describing the function $f(z) = \exp(a \cdot z)$ is different from 0, then $|f(z)|$ attains arbitrarily large values either for $z \to \infty$ (if $\mathrm{Re}(a) > 0$), or for $z \to -\infty$ (if $\mathrm{Re}(a) > 0$). In both cases, the mathematical expectations may become infinite, thus making the above formula meaningless. So, to make this formula always meaningful, we should only consider coefficients a with $\mathrm{Re}(a) = 0$, i.e., purely imaginary coefficients $a = i \cdot \omega$ for real values ω. For such a, all the above expectations are of the form $E[\exp(i \cdot z)]$ for some real number z. This expression is called a *characteristic function* of the probability distribution and denoted by $\chi(z)$. If we use this notation, then the above formula becomes

$$\chi(\omega \cdot \Delta) = \chi(\omega \cdot c_1 \cdot \Delta_1) \cdot \ldots \cdot \chi(\omega \cdot c_n \cdot \Delta_n).$$

For $n = 1$, $\Delta_1 = 1$, and $c_1 = -1$, the formula (4) leads to $\Delta = 1$, so we can conclude that $\chi(\omega) = \chi(-\omega)$, i.e., that the value $\chi(\omega)$ depends only on the absolute value of ω and not on it sign: $\chi(\omega) = \chi(|\omega|)$.

For $n = 2$, $\Delta_1 = \Delta_2 = 1$, $c_1 > 0$, and $c_2 > 0$, we have $\Delta = c_1 + c_2$, so we can conclude that $\chi(c_1 + c_2) = \chi(c_1) \cdot \chi(c_2)$. It is known that the only measurable function which satisfies this functional equation is $\chi(\omega) = \exp(b \cdot \omega)$ for some real value b [1]. (The proof is that χ's logarithm is an additive function, and it has been, basically, known since Cauchy that the only measurable additive functions are linear ones.) Hence, for $\omega > 0$, we have $\chi(\omega) = \exp(b \cdot \omega)$, and for general ω, we have $\chi(\omega) = \exp(b \cdot |\omega|)$. Since $\chi(\omega)$ is defined as a mathematical expectation of the expression $\exp(i \cdot \omega \cdot \alpha)$ whose absolute value is 1, the absolute value of $\chi(\omega)$ cannot exceed 1, thus $b < 0$. For simplicity, we will take $b = -1$ (but any other value will also work).

By definition, the characteristic function is a Fourier transform of a probability density function $\rho(z)$; thus, we can reconstruct the probability density by taking the inverse Fourier transform of $\chi(\omega) = \exp(-|\omega|)$. As a result, we get the probability density $\rho(z) = 1/(\pi \cdot (z^2 + 1))$ called *Cauchy distribution*.

Towards Implementation of the Main Idea. In our simulations, we multiply variables α distributed according to this distribution by positive real numbers Δ. If α is distributed according to this distribution, then the probability density for $\delta = \Delta \cdot \alpha$ is equal to

$$\rho(z) = \frac{\Delta}{\pi \cdot (z^2 + \Delta^2)};$$

This more general formula is also called *Cauchy distribution*. The value Δ will be called the *scale parameter* of this distribution, or simply a *parameter*, for short.

The property that we aimed for (and that we will use) is that if $z_1, \ldots, z_n$ are independent random variables, and each of z_i is distributed according to the Cauchy law with parameter Δ_i, then their linear combination $z = c_1 \cdot z_1 + \ldots + c_n \cdot z_n$ is also distributed according to a Cauchy law, with a scale parameter $\Delta = |c_1| \cdot \Delta_1 + \ldots + |c_n| \cdot \Delta_n$.

Therefore, if we take random variables δ_i which are Cauchy distributed with parameters Δ_i, then the value

$$c = f(\tilde{x}_1 + \delta_1, \ldots, \tilde{x}_n + \delta_n) - f(\tilde{x}_1, \ldots, \tilde{x}_n) = c_1 \cdot \delta_1 + \ldots + c_n \cdot \delta_n$$

is Cauchy distributed with the desired parameter (4). So, repeating this experiment N times, we get N values $c^{(1)}, \ldots, c^{(N)}$ which are Cauchy distributed with the unknown parameter, and from them we can estimate Δ.

The bigger N, the better estimates we get.

There are two questions to be solved:

- how to simulate the Cauchy distribution;

- how to estimate the parameter Δ of this distribution from a finite sample.

Simulation can be based on the functional transformation of uniformly distributed sample values: $\delta_i = \Delta_i \cdot \tan(\pi \cdot (r_i - 0.5))$, where r_i is uniformly distributed on the interval $[0, 1]$.

In order to estimate σ we can apply the Maximum Likelihood Method $\rho(d^1) \cdot \rho(d^2) \cdot \ldots \cdot \rho(d^n) \to \max$, where $\rho(z)$ is a Cauchy distribution density with the unknown Δ. When we substitute the above-given formula for $\rho(z)$ and equate the derivative of the product with respect to Δ to 0 (since it is a maximum), we get an equation

$$\frac{1}{1 + \left(\frac{c^{(1)}}{\Delta}\right)^2} + \ldots + \frac{1}{1 + \left(\frac{c^{(N)}}{\Delta}\right)^2} = \frac{N}{2}. \tag{9}$$

The left-hand side of (9) is an increasing function that is equal to $0(< N/2)$ for $\Delta = 0$ and $> N/2$ for $\Delta = \max \left|c^{(k)}\right|$; therefore the solution to the equation (9) can be found by applying a bisection method to the interval $\left[0, \max \left|c^{(k)}\right|\right]$.

So, we arrive at the following algorithm:

Algorithm.   For $k = 1, 2, \ldots, N$, repeat the following:

- use the random number generator to compute n numbers $r_i^{(k)}$, $i = 1, 2, \ldots, n$, that are uniformly distributed on the interval $[0, 1]$;

- compute $\delta_i^{(k)} = \Delta_i \cdot \tan(\pi \cdot (r_i^{(k)} - 0.5))$;

- compute the length $\delta^{(k)} = \left\|\vec{\delta}^{(k)}\right\|$ of the vector $\vec{\delta}^{(k)} = (\delta_1^{(k)}, \ldots, \delta_n^{(k)})$;

- compute the normalized coefficient $K_{\text{norm}}^{(k)} = \delta^{(k)}/\delta_0$;

- compute the auxiliary vector $\vec{\beta}^{(k)} = \vec{\delta}^{(k)}/K_{\text{norm}}^{(k)}$ with components $\beta_i^{(k)} = \delta_i^{(k)}/K_{\text{norm}}^{(k)}$;

- substitute $\tilde{x}_i + \beta_i^{(k)}$ into the program f and compute

$$c^{(k)} = \frac{\delta^{(k)}}{\delta_0} \cdot \left(f\left(\tilde{x}_1 + \beta_1^{(k)}, \ldots, \tilde{x}_n + \beta_n^{(k)}\right) - \tilde{y}\right);$$

- compute Δ by applying the bisection method to solve the equation (9).

Philosophical Comment: Sometimes, Distortion of Simulated Phenomenon Makes Simulation More Efficient.   The use of Cauchy distribution in the above algorithm may seem somewhat counter-intuitive (see, e.g., [36, 49]). Indeed, in the interval setting, we do not know the exact probability distribution of each error Δ_i, but we do know that each error Δ_i belongs to the corresponding interval $[-\Delta_i, \Delta_i]$, so the actual (unknown) probability distribution for Δ_i must be located on this interval with probability 1. So, at first glance, if we want to design a simulation-type technique for computing Δ, we should use one of such possible distributions in our simulations. Instead, we use a Cauchy distribution for which the probability to be outside the interval $[-\Delta_i, \Delta_i]$ is non-zero. In other words, in order to make the simulations work, in these simulations, we *distort* the actual distributions.

At first glance, it may therefore seem natural to use, in our simulations, instead of n independent variables distributed according to Cauchy distribution, n independent variables δ_i distributed according to some distributions which are actually located on the interval $[-\Delta_i, \Delta_i]$. It is sufficient to select a distribution corresponding to $\Delta_i = 1$; let a and σ denote the average and standard deviation of this variable. Then, by scaling (namely, by multiplying by Δ_i), we can get a distribution corresponding to an arbitrary Δ_i. In this case, for each variable δ_i, its average is equal to $\Delta_i \cdot a$, and its standard deviation is equal to $\Delta_i \cdot \sigma$.

As a result of each simulation, we get the value $c_1 \cdot \delta_1 + \ldots + c_n \cdot \delta_n$. For large n, we can apply the limit theorem to this sum and conclude that this value is approximately normally distributed, with an average $\sum c_i \cdot \Delta_i$ and the standard deviation $\sqrt{\sum c_i^2 \cdot \Delta_i^2}$. The larger n, the closer this resulting distribution to normal. It is known that a normal distribution is uniquely determined by its first two moments; hence, for large n, the only information that we will be able to extract from the simulation results are the average and the standard deviation of the resulting distribution. From these two sums $\sum c_i \cdot \Delta_i$ and $\sum c_i^2 \cdot \Delta_i^2$, we cannot uniquely determine the desired value $\sum |c_i| \cdot \Delta_i$. Thus, we cannot use un-distorted simulation, and *distortion is inevitable*.

A general conclusion is: *In simulation, sometimes distorting the simulated process leads to a faster simulation-based algorithm.*

At a more general level, the advantages of simulations with distortions over accurate simulations may explain:

- why an artistic (somewhat geometrically distorted) portrait often captures our impression of a person much better than a (geometrically correct) photo, and

- why, in spite of humans' high optical abilities, we sometimes (as in optical illusions) distort the image that we are trying to reproduce.

When Is This Randomized Algorithm Better Than Deterministic Numerical Differentiation. To determine the parameter Δ, we use the maximum likelihood method. It is known that the error of this method is asymptotically normally distributed, with 0 average and standard deviation $1/\sqrt{N \cdot I}$, where I is Fisher's information:

$$I = \int_{-\infty}^{\infty} \frac{1}{\rho} \cdot \left(\frac{\partial \rho}{\partial \Delta} \right)^2 \, dz.$$

For Cauchy probability density $\rho(z)$, we have $I = 1/(2\Delta^2)$, so the error of the above randomized algorithm is asymptotically normally distributed, with a standard deviation $\sigma_e \sim \Delta \cdot \sqrt{2/N}$. Thus, if we use a "two sigma" bound, we conclude that with probability 95%, this algorithm leads to an estimate for Δ which differs from the actual value of Δ by $\leq 2\sigma_e = 2\Delta \cdot \sqrt{2/N}$. So, if we want to achieve a 20% accuracy in the error estimation, we must use the smallest N for which $2\sigma_e = 2\Delta \cdot \sqrt{2/N} \leq 0.2 \cdot \Delta$, i.e., to select $N_f = N = 200$.

When it is sufficient to have a standard deviation of 20% (i.e., to have a "two sigma" guarantee of 40%), we need only $N = 50$ calls to f. For $n \approx 10^3$, both values N_f are much smaller than $N_f = n$ required for numerical differentiation.

So, if we have to choose between the (deterministic) numerical differentiation and the randomized Monte-Carlo algorithm, we must select:

- a deterministic algorithm when the number of variables n satisfies the inequality $n \leq N_0$ (where $N_0 \approx 200$), and

- a randomized algorithm if $n \geq N_0$.

A natural question is: is this the optimal choice? In the interval setting, in contrast to the statistical setting, we do not know the optimal algorithm. However, we will still be able to conclude that the answer to the above question is: No, for many different values of n, it is possible to find a new algorithm which is better than both above algorithms. This algorithm was first announced in [45].

3.2   MONTE-CARLO SIMULATION FOR INTERVAL SETTING: A NEW ALGORITHM

Main Idea Behind the New Algorithm.   In the interval setting, we have the following error estimation algorithms for indirect measurements:

- we have a (deterministic) *numerical differentiation* algorithm A_n which requires $N_f = n$ calls to a black-box program f and returns the exact result Δ (with 0 standard deviation);

- for every positive integer $N < n$, we have a (randomized) Cauchy-based algorithm A_N which requires N calls to a black-box program f, and which returns an approximate estimate whose standard deviation is $\sim \sqrt{2/N} \cdot \Delta$.

The larger the number N_f of calls to f, the better the estimate of Δ, until we reach $N_f = n$ for which we can get the exact estimate. In many real-life situations, as we have mentioned earlier, we cannot afford n calls to f, so we must use approximate approximate estimates.

Let $\bar{N}$ be the average number of calls which we can afford. Then, instead of simply using the randomized algorithm corresponding to this number of calls, we can use the following *combined* algorithm:

- select probabilities $p_1, \ldots, p_n$ for which $\sum p_i = 1$; and

- with probability p_i, use the algorithm A_i.

The probabilities p_i serve as parameters for the new algorithm. Which values of these parameters should we choose to get the best estimate?

Optimal Choice of Parameters of the New Algorithm. We must select the parameters p_i of the new algorithm so as to minimize the standard deviation (or, equivalently, the variance) of the resulting algorithm under the condition that the average number of calls is equal to a given number $\bar{N}$.

The variance V of an algorithm A_i is equal to $2\Delta^2/i$ for $i < n$ and to 0 for $i = n$; the number of calls for A_i is equal to i. Thus, the above optimization problem takes the following form:

$$V = \sum_{i=1}^{n-1} p_i \cdot \frac{2\Delta^2}{i} \to \min_{p_1, \ldots, p_n} \tag{10}$$

under the conditions that

$$p_i \geq 0; \quad \sum_{i=1}^{n} p_i = 1; \quad \text{and} \quad \sum_{i=1}^{n} p_i \cdot i = \bar{N}. \tag{11}$$

The solution to this problem is given by the following theorem:

Theorem 2. *Let $\bar{n} = \lfloor \bar{N} \rfloor$ denote the integer part of $\bar{N}$, and $\theta = \bar{N} - \bar{n}$ its fractional part. Then, the solution to the problem* (10), (11) *is as follows:*

- *When $\bar{N} \leq n/2$, we have $p_{\bar{n}} = 1 - \theta$, $p_{\bar{n}+1} = \theta$, and $p_i = 0$ for all other i. In this case,*

$$V = 2\Delta^2 \cdot \left(\frac{1 - \theta}{\bar{n}} + \frac{\theta}{\bar{n} + 1} \right).$$

 (In particular, when $\bar{N}$ is an integer, we have $p_{\bar{N}} = 1$ and $p_i = 0$ for all other i.)

- *When $\bar{N} > n/2$, we have*

$$p_{\lfloor n/2 \rfloor} = \frac{n - \bar{N}}{\lceil n/2 \rceil}, \quad p_n = \frac{\bar{N} - \lfloor n/2 \rfloor}{\lceil n/2 \rceil},$$

 and $p_i = 0$ for all other i. In this case,

$$V = \frac{2\Delta^2 \cdot (n - \bar{N})}{\lfloor n/2 \rfloor \cdot \lceil n/2 \rceil}.$$

Comment. In slightly less formal terms, we can reformulate this theorem as follows:

- when $\bar{N} \leq n/2$, combination does not help much, but

- when $\bar{N} > n/2$, it is better, instead of using a randomized algorithm $A_{\bar{N}}$ with $\bar{N}$ calls to f, to "flip a coin" and use either a randomized algorithm $A_{\lceil n/2 \rceil}$ with $\approx n/2$ calls, or numerical differentiation A_n.

For $\bar{N} > n/2$, the combined algorithm is indeed better: e.g., when n is even and $\bar{N} = n - 1$, the combined algorithm has a variance $8\Delta^2/n^2$, while the randomized algorithm A_{n-1} has a variance $2\Delta^2/(n-1) \sim \Delta^2/n$ which is $\approx n/4$ times larger. This algorithm may be not optimal; finding an optimal algorithm for interval setting is an important *open problem*.

4. NECESSITY AND POSSIBILITY OF PARALLELIZATION

Parallelization Is Necessary. The most time-consuming part of the above algorithms is calling the program f.

Even for the optimal algorithm corresponding to statistical setting, when the number of variables n is large, we need at least 50 calls to f. If each call requires a minute, the resulting time takes about an hour, which may be too long for on-line results.

Since this algorithm is optimal, we cannot decrease the number of calls, so the only way to decrease the total computation time is to run these calls *in parallel*.

Parallelization Is Possible. Not every algorithm can be easily parallelized. In principle, we could have algorithms for which the next vector $\vec{\delta}$ depends on the results of the previous call (and our general definitions allow such algorithms).

Luckily to us, however, for the above optimal algorithm (and for all other algorithms described above), the vectors $\vec{\delta}$ do not depend on the results of the previous calls. Thus, we can make all the calls in parallel (see, e.g., [6, 7, 9, 14, 15, 16, 39, 40, 41, 42, 47, 63]). If we do not have the actual parallel machine, we can use several processors connected into a net [7, 9, 47].

The more processors we have, the less time the resulting computation will take. If we have as many processors as the required number of calls, then the time needed to estimate the error of indirect measurement becomes equal to the time of a single call, i.e., to the time necessary to compute the result $\tilde{y}$ of this indirect measurement. Thus, if we have enough processors working in parallel, we can compute the result of the indirect measurement *and* estimate its error during the same time that it normally takes just to compute the result.

In particular, if the result $\tilde{y}$ of indirect measurement can be computed in real time, we can estimate the error of this result in real time as well.

Optimal Choice of Parallel Architecture. The efficiency of parallelization depends on the exact computer architecture. For the above Monte-Carlo-type algorithms, asymptotically optimal parallel architecture is described in [63].

Practical Applications. The above parallelization techniques have been applied to real-life problem: to expert systems [14, 15], to geophysics and petroleum engineering [9, 16, 47], etc.

5. ERROR ESTIMATION FOR INDIRECT MEASUREMENT: MORE COMPLICATED SETTINGS

In the above text, we considered typical settings, when:

- we either know that the errors Δx_i are independent (statistical setting), or we have no information about probabilities at all (interval setting), and

- we have a program which computes $f(x_1, \ldots, x_n)$ exactly.

In real life, in addition to these typical settings, we sometimes encounter more complicated situations. In this section, we describe how the above algorithms can be modified to cover such settings.

5.1 POSSIBLE CORRELATIONS IN STATISTICAL SETTING

In the above text, we assumed that the errors Δx_i are independent random variables with 0 average and known standard deviations σ_i. In real life applications, some of the measurement errors may be caused by the same physical phenomenon and thus, may be correlated. It is therefore desirable to consider a more general setting in which about some variables are known to be independent from each other, while others may be correlated.

When all the variables Δx_i are independent, the standard deviation σ of the error Δy (defined by the formula (1)) is uniquely determined by the known values σ_i (formula (3)). In a more general setting, the actual value of σ may also depend on the correlation between Δx_i; if we do not know this correlation, we can only determine the *upper bound* $\tilde{\sigma}$ for the standard deviation σ.

We can describe the general information about independence between n variables by an (undirected) *graph G* if we take n vertices corresponding to n variables and connect two vertices by an edge if and only if the corresponding variables are known to be independent. In Section 2, we considered the setting in which this graph G is *complete*. As another extreme case, we can consider the case when the graph is *fully disconnected*, i.e., when we have no information about independence. In this case, the largest possible value $\tilde{\sigma}$ of the standard deviation σ is attained when all the variables are fully correlated, and it is given by a formula

$$\tilde{\sigma} = |c_1| \cdot \sigma_1 + \ldots + |c_n| \cdot \sigma_n. \tag{4a}$$

This formula is identical to the formula (4) from interval setting, and thus, all the above-described interval-setting algorithms can be also used to estimate $\tilde{\sigma}$.

In real life, in addition to these two extreme cases, we have many intermediate situations. In practical problems, usually, we can divide the variables Δx_i into several classes of related variables so that within each class, variables are related and thus may be dependent, but variables belonging to different classes are mutually independent. In graph terms, such situations can be described by *hierarchical* graphs, in which the vertices of a graph G can be divided into subgraphs $G_1, \ldots, G_k$ so that within each graph G_i there are no edges, but each two vertices belonging to different sets are connected by an edge. In this case, we can find the desired estimate $\tilde{\sigma}$ as follows:

$$\tilde{\sigma} = \sqrt{\left(\sigma^{(1)}\right)^2 + \ldots + \left(\sigma^{(k)}\right)^2},$$

where

$$\sigma^{(j)} = \sum_{i \in G_j} |c_i| \cdot \sigma_i.$$

These formulas describe a two-level hierarchy, in which we have subgraphs $G_1, \ldots, G_k$ (level 1) and the graph itself (level 2). These formulas can be naturally generalized to a *multi-level* hierarchy. Such a hierarchy can be described by a *tree* of clusters (subsets of the graph):

- the root of this tree is the graph G itself; this cluster corresponds to the highest level L of the hierarchy;

- its direct children are the clusters $G_1, \ldots, G_k$ of the next level $L - 1$;

- etc., until we reach the leaves of the tree – vertices – "clusters" of the lowest possible level 0.

The intuition behind this hierarchy is that vertices from a node of level 1 may be dependent; clusters of level 1 belonging to a cluster of level 2 are independent; clusters of level 2 belonging to a single cluster of level 3 may be dependent, etc. In other words, to determine whether two vertices are connected or not, we must find the smallest cluster which contains both;

- if this cluster is on the even level, these vertices are connected;

- if this cluster is on the odd level, the vertices are not connected.

For such a hierarchical graph, we can estimate the desired bound $\tilde{\sigma}$ corresponding to the largest cluster by first estimating similar bounds $\sigma(H)$ for the standard deviation of the sum $\sum_{i \in G} c_i \cdot \Delta x_i$ corresponding first to clusters of level 0, then to clusters of level 1, etc.:

- For clusters of level 0 (i.e., for clusters $H = \{i\}$), we have $\sigma(\{i\}) = |c_i| \cdot \sigma_i$.

- If we have already computed $\sigma(H)$ for all clusters of levels $0, \dots, k-1$, then, to compute $\sigma(H)$ for clusters of level k, we must do the following:

  - for odd k, we must use the formula

  $$\sigma(H) = \sum_{H_i \subset H} \sigma(H_i),$$

  where the sum is taken over all subclusters of level $k - 1$;
  - for even k, we must use the formula

  $$\sigma(H) = \sqrt{\sum_{H_i \subset H} \sigma^2(H_i)}.$$

This generalized situations covers a lot of practical situations, but sometimes, the graph has a more complicated structure. For two variables, they are either connected or not, so nothing more complicated can happen. But already for three variables, we can have a non-hierarchical situation when Δx_1 and Δx_2 are independent, Δx_2 and Δx_3 are independent, but we have no information about the dependence between Δx_1 and Δx_3. This particular 3-variable situation has an explicit solution: $\tilde{\sigma} = \sqrt{(\sigma^{(1)} + \sigma^{(3)})^2 + (\sigma^{(2)})^2}$, but in general, we do not know of any efficient algorithm which would work for an arbitrary graph.

It is known that normally distributed random variables can be represented as vectors in multi-D space, so that linear combinations turn into linear combinations, and a length of a vector is equal to its standard deviation. Due to this representation, the above *open problem* can be reformulated in geometric terms:

- we have a graph G with n nodes, and we have n positive real numbers $\sigma^{(1)}, \dots, \sigma^{(n)}$;

- we want to find the largest possible length $\tilde{\sigma}$ of a vector $\vec{x} = \vec{x}^{(1)} + \dots + \vec{x}^{(n)}$ for which the following two conditions hold:

  - for every i, $\left\| \vec{x}^{(i)} \right\| = \sigma^{(i)}$, and

  - if nodes i and j are connected in the graph G, then the vectors $\vec{x}^{(i)}$ and $\vec{x}^{(j)}$ are orthogonal: $\vec{x}^{(i)} \cdot \vec{x}^{(j)} = 0$.

5.2 POSSIBLE CORRELATIONS IN INTERVAL SETTING

Formulation of the Problem. In the interval setting, in addition to the upper bounds Δ_i on each of the errors Δx_i, we may have some additional information about the relation between the errors Δx_i: e.g., in addition to knowing that $|\Delta x_1| \leq 0.1$ and $|\Delta x_2| \leq 0.1$, we may also know that the sum $\Delta x_1 + \Delta x_2$ of these two errors should also not exceed 0.1. This additional information can be easily reformulated in geometric terms:

- in the standard interval setting, the possible values of the vector $\vec{\Delta} = (\Delta x_1, \ldots, \Delta x_n)$ form a *box* (parallelepiped) $[-\Delta_1, \Delta_1] \times \ldots \times [-\Delta_n, \Delta_n] \subseteq R^n$;

- when we have an additional information about the errors, not all points from this box are possible; as a result, the set G of possible values of the vector $\vec{\Delta}$ can have a more complicated shape than a box.

In this more complicated setting, we also want to be able, given a black-box program $f(x_1, \ldots, x_n)$ and n values $\tilde{x}_1, \ldots, \tilde{x}_n$, to determine the interval of possible values of Δy (determined by the formula (1)).

In more precise terms, we are interesting in computing the endpoints of the interval $[\Delta_-, \Delta_+] = [f_-(\vec{c}), f_+(\vec{c})]$, where we denoted

$$f_-(\vec{c}) = \min_{\vec{x} \in G} \vec{c} \cdot \vec{x}, \quad f_+(\vec{c}) = \max_{\vec{x} \in G} \vec{c} \cdot \vec{x}.$$

Solution: Main Idea. We will show that this problem can be solved by an appropriate modification of the Cauchy-distribution based randomized algorithm described in Section 3.1 (this modification was first described in [67] and announced in [49]). The main difference will be that instead of using n independent random variables δ_i, we will now use an appropriate joint distribution for the vector $\vec{\delta}$ in which components are not necessarily independent.

Specifically, we will be using an n-dimensional distribution for which the probability density $\rho(\vec{x})$ is equal to the inverse Fourier transform of the complex function $\exp(i \cdot f_a(\vec{\omega}) - f_s(\vec{\omega}))$, where $f_a(\vec{\omega})$ and $f_s(\vec{\omega})$ are, correspondingly, the antisymmetric and symmetric parts of the above functions $f_-(\vec{\omega})$ and $f_+(\vec{\omega})$:

$$f_s(\vec{\omega}) = \frac{1}{2} \cdot (f_+(\vec{\omega}) + f_-(\vec{\omega})); \quad f_a(\vec{\omega}) = \frac{1}{2} \cdot (f_+(\vec{\omega}) - f_-(\vec{\omega})).$$

The resulting algorithm consists of two stages:

- first, a *preliminary stage*: when we know the set G, but before we know anything about the function f, we prepare the random number generator to generate numbers distributed according to the probability density $\rho(\vec{c})$;

- then, the *main stage*: for any given black-box program f, we use the random number generator constructed on the preliminary stage to estimate the desired interval $[f_-(\vec{c}), f_+(\vec{c})]$.

Algorithm: Preliminary Stage. To simulate the desired distribution, we first do some preliminary computations:

- First, we compute the values $f_-(\vec{c})$ and $f_+(\vec{c})$ for several values $\vec{c}$.

- Then, we compute the values $f_s(\vec{c})$ and $f_a(\vec{c})$.

- After that, we compute the inverse Fourier transform $\rho(\vec{x})$ of the function $\exp(i \cdot f_a(\vec{\omega}) - f_s(\vec{\omega}))$.

- The density $\rho(\vec{x})$ tends to 0 as $\|\vec{x}\| \to \infty$, so there exists a number D such that for $\|\vec{x}\| > D$, the value of $\rho(\vec{x})$ is so small that we can safely assume that it is equal to 0; let us compute such a D.

- Finally, we compute the maximum $s = \sup \rho(\vec{x})$ of the density $\rho(\vec{x})$.

Now, we are ready to design the desired random vector generator. Namely, to get a vector $\vec{\delta}$ with the desired probability density function $\rho(\vec{x})$:

- we run $n+1$ times a standard random number generator which produced random numbers $r_1, \ldots, r_n, r_{n+1}$ uniformly distributed on the interval $[0, 1]$;

- then, we form a vector $\vec{x} = (x_1, \ldots, x_n)$ with components $x_i = (2r_i - 1) \cdot D$, and check whether $r_{n+1} \leq \rho(\vec{x})/s$;

- if this inequality holds, we return $\vec{\delta} = \vec{x}$, otherwise, we generate r_i anew, etc.

One can easily see that for this algorithm, the probability of a point $\vec{x} \in [-D, D]^n$ to be produced is proportional to $\rho(\vec{x})$, and thus, we get the desired distribution.

Algorithm: Main Stage. For $k = 1, 2, \ldots, N$, repeat the following:

- use the random number generator designed on the preliminary stage to compute a random vector $\vec{\delta}^{(k)} = (\delta_1^{(k)}, \ldots, \delta_n^{(k)})$;

- compute the length $\delta^{(k)} = \left\| \vec{\delta}^{(k)} \right\|$ of the vector $\vec{\delta}^{(k)}$;

- compute the normalized coefficient $K_{\text{norm}}^{(k)} = \delta^{(k)}/\delta_0$;

- compute the auxiliary vector $\vec{\beta}^{(k)} = \vec{\delta}^{(k)}/K^{(k)}_{\text{norm}}$ with components $\beta_i^{(k)} = \delta_i^{(k)}/K^{(k)}_{\text{norm}}$;

- substitute $\tilde{x}_i + \beta_i^{(k)}$ into the program f and compute

$$c^{(k)} = \frac{\delta^{(k)}}{\delta_0} \cdot \left(f\left(\tilde{x}_1 + \beta_1^{(k)}, \ldots, \tilde{x}_n + \beta_n^{(k)}\right) - \tilde{y}\right);$$

- compute the average

$$\Delta_a = \frac{1}{N} \cdot \sum_{k=1}^{N} c^{(k)};$$

- compute Δ_s by applying the bisection method to solve the equation

$$\frac{1}{1 + \left(\frac{c^{(1)} - \Delta_a}{\Delta_s}\right)^2} + \ldots + \frac{1}{1 + \left(\frac{c^{(N)} - \Delta_a}{\Delta_s}\right)^2} = \frac{N}{2}; \qquad (9a)$$

- compute $\Delta_- = \Delta_s - \Delta_a$ and $\Delta_+ = \Delta_s + \Delta_a$.

Comment. The justification of this algorithm is given in the Proofs section.

Special Case When the Set G is an Ellipsoid.   An important particular case is when the set G is an ellipsoid with a center at $\vec{0}$. Ellipsoids are often used to describe uncertainty domains; see, e.g., [4, 11, 12, 22, 24, 51, 57, 58, 60, 61]. Moreover, it has been shown, both experimentally [11, 12] and theoretically [23], that for many practical problems, ellipsoids are an *optimal* family for describing domains.

For ellipsoids, we can describe the *optimal* error estimation algorithm (the idea of this algorithm was first partly presented in [35]).

Let us start with the case when G is a *sphere* of a given radius r, i.e., the set of all the vectors $\vec{\Delta} = (\Delta x_1, \ldots, \Delta x_n)$ for which $\left\|\vec{\Delta}\right\| \leq r$. We want to find the maximum and minimum of the scalar product $\vec{c} \cdot \vec{\Delta}$ under this inequality. Geometrically, the scalar product is equal to $\|\vec{c}\| \cdot \left\|\vec{\Delta}\right\| \cdot \cos(\theta)$, where θ is an angle between the two vectors, so maximum and minimum are attained, correspondingly, when $\cos(\theta) = 1$ or $\cos(\theta) = -1$, i.e., when the vectors are parallel or anti-parallel. Thus, the maximum is this attained for $\vec{\Delta} = \vec{c} \cdot (r/\|\vec{c}\|)$, and the resulting maximum is equal to $\vec{c} \cdot \vec{\Delta} = \|\vec{c}\| \cdot r$. Similarly, the minimum is equal to $-\|\vec{c}\| \cdot r$.

Thus, to find both maximum and minimum, it is sufficient to estimate the length $\|\vec{c}\|$ of the (unknown) gradient vector $\vec{c}$. This is exactly the problem that we solved in statistical setting, a problem for which we even produced the optimal algorithm. Thus, we can use that algorithm here as well.

This idea can be naturally generalized to a general *non-spherical* ellipsoid. Namely, for each ellipsoid, one can easily find new coordinates $x_i' = \sum_j a_{ij} \cdot x_j$ in which this ellipsoid becomes a sphere. By inverting the matrix a_{ij}, we can get the expression of the old coordinates x_i in terms of the new ones: $x_i = \sum_j b_{ij} \cdot x_j'$, which enables us to re-describe the function f in terms of the new coordinates x_i': $f(x_1, \ldots, x_n) = g(x_1', \ldots, x_n')$, where

$$g(x_1', \ldots, x_n') = f\left(\sum_{j=1}^{n} b_{1j} \cdot x_j', \ldots, \sum_{j=1}^{n} b_{nj} \cdot x_j'\right).$$

For the new function g, the corresponding region G is a sphere, so, as we have mentioned, we can solve the problem of finding the interval of possible values for Δy by using the smallest possible number of calls to g. According to the above explicit formula for g, every call to g consists of a single call to f (plus a few simple computations); thus, we still get the optimal number of calls.

5.3 A PRACTICALLY USEFUL CASE OF KNOWN CORRELATION

Description of the Case: Determining Parameters of a Known Model. In this paper, we consider situations in which the program $f(x_1, \ldots, x_n)$ implements a complex algorithm which is given as a black box. In many practical applications, we have an explicit parameterized model, with unknown parameters $y_1, \ldots, y_m$, which predicts the values of the measured quantities $x_1, \ldots, x_n$ as $x_i = M_i(y_1, \ldots, y_m)$. In these applications, one of the main goals of data processing is to find the values of these parameters which fit with the known measurement results $\tilde{x}_i$. Usually, the functions M_i which represent these models are either explicitly given or rather easy to compute.

When the functions M_i are highly non-linear, the program f – which solves these equations and actually finds the values y_i – can be really complicated, and thus, require a lot of computation time (much more than computing $M_i(y_1, \ldots, y_m)$ for known values y_j). If we use the above algorithms to estimate the error of the corresponding indirect measurement of y_i, we thus have to make several calls to f and hence, further increase the computation time.

In This Case, It Is Possible to Estimate the Error of Indirect Measurements Without Any Extra Calls to f. Let us show that in this case, we can estimates the errors of indirect measurements much faster than in the general situation: namely, we can do it without any extra calls to the time-consuming program f.

Indeed, since we assumed that the errors Δx_i and Δy_j are small, we can linearize the above system of equations, and get a linear system

$$\Delta x_i = \sum_j M_{i,j} \cdot \Delta x_j.$$

Here, the derivatives $M_{i,j}$ of the functions M_i can be computed, e.g., by applying numerical differentiation to the functions M_i. Since computing M_i is much faster than computing f (i.e., than solving the system of equations), the computation time necessary for computing these partial derivatives is much smaller than a single call to f.

Inverting the matrix $M_{i,j}$, we can explicitly express Δy_j in terms of Δx_i – as $\Delta y_j = \sum_i c_i^{(j)} \cdot \Delta x_i$ with known coefficients $c_i^{(j)}$. Then, we can use the corresponding formula (3) or (4) to estimate the error of indirect measurement of y_j.

This method was successfully used in *pavement engineering* [10, 21].

5.4   WHAT IF THE BLACK BOX PROGRAM IS ITSELF A RANDOMIZED ALGORITHM

Formulation of the Problem. In the previous text, we assumed that we have a black-box program which computes $f(x_1, \ldots, x_n)$ *exactly*. In many practical cases, the black-box program implements a *randomized* algorithm which, instead of computing f exactly, returns an *approximate* value $\tilde{f}$ of f. In most such cases, in good accordance with the central limit theorem, the distribution of the corresponding error $\Delta f = \tilde{f} - f(x_1, \ldots, x_n)$ is almost normal, so for all practical purposes, we can assume it to be normal. We can also assume that the estimate $\tilde{f}$ is unbiased, i.e., that the average approximation error $E[\Delta f]$ is 0.

To get a more accurate estimate, we can run the original black-box (randomized) program several (K) times, get K results $\tilde{f}^{(1)}, \ldots, \tilde{f}^{(K)}$, and compute their arithmetic average $f_{av} = (\tilde{f}^{(1)} + \ldots + \tilde{f}^{(K)})/K \approx f(x_1, \ldots, x_n)$. As a result of this averaging, the random error component of Δf will decrease $\sqrt{K}$ times. For example, if we want to decrease it 10 times, we must use $K = 100$.

We can also use these same values $\tilde{f}^{(1)}, \ldots, \tilde{f}^{(K)}$ to estimate the standard deviation σ_f of the original Δf as

$$\sigma_f = \sqrt{\frac{1}{K-1} \cdot \sum_{k=1}^{K} \left(\tilde{f}^{(k)} - f_{av} \right)^2}.$$

To estimate the error of indirect measurement, we can, in principle, apply one of the above black-box-oriented algorithms directly: namely, every time the algorithm asks us to apply the function f to a vector $(\tilde{x}_1 + \delta_1, \ldots, \tilde{x}_n + \delta_n)$, we apply the black-box procedure K times and do the averaging.

The problem with this direct application is that in the above algorithms, we already had quite a few calls to f (for large n, 50 or 200), and if each call requires quite a few (e.g., $K = 100$) applications of the black-box program $\tilde{f}$, then we have a lot of applications of the black-box program which take a lot of time.

It turns out that if the function f is smooth enough so that the quadratic terms can be neglected for reasonable δ_i (i.e., that δ_0 is large enough), then we can drastically decrease the number of calls to a black box. We will describe the corresponding modified algorithms for statistical and interval settings.

Algorithm for Statistical Setting: Motivation. In the standard Monte-Carlo simulation for statistical setting, we do the following for $k = 1, \dots, N$:

- first, we simulate random values $\delta_i^{(k)}$ which are independently normally distributed with 0 average and standard deviation σ_i;

- then, we apply the black box program (in our case, $\tilde{f}$) to the vector $\vec{\tilde{x}} + \vec{\delta}$ and compute the value

$$c^{(k)} = \tilde{f}\left(\tilde{x}_1 + \delta_1^{(k)}, \dots, \tilde{x}_n + \delta_n^{(k)}\right) - \tilde{y}.$$

We know that

$$\tilde{f}\left(\tilde{x}_1 + \delta_1^{(k)}, \dots, \tilde{x}_n + \delta_n^{(k)}\right) = f\left(\tilde{x}_1 + \delta_1^{(k)}, \dots, \tilde{x}_n + \delta_n^{(k)}\right) + \Delta f^{(k)},$$

where $\Delta f^{(k)}$ is a normally distributed random variable with 0 average and standard deviation σ_f. We also know that

$$f\left(\tilde{x}_1 + \delta_1^{(k)}, \dots, \tilde{x}_n + \delta_n^{(k)}\right) = \tilde{y} + c_1 \cdot \delta_1^{(k)} + \dots + c_n \cdot \delta_n^{(k)}.$$

Therefore, $c^{(k)} = c_1 \cdot \delta_1^{(k)} + \dots + c_n \cdot \delta_n^{(k)} + \Delta f^{(k)}$ is a linear combination of independent normally distributed random variables. Hence, the standard deviation $\tilde{\sigma}$ of $c^{(k)}$ is equal to

$$\tilde{\sigma} = \sqrt{c_1^2 \cdot \sigma_1^2 + \dots + c_n^2 \cdot \sigma_n^2 + \sigma_f^2},$$

i.e., to $\sqrt{\sigma^2 + \sigma_f^2}$. Thus, we can estimate $\tilde{\sigma}$ by using standard statistical techniques, and then determine σ as $\sigma = \sqrt{(\tilde{\sigma})^2 - \sigma_f^2}$. As a result, we arrive at the following algorithm:

Algorithm for Statistical Setting. For $k = 1, \ldots, N$, repeat the following:

- use a random number generator to compute n numbers $\alpha_i^{(k)}$, $i = 1, 2, \ldots, n$, that are normally distributed with 0 average and standard deviation 1;

- compute $\delta_i^{(k)} = \sigma_i \cdot \alpha_i^{(k)}$;

- substitute $\tilde{x}_i + \delta_i^{(k)}$ into the black box $\tilde{f}$ and compute

$$c^{(k)} = \tilde{f}\left(\tilde{x}_1 + \delta_1^{(k)}, \ldots, \tilde{x}_n + \delta_n^{(k)}\right) - \tilde{y};$$

- compute

$$\sigma = \sqrt{\frac{1}{N-1} \cdot \sum_{k=1}^{N} \left(c^{(k)}\right)^2 - \sigma_f^2}.$$

Comment. If $\tilde{f}$ provides a biased estimate, with non-zero $E[\Delta f]$, then this algorithm still works, because when we compute the differences $c^{(k)}$, the biases cancel each other.

Algorithm for Interval Setting: Motivation. If we apply the interval-setting algorithm based on Cauchy distribution, then we conclude that the random variable $c^{(k)}$ is a sum of two independent random components: a component $\vec{c} \cdot \vec{\delta}$ which is Cauchy distributed with the desired parameter Δ and a normally distributed component Δf with 0 average and standard deviation σ_f.

Since these components are independent, the characteristic function $E(\exp(i \cdot \omega \cdot c)]$ of this sum is equal to the product of the corresponding characteristic functions, i.e., to $\chi(\omega) = \exp(-|\omega| \cdot \Delta - \omega^2 \cdot \sigma_f^2)$. Hence, to determine Δ, we can estimate $\chi(\omega)$, compute its negative logarithm, and then compute Δ (see the formula below).

Since the value $\chi(\omega)$ is real, it is sufficient to consider only the real part $\cos(\ldots)$ of the complex exponent $\exp(i \cdot \ldots)$. Thus, we arrive at the following algorithm:

Algorithm for Interval Setting. For $k = 1, 2, \ldots, N$, repeat the following:

- use a random number generator to compute n numbers $r_i^{(k)}$, $i = 1, 2, \ldots, n$, that are uniformly distributed on the interval $[0, 1]$;

- compute $\delta_i^{(k)} = \Delta_i \cdot \tan(\pi \cdot (r_i^{(k)} - 0.5))$;

- substitute $\tilde{x}_i + \delta_i^{(k)}$ into the black box $\tilde{f}$ and compute

$$c^{(k)} = \tilde{f}\left(\tilde{x}_1 + \delta_1^{(k)}, \ldots, \tilde{x}_n + \delta_n^{(k)}\right) - \tilde{y};$$

- for a real number $\omega > 0$, compute

$$\chi(\omega) = \frac{1}{N} \cdot \sum_{k=1}^{N} \cos\left(\omega \cdot c^{(k)}\right);$$

- compute $\Delta = -\dfrac{\ln(\chi(\omega))}{\omega} - \sigma_f^2 \cdot \dfrac{\omega}{2}.$

Comment. Instead of a *statistical* information about the error Δf, we sometimes only have an upper bound for Δf, i.e., an *interval* of possible values for f. In this case, the exact error estimation for indirect measurements becomes exponentially hard [38].

6. ERROR ESTIMATION ALGORITHMS CAN BE APPLIED TO OTHER PRACTICAL PROBLEMS

In the previous sections, we mentioned that the above black-box-oriented algorithms have been actually used to estimate the error for actual indirect measurements. Let us give two examples in which these same algorithms can be also used in solving other practical problems.

6.1 TOLERANCE ANALYSIS

Main Problem of Tolerance Analysis. The shape of a manufactured object is provided by several manufacturing operations. Each operation produces a simple geometric form, and each operation is not perfectly accurate: the resulting characteristics x_i of a shape (position of a hole, radius of a circle, size of an line, angle, etc) are maintained only within a certain limit $[\tilde{x}_i - \Delta_i, \tilde{x}_i + \Delta_i]$. The ideal value of the characteristic $\tilde{x}_i$ is called its *nominal value*, and Δ_i is called *tolerance.*

When after several operations, we assemble the parts into the final product, we want to know the geometric characteristics y of the resulting shape. For the ideal case when all manufacturing operations are precise, the corresponding value $\tilde{y}$ of the desired characteristic (it is called a *nominal* value of y) have been computed during design. So, we are only interested in the interval of possible deviations $\Delta y = \tilde{y} - y$ of this characteristic from the nominal one. Computation of such an interval is called *tolerance analysis* [2, 3, 55, 59, 64].

Tolerances are small, so we can usually safely neglect quadratic terms and consider a linearized problem.

From the physical viewpoint, this situation is slightly different from our main problem:

- we do not actually measure x_i; we just rely on the nominal values that have been supplied to us; and

- we do not actually calculate y; we simply use the value that has been pre-calculated during the design.

However, mathematically, we have exactly the same problem as with indirect measurements.

Inverse Problem of Tolerance Analysis. If after estimating the interval of possible values for y, we find out that all these values lie within the desired tolerance interval, then the analyzed manufacturing process is thus verified. However, sometimes, the interval of possible values of y turns out to be too wide. In this case, we are interested in finding the parameters of the manufacturing process that need to be improved.

In principle, this problem can be solved by computing the contribution of each manufacturing parameter. However, if there are many parameters (as, e.g., in VLSI manufacturing), then this algorithm takes too long. It turns out that the ideas of Cauchy-based simulation can lead to a faster algorithm [5].

6.2   PROCESSING EXPERT ESTIMATES

The Need for Expert Estimates. In the interval setting, the above algorithms enable us to compute the upper bound Δ on the error Δy of indirect measurement. This upper bound is based on the guaranteed upper bounds Δ_i for the errors Δx_i of the corresponding direct measurements. These upper bounds must be guaranteed, therefore, they are based on the worst-case estimations. In many cases, the actual error will be much lower. E.g., if in 99% of all measurements the error was ≤ 0.01 V, and in only one case out of 100 it was equal to 0.05 V, then we can only guarantee the error ≤ 0.05 V.

In some cases, the manufacturer also provides us with the probabilities of different values of error; in these cases, we can also estimate probabilities of different values of Δy. But to obtain these probabilities, we must undertake many experiments. The necessary statistical analysis of a sensor is often very time- and cost-consuming, and is, therefore, often avoided.

In many of such cases, when we do not have statistically justified probabilities (because we did not make enough experiments), we often have an opinion of the experts who designed and tested the sensors about what the errors can be. For example, in the above-described case, the guaranteed estimate for a sensor is 0.05 V, but since an expert who designed this sensor knows that usually (in 99 cases out of 100) the error was ≤ 0.01, he can also say that usually, this error is of order ≤ 0.01 V.

He cannot guarantee that, because it is not a precise statement (e.g., what does "of order" mean?). However, this is an additional information about the errors of Δx_i, and it is desirable to use this information in estimating the error Δy of an indirect measurement.

How to Formalize Expert Estimates. We are interested in formalizing a statement "the error e is (usually) of order $\leq \sigma$". This is not a precise statement, therefore, for given e and σ, we cannot say whether it is exactly true or not. If $e \leq \sigma$, this is true; if $e = 1.5\sigma$, this is probably true. If $e = k\sigma$ for some $k > 0$, then, the bigger k, the smaller is our degree of belief that this particular e satisfies this statement.

This degree of belief is usually expressed by numbers from the interval [0,1] (0 corresponds to no belief, 1 to absolute belief, and values between 0 and 1 correspond to intermediate degrees of belief). This idea was first proposed by L. Zadeh [66] under the name of *fuzzy set theory*. There are many different procedures that enable to obtain these numerical values (see, e.g., [19, 32]). If we apply any of these procedures, then for every e, we get the degree of belief $\mu(e)$ that this e satisfies the above statement. In mathematical terms, we get a function from the set of all real numbers to the interval [0,1]. This function is called a *membership function*. So, for each σ, we get a separate membership function. How are these functions related? First, let's notice that $e \leq \sigma$ if and only if $e/\sigma \leq 1$. Therefore, if we know that e is usually of order $\leq \sigma$, then we can conclude that e/σ is usually of order ≤ 1, and vice versa. In other words, for every e and σ, our degree of belief that e is of order $\leq \sigma$ is equal to the degree of belief that e/σ is of order ≤ 1. Therefore, it is sufficient to describe only one membership function $\mu_0(e)$ that describes degrees of belief for $\sigma = 1$. Then, for an arbitrary σ, the degree of belief that e is of order $\leq \sigma$, is equal to $\mu_0(e/\sigma)$.

What are the properties of the function $\mu_0(e)$?

- The value e is of order ≤ 1 if and only if $-e$ is of order ≤ 1; therefore, $\mu_0(-e) = \mu_0(e)$ (i.e., μ_0 is an even function).

- If we consider only positive values of e, then the bigger e, the smaller is our degree of belief that e is of order ≤ 1. In other words, for positive e, μ_0 is a *non-increasing function*.

Because of these two properties, μ_0 is non-decreasing for $e < 0$, and it is non-increasing for $e > 0$. For such functions μ_0, membership functions $\mu_0(e/\sigma)$ were called *LR-numbers* by D. Dubois and H. Prade who introduced them in [17, 18]. They also described how to apply arithmetic operations to these "fuzzy numbers". For detailed expositions of LR-numbers, see [19], Section II.2.B; [30], Section 1.9; [20], Chapter 2 and Section 3.2.4; [69], Chapter 5.

Let's describe their main formulas in application to our problem.

How to Describe a Resulting Expert Estimate For Δy? Extension Principle For LR-Numbers: General Idea. For each i, we know (from the experts) that Δx_i is of order $\leq \sigma_i$. How to compute the membership function that correspond to Δy?

The value Δy is possible if and only if there exists the real numbers Δx_1, ..., Δx_n such that Δx_1 is of order $\leq \sigma_1$, ..., Δx_n is of order $\leq \sigma_n$, and after applying formula (1) to these Δx_i, we get this particular value of Δy. In other words, Δy is possible if and only if:

$$\bigvee_{\Delta x_1, ..., \Delta x_n} [(\Delta x_1 \text{ is of order } \leq \sigma_1) \& \ldots \& (\Delta x_n \text{ is of order } \leq \sigma_n) \&$$

$$(\Delta y = c_1 \cdot \Delta x_1 + \ldots + c_n \cdot \Delta x_n)].$$

Here, $\vee$ means "there exists", which is the same as "or" (i.e., either for one set of values Δx_i, or for some other set, the statement in the square brackets must be true).

We know the degrees of the component statements: for "Δx_i is of order $\leq \sigma_i$", the degree of belief is equal to $\mu_0(\Delta x_i/\sigma_i)$. For "$\Delta y = c_1 \cdot \Delta x_1 + \ldots + c + n \cdot \Delta x_n$", the degree of belief is equal to 1 if and only if this statement is true, else it is equal to 0.

So, to find the membership function for Δy, it is sufficient to find out how to estimate our degree of belief $t(A \& B)$ $(t(A \vee B))$ of a composite statement (obtained by applying "and" and "or" to the component ones A and B) if we know the degrees of beliefs $t(A)$ and $t(B)$ of the component statements.

Which &- and $\vee$-Operations Should We Choose. Experiments with expert estimates showed [29, 53, 68] that, on average, the following estimates describe the human reasoning the best: $\min(t(A), t(B))$ for $t(A \& B)$ and $\max(t(A), t(B))$ for $t(A \vee B)$. Second best fits are $t(A) \cdot t(B)$ for $t(A \& B)$ and $t(A) + t(B) - t(A) \cdot t(B)$ for $t(A \vee B)$.

Thus, we have two possibilities for each operation, and hence we must consider four possible combinations of these operations. Let's show that two of them are meaningless for our problem. Namely, we will show that $t(A) + t(B) - t(A) \cdot t(B)$ is not applicable to it.

Indeed, we are applying "or" to infinitely many statements that correspond to different tuples Δx_i. The $\vee$-operation $a, b \rightarrow a + b - a \cdot b$ has the following property: when we apply it to several statements with equal degrees of belief, then the resulting degree of belief increases. If we combine two statements, we go from $a = t(A)$ to $2a - a^2 = 1 - (1 - a)^2$; if we combine three statements, we get $1 - (1 - a)^3$, and when we combine k statements, we get $1 - (1 - a)^k$. When $k \rightarrow \infty$, the result of this combination tends to 1. So, if we apply this operation to describe the membership function for Δy, we will get a function whose value is 1 for all Δy, which is meaningless.

Therefore, for our problem, the only possible $\vee$-operation is $\max(t(A), t(B))$. As a result, we are left with two cases:

- $t(A \vee B) \approx \max(t(A), t(B))$ and $t(A \& B) \approx \min(t(A), t(B))$;

- $t(A \vee B) \approx \max(t(A), t(B))$, $t(A \& B) \approx t(A)t(B)$.

Let's consider these two cases.

Extension Principle For the Case When min is Used as an &-Operation.
If in the above definition of the term "Δy is possible", we formalize "and" as min, and "or" as max, then we will obtain the following expression for the desired membership function $\mu(\Delta y)$ (i.e., for the degree of belief $\mu(\Delta y)$ that Δy is possible):

$$\mu(\Delta y) = \max_{\Delta x_1, \ldots, \Delta x_n} \left[\min \left(\mu_0 \left(\frac{\Delta x_1}{\sigma_1} \right), \ldots, \mu_0 \left(\frac{\Delta x_n}{\sigma_n} \right) \right) \right],$$

and maximum is taken over all values Δx_i for which $\Delta y = c_1 \cdot \Delta x_1 + \ldots + c_n \cdot \Delta x_n$ (if $\Delta y \neq c_1 \cdot \Delta x_1 + \ldots + c_n \cdot \Delta x_n$, then the degree of belief in this equality is 0, hence the minimum of $n + 1$ terms is 0, so these values need not be considered at all).

Comment. This formula is called an *extension principle*. It was proposed by Zadeh in his pioneer paper [66].

In [18], it was shown that $\mu(\Delta y) = \mu_0(\Delta y/\sigma)$, where $\sigma = |c_1| \cdot \sigma_1 + \ldots + |c_n| \cdot \sigma_n$. To describe this membership function, it is therefore sufficient to compute the value σ. The formula relating σ with σ_i is exactly the formula (4). Thus, *we can use the above algorithms for interval setting to find the expert estimate for Δy.*

Extension Principle for the Case When a Product Is Used as an &-Operation. If in the above definition of the term "Δy is possible", we formalize "and" as a product, and "or" as max, then we will obtain the following expression for the desired membership function $\mu(\Delta y)$ (i.e., for the degree of belief $\mu(\Delta y)$ that Δy is possible):

$$\mu(\Delta y) = \max_{\Delta x_1, \ldots, \Delta x_n} \left[\mu_0 \left(\frac{\Delta x_1}{\sigma_1} \right) \cdot \ldots \cdot \mu_0 \left(\frac{\Delta x_n}{\sigma_n} \right) \right],$$

where maximum is taken over all values Δx_i for which $\Delta y = c_1 \cdot \Delta x_1 + \ldots + c_n \cdot \Delta x_n$.

In this case, the result of applying extension principle essentially depends on what exactly membership function we use. Dubois and Prade considered a Gaussian membership function $\mu_0(e) = \exp(-\beta \cdot e^2)$ for some constant $\beta > 0$ (additional arguments in favor of choosing a Gaussian membership function are presented in [46]). For this function, they showed that the extension principle leads to the membership function $\mu_0(\Delta y/\sigma)$, where σ is determined by the formula (3). This is exactly the same formula as for statistical setting, so *we can use the above algorithms for interval setting to find the expert estimate for Δy.*

7. PROOFS

7.1 PROOF OF THEOREM 1

The Main Idea of the Proof. The original problem is rotation-invariant. Therefore, if we have an (n, N)-algorithm for solving this problem which is not rotation-invariant, we can design a new algorithm if we first apply a random rotation and then apply this algorithm to the rotated problem. The resulting new algorithm is rotation-invariant, and because the function (inaccuracy) that we want to minimize is assumed to be convex, additional randomization can only decrease its expected value, and thus, can only improve the algorithm. This, when we are looking for an optimal algorithm, we only need to consider rotation-invariant algorithms. We will show, step-by-step, that the only rotation-invariant algorithm is the algorithm described in Theorem 1.

Comment. For a general overview of how the general ideas of symmetry and invariance help in solving optimization problems in computer-related areas, see, e.g., [50].

When Looking for an Optimal Algorithm, We Can Always Assume That the Vectors $\vec{\delta}^{(k)}$ Produced by an Algorithm Form an Orthonormal Basis. Indeed, let us show that an arbitrary (n, N)-algorithm can be modified, without changing its inaccuracy, into a new algorithm which only deals with the orthonormal bases.

Ultimately, whenever the original algorithm produces a sequence of vectors $\vec{\delta}^{(1)}, \ldots, \vec{\delta}^{(N)}$, the new algorithm will produce the Gram-Schmidt orthogonalization $\vec{e}^{(1)}, \ldots, \vec{e}^{(N)}$ of this sequence of vectors. To be more precise, a new algorithm follows the original one, with the following modifications:

- On the 1st step, when the original algorithm produced a vector $\vec{\delta}^{(1)}$ and wants to compute the value $c^{(1)} = \vec{c} \cdot \vec{\delta}^{(1)}$, the new algorithm first produces the unit vector $\vec{e}^{(1)}$ by applying the first orthogonalization step (7), then computes a new scalar product $c_{\text{new}}^{(1)} = \vec{c} \cdot \vec{e}^{(1)}$, and then reconstructs $c^{(1)}$ as $c^{(1)} = c_{\text{new}}^{(1)} \cdot \left\| \vec{\delta}^{(1)} \right\|$.

- For each $k \geq 2$, on k-th step, when the original algorithm produced a vector $\vec{\delta}^{(k)}$, and wants to compute the value $c^{(k)} = \vec{c} \cdot \vec{\delta}^{(k)}$, the new algorithm first performs the k-th step (8) of the orthogonalization procedure, i.e., produces a vector $\vec{e}^{(k)}$, then computes $c_{\text{new}}^{(k)} = \vec{c} \cdot \vec{e}^{(k)}$, and reconstructs the desired value as

$$c^{(k)} = \left(\vec{\delta}^{(k)} \cdot \vec{e}^{(1)} \right) \cdot c_{\text{new}}^{(1)} + \ldots + \left(\vec{\delta}^{(k)} \cdot \vec{e}^{(k)} \right) \cdot c_{\text{new}}^{(k)}.$$

The possibility of this reconstruction follows from the fact that the vectors $\vec{e}^{(i)}$, $1 \leq i \leq k$, form an orthonormal basis in a space which includes

$\vec{\delta}^{(k)}$ and therefore,

$$\vec{\delta}^{(k)} = \sum_{i=1}^{k} \left(\vec{\delta}^{(k)} \cdot \vec{e}^{(i)} \right) \cdot \vec{e}^{(i)};$$

multiplying both sides of this equality by $\vec{c}$ and taking into consideration that $\vec{c} \cdot \vec{e}^{(i)} = c_{\text{new}}^{(i)}$, we get the above formula.

The final result of the new algorithm is exactly the same as for the old one, so the inaccuracy of the new algorithm is the same as for the old one. On each step, we replace each call to f (i.e., each computation of a scalar product) by exactly one call (computation); thus, the total number of calls (N) remains the same. However, in the new algorithm, we only compute the scalar product for a sequence of vectors which forms an orthonormal basis.

In view of this transformation, we can, without losing generality, assume that an (n, N)-algorithm only uses orthonormal bases, i.e., that with probability 1, in each trajectory, the vectors $\vec{\delta}^{(1)}, \ldots, \vec{\delta}^{(N)}$ form an orthonormal basis.

A Known Result: There Is Only One Rotation-Invariant Probability Measure on the Set of All Bases. We want to generate random orthonormal bases. Since the problem is rotation-invariant, it is reasonable to look for rotation-invariant probability measures on the set of all such bases.

In our further proof, we will use the fact that there is only one such measure. This result is known, but, for completeness, we will describe the main idea of the proof. To describe the probability measure on the set of all N-element orthonormal bases, we must describe: the probabilities of different values of the first vector $p\left(\vec{e}^{(1)}\right)$, the conditional probabilities of different values of the second vector given the first vector $p\left(\vec{e}^{(2)} \mid \vec{e}^{(1)}\right)$, and, for all $k = 3, \ldots, N$, the conditional probabilities $p\left(\vec{e}^{(k)} \mid \vec{e}^{(1)}, \ldots, \vec{e}^{(k-1)}\right)$.

For $k = 1$, all possible vectors $\vec{e}^{(1)}$ form a unit sphere (since the bases are orthonormal). On the unit sphere, there is a only one rotation-invariant distribution: uniform distribution on this sphere, for which the probability measure is equal to the normalized Lebesgue measure on this sphere. Similarly, when the vectors $\vec{e}^{(1)}, \ldots, \vec{e}^{(k-1)})$ are fixed, then, due to orthonormality, $\vec{e}^{(k)}$ must be a unit vector orthogonal to all the previous vectors $\vec{e}^{(i)}$, $1 \le i \le k - 1$. Thus, all such vectors form a unit sphere in a linear subspace of R^n, namely, on a space $\left[\text{Lin}\left(\vec{e}^{(1)}, \ldots, \vec{e}^{(k-1)}\right)\right]^{\perp}$ which is formed by all vectors orthogonal to all $k - 1$ previous vectors. Again, on a unit sphere, there is only one rotation-invariant probability distribution, for which the probability measure is equal to the normalized Lebesgue measure on the corresponding linear subspace.

The uniqueness is proven.

The Algorithm from Section 2.3 Produces a Basis Which Is Random With Respect to a Rotation-Invariant Distribution. Indeed, according to the construction from Section 2.3, each component a_i of each vector $\vec{a} = (a_1, \ldots, a_n)$ is an independent Gaussian variable with 0 average and standard deviation 1. Thus, the corresponding probability density for each component a_i is equal to const $\cdot \exp(-a_i^2/2)$, and the resulting probability density for each vector $\vec{a}$ is equal to the product of these densities: const $\cdot \exp(-a_1^2 - \ldots - a_n^2) = $ const $\cdot \exp\left(-\|\vec{a}\|^2/2\right)$. This probability density depends only on the length $\|\vec{a}\|$ of the vector $\vec{a}$, and is, therefore, rotation-invariant.

Since the distribution of each of the vectors $\vec{a}^{(k)}$ is rotation-invariant, and these vectors are, by construction, statistically independent, the resulting distribution on the set of all tuples $(\vec{a}^{(1)}, \ldots, \vec{a}^{(N)})$ is also rotation-invariant. Since Gram-Schmidt orthogonalization is also a rotation-invariant procedure, the resulting probability distribution on the set of all the bases is also rotation-invariant. The statement is proven.

When Looking for an Optimal Algorithm, We Can Always Assume That the Vectors $\vec{\delta}^{(k)}$ Form an Orthonormal Basis Which Is Random With Respect to a Rotation-Invariant Distribution. Indeed, let us show that an arbitrary (n, N)-algorithm can be modified, without worsening its inaccuracy, into a new algorithm in which orthonormal bases are random with respect to the rotation-invariant distribution.

To produce this new algorithm, we will use a *random rotation* T, which can be defined as a linear transformation that turns the coordinate unit vectors $\vec{b}^{(1)} = (1, 0, \ldots, 0), \ldots, \vec{b}^{(n)} = (0, \ldots, 0, 1)$ into elements of an n-element random orthonormal basis $\vec{e}^{(1)}, \ldots, \vec{e}^{(n)}$. Due to linearity, this rotation turns every vector $\vec{x} = (x_1, \ldots, x_n)$ into a vector $T(\vec{x}) = x_1 \cdot \vec{e}^{(1)} + \ldots + x_n \cdot \vec{e}^{(n)}$.

Now that a random rotation is defined, we can describe the new (n, N)-algorithm. First, we pick a random rotation T. Then, we follow the old algorithm step-by-step. Every time the old algorithm generates a vector $\vec{\delta}^{(k)}$ for computing $c^{(k)} = \vec{c} \cdot \vec{\delta}^{(k)}$, in the new algorithm, we call f with the *new* vector $\vec{\delta}_{\text{new}}^{(k)} = T\left(\vec{\delta}^{(k)}\right)$, i.e., compute $c_{\text{new}}^{(k)} = \vec{c} \cdot T\left(\vec{\delta}^{(k)}\right)$.

Let us show that for the new algorithm, the inaccuracy can only become smaller. Indeed, since the scalar product is rotation-invariant, the new value of $c_{\text{new}}^{(k)}$, which we obtained by combining the new (rotated) vector $\vec{\delta}_{\text{new}}^{(k)}$ with the old vector $\vec{c}$, is equal to the result $T^{-1}(\vec{c}) \cdot \vec{\delta}^{(k)}$ of combining the old (unrotated) vector $\vec{\delta}^{(k)}$ with the new vector $\vec{c}_{\text{new}} = T^{-1}(\vec{c})$ (which is obtained from $\vec{c}$ by the reverse rotation).

If we fix T, the trajectory of an old algorithm for a vector $\vec{c}$ is a trajectory of the new algorithm for the rotated vector $\vec{c}_{\text{new}} = T^{-1}(\vec{c})$. Thus, when T

is fixed, the new algorithm has the same inaccuracies as the old one, except that the new algorithm has them for different vectors $\vec{c}$. So, if we had fixed T, the maximum over $\vec{c}$ would be exactly the same for the old and for the new algorithms.

In the actual new algorithm, there is a new randomization step: instead of fixing T, we choose T according to the (rotation-invariant) random distribution. As a result, for each vector $\vec{c}$, the actual average for a new algorithm is equal to the inaccuracy averaged over different $\vec{c}$ (to be more precise, averaged over all vectors $T^{-1}(\vec{c})$ for a random rotation $\vec{c}$). The average cannot exceed the maximum, and therefore, for every $\vec{c}$, this average (i.e., the inaccuracy of the new algorithm) cannot exceed the worst-case inaccuracy of the old algorithm. Thus, the inaccuracy of the new algorithm is indeed smaller than or equal to the inaccuracy of the old one.

To complete the proof of this statement, it is sufficient to show that in the new algorithm, the resulting probability distribution of the bases $\vec{\delta}_{new}^{(k)}$ is indeed rotation-invariant. Indeed, if we apply an additional rotation S to all bases, it is equivalent to changing from T to the composition $S \circ T$ of the rotations, i.e., equivalently, to replacing a random orthonormal basis $\vec{e}^{(1)}, \ldots, \vec{e}^{(n)}$ in the definition of the rotation T by a rotated basis $S(\vec{e}^{(1)}), \ldots, S(\vec{e}^{(n)})$. The distribution on the set of all bases is rotation-invariant; thus, applying a fixed rotation S to all elements of all the bases should not change the probabilities. Therefore, the resulting distribution of the bases $\vec{\delta}_{new}^{(k)}$ is also rotation-invariant. The statement is proven.

First Preliminary Conclusion. Thus, when we are looking for an optimal algorithm, we can always assume that the bases are distributed according to the rotation-invariant distribution. The only difference between different (n, N)-algorithms of this type is how they estimate c based on these bases and on the results $c^{(k)}$ of multiplying $\vec{c}$ and the vectors from these bases. Let us therefore analyze which estimations lead to smaller inaccuracy.

When Looking for an Optimal Algorithm, We Can Always Assume that the Final Estimate $\tilde{e}$ Is Deterministic. In our general definitions, we assumed that after we get all the values $c^{(k)}$, we may still need to use some (Monte-Carlo) randomization to produce the final estimate $\tilde{c}$. Let us show that if this is the case, then we can replace this randomized estimate $\tilde{c}$ by its deterministic mathematical expectation $\tilde{c}_{new} = E(\tilde{c})$ and not increase the algorithm's inaccuracy. (To be more precise, by a mathematical expectation, we mean a *conditional* mathematical expectation, under the condition that the values $c^{(k)}$ and $\vec{\delta}^{(k)}$ are fixed.)

Indeed, since the inaccuracy function $\Phi(\tilde{c}, c)$ is convex in $\tilde{c}$, and for convex functions, $\Phi(E(z)) \leq E(\Phi(z))$, we have $\Phi(\tilde{c}_{new}, c) = \Phi(E(\tilde{c}_{old}), c) \leq$

$E(\Phi(\tilde{c}_{old}, c))$. Therefore, the inaccuracy of the new algorithm (with the deterministic last step) cannot exceed the inaccuracy of the old algorithm (with possibly stochastic last step). The statement is proven.

When Looking for an Optimal Algorithm, We Can Assume That the Final Estimation Depends Only On $c^{(k)}$ And Not on the Vectors $\vec{\delta}^{(k)}$. Indeed, let us take an arbitrary (n, N)-algorithm which have passed through all our previous modifications. In particular, for this algorithm, the final step is deterministic, i.e., $\tilde{c} = \tilde{e}\left(c^{(1)}, \ldots, c^{(N)}, \vec{\delta}^{(1)}, \ldots, \vec{\delta}^{(N)}\right)$ for some function $\tilde{e}$.

For an arbitrary rotation T, we can consider an alternative function

$$\tilde{e}_T\left(c^{(1)}, \ldots, c^{(N)}, \vec{\delta}^{(1)}, \ldots, \vec{\delta}^{(N)}\right) =$$

$$\tilde{e}\left(c^{(1)}, \ldots, c^{(N)}, T\left(\vec{\delta}^{(1)}\right), \ldots, T\left(\vec{\delta}^{(N)}\right)\right).$$

The use of this function is equivalent to using the new (rotated) basis $T\left(\vec{\delta}^{(k)}\right)$ *and* the new (rotated) unknown vector $\tilde{c}_{new} = T^{-1}(\tilde{c})$.

Since we consider an algorithm which has already passed through our improvements, the bases $\vec{\delta}^{(k)}$ are simply randomly distributed in a rotation-invariant way, so adding a rotation T should not change the distribution of these bases. For the same reason, the quality of the new algorithm should not change is we replace $\vec{c}$ by its rotation. Therefore, the new algorithm should have the same inaccuracy as the old one. This is true for all rotations T, i.e., for every T, we can take $\tilde{c} = \tilde{e}_T$ and get the same inaccuracy.

Now, we are ready for the final step in the proof of this statement: we can take an *average* of such estimates for all possible rotations T (average over the rotation-invariant measure on all T):

$$\tilde{c}_{new} = E_T\left[\tilde{e}_T\left(c^{(1)}, \ldots, c^{(N)}, \vec{\delta}^{(1)}, \ldots, \vec{\delta}^{(N)}\right)\right] =$$

$$E_T\left[\tilde{e}\left(c^{(1)}, \ldots, c^{(N)}, T\left(\vec{\delta}^{(1)}\right), \ldots, T\left(\vec{\delta}^{(N)}\right)\right)\right].$$

Due to convexity, this averaging can only decrease the inaccuracy of the original algorithm. Since the basis is randomly distributed according to a rotation-invariant distribution, averaging over all T is equivalent to averaging over all possible bases. Hence, this averaging leads to an estimate which does not depend on the basis vectors at all. The statement is proven.

Second Preliminary Conclusion. Thus, when we are looking for optimal algorithms, it is sufficient to consider final estimates of the type $\tilde{c} = \tilde{e}\left(c^{(1)}, \ldots, c^{(N)}\right)$. For the following arguments, it is convenient to combine the N values $c^{(1)}, \ldots, c^{(N)}$ into a single N-dimensional vector **c**. (We will denote N-dimensional vectors differently to avoid confusion with n-dimensional ones.) In these terms, $\tilde{c} = \tilde{e}(\mathbf{c})$.

When Looking for an Optimal Algorithm, We Can Assume That the Final Estimate Only Depends on $\sum (c^{(k)})^2$. This statement can be proved in a similar way, with the only difference that instead of arbitrary rotations in a n-dimensional space, we can consider only rotations within the N-dimensional space generated by the vectors from the basis $\vec{\delta}^{(k)}$, $1 \leq k \leq N$.

Indeed, if we rotate the basis by such a rotation **T**, we get a new algorithm, which is equivalent to the old one (because additional rotation does not change the rotation-invariant distribution), but for which the new vectors $\vec{\delta}_{\text{new}}^{(k)} = \sum_l r_{kl} \cdot \vec{\delta}^{(l)}$, where r_{kl} is an $N \times N$ rotation matrix **T** in a N-dimensional space. For this new basis, the new values $c_{\text{new}}^{(k)} = \vec{c} \cdot \vec{\delta}_{\text{new}}^{(k)}$ are obtained by the same rotation from the old value $c^{(k)}$: $c_{\text{new}}^{(k)} = \sum_l r_{kl} \cdot c_{\text{old}}^{(l)}$. In other words, the new N-dimensional vector $\mathbf{c}_{\text{new}}$ is obtained from the old vector by a rotation **T**: $\mathbf{c}_{\text{new}} = \mathbf{T}(\mathbf{c}_{\text{old}})$. Thus, instead of the old estimate $\tilde{c} = \tilde{e}(\mathbf{c})$, we can, for every $(N \times N)$-rotation matrix **T**, consider a new estimate $\tilde{c} = \tilde{e}_{\mathbf{T}}(\mathbf{c}) = \tilde{e}(\mathbf{T}(\mathbf{c}))$.

Now, we are ready for the final step in the proof of this statement: we can take an *average* of such estimates for all possible N-dimensional rotations **T** (average over the rotation-invariant measure on all T): $\tilde{c}_{\text{new}} = E_{\mathbf{T}} [\tilde{e}_{\mathbf{T}}(\mathbf{c})] = E_{\mathbf{T}} [\tilde{e}(\mathbf{T}(\mathbf{c}))]$. Due to convexity, this averaging can only decrease the inaccuracy of the original algorithm. Since the rotation **T** is randomly distributed according to a rotation-invariant distribution, the vector $\mathbf{T}(\mathbf{c})$ is rotation-invariant distributed; all such vectors have the same length as the original N-dimensional vector **c**. Therefore, the vector $\mathbf{T}(\mathbf{c})$ is uniformly distributed over a sphere of radius $\|\mathbf{c}\| = \sqrt{\sum (c^{(k)})^2}$. Hence, this averaging leads to an estimate which does not depend on the actual values of the components of the vector **c**, but only depends on this length.

The statement is proven, and so is the theorem.

7.2 PROOF OF THEOREM 2

Main Idea of the Proof. The problem of minimizing (10) under the condition (11) is equivalent to minimizing

$$J = \sum_{i=1}^{n-1} \frac{p_i}{i} \to \min_{p_1, \ldots, p_n} \tag{10a}$$

under the same condition (11). The optimization problem (10a), (11) is a linear programming problem, and it is known that for such problems, the optimum is always attained at one of the vertices. Therefore, to find the solution to our problem, it is sufficient to enumerate all such vertices, i.e., in this case, all combinations p_i in which at most two probabilities p_i are different from 0. Without losing generality, we can therefore assume that $p_i = 0$ except for two values p_m and p_M, $m < M$, which may be different from 0. From the conditions (11), we conclude that $p_m + p_M = 1$ and that $p_m \cdot m + p_M \cdot M = \bar{N}$. From the first of these equations, we conclude that $p_M = 1 - p_m$. Substituting this expression into the second equation, we conclude that $p_m \cdot m + (1 - p_m) \cdot M = \bar{N}$ and hence,

$$p_m = \frac{M - \bar{N}}{M - m} \quad \text{and} \quad p_M = 1 - p_m = \frac{\bar{N} - m}{M - m}.$$

Since the probabilities have to be positive and $M > m$, we conclude that $\bar{N} \leq M$. Similarly, from the fact that $p_M \geq 0$, we can conclude that $m \leq \bar{N}$, so $m \leq \bar{N} \leq M$.

For these values p_m and p_M, the value of the expression (10a) depends on whether $M = n$ or $M < n$. Let us find the find for each of these cases, and then find the actual minimum by comparing the resulting two minima.

First Case: $M < n$.   If $M < n$, then the minimized expression (10a) takes the form

$$J = \frac{p_m}{m} + \frac{p_M}{M} = \frac{p_m}{m} + \frac{1 - p_m}{M} = p_m \cdot \left(\frac{1}{m} - \frac{1}{M} \right) + \frac{1}{M} =$$

$$\frac{p_m \cdot (M - m)}{m \cdot M} + \frac{1}{M}.$$

Substituting the above expression for p_m, we conclude that

$$J = \frac{M - \bar{N}}{M \cdot m} + \frac{1}{M}.$$

For a fixed M, this expression is a decreasing function of m; thus, its minimum is attained when m takes the largest possible value. Since m can only take values from 1 to $\bar{N}$, we therefore conclude that $m = \lfloor \bar{N} \rfloor$.

Similarly, we can conclude that

$$J = \frac{p_m}{m} + \frac{p_M}{M} = \frac{1 - p_M}{m} + \frac{p_M}{M} = p_M \cdot \left(\frac{1}{M} - \frac{1}{m} \right) + \frac{1}{m} =$$

$$-\frac{p_M \cdot (M - m)}{m \cdot M} + \frac{1}{m}.$$

Substituting the above expression for p_M, we conclude that

$$J = -\frac{\bar{N} - m}{M \cdot m} + \frac{1}{m}.$$

For a fixed m, this expression is an increasing function of M; thus, its minimum is attained when M takes the smallest possible value. Since M can only take values from $\bar{N}$ to n, we therefore conclude that $M = \lceil \bar{N} \rceil$.

Second Case: $M = n$. In this case, the minimized expression (10a) takes the form

$$J = \frac{p_m}{m} = \frac{n - \bar{N}}{(n - m) \cdot m}.$$

Hence, its minimum is attained when the value $(n - m) \cdot m$ is the largest possible among the values $m \leq \bar{N}$.

The function $(n - m) \cdot m$ is increasing until $m = n/2$ and then decreases. So:

- if $\bar{N} > n/2$, the maximum of this function is attained at $m = \lfloor n/2 \rfloor$, and

- if $\bar{N} \leq n/2$, its maximum is attained when $m = \lfloor \bar{N} \rfloor$.

Comparing the Two Cases: General Idea. To find the optimal values of m and M, we must compare the optimal values corresponding to the two cases $M < n$ and $M = n$. Since in the second case, we get different values of m depending on whether $\bar{N} > n/2$ or $\bar{N} \leq n/2$, let us consider these two situations separately.

Within each two situations, the formulas will be slightly different depending on whether $\bar{N}$ is an integer or not. In this proof, we will describe the generic situation when $\bar{N}$ is not an integer. In this generic case, we have $\lceil \bar{N} \rceil = \bar{n} + 1$.

Situations when $\bar{N}$ is an integer can be either considered separately (and similarly), or can be obtained from the generic situation as a limit case, when $\bar{N}$ tends to an integer.

Comparing the Two Cases When $\bar{N} > n/2$. Let us first show that when $\bar{N} > n/2$, then $M = n$ leads to a better estimate. In this proof, we will consider two subcases: when $\bar{N} > \lceil n/2 \rceil$ and when $n/2 < \bar{N} \leq \lceil n/2 \rceil$.

In the first subcase, when $M < n$, we get $m = \lfloor \bar{N} \rfloor = \bar{n}, M = \lceil \bar{N} \rceil = \bar{n} + 1$, $p_m = 1 - \theta, p_M = \theta$, and the optimized expression J takes the value

$$J_< = \frac{1 - \theta}{\bar{n}} + \frac{\theta}{\bar{n} + 1}.$$

When $M = n$, we get $m = \lfloor n/2 \rfloor$, so

$$p_m = \frac{n - (\bar{n} + \theta)}{n - \lfloor n/2 \rfloor},$$

and the corresponding value of J is equal to

$$J_= = \frac{n - (\bar{n} + \theta)}{(n - \lfloor n/2 \rfloor) \cdot \lfloor n/2 \rfloor}.$$

We want to prove that $M = n$ leads to a better solution, i.e., that $J_< \geq J_=$. We will prove it in two steps: first, we will prove that $J_< \geq J_0$, where by J_0, we denoted $J_0 = 1/(\bar{n} + \theta)$, and then, we will prove that $J_0 \geq J_=$.

The inequality $J_< \geq J_0$ is equivalent to

$$\frac{1 - \theta}{\bar{n}} + \frac{\theta}{\bar{n} + 1} \geq \frac{1}{\bar{n} + \theta}.$$

Multiplying both sides of this desired inequality by the product of all three denominators, we get an equivalent inequality:

$$(1 - \theta) \cdot (\bar{n} + \theta) \cdot (\bar{n} + 1) + \theta \cdot \bar{n} \cdot (\bar{n} + \theta) \geq \bar{n} \cdot (\bar{n} + 1).$$

If we perform all the multiplications and group together terms proportional to $\bar{n}^2$, to $\bar{n}$, and to 1, we get an equivalent inequality

$$[(1 - \theta) + \theta] \cdot \bar{n}^2 + [(1 - \theta^2) + \theta^2] \cdot \bar{n} + (1 - \theta) \cdot \theta \geq \bar{n}^2 + n.$$

Terms proportional to $\bar{n}^2$ and to $\bar{n}$ cancel each other, so we get the equivalent inequality $(1 - \theta) \cdot \theta \geq 0$ which is true for every $\theta \in [0, 1]$. So, $J_< \geq J_0$.

Let us now prove the second auxiliary inequality $J_0 \geq J_=$. Since $n - \lfloor n/2 \rfloor = \lceil n/2 \rceil$, the inequality $J_0 \geq J_=$ can be reformulated in the following equivalent form:

$$\frac{1}{\bar{n} + \theta} \geq \frac{n - (\bar{n} + \theta)}{\lceil n/2 \rceil \cdot (n - \lceil n/2 \rceil)}.$$

Multiplying both sides by the common denominator, we get an equivalent inequality

$$\lceil n/2 \rceil \cdot (n - \lceil n/2 \rceil) \geq (\bar{n} + \theta) \cdot (n - (\bar{n} + \theta)).$$

This inequality is true because, as we mentioned earlier, the function $z \cdot (n - z)$ is decreasing for $z \geq n/2$, and we are considering the case when $\bar{N} = \bar{n} + \theta \geq \lceil n/2 \rceil$. Thus, $J_0 \geq J_=$ and hence, $J_< \geq J_=$, i.e., the solution corresponding to the case $M = n$ is indeed optimal.

To complete the proof for the case $\bar{N} > n/2$, we must now consider the situation when $n/2 < \bar{N} \leq \lceil n/2 \rceil$. This is only possible if $n/2$ is not an integer; then, $\lfloor n/2 \rfloor = \lfloor \bar{N} \rfloor = \bar{n}$, so $\lceil n/2 \rceil = \bar{n} + 1$, and $n = 2\bar{n} + 1$. In this case, the desired inequality $J_< \geq J_=$ takes the following form:

$$\frac{1 - \theta}{\bar{n}} + \frac{\theta}{\bar{n} + 1} \geq \frac{\bar{n} + 1 - \theta}{\bar{n} \cdot (\bar{n} + 1)}.$$

722

Multiplying both sides by $\bar{n} \cdot (\bar{n} + 1)$ and performing all the multiplications, we get the equality, so, in this case, we also have $J_< \geq J_=$.

Thus, when $\bar{N} > n/2$, the solution corresponding to the case $M = n$ is indeed optimal.

Comparing the Two Cases When $\bar{N} \leq n/2$. Let us now prove that when $\bar{N} \leq n/2$, then $M < n$ leads to a better estimate, i.e., that $J_< \leq J_=$. Here, $J_<$ is described by the same expression as in the case $\bar{N} > n/2$, but the expression for $J_=$ is now different, because for $M = n$, we now have $m = \bar{n}$ and hence

$$J_= = \frac{n - (\bar{n} + \theta)}{\bar{n} \cdot (n - \bar{n})}.$$

In this case, the desired inequality $J_< \leq J_+$ takes the following form:

$$\frac{1 - \theta}{\bar{n}} + \frac{\theta}{\bar{n} + 1} \leq \frac{n - (\bar{n} + \theta)}{\bar{n} \cdot (n - \bar{n})}.$$

Reducing the left-hand side to the common denominator, we get the equivalent inequality

$$\frac{\bar{n} + 1 - \theta}{\bar{n} \cdot (\bar{n} + 1)} \leq \frac{n - (\bar{n} + \theta)}{\bar{n} \cdot (n - \bar{n})}.$$

Multiplying both sides by the common denominator, we get an equivalent inequality

$$(\bar{n} + 1 - \theta) \cdot (n - \bar{n}) \leq (\bar{n} + 1) \cdot (n - \bar{n} - \theta).$$

If we perform all the multiplications and cancel equal terms on both sides of the inequality, we get an equivalent inequality $-\theta \cdot (n - \bar{n}) \leq -\theta \cdot \bar{n}$, which, in its turn, is equivalent to $n - \bar{n} \geq \bar{n}$, i.e., to $2\bar{n} \leq n$ and to $\bar{n} \leq n/2$. Since we consider the case $\bar{N} \leq n/2$ and $\bar{n} = \lfloor \bar{N} \rfloor \leq \bar{N}$, therefore, indeed $\bar{n} \leq n/2$, and hence, $J_< \leq J_=$.

Thus, when $\bar{N} \leq n/2$, the solution corresponding to the case $M < n$ is indeed optimal. The theorem is proven.

7.3   CORRELATIONS IN INTERVAL SETTING: JUSTIFICATION OF THE ALGORITHM

Let us show that the values $c^{(k)}$ are distributed according to Cauchy law with the parameter $\Delta_s = f_s(\vec{c})$ and the average $\Delta_a = f_a(\vec{c})$, i.e., that its probability density is equal to

$$\rho(z) = \frac{\Delta_s}{\pi \cdot ((z - \Delta_a)^2 + \Delta_s^2)}.$$

Indeed, by construction, $c = \vec{c} \cdot \vec{\delta}$, where δ is distributed according to the probability density $\rho(\vec{x})$. Thus, the characteristic function

$$\chi(\omega) = E[\exp(i \cdot \omega \cdot c)]$$

of this distribution is equal to $\chi(\omega) = E[\exp(\mathrm{i} \cdot \omega \cdot \vec{c} \cdot \vec{\delta})]$, i.e., $\chi(\omega) = E[\exp(\mathrm{i} \cdot \vec{\omega} \cdot \vec{\delta})]$, where we denoted $\vec{\omega} = \omega \cdot \vec{c}$.

By definition, the characteristic function $E[\exp(\mathrm{i} \cdot \vec{\omega} \cdot \vec{\delta})]$ is equal to the Fourier transform of the probability density function $\rho(\vec{x})$. Since this density function was defined as an inverse Fourier transform of the expression

$$\exp(\mathrm{i} \cdot f_a(\vec{\omega}) - f_s(\vec{\omega})),$$

we thus conclude that $E[\exp(\mathrm{i} \cdot \vec{\omega} \cdot \vec{\delta})] = \exp(\mathrm{i} \cdot f_a(\vec{\omega}) - f_s(\vec{\omega}))$. Substituting the expression $\vec{\omega} = \omega \cdot \vec{c}$ into this formula, we conclude that

$$\chi(\omega) = \exp(\mathrm{i} \cdot f_a(\omega \cdot \vec{c}) - f_s(\omega \cdot \vec{c})).$$

From the definition of the functions f_- and f_+, we can easily conclude that

- for $\omega > 0$, we have $f_-(\omega \cdot \vec{c}) = \omega \cdot f_-(\vec{c})$ and $f_+(\omega \cdot \vec{c}) = \omega \cdot f_+(\vec{c})$;

- for $\omega < 0$, we have $f_-(\omega \cdot \vec{c}) = \omega \cdot f_+(\vec{c})$ and $f_+(\omega \cdot \vec{c}) = \omega \cdot f_-(\vec{c})$.

Thus, for f_a and f_s, we have, for arbitrary real ω, that $f_a(\omega \cdot \vec{c}) = \omega \cdot f_a(\vec{c})$ and $f_s(\omega \cdot \vec{c}) = |\omega| \cdot f_s(\vec{c})$. Hence, $\chi(\omega) = \exp(\mathrm{i} \cdot \omega \cdot f_a(\vec{c}) - |\omega| \cdot f_s(\vec{c}))$.

This is a known characteristic function of Cauchy distribution with an average $f_a(\vec{c})$ and a parameter $f_s(\vec{c})$. Thus, we can estimate $f_a(\vec{c})$ as the average Δ_a of the sample values $c^{(k)}$. For the values $c^{(k)} - \Delta_a \approx c^{(k)} - f_a(\vec{c})$, the average is 0, so we can use the above maximum likelihood method to determine the parameter $f_s(\vec{c})$ of this distribution.

From the resulting estimates $\Delta_a \approx f_a(\vec{c})$ and $\Delta_s \approx f_s(\vec{c})$, we can conclude that $\Delta_- = \Delta_s - \Delta_a \approx f_s(\vec{c}) - f_a(\vec{c}) = f_-(\vec{c})$ and similarly that $\Delta_+ \approx f_+(\vec{c})$. The justification is completed.

Acknowledgments

This work was supported in part by NASA under cooperative agreement NCC5-209, by NSF grants No. DUE-9750858 and CDA-9522207, by United Space Alliance, grant No. NAS 9-20000 (PWO C0C67713A6), by the Future Aerospace Science and Technology Program (FAST) Center for Structural Integrity of Aerospace Systems, effort sponsored by the Air Force Office of Scientific Research, Air Force Materiel Command, USAF, under grant number F49620-95-1-0518, and by the National Security Agency under Grant No. MDA904-98-1-0561.

The authors are thankful to the editors for their invitation, and to Chitta Baral and to R. Baker Kearfott for helpful discussions.

724

References

[1] J. Aczel, *Lectures on functional equations and their applications*, Academic Press, New York, London, 1966.

[2] S. Al Wakil, *Processes and design for manufacturing*, Prentice-Hall, Englewood Cliffs, NJ, 1989.

[3] ANSI (American National Standards Institute), *Standard* ANSI Y14.5M-1982, Industrial Press, N.Y., 1982.

[4] G. Belforte and B. Bona, "An improved parameter identification algorithm for signal with unknown-but-bounded errors", *Proc. 7th IFAC Symposium on Identification and Parameter Estimation*, York, U.K., 1985.

[5] M. Beltran and V. Kreinovich, "How To Find Input Variables Whose Influence On The Result Is The Largest, or, How To Detect Defective Stages In VLSI Manufacturing?", *Reliable Computing*, 1995, Supplement (Extended Abstracts of APIC'95: International Workshop on Applications of Interval Computations, El Paso, TX, Febr. 23–25, 1995), pp. 34–37.

[6] A. Bernat, L. Cortes, V. Kreinovich, K. Villaverde, "Intelligent parallel simulation – a key to intractable problems of information processing." *Proceedings of the Twenty-Third Annual Pittsburgh Conference on Modelling and Simulation*, Pittsburgh, PA, 1992, Vol. 2, pp. 959–969.

[7] A. Bernat, E. Villa, K. Bhamidipati, V. Kreinovich, "Parallel interval computations as a background problem: when processors come and go", *International Conference on Interval and Computer-Algebraic Methods in Science and Engineering (Interval'94), St. Petersburg, Russia, March 7-10, 1994, Abstracts*, pp. 51–53.

[8] M. Berz, C. Bischof, G. Corliss, and A. Griewank, *Computational differentiation: techniques, applications, and tools*, SIAM, Philadelphia, 1996.

[9] K. Bhamidipati, "PVM estimates errors caused by imprecise data", In: *Proceedings of the 1994 PVM Users' Group Meeting, Oak Ridge, TN, May 19–20*, Center for Research on Parallel Computations, Session 2A, 1994.

[10] C.-C. Chang, *Fast Algorithm That Estimates the Precision of Indirect Measurements*, Master Project, Computer Science Dept., University of Texas at El Paso, 1992.

[11] F. L. Chernousko, *Estimation of the phase space of dynamic systems*, Nauka publ., Moscow, 1988 (in Russian).

[12] F. L. Chernousko, *State estimation for dynamic systems*, CRC Press, Boca Raton, FL, 1994.

[13] A. A. Clifford, *Multivariate error analysis*, J. Wiley & Sons, N.Y., 1973.

[14] L. A. Cortes, "How to design an expert system that for a given query Q, computes the interval of possible values of probability $p(Q)$ that Q is true", *Abstracts for a Workshop on Interval Methods in Artificial Intelligence, International Conference on Numerical Analysis with Automatic Result Verification: Mathematics, Application and Software, February 25–March 1, 1993*, Lafayette, LA, 1993, p. 11.

[15] L. A. Cortes, *Calculating belief probabilities and intervals*, Master Thesis, Department of Computer Science, University of Texas at El Paso, May 1994.

[16] D. I. Doser, K. D. Crain, M. R. Baker, V. Kreinovich, and M. C. Gerstenberger "Estimating uncertainties for geophysical tomography", *Reliable Computing*, 1998, Vol. 4, No. 3, pp. 241–268.

[17] D. Dubois and H. Prade, "Operations on fuzzy numbers", *International Journal of Systems Science*, 1978, Vol. 9, pp. 613–626.

[18] D. Dubois and H. Prade, "Fuzzy real algebra: some results", *Fuzzy Sets and Systems*, 1979, Vol. 2, pp. 327–348.

[19] D. Dubois and H. Prade, *Fuzzy sets and systems: theory and applications*, Academic Press, N.Y., London, 1980.

[20] D. Dubois and H. Prade, *Possibility theory. An approach to computerized processing of uncertainty*, Plenum Press, N.Y. and London, 1988.

[21] C. Ferregut, S. Nazarian, K. Vennalganti, C.-C. Chang, and V. Kreinovich, "Fast Error Estimates For Indirect Measurements: Applications To Pavement Engineering", *Reliable Computing*, 1996, Vol. 2, No. 3, pp. 219–228.

[22] A. F. Filippov, "Ellipsoidal estimates for a solution of a system of differential equations", *Interval Computations*, 1992, No. 2(4), pp. 6–17.

[23] A. Finkelstein, O. Kosheleva, and V. Kreinovich, "Astrogeometry, error estimation, and other applications of set-valued analysis", *ACM SIGNUM Newsletter*, 1996, Vol. 31, No. 4, pp. 3–25.

[24] E. Fogel and Y. F. Huang, "On the value of information in system identification. Bounded noise case", *Automatica*, 1982, Vol. 18, pp. 229–238.

[25] W. A. Fuller, *Measurement error models*, J. Wiley & Sons, New York, 1987.

[26] C. F. Gerald and P. O. Wheatley, *Applied Numerical Analysis*, Addison-Wesley, Reading, MA, 1994.

[27] A. Griewank, *Evaluating derivatives: Principles and techniques of algorithmic differentiation*, SIAM, Philadelphia, 2000.

[28] H. G. Hecht, *Mathematics in chemistry. An introduction to modern methods*, Prentice Hall, Englewood Cliffs, NJ, 1990.

726

[29] H. M. Hersch and A. Caramazza, "A fuzzy-set approach to modifiers and vagueness in natural languages", *J. Exp. Psychol.: General*, 1976, Vol. 105, pp. 254–276.

[30] A. Kauffman and M. M. Gupta, *Introduction to fuzzy arithmetic: Theory and applications,* Van Nostrand, N.Y., 1985.

[31] R. B. Kearfott and V. Kreinovich (eds.), *Applications of interval computations,* Kluwer, Dordrecht, 1996.

[32] G. Klir and B. Yuan, *Fuzzy sets and fuzzy logic: theory and applications,* Prentice Hall, Upper Saddle River. NJ, 1995.

[33] E. Koltik, V. G. Dmitriev, N. A. Zheludeva, and V. Kreinovich, "An optimal method for estimating a random error component," *Investigations in Error Estimation,* Proceedings of the Mendeleev Metrological Institute, Leningrad, pp. 36–41, 1986 (in Russian).

[34] V. Ya. Kreinovich, "A General Approach to Analysis of Uncertainty in Measurements", *Proceedings of the the 3-rd USSR National Symposium on Theoretical Metrology*, Leningrad, Mendeleev Metrology Institute (VNIIM), 1986, pp. 187–188 (in Russian).

[35] V. Kreinovich, *A method of error estimation for indirect measurements for the case when the set of values of input variables forms an ellipsoid,* Center for New Information Technology "Informatika", Leningrad, 1989 (in Russian).

[36] V. Kreinovich, *In simulation modeling, sometimes simulation with distortions is useful,* Center for New Information Technology "Informatika", Leningrad, 1989 (in Russian).

[37] V. Kreinovich, "A simplified version of the tomography problem can help to estimate the errors of indirect measurements", In: A. Mohamad-Djafari (ed.), *Bayesian Inference for Inverse Problems*, Proceedings of the SPIE/International Society for Optical Engineering, Vol. 3459, San Diego, CA, 1998, pp. 106–115.

[38] V. Kreinovich, "Error estimation for indirect measurements is exponentially hard," *Neural, Parallel, and Scientific Computations,* 1994, Vol. 2, No. 2, pp. 225–234.

[39] V. Kreinovich and A. Bernat, "Parallel algorithms for interval computations: an introduction", *Interval Computations*, 1994, No. 3, pp. 6–62.

[40] V. Kreinovich, A. Bernat, E. Villa and Y. Mariscal, "Parallel computers estimate errors caused by imprecise data", *Proceedings of the Fourth ISMM (International Society on Mini and Micro Computers) International Conference on Parallel and Distributed Computing and Systems*, Washington, 1991, Vol. 1, pp. 386–390.

[41] V. Kreinovich, A. Bernat, E. Villa, and Y. Mariscal, "Parallel computers estimate errors caused by imprecise data", *Interval Computations*, 1991, Vol. 2, pp. 21–46.

[42] V. Kreinovich, A. P. Bernat, E. Villa, and Y. Mariscal, "Parallel computers estimate errors caused by imprecise data", *Technical Papers of the the Society of Mexican American Engineers and Scientists 1992 National Symposium*, San Antonio, Texas, April 1992, pp. 192–199.

[43] V. Kreinovich, A. Lakeyev, J. Rohn, and P. Kahl, *Computational complexity and feasibility of data processing and interval computations*, Kluwer, Dordrecht, 1998.

[44] V. Kreinovich, M. I. Pavlovich, "Error estimate of the result of indirect measurements by using a calculational experiment", *Measurement Techniques*, 1985, Vol. 28, No. 3, pp. 201–205.

[45] V. Kreinovich, S. A. Starks, and R. Trejo, "Automatic Differentiation or Monte-Carlo Methods: Which is Better for Error Estimation?", *Abstracts of the SIAM Annual Meeting*, Toronto, July 13–17, 1998, p. 51.

[46] V. Kreinovich, C. Quintana, L. Reznik, "Gaussian membership functions are most adequate in representing uncertainty in measurements", *Proceedings of NAFIPS'92: North American Fuzzy Information Processing Society Conference*, Puerto Vallarta, Mexico, December 15–17, 1992, NASA Johnson Space Center, Houston, TX, 1992, Vol. II, pp. 618–624.

[47] D. Morgenstein and J. Murphy, "An application of parallel interval techniques to geophysics", *Reliable Computing*, 1995, Supplement (Extended Abstracts of APIC'95: International Workshop on Applications of Interval Computations, El Paso, TX, Febr. 23–25, 1995), p. 155

[48] *MU 25.750 – 85. Methods of calibration, estimation, and testing for metrological characteristics of information processing algorithms*, Industrial Standard, Leningrad, VNIIEP, 1985 In Russian).

[49] S. Nesterov and V. Kreinovich, "The worse, the better: a survey of paradoxical computational complexity of interval computations", In: M. A. Campos (ed.), *Abstracts of the II Workshop on Computer Arithmetic, Interval and Symbolic Computation (WAI'96)*, Recife, Pernambuco, Brazil, August 7-8, 1996, pp. 61A–63A.

[50] H. T. Nguyen and V. Kreinovich, *Applications of continuous mathematics to computer science*, Kluwer, Dordrecht, 1997.

[51] J. P. Norton, "Identification and application of bounded parameter models", *Proc. 7th IFAC Symposium on Identification and Parameter Estimation*, York, U.K., 1985.

[52] P. V. Novitskii and I. A. Zograph, *Estimating the measurement errors*, Energoatomizdat, Leningrad, 1991 (in Russian).

[53] G. C. Oden, "Integration of fuzzy logical information", *Journal of Experimental Psychology: Human Perception Perform.*, 1977, Vol. 3, No. 4, pp. 565–575.

[54] A. I. Orlov, "How often are the observations normal?", *Industrial Laboratory*, 1991, Vol. 57, No. 7, pp. 770–772.

[55] D. E. Punccohar, *Interpretation of geometric dimensioning and tolerancing*, Society of Manufacturing Engineers (SME) Press, MI, 1990.

[56] S. Rabinovich, *Measurement errors: theory and practice*, American Institute of Physics, N.Y., 1993.

[57] F. C. Schweppe, "Recursive state estimation: unknown but bounded errors and system inputs", *IEEE Transactions on Automatic Control*, 1968, Vol. 13, p. 22.

[58] F. C. Schweppe, *Uncertain dynamic systems*, Prentice Hall, Englewood Cliffs, NJ, 1973.

[59] S. G. Shina, *Concurrent engineering and design for manufacturing of electronics products*, Van Nostrand Reinhold, N.Y., 1991.

[60] S. T. Soltanov, "Asymptotic of the function of the outer estimation ellipsoid for a linear singularly perturbed controlled system", In: S. P. Shary and Yu. I. Shokin (eds.), *Interval Analysis*, Krasnoyarsk, Academy of Sciences Computing Center, Publication No. 17, 1990, pp. 35–40 (in Russian).

[61] G. S. Utyubaev, "On the ellipsoid method for a system of linear differential equations", In: S. P. Shary (ed.), *Interval Analysis*, Krasnoyarsk, Academy of Sciences Computing Center, Publication No. 16, 1990, pp. 29–32 (in Russian).

[62] S. A. Vavasis, *Nonlinear optimization: complexity issues*, Oxford University Press, N.Y., 1991.

[63] E. Villa, A. Bernat, and V. Kreinovich, "Estimating errors of indirect measurement on realistic parallel machines: routings on 2-D and 3-D meshes that are nearly optimal", *Interval Computations*, 1993, No. 4, pp. 154–175.

[64] O. R. Wade, *Tolerance control in design and manufacturing*, Industrial Press, N.Y., 1989.

[65] Website on interval computations: http://cs.utep.edu/interval-comp.

[66] L. Zadeh, "Fuzzy sets", *Information and control*, 1965, Vol. 8, pp. 338–353.

[67] N. A. Zheludeva and V. Kreinovich, *A method of error estimation for indirect measurements for the case when the set of values of input variables forms a domain*, Center for New Information Technology "Informatika", Leningrad, 1989 (in Russian).

[68] H. J. Zimmerman, "Results of empirical studies in fuzzy set theory". In: G. Klir, ed., *Applied General System Research*, Plenum, New York, 1978, pp. 303–312.

[69] H. J. Zimmerman, *Fuzzy set theory and its applications*, Kluwer, Dordrecht, 1985.

Chapter 18

DERANDOMIZATION IN COMBINATORIAL OPTIMIZATION

Anand Srivastav

Mathematisches Seminar;
Universität zu Kiel;
Ludewig-Meyn-Strasse 4, D-24098 Kiel, Germany
asr@numerik.uni-kiel.de

1.   INTRODUCTION

Over the last 20 years randomized algorithms have advanced the design of algorithms in algorithmic discrete mathematics and theoretical computer science. Many notoriously hard algorithmic problems have been solved with randomized algorithms or probabilistic methods, either optimally or in a satisfactory approximative way. Among the beautiful examples of the power of randomization are

- Prime number testing and factoring algorithms together with their application in coding theory and cryptography.

- The notion of Vapnik-Chervonenkis dimension (VC-dimension) and the design of efficient algorithms for geometrical problems in bounded VC-dimension.

- Rapidly mixing Markoff chains and approximative solutions of # P-complete problems, e.g. volume computation and counting of discrete structures.

- Randomized algorithms in combinatorial optimization.

- The concept of k-wise independence and the design of parallel randomized algorithms.

- The theory of probabilistically checkable proofs and non-approximability results for combinatorial optimization problems.

731

S. Rajasekaran et al (eds.), Handbook of Randomized Computing, Volume 2, pp, 731–842.
© 2001 *Kluwer Academic Publishers.*

732

With the evident success of randomized algorithms a natural question arises:

"Can a randomized algorithm or probabilistic existence result be tranformed into an efficient, i.e. polynomial-time, deterministic algorithm, or in other words, can it be de-randomized ? "

A key to answering this question is a technique, due to Erdös and Selfridge (1973) and Spencer (1987), known as the the method of conditional probabilities.

The paradigm of parallel computation motivated the search for parallel counterparts of algorithms designed with the conditional probability method. It turned out that random variables with only limited independence correspond to "small" sample spaces. Formulating a problem over such a sample space, if possible, sometimes helps to derive a parallel algorithm. These two derandomization techniques, the conditional probability method and the concept of limited independence, form the fundament for the design of deterministic, polynomial-time algorithms:

- In computational geometry, derandomization can be considered as successful. The perhaps most prominent example is Chazelle's computation of the convex hull of n points in fixed dimension d with an optimal running-time algorithm (Chazelle [63]).

- In combinatorial optimization, the derandomized approximation algorithms for 0/1-integer programming problems of packing type (Raghavan [269]) and the max-cut problem (Goemans, Williamson [127]) opened the gate to new approximation algorithms

- The parallel algorithms of Karp, Widgerson [181] and Luby [211] for the maximal independent set problem using constant independence, the extension of the conditional probability method to $\log^c n$-wise independent random variables due to Berger and Rompel [43] and Motwani, Naor, Naor [239] and the concept of almost k-wise independence introduced by Naor & Naor [245]J are milestones in the design of parallel derandomized algorithms.

In this article we will survey developments in the design of derandomized algorithm in combinatorial optimization and some related fields.

Acknowledgement. I thank Nitin Ahuja, Andreas Baltz, Benjamin Doerr, Marian Margraf and Tomasz Schoen for all their help in preparing this article. I thank Jiří Matoušek for his helpful comments on preliminary versions of this article.

2.   PRELIMINARIES

Graphs and Hypergraphs.   We use the standard notation of graphs and hypergraphs. A graph $G = (V, E)$ is a pair of a finite set V (the set of vertices or nodes) and a subset $E \subseteq \binom{V}{2}$, where $\binom{V}{2}$ denotes the set of all $2-$element subsets of V. The elements of E are called edges. A hypergraph or set system $\mathcal{H} = (V, \mathcal{E})$ is a pair of a finite set V and a subset $\mathcal{E} \subseteq \mathcal{P}(V)$ of the powerset $\mathcal{P}(V)$. The elements of $\mathcal{E}$ are called hyperedges. The degree of a vertex $v \in V$ in $\mathcal{H}$, denoted by $\deg(v)$ or $d(v)$ is the number of hyperedges containing v, and $\deg(\mathcal{H}) = \max_{v \in V} d(v)$ is the (vertex-)degree of $\mathcal{H}$. For a pair of vertices $u, v \in V$, $\mathrm{codeg}(u, v)$ is the co-degree of u and v, and is the number of edges containing both u and v, and $\mathrm{codeg}(\mathcal{H})$ is the maximum over all $\mathrm{codeg}(u, v)$. $\mathcal{H}$ is called $r-$regular resp. $s-$uniform, if $\deg(v) = r$ for all $v \in V$ resp. $|E| = s$ for all $E \in \mathcal{E}$. It is convenient to order the vertices and hyperedges, $V = \{v_1, \cdots, v_n\}$ and $\mathcal{E} = \{E_1, \cdots, E_m\}$, and to identify vertices and edges with their indices. The vertex-hyperedge incidence matrix of a hypergraph $\mathcal{H} = (V, \mathcal{E})$, with $V = \{v_1, \cdots, v_n\}$ and $\mathcal{E} = \{E_1, \cdots, E_m\}$, is a matrix $A = (a_{ij}) \in \{0, 1\}^{n \times m}$, where $a_{ij} = 1$ if $v_i \in E_j$, and 0 else. Sometimes the hyperedge-vertex incidence matrix A^T is used.

For a modern treatment of graph theory, we refer to the books of Berge [42], Bollobás [53], Diestel [89] and West [West96].

Large Deviations.   Throughout this article we consider only finite probability spaces $(\Omega, \mathbb{P})$, where Ω is a finite set and $\mathbb{P}$ is a probability measure with respect to the powerset $\mathcal{P}(\Omega)$ as the sigma field. Let $u_1, \ldots, u_n$ and $v_1, \ldots, v_n$ be integers, and let $X_1, \ldots, X_n$ be mutually independent (briefly independent) random variables, where X_j takes the values u_j and v_j, $1 \leq j \leq n$ and

$$\mathbb{P}[X_j = u_j] = p_j, \ \mathbb{P}[X_j = v_j] = 1 - p_j$$

for real probabilities p_j for all $1 \leq j \leq n$. For $1 \leq j \leq n$ let w_j denote rational weights with

$$0 \leq w_j \leq 1$$

and let

$$\psi = \sum_{j=1}^{n} w_j X_j$$

be the weighted sum. For $u_j = 1$, $v_j = 0$, $w_j = 1$ and $p_j = p$ for all $j = 1, \ldots, n$, $\psi = \sum_{j=1}^{n} X_j$ is the well-known binomially distributed random variable with mean np. The inequalities given below can be found in the books of Alon, Spencer and Erdös [18], Habib, McDiarmid, Ramirez-Alfonsin and Reed [141], and Janson, Łuczak, Ruciński [163].

Theorem 2.1 (Markov Inequality) *Let* $(\Omega, \mathbb{P})$ *be a probability space and* $X : \Omega \longrightarrow \mathbb{R}^+$ *a random variable with expectation* $\mathbb{E}(X) < \infty$. *Then for any* $\lambda \in \mathbb{R}^+$

$$\mathbb{P}[X \geq \lambda] \leq \frac{\mathbb{E}(X)}{\lambda}.$$

An sharper bound is the well-known inequaltity of Chebyshev:

Theorem 2.2 (Chebyshev Inequality) *Let* $(\Omega, \mathbb{P})$ *be a probability space and* $X : \Omega \longrightarrow \mathbb{R}$ *a random variable with expectation* $\mathbb{E}(X)$ *and variance Var(X).* *Then for any* $\lambda \in \mathbb{R}^+$

$$\mathbb{P}[|X - \mathbb{E}(X)| > \lambda \sqrt{Var(X)}] \leq \frac{1}{\lambda^2}.$$

The following basic large deviation inequality is implicitly given in Chernoff [71] in the Binomial case. In explicit form it can be found in Okamoto [252]. Its generalization to arbitrary weight is due to Hoeffding [155]:

Theorem 2.3 (Hoeffding 1963) *Let* $u_j = 1, v_j = 0, 0 \leq w_j \leq 1, 0 \leq p_j \leq 1$ *for all* $j = 1, \ldots, n$ *and let* $\lambda > 0$. *Then*

(a) $\mathbb{P}(\psi > \mathbb{E}(\psi) + \lambda) \leq \exp(-\frac{2\lambda^2}{n})$

(b) $\mathbb{P}(\psi < \mathbb{E}(\psi) - \lambda) \leq \exp(-\frac{2\lambda^2}{n})$.

In the literature Theorem 2.3 is well known as the Chernoff bound. For k-wise independent random variables similar bounds can be found in the article of Schmidt, Siegel and Srinivasan [285]. For small expectations, i.e $\mathbb{E}(\psi) \leq \frac{n}{6}$, the following inequalities due to Angluin and Valiant [24], give sharper bounds than Chernoff's inequality.

Theorem 2.4 (Angluin, Valiant 1979) *Let* $u_j = 1, v_j = 0, 0 \leq w_j \leq 1, 0 \leq p_j \leq 1$ *for all* $j = 1, \ldots, n$ *and let* $0 < \beta \leq 1$. *Then*

(a) $\mathbb{P}(\psi > \mathbb{E}(\psi)(1 + \beta)) \leq \exp(-\frac{\beta^2 \mathbb{E}(\psi)}{3})$

(b) $\mathbb{P}(\psi < \mathbb{E}(\psi)(1 - \beta)) \leq \exp(-\frac{\beta^2 \mathbb{E}(\psi)}{2})$.

For random variables with zero expectation there are two inequalities which can be found in the book of Alon, Spencer and Erdös ([18], Appendix). The first inequality goes back to Hoeffding [155], while the second inequality is due to Alon and Spencer [18].

Theorem 2.5 (Hoeffding 1963) *Let* $u_j = 1, v_j = -1, w_j = 1, 0 \leq p_j \leq 1$ *for all* $j = 1, \ldots, n$. *For* $\lambda > 0$ *we have*

(a) $\mathbb{P}(\psi > \lambda) \leq \exp(-\frac{\lambda^2}{2n})$

(b) $\mathbb{P}(\psi < -\lambda) \leq \exp(-\frac{\lambda^2}{2n})$.

Alon and Spencer improved the Hoeffding bound $e^{-\frac{2\lambda^2}{n}}$ replacing n by $pn = p_1 + \cdots + p_n$.

Theorem 2.6 (Alon, Spencer 1992) *Let $u_j = 1 - p_j$, $v_j = -p_j$, $w_j = 1$, $0 \leq p_j \leq 1$ for all $j = 1, \ldots n$ and let $\lambda > 0$. Set $p = \frac{1}{n}(p_1 + \cdots + p_n)$. Then*

(a) $\mathbb{P}(\psi > \lambda) \leq \exp(-\frac{\lambda^2}{2pn} + \frac{\lambda^3}{2(pn)^2})$

(b) $\mathbb{P}(\psi < -\lambda) \leq \exp(-\frac{\lambda^2}{2pn})$.

∎

3. SEQUENTIAL DERANDOMIZATION

The Method of Conditional Probabilities. Let N and n be non-negative integers and let $[N]$ resp. $[n]$ denote the set $\{0, \cdots, N\}$ resp. $\{1, \cdots, n\}$. We consider the probability space $(\Omega, \mathbb{P})$ where $\Omega = [N]^n$, the powerset $\mathcal{P}(\Omega)$ is the σ−field and $\mathbb{P}$ is a probability measure on Ω. Let $E \subset \Omega$ be an event with $\mathbb{P}(E) > 0$ and let E^c denote the complement of E. For $y \in \Omega$, $y = (y_1, \cdots, y_n)^T$ and $\omega_1, \cdots, \omega_l \in [N]$, $1 \leq l \leq n$,

$$\mathbb{P}[E^c | \omega_1, \cdots, \omega_l] := \mathbb{P}[E^c | y_1 = \omega_1, \cdots, y_l = \omega_l]$$

is the conditional probability of E^c under the condition that the first l components $y_1, \cdots, y_l$ of the vector y are $\omega_1, \cdots, \omega_l$. The following simple procedure constructs a vector in E.

Algorithm CONDPROB

Input: An event $E \subset \Omega$ with $\mathbb{P}(E) > 0$.

Output: A vector $x \in E$.

1. Choose x_1 as the miminizer of the function $\omega \mapsto \mathbb{P}[E^c | y_1 = \omega], \omega \in [N]$.

2. For $l = 2, \ldots, n$ do:
   If $x_1, \ldots, x_{l-1} \in [N]$ have been selected, set $y_l = x_l$ where x_l minimizes the function $\omega \mapsto \mathbb{P}[E^c | x_1, \ldots, x_{l-1}, y_l = \omega], \omega \in [N]$.

∎

A similar algorithm works for conditional expectations.

Algorithm CONDEXP

Input: A function $F : \Omega \mapsto \mathbb{Q}$.

Output: A vector $x \in \Omega$ with $F(x) \geq \mathbb{E}(F)$.

1. Choose x_1 as the miminizer of the function $\omega \mapsto \mathbb{E}[F|y_1 = \omega], \omega \in [N]$.

2. For $l = 2, \ldots, n$ do:
   If $x_1, \ldots, x_{l-1} \in [N]$ have been selected, set $y_l = x_l$ where x_l maximizes the function $\omega \mapsto \mathbb{E}[F|x_1, \ldots, x_{l-1}, y_l = \omega], \omega \in [N]$.

■

The striking observation is

Proposition 3.1 *The algorithms* CONDPROB *and* CONDEXP *are correct.*

Proof. Let $x_1, \cdots, x_{l-1} \in [N]$, $1 \leq l \leq n$. Conditional probabilities can be written as a convex combination:

$$\mathbb{P}[E^c|x_1, \cdots, x_{l-1}] = \sum_{\omega \in [N]} \mathbb{P}[y_l = \omega] \cdot \mathbb{P}[E^c|x_1, \cdots, x_{l-1}, y_l = \omega]. \quad (3.1)$$

By the choice of the x_l's and the assumption $\mathbb{P}[E] > 0$,

$$1 > \mathbb{P}[E^c] \geq \mathbb{P}[E^c|x_1] \geq \cdots \geq \mathbb{P}[E^c|x_1, \cdots, x_n] \in \{0, 1\},$$

so $\mathbb{P}[E^c|x_1, \cdots, x_n] = 0$ and $x \in E$. For the correctness proof of the algorithm CONDEXP a simular argument works. ■

Efficiency of these algorithms depends on the efficient computation of the conditional probabilities resp. expectations. In general, it seems to be hopeless to compute conditional probabilities directly. But for the purpose of derandomization it suffices to compute upper bounds for the conditional probabilities, which play the role of the conditional probabilities. Such upper bounds have been introduced by Spencer [299] in the hyperbolic cosine algorithm, and later defined in a rigorous way as so called pessimistic estimators by Raghavan [269]. We give a little modified definition, covering multivalued random variables.

Definition 3.2 (Pessimistic Estimator) *Let $\mathcal{U}$ be a family of functions U_l : $[N]^l \mapsto \mathbb{Q}$, $l \in [n]$ containing a constant function U_0 . Let $(\Omega, \mathbb{P})$ be a probability space and $E \subset \Omega$ an event with $\mathbb{P}[E] \geq \delta$ for some $0 < \delta < 1$. $\mathcal{U}$ is called a pessimistic estimator for the event E, if for each $l \in [n]$ the following conditions are satisfied:*

(a) $\mathbb{P}[E^c|\omega_1, \ldots, \omega_l] \leq U_l(\omega_1, \ldots, \omega_l)$ *for all $\omega_1, \ldots, \omega_l \in [N]$.*

(b) *Given $\omega_1, \ldots, \omega_{l-1}$ there exists an $\omega_l \in [N]$ such that*
$$U_l(\omega_1, \ldots, \omega_l) \leq U_{l-1}(\omega_1, \ldots, \omega_{l-1}).$$

(c) $U_0 \leq 1 - \delta$

(d) *Each value $U_l(\omega_1, \ldots, \omega_l)$ can be computed in time polynomially bounded in n, N and $\log \frac{1}{\delta}$.*

If only conditions (a)–(c) are satisfied, let us call $\mathcal{U}$ a *weak pessimistic estimator*. Note that condition (b) is automatically satisfied for "convex" functions U_l: We call a family $\mathcal{U}$ as in Definition 3.2 convex, if there are $\mu_j > 0$ with $\mu_1 + \ldots + \mu_N = 1$ such that

$$\sum_{j=1}^{N} \mu_j U_l(\omega_1, \ldots, \omega_{l-1}, j) \leq U_{l-1}(\omega_1, \ldots, \omega_{l-1}). \qquad (3.2)$$

This is nothing else than the convex decomposition property of conditional probabilities stated in (3.1). With a pessimistic estimator we have a polynomial-time implementation of the algorithm CONDPROB.

Algorithm DERAND

Input: An event $E \subset \Omega$ with $\mathbb{P}[E] \geq \delta > 0$ and a pessimistic estimator $\mathcal{U}$ for E.

Output: A vector $x \in E$.

Algorithm For $l = 0, \ldots, n - 1$ do:
If $x_1, \ldots, x_{l-1} \in [N]$ have been selected, choose $x_l \in [N]$ as the minimizer of the function $\omega \to U_l(x_1, \ldots, x_{l-1}, \omega)$, $\omega \in [N]$. ∎

The pessimistic estimator-version of Proposition 3.1 is:

Proposition 3.3 *The algorithm* DERAND *runs in polynomial time and $x \in E$.*

Proof. Since $\mathcal{U}$ is a pessimistic estimator, each $U_l(x_1, \cdots, x_{l-1}, \omega)$, $\omega \in [N]$, can be computed in polynomial time, thus the minimizer x_l can computed in polynomial time. The vector $x = (x_1, \cdots, x_n)$ satisfies

$$\mathbb{P}[E^c | x_1, \ldots, x_n] \leq U_n(x_1, \ldots, x_n) \quad \leq \quad U_{n-1}(x_1, \ldots, x_{n-1})$$

$$\vdots$$

$$\leq \quad U_0 \leq 1 - \delta < 1,$$

so $\mathbb{P}[E^c | x_1, \ldots, x_n] = 0$, and $x \in E$. ∎

The correctness of an randomized algorithm is often proved with large deviation inequalities for sums of random variables. When the random variables are independent, or form a martingale, pessimistic estimators can be constructed. Let us examine the case of independent 0/1-random variables more closely.

We are given n independent $0/1$ random variables $X_1, \ldots, X_n$ where $\mathbb{P}[X_j = 1] = p_j$ and $\mathbb{P}[X_j = 0] = 1 - p_j$ for some rational numbers $0 \leq p_j \leq 1$. For $1 \leq i \leq m$, $1 \leq j \leq n$ let w_{ij} be rational numbers with $0 \leq w_{ij} \leq 1$ and denote by ψ_i the random variables $\psi_i = \sum_{j=1}^{n} w_{ij} X_j$. Now let $E_i^{(+)}$ be the event "$\psi_i \leq \mathbb{E}(\psi_i)(1 + \beta_i)$" and let $E_i^{(-)}$ denote the event "$\psi_i \geq \mathbb{E}(\psi_i)(1 - \beta_i)$". Furthermore, set $E = \bigcap_{i=1}^{m} E_i$ where E_i is either $E_i^{(+)}$ or $E_i^{(-)}$. Let $f(\beta_i) := \exp(-\frac{\beta_i^2 \mathbb{E}(\psi)}{d})$ with $d \in \{2, 3\}$, $0 < \beta_i < 1$ and suppose that for some $0 < \delta < 1$ the inequality

$$\sum_{i=1}^{m} f(\beta_i) \leq 1 - \delta \tag{3.3}$$

is satisfied. By the Angluin-Valiant inequality (Theorem 2.4), $\mathbb{P}(\bigcap_{i=1}^{m} E_i) \geq \delta$, hence $\bigcap_{i=1}^{m} E_i$ is not empty. Derandomization asks for the construction of a vector $x \in \bigcap_{i=1}^{m} E_i$ in deterministic time bounded by a polynomial in n, m and $\log \frac{1}{\delta}$. The following theorem proved by Srivastav and Stangier [309], called the algorithm version of the Angluin-Valiant inequality, is an implementation of the general algorithm DERAND for events estimated by the Angluin-Valiant bound.

Theorem 3.4 (Srivastav, Stangier 1996) *Let* $0 < \delta < 1$ *and* $E_1, \ldots, E_m$ *be events such that (3.3) is satisfied. Then* $\mathbb{P}(\bigcap_{i=1}^{m} E_i) \geq \delta$ *and a vector* $x \in \bigcap_{i=1}^{m} E_i$ *can be constructed in time* $\mathcal{O}\left(mn^2 \log \frac{mn}{\delta}\right)$.

Under the assumptions that real numbers and the exponential function can be computed exactly, Theorem 3.4 can be proved as in the article of Raghavan [269] by simulating the proof of the Angluin-Valiant inequality in an algorithmic way, and a time complexity of $O(mn^2)$ can be fixed. We will demonstrate this proof soon.

More effort is needed when we use finite machine models and the RAM model of computation, where the number of elementary arithmetic operations is counted. For the proof in this situation Srivastav and Stangier consider the first step in the proof of the Angluin-Valiant bound where the Markov inequality is applied. The so obtained upper bounds are essentially the moment generating functions $\mathbb{E}(e^{t_i \psi_i})$ for some $t_i > 0$. For rational weights w_{ij} and optimal choice of t_i the moment generating functions are transcendental. Taking low degree Taylor polynomials as approximations a pessimistic estimator can be constructed. The degree of the Taylor polynomials influences the running time and inreases the $\mathcal{O}(mn^2)$−time bound to $O\left(mn^2 \log \frac{mn}{\delta}\right)$.

For technical simplification we define the sum $\bigoplus_{i=1}^{m} \mathcal{U}^{(i)}$ of m families $\mathcal{U}^{(i)}$ as the set of functions U_l with $U_l := \sum_{i=1}^{m} U_l^{(i)}$, where $U_l^{(i)} \in \mathcal{U}^{(i)}$, $l \in [n]$. From the definition of convexity (3.2) it is straightforward to prove:

Lemma 3.5 *Let $\mathcal{U}^{(i)}$ be a weak pessimistic estimator for an event $E_i \subseteq \Omega$, $i = 1, \ldots, m$ such that $\sum_{i=1}^{m} U_0^{(i)} \leq 1 - \delta$ for some $\delta > 0$. If $\bigoplus_{i=1}^{m} \mathcal{U}^{(i)}$ is convex, then $\bigoplus_{i=1}^{m} \mathcal{U}^{(i)}$ is a weak pessimistic estimator for $\bigcap_{i=1}^{m} E_i$.*

Proof of Theorem 3.4. For simplicity we assume $m = 2$. Let E_1, E_2 be the events

$$\text{``}\psi_1 \leq \mathbb{E}(\psi_1) + \lambda_1\text{''} \quad \text{and} \quad \text{``}\psi_2 \geq \mathbb{E}(\psi_2) - \lambda_2\text{''}.$$

We construct a pessimistic estimator for the event $E_1 \cap E_2$ and find with the algorithm DERAND an $y \in E_1 \cap E_2$. Let $\sigma_1 = +1, \sigma_2 = -1$ and $y_1, \cdots, y_l \in \{0, 1\}$. For $t_i > 0$ we define

$$V_l^{(i)}(y_1, \cdots, y_l) := e^{-t_i \lambda_i} \mathbb{E}\left(e^{\sigma_i t_i \psi_i} | X_1 = y_1, \cdots, X_l = y_l\right), \qquad (3.4)$$

$1 \leq l \leq n$, and

$$V_0^{(i)} := e^{-t_i \lambda_i} \mathbb{E}\left(e^{\sigma_i t_i \psi_i}\right). \qquad (3.5)$$

By Markov's inequality (Theorem 2.1)

$$\mathbb{E}\left[E_i^c | X_1 = y_1, \cdots, X_l = y_l\right] \leq V_l^{(i)}(y_1, \cdots, y_l), \qquad (3.6)$$

and an optimal choice of t_i leads to the Angluin-Valiant bound.

$$\mathbb{P}[E_i^c] \leq V_0^{(i)} \leq f(\beta_i). \qquad (3.7)$$

Put $\mathcal{V}^{(i)} := (V_l^{(i)})_{l=0}^{n}$. Let us show that $\mathcal{V}^{(i)}$ is a pessimistic estimator for the event E_i, $i = 1, 2$:

- By (3), condition a) of Definition 3.2 is satisfied.

- By (4) condition c) of Definition 3.2 is true.

- Since the $V_l^{(i)}(y_1, \cdots, y_l)$ are conditional expectations, they can be written as convex sums, thus condition b) is valid.

- For an efficient computation of the $V_l^{(i)}(y_1, \cdots, y_l)$ observe that

$$
\begin{aligned}
V_l^{(i)}(y_1, \cdots, y_l) &= e^{-t_i \lambda_i} \mathbb{E}\left[e^{\sigma_i t_i \psi_i} | X_1 = y_1, \cdots, X_l = y_l \right] \\
&= e^{-t_i \lambda_i} \mathbb{E}\left[e^{\sigma_i t_i \sum_{j=1}^n w_{ij} X_j} | X_1 = y_1, \cdots, X_l = y_l \right] \\
&= e^{-t_i \lambda_i} e^{\sigma_i t_i \sum_{j=1}^l w_{ij} y_j} \mathbb{E}\left[e^{\sigma_i t_i \sum_{j=l+1}^n w_{ij} X_j} \right] \\
&= e^{-t_i \lambda_i} e^{\sigma_i t_i \sum_{j=1}^l w_{ij} y_j} \mathbb{E}\left[\prod_{j=l+1}^n e^{\sigma_i t_i w_{ij} X_j} \right] \\
&= e^{-t_i \lambda_i} e^{\sigma_i t_i \sum_{j=1}^l w_{ij} y_j} \prod_{j=l+1}^n \mathbb{E}\left[e^{\sigma_i t_i w_{ij} X_j} \right] \\
&= e^{-t_i \lambda_i} e^{\sigma_i t_i \sum_{j=1}^l w_{ij} y_j} \prod_{j=l+1}^n \left[p_j e^{\sigma_i t_i w_{ij}} + 1 - p_j \right].
\end{aligned}
$$

Assuming that the exponential function can be computed exactly, we can compute $V_l^{(i)}(y_1, \cdots, y_l)$ with $\mathcal{O}(n)$ arithmetic operations.

Finally, since $\mathcal{V}^{(1)} \oplus \mathcal{V}^{(2)}$ is convex, $\mathcal{V}^{(1)} \oplus \mathcal{V}^{(2)}$ is a pessimistic estimator for the event $E_1 \cap E_2$ according to Lemma 3.4. The running time of the algorithm DERAND is $\mathcal{O}(mn^2)$ where the $\mathcal{O}(n)$−term in the running time is due to the computation of each $V_l^{(i)}(y_1, \cdots, y_l)$, $l = 0, \cdots, n$. For a polynomial-time implementation in the Turing machine model, we must approximate all real functions. As remarked before, this increases the time bound to $\mathcal{O}(mn^2 \log \frac{mn}{\delta})$.

∎

Remark 3.6 An algorithmic version of the Chernoff /Hoeffding resp. Alon-Spencer inequality can be derived in the same way as the algorithmic Angluin-Valiant inequality, with the same time bound of $\mathcal{O}(mn^2 \log \frac{mn}{\delta})$. We will use them in the following frequently. These results can be found in Srivastav, Stangier ([309], Theorem 2.11 and 2.12).

BIBLIOGRAPHY AND REMARKS. The derandomization technique of conditional probabilities resp. conditional expectations was invented by Erdös and Selfridge [101]. In implicit form this method has been used to analyse a greedy algorithm for the maximum cut problem (MAXCUT) by Sahni and Gonzales [134]. The explicite form of the conditional probability method was fixed in the context of discrepancy theory by Beck and Fiala [39] and Spencer [299] ("the hyperbolic cosine algorithm"). According to Raghavan's [269] notation of pessimistic estimators the hyperbolic cosine algorithm provides a pessimistic estimator for combinatorial discrepancies. Raghavan's concept of randomized rounding and derandomization with the pessimistic estimator technique had a strong impact in the design of approximation algorithm in combinatorial optimization and computational geometry (for derandomization in computational geometry see the excellent survey article of Matoušek [228]). The polynomial-time implementation of the pessimistic estimator method for large deviation inequalities from the Chernoff-Hoeffding family are treated in Srivastav, Stangier [309] and Srivastav [305]. For sum of dependent random variables Srivastav [305] gave an algorithmic version of Azumas martingale inequality.

For applications of derandomization to two-person games we refer to the papers of Spencer [297], Spencer and Winkler [301], Deng and Mahajan [86] and the book of Du and Hwang [92]. Interestingly, derandomization in games was already used by Romanovskij in 1964 [277]. Recent work on derandomization in games theory has been done by Vovk [321]. For derandomization in combinatorial design theory we refer to the survey article of Gopalakrishnan and Stinson [138].

4. PARALLEL DERANDOMIZATION AND K-WISE INDEPENDENCE

The conditional probability method can be viewed as a binary search for a "good" vector in the sample space Ω. If the size of Ω is polynomial in N and n, an exhausitive search is already a polynomial-time algorithm. Small sample spaces correspond to random variables with limited rather than full independence. An interesting aspect of limited independence is that it is applicable not only for derandomization, but also for parallelization.

4.1 Small Sample Spaces.

Let $\Omega = \{0,1\}^n$, $n \in \mathbb{N}$, $\mathcal{P}(\Omega)$ the powerset of Ω and $\mathbb{P}$ a probability measure on Ω with respect to $\mathcal{P}(\Omega)$ as the $\sigma-$field. Throughout this section let $X_1, \cdots, X_n : \Omega \longrightarrow \{0,1\}$ be uniformly distributed $0/1-$random variables, i.e. $\mathbb{P}[X_i = 1] = \mathbb{P}[X_i = 0] = \frac{1}{2}$ for all $i = 1, \cdots, n$. For $\omega \in \Omega$ we identify $X_i(\omega)$ with ω_i, the $i-$th component of ω.

Definition 4.1 (k-wise Independence) $X_1, \cdots, X_n$ are $k-$wise independent if for any $\alpha \in \{0,1\}^k$ and any choice of k variables $X_{i_1}, \cdots, X_{i_k}$, $1 \leq i_1 < i_2 < \cdots < i_k < n$,

$$\mathbb{P}\left[(X_{i_1}, \cdots, X_{i_k}) = \alpha\right] - 2^{-k} = 0. \tag{4.1}$$

A motivation for $k-$wise independence from the computational point of view is the following. If $X_1, \cdots, X_n$ are independent, then the generation of a vector $(X_1, \cdots, X_n) = (\omega_1, \cdots, \omega_n)$ requires n bits (n independent Bernoulli trials). Now suppose that there is a $n \times l-$matrix B such that

$$BY = X,$$

where $Y = (Y_1, \cdots, Y_l)^T$, $X = (X_1, \cdots, X_n)^T$, the $Y_1, \cdots, Y_l$ are independent and the $X_1, \cdots, X_n$ are $k-$wise independent. If l is smaller than n, we need only $l < n$ random bits in order to generate $X_1, \cdots, X_n$. In other words, $\Omega = \{0,1\}^l$ instead of $\Omega = \{0,1\}^n$ is the sample space corresponding to the $X_1, \cdots, X_n$.

Alon, Babai and Itai [9] showed that k-wise independent random variables can be constructed from mutually independent random variables using a matrix over $GF(2)$. Their construction goes as follows.

Let $n = 2^{n'} - 1$ for some $n' \in \mathbb{N}$, $n' \geq 2$. For simplicity we assume k to be odd, i. e. $k = 2k' + 1$ for some $k' \in \mathbb{N}$. A representation of $GF(2^{n'})$ as a n'-dimensional algebra over $GF(2)$ can be explicitly constructed in polynomial time [206]. Let $b_1, \dots, b_n$ be the n non-zero elements of $GF(2^{n'})$ in such a representation and let $A = (a_{ij})$ be the following $n \times \frac{k-1}{2}$ matrix over $GF(2^{n'})$

$$A = \begin{bmatrix} 1 & b_1 & b_1^3 & \cdots & b_1^{k-2} \\ 1 & b_2 & b_2^3 & & b_2^{k-2} \\ 1 & b_3 & b_3^3 & & b_3^{k-2} \\ \vdots & \vdots & \vdots & \vdots & \vdots \\ 1 & b_n & b_n^3 & & b_n^{k-2} \end{bmatrix}. \tag{4.2}$$

A can be viewed as a $n \times l$ matrix over $GF(2)$ with $l = 1 + k' \cdot n' = 1 + \frac{k-1}{2} \log(n+1) = \mathcal{O}(k \log n)$. In coding theory A is well-known as the parity check matrix of binary BCH codes [220].

Theorem 4.2 (Alon, Babai, Itai 1996) *Let* $B := A^T$, *where* A *is the* $n \times l$ *matrix over* $GF(2)$ *as in* (4.2). *If* $Y_1, \dots, Y_l$ *are uniformly distributed, independent* $0/1-$*random variables, then* $X_1, \dots, X_n$, *where* $X_i = (BY)_i$ *and* $Y = (Y_1, \dots, Y_l)^T$, *are* k-*wise independent, uniformly distributed random variables.*

Proof. Choose $J \subseteq \{1, \dots, n\}$ with $|J| = k$. Remember that from the definition of A, $l = 1 + \frac{k-1}{2} \log(n+1)$. Let $x \in \{0,1\}^k$ be an arbitrarily choosen but fixed vector. Set $X_J = (X_j)_{j \in J}$. For $k-$wise independence we must show $\mathbb{P}[X_J = x] = \frac{1}{2^k}$. Let A_J be the submatrix of A with row indices from J and set $B_J = A_J^T$. B_J is a $l \times k$ matrix over $GF(2)$. Every set of k row vectors of A is linearly independent over $GF(2)$ [220], thus the columns of B_J are linearly independent over $GF(2)$. Since the dimension of the vector space $\{0,1\}^l$ over $GF(2)$ is $l \geq k$, we can supplement the k columns of B_J to l linearly independent vectors over $GF(2)$ in $\{0,1\}^l$, for example with Steinitz's theorem in Linear Algebra. Let C be the $l \times l$ matrix extending B_J in this way. Then C is non-singular. Put $\Omega_x = \{x' \in \{0,1\}^l, x_i' = x_i \text{ for } i = 1, \dots, k\}$.

Obviously $|\Omega_x| = 2^{l-k}$, and with $Y := (Y_1, \cdots, Y_l)^T$ we have

$$
\begin{aligned}
\mathbb{P}[X_J = x] &= \mathbb{P}[BY = x] \\
&= \mathbb{P}[CY \in \Omega_x] \\
&= \sum_{x' \in \Omega_x} \mathbb{P}[CY = x'] \\
&= \sum_{x' \in \Omega_x} \mathbb{P}[Y = C^{-1}x'] \\
&= \sum_{x' \in \Omega_x} \frac{1}{2^l} \qquad \text{(the } Y_i\text{'s are independent !)} \\
&= \frac{|\Omega_x|}{2^l} = \frac{2^{l-k}}{2^l} = 2^{-k}.
\end{aligned}
$$

∎

How large is the sample space corresponding to n, k-wise independent $0/1$ random variables? The random variables X_i constructed in Theorem 4.2 can be viewed as mappings $X_i : \{0,1\}^l \mapsto \{0,1\}$ with $X_i\omega = (B\omega)_i$, $\omega \in \{0,1\}^l$ and the sample space corresponding to the $X_1, \ldots, X_n$ is $\{0,1\}^l$. According to Theorem 4.2 it's size is $2^l = \mathcal{O}((n+1)^{\lfloor \frac{k}{2} \rfloor})$. In view of the lower bound $\Omega(n^k)$ for sample spaces of certain k−wise independent random variables given by Chor, Friedmann, Goldreich, Hastad, Rudich and Smolensky [72] this is best possible up to constants in the exponent. Note that the sample space constructed by Alon, Babai and Itai is polynomial only if k is constant. In applications k is often not constant, but $k = \mathcal{O}(\log^c n)$ with some constant $c > 0$. Derandomization in such sample spaces can be done combining $\mathcal{O}(\log^c n)$−wise random variables with the conditional probability method (section 4.2). A significant reduction of the size of the sample space was achieved by Naor and Naor [245]. The heart of the construction of Naor and Naor is the notation of almost k−wise independent random variables. Almost k−wise independence is an approximate version of equation (4.1):

Definition 4.3 (Almost k−wise Independence)
Let $\varepsilon, \delta > 0$ and $1 \le k \le n$, $n \in \mathbb{N}$.

(i) *$X_1, \cdots, X_n$ are (ε, k)−independent, if for any $\alpha \in \{0,1\}^k$ and any choice of indices $1 \le i_1 < i_2 < \cdots < i_k \le n$, we have*

$$
\left| \mathbb{P}\left[(X_{i_1}, \cdots, X_{i_k}) = \alpha\right] - 2^{-k} \right| \le \varepsilon.
$$

744

(ii) $X_1, \cdots, X_n$ are δ-away from k-wise independence, if for any $\alpha \in \{0,1\}^k$ and any choice of indices $1 \leq i_1 < i_2 < \cdots < i_k \leq n$, we have

$$\sum_{\alpha \in \{0,1\}^k} \left| \mathbb{P}\left[(X_{i_1}, \cdots, X_{i_k}) = \alpha\right] - 2^{-k} \right| \leq \delta.$$

(ε, k)-independency measures the deviation from the uniform distribution in the maximum norm, while definition 4.2 (ii) describes statistical closeness to the uniform distribution in the L^1-norm.

Observe, that (ε, k)-independent random variables are at most $2^k \varepsilon$-away from k-wise independence, whereas if they are δ-away from k-wise independence, they are (δ, k)-independent. For a set $S \subseteq \{1, \cdots, n\}$ let $X_S := \sum_{i \in S} X_i$ and put $X := \sum_{i=1}^n X_i$. The number X_S mod 2 is 0 iff X_S is even and 1 else. It can be viewed as the parity of S. If the $X_i's$ are independent, and $\mathbb{P}[X_i = 1] = \mathbb{P}[X_i = 0]$ for all $i = 1, \cdots, n$, then

$$\mathbb{P}[X_s \bmod 2 = 1] = \mathbb{P}[X_s \bmod 2 = 0]. \tag{4.3}$$

An approximative version of (4.3) leads to the concept of ε-biased random variables (Vazirani [318], Naor and Naor [245] and Peralta [257]). The bias of S is

$$\text{bias}(S) := \left| \mathbb{P}\left[X_s \bmod 2 = 0\right] - \mathbb{P}\left[X_s \bmod 2 = 1\right] \right|. \tag{4.4}$$

So for independent uniform random variables, $\text{bias}(S) = 0$ according to (4.3).

Definition 4.4 (ε-Bias)

(i) $X_1, \cdots, X_n$ are ε-biased, if $\text{bias}(S) \leq \varepsilon$ for all $S \subseteq \{1, \cdots, n\}$.

(ii) $X_1, \cdots, X_n$ are k-wise ε-biased, if $\text{bias}(S) \leq \varepsilon$ for all $S \subseteq \{1, \cdots, n\}$, $|S| \leq k$.

k-wise ε-biased random variables and almost k-wise independence are closely related. To see this, let $D : \Omega \longrightarrow [0, 1]$ be the probability distribution induced by the measure $\mathbb{P}$, so $D(\omega) = \mathbb{P}(\{\omega\}) := \mathbb{P}[(X_1 \ldots, X_n) = \omega]$, and let U be the uniform distribution, i.e. $U(\omega) = 2^{-n}$ for all $\omega \in \Omega$. The variation distance $\|D - U\|$ between D and U is the L^1-Norm of $D - U$,

$$\|D - U\| := \sum_{\omega \in \Omega} |D(\omega) - U(\omega)| = \sum_{\omega \in \Omega} |D(\omega) - 2^{-n}|.$$

$\|D - U\|$ is a measure for the distance of D from the uniform distribution. Let $D(S)$ resp. $U(S)$ be the restriction of D resp. U to a subset $S \subseteq \{1, \cdots, n\}$.

Definition 4.5 $X_1, \ldots, X_n$ *are* $k-$*wise* $\delta-$*dependent, if for all subsets* $S \subseteq \{1, \ldots, n\}$ *with* $|S| \leq k$,

$$\|D(S) - U(S)\| \leq \delta.$$

Note that if the $X_1, \cdots, X_n$ are $k-$wise $\delta-$dependent, then they are $\delta-$away from $k-$wise independency. Taking the Fourier transform $\widehat{D} - U$, Diaconis and Shahashahani [87] proved

$$\|D - U\|^2 = \|\widehat{D - U}\|^2 \leq \sum_{S \subseteq \{1, \cdots, n\}} \text{bias}_D(S).$$

This inequality immediately implies:

Corollary 4.6 *If* $X_1, \cdots, X_n$ *are* $\delta-$*biased, then they are* $k-$*wise* $2^{k/2}\delta-$*dependent and* $(2^{k/2}\delta, k)-$*independent.*

In conclusion, $\varepsilon-$biased random variables for small $\varepsilon > 0$ should behave as $k-$wise independent variables. For derandomization the hope is to replace $k-$wise independence by the weaker notion of almost $k-$wise independence and to obtain an even smaller sample space. Naor and Naor [245] proved that $\varepsilon-$biased random variables can be constructed with only "few" random bits:

Theorem 4.7 (Naor, Naor 1993) *Let* $\varepsilon > 0$ *and* $n \in \mathbb{N}$, $k \leq n$.

  (i) *Uniformly distributed* $0/1-$*valued random variables* $X_1, \cdots, X_n$ *which are* $\varepsilon-$*biased can be constructed using* $\mathcal{O}(\log n + \log \frac{1}{\varepsilon})$ *random bits, and the size of the corresponding sample space is* $2^{\mathcal{O}(\log n + \log \frac{1}{\varepsilon})} = (\frac{n}{\varepsilon})^c$ *for a constant* $c > 0$.

  (ii) *Uniformly distributed* $0/1-$*valued random variables* $X_1, \cdots, X_n$ *which are* $k-$*wise* $\varepsilon-$*dependent can be constructed using* $\mathcal{O}(\log \log n + k + \log \frac{1}{\varepsilon})$ *random bits. The size of the corresponding sample space is* $\left(\frac{2^k \log n}{\varepsilon}\right)^{\mathcal{O}(1)}$.

The constant c in Theorem 4.7 (i) depends on the expansion rate of an expander graph and is sligthly larger than 4 for the asymptotically optimal expanders of Lubotzky, Phillips and Sarnak [210]. For polynomially large $1/\varepsilon$, i.e. $1/\varepsilon = \mathcal{O}(poly(n))$, the size of the sample space is polynomial in n. In applications of the method of almost $k-$wise independence one would like to have k large and ε small while the size of the sample space should remain polynomial. The bound $\left(\frac{2^k \log n}{\varepsilon}\right)^{\mathcal{O}(1)}$ limits the growth of k, in fact for $1/\varepsilon = \mathcal{O}(poly(n))$, k must be about $\mathcal{O}(\log(1/\varepsilon)) = \mathcal{O}(\log n)$. Seminal papers managed to reduce the size of sample spaces corresponding to $\varepsilon-$biased random variables. (Azar, Motwani, Naor [27], Even, Goldreich, Luby, Nisan, Veliv̌ković [103] and Alon, Goldreich, Håstad, Peralta [12]). Alon, Goldreich, Håstad and Peralta achieved a size of roughly $(n/\varepsilon)^2$ for ε−biased random variables. In particular, they generate n random variables which are ε−away from k−wise independence with only $(2+o(1))(\log\log n + k/2 + \log k + \log\frac{1}{\varepsilon})$ random bits. This beats the bound of Naor & Naor as long as $\varepsilon < 1/(k\log n)$. There are two critical aspects in all of these results.

1. The random variables $X_1, \cdots, X_n$ are 0/1-valued and uniform, i.e. $\mathbb{P}[X_i = 1] = \mathbb{P}[X_i = 0] = 1/2$ for all i. In applications, this may not be the case.

2. The usual strategy for derandomization using k−wise independence is to construct a small sample space *a priori* and to simulate a randomized algorithm for a specific problem in such a space, if possible. Thus the sample space is choosen independently of the problem!

Schulman [286] gave an interesting, different approach for problem 2: He observed that for concrete problems only certain sets of d random variables, so called d−neighborhoods, among the n random variables need to be independent. Thus the choice of the magnitude of independence is *driven by the problem*. Koller and Megiddo [190] and Karger and Koller [169] further developed Schulmans approach covering also multivalued, non-uniformly distributed random variables. Koller and Megiddo showed that the sample space of k−wise independent random variables $X_1, \cdots, X_n$ with non-uniform probabilities $\mathbb{P}[X_i = 1] = p_i$, $1 \leq i \leq n$, correspond to a sample space of size at most $m(n,k) = \binom{n}{k} + \binom{n}{k-1} + \cdots + \binom{n}{0}$. Karloff and Mansour [172] showed the existence of $p_1, \cdots, p_n$ such that the size of any k−wise independent 0/1−probability space over $p_1, \cdots, p_n$ is at least $m(n,k)$. An interesting connection between small hitting sets for combinatorial rectangles in high dimension and the construction of a small sample spaces for general multivalued random variables is given in the paper of Linial, Luby, Saks and Zuckerman [205].

4.2 Parallelization.

In this section we show the algorithmic impact of k-wise independence, in particular it's power for parallelizing algorithms. Berger, Rompel [43] and Motwani, Naor & Naor [239] gave a parallelization of the conditional probability method using k-wise independence. For convenience, we formulate the derandomization problem in the following setting. Let $\mathbb{Z}_2^k$ denote the k-dimensional vector space over $GF(2)$. We describe a class of functions $F : \mathbb{Z}_2^n \to \mathbb{Q}$ for which the derandomization problem can be solved in parallel. Let $N = N(n, m, k)$ be an integer valued function. At the moment we do not assume that N is polynomially bounded in m and n. Let F be of the form:

$$F(x_1, \dots, x_n) = \sum_{i=1}^{n} \alpha_i f_i(x_{i_1}, \dots, x_{i_k}) \qquad (4.5)$$

with $\alpha_i \in \mathbb{Q}$ and $f_i(x_{i_1}, \dots, x_{i_k}) = (-1)^{\sum_{j=1}^{k} x_{i_j}}$.

Theorem 4.8 *Let $k, n \in \mathbb{N}$, $k \le n$, and $X_1, \dots, X_n$ be k-wise indepen-dent $0/1$ uniformly distributed random variables as in Theorem 4.2. Let $F : \mathbb{Z}_2^n \to \mathbb{Q}$ be a function as in (4.5). With $\mathcal{O}(N)$ parallel processors we can construct $x_0 \in \{0, 1\}^n$ in $\mathcal{O}(k \log n \log N)$-time such that $F(x_0) \ge \mathbb{E}(F(X_1, \dots, X_n))$.*

Proof. The random variables $X_1, \dots, X_n$ by definition have the form $X_i = (BY)_i$, where $Y = (Y_1, \dots, Y_l)^T$, the Y_i's are uniformly distributed $0/1$ random variables, $l = 1 + \frac{k-1}{2} \log(n+1)$, $B = A^T$ and A is the $n \times l$ parity check matrix of BCH codes as in (4.2). Therefore it suffices to give assignments for the Y_i's. The conditional probability method then goes as follows:

Suppose that for some $1 \le t \le l$ we have computed the assignments

$$Y_1 = y_1, \dots, Y_{t-1} = y_{t-1}.$$

Then we choose for Y_t the value $y_t \in \{0, 1\}$ which maximizes the function

$$w \to \mathbb{E}(f(X_1, \dots, X_n) \mid Y_1 = y_1, \dots, Y_{t-1} = y_{t-1}, Y_t = w). \qquad (4.6)$$

Obviously, after l steps we get $(Y_1, \cdots, Y_l) = y$ for some $y \in \{0, 1\}^l$. Thus $x_0 := By$ is a solution according to the correctness of the algorithm CONDEXP in Section 3. We are done, if the conditional expectations

$$\mathbb{E}(F(X_1, \dots, X_n) \mid Y_1 = y_1, \dots, Y_{t-1} = y_{t-1}, Y_t = y_t)$$

can be computed within the claimed time and space bounds. By linearity of expectation, it is sufficient to compute for each $i = 1, \dots, N$

$$\mathbb{E}(f_i(X_{i_1}, \dots, X_{i_k}) \mid Y_1 = y_1, \dots, Y_{t-1} = y_{t-1}, Y_t = y_t). \qquad (4.7)$$

Let B_i be the i-th row of B and put for $1 \le t \le k$, $b := \sum_{j=1}^{k} B_{i_j}$. Let $\langle b, Y \rangle$ denote the inner product of b and Y. Then $f_i(X_{i_1}, \dots, X_{i_k}) = (-1)^{\langle b, Y \rangle}$. Let s be the last position in the vector b which contains a 1. To shorten notation put $\vec{Y}_t = (Y_1, \dots, Y_t)$ and $\vec{y}_t = (y_1, \dots, y_t)$. Now

$$\mathbb{E}((-1)^{\langle b, Y \rangle} \mid \vec{Y}_t = \vec{y}_t) = \begin{cases} (-1)^{\sum_{j=1}^{s} b_j y_j} & \text{if } t \ge s \\ \\ 0 & \text{if } t < s. \end{cases} \qquad (4.8)$$

(4.8) follows from the following observations. If $t < s$ then

$$\langle b, Y \rangle = \underbrace{\sum_{j=1}^{t} b_j y_j}_{=:\alpha} + \underbrace{\sum_{j=t+1}^{s} b_j Y_j}_{=:\beta}$$

Hence $\mathbb{E}\left((-1)^{\alpha+\beta}\right) = (-1)^{\alpha} \mathbb{E}\left((-1)^{\beta}\right) = (-1)^{\alpha} \cdot 0$, because the Y_j's are independent and uniformly distributed. If $t > s$, then obviously $\langle b, Y \rangle = \sum_{j=1}^{s} b_j Y_j$, due to condition $\vec{Y_t} = \vec{y_t}$. In other words, for fix i and t we can compute $\mathbb{E}(f_i \mid \vec{Y_t} = y_t)$ in constant time. The total running time is computed as follows: we assign to each $f_i(X_{i_1}, \ldots, X_{i_k})$ one processor, so we have N processors in total. In the t-th step, $1 \le t \le l$, we can compute the sum $\sum_{i=1}^{N} \mathbb{E}(f_i)$ in $\mathcal{O}(\log N)$-time using N parallel processors. Summing up over the l steps we get a total running time of $\mathcal{O}(l \log N)$.　■

Theorem 4.8 gives a parallel ($\mathcal{NC}$) derandomized algorithm, provided $N(n, m, k)$ is polynomially bounded in n and m. It can be extended to uniformly distributed multivalued variables ([43], [310]).

BIBLIOGRAPHY AND REMARKS. In probability theory, k−wise independence was already used by Joffe [166]. This concept swapped into combinatorics and the design of algorithms in the years 1985/86, where the fundamental papers of Karp and Widgerson [181] and Luby [211] on 2−wise independence, the work of Alon, Babai and Itai on k−wise independence [9] and the lower bound proof for the size of k−wise independent sample space by Chor et al. [72] were published. Derandomization of space bounded computations is treated by Armoni [25] and Saks [281]. A survey on parallel derandomization techniques is given in the paper of Han [147].

In computational geometry, k−wise independence was applied by Berger et al. [44] and later by Goodrich [135, 136, 137], and by Amato et al. [20]. For the parallelization of derandomized geometric algorithms Mulmuley [242] extended earlier work of Karloff and Raghavan [173] on limiting random resources using bounded independence distributions and demonstrated that a polylogarithmetric number of random bits is sufficient for guaranteeing a good expected performance of many randomized incremental algorithms.

Corollary 4.6 is due to U. Vazirani (PhD-Thesis 1986, [318], see also the papers of Vazirani, Vazirani [319] and Chor et al [72]).

The method of k−wise independence has been applied sucessfully to reduce or remove randomness from probabilistic constructions and from algorithms in various fields, like hashing, pseudo-random generators, one-way functions, circuit and communication complexity and Boolean matrix multiplication.

Many of these aspects are treated in the lecture notes of Luby and Widgerson [214].

Applications to learning algorithms are discussed in [246]. Sitharam and Straney [295] applied derandomization to learn Boolean functions.

Yao [329] introduced the concept of a pseudo-random generator. An excellent book on pseudo-random generators and applications to cryptography (and other areas) are the lectures of Luby [212]. Blum and Micali [51] showed that the problem of constructing a pseudo-random generator in based on the concept of computational indistinguishability introduced by Goldwasser and

Micali [132]. Pseudo-random generators have been designed by Karp, Pippenger, Sipser [179], Ajtai, Komlos, Szemeredi [5], Chor and Goldreich [73], Nissan [249, 250], Håstad, Impagliazzo, Zuckerman [161], Impagliazzo, Levin, Luby [151], Sipser [294], Håstad, Impagliazzo, Levin, Luby [150] and Impagliazzo, Widgerson [160] and Andreev, Clementi and Rolim [23], to name same of the researchers.

Santha [282] showed how to sample with a small number of random points. Feder, Kushilevitz and Naor [106] applied almost $k-$wise independence to amortize the communication complexity. Application to hashing and Boolean matrix multiplication were discussed by Alon, Galil, Margalit, Naor [11], Alon, Galil [10], Alon, Naor [17] and Naor, Naor [245]. Approximation of DNF via derandomization was carried out by Luby and Velickovic [216] and approximate counting of depth-2 circuits is shown by Luby, Widgerson and Velickovic [215].

Blum et al. [50] applied hash functions to authenticate memories. An interesting technique for deterministic construction of small sample spaces for general multivalue random variables via the construction of point sequences with a discrepancy resp. $\varepsilon-$net property was introduced by Linial, Luby, Saks and Zuckerman [205].

5. HYPERGRAPH-2-COLORING AND DISCREPANCIES

One root for derandomization certainly is the application of the probabilistic method to combinatorial discrepancy theory. Combinatorial discrepancy theory deals with the problem of partitioning the vertices of a hypergraph in such a way that all hyperedges are split into about equal parts by the partition classes. Discrepancy measures the deviation of an optimal partition from an ideal one, that is one where all edges contains the same number of vertices in any partition class. An introduction to combinatorial discrepancy theory is given in the book of Alon, Spencer and Erdös [18] and the survey article of Beck and Sós [40]. Excellent sources covering many aspects (and proofs) of combinatorial, geometric as well as classical discrepancy theory are the books of Matoušek [229] and Chazelle [64] for the connection of derandomization and discrepancy theory. For an extension of 2-color discrepancy theory to c-colors see Doerr and Srivastav [90].

5.1 Sequential Algorithms for General Hypergraphs.

Let $\mathcal{H} = (X, \mathcal{E})$ be a hypergraph and let $\chi : X \rightarrow \{-1, +1\}$ be a function. Identifying -1 and $+1$ with colors, say red and blue, χ is a *2-coloring* of X. The sets $\chi^{-1}(-1)$ and $\chi^{-1}(+1)$ build the partition of X induced by χ. The imbalance of a hyperedge $E \in \mathcal{E}$ with respect to χ can be expressed by $\chi(E) := \sum_{x \in E} \chi(x)$. The *discrepancy* of $\mathcal{H}$ with respect to χ is

$$\mathrm{disc}(\mathcal{H}, \chi) := \max_{E \in \mathcal{E}} |\chi(E)|,$$

and the *discrepancy* of $\mathcal{H}$ is

$$\mathrm{disc}(\mathcal{H}) := \min_{\chi : X \rightarrow \{-1, +1\}} \mathrm{disc}(\mathcal{H}, \chi).$$

Theorem 5.1 (Spencer 1987) *A 2–coloring of X with $\mathrm{disc}(\chi) \leq \sqrt{2n \log 4m}$ can be constructed in $\mathcal{O}(mn^2 \log(mn))$-time.*

Proof. Put $\alpha := \sqrt{2n \log 4m}$. Let $\chi_1, \ldots, \chi_n$ be independent random variables defined by $\mathbb{P}[\chi_j = +1] = \mathbb{P}[\chi_j = -1] = \frac{1}{2}$ for all j. The vector $\chi = (\chi_1, \ldots, \chi_n)$ is a random 2-coloring of X. For $1 \leq i \leq m$ let E_i be the hyperedges in $\mathcal{E}$ and F_i be the events „$|\chi(E_i)| \leq \alpha$". From Hoeffdings inequality (Theorem 2.5) we infer $\mathbb{P}[F_i^c] \leq 1/(2m)$, thus $\mathbb{P}[\bigcup_{i=1}^m F_i^c] \leq 1/2$, and a coloring with discrepancy at most α exists. With the algorithmic version of the Chernoff-Hoeffding inequality [309] it can be constructed in $\mathcal{O}(mn^2 \log(mn))$-time. ∎

The original proof of Spencer is done with the hyperbolic cosine algorithm instead of the algorithmic Chernoff-Hoeffding bound.

It is known that the discrepancy bound $\mathcal{O}(\sqrt{n \ln m})$ is not optimal. The celebrated "six-standard-deviation" theorem of Spencer [298] proves the existence of a 2-coloring with discrepancy at most $6\sqrt{n}$ (for $m = n$). In general, the bound $\mathcal{O}(\sqrt{n})$ for discrepancy is optimal up to a constant factor, since the discrepancy of the set system induced by an $n \times n$ Hadamard matrix is at least $\frac{1}{2}\sqrt{n}$ (see [18]). It is a challenging open problem to give a randomized or deterministic polynomial time algorithm finding such a coloring. The main hinderance to the transformation of this existence result into an algorithm is the use of the pigeonhole principle in the proof.

5.2 Parallelization.

We proceed to the parallelization of the discrepancy algorithm in Theorem 5.1. Berger/Rompel [43] and Motwani, Naor, Naor [239] presented in 1989 the first parallel algorithm for this problem. The key for the success of their approach is the computation of k-th moments of the discrepancy function over a k-wise independent distribution.

Let $A = (a_{ij})$ be the edge-vertex incidence matrix of $\mathcal{H}$. Let $X_1, \ldots, X_n$ be $(-1/1)$–random variables with $\mathbb{P}[X_j = +1] = \mathbb{P}[X_j = -1] = \frac{1}{2}$ and set

$$\psi_i = \sum_{j=1}^n a_{ij} X_j, \quad \Delta_i = \sum_{j=1}^n a_{ij}, \quad \Delta = \max_{1 \leq i \leq m} \Delta_i$$

for each $i = 1, \ldots, m$. We assume that the $X_1, \ldots, X_n$'s are k-wise independent. Let $I = \{1, \ldots, n\}$ and let I^k be the k-product of I. For every multiindex $\alpha \in I^k, \alpha = (j_1, \ldots, j_k)$, we define

$$a_{i\alpha} := \prod_{l=1}^k a_{ij_l} \quad \text{and} \quad \phi_\alpha := \prod_{l=1}^k X_{j_l}.$$

Let $\mathbb{E}_{k-wise}(\psi_i^k)$ be the expectation with respect to the probability measure induced by the k-wise independent random variables $X_1, \cdots, X_n$. We have

$\psi_i^k = \sum_{\alpha \in I^k} a_{i\alpha}\phi_\alpha$, hence $\mathbb{E}_{k-wise}(\psi_i^k) = \mathbb{E}(\psi_i^k)$, so

$$
\begin{aligned}
\mathbb{E}_{k-wise}(\psi_i^k) = \mathbb{E}(\psi_i^k) &= \int_0^\infty \mathbb{P}[|\psi_i|^k > x]dx \\
&= \int_0^\infty \mathbb{P}[|\psi_i| > x^{\frac{1}{k}}]dx \\
&\leq \int_0^\infty 2 \cdot \exp\left(-\frac{x^{\frac{2}{k}}}{2\Delta}\right) dx \quad \text{(Theorem 2.5)} \\
&= 2(\frac{k}{2})!(2\Delta)^{\frac{k}{2}} \\
&\leq 2(k\Delta)^{\frac{k}{2}},
\end{aligned}
$$

Substituting $k = 2\left\lceil \frac{\log 2m}{2\varepsilon \log \Delta} \right\rceil$ and assumming $\varepsilon \geq \frac{1}{\log \Delta}$ we obtain

$$
\left(\mathbb{E}_{k-wise}(\sum_{i=1}^m \psi_i^k)\right)^{\frac{1}{k}} \leq \Delta^{\frac{1}{2}+\varepsilon}\sqrt{\log 2m}. \tag{5.1}
$$

Theorem 5.2 (Berger, Rompel 1989, Motwani, Naor, Naor 1989) *Let $\mathcal{H} = (X, \mathcal{E})$ be a hypergraph with $|X| = n$, $|\mathcal{E}| = m$ and $|E| \leq \Delta$ for all $E \in \mathcal{E}$. For every $\frac{1}{\log \Delta} \leq \varepsilon \leq 1$ a 2-coloring χ of X with*

$$
disc(\chi) \leq \Delta^{\frac{1}{2}+\varepsilon}\sqrt{\log 2m}
$$

can be constructed with $\mathcal{O}(n^2 m^{1+\frac{1}{\varepsilon}})$ parallel processors in $\mathcal{O}(\log n \log^2 m)$-time.

Proof. We invoke Theorem 4.8. Choose k and ε as in (5.1). Let $z \in \{0,1\}^n$. Each X_j has the form $X_j(z) = (-1)^{z_j}$. We regard ψ_i as a function of z, $\psi_i(z) = \sum_{j=1}^n a_{ij}X_j(z)$. For every multiindex $\alpha \in I^k$, $\alpha = (\alpha_1, \dots, \alpha_n)$, we have $\phi_\alpha(z) = (-1)^{\sum_{j=1}^k z_{\alpha_j}}$. Therefore

$$
\sum_{i=1}^m \psi_i^k(z) = \sum_{i=1}^m \sum_{\alpha \in I^k} a_{i\alpha}\phi_\alpha(z) = \sum_{i=1}^m \sum_{\alpha \in I^k} a_{i\alpha}(-1)^{\sum_{j=1}^k z_{\alpha_j}}. \tag{5.2}
$$

We define the right hand side of (5.2) as the function F in Theorem 4.8. The number $N = N(n, m, k)$ of terms in (5.2) is $N = m \cdot n^k = n^2 \cdot m^{\frac{1}{\varepsilon}+1}$. By Theorem 4.8 we can construct $y \in \{0,1\}^n$ using $\mathcal{O}(N) = \mathcal{O}(n^2 \cdot m^{1+\frac{1}{\varepsilon}})$ parallel processors in $\mathcal{O}(k \log n \log N) = \mathcal{O}(\log^2 m \log n)$ time such that $F(y) \leq \mathbb{E}(F(X_1, \dots, X_n))$. The vector $x = (x_1, \dots, x_n)$, where $x_i =$

$(-1)^{y_i}$, defines a 2-coloring χ. Its discrepancy is

$$
\begin{aligned}
\mathrm{disc}(\mathcal{H}, \chi) = \|Ax\|_\infty &\leq \left(\sum_{i=1}^{m} |(Ax)_i|^k \right)^{\frac{1}{k}} \\
&= \left(\sum_{i=1}^{m} [(Ax)_i]^k \right)^{\frac{1}{k}} \qquad \text{(because } k \text{ is even)} \\
&= (F(y))^{\frac{1}{k}} \leq \mathbb{E}(F)^{\frac{1}{k}} \leq \Delta^{\frac{1}{2}+\varepsilon} \sqrt{\log 2m} \quad \text{(by 5.1)}.
\end{aligned}
$$

∎

Note that the proof of Theorem 5.2 can be done also via almost k−wise independence and yields little better bounds for the running time and space requirement [245].

5.3 Bounded VC-Dimension.

A very interesting situation arises, when the hypergraph is sparse and moreover its sparsity is hereditary. The fundamental notion of VC-dimension due to Vapnik and Chervonenkis [317] captures hereditary sparsity. Let $\mathcal{H} = (X, \mathcal{E})$ be a hypergraph and $A \subseteq X$. $\mathcal{E}/A$ is the set of all edges of the form $E \cap A$, $E \in \mathcal{E}$ and $\mathcal{H}_A := (A, \mathcal{E}/A)$ is the hypergraph induced by A. By 2^A we denote the powerset of A.

Definition 5.3 Let $\mathcal{H} = (X, \mathcal{E})$ be a hypergraph and $Y \subseteq X$. Y is called shattered, if $\mathcal{H}_Y := (Y, 2^Y)$. The VC - dimension (Vapnik-Chervonenkis dimension) of $\mathcal{H}$ is the maximal cardinality of a shattered subset of X.

Sauer (1972), and independently Perles and Shelah, and, in a slightly weaker form Vapnik and Chervonenkis [317], showed

Proposition 5.4 If $\mathcal{H} = (X, \mathcal{E})$ has VC-dimension d, then $|\mathcal{E}| \leq 1 + \binom{n}{1} + \cdots + \binom{n}{d-1} + \binom{n}{d} < n^d$ for sufficiently large n.

Hypergraphs arising in computational geometry often have a small, constant VC-dimension leading to optimal discrepancy bounds (as we will see soon). Closely related to the VC-dimension are the shatter functions. The hypergraph $\mathcal{H}^* = (\mathcal{E}, X)$, where the edges are points and an element $x \in X$ is identified with the set $\mathcal{E}(x) := \{E \in \mathcal{E}; x \in E\}$ (the edges incident in x) is called the dual hypergraph of $\mathcal{H}$.

Definition 5.5 (Shatter Functions)
The function $\pi_\mathcal{H} : \{0, \dots, |X|\} \to \mathbb{N}, n \mapsto \max\{|\mathcal{E}/A|; A \subseteq X, |A| = n\}$ is the primal shatter function of $\mathcal{H}$. The dual shatter function $\pi_\mathcal{H}^*$ is the primal shatter function of the dual hypergraph.

Note that these definitions can be directly generalized to infinite sets X and $\mathcal{E}$.

For many hypergraphs in computational geometry, the shatter functions are polynomially bounded, i.e. $\pi_{\mathcal{H}}(n) \leq cn^l$ resp. $\pi_{\mathcal{H}}^*(n) \leq c'n^{l'}$ for some $l, l' \in \mathbb{N}_0$.

The reader may observe that if $X = \mathbb{R}^2$ and $\mathcal{E}$ ist the set of all closed halfplanes then $\pi_{\mathcal{H}}(n) \leq cn^2$, $c > 0$ a constant. If $X = \mathbb{R}^d$ and $\mathcal{E}$ ist the set of all balls then $\pi_{\mathcal{H}}(n) \leq cn^{d+1}$, $c > 0$ a constant.

If the VC-dimension of $\mathcal{H}$ is d,

$$\pi_{\mathcal{H}}(n) \leq \binom{n}{0} + \binom{n}{1} + \cdots + \binom{n}{d} = \Theta(n^d) \tag{5.3}$$

(see [18]).

The discrepancy of hypergraphs with bounded VC-dimension is considerably smaller than Spencer's $\mathcal{O}(\sqrt{n})$ bound [233]:

Theorem 5.6 (Matoušek, Welzl, Wernisch 1991)
Let $\mathcal{H}$ be a hypergraph on n points and d, $C_1, C_2 > 0$ be constants.

(a) *If $\pi_{\mathcal{H}}(m) \leq C_1 m^d$ for all $m \leq n$, then the discrepancy of $\mathcal{H}$ is bounded by*

$$\mathcal{O}(n^{\frac{1}{2}-\frac{1}{2d}}(\log n)^{1+\frac{1}{2d}}), \text{ if } d > 1, \text{ and } \mathcal{O}(\log^{\frac{5}{2}} n), \text{ if } d = 1.$$

(b) *If $\pi_{\mathcal{H}}^*(m) \leq C_2 m^d$ for all $m \leq n$ then the discrepancy of $\mathcal{H}$ is bounded by $\mathcal{O}(n^{\frac{1}{2}-\frac{1}{2d}}\log n)$, if $d > 1$ and by $\mathcal{O}(\log^{\frac{3}{2}} n)$, if $d = 1$.*

For the hypergraph with a point set $X \subseteq \mathbb{R}^d$, $|X| = n$ and edges as sets obtained as intersections of X with balls in $\mathbb{R}^d$, the dual-shatter function bound can be improved to $\mathcal{O}(n^{\frac{1}{2}-\frac{1}{2d}}\sqrt{\log n})$ [233]. For $d = 2$, this is $\mathcal{O}(n^{\frac{1}{4}}\sqrt{\log n})$, and the best known lower bound is $\Omega(n^{\frac{1}{4}})$.

It is an interesting open problem whether the $\sqrt{\log n}$ factor can be improved or even removed.

In the primal-shatter function case (Theorem 5.6 (a)), for $d > 1$ Matoušek [227] removed the $\log$ −factor and showed that this bound is best possible up to a multiplicative constant:

Theorem 5.7 (Matoušek 1995) *Let $\mathcal{H}$ be as in Theorem 5.6 (a). Then for $d > 1$, $\mathrm{disc}(\mathcal{H}) \leq C'n^{\frac{1}{2}-\frac{1}{2d}}$, where the constant C' depends on the constant C_1 and d.*

The proof of Theorem 5.6 (a) is based on a lemma of Beck [38] which invokes the pigeonhole principle in an exponentially large space (as the proof of Spencer's "six-standard-diviation" theorem). Presently, no constructive and efficient method is known to circumvent the pigeonhole principle.

But interestingly, the proof of the discrepancy bound via the dual shatter function can be turned into a polynomial-time algorithm. The proof is based on results of Chazelle and Welzl [327] on spanning paths with a low crossing number. A spanning path P on X is a linear ordering $x_1, x_2, \cdots, x_n$ of the points of X. Let us call $\{x_1, x_2\}, \{x_2, x_3\}, \cdots$ the edges of P. A hyperedge $E \in \mathcal{E}$ crosses a path-edge $\{x_i, x_{i+1}\}$ iff $|E \cap \{x_i, x_{i+1}\}| = 1$. The crossing number of P is the maximum number of edges crossed by an hyperedge, over all $E \in \mathcal{E}$.

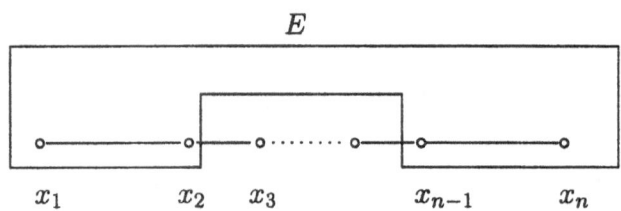

$$E$$

$$x_1 \qquad x_2 \quad x_3 \qquad\qquad x_{n-1} \qquad x_n$$

Figure 18.1   E with crossing number 2

Chazelle and Welzl showed that for hypergraphs with dual shatter function $\pi_{\mathcal{H}}^*(m) = \mathcal{O}(m^d)$, there exists a spanning path on X with crossing number $\mathcal{O}(n^{1-1/d} \log n)$ if $d > 1$ and $\mathcal{O}(\log^2 n)$, if $d = 1$. Moreover, such a path can be constructed in polynomial time. The striking observation is that a specific random 2−coloring χ of a hypergraph with crossing number κ has discrepancy $\mathcal{O}(\sqrt{\kappa \log |\mathcal{E}|})$ with high probability and such a coloring can be constructed via derandomization [233]:

Theorem 5.8 (Matoušek, Welzl, Wernisch 1991) *If $\pi_{\mathcal{H}}^*(m) = \mathcal{O}(m^d)$, then a 2−coloring χ of $\mathcal{H}$ with discrepancy $\mathcal{O}(n^{\frac{1}{2}-\frac{1}{2d}} \log n)$ can be constructed in time polynomial in $|X|$ and $|\mathcal{E}|$.*

Proof. Let $P = x_1, \cdots, x_n$ be a spanning path on X with crossing number κ and let M be the matching $\{\{x_1, x_2\}, \{x_3, x_4\}, \cdots, \{x_{n-1}, x_n\}\}$. Let $\mathcal{C}$ be the set of 2−colorings χ with the property $\chi(\{x, y\}) = 0$ for any pair $\{x, y\} \in M$. For $E \in \mathcal{E}$ let $M_E \subseteq X$ be the union over the set of edges in M crossed by E. Let us calculate $\chi(E)$: If $\{x, y\} \in M$ is not crossed by E, then $x, y \in E$. But $\{x, y\} \in M$ means $\chi(\{x, y\}) = 0$, and x, y do not contribute to $\chi(E)$. So only points in crossed edges of M constribute to $\chi(E)$. Thus $\chi(E) = \chi(E \cap M_E)$, and because $|E \cap M_E| \leq \kappa$, we have $|\chi(E)| \leq \kappa$. Among the 2−colorings of $\mathcal{C}$ there must be one with a much smaller discrepancy. Indeed, we may generate a random 2−coloring $\chi \in \mathcal{C}$, for example by choosing the values

$\chi(x_1), \chi(x_3), \cdots \in \{-1, +1\}$ by a sequence of independent Bernoulli trials. Putting $\chi(x_2) = -\chi(x_1)$, $\chi(x_4) = -\chi(x_3)$ etc., we get $\chi \in \mathcal{C}$. The Hoeffding-bound (Theorem 2.5) implies for $\lambda = \sqrt{2\ln(4|\mathcal{E}|)}$

$$\mathbb{P}\left[|\chi(E)| > \lambda\sqrt{\kappa}\right] \leq \mathbb{P}\left[|\chi(E \cap M_E)| > \lambda\sqrt{|E \cap M_E|}\right]$$

$$\leq 2\exp\left(-\frac{\lambda^2|E \cap M_E|}{2|E \cap M_E|}\right) = 1/(2|\mathcal{E}|).$$

Thus

$$\mathbb{P}[|\chi(E)| > \lambda\sqrt{\kappa} \text{ for } \underline{\text{some}} \ E \in \mathcal{E}] \leq \sum_{E \in \mathcal{E}} \mathbb{P}[\chi(E)| > \lambda\sqrt{\kappa}]$$

$$\leq |\mathcal{E}| \cdot \frac{1}{2|\mathcal{E}|} = \frac{1}{2},$$

so

$$\mathbb{P}[|\chi(E)| \leq \lambda\sqrt{\kappa} \text{ for } \underline{\text{all}} \ E \in \mathcal{E}] > \frac{1}{2}.$$

Finally, with the algorithmic version of the Chernoff-Hoeffding-bound (Remark 3.5), a desired coloring can be constructed in $\mathcal{O}(n^2|\mathcal{E}|\log(|\mathcal{E}|n))$ time. ∎

5.4 Geometric Discrepancies

Geometric discrepancy theory, also called the theory of irregularities of distributions, studies the deviation of point distributions in $\mathbb{R}^d$ from a perfectly uniform distribution. Discrepancy theory has its roots in the work of van der Corput [81, 82], and has grown to a mathematical theory with many far reaching applications, e.g. to numerical integration in high dimension, probability theory, ergodic theory, Ramsey theory and communication complexity. We refer the interested reader to the books of Hlawka [153], Beck and Chen [38], Niederreiter [248], the handbook article of Beck and Sós [40] and the recent books of Matoušek [229] and Chazelle [64].

Some of the best bounds for the discrepancy of point distributions with respect to axis-parallel rectangles have been achieved with probabilistic methods. An exciting, recent breakthrough is the deterministic construction of optimal point distributions by Chen and Skriganov [70] resolving a conjecture of Roth which has been open for more than 40 years.

Let $U^d = [0, 1]^d$ the unit cube in $\mathbb{R}^d$. For $x \in U^d$ let C_x by the axis-parallel cube with corner x and let $\mathcal{C}_d$ be the set of all such cubes C_x. We consider a point distribution P of n points in U^d.

The discrepancy of P with respect to C_x is

$$\mathcal{D}(P, C_x) := ||P \cap C_x| - n \cdot \mathrm{vol}(C_x)|,$$

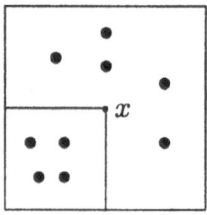

Figure 18.2 $U^2, n = 9$

the discrepancy (or L^∞−discrepancy) of P with respect to the set system C_d is

$$\mathcal{D}(P, C_d) := \max_{x \in U^d} \mathcal{D}(P, C_x),$$

and the L^2−discrepancy of P is

$$\mathcal{D}_2(P, C_d) := \sqrt{\int_{x \in U^d} \mathcal{D}(P, C_x)^2 d\lambda(x)},$$

where λ is the Lesbeque measure in U^d. The discrepancy resp. L^2−discrepancy of C_d for an n point distribution is

$$\mathcal{D}(n, C_d) := \inf_{P \subseteq U^d, |P|=n} \mathcal{D}(P, C_d) \quad \text{resp.} \quad \mathcal{D}_2(n, C_d) := \inf_{P \subseteq U^d, |P|=n} \mathcal{D}_2(P, C_d).$$

A celebrated theorem of Roth [278] states:

Theorem 5.9 (Roth 1954) $c(\log n)^{\frac{d-1}{2}} \leq \mathcal{D}_2(n, C_d) \leq C(\log n)^{\frac{d-1}{2}}$ *for constants* $c, C > 0$ *depending only on* d.

The lower bound is proved with Fourier analysis while the existence of a point distribution satisfying the upper bound is shown with the probabilistic method. Explicit constructions of such a point distribution were known only for $d = 2$. With methods from number theory, coding theory and Fourier analysis, Chen and Skriganov [70] gave an explicit construction for $d \geq 2$.

Theorem 5.10 (Chen, Skriganov 1999) *Let* $p \geq 2d^2$ *be a prime. For every* $n > 1$ *a distribution* $P \subseteq U^d$, $|P| = n$, *can be constructed explicitly to satisfy the inequality* $\mathcal{D}(P, C_d) < 2^{d+1} p^{2d} (\log n + 2d + 1)^{\frac{1}{2}(d-1)}$.

Remark 5.11 The construction of Chen and Skriganov does not use probabilistic arguments, but it is not known whether this construction can be done in time polynomial in n (for fixed d).

BIBLIOGRAPHY AND REMARKS: Beck and Fiala [39] showed for a general hypergraph $\mathcal{H}$, disc$(\mathcal{H}) \leq 2 \deg(\mathcal{H})$. A long-standing conjecture of great interest is whether a bound of

$\mathcal{O}(\sqrt{\deg(\mathcal{H})})$ is valid. Beck [36] showed for any hypergraph $\mathcal{H}$ of degree t, $\text{disc}(\mathcal{H}) = \mathcal{O}(\sqrt{t}\ln t \ln n)$. Srinivasan [303] improved the bound of Beck to $\mathcal{O}(\sqrt{t}\ln n)$ and gave better bounds for the lattice approximation problem (see Beck and Fiala [39] and Raghavan [269] for bounds derived with Chernoff-Hoeffding type inequalities). The best known bound $\mathcal{O}(\sqrt{t}\log n)$ is due to Banaszczyk [31]. The quadratic lattice approximation problem, a generalization of the (linear) lattice approximation problem, has been studied by Srivastav and Stangier [307], and a derandomized algorithm using Azuma's martingale was obtained.

Among hypergraphs of special interest is the hypergraph of arithmetic progressions, where $[N] := \{1, \ldots, N\}$ is the node set and all arithmetic progressions in $[N]$ are the hyperedges. For the hypergraph AP of arithmetic progressions in the first N integers Roth [279] proved $\text{disc}(AP) \geq cN^{\frac{1}{4}}$. In 1994, Matoušek and Spencer [232] showed $\text{disc}(AP) \leq c'N^{\frac{1}{4}}$, resolving the problem ($c, c' > 0$ are constants). For the hypergraph of cartesian products of arithmetic progressions AP^d in $[N]^d$ Doerr, Srivastav and Wehr [91] proved $\text{disc}(AP^d) = \Theta(N^{\frac{d}{4}})$, with constants depending only on d. For extension of these results to multicolor discrepancy we refer to Doerr, Srivastav [90]. The c−color discrepancy, $c \geq 2$, of a hypergraph on n points with n hyperedges is at most $\mathcal{O}(\sqrt{\frac{n}{c}\ln c})$, and there are examples of hypergraphs with c−color discrepancy at least $\mathcal{O}(\sqrt{\frac{n}{c}})$. For the hypergraph of arithmetic progressions over $[N]$ the c−color discrepancy is at most $\mathcal{O}(c^{-0.16}\sqrt[4]{N})$, and is at least $\mathcal{O}(c^{-0.5}\sqrt[4]{N})$.

Another problem dealing with cartesian products was studied by Matoušek [230], who proved -answering a question of Beck and Chen- that the discrepancy of the family of cartesian products of circular discs in the plane in $\mathbb{R}^4$ is $\mathcal{O}(n^{\frac{1}{4}+\varepsilon})$ for an arbitrarily small constant $\varepsilon > 0$, thus it is essentially the same as for circular discs in the plane.

6. HYPERGRAPH-2-COLORING AND LOVÁSZ-LOCAL-LEMMA

Another aspect of hypergraph 2−coloring is related to Ramsey theory. Let $\mathcal{H} = (X, \mathcal{E})$ be a n-uniform hypergraph. Let the degree of an edge $E \in \mathcal{E}$ be the number of edges intersecting $\mathcal{H}$ and let the edge-degree of the hypergraph be the maximum edge degree. As usual, the size of $\mathcal{H}$ is the number of edges and let us assume that the size is M. A 2−coloring of X is called non-monochromatic, if no hyperedge is monochromatic with respect to the coloring. We say that $\mathcal{H}$ has property $\mathcal{B}$ if there is a 2-coloring of X such that no hyperedge is monochromatic. In 1975, Erdös and Lovász [100] showed that a n-uniform hypergraph of degree at most 2^{n-3} has property $\mathcal{B}$. The remarkable conclusion of this theorem is that without any assumption on the size of the hypergraph an only local condition suffices to guarantee property $\mathcal{B}$. The key in the proof is a sieve method called the Lovász-Local-Lemma (LLL), which has many applications in combinatorial optimization (see sections 6 and 7) and in combinatorics, for example to k-coloring of real numbers, [100], linear and star arboricity of graphs [7] [15], acyclic colorings of graphs [16], and construction of edge disjoint paths in expander graphs [57].

Theorem 6.1 (Lovász Local Lemma; Symmetric Case) *Let $A_1, \ldots, A_n$ be events in a probability space where each A_i depends on at most d other events*

A_j. Let $\Pr[A_i] \leq p$ for all $1 \leq i \leq n$. If $ep(d+1) \leq 1$, then

$$\Pr\left(\bigwedge_{i=1}^{n} A_i^c\right) \geq (1 - \frac{1}{d+1})^d.$$

The dependency graph of the events $A_1, \ldots, A_n$ is the graph with vertices $A_1, \ldots A_n$ and edges $\{A_i, A_j\}$ iff A_i depends on A_j. In the same manner the *dependency graph* $\mathcal{D}(\mathcal{H})$ of a hypergraph $\mathcal{H} = (X, \mathcal{E})$ is defined: The hyperedges $E \in \mathcal{E}$ are the vertices of $\mathcal{D}(\mathcal{H})$, and for $E, F \in \mathcal{E}$, $\{E, F\}$ forms an edge of $\mathcal{D}(\mathcal{H})$ if $E \cap F \neq \emptyset$. The notation of dependency graphs will be usefull later.

For property $\mathcal{B}$ we get:

Corollary 6.2 *A hypergraph* $\mathcal{H} = (X, \mathcal{E})$ *with maximum degree* $d \leq 2^{n-3}$ *in which each edge has at least* n *vertices,* $n \geq 5$ *has property* $\mathcal{B}$.

Proof. Randomly and independently color the nodes of $\mathcal{H}$ with two colors, say red and blue, with probability $1/2$. Let us denote this random 2-coloring by χ. For an edge $E \in \mathcal{E}$ let A_E be the event that E is monochromatic with respect to χ. Put $p = 2^{-n+1}$. We have $\mathbb{P}[A_E] = 2 \cdot 2^{-|E|} \leq p$. Furthermore $ep(d+1) \leq e2^{-n+1}(2^{n-3} + 1) \leq \frac{5e}{16} < 1$ for $n \geq 5$. By the Local-Lemma (Theorem 6.1)

$$\mathbb{P}\left[\bigwedge_{E \in \mathcal{E}} A_E^c\right] \geq \underbrace{(1 - \frac{1}{d+1})^d}_{\approx \theta 2^n \text{ for some } \theta \in (0,1)} > 0.$$

Hence with non-zero probability, χ is a non-monochromatic 2-coloring, and $\mathcal{H}$ has property $\mathcal{B}$. ∎

Further results on the analysis of hypergraph coloring with the Local-Lemma are discussed in the article of McDiarmid [235]. Here we proceed to the derandomization of Corollary 6.2.

But it may happen that the probability of χ being non-monochromatic is only exponentially smal.Thus an attempt of finding a non-monochromatic 2-coloring is „the search for a needle in a haystick" (Beck 1991). For general hypergraphs this is even hopeless, since the problem to decide whether a hypergraph has property $\mathcal{B}$ is $\mathcal{NP}$-complete (Lovász [100]). In 1991 Beck [37] gave the first algorithmic version of the Local-Lemma and demonstrated his method – which is an ingenious interplay between global arguments (hypergraph size) and local considerations (hypergraph degree) – for the property $\mathcal{B}$ problem:

Theorem 6.3 (Beck 1991) *Let* $\mathcal{H} = (X, \mathcal{E})$ *be a* n-*uniform hypergraph with* $M := |\mathcal{E}|$ *and degree at most* d. *Suppose that* $d \leq 2^{\frac{n}{48}}$ *and* n *and* M *to be* *sufficiently large, i.e. $n \geq 400$ and $\log M \geq 2^{10}$. Then a non-monochromatic 2−coloring of $\mathcal{H}$ can be constructed in $\mathcal{O}(M^{const.})$ time.*

Alon [8] translated Beck's idea into a probabilistic setting, gave a probabilistic algorithm and derandomized the probabilistic algorithm with the method of almost k-wise independence for hypergraphs of degree at most $2^{\frac{n}{500}}$.

Let us give a sketch of the proof of Theorem 6.3. Before we describe Beck's algorithm let us briefly fix a class of hypergraphs for which a two coloring with property $\mathcal{B}$ can be constructed directly with the conditional probability method. This construction will be used as a subprocedure in Beck's algorithm.

Theorem 6.4 (Basic Coloring Theorem)
Let $\mathcal{H} = (X, \mathcal{E})$ be a hypergraph. Suppose that an integer $n \in \mathbb{N}$ exists such that $|E| \geq n$ for all $E \in \mathcal{E}$ and $|\mathcal{E}| \leq 2^{\frac{n}{2}}$. Then a non–monochromatic 2-coloring of $\mathcal{H}$ can be constructed in $\mathcal{O}(|\mathcal{E}||X|^2 \log(|\mathcal{E}||X|))$ time.

Proof. Let χ be a random 2-coloring of X, i.e. $\mathbb{P}[\chi(i) = 1] = \mathbb{P}[\chi(i) = 0] = \frac{1}{2}$ for all $i \in X$ independently. For $E \in \mathcal{E}$ we define $\chi(E) := \sum_{i \in E} \chi(i)$. Let A_E^0 resp. A_E^1 be the events "$\chi(E) \leq 0$" resp. "$\chi(E) > n$". By Chernoff's inequality (Theorem 2.3)

$$\mathbb{P}[A_E^1] \leq \exp(-\frac{2(|E|/2)^2}{|E|}) = \exp(-\frac{|E|}{2}) \leq e^{-\frac{n}{2}},$$

and $\mathbb{P}[A_E^0] \leq e^{-\frac{n}{2}}$. Thus $\mathbb{P}[(\bigvee_{E \in \mathcal{E}} A_E^1) \vee A_E^0] \leq (|\mathcal{E}| \cdot e^{-\frac{n}{2}} + e^{-\frac{n}{2}}) < \frac{1}{2}$ for $n \geq 10$. Hence with probability at least $1/2$ χ is not monochromatic, and $\mathcal{H}$ has property $\mathcal{B}$. With the algorithmic Chernoff-Hoeffding inequality (Remark 3.5) we can construct such a χ within the claimed time bound. ∎

The deterministic algorithm of Beck starts with a partial coloring and iterates until a complete coloring is obtained, a concept widely applied in the theory of 2−coloring of hypergraphs. A partial coloring of $\mathcal{H} = (X, \mathcal{E})$ is a mapping $\chi : X \mapsto \{-1, 0, 1\}$, and we identify $1, -1$ with colors, say red and blue, and 0 with a non-color. Let a partial coloring be given and let $\gamma \geq 1$. An edge $E \in \mathcal{E}$ is called *dangerous*, with respect to this partial coloring if it has $\lceil n/\gamma \rceil$ points in one color class. $\lceil n/\gamma \rceil$ is called a treshold value. Dangerous edges tend to become monochromatic and that is the reason why they are "dangerous" for property $\mathcal{B}$.

In the following we describe Beck's algorithm along with its analysis, leading to the proof of Theorem 6.3. For simplicity we assume n to be divisible by 2, 4 and 6 and consider in Theorem 6.3 the stronger degree condition $d \leq 2^{\frac{n}{96}}$.

Algorithm $\mathcal{H}-$COLOR

1. *First Pass.* Treshold value is $n/2$.

a) *Partial Coloring.* Color the points of X with the basic coloring algorithm sequentially in the following way: If we reach an uncolored point $x \in X$, and some $E \in \mathcal{E}$ containing x is dangerous, then we do not color x and proceed to the next uncolored point in X. After the first coloring pass all points of X are either colored or uncolored. Let $S \subseteq X$ be the set of the colored points.

b) *Truncation.* All bicolored edges never will become monochromatic, and we may remove them from $\mathcal{E}$. Furthermore, we remove the colored points from X, so consider $X \backslash S$. Let $\mathcal{H}^{(1)} = (X^{(1)}, \mathcal{E}^{(1)})$ be the so obtained new hypergraph (where $X^{(1)} = X \backslash S$, $\mathcal{E}^{(1)} = (\mathcal{E} \backslash \{\text{bichromatic edges}\})|_{X^{(1)}}$). Observe that every edge in $\mathcal{H}^{(1)}$ has at least $n/2$ points.

Before we proceed with the algorithm let us examine $\mathcal{H}^{(1)}$ more closely. We expect that $\mathcal{H}^{(1)}$ is sparser than $\mathcal{H}$ in the sense that it has fewer vertices and edges. But not only that. $\mathcal{H}^{(1)}$ has a less congested dependency structure compared with $\mathcal{H}$. The whole analysis is based on the following fundamental observation about the structure of the truncated hypergraph.

LEMMA (Main Lemma) *Every (connected) component of $\mathcal{D}(\mathcal{H}^{(1)})$ has size at most $\frac{\beta \log M}{n} \cdot 2^{4\alpha n}$, where α, β are constants choosen such that*

$$1 + 6\alpha\beta + 8\beta/n \leq \beta/2.$$

The proof of this lemma is the heart of Beck's analysis.
Now we can proceed to the coloring step.

c) *Colorability Test.* By the main lemma, taking $\alpha = 1/48$ and $\beta = 4$, $\mathcal{D}(\mathcal{H}^{(1)})$ (and therefore $\mathcal{H}^{(1)}$) falls apart in components, say $C_1, \ldots, C_r$ of size at most

$$f_1 := \frac{4 \log M}{n} \cdot 2^{\frac{n}{12}}.$$

Let $C \in \{C_1, \ldots, C_r\}$ and $\mathcal{H}_C^{(1)} = (X_{1,C}, \mathcal{E}_C^{(1)})$ the subhypergraph of $\mathcal{H}^{(1)}$ having only the hyperedges from C.

Case 1: $f_1 < 2^{\frac{n}{4}}$. Then $\mathcal{E}_C^{(1)} \leq f_1 \leq 2^{\frac{n}{4}-1}$, and since every hyperedge from $\mathcal{E}_C^{(1)}$ has at least $n/2$ vertices, by Theorem 6.4 we can find a non-monochromatic coloring for all $\mathcal{H}_C^{(1)}, C \in \{C_1, \ldots, C_r\}$

Case 2: $f_1 \geq 2^{\frac{n}{4}}$. In this case the size of the connected components of $\mathcal{D}(\mathcal{H}^{(1)})$ is even smaller: $\frac{4 \log M}{n} \cdot 2^{\frac{n}{12}} \geq 2^{\frac{n}{4}}$ implies $\frac{4 \log M}{n} \geq 2^{\frac{n}{6}}$. Hence

$$|\mathcal{E}_C^{(1)}| \leq \frac{4 \log M}{n} \left(\frac{4 \log M}{n} \right)^{\frac{1}{2}} = \left(\frac{4 \log M}{n} \right)^{\frac{3}{2}} \leq (\log M)^{\frac{3}{2}}$$

for $n \geq 4$.

We enter the second pass of the algorithm.

2. Second Pass. Treshold values is $\frac{n}{4}$. With $\mathcal{H}^{(1)}$ we go through steps a) and b) of the first pass. Let $\mathcal{H}^{(2)} = (X^{(2)}, \mathcal{E}^{(2)})$ the hypergraph after the truncation step.

Let us consider $\mathcal{D}(\mathcal{H}^{(2)})$. We apply the main lemma to $\mathcal{D}(\mathcal{H}^{(2)})$ choosing $\alpha = \frac{1}{96}$ and $\beta = 6$: Every connected component of $\mathcal{D}(\mathcal{H}^{(2)})$ has size at most

$$\frac{6 \log[(\log M)^{\frac{3}{2}}]}{n/2} \cdot 2^{\frac{n}{24}} = \underbrace{\frac{18 \log \log M}{n} \cdot 2^{\frac{n}{24}}}_{:=f_2}.$$

d) *Colorability Test.* Each hypergraph of $\mathcal{D}(\mathcal{H}^{(2)})$ has size at least $n/2 - n/4 \geq n/6$.

Case 1: If $f_2 < 2^{\frac{n}{12}} = 2^{\frac{n/6}{2}}$, then we can color all points of $\mathcal{H}^{(2)}$ with Theorem 6.4.

Case 2: If $f_2 \geq 2^{\frac{n}{12}}$, then $\frac{18 \log \log M}{n} \geq 2^{\frac{n}{24}}$, so the size of the components of $\mathcal{D}(\mathcal{H}^{(2)})$ is at most

$$\frac{18 \log \log M}{n} \cdot 2^{\frac{n}{24}} \leq \left(\frac{18 \log \log M}{n} \right)^2 \leq \frac{\log M}{n}, \tag{6.1}$$

since $n \geq 400$ and $\log M \geq 2^{10}$.

Brute-Force Coloring.

Let C be a component of $\mathcal{D}(\mathcal{H}^{(2)})$, $\mathcal{E}_C^{(2)}$ be the set of all hyperedges corresponding to C and let $X_C^{(2)} \subseteq X^{(2)}$ be the set of all points from the hyperedges of $\mathcal{E}_C^{(2)}$. By (6.1) we have

$$|X_C^{(2)}| \leq n \cdot \frac{\log M}{n} = \log M \tag{6.2}$$

Hence, the number of 2-colorings of $\mathcal{H}_C^{(2)}$ is $2^{\mathcal{O}(\log M)} = M^{\mathcal{O}(1)}$. Now we can test all these colorings in polynomial-time $M^{\mathcal{O}(1)}$ whether or not there is a non-monochromatic coloring among them. But is there such a coloring? Fortunately, the Local-Lemma (resp. Corollary 6.2) proves the existence of at least one non-monochromatic coloring, because

$$\deg(\mathcal{H}^{(2)}) \leq \deg(\mathcal{H}) \leq 2^{\frac{n}{96}} \leq 2^{\frac{n}{4}-3}.$$

This finishes the proof of Theorem 6.3. ∎

Molloy and Reed [238] generalized Beck's algorithm and gave new applications. Their main result can be formulated as follows.

Let $f_1, \cdots, f_m$ be independent random variables where each f_i takes values in some domain of cardinality at most γ. Suppose furthemore that we are given m "bad" events $A_1, \cdots, A_m$ where A_i is determined by the random variables in some set $F_i \subseteq \{f_1, \cdots, f_m\}$, and let $w \geq \max_{i=1,\ldots,m} |F_i|$. We say, A_1 depends on A_j iff $F_i \cap F_j \neq \emptyset$. Let d_i be the number of A_j's on which A_i depends and let

$$d \geq \max_{1 \leq i \leq m} d_i \quad \text{and} \quad p \geq \max_{1 \leq i \leq m} \mathbb{P}[E_i].$$

Some assumption on the computation time of the probabilities are necessary. Let t_1 be the time to carry out the random trial f_i, $1 \leq i \leq m$ and let t_2 the time to compute the conditional probabilities $\mathbb{P}[A_i | f_{j_1} = w_1, \cdots, f_{j_k} = w_k]$ for $f_{j_1}, \cdots, f_{j_k} \in F_i$ and $w_1, \cdots, w_k$ in the domains of the $f_{j_1}, \cdots, f_{j_k}$.

Theorem 6.5 (Molloy, Reed 1998) *If $pd^9 < 1/8$, then an assignment of all the X_i can be found with a randomized $\mathcal{O}(md(t_1 + t_2) + m\gamma^{wd} \log \log m) - time$ algorithm such that* $\mathbb{P}\left[\bigcap_{i=1}^{n} A_i^c\right] > 0.$

Roughly speaking the result says that whenever the weaker condition $8pd^9 \leq 1$ instead of the Local Lemma condition $4pd \leq 1$ holds, an at least randomized algorithmic version of the Local Lemma (in a more general framework) can be given. The proof is based on a variant of Beck's algorithm.

Among the various striking applications of the Local Lemma is certainly acyclic edge coloring. A proper edge coloring of a graph is *acyclic* if the union of two color classes is a forest in the graph. Alon [8] proved:

Theorem 6.6 (Alon 1991) *If G is a graph with maximum vertex degree Δ, then G has an acyclic edge coloring using at most 16Δ colors.*

Molloy and Reed showed with their method how this result can made constructive.

Another aspect of property $\mathcal{B}$ was studied by Beck [35] who proved that a n−uniform hypergraph with at most $n^{1/3-o(1)}2^n$ hyperedges has property $\mathcal{B}$. Recently Radhakrisnan and Srinivasan [268] gave a remarkable improvement.

Theorem 6.7 (Radhakrisnan, Srinivasan 2000) *Let* $\mathcal{H} = (X, \mathcal{E})$ *be a* n−*uniform hypergraph with at most* $0.7\sqrt{\frac{n}{\ln n}}2^n$ *hyperedges.* $\mathcal{H}$ *has property B and a non-monochromatic* 2−*coloring of* $\mathcal{H}$ *can be found in* $\mathcal{O}(n)$−*deterministic time using* $2^{\Theta(n)}$ *parallel processors.*

BIBLIOGRAPHY AND REMARKS. For a proof of the Lovász-Local-Lemma and several applications in combinatorics we refer to the book of Alon, Spencer and Erdös [18]. Alon [8] also considered the following generalization of the two coloring problem: for a $\gamma \in [\frac{1}{n}, \frac{1}{2}]$ the problem is to find a two coloring so that in each edge there are at least $\lfloor \gamma n \rfloor$ vertices in each color. For hypergraphs with degree at most $2^{\frac{n}{500}}$ and $\gamma = \frac{1}{1000}$ he solved this problem with a parallel algorithm. In [305] a generalized version of Beck's algorithms is used to construct a $\gamma = \frac{1}{1000}$ coloring for hypergraphs with degree at most $2^{\frac{n}{220}}$. Molloy and Reed observed that their algorithm runs no longer in polynomial-time, if γ, the size of the domains of the random variables is large, say $\gamma = (m + n)^c$, $c > 0$ a constant. Here Leighton, Rao and Srinivasan [201] gave constructive and efficient extensions and covered new applications to the disjoint path problem in expander graphs, hypergraphs partitioning and routing with low congestion. Applications of the Lovász-Local-Lemma to integer programming resp. scheduling were given by Srinivasan [302], resp. Leighton, Maggs and Rao [198] and Ahuja, Srivastav [3]. Some of these achievements will be discussed in section 8.

7. PACKING AND COVERING INTEGER PROGRAMS

Integer programming plays a central role in combinatorial optimization. To solve an integer program in an approximative way, the primal-dual method and the randomized rounding technique turned out to be successfull. These methods have been applied to several $\mathcal{NP}$-hard problems, and have exhibited the best known approximation guarantees. In this section we discuss the design of approximation algorithms for integer programs of packing and covering type based on randomized rounding and derandomization. Let us consider the following integer linear programs.

PACKING INTEGER PROGRAM (PIP)   COVERING INTEGER PROGRAM (CIP)

$$\max c^T x \qquad\qquad \min d^T y$$
$$Ax \le b, x \in \mathbb{Z}_+^n \qquad\qquad By \ge a, y \in \mathbb{Z}_+^m$$

where $a \in \mathbb{Q}^n, b \in \mathbb{Q}^m, c \in [0,1]^n \cap \mathbb{Q}^n, d \in [0,1]^m \cap \mathbb{Q}^m, A = (a_{ij}) \in [0,1]^{m \times n}$ resp. $B = (b_{ij}) \in [0,1]^{n \times m}$, where $a_{ij}, b_{ij} \in \mathbb{Q}$ for all i, j. By OPT we denote the value of an optimal solution to PIP, resp. CIP and by OPT* the value of an optimal solution to the relaxations $x \in \mathbb{Q}_+^n$ resp. $y \in \mathbb{Q}_+^m$. Interesting examples of these integer programs are the matching problem in hypergraphs, the vertex cover problem (also known as the hitting set problem) and the set cover problem in hypergraphs.

Matching Problem.

Let $\mathcal{H} = (V, \mathcal{E})$ be a hypergraph with $|V| = n$, $|\mathcal{E}| = m$, rational hyperedge weights $c_j \geq 0$, $j = 1, ..., m$. A *matching* or *packing* is a subset of hyperedges $\mathcal{E}_0 \subseteq \mathcal{E}$ such that no vertex is contained in more than one hyperedge from $\mathcal{E}_0$. The objective is to find a matching $\mathcal{E}_0 \subseteq \mathcal{E}$ with maximum weight $c(\mathcal{E}_0) = \sum_{j \in \mathcal{E}_0} c_j$.

Set-Cover Problem.

Given a hypergraph $\mathcal{H} = (V, \mathcal{E})$, $|V| = n, |\mathcal{E}| = m$, with edge weights $d : \mathcal{E} \mapsto \mathbb{Q}^+$, a *set cover* is a set of hyperedges $C \subset \mathcal{E}$ such that $V = \bigcup_{E \in C} E$. The goal is to find a set cover C of minimum weight $d(C) = \sum_{j \in C} d_j$.

Vertex Cover or Hitting Set Problem.

Let $\mathcal{H} = (V, \mathcal{E})$ be a hypergraph with $|V| = n$, $|\mathcal{E}| = m$, and node weights $w : V \mapsto \mathbb{Q}^+$. A *vertex cover* or *hitting set* is a set of nodes $T \subset V$ such that $E \cap T \neq \emptyset$ for all edges $E \in \mathcal{E}$. Here the objective is to find a vertex cover T of minimum weight $w(T) = \sum_{i \in T} w_i$.

If all weights are equal to 1, we speak of the unweighted version of these problems.

Note that the set covering problem in the primal hypergraph $(V, \mathcal{E})$ is equivalent to the vertex covering problem in the dual hypergraph $(\mathcal{E}, V)$. The integer linear programing formulation of these problems is easily stated using the vertex-edge incidence matrix $A = (a_{ij}) \in \{0,1\}^{n \times m}$ of $\mathcal{H}$, where $a_{ij} = 1$ if vertex i is contained in hyperedge j, and 0 otherwise. We consider a generalization with a parameter $k \in \mathbb{N}$.

k − Matching	k − Vertex Cover	k − Set Cover
$\max c^T x$	$\min w^T y$	$\min d^T z$
$Ax \leq k, x \in \{0,1\}^m$	$A^T y \geq k, y \in \{0,1\}^n$	$Az \geq k, z \in \{0,1\}^m$.

In the k—matching problem, we look for a set of hyperedges where each vertex is contained in at most k of the chosen hyperedges. In the k—vertex covering problem we want to find a set of vertices such that every hyperedge is incident in at least k of the chosen vertices, and in the k—set covering problem we seek for a set of hyperedges such that every vertex is contained in at least k of the chosen hyperedges.

The problems with $k = 1$ are the matching, vertex cover resp. set cover problems in hypergraphs. Let $\nu_k(\mathcal{H})$, $\tau_k(\mathcal{H})$ and $s_k(\mathcal{H})$ denote the value of an optimal integer solution to the $k-$matching resp. $k-$vertex cover resp. $k-$set cover problem, and let $\nu_k^*(\mathcal{H})$, $\tau_k^*(\mathcal{H})$ and $s_k^*(\mathcal{H})$ denote the value of an optimal solution to their LP-relaxations, where $x \in [0,1]^m$ resp. $y \in [0,1]^n$ resp. $z \in [0,1]^m$. For $k = 1$ we write $\nu(\mathcal{H})$, $\tau(\mathcal{H})$ and $s(\mathcal{H})$ etc.. $\nu(\mathcal{H})$ is known as the *matching number*, τ is the *blocking number* or the *hitting set number*. There are two questions concerning packing and covering integer programming which have been studied thoroughly.

1) For which instances can a polynomial time approximation algorithm be constructed ?

2) What is the relationship between the integral and fractional optima ?

Early work on the second question has been done by Faber and Lovász [105], Lovász [207] and Füredi [116]. Lovász [207] proved for unweighted vertex covering the inequality $\nu(\mathcal{H}) \le (1+\log n)\nu^*(\mathcal{H})$. This inequality was generalized by Kuzyurin [195] to covering integer programs with non-negative integer data. Kuzyurin's result implies in particular for the weighted k-vertex cover-problem with non-negative integer weights, $\nu_k(\mathcal{H}) \le (1 + \log(kn))\nu_k^*(\mathcal{H})$. In the unweighted case of vertex covering, Aharoni, Erdös and Linial [2] improved on the bound of Lovász and obtained a tight bound for the ratio of the fractional and integral matching number, $\frac{\nu(\mathcal{H})}{\nu^*(\mathcal{H})} \ge \frac{\nu^*(\mathcal{H})}{n}$. Füredi, Kahn and Seymour [117] confirmed a conjecture of Füredi [116] and showed for the weighted matching problem with a uniform or intersecting hypergraph $\mathcal{H}$, or constant edge weights, the existence of a set of matching edges $\mathcal{M} \subseteq \mathcal{E}$ obeying the inequality $\nu^*(\mathcal{H}) \le \sum_{E \in \mathcal{M}}(|E| - 1 + \frac{1}{|E|})c(E)$. For hypergraphs with bounded VC-dimension a much smaller vertex cover can be obtained as it was pointed out by Ding, Seymour and Winkler [88].

7.1 Packing Integer Programs.

Linear Programming Techniques. In a pioneering article Raghavan and Thompson [270] gave a randomized approximation algorithm for the $k-$matching problem in the unweighted case which finds a $k-$matching such that

$$|\mathcal{M}| \ge (1 - \delta)\nu_k(\mathcal{H}) - O(\sqrt{(1 - \delta)\nu_k^*(\mathcal{H})}),$$

with some $\delta \in (0, \frac{1}{2})$, provided that $k \ge 6 \ln n$. Raghavan [269] transformed this probabilistic result into a deterministic algorithm with nearly the same approximation guarantee with the pessimistic estimator technique. These papers can be considerd as the historical source for randomized rounding in integer linear programming.

During the last decade, randomized rounding turned out to be a powerful technique in the design of approximation algorithms. In all its variants it consists of two main steps.

Let us consider an instance of (PIP) with $0/1$ variables: $x \in \{0,1\}^n$.
First, the integer linear program is relaxed to a linear program (LP) by allowing fractional solutions, i.e. $0 \le x_i \le 1$ for all variables x_i. The crucial point is that an optimal fractional solution $y \in [0,1]^n$ can be found in polynomial time, for example with the ellipsoid algorithm [140] or Karmarkar's algorithm [175]. *Secondly*, since $0 \le y_i \le 1$ for all $i = 1, \ldots, n$, the y_i serve as probabilities to "round" the fractional solution to an integer one: Independently we set $x_i = 1$ with probability y_i and $x_i = 0$ with probability $1 - y_i$ for all i, thus creating a sequence of independent Bernoulli trials.

Observe that the expectation $\mathbb{E}(x_i)$ is y_i. By linearity of expectation this yields $\mathbb{E}(c^T x) = c^T y$ and $\mathbb{E}(Ax) = Ay$. Since y is an optimal fractional solution, the crucial observation is $\mathbb{E}(c^T x) = c^T y = \text{OPT}^* \ge \text{OPT}$ and $\mathbb{E}(Ax) = Ay \le b$, in other words, in *expectation* x is an optimal solution for PIP. Of course, in *distribution* x will be off the expectation. But the hope is that it should not be too far away from the expected, i.e. the fractional optimal solution. So we wish to prove inequalities of the form

$$c^T x \ge (1 - \varepsilon)\text{OPT}^* \quad \text{and} \quad Ax \le (1 + \varepsilon)b,$$

which say that $c^T x$ is close to the optimum and Ax is not too far outside the feasible region.

Such bounds can be derived with the help of large deviation inequalities for sums of independent random variables. In fact, scaling the rounding probability y_i down to $(1 - \varepsilon)y_i$ we even can enforce feasibility, i.e. $Ax \le b$, with high probability.

Derandomization with the algorithmic Angluin-Valiant inequality immediately gives [309]:

Theorem 7.1 (Srivastav, Stangier 1996) *Let $0 < \varepsilon < 1$ and $b_i \ge \frac{12}{\varepsilon^2} \log(2m)$ for all $i = 1, \ldots, m$. An integral solution $x \in \{0,1\}^n$ for PIP with $0/1$ variables satisfying $c^T x \ge (1 - \varepsilon)\text{OPT}^*$ can be constructed in $\mathcal{O}(t_{LP} + mn^2 \log(mn))-$ time, where t_{LP} is the time to solve the LP-relaxation.*

Proof. Let $y \in [0,1]^n$ be an optimal fractional solution for PIP, constructed with a polynomial-time LP-solver (for example Karmarkars algorithm). For technical reasons we consider the unweighted case where $c_j = 1$ for all $j = 1, \ldots, n$.
Randomized Rounding: Let $X_1, \ldots, X_n$ be mutually independent random variables with

$$\mathbb{P}[X_j = 1] = (1 - \varepsilon/2)y_j \quad \text{and} \quad \mathbb{P}[X_j = 0] = 1 - (1 - \varepsilon/2)y_j,$$

for $j = 1, \ldots, n$. Let $x \in \{0,1\}^n$ be the outcome of the $X_1, \ldots, X_n$. For $i = 1, \ldots, m$ let E_i be the events "$(Ax)_i \leq b_i$" and E_0 be the event "$c^T x \geq (1 - \varepsilon)\text{OPT}^*$". With the Angluin-Valiant inequality it is not difficult to verify that $\mathbb{P}[E_i^c] \leq 1/(2m)$ for all $i = 1, \ldots, m$ and $\mathbb{P}[E_0^c] \leq 1/4$. Thus

$$\mathbb{P}\left[\bigcup E_i^c\right] \leq \sum_{i=1}^{m} \mathbb{P}[E_i^c] \leq m\left(\frac{1}{2m}\right) = \frac{1}{2}, \tag{7.1}$$

hence with probability at least $1/4$,

$$Ax \leq b \quad \text{and} \quad c^T x \geq (1 - \varepsilon)\text{OPT}^*.$$

Derandomization: Since $\mathbb{P}\left[\bigcup E_i^c\right] \leq 3/4$, we can invoke the algorithmic version of the Angluin-Valiant inequality with $\delta = 1/4$ to construct a vector $x \in \bigcap_{i=0}^{m} E_i$ in $\mathcal{O}(mn^2 \log(mn))$ deterministic time. ∎

For the $k-$matching problem we obtain a $k-$matching $\mathcal{M} \subseteq \mathcal{E}$ with $|\mathcal{M}| \geq (1-\varepsilon)\nu_k^*(\mathcal{H})$ whenever $k = \Omega\left(\frac{\log n}{\varepsilon^2}\right)$. Note that Theorem 7.1 can be extended to the general case of non-negative integer variables $x \in \mathbb{Z}_+^n$ [309].

Unfortunately, the above discussed results are restricted to PIPs with large right hand side, $b_i = \Omega(\log m)$ for all $i = 1, \ldots, m$ This is due to the "naive" estimate (7.1)

$$\mathbb{P}\left[\bigcup_{i=0}^{m} E_i^c\right] \leq \sum_{i=0}^{m} \mathbb{P}[E_i^c]. \tag{7.2}$$

A major improvement has been obtained by Srinivasan [304] leading to approximation guarantees also for a small right hand side. The main observation is that the events $E_1, \ldots, E_m$ are positively correlated: This means, for every subset $S \subseteq \{1, \ldots, m\}$,

$$\mathbb{P}\left[\bigcap_{i \in S} E_i\right] \geq \prod_{i \in S} \mathbb{P}[E_i]. \tag{7.3}$$

For the proof of (7.3) Srinivasan invoked the FKG-inequality (Fortuin, Kasteleyn and Ginibre, see [18]). Let us explain the intuitive meaning of positive correlation. For $m = 2$, (7.3) reads as

$$\mathbb{P}[E_2] \leq \frac{\mathbb{P}[E_1 \cap E_2]}{\mathbb{P}[E_1]} = \mathbb{P}[E_2 | E_1],$$

saying that the probability of event E_2 under the condition that E_1 hold is at least the probability of E_2. In other words, whenever some constraints $(Ax)_i \leq b_i$

are satisfied, they boost the chance that other constraints are satisfied, too. The improvement over the naive bound is now

$$\mathbb{P}\left[\bigcup_{i=1}^{m} E_i^c\right] = 1 - \mathbb{P}\left[\bigcap_{i=1}^{m} E_i\right] \leq 1 - \prod_{i=1}^{m} \mathbb{P}[E_i]. \qquad (7.4)$$

Observe that the right hand side in (7.4) is at most $\sum_{i=1}^{m} \mathbb{P}[E_i^c]$. Unfortunately, the terms $\prod_{i=1}^{m} \mathbb{P}[E_i]$ can be extremely, i.e. exponentially small. In such cases randomized rounding analyzed with bound (7.4) would give a very poor success probability. The surprising result of Srinivasan is that derandomization via the pessimistic estimator technique leads to a polynomial-time deterministic algorithm even in this situation.

In the following theorem a packing integer program (PIP2) is considered, where $x \in \mathbb{Z}_+^n$, $b \in [1, \infty)^m$, $c \in [0, 1)^n$ with $\max_{1 \leq j \leq n} c_j = 1$.

Theorem 7.2 (Srinivasan 1995, 1999) *Let* $\gamma := (\min_{1 \leq i \leq m} b_i) - 1$ *and let* y^* *be an optimal solution for the LP relaxation of PIP2. A solution* $x \in \mathbb{Z}_+^n$ *for PIP2 satisfying* $c^T x \geq \alpha y^* (y^*/m)^{1/\gamma}$, *where* $\alpha > 0$ *is a constant, can be constructed in polynomial time.*

This approximation is good if $(y^*/m)^{1/\gamma} = 2^{-\frac{1}{\gamma}(\log m - \log y^*)}$ tends to 1. For $y^* = \mathcal{O}(m)$ and constant γ, $(y^*/m)^{1/\gamma}$ is constant, while for large γ, i.e. $\gamma = \Omega(\log m)$, it is $o(1)$, and does not beat the approximation guarantee of basic randomized rounding (Theorem 7.2). Srinivasan improved the approximation guarantee further by applying Janons's inequality and the Lovász-Local-Lemma [302]. However, no efficient algorithm achieving these improvements is known. Parallel derandomized counterparts of Theorem 7.2 have been given by Alon and Srinivasan [19].

The Rödl-Nibble. A variant of the matching problem in hypergraphs is the task to find a matching which covers at least $(1-\varepsilon)n$ nodes for some $\varepsilon > 0$, a so called ε−*near-perfect matching*. For a hypergraph $\mathcal{H}$ let $\mathcal{U}(\mathcal{H})$ be the minimum number of vertices not covered by a matching. We recall that the codegree, $\text{codeg}(\mathcal{H})$, of $\mathcal{H}$ is the maximum number of edges having two points in common. The main impact on this version of the hypergraph matching problem came from combinatorics, namely the confirmation of the Erdös-Hanani-Conjecture by Rödl [275]. To formulate the conjecture we need the notion of *partial Steiner Systems*. A partial Steiner system $S(t, r, m)$ is an r−uniform hypergraph on m vertices such that every set of t vertices is contained in at most one hyperedge.

Erdös-Hanani-Conjecture ([99]): *Let* t *and* r *be fixed and let* $m \to \infty$. *Then there is a* $S(t, r, m)$ *partial Steiner tree system with* $(1 - o(1))\binom{m}{t}/\binom{r}{t}$ *edges.*

This conjecture was confirmed by Rödl 1985 in a pioneering paper which is the origin of the so-called semi-random method. The proof-technique (resp. algorithm) in this paper is today known as the "Rödl-Nibble". Later, the Nibble-technique was successfully used to show the existence of large near-perfect matchings, whenever the hypergraph is sufficiently regular and has a negligible codegree. Pippenger and Spencer [261] and Frankl and Rödl [113] proved:

Theorem 7.3 *Let $\mathcal{H}$ be an r—uniform and D—regular hypergraph on n vertices. Assume that $codeg(\mathcal{H}) = C = o(D)$. Then there is an $o(n)$—near perfect matching, i.e. $\mathcal{U}(\mathcal{H}) \leq o(n)$.*

In this result the dependency of the $o(n)$ term on C and D is not clear. Progress in this direction was made by Kostochka and Rödl [KR97] and Alon, Kim and Spencer [AKS97]. The presently best result is due to V. H. Vu [322] who proved that

$$\mathcal{U}(\mathcal{H}) \leq \mathcal{O}\left(n \left(\frac{C}{D}\right)^{\frac{1}{r-1}} \log^c D \right)$$

where $c > 0$ is a constant. Not only this result is interesting, but also the powerful martingale inequalities introduced in [322] are valuable. A survey on this new martingale results and their applications is given in the recent paper of Vu [323].

It is not known whether a matching with at most $\mathcal{O}(n(\frac{C}{D})^{\frac{1}{r-1}} \log^c D)$ unmatched vertices can be constructed in deterministic polynomial time. But as shown by Grable [139], the Nibble-algorithm can be derandomized, even in parallel, leading to an parallel algorithm for the Pippenger/Spencer resp. the Frankl/Rödl theorem (Theorem 7.3).

BIBLIOGRAPHY AND REMARKS. A generalization of Theorem 7.1 resp. Theorem 7.2 to general packing integer programs can be found in Srivastav, Stangier [309], resp. Srinivasan [302]. In [122], Garg and Könemann gave a combinatorial algorithm for PIP with the primal-dual method and obtained roughly the same approximation guarantee as in Theorem 7.1. In [126] Goemans, Mihail, Vazirani, and Williamson presented deterministic, polynomial-time approximation algorithms for a large class of packing and network problems with the primal-dual method.

7.2 Covering Integer Programs.

A comprehensive discussion of approximation algorithms for vertex covering and set covering based on primal-dual linear programming methods is given in the book of Hochbaum [154], section 3. In this section we focus on randomized methods and derandomization. For the unweighted set covering problem

Johnson [167] and Lovász [207] showed that the greedy heuristic achieves a $H(n) \leq \log(n) + 1$ factor approximation, where $H(n) = \sum_{i=1}^{n} \frac{1}{i}$. Chvatal [74] extended this result to arbitary non-negativ edge weights. Lund and Yannakakis [219] proved for set covering that for any $\alpha < \frac{1}{4}$ the existence of a polynomial-time $(\alpha \log n)$-factor approximation algorithm would imply that $\mathcal{NP}$ has superpolynomial, i.e. $n^{\mathcal{O}(poly(\log n))}$ deterministic algorithms. In a breakthrough paper Feige [107] showed that $(1 + o(1)) \ln n$ is a treshold below which set cover cannot be approximated in polynomial time, unless $\mathcal{NP}$ has $n^{\mathcal{O}(\log \log n)}$ deterministic algorithms. This closed the gap to the greedy approximation factor of $1 + \log n = (1 + o(1)) \ln n$.

For the vertex cover problem the result of Feige implies that $\log m$ is essentially the best possible approximation factor. But for special hypergraph it is possible to break the $\log m$−barrier. Most interesting are the hypergraphs with constant VC-dimension d (see section 5.3 for a definition).

But first let us start with a randomized rounding approach for the general vertex covering problem due to Bertsimas and Vohra [48].

Let $\mathcal{H} = (V, \mathcal{E})$ a hypergraph with $|V| = n$, $|\mathcal{E}| = m$, and let as usual $d_i = \deg(v_i)$ be the degree of vertex $v_i \in V$ and $D = \max_{1 \leq i \leq n} d_i$ the maximum vertex degree of $\mathcal{H}$. For $k \in \mathbb{N}$ set $f(x) := 1 - (1 - x)^k$. Let $y \in [0, 1]^n$ be an optimal solution to the LP-relaxation of the vertex cover problem. The randomized rounding algorithm of Bertsimas and Vohra is:

Algorithm VERTEXCOVER1

- With probability $f(y_j)$ set $x_j = 1$ for all $j = 1, \ldots, n$, independently.

- Output $x = (x_1, \ldots, x_n) \in \{0, 1\}^n$. ∎

This algorithm can be interpreted as follows: for each $j \in V$ we k times flip a coin that has probability y_j of giving heads. Now, $f(y_j)$ is the probability that at least one outcome is heads.

Theorem 7.4 (Bertsimas, Vohra 1996) *Put $W(z) := w^T z$, $z \in [0, 1]^n$.*

(i) $\mathbb{E}(W|x \text{ is feasible}) \leq (1 - 1/D)^{-D} \log D \; \tau^*(\mathcal{H})$.

(ii) In deterministic polynomial-time a feasible vertex covering vector x can be constructed such that $W(x) \leq (1 - 1/m)^{-m} \log m \; \tau^*(\mathcal{H})$.

Proof. (i) Let A_i be the event that constraint i is violated by x and put $B_i = A_i^c$. Then, $F = \bigcap_{i=1}^{m} B_i$ is the event that x is feasible. Now it is an easy exercise to prove

$$\mathbb{P}[A_i] = \mathbb{P}[\sum_{j=1}^{n} a_{ij} x_j = 0] \;=\; \mathbb{P}[\sum_{j \in E_i} x_j = 0]$$

$$= \prod_{j \in E_i} \mathbb{P}[x_j = 0]$$

$$\leq \; e^{-k}. \tag{7.5}$$

Furthermore,

$$\mathbb{E}(W|F) = \sum_{j=1}^{n} w_j \mathbb{P}[x_j = 1|F] = \sum_{j=1}^{n} w_j \frac{\mathbb{P}[F|x_j = 1]}{\mathbb{P}[F]} \mathbb{P}[x_j = 1]. \tag{7.6}$$

Claim: $\frac{\mathbb{P}[F|x_j=1]}{\mathbb{P}[F]} \leq (1 - e^{-k})^{-d_j}.$

The claim proves part (a), since (7.6) implies

$$\mathbb{E}(W|F) \;\leq\; \sum_{j=1}^{n} w_j (1 - e^{-k})^{-|d_j|} (1 - (1 - y_j)^k)$$

$$\leq\; k(1 - e^{-k})^{-D} \underbrace{\sum_{j=1}^{n} w_j y_j}_{=: \tau^*(\mathcal{H})}$$

$$\leq\; \left(1 - \frac{1}{D}\right)^{-D} \log D \; \tau^*(\mathcal{H}) \quad \text{(taking } k = \log D\text{)}.$$

The proof of the claim is the main difficulty. As in the previously discussed result of Srinivasan (Theorem 7.2), positive correlation plays the main role. Let $\mathcal{E}_j \subseteq \mathcal{E}$ be the set of edges containing vertex v_j. Then

$$\frac{\mathbb{P}[F|x_j = 1]}{\mathbb{P}[F]} = \frac{\mathbb{P}[\bigcap_{i \notin \mathcal{E}_j} B_i]}{\mathbb{P}[\bigcap_{i=1}^{m} B_i]}. \tag{7.7}$$

But the $B_1, \ldots, B_m$ are positive correlated, thus by definition the inequality

$$\mathbb{P}[\bigcap_{i=1}^{m} B_i] \geq \mathbb{P}[\bigcap_{i \in I} B_i] \mathbb{P}[\bigcap_{i \notin I} B_i] \tag{7.8}$$

holds for any subset $I \subseteq \{1, \ldots, m\}$ (this can be proved directly (with some effort), or we may use the FKG-inequality [18]).

By (7.8), (7.7) becomes (with $I := \mathcal{E}_j$)

$$
\begin{aligned}
\frac{\mathbb{P}[F|x_j = 1]}{\mathbb{P}[F]} &\leq \frac{1}{\mathbb{P}[\bigcap_{i \in \mathcal{E}_j} B_i]} \\
&\leq \frac{1}{\prod_{i \in \mathcal{E}_j} \mathbb{P}[B_i]} \quad \text{(use (7.8))} \\
&\leq \frac{1}{\prod_{i \in \mathcal{E}_j}(1 - \mathbb{P}[A_i])} \\
&\leq (1 - e^{-k})^{-d_j} \quad \text{with } d_j = |\mathcal{E}_j| \text{ and (7.5)},
\end{aligned}
$$

and the claim is proved. In particular (7.8) implies

$$
\mathbb{P}[F] = \mathbb{P}[\bigcap_{i=1}^{m} B_i] \geq \prod_{i=1}^{n} \mathbb{P}[B_i] = \prod_{i=1}^{m}(1 - \mathbb{P}[A_i]) \geq (1 - e^{-k})^m. \quad (7.9)
$$

(ii) We derandomize this result computing conditional expectations. The trick is to define an appropriate potential function ("pessimistic estimator") $\Phi(z_1, \ldots, z_n)$, $z_i \in \{0, 1\}$, $i = 1, \ldots, n$, with the property that whenever $\Phi(z_1, \ldots, z_n) \leq M$ for a certain number $M > 0$, then $z = (z_1, \ldots, z_n)$ is a feasible solution for the vertex covering problem. M will be specified later. Put

$$
\Phi(z_1, \ldots, z_n) := \sum_{j=1}^{n} w_j z_j + M \cdot G(z)
$$

where $G(z) = 1$ if $\sum_{j=1}^{n} a_{ij} z_j = 0$ for some i and $G(z) = 0$ otherwise. Note that any feasible $z \in \{0, 1\}^n$ satisfies $\Phi(z) \leq M$ while any infeasible $z \in \{0, 1\}^n$ satisfies $\Phi(z) > M$. Furthermore,

$$
\mathbb{E}(G) = \mathbb{P}[z \text{ is not feasible}] = 1 - \mathbb{P}[F].
$$

Let k be a non-negative integer, also to be specified later. Let $X_1, \ldots, X_n$ be independent 0/1 random variables, defined by $X_j = 1$ with probability $f(y_j)$ and $X_j = 0$ with probability $1 - f(y_j)$. Elementary calculus gives

$$
f(y_j) \leq k y_j. \quad (7.10)
$$

Then, with the algorithm CONDEXP we can find an $x \in \{0, 1\}^n$ such that

$$
\begin{aligned}
\Phi(x) &\leq \mathbb{E}(\Phi(X_1, \ldots, X_n)) \\
&= \sum_{j=1}^{n} w_j f(y_j) + M(1 - \mathbb{P}(F)) \\
&\leq k W(y) + M(1 - (1 - e^{-k})^m) \quad \text{by (7.9) and (7.10).}
\end{aligned}
$$

Taking $k := \log m$ and $M := (1 - 1/m)^{-m} (\log m) W(y)$ the right hand side of the last inequality is at most M, hence x is feasible. ∎

Srinivasan [304] obtained improved approximation guaranties using positive correlations for general covering integer programs (CIP).

Theorem 7.5 (Srinivasan 1995, 1999) *Let $\mathcal{H} = (V, \mathcal{E})$ be a hypergraph with $|V| = n$ and $|\mathcal{E}| = m$. For the unweighted set covering problem a set cover $\mathcal{C} \subseteq \mathcal{E}$ satisfying*

$$|\mathcal{C}| \leq \left(\ln \left(\frac{n}{\tau^*(\mathcal{H})} \right) + \mathcal{O} \left(\ln \ln \left(\frac{n}{\tau^*(\mathcal{H})} \right) \right) + O(1) \right) \tau^*(\mathcal{H})$$

can be found in polynomial time.

The leading term in the approximation factor is $\ln \left(\frac{n}{\tau^*(\mathcal{H})} \right)$ which clearly improves on the greedy approximation factor of $\ln(n)$. Further improvements resp. parallelization are discussed in the articles of Srinivasan [302] resp. Alon, Srinivasan [19].

If the VC-dimension of the hypergraph $\mathcal{H} = (V, \mathcal{E})$ is bounded by d, then we know that the number of hyperedges is polynomially small, i.e. $|\mathcal{E}| = \mathcal{O}(n^d)$. In such a situation Brönnimann and Goodrich [59] and Lovász [209] were able to improve the approximation factor for the vertex cover problem of $\mathcal{O}(\log m)$ to $\mathcal{O}(d \log(d\tau(\mathcal{H})))$ resp. $\mathcal{O}(d \log(d\tau^*(\mathcal{H})))$.

The Brönnimann-Goodrich Algorithm.
In section 11 we define an $\varepsilon-$net for a hypergraph $\mathcal{H} = (V, \mathcal{E})$ as a subset $V_0 \subseteq V$ such that $|V_0 \cap E| \geq 1$ for all "$\varepsilon-$large" hyperedges, i.e. for all $E \in \mathcal{E}$ such that $|E| \geq \varepsilon|V|$. Thus a vertex cover for $\mathcal{H}$ is an $\varepsilon-$net for any $\varepsilon > 0$. The algorithm of Brönnimann and Goodrich is based on the computation of ε-nets with node weights. Matoušek [225] showed for hypergraphs with VC-dimension d for which a subsystem oracle of degree d exists [1] that a $(1/r)$-net of size $\mathcal{O}(dr \log(dr))$ can be computed in $\mathcal{O}(nr^d \log^d(dr))$ time.

Algorithm VERTEXCOVER2
a) Such an $\varepsilon-$net can serve as a start solution for the vertex covering problem.
b) If this net is already a vertex cover, we stop. Otherwise, if some edge $E \in \mathcal{E}$ is not hit by the net, Brönnimann and Goodrich double the weight of the nodes of E and iterate until a vertex cover is found.

Theorem 7.6 (Brönnimann Goodrich 1994) *In time $\mathcal{O}(n^{2+d} \log^{2+d} n)$ a vertex cover of size $\mathcal{O}(\tau(\mathcal{H}) d \log(d\tau(\mathcal{H})))$ can be constructed.*

The Lovász Algorithm.
The algorithm of Lovász [209] is a random greedy heuristic based on randomized rounding. The crux is a better analysis for hypergraphs with fixed VC-dimension d.

Algorithm VERTEXCOVER3

(a) Solve the LP associated to the vertex covering problem. Let $p = (p_1, \ldots, p_n)$ be an optimal fractional solution vector.

(b) Generate nodes from the probability distribution $p = (p_1, \ldots, p_n)$ randomly and independently until all edges are hit. Let Y be this set of nodes. ∎

Without any assumptions on the hypergraph it can be proved that the expected size of Y is $\mathcal{O}(\tau^*(\mathcal{H})\log m)$. But in bounded VC-dimension this can be improved considerably.

Theorem 7.7 (Lovász 1995) *If $\mathcal{H} = (V, \mathcal{E})$ has VC-dimension d, then*
$$\mathbb{E}(|Y|) \leq 16d\tau^*(\mathcal{H})\log(d\tau^*(\mathcal{H})).$$

Presently this is the best approximation algorithm for hypergraphs with bounded VC-dimension. It would be interesting to design a derandomized counterpart of this algorithm.

BIBLIOGRAPHY AND REMARKS. Parallel approximation algorithms for the set covering problem within a factor of $\mathcal{O}(\log n)$ resp. $\mathcal{O}(H(n))$ have been presented by Berger, Rompel and Shor [44] and Rajagopalan and Vazirani [272]. An interesting construction of a small hitting set (vertex cover) for combinatorial rectangles in high dimension can be found in the paper of Linial, Luby, Saks and Zuckerman [205].

8. SCHEDULING

In the last years randomized and derandomized algorithms could be designed for several $\mathcal{NP}$-hard scheduling problems leading to the presently best approximation guarantees. Among them are scheduling with release dates, flow shop scheduling, job shop scheduling and resource constrained scheduling.

8.1 Scheduling with Release Dates.

Scheduling with release dates is the following problem. We are given n jobs $J_1, \cdots, J_n$ that should be processed on a single machine. Job J_j has a release time r_j (that is the time, when job j becomes avalable), it occupies the machine for p_j time units and has a weight w_j. The goal is to schedule all jobs respecting their release times so that the weighted sum $\sum_{j=1}^{n} w_j C_j$ is minimum, where C_j is the completion time of job J_j in the schedule. Let OPT be the minimum completion time. Several linear programming relaxations have been studied for this problem. A key observation is that the choice of the relaxation influences the analysis in a crucial way. The first deterministic constant factor approximation guarantee was given by Phillips, Stein and Wein [260]. Their algorithm yields a $(16 + \varepsilon)-$approximation. Hall, Schulz, Shmoys and

Wein [144] improved the approximation factor to 4, and Schulz [287] further improved it to 3. Chakrabarti, Phillips Schulz, Stein and Wein [62] obtained a factor of 2.443. The presently best performance guarantee is due to Goemans [125] who gave a derandomized 2−approximation algorithm. Its randomized version is stated as follows.

A *preemptive* schedule is a schedule in which we may process a part of a job, say αp_j units, then interrupt and continue later, while in a *non-preemptive* schedule the whole job must be processed without any interruption. It is clear that preemptive scheduling is a relaxation of non-preemptive scheduling. Goemans' algorithm first constructs a preemptive schedule, and following an idea of Phillips, Stein and Wein [260], -fixes a non-preemptive schedule by ordering the jobs driven by the preemptive schedule.

Goemans' algorithm consists of two steps:

1. *Preemptive Schedule.* Consider for example the time axis with the release times r_j. At time r_{j_1} we schedule job j_1. If j_1 is finished before release

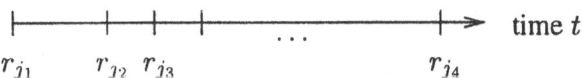

time r_{j_2}, fine, if not, we process at time r_{j_2} the job with the largest ratio w_j/p_j. This means, if $\frac{w_{j_1}}{p_{j_1}} > \frac{w_{j_2}}{p_{j_2}}$ we continue with j_1 otherwise the processing of j_1 will be interrupted, and j_2 starts. Henceforth, at any release time r_j we insert the job with the largest ratio w_j/p_j among the available jobs.

2. *Algorithm A_α.*
For $\alpha \in [0,1]$ let $t_j(\alpha)$ be the time at which a fraction of αp_j of job j has been scheduled in the preemptive schedule and let $\pi(1), \cdots, \pi(n)$ be an ordering such that $t_{\pi(1)}(\alpha) \leq \cdots \leq t_{\pi(n)}(\alpha)$. $t_j(\alpha)$ is called the α−point of job j. We schedule the jobs along this ordering as early as possible. Let C_α denote the weight of the schedule generated by A_α.

The randomized scheduling algorithm now is:

Algorithm RANDOM(α).

1. Generate a preemtive schedule.

2. Choose $\alpha \in [0,1]$ uniformly at random and call Algorithm A_α.

Thus, C_α is a random variable. The heart of the work of Goemans is the (sophisticated) proof that $\mathbb{E}(C_\alpha) \leq 2 \cdot \text{OPT}$. Here certain LP-relaxations of the problem are involved.

In order to derandomize RANDOM(α), we must find an $\alpha \in [0,1]$ with $C_\alpha \leq \mathbb{E}(C_\alpha)$. Since α is a real number, there are infinitely many choices for α, and moreover, so far no nice expression for $\mathbb{E}(C_\alpha)$ as a function of α is on hand. But the problem does not really possess a continous character. In fact, Goemans observed that among all choices of α, A_α can generate only n different schedules and these schedules can be computed in polynomial-time. Among them we pick one with the best weight, and derandomization is done:

Theorem 8.1 (Goemans 1997) *For the scheduling with release dates a schedule of weight at most* $2 \cdot \text{OPT}$ *can be computed in* $\mathcal{O}(n^2)-$ *time.*

8.2 Shop Scheduling and the Lovász-Local-Lemma.

An instance of the shop scheduling problem consists of

- n jobs $J_1, \dots, J_n$

- m machines $M_1, \dots, M_m$

- K operations $O_1, \dots, O_K$.

Each operation O_k belongs to some job J_j and must be processed on a specific machine M_i. The processing of the operations cannot overlap in time.

Open shop is a shop problem in which the operations of a job can be processed in any order. In a *job shop* problem they must be processed in a job-dependent order. A *flow shop* problem is a special job shop problem in which each job has exactly m operations, one for every machine, and the order in which the operations have to be processed is the same for all jobs. The following notations are needed:

- p_k is the processing time of operation O_k, and $p_{\max} := \max_k p_k$.

- P_j is the total processing time of job J and $P_{max} := \max_j P_j$ is the maximum job-load.

- Π_i is the total processing time that must take place on machine M_i and $\Pi_{max} := \max_i \Pi_i$ is the maximum machine-load.

- μ_j is the number of operations for job J_j, and $\mu := \max_j \mu_j$.

The goal is to minimize the makespan of the schedule, defined as the time at which all operations are completely processed. All of these problems are $\mathcal{NP}-$hard. For a detailed discussion of approximation algorithms for scheduling we refer to the survey article of Hall [143]. Here we only review the impact of probabilistic methods on shop problems. Let us consider the job shop problem in which all operations have unit length and each job has at most one operation per machine. One of the most remarkable and surprising results for job shop was proved by Leighton, Maggs and Rao [198].

Theorem 8.2 (Leighton, Maggs, Rao 1994) *There exists a schedule whose makespan is at most* $\mathcal{O}(\Pi_{max} + P_{max})$.

Proof Ideas. The proof relies on the Lovasz-Local-Lemma (Theorem 6.1). We discuss some ingredients of the proof. One important concept is the notion of *random delays*. Let us schedule the jobs in a greedy manner independently of each other, where every job starts at time zero and is processed until completion. Such a schedule certainly will be infeasible, violating the non-interference condition for the operations etc.. Now we choose for every job's greedy schedule a time $t \in \{0, 1, 2, \dots, \Delta\}$ and delay it by t. Let us call such a schedule a greedy *pseudo-schedule* with $\{0, \Delta\}$-delay. Note that the makespan of such a delayed schedule is at most $P_{max} + \Delta$. Randomization comes in by selecting t *randomly* from $\{0, \dots, \Delta\}$.

We expect that if Δ is large then there should be enough "room" left in a random $\{0, \dots, \Delta\}$-delayed schedule, so it might be unlikely that two operations will run simultaneously on the same machine.

The key step in the proof of Leighton, Maggs and Rao is the proof that the number of operations performed on any machine in a random delayed schedule can be bounded by $\log \Pi_{max}$, provided that $\Delta = \mathcal{O}(\Pi_{max})$. Let us give at least an indication why this is true.

For simplicity, we assume $\mu = 1$ and $\Pi_{max} \geq P_{max}$. Let us consider a long time interval I of length T in a random $\{0, \alpha\Pi_{max}\}$-delayed schedule and let C be the largest number of operations performed on any of the machines in I. We would like to choose T in a way that:

(a) $C \leq T$,

(b) T is small,

(c) α is a small constant.

Of course, if α is large we do not expect to have problems in order to meet (a) and (b) simultaneously. The Local-Lemma proves α and T can be ept *simultaneously* small: Let A_i be the event that more than T operations get assigned to machine M_i, for some interval I of length T. Thus the good event is $\bigwedge_{i=1}^{m} A_i^c$ and we wish to bound its probability away from 0. For any machine M_l that does not have a job in common with machine M_i, the event A_l does not depend on A_i. Bounding all of these dependencies by $d = \Pi_{max} P_{max}$ one can show that

$$\mathbb{P}[A_i] \leq (1 + \alpha)\Pi_{max}^2 \left(\frac{e}{\alpha}\right)^{\Pi_{max}}.$$

Then the assumption of the Local-Lemma, $4pd \leq 1$, holds for an even *constant* $\alpha > 0$ provided that $T \geq \log P_{max}$. In conclusion there exists a greedy pseudo-schedule with $\{0, \alpha\Pi_{max}\}$-delays, $\alpha > 0$ constant, with at most $\log(\Pi_{max})$ operations on each machine M_i in each time interval of length $T = \log(\Pi_{max})$. ∎

For general job-shop scheduling Shmoys, Stein and Wein [293] gave a deterministic polynomial-time algorithm which delivers a schedule of makespan $\mathcal{O}(\rho \cdot (\Pi_{\max} + P_{\max}))$ with

$$\rho = \frac{\log(m\mu)}{\log\log(m\mu)} \log(\min\{m\mu, p_{\max}\}).$$

The presently best algorithm for job-shop is due to Goldberg, Paterson, Srinivasan, Sweedyk [128], improving the above by a factor of $1/(\log\log(m\mu))$.

Theorem 8.3 (Goldberg, Paterson, Srinivasan, Sweedyk 1997) *There is a derandomized $\rho-$approximation for job-shop scheduling with*

$$\rho = \frac{\log(m\mu)}{(\log\log(m\mu))^2} \log(\min\{m\mu, p_{\max}\}).$$

For $m = n$ it can be parallelized.

8.3 Resource Constraind Scheduling.

An instance of resource constand scheduling consist of:

- a set $\mathcal{J} = \{J_1, \ldots, J_n\}$ of independent jobs. Each job J_j needs one time unit for its completion and cannot be scheduled before its start time r_j, $r_j \in \mathbb{N}$.

- a set $\mathcal{P} = \{P_1, \ldots, P_m\}$ of identical processors. Each job needs one processor.

- a set $\mathcal{R} = \{R_1, \ldots, R_s\}$ of limited resources. This means that at any time all resources are available, but the available amount of each resource R_i is bounded by $b_i \in \mathbb{N}$, $i = 1, \ldots, s$.

- For $i = 1, \ldots, s$, $j = 1 \ldots, n$ let $R_i(j)$ be 0/1 resource requirements saying that every job T_j needs $R_i(j)$ units of resource R_i during its processing time. For a job $T_j \in \mathcal{T}$ and a time $z \in \mathbb{N}$ let x_{jz} be the 0/1 variable which is 1 iff job T_j is scheduled at time z.

- Given a valid schedule let C_{max} be the latest completion time defined by $C_{max} = \max\{z \mid x_{jz} > 0, \ j = 1, \ldots, n\}$.

The combinatorial optimization problem is to find a schedule, that is a 0/1 assignment for all varibles x_{jz}, subject to the start time, processor and resource constraints such that $\sum_{z \in \mathbb{N}} x_{jz} = 1$ for all jobs T_j and C_{max} is minimum. Let C_{opt} denote this minimum. [2]

A fractional schedule is an assignment of each x_{jz} to a rational number in the closed interval $[0, 1]$ subject to the start times, processor and resource constraints so that $\sum_{z \in \mathbb{N}} x_{jz} = 1$ for all jobs T_j and C_{max} is minimum. Let C denote this minimum. We call C_{opt} the (integral) optimal and C the fractional optimal schedule.

The integral problem is $\mathcal{NP}$-hard in the strong sense, even if $r_j = 0$ for all $j = 1, \ldots, n$, $s = 1$ and $m = 3$ [119], while the fractional problem can be solved by linear programming in polynomial time. An interesting special case of resource constrained scheduling is the following generalized version of the multidimensional bin packing problem. Bin Packing Problem $BIN(\vec{l}, d)$ Let $d, n, l_i \in \mathbb{N}$, $i = 1, \ldots, d$, and let $\vec{l} = (l_1, \ldots, l_d)$. Given vectors $\vec{v}_1, \ldots, \vec{v}_n \in [0, 1]^d$, pack [3] all vectors in a minimum number of bins such that in each bin B and for each coordinate i, $i = 1, \ldots, d$, $\sum_{\vec{v}_j \in B} v_{ij} \leq l_i$. Define $L_R = \lceil \max_{1 \leq i \leq d} \frac{1}{l_i} \sum_{j=1}^{n} v_{ij} \rceil$. (Observe that L_R is the minimum number of bins, if fractional packing is allowed.) $BIN(1, d)$ is the multidimensional bin packing problem, and $BIN(1, 1)$ is the classical bin packing problem.

Polynomial-time approximation algorithms for resource constrained scheduling with zero start times (problem class $P|\text{res} \cdots, r_j = 0, p_j = 1|C_{\max}$) are due to Garey, Graham, Johnson, Yao [119] and Röck and Schmidt [274]. Garey et al. constructed with the First-Fit-Decreasing heuristic a schedule of length C_{FFD} which asymptotically is a $(s + \frac{1}{3})$-factor approximation, i.e. there is a non-negative integer C_0 such that $C_{FFD} \leq C_{opt}(s + \frac{1}{3})$ for all instances with $C_{opt} \geq C_0$. De la Vega and Lueker [320] improved this result presenting for every $\varepsilon > 0$ a linear-time algorithm which achieves an asymptotic approximation factor of $s + \varepsilon$. Röck and Schmidt showed, employing the polynomial-time solvability of the simpler problem with two processors [4] an $\lceil \frac{m}{2} \rceil$-factor polynomial-time approximation algorithm. Thus for problems with small optimal schedules or many resource constraints resp. processors these algorithms have a weak performance. Note that all these results are based on the assumption that the start-times of all jobs are zero. For example, Röck and Schmidt's algorithm cannot be used, when non-zero start-times are given. [5] With randomized rounding and derandomization an approximation guarantee independent of the number of resources can by proved [310]:

Theorem 8.4 (Srivastav, Stangier 1997) *Let $\varepsilon > 0$ with $(1/\varepsilon) \in \mathbb{N}$. For the resource constrained scheduling problem with start times a valid integral schedule of size at most $\lceil (1 + \varepsilon)C \rceil$ can be found in polynomial time, provided that $m \geq \frac{3(1+\varepsilon)}{\varepsilon^2} \lceil \log(8C) \rceil$ and $b_i \geq \frac{3(1+\varepsilon)}{\varepsilon^2} \lceil \log(8Cs) \rceil$ for all $i = 1, \ldots, s$.*

For the proof we need the following result for integral solvability of a linear inequallity system which is fractionally solvable. Its proof can be derived with a variant of the algorithmic version of the Angluin-Valiant inequality [310]. Let $\widehat{s}, n, N$ be non-negative integers. Let $A = (a_{ij})$ be a $\widehat{s} \times n$, 0/1 matrix and let $b \in \mathbb{Q}_+^{\widehat{s}}$. For a vector $x_j \in \{0,1\}^N$, $j = 1, \ldots, n$, and for $k = 1, \ldots, N$ let x_{jk} be its k-th component. Let $x^{(k)}$ be the vector of all the k-th components, i.e. $x^{(k)} = (x_{1k}, \ldots, x_{nk})$. Let $r_j \in \{1, \ldots, N\}$ be numbers which will later play the role of start times. Consider the following system of linear inequalities

$$
\text{(IS)} \quad
\begin{aligned}
Ax^{(k)} &\leq b \quad \forall k = 1, \ldots, N \\
\|x_j\|_1 &= 1 \quad \forall j = 1, \ldots, n \\
x_j &\in \{0,1\}^N \quad \forall j = 1, \ldots, n \\
x_{jk} &= 0 \quad \forall k < r_j \ \forall j = 1, \ldots, n \\
x^{(k)} &= (x_{1k}, \ldots, x_{nk}) \quad \forall k = 1, \ldots, N.
\end{aligned}
\tag{8.1}
$$

We will later show how resource constrained scheduling can be modelled in such a way. Let us define the ε-version of (IS) for parameters $0 < \varepsilon \leq 1$ by

$$
\text{IS}(\varepsilon) \quad
\begin{aligned}
Ax^{(k)} &\leq (1+\varepsilon)^{-1} b \quad \forall k = 1, \ldots, N \\
\|x_j\|_1 &= 1 \quad \forall j = 1, \ldots, n \\
x_j &\in \{0,1\}^N \quad \forall j = 1, \ldots, n \\
x_{jk} &= 0 \quad \forall k < r_j \ \forall j = 1, \ldots, n \\
x^{(k)} &= (x_{1k}, \ldots, x_{nk})^T \quad \forall k = 1, \ldots, N.
\end{aligned}
\tag{8.2}
$$

We further need the following parameters. Let s_1 be an integer with $s_1 \leq \widehat{s}$ and define

$$
b_{\varepsilon,1} = \frac{3(1+\varepsilon)}{\varepsilon^2} \log(4s_1 N) \text{ and } b_{\varepsilon,2} = \frac{3(1+\varepsilon)}{\varepsilon^2} \log(4(\widehat{s} - s_1)N). \tag{8.3}
$$

(IS) can be solved, if a fractional solution to IS(ε) is given:

Theorem 8.5 (Srivastav, Stangier 1997) *Let $0 < \varepsilon \leq 1$. Let s_1 and $b_{\varepsilon,i}$ ($i = 1, 2$) be as in (8.3). Suppose that $b_i \geq b_{\varepsilon,1}$ for all $i = 1, \ldots, s_1$ and $b_i \geq b_{\varepsilon,2}$ for all $i > s_1$. If $u = (u_1, \ldots, u_n)^T$ with $u_j \in [0,1]^N$ is a fractional solution for the ε-inequality system IS(ε), then a vector $x = (x_1, \ldots, x_n)^T$ with $x_j \in \{0,1\}^N$ satisfying system (IS) can be constructed in $O(N\widehat{s}n^2 \log(N\widehat{s}n))$ time.*

Proof Theorem 8.4. The randomized algorithm consists of 3 steps. First, we compute a fractional schedule and the number C. Secondly, the fractional schedule is enlarged to $\lceil (1+\varepsilon) \rceil C$. In the enlarged schedule a fraction $\tilde{x}_{jz}$ of job j is assigned to time z for all j and z.

Finally, in the enlarged schedule the jobs are assigned to time randomly and independently with probabilities roughly proportional to the $\tilde{x}_{jz}$. The $\tilde{x}_{jz}$ are

computed with the following algorithm.
The computations are done with the following algorithm.

Algorithm RANDOMSCHEDULE

Step 1: Let us assume that $C_{\text{opt}} \leq n$. (This assumption can be made w.l.o.g.). Start with an integer $\tilde{C} \leq n$ and check whether the LP

$$
\begin{aligned}
\sum_{j=1}^{n} R_i(j) x_{jz} &\leq b_i & &\forall R_i \in \mathcal{R}, \\
& & &z \in \{1, \dots, n\} \\
\sum_{z=1}^{n} x_{jz} &= 1 & &\forall J_j \in \mathcal{J} \\
x_{jz} &= 0 & &\forall J_j \in \mathcal{J}, z < r_j \text{ and} \\
& & &\forall J_j \in \mathcal{J}, z > \tilde{C} \\
x_{jz} &\in [0,1] & &\forall J_j \in \mathcal{J} \; \forall z \in \{1, \dots, n\}
\end{aligned}
\tag{8.4}
$$

has a solution. Using binary search we can find C along with fractional assignments $(\tilde{x}_{jz})$ solving at most $\log n$ such LPs. Hence C can be computed in polynomial-time, with standard polynomial-time LP algorithms.

Step 2: Consider the time interval $\{1, \dots, \lceil (1+\varepsilon)C \rceil \}$ and put $\delta = \frac{1}{1+\varepsilon}$ and $\alpha = \varepsilon \delta \lceil \varepsilon C \rceil$. Set

$$
\widehat{x}_{jl} := \begin{cases} \delta \tilde{x}_{jl} & \text{for } l \in \{1, \cdots, C\} \\ \sum_{t=1}^{C} \alpha \tilde{x}_{jt} & \text{for } l \in \{C+1, \cdots, C + \lceil \varepsilon C \rceil \}. \end{cases}
$$

Step 3:
Schedule the jobs at times selected by the following randomized procedure:

(a) Cast n mutually independent dice each having $N = \lceil (1+\varepsilon)C \rceil$ faces where the z-th face of the j-th die corresponding to job j appears with probability $\widehat{x}_{jz}$. (The faces stand for the scheduling times)

(b) Schedule for each $j = 1, \dots, n$ the jobs J_j at the time selected in step (a).

It is straightforward to prove that the $\widehat{x}_{jl}$ define a valid fractional schedule with makespan $\lceil (1+\varepsilon)C \rceil$. In order to find an integral schedule we invoke the rounding lemma for inequaltiy systems. As above let $N = \lceil (1+\varepsilon)C \rceil$. Consider the inequality system (IS) where $A = (R_i(j))_{ij}$, $(R_i(j))_{ij}$ is the $(s+1) \times n$ resource requirement matrix where the resource R_{s+1} represents the processors with requirements 1 for all jobs and bound $b_{s+1} = m$. The assertion of the theorem is equivalent to the problem of finding a solution to (IS). The vectors $\widehat{x}_1, \dots, \widehat{x}_n$ with $\widehat{x}_j \in [0,1]^N$ and $\sum_{k=1}^{N} \widehat{x}_{jk} = 1$ form a fractional solution to the ε-inequality system IS(ε). We apply the rounding theorem (Theorem 8.5) with $\hat{s}_h = s + 1$ and $s_1 = s$: Since

$$b_i \geq \frac{3(1+\varepsilon)}{\varepsilon^2} \lceil \log(8Cs)) \rceil \geq \frac{3(1+\varepsilon)}{\varepsilon^2} \lceil \log(4s_1 N) \rceil \ \forall i = 1, \ldots, s$$

and

$$m = b_{s+1} \geq \frac{3(1+\varepsilon)}{\varepsilon^2} \lceil \log(8C)) \geq \frac{3(1+\varepsilon)}{\varepsilon^2} \lceil \log(4N) \rceil,$$

the hypothesis of Theorem 8.5 are satisfied and we can find in deterministic polynomial time a solution to the inequality system (IS). ∎

Interestingly the approximation guarantee of Theorem 8.4 is best possible:

Theorem 8.6 (Srivastav, Stangier 1997) *Under the assumption that there exists a fractional schedule of $C \geq 3$, C fixed, and an integral schedule of size $C + 1$, $b_i = \Omega(\log(Cs))$, $R_i(j) \in \{0, 1\}$ for all $i = 1, \cdots, s$, $j = 1, \cdots, n$ is it $\mathcal{NP}$−complete to decide wether or not there exists an integral schedule of size C.*

Proof. We give the basic argument for $b_i = 1$ for all i. (the main work is its extension to large bounds $b_i = \Omega(\log(Cs))$). We use a reduction to the $\mathcal{NP}$−comlete problem of determining the chromatic index of graph. Let $G = (V, E)$ be a graph with $|V| = \nu$, $|E| = \mu$ and $\deg(v) \leq \Delta$ for all $v \in V$. We construct an instance of resource constrained scheduling as follows. Introduce for every edge $e \in E$ exactly one job J_e and consider $\mu = |E|$ identical processors. Let us freely call the edges jobs and vice versa. For every node $v \in V$ define a resource R_v with bound 1 and resource/job requirements

$$R_v(e) = \begin{cases} 1 & \text{if } v \in e \\ 0 & \text{if } v \notin e. \end{cases}$$

It is straightforward to verify that there exists a coloring that uses Δ colours if and only if there is a feasible integral schedule of size Δ. Furthermore, there is a fractional schedule of size $C = \Delta$: Simply set $x_{ez} = \frac{1}{\Delta}$ for all $z = 1, \ldots, \Delta$. ∎

BIBLIOGRAPHY AND REMARKS: For a concise introduction to approximation algorithms for scheduling problems we refer to the survey article of L. Hall [143].

Scheduling with Release Dates. Polyhedral aspects of the problem have been treated by Queyranne [265], Queyranne and Schulz [266] and Goemans [125]. In case of all weights being 1, a 2−factor approximation algorithm was already presented by Pillips, Stein and Wein [260]. In 1997 Chekuri et al. [69] obtained in this case the improved factor of $e/(e-1)$. The algorithm of Goemans applies also to the on-line case, but there derandomization is not known.

Job-Shop Scheduling. For general job-shop scheduling Shmoys, Stein and Wein [293] gave a deterministic polynomial-time ρ−approximation algorithm with $\rho = \frac{\log(m\mu)}{\log\log(m\mu)} \log(\min\{m\mu, p_{\max}\})$, Goldberg, Paterson, Srinivasan, Sweedyk [128] gave a derandomized algorithm improving ρ by a factor of $1/(\log \log(m\mu))$ and could parallelize it for $m = n$.

Resource Constrained Scheduling. Applying Berger/Rompel's [43] extension of the method of $\log^c n$-wise independence to multivalued random variables, a parallelization has been given in [310]. For $\tau \geq \frac{1}{\log n}$ there is an NC-algorithm which guarantees for every constant $\alpha > 1$, a $2^{\lceil \log \alpha C \rceil}/C \leq 2\alpha$-factor approximation, under the conditions $m, b_i \geq \alpha(\alpha - 1)^{-1} n^{\frac{1}{2}+\tau} \lceil \log 3n(s+1) \rceil^{1/2}$ for all $i = 1, \ldots, s$. Ahuja and Srivastav [3] obtained improved results with the Lovász Local Lemma.

For scheduling of unrelated parallel machines results of similar flavour have been achieved by Lenstra, Shmoys and Tardos [203] and Lin and Vitter [204]. Lenstra, Shmoys and Tardos [203] gave a 2-factor approximation algorithm for the problem of scheduling independent jobs with different processing times on *unrelated* processors and also proved that there is no ρ-approximation algorithm for $\rho < 1.5$, unless $P = NP$. Lin and Vitter [204] considered the generalized assignment problem and the problem of scheduling of unrelated parallel machines. For the generalized assignment problem with resource constraint vector b they could show for every $\varepsilon > 0$ an $1 + \varepsilon$ approximation of the minimum assignment cost, which if feasible within the enlarged packing constraint $(2 + \frac{1}{\varepsilon})b$.

9.   MULTICOMMODITY FLOWS AND ROUTING IN VLSI-LAYOUT

The randomized rounding scheme has shown to be a successful method for finding integral multicommodity flows in graphs and routing in VLSI chips.

9.1 Multicommodity Flows.

A central problem in combinatorial optimization is the multicommodity flow problem which is a generalization of the well known maximum flow problem. An instance of the multicommodity flow problem is a graph $G = (V, E)$ (the supply graph) with $|V| = n, |E| = m$, and a graph $H = (T, D)$ (the demand graph) with terminal set $T \subseteq V$, $|T| = 2k$ and commodity set $D = \{(s_1, t_1), \ldots, (s_k, t_k)\}$ where $s_i, t_i \in T$. $(s_i, t_i) \in D$ are the k source-sink pairs, also called demand edges or commodities. For each commodity $d = (s, t) \in D$ let σ_d be an orientation of G forming the directed graph (V, A_d) and let $F(d)$ be an integral (s, t)-flow in (V, A_d). Then the $|D|$-tuple of flows $F = (F(d))_{d \in D}$ is called an integral multicommodity flow. It is a $0/1$ multicommodity flow, if all flows are either 0 or 1. Given a capacity function $c : E \mapsto \mathbb{Z}_+$ and a demand function $r : D \mapsto \mathbb{Z}_+$ the multicommodity flow is feasible subject to c, if for each edge $e \in E$ the sum of the flows through e (in both directions) is at most $c(e)$, and is feasible subject to r, if for each demand edge $d \in D$ the d-th flow value $f(d)$ is at least $r(d)$. The different multicommodity flow problems considered in this paper are:

Definition 9.1   *(i) (Specified Demands) Given (G, H, c, r), find an integral multicommodity flow, subject to c and r, if possible.*

(ii) (Maximum Integral Problem) *Given (G, H, c), find an integral multi-commodity flow subject to c with maximum total flow value, denoted by f_{opt}.*

(iii) (Maximum 0/1 Problem) *Given (G, H, c), find a 0/1 multicommodity flow subject to c with maximum total flow value, denoted by f_{opt}.*

All these problems are $\mathcal{NP}$—hard.

Approximating Maximum Flows. The design of derandomized algorithms for the integral multicommodity flow problem started with the first application of randomized rounding to the $0/1$—multicommodity flow problem by Raghavan and Thomson [270]. Raghavan and Thomson considered two variants of the problem: the first one is exactly the $0/1$—multiflow problem as in Definition 9.1 *(iii)*, while in the second problem we wish to route all the k—commodities keeping the capacity violation as small as possible. In this variant, we wish to find $0/1$—flows f_i for all $1 \leq i \leq k$ such that $\sum_{i=1}^{n} f_i(e) \leq C$, $e \in E$ and C is minimum. Let C_{opt} be the optimal integral solution, and C^* the optimal fractional solution for the linear programming relaxation $f_i(e) \in [0,1]$, $1 \leq i \leq k$ and $e \in E$.

Theorem 9.2 (Raghavan, Thompson 1987) *Let $0 < \varepsilon < 1$ and suppose that $c(e) \geq 2 \ln m$ for all $e \in E$. With probability at least $1 - \varepsilon$ all k commodities can be routed within $C^* + (3C^* \ln(m/\varepsilon))^{1/2}$.*

For the $0/1$—maximization problem Raghavan [269] derived a deterministic algorithm with the pessimistic estimator technique.

Theorem 9.3 (Raghavan 1988) *For the $0/1$—maximum multicommodity flow problem a flow with total value $f \geq \alpha f_R$ can be constructed using pessimistic estimators provided that $c(e) \geq \beta \ln m$ for all $e \in E$, where $\beta > 0$ and $0 < \alpha < 1$ are constant.*

The computation of the constant α is quite involved, and it appears as the solution of a functional equation ([269], Equation 1.12). Srivastav and Stangier observed that Raghavans approach to the integral multicommolity flow problem can be simplified using the Angluin-Valiant inequality, and more important, a polynomial running time can be fixed with the algorithmic version of the Angluin-Valiant bound. As we will see, concise approximation results for integer multicommodity flow problems can be derived in the framework of integer solvability of systems of linear inequalities and equations in an unified way.

The randomized approximation result we would like to obtain in a deterministic way is given by the following theorem of Motwani, Naor and Raghavan [240] for 0/1-maximum multiflows.

Theorem 9.4 (Motwani, Naor, Raghavan 1995) *Let* $\varepsilon \geq 0.62$. *For the 0/1-maximum multiflow problem a 0/1-flow of total value* f *can be contructed in polynomial time such that* $f \geq (1 - \varepsilon)^2 f_{opt}$ *holds with probability at least* $1 - 1/m - \exp(-0.38\varepsilon^2 f_{opt})$, *provided that the edge capacities are large, i.e.* $c \in \Omega(\log m)$.

We show a derandomized „counterpart" of Theorem 9.4, covering general integer flows. The integer maximum multiflow problem is equivalent to the following integer linear program.

(IP-Flow) $\quad \max \sum_{i=1}^{k} \sum_{\{v \in V; (s_i, v) \in E\}} (f_{s_i v}^{(i)} - f_{v s_i}^{(i)})$
$\quad\quad$ such that:

$$\sum_{\substack{\{v \in V \setminus \{s_i, t_i\}; \\ (u,v) \in E\}}} f_{uv}^{(i)} = \sum_{\substack{\{v \in V \setminus \{s_i, t_i\}; \\ (u,v) \in E\}}} f_{vu}^{(i)} \quad \forall i, \forall u \in V - \{s_i, t_i\}$$

$$\sum_{i=1}^{k} f_{uv}^{(i)} + f_{vu}^{(i)} \leq c(u, v) \quad \forall (u, v) \in E$$

$$f_{vu}^{(i)} \in \mathbb{Z}_+ \quad \forall i, \forall (u, v) \in E.$$

It is a standard and convenient approach to transform an optimal fractional solution to this integer linear program into a flow- path formulation. Having solved the LP relaxation of (IP-Flow), where $f_{vu}^{(i)} \in \mathbb{Q}_+$ for all i and v, u, we can construct with standard algorithms ([223], [270]) in polynomial-time for each commodity $d = (s, t) \in D$ a set of (s, t)-flow paths Γ_d with $|\Gamma_d| \leq m$. Each such path has a fractional path value $\lambda(P) \in \mathbb{Q}_+$. The following conditions making sure that each flow is conveyed through the flow-paths without violating the capacity bounds:

(a) (Capacity Condition) For each $e \in E$, $\sum_{P, e \in P} \lambda(P) \leq c(e)$.

(b) (Demand Condition) For each $d \in D$, $\sum_{d \in D} \sum_{P \in \Gamma_d} \lambda(P) = f_R$.

Let $\Gamma = \bigcup_{d \in D} \Gamma_d$ be the set of all flow paths. Note that a path in our graph G may occur as many times in Γ as it is a flow path for some commodities, thus Γ is a multiset.

The idea in the randomized algorithm behind the following theorem is to pick for every commodity d a flow-path randomly and independently.

Theorem 9.5 (Srivastav, Stangier 1993) *Let* (G, H, c) *be an instance of the maximum integral multicommodity flow problem. Let* $0 < \varepsilon \leq 1$ *and suppose that* $c(e) \geq \frac{6(2-\varepsilon)}{\varepsilon^2} \lceil \log(2m) \rceil$ *for all* $e \in E$. *An integral multicommodity flow with total value* $f \geq (1 - \varepsilon) f_R \geq (1 - \varepsilon) f_{opt}$ *can be contructed in* $\mathcal{O}(t_{LP} + (1/\varepsilon)^4 (k^2 m^5 \log^2 m \log(m/\varepsilon))$ *time, where* t_{LP} *is the time to solve the LP relaxation of (IP-Flow).*

For the proof we formulate the problem in terms of an inequality system studied in [306]. Let A be a rational $m \times n$ matrix, C a rational $\ell \times n$ matrix with

$0 \leq a_{ij} \leq 1$ and $0 \leq c_{ij} \leq 1$. Let $b \in \mathbb{Z}_+^m$ and $u \in \mathbb{Z}_+^\ell$. We consider the following inequality System.

$$\text{(IIS)} \qquad \left(\begin{smallmatrix} A \\ -C \end{smallmatrix}\right) x \leq \left(\begin{smallmatrix} b \\ -u \end{smallmatrix}\right), \; x \in \mathbb{Z}_+^n. \tag{9.1}$$

Note that the decision version of packing integer programming is a special case of (IIS). For $0 \leq \varepsilon < 1$ the ε-relaxation of (IIS) is:

$$\text{IIS}(\varepsilon) \qquad \left(\begin{smallmatrix} A \\ -C \end{smallmatrix}\right) x \leq \left(\begin{smallmatrix} b \\ -(1-\varepsilon)u \end{smallmatrix}\right), \; x \in \mathbb{Z}_+^n. \tag{9.2}$$

$\text{IIS}(\varepsilon)$ is still an integer inequality system. An integral solution to $\text{IIS}(\varepsilon)$ can be obtained from a fractional solution to (IIS) by randomly rounding the components of the fractional solution [306].

Theorem 9.6 (*Srivastav, Stangier 2000*) *Let* $0 < \varepsilon \leq 1$ *and suppose that* $b_i \geq \lceil \frac{6(2-\varepsilon)}{\varepsilon^2} \rceil \lceil \log(2m) \rceil$ *for all* $i = 1, \ldots, m$ *and* $u_j \geq \frac{16}{\varepsilon^2} \lceil \log(2\ell) \rceil$ *for all* $j = 1, \ldots, \ell$. *Let* $y \in \mathbb{Q}_+^n$ *be a fractional solution to (IIS). Then an integral solution* $x \in \mathbb{Z}_+^n$ *to* $\text{IIS}(\varepsilon)$ *can be constructed in* $O((1/\varepsilon)^4 (t^3 n^2 \log^2 t \log(nt/\varepsilon)))$ *time, where* $t = \max(m, l)$.

Proof of Theorem 9.5. First we solve the LP-relaxation and construct the corresponding fractional flow path vector $y = (\lambda(P))_{P \in \Gamma}$. Let $A = (a_{eP})_{e \in E, P \in \Gamma}$ be the edge-path incidence matrix and let $\mathbb{1} \in \mathbb{Q}^{|\Gamma|}$ be the vector whose components are 1. Let us identify c with the vector $(c(e))_{e \in E}$, and fix any ordering for the edges of E such that the i-th row of A and the i-th component of c are indexed by the same edge $e \in E$. Consider the system

$$\begin{pmatrix} A \\ -\mathbb{1}^T \end{pmatrix} z \leq \begin{pmatrix} c \\ -(1-\varepsilon) f_R \end{pmatrix}, \; z \in \mathbb{Z}_+^{|\Gamma|}. \tag{9.3}$$

The theorem is proved, if we can find a solution of (9.3) in polynomial time. Since y is a fractional solution for the system

$$\begin{pmatrix} A \\ -\mathbb{1}^T \end{pmatrix} z \leq \begin{pmatrix} c \\ -u f_R \end{pmatrix}, \; z \in \mathbb{Z}_+^{|\Gamma|},$$

we invoke Theorem 9.6 and can solve (9.3) within the claimed time bound. ∎

For the maximum 0/1-multiflow problem a similar approximation result can be proved.

Approximating Specified Demands.. Let us start with an interesting result of Korach and Penn [191]. We say, the cut-condition is satisfied, if for every cut-set $\delta S \subset E$ the demand of the cut is at most the capacity of the cut:

$$\sum_{(s,t) \in D, \, \delta S \text{ separates } s,t} r(s,t) \leq \sum_{e \in \delta S} c(e).$$

Theorem 9.7 (Korach, Penn 1988) *Let (G, H, c, r) be an instance of the multicommodity flow problem with specified demands. If $G \cup H$ is planar and if the cut condition is satisfied, then the reduced multicommodity flow problem $(G, H, c, r - 1)$ can be solved in polynomial time.*

It is quite surpising that the reduction of one unit from every demand makes the integer problem solvable. The theorem of Korach and Penn motivates the following notion of integrality gaps. For a specified demand problem (G, H, c, r) let $\kappa_I : D \mapsto \mathbb{Z}_+$ and $\kappa_R : D \mapsto \mathbb{Q}_+$ be functions with the property that the reduced problems $(G, H, c, r - \kappa_I)$ resp. $(G, H, c, r - \kappa_R))$ admit an integral resp. fractional solution, and the maximum norms $\|\kappa_I\|_\infty$ resp. $\|\kappa_R\|_\infty$ are minimum. We call the numbers $\|\kappa_I\|_\infty$ resp. $\|\kappa_R\|_\infty$ the integral resp. fractional demand gap, and call the difference $\|\kappa_I\|_\infty - \|\kappa_R\|_\infty$ the integrality gap for the problem (G, H, c, r).

Since instances with planar $G \cup H$ and cut-condition are fractionally solvable ([140], Theorem 8.6.6), by the Korach/Penn theorem the integrality gap is at most one. Using a construction suggested by Pfeiffer [259] we show that for planar G and H, but nonplanar $G \cup H$ the integrality gap is unbounded, thus planarity of $G \cup H$ is essential in order to bound the integrality gap.

Proposition 9.8 *For every odd integer $C \geq 1$ and for every non-negative integer K there is a fractionally solvable multicommodity flow problem (G, H, c, r) with $r = r(C, K)$ and $c(e) \geq C$ for all $e \in E$ such that $(G, H, c, r - K)$ has no integral solution.*

Proof. Let $C = 2\alpha + 1$, $\alpha \in \mathbb{Z}_+$. Construct a $(2K + 1) \times (2K + 1)$ grid (as shown in Figure 18.3), where every node of the grid is replaced by a C_4 and each edge has capacity C. Let the supply graph G be this grid and let (s, t) and (s', t') be commodities with demands $r(s, t) = r(s', t') = (2K + 1)C$. The demands can be satisfied by routing the commodities half-integrally, thus the problem is fractionally solvable. Assume for a moment that $(G, H, c, r - K)$ has an integral solution and let F resp. F' such (s, t) resp. (s', t')-flows. Then at most K units of F resp. F' can be routed "around" the grid using edges incident to s or t resp. s' or t'. So for both F and F' there is a F-saturated (s, t)-path P resp. F'-saturated (s', t')-path P' through the grid. But P and P' must cross in some C_4, which is impossible because P and P' are saturated (see figure 18.3). ∎

For the general integer multiflow problem with specified demands derandomization yields:

Theorem 9.9 (Srivastav, Stangier 1993) *Let (G, H, c, r) be a multicommodity flow problem and $0 < \varepsilon \leq 1$. Suppose that $c(e) \geq \frac{6(2-\varepsilon)}{\varepsilon^2} \lceil \log(2m) \rceil$ for all $e \in E$ and $r(d) - \kappa_R(d) \geq \frac{16}{\varepsilon^2} \lceil \log(2k) \rceil$ for all $d \in D$. Then in polynomial*

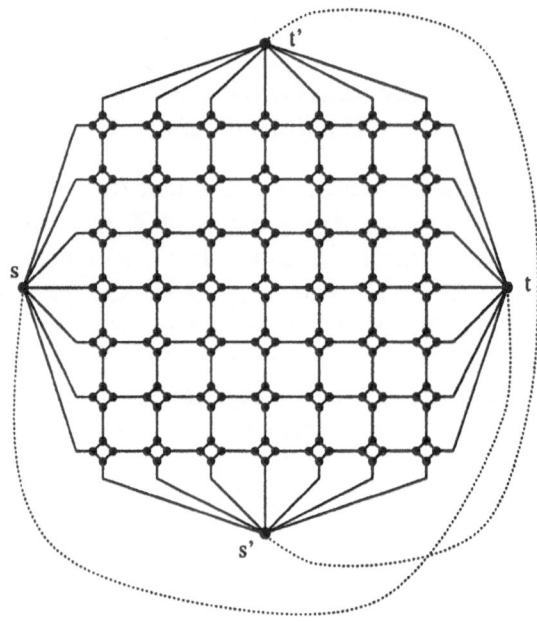

Figure 18.3 The construction with $K = 3$

time an integral multicommodity flow can be constructed such that for all $d \in D$
we have $f(d) \geq (1 - \varepsilon)(r(d) - \kappa_R(d)) \geq (1 - \varepsilon)(r(d) - \kappa_I(d))$.

BIBLIOGRAPHY AND REMARKS.

Maximum Flows: When the union of supply and demand graph is planar and there are only
two commodities then the problem can be solved in $O(n\sqrt{\log n})$ time [192]. In case that
the supply graph is a tree, Garg, Vazirani and Yannakakis [123] gave (with the primal-dual
method) a polynomial-time 1/2-factor approximation and showed the $MAXSNP$-hardness
even in this case. For densly embedded and nearly-Eulerian supply graphs (which includes
a two-dimensional mesh) the maximum edge-disjoint path problem admits a polynomial-time
contant factor a pproximation algorithm (Kleinberg/Tardos [187]). Garg and Könemann [122]
gave the presently fastest deterministic approximation algorithm which for every $\varepsilon > 0$ and
large capacities (i.e. $c(e) \in \Omega(\frac{\ln((1+\varepsilon)m)}{\varepsilon \ln(1+\varepsilon)})$ for all edges $e \in E$) routes at least an $(1 - \varepsilon)^2$
fraction of the maximum integral flow in $\mathcal{O}(\frac{km^2 \log m \log n}{\varepsilon \log(1+\varepsilon)})$ time.

Specified Demands: For planar supply graphs and a fixed number of commodities Sebö [288]
showed that the problem is solvable in polynomial time. For graphs with unit edge capacities
and unit demands the integral multicommodit y flow problem with specified demands is the
edge-disjoint path problem. For a planar and Eulerian supply graph with terminals on the
boundary, Okamura and Seymour [253] gave a polynomial-time algorithm. Wagner and Weihe
[325] showed even a linear running time algorithm. The problem with integer capacities
can be solved in polynomial-time, too, if a modified evenness condition holds (see [325]
for a discussion). Korach and Penn [191] Connections of the specified demand problem to
discrepancies of hypergraphs were discussed in [311], where it is proved that a fractionally

solvable multicommodity flow problem with specified demands (G, H, c, r) admits an integral solution for the reduced problem $(G, H, c + 6\sqrt{m+k}, r - 6\sqrt{m+k})$.

9.2 Routing in VLSI-Layout

The layout of VLSI logic chips, so called physical design, faces a number of $\mathcal{NP}$—hard problems from combinatorial optimization. Among them two problems play a major role: the placement problem where the gates of a VLSI chip have to be placed on the feasible chip area and the routing (or wiring) problem that has the task to realize a wiring essentially minimizing the total wire length (which directly influences the cycle time of the VLSI processor).

Obviously both problems are intertwined with each other. Due to the enormous complexity of such problems in real-world applications where several hundred thousand of gates must be placed, and nets connecting them with several million pins must be routed, a hierachial approach is widely used. First, a placement is computed driven by a global routing and then the final and local routing is carried out. Here we focus on the global routing problem. For a first insight in the nature of routing problems it might be sufficient to discuss routing in a 2—dimensional array of gates, called a gate-array. Following Raghavan and Thompson [271] a regular array is a two dimensional $m \times n$ lattice $L = L(V, E)$ where V is the set of lattice nodes and E is the set of edges connecting two neighboring lattice points. The lattice nodes represent the gates and the edges are the wiring channels through which the gates must be connected.

A net N is a subset of nodes, $N \subseteq V$. Let R be the set of all nets with $r = |R|$. A connection between two nodes u and v is one of the simple paths P_1, P_2 :

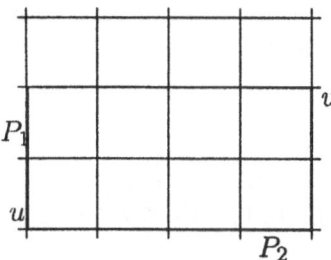

A routing for a net is a tree in the grid where with the net-nodes as it leaves. Let S be a set of trees for the nets in R. We call S a routing for R. The width of an edges $e \in E$, noted $w(e)$ is the maximum number of trees in S containing edge, and the width $W(S)$ is the maximum edge width among all edges.

Problem GLOBALROUTING.

Given (L, R) find a routing S for R with minimum width $W(S)$.

The leading idea of Raghavan and Thompson is to model the global routing problem as a multiterminal, multicommodity flow problem. For simplicity let us consider only 3 terminal nets, so $|N| = 3$ for all $N \in R$. A routing tree T for a net $N_i = \{t_{i1}, t_{i2}, t_{i3}\}$ must have a point $v_j \in V$ to which the t_{i1}, t_{i2} and t_{i3} are connected by simple paths:

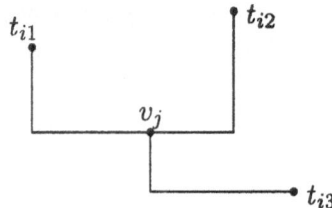

s_{ij} is called a Steiner point, and T is a so called Steiner tree for t_{i1}, t_{i2} and t_{i3} (see bibliography for the notation of Steiner trees). Let $s_{ij} \in \{0, 1\}$ denote a variable which is 1 if v_j is a Steiner point for net N_i and zero else.

The multiterminal, multicommodity flow problem for (L, R) is the multicommodity flow problem in which s_{ij} unit of an integral flow f_i must be transported from each of the terminal nodes t_{i1}, t_{i2} and t_{i3} to node v_j such that $\sum_{i=1}^{r} f_i(e)$ is minimum over all $e \in E$ ($f_i(e) \in \{0, 1\}$ is the f_i–flow through $e \in E$).

Note that this problem is equivalent to the global routing problem. An integer linear programming formulation of the multiterminal, multicommodity flow problem is the following:

$$
\begin{array}{ll}
& \min W \\
\text{(IP)} & \sum_{j=1}^{mn} s_{ij} = 1, \qquad \forall 1 \leq i \leq r \\
& \sum_{i=1}^{r} f_i(e) \leq W, \quad \forall e \in E, \\
& f_i(e), s_{ij} \in \{0, 1\} \quad \forall 1 \leq i \leq r,\, 1 \leq j \leq mn,\, e \in E
\end{array}
$$

(and f_i is an integral multiterminal, multicommodity flow for net N_i).

Let (LP) be the linear programming relaxation of (IP), where $f_i(e), s_{ij} \in [0, 1]$ for all i, j, and the f_i's are real multiterminal, multicommodity flows (satisfying all conservation laws). Let $\hat{W}$ be an optimal solution to (LP). The algorithm proceed in two stages: First, the Steiner points are picked. Then for every net N_i, a $0/1$–flow f_i is realized, leading to a routing tree for net f_i with the Steiner points choosen in the first step. With the pessimistic estimator technique the following theorem can be proved.

Theorem 9.10 (Raghavan, Thompson 1991) *If* $\hat{W} > \ln|E|$, *then a routing S for* (L, R) *can be found in polynomial time such that*

$$W(S) \leq \hat{W} + (e - 1)\sqrt{\hat{W} \ln|E|}.$$

Thus, the width $W(S)$ of the solution is $(1 + \mathcal{O}(\sqrt{\frac{\ln|E|}{\hat{W}}}))$–approximation of the optimal fractional width. This approach can be extended to nets with more than 3 terminals, but then the choices for the Steiner points for a net grow exponentially in the number of nodes in the net. Similar methods based on randomized rounding have been applied to other global routing problems by Ng, Raghavan and Thompson [247]. As pointed out by Lengauer [202], the randomized rounding/derandomization scheme applies sucessfully to unconstrained global routing problems, for example where no bounds for the channel capacities are given. In real-world applications constraints are not avoidable. Presently, the most effective routers for VLSI chips, handling more than 5 million transistors, are based on efficient Steiner tree heuristics and placement algorithms using quadratic programming (see [193], [324] and [188] to quote only a few references).

BIBLIOGRAPHY AND REMARKS.

Routing in VSLI-Design. A comprehensive presentation of combinatorial algorithm in VSLI-design is given in the book of Lengauer [202]. For more recent work we refer to the proceedings of the annual design autotation conferences. Derandomized algorithms for packet routing and sorting in multidimensional lattices were presented by Kaufmann, Sibeyn and Suel [182].

Steiner Trees. For a graph $G = (V, E)$ and a set $T \subseteq V$ called terminal set, a subgraph $S = (V(S), E(S))$ of G is called Steiner tree for T, if S is a tree in G with $T \subseteq V(S)$ and each leave of S is a terminal. The nodes in the set $V(S) \setminus T$ are called Steiner points (or nodes). For a cost function $l : E \rightarrow \mathbb{Q}_0^+$, $l(S) = \sum_{e \in E(S)} l(e)$ is the length of the Steiner tree S. A Steiner tree of minimum length is a *Steiner minimal tree* for T and the Steiner problem in graphs is o find a Steiner minimal tree for a given set of terminals T.. The Steiner problem in graphs is $\mathcal{NP}$-hard. In the last years a race for finding the best polynomial time approximation algorithms for the Steiner problem in graphs started initiated by the paper of Zelikovsky [330] who improved the greedy approximation factor of 2 (due to Takahasi and Matsuyama [314]) to 1.834. The presently best approximation factor of 1.550 is due to Robins and Zelikovsky [273]. For a history and recent developments for approximation algorithms for the Steiner tree problem we refer the reader to the article of Hougardy and Prömel [159].

10. INDEPENDENT SETS IN HYPERGRAPHS

How many queens can be placed on a chessboard so that no queen can capture another queen? How many points can be picked from n-dimensional Euclidean space such that all angles determined by three points are strictly less then $\frac{\pi}{2}$? How large is a maximum subset of $\{1, \dots, n\}$ in which all pairwise sums are distinct? Extremal selection problems like these are equivalent to the problem of determining the independence number of a hypergraph.

Definition 10.1 *Let* $\mathcal{H} = (V, \mathcal{E})$ *be a hypergraph. A subset I of V is called independent if no edge is entirely contained in I, that is if $\mathcal{P}(I) \cap \mathcal{E} = \emptyset$. The independence number $\alpha(\mathcal{H})$ is the cardinality of a maximum independent set.*

Both, calculating $\alpha(\mathcal{H})$ and finding an independent set of size $\alpha(\mathcal{H})$ are $\mathcal{NP}$-hard problems [120]. We refer to the latter as the IS problem. On the other hand, the *maximal independent set problem* (briefly, MIS) which seeks for an inclusion-maximal independent set, can be solved in polynomial time. The design of parallel algorithms for this problem and has attracted a wide interest, and in same sence it has served as a "benchmark" problem for parallel derandomization via k-wise independence.

10.1 Maximum Independent Sets.

Since the calculation of $\alpha(\mathcal{H})$ is $\mathcal{NP}$-hard, we are interested in good approximation results. The celebrated theorem of P. Turán [316] on the independence number of graphs states $\alpha(\mathcal{H}) \geq \frac{|V|^2}{2|\mathcal{E}|+|V|}$. In the general case of $k+1$-uniform hypergraphs two probabilistic strategies of finding large independent sets were successful. The *random sampling* approach is based on choosing a random set $S \subseteq V$ and deleting one element from each edge in S. The expected size of the resulting set yields a lower bound for $\alpha(\mathcal{H})$ in terms of $|V|$ and $|\mathcal{E}|$. The second strategy relies on the fact that our problem can be formulated as an integer *linear program*. Analyzing a randomized rounding procedure gives a lower bound with respect to the optimal solution to the relaxed program.

10.1.1 Basic Random Sampling.

Let $\mathcal{H} = (V, \mathcal{E})$ be a hypergraph with $n := |V|$ and $m := |\mathcal{E}|$. We call $\mathcal{H}$ $k+1$-*uniform* if $|E| = k+1$ for all $E \in \mathcal{E}$. The *average degree* of a $k+1$-uniform hypergraph is $t^k := \frac{(k+1)m}{n}$, the average number of edges incident in each vertex. As a warming-up exercise we prove the following result of Spencer [18].

Theorem 10.2 *Let* $\mathcal{H} = (V, \mathcal{E})$ *be a $k+1$-uniform hypergraph with average degree $t^k \geq 1$. Then*

$$\alpha(\mathcal{H}) \geq \left(1 - \frac{1}{k+1}\right) \frac{n}{t},$$

and an independent set of at least that size can be found in time $\mathcal{O}(m)$.

Proof. Let $S \subseteq V$ be obtained by picking each vertex independently with probability $p := t^{-1}$. Deleting one vertex from every edge in S yields an independent set of size $|S| - |\mathcal{P}(S) \cap \mathcal{E}|$. Hence

$$\alpha(\mathcal{H}) \geq \mathbb{E}[|S| - |\mathcal{P}(S) \cap \mathcal{E}|] = np - m \cdot p^{k+1} = \left(1 - \frac{1}{k+1}\right) \frac{n}{t}.$$

In order to find an independent set we derandomize this procedure. To simplify notations we assume $V = \{1, \ldots, n\}$. Define a function $f : [0,1]^n \longrightarrow \mathbb{R}$ by

$$f(x_1, \ldots, x_n) := \sum_{i=1}^{n} x_i - \sum_{E \in \mathcal{E}} \prod_{i \in E} x_i \text{ for all } (x_1, \ldots, x_n) \in [0,1]^n.$$

If each vertex i is picked independently with probability p_i, we have $\mathbb{E}(|S| - |\mathcal{P}(S) \cap \mathcal{E}|) = f(p_1, \ldots, p_n)$. Since for all $i \in V$

$$
\begin{aligned}
f(x_1, \ldots, x_i, \ldots, x_n) &= x_i \cdot f(x_1, \ldots, 1, \ldots, x_n) \\
&+ (1 - x_i) \cdot f(x_1, \ldots, 0, \ldots, x_n)
\end{aligned}
$$

we can select $p_i \in \{0, 1\}$ in order to maximize f. By successively setting

$$
p_i := \begin{cases}
1, & \text{if } f(p_1, \ldots, p_{i-1}, 1, t^{-1}, \ldots, t^{-1}) \\
& \quad > f(p_1, \ldots, p_{i-1}, t^{-1}, \ldots, t^{-1}) \\
\\
0, & \text{if } f(p_1, \ldots, p_{i-1}, 0, t^{-1}, \ldots, t^{-1}) \\
& \quad \leq f(p_1, \ldots, p_{i-1}, t^{-1}, \ldots, t^{-1}).
\end{cases}
$$

we ensure that $I := \{i \in V \mid p_i = 1\}$ does not contain any edges. Hence I is an independent set of size

$$\sum_{i=1}^{n} p_i = f(p_1, \ldots, p_n) \geq f(t^{-1}, \ldots, t^{-1}) = \left(1 - \frac{1}{k+1}\right) \frac{n}{t}.$$

■

For certain dense hypergraphs – like unions of $k + 1$-cliques – the bound of Theorem 10.2 is sharp [47]. In the case of hypergraphs containing no small cycles a stronger result is possible by iterating the random sampling procedure.

10.1.2 Iterated Random Sampling. Let $\mathcal{H} = (V, \mathcal{E})$ be a $k + 1$-uniform hypergraph. A cycle of length $l \geq 2$ (briefly l-cycle) is a sequence $(v_1, E_1, v_2, E_2, v_3, \ldots, v_l, E_l, v_1)$ such that

- $E_1, E_2, \ldots, E_l$ are distinct edges of $\mathcal{H}$,
- $v_1, v_2, \ldots, v_l$ are distinct vertices of $\mathcal{H}$,
- $v_i, v_{i+1} \in E_i$ $(i = 1, 2, \ldots, l - 1)$,
- $v_l, v_1 \in E_l$.

We say, $\mathcal{H}$ is *uncrowded* if $\mathcal{E}$ contains no cycles of length 2,3, or 4.
Ajtai, Komlós, Pintz, Spencer, and Szemerédi proved the following remarkable result [4].

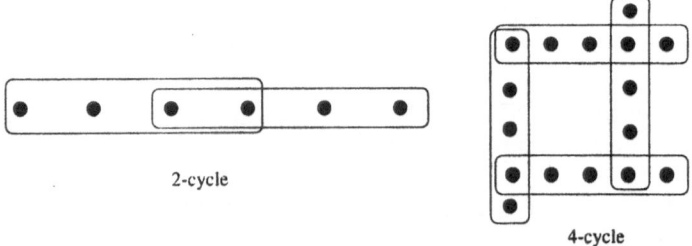

2-cycle

4-cycle

Theorem 10.3 (Ajtai, Komlós, Pintz, Spencer, Szemerédi 1982) *Let $\mathcal{H} = (V, \mathcal{E})$ be an uncrowded $k + 1$-uniform hypergraph with average degree t^k where $t \longrightarrow \infty$ as $n \longrightarrow \infty$. Then $\mathcal{H}$ contains an independent set of size at least $\Omega\left(\frac{n}{t}(\ln t)^{1/k}\right)$.*

This improves on the random sampling bound by the factor of $(\ln t)^{1/k}$. The proof relies on the analysis of the following randomized algorithm:

Algorithm: MAXIMUM-INDEPENDENT-SET

1. Start with $s := 1$, $\mathcal{H}_1 = (V_1, \mathcal{E}_1) := \mathcal{H}$.

2. Let t_s^k be the average degree in $\mathcal{H}_s$, and choose a random subset C_s of V_s by independently selecting each vertex with probability t_s^{-1}.

3. Let D_s denote the set of vertices v such that all the neighbours of v (i.e. the vertices sharing an edge with v) are chosen for C_s. Remove these "dangerous" vertices to get an independent set $I_s := C_s \setminus D_s$.

4. Define $\mathcal{H}_{s+1} := (V_{s+1}, \mathcal{E}_{s+1})$, where $V_{s+1} := (V \setminus C_s) \setminus D_s$, and $\mathcal{E}_{s+1} := \{E \cap V_{s+1} \mid E \in \mathcal{E}_s\}$. If $|V_{s+1}|$ is sufficiently large update $s := s + 1$ and go back to step 2. Otherwise output $I := \bigcup_{j=1}^{s} I_j$. ∎

Note that since no vertex adjacent to I_s is in V_{s+1}, the independent sets $I_1, I_2, \ldots$ combine to a large independent set I. A detailed analysis involves considerable technical effort, so we present only a rough sketch of the proof.

Sketch of proof of Theorem 10.3. For simplicity let us assume that $\mathcal{H}_s$ is regular with degree t_s^k, i.e. every vertex x is contained in t_s^k edges. Since $\mathcal{H}$ does not contain 2-cycles, these edges intersect in x only, and the events "$E \setminus \{x\} \subseteq C_s$" are independent for each edge E containing x. Thus for every $x \in V$:

$$\mathbb{P}[x \notin D_s] = \prod_{\substack{E \in \mathcal{E}_s \\ \text{with } x \in E}} \mathbb{P}[E \setminus \{x\} \nsubseteq C_s] = \left(1 - \frac{1}{t_s^k}\right)^{t_s^k} \sim e^{-1}.$$

The events "$x \in D_s$", "$x \in C_s$" are independent, so

$$
\begin{aligned}
\mathbb{P}[x \in I_s] &= \mathbb{P}[x \notin D_s]\mathbb{P}[x \in C_s] \sim \frac{1}{t_s \cdot e} \\
\mathbb{P}[x \in V_{s+1}] &= \mathbb{P}[x \notin D_s]\mathbb{P}[x \notin C_s] \sim \left(1 - \frac{1}{t_s}\right) e^{-1} \sim e^{-1}.
\end{aligned}
$$

Uncrowdedness guarantees that only few vertices have a common neighbour. Thus for almost all $x, y \in V$ the events "$x \in I_s$", "$y \in I_s$" as well as the events "$x \in V_{s+1}$", "$y \in V_{s+1}$" are independent. From Chebyshev's inequality we infer ($e := \exp(1)$)

$$
\mathbb{P}[|I_s| - \mathbb{E}[I_s]| \geq 0.01\mathbb{E}[|I_s|]] \leq \frac{net_s^{-1}(1 - et_s^{-1})}{(0.01net_s^{-1})^2} < \frac{10000t_s}{ne} \overset{n \to \infty}{\longrightarrow} 0,
$$

so

$$
|I_s| \sim \frac{n}{et_s} \text{ and similarly } |V_{s+1}| \sim \frac{n}{e}.
$$

Moreover, it can be shown that by passing from $\mathcal{H}_s$ to $\mathcal{H}_{s+1}$ the degree of each vertex decreases by a factor of e^{-k}. After roughly $\ln t$ iterations the average degree t (and consequently $|V_{\ln t}|$) is very small. For $k = 1$ the algorithm terminates with an independent set of size $\sum_{i=1}^{\ln t} \frac{n}{t}e^{-i} = \Omega\left(\frac{n}{t}\ln t\right)$. If $k > 1$ we must replace $\ln t$ with its k-th root. This loss is due to the fact that we have to cope with $\mathcal{H}_s$ being no longer k-uniform after the first iteration. ∎

10.1.3 Derandomization.

The probabilistic proof of Theorem 10.3 relies heavily on Chebyshev's inequality. It has been derandomized by Fundia [118] with the method of conditional expectations.

Theorem 10.4 (Fundia 1996) *Under the hypothesis of Theorem 10.3 there is an algorithm that finds an independent subset of size* $\Omega\left(\frac{n}{t}(\ln t)^{1/k}\right)$ *in time* $\mathcal{O}(n^3 t^{6k} \ln t)$.

Sketch of Proof. The main difficulty is to handle more than one random variable simultaneously. Consider for example the variables I_s, $s = 1, \ldots, [\ln t]$. The method of conditional probabilities cannot be applied directly to guarantee $I_s \geq \frac{n}{et_s}$ for all s, since $\mathbb{P}[I_s < \frac{n}{et_s}]$ cannot be computed efficiently. Fundia overcame this problem by applying the following modification of the pessimistic estimator technique.

A nonnegative random variable Z defined on a finite probability space $(\Omega, \mathcal{P}(\Omega), \mathbb{P})$ is a pessimistic estimator for an event A, if $\mathbb{E}(Z) < 1$ and

$\mathbb{P}[A \mid U] \leq \mathbb{E}[Z \mid U]$ for any $U \subseteq \Omega$ with $\mathbb{P}[U] > 0$.

Let A_s denote the event "$I_s < \frac{n}{et_s}$" and let $A := \bigcup_{s=1}^{[\ln t]} A_s$. A is the event "there is an $i \in \{1, \ldots, \ln t\}$ such that $I_s < \frac{n}{et_s}$". Now, if Z is a pessimistic estimator for A we can find a set C_s such that each I_s is of size at least $\frac{n}{et_s}$ by minimizing $\mathbb{E}[Z \mid C_s$ is chosen$]$. After n conditioning steps we get $\{0, 1\} \ni \mathbb{P}[A \mid C_s$ is chosen$] \leq \mathbb{E}[Z \mid C_s$ is chosen$] \leq \mathbb{E}[Z] < 1$, hence $\mathbb{P}[A \mid C_s$ is chosen$] = 0$. Fundia observed that a pessimistic estimator for an event A of the form "there is an $i \in \{1, \ldots, l\}$ such that $A_i < a_i$" is given by the random variable $Z := \sum_i Z_i$, where

$$Z_i = \frac{\mu_i^2 - 2A_i\mu_i + A_i^2}{(\mu_i - a_i)^2} \tag{10.1}$$

for arbitrary $\mu_i < a_i$. Equation (10.1) can be seen as as a generalization of Chebyshev's inequality since after replacing μ_i by $\mathbb{E}[A_i]$ and taking expectations the right hand side of (10.1) becomes $\frac{\text{Var}[A_i]}{(\mathbb{E}[A_i] - a_i)^2}$. ∎

Bertram-Kretzberg and Lefmann [47] improved Fundia's result. They gave an algorithmic version of a theorem of Duke, Lefmann, and Rödl [93] who observed that hypergraphs with a small number of 2-cycles contain large uncrowded subhypergraphs. Using the notation $s_{2,j}(\mathcal{H})$ for the number of 2-cycles $\{E, E'\}$ with $|E \cap E'| = j$, Theorem 10.3 can be strengthened as follows.

Theorem 10.5 (Bertram-Kretzberg, Lefmann 1998) *Let $k \geq 2$ be a fixed integer. Let $\mathcal{H} = (V, \mathcal{E})$ be a $k + 1$-uniform hypergraph on n vertices with average degree at most t^k where $t \longrightarrow \infty$ with $n \longrightarrow \infty$. If $s_{2,j}(\mathcal{H}) \leq cn \cdot t^{2k+1-j-\gamma}$ for $j \in \{2, 3, \ldots, k\}$ and some constants $c, \gamma > 0$, then one can find for every $\delta > 0$ in time $\mathcal{O}\left(n \cdot t^k + \sum_{j=2}^k s_{2,j}(\mathcal{H}) + \frac{n^3}{t^{3-\delta}}\right)$ an independent set of size at least $\Omega\left(\frac{n}{t}(\ln t)^{1/k}\right)$.*

Related results can be found in the work of Hofmeister and Lefmann [156].

10.2 An Application: Sidon Sets.

An interesting example of independent sets are Sidon sets which play an important role in additive number theory [102] and functional analysis [280]. We can apply Theorem 10.5 to prove the existence of large Sidon sets in an arbitrary abelian group G of order n. $S \subseteq G$ is a Sidon set if all pairwise sums $s_1 + s_2$ ($s_1, s_2 \in S$, $s_1 \neq s_2$) are distinct. S cannot have more than $\mathcal{O}(\sqrt{n})$ elements, since the Sidon property implies $\binom{|S|}{2} \leq n$. While a well-known theorem of Erdös and Turán [102] states that maximum sized Sidon sets in the integers $\{1, \ldots, n\}$ have $\sqrt{n} + O(n^{1/4})$ elements, tight bounds are unknown for the general case of abelian groups. Thiele [315] observed that for a random subset

$S \subseteq G$ the probability of being Sidon tends to 1 if $|S| = o(\sqrt[4]{n})$ and to 0 if $|S| = \omega(\sqrt[4]{n})$ for $n \longrightarrow \infty$. We prove the following result of Alon, Lefmann, and Rödl [13].

Proposition 10.6 *Let G be an abelian group of order n, then G contains a Sidon set with at least $\Omega(\sqrt[3]{n \ln n})$ elements.*

Proof. Consider the 4-uniform hypergraph $\mathcal{H} = (V, \mathcal{E})$, where $V := G$ and $\mathcal{E} := \{\{g_1, g_2, g_3, g_4\} \subseteq G \mid |\{g_1, g_2, g_3, g_4\}| = 4, \ g_1 + g_2 = g_3 + g_4\}$. Note that an independent set in $\mathcal{H}$ is a Sidon set in G and vice versa.
Each $E \in \mathcal{E}$ is the union of two sets $\{g_1, g_2\}, \{g_3, g_4\} \subseteq V$ satisfying $g_1 + g_2 = g_3 + g_4$. There are n possible values for the sums $g_1 + g_2$, and each sum has roughly $\lfloor \frac{n}{2} \rfloor$ representations. Hence

$$|\mathcal{E}| \approx n \cdot \binom{\lfloor \frac{n}{2} \rfloor}{2} \approx \frac{n^3}{8} \text{ and } t = \sqrt[3]{\frac{4|\mathcal{E}|}{|V|}} \approx \sqrt[3]{\frac{n^2}{2}}.$$

To bound $s_{2,3}(\mathcal{H})$ and $s_{2,2}(\mathcal{H})$ we fix an edge $E \in \mathcal{E}$ and estimate the number of edges $E', E'' \in \mathcal{E}$ with $|E \cap E'| = 3, |E \cap E''| = 2$.
There are $\binom{4}{3}$ choices for $E \cap E'$ and $\binom{4}{2}$ choices for $E \cap E''$. $E \cap E''$ can be extended to a three element set $\not\subseteq E$ in $n - 4$ ways. Each three element set determines $\binom{3}{2}$ sums. Summing over all $E \in \mathcal{E}$ we count each 2-cycle at least twice, thus

$$
\begin{aligned}
s_{2,3}(\mathcal{H}) &\leq & \binom{4}{3} \cdot \binom{3}{2} \cdot \frac{|\mathcal{E}|}{2} &= 6|\mathcal{E}| &< \tfrac{3}{4}n^3 < 4nt^{4-\frac{1}{2}} \\
s_{2,2}(\mathcal{H}) &\leq \binom{4}{2} \cdot (n-4) \cdot \binom{3}{2} \cdot \frac{|\mathcal{E}|}{2} &= 9|\mathcal{E}|(n-4) &< \tfrac{9}{8}n^4 < 4nt^{5-\frac{1}{2}}.
\end{aligned}
$$

The conditions of Theorem 10.5 are satisfied and the claim follows. ∎

By applying Theorem 10.5 to an appropriate random subhypergraph, Bertram-Kretzberg and Lefmann [47] showed that Sidon sets of size $\Omega(\sqrt[3]{n \ln n})$ in arbitrary groups can be computed in time $\mathcal{O}(n^3)$.

10.3 LP-based Bounds.

If k is larger than $\ln n$, iterated random sampling can no longer ensure a better bound for $\alpha(\mathcal{H})$ than the basic sampling method. On the other hand, for large k the basic randomized rounding approach gives an interesting result. Assume $\mathcal{H} = (V, \mathcal{E})$ to be $k + 1$-uniform with vertices $v_1, \dots, v_n$ and edges $E_1, \dots, E_m$. Recall that the incidence matrix $A = (a_{ij})_{m \times n}$ of $\mathcal{H}$ is defined by $a_{ij} := 1$, if $v_j \in E_i$ and $a_{ij} := 0$ otherwise. The independent set problem is equivalent to the following integer program. Let $c := (1, \dots, 1)^T \in \mathbb{R}^n$.

$$\max\{c^T x \mid A^T x \leq k, \ x \in \{0, 1\}^n\}$$

The relaxation $x \in [0,1]^n$ yields a linear program which can be solved in polynomial time.

Proposition 10.7 *Let OPT* be the optimal value of the objective function of the relaxed linear program. If $k = \Omega\left(\frac{\ln m}{\varepsilon^2}\right)$, then $\mathcal{H}$ contains an independent set with at least $(1 - \varepsilon)OPT^*$ elements, and such a set can be constructed in polynomial time.*

Proof. Let $x_1^*, \dots, x_n^*$ be an optimal solution to the relaxed linear program. Let $(x_1, \dots, x_n)$ be obtained by independently rounding each x_i^* to 1 with probability $\left(1 - \frac{\varepsilon}{2}\right) x_i^*$ and to 0 with probability $1 - \left(1 - \frac{\varepsilon}{2}\right) x_i^*$. If the events

$$B_0 \ : \ \sum_{i=1}^n x_i \geq (1 - \varepsilon)\text{OPT}^*$$

and

$$B_j \ : \ \sum_{i=1}^n a_{ij}x_j \leq k, \ j \in \{1, \dots, m\}$$

simultaneously have positive probability, we know that $\mathcal{H}$ contains an independent set of size at least $(1 - \varepsilon)\text{OPT}^*$. Through randomized rounding, the x_i have the nice property that their expected value is x_i^* (scaled down by $\left(1 - \frac{\varepsilon}{2}\right)$ to boost the probability for the B_j). We can now use the Angluin-Valiant inequality (Theorem 2.4) to bound the probability of the negative events $\overline{B_l}$:

$\mathbb{P}(\overline{B_l}) \leq e^{-\frac{\varepsilon^2 k}{12}}$ for all $l \in \{0, \dots, m\}$. Hence

$$\mathbb{P}(\bigwedge_{l=0}^m B_l) \geq 1 - (m+1) \cdot e^{-\frac{\varepsilon^2 k}{12}} > 0 \quad \text{if} \quad k > \frac{12}{\varepsilon^2} \ln(m+1).$$

Finally, the derandomization is easily carried out with the algorithmic Angluin-Valiant inequality (Theorem 3.4). ∎

It is not difficult to prove that an uncrowded $\mathcal{H}$ cannot have more than

$$\frac{n}{2(k+1)} \left(1 - k + \sqrt{4n + k^2 - 2k - 3}\right) \sim n\frac{\sqrt{n}}{k}$$

edges, and it cannot have more than $\frac{\binom{n}{2}}{\binom{k+1}{2}} \sim \frac{n^2}{k^2}$ edges when being free of 2-cycles. Thus for hypergraphs as in Theorem 10.3 or Theorem 10.5 the above bound is already interesting if $k = \Omega(\ln n)$.

Further improvements can be obtained by involving positive correlation among the events B_j:

$$\mathbb{P}(\bigwedge_{l=0}^m B_l) \geq 1 - \prod_{l=1}^m (1 - \mathbb{P}(\overline{B_l})) + \mathbb{P}(\overline{B_0}). \tag{10.2}$$

Srinivasan's [304] analysis of randomized rounding for integer packing and covering programs using positive correlations yields a deterministic polynomial-time algorithm for computing an independent set of size $\Omega\left(\text{OPT}^* \cdot \sqrt[k]{\frac{\text{OPT}^*}{m}}\right)$. By applying Janson's inequality and an extended version of the Lovász Local Lemma, Srinivasan [302] obtained a randomized approximation guarantee of $\Omega\left(\frac{\text{OPT}^*}{\sqrt[k]{\Delta}}\right)$, where Δ denotes the maximum degree $\max_{v \in V} |\{E \in \mathcal{E} \mid v \in E\}|$.

10.4 Approximation Ratios and Inapproximability Results.

Let I_n be the set of all hypergraphs on n vertices. We denote by $A(\mathcal{H})$ the cardinality of an independent set found by an algorithm A and define

$$R_A(n, C_n) := \max_{\mathcal{H} \in C_n} \frac{\alpha(\mathcal{H})}{A(\mathcal{H})}$$

to be the *approximation ratio* of the algorithm A on a subset $C_n \subseteq I_n$.

We say that the problem IS of finding a maximum independent set for a class C_n of hypergraphs from I_n is *approximable within* $\mathcal{O}(p)$ for some expression p, if there is a polynomial time algorithm A such that $R_A(n, C_n) = \mathcal{O}(p)$. Theorem 10.2 ensures that IS is approximable within $\mathcal{O}(t)$ for $k + 1$-uniform hypergraphs on n vertices with average degree t^k. The best result in terms of the number of vertices is due to Halldórsson[146], who proved approximability within $\mathcal{O}\left(\frac{n}{\log n}\right)$. Concerning inapproximability, Hofmeister and Lefmann [157] generalized a result of Håstad[148] for graphs by showing that IS cannot be approximated within a factor of $\mathcal{O}(n^{1-\varepsilon})$ for any $\varepsilon > 0$, unless $\mathcal{NP}=\mathcal{ZPP}$ (i.e. unless $\mathcal{NP}$-hard problems have randomized polynomial algorithms).

10.5 Maximal Independent Sets in Graphs.

A *maximal independent set* of vertices of a graph has the property that no vertex can be added without creating an edge. The problem MIS of finding such a set can be solved by means of a very simple linear-time algorithm.

Algorithm LFMIS
$I := \emptyset$
For $i := 1$ to n do
if vertex i is not adjacent to any vertex in I then add vertex i to I
Output I ∎

Can we do substantially better using a polynomial bounded number of processors? Cook [80] proved that deciding whether a vertex belongs to the lexicographically first maximal independent set is logspace-complete in $\mathcal{P}$. Thus there is no hope to parallelize LFMIS. Yet, the MIS problem is not inherently sequential. Karp and Wigderson [181] were the first to present a polylogarithmic algorithm, proving that MIS $\in \mathcal{NC}^4$. Alon, Babai, and Itai [9], and, independently, Luby [211] found randomized algorithms with expected running-time $\mathcal{O}(\log^2 n)$ on $\mathcal{O}(m)$ processors. Moreover, Luby was able to remove the randomness, showing that $\mathcal{O}(n^2 m)$ processors can solve MIS in $\mathcal{O}(\log n)$ time, and hence MIS $\in \mathcal{NC}^2$. The key for parallelizing MIS is to successively add not a *single* vertex to the independent set under construction but a *whole* independent set S.

Algorithm: PARALLELMIS(High Level)
$I := \emptyset$
while $V(\mathcal{H}) \neq \emptyset$ do
begin
select $S \subseteq V(\mathcal{H})$ which is independent in $\mathcal{H}$
$I := I \cup S$
update $\mathcal{H} := \mathcal{H} \setminus (I \cup N(I))$, where $N(I)$ denotes the set of neighbours of I
end
Output I ∎

How do we choose S? Luby's idea is to include each vertex v independently with probability $\frac{1}{\deg(v)}$ and delete from every edge in S the vertex with smallest degree. The select step can be implemented using $\mathcal{O}(m)$ processors and takes time $\mathcal{O}(\log n)$ in each run. It takes an expected number of $\mathcal{O}(\log n)$ executions of the while loop for the algorithm to terminate, so the overall running-time is $\mathcal{O}(\log^2 n)$. Analyzing the algorithm, Luby observed that it only requires the random insertions of vertices into S to be *pairwise* independent. Using the construction described in section 4 we obtain a sample space of size $\mathcal{O}(n^2)$. Now exhaustive search of this sample space can replace the random select step.

BIBLIOGRAPHY AND REMARKS.

Maximum Independent Sets. Boppana and Halldórsson[55] proved that INDEPENDENTSET for general graphs is approximable within $\mathcal{O}\left(\frac{n}{\ln^2 n}\right)$. Various special cases have been studied. Results include a PTAS for IS in planar graphs[29] and an approximation algorithm with ratio $\mathcal{O}(\Delta)$ [45] as well as an $\mathcal{APX}$-completeness proof for IS in graphs with degree bounded by $\Delta \geq 3$ [256],[46]. For an account of these special cases see the survey article of Halldórsson [145] and the compendium of Crescenzi and Kann[83].
For a K_r-free graph $\mathcal{H}$ with average degree $t = \frac{2m}{n}$ Shearer [291],[292] proved $\alpha(\mathcal{H}) = \Omega(\frac{t \ln t - t + 1}{(t-1)^2}n)$ if $r = 3$ and $\alpha(\mathcal{H}) = \Omega(c_r \frac{\ln t}{t \ln \ln t}n)$ if $r \geq 4$. Bollobás [52] observed that Shearer's theorem for $r = 3$ can be proved via analyzing a random greedy algorithm. For a graph $\mathcal{H}$ and a vertex $v \in V(\mathcal{H})$ let $\mathcal{H} \setminus v$ denote the subgraph of $\mathcal{H}$ obtained by deleting v and all incident edges. The algorithm RANDOM REMOVAL produces an independent set by randomly choosing $v \in V(\mathcal{H})$ and setting $\mathcal{H} := \mathcal{H} \setminus v$ until $\mathcal{E}(\mathcal{H}) = \emptyset$. If we do not select v at random but choose a vertex with maximum degree instead, RANDOM REMOVAL turns into the so-called MAX algorithm. Let $F(\mathcal{H}) := \sum_{v \in V(\mathcal{H})} \frac{1}{\deg(v)+1}$, where $\deg(v)$ denotes the number of edges containing v. If x is a vertex of maximum degree Δ and $N(v)$ denotes the set of vertices adjacent to x, then $F(\mathcal{H} \setminus x) - F(\mathcal{H}) = \sum_{v \in N(x)} (\frac{1}{\deg(v)} - \frac{1}{\deg(v)+1}) - \frac{1}{\Delta+1} \geq \Delta \cdot \frac{1}{\Delta(\Delta+1)} - \frac{1}{\Delta+1} = 0$, so F is nondecreasing during the MAX algorithm. At termination point $\mathcal{H}$ consists of isolated vertices only, and $F(\mathcal{H})$ simply counts the vertices of an independent set. By comparing the initial and the final value of F we obtain the bound of of Caro [60] and Wei [326],

$\alpha(\mathcal{H}) \geq \sum_{v \in V} \frac{1}{\deg(v)+1}$.

Caro and Tuza [61] generalized this result to $k + 1$-uniform hypergraphs by showing that $\alpha(\mathcal{H}) \geq \sum_{v \in V} f(\deg(v))$, where $f(x) := \prod_{i=1}^{x} \left(1 - \frac{1}{ik+1}\right)$. Note that graphs with identical degree sequences can considerably differ in their independence number. (Take, for example, $\mathcal{H}_1$ as the complete bipartite graph $K_{l,l}$ and $\mathcal{H}_2$ as the union of two cliques of size l joined by a perfect matching, then $\alpha(\mathcal{H}_1) = l$ while $\alpha(\mathcal{H}_2) = 2$.) For results on the independence number of random graphs see Bollobás [52]. Note also that Goldberg and T.Spencer[131] have given an efficient parallel algorithm for computing an independent set in graphs of size at least the Turán bound. Applications of Sidon sets to problems in additive number theory are given in the paper of Baltz, Schoen and Srivastav [30].

Maximal Independent Sets. Karp, Upfal, and Wigderson [180] developed a randomized algorithm for finding a maximal independent set in an independence system. Their algorithm can be adapted to compute a MIS in a general hypergraph in time $\mathcal{O}(\sqrt{n}(\log n + \log m))$ on $m \cdot n$ processors. Faster parallel algorithms have been described by Goldberg and T.Spencer [129],[130]. Dahlhaus and Karpiński [84], and, independently, Kelsen [183] have given efficient $\mathcal{NC}$-algorithms for computing MIS in 3-uniform hypergraphs (journal version in [85]). Beame and Luby [34] studied the MIS problem for hypergraphs, where the size of each edge is bounded by a fixed absolute constant c. They presented a generalized version of the randomized PARALLELMIS-algorithm running on $\mathcal{O}(n + c)$ processors with running time polynomial in $\log(n + c)$, as verified by Kelsen [184]. They also gave a similar algorithm for the general case of k-uniform hypergraphs, and conjectured a running-time polynomial in $\log(n + m)$ on $\mathcal{O}(n \cdot m)$ processors. Luczak and Szymańska [218] observed that a linear hypergraph, i.e. a hypergraph in which each pair of edges has at most one vertex in common, contains a large subhyperpgraph without vertices of large degree. They proved that the randomized algorithm of Beame and Luby finds a maximal independent set in a linear hypergraph in polylogarithmic expected running-time if a preprocessing step is added to make the algorithm perform on such an "equitable" subhypergraph. A derandomized version of their result has been given in the Ph.D. thesis of Edita Szymańska [313].

11. SEMIDEFINITE PROGRAMMING

The natural relaxation of an integer linear program is a linear program. But for some $\mathcal{NP}$-hard problems it is most natural to model them as a quadratic integer optimization problem of the form

$$(\text{QP}) \quad \begin{array}{ll} \min & x^T Q x + c^T x \\ \text{s.t.} & Ax = b, \ x \in \{0, 1\}^n, \end{array}$$

where $Q \in \mathbb{Q}^{n \times n}$, $A \in \mathbb{Q}^{m \times n}$, and $b \in \mathbb{Q}^m$. A straightforward relaxation of (QP) is $x \in \mathbb{R}^n$ and $\|x\|_2 \leq 1$. However, if the objective function is non-convex even the relaxed minimization problem becomes $\mathcal{NP}$-hard [244]. Therefore it is important to choose a relaxation which can be solved in polynomial time. Semidefinite relaxations of quadratic optimization problems arising from $\mathcal{NP}$-hard graph-theoretical problems turned out to be most promising.

11.1 Maximum Cuts in Graphs.

In the previous sections we considered optimization problems which have a nice representation as an integer linear program. But for cut-problems, for example, the corresponding integer linear programs involve dependencies among the variables and therefore can hardly be analyzed by randomized rounding techniques. Boppana [54] and Poljak and Rendl [262], [263], [264] observed that cut-problems can be represented as semidefinite non-linear programs in a natural way. They proposed the first rounding algorithms for the graph bisection and the max-cut problem, solving a relaxation of the non-linear program and rounding the fractional solution in a certain way. The striking new idea of Goemans and Williamson [127] in this context is to round the fractional solution of a non-linear semidefinite program by randomly selecting a hyperplane and then to derandomize.

With this technique Goemans and Williamson improved the previously known approximation factor of $(\frac{1}{2} + o(1))$ for the max-cut problem – which stood for 20 years – to 0.87856.

Problem MAXCUT: Let $G = (V, E)$ be a graph, $|V| = n$ and $|E| = m$. Let $w : E \to \mathbb{Q}^+$ be edge weights. A cut is a partition (S, S^c) of V. Let δS denote the edges between S and S^c. For vertices i, j let w_{ij} be edge weights with the convention that $w_{ij} = 0$, if (i, j) is not an edge. Let $w(\delta S)$ be the sum of the weights of edges in δS. The MAXCUT-problem is the problem of finding a cut of maximum weight.

The MAXCUT-problem is $\mathcal{NP}$–hard and $\mathcal{MAX} - \mathcal{SNP}$–hard [256]. Håstad [149] proved that it is even $\mathcal{NP}$–hard to approximate MAXCUT within a factor better than $16/17$. For planar graphs [142] and K_5-free graphs [32] the MAXCUT problem can be solved in polynomial time. Gonzales and Sahni [134] presented a 0.5-factor approximation for MAXCUT.

As a warming-up exercise, let us show the 0.5-approximation with the method of conditional expectations. Consider the function $F : [-1, 1]^n \to \mathbb{R}$ with $F(x_1, \dots, x_n) := \frac{1}{2} \sum_{i<j} w_{ij}(1 - x_i x_j)$. For a vector $x \in \{-1, 1\}^n$, the set $S := \{i \in V, x_i = 1\}$ determines a cut in V and $F(x) = w(\delta S)$. Let $X_1, \dots, X_n$ be $-1, +1$-valued independent random variables with $\mathbb{P}[X_i = 1] = \frac{1}{2} = \mathbb{P}[X_i = -1]$ for all i. By X we denote the vector $(X_1, \dots, X_n)^T$.

Theorem 11.1 (Sahni, Gonzales 1976) *Let OPT be the weight of a maximum cut.*

(i) $\mathbb{E}(F(X)) \geq \frac{OPT}{2}$.

(ii) *A vector* $x \in \{-1, 1\}^n$ *satisfying* $F(x) \geq \frac{OPT}{2}$ *can be constructed in* $O(n^3)$ *time.*

Proof. (i)

$$
\begin{aligned}
\mathbb{E}(F(X)) &= \frac{1}{2} \sum_{i<j} w_{ij}(1 - \mathbb{E}(X_i X_j)) \\
&= \frac{1}{2} \sum_{i<j} w_{ij}(1 - \mathbb{E}(X_i)\mathbb{E}(X_j)) \\
&= \frac{1}{2} \sum_{i<j} w_{ij} \\
&= \frac{w(E)}{2} \geq \frac{OPT}{2}
\end{aligned}
$$

(ii) Directly apply the method of conditional expectations. ∎

11.1.1 The Randomized Algorithm.

Let us look more closely at the approach of Goemans and Williamson. The MAXCUT problem is equivalent to the following quadratic program

$$
\text{(QP)} \qquad \begin{aligned} &\max \tfrac{1}{2} \sum_{i<j} w_{ij}(1 - y_i \cdot y_j) \\ &y_i \in \{-1, 1\} \quad \forall i \in V. \end{aligned}
$$

We consider the following relaxation on the n-dimensional unit sphere S_n:

$$
\text{(RQP)} \qquad \begin{aligned} &\max \tfrac{1}{2} \sum_{i<j} w_{ij}(1 - v_i \cdot v_j) \\ &v_i \in S_n \quad \forall i \in V. \end{aligned}
$$

A similar relaxation was introduced by Boppana [54] for the graph bisection problem. The randomized algorithm is:

Algorithm RANDOMMAXCUT

(i) Solve (RQP). Let v_i $(i = 1, \ldots, n)$ be an optimal set of vectors.

(ii) Let r be a random vector uniformly distributed on S_n.

(iii) Output $S = \{i;\ v_i \cdot r \geq 0\}$. ∎

In other words we "round" the fractional solution by separating the v_i's with a random hyperplane H through the origin with r as its normal.

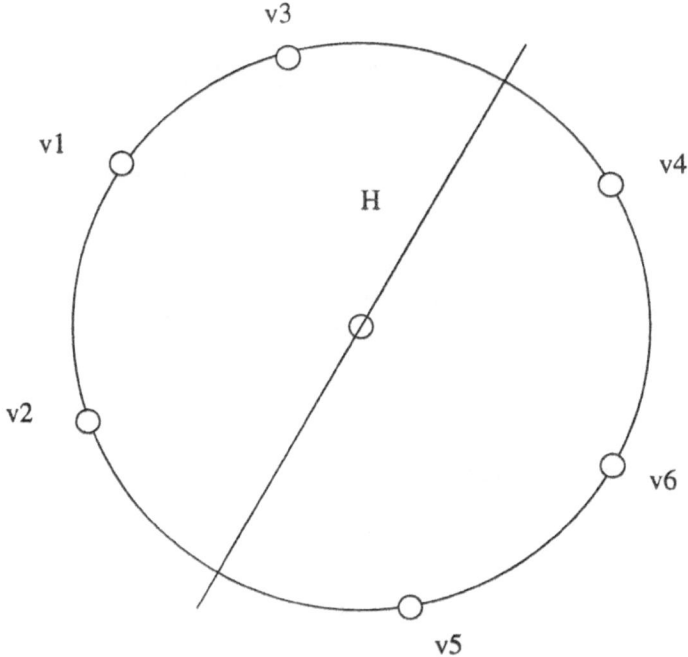

Let Z_R denote the optimal solution of (RQP) and let Z_{OPT} be the optimal value of the maximum cut.

Theorem 11.2 (Goemans, Williamson 1994) *Let W be the weight of the random cut δS. Then $\mathbb{E}(W) \geq \alpha Z_R \geq \alpha Z_{\mathrm{OPT}}$, where $\alpha \geq 0.87856$.*

Proof. Let H be the random hyperplane through the origin with normal r.

By symmetry, $\mathbb{P}[H$ separates v_i and $v_j] = 2\mathbb{P}[v_i \cdot r \geq 0, v_j \cdot r < 0]$. The probability that H separates v_i and v_j is directly proportional to the angle $\theta_{ij} := \arccos(v_i v_j)$.

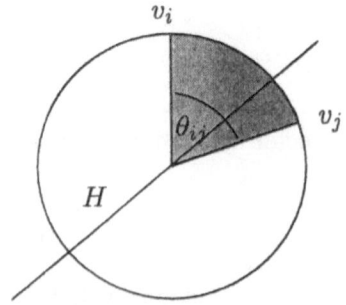

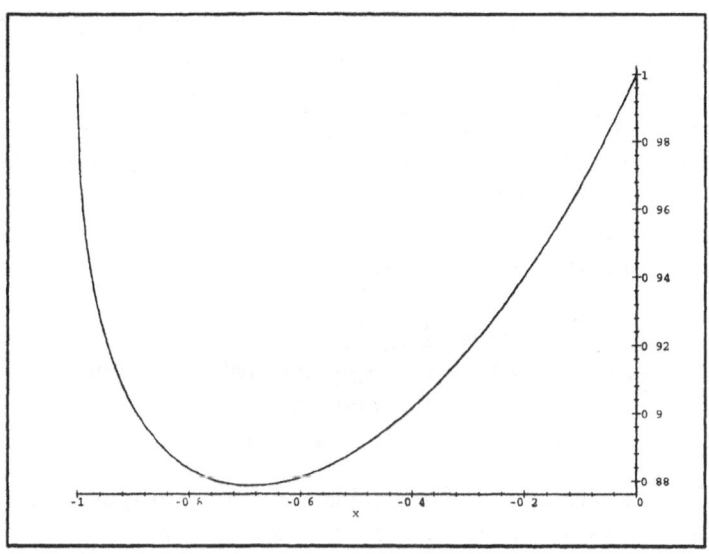

Figure 18.4   Calculating α; $x = \cos\phi$

In fact, the set $\{r \in \mathbb{R}^n; v_i \cdot r \geq 0, v_j \cdot r < 0\}$ is the intersection of two halfplanes with dihedral angle θ_{ij}. Its intersection with S_n is a spherical digon of measure $\frac{\theta_{ij}}{2\pi}$, so $\mathbb{P}[v_i \cdot r \geq 0, v_j \cdot r < 0] = \frac{\theta_{ij}}{2\pi}$, and we have

$$
\begin{aligned}
\mathbb{E}(W) &= \sum_{i<j} w_{ij}\mathbb{P}[H \text{ separates } v_i \text{ and } v_j] \\
&= \sum_{i<j} w_{ij}\frac{\theta_{ij}}{\pi} \qquad\qquad\qquad\qquad (11.1) \\
&= \sum_{i<j} w_{ij}\frac{2}{\pi}\cdot\frac{\theta_{ij}}{1-\cos\theta_{ij}}\cdot\frac{1-\cos\theta_{ij}}{2}
\end{aligned}
$$

For $\alpha := \min\{\frac{2}{\pi}\cdot\frac{\phi}{1-\cos\phi}; 0 < \phi < \pi\}$ we obtain using elementary calculus that $\alpha \geq 0.87856$ (see figure 11.1). Hence,

$$
\begin{aligned}
\mathbb{E}(W) &\geq \alpha\sum_{i<j} w_{ij}\frac{1-\cos\theta_{ij}}{2} \\
&= \alpha\sum_{i<j} w_{ij}\frac{1-v_iv_j}{2} \quad \text{(with } v_iv_j = \cos\theta_{ij}\text{)} \\
&= \alpha Z_R.
\end{aligned}
$$

Before we proceed to derandomization, let us show why the program (RQP) can be solved in polynomial time. Let $v_1, \ldots, v_n$ be a solution to (RQP). If we define a matrix $Y = (y_{ij})$ as the Gram matrix of the vectors $v_1, \ldots, v_n$ (so $y_{ij} := v_i \cdot v_j = v_i^T v_j$) then we have $y_{ii} = 1$ for all i and Y is positive semidefinite. On the other hand, for each positive semidefinite matrix Y with $y_{ii} = 1$ for all i, we can find with the Cholesky decomposition orthonormal vectors $v_1, \ldots, v_n$ such that $y_{ij} = v_i^T v_j$. The set $Q := \{Y; Y$ positive semidefinite, $y_{ii} = 1,\ Y$ symmetric$\}$ is convex, thus (RQP) is equivalent to the problem

$$
\text{(SDP)} \quad
\begin{aligned}
&\max \quad \sum_{(i,j) \in E} \frac{1 - y_{ij}}{2} \\
&\text{s.t.} \quad Y = (y_{ij}) \text{ positive semidefinite,} \\
&\qquad\ \ Y \text{ symmetric,} \\
&\qquad\ \ y_{ii} = 1 \text{ for } i = 1, \ldots, n.
\end{aligned}
$$

Since (SDP) is a special case of the problem of finding the global maximum of a linear function over a convex set, known as a semidefinite program, it can be solved in polynomial time, for example with the ellipsoid method [140].

11.1.2 Derandomization.

In principle, with the method of conditional expectations we can construct a cut S of weight $W(\delta S) \geq \mathbb{E}(W)$, *if* there is a sequential procedure to determine the hyperplane H, or equivalently to construct the vector r. But it is not clear at all, how a vector r uniformly distributed on the sphere can be constructed deterministically.

Recall, that in the algorithms of Sahni and Gonzales we had a much simpler situation since the components of the vector X were binomially distributed.

Goemans and Williamson argue as follows: Let $r = (r_1, \ldots, r_n)^T$. First we determine r_n and r_{n-1}, then proceed to find r_{n-2}, r_{n-3}, and so on. In polar coordinates

$$
r_n = \gamma_n \cos \theta_n \qquad \text{and} \qquad r_{n-1} = \gamma_n \sin \theta_n,
$$

with $\theta_n \in \left[-\frac{\pi}{2}, \frac{\pi}{2}\right]$ and $\gamma_n = \sqrt{r_n^2 + r_{n-1}^2}$. So θ_n and γ_n are random variables and it is known (see [189]) that they are independent. For a set of n vectors $z_1, \ldots, z_n \in \mathbb{R}^n$ and unit edge weights let

$$
f(z_1, \ldots, z_n) := \sum_{i < j} \frac{\arccos(z_i \cdot z_j)}{\pi}.
$$

According to the notation $\mathbb{E}(W) = f(v_1, \ldots, v_n)$ we put $f_{\theta_n}(v_1, \ldots, v_n) := \mathbb{E}(W|\theta_n)$, where $\mathbb{E}(W|\theta_n)$ is the expected weight under the condition that the angle θ_n is given. Then

$$
\begin{aligned}
f(v_1, \ldots, v_n) &= \mathbb{E}(W) = \mathbb{E}(\mathbb{E}(W|\theta)) \\
&= \frac{1}{\pi} \int_{-\frac{\pi}{2}}^{\frac{\pi}{2}} f_\theta(v_1, \ldots, v_n) d\theta \\
&\leq \max_{-\frac{\pi}{2} \leq \theta \leq \frac{\pi}{2}} f_\theta(v_1, \ldots, v_n).
\end{aligned}
\tag{11.2}
$$

In the first step we compute for θ_n a value, say β_1, which maximizes $f_{\theta_n}(v_1, \ldots, v_n)$. (For the moment, we assume that an appropriate discretization of $\left[-\frac{\pi}{2}, \frac{\pi}{2}\right]$ is given, so that this computation can be done in finite time.) Next, the iteration step must be defined. We would like to find vectors, say $v_1^{(1)}, \ldots, v_n^{(1)} \in \mathbb{R}^{n-1}$ and $r^{(1)} \in \mathbb{R}^{n-1}$ such that

$$
f_{\theta_n}(v_1, \ldots, v_n) = f\left(v_1^{(1)}, \ldots, v_n^{(1)}\right).
$$

Now, as above, compute θ_{n-1} and $v_1^{(2)}, \ldots, v_n^{(2)}$. Since the dimension of the vectors $v_j^{(i)}$ for $j = 1, \ldots, n$ in each step i goes down by one, finally we end up with scalars, say $y_1, \ldots, y_n$. If in addition all vectors computed have norm 1, then $y_i \in \{-1, 1\}^n$, hence $f(y_1, \ldots, y_n)$ is the value of the cut δS, where $S = \{i; y_i = 1\}$. Because in each step a maximization is performed

$$
\begin{aligned}
\mathbb{E}(W) &= f(v_1, \ldots, v_n) \\
&\leq f\left(v_1^{(1)}, \ldots, v_n^{(1)}\right) \\
&\vdots \\
&\leq f(y_1, \ldots, y_n) = W(\delta S),
\end{aligned}
$$

and we are done.

So, neglecting numerical problems with the method, the main problem is to find $v_1^{(1)}, \ldots, v_n^{(1)}$ such that

$$
f_{\theta_n}(v_1, \ldots, v_n) = f\left(v_1^{(1)}, \ldots, v_n^{(1)}\right).
\tag{11.3}
$$

Goemans and Williamson define the $v_i^{(1)}$'s as follows: (remember that we have chosen $\theta_n = \beta_1$). Let $v_{i,n} = t \cos \gamma_n$ and $v_{i,n-1} = t \sin \gamma_n$ and define $w = t \cos(\beta_1 - \gamma_n)$. Thus w depends on the derandomized components of r and the vectors $v_{i,n}$ and $v_{i,n-1}$. Put

$$
\bar{r}^{(1)} := \begin{pmatrix} r_1 \\ \vdots \\ r_{n-2} \\ X_n \end{pmatrix}
\quad \text{and} \quad
\bar{v}_i^{(1)} := \begin{pmatrix} v_{i,1} \\ \vdots \\ v_{i,n-2} \\ w \end{pmatrix},
$$

and let

$$r^{(1)} := \frac{\bar{r}^{(1)}}{\|\bar{r}^{(1)}\|} \quad \text{and} \quad v_i^{(1)} := \frac{\bar{v}_i^{(1)}}{\|\bar{v}_i^{(1)}\|}.$$

Under the assumption that γ_n is normally distributed as a function of θ, (11.3) can be proved. Unfortunately, this assumption is wrong! As pointed out by Mahajan and Ramesh [221], γ_n is "half-gamma" distributed and there is a counter example for (11.3).

The direction in which one has to look for a correct derandomization procedure is indicated by the proof of Theorem 11.2. We already know that

$$\mathbb{E}(W) = \sum_{i<j} \mathbb{P}[sgn(r \cdot v_i) \neq sgn(r \cdot v_j)].$$

Instead of reducing the dimension of the vectors v_i, Mahajan and Ramesh sequentially determine the components of r. Finally, they get a vector s. Why does this procedure work? Define for $R \in \mathbb{R}^n$ the function $g_R(v_1, \dots, v_n)$ by

$$g_R(v_1, \dots, v_n) = \sum_{i<j} \mathbb{P}[sgn(R \cdot v_i) \neq sgn(R \cdot v_j)].$$

Suppose that we are entering step i and the first $i-1$ components of the normal vector r have been chosen to, say $s_1, \dots, s_{i-1}$. Now s_i is the value for r_i which maximizes the function

$$t \mapsto g_r(v_1, \dots, v_n | r_1 = s_1, \dots, r_{i-1} = s_{i-1}, r_i = t).$$

Actually, this maximum cannot be computed exactly, but it can be computed approximately in polynomial time within an error of at most $\varepsilon \mathbb{E}(W)$, for any given fixed $\varepsilon > 0$. Now, because $g_s(v_1, \dots, v_n)$ is nothing else but the weight of the cut induced by the hyperplane through the origin with normal vector s, we have

$$g_s(v_1, \dots, v_n) \geq \mathbb{E}(W)(1 - n\varepsilon).$$

Taking a sufficiently small ε, we are done:

Theorem 11.3 (Mahajan and Ramesh 1995) *A cut with value at least* $0.87856 Z_R$ *can be computed in polynomial time.*

Note that a kind of parallel counterpart of the derandomized computation of maximum cuts has been obtained by Pantziou, Spirakis, and Zaroliagis [254].

11.2 Approximate Graph Coloring.

A legal vertex coloring of a graph $G = (V, E)$ is an assignment of colors to its vertices such that no two adjacent vertices receive the same color. G is

k-colorable if there exists a legal vertex coloring using not more than k colors. It is well-known that determining the chromatic number χ of a graph (i.e. the minimum number of colors needed for a legal vertex coloring) is $\mathcal{NP}$-hard [120]. Lund and Yannakakis [219] showed that even approximating χ within a factor of n^ε, for a small constant $\varepsilon > 0$, implies $\mathcal{P} = \mathcal{NP}$. Another hardness result due to Feige and Kilian [111] states that it is impossible to approximate the chromatic number within $n^{1-\delta}$, for any $\delta > 0$, unless $\mathcal{P} = \mathcal{RP}$. Khanna, Linial, and Safra [186] have shown that it is $\mathcal{NP}$-hard to find a 4-coloring of a graph G that is known to be 3-colorable. Following Karger, Motwani, and Sudan [171] let us see how semidefinite programming and randomized rounding can serve to handle the problem of coloring a k-colorable graph in polynomial time with almost as few colors as possible.

A Vector Relaxation

Let $G = (V, E)$ be a k-colorable graph with maximum degree Δ. Relaxing our original problem we ask for a vector k-coloring of G, i.e. an assignment φ of unit vectors to the vertices of G such that the inner product $\varphi(v) \cdot \varphi(w) \leq -\frac{1}{k-1}$ whenever $v, w \in V$ are adjacent. The vector coloring represents a hypothetical "fractional k-coloring" which we will "round" to a legal coloring. Observe that every k-colorable graph is vector k-colorable. On the other hand, there are graphs with chromatic number $n^{\Omega(1)}$ which have bounded vector chromatic number [171].

Let us consider the following semidefinite program:

$$
\begin{aligned}
\min \quad & \alpha \\
\text{s.t.} \quad & M = (m_{ij}) \text{ positive semidefinite,} \\
& M \text{ symmetric,} \\
& m_{ij} \leq \alpha \text{ if } \{i, j\} \in E, \\
& m_{ii} \doteq 1 \text{ for } i = 1, \dots, n.
\end{aligned}
$$

Given a solution M, we can apply the Cholesky decomposition to find a matrix U such that $UU^T = M$. It is not hard to see that the rows of U are vectors forming a vector k-coloring.

Rounding via Random Hyperplane Partitions.

Let us focus on the case $k = 3$. Our crucial task is to algorithmically transform the vector coloring into an almost legal "semicoloring". We define a k-semicoloring of G as a legal assignment of k colors to at least half of the vertices of G. An approximate coloring can be obtained from a semicoloring algorithm in a very natural way:

- semicolor G

- let S be the set of uncolored vertices

- keep the colors of vertices in $V \setminus S$ and semicolor the subgraph induced by S using a new set of colors.

- iterate this procedure until G is completely colored.

Karger, Motwani, and Sudan [171] proved that this method increases the number of colors used by the semicoloring algorithm by no more than a constant factor, if the semicoloring algorithm consumes a polynomial number of colors. Such an algorithm is given by the following extension of the random hyperplane method.

Let $\varphi : V \to \mathbb{R}^n$ denote the vector coloring. Independently choose r random hyperplanes through the origin to partition $\mathbb{R}^n$ into at most 2^r distinctly colored regions and assign to each vertex i the color of the region its vector $\varphi(i)$ lies in. Uncolor one vertex of each edge $\{i, j\}$ that is not "cut" by one of the r hyperplanes such that $\varphi(i)$ and $\varphi(j)$ fall into distinct regions.

Putting $r := 2 + \lceil \log_3 \Delta \rceil$ Markov's inequality guarantees that with probabil-

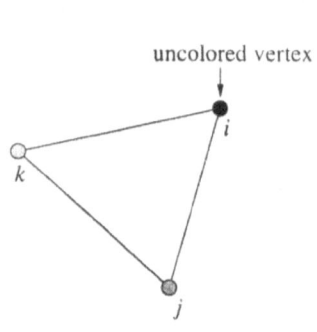

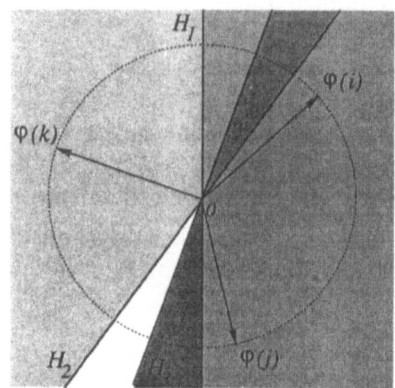

Figure 18.5 Colored Regions

ity at least $\frac{1}{2}$ we get a set of more than $\frac{3n}{4}$ legally colored vertices and hence a semicoloring. Thus we have:

Theorem 11.4 *A 3-colorable graph with n vertices and maximum vertex degree Δ can be colored using $\mathcal{O}(\Delta^{\log_3 2}) \leq \mathcal{O}(n^{0.631})$ colors in polynomial randomized time.*

This result can be improved by a technique of Wigderson [328]. The idea is to bring the maximum degree below a threshold value δ at the cost of at most $\frac{2n}{\delta}$ colors by successively 2-coloring and removing the neighbors of each vertex with degree $\geq \delta$. The optimal choice of δ is around $n^{0.631}$ which implies a coloring using $\mathcal{O}(n^{0.387})$ colors.

Rounding via Vector Projections.

By using a different rounding strategy we get an improved result which applies to general k-colorable graphs.

Theorem 11.5 (Karger, Motwani, Sudan 1998) *Any k-colorable graph on n vertices with maximum degree Δ can be colored, in probabilistic polynomial time, using* $\min\{\mathcal{O}(\Delta^{1-\frac{2}{k}}\sqrt{\ln\Delta}\log n), \mathcal{O}(n^{1-\frac{3}{k+1}}\sqrt{n})\}$ *colors.*

Proof. Again, we start with a vector k-coloring φ. Choose a random n-dimensional vector r from the n-dimensional standard normal distribution. Fix a parameter $c := \sqrt{2\frac{k-2}{k}\ln\Delta}$ and consider the subgraph H induced by the vertices v with $\varphi(v) \cdot r \geq c$. We can think of these vertices as being "captured" by the random projection. Intuitively, since the vectors assigned to adjacent vertices point away from each other, H is likely to contain few edges only. An analysis shows that in fact deleting all the edges from S yields an independent set of size $\Omega(\frac{n}{\Delta^{1-\frac{2}{k}}\sqrt{\ln\Delta}})$ with high probability. We can now assign one color to the vertices in this independent set and iterate on the remaining graph to get an $\mathcal{O}(\Delta^{1-\frac{2}{k}}\sqrt{\ln\Delta})$ coloring. Application of Wigderson's technique concludes the proof. ∎

Note that the algorithm of Karger, Motwani, and Sudan has been derandomized by Mahajan and Ramesh [221].

Bibliography and Remarks.

Maximum Cuts. The work of Goemans and Wiliamson has initiated intense research activities, and brought a new direction in the design of approximation algorithms in combinatorial optimization. As the derandomization is technically difficult already for the MaxCut problem, most of the results obtained by the semidefinite programming technique are randomized approximation algorithms. For MaxCut the presently best approximation factor which is slightly better than 0.87856 is due to Feige, Karpinski and Langberg [110]. For a more comprehensive discussion of semidefinite programming techniques for MaxCut and related problems, like Max-k-Sat, $k = 2, 3, 4$ Max-DiCut and Max-Not-All-Equal-3-Sat we refer to the STOC'99 article of Zwick [331], and the survey article of Feige [108]. Zwick [331] gave a derandomized approximation algorithm improving the Goemans/Williamson factor. The basic random hyperplane method has been further developed to handle semidefinite programs with constraints. Jerum and Frieze [114] analyzed the MaxBisection and the Max-k-Cut problem. For the Max-3-Sat problem, Håstad [149] showed that there cannot exist a polynomial time ρ-approximation algorithm with $\rho > \frac{7}{8}$. Karloff and Zwick [174] gave a simple algorithm (with an involved analysis) for which the $\frac{7}{8}$-approximation guarantee is conjectured. Andersson [22] improved the Max-Cut approximation further and obtained a $(1 - \frac{1}{k} + \Theta(k^{-3}))$-approximation.

Dense Subgraphs. For the DenseSubgraph problem, which given a graph on n vertices and an integer $k \leq n$ seeks for an induced subgraph on k vertices with maximum number of edges, randomized approximation algorithms have been given by Goemans [124], Feige and Seltser [112] with linear randomized rounding, and via semidefinite programming for $k = \Omega(n)$ by

Srivastav and Wolf [312] and Langberg [197]. Extending techniques of [312] Langberg proved the presently best approximation factor which is slightly larger than $1/c$ for $k = n/c, c \geq 2$.

This list is by no means complete! There might be more than about 100 papers in the area of semidefinite programming driven approximation algorithms. Here Feige [108] and Zwick [331] give a more comprehensive list of references.

Minimum Cuts. Though the minimum cut problem can be solved in polynomial time, it is an interesting problem to derive parallel algorithms for the MINCUT problem. Karger and Motwani [170] gave a parallel $(2 + \varepsilon)$-approximation algorithm for the minimum cut problem via pairwise independence and derandomization.

12. ε-APPROXIMATIONS AND LINEAR PROGRAMMING

The theory of derandomization has been quite successful in the design of polynomial-time algorithms in computational geometry. An illuminating example is the problem of computing the convex hull of n points in $\mathbb{R}^d$ in optimal running time of $\mathcal{O}(n^{\lfloor \frac{d}{2} \rfloor})$ by Chazelle (1993). In 1988 Clarkson and Shor [79] gave the first randomized algorithm with optimal running time and five years later Chazelle [63] published the derandomized version, which is a tour d'horizon through probabilistic methods in computational geometry. In almost all state-of-the-art derandomized algorithms in computational geometry the efficient computation of ε-nets, a concept introduced by Haussler and Welzl [152], plays a central role. For example, for linear programming in fixed dimension an elegant derandomized algorithm using ε-approximations has been presented by Chazelle and Matoušek [67].

12.1 ε-Approximations.

In computational geometry the notion of range spaces is widely used, while in combinatorics the same object is called a set system or hypergraph. A range space is a pair $(X, \mathcal{R})$, where X is a set of points, the ground set, and $\mathcal{R}$ is a set of subsets of X. The elements of $\mathcal{R}$ are called ranges. X and $\mathcal{R}$ may be finite or infinite, for example take $X = \mathbb{R}^2$ and $\mathcal{R}$ as the set of all halfspaces. In this section we assume that X itself is a range, i.e. $X \in \mathcal{R}$. For $Y \subseteq X$ let $\mathcal{R}_Y = \{R \cap Y \mid R \in \mathcal{R}\}$ and $(Y, \mathcal{R}_Y)$ the range space induced by Y.

Definition 12.1 (Haussler, Welzl 1987) *Let $(X, \mathcal{R})$ be a range space. A subset $A \subseteq X$ is called ε - net for $(X, \mathcal{R})$, if for all $R \in \mathcal{R}$ with $|R| > \varepsilon |X|$, $A \cap R \neq \emptyset$.*

Presently, the only way to compute ε-nets deterministically is via ε-approximations. ε-approximations are special ε-nets with nice properties not shared by arbitrary ε-nets: for example ε-approximations are stable under divide-and-conque arguments.

Definition 12.2 *Let $(X, \mathcal{R})$ be a range space. A subset $A \subseteq X$ is called an ε - approximation for $(X, \mathcal{R})$, if for all $R \in \mathcal{R}$*

$$\left| \frac{|A \cap R|}{|A|} - \frac{|R|}{|X|} \right| \leq \varepsilon.$$

In the following let $(X, \mathcal{R})$ be a range space with $|X| = n$ and $|\mathcal{R}| = m$. Matoušek, Welzl and Wernisch [233] observed that a 2-coloring with small discrepancy gives a small ε-approximation. Indeed, let $\chi : X \to \{-1, +1\}$ be a coloring with $\text{disc}(\varkappa, \chi) = \delta$, and say $A' = \chi^{-1}(1)$ is larger than $\chi^{-1}(-1)$ (see figure below where $X = A \cup B \cup C$ and $A' = A \cup B$). Thus $A' \geq \lceil \frac{n}{2} \rceil$, and on removing some points from A' we obtain a set A of size $\lceil \frac{n}{2} \rceil$.

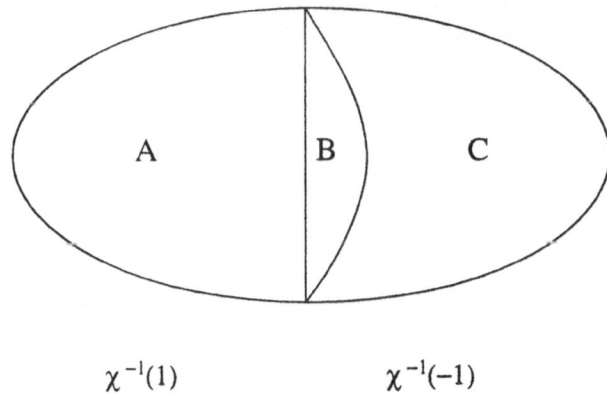

$$\chi^{-1}(1) \qquad\qquad\qquad \chi^{-1}(-1)$$

It is not difficult to show

Proposition 12.3 *A is a $\frac{2\delta}{n}$-approximation.*

For $m = n$ the six-standard-deviation result of Spencer [298] ensures the existence of a 2-coloring with disrepancy $6\sqrt{n}$. By Proposition 12.3, with $r = \frac{\sqrt{n}}{12}$, there is a $(1/r)$-approximation of size $\mathcal{O}(r^2)$. And, since the discrepancy of the set system associated to a Hadamard matrix is $\Omega(\sqrt{n})$ [297], $\Omega(r^2)$ is the lower bound for the size of $(1/r)$-approximations. But for arbitrary range spaces the algorithmically reachable size presently is only $\mathcal{O}(r^2 \log m)$. Let us first describe an algorithm for the computation of a $(1/r)$-approximation of size $\mathcal{O}(r^2 \log m)$ in deterministic polynomial time. The probabilistic "algorithm" is very simple:

Algorithm ε-GENERATE.

Input: Let $r \geq 1$, $s \geq 8r^2(\lceil \log 4m \rceil + 4)$ and $p = \frac{s}{n}$.

Output: A set $A \subset X$ of size s.

- Choose a set $A' \subset X$ performing a sequence of Bernoulli trials, where a point from X is chosen with probability p, independently.

- If $|A'| = s$ then $A := A'$, otherwise remove (resp. add) points from (resp. to) A' such that the resulting set has size s. This set is A.

∎

The reader is invited to show with the Alon-Spencer inequality (Theorem 2.6):

Proposition 12.4 *A is a $(1/r)$-approximation for $(X, \mathcal{R})$ of size s with probability at least $\frac{1}{2}$.*

Its derandomized version has been derived by Matoušek [225].

Theorem 12.5 (Matoušek 1991) *Let $(X, \mathcal{R})$ be a range space with $|X| = n$ and $|\mathcal{R}| = m$. For every $r \geq 1$ and $s \geq 8(\lceil \log(4m) \rceil + 4)$ an $(1/r)$-approximation A of size $|A| = s$ can be computed in $\mathcal{O}(n^2 + \sum_{R \in \mathcal{R}} |R|)$ time.*

Matoušek showed a time bound of $\mathcal{O}(n^2 + mn)$. A close inspection of Matoušek proof reveals the improved time bound $\mathcal{O}(n^2 + \sum_{R \in \mathcal{R}} |R|)$, which can also be derived by an algorithmic version of the Alon-Spencer inequality [305].

A remarkable observation is that in bounded VC-dimension the size of the approximation can be choosen independently of the size of the range space. From Theorem 5.8 and Proposition 12.3 it follows that an $(1/r)$-approximation of size $\mathcal{O}(r^2 \log r)$ can be computed in polynomial time, provided the dual shatter function is bounded by $\mathcal{O}(m^d)$, while in the case of bounded primal shatter function we only have an existence result for such an $(1/r)$-approximation due to Theorem 5.6. Matoušek showed that even in this case a $(1/r)$-approximation of size $\mathcal{O}(r^2 \log r)$ can be efficiently computed by a divide-and-conquer strategy: First the ground set is partitioned in small sets, then approximations in the set system induced by the partitions are computed, and finally some of the partitions and their approximations are combined. A computational problem needs our attention. For a subset $Y \subseteq X$ we would like to list all sets $\mathcal{R}_Y$ efficiently. Such a procedure is called a subsystem oracle. Note that $|\mathcal{R}_Y|$ is bounded by $\pi_{\mathcal{R}}(|Y|)$. A subsystem oracle of dimension d is a procedure which lists all the sets of $\mathcal{R}_Y$ in $\mathcal{O}(|Y|^{d+1})$ time.

Theorem 12.6 (Matoušek 1995, 1990) *Let $(X, \mathcal{R})$ be a set system with a subsystem oracle of dimension d and primal shatter function $\pi_{\mathcal{R}}(n) = \mathcal{O}(n^d)$. For every $1 \leq r \leq n$ an $(1/r)$-approximation of size $\mathcal{O}(r^2 \log r)$ can be constructed in $\mathcal{O}(n(r^2 \log r)^d)$ time.*

Since the divide-and-conquer strategy per se is accessible for parallelization, parallel counterparts of the above presented algorithms can be given (see *Bibliography and Remarks*). The presently most efficient parallel algorithm is due to Goodrich [137]. It has been derived by using a slightly different version of ε-approximation: a set $A \subseteq X$ is a δ -relative ε-approximation for $(X, \mathcal{R})$ if

$$\left| \frac{|R \cap A|}{|A|} - \frac{|R|}{|X|} \right| \leq \delta \frac{|R|}{|X|} + \varepsilon \tag{12.1}$$

for all $R \in \mathcal{R}$.

Theorem 12.7 (Goodrich 1996) *Let* $(X, \mathcal{R})$ *be a range space with VC-dimension* d, $d > 0$ *is some constant, and let* $|X| = n$. *Also let* $1 < r < n$ *be a given parameter and let* $\varepsilon > 0$ *be any fixed constant. Then, in the CRCW PRAM model, for some constant* $c > 0$ *a* $(\log \log n)^{-b}$-*relative* $(1/r)$-*approximation of* $(X, \mathcal{R})$ *of size* $\Theta(r^{2+\varepsilon})$ *resp.* $\Theta(r^2 \log r)$ *can be constructed in* $\mathcal{O}(1)$ *time using* $\mathcal{O}((nr)^c)$ *resp.* $\mathcal{O}((nr)^{c \log r})$ *parallel processors.*

12.2 Linear Programming in Fixed Dimension.

Linear programming, in small dimensions, is one area where the advantages of randomization, namely speed and simplicity, are clearly evident. In fact the best known deterministic algorithms for linear programming in fixed dimension have been obtained by derandomization. Let $A \in \mathbb{Q}^{n \times d}$, $b \in \mathbb{Q}^n$ and $c \in \mathbb{R}^d$. The linear programming problem (minimization form) in dimension d is to minimize $c^T x$ such that $Ax \leq b$ where $x = (x_1, x_2, \ldots, x_d)^T \in \mathbb{Q}^d$. Throughout this section we will assume d to be the number of variables, H to be the set of the given constraints with $|H| = n$. Most of the algorithms in this field are based on two techniques known as random sampling and incremental construction. We first give a variant of the algorithm first proposed by Clarkson(see [76],[77] and [78]) and then present a derandomized version by Chazelle and Matoušek [67]. For incremental algorithms we refer to [243]. Note that the algorithms presented here treat operands as real numbers on which an arithmetic operation can be performed in constant time. Any deviation from this model will be explicitly stated. To keep the analysis simple we make the following assumptions:

   (i) The polyhedron given by H is nonempty and bounded.

   (ii) The objective is to minimize $c^T x = x_1$.

   (iii) The minimum occurs at a unique vertex of the polyhedron.

   (iv) Each vertex of the polyhedron is defined by exactly d constraints.

It is known that these assumptions can be removed by using standard techniques [77] [289].

Clarkson's Algorithm and Derandomization.

A few definitions are needed before we describe the algorithms. For a subset $S \subseteq H$ let $\mathsf{LP}(S)$ be the optimum value of the objective function subject to S. A subset $B \subseteq H$ is called *basis* if $\mathsf{LP}(B) > -\infty$ and $\mathsf{LP}(\hat{B}) < \mathsf{LP}(B)$ for any $\hat{B} \subset B$. By assumption 1 a basis exists. The *basis of a subset* $G \subseteq H$, denoted by $\mathsf{B}(G)$, is a minimal subset $B \subseteq G$ such that $\mathsf{LP}(B) = \mathsf{LP}(G)$. Observe that $|\mathsf{B}(H)| = d$ because of the assumptions 1, 3 and 4. For a subset $S \subseteq H$, we say a constraint $h \in H$ *violates* $\mathsf{LP}(S)$, if $\mathsf{LP}(S \cup \{h\}) > \mathsf{LP}(S)$. The first algorithm RANDLP uses random sampling to collect $\mathsf{B}(H)$ and discard redundant constraints efficiently. The algorithm takes H as input, performs at most d phases to construct $\mathsf{B}(H)$ and returns $\mathsf{LP}(\mathsf{B}(H))$.

Algorithm RANDLP

1 If $n \leq 9d^2$ then use the simplex algorithm to find $\mathsf{LP}(H)$, else

   1.1 $V = H$; $S = \emptyset$.

   1.2 Perform the following steps until $V = \emptyset$

      1.2.1 Choose $R \subset H\backslash S$ at random such that $|R| = r = \min\{2d\sqrt{n}\log n, |H\backslash S|\}$.

      1.2.2 $opt = $ RANDLP$(R \cup S)$.

      1.2.3 $V = \{h \in H \mid opt$ violates $h\}$.

      1.2.4 If $|V| \leq \sqrt{n}$ then $S = S \cup V$.

   1.3 return opt.

It is easy to see that after a run through steps 1.2.1 to 1.2.4, if $V \neq \emptyset$ then V may not contain all of $\mathsf{B}(H)$ but it will surely contain at least one constraint of $\mathsf{B}(H)$ that is not already in S. It is also known that the running time of the simplex algorithm is $\mathcal{O}(d^{\frac{d}{2}+\mathcal{O}(1)})$ if the number of constraints is at most $9d^2$. We will see that with high probability V will contain at most $\sqrt{n}$ constraints.

Theorem 12.8 *The linear programming problem with n constraints in $\mathbb{R}^d$ can be solved in* $(\log_d n)^{\log d} \mathcal{O}(d^{\frac{d}{2}+\mathcal{O}(1)}) + \mathcal{O}(d^2 n)$ *expected time.*

Proof. Suppose we ran the algorithm on a linear programming problem with n constraints and in some phase we obtained $opt = v$ such that $|V| > \sqrt{n}$. So we repeat the steps and again choose constraints from $H\backslash S$ at random. The probability that a constraint $h \in H\backslash S$, chosen at random, lies in V is at least $|V|/n$. Since we choose r constraints again, the probability that none of these constraints lies in V is bounded from above by

$$\left(1 - \frac{|V|}{n}\right)^r \leq \left(1 - \frac{\sqrt{n}}{n}\right)^r = \left(1 - \frac{1}{\sqrt{n}}\right)^{2d\sqrt{n}\log n} = \mathcal{O}\left(\frac{1}{n^{2d}}\right). \quad (12.2)$$

Since there are at most n^d vertices of the polyhedron given by H, the probability of choosing a vertex v for which $|V| > \sqrt{n}$ is at most $\frac{1}{n^d}$. Thus, the probability that $|V| > \sqrt{n}$ is at most 0.5. This implies that steps 1.2.1 to 1.2.4 will be repeated at most twice between successive addition to the set S. After i phases the size of S is at most $i\sqrt{n}$ and the size of R is $2d\sqrt{n}\log n$. Thus, at any step the number of constraints in the recursive calls is at most $3d\sqrt{n}\log n$. Notice that for each of the d phases we perform at most n constraint violation checks at the cost of $\mathcal{O}(d)$ for each check. Using all the facts obtained above we get the following recurrence for $T(n)$, the maximum expected running time of RANDLP:

$$T(n) = \mathcal{O}(d^{\frac{d}{2}+\mathcal{O}(1)}) \quad \text{if } n \le 9d^2 \tag{12.3}$$

$$T(n) \le 2d\,T(3d\sqrt{n}\log n) + \mathcal{O}(d^2 n) \quad \text{if } n > 9d^2 \tag{12.4}$$

Resolving this recurrence yields $T(n) \le (\log_d n)^{\log d}\,\mathcal{O}(d^{\frac{d}{2}+\mathcal{O}(1)}) + \mathcal{O}(d^2 n)$. ∎

We proceed to derandomization. To every linear programming problem we associate a set system $(H, \mathcal{R})$ as follows. For every basis $B \subseteq H$, let $V(B)$ be the set of constraints violated by $\text{LP}(B)$. Let $\mathcal{R} = \{V(B) \mid B \text{ is a basis in } H\}$.

The main idea is to use a $(1/r)$-approximation for $(H, \mathcal{R})$ instead of picking a random sample R everytime. Since this $(1/r)$-approximation contains some constraints from every $V(B) \in \mathcal{R}$ we are able to bound the number of constraints violated in every phase of the algorithm. For the computation of a $(1/r)$-approximation with the algorithm of Matoušek (Theorem 12.6) we need to list all elements from $\mathcal{R}$, i.e. a a subsytem oracle of dimension d is required.

We are now ready to state the algorithm DETLP which takes H as input and returns $\text{LP}(H)$.

Algorithm DETLP

1 If $n \le Cd^4 \log d$ ($C > 0$ is a constant) then use the simplex algorithm to find $\text{LP}(H)$, else

    1.1 Compute $\widehat{R}$, a $(1/r)$-approximation for $(H, \mathcal{R})$ where $r = 4d^2$, of size $\mathcal{O}(dr^2 \log(dr))$ with Matoušek's algorithm (Theorem 12.6) in $\mathcal{O}(n\,d^{3d} r^{2d} \log^d(dr))$ time.

    1.2 $V = H$; $S = \emptyset$.

    1.3 Perform the following steps until $V = \emptyset$

        1.3.1 $opt = \text{DETLP}(\widehat{R} \cup S)$.

        1.3.2 $V = \{h \in H \mid opt \text{ violates } h\}$.

        1.3.3 $H = H \backslash V$; $S = S \cup V$;

`1.4 return` *opt*.

Theorem 12.9 (Chazelle, Matoušek 1996) *The linear programming problem with n constraints in $\mathbb{R}^d$ can be solved in $\mathcal{O}(n\,d^{7d}\log^d d)$ deterministic time.*

Proof. Again observe that if $V \neq \emptyset$ then V will surely contain at least one constraint of $\mathsf{B}(H)$ that is not already in S. Since $|\mathsf{B}(H)| = d$ (assumptions 1,3 and 4) the algorithm terminates after at most d passes through steps 1.3.1 to 1.3.3. The fact that $\widehat{R}$ is a $(1/r)$-approximation, thus is also a $(1/r)$-net for $(H, \mathcal{R})$, implies that any value of variable *opt* which violates more than $\frac{n}{r}$ constraints of H also violates at least one constraint of $\widehat{R}$. But no constraint of $\widehat{R}$ is violated in the algorithm which implies that at the end $|S| \leq \frac{dn}{r} = \frac{n}{4d}$. According to the theorem stated above, constructing $\widehat{R}$ takes at most $\mathcal{O}(n\,d^{7d}\log^d d)$ time and $|\widehat{R}| = \mathcal{O}(d^5 \log d)$. Thus, the algorithm DETLP is recursively called with no more than $\mathcal{O}(d^5 \log d) + \frac{n}{4d}$ constraints in each of the d phases. If, in step 1 we choose the constant C appropriately, then we can achieve $\mathcal{O}(d^5 \log d) + \frac{n}{4d} < \frac{n}{2d}$. The total cost of violation checks is $\mathcal{O}(d^2 n)$ which can be neglected. We have the following recurrence for $T(n)$, the worst case running time of DETLP:

$$T(n) = \mathcal{O}\left(\binom{Cd^4 \log d}{\lceil \frac{d}{2} \rceil}\right) \quad \text{if } n \leq Cd^4 \log d \tag{12.5}$$

$$T(n) \leq dT\left(\frac{n}{2d}\right) + \mathcal{O}(n\,d^{7d}\log^d d) \quad \text{if } n > Cd^4 \log d. \tag{12.6}$$

This recurence gives $T(n) \leq \mathcal{O}(n\,d^{7d}\log^d d)$.

All we need to show is the construction of the subsystem oracle of dimension d. Recall the four assumptions that we made in the beginning. Since the VC-dimension of $(H, \mathcal{R})$ is d by assumption 5, the primal shatter function of such set systems is bounded by $\sum_{i=0}^{d} \binom{m}{i}$ (section 4.3). A vertex P defined by d constraints of H violates a constraint $h \in H$ if the halfplane corresponding to h intersects the horizontal semiline starting from P and travelling rightwards (by assumption 2). In the figure given above FJRSKF is the feasible region and the vertex P violates constraints EX and GY. Thus, any set of $\mathcal{R}$ corresponds to a set of halfplanes that intersect a horizontal semiline travelling to the right. Now assume that we are given a set A of m constraints and we are looking for solutions of the linear programming problem inside the bounding box (ABCD here). Let $\tilde{A}$ be the corresponding set of m halfplanes. If a vertex P violates k constraints of A, then the semiline from P passes through the region defined by the k corresponding halfplanes from $\tilde{A}$ and the halfplanes defining the bounding box. This is the region EFGCBE in the figure. Every set of $\mathcal{R}_A$ occurs in some such region. Since, the primal shatter function is polynomially bounded, it implies that we do limited work (see [67]) in searching such regions. ∎

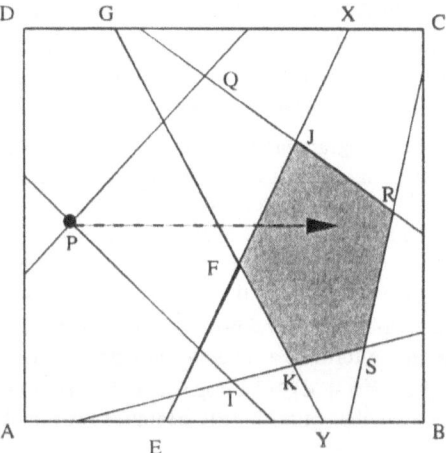

A lot of work has been done in designing parallel algorithms for linear programming. In general, it is hopeless to obtain an efficient parallel algorithm for linear programming, since even approximating the optimal value of a linear program up to a constant factor is $\mathcal{P}$–complete [162]. Thus the hope is to find efficient parallel implimentations in small dimensions. Alon and Megiddo [14] gave a probabilistic parallel algorithm (in the CRCW PRAM model) which solves the given linear programming problem with $\mathcal{O}(n)$ processors with high probability. In [6] they state a deterministic algorithm. The idea here is first to project the constraints in a plane, throw away some of the constraints using techniques of Dyer [94] and then invoke expander graphs to eliminate some more redundant constraints before computing the entire convex hull in parallel (on CRCW PRAM).

Theorem 12.10 (Ajtai, Megiddo 1992) *The linear programming problem with d variables and n linear inequalities can be solved deterministically by n parallel processors in $\mathcal{O}((\log \log n)^d)$ time.*

BIBLIOGRAPHY AND REMARKS. *ε-Approximations.* The first polynomial-time deterministic algorithms for the computation of ε-approximations and ε-nets for set systems with polynomially bounded shatter function were given by Matoušek [225] (see also [67]) and by Matoušek, Welzl and Wernisch [233]. For papers leading to Matoušek's results in [225] we refer to [224], [226], Agarwal [1] and Chazelle and Friedman [65]. Brönnimann, Chazelle and Matoušek [58] modify the algorithm to compute sensitive ε-approximations. They give an algorithm that computes a sensitive $(1/r)$-approximation of size $\mathcal{O}(dr^2 \log(dr))$ in $\mathcal{O}(d)^{3d} r^{2d} \log^d(dr)|X|$ time, and a $(1/r)$-net of size $\mathcal{O}(dr \log(dr))$ in $\mathcal{O}(d)^{3d} r^d \log^d(dr)|X|$ time.
The ideas of Chazelle and Matoušek [67] have been used to develop parallel algorithms for computing ε-nets and ε-approximations. Efficient parallel implementations have been given by Mahajan et al. [222] and by Goodrich [135]. Mahajan et al. [222] use the techniques of Nissan [250] and Karger and Koller [169] to get a $(1/r)$-approximation of size $\mathcal{O}(r^2 \log r)$ in NC, while the size of the previous best approximation was $\mathcal{O}(r^{2+\delta})$. Goodrich [135] presented

algorithms for the EREW PRAM model that can compute an $(1/r)$-approximation of size $\mathcal{O}(r^{2+\delta})$ and $\mathcal{O}(n^{\delta}r^2)$ in time $\mathcal{O}(\frac{\log^2 n}{\log r})$ and $\mathcal{O}(\log n)$ respectively, both with $\mathcal{O}(nr^c)$ work (c is a constant). Goodrich showed fast parallel (see Theorem 12.7) algorithms for computing $(1/r)$-approximations in the CRCW PRAM model.

Linear Programming. Linear programming is a problem where the dependence of the running time on the dimension has been closely studied. Many resaerchers considered the fine-tuning of different techniques to obtain bounds for sequential algorithms with a *sub-exponential* dependence of the running time on the dimension d. Polynomial time deterministic algorithms were given by Khachiyan [185] and Karmarkar [175] but the number of arithmetic operations they perform depends on the precision of the numbers describing the input. In a landmark paper Megiddo [236] gave the first deterministic algorithm with $\mathcal{O}(n2^{2^d})$ running time. The next breakthrough came in the form of a randomized algorithm from Clarkson [77] with the expected running time of $\mathcal{O}(d^2 n + d^{\frac{d}{2}+\mathcal{O}(1)}\log n)$. Kalai [168] presented a randomized algorithm with the expected number of arithmetic operations required bounded from above by $\min\{n^K\sqrt{\frac{d}{\log d}}, n2^K\sqrt{(n-d)\log d}\}$ where K is an absolute constant. Matoušek, Sharir and Welzl [231] have also developed randomized algorithms with sub-exponential dependency on both n and d. For more algorithms on linear programming we refer to the articles of Dyer and Frieze [96] and R. Seidel [289]. The parallel algorithm of Goodrich [137] runs on CRCW PRAM and EREW PRAM respectively. It is based on ideas of Dyer [95] and the parallel computation of ε-nets.

Notes

1. an oracle which given a set $Y \subseteq V$ lists all distinct $e \cap Y, e \in E$ in $\mathcal{O}(|Y|^d)$ time.

2. According to the standard notation of scheduling problems the integral problem can be formalized as $P|\text{res} \cdot \cdot 1, r_j, p_j = 1|C_{\max}$. This notation means: the number of identical processors is part of the input ($P|$), resources are involved (res), the number of resources and the amount of every resource are part of the input (res $\cdot \cdot$), every job needs at most 1 unit of a resource (res $\cdot \cdot 1$), start times are involved (r_j), the processing time of all jobs is equal 1 ($p_j = 1$) and the optimization problem is to schedule the jobs as soon as possible ($|C_{max}$).

3. Packing simply means to find a partitioning of vectors in a minimum number of sets so that the vectors in each partition (= bin) satisfy the upper bound conditions of the definition.

4. problem class $P2|\text{res} \cdot \cdot \cdot, r_j = 0, p_j = 1|C_{\max}$

5. this is due to the $\mathcal{NP}$-completeness of the problem $P2|\text{res} \cdot \cdot 1, r_j, p_j = 1|C_{\max}$.

References

[1] P. K. Agarwal. *Partitioning arrangements of lines: I. An efficient deterministic algorithm.* Discrete Computational Geometry **5** (1990), 449–483.

[2] R. Aharoni, P. Erdös, N. Linial. *Optima of dual integer linear programs.* Combinatorica **8**:1 (1988), pp. 13-20. Preliminary Version: *Dual integer linear programs and the relationship between their optima.* in Proceedings of the 17th Annual ACM Symposium on the Theory of Computing, New York 1985, (STOC '85), pp. 476–483.

[3] N. Ahuja, A. Srivastav. *Resource constrained scheduling and the Lovász Local Lemma.* Preprint, Mathematisches Seminar, Universität Kiel, 2000.

[4] M. Ajtai, J. Komlós, J. Pintz, J. Spencer, E. Szemerédi. *Extremal uncrowded hypergraphs.* Journal of Combinatorial Theory Ser. A **32** (1982), 321–335.

[5] M. Ajtai, J. Komlós, E. Szemeredi. *Deterministic simulation in LOGSPACE.* Proceedings of the 19th Annual ACM Symposium on the Theory of Computing 1987, (STOC '87), pp. 132-140.

[6] M. Ajtai and N.Megiddo. *A deterministic poly*(log log n) *time n-processor algorithm for linear programming in fixed dimension.* SIAM J. Computing **25**:6 (1996), pp. 1171-1195. Preliminary Version in: Proc. 24th Annual ACM Symposium on the Theory of Computing 1992, (STOC '92), pp. 327–338.

[7] N. Alon. *The linear arboricity of graphs.* Israel J. Math. **62** (1988), 311–325.

[8] N. Alon. *A parallel algorithmic version of the Local Lemma.* Random Structures and Algorithms 2:4 (1991), 367–378.

[9] N.Alon, L Babai, A. Itai. *A fast and simple randomized algorithm for the maximal independent set problem.* Journal of Algorithms **7** (1986), 567–583.

[10] N. Alon, Z. Galil. *On the exponent of all pairs shortest path problem.* J. Computer System Sciences **54**:2 (1997), pp. 255-262. Preliminary Version in: Proceedings of the 32nd IEEE Symposium on Foundations of Computer Science 1991, (FOCS '91), pp. 569–575.

[11] N. Alon, Z. Galil, O. Margalit, M. Naor. *Witnesses for matrix multiplication and for shortest paths.* Proceedings of the 33rd IEEE-Symposium on Foundations of Computer Science 1992, (FOCS '92), pp. 417-426.

[12] N. Alon, O. Goldreich, J. Håstad, R. Peralta. *Simple constructions for almost k-wise independent random variables.* Random Structures and Algorithms **3**:3 (1992), 289–304.

[AKS97] N. Alon, J. H. Kim and J. Spencer. *Nearly perfect matching in regular simple hypergraphs.* Israel J. Math. **100** (1997), 171-187.

[13] N. Alon, H. Lefmann, V. Rödl. *On an anti-Ramsey type result.* Colloquia Mathematica Societatis János Bolyai **60**, *Sets, Graphs and Numbers.* 1991, pp. 9–22.

[14] N. Alon, N. Megiddo. *Parallel linear programming in fixed dimension almost surely in constant time.* J. Assoc. Comput. Mach. **41**:2 (1994), 422-434. Preliminary Version in: Proceedings of the 31st IEEE Symposium on Foundations of Computer Science 1990, (FOCS '90), pp. 574–582.

[15] N. Alon, C. McDiarmid, B. Reed. *Star arboricity.* Combinatorica **12**:4 (1992), 375-380.

[16] N. Alon, C. McDiarmid, B. Reed. *Acyclic coloring of graphs.* Random Structures and Algorithms 2:3 (1991), 277–288.

[17] N. Alon, M. Naor. *Derandomization, witnesses for boolean matrix multiplication and construction of perfect hash functions.* Algorithmica **16**:4-5 (1996), 434-449. Preliminary Version in: Proceedings of the 33rd IEEE Symposium on Foundations of Computer Science 1992, (FOCS '92).

[18] N. Alon, J. Spencer, P. Erdös. *The probabilistic method.* John Wiley & Sons, Inc. 1992.

[19] N. Alon, A. Srinivasan. *Improved parallel approximation of a class of linear programming problems.* Algorithmica **17** (1997), 449–462.

[20] N. M. Amato, M. T. Goodrich, E. A. Ramos. *Parallel algorithms for higher dimensional convex hulls.* In Proceedings of the 35th IEEE Symposium on Foundation of Computer Science 1994, (FOCS'94), pp. 683 – 694.

[21] N. M. Amato, M. T. Goodrich, E. A. Ramos. *Computing faces in segment and simplex arrangements.* Proceedings of the 27th annual ACM symposium on the theory of computing (STOC). Las Vegas, NV, USA, May 29 - June 1, 1995. New York, NY: ACM, 672-682 (1995).

[22] G. Andersson. *An approximation algorithm for max-p-section.* Preprint 1998.

[23] A. F. Andreev, A. E. F. Clementi, J. D. P. Rolim. *A new general derandomization method.* J.ACM **45**:1 (1998), 179–213.

[24] D. Angluin, L.G. Valiant. *Fast probabilistic algorithms for Hamiltonion circuits and matchings.* J. Computer System Sciences **18** (1979), 155–193.

[25] R. Armoni. *On the derandomization of space-bounded computations.* in ; Proceedings of the 2nd international workshop on Randomization and approximation techniques in computer science (RANDOM'98). Barcelona, Spain, October 8-10, 1998. Berlin: Springer. Lect. Notes Comput. Sci. **1518** (1998), pp. 47-59.

[26] Arora S. Arora, C. Lund, R. Motwani, M. Sudan, M. Szegedy. *Proof verification and the intractability of approximation algorithms.* ACM **45** (1998), 501-555. Preliminary Version in: Proceedings of the 33rd Annual IEEE Symposium on the Foundations of Computer Science, Pittsburgh, Pennsylvania 1992, (FOCS '92), pp. 14–23.

[27] Y. Azar, R. Motwani, J. Naor. *Approximating probability distributions using small sample spaces.* Combinatorica **18**:2 (1998), 151-171.

[28] K. Azuma. *Weighted sums of certain dependent variables.* Tohoku Math. Journ. **3** (1967), 357–367.

[29] B. S. Baker. *Approximation algorithms for $\mathcal{N}\mathcal{P}$-complete problems on planar graphs.* J.ACM **41** (1994), 153–180.

[30] A. Baltz, T. Schoen, A. Srivastav. *Probabilistic construction of small sum-free sets via large Sidon sets.* To appear in Colloq. Math. 2000. (Preliminary Version in: Proceedings of 3rd. International Workshop on Randomization and Approximation Techniques in Computer Science, Berkeley, California, Springer LNCS 1671, pp. 138 – 143.)

[31] W. Banaszczyk. *Balancing vectors and Gaussian measure of n-dimensional convex bodies.* Random Structures & Algorithms **12** (1998), 351-360.

[32] F. Barahona, M.Grötschel, M.Jünger, G.Reinelt. *An application of combinatorical optimization to statistical physics and circuit layout design.* Operations Research **36** (1988), 493–513.

[33] H. Bauer. *Probability theory and elements of measure theory.* Academic Press, 1981.

[34] P. Beame, M. Luby. *Parallel search for maximal independence given minimal dependence.* Proceedings of the 1st Annual ACM-SIAM Symposium on Discrete Algorithms 1990, (SODA '90), pp. 212–218.

[35] J. Beck. *On 3-chromatic hypergraphs.* Discrete Math. **24** (1978), 127–137.

[36] J.Beck. *Roth's estimate of the discrepancy of integer sequences is nearly sharp.* Combinatorica **1**:4 (1981), 319–325.

[37] J. Beck. *An algorithmic approach to the Lovász Local Lemma I.* Random Structures & Algorithms **2**:4 (1991), 343–365.

[38] J. Beck, W. Chen. *Irregularities of distribution.* Cambridge University Press, 1987.

[39] J.Beck, T. Fiala. *Integer-making theorems.* Disc.Appl.Math. **3** (1981), 1–8.

[40] J. Beck, V. Sós. *Discrepancy theory.* Handbook of Combinatorics, Vol. II, eds.: R.L. Graham, M. Grötschel, L. Lovász; North-Holland, Cambridge 1995, pp. 1405–1446.

[41] J. Beck, J. Spencer. *Balancing matrices with line shifts.* Combinatorica **3**:3 and 4 (1983), 299–304.

[42] C. Berge. *Graphs and Hypergraphs.* North Holland, Amsterdam 1973.

[43] B. Berger, J. Rompel. *Simulating $(log^c n)$-wise independence in NC.* J.ACM **38**:4 (1991), 1026–1046. Preliminary Version in: Proceedings of the IEEE Symposium on the Foundation of Comuter Science (FOCS) 1989.

[44] B. Berger, J. Rompel, P. Shor. *Efficient NC algorithms for set cover with applications to learning and geometry.* J. Computer System Science **49** (1994), 454-477. Preliminary Version in: Proceedings of the 30th Annual IEEE Symposium on the Foundations of Computer Science 1989, (FOCS '89), pp. 54–59.

[45] P. Berman, M. Fürer. *Approximating maximum independent set in bounded degree graphs.* Proceedings of the 5th Annual ACM-SIAM Symposium on Discrete Algorithms 1994, (SODA'94), pp. 365–371.

[46] P. Berman, T. Fujito. *Approximating independent sets in degree 3 graphs.* Proceedings of the 4th Workshop on Algorithms and Data Structures,

824

Springer Lecture Notes in Computer Science (LNCS) 955, 1995, pp. 449–460.

[47] C. Bertram-Kretzberg, H. Lefmann. *The algorithmic aspects of uncrowded hypergraphs.* SIAM J. Computing **29**:1 (1998), 201–230.

[48] D. Bertsimas, R. Vohra. *Rounding algorithms for covering problems.* Math. Program. **80A** (1998), 63-89.

[49] J. Błazewicz, K. Ecker, G. Schmidt, J. Węglarz. *Scheduling in computer and maufacturing systems.* Springer-Verlag, Berlin 1993.

[50] M. Blum, W. Evans, P. Gemmell, S. Kannan and M. Naor. *Checking the correctness of memories.* Proceedings of the 32nd IEEE Symposium on the Foundations of Computer Science 1991, (FOCS '91), pp. 90–99.

[51] M. Blum, S. Micali. *How to generate cryptographically strong sequences of pseudo-random bits.* SIAM J. Computing **13** (1984), 850–864.

[52] B. Bollobás. *Random Graphs.* Academic Press, London 1985.

[53] B. Bollobás. *Modern Graph Theory.* Springer, New York 1998.

[54] R. Boppana. *Eigenvalues and graph bisection: An average case analysis.* Procedings of the 28th Annual IEEE Symposium on the Foundations of Computer Science 1987, (FOCS '87), pp. 39–46.

[55] R. Boppana, M. M. Halldórsson. *Approximating maximum independent set by excluding subgraphs.* Bit **32** (1992), pp. 180–196.

[56] R. P. Brent. *Fast multiple-precision evaluation of elementary functions.* J.ACM **23** (1976), 242–251.

[57] A. Z. Broder, A. M. Frieze, E. Upfal. *Existence and construction of edge disjoint paths on expander graphs.* SIAM J. Computing **23**:5 (1994), 976-989. Preliminary Version in: Proceedings of the 24th Annual ACM Symposium on the Theory of Computing 1992, (STOC '92), pp. 140–149.

[58] H. Brönnimann, B. Chazelle, J. Matoušek. *Product range spaces, sensitive sampling, and derandomization.* SIAM J. Computing **28**:5, (1999), 1552-1575.

[59] H. Brönnimann, M. T. Goodrich. *Almost optimal set covers in finite VC-dimension.* Discrete & Computational Geometry **14**:4 (1995), 463-479. Preliminary Version in: Proceedings of the 10th ACM Symposium on Computational Geometry 1994, pp. 293–302.

[60] Y. Caro. *New results on the independence number.* Technical report, Tel-Aviv University, (1979).

[61] Y. Caro, Z. Tuza. *Improved lower bounds on k-independence.* Journal of Graph Theory **15** (1991), 99–107.

[62] S. Chakrabarti, C. Phillips, A.S. Schulz, D.B. Shmoys, C. Stein, J. Wein. *Improved algorithms for min-sum criteria.* Proceedings of the 23rd International Colloquium on Automata, Languages and Processing, 1996.

[63] B. Chazelle. *An optimal convex hull algorithm in any fixed dimension.* Disc. Comput. Geom. **10** (1993), 377–409.

[64] B. Chazelle. *The Discrepancy Method: Randomness and Complexity.* Cambridge University Press 2000.

[65] B. Chazelle, J. Friedman. *A deterministic view of random sampling and its use in geometry.* Combinatorica **10**:3 (1990), 229–249.

[66] B. Chazelle, L. J. Guibas, D. T. Lee. *The power of geometric duality.* BIT **25** (1985), 76–90.

[67] B. Chazelle, J. Matoušek. *On linear-time deterministic algorithms for optimization problems in fixed dimension.* J. Algorithms **21** (1996), 116-132.

[68] B. Chazelle, J. Matoušek. *Derandomizing an output-sensitive convex hull algorithm in three dimensions.* Computational Geometry: Theor. Appl. **5** (1994), 27–32.

[69] C. Chekuri, R. Motwani, B. Natarajan, C. Stein. *Approximation techniques for average completion time scheduling.* Proceedings of the 7th Annual ACM-SIAM Symposium on Discrete Algorithms 1997, (SODA '97).

[70] W.W. L. Chen, M.M. Skriganov. *Explicite construction in the classical mean squares problem in irregularities of point distribution.* Preprint, Department of Mathematics, Macquarie University, Sydney, 1999.

[71] H. Chernoff. *A measure of asymptotic efficiency for test of a hypothesis based on the sum of observation.* Ann. Math. Stat. **23** (1952), 493–509.

[72] B. Chor, J. Freidmann, O. Goldreich, J. Hastad, S. Rudich, R. Smolensky. *The bit extraction problem and t-resilient functions.* Proceedings of the 26th IEEE Annual Symposium on the Foundations of Computer Science 1985, (FOCS '85), pp. 396–407.

[73] B. Chor, O. Goldreich. *On the power of two-point sampling.* Journal of Complexity **5** (1989), 96–106.

[74] V. Chvatal. *A greedy heuristic for the set covering problem.* Math. Oper. Res. **4** (1979), 233–235.

[75] V. Chvatal. *The tail of the hypergeometric distribution.* Discrete Math. **25** (1979), 285–287.

[76] K.L. Clarkson. *New applications of random sampling in computational geometry.* Discrete & Computational Geometry **2** (1987), 195–222.

[77] K.L. Clarkson. *Las Vegas algorithm for linear programming when the dimension is small.* J. ACM **42**:2 (1995), 488-499. Preliminary Version in: Proceedings of the 29th Annual IEEE Symposium on the Foundations of Computer Science 1988, (FOCS '88), pp. 452–457.

[78] K.L. Clarkson, P. W. Shor. *Applications of random sampling in computational geometry II.* Discrete & Computational Geometry **4**:5 (1989), 423-432. Preliminary Version in: Proceedings of the 4th Annual ACM Symposium on Computational Geometry 1988, pp. 1–11.

826

[79] K. L. Clarkson, P. W. Shor. *Applications of random sampling in computational geometry II.* Discrete & Computational Geometry **4** (1989), 423–432.

[80] S. Cook. *A taxonomy of problems with fast parallel algorithms.* Information and Control **64**:1–3, (1985), Academic Press, New York.

[81] J.G van der Corput. *Verteilungsfunktionen I.* Akad. Wetensch. Amsterdam, Proc., **38** (1935), 813-821.

[82] J.G van der Corput. *Verteilungsfunktionen II.* Akad. Wetensch. Amsterdam, Proc., **38** (1935), 1058-1066.

[83] P. Crescenzi, V. Kann. *A Compendium of $\mathcal{NP}$ optimization problems.* Technical Report SI/RR-95/02 (1995), Department of Computer Science, University of Rome "La Sapienza".

[84] E. Dahlhaus, M. Karpiński. *An efficient algorithm for the 3MIS problem.* Technical Report TR-89-052 (1989), International Computer Science Institute, Berkeley, CA.

[85] E. Dahlhaus, M. Karpiński, P. Kelsen. *An efficient parallel algorithm for computing a maximal independent set in a hypergraph of dimension 3.* Inf. Process. Lett. **42:6** (1992), 309-313.

[86] X. Deng, S. Mahagan. *The cost of derandomization: computability or competitation.* SIAM J. Computing **26** (1997), 786-802.

[87] P. Diaconis. *Group representations in probability and statistics.* IMS Lecture Notes, Monograph Series, Volume 11, Hayward, CA, 1988.

[88] G.-L. Ding, P. D. Seymour, P. Winkler. *Bounding the vertex cover number of a hypergraph.* Combinatorics, Probability and Computing **14** (1994), 23 – 34.

[89] R. Diestel. *Graph Theory.* Springer, New York 1997.

[90] B. Doerr, A. Srivastav. *Approximation of multicolor discrepancy.* in: Proceedings of the 2nd International Workshop on Approximation Algorithms for Combinatorial Optimization Problems, Berkeley, Springer LNCS 1671, pp. 39 – 50.

[91] B. Doerr, A. Srivastav, P. Wehr. *The discrepancy of cartesian products of arithmetric progressions.* Preprint, Mathematisches Seminar, Universität Kiel, 1999.

[92] D.-Z. Du, F.K.Hwang. *Combinatorial group testing and its application.* Series on Applied Mathematics, Volume 3, World Sciencific Publishing Co. Pte. Ltd., Singapore, 1993.

[93] R. A. Duke, H. Lefmann, V. Rödl. *On uncrowded hypergraphs.* Random Structures & Algorithms **6** (1995), 209–212.

[94] M. Dyer. *Linear time algorithms for two and three variable linear prog rams.* SIAM J. Computing **13** (1984), 31–45.

[95] M. Dyer. *A parallel algorithm for linear programming in fixed dimension.* in: Proceedings of the 11th ACM Symposium on Computational Geometry 1995, pp. 345–349.

[96] M. Dyer and A.M. Frieze. *A randomized algorithm for fixed-dimensional linear programming.* Mathematical Programming **44** (1989) 203-212.

[97] M. E. Dyer, A. M. Frieze. *The solution of some random NP-hard problems in polynomial expected time.* Journal of Algorithms **10** (1989), 451-489.

[98] P. Erdös. *Some remarks on the theory of graphs.* Bulletin of the American Mathematics Society **53**, 292–294.

[99] P. Erdös, H. Hanani. *On a limit theorem in combinatorial analysis.* Publ. Math. Debrecen **10** (1963), 10–13.

[100] P. Erdös, L. Lovász. *Problems and results on 3 - chromatic hypergraphs and some related questions.* Infinite and finite sets, A. Hajnal et al., Colloq. Math. Soc. J. Bolyai, Vol. II. (1975), 609–627.

[101] P. Erdös, J.L. Selfridge. *On a combinatorial game.* J. Combinatorial Theory, Ser.A **14** (1973), 298–301.

[102] P. Erdös, P. Turán. *On a problem of Sidon in additive number theory and on some related problems.* J. London Math. Soc. **16** (1941), 212-215.

[103] G. Even, O. Goldreich, M. Luby, N. Nisan, B. Veliv̆ković. *Efficient approximation of product distributions.* Random Structures & Algorithms **13**:1 (1998), 1-16. Preliminary Version *Approximation of general independent distributions.* in: Proceedings of the 24th Annual ACM Symposium on Theory of Computing 1992, (STOC '92), pp. 10–16.

[104] S. Even, A. Itai, A. Shamir. *On the complexity of timetable and multicommodity flow problems.* SIAM J. Computing **5**:4 (1976), 691–703.

[105] V. Faber, L. Lovász. *Problem 18* Lecture Notes in Mathematics 411, 1974, p.284, C. Berge and D.K. Ray-Chaudhri(eds.), Springer Verlag, Berlin/New York.

[106] T. Feder, E. Kushilevitz, M. Naor. *Amortized communication complexity.* SIAM J. Computing **24**:4 (1995), 736-750. Preliminary Version in: Proceedings of the 32nd IEEE Symposium on the Foundation of Computer Science 1991, (FOCS91), pp. 239 – 248.

[107] U. Feige. *A treshold of* ln *n for approximating set cover.* in: Proceedings of the 28th Annual ACM Symposium on the Theory of Computing 1996, (STOC '96), pp. 314–318.

[108] U. Feige. *Randomized rounding for semidefinite programs - Variations of the* MAXCUT *example.* in: Proceedings of the 2nd International Workshop on Approximation Algorithms for Combinatorial Optimization Problems, Berkeley 1999, Springer LNCS 1671, pp. 189–196.

[109] U. Feige, M.X. Goemans. *Approximating the value of two prover proof systems, with applications to* MAX 2SAT *and* MAX DICUT. Proceedings

of the 3rd Israel Symposium on the Theory of Computing and Systems 1995, pp. 182–189.

[110] U. Feige, M. Karpinski, Michael Langberg. *Improved Approximation of MAX-CUT on Graphs of Bounded Degree.* Electronic Colloquium on Computational Complexity **21** (2000).

[111] U. Feige, J. Kilian. *Zero knowledge and chromatic number.* J. Computer System Sciences **57**:2 (1998), 187-199. Preliminary Version in: Proceedings of the 11th Annual Conference on Structure in Complexity Theory 1996.

[112] U. Feige and M. Seltser. *On the densest k-subgraph problem.* Technical report, Department of Applied Mathematics and Computer Science, The Weizmann Institute, Rehovot, September 1997.

[113] P. Frankl, V. Rödl. *Near perfect coverings in graphs and hypergraphs.* European Journal of Combinatorics **6** (1985), 317–326.

[114] A. Frieze, M. Jerrum. *Improved approximation algorithms for* MAX k-CUT *and* MAX BISECTION. Algorithmica **18** (1997), 67–81.

[115] A.M. Frieze, R. M. Karp, B. Reed. *When is the assignment bound tight for the asymmetric traveling-salesman problem.* SIAM J. Computing **24**:3 (1995), 484-493. Preliminary Version: Preprint 1992, Research Institute for Discrete Mathematics, University of Bonn.

[116] Z. Füredi. *Maximum degree and fractional matchings in uniform hypergraphs.* Combinatorica **1** (1981), 155–162.

[117] Z. Füredi, J. Kahn, P.D. Seymour. *On the fractional matching polytope of a hypergraph.* Combinatorica **13**:2 (1993), 167–180.

[118] A. D. Fundia. *Derandomizing Chebyshev's Inequality to find Independent Sets in Uncrowded Hypergraphs.* Random Structures & Algorithms **8** (1996), 131–147.

[119] M. R. Garey, R. L. Graham, D. S. Johnson, A.C.-C. Yao. *Resource constrained scheduling as generalized bin packing.* J. Combinatorial Theory Ser. A **21** (1976), 257–298.

[120] M. R. Garey, D. S. Johnson. *Computers and Intractability.* W. H. Freeman and Company, New York (1979).

[121] G. R. Garey, D. S. Johnson, L. Stockmeyer. *Some simplified NP-complete graph problems.* Theoretical Computer Science **1** (1967), 237–267.

[122] N. Garg, J. Könemann. *Faster and simpler algorithms for multicommodity flow and other fractional packing problems.* Technical Report, Max-Planck-Institut für Informatik, Saarbrücken, Germany (1997).

[123] N. Garg, V.V. Vazirani, M. Yannakakis. *Primal-Dual algorithms for integral flow and multicut in trees, with applications to matching and set cover.* Algorithmica **18** (1997), 3–20.

[124] M.X. Goemans. *A supermodular relaxation for scheduling with release dates.* Proceedings of the 5th MPS Integer Programming and Combinatorial Optimization Conference 1996, pp. 288-301.

[125] M.X. Goemans. *Improved approximation algorithms for scheduling with release dates.* Proceedings of the 8th Annual ACM-SIAM Symposium on Discrete Algorithms 1997, (SODA '97), pp. 591–598.

[126] M.X. Goemans, M. Mihail, V. Vazirani, D. Williamson. *A primal-dual approximation algorithm for generalized Steiner network problems.* Combinatorica **15:3** (1995), 435-454. Preliminary Version in: Proceedings of the 25th Annual ACM Symposium on the Theory of Computing 1993, (STOC '93), pp. 708–717.

[127] M.X. Goemans, D. Williamson. *Improved approximation algorithms for maximum cut and satisfiability problems using semidefinite programming.* J. ACM **42**:6 (1995), 1115-1145. Preliminary Version *.878-approximation algorithms for MAX CUT and MAX 2SAT.* in: Proceedings of the 26th Annual ACM Symposium on the Theory of Computing 1994, (STOC '94), pp. 422–431.

[128] L. M. Goldberg, M. Paterson, A. Srinivasan, E. Sweedyk. *Better approximation guarantees for job-shop scheduling.* Accepted for publication in SIAM Journal on Discrete Mathematics. Preliminary Version in: Proceedings of the 7th ACM-SIAM Symposium on Discrete Algorithms, 1997 (SODA'97), pp. 599 – 608.

[129] M. Goldberg, T. Spencer. *A new parallel algorithm for the maximum independent set problem.* SIAM J. Computing, **18** (1989), 419–427.

[130] M. Goldberg, T. Spencer. *Constructing a maximal independent set in parallel.* SIAM J. Discrete Math. **2** (1989), 322–328.

[131] M. Goldberg, T. Spencer. *An efficient parallel algorithm that finds independent sets of guaranteed size.* SIAM J. Discrete Math. **6** (1993), 443–459.

[132] S. Goldwasser, S. Micali. *Probabilistic Encryption.* J. of Computer and System Sciences **28** (1984), 270–299. Preliminary version in Proceedings STOCS'82, pp. 365-377.

[133] G. H. Golub, C.F. van Loan. *Matrix Computations.* The John Hopkins University Press, Baltimore, London (1993).

[134] T. Gonzales, S. Sahni. *P-complete approximation.* J. ACM **23** (1976), 555–565.

[135] M. T. Goodrich. *Geometric partitioning made easier, even in parallel.* Proceedings of the 9th Annual ACM Symposium on Computational Geometry 1993, pp. 73–82.

[136] M. T. Goodrich. *Constructing arrangements optimally in parallel.* Discrete & Computational Geometry **9** (1993), 371–385.

830

[137] M. T. Goodrich. *Fixed-dimensional parallel linear programming via relative ε-approximations*. Proceedings of the 7th Annual ACM-SIAM Symposium on Discrete Algorithms 1996, (SODA '96), pp. 132–141.

[138] K. Gopalakrishnan; D.R. Stinson. *Derandomization*. in: Colbourn, Charles J. (ed.) et al., The CRC handbook of combinatorial designs. Boca Raton, CA: CRC Press. CRC Press Series on Discrete Mathematics and its Applications. pp. 558-560, 1996.

[139] D.A. Grable. *Nearly-perfect hypergraph packing is in NC*. Information Processing Letters **60**:6 (1996), 295–299.

[140] M. Grötschel, L. Lovász, A. Schrijver. *Geometric algorithms and combinatorial optimization*. Springer-Verlag 1988.

[141] M. Habib, C. McDiarmid, J. Ramirez-Alfonsin and B. Reed. *Probabilistic methods for algorithmic discrete mathematics*. Springer Series *Algorithms and Combinatorics, Vol. 16*, Springer Verlag, 1998.

[142] F. Hadlock. *Finding a maximun cut of a planar graph in polynomial time*. SIAM J. Computing **4** (1975), 221–225.

[143] L.A. Hall. *Approximation algorithms for scheduling*. Approximation algorithms for NP-hard problems, D.S.Hochbaum(ed.), PWS Publishing Company, Boston 1997, pp. 1–46.

[144] L.A. Hall, A.S. Schulz, D.B. Shmoys, J. Wein. *Scheduling to minimize average completion time: Off-line and on-line approximation algoritms*. Math. Oper. Res. **22**:3 (1997), 513-544.

[145] M. M. Halldórsson. *Approximations of independent sets in graphs*. in: Proceedings of the 2nd International Workshop on Approximation Algorithms for Combinatorial Optimization Problems, Berkeley, Springer LNCS 1671, pp. 1–14.

[146] M. M. Halldórsson. *Approximations of independent set variants and hereditary subset problems*. COCON'99, Springer Lecture Notes in Computer Science (LNCS) 1627, 1999 pp. 261-270.

[147] Y. Han. *Parallel derandomization techniques*. in: Advances in Parallel Algorithms, eds.: L. Kronsjö, D. Shumsheruddin, Blackwell Sci. Publ., Oxford, 1992, pp. 368 – 395.

[148] J. Håstad. *Clique is hard to approximate within $n^{1-\varepsilon}$*. accepted for publication in Acta Mathematica. Preliminary Version in: Proceedings of the 37th IEEE Symposium on Foundations of Computer Science 1996, (FOCS '96), pp. 627-636.

[149] J. Håstad. *Some optimal inapproximability results*. Proceedings of the 28th Annual ACM Symposium on the Theory of Computing, El Paso, Texas 1997 (STOC '97), pp. 1–10.

[150] J. Håstad, R. Impagliazzo, L. Levin, M. Luby. *Construction of a pseudorandom generator from any one-way function*. ICSI Technical Report No. 91-068, 1997.

[151] J.Håstad, R. Impagliazzo, L. Levin, M. Luby. *A pseudo-random generator from any one-way function.* SIAM J. Computing **28**:4 (1999), 1364-1396.

[152] D. Haussler, E. Welzl. ε - *nets and simplex range queries.* Discrete & Computational Geometry **2** (1987), 127–151.

[153] E. Hlawka. *Theorie der Gleichverteilung.* Bibliographisches Inst., Mannheim, Wien, 1979.

[154] D.S. Hochbaum. *Approximation algorithms for $\mathcal{N}P$–hard problems.* D.S. Hochbaum, ed., PSW Publishing Company, Boston, 1997.

[155] W. Hoeffding. *Probability inequalities for sums of bounded random variables.* J. Amer. Statist. Assoc. **58** (1963), 13–30.

[156] T. Hofmeister, H. Lefmann. *Derandomization for sparse approximations and independent sets.* Springer Lecture Notes in Computer Science (LNCS) **969** (1995), 201–210.

[157] T. Hofmeister, H. Lefmann. *Approximating maximum independent sets in uniform hypergraphs.* Springer Lecture Notes in Computer Science (LNCS) **1450** (1998), 562–570.

[158] I. Holyer. *The NP-copmpleteness of edge coloring.* SIAM J. Computing **10**:4 (1981), 718–720.

[159] S. Hougardy, H.J. Prömel. *A 1.598 approximation algorithm for the Steiner problem in graphs.* Proceedings of the 10th Annual ACM-SIAM Symposium on Discrete Algorithms 1998, (SODA '98), pp. 448–453.

[160] R. Impagliazzo, A. Wigderson. *P=BPP unless E has sub-exponential circuits, derandomizing the XOR Lemma.* Proceedings of the 29th Annual ACM Symposium on the Theory of Computing 1997, (STOC '97).

[161] R. Impagliazzo, D. Zuckerman. *How to recycle random bits.* Proceedings of the 31st Annual IEEE Symposium on the Foundations of Computer Science 1990, (FOCS '89), pp. 248–253.

[162] J. JaJa. *An introduction to parallel algorithms.* Addison-Wesley Publishing Company, Reading, Massachusetts, 1992.

[163] S. Janson, T. Łuczak, A. Ruciński. *Random Graphs.* Wiley-Interscience Series in Discrete Mathematics and Otimization, John Wiley & Sons, Inc. New York Toronto 2000.

[164] M. R. Jerrum, A. J. Sinclair. *Approximating the permanent.* SIAM. J. Computing **18** (1989), 1149–1178.

[165] M. Jerrum, G.B. Sorkin. *The Metropolis algorithm for graph bisection.* Discrete Appl. Math. **82**:1-3 (1998), 155-175. Preliminary Version *Simulated annealing for graph bisection.* in: Proceedings of the 34th IEEE Annual Symposium on the Foundation of Computer Science 1993, Palo Alto, California, (FOCS '93), pp. 94–103.

[166] A. Joffe. *On a sequence of almost deterministic k-independent random variables.* Annals of Probability **2** (1974), 161-162. Preliminary Version in: Proceedings of the AMS **29** (1971), 381–382.

[167] D. S. Johnson. *Approximation algorithms for combinatorial problems.* J. Computer System Sciences **9** (1974), 256–278.

[168] G. Kalai. *Linear programming, the simplex algorithm and simple polytopes.* Math. Programming **79B**:1-3 (1997), 217-233. Preliminary Version *A sub-exponential randomized simplex algorithm.* in: Proceedings of the 24th Annual ACM Symposium on the Theory of Computing 1992, (STOC '92), pp. 475–482.

[169] D. Karger, D. Koller. *(De)randomized construction of small sample spaces in NC.* J. Computer System Sciences **55**:3 (1997) 402-413. Preliminary Version in: Proceedings of the 35th IEEE Symposium on the Foundations of Computer Science 1994, (FOCS '94), pp. 252–263.

[170] D. Karger, R. Motwani. *Derandomization through approximation: an NC-algorithm for minimum cuts.* SIAM J. Computing **26**:1 (1997) 255-272. Preliminary Version in: Proceedings of the 26th Annual ACM Symposium on the Theory of Computing 1994, (STOC '94), pp. 497– 506.

[171] D. Karger, R. Motwani, M. Sudan. *Approximate Graph Coloring by Semidefinite Programming.* J. ACM **45** (1998), 246–265.

[172] H. Karloff, Y. Mansour. *On construction of $k-wise$ independent random variables.* Combinatorica **17**:1 (1997) 91-107. Preliminary Version in: Proceedings of the 26th ACM Symposium on the Theory of Computing 1994, (STOC '94), pp. 564–573.

[173] H. Karloff, P. Raghavan. *Randomized algorithm and pseudorandom numbers.* J. ACM **40**:3 (1993), pp. 454-476.

[174] H. Karloff, U. Zwick. *A 7/8-approximation for* MAX-3-SAT*?* Proceedings of the 38rd Annual IEEE Symposium on Foundations of Computer Science, Miami Beach, Florida 1997, (FOCS '97), pp. 406–415.

[175] N. Karmarkar. *A new polynomial time algorithm for linear programming.* Combinatorica **4** (1984), 373–395.

[176] R. M. Karp. *Reducibility among combinatorial problems. Complexity of Computer Computations,* R.E. Miller and J.W. Thatcher (eds.), Plenum Press, New York 1972, 85–103.

[177] R. M. Karp. *An introduction to randomized algorithms.* Discrete Appl Math. **34** (1991), 165–201.

[178] R. M. Karp, F. T. Leighton, R. L. Rivest, C. D. Thompson, U. V. Vazirani, V. V. Vazirani. *Global wire routing in two-dimensional arrays.* Algorithmica **2** (1987) 113-129. Preliminary Version in: Proceedings of the 24th IEEE Annual Symposium on the Foundation of Computer Science 1983, (FOCS '83), pp. 453–459.

[179] R. Karp, N. Pippenger, M. Sipser. appaears in M. Sipser, *Expanders, Randomness, or Time versus Space.* in: Proceedings of the First Annual Conference on Structure in Complexity Theory, 1986, pp. 325-329.

[180] R. M. Karp, E. Upfal, A. Wigderson. *The complexity of parallel search.* J. Computer and System Sciences **36** (1988), 225–253.

[181] R. M. Karp, A. Wigderson. *A fast parallel algorithm for the maximal independent set problem.* J. ACM **32** (1985), 762–773. Preliminary Version in: Proceedings of the 16th ACM Symposium on Theory of Computing 1984, (STOC '84), pp. 266–272.

[182] M. Kaufmann, J.F. Sibeyn, T. Suel. *Derandomizing algorithms for routing and sorting on meshes.* Proceedings of the 5th Annual ACM-SIAM Symposium on Discrete Algorithms 1994, (SODA '94), pp. 669–679.

[183] P. Kelsen. *An efficient parallel algorithm for finding an mis in hypergraphs of dimension 3.* Manuscript, Department of Computer Sciences, University of Texas, Austin, TX, 1990.

[184] P. Kelsen. *On the parallel complexity of computing a maximal independent set in a hypergraph.* in: Proceedings of the 24th ACM Symposium on the Thoery of Computing 1992, (STOC '92), pp. 339–350.

[185] L.G. Khachiyan. *Polynomial algorithm in linear programming.* U.S.S.R Comput. Math. and Math. Phy. **20** (1980), 53–72.

[186] S. Khanna, N. Linial, S. Safra. *On the Hardness of Approximating the Chromatic Number.* in: Proceedings of the 2nd Israeli Symposium on Theory and Computing Systems (1992), 250–260.

[187] J. Kleinberg, E. Tardos. *Routing with low congestion in densely embedded graphs.* Preprint 1996.

[188] J.M. Kleinhans, G.Sigl, F.M. Johannes. *Sea-of-gates placement by simultaneous quadratic programming combined with improved partitioning.* Proceedings of VLSI'89, Elsevier Science Publishers B.V., Amsterdam, The Netherlands, 1989.

[189] D.E. Knuth. *Seminumerical algorithms.* Volume 2 of The Art of Computer Programming, Addison-Wesley, Reading, MA, second edition, 1981.

[190] D. Koller and N. Megiddo. *Constructing small sample spaces satisfying given constraints.* SIAM J. Discrete Math. **7**, No.2 (1994), 260-274. Preliminary Version in: Proceedings of the 25th ACM Symposium on Theory of Computing 1993, (STOC '93), pp. 268–277.

[191] E. Korach, M. Penn. *Tight integral duality gap in the chinese postman problem* Technical report, Computer Science Department, Israel Institute of Technology, Haifa, Revised Version, December 1989.

[192] E. Korach, M. Penn. *A fast algorithm for maximum integral two-commodity flow in planar graphs.* Discrete Appl. Math. **47** (1993), 77–83.

[193] B. Korte, L. Lovász, H.J. Prömel, A. Schrijver (eds.). *Paths, flows, and VLSI-layout.* Springer 1990

834

[KR97] A. Kostochka and V. Rödl. *Partial Steiner systems and matchings in hypergraphs.* Random Struc. and Algorithms **13** (1997), 335-347.

[194] K. L. Krause, V. Y. Shen, H .D. Schwetmann. *Analysis of several job-scheduling algorithms for a model of multiprogramming computer systems.* J. ACM **22** (1975), 522–550. Erratum: JACM **24** (1977), p. 527.

[195] N.N. Kuzyurin. *On the relationship between the optima of linear and integer programming.* Discrete Math. Appl. **2**:3 (1992), 305–311.

[196] R.E. Ladner. *On the structure of polynomial time reducibility.* J. ACM **22** (1975), 155–171.

[197] M. Langberg. *Approximation algorithms for maximization problems arising in graph partitioning.* M.Sc. Thesis, Weizmann Institute of Science, Rehovot, Israel, December 1998.

[198] T. Leighton, B. Maggs, S. Rao. *Packet routing and job-shop scheduling in O(Congestion+Dilation) steps.* Combinatorica **14** (1994), 167–186.

[199] T. Leighton, F. Makedon, S. Plotkin, C. Stein, E. Tardos, S. Tragondas. *Fast approximation algorithms for multicommodity flow problems.* J. Computing System Sciences **50**, No.2 (1995), 228-243. Preliminary Version in: Proceedings of the 23rd Annual ACM Symposium on the Theory of Computing 1991, (STOC '91), pp. 101–111.

[200] T. Leighton, S. Rao. *An approximate max-flow min-cut theorem for uniform multicommodity flow problems with applications to approximation algorithms.* Proceedings of the 29th Annual IEEE Symposium on the Foundation of Computer Science, Los Alamitos 1988, (FOCS '88), pp. 422–431.

[201] T. Leighton, S. Rao, A. Srinivasan. *New algorithmic aspects of the local lemma with applications to routing and partitioning.* in: Proceedings of the 10th ACM-SIAM Symposium on Discrete Algorithms, 1999, (SODA'99), pp. 643 – 652.

[202] C.T. Lengauer. *Combinatorial algorithms for integrated circuit layout.* Wiley-Teubner 1990.

[203] J. K. Lenstra, D. B. Shmoys, E. Tardos. *Approximating algorithms for scheduling unrelated parallel machines.* Math. Programming **46** (1990), 259–271.

[204] J.-H. Lin, J. S. Vitter. *ε-approximations with minimum packing constraint violation.* Proceedings 24th Annual ACM Symposium on the Theory of Computation 1992, Victoria, B.C., Canada, (STOC '92), pp. 771–782.

[205] N. Linial, M. Luby, M. Saks, D. Zuckerman. *Efficient construction of a small hitting set for combinatorial rectangles in high dimension.* Combinatorica **17**, No.2 (1997), 215-234. Preliminary Version in: Proceedings of the 25th ACM Symposium on the Theory of Compututing 1993, (STOC '93), pp. 258–267.

[206] J. H. van Lint. *Introduction to coding theory.* Springer Verlag New York, Heidelberg, Berlin 1982.

[207] L. Lovász. *On the ratio of optimal integral and fractional covers.* Discrete Mathematics **13** (1975), 383–390.

[208] L. Lovász. *Combinatorial problems and exercises.* North Holland and Akadémia Kiadó, Amsterdam 1979.

[209] L. Lovász. *Combinatorial Optimization.* in: Papers from the DIMACS Speecial Year (ed. W. Cook, L. Lovász, P. Seymour). DIMACS Series in Discrete Mathematics and Combinatorial Optimization **20** (1995), Amer. Math. Soc. Providence, pp. 153 – 179.

[210] A. Lubotzky, R. Phillips, P. Sarnak. *Ramanujan Graphs.* Combinatorica **8**:3 (1988), 261–277.

[211] M. Luby. *A simple parallel algorithm for the maximal independent set problem.* SIAM J. Computing **15** (1986), 1036–1053.

[212] M. Luby. *Pseudorandomness and cryptographic applications.* Princeton Computer Science Notes, Princeton Unviversity Press, New Jersey 1996.

[213] M. Luby, N. Nisan. *A parallel approximation algorithm for positive linear programming.* in: Proceedings of the 24th ACM Symposium on the Theory of Computing 1993, (STOC '93), pp. 448–457.

[214] M. Luby, A. Wigderson. *Pairwise independence and derandomization.* International Computer Science Institute, UC Berkeley, TR-95-035, July 1995.

[215] M. Luby, A. Wigderson, B. Velickovic. *Deterministic approximate counting of depth−2 circuits.* in: Proceedings of the Second Israel Symposium on Theory of Computing and Systems 1993, pp.18-24.

[216] M. Luby, B. Velickovic. *On deterministic approximation of DNF.* Algorithmica **16**, No. 4/5 (1996), pp.415-433.

[217] T. Łuczak. *A note on the sharp concentration of the chromatic number of random graphs.* Combinatorica **11**:3 (1991), 295–297.

[218] T. Łuczak, E. Szymańska. *A parallel randomized algorithm for finding a maximal independent set in a linear hypergraph.* J. Algorithms **25**:2 (1997), 311–320

[219] C. Lund, M. Yannakakis. *On the hardness of approximating minimization problems.* J. ACM **41**, No.5 (1994), 960-981 Preliminary Version in: Proceedings of the 25th ACM Symposium on the Theory of Computing 1993, (STOC '93), pp. 286–293.

[220] F.J. MacWilliams, N.J.A. Sloane. *The theory of error correcting codes.* North Holland, Amsterdam 1977.

[221] S. Mahajan and H. Ramesh. *Derandomizing approximation algorithms based on semidefinite programming.* SIAM J. Computing **28**:5 (1999), 1641-1663. Preliminary Version *Derandomizing semidefinite program-*

836

ming based approximation algorithms. in Proceedings of the 36th Annual IEEE Symposium on Foundations of Computer Science (1995), 162-169.

[222] S. Mahajan, E. A. Ramos and K. V. Subrahmanyam. *Solving some discrepancy problems in NC.* Manuscript 1997.

[223] V. M. Malhotra, M. P. Kumar, S. N. Maheshwari. *An $O(|V|^2)$-algorithm for finding maximum flows in networks.* Information Processing Letters **7** (1978), 277–278.

[224] J. Matoušek. *Construction of ε-nets.* Discrete Computational Geometry **5** (1990), 427–448.

[225] J. Matoušek. *Approximations and optimal geometric divide-and-conquer.* J. Computer and System Sciences **50** (1995), 203-208.

[226] J. Matoušek. *Cutting hyperplane arrangement.* Discrete Computational Geometry **6** (1991), 385–406.

[227] J. Matoušek. *Tight upper bounds for the discrepancy of halfspaces.* Discrete & Computational Geometry **13** (1995), pp. 593-601.

[228] J. Matoušek. *Derandomization in computational geometry.* Hanbook of Computational Geometry. Elsevier 2000, pp. 559-595.

[229] J. Matoušek. *Geometric Discrepancy.* Springer Verlag, Heidelberg, New York, 1999.

[230] J. Matoušek. *On the discrepancy for cartesian products.* J. London Math. Soc. **61**:2 (2000), 737-747.

[231] J. Matoušek, M. Sharir and E. Welzl. *A sub-exponential bound for linear programming.* Algorithmica **16**:4-5, (1996), 498-516. Preliminary version in: Proceedings of the 8th ACM Symposium on Computational Geometry 1992, pp. 1–8.

[232] J. Matoušek, J. Spencer. *Discrepancy of arithmetic progressions.* J. Amer. Math. Soc. **9** (1996) 195–204.

[233] J. Matoušek, E. Welzl, L. Wernisch. *Discrepancy and ε-approximations for bounded VC-dimension.* Combinatorica **13**:4, (1991), 455-466.

[234] C. McDiarmid. *On the method of bounded differences.* Surveys in Combinatorics, 1989. J. Siemons, Ed.: London Math. Soc. Lectures Notes, Series 141, Cambridge University Press, Cambridge, England 1989.

[235] C. McDiarmid. *Hypergraph coloring and the Lov'asz Local Lemma.* Discrete Math. **167/168** (1997), 481-486.

[236] N. Megiddo. *Linear programming in linear time when the dimension is fixed.* J. ACM **31** (1984), 114–127.

[237] K. Mehlhorn. *Data structures and algorithms 1: Sorting and searching.* Sringer-Verlag 1984.

[238] M. Molloy, B. Reed. *Further algorithmic aspects of the Local Lemma.* in: Proceedings of the 30th ACM-Symposium on the Theory of Computing, 1998, (STOC'98), pp. 524 – 530.

[239] R. Motwani, J. Naor, M. Naor. *The probabilistic method yields deterministic parallel algorithms.* J. Computer and System Sciences **49** (1994), 478-516. Preliminary Version in: Proceedings of the 30th Annual IEEE Symposium on Foundations of Computer Science, 1989, pp. 8-13.

[240] R. Motwani, J. Naor, P. Raghavan. *Randomized approximation algorithms in combinatorial optimization.* Approximation algorithms for NP-hard problems, D. S. Hochbaum (ed.), PWS Publishers 1995, pp. 447-481.

[241] R. Motwani, P. Raghavan. *Randomized algorithms.* Cambridge University Press, USA, 1995.

[242] K. Mulmuley. *Randomized geometric algorithms and pseudo-random generators.* Algorithmica **16** (1996), 450-463. Preliminary Version in Proceedings of the 33rd Annual IEEE Symposium on the Foundations of Computer Science 1992, (FOCS '92), pp. 90–100.

[243] K. Mulmuley. *Computational Geometry, An introduction through randomized algorithms* Prentice Hall, Englewood Cliffs, NJ 07632, 1994.

[244] M. G. Murty, S. N. Kabadi. *Some NP-complete problems in quadratic and non-linear programming.* Math. Programming **39** (1987), 117–129.

[245] J. Naor, M. Naor. *Small bias probability spaces: Efficient constructions and applications.* SIAM J. Computing **22**:4 (1993), 838–856. Preliminary Version in Proceedings of the 22nd Annual ACM Symposium on the Theory of Computing 1990, (STOC '90), pp. 213–223.

[246] M. Naor, L. J. Schulman, A. Srinivasan. *Splitters and near optimal derandomization.* Proceedings of the 36th Annual IEEE Symposium on the Foundations of Computer Science 1995, (FOCS '95), pp. 182–191.

[247] A. P.-C. Ng, P. Raghavan, C. D. Thompson. *Experimental results for a linear program global router.* Computers and Artificial Intelligence **6**:3 (1987), 229–242.

[248] H. Niederreiter. *Random number generation and quasi-monte-carlo methods.* SIAM Conf. Ser. App. Math. **63**, Philadelphia, 1992.

[249] N. Nisan. *Pseudorandom bits for constant depth circuits.* Combinatorica **1** (1991), 63–70.

[250] N. Nisan. *Pseudorandom generators for space-bounded computation.* Combinatorica **12** (1992), 449–461.

[251] N. Nisan. *RL⊆SC.* Comput. Complexity **4**:1 (1994), 1-11. Preliminary Version in: Proceedings of the 23rd Annual ACM Symposium on the Theory of Computing 1992, (STOC '92), pp. 19–623.

[252] M. Okamoto. *Some inequalities relating to the partial sum of binomial probabilities.* Ann. Inst. Statist. Math. **10** (1958), pp. 29-35.

[253] H. Okamura, P. D. Seymour. *Multicommodity flows in planar graphs.* J. Combinatorial Theory Ser. B **31** (1981), 75–81.

[254] G. Pantziou, P. Spirakis, C. Zaroliagis. *Fast parallel approximations of the maximum weighted cut problem through derandomization.* Foundations

838

of software technology and theoretical computer science, Proc. 9th Conf., Bangalore/India 1989, Lect. Notes Comput. Sci. **405** (1989), 20-29.

[255] C. H. Papadimitriou, K. Steiglitz. *Combinatorial optimization: Algorithms and complexity.* Prentice-Hall, Englewood Cliffs NJ, 1982.

[256] C. H. Papadimitriou, M. Yannakakis. *Optimization, approximation, and complexity classes.* J. Computer and System Sciences **43** (1991), 425–440.

[257] R. Peralta. *On the randomness complexity of algorithm.* CS Res. Report TR90-1, Universit of Wisconsin, Milwaukee, WI.

[258] F. Pfeiffer. *Zur Komplexität des disjunkte-Wege-Problems.* Dissertation, Forschungsinstitut für Diskrete Mathematik , Universität Bonn, 1991.

[259] F. Pfeiffer. Personal Communication (1992).

[260] C. Phillips, C. Stein, J. Wein. *Scheduling jobs that arrive over time.* Proceedings of the 4th Workshop on Algorithms and Data Structures, Springer Lecture Notes on Computer Science (LNCS) 955, 1995, 86–97.

[261] N. Pippenger, J. Spencer. *Asymptotic behavior of the chromatic index for hypergraphs.* J. Combinatorial Theory, Ser. A **51** (1989), 24–42.

[262] S. Poljak, F. Rendl. *Solving the max-cut problem using eigenvalues.* Report OR-91735, Forschungsinstitut für Diskrete Mathematik, Universität Bonn, 1991.

[263] S. Poljak, F. Rendl. *Nonpolyhedral relaxations of graph-bisection problems.* DIMACS Technical Report 92-55, 1992.

[264] S. Poljak, F. Rendl. *Computational experiments with node and edge relaxations of the max-cut problem.* Report 266, Technische Universität Graz, 1993.

[265] M.Queyranne. *Structure of a simple scheduling polyhedron.* Math. Programming **58** (1993), 263–285.

[266] M.Queyranne, A.S. Schulz. *Polyhedral approaches to machine scheduling.* Preprint 408/1994, Department of Mathematics, Technical Univercity of Berlin, Berlin, Germany 1994.

[267] M.O. Rabin. *Probabilistic algorithms.* J.F. Traub (ed.), *Algorithms and Complexity*, Academic Press, New York (1976), 21–39.

[268] J. Radhakrisnan, A. Srinivasan. *Improved bounds and algorithms for hypergraoh 2−coloring.* Random Structures & Algorithms **16**:1 (2000), 4–32.

[269] P. Raghavan. *Probabilistic construction of deterministic algorithms: approximating packing integer programs.* J. Computer System Sciences **37** (1988), 130–143.

[270] P. Raghavan, C. D. Thompson. *Randomized rounding: a technique for provably good algorithms and algorithmic proofs.* Combinatorica **7**:4 (1987), 365–374.

[271] P. Raghavan, C. D. Thompson. *Multiterminal global routing: a deterministic approximation scheme.* Algorithmica **6** (1991), 73–82.

[272] S. Rajagopalan, V. V. Vazirani. *Primal-dual RNC approximation algorithms for (multi)-set (multi)-cover and covering integer programs.* SIAM J. Computing **28**:2 (1998), 525-540. Preliminary Version in: Proceedings of the 33rd Annual IEEE Symposium on the Foundations of Computer Science 1993, (FOCS '93), pp. 322–331.

[273] G. Robins, A. Zelikovsky. *Improved Steiner tree approximation in graphs.* in: Proceedings of the 11th annual ACM-SIAM symposium on Discrete algorithms (SODA'00), pp. 770-779.

[274] H. Röck, G. Schmidt. *Machine aggregation heuristics in shop scheduling.* Methods Oper. Res. **45** (1983), 303–314.

[275] V. Rödl. *On a packing and covering problem.* European Journal of Combinatorics **5** (1985), 69–78.

[276] V. Rödl, L. Thoma. Talk at *The Seventh International Conference on Random Structures and Algorithms.* Atlanta, May 1995.

[277] I.V. Romanovskij. *Derandomization of optimal strategies in antagonistic games with bluffing* (English of Russian original) Sov. Phys., Dokl. **9** (1965), pp. 630-632 ; translation from Dokl. Akad. Nauk SSSR **157** (1964), pp. 1066-1068.

[278] K.F. Roth. *On irregularities of distribution.* Mathematika **1** (1954), 73–79.

[279] K.F. Roth. *Remark concerning integer sequences.* Acta Arithmetica **9** (1964), 257-260.

[280] W. Rudin. *Functional analysis.* McGraw-Hill Series in Higher Mathematics. New York etc.: McGraw-Hill Book Comp. XIII, (1973).

[281] M. Saks. *Randomization and derandomization in space-bounded computation.* Eleventh Annual IEEE Conference on Computational Complexity, pp. 128-149, Philadelphia, Pennsylvania, 24-27 May 1996.

[282] M. Santha. *On using deterministic functions to reduce randomness in probabilistic algorithms.* Information and Computation **74** (1987), 241 – 249.

[283] H. Saran, V. V. Vazirani. *Finding k-cuts within twice the optimal.* SIAM J. Computing **24**:1 (1995), 101-108. Preliminary Version in: Proceedings of the 33rd Annual IEEE Symposium on the Foundation of Computer Science 1992, (FOCS '92), pp. 743–751.

[284] T. J. Schaefer. *The complexity of satisfiability problems.* Proceedings of the 10th Annual ACM Symposium on the Theory of Computing, New York 1978, (STOC '78), pp. 216–226.

[285] J. P. Schmidt, A. Siegel, A. Srinivasan. *Chernoff-Hoeffding bounds for applications with limited independence.* SIAM J. Discrete Math. **8**:2 (1995), 223-250. Preliminary Version in: Proceedings of the fourth ACM-SIAM Symposium on Discrete Algorithms 1993, (SODA '93), pp. 331–340

[286] L. Schulman. *Sample spaces uniform on neighborhoods*. Proceedings of 24th Annual ACM Symposium on Theory of Computing 1992, (STOC '92), pp. 17–25.

[287] A.S. Schulz. *Scheduling to minimize total weighted completion time: Performance guarantees of LP-based heuristics and lower bounds*. Proceedings of the 5th MPS Integer Programming and Combinatorial Optimization Conference, 1996, pp. 301–315.

[288] A. Sebo. *Integer plane multicommodity flows with a bounded number of demands*. Tech. Report. OR-885340, Research Institute for Discrete Mathematics, University of Bonn, Germany 1988.

[289] R. Seidel. *Small dimensional linear programming and convex hulls made easy*. Discrete & Computational Geometry **6**:5 (1991), 423–434.

[290] E. Shamir, J. Spencer. *Sharp concentration of the chromatic number of random graphs $G_{n,p}$*. Combinatorica **7** (1987), 121–129.

[291] J. B. Shearer. *A Note on the independence Number of Triangle-free Graphs*. Discrete Mathematics **46** (1983), 83–87.

[292] J. B. Shearer. *On the independence number of sparse graphs*. Random Structures & Algorithms **5** (1995), 269–271.

[293] D.B. Shmoys, C. Stein, J. Wein. *Improved approximation algorithms for shop scheduling problems*. SIAM J. Computing **23** (1994), 617–632.

[294] M. Sipser. *Expanders, randomness, or time versus space*. J. Computer & Systems Sci. **36** (1988), pp. 379–383.

[295] M. Sitharam, T. Straney. *Derandomized learning of Boolean functions*. Li, Ming (ed.) et al., Algorithmic learning theory. in: Proceedings of the 8th international workshop, ALT '97, Sendai, Japan, October 6–8, 1997, Lecture Notes in Computer Science **1316** (1997), pp. 100-115.

[296] R. Solovay, V. Strassen. *A fast Monte Carlo test for primality*. SIAM J. Computing **6** (1977), 84–85. (erratum: **7** (1978), 118).

[297] J. Spencer. *Guess number-with lying*. Math. Mag. **57** (1984), 105-108.

[298] J. Spencer. *Six standard deviation suffice*. TAMS **289** (1985), 679–706.

[299] J. Spencer. *Ten lectures on the probabilistic method*. SIAM, Philadelphia 1987.

[300] J. Spencer. *Searching game with a fixed number of lias*. Theoretical Computer Science **95** (1992), 307-322.

[301] J. Spencer, P. Winkler. *Three thresholds for a liar*. Combinatorics, Probability and Computing **1**, No.1 (1992), 81-93.

[302] A. Srinivasan. *An extension of the Lovász Local Lemma, and its application to integer programming*. Proceedings of the 7th Annual ACM-SIAM Symposium on Discrete Algorithms 1996, (SODA '96), pp. 6–15.

[303] A. Srinivasan. *Improving the discrepancy bound for sparse matrices: better approximation for sparse lattice approximation problems*. Proceed-

ings of the 7th ACM/SIAM Symposium on Discrete Algorithms 1997, (SODA '97), pp. 692-701.

[304] A. Srinivasan. *Improved approximation guarantees for packing and covering integer programs.* SIAM J. Computing **29** (1999), 648–670.

[305] A. Srivastav. *Derandomized algorithms in combinatorial optimization* Habilitation-Thesis. Freie Universität zu Berlin, 1995, 180 pages.

[306] A. Srivastav, P. Stangier. *Integer multicommodity flows with reduced demands.* Lecture Notes in Computer Science (LNCS) 726, 1993, 360–372, Springer Verlag. (Proceedings of the 1st Annual European Symposium on Algorithms (ESA'93), Bonn/Bad Honnef 1993, ed.: T. Lengauer)

[307] A. Srivastav, P. Stangier. *On quadratic lattice approximations.* Lecture Notes in Computer Science (LNCS) 762, 1993, 176–184, Springer Verlag. (Proceedings of the 4th International Symposium on Algorithms and Computation (ISAAC'93), Hong-Kong, eds.: K.W. Ng, P. Raghavan, N.V. Balasubramanian)

[308] A. Srivastav, P. Stangier. *Weighted fractional and integral k-matching in hypergraphs.* Discrete Appl. Math. **57** (1995), 255–269.

[309] A. Srivastav, P. Stangier. *Algorithmic Chernoff-Hoeffding inequalties in integer programming.* Random Structures & Algorithms **8**:1 (1996), 27–58.

[310] A. Srivastav, P. Stangier. *Tight aproximations for resource constrained scheduling and bin packing.* Discrete Appl. Math. **79** (1997), 223–245.

[311] A. Srivastav, P. Stangier. *Complexity, representation and approximation of integral multicommodity flows.* Discrete Appl. Math. **99** (2000), 183–208.

[312] A. Srivastav, K. Wolf. *Finding dense subgraphs with semidefinite programming.* in: Proceedings of the 1st International Workshop on Approximation Algorithms for Combinatorial Optimization Problems, July 1998, Aalborg, Denmark. K. Jansen and J. Rolim (Eds.), Lecture Notes in Computer Science (LNCS) 1444, 1998, pp. 181–193, Springer-Verlag.

[313] E. Szymańska. Ph. D. thesis, Adam Mickiewicz University, Poznán, Poland, 1998.

[314] H. Takahashi, A. Matsuyama. *An approximative solution for the Steiner problem in graphs.* Math. Japonica **24** (1980), 573–577.

[315] T. Thiele. *Geometric Selection Problems and Hypergraphs.* Ph. D. thesis, Freie Universität zu Berlin, 1995.

[316] P. Turán. *On an extremal problem in graph theory.* Matematicko Fizicki Lapok **48** (1941), 436–452 (See, for instance, N. Alon, J. H. Spencer, P. Erdös. *The Probabilistic Method.* John Wiley & Sons, Inc., 1992, 81–82.)

[317] V.N. Vapnik, A. Ya. Chervonenkis. *On the uniform convergence of relative frequencies of events to their probabilities.* Theory Probab. Appl. **16** (1971), 264–280.

[318] U. V. Vazirani. *Randomness, adversaries and computation*. Thesis, University of California, Berkeley, CA, 1986.

[319] U. V. Vazirani and V. V. Vazirani. *Efficient and secure pseudo-random number generation*. Proceedings of the 25th Annual IEEE Symposium on Foundations of Computer Science 1984, (FOCS '84), pp. 458–463.

[320] W. F. de la Vega, C. S. Luecker. *Bin packing can be solved in within $1 - \varepsilon$ in linear time*. Combinatorica **1** (1981), 349–355. bibitemvizing V. G. Vizing. *On an estimate of the chromatic class of a p-graph*. (Russian), Diskret. Analiz. **3** (1964), 25–30.

[321] V. Vovk. *Derandomizing stochastic prediction strategies*. Mach. Learn. **35**, No.3 (1999), pp. 247-282.

[322] V. H. Vu. *New bounds on nearly perfect matchings in hypergraphs: higher codegrees do help*. Random Structures & Algorithms **17** (2000), 29-63.

[323] V. H. Vu. *Concentration of non-Lipschitz functions and applications*. preprint 2000.

[324] J. Vygen. *Algorithms for large-scale flat placement*. Proceedings of the 34th ACM Design Automation Conference 1997, (DAC '97), pp. 746–751.

[325] D. Wagner, K. Weihe. *A linear-time algorithm for edge-disjoint paths in planar graphs*. Combinatorica **15**:1 (1995), 135–150.

[326] V. K. Wei. *A lower bound on the stability number of a simple graph*. Technical report 81-11217-9, Bell Laboratories, 1981.

[327] E. Welzl. *Partition trees for triangle counting and other range searching problems*. Proceedings of the 4th ACM Symposium on Computational Geometry, 1988, pp. 23–33.

[West96] D. B. West. *Introduction to Graph Theory*. Prentice-Hall, Inc. Simon & Schuster 1996.

[328] A. Wigderson. *Improving the performance guarantee for approximate graph coloring*. J. ACM **30** (1983), 729–735.

[329] A. Yao. *Theory and Applications of Trapdoor Functions*. Proceedings of the 23th Annual IEEE Symposium on Foundations of Computer Science 1982, (FOCS '82), pp. 80–91.

[330] A. Zelikovsky. *An 11/6-approximation algorithm for the network Steiner problem*. Algorithmica **9** (1993), 463–470.

[331] U. Zwick. *Outward rotations: a tool for rounding solutions of semidefinite programming relaxations with applications to MAXCUT and other problems*. Proceedings of the 31. ACM Symposium on Theory of Computing 1999, (STOC '99), pp. 679–687.

Chapter 19

DERANDOMIZING COMPLEXITY CLASSES

Peter Bro Miltersen
BRICS
University of Aarhus
bromille@brics.dk

1.   INTRODUCTION

Consider a randomized algorithm, such as the *Rabin primality test* of Rabin, 1976. This algorithm is given as input an integer n, and, as auxiliary input, a sequence of *coin tosses*, i.e., a vector of unbiased, independent random bits. The test gives as output either "n is a prime" or "n is not a prime". For every input n, the probability that the output is a false statement is bounded from above by a small constant, say, $\frac{1}{4}$. The test runs in time polynomial in the size of the binary representation of n. In particular, the number r of random bits needed is polynomial in $\log n$.

Randomized algorithms such as the Rabin primality test are immensely useful, also in practice. An obstacle for using them is that they are not very suited for implementation on conventional hardware because of the auxiliary input, a sequence of unbiased, independent random bits. If the reader has never before considered the task of obtaining such a sequence, or even just a *single* unbiased random bit, he is invited to spend a few minutes pondering about possible ways of performing this task. It is certainly a non-trivial task, an expensive one (compared to other atomic operations, such as taking the AND of two bits), and, in many settings where no special hardware is available, arguably even an impossible one. The fact of life we face is that *independent random bits are scarce*. Thus, we want to use as few as possible. In this chapter, we describe theoretical techniques for saving on independent random bits. In particular, we ask the following three related questions.

1. It is well known that if we are unhappy about the error probability of a randomized algorithm, we can decrease it at an exponential

S. Rajasekaran et al (eds.), Handbook of Randomized Computing, Volume 2, pp, 843–941.

rate by running the algorithm a number of times. Running it k times increases the random bits used by a factor of k. Can we save on the number of random bits used to achieve a certain error probability?

2. Independent random unbiased bits are hard to obtain, but we may be able to get our hands on a "messier" source of randomness, such as a bit sequence $x \in \{0,1\}^r$ sampled from a source we don't quite know how to model, but which presumably "contains some randomness". Can we simulate our algorithm with such a bit sequence instead of unbiased independent bits?

3. Can we replace the randomized algorithm with a deterministic one, thus avoiding using randomness altogether?

The purpose of this chapter is to give an account of the state-of-the-art answers to these three questions, as of around the turn of the millennium[1]. Before making the questions more precise, there are three general qualifications we should make.

First, we concentrate on randomized algorithms solving *decision problems*, with a small probability of error on every instance, such as the Rabin primality test. The important property is that there is a single well defined answer we are trying to obtain and the probability that we will *not* obtain it is small[2]. There are other ways in which an algorithm may use randomness and we won't be considering derandomizing those. For instance, we may want an algorithm to actually produce an output distributed according to some distribution, i.e., we may want the output to *contain* some randomness. Clearly, such an algorithm could never be completely derandomized[3]. As another example, there are randomized algorithms which generate, with high probability, a particular kind of combinatorial object. For instance, a randomized algorithm for generating, with very high probability, the truth table of a Boolean function $\{0,1\}^n \rightarrow \{0,1\}$ with circuit complexity $(1 - o(1))\frac{2^n}{n}$ (i.e., almost maximum possible) is by picking each of the 2^n entries in the truth table independently at random. Clearly, we'll get a different truth table every time we run the algorithm, so we shall not attempt to derandomize this algorithm either (but the example is somewhat relevant to our techniques after all, as we shall see).

Second, we shall care much more about the number of random bits used than about other computational resources, such as time. Of course, if we do not care about time at all, everything is trivialised. Indeed, we can deterministically simulate a randomized algorithm using r random bits by running it on all possible 2^r bit sequences. Thus, an exponential slowdown is uninteresting, but we shall routinely accept any polynomial slowdown of our algorithm to save, say, a constant factor on the random bits used to achieve a certain error probability. We shall even consider a slowdown which is only just non-trivial, i.e., slightly subexponential, as interesting (though of course less interesting than a polynomial slowdown). Thus, we take a very fine-grained point of view when it comes to randomness, but a very coarse-grained point of view when it comes to time. Though this point of view of course has a bearing of the practicality of the results obtained (an issue we do not address in this chapter), it seems the right point of view from a scientific perspective: To get a basic understanding of the power of one resource, in this case randomness, we have to consider other resources as of secondary importance, to avoid cluttering up the big picture. After a basic understanding has been obtained, a more complete picture of the finer trade-offs between time and randomness can be studied (and such a picture is indeed emerging in current literature).

Third, we shall concentrate on *generic*, general-purpose derandomization, i.e., techniques that will work on *any* randomized algorithm, *without any knowledge necessary about how the algorithm operates*, except, possibly, some bound on the computational complexity of the algorithm, such as its time or space bound (hence, the title of the chapter). For example, we should be able to give good answers to the three questions above for, say, Rabin's primality algorithm, based *only* on the account of the algorithm given above. An alternative and very important strategy for derandomization which we will not cover, is to use all we know about the algorithm and the analysis leading to our estimate on its probability of error in order to answer the three questions above, for the particular algorithm in question. For the specific case of Rabin's primality test, there are reasonable (but unproven) number theoretic assumptions that allow a complete derandomization. The more general tools available for special-purpose derandomization may or may not work for a given randomized algorithm and usually require non-trivial adaptation to each particular case. Important cases of such special-purpose techniques are the techniques of *conditional probabilities* and *small sample spaces*[4]. We refer to Motwani and Raghavan, 1995, for an exposition of these techniques.

With these points in mind, let us define the three basic tasks a bit more precisely and state which results we shall cover about them in this chapter.

Randomness-efficient amplification.   Suppose we have a randomized polynomial algorithm using r bits to obtain an error probability of

$\frac{1}{10}$. Suppose we want a smaller error probability than this. By repeating the algorithm k times and taking the majority of the answers, we can reduce the error probability to $2^{-O(k)}$ using rk random bits. We can also think of it as *amplifying* the success probability to $1 - 2^{-O(k)}$. If k is polynomial, the resulting algorithm is a polynomial algorithm. However, if random bits are very expensive, using a factor of k more may be unacceptable. The task of randomness-efficient amplification is to *obtain the same probability of error using fewer random bits*, while still preserving the efficiency.

It is not at all obvious that this task is possible at all. However, we shall see that it is, using no unproven assumptions. In particular we shall discuss the following simulations, all slowing down the randomized algorithm at most polynomially:

1. We can obtain an error probability of $\frac{1}{r^k}$ for any desired constant k using r bits, i.e. using *no* additional randomness.

2. We can obtain an error probability of 2^{-R} for any desired R, using (a) $r + O(R)$ random bits, and even (b) $(1+\epsilon)(r + R)$ random bits, for any desired constant $\epsilon > 0$.

3. Given any constant $\epsilon > 0$, we can, using R random bits, where R is a certain polynomial in r, obtain an error probability of 2^{-R+R^ϵ}.

4. If the algorithm has one-sided error (i.e., has error probability 0 on negative instances), we can, using R random bits for any parameter $R \geq r$, obtain an error probability of $2^{-R+r+\text{polylog } R}$.

Note that the flavor of the last two results is that *each* random bit used reduces the error probability of the algorithm by a factor of $2 - o(1)$. Since the error probability of an algorithm using R random bits must be an integer multiplum of 2^{-R}, one cannot do much better without derandomizing the algorithm completely.

Computing with weak random sources. Again, suppose we have a randomized algorithm using r random bits to achieve an error probability of $\frac{1}{10}$. Suppose that we are happy with the error probability, but at a loss on how to supply the r unbiased, independent coin tosses needed. How do we construct hardware to flip coins? Many physical systems can reasonably be assumed to have a behavior exhibiting some randomness. A radioactive source is an obvious example. We can consider making a measurement of such a system, encoding the measurement as a bit string of length r, and using this bit string as the sequence of coin tosses to our algorithm. However, it is not very likely that we can easily find and build into our hardware a physical system where such a direct approach would give us independent, unbiased coin flips. It is far more likely that we would not able to state the exact distribution of the output at all. A source that is assumed to contain some randomness but with an unknown distribution is called a *weak random source*. Our objective is to simulate our randomized algorithm *using the weak random source instead of independent, unbiased coin flips*, while preserving (roughly) the efficiency and (precisely) the error probability of our algorithm.

The above task needs to be formalized before we can solve it. Obviously, we need some measure on the amount of randomness contained in a source, to be able to prove a theorem stating that any source containing a certain amount of randomness can be used to simulate our algorithm.

One classical measure for the amount of randomness contained in a source is the *Shannon entropy* of the source. Thus, it would be nice to prove a theorem of the form: Any source with a Shannon entropy above a certain threshold can be used to simulate our r-bit algorithm. However, we cannot obtain such a theorem as a source can have high Shannon entropy and yet only exhibit randomness with low probability. For instance, consider the following source. Let R be a value much bigger than r. With probability $\frac{99}{100}$, the source produces the value $(0, 0, \ldots, 0) \in \{0, 1\}^R$. With probability $\frac{1}{100}$, the source produces a uniformly chosen value in $\{0, 1\}^R$. Then, the source has Shannon entropy more than $\frac{R}{100}$. Thus, by letting R grow, we can make a source with Shannon entropy exceeding r as much as we want. However, it does not seem likely that we can use it to simulate our randomized algorithm with success probability $\frac{9}{10}$, as we only get something containing randomness at all with probability $\frac{1}{100}$! Indeed, if we were able to do so, we would also be able to derandomize our algorithm completely.

A different measure for the amount of randomness contained in a source turns out to be much more useful. We say that a source has *min-entropy* at least t if any outcome of the source has probability at most 2^{-t}. Thus, a high min-entropy source is a source where no particular outcome is very likely to occur. It is easy to see that the min-entropy of a source is smaller than its Shannon entropy, i.e, a high min-entropy source is a stricter notion than a high Shannon entropy source. In particular, the problem described above is avoided, and it turns out that we *can* simulate our randomized algorithm using any such source.

5. For any constant $\epsilon > 0$ and any randomized polynomial algorithm A using r bits to obtain an error probability of $\frac{1}{3}$, there is a randomized polynomial algorithm A' which, instead of r random bits,

takes as auxiliary input $y \in \{0,1\}^R$, where R is polynomial in r. If y is sampled from any R-bit source with min-entropy at least R^ϵ, the error probability of the original algorithm is maintained.

Deterministic simulation. Finally, consider the task of removing the randomness requirement of our randomized algorithm all together, simulating it with an efficient deterministic one. Two (oddly contradictory) points make this task somewhat different from the two previously mentioned ones. First, *in the most general setting we will only be able to do it under an unproven assumption*. Indeed, it is at the time of writing an open problem if every problem solvable in exponential time can be solved by an efficient randomized algorithm. If this is the case, no general subexponential derandomization is possible, as this would imply that deterministic exponential time equals deterministic subexponential time, contradicting the time hierarchy theorem. Second, *deterministic simulation is being done all the time in the real world and is part of routine software engineering*. Indeed, lacking a source of truly random bits, most implementations of randomized algorithms currently running on conventional computers are replacing this source with a deterministically generated sequence of bits in some ad hoc way, e.g., with the output of the "pseudorandom generator" built into the C language. Of course, doing this without further analysis changes a randomized algorithm with a well analyzed error probability, such as the Rabin primality test, into a heuristic which may, conceivably, fail completely and give incorrect answers on all inputs.

What we shall see is that it *is* possible to construct pseudorandom generators (different from the one built into the C language) which makes a variation of the above scheme a correct simulation, under a reasonable complexity theoretic assumption.

In particular, we shall see that a sufficient condition is that "truth tables of hard Boolean functions can be generated efficiently deterministically" and prove the following theorems.

6. *If* there is an algorithm which on input n runs in time $2^{n^{O(1)}}$ and generates the truth table of a Boolean function $g : \{0,1\}^n \to \{0,1\}$ with circuit complexity at least $s(n)$, for some superpolynomial function s, *then* any decision problem that can be solved by a randomized polynomial time algorithm can also be solved by a deterministic one running in subexponential time.

7. *If* there is a constant $\epsilon > 0$ and an algorithm which on input n runs in time $2^{O(n)}$ and generates the truth table of a Boolean function $g : \{0,1\}^n \to \{0,1\}$ with circuit complexity at least $2^{\epsilon n}$

then any decision problem that can be solved by a randomized polynomial time algorithm can also be solved by a deterministic one in polynomial time.

While the two statements above are conditional, the conditions seem quite likely to be true, especially the condition of Statement 6. Note that since we have time more than $2^{O(n)}$ to generate the truth table for $g : \{0,1\}^n \to \{0,1\}$, we can, say, let g be the truth table of an **NP**-complete problem such as SATISFIABILITY, i.e., let $g(x) = 1$ if and only if x is the encoding of a Boolean formula with a satisfying assignment. While it has not been proved that g such defined does not have circuits of size $2^{o(\frac{n}{\log n})}$ (this would imply $\mathbf{P} \neq \mathbf{NP}$), no such circuits are known.

If we move to classes of algorithms defined by a space constraint, rather than a time constraint, some unconditional non-trivial deterministic simulation results are possible (the space measure is the standard one of complexity theory, i.e., we do not count the space taken up by the input in the space bound of the machine).

8. A decision problem solved by a randomized algorithm using space $O(\log n)$ and $O(\text{polylog } n)$ random bits on inputs of length n can be solved by a deterministic algorithm running in space $O(\log n)$.

9. A decision problem solved by a randomized algorithm using space $O(\log n)$ can be solved by a deterministic algorithm running in space $O((\log n)^2)$ and in polynomial time.

10. A decision problem solved by a randomized algorithm using space $O(\log n)$ can be solved by a deterministic algorithm running in space $O((\log n)^{\frac{3}{2}})$.

1.1 OVERVIEW OF CHAPTER

The results 1-10 stated above are landmark results in the theory of general derandomization. The main purpose of this chapter is to give almost (but not quite) self-contained proofs of many (but not all) of them, concentrating on those results that have not been covered in previous surveys and texts, such as Alon and Spencer, 1992, Motwani and Raghavan, 1995, Saks, 1996, Clementi et al., 1998, Goldreich, 1999, and Nisan and Ta-Shma, 1999. Though all of the results have an associated single paper where the result first appeared, many are the culmination of years of work appearing in many papers. The proof we want to give for each result is not always the original proof, nor is our sequence of proofs the same as the historical sequence, as many important connections were only realised recently.

In Section 2 (*Preliminaries*), we discuss three preliminary issues before starting the main story. The first of these is basic computational complexity theory, especially that concerned with randomized complexity classes. There is no hope of giving a comprehensive overview here, so we only explain what "extra" basics we need, apart from standard texts, such as Papadimitriou, 1994. The second is universal hashing and the construction of small and efficient families of universal hash functions. The third is the theory of error correcting codes. Both universal hashing and error correcting codes play crucial parts in many places of our main story and are thus well worth singling out.

A main key to solving all the three derandomization tasks turns out to be the concepts of *dispersers* and *extractors*. In Section 3 (*Dispersers and extractors*) we define these concepts and show how such objects solve the tasks of randomness-efficient amplification and computing using weak random sources. In Section 4 (*Constructions of dispersers and extractors*), we then show how to build these objects, leading to Statements 1-5 stated above. Here, we also introduce *Nisan-Wigderson designs*, an important tool in many of the proofs in the chapter.

The rest of the chapter is concerned with deterministic simulation. In Section 5 (*Hitting set and pseudorandom generators*), we introduce these two concepts that are the keys to all known deterministic simulation results. The most important kind of hitting set and pseudorandom generators, *hardness based generators*, will be shown to be special cases of dispersers and extractors. In Section 6, (*A low end generator*) and Section 7 (*A high end generator*) we prove Statement 6 and 7. As important lemmas, we prove in Section 6 *Yao's XOR lemma* and in Section 7 *Impagliazzo's hardcore lemma*. In Section 8 (*The compression approach to deterministic simulation*), we present a different approach to deterministic simulation that can be used to obtain similar results somewhat more easily and in Section 9 (*Other hardness based deterministic simulation results*), we survey which other results that can be proved along the lines of the proofs of Section 6, 7 and 8. Finally, in Section 10 (*Unconditional derandomization of space-bounded computation*) we present the unconditional deterministic simulation results for randomized space bounded computation, proving Statements 8-10. Again, extractors turn out to be key components.

We use standard mathematical notation in the theory of computation tradition. For instance, log will denote $\log_2$, the size of a finite set A will be denoted $|A|$ and the density of a set $A \subset \{0,1\}^n$, i.e., $|A|/2^n$ will be denoted $\mu(A)$. Though many theorems will be given fairly detailed proofs, we shall, in an attempt to improve readability, generally ignore round-off errors, i.e., sometimes tacitly assume expressions such as $\log n$

to be integers, even though we cannot strictly speaking assume that n is a power of two, without loss of generality. Fixing the proofs to hold in general is usually straightforward, amounting to adding $\lfloor \cdot \rfloor$ and $\lceil \cdot \rceil$ as appropriately, but sometimes requires slightly more work.

1.2   NOTES

The first paper to point out that randomness-efficient amplification is possible was Karp et al., 1986. The notion of computing with weak random sources are due to Santha and Vazirani, 1986. The observation that the min-entropy of a source is the right measure to consider is due to Zuckerman, 1990. The heuristic idea of using pseudorandom generators to simulate randomized algorithm has a long history. See Knuth, 1981, for an excellent survey of this. Blum and Micali, 1984, and Yao, 1982, were first to find ways of putting pseudorandom generators on more rigorous ground.

The results labelled 1 to 10 above first appeared in Karp et al., 1986 (result 1 for one-sided error algorithms), Cohen and Wigderson, 1989 (result 1 for two-sided error algorithms), Impagliazzo and Zuckerman, 1989, and Cohen and Wigderson, 1989 (result 2a), Zuckerman, 1996 (result 2b), Saks et al., 1998 (results 3 and 5 for one-sided error algorithms), Andreev et al., 1999 (results 3 and 5), Ta-Shma, 1998 (result 4), Babai et al., 1991a (result 6), Impagliazzo and Wigderson, 1997 (result 7), Nisan and Zuckerman, 1996 (result 8), Nisan, 1994 (result 9), and Saks and Zhou, 1999 (result 10). As the papers cited here are journal papers (where available), the above dates do not necessarily say much about the dates of the results. However, all the results are from the period 1986-1998, though some of the proofs presented in this chapter are more recent.

2.   PRELIMINARIES

2.1   COMPLEXITY THEORY

We assume that the reader is familiar with standard notation for complexity classes (a good textbook is Papadimitriou, 1994).

In particular, **TIME**$(t(n))$ where $t(n)$ is a proper time bound, is the class of languages decided by a deterministic Turing machine running in time $O(t(n))$ for inputs of length n, **P** is $\cup_{k \geq 1}$**TIME**(n^k), **E** is $\cup_{k \geq 1}$ **TIME**(2^{kn}) and **EXP** is $\cup_{k \geq 1}$**TIME** (2^{n^k}), while **SUBEXP** is $\cap_{\epsilon > 0}$**TIME**(2^{n^ϵ}). Also, **SPACE**$(s(n))$ where $s(n)$ is a proper space bound, is the class of languages decided by a deterministic Turing machine using space $O(s(n))$ for inputs of length n, **L** is **SPACE**$(\log n)$, **LINSPACE** is **SPACE**(n) and **PSPACE** is **SPACE**$(n^{O(1)})$.

Apart from the uniform Turing machine model, we shall also consider the non-uniform circuit model. Unlike a Turing machine, a circuit only decides a finite language $L \subseteq \{0,1\}^n$ for a particular n. We say that a circuit C with n Boolean inputs and one output decides $L \subseteq \{0,1\}^n$ if $C(x) = 1$ if and only if $x \in L$. Turing machines can be simulated by circuits with a polynomial overhead. In particular, if a language L is in **P**, the circuit complexity of $L \cap \{0,1\}^n$ is polynomial in n. The class of languages L for which the circuit complexity of $L \cap \{0,1\}^n$ is polynomial in n is called **P/poly**. For all languages L, the circuit complexity of $L \cap \{0,1\}^n$ is at most exponential. More precisely, if we by $M_{n,m}$ denote the biggest possible circuit complexity of functions from $\{0,1\}^n$ to $\{0,1\}^m$, we have the Shannon-Lupanov bound of Shannon, 1949; Lupanov, 1965,

Theorem 1 $M_{n,m} = (1 + o(1))\frac{2^n m}{n + \log m}$.

A family of circuits C_n of size at most $s(n)$ is said to be a *uniform* family of size $s(n)$ if there is a Turing machine operating in space $O(\log s(n))$ which on input n outputs a description of C_n.

A *nondeterministic* Boolean circuit C contains, in addition to AND, OR and NOT gates, *choice*-gates of fan-in 0. The circuit evaluates to 1 on an input x, and we say that $C(x) = 1$, if there is *some* assignment of truth values to the choice-gates that makes the circuit evaluate to 1.

All complexity classes defined by a time bound or circuit size bound have *relativized* versions, where we assume "free" access to an oracle deciding a language L. In particular, **TIME**$[L](t(n))$ is the class of languages decided in time $t(n)$ with an oracle for L and **P**$[L]$, **EXP**$[L]$, etc. are defined accordingly. If **C** is a complexity class, **TIME**$[\mathbf{C}](t(n))$ is $\cup_{L \in \mathbf{C}}$ **TIME**$[L](t(n))$ and **P**$[\mathbf{C}]$, **EXP**$[\mathbf{C}]$, etc. are defined accordingly. A relativized circuit with oracle L (an L-circuit) contains L-gates deciding membership in L. When we consider the size of L-circuits, an L-gate with m inputs is charged size $m - 1$.

A complexity theoretic theorem and/or proof is said to *relativize* if it holds for relativized complexity classes with any fixed oracle L. Most complexity theoretic theorems and proofs do relativize but there are exceptions.

The complexity classes relevant for understanding time-efficient randomized computation are **RP** and **BPP**. We present the basic facts about these classes in somewhat more detail.

A language L is in **RP** if there is a randomized polynomial time Turing machine M so that for all $x \in L, M$ accepts with probability at least $\frac{1}{2}$ and for all $x \notin L$, M accepts with probability 0. Such a randomized Turing machine is said to have *one-sided error*. An example of a language in **RP** not known to be in **P** is NON-PRIMALITY, the language of integers which are not primes. Rabin's primality test is an **RP**-machine for this language. If, actually, for all $x \in L, M$ accepts x with probability at least $1 - \epsilon$, we say that M has error probability at most ϵ. If we have an **RP**-machine M for a language L with error probability at most ϵ, we can get an **RP**-machine for a language L with error probability at most ϵ^k, by doing k independent runs of M and accepting if at least one of the runs accepts.

A language L is in **BPP** if there is a randomized polynomial time Turing machine M so that for all $x \in L, M$ accepts with probability at least $\frac{2}{3}$ and for all $x \notin L$, M accepts with probability at most $\frac{1}{3}$. Such a randomized Turing machine is said to have *two-sided error*. If, actually, for all $x \in L (x \notin L), M$ accepts (rejects) x with probability at least $1 - \epsilon$, we say that M has error probability at most ϵ. If we have a **BPP**-machine M for a language L with error probability at most $\epsilon \leq \frac{1}{3}$, we can get a **BPP**-machine with error probability $\epsilon^{O(k)}$, by doing k independent runs of M and accepting if most of the runs accepts. This is an easy corollary of the *Chernoff bound*, a proof of which can be found, for instance, in Motwani and Raghavan, 1995:

Lemma 1 *Let* $X_1, X_2, \ldots, X_m$ *be independent 0-1 random variables and let* $X = \sum X_i$. *Let* $\mu = \mathrm{E}[X]$. *Then, for any* $\delta > 0$, $\Pr[X > (1 + \delta)\mu] < [\frac{e^\delta}{(1+\delta)^{1+\delta}}]^\mu$ *and* $\Pr[X < (1 - \delta)\mu] < e^{\frac{-\mu\delta^2}{2}}$

An immediate consequence of Lemma 1 is that the error probability of a **BPP**-machine can be made smaller than 2^{-n} on instances of length n, while preserving polynomial efficiency. This means that there is one particular coin toss sequence which will be good for all inputs of length n. As the Turing machine can be simulated by a circuit with polynomial overhead and the random coin toss sequence can be frozen and build into the circuit, we have that **BPP** $\subseteq$ **P/poly**, a fact first noted by Adleman, 1978.

Clearly **P** $\subseteq$ **RP** $\subseteq$ **BPP** $\subseteq$ **EXP**. We know that **P** $\neq$ **EXP** by the time hierarchy theorem, but at present, we don't know that **RP** $\neq$ **EXP**, though many people suspect **P** = **BPP**. In complexity theory, we often seek to understand a class through its *complete* problems, using, say, logspace reductions. For instance, to prove that **P=NP**, it is enough to find an efficient algorithm for a particular **NP**-complete problem. Similarly, if we wish to prove that **P=BPP**, it would seem like a good concrete strategy to first find a **BPP**-complete problem and focus on finding an efficient algorithm for that. Unfortunately, the classes

BPP and **RP** do *not* seem to have complete problems. This may seem somewhat surprising at first, as we would expect a complexity class to have at least the corresponding machine simulation problem as a complete problem. For instance, the problem NONDETERMINISTIC TURING MACHINE SIMULATION given by the language $\{\langle M, x\rangle | M$ is a nondeterministic Turing machine accepting x in at most $|x|^2$ steps $\}$ is complete for **NP**. The corresponding problem for two-sided error randomized computation would be RANDOMIZED TURING MACHINE SIMULATION, given by the language $\{\langle M, x\rangle | M$ is a two-sided error randomized Turing machine accepting x in at most $|x|^2$ steps $\}$. While this problem is *hard* for **BPP**, it is not *complete* for **BPP**, as it is not *in* **BPP**. The problem is that while being a nondeterministic Turing machine is a *syntactic* property, being a randomized one is a *semantic* one: Deciding whether a given Turing machine is a proper two-sided error randomized machine, i.e., whether it on every input accepts either with probability at least $\frac{2}{3}$ or with probability at most $\frac{1}{3}$, is not even decidable.

This somewhat annoying problem is really only an artifact of our definitions. In particular, it is caused by the decision to model all computational problems in terms of *languages*. The situation is much nicer if we extend the kind of problems we look at to include *partial languages* as well. Partial languages are also known in complexity theory as *promise problems*. A partial language or promise problem is simply a language $L \subseteq U$, where U is some subset of Σ^*. A Turing machine solves a promise problem L if it accepts all instances of L and rejects all instances of $U - L$. On every instance $x \notin U$, the machine can have arbitrary behavior. The trick is that the Turing machine does not have to be able to decide membership in U or even "know what U is". Informally speaking, we *promise* that we will only give the machine inputs from U. We can extend each complexity classes **C** considered above to a class of promise problems by changing its definition in the obvious way. This "promise-version" of a class **C** will be denoted **pC**, with **p** meaning either "promise" or "partial"[5].

For "syntactic classes" such as **P** or **NP**, there is not much reason to consider the promise-versions of the classes, as they behave exactly like the original classes in all important respects. For instance, **P**=**NP** if and only if **pP**=**pNP**. For "semantic classes" such as **RP** or **BPP**, the promise-versions are much better behaved than the original classes. For instance, the promise problem RANDOMIZED TURING MACHINE SIMULATION given by the partial language $\{\langle M, x\rangle | M$ is a two-sided error randomized Turing machine accepting x in at most $|x|^2$ steps $\}$ defined on $U = \{\langle M, x \rangle | M$ is a two-sided error randomized Turing machine $\}$ *is* complete for **pBPP**, as is easily seen.

The problems above are somewhat artificial. The classes **pBPP** and **pRP** also have natural complete problems, i.e. problems not mentioning simulating Turing machines.

Let CIRCUIT ACCEPTANCE PROBABILITY ESTIMATION be the following promise problem. Given a Boolean circuit C of size n with n inputs and one output with a promise that *either* (a) $\Pr[C(x) = 1] \leq \frac{1}{3}$ *or* (b) $\Pr[C(x) = 1] \geq \frac{2}{3}$ for a randomly chosen $x \in \{0,1\}^n$, we should tell which of (a) or (b) is the case.

Let ONE-SIDED CIRCUIT ACCEPTANCE PROBABILITY ESTIMATION be the following promise problem. Given a Boolean circuit C of size n with n inputs and one output with a promise that *either* (a) $\Pr[C(x) = 1] = 0$ *or* (b) $\Pr[C(x) = 1] \geq \frac{1}{2}$ for a randomly chosen $x \in \{0,1\}^n$, we should tell if (b) is the right case.

Lemma 2 *CIRCUIT ACCEPTANCE PROBABILITY ESTIMATION is complete for* **pBPP** *and ONE-SIDED CIRCUIT ACCEPTANCE PROBABILITY ESTIMATION is complete for* **pRP**.

Proof We only describe the **pBPP**-case. The **pRP**-case is similar.

First, CIRCUIT ACCEPTANCE PROBABILITY ESTIMATION is *in* **pBPP**: If the promise is kept we can see which of two possibilities is the case by randomly picking a sufficient number of assignments and seeing if the majority is accepted. The error probability can be made smaller than any desired constant.

Second, let any partial language L in **pBPP** be given and let M be a **pBPP**-machine for L with running time $p(n)$ on inputs of length n. Given an input length n, we can in space $O(\log n)$ build a circuit C of size $p(n)^{O(1)}$ with two inputs, the real input x, and a random bit sequence y, so that M accepts x when the random bit sequence is y if and only if $C(x, y) = 1$. Now, given an input x, we can build it into the circuit C, i.e. we let C_x be the circuit given by $C_x(y) = C(x, y)$. If x is in the domain of definition of L, C_x is a yes-instance of CIRCUIT ACCEPTANCE PROBABILITY ESTIMATION if and only if $x \in L$. Similarly, C_x is a no-instance of CIRCUIT ACCEPTANCE PROBABILITY ESTIMATION if and only if $x \notin L$. This completes the description of the reduction, except that the circuit constructed may have a size different from its number of inputs. To get a circuit with the same number of inputs as its size, we either add a number of "dummy inputs" that the circuit does not read, or add a number of "dummy gates" to the circuit that do not contribute to the output. This does not change the acceptance probability of the circuit. ∎

Most known randomized algorithms for natural decision problems have one-sided error. Also, technically, derandomizing one-sided error algorithms is easier than derandomizing two-sided error algorithms. Thus, it seems to be a good strategy to focus on derandomizing one-sided error algorithms. It can even be formally shown that this is sometimes a sufficient strategy, even if we are ultimately interested in two-sided error algorithms. This is a corollary of the following theorem, implicit in the work of Sipser, 1983, and Lautemann, 1983, and made explicit by Buhrman and Fortnow, 1999.

Theorem 2 pBPP = pRP[pRP].

Proof It is easy to see that **pBPP = pBPP[pBPP]**. Thus, **pRP[pRP]** $\subseteq$ **pBPP**.

For the other direction, take a problem $L \subseteq U$ in **pBPP** by a randomized machine M. Assume that M uses $r(n)$ random bits on inputs of length n. By repeating the machine $O(r(n))$ times and taking majority of answers, we obtain a machine M' using $R = R(n) = O(r(n)^2)$ random bits and with error probability $2^{-O(r(n))} = 2^{-O(R^{1/2})}$.

Fix a particular $x \in U$. Let $A \subseteq \{0,1\}^R$ be the set of coin toss sequences that make M' accept on x. Either the density of A is greater than $\geq 1 - 2^{-O(R^{1/2})}$, in which case $x \in L$ or smaller than $2^{-O(R^{1/2})}$, in which case $x \notin L$.

Given A, and a vector $y \in \{0,1\}^R$, let $A \oplus y = \{x \oplus y | x \in A\}$, where $\oplus$ is bitwise XOR. Consider picking $y_1, y_2, \ldots, y_R$ at random. Let $A' = (A \oplus y_1) \cup (A \oplus y_2) \cup \cdots \cup (A \oplus y_R)$.

We claim that if $x \in L$, with very high probability, over the choice of $y_1, y_2, \ldots, y_R$, we have $A' = \{0,1\}^R$. On the other hand, if $x \notin L$, we immediately have that the density of A' is at most $R2^{-O(R^{1/2})}$, no matter which y_i's are chosen.

Suppose the claim is true. Then we show that we have the desired **pRP[pRP]** algorithm for L. Given $y_1, y_2, \ldots, y_R$, we can easily construct a small circuit deciding membership in A': The circuit checks, on input z if there is a y_i so that the coin toss sequence $y_i \oplus z$ makes M' accept x. Given a random $y_1, y_2, \ldots, y_R$ we have that with very high probability either A' is very small or everything. Thus, if we negate the circuit for A', we have a circuit which either accepts with probability 0 or with very high probability. Deciding which is the case is in **pRP** by Lemma 2, in particular, we can take a **pRP** oracle O solving it. Then the following is a **pRP[O]** algorithm for L: Pick $y_1, y_2, \ldots, y_l$ at random, compute the circuit for A' and ask the oracle O if A' is small or everything. If the oracle says that A' is small, we say that $x \notin L$. Otherwise we say that $x \in L$. Note that $x \notin L$, then no matter which random

$y_1, y_2, \ldots, y_l$ we pick, A' will be small, so we always keep our promise to the oracle and thus always return the right answer. On the other hand, if $x \in L$, A' will be large with high probability, so we keep our promise to the oracle with high probability, and thus answer correctly with high probability. Thus, our algorithm is indeed an $\mathbf{pRP}[O]$ algorithm.

We now prove the claim. Assume that the density of A is at least $1 - 2^{-O(R^{1/2})}$. Fix a z in $\{0,1\}^R$ and consider the probability that it is not included in A'. Note that $z \in A \oplus y_i$ if and only if $z \oplus y_i$ is in A and $z \oplus y_i$ is uniformly distributed in $\{0,1\}^R$ and mutually independent of $z \oplus y_j$ for $j \neq i$. Thus, the probability that $z \notin A \oplus y_i$ is at most $2^{-O(R^{1/2})}$ and the probability that $z \notin A'$ is at most $(2^{-O(R^{1/2})})^R = 2^{-O(R^{3/2})}$ Thus, the probability that *some* z is not in A' is at most $2^{-O(R^{3/2})+R} = 2^{-O(R^{3/2})}$. ∎

As $\mathbf{pP}[\mathbf{pP}] = \mathbf{pP}$, we have the following corollary.

Corollary 1 *If* $\mathbf{pRP}{=}\mathbf{pP}$, *then* $\mathbf{pBPP}{=}\mathbf{pP}$ *and in particular,* $\mathbf{BPP} = \mathbf{P}$.

Thus, if our aim is to prove that $\mathbf{pBPP}{=}\mathbf{pP}$, it is sufficient to prove the, a priori, easier statement $\mathbf{pRP}{=}\mathbf{pP}$. Corollary 1 is only known for promise problems. It is not known if $\mathbf{RP}{=}\mathbf{P}$ implies $\mathbf{BPP}{=}\mathbf{P}$. In fact, Buhrman and Fortnow show that there are oracles relative to which this implication fails. One should be aware that Corollary 1 does not mean that we can answer any question about $\mathbf{pBPP}$ by considering the corresponding question for $\mathbf{pRP}$. For instance, the statement "$\mathbf{pRP} \subseteq \mathbf{pSUBEXP}$ implies $\mathbf{pBPP} \subseteq \mathbf{pSUBEXP}$" has not been proven and its validity is not currently known (it does not even seem to be known if there is an oracle relative to which it fails). Also, the corollary is only relevant for deterministic simulation and not for randomness-efficient amplification nor for computing with weak random sources.

We now move to discuss classes defined by randomized computation with a bound on space, rather than a bound on time. There are certain subtleties and traps that one should be aware of when defining randomized space bounded computation. We refer to Saks, 1996, for an excellent exposition on this. We say that a language L is in $\mathbf{BP_H SPACE}(s(n))$ if there is a randomized machine M using space $O(s(n))$ on inputs of length n *and* with every computation eventually halting, so that for all $x \in L$, M accepts x with probability at least $\frac{2}{3}$ and for all $x \notin L$, M accepts x with probability at most $\frac{1}{3}$. We let $\mathbf{BP_H L}$ be $\mathbf{BP_H SPACE}(\log n)$. Simon, 1981, showed that $\mathbf{BP_H SPACE}(\log n)$ is included in $\mathbf{P}$ as well as $\mathbf{SPACE}(\log^2 n)$. An example of a language in $\mathbf{BP_H L}$ which is not known to be in **L** is UNDIRECTED GRAPH CONNECTIVITY, by the result of Aleliunas et al.. 1979.

We saw that estimating the acceptance probability of a Boolean circuit is complete for **pBPP**. We can get a complete problem for **pBP$_H$L** by replacing "circuit" with a simpler computational device. Let AUTOMATON ACCEPTANCE PROBABILITY ESTIMATION be the following promise problem. Given a deterministic finite automaton A over the alphabet $\{0,1\}$ and with n states, and with a single accepting state and with a promise that *either* (a) $\Pr[A \text{ accepts } x] \leq \frac{1}{3}$ *or* (b) $\Pr[A \text{ accepts } x] \geq \frac{2}{3}$ for a randomly chosen $x \in \{0,1\}^n$, we should tell which of (a) or (b) is the case.

Lemma 3 *AUTOMATON ACCEPTANCE PROBABILITY ESTIMATION is complete for* **pBP$_H$L**.

Proof First, AUTOMATON ACCEPTANCE PROBABILITY ESTIMATION is *in* **pBP$_H$L**: If the promise is kept we can see which of the two possibilities is the case by randomly picking a sufficient number of assignments and seeing if the majority is accepted (simulating the automaton on a particular input is easily done in logspace).

Second, let any partial language L in **pBP$_H$L** be given and let M be a **pBP$_H$L**-machine for L using space $O(\log n)$ and with running time $p(n) = n^{O(1)}$ on inputs of length n. We can without loss of generality assume that M has a single accepting and halting configuration and a single rejecting and halting configuration. We modify the machine so that once it reaches one of these two configurations, it stays there rather than halting. Given an input x of length n, we can in space $O(\log n)$ build x into the machine, getting a finite state automaton with $s = n^{O(1)}$ states. The input of the automaton is the sequence of coin tosses made by the original machine. Now $x \in L$, if and only if this automaton accepts a random input of length s with probability more than $\frac{2}{3}$ and $x \notin L$, if and only if this automaton accepts a random input of length s with probability less than $\frac{1}{3}$. ∎

An arguably even more natural complete problem for **pBP$_H$L** is APPROXIMATE STOCHASTIC MATRIX EXPONENTIATION: Given an $n \times n$ stochastic matrix M with entries taken from $0, 1, \frac{1}{2}$ and indices i, j, with a promise that $(M^n)_{ij}$ is either at least $\frac{2}{3}$ or at most $\frac{1}{3}$, tell which of the two is the case.

Lemma 4 *APPROXIMATE STOCHASTIC MATRIX EXPONENTIATION is complete for* **pBP$_H$L**.

Proof We present two reductions, one in each direction, between AP-PROXIMATE STOCHASTIC MATRIX EXPONENTIATION and AUTOMATON ACCEPTANCE PROBABILITY ESTIMATION.

Given an instance M, i, j of APPROXIMATE STOCHASTIC MATRIX EXPONENTIATION, where M is an $n \times n$ matrix, we can build a finite automaton A with n states $\{1, 2, \ldots, n\}$. The transition function δ of the automaton is defined as follows. Each row r of M either contains a single 1 in one entry (say, s) and the rest of the entries 0, or it contains two $\frac{1}{2}$'s (say, in entries $s_1 < s_2$) and the rest 0. In the first case, we let $\delta(i, 0) = \delta(i, 1) = j$. In the second case, we let $\delta(i, 0) = s_1$ and $\delta(i, 1) = s_2$. We let the start state of the automaton be i and the single accepting state be j. It is easy to see that the probability that the automaton so constructed accepts a random input $x \in \{0, 1\}^n$ is exactly $(M^n)_{ij}$, so the reduction is valid.

For the other direction, given an instance A of AUTOMATON AC-CEPTANCE PROBABILITY ESTIMATION with n states and labelled $1, 2, \ldots, n$ with start state 1 and single accepting state n, we can build an $n \times n$ matrix M with M_{rs} being the probability A goes from state r to state s in a single step. Then $M, 1, n$ is an instance of APPROX-IMATE STOCHASTIC MATRIX EXPONENTIATION and the reduction is easily seen to be valid. ∎

2.2 UNIVERSAL HASHING

The notion of a hash function is a 1950's notion, originating from the programming of space-efficient lookup tables. Originally, the concept was heuristic rather than rigorous: A hash function was a single fixed function that was assumed, in all respects, to behave like a random function. This is of course an impossibility and thus somewhat unsatisfactory from a theoretical point of view. The "modern" point of view was introduced in the seminal paper of Carter and Wegman, 1979. In modern terminology, we rarely speak about single hash functions in isolation, but only about families of hash functions. What we aim for is a small family of functions so that if one function is picked uniformly at random from the family, certain *specific* events occur with roughly the same probability as they would have, had the functions been picked uniformly at random from the space of all functions.

An important event is a *collision* between two elements x, y introduced by the hash function h, i.e., the event that $h(x) = h(y)$. Concentrating on such events leads to Carter and Wegman's definition of a universal class of hash functions. To be precise, let $\mathcal{H}$ be a family of functions with each $h \in \mathcal{H}$ mapping $\{0, 1\}^n$ to $\{0, 1\}^m$. We say that $\mathcal{H}$ is a *universal family of hash functions.* if, for all fixed $x, y \in \{0,1\}^n$, $x \neq y$ implies that if h is picked uniformly at random from $\mathcal{H}$, $\Pr[h(x) = h(y)] \leq 2^{-m}$.

We say that $\mathcal{H}$ is *strongly universal* if, for all fixed $x, y \in \{0,1\}^n$ and $a, b \in \{0,1\}^m$, we have that $x \neq y$ implies that if h is picked uniformly at random from $\mathcal{H}$, $\Pr[h(x) = a \wedge h(y) = b] = 2^{-2m}$.

A reformulation of the notion of a strongly universal family of hash functions is the notion of a pairwise independent sample space: The definition of strong universality is equivalent to stating that if h is picked uniformly at random from $\mathcal{H}$, then all random variables in the set $\{h(x)\}_{x \in \{0,1\}^n}$ are uniformly distributed and pairwise independent. Thus, constructing a small strongly universal family of hash functions mapping $\{0,1\}^n$ to $\{0,1\}^m$ is equivalent to constructing a small uniform probability space with 2^n associated random variables, each uniformly distributed in $\{0,1\}^m$ and pairwise independent.

Sometimes more than pairwise independence is needed. Given an integer $k \geq 2$, a family $\mathcal{H}$ is a k-*wise* independent family of hash functions mapping $\{0,1\}^n$ to $\{0,1\}^m$ if the random variables in the set $\{h(x)\}_{x \in \{0,1\}^n}$ are k-wise independent and uniform on $\{0,1\}^m$ when h is picked uniformly at random from $\mathcal{H}$.

Finally we say that $\mathcal{H}$ is k-*weakly universal* if for each $h \in \mathcal{H}$, if, for all fixed $x, y \in \{0,1\}^n$, $x \neq y$ implies that if h is picked uniformly at random from $\mathcal{H}$, $\Pr[h(x) = h(y)] \leq k2^{-m}$.

We shall need small, efficiently computable families of universal hash functions. A straightforward way of constructing such families is by using the properties of arithmetic over finite fields. In particular, if $\mathbf{F}$ is a finite field then the family of functions $\mathcal{H} = \{h_{a,b} : \mathbf{F} \to \mathbf{F} | a, b \in \mathbf{F}, h_{a,b}(x) = ax + b\}$ is strongly universal as is easily seen. If $|\mathbf{F}| = 2^n$, we can identify $|\mathbf{F}|$ with $\{0,1\}^n$. This identification can be made so that arithmetic is polynomial time and polylog space efficient, using a construction of Shoup, 1994. If we desire a universal family mapping $\{0,1\}^n$ to $\{0,1\}^m$ for $n \neq m$, we can easily modify the family above: If $n > m$, we take the family above and throw away $n - l$ bits of the outputs of the functions. If $n < m$, we take a family mapping $\{0,1\}^m$ to $\{0,1\}^m$ and take the restriction of every function in the family to a subdomain of size 2^n.

The construction easily generalizes to k-wise independent families: We take $\mathcal{H} = \{h_{a_0, a_1, \ldots, a_{k-1}} : \mathbf{F} \to \mathbf{F} \mid a_0, a_1, \ldots, a_{k-1} \in \mathbf{F}, h_{a_0, a_1, \ldots, a_{k-1}}(x) = a_0 + a_1 x + a_2 x^2 + \ldots a_{k-1} x^{k-1}\}$.

Summing up, we have

Lemma 5 *For every k, and every n, m, there is a k-wise independent family $\mathcal{H}$ of size at most $2^{k \max(n,m)}$ mapping $\{0,1\}^n$ to $\{0,1\}^m$, with* *each member in $\mathcal{H}$ described as a bit string of length $k\max(n,m)$, so that members in $\mathcal{H}$ can be evaluated in polynomial time and linear space.*

Sometimes we need a bit better performance than just polynomial time and linear space. For pairwise independence, we can easily do somewhat better: For any integer n, and $x, y \in \mathrm{GF}(2)^n$, the wrapped convolution $x \circ y \in \mathrm{GF}(2)^n$ of x and y is given by $(x \circ y)_i = \sum_{j=1}^n x_j y_{(i-j)\bmod n}$. Impagliazzo and Zuckerman, 1989, and Mansour et al., 1993, observed that $\mathcal{H} = \{h_{a,b} : \{0,1\}^n \rightarrow \{0,1\}^n | a \in \mathbf{F}, b \in \mathbf{F}, h_{a,b}(x) = a \circ x + b\}$ is strongly universal and, for $m \leq n$ $\mathcal{H} = \{h_a : \{0,1\}^n \rightarrow \{0,1\}^m | a \in \mathbf{F} - \{0\}, h_a(x) = \text{first } m \text{ bits of } a \circ x\}$ is universal. Unlike the family based on finite field, we can easily compute this family in logspace. Furthermore, wrapped Boolean convolution has nearly linear sized circuits by the classical result of Schönhage and Strassen, 1971. Summing up, we have

Lemma 6 *For every $m \leq n$, there is a strongly universal logspace computable family of size 2^{2n} mapping $\{0,1\}^n$ to $\{0,1\}^n$ with circuits of size $O(n\log n\log\log n)$, and a universal logspace computable family of size 2^n mapping $\{0,1\}^n$ to $\{0,1\}^m$ with circuits of size $O(n\log n\log\log n)$.*

Dietzfelbinger, 1996, presented a similar construction, yielding a k-wise independent family, for any $k \geq 2$ (his family is based on integer multiplication, rather than wrapped convolution). Thus, even k-wise independence can be done in logspace and with almost linear sized circuits. We shall not need this fact, however.

It is an interesting open problem if there is a strongly universal family with linear sized circuits. However, Andreev et al., 1997, noted that if we only want universality and not strong universality, and furthermore want the size of the co-domain to be only slightly smaller than the size of the domain, linear sized circuits are possible.

Lemma 7 *For any $n \geq 1$, there is a universal family $\mathcal{H}$ mapping $\{0,1\}^n$ to $\{0,1\}^{n-1}$ with the following properties.*

- *Each $h \in \mathcal{H}$ is two-to-one, i.e., each $y \in \{0,1\}^{n-1}$ has exactly two values x_1, x_2 in $\{0,1\}^n$, so that $h(x_1) = h(x_2) = y$.*

- *For each $h \in \mathcal{H}$, h and h^{-1} have linear sized circuits (the latter circuit returning two values).*

Proof Let h be given by $\mathcal{H} = \{h_a\}_{a \in \{0,1\}^n, a \neq 0}$ where h_a is the natural group homomorphism $h_a : \mathrm{GF}(2)^n \rightarrow \mathrm{GF}(2)^n/(a)$. Obviously, each member of $\mathcal{H}$ is two-to-one, and it is easily checked that $\mathcal{H}$ is universal.

Thus, we just need to find a representation of $GF(2)^n/(a)$ as bit vectors of length $n-1$ so that h and h^{-1} have linear circuits.

The representation we choose is as follows. Let i be any index in $\{1,\dots,n\}$ for which $a_i \neq 0$. Let $\tilde{x}$ be a member of $GF(2)^n/(a)$. We can view $\tilde{x}$ as a two-element subset of $GF(2)^n$ and may note that exactly one element of the subset, say x, has $x_i = 0$. We represent $\tilde{x}$ as the bit string x with the i'th bit removed. It is easily checked that with this representation, h_a and h_a^{-1} have linear sized circuits and we are done. ∎

When considering weakly universal family, we clearly want to gain something compared to the bounds above. It turns out that we can make the size of weakly universal families much smaller than the size of universal families. A simple way to construct a small k-weakly universal family for an integer $k \geq 2$ is to use univariate polynomials over a finite field again, but switch the roles of the hash function and the value to be hashed. To be precise, if we take $\mathcal{H} = \{h_x : \mathbf{F}^k \rightarrow \mathbf{F} \mid x \in \mathbf{F}, h_x(a_0, a_1, \dots, a_k) = a_0 + a_1 x + a_2 x^2 + \dots a_{k-1} x^{k-1}\}$, we get a k-weakly universal family, so we have

Lemma 8 *For every k, and every n, there is a k-weakly universal family $\mathcal{H}$ of size 2^n mapping $\{0,1\}^{kn}$ to $\{0,1\}^n$ that can be evaluated in polynomial time and linear space.*

By a somewhat more involved construction one can get small k-weakly universal families of comparable size with k very close to 1. Goldreich and Wigderson, 1994, construct such a family which we state without proof.

Lemma 9 *For every $\epsilon > 0$, and every n, m, there is a $(1 + \epsilon)$-weakly universal family $\mathcal{H}$ of size $(n2^m/\epsilon)^{O(1)}$ mapping $\{0,1\}^n$ to $\{0,1\}^m$ that can be evaluated in polynomial time and linear space.*

2.3   ERROR CORRECTING CODES

The original motivation of the theory of error correcting codes is data transmission over a noisy channel. We consider the following setup. A *sender* wants to transmit a bit string x to a *receiver*. However, a *meddler* is allowed to introduce a limited amount of noise in any message sent from the sender to the receiver by flipping a certain fraction of the bits in the message. How does the sender transmit the message reliably?

The solution is to use an *error correcting code*. Though we are mainly interested in Boolean codes, we need the concept of a code over an arbitrary set in order to construct good Boolean codes. Let n, m be integers with $n < m$ and let S be a set of size at least 2. An *error correcting* code over S with *relative minimum distance* δ is a map $e : S^n \to S^m$, so that for any $x \neq y$, the fraction of entries where $e(x)$ and $e(y)$ differ (i.e., the relative Hamming distance between $e(x)$ and $e(y)$), is at least δ. The set $\{e(x) | x \in S^n\}$ is the set of *code words* of e. The *rate* of the code is the value n/m. For the applications we consider, it is often more convenient to speak of the *blowup* of the code rather than the rate. The blowup is simply the value m, as a function of n. For example, if $m = n^2$, we speak about a code with quadratic blowup. If $S = \{0, 1\}$, the code is a *Boolean* code. If the original message x is a prefix of the encoded message $e(x)$, i.e. $e(x) = x \circ e'(x)$ for some map e', we say that the code is *systematic*. If S is a field and the set of code words form an n-dimensional subspace of the vector space over S of dimension m, we say that the code is *linear*. All codes considered in this chapter will be linear.

Let us see that error correcting codes solve the data transmission problem. Let e be a Boolean error correcting code with relative minimum distance δ. Given $x \in \{0, 1\}^n$, if the sender computes $\tilde{x} = e(x)$ and the meddler flips less than a $\delta/2$ fraction of the bits of $\tilde{x}$, and transmits the result x^* to the receiver, the receiver can reconstruct the original message as the value x minimizing the Hamming distance between $e(x)$ and x^*. This procedure is called *maximum likelihood decoding*. Note that it as stated is an exponential procedure. Sometimes we can find more efficient procedures.

Coding theory is concerned with the construction of codes with small blowup and large relative minimum distance. In the traditional applications of coding theory, often a linear blowup $n \to O(n)$ or even $n \to n + o(n)$ is required and lots of effort is spent constructing such codes. For the applications to derandomizing complexity classes, it is sufficient to consider codes with polynomial blowup. Such codes are rather easily constructed, as we see next.

Let $\mathbf{F}$ be a finite field of size $m > n$. The *Reed-Solomon* code over $\mathbf{F}$ is the code $e_{\mathrm{RS}} : \mathbf{F}^n \to \mathbf{F}^m$ defined as follows. Let $i_1, i_2, \ldots, i_n$ be arbitrary distinct members of $\mathbf{F}$. Given $x \in \mathbf{F}^n$, let p be the unique polynomial of degree at most $n - 1$, so that $p(i_j) = x_j$ for all $j \in \{1, \ldots, n\}$. Let $e_{\mathrm{RS}}(x)$ be $(p(y))_{y \in \mathbf{F}}$, i.e., the enumeration of p on all members of $\mathbf{F}$. As every two distinct polynomials of degree at most $n - 1$ agree on at most $n - 1$ points, we find that the Reed-Solomon code has relative minimum distance $\frac{m-n+1}{m}$. Thus, if we let $m = n^c$ for a constant c, we get a code with polynomial blowup and relative minimum distance close to 1.

The Reed-Solomon code has an efficient maximum likelihood decoding procedure. More generally, such a decoding procedure exists even if we modify the code so that the encoding of x is not the enumeration of p over *all* members of $\mathbf{F}$ but only on some subset containing $i_1, i_2, \ldots, i_n$. This was shown by Berlekamp and Welch, 1986. We state the existence of this procedure as a lemma.

Lemma 10 *There is a polynomial procedure, which, given as input a subset $M \subseteq \mathbf{F}$ of size m and the tabulation of a function $f : M \to \mathbf{F}$ which disagree with some polynomial p of degree at most d on less than a $\frac{1}{2} - \frac{d}{2m}$ fraction of M, outputs the unique such polynomial p.*

The Reed-Solomon code is a great code. Its only problem is that it is not Boolean. For Boolean codes, the *Plotkin bound* states that any Boolean code with relative minimum distance $d > \frac{1}{2}$ has at most $\frac{2d}{2d-1}$ code words, i.e., if d is bounded away from $\frac{1}{2}$ by a constant, the number of code words is constant. Thus, for asymptotically non-trivial Boolean codes, the best relative minimum distance we can hope for is $\frac{1}{2}$. The *Hadamard* code achieves this. Given any integer value l, the Hadamard code $e_H : \{0,1\}^l \to \{0,1\}^{2^l}$ is defined $e_H(x) = (\langle x, y \rangle)_{y \in \{0,1\}^l}$, i.e., we encode x as the inner products of x with all possible vectors y. It is easy to see that the relative minimum distance of the Hadamard code is $\frac{1}{2}$: Given fixed vectors $x_1 \neq x_2$, the probability that a randomly chosen vector y has $\langle x_1 - x_2, y \rangle = 0$ is exactly $\frac{1}{2}$. Thus, the Hadamard code has exponential blowup but a good minimum distance.

To get a Boolean code with satisfactory blowup and relative minimum distance, we introduce the concept of *code concatenation*, which we define by the example relevant to us: Given the Reed-Solomon code $e_{RS} : \mathbf{F}^n \to \mathbf{F}^m$ for a field of size m, where $m = 2^l$ take any encoding c of field elements as bit vectors of length l, and consider the Hadamard code e_H with parameter l. Now we define the concatenation of the Reed-Solomon code with the Hadamard code $e_{RS} \circ e_H$ as the Boolean code $e_{RS} \circ e_H : \{0;1\}^n \to \{0,1\}^{m2^l}$, defined by

$$e_{RS} \circ e_H(x) = e_H(c(y_1)) \circ e_H(c(y_2)) \circ \cdots \circ e_H(c(y_m))$$

where $y = y_1 y_2 \ldots y_m = e_{RS}(x)$ and $\circ$ denotes concatenation of strings. In short, we take the Reed-Solomon code of x (viewing the bits of x as the field elements 0 and 1 of $\mathbf{F}$) and encode each of the field elements of the resulting code word in the Hadamard code.

Lemma 11 *Given any parameters $n \leq m$ with m a power of two, the concatenation of the Reed-Solomon code with the Hadamard code $e = e_{RS} \circ e_H : \{0,1\}^n \to \{0,1\}^{m^2}$ is a Boolean code with minimum distance $\frac{1}{2} - \frac{n-1}{2m}$.*

Proof Take any two different code words of e. The corresponding Reed-Solomon code words differ on at least $m - n + 1$ entries, i.e., a $\frac{m-n+1}{m}$ fraction of all the entries. Where the Reed-Solomon code words differ, the Boolean Hadamard encoding of the corresponding field element differ on half the entries. Thus, the code word of e differ on at least a $\frac{m-n+1}{m} \cdot \frac{1}{2}$ fraction of the bits. ∎

The code e of Lemma 11 will satisfy all the "generic" coding needs of this chapter. In fact this is the case for the following specific setting of the parameters. Note that the code can be made systematic (so that x is a prefix of $e(x)$) by an appropriate renumbering of the bit positions of the code words.

Lemma 12 *For n a power of two, there is a polynomial time computable, linear, systematic code $e : \{0,1\}^n \to \{0,1\}^{n^{10}}$ with relative minimum distance at least $\frac{1}{2} - \frac{1}{n^4}$.*

If an error correcting code has minimum distance d, and we are given a vector which has distance less than $d/2$ to a code word, maximum likelihood decoding finds this code word. Thus, if $d = \frac{1}{2} - \epsilon$, we can correct up to a $\frac{1}{4} - \epsilon$ fraction of errors. For some applications we shall see, this correction distance is not good enough. For such applications we can consider *list decoding*: If we are content with finding not one original code word, but a small list of possibilities, we can do much better. The following lemma, due to Bellare et al., 1998, Lemma A.1, and stated here without proof (the proof is a clever, but elementary, counting argument), expresses this fact.

Lemma 13 *Let $0 < \alpha < \frac{1}{2}$ and let a Boolean code of relative minimum distance $\frac{1}{2} - \alpha$ be given. The number of code words within distance $\frac{1}{2} - \alpha^{\frac{1}{2}}$ from any given string is at most $\frac{1}{2\alpha}$.*

For the special case of the code of Lemma 12, we get

Lemma 14 *Any Hamming ball in $\{0,1\}^{n^{10}}$ of relative radius $\frac{1}{2} - \frac{1}{n^2}$ contains at most $\frac{n^4}{2}$ code words of e, where e is the code of Lemma 12.*

Finally, we shall need a connection between error correcting codes and universal hashing. In Miltersen, 1998, it was observed that there is a very simple weakly universal family of hash functions for the set of code words for an error correcting code. Let the family $\mathcal{H}$ of *projections* be the family of functions of the form $h_S : \{0,1\}^m \to \{0,1\}^l$, where S is an l-subset of $\{1, 2, \ldots, m\}$ and where $h(x)$ is the concatenation of the bits of x, indexed by S. For example, $h_{\{2,3,5\}}(011000101) = 110$. Unlike the hash functions of the previous subsection, projections are trivial to compute and intuitively, "does not mess up the data". Still, we have the following lemma.

Lemma 15 *Let $C \subseteq \{0,1\}^m$ be the set of code words of an error correcting code of relative minimum distance $\frac{1}{2} - \epsilon$. The projection family $\mathcal{H}$ is a $(1 + 2\epsilon)^l$-weakly universal family from C to $\{0,1\}^l$.*

Proof Let $\delta = \frac{1}{2} - \epsilon$. Given $x, y \in \{0,1\}^m$, we must show that a random element h_S of $\mathcal{H}$ has $h_S(x) = h_S(y)$ with probability at most $(1 + 2\epsilon)^l 2^{-l} = (1 - \delta)^l$.

We can pick a random element $h_S \in \mathcal{H}$ by picking each element of S at random, one at a time, without replacement. The final hash function h_S will have $h_S(x) \neq h_S(y)$ if at least one of the bits chosen is a bit where x and y differ. Suppose we have picked elements $i_1, i_2, \ldots, i_j$ of S and have not separated x and y. We pick i_{j+1} at random from $\{1, \ldots, m\} - \{i_1, i_2, \ldots, i_j\}$. We can bound from below the conditional probability that the inclusion of i_{j+1} will separate x and y, by $\frac{\delta m}{m-j} \geq \delta$. Thus, the probability that none of the chosen elements in the final set S will separate x and y is at most $(1 - \delta)^l$, as desired. ∎

3.   DISPERSERS AND EXTRACTORS

3.1   DISPERSERS

Consider the following, purely combinatorial, two-player game. Let N be a positive integer and let $\epsilon > 0$. The first player, Alice, secretly chooses at least $(1 - \epsilon)N$ cards from a deck of N cards and puts a mark on the face of the cards chosen. Then she places all the cards in the deck, face down, before the second player, Bob. Now Bob must turn over at most d cards. If a mark is revealed, Bob wins the game.

If $d \leq \epsilon N$, no *deterministic* strategy for Bob can be good: If Alice knows the strategy, she can simply mark a set of cards she knows that Bob will not choose and Bob loses. On the other hand, a good and simple *randomized* strategy for Bob would be to choose the d cards to turn over randomly, without replacement, yielding a loosing probability for Bob which can be bounded from above by ϵ^d.

For choosing d cards independently at random Bob needs to use approximately $d \log N$ random bits. We now add a twist to the game: We want Bob to adhere to a randomized strategy *using much fewer random bits*. Formally, we define an *R-bit strategy* for Bob to be a map ϕ from $\{0,1\}^R$ to the set of subsets of $\{1, \ldots, N\}$ of size at most d. Bob executes the strategy by picking $x \in \{0,1\}^R$ uniformly at random and turning over the cards of $\phi(x)$.

The general task we now face is, for given values of N, ϵ, R, and d to find an R-bit strategy minimizing Bob's chance of loosing the game, or at least making this chance strictly smaller than some value p. Let us (temporarily, until Proposition 1) denote such a strategy an (N, ϵ, R, d, p)-*strategy*.

It turns out that we can conveniently rephrase the notion of a strategy in the language of bipartite graphs.

Definition 1 *A bipartite graph on vertex sets U and V and with edges $E \subseteq U \times V$ is called an $(1 - \epsilon)$-hitting disperser with threshold T if, for all subsets S of U of size $|S| \geq T$, we have that the size of the set of neighbors of S in V, $|\Gamma(S)|$, is more than $\epsilon|V|$.*

Given an R-bit strategy ϕ, we can form a bipartite graph with $U = \{0,1\}^R$ and $V = \{1, \ldots, N\}$ and edges defined by $\Gamma(x) = \phi(x)$ for $x \in U$, where $\Gamma(x)$ is the set of neighbors in V of x. We now have

Proposition 1 *A strategy is an (N, ϵ, R, d, p)-strategy if and only if the corresponding graph is a $(1 - \epsilon)$-hitting disperser with threshold $p|U|$.*

Proof First, we show that the graph of a good strategy is a good disperser. So, let the graph of an (N, ϵ, R, d, p)-strategy be given. We must show that in the corresponding graph, every U-subset of size at least $p|U|$ has more than $\epsilon|V|$ neighbors. Suppose not. Then some U-subset S of size at least $p|U|$ has at most $\epsilon|V|$ neighbors. This means that there is a subset S' of $|V|$ of size at least $(1 - \epsilon)|V|$ and no edges between S and S'. Now we play the game between Alice and Bob with S' being the set of marked cards. By definition of an (N, ϵ, R, d, p)-strategy, if Bob chooses a random vertex $x \in U$, the probability that the set of neighbors of x does not intersect S' is less than p. Thus, this is the case for less than $p|U|$ vertices x in U. On the other hand, it is the case for all the vertices of S, and $|S| \geq p|U|$, a contradiction.

Next, we show that a good disperser defines the graph of a good strategy. So let G be a $(1 - \epsilon)$-hitting disperser on $U \times V$ with threshold $p|U|$ and appropriate degree and size of U and V. We want to show that if we take any subset S' of V of size at least $(1 - \epsilon)|V|$ and a random vertex x of U, the probability that S' does not contain a neighbor of x is less than p. So suppose it is at least p, and let S be the subset of U which is not neighboring S'. As $|S| \geq p|U|$, by the disperser-property, we have that the set of neighbors of S has size more than $\epsilon|V|$. But this is impossible, as $|S'| \geq (1 - \epsilon)|V|$. ∎

It is sometimes convenient to describe a disperser as a *map* $D :$ $U \times \{1, 2, \ldots, d\} \to V$ instead of a bipartite graph, with the natural correspondence between maps and graphs in mind, i.e., $D(x, i)$ is the i'th neighbor of x, relative to some arbitrary enumeration of its neighbors.

It is immediate how we can use strategies developed for the card game to solve the randomness-efficient amplification problem for a one-sided error randomized algorithm deciding a language L: If we have a randomized algorithm for L using r random bits on instances of a certain size and with error probability at most ϵ, we let $N = 2^r$, and let the cards of Alice be the possible sequences of r random coin tosses. For a positive instance $x \in L$, we let the marked cards be the coin toss sequences leading to acceptance of x. Now we get the following randomized algorithm deciding L and using R random bits: Take a random element $y \in \{0, 1\}^R$. Compute the set $\phi(y)$. Execute the original algorithm on input x with the set of coin toss sequences in $\phi(x)$. If one of them leads to acceptance, we accept, otherwise we reject. The error probability of our algorithm is at most Bob's risk of loosing the game.

To solve the amplification problem in the way we described above while maintaining efficiency, we need the computation of $\phi(y)$, given y to be efficient. The notion of an efficient strategy, translated into the language of graphs, becomes the notion of an *explicit disperser*.

Definition 2 *A family of dispersers $E_r \subseteq U_r \times V_r$ with $|V_r| = 2^r$ is* explicit, *if there is an algorithm which on input r and $x \in U_r$ enumerates the neighbors of x in V_r, in time polynomial in $r, \log |U_r|$ and the number of such vertices.*

The discussion above yields the following "master lemma".

Lemma 16 *If there is an explicit family of $\frac{1}{2}$-hitting dispersers $E_r \subseteq U_r \times V_r$, $r = 1, 2, \ldots$ with $|V_r| = 2^r$, $|U_r| = 2^{R(r)}$, degree $d(r)$ for each vertex in U_r, and threshold $T(r)$, then a one-sided error algorithm running in time t, using r random bits to achieve error probability $\frac{1}{2}$ can be converted to a one-sided error algorithm, running in time $(R(r) + d(r) + t)^{O(1)}$ using $R(r)$ random bits to achieve error probability $T(r)/2^{R(r)}$.*

We will next show that, perhaps somewhat surprisingly, dispersers not only immediately solve the amplification problem, but also the weak random source problem, for one-sided error algorithms.

Lemma 17 *If there is an explicit family of $\frac{1}{2}$-hitting dispersers $E_r \subseteq U_r \times V_r$, $r = 1, 2, \ldots$ with $|V_r| = 2^r$, $|U_r| = 2^{R(r)}$, degree $d(r)$ for each* *vertex in U_r, and threshold $T(r)$, then a one-sided error algorithm running in time t, using r random bits to achieve a error probability $\frac{1}{2}$ can be simulated by an algorithm running in time $(R(r)+d(r)+t)^{O(1)}$ and using any $R(r)$-bit weak random source with min-entropy at least $\log T(r) + k$ to achieve error probability 2^{-k}.*

Proof By lemma 16, the disperser enables us to reduce the error probability of the algorithm to $T(r)/2^{R(r)}$, using $R(r)$ unbiased random bits. Now consider just giving this modified algorithm a sample of the weak random source instead of the coin toss sequence. As the error probability is at most $T(r)/2^{R(r)}$, at most $T(r)$ specific coin toss sequences make the algorithm give the wrong output. As the source has min-entropy at least $\log T(r) + k$, each of these bad sequences will be given with probability at most $\frac{1}{2^k T(r)}$. Thus, the probability of getting a bad sequence is at most 2^{-k}. This is also a bound on the error probability of the modified algorithm using the weak random source. ∎

3.2   EXTRACTORS

We shall next see what the corresponding definitions would look like for two-sided error algorithms. We should consider a modified card game between Alice and Bob. Again, Alice marks at least a $1 - \epsilon$ fraction of the cards, but now, it is not sufficient for Bob that one of the cards he turns over is marked: *More than half of them should be marked.* By the Chernoff bound, choosing cards uniformly at random is a good strategy for Bob. Again, we want to find strategies using fewer random bits. We can use a strategy for Bob to do randomness-efficient amplification for two-sided error algorithms, as the strategy would also enable Bob to distinguish between a situation where Alice marks at least a $1-\epsilon$ fraction of the cards from a situation where she marks at most an ϵ fraction of the cards. As we did above, we can turn the definition of such a strategy into a graph theoretic property of bipartite graphs. This would lead to the definition of what is known as a *majority disperser*.

While explicit majority dispersers are sufficient for both randomness-efficient amplification and computing with weak random sources, we choose in this chapter to focus on a somewhat stronger property, leading to the graph theoretic notion of an *extractor*. An extractor is also a majority disperser and a disperser, but extractors have many other applications as well. In particular, they are better suited than dispersers and majority dispersers to being used as building blocks in other constructions.

Before defining extractors, let us introduce and analyze a closeness measure for probability distributions. We say that two distributions $\mathcal{D}_1$ and $\mathcal{D}_2$ (defined on the same domain) are ϵ-close if their L_1-distance $\|\mathcal{D}_1 - \mathcal{D}_2\|_1$ is at most ϵ, i.e., if the sum, over all possible outcomes, of the distance between the two probabilities of the outcome according to the two distributions, is at most ϵ. The following lemma is easily proved and very useful.

Lemma 18 *Two distributions $\mathcal{D}_1$ and $\mathcal{D}_2$ on domain S are ϵ-close if and only if, for any event $T \subseteq S$, $|\Pr_{x \in \mathcal{D}_1}[x \in T] - \Pr_{x \in \mathcal{D}_2}[x \in T]| \leq \epsilon/2$*

In Lemma 18, one should think of T as a *statistical test*, trying to distinguish between $x \in \mathcal{D}_1$ and $x \in \mathcal{D}_2$. Thus, we also say that $\mathcal{D}_1$ and $\mathcal{D}_2$ are *statistically close*.

Statistical closeness is preserved under deterministic and random transformations.

Lemma 19 *If x_1 is a random variable with distribution $\mathcal{D}_1$ and x_2 is a random variable with distribution $\mathcal{D}_2$ and $\mathcal{D}_1$ and $\mathcal{D}_2$ are ϵ-close, then for any function f, $f(x_1)$ and $f(x_2)$ have distributions which are ϵ-close. Furthermore if y is a random variable independent of x_1 and x_2, $f(x_1, y)$ and $f(x_2, y)$ have distributions which are ϵ-close.*

We present two lemmas which are useful for proving closeness when one of the two distributions is the uniform distribution.

Let $\mathcal{D}$ be a probability distribution. The *collision probability* of $\mathcal{D}$ is the probability that two independent samples from $\mathcal{D}$ are identical. The first lemma is due to Impagliazzo et al., 1989.

Lemma 20 *Let $\mathcal{D}$ be a probability distribution on a domain of size s. If the collision probability of $\mathcal{D}$ is at most $(1 + \epsilon)/s$, then $\mathcal{D}$ is $\sqrt{\epsilon}$-close to uniform.*

Proof Let $v(x) = \mathcal{D}(x) - 1/s$. Then, if we view v as a real vector of length s, we want to bound the one-norm of v, as this is the distance between $\mathcal{D}$ and the uniform distribution.

Viewing $\mathcal{D}$ as a vector, the collision probability of $\mathcal{D}$ is $\langle \mathcal{D}, \mathcal{D} \rangle$, so we have $\langle \mathcal{D}, \mathcal{D} \rangle \leq (1 + \epsilon)/s$. But $\langle \mathcal{D}, \mathcal{D} \rangle = \langle v + 1/s, v + 1/s \rangle = \langle v, v \rangle + \langle 1/s, 1/s \rangle + 2 \langle v, 1/s \rangle = \langle v, v \rangle + 1/s \leq (1 + \epsilon)/s$, so $\langle v, v \rangle \leq \epsilon/s$. By the Cauchy-Schwartz inequality $\|v\|_1 \leq \sqrt{s} \sqrt{\langle v, v \rangle} \leq \sqrt{\epsilon}$. ∎

The second lemma is due to Trevisan, 1999, building on Yao, 1982. Here, we consider a probability distribution $\mathcal{D}$ on $\{0, 1\}^m$, rather than an arbitrary domain and take the structure of $\{0, 1\}^m$ into account. We say that a probability distribution $\mathcal{D}$ on $\{0,1\}^m$ is ϵ-*predictable* if for some i between 1 and m, there is a function (the *predictor*) $f : \{0,1\}^{i-1} \to \{0,1\}$ which predicts the i'th bit of $x \in \mathcal{D}$ well, given the previous bits, in the following sense. If $x \in \{0,1\}^n$ is chosen randomly according to $\mathcal{D}$, then

$$\Pr[f(x_{1..(i-1)}) = x_i] \geq \frac{1}{2} + \epsilon.$$

The value ϵ is called the *advantage* of the predictor.

Lemma 21 *If a distribution $\mathcal{D}$ on $\{0,1\}^m$ is not ϵ-close to uniform, then it is $\frac{\epsilon}{2m}$-predictable.*

Proof We consider the $m+1$ distributions $\mathcal{D}_0, \ldots, \mathcal{D}_m$, defined as follows. To pick a random element of $\mathcal{D}_i$, we pick $x \in \{0,1\}^m$ according to $\mathcal{D}$ and $y \in \{0,1\}^m$ according to the uniform distribution $\mathcal{U}$ and return $x_1, \ldots, x_i, y_{i+1}, \ldots, y_m$. Thus, $\mathcal{D}_0$ is the uniform distribution, $\mathcal{D}_m$ is $\mathcal{D}$, and the $\mathcal{D}_i$'s in between are *hybrids* of the uniform distribution and $\mathcal{D}$.

If $\| \mathcal{D} - \mathcal{U} \|_1 > \epsilon$, then, by the triangle inequality, there must be i so that $\| \mathcal{D}_{i+1} - \mathcal{D}_i \|_1 > \epsilon/m$. Let $\mathcal{D}'_{i+1}$ and $\mathcal{D}'_i$ be the distribution induced by $\mathcal{D}_{i+1}$ and $\mathcal{D}_i$ by removing all entries in the vectors except the first $i+1$ ones. Clearly, $\| \mathcal{D}'_{i+1} - \mathcal{D}'_i \|_1 = \| \mathcal{D}_{i+1} - \mathcal{D}_i \|_1 > \epsilon/m$. Also note that we can sample $\mathcal{D}'_i$ by sampling $\mathcal{D}'_{i+1}$ and replacing the last bit of the sample with an unbiased random bit.

Now, a predictor $f : \{0,1\}^i \to \{0,1\}$ is given as follows: Given input $y \in \{0,1\}^i$, the value 0 is predicted if for random $x \in \{0,1\}^{i+1}$ according to $\mathcal{D}'_{i+1}$, $\Pr[x_{i+1} = 0 | x_1 = y_1 \wedge x_2 = y_2 \wedge \ldots \wedge x_i = y_i]$ is bigger than $\frac{1}{2}$. Otherwise 1 is predicted. A calculation shows that the advantage of the predictor is exactly $\frac{1}{2} \| \mathcal{D}'_{i+1} - \mathcal{D}'_i \|_1$. ∎

Having introduced the notion of closeness of distributions, we are ready to define an extractor.

Definition 3 *A bipartite (multi-)graph on vertex sets U and V and with edges $E \subseteq U \times V$ so that every vertex x in U has exactly d neighbors is called an* extractor *with error ϵ and min-entropy threshold t if the following holds: If $x \in U$ is chosen according to a any distribution with min-entropy at least t, and a vertex $y \in V$ is then chosen uniformly at random from the neighbors of x, the resulting distribution of $y \in V$ is ϵ-close to uniform.*

The right way to think of an extractor is as a *randomness-refiner*, taking as input a weak random input (i.e., $x \in U$) and a short uniformly random *seed* (i.e., a random outgoing edge from x) and outputs a value, nearly uniformly distributed (i.e., y). The extractor *extracts* the min-entropy from the source U, using $\log d$ truly random bits, and producing $\log |V|$ bits of pure randomness. We shall often describe an extractor as a *map* $E : U \times \{1, 2, \ldots, d\} \to V$ instead of a bipartite graph (with the natural correspondence between maps and graphs in mind).

It is easy to see that extractors are also dispersers.

Proposition 2 *An extractor with error strictly less than ϵ and min-entropy threshold t is also a $\frac{\epsilon}{2}$-hitting disperser with threshold 2^t.*

Furthermore, extractors are also majority dispersers. This means that they solve the randomness-efficient amplification and weak random source problems for two-sided error algorithms.

Lemma 22 *If there is an explicit family of extractors $E_r \subseteq U_r \times V_r$, $r = 1, 2, \ldots$ with $|V_r| = 2^r$, $|U_r| = 2^{R(r)}$, degree $d(r)$, min-entropy threshold $t(r)$, and error less than $\frac{1}{3}$, then a two-sided error algorithm running in time T, using r random bits to achieve a error probability of $\frac{1}{3}$ can be converted to a two-sided error algorithm, running in time $(R(r) + d(r) + T)^{O(1)}$ using $R(r)$ random bits to achieve error probability $2^{t(r)-R(r)}$.*

Proof The simulating algorithm uses its $R(r)$ bit input as an index to a node $x \in U_r$. The neighbors of x in V_n is $y_1, y_2, \ldots, y_{d(r)} \in \{0,1\}^r$. The simulating algorithm simulates the original algorithm on coin toss sequences $y_1, y_2, \ldots, y_r$ and accepts if the majority of these computations accept. We now show that this has the desired probability of error. Suppose without loss of generality that the right behavior is "accept", i.e., that at least a $\frac{2}{3}$ fraction of V_r corresponds to accepting computations of the original algorithm. We need to show that the error probability is less than $2^{t(r)-R(r)}$, i.e., that less than $2^{t(r)}$ choices of x causes reject. Suppose that the set of choices of x leading to reject has size at least $2^{t(r)}$. Then the uniform distribution on these choices has min-entropy at least $t(r)$, and thus picking a random one of the choices x and a random neighbor y of x is $\frac{1}{3}$- close to the uniform distribution on V_r. But now consider the statistical test $T \subset V_r$ consisting of all accepting coin toss sequences of the original algorithm. The probability that a uniform sample from V_r is in T is at least $\frac{2}{3}$. Thus, by Lemma 18 the probability that y generated as above is in T is at more than $\frac{1}{2}$. But this is a contradiction, as for *each* of the choices of x leading to reject, the majority of neighbors are not in T. ∎

By a similar proof, we have the corresponding result for computing with weak random sources.

Lemma 23 *If there is an explicit family of extractors $E_r \subseteq U_r \times V_r$, $r = 1, 2, \ldots$ with $|V_r| = 2^r$. $|U_r| = 2^{R(r)}$, degree $d(r)$, min-entropy threshold $t(r)$, and error less than $\frac{1}{3}$, then a two-sided error algorithm running in time T, using r random bits to achieve an error probability $\frac{1}{3}$ can be simulated by an algorithm running in time $(R(r) + d(r) + T)^{O(1)}$ and using any $R(r)$-bit weak random source with min-entropy at least $t(r) + k$ to achieve an error probability of 2^{-k}.*

3.3   NOTES

The notion of dispersers and majority disperser and the observation that they capture the problems of randomness efficient amplification and computing with weak random sources is due to Cohen and Wigderson, 1989. Even earlier, Sipser, 1988, had used a disperser graph for randomness-efficient amplification, without giving his graph a name. The stronger notion of extractors and the realization of their importance is due to Nisan and Zuckerman, 1996, though extractors were implicitly used in constructions beginning with Impagliazzo et al., 1989.

We have shown that dispersers can be used for randomness efficient amplification for one-sided error algorithm and extractors for randomness efficient amplification for two-sided error algorithms. A clever construction by Andreev et al., 1999 shows that dispersers can be used for randomness efficient amplification for two-sided error algorithms as well.

4.   CONSTRUCTIONS OF DISPERSERS AND EXTRACTORS

There are essentially three known ways of constructing explicit dispersers and extractors. The first is based on explicit expander constructions. The second is based on iterated universal hashing. The third is based on error correcting codes and Nisan-Wigderson designs. We shall describe all three ways in the next three subsections, but only the technique using error correcting codes and Nisan-Wigderson designs will be described in full detail. This technique was only discovered recently, by Trevisan, 1999. It is simpler than the other two techniques, and for some important settings of the parameters it is the only known technique that will work.

4.1   EXPANDER BASED DISPERSERS

The explicit expander constructions of Margulis, 1973, Gabber and Galil, 1981, and Lubotzky et al., 1986, form an important way of building explicit dispersers and extractors. Historically, it was also the first way.

Let $G = (V, E)$ be an undirected d-regular graph. We say that G is a c-expander, for $c > 0$, if for any subset V_1 of V with $|V_1| \leq |V|/2$, we have $|V_1 \cup \Gamma(V_1)| \geq (1+c)|V_1|$, where $\Gamma(V_1)$ are vertices adjacent to some vertex in V_1.

In the papers mentioned above, families of explicit expanders with $|V| \to \infty$ and the parameters $c > 0$ and d being constants were constructed. The definitions of the expanders are generally not very involved. Particularly easy to describe is the 5-regular Gabber-Galil expander (V, E) with size $V = 2m^2$, for any desired integer m. Identifying V with $\{0, 1\} \times (\mathbf{Z}/m\mathbf{Z})^2$, each vertex (b, x, y) has edges to $(1 - b, x, y)$, $(1 - b, x, x + y)$, $(1 - b, x, x + y - 1)$, $(1 - b, x + y, y)$ and $(1 - b, x + y + 1, y)$. The graph is a c-expander for a small constant $c > 0$.

The proofs that the graphs defined are in fact expanders are, on the other hand, very involved, generally proceeding as follows. First, one appeals to a correspondence between the expansion properties of a graph and the second largest eigenvalue of its adjacency matrix discovered and described by Tanner, 1984, Alon and Milman, 1985, and Alon, 1986, and then one analyzes the eigenvalues of the adjacency matrix of the graph described above using various tools, such as Fourier analysis and number theory. The details are most certainly beyond the scope of this chapter!

The definition of an expander has the flavor: *Any set of vertices has "many" neighbors.* Note that the definition of a disperser has the same flavor (though with a different interpretation of "many" and a restriction of "any"), so it is not too surprising that we can use expanders to construct dispersers.

There are two known ways of constructing a disperser (U, V, E) from an expander $G = (V', E')$.

The first construction, due to Karp et al., 1986, is direct and very easy (once the expansion property of the graph is known). We let $U = V = V'$ and, for some parameter $l \geq 1$, we let $E = \{(u, v)|$ there is a path of length at most l from u to v in $G\}$. It follows directly from the definitions that, for any $\epsilon > 0$, (U, V, E) is an $(\frac{1}{2} + \epsilon)$-hitting disperser with threshold $\frac{1}{2}(1 + c)^{-l}$ and degree at most d^{l+1}.

Using this construction on, for instance, the Gabber-Galil expanders, we get the following theorem.

Theorem 3 *For any desired constant k, there is an explicit $\frac{1}{2}$-hitting disperser $E_r \subset V_r \times V_r$ with $|V_r| = 2^r$, degree $r^{O(1)}$ and threshold $2^r/r^k$.*

Actually, when proving the theorem, we can only use the Gabber-Galil expander if r is odd, as we then have that 2^r is of the form $2m^2$. However, we can always reduce the number of random bits used by an additive constant, by trying all possibile settings of the bits removed. Combining Theorem 3 with Lemma 16, we get Statement 1 of the introduction (for one-sided error). A slightly more involved argument gives majority dispersers and extractors with similar parameters, establishing Statement 1 for two-sided error algorithms. We refer to Cohen and Wigderson, 1989, Alon and Spencer, 1992, and Goldreich and Wigderson, 1994 for details. In particular, the extracting property of the power of an expander graph follows from the *expander mixing lemma* appearing in Goldreich and Wigderson, 1994 as Lemma 2.2.

The second construction gives us the possibility of varying the size of U. For each of the vertices of V', we label the outgoing edges $1, 2, \ldots, d$ in an arbitrary way. Thus, each edge gets labelled twice, once for each "direction", but the two labels do not have to agree. Then, for some parameter l, we let $U = V' \times \{1, \ldots, d\}^l$ and we let $V = V'$. Now we let the edges of the disperser be $E = \{((u, e_1, e_2, \ldots, e_l), v) |$ The walk starting in u and following edges labelled $e_1, e_2, \ldots, e_l$ (in that order) encounters v at some point$\}$.

Thus, the disperser defined has $|U| = |V| d^l$ and degree l. It was shown by Cohen and Wigderson, 1989, and Impagliazzo and Zuckerman, 1989, (based on earlier work of Ajtai et al., 1987), that it is a $\frac{1}{2}$-hitting disperser with threshold $2^{-\Omega(l)} |U|$, and a majority disperser with similar parameters. This gives us Statement 2(a) of the introduction.

Unlike the simpler way of using expanders to construct dispersers, there is no known proof of these results based directly on the expansion property of the graph. The only known proofs uses again the relationship between the expansion of a graph and the second largest eigenvalue of its adjacency matrix. Once this relationship is known, the proof is clever, but not difficult. A good exposition can be found in the text of Alon and Spencer, 1992.

4.2 CONSTRUCTIONS BASED ON UNIVERSAL HASHING

The second way of constructing dispersers and extractors is using universal families of hash functions. We first describe a simple disperser construction of Chor and Goldreich, 1989.

Given parameters $V = \{1, \ldots, m\}$ and $\epsilon > 0$. Let D be a set of size more than $\lceil 1/\epsilon \rceil$. Let $\mathcal{H}$ be an explicit strongly universal family of hash functions mapping D to V with $|\mathcal{H}| = (|D||V|)^{O(1)}$. We let $U = \mathcal{H}$ and let E be given by $\{(h, h(x)) | h \in U, x \in D\}$. Thus, each vertex of U corresponds to a hash function and each outgoing edge from such a vertex corresponds to applying this hash function to a particular value.

We shall show that this is a $\frac{1}{2}$-hitting disperser with threshold ϵ. This, combined with Lemma 16 means that it can be used to reduce the error probability of any one-sided error randomized algorithm using r bits to achieve error probability $\frac{1}{2}$ to at most $1/r^k$, for any constant k with only a polynomial blow up in the running time and using only a constant factor more random bits. We saw in the last subsection that we can use expanders to achieve the same error probability using *no* extra random bits, but the construction here is much more elementary.

Theorem 4 *For any value of $|V|$ and any $\epsilon > 0$, the above is an explicit $\frac{1}{2}$-hitting disperser with $|U| = (|V|/\epsilon)^{O(1)}$, degree $O(1/\epsilon)$, and threshold $\epsilon|U|$.*

Proof We need to show that for any subset S of U of size at least $\epsilon|U|$, we have that $|\Gamma(S)|$ is greater than $|V|/2$. Equivalently, we must show that if we take any subset V' of V of size at least $|V|/2$ and a randomly chosen vertex h of U, the probability that there is an edge between h and V' is more than $1 - \epsilon$.

For fixed V' and $i \in D$, define a random variable Y_i which is 1 if $h(i) \in V'$. Thus, there is an edge between h and V' if and only if $\sum_{i \in D} Y_i > 0$. The expectation of Y_i is $E[Y_i] = |V'|/|V| \geq \frac{1}{2}$. Thus the variance of Y_i is $V[Y_i] = E[Y_i](1 - E[Y_i]) \leq 1/4$. By linearity of expectation, $E[\sum_i Y_i] \geq |D|/2$ and as the variables Y_i are pairwise independent, we have $V[\sum Y_i] = \sum V[Y_i] \leq |D|/4$.

Thus, $\Pr[$ There is *not* an edge between h and V' $] = \Pr[\sum_{i \in D} Y_i = 0]$
$= \Pr[\sum_{i \in D} Y_i \leq E[\sum_{i \in D} Y_i] - |D|/2] \leq (|D|/4)/((|D|/2)^2) = 1/|D| < \epsilon$,
where the first inequality is Chebyshev's, stating that if X is a random variable with expectation μ and variance σ^2, then for any real number $t > 0$, $\Pr[|X - \mu| \geq t] \leq \sigma^2/t^2$. ■

A similar argument establishes that the same construction yields a majority disperser. Thus, it can also be used as an amplifier for two-sided error algorithms.

In the disperser construction of Chor and Goldreich, each hash function in the strongly universal family corresponds to a vertex in U and the elements we apply them to corresponds to outgoing edges from U. By switching the two roles, we get an extractor, as stated in the *leftover hash lemma* by Impagliazzo et al., 1989.

Lemma 24 *Let $\mathcal{H}$ be a universal family of hash functions of size 2^n indexed by and identified with $\{0,1\}^n$, mapping $\{0,1\}^n$ to $\{0,1\}^l$. Let $E : \{0,1\}^n \times \{0,1\}^n \to \{0,1\}^{n+l}$ be defined by $E(x,h) = (h(x), h)$. Then, for any a, E is an extractor with min-entropy threshold $l + a$ and error at most $2^{-a/2}$.*

Proof Let $x_1, x_2 \in \{0,1\}^n$ be independent samples from a source with min-entropy at least $l + a$ and let $h_1, h_2 \in \{0,1\}^s$ be independent uniform samples. By Lemma 20, we just need to show that the collision probability $\Pr[E(x_1, h_1) = E(x_2, h_2)]$ is very close to $1/2^{n+l}$. Note that $\Pr[E(x_1, h_1) = E(x_2, h_2)] = \Pr[h_1 = h_2 \wedge h_1(x_1) = h_1(x_2)] \leq \Pr[h_1 = h_2 \wedge x_1 = x_2] + \Pr[h_1 = h_2 \wedge h_1(x_1) = h_1(x_2) | x_1 \neq x_2]$. As $\mathcal{H}$ is universal, the latter probability can be bounded from above by $2^{-n} \cdot 2^{-l}$, i.e., exactly $1/2^{n+l}$. Thus, we just need to bound the former, which is equal to $2^{-n} \Pr[x_1 = x_2] \leq 2^{-n} 2^{-(l+a)}$, as x_1, x_2 were from a source of min-entropy at least $l + a$. Thus, the collision probability is at most $(1 + 2^{-a}) 2^{-(n+l)}$, and, by Lemma 20, $E(x_1, h_1)$ is $2^{-a/2}$ close to uniform. ∎

The lemma is called the leftover hash lemma, because h appears both in the input and the output of the extractor. Thus, h is "recycled".

The leftover hash lemma does not immediately define an extractor useful for deterministic amplification or computing with weak random sources because of the high degree: Viewing the extractor as a graph, each vertex in U has 2^n neighbors and the corresponding algorithm would thus run in exponential time.

However, the extractor can be used as a component in the construction of dispersers and extractors with parameters that are directly useful for randomness-efficient amplification.

The most simple way of doing this was described by Impagliazzo and Zuckerman, 1989, who called their technique *recycling of random bits*. They showed how to use the extractor of the leftover hash lemma to get, given any size of V, a $\frac{1}{2}$-hitting disperser $E \subseteq U \times V$ with $|U| = |V|^{O(1)}$, and degree $(\log |V|)^{O(1)}$ with threshold $2^{-\Omega(\sqrt{\log |V|})}|U|$.

Thus, it can be used to convert any **RP** algorithm using r random bits to achieve an error probability of $\frac{1}{3}$ to an **RP** algorithm using $O(r)$ random bits to achieve an error probability of $2^{-\Omega(\sqrt{r})}$. Note that this is not as good as the expander based disperser of the last subsection, but a self-contained proof that it works is much easier (in particular, it can be given here).

Let $U = \{0,1\}^{3n}$ and $V = \{0,1\}^n$. Let $E : \{0,1\}^n \times \{0,1\}^n \to \{0,1\}^{2n-\sqrt{n}}$ be the extractor with error $2^{-(\sqrt{n}-1)/2}$ and min-entropy threshold $n - 1$ of Lemma 24 (i.e., we set $l = n - \sqrt{n}$, $a = \sqrt{n} - 1$).

The disperser D is defined as follows. Given $x \in U = \{0,1\}^{3n}$ split x into two strings x_1 and y_1 of length n, and $\sqrt{n}$ strings $s_1, s_2, \ldots, s_{\sqrt{n}}$, each of length $\sqrt{n}$. Then recursively define $(x_2, y_2), (x_3, y_3), \ldots (x_{\sqrt{n}}, y_{\sqrt{n}})$, with $x_i, y_i \in \{0,1\}^n$ by $(x_{i+1}, y_{i+1}) = E(x_i, y_i) \circ s_i$, where $\circ$ denotes concatenation. The neighbors of x in V should then be $x_1, x_2, \ldots, x_{\sqrt{n}}$.

The proof that this works is very typical of the way extractors are used in other constructions. The intuition is the following. We are trying to hit a set S, using the uniformly random, but not independent, elements $x_1, x_2, \ldots, x_{\sqrt{n}}$. Suppose we have already tried $x_1, x_2, \ldots, x_i$, without hitting S. Then the conditional distribution of x_i given *this* still contains a lot of min-entropy. Letting $(x_{i+1}, y_{i+1}) = E(x_i, y_i)$ makes the conditional distribution of x_{i+1}, given that we did not yet hit S, close to uniform, and thus our chance of hitting S with x_{i+1} is almost as good, as if x_{i+1} had been independent from $x_1, x_2, \ldots, x_i$.

In the formal proof, we have to be a bit more careful.

Theorem 5 *The graph defined above is a $\frac{1}{2}$-hitting disperser with threshold $2^{-\Omega(\sqrt{n})} 2^{3n}$.*

Proof We should show that if S is a subset of V of density at least $\frac{1}{2}$, then for random x, the probability that $x_1, x_2, \ldots, x_{\sqrt{n}}$ does not intersect S at most $2^{-O(\sqrt{n})}$. We assume without loss of generality that the density of S is exactly $\frac{1}{2}$.

Let $b_1, b_2, \ldots, b_{\sqrt{n}}$ be indicator variables, with b_i indicating if $x_i \in S$.

Let $\mathcal{D}_i$ be the distribution on $\{0,1\}^{i+2n}$, induced by the variables $b_1, b_2, \ldots, b_i, x_{i+1}, y_{i+1}$. Let $\mathcal{U}_i$ be the uniform distribution on $\{0,1\}^{i+2n}$. Given any distribution $\mathcal{D}'$ on $\{0,1\}^{i-1+2n}$, let extend($\mathcal{D}'$) be the distribution on $\{0,1\}^{i+2n}$ obtained as follows: Take a sample $(j_1, j_2, \ldots, j_{i-1}, u_i, v_i)$ from $\mathcal{D}'$, let $j_i = 1_{u_i \in S}$ and $(u_{i+1}, v_{i+1}) = E(u_i, v_i) \circ s$, where s is a vector of $\sqrt{n}$ unbiased independent bits, and output the vector $(j_1, j_2, \ldots, j_i, u_{i+1}, v_{i+1})$. Thus, we have that $\mathcal{D}_i = \text{extend}(\mathcal{D}_{i-1})$.

We'll show by induction that $\|\mathcal{U}_i - \mathcal{D}_i\|_1 \leq i 2^{-\frac{\sqrt{n}-1}{2}}$.

Note that $\|\mathcal{U}_{\sqrt{n}} - \mathcal{D}_{\sqrt{n}}\|_1 \leq \sqrt{n} 2^{-\frac{\sqrt{n}-1}{2}}$ implies the statement of the lemma. Indeed, by Lemma 18 we have

$\Pr[\forall i, x_i \notin S] \leq 2^{-\sqrt{n}} + \frac{1}{2}\sqrt{n} 2^{-\frac{\sqrt{n}-1}{2}} = 2^{-\Omega(\sqrt{n})}$.

For $i = 0$, the statement is true, as $U_0 = D_0$.

Suppose that the statement holds for $i - 1$, i.e. that the distribution of the bit vector $b_1, b_2, \ldots b_{i-1}, x_i, y_i$ is $(i-1)2^{-\frac{\sqrt{n}-1}{2}}$-close to the uniform distribution $\mathcal{U}_{i-1}$.

We will bound $\|\mathcal{D}_i - \mathcal{U}_i\|_1 = \|\text{extend}(\mathcal{D}_{i-1}) - \mathcal{U}_i\|_1$ from above by $\|\text{extend}(\mathcal{D}_{i-1}) - \text{extend}(\mathcal{U}_{i-1})\|_1 + \|\text{extend}(\mathcal{U}_{i-1}) - \mathcal{U}_i\|_1$.

By Lemma 19, we can bound $\|\text{extend}(\mathcal{D}_{i-1}) - \text{extend}(\mathcal{U}_{i-1})\|_1$ from above by $\|\mathcal{D}_{i-1} - \mathcal{U}_{i-1}\|_1$ which is at most $(i-1)2^{-\frac{\sqrt{n}-1}{2}}$ by the induction hypothesis. Thus we only have to bound $\|\text{extend}(\mathcal{U}_{i-1}) - \mathcal{U}_i\|_1$.

Now, let $u_1, u_2, \ldots u_{i-1}, x_i', y_i'$ be a sample from $\mathcal{U}_{i-1}$. Letting $b_i' = 1_{x_i' \in S}$ and $(x_{i+1}', y_{i+1}') = E(x_i, y_i) \circ s_i$, we should prove that the distribution of $u_1, u_2, \ldots, u_{i-1}$ b'_i, x'_{i+1}, y'_{i+1} is close to uniform. The bits $u_1, u_2, \ldots, u_{i-1}$ are uniformly distributed and independent of what comes after. The i'th bit, b'_i is unbiased. as the density of S is $\frac{1}{2}$. Also, for any setting of $b'_i \in \{0, 1\}$, the conditional distribution of x'_i given b'_i, loses at most 1 bit of min-entropy compared to the unconditional distribution of x'_i, as the conditional probability of any outcome, conditioned by an event of probability $\frac{1}{2}$ is at most a factor of two bigger than the unconditional probability. So, the conditional min-entropy of x'_i, given b'_i, for any outcome of b'_i, is at least $n - 1$. By the extractor property of E, the conditional distribution of $(x'_{i+1}, y'_{i+1}) = E(x'_i, y'_i)$, given the value of b'_i, is $2^{-\frac{\sqrt{n}-1}{2}}$- close to uniform. We conclude that $(u_1, u_2, \ldots, u_{i-1}, b'_i, x'_{i+1}, y'_{i+1})$ is $2^{-\frac{\sqrt{n}-1}{2}}$-close to uniform.

Thus, $\|\mathcal{D}_i - \mathcal{U}_i\|_1 \leq (i-1)2^{-\frac{\sqrt{n}-1}{2}} + 2^{-\frac{\sqrt{n}-1}{2}} = i2^{-\frac{\sqrt{n}-1}{2}}$, and we are done. ∎

The above constructions were relatively straightforward. Following them, in a series of papers, Zuckerman, 1990, Zuckerman, 1991, Nisan and Zuckerman, 1996, Wigderson and Zuckerman, 1999, Goldreich and Wigderson, 1994, Srinivasan and Zuckerman, 1999, Saks et al., 1998, Zuckerman, 1996, Ta-Shma, 1996, Ta-Shma, 1998, expanded and refined the ideas. It is beyond the scope of this chapter to give a detailed account here. An excellent survey of this line of research is given by Nisan and Ta-Shma, 1999. We merely state some of the important ingredients and then state the landmark results obtained. Many of the results can also be obtained using the technique using error correcting codes of the next subsection, but there are exceptions.

One important ingredient is the family of hash function used. First, note that the proof of the leftover hash lemma works, even if the family used is only $(1 + \epsilon)$-weakly universal, for some small $\epsilon > 0$. Thus, we can use the family of Lemma 9 instead of the family based on arithmetic over a field. This won't do us any good in the construction of Impagliazzo and Zuckerman described above, but in other settings, where the parameters are different, it will.

Another important ingredient is the notion of *extractor composition* by Wigderson and Zuckerman, 1999: If E_1 and E_2 are extractors with certain parameters, then E defined by $E(x, y_1 \circ y_2) = E_1(x, y_1) \circ E_2(x, y_2)$ is also an extractor. The proof sketch of this is somewhat similar to the proof of Theorem 5: If x has high min-entropy, then $E_1(x, y_1)$ is close to uniform. Furthermore, if a is a particular outcome of $E_1(x, y_1)$ with high probability, then the min-entropy of x given that $E_1(x, y_1) \circ y_2 = a$ is still high, so even given this event, $E_2(x, y_2)$ has distribution close to uniform.

The error introduced by outcomes a of low probability is low, so we can safely deal with these outcomes and conclude that $E_1(x, y_1) \circ E_2(x, y_2)$ is close to uniform.

A third important ingredient is the notion of a *blockwise source*. A blockwise source is a source $(X_1, X_2, \ldots, X_k)$ on $S_1 \times S_2 \times \cdots \times S_k$, for some value k and some sets S_i, so that the conditional marginal distribution of X_i, conditioned on any particular outcome of $X_1, X_2, \ldots, X_{i-1}$, still has high min-entropy. Nisan and Zuckerman, 1996, show that by picking out subsets of bits from a weak random source on $\{0, 1\}^R$ randomly, one can obtain a blockwise source. The bits can be taken out randomness-efficiently by using k-wise independent families of hash functions (similar to the use of pairwise independence in Theorem 4 - intuitively, one is trying to *hit* some *really* random bits in the weak random sample). Saks et al., 1998 describes a different way of obtaining a blockwise source, based on splitting the input in blocks in an "adaptive" way. This technique was also used by Ta-Shma, 1996 and Ta-Shma, 1998.

Having obtained a block-wise source $(X_1, X_2, \ldots, X_k)$, one can extract a value Y_0 with close to uniform distribution from it, in the following iterative way, using extractors of appropriate parameters: Given a truly random seed y_k, one lets $y_{k-1} = E(X_k, y_k)$, $y_{k-2} = E(X_{k-1}, y_{k-1})$, $\ldots$, $y_0 = E(X_1, y_1)$. Again, the proof that this works is similar to the proof of Theorem 5. It is, however, rather fragile: It is important that one does the construction "backwards", starting with X_k rather than X_1.

Finally, Ta-Shma, 1996, introduced the notion of a *merger*, an important component in the latest constructions: Loosely speaking, a merger is a deterministic algorithm which, given a number of sources, an unknown one of them uniform, outputs a nearly uniform distribution.

Best results obtained using these techniques include the following, the first by Zuckerman, 1996.

Theorem 6 *For any constants $0 < \alpha < \delta < 1$, there is an explicit extractor E for all parameters $n \geq 1, \epsilon > 0$, with*

$$E : \{0, 1\}^n \times \{0, 1\}^{O(\log n + \log \epsilon^{-1})} \to \{0, 1\}^{(\delta - \alpha)n},$$

error ϵ, and min-entropy threshold δn.

The theorem of Zuckerman, combined with Lemma 22 implies Statement 2(b) of the introduction.

The second by Ta-Shma, 1996.

Theorem 7 *There is an explicit extractor E for all parameters n, m, ϵ with $E : \{0, 1\}^n \times \{0, 1\}^{(\log n + \log \epsilon^{-1})^{O(1)}} \to \{0, 1\}^m$, error ϵ, and min-entropy threshold m.*

Note that Ta-Shma's extractor extracts *all* the min-entropy from the source: The min-entropy threshold equals the number bits of the output domain. On the other hand, the degree of the extractor is $2^{\text{polylog}n}$, so using the extractor for deterministic amplification would induce a superpolynomial slowdown.

The third by Ta-Shma, 1998.

Theorem 8 *There is an explicit ϵ-hitting disperser D for all parameters n, m, ϵ with $D : \{0,1\}^n \times \{0,1\}^{O(\log n + \log \epsilon^{-1})} \to \{0,1\}^{m-\text{polylog}n}$, with threshold 2^m.*

Combined with Lemma 16, this theorem implies Statement 4 of the introduction.

Missing from the above theorems is an extractor, capable of extracting randomness from a source of sublinear min-entropy, using a seed of logarithmic length. It is not known how to obtain such an extractor using iterated universal hashing, but it can be obtained using Trevisan's technique of error correcting codes, to be explained in the next subsection.

4.3 THE TREVISAN EXTRACTOR

In this subsection, we shall give a detailed account of the extractor of Trevisan, 1999, proving the following theorem.

Theorem 9 *For any constants $k > c > 1$ there is an explicit extractor for any parameter r with $E : \{0,1\}^{r^k} \times \{0,1\}^{O(\log r + \log \epsilon^{-1})} \to \{0,1\}^r$, with error ϵ and min-entropy threshold r^c.*

For the presentation, we shall be content with proving the theorem for $c = 3$, rather than c arbitrarily close to one, but with k being as large as desired. We shall also deal with constant error $\epsilon = \frac{1}{10}$ only. Still, combining the theorem thus proved with Lemma 22, we get Statement 3 of the introduction, and combining it with Lemma 23, we get Statement 5 of the introduction. Also, this version of the theorem is enough for the important applications to pseudorandom generators we describe later.

Trevisan's extractor construction was a major breakthrough. Curiously, the entire construction was in some sense already known when Trevisan found it, as a *hardness based pseudorandom generator* with similar parameters was already constructed by Impagliazzo and Wigderson, 1997. In Section 5, we introduce this concept and show that it is a special case of an extractor. Nobody realized this, however, until Trevisan's work. Trevisan essentially took the generator of Impagliazzo and Wigderson, 1997, and threw away what was not needed to keep the extractor property.

The construction has two parts: The use of error correcting codes and the use of Nisan-Wigderson designs. We explain each component separately.

Min-entropy and source coding. To prove that a map $E : U \times A \to V$ is an extractor, we have to prove that if we pick $x \in U$ distributed according to $\mathcal{D}$, where $\mathcal{D}$ is a distribution of high min-entropy and $y \in A$ is uniformly distributed, then $E(x, y)$ has a distribution which is close to uniform. If we make such a proof by way of contradiction, we would assume that $E(x, y)$ is *not* close to uniform and then prove that $\mathcal{D}$ does not have high min-entropy. Thus, it is useful to find strategies for proving that a distribution does not have high entropy. Concepts from *source coding* turn out to be useful for this.

A *lossless source code* for $\{0, 1\}^m$ is a pair of maps, the *encoder* $c : \{0, 1\}^m \to \{0, 1\}^*$ and the *decoder* $d : \{0, 1\}^* \to \{0, 1\}^m$ so that $\forall x \in \{0, 1\}^m, d(c(x)) = x$. The set $\{c(x) | x \in \{0, 1\}^m\}$ is the set of *code words*. A *prefix* code is a lossless source code with the extra requirement that no code word is a prefix of another code word. A *fixed length* code has the extra requirement that all code words have the same length.

Let $\mathcal{D}$ be a distribution on $\{0, 1\}^m$. Shannon's lossless source coding theorem states that any prefix code $c : \{0, 1\}^m \to \{0, 1\}^*$ must have $E_{x \in \mathcal{D}}(|c(x)|) \geq H(\mathcal{D})$. Thus we can prove that a distribution does not have high *Shannon* entropy by providing a source code for the distribution with small average code length.

We can attempt to adopt a similar strategy to prove that a distribution does not have high min-entropy using the following simple observation: If a distribution $\mathcal{D}$ has min-entropy t, then any *fixed length* code for $\mathcal{D}$ must have code length at least t. Thus we can prove that a distribution does not have high min-entropy by providing a source code for the distribution with small worst case code length. While valid, this strategy is not generally useful as it can only be used in very special cases.

To get a more useful strategy, we have to consider *lossy source codes*. A *lossy* fixed length source code with *rate* l is a pair of maps, the encoder $c : \{0, 1\}^m \to \{0, 1\}^l$ and the decoder $d : \{0, 1\}^l \to \{0, 1\}^m$. Now there is no requirement that $\forall x \in \{0, 1\}^m d(c(x)) = x$. Instead, we refer to $d(c(x))$ as the *quantized* value of x and, given a distribution $\mathcal{D}$ on $\{0, 1\}^m$, we define the *distortion* of the source code relative to $\mathcal{D}$ as the probability $\Pr[d(c(x))_i \neq x_i]$ where x is chosen randomly according to $\mathcal{D}$ and i is chosen uniformly at random in $\{1, \dots, m\}$.

We cannot directly relate the min-entropy of a distribution to the rate-distortion performance of a lossy fixed length code for the distribution, as some distributions with high entropy or high min-entropy admit lossy codes with low rate and distortion (indeed, industrial lossy data compression, such as the JPEG standard, is based on this fact). However, if we first encode our distribution in a good error correcting code, and consider source codes for the encoded distribution, high min-entropy of the original distribution (and thus for the encoded distribution) *does* imply that we cannot have low rate and distortion simultaneously, *even if we by "low distortion" mean a distortion just below the trivially obtainable distortion level of* $\frac{1}{2}$. This simple but crucial fact is reflected in the following lemma.

Lemma 25 *Let* $e : \{0,1\}^m \to \{0,1\}^{m^{10}}$ *be the error correcting code of Lemma 12. Let* $\mathcal{D}$ *be a distribution on* $\{0,1\}^m$ *with min-entropy at least* t *and let* $e(\mathcal{D})$ *be the induced distribution on* $\{0,1\}^{m^{10}}$. *Then any lossy fixed length source code for* $\{0,1\}^{m^{10}}$ *with distortion less than* $\frac{1}{2} - \frac{2}{m^2}$ *relative to* $e(\mathcal{D})$, *has rate at least* $t - 6 \log m$.

Proof Let the source encoder be c and the decoder d. Consider the event G that a randomly chosen sample y of $e(\mathcal{D})$ has Hamming distance smaller than $\frac{1}{2} - \frac{1}{m^2}$ to its quantized value $d(c(y))$. We have that $\Pr[y \in G] \geq \frac{1}{m^2}$, for otherwise the distortion of the source code would be more than $(1 - \frac{1}{m^2})(\frac{1}{2} - \frac{1}{m^2}) > \frac{1}{2} - \frac{2}{m^2}$.

Assume that the rate of the source code is l. Thus, there are at most 2^l source code words. Let w be the most popular value of $c(y)$, given that $y \in G$. Let G' be the event $y \in G \wedge c(y) = w$. For a randomly chosen y of $e(\mathcal{D})$ we have that $\Pr[y \in G'] \geq \frac{1}{2^l m^2}$.

On the other hand, all $y \in G$ are e-code words and if $c(y) = w$, y is in the Hamming ball of radius $\frac{1}{2} - \frac{1}{m^2}$ with center $d(w)$. Thus, by Lemma 14, there are at most $m^4/2$ outcomes of G', and since $\mathcal{D}$ and thus $e(\mathcal{D})$ has min-entropy t, each outcome has probability at most 2^{-t}. Thus $\Pr[G'] \leq \frac{m^4}{2} 2^{-t}$.

Combining the two inequalities, we have $\frac{1}{2^l m^2} \leq \frac{m^4}{2} 2^{-t}$, so $l \geq t - 6 \log m$. ∎

In the extractor construction to follow, the first step of the extractor (viewed as a function) will be to apply an error correcting code to its first input, i.e. the output of the weak random source. Then another transformation, to be outlined next will follow. We will prove the extractor construction correct by showing that if a distribution does *not* yield a distribution ϵ-close to uniform when sent through the extractor, we can make a low rate lossy source code with distortion bounded away from $\frac{1}{2}$ for the encoded distribution.

Nisan-Wigderson designs. A Nisan-Wigderson design, introduced in Nisan and Wigderson, 1994, is a set system with small pairwise intersections. More precisely, an (m, n, s, l)-Nisan-Wigderson design is a system of m sets $S_1, S_2, \ldots, S_m$, each an s-subset of $\{1, 2, \ldots, n\}$, so that for $i \neq j, |S_i \cap S_j| \leq l$.

Explicit Nisan-Wigderson designs can be constructed using weakly universal families of hash functions.

Lemma 26 *For any parameters m, n which are powers of two, there is an explicit Nisan-Wigderson design with parameters (n^m, n^2, n, m).*

Proof Let $\mathcal{H}$ be the explicit m-weakly universal family of Lemma 8 of size n mapping $\{1, 2, \ldots, n\}^m \to \{1, 2, \ldots, n\}$, and let the system consist of the sets $S_x = \{(h, h(x)) | h \in \mathcal{H}\}, x \in \{1, 2, \ldots, n\}^m$. ∎

This construction is essentially the best explicit one known, if we define an explicit set system to be a set system where the corresponding bipartite graph is explicit. Unfortunately, it does not quite give us the parameters necessary for the application at hand. Fortunately, we do not really need explicit designs, but only polynomial time constructible ones.

Lemma 27 *For any constant $c \geq 1$, there is a deterministic algorithm which, given as input m, outputs an $(m, O(\log m), c \log m, \log m)$-Nisan-Wigderson design, using time polynomial in m.*

Proof The algorithm constructs the set system on a domain of size $100c^2 \log m$. The system is constructed in a "greedy" way by picking one set at a time. Suppose we have already picked $S_1, S_2, \ldots, S_i$ of the correct size and with the correct bound on the size of the pairwise intersections. If we can prove that there *exists* a set S_{i+1} we can add to the system without violating the constraints, we are done, as the algorithm can exhaustively search for such a set in time polynomial in m, since $\binom{100c^2 \log m}{c \log m} = m^{O(1)}$, as c is constant.

To show the existence of an appropriate set S_{i+1}, we pick randomly, with replacement, $2c \log m$ members of $1, 2, \ldots, 100c^2 \log m$. Let the resulting multi-set be denoted S. Lemma 1 implies that

1. With very high probability, the multi-set S has at least $c \log m$ different members.

2. With very high probability, the size of the intersection of S with any of the previously chosen sets, is less than $\log m$.

Thus, the desired set S_{i+1} can be obtained by taking an appropriate subset of S. ∎

A design constructed using weakly universal families would have domain size $O((\log m)^2)$, rather than $O(\log m)$.

Putting things together. The Trevisan extractor is most conveniently described as an efficiently computable map $E : U \times A \to V$ with $U = \{0,1\}^{r^k}$, $V = \{0,1\}^r$, and $A = \{0,1\}^{c \log r}$. Here, k is any constant greater than 5 and c is a constant dependent on k. For sufficiently large r, we shall prove the construction to be an extractor with error at most $\frac{1}{10}$ and min-entropy threshold at least r^3, thus proving the special case of Theorem 9 outlined below the statement of the theorem.

We describe the operation of E, viewed as an algorithm. First, given the parameter r, the extractor uses Lemma 27 to construct an $(r, c \log r, 10k \log r, \log r)$-Nisan-Wigderson design $S_1, S_2, \ldots, S_r$, where c is a constant depending on k. This constant c is also the constant c in the definition of the set of seeds A.

Now, given the first input $x \in U$, the extractor computes the value $e(x) \in \{0,1\}^{r^{10k}}$. For notational convenience, we represent the sequence $e(x)$ by its characteristic function $g : \{0,1\}^{10k \log r} \to \{0,1\}$.

Then, given the second input $y \in A$, the final output $E(x,y)$ is computed as follows. Each set in the Nisan-Wigderson design is a subset of $\{1, \ldots, c \log r\}$. View y as a bit sequence of length $c \log r$, and for S being a set in the design, let y_S be the subsequence consisting of the bits indexed by S. For example, if $v = 10111001011$ and $S = \{2,4,8\}$, $y_S = 011$. Now, we give as output $E(x,y) = g(y_{S_1})g(y_{S_2}) \cdots g(y_{S_r})$.

We claim that the map E so defined is an extractor with min-entropy threshold r^3 and error at most $\frac{1}{10}$.

To see this, let $\mathcal{D}$ be some distribution on $U = \{0,1\}^{r^k}$ with min-entropy at least r^3. Pick $x \in \mathcal{D}$ at random and $y \in A$ uniformly at random. We must show that $E(x,y) \in \{0,1\}^r$ has a distribution which is $\frac{1}{10}$-close to the uniform distribution on $\{0,1\}^r$. Suppose not. Then, by Lemma 21, there is a predictor with advantage $\epsilon = \frac{1}{20r}$ for one of the bits of $E(x,y)$, say the i'th bit, given the previous bits. We will use this predictor to construct a lossy source code for $c(x)$ with a better distortion than possible by Lemma 25, and thus reach a contradiction. The i'th bit of $E(x,y)$ is $g(y_{S_i})$. Our predictor predicts $g(y_{S_i})$ on input $g(y_{S_1}), g(y_{S_2}), \ldots, g(y_{S_{i-1}})$ with advantage ϵ when x is chosen from $\mathcal{D}$ and y is chosen uniformly at random.

It does not seem immediately useful to be able to predict $g(y_{S_i})$ on input $g(y_{S_1}), g(y_{S_2}), \ldots, g(y_{S_{i-1}})$. Suppose, on the other hand, that we could predict $g(y_{S_i})$ from y_{S_i} with a positive advantage. Then we could reconstruct the entire vector $e(x)$ (for which g is the characteristic function) by cycling through all possible settings of y_{S_i}. The advantage would give us a reconstruction of $e(x)$ with a distortion bounded away from $\frac{1}{2}$.

To accomplish this, we *freeze* the bits of $y_{S_i^c}$ (i.e., all bits except the bits of y_{S_i}) to fixed values while preserving the advantage of our predictor for a random setting of y_{S_i}. Since our predictor predicts $g(y_{S_i})$ from $g(y_{S_1}), g(y_{S_2}), \ldots, g(y_{S_{i-1}})$ rather than y_{S_i}, our strategy will be to *compute* $g(y_{S_1}), \ldots, g(y_{S_{i-1}})$ given y_{S_i}. As $S_1 \cap S_i$ only contains $\log r$ bit positions and the bits of $y_{S_1 \setminus S_i}$ have been frozen, then for any fixed x (and thus fixed g), we can *tabulate* the value of $g(y_{S_1})$ as a function of $y_{S_1 \cap S_i}$. The table can be represented as a bit string of $2^{\log r} = r$ bits. Similarly, we can tabulate the value of $g(y_{S_j})$ as a function of $y_{S_j \cap S_i}$ for each $j = 2, \ldots, i - 1$.

The value i in binary representation and the concatenation of these $i - 1$ tables defines our source code word $c(e(x))$ for the value $e(x)$. We pad the code word with zeros to make it a fixed length code. The rate of the source code so defined is $r^2 + O(\log r)$.

To decode a code word $y = c(e(x))$ (without knowing $e(x)$), we do the following. We cycle through all possible values of y_{S_i}, and use the tables of the code word to look up the corresponding values of $g(y_{S_1}), \ldots, g(y_{S_{i-1}})$. These we feed into the predictor, to get a prediction for $g(y_{S_i})$, i.e., for the bit of $e(x)$ indexed by y_{S_i}. Concatenating all the bits predicted gives us $d(y) = d(c(e(x)))$. As the predictor has advantage ϵ, we see that for random $x \in \mathcal{D}$ and random i, the probability that $d(c(e(x)))_i \neq e(x)_i$ is at most $\frac{1}{2} - \epsilon$, i.e. the distortion of the source code we have defined is at most $\frac{1}{2} - \epsilon = \frac{1}{2} - \frac{1}{20r}$.

By Lemma 25, the rate of the source code must be at least $r^3 - O(\log r)$. This is a contradiction, as we already established the rate to be $r^2 + O(\log r)$. Thus, our assumption that the distribution of $E(x, y)$ is not ϵ-close to uniform must be false, and we are done.

4.4 TOWARDS OPTIMAL EXTRACTORS

The extractor construction technique of Trevisan was further refined by Raz et al., 1999b, Raz et al., 1999a, Impagliazzo et al., 1999, and Impagliazzo et al., 2000.

Some of the ingredients in the improvements of Raz et al., 1999b are the following:

First, a weakening of the notion of Nisan-Wigderson design is introduced. A *weak* Nisan-Wigderson designs is defined similarly to a Nisan-Wigderson design, except that, for each set S in the design, one demands a bound on $\sum_T 2^{|S \cap T|}$ rather than $\max_T |S \cap T|$, where the sum and max are over the other sets in the design. As is apparent in the proof of the correctness of the Trevisan extractor, it is enough to bound the first quantity. This makes better parameters possible.

Second, by using a special error correcting code, rather than a "generic" error correcting code, Raz et al., 1999b showed that the tabulation of the functions $g(S_j)$ used in the source code can be replaced with a more succinct representation, making the source code shorter.

Third, an improved version of extractor composition is introduced, where two extractors E_1, E_2 are composed as $E(x, y_1 \circ y_2) = E_1(x, y_1) \circ E_2(x \circ y_1, y_2)$ rather than $E(x, y_1 \circ y_2) = E_1(x, y_1) \circ E_2(x, y_2)$, as done by Wigderson and Zuckerman, 1999.

These improvement makes it possible, for instance, to match the parameters of the extractor of Theorem 7.

To further describe the improvement obtained, one should first note that in the extractor constructions stated so far, we have only considered rough tradeoffs between the five parameters (U, V, the degree d, the error ϵ, and the min-entropy threshold). When looking at the finer tradeoffs possible, for extractors extracting all or almost all the min-entropy of a source, the notion of *entropy loss* of the extractor becomes relevant. The entropy loss of an extractor is the difference between the number of almost unbiased bits produced by the extractor and the sum of its min-entropy threshold and the number of truly random bits given as input, i.e. $\log d$. Radhakrishnan and Ta-Shma, 1997, prove that the entropy loss of any extractor must be at least $\log \epsilon^{-1}$. Unlike any previous extractors, an extractor of Raz et al., 1999b matches this bound within an additive constant, with d being quasipolynomial.

Further improvements and refinements were made in Raz et al., 1999a, Impagliazzo et al., 1999, and Impagliazzo et al., 2000. It is beyond the scope of this chapter to discuss the various tradeoffs obtained. Also, presumably any information given would quickly be obsolete, as the possible combinations of the available techniques do not seem to have been fully explored yet.

We shall here just mention that it is, at the time of writing, not known how to match the parameters of Theorem 6 or Theorem 8 using error correcting codes and designs.

The ultimate goal and the most important open problem is to construct, for parameters $r < R$ and $\epsilon > 0$ an explicit extractor $E : \{0,1\}^R \times \{0,1\}^{O(\log R + \log \epsilon^{-1})} \to \{0,1\}^r$ with error ϵ and min-entropy threshold r. Such an extractor would have as consequence that for any parameter $R > r$, a randomized polynomial algorithm using r random bits could be simulated with polynomial overhead, using R random bits, and achieving an error probability of 2^{-R+r}. Note that Theorem 7 (or the alternative construction given by Raz et al., 1999b) makes such a simulation possible if we allow a quasipolynomial, rather than polynomial, overhead in time.

5. HITTING SET AND PSEUDORANDOM GENERATORS

For the rest of this chapter, we shall deal with the third derandomization task: Simulating randomized algorithms deterministically with a non-trivial overhead in time. As mentioned earlier, it is, at the time of writing, an open problem if **BPP=EXP**. Thus, we can only hope for conditional results, such as reasonable conditions implying **BPP=P**, or weaker statements such as **BPP ⊆ SUBEXP**.

All known proofs of such implications proceed by the construction of either *hitting set generators* or *pseudorandom generators*.

To motivate these concepts, let us go back to the card game between Alice and Bob of Section 3.1, used to define dispersers, and try to make it match the problem of deterministic simulation. As in the case of randomness-efficient amplification, each card of Alice corresponds to the coin toss sequence in a particular computation of a one-sided error algorithm with error probability bounded by ϵ. The marked cards correspond to accepting computations. To efficiently derandomize the randomized algorithm using the card game, it seems that we would need a *deterministic* efficient strategy for Bob, i.e., a small set of cards to turn over, with a guarantee that no matter how Alice marked her cards, one of the marked cards would be turned over. But as we pointed out when we first discussed the card game, such strategies do not exist, and we seem to be stuck.

The key observation that makes progress possible is the fact that for the intended application, the set of marked cards is *not* an arbitrary set of the given size: *it is the set of coin toss sequences that make an efficient randomized algorithm accept a certain input*. Thus, we should consider a modified card game, where Alice must choose her set of marked cards from a quite restricted class C of possibilities. This changes the game completely! Now it is very possible that a good deterministic strategy for Bob exists. Such a strategy we call a *hitting set* for C.

Definition 4 *Let C be a set system over a finite universe. A $(1 - \epsilon)$-hitting set for C is a set H so that $H \cap S \neq \emptyset$ for every $S \in C$ of density at least $1 - \epsilon$.*

To see which class C to consider for the task at hand, we recall Lemma 2: To derandomize **RP**, it is sufficient to be able to find satisfying assignments to circuits $C : \{0, 1\}^r \to \{0, 1\}$ with acceptance probability at least $\frac{1}{2}$. Thus, we should define $\mathcal{C}$ to be the family of subsets of $\{0,1\}^r$ recognized by circuits of size r and consider $\frac{1}{2}$-hitting sets for $\mathcal{C}$. Abusing terminology slightly, we shall refer to such a hitting set as a $\frac{1}{2}$-hitting set "for circuits of size r" rather than as a $\frac{1}{2}$-hitting set "for the sets decided by circuits of size r". To see if an instance of CIRCUIT ACCEPTANCE PROBABILITY ESTIMATION satisfies at least half its assignments, we simply check if one of the elements of the hitting set satisfies the circuit. For this to be an efficient deterministic procedure, we need to be able to efficiently generate, on input r a small hitting set for circuits of size r. This leads us to the notion of a *hitting set generator*.

Definition 5 *A $(1 - \epsilon)$-hitting set generator for $\mathcal{C}$ is a deterministic algorithm which on input r outputs a $(1 - \epsilon)$-hitting set $H_r \subseteq \{0,1\}^r$ for $\mathcal{C}$.*

The discussion above is encapsulated in the following Lemma.

Lemma 28 *If a $\frac{1}{2}$-hitting set generator for circuits of size r, running in time $t(r)$ exists, then* **RP** *is contained in* **TIME**$(t(n^{O(1)})n^{O(1)})$.

In particular, if a hitting set generator running in polynomial time exists, then **RP=P**.

It is not so easy to directly construct a generator with the parameters of Lemma 28, even making reasonable unproven assumptions. To make the task of eliminating the use of random bits easier, it seems a good strategy to first try to use as few as possible. Thus, the work we have already done on randomness-efficient amplification should be very useful to us. That this intuition is true, is expressed in the following lemma.

Given an integer q and a real value $\epsilon > 0$, the problem (q, ϵ)-STRONG ONE-SIDED CIRCUIT ACCEPTANCE PROBABILITY ESTIMATION is defined as follows: Given a Boolean circuit C of size n^q with n inputs and one output, with a promise that *either* (a) $\Pr[C(x) = 1] = 0$ *or* (b) $\Pr[C(x) = 1] \geq 1 - 2^{-n+n^\epsilon}$ for a randomly chosen $x \in \{0,1\}^n$ we should tell which of (a) or (b) is the case.

Lemma 29 *For any constant $\epsilon > 0$, there is a constant q, so that (q, ϵ)-STRONG ONE-SIDED CIRCUIT ACCEPTANCE PROBABIL-ITY ESTIMATION is complete for* **pRP**.

Proof Given an **pRP** problem solved by a machine M, we reduce it to (q, ϵ)-STRONG ONE-SIDED CIRCUIT ACCEPTANCE PROBABIL-ITY ESTIMATION as follows. First, using Theorem 9, combined with Lemma 22, convert M to a polynomial machine M' using $R(n) > n$ random bits and achieving error probability $2^{-R(n)+R(n)^\epsilon}$. Then, as in the proof of Lemma 2, given an input x, convert M' to a circuit for instances of length $|x|$ with two inputs, the real input x and the $R(|x|)$ random bits and build the input x into the circuit, leaving only the random bits as input. The circuit has size $R(|x|)^q$, for some constant q depending on ϵ. Note that unlike the proof of Lemma 2, we cannot make the circuit linear size by adding dummy inputs, as we now measure the error probability as a function of the number of random bits. ∎

Using Lemma 29 instead of Lemma 2, we get the following variation of Lemma 28.

Lemma 30 *For any constant $\epsilon > 0$, there is a constant $q \geq 1$, so that if a $(1 - 2^{-r+r^\epsilon})$-hitting set generator for circuits $C : \{0,1\}^r \to \{0,1\}$ of size r^q exists, running in time $t(r)$, then* **RP** *is contained in* **TIME**$(t(n^{O(1)})n^{O(1)})$.

Thus, we have shown that we only have to construct a $(1 - 2^{-r+r^\epsilon})$-hitting set generator, rather than a $\frac{1}{2}$-hitting set generator to derandomize **RP**. A slightly different way of seeing this is by using the following lemma, which states that we can simply convert a $(1 - 2^{-r+r^\epsilon})$-hitting set into a $\frac{1}{2}$-hitting set.

Lemma 31 *For any constant $\epsilon > 0$, there are constants $q \geq 1$ and $\delta > 0$ so that the following holds. There is a polynomial time procedure which, on input H where H is $(1 - 2^{-r+r^\epsilon})$-hitting set in $\{0,1\}^r$ for circuits of size r^q, outputs a $\frac{1}{2}$-hitting set in $\{0,1\}^{r'}$ for circuits of size r', where $r' = r^\delta$.*

The polynomial time procedure of Lemma 31 does the following: It interprets H as a subset of U, where $E \subseteq U \times V$ is the explicit extractor (viewed as a disperser) of Theorem 9 and outputs the neighbors of H in V. We omit the details, as they are essentially just another way of looking at lemma 29 and its proof.

Lemmas 28 and 30 hold with **RP** replaced with **pRP**. Thus, Theorem 2 implies that if a hitting set generator running in polynomial time exists, then also **BPP=P**. On the other hand, Theorem 2 does not give us a more general theorem of the form: "If a hitting set generator running in time $t(r)$ exists, then **BPP** is contained in **TIME**$(t(n^{O(1)})n^{O(1)})$". The proof of Theorem 2 would yield only (roughly) that **BPP** is contained in **TIME**$(t(t(n^{O(1)}))n^{O(1)})$. However, Andreev et al., 1998, showed by a different proof that the stronger conclusion does in fact hold. The proof was simplified significantly by Goldreich and Wigderson, 1999, the latter proof using extractors.

A more direct way of derandomizing **BPP** is to consider a stronger generator than a hitting set generator. The definition of a $(1 - \epsilon)$-hitting set generator is strongly analogous to the definition of a $(1 - \epsilon)$-hitting explicit disperser: In the disperser case, we want to hit each subset of V of density at least $1 - \epsilon$ with high probability. In the hitting set generator case, we want to hit every subset of V of density at least $1 - \epsilon$, recognized by a small circuit. To derandomize two-sided error algorithms, we want to make a definition similarly analogous to the definition of an extractor. To get the correct definition, it is useful to consider Lemma 18. An extractor produces a distribution which looks, to every statistical test, almost as the uniform distribution. Analogously, a *pseudorandom generator* produces a distribution which looks, to every statistical test given by a small circuit, almost as the uniform distribution. Now we get the following definitions, analogous to the definitions of hitting sets and hitting set generators.

Definition 6 *If C is a set system over $\{0,1\}^r$, a pseudorandom set for C with error ϵ is a (multi-)set P in $\{0,1\}^r$ so that for every set $S \in C$, $|\Pr_{x \in P}[x \in S] - \Pr_{x \in \{0,1\}^r}[x \in S]| \leq \frac{\epsilon}{2}$.*

We use the terminology "A pseudorandom set with error ϵ for circuits of size r" a short hand for "A pseudorandom set with error ϵ for the subsets of $\{0,1\}^r$, decided by circuits of size r".

A pseudorandom generator with error ϵ for circuits of size r is a deterministic algorithm which on input r, outputs a pseudorandom set in $\{0,1\}^r$ for circuits of size r.

Now, Lemma 2 immediately yields

Lemma 32 *If a pseudorandom generator with error less than $\frac{1}{3}$ for circuits of size r, running in time $t(r)$ exists, then **BPP** is contained in* $\mathbf{TIME}(t(n^{O(1)})n^{O(1)})$.

As mentioned, it is not known how to make efficient hitting set or pseudorandom generators without appealing to unproven assumptions. In the rest of this section, we discuss three ways of constructing hitting set and pseudorandom generators based on different assumptions.

5.1   CRYPTOGRAPHIC PSEUDORANDOM GENERATORS

The first constructions of pseudorandom generators, done by Blum and Micali, 1984, and Yao, 1982, were intended primarily for cryptography, with derandomization as an interesting secondary application. This means that the parameters of interest were somewhat different.

Recall that we want our pseudorandom generator to efficiently output, on input r, a pseudorandom set for circuits of size r. If the set has size $d(r)$, we should of course be allowed time at least $d(r)$ to output the set. For cryptographic applications it turns out that we want something stronger. We want the generator, on input r and an index $i \in \{1, 2, \ldots, d(r)\}$ to output the i'th member of the pseudorandom set (relative to some arbitrary enumeration of the set) in time polynomial in the length of i, i.e., $\log d(r)$. Thus, we should be able to *sample* from the set very efficiently. This is not necessary for deterministic simulation, as we intend to run our original randomized algorithm on all coin toss sequences in the set (or use them as assignments to the circuit of Lemma 2). Furthermore, in the cryptographic case, we want the set to be pseudorandom for circuits of size r^k, for all constant k, simultaneously, rather than just circuits of size r. In particular, the sampling procedure will have shorter time to produce a member of the set that the size of the circuits the resulting distribution should be pseudorandom for.

Clearly, the cryptographic requirements are much more restrictive than the requirements for derandomization and it is only known how to construct cryptographic generators under much harsher assumptions than the assumptions we are aiming for (such as the assumptions of Statement 6 and 7 in the introduction). Blum and Micali, 1984, and Yao, 1982, constructed cryptographic generators under average case hardness assumptions about specific number theoretic functions (discrete log and factoring). Later, Goldreich and Levin, 1989 showed that such generators can be constructed under the assumption that a *cryptographic one-way permutation* exists, and Impagliazzo et al., 1989, showed that they can be constructed under the weaker assumption that a cryptographic one-way *function* exists. This latter condition can also be shown to be necessary. A cryptographic one-way function is a polynomial time computable function f mapping $\{0,1\}^*$ to $\{0,1\}^*$, with $|f(x)| \geq |x|$ for all x, so that for any polynomial time computable function, g does a very bad job inverting f. More precisely, for any polynomial time function g, for a random $x \in \{0,1\}^n$, the probability that $g(f(x))$ is an inverse of $f(x)$, i.e., the probability that $f(g(f(x))) = f(x)$, is negligible, i.e., smaller than the inverse of any polynomial in n.

Impagliazzo et al., 1989, building on Blum and Micali, 1984, Yao, 1982, and Goldreich and Levin, 1989, prove the following theorem.

Theorem 10 *If a cryptographic one-way function exists, then* **BPP** *is included in* **SUBEXP**.

Today, this theorem is mainly of historical interest, as we can get a (provably) stronger statement using the hardness based generators to be described below. On the other hand, there is no known way of using the constructions we shall concentrate on in this chapter directly for cryptography, so cryptographic pseudorandom generators remain extremely important for such applications.

5.2   SIPSER'S HITTING SET GENERATOR

Sipser proved the existence of a hitting set generator using the assumption that some computations need almost as much space as time.

Theorem 11 *There is a constant $\delta > 0$ so that the following holds. If, for some constant $c \geq 1$, there is a language L in* **TIME**(2^{cn}) *so that any Turing machine for L must use space at least $2^{(c-\delta)n}$ on all sufficiently large input lengths n, then* **pP** $=$ **pRP** *(and hence* **P** $=$ **BPP**).

Proof By Lemma 30, it is enough to show that there is a generator which on input r runs in time polynomial in r and outputs a set $H_r \subseteq \{0,1\}^r$ which is a $(1 - 2^{-r+r^{1/2}})$-hitting set for circuit of size r^q, for a certain constant q. The value of δ for which we will show the theorem will be $\delta = \frac{1}{5q}$.

Let L be a language in **TIME**(2^{cn}) and let M be a Turing machine witnessing this fact. We can assume that the work tapes of M are all over the Boolean alphabet $\{0,1\}$, and that each work tape has length at most 2^{cn}. Given a parameter r, we split each work tape into $2^{cn}/r$ blocks of length r. Given an input x to M, during the computation of M, each block of each work tape will attain a sequence of configurations, each configuration being a bit vector in $\{0,1\}^r$. Let H_x be the union of all these configurations, taken over all blocks of all the work tapes. Let $H_{n,r}$ be the union of all sets H_x, for all $x \in \{0,1\}^n$.

Now consider the set $H_r := H_{2q \log r, r}$. Observe that the set can be generated in polynomial time (in r). We claim that, for sufficiently large r, this is a $(1 - 2^{-r+r^{1/2}})$-hitting set in $\{0,1\}^r$ for circuits of size r^q.

Suppose not. Then there are arbitrarily large values of r and circuits $C_r : \{0,1\}^r \to \{0,1\}$ of size r^q with $|Z(C_r)| \leq 2^{r^{1/2}}$ and $H_r \subseteq Z(C_r)$, where $Z(C_r) = \{x | C_r(x) = 0\}$.

The crucial observation is that we can now define a *compression* function $c : H_r \to \{0,1\}^{r^{1/2}}$ and a decompressor $d : \{0,1\}^{r^{1/2}} \to \{0,1\}^r$, so that $d(c(x)) = x$ for every $x \in H_r$ by letting $c(x)$ be the *rank* of x in $Z(C_r)$. If C_r is given as advice, the functions c and d can be computed in space $O(r^q)$, as this is the size of the circuit C_r.

We now describe a simulation of M on inputs of length $n = 2q \log r$ using space $2^{(c-\delta)n}$, contradicting the assumption about L. We first assume that the machine is given as advice the circuit C_r on a special tape. At any given time, the simulating machine M' stores on its work tapes the compressed values $c(y)$ for each block y on each work tape of M, except for those blocks containing a tape head of M. These are stored in uncompressed, "verbatim" form. Whenever a tape head of M is to be moved from one block u to another v, M' simulates this by compressing u, replacing it with $c(u)$ and decompressing the value $c(v)$ stored on its tapes, replacing it with v. Thus, the entire computation of M can be simulated.

The space use is $O(r^q)$ for the compression/decompression operations, $O(r)$ for the uncompressed blocks and $O(r^{\frac{1}{2}} \cdot 2^{cn}/r)$ for the compressed ones, i.e., a total of $O(r^q + 2^{cn}/r^{\frac{1}{2}})$ which is less than $2^{(c-\delta)n}$, as desired.

We now just need to show that we don't need the advice C_r. This follows from the fact that the above simulation can be done using *any* circuit $C : \{0,1\}^r \to \{0,1\}$ instead of C_r and will work correctly if we have $|Z(C)| \le 2^{r^{\frac{1}{2}}}$ and any block of the computation turns out to be a zero of C. We can test for these properties when we do the simulation and abort the simulation if one of them fails. Thus to simulate M without the advice, we try the simulation of *all* possible C and if they all fail, we do a naive simulation. For each r where H_r is not a hitting set, our simulation works in space $2^{(c-\delta)n}$, as it will eventually find C_r or another circuit that works. This contradicts the assumption about L. $\blacksquare$

We can optimize the value of δ by using the best available dispersers to get as good a tradeoff between ϵ and q as possible in Lemma 29. The value $\delta = 0.01$ is certainly possible by the above argument. However, there is not much reason to do so, as we can obtain a stronger theorem as a corollary to the hardness based generator of Impagliazzo and Wigderson, described in the next section.

5.3   HARDNESS BASED GENERATORS

Hardness based generators are the most important kind of hitting set and pseudorandom generators for applications in derandomization. The next two sections of the chapters will be devoted to constructing them. In Section 6, we prove the following Theorem, first proved by Babai et al., 1991a, building on Nisan and Wigderson, 1994.

Theorem 12 *If there is a language L in* **EXP** *so that, for all n, the circuit complexity of $L \cap \{0,1\}^n$ is $n^{\omega(1)}$, then* **BPP** $\subseteq$ **SUBEXP**.

This establishes Statement 6 of the introduction: Note that the assumption stating that a language in **EXP** of superpolynomial circuit complexity for all n exists is just a reformulation of the assumption that the truth table of a Boolean function on n inputs with superpolynomial circuit complexity can be generated in time $2^{n^{O(1)}}$.

Also note that the assumption of the theorem is *not* equivalent to "**EXP** $\not\subseteq$ **P/poly**"., i.e., to the statement that **EXP** does not have polynomial sized circuits. Indeed it is conceivable that any language in **EXP** has polynomial sized circuits, except on some "sparse" infinite sequence of input lengths. If this is the case, the statement "**EXP** does not have polynomial sized circuits" could be true, but the assumption of theorem 12 would be false. However, the proof of Theorem 12 also yields a proof of the following theorem.

Theorem 13 *If* **EXP** *does not have polynomial circuits then* **BPP**$\subseteq$ $\cap_{\epsilon>0}$**TIME**$_{i.o}(2^{n^{\epsilon}})$

Here, **TIME**$_{i.o}(t)$ is the class of languages decided by a deterministic Turing machine running in time $t(n)$ and deciding the correct language for infinitely many input lengths n.

The theorems above provide a relatively weak conclusion from a relatively weak assumption, and are often referred to as *low end* simulation theorems. A *high end* simulation theorem is the theorem of Impagliazzo and Wigderson, 1997, which we shall prove in Section 7.

Theorem 14 *If there is a language L in* **E** *and an $\epsilon > 0$, so that for all n, the circuit complexity of $L \cap \{0,1\}^n$ is at least $2^{\epsilon n}$, then* **BPP** $=$ **P**.

This yields Statement 7 of the introduction, as the assumption of the theorem is, again, just a reformulation of the assumption of that statement.

The proofs of the theorems above work by constructing *hardness based* pseudorandom or hitting set generators. Such a generator works as follows. First the generator efficiently generates the truth table of a Boolean function f with big circuit complexity. This is possible by the assumption of the theorem. Then, using a series of transformation, the generator transforms this truth table into a hitting set or pseudorandom set for circuits of the appropriate size.

In Trevisan, 1999, it was observed that it is useful to think of the truth table as an extra input to the generator, rather than being generated inside the generator. Thus we define:

Definition 7 *A hardness based $(1 - \epsilon)$-hitting set generator with hardness threshold $s(\cdot)$ is an algorithm which takes as input the truth table of a Boolean function $f : \{0,1\}^{\log n} \to \{0,1\}$ (i.e. n bits), with a promise* *that the circuit complexity of f is at least $s(\log n)$ and outputs a $(1 - \epsilon)$-hitting set $H \subseteq \{0,1\}^{n'}$ for circuits of size n', for some n' depending on n.*

A similar definition can be made for the case of a hardness based pseudorandom generator. The crucial observation of Trevisan, 1999 is that when we look at it this way, the syntactic similarity to the definitions of dispersers and extractors are immediate. Indeed, given a hardness based hitting set generator, we can define a graph on $U \times V$, where $U = \{0,1\}^n$ and $V = \{0,1\}^{n'}$ by putting an edge between $x \in \{0,1\}^n$ and $y \in \{0,1\}^{n'}$, if y is a member of the set produced by the generator if it is given as input the truth table x. Now, this graph has the property that if we take a *fixed* vertex in U with big circuit complexity, its set of neighbors intersects any set $S \subseteq V$ of density $1 - \epsilon$ which is recognized by a small circuit. Recall that the disperser property states that for any set $S \subseteq V$ of density $1 - \epsilon$, this will be the case for *the vast majority of vertices of U*, but not any particular fixed one. On the other hand, the disperser does not require S to be recognized by a small circuit. Thus, it seems that we are getting something more from the disperser (we are hitting all sets), by paying something more (we pick a random vertex from U, rather than a fixed one with a particular property).

We now show that the similarity is not merely syntactic: Hardness based hitting set generators *are* dispersers, if the proof that they work *relativizes*, i.e., holds under any oracle.

Definition 8 *A relativizable hardness based $(1 - \epsilon)$-hitting set generator with hardness threshold $s(\cdot)$ is an algorithm which takes as input the truth table of a Boolean function $f : \{0,1\}^{\log n} \to \{0,1\}$ (i.e. n bits), with a promise that the circuit complexity of f is at least $s(\log n)$ (with respect to some arbitrary oracle A) and outputs a $(1 - \epsilon)$-hitting set $H \subseteq \{0,1\}^{n'}$ for circuits of size n' (containing oracles for A), for some n' depending on n.*

The hardness based generators we shall construct will be relativizable, as they use no special properties of Boolean circuits that are not satisfied by oracle circuits as well. The reason they are dispersers is, intuitively, that if a construction relativizes, it cannot use any particular property of "big sets recognized by small circuits" as opposed to just "big sets". Thus, even though the hitting set generator as opposed to a disperser allows us to just hit the big sets recognized by small circuits, we can only take advantage of this to a limited extend in the proof of the correctness of the generator. Formally, we have the following theorem.

Theorem 15 *A relativizable hardness-based* $(1-\epsilon)$-*hitting set generator with hardness threshold* $s(\cdot)$ *is, viewed as a graph, an explicit* $(1 - \epsilon)$-*hitting disperser on* $\{0,1\}^n \times \{0,1\}^{n'}$ *with threshold* $2^{2s(\log n)\log s(\log n)}$.

Proof We need to prove that given any subset S of U with $|S| \geq 2^{2s(\log n)\log s(\log n)}$, more than ϵn of the vertices of V are adjacent to S. Suppose not. Let S be a set for which this is not the case, and let A be the non-neighbors of S, so we have $|A| \geq (1 - \epsilon)|V|$. Viewed as a subset of $\{0,1\}^{\log |V|}$, we can use A as an oracle and consider circuit complexity relative to A. By Shannon's counting argument, viewed as truth tables for Boolean functions on $\log |U|$ variables, at least one of the members of S must have oracle circuit complexity with oracle gates for A at least $\frac{1}{2}\log|S|/\log\log|S| > s(\log n)$. Let this element of S be denoted a. Thus, by the hitting set generator property, the vertices in V adjacent to a will intersect every set in V which is the characteristic (accepted) set of an oracle circuit with oracle gates for A of size at most n *and* has size at least $(1 - \epsilon)|V|$.

But then consider the oracle circuit defined by $x \rightarrow A(x)$. It has size n, density at least $1 - \epsilon$ and the neighbors of a do not intersect its characteristic set (as this set is the non-neighbors of S). A contradiction. ∎

By an analogous proof, one can show that relativizable hardness based pseudorandom generators are extractors. Thus, we cannot build these generators without building dispersers and extractors as well.

This leads to two strategies for building the generators: We can either refine the disperser/extractor constructions, making them obtain the stronger generator properties. This is the strategy used in Section 6 where we shall present a proof of Theorem 12, based on extending the extractor construction of Trevisan (as we mentioned earlier, historically, it was the other way around). Or, alternatively, we can build dispersers or extractors with the appropriate parameters *into* the constructions, using them as "black boxes" and then try to "add on" the stronger generator property. This is the strategy used in Section 7 where we shall present a proof of Theorem 14. It is also very close to the way we proved Theorem 11, even though the generator of that theorem was not hardness-based.

First however, we shall note that Theorems 12 and 14 give stronger results that either the cryptographic generators or Sipser's generator. Indeed, Theorem 10 can also be obtained as a corollary of Theorem 12, as the assumption of a cryptographic one-way function easily yields the appropriate language in **EXP**. Also, as corollary of Theorem 14, we get the following stronger version of Theorem 11.

Corollary 2 *If, for any constant $\epsilon > 0$,there is a language L in* **E** *so that any Turing machine for L must use space at least $2^{\epsilon n}$ on all sufficiently large input lengths n, then* **P**=**BPP**.

Proof We show that the assumption of Corollary 2 implies the assumption of Theorem 14. So let a (without loss of generality, Boolean) language L in **TIME** (2^{cn}) be given so that any machine for L must use space $2^{\epsilon n}$. We shall construct a language L' in **E** so that $L' \cap \{0,1\}^n$ has no circuit of size $2^{\frac{\epsilon}{10c}n}$.

Given $L \in$ **E** by a one-tape Turing machine M, we define the language $L' = \{\langle x,i,j,\sigma,b,\tau\rangle|$ the contents of the i'th cell at time j during the computation of M on x is σ and b is a Boolean which is true if the tape head is on cell i at time j. Furthermore, if b is true, then the state of the machine at time j is $\tau\}$. Then L' is *locally checkable*, i.e., we can tell if circuit C decides $L' \cap \{0,1\}^n$ for some input length n correctly if an exponential number of easily generated local conditions of the form $C(y) \Leftrightarrow C(y_1) \wedge C(y_2) \ldots \wedge C(y_l)$, where l is a constant, are satisfied. This means that we can check if a certain circuit C of size $2^{\frac{\epsilon}{10c}n}$ computes $L' \cap \{0,1\}^n$, using space linear in the description of the circuit. Now suppose such a circuit actually exists, for infinitely many n. We show how to evaluate L space-efficiently on infinitely many input lengths. On input $x \in \{0,1\}^n$, we reduce the question $x \in L$? to a question $x' \in L'$? for x' of length $n' \leq 10nc$. Now we do an exhaustive search for a circuit of size $2^{\frac{\epsilon}{10c}n'}$ deciding L', testing each possibility as described above. If we find one, we use it to decide $x' \in L'$ and we are done. If we fail to find a correct circuit, we run the original machine M to decide $x \in L$. As we will find a circuit for infinitely many input lengths, we have an algorithm in space $2^{\epsilon n}$ for infinitely many n, contradicting the assumption on L. ∎

Even though we can get the stronger statement of Corollary 2, a self-contained proof, including the proof of Theorem 14 would be much harder than the proof of Theorem 11 given before. To get the easiest known self-contained proof of Corollary 2, we can use Theorem 20 to be described later, instead of Theorem 14.

6. A LOW END GENERATOR

In this section, we prove Theorem 12. For this, we describe, for any constant $k \geq 1$, a procedure running in time $2^{O(n^6)}$, converting the truth table of a function $f : \{0,1\}^n \to \{0,1\}$ with no circuits of size $s(n)$, for a superpolynomial function s, into a pseudorandom set in $\{0,1\}^{n'}$ with small error, for circuits of size n' with $n' = n^k$. The construction relativizes. By the discussion of Section 5.3, this means that the procedure also defines an explicit extractor $E : U \times A \to V$ with $|U| = 2^{2^n}$, $|V| = 2^{n^k}$, $|A| = 2^{O(n^c)}$ and min-entropy threshold $s(n)^{1+o(1)}$. While we have not previously explicitly described an extractor construction with these parameters, having concentrated on extractors with polynomial degree, it is not difficult to see that by setting the parameters right in the Trevisan extractor, we can obtain such an extractor.

The actual construction of the pseudorandom generator follows the steps of the Trevisan extractor quite closely.

Recall that the Trevisan extractor performs two operations on its inputs $x \in U$ and $y \in A$, where x is randomly chosen with a distribution with min-entropy at least t, and y is uniformly chosen. First, it applies an error correcting code e to x, obtaining $e(x)$. If x has min-entropy at least t, any lossy source code for $e(x)$ with distortion bounded away from $\frac{1}{2}$ must have rate close to t. Second, the value y defines, through Nisan-Wigderson designs, a number of bit positions of $e(x)$ which are taken out and concatenated. This yields a value $E(x, y)$ with distribution close to uniform.

The pseudorandom generator goes through two completely analogous steps. Instead of starting with a value x sampled from a distribution on m bits with min-entropy at least t, we start with a *computational analogue* of this: The truth table of a function $f : \{0,1\}^n \to \{0,1\}$ with n roughly $\log m$ and with circuit complexity at least (roughly) t. Intuitively, if we think of a distribution with min-entropy at least t simply as a distribution with no lossless fixed length source code with rate t, we see that the relevant computational analogue of a lossless source code is a circuit. Then, rather than generating a value $e(x)$ distributed so that any lossy source code with rate roughly t has distortion close to one half, we generate a computational analogue of this: *The truth table of a Boolean function $\tilde{f}$ so that any circuit C of size close to t attempting to compute $\tilde{f}$ must disagree with $\tilde{f}$ on a fraction of inputs close to $\frac{1}{2}$.* Finally, we use the second input y to pick out, through Nisan-Wigderson designs, a number of inputs on which to evaluate $\tilde{f}$. As in the extractor case, we concatenate the outputs. This yields a value $G(x, y)$. The values $\{G(x, y) | y \in A\}$ now yield a computational analogue of a distribution ϵ-close to random: an ϵ-pseudorandom set for small circuits.

In the next subsections we formalize the above intuitive description. The second step causes the fewer problems and will be dealt with first, in Subsection 6.1. Essentially, we can just redo the corresponding information theoretic proof replacing "low rate source codes" with "small circuits" and "predictors" with "predictors with small circuits". To improve readability, we build the generator of the original truth table into our generators, rather than regard the truth table as an extra input. Thus, the last step of the construction will be given by the following lemma.

Lemma 33 *Let $\tilde{s}(n)$ be a superpolynomial function and let constant $c \geq 2$. If there is a procedure running in time $2^{O(n^c)}$ which on input n outputs the truth table of a function $\tilde{f} : \{0,1\}^n \to \{0,1\}$ so that no circuit of size $\tilde{s}(n)$ agrees with $\tilde{f}$ on more than a $\frac{1}{2} + \tilde{s}(n)^{-1}$ fraction of the inputs, then, for any constant $\epsilon > 0$, there is procedure running in time $2^{O(n^\epsilon)}$ which on input n outputs a pseudorandom set for circuits of size n with error $\frac{1}{10}$, and thus, $\mathbf{BPP}$ is in $\mathbf{TIME}(2^{O(n^\epsilon)})$.*

To successfully implement the first step, we need to do more work. While the key is still the use of error correcting codes, it turns out that generic error correcting codes will not do the trick for us. We need a special kind of error correcting codes, not traditionally studied in coding theory: *locally decodable codes*. Locally decodable error correcting codes will enable us to transform a function f with no circuit of size t to a function f' so that all circuits of size roughly t must disagree with f' on a non-trivial fraction of the inputs. Thus, in Section 6.2, we prove the following lemma.

Lemma 34 *Let $s(n)$ be a superpolynomial function and let constant $c \geq 2$. If there is a procedure running in time $2^{O(n^c)}$ which on input n outputs the truth table of a function $f : \{0,1\}^n \to \{0,1\}$ with no circuits of size $s(n)$ then there is a procedure running in time $2^{O(n^c)}$ which on input n outputs the truth table of a function $f' : \{0,1\}^n \to \{0,1\}$ so that any circuit of size $s'(n)$ disagrees with f' on at least a $1/n^2$ fraction of the inputs, where $s'(n)$ is another superpolynomial function.*

Unfortunately, locally decodable codes will not give us what we really desire to get in a position to apply Lemma 33: A function $\tilde{f}$ so that all small circuits disagree with $\tilde{f}$ on almost half the inputs. Instead we convert, in Section 6.3, our function f' to such a function $\tilde{f}$ using a different, non-coding theoretic, construction: If $f' : \{0,1\}^n \to \{0,1\}$, we let $\tilde{f} : \{0,1\}^{n^5} \to \{0,1\}$ be given by $\tilde{f}(x_1, x_2, \ldots, x_{n^4}) = f'(x_1) \oplus f'(x_2) \oplus \cdots \oplus f'(x_{n^4})$, where $\oplus$ denotes Boolean XOR, and each $x_i \in \{0,1\}^n$. A lemma, *Yao's XOR lemma* will tells us that this construction has the desired property. Thus, we can bridge the gap between Lemma 34 and Lemma 33, proving the following lemma.

Lemma 35 *Let $s'(n)$ be a superpolynomial function and let $c \geq 2$. If there is a procedure running in time 2^{n^c} which on input n outputs the truth table of a function $f' : \{0,1\}^n \to \{0,1\}$ so that all circuits of size*

$s'(n)$ *disagrees with* f' *on at least a* $1/n^2$ *fraction of the inputs, then there is a procedure running in time* $2^{O(n^c)}$ *which on input* n *outputs the truth table of a function* $\tilde{f} : \{0,1\}^n \to \{0,1\}$ *so that no circuit of size* $\tilde{s}(n)$ *agrees with* $\tilde{f}$ *on more than a* $\frac{1}{2} + 1/\tilde{s}(n)$ *of the inputs, where* $\tilde{s}(n)$ *is another superpolynomial function.*

Lemma 34, Lemma 35 and Lemma 33 together proves Theorem 12. Thus, the construction we give has 3 steps rather than 2. It is possible to do the construction without the middle step by using a different, more relaxed, kind of code than locally decodable codes, essentially capturing "local list decodability". This was done by Sudan et al., 1999. We present the three step proof here, as it arguably gives the easiest self-contained proof of Theorem 12. It was also the original proof given by Nisan and Wigderson, 1994, and Babai et al., 1991a.

6.1   UNPREDICTABILITY AND PSEUDORANDOMNESS

In this subsection, we prove Lemma 33. For this, we need a computational analogue of Lemma 21. This lemma was proved before Lemma 21, by Yao, 1982.

Lemma 36 *Given* $\epsilon > 0$. *If a set* $S \subseteq \{0,1\}^m$ *is not* ϵ*-pseudorandom for circuits of size* m, *then the uniform distribution* $\mathcal{D}$ *on* S *is* $\frac{\epsilon}{2m}$*-predictable, with the predictor given by a circuit of size* m.

Proof We mimic the proof of Lemma 21 closely. We consider the $m + 1$ distributions $\mathcal{D}_0, \dots, \mathcal{D}_m$, defined as follows. To pick a random element of $\mathcal{D}_i$, we pick $x \in \{0,1\}^m$ according to $\mathcal{D}$ and $y \in \{0,1\}^m$ according to the uniform distribution $\mathcal{U}$ and return $x_1, \dots, x_i, y_{i+1}, \dots, y_m$. Thus, $\mathcal{D}_0$ is the uniform distribution, $\mathcal{D}_m$ is $\mathcal{D}$, and the $\mathcal{D}_i$'s in between are hybrids of the uniform distribution and $\mathcal{D}$.

By definition, if S is not pseudorandom for circuits of size m, we have a circuit C of size m, so that $|\Pr_{x \in S}[C(x) = 1] - \Pr_{x \in \{0,1\}^m}[C(x) = 1]| > \frac{\epsilon}{2}$, i.e. $|\Pr_{x \in \mathcal{D}_m}[C(x) = 1] - \Pr_{x \in \mathcal{D}_0}[C(x) = 1]| > \frac{\epsilon}{2}$. By the triangle inequality, there must be i so that $|\Pr_{x \in \mathcal{D}_{i+1}}[C(x) = 1] - \Pr_{x \in \mathcal{D}_i}[C(x) = 1]| > \frac{\epsilon}{2m}$.

We now define a randomized predictor f for the $i+1$'st bit with good (expected) advantage. On input $x_1 x_2 \dots x_i$ generate uniform random bits $r_{i+1}, \dots, r_m$. Thus, $\tilde{x} = x_1 x_2 \dots x_i r_{i+1} \dots r_m$ is a sample from $\mathcal{D}_{i+1}$. Now, if $C(\tilde{x}) = 1$, predict x_{i+1} to be r_{i+1}, otherwise predict x_{i+1} to be $\neg r_{i+1}$. The expected advantage of the predictor is at least $\frac{\epsilon}{2m}$. We can freeze the random decisions of the predictor to get a predictor with the same advantage, given by a circuit of size m. ∎

For a much more detailed coverage of the computational version of the hybrid technique, see Goldreich, 1999.

Now we are ready to prove Lemma 33. Let the parameter $\epsilon > 0$ be fixed. Given input n, we describe a procedure for generating a pseudo-random set for circuits of size n, using time $2^{O(n^\epsilon)}$.

We let $\epsilon' = \epsilon/c$, where c is the constant in the assumption of the Lemma. Then, using this assumption, we generate in time $2^{O(n^\epsilon)}$ the truth table of a function $f : \{0,1\}^{n^{\epsilon'}} \to \{0,1\}$ so that any circuit of size $s(n^{\epsilon'})$ agrees with f on at most a $\frac{1}{2} + \frac{1}{s(n^{\epsilon'})}$ fraction of the inputs.

Now, use Lemma 26 to construct an $(n, n^{2\epsilon'}, n^{\epsilon'}, \log n)$ Nisan-Wigderson design $S_1, S_2, \ldots, S_n$. The pseudorandom set for circuits of size n is defined to be $P = \{v_y\}_{y \in \{0,1\}^{n^{2\epsilon'}}}$, with v_y defined as follows: View y as a bit sequence of length $n^{2\epsilon'}$, and for S being a set in the design, let y_S by the subsequence consisting of the bits indexed by S. Then $v_y = f(y_{S_1})f(y_{S_2})\ldots f(y_{S_n})$.

Suppose this is not a pseudorandom set for circuits of size n with error at most $\frac{1}{10}$. Then, if y is chosen at random, by Lemma 36, there is a predictor with advantage $\frac{1}{20n}$, given by a circuit of size n, for one of the bits of v_y, say the i'th bit $f(y_{S_i})$, given the previous bits.

Freeze the bits of $y_{S_i^c}$ (i.e., all bits except the bits of y_{S_i}) to fixed values while preserving the advantage of the predictor for a random setting of y_{S_i}. As $S_1 \cap S_i$ only contains $\log n$ bit positions and the bits of $y_{S_1 \setminus S_i}$ have been frozen, we can tabulate the value of $f(y_{S_1})$ as a function of $y_{S_1 \cap S_i}$ using a polynomial size table. Similarly, we can tabulate the value of $f(y_{S_j})$ as a function of $y_{S_j \cap S_i}$ for each $j = 2, \ldots, i-1$.

Now we have the following polynomial circuit approximating $f(y)$. The circuit has the tables above given as advice. On input y, the circuit looks up $f(y_{S_1}), \ldots, f(y_{S_{i-1}})$ in the tables above, These values are then fed into the predictor, to get a prediction for $f(y_{S_i})$ which we give as output. By the advantage of the predictor, the circuit is correct on random input with probability at least $\frac{1}{2} + \frac{1}{2n}$. The circuit has polynomial size. This contradicts the assumption on f. Thus, we have proved Lemma 33.

6.2   LOCALLY DECODABLE CODES

To introduce locally decodable error correcting codes, let us first recall the motivation for classical error correcting codes presented in Section 2.3: A sender wants to transmit a bit string x to a receiver reliably, even if a meddler is allowed to introduce a limited amount of noise in any message sent. Encoding x in an error correcting codes gave a satisfactory solution to this problem. Now consider a setting where the string x is rather long, say, an encoding of the Encyclopedia Britannica, and the receiver is really only interested in a small piece of x, say, the piece about the nesting habits of night owls. Thus, the receiver does not really want to read and decode the entire received string x^*. Ideally, he wants to read only a piece of x^* comparable in size to the piece about night owls in the original string x. Informally, a *locally decodable code* is an error correcting code that allows him to achieve this goal.

At first sight, it seems that locally decodable codes cannot exist for interesting settings of the parameters. In general, we would want the receiver to look at much fewer bits than the meddler is allowed to mix up. Thus, if the meddler knows which piece the receiver is interested in (and we assume that he does), the meddler will be able to mess up the communication completely by replacing all the bits the receiver is going to look at by random bits. Indeed, this impossibility proof is correct, but only if we assume that the receiver will choose which bits to look at *deterministically*. If we allow the receiver to pick the bits to look at using a *randomized* strategy and also allow him to retrieve an incorrect message piece with some small probability, locally decodable codes do exist, as we shall see in this section. First, we give a formal definition of a locally decodable error correcting code.

Definition 9 *A family of error correcting codes* $e : \{0,1\}^n \to \{0,1\}^m$ *is locally decodable with error* ϵ *and relative correction distance* δ *if there is an efficient randomized algorithm, the decoding algorithm, which is given as input an index* $i \in \{1, \ldots, n\}$, *and, in a random access memory, a string* x^*, *so that the relative distance between* x^* *and* $\tilde{x} = e(x)$ *is at most* δ, *for some string* $x \in \{0,1\}^n$. *The algorithm runs in time polynomial in* $\log n$ *and returns a bit* u, *so that, with probability at least* $1 - \epsilon$, *over the random coin flips of the algorithm, we have* $u = x_i$.

Our intended application for a locally decodable code $e : \{0,1\}^n \to \{0,1\}^m$ is to convert the truth table of a function $f : \{0,1\}^{\log n} \to \{0,1\}$ with no small circuits to the truth table of a function $f : \{0,1\}^{\log m} \to \{0,1\}$ which cannot be approximated well by small circuits. We now verify that locally decodable codes indeed do this for us.

Lemma 37 *Let* $e : \{0,1\}^n \to \{0,1\}^m$ *be a locally decodable error correcting code with error less than* $1/n$ *and relative correction distance* δ. *Then for some constant* k, *the following is true: If* x *is the truth table of a function* f *on* $\log n$ *variables with no circuits of size* s, *then* $e(x)$ *is the truth table of a function* f' *on* $\log m$ *variables so that any circuit of size* $s/(\log n)^k$ *attempting to compute* f' *must make errors on more than a* δ *fraction of the inputs.*

Proof Let C' be a circuit of size s' computing f' incorrectly on at most a δ fraction of the inputs. The truth table of C', viewed as a bit string of length m differs from the truth table of f' on at most a δ fraction of the inputs. Thus, the local decoding procedure applied to the truth table of C' will reliably retrieve any bit of the truth table of f, i.e., it will evaluate f on any desired input with an error probability of at most $1/n$. Individual bits of the truth table of C' can be obtained by evaluating C'. Thus, by building C' into the local decoding procedure, we get a randomized circuit for f with an error probability of less than $1/n$ and of size $(\log n)^k s'$ for some constant k, as the time complexity of the procedure is polynomial in $\log n$. As the error probability is smaller than $1/n$ and there are only $2^{\log n} = n$ different inputs to give the circuit, we can freeze the random bits to get a deterministic circuit for f of the same size. Thus $(\log n)^k s' > s$ or $s' > s/(\log n)^k$, as desired. ∎

We present two locally decodable codes. The first one, due to Babai et al., 1991b, is very simple, but has a rather bad blowup: It expands n bits to $2^{O(\log n \log \log n)}$ bits, i.e., the blowup is slightly superpolynomial. Also, it has a subconstant relative correction distance. It is, however, good enough to be used to prove Lemma 34 and we can give a self-contained proof of its local decodability without appealing to any non-trivial coding theory.

The Babai-Fortnow-Lund code (for shorts, the BFL-code) for $x \in \{0,1\}^n$ is defined as follows. Given $x \in \{0,1\}^n$, identify x with its characteristic function $x : \{0,1\}^{\log n} \to \{0,1\}$. Let $\mathbf{F}$ be a finite field of size roughly $(\log n)^2$. Let a *multilinear polynomial* over $\mathbf{F}$ be a polynomial where each variable occurs with degree at most 1 in each term. It is easy to see that there is a unique $\log n$-variate multilinear polynomial $p : \mathbf{F}^{\log n} \to \mathbf{F}$ so that p agrees with x on $\{0,1\}^{\log n}$. This polynomial is called the *multilinear extension* of x. Our code for x is the enumeration of all values of p, i.e., $\tilde{x} = (p(y))_{y \in \mathbf{F}^{\log n}}$. To make the code Boolean, we simply replace each field element $\mathbf{F}$ by an arbitrary Boolean encoding using $O(\log \log n)$ bits.

First we observe that the rate of the code is indeed $n \to 2^{O(\log n \log \log n)}$. Now we prove that the code is locally decodable.

Theorem 16 *The BFL-code is locally decodable with error $1/n$ and relative correction distance $1/(\log n)^2$.*

Proof Assume that we are a given a string x^* with distance at most $1/(\log n)^2$ to a BFL-code word $\tilde{x}$. Let $p : \mathbf{F}^{\log n} \to \mathbf{F}$ be the corresponding multilinear map. With random access to $\tilde{x}$, we can, for a given v, read off the value of $p(v)$ by reading $O(\log \log n)$ bits of $\tilde{x}$. If v is *randomly* chosen and we have random access to x^* instead of $\tilde{x}$, the probability that the same procedure does *not* give us the correct value of $p(v)$ is at most $O(\log\log n/(\log n)^2)$.

However, to prove locally decodability, we have to show that given some specific, i.e., *non-random*, value of v, we can retrieve the correct value of $p(v)$ with high probability, where the probability is over the choices of our decoding algorithm only. To achieve this, we reduce the non-random case to the random case as follows:

Let $i_1, i_2, \ldots, i_{\log n+1}$ be arbitrary non-zero members of $\mathbf{F}$. Given v, pick $u \in \mathbf{F}^{\log n}$ at random and read off from x^* the values $p(v+i_1 u), p(v+i_2 u), \ldots, p(v+i_{\log n+1} u)$, or, more precisely, what *would* have been these values, had x^* been uncontaminated. However, as each of $v + iu$ for $i \in \{1, \ldots, \log n + 1\}$ are uniformly chosen in $\mathbf{F}^{\log n}$ (though not independent) we have by the union bound that all of the retrieved values are correct with probability at least $1 - O((\log n + 1)(\log\log n/(\log n)^2)) = 1 - O(\log\log n/\log n)$. As p is multilinear, the map $f(t) = p(v + ut)$ is a univariate polynomial of degree at most $\log n$. Given that the retrieved values are all correct, we have found $f(i_1), f(i_2), \ldots, f(i_{\log n+1})$ and can uniquely interpolate the polynomial f. Then, we evaluate $f(0)$ which is the desired value $p(v)$.

The above procedure has error probability at most $O(\log\log n/\log n)$. To reduce the error probability to $1/n$, we repeat it $O(\log n)$ times. ∎

Combining the BFL code with Lemma 37, we have a proof of Lemma 34: Using the assumption of the Lemma, we generate the truth table of a Boolean function with no small circuits, and return the BFL-encoding of this truth table. A slight technical problem is that to prove Lemma 34, we need, on any input n, to produce a truth table of length 2^n and this may not match the length of a BFL-codeword, for any setting of the parameters. A simple way to solve this which is easily seen to work if the parameters are set carefully, is to generate a BFL-codeword of length a power of two close to, but not exceeding the desired length 2^n, and concatenating it with itself a sufficient number of times to make the total length exactly 2^n.

For later use in the proof of Theorem 14 in Section 7, we shall also present a better locally decodable code due to Babai et al., 1991a which has only polynomial blowup $n \to n^{O(1)}$ and constant relative correction distance. Though similar to the above BFL-code, the decoding procedure of the BFLS-code is somewhat more complicated and uses as subroutine the Berlekamp-Welch decoder, presented in Section 2.3 as Lemma 10.

To define the BFLS-code, we first define a code $e' : \mathbf{F}^n \to \mathbf{F}^{n'}$, where $\mathbf{F}$ is a finite field of size roughly $(\log n)^4$. The final BFLS-code code is then the concatenated code $e'' : \{0,1\}^n \to \{0,1\}^{n'r}$. defined by $e''(x) = e' \circ e(x) = e(e'(x)_1) \circ e(e'(x)_2) \circ \cdots \circ e(e'(x)_{m'})$, where e is the code of Lemma 12. Note that the BFL-code is really a concatenated code too, but with a trivial encoding as the inner code.

The outer code e' can be seen both as a generalization of the multi-linear extension to higher degree polynomials and as a generalization of the Reed-Solomon code to higher dimensions. It is called the *low degree extension* and is defined as follows.

Pick a subset S of $\mathbf{F}$ of size $\log n$ and identify the set $\{1, \dots, n\}$ with the set $S^{\log n / \log \log n}$ in some arbitrary way. Thus, given a message $x \in \mathbf{F}^n$, we can view x as a map $x : S^l \to \{0,1\}$, where $l = \log n / \log \log n$. There is a unique l-variate polynomial p of degree at most $|S| - 1$ in each variable agreeing with f on S^l. The low degree extension $e'(x)$ is the enumeration of all values of this map, i.e. the vector $(p(v))_{v \in \mathbf{F}^l}$.

The sole advantage we get from using the low degree extension over the multilinear extension is the fact that it only blows up the size of its input by a polynomial. Indeed, the number of field elements in a code word is roughly $((\log n)^4)^{\log n / \log \log n} = n^4$.

Now we define the BFLS-code as the concatenation of the low degree extension with the code e of Lemma 12. Observe that the blowup of the code is indeed polynomial.

Theorem 17 *The BFLS-code is locally decodable with error $1/n$ and relative correction distance $\frac{1}{20}$.*

Proof Assume that we are a given a string x^* with distance at most 0.05 to a BFLS-code word $\tilde{x}$. Let $p : \mathbf{F}^l \to \mathbf{F}$ be the corresponding low degree polynomial. With random access to $\tilde{x}$, we can, for a given v, read off the value of $\tilde{p}(v)$ by reading $O(\log \log n)$ bits of $\tilde{x}$. If we have random access to x^* instead, a value so found may be incorrect. We can at least replace a value which is not an encoding of a field element by the closest valid encoding (an exhaustive search through all field elements is feasible, as $|\mathbf{F}|$ is only polylog n). Then, for a *randomly* chosen v, the probability that this procedure does *not* give us the correct value of $p(v)$ is less than 0.24. This is because less than a 0.24 fraction of the encodings of field elements can have more than a 0.24 fraction of the bits flipped - otherwise the relative distance between x^* and $\tilde{x}$ would be more than $(0.24)(0.24) > 0.05$.

Again, we reduce the non-random case to the random one in a similar way as before. Let $i_1, i_2, \dots, i_{(\log n)^3}$ be arbitrary non-zero members of $\mathbf{F}$. Given v, pick $x \in \mathbf{F}^l$ at random and read off from x^* the values $p(v +$

$i_1 x), p(v+i_2 x), \ldots, p(v+i_{(\log n)^3} x)$, or rather, what *would* have been these values, had x^* been uncontaminated. However, as each of $v + ix$ for $i \in \{1, \ldots, (\log n)^3\}$ are uniformly chosen in $\mathbf{F}^l$ (though not independent), each is incorrect with probability at most 0.24, so the expected fraction of incorrect values is at most 0.24. Thus, the probability that more than a 0.501 fraction of values are correct is at least 0.501 (as $0.499 \cdot 0.499 >$ 0.24). As p is a polynomial with a degree of at most $\log n$ in each variable, the map $f(t) = p(v + xt)$ is a univariate polynomial of degree at most $(\log n)^2 / \log \log n$. Given that at least a 0.501 fraction of the retrieved values are correct, and n is sufficiently large, as we can assume without loss of generality, we can use the Berlekamp-Welch procedure of Lemma 10 to reconstruct this polynomial. Then, we evaluate $f(0)$ which is the desired value $p(v)$.

The above procedure gives us the correct answer with probability at least 0.501. To get the desired error probability, we repeat the procedure a number of times and take the most common answer. ∎

Locally decodable codes form an interesting topic in their own right. An important open problem is whether there are locally decodable codes with constant correction distance and with only linear blowup (rather than polynomial blowup). By doing the above construction more carefully, the blowup can be made $n^{1+\epsilon}$, for any desired constant $\epsilon > 0$. This is essentially the best upper bound currently known for locally decodable codes.

The BFLS code and Lemma 37 gives us the following improvement of Lemma 34 which will be used in the proof of the Impagliazzo-Wigderson theorem.

Lemma 38 *Given a constant $\epsilon > 0$. If there is a procedure which runs in time $2^{O(n)}$ and on input n generates a truth table of some Boolean function $f : \{0,1\}^n \to \{0,1\}$ with circuit complexity at least $2^{\epsilon n}$, then there is a constant $\epsilon' > 0$ and a procedure which runs in time $2^{O(n)}$ and on input n generates a truth table of some Boolean function $f' : \{0,1\}^n \to \{0,1\}$ so that any Boolean circuit of size $2^{\epsilon' n}$ fails to compute f' on at least a $\frac{1}{20}$ fraction of the inputs.*

6.3   THE XOR LEMMA

In this subsection we prove Lemma 35, thus completing the proof of Theorem 12. The statement of the lemma is a corollary of *Yao's XOR lemma*. For convenience, in the statement and proof of this lemma, we consider Boolean functions mapping $\{0,1\}^n$ to $\{-1,1\}$ rather than $\{0,1\}$. Then Boolean XOR is just integer multiplication. Furthermore, if we define the *correlation* of two functions f_1 and f_2 as the value $2^{-n} \sum_{x \in \{0,1\}^n} f_1(x) f_2(x)$, we see that the correlation of f_1 and f_2 is δ, if and only if the probability over random input x that $f_1(x) = f_2(x)$, is $\frac{1}{2} + \frac{\delta}{2}$.

Yao's XOR-lemma, of Yao, 1982, is the following lemma.

Lemma 39 *Let* $f : \{0,1\}^n \to \{-1,1\}$ *and assume that all circuits of size at most* s *have correlation at most* δ *with* f. *Now define for any parameter* $k \geq 1$

$$f^{(k)} : \{0,1\}^{kn} \to \{-1,1\}$$
$$f^{(k)}(x_1, x_2, \ldots, x_k) = f(x_1) f(x_2) \ldots f(x_k),$$

where each $x_i \in \{0,1\}^n$. *Then all circuits of size at most*

$$\frac{s - 10(\frac{n}{\epsilon})^{10}}{\left(\frac{n}{\epsilon}\right)^{10}}$$

have correlation at most $\delta^k + k\epsilon$ *with* $f^{(k)}$ *(for any parameter* $\epsilon > 0$).

Before proving the Lemma, let us see that it implies Lemma 35. To generate the desired truth table, we generate a truth table f' using the assumption of Lemma 35 and let $\tilde{f} : \{0,1\}^{n^5} \to \{0,1\}$ be defined by $\tilde{f}(x_1, x_2, \ldots, x_{n^4}) = f'(x_1) \oplus f'(x_2) \oplus \cdots \oplus f'(x_{n^4})$. As in the proof of Lemma 34 in Subsection 6.2, we may not be able to match exactly the size of the truth table we are supposed to generate, but as we did there, we can solve this by simply concatenating a number of copies of the generated truth table.

The proof of the XOR lemma we shall present is due to Levin, 1987 (for much more information about the XOR lemma and other proofs, see Goldreich et al., 1995). It is a proof by induction in k, with the induction step being handled by the following lemma, the *isolation lemma*.

Lemma 40 *Let* $f_1 : \{0,1\}^{n_1} \to \{-1,1\}$ *and* $f_2 : \{0,1\}^{n_2} \to \{-1,1\}$. *Assume that all circuits of size at most* s_1 *have correlation at most* δ_1 *with* f_1 *and that all circuits of size at most* s_2 *have correlation at most* δ_2 *with* f_2. *Then all circuits of size at most*

$$s = \min\left(\frac{s_1 - 10(\frac{n_1}{\epsilon})^{10}}{\left(\frac{n_1}{\epsilon}\right)^{10}}, s_2\right)$$

have correlation at most $\delta = \delta_1 \delta_2 + \epsilon$ *with* $f(x,y) = f_1(x) f_2(y)$.

Proof Assume to the contrary that a circuit C of size s has correlation more than δ with f.

Then, if we define $T(x) = \mathrm{E}_y[C(x,y)f_2(y)]$, we have

$$\delta < \mathrm{E}_{x,y}[C(x,y)f(x,y)] = \mathrm{E}_{x,y}[C(x,y)f_1(x)f_2(y)] = \mathrm{E}_x[f_1(x)T(x)].$$

Assume that $\exists x : T(x) > \delta_2$. Then the circuit $y \to C(x,y)$ has a correlation with f_2 better than allowed by the lemma hypothesis. Similarly, assume that $\exists x : T(x) < -\delta_2$. Then the circuit $y \to \neg C(x,y)$ has a correlation with f_2 better than allowed by the lemma hypothesis.

Hence, $\forall x : \frac{T(x)}{\delta_2} \in [-1; 1]$. If the function $T(x)/\delta_2$ was Boolean and computable by a small circuit, it would contradict the assumption about f_1. Our strategy is to modify the function so that this will be the case.

Let $(y_i, f_2(y_i))_{i=1,2,\dots}$ be a sequence of randomly generated samples, with y_i uniformly chosen in $\{0, 1\}^{n_2}$.

Define

$$\tilde{T}(x) = \sum_{i=1}^{(\frac{n_1}{\epsilon})^{10}} f_2(y_i)C(x, y_i)$$

With the random sequence given as advice, $\tilde{T}$ is computable by a circuit of size at most $s(n_1/\epsilon)^{10} + 5(n_1/\epsilon)^{10} \leq s_1 - 5(n_1/\epsilon)^{10}$. Furthermore, by the Chernoff bound of Lemma 1 we have that for fixed x,

$$\Pr[|\tilde{T}(x) - T(x)| > \epsilon] < 2^{-2n_1}.$$

So with high probability $|\tilde{T}(x) - T(x)| \leq \epsilon$ holds for all x. We now freeze the choice of random pairs so that this is in fact the case. Now consider the value $\frac{\tilde{T}(x)}{\delta_2}$. Let $\Delta(x) = \tilde{T}(x) - T(x)$.

$$
\begin{aligned}
\mathrm{E}_x\left(f_1(x)\frac{\tilde{T}(x)}{\delta_2}\right) &= \mathrm{E}_x\left(f_1(x)\frac{T(x)+\Delta(x)}{\delta_2}\right) \\
&= \mathrm{E}_x\left(\frac{f_1(x)T(x)}{\delta_2}\right) + \mathrm{E}_x\left(\frac{f_1(x)\Delta(x)}{\delta_2}\right) \\
&\geq \frac{\delta_1\delta_2+\epsilon}{\delta_2} - \frac{\epsilon}{\delta_2} \\
&= \delta_1.
\end{aligned}
$$

We thus have the desired violation of correlation, but the function we have defined is not Boolean. We now round the values $\tilde{T}(x)$ to values in $\{-1, 1\}$ in a way that preserves correlation. Consider a randomized circuit doing the following on input x: Compute $\tilde{T}(x)$ and with probability $(\frac{\tilde{T}(x)}{\delta_2} + 1)/2$ output 1, otherwise output -1. This circuit has the same expected correlation with f_1 as $\tilde{T}/\delta_2$. Now freeze the random choices made, while preserving the correlation. We now have a deterministic circuit of size at most $s_1 - 5(n_1/\epsilon)^{10} + 5n_1^2 \leq s_1$ with correlation at least δ_1 with f_1, contradicting the assumption about f_1.   ∎

Proof (of Lemma 39) By induction in k using the isolation lemma.

Induction step (assume true for $k-1$): Let $f_1 = f$ and $f_2 = f^{(k-1)}$, $s_1 = s$ and $s_2 = \frac{s - 10(\frac{n}{\epsilon})^{10}}{(\frac{n}{\epsilon})^{10}}$. We conclude by Lemma 40 that all circuits of size at most $\frac{s - 10(\frac{n}{\epsilon})^{10}}{(\frac{n}{\epsilon})^{10}}$ will have correlation at most

$$\delta\left(\delta^{k-1} + (k-1)\epsilon\right) + \epsilon = \delta^k + \delta(k-1)\epsilon + \epsilon \le \delta^k + k\epsilon$$

with $f^{(k)}$. ∎

7. A HIGH END GENERATOR

In this section, we prove Theorem 14. By now, there are several known ways of doing this. First observe that if one sets the parameters of the proof in Section 6 carefully, and in particular replaces Lemma 34 with Lemma 38 one gets very close to obtaining a proof of Theorem 14. In fact, one can prove the theorem with the conclusion **BPP = P** replaced with **BPP ⊆ TIME**$(2^{\mathrm{polylog} n})$. The only step in the proof causing the superpolynomial blowup is Lemma 35, the XOR lemma. To transform a truth table of a function so that all small circuits fails to compute the function on a constant fraction of the inputs into a truth table of a function so that all small circuits fail to compute the function on almost half the inputs, the XOR lemma as stated needs to blowup the size of the truth table quasipolynomially.

In the first proof of Theorem 14, by Impagliazzo and Wigderson, 1997, the use of the XOR lemma was replaced with a modified XOR lemma, where the function $f^{(k)}$ is not defined on *all* k-tuples of inputs, but only on a carefully selected subset. This makes a polynomial blowup possible. The proof is technically involved.

Two simpler proofs were given by Sudan et al., 1999. The second of the proofs consists of extending the notion of locally decodable codes to "locally list-decodable codes", thus eliminating the use of the second step altogether and obtaining a proof very close to the proof of Theorem 9.

The proof we shall present in this section is a further simplification of the first proof presented by Sudan et al., 1999. The proof is based on the second strategy for constructing hardness based generators we outlined in Section 5.3: We use the extractor of Theorem 9 as a black box in the construction and try to "add on" the generator property. Following this approach, Sudan et al., 1999 obtain a pseudorandom generator. The simpler version we present here only gives a hitting set generator.

This makes it possible to eliminate one of the steps performed in the construction of Sudan et al., 1999.

In Subsection 7.1, we prove an important lemma for the construction, Impagliazzo's hardcore lemma. In Subsection 7.2, we present the proof of Theorem 14.

7.1   THE HARDCORE LEMMA

Impagliazzo's hardcore lemma, due to Impagliazzo, 1995, states that any function f which disagrees with any small circuit on a constant fraction of the inputs, has a *hardcore*. Informally speaking, the hardcore is a set of hard instances of the function. More precisely, the hardcore is a set of inputs H of constant density so that any small circuit must fail to compute f on almost half the instances of H. The proof we present is attributed to Nisan in Impagliazzo, 1995. For much more information about the hardcore lemma and other proofs, see this reference, and Klivans and Servedio, 1999.

Lemma 41 *Let $0 < \epsilon < \frac{1}{10}$ be a fixed constant and let n be sufficiently large. Given a function $f : \{0,1\}^n \rightarrow \{0,1\}$ so that all circuits of size $2^{\epsilon n}$ compute f incorrectly on at least a $\frac{1}{20}$ fraction of the inputs. Then there is a set $H \subseteq \{0,1\}^n$ of density at least $\frac{1}{25}$, so that all circuits of size $2^{(\epsilon/2)n}$ compute f incorrectly on at least a $\frac{1}{2} - \frac{1}{s(n)}$ fraction of H, where s is a superpolynomial function. The set H is called a hardcore for f.*

Proof We consider the following two-player zero-sum game between Alice and Bob. Alice chooses a circuit C of size $2^{(\epsilon/2)n}$. Simultaneously, Bob chooses a subset S of $\{0,1\}^n$ of density at least $\frac{1}{20}$. The payoff of Alice is the probability, over a uniformly chosen input $x \in S$, that $C(x) = f(x)$. The payoff of Bob is this value negated.

Intuitively, a good deterministic move for Bob would be a hardcore set. On the other hand, it is easy to see that Alice does not have a good deterministic move: Whatever circuit Alice chooses, Bob can choose a set, so that the payoff of Alice is 0.

According to the min-max theorem of game theory, the game has a value, say v, so that there is a *randomized* strategy for Alice ensuring that her expected payoff is at least v no matter what Bob does, and there is *randomized* strategy for Bob ensuring that his expected payoff is at least $-v$, no matter what Alice does.

Our aim is to show that if $v > \frac{1}{2} + n^{-\log n}$, then there is a circuit of size $2^{\epsilon n}$ computing f incorrectly on less than a $\frac{1}{20}$ fraction of the inputs, contradicting the assumption of the lemma. On the other hand, we'll show that if $v \leq \frac{1}{2} + n^{-\log n}$, Bob's optimal randomized strategy can be made deterministic, yielding a hardcore set of the desired parameters.

First assume $v > \frac{1}{2} + n^{-\log n}$ and consider the optimal strategy of Alice. The strategy is a probability distribution over circuits of size $2^{(\epsilon/2)n}$. We must derive from it a single, deterministic circuit $\tilde{C}$ of size at most $2^{\epsilon n}$ computing f "too well". We construct this deterministic circuit probabilistically: We sample $2^{(\epsilon/4)n}$ circuits from the distribution of the optimal strategy of Alice and take their majority. This is the circuit $\tilde{C}$. We now argue that it, with high probability, has the correct property.

To see this, we partition $x \in \{0,1\}^n$ in two parts, G and B. G is the set of strings which are good for the strategy of Alice: $x \in G$ if and only if a random circuit C according to her strategy has $C(x) = f(x)$ with probability at least $\frac{1}{2} + n^{-\log n}$. B is the set of strings which are bad for the strategy of Alice, i.e., the rest of the strings. We argue that $\tilde{C}$ does well on G and that B is so small that it doesn't matter how well $\tilde{C}$ does on B.

First we argue that B is small, in particular that the density of B is less than $\frac{1}{20}$: Take the $\frac{2^n}{20}$ strings in $\{0,1\}^n$ on which the strategy of Alice does worst. If Bob plays this set, the expected payoff for Alice is still at least $\frac{1}{2} + n^{-\log n}$. Thus the set contains some element *not* in B and hence *all* of B as a proper subset.

Next fix any $x \in G$. By the definition of G, a random circuit according to the strategy of Alice answers correctly on x with probability at least $\frac{1}{2} + n^{-\log n}$. By the Chernoff bound of Lemma 1, the probability that $\tilde{C}$ is correct on x is overwhelming, certainly, it is bigger than $1 - 2^{-n}$. Thus, we can fix a choice of $\tilde{C}$ which is correct on all members of G. This circuit has size at most $2^{\epsilon n}$ and computes f except on B, a set of density less than $\frac{1}{20}$, contradicting the assumption about f. Thus the value of the game is less than $\frac{1}{2} + n^{-\log n}$.

Now consider an optimal strategy of Bob. Consider the distribution $\mathcal{D}$ on $\{0,1\}^n$ defined by picking a random set according to the strategy of Bob and then a random element of the set. For a randomly chosen element according to $\mathcal{D}$, any fixed circuit of Alice answers correctly with probability less than $\frac{1}{2} + n^{-\log n}$. After the set has been chosen, any particular element has probability at most $(2^n/20)^{-1}$ of being chosen, so $\mathcal{D}$ assigns probability at most $20/2^n$ to any particular element.

We now construct the desired hardcore set H probabilistically. For each $x \in \{0,1\}^n$, we make an independent decision about whether to include $x \in H$. We include x with probability $2^n \mathcal{D}(x)/20$, which is at most 1, by the above analyses.

By the Chernoff bound of Lemma 1, the probability that H has density somewhere in the interval $\frac{1}{20} \pm n^{-\log n}$ is overwhelming. Given any fixed particular circuit C of Alice, we have $\sum_{x \in \{0,1\}^n} \mathcal{D}(x)\tau_C(x) \leq \frac{1}{2} + n^{-\log n}$, where $\tau_C(x)$ is 1 if $C(x) = f(x)$ and 0 otherwise. Thus, $\sum_{x,\tau_C(x)=1} \mathcal{D}(x) \leq \frac{1}{2} + n^{-\log n}$. Therefore, the probability that the density of the set $H \cap \{x|\tau_C(x) = 1\}$ is at most $(\frac{1}{2} + n^{-\log n})(\frac{1}{20} + n^{-\log n})$ is overwhelming, certainly bigger than $1 - 2^{2^{-n/5}}$. Therefore the probability that the density of $H \cap \{x|\tau_C(x) = 1\}$ is smaller than this value for *all* circuits C of Alice is also overwhelming. In particular, we can fix a set H with these properties. This is the hardcore set. ∎

7.2   THE GENERATOR

We now prove theorem 14. By Lemma 30, it is enough to show that the assumption of the theorem implies that there is a generator which on input r runs in time polynomial in r and outputs a set $W \subseteq \{0,1\}^r$ which is a $(1 - 2^{-r+r^{1/2}})$-hitting set for circuits of size r^q, for a certain constant q, which we assume, without loss of generality, to be bigger than 2.

The assumption of the theorem allows us, on input n, to generate the truth table of a function $f : \{0,1\}^n \to \{0,1\}$ with no circuits of size $2^{\epsilon n}$, in polynomial time in the size of the truth table. By Lemma 38, we can convert this generator into a generator which on input n gives us the truth table of a function $f : \{0,1\}^n \to \{0,1\}$ so that any circuit of size $2^{\epsilon' n}$ disagrees with f on at least a $\frac{1}{20}$ fraction of the input.

We now define the hitting set generator. Given the parameter r, the generator uses Lemma 27 to construct an $(r, c \log r, 4q \log r/\epsilon', \log r)$-Nisan-Wigderson design $S_1, S_2, \ldots, S_r$, where c is a constant depending on q and ϵ'. Letting $n = 4q \log r/\epsilon'$, the generator builds the truth table of a function $f : \{0,1\}^n \to \{0,1\}$ so that any circuit of size $2^{\epsilon' n}$ disagrees with f on at least a $\frac{1}{20}$ fraction of the inputs. The time used is polynomial in 2^n, i.e., polynomial in r. Now, the generator outputs the set

$$W = \{f(y_{S_1})f(y_{S_2}) \cdots f(y_{S_r}) | y \in \{0,1\}^{c \log r}\}.$$

We claim this is the desired hitting set.

Assume, to the contrary that $C : \{0,1\}^r \to \{0,1\}$ is a circuit of size r^q, so that if $Z = \{x \in \{0,1\}^r | C(x) = 0\}$, we have $|Z| \leq 2^{r^{1/2}}$, and $W \subseteq Z$.

By Lemma 41, f has a hardcore $H \subset \{0,1\}^n$ of density at least $\frac{1}{25}$, so that all circuits of size $2^{(\epsilon'/2)n} = r^{2q}$ fail to compute f on a $\frac{1}{2} - \frac{1}{s(n)}$ fraction of H, where s is a superpolynomial function.

Let $\mathcal{D}$ be the uniform distribution on W and consider the following derived distributions $\mathcal{D}_0, \mathcal{D}_1, \dots, \mathcal{D}_r$, defined by the following sampling procedures.

A random sample from $\mathcal{D}_i$ is obtained by picking a random sample x from $\mathcal{D}$, together with the "history" of x, i.e., the seed $y \in \{0,1\}^{c \log r}$ so that $x = f(y_{S_1})f(y_{S_2})\dots f(y_{S_r})$. Looking at y, we identify those bits among $x_{i+1}, \dots, x_r$, i.e., among the last $r - i$ bits of x, that are the results of applying the function f to elements of the hardcore H. *Each of these "hardcore bits" we replace with an unbiased, independent random bit.*

Note that $\mathcal{D}_r = \mathcal{D}$ and $\mathcal{D}_0$ is the distribution resulting from $\mathcal{D}$ when we replace every "hardcore bit" with a randomly chosen bit.

We immediately have that for a randomly chosen sample $x \in \mathcal{D}_r$, $\Pr[C(x) = 1] = 0$.

We also have that for a randomly chosen sample $x \in \mathcal{D}_0$, the probability of $\Pr[C(x) = 1] \geq \frac{1}{100}$. This can be seen as follows: The procedure for sampling $x \in \mathcal{D}_0$, first samples $x' \in \mathcal{D}$. As the hardcore H has density at least $\frac{1}{25}$ in $\{0,1\}^{m'}$, a randomly chosen bit in such a random sample x' will be a "hardcore bit" bit with probability at least $\frac{1}{25}$. Thus, with probability at least $\frac{1}{50}$, more than a $\frac{1}{50}$ fraction of the bits of the sample are hardcore bits. Call such a sample *good*. Then, if we fix our attention to a single good sample, the conditional distribution we get when we replace the hardcore bits of the sample with randomly chosen bits, is the uniform distribution on a subset of $\{0,1\}^r$ of size at least $2^{r/50}$. At most $2^{r^{1/2}}$ of these make C reject. Thus, the conditional probability that $C(x) = 1$, given that x comes from a good sample, is $1 - o(1)$, so the probability that $C(x) = 1$ for a randomly chosen sample $x \in \mathcal{D}_0$, is at least $(1 - o(1))\frac{1}{50} \geq \frac{1}{100}$.

As $\Pr[C(x) = 1] = 0$ if x is chosen at random from $\mathcal{D}_r$ and $\Pr[C(x) = 1] \geq 1/100$ if x is chosen at random from $\mathcal{D}_0$, the triangle inequality gives us an i, so that

$$\Pr_{x \in \mathcal{D}_i}[C(x) = 1] - \Pr_{x \in \mathcal{D}_{i+1}}[C(x) = 1] \geq \frac{1}{100r}$$

Note that a sample from $\mathcal{D}_i$ can be obtained from a sample from $\mathcal{D}_{i+1}$ by replacing the $i + 1$'st bit with a random bit if it is a hardcore bit.

Samples from $\mathcal{D}_i$ and $\mathcal{D}_{i+1}$ are generated from a seed y. Let $\mathcal{D}'_i$ and $\mathcal{D}'_{i+1}$ be the distributions from $\mathcal{D}_i$ and $\mathcal{D}_{i+1}$ obtained by freezing all bits of y, *except* the bits of S_{i+1} to fixed values, while preserving

$$\Pr_{x \in \mathcal{D}'_i}[C(x) = 1] - \Pr_{x \in \mathcal{D}'_{i+1}}[C(x) = 1] \geq \frac{1}{100r}$$

As $\mathcal{D}'_i$ and $\mathcal{D}'_{i+1}$ only differ when $y_{S_{i+1}}$ is a hardcore bit, we also have,

$$\Pr_{x \in \mathcal{D}'_i}[C(x) = 1 | y_{S_{i+1}} \in H] - \Pr_{x \in \mathcal{D}'_{i+1}}[C(x) = 1 | y_{S_{i+1}} \in H] \geq \frac{1}{100r}.$$

Now, we describe a small circuit C' for f which will compute f correctly on significantly more than half its hardcore, and thus we shall reach a contradiction.

In the description and analysis of the circuit C', we shall assume that the input $z \in \{0,1\}^r$ given to the circuit is indeed a member of the hardcore. What happens on other inputs is irrelevant.

The circuit C' generates, on input $z \in H$, a random value x according to the distribution $\mathcal{D}'_{i+1}$, conditioned by $y_{S_{i+1}} = z$, where $y \in \{0,1\}^{c \log r}$ is the seed leading to x. We claim that this can be done by a circuit of size $O(r^3)$. This follows from the fact that each set S_j with $j \neq i$ intersects S_{i+1} in at most $\log n$ places and we can thus tabulate the value of $f(y_{S_j})$ as a function of $y_{S_j \cap S_{i+1}}$. We can also tabulate the Boolean predicate $y_{S_j} \in H$? as a function of $y_{S_j \cap S_{i+1}}$. The total size of these tables are $O(r^2)$ and a random value x according to $\mathcal{D}'_{i+1}$, conditioned by $y_{S_{i+1}} = z$, is easily generated from them. Now our circuit checks if $C(x) = 1$. If this is the case, it outputs $\neg x_{i+1}$. Otherwise it outputs x_{i+1}. If z is randomly chosen from the hardcore, we have that $C'(z) = f(z)$ with probability at least $\frac{1}{2} + \frac{1}{100r}$, as $\Pr_{x \in \mathcal{D}'_i}[C(x) = 1 | y_{S_{i+1}} \in H] - \Pr_{x \in \mathcal{D}'_{i+1}}[C(x) = 1 | y_{S_{i+1}} \in H] \geq \frac{1}{100r}$. We can freeze the random choices made by C' to get a deterministic circuit with the same property. This contradicts H being a hardcore of f.

Thus, our assumption that W did not hit C was false, so W is a hitting set with the desired parameters. This completes the proof of Theorem 14.

8. THE COMPRESSION APPROACH TO GENERATOR CONSTRUCTION

In this section, we describe an alternative approach to hardness-based hitting set generators, developed in Andreev et al., 1997, and Miltersen and Vinodchandran, 1999. It is somewhat simpler than the approach using locally decodable codes and Nisan-Wigderson designs and yields comparable results.

The first generator, due to Andreev et al., 1997, takes as input a table of a *multi-output* Boolean function with *maximum* circuit complexity and produces a $(1 - \frac{1}{r^3})$-hitting set for circuits of size r. Such a hitting set derandomizes **pRP** and hence **BPP**. Furthermore, we do not have to rely on explicit dispersers and Lemma 29 to see that: An error probability of $1/r^3$, using r random bits, can be obtained by simple repetition, so a simple modification of Lemma 2 will do.

Since the requirement for the hardness of the function that is given as input is so harsh, we see that it does *not* follow from the techniques of Section 5 that the generator is a disperser with non-trivial parameters. Indeed it isn't. A fortunate side-effect of this is that the correctness of the generator can be given a completely elementary one-page self-contained proof, unlike any of the other hitting set generators we have seen.

Theorem 18 *If there is an algorithm which on input n, m outputs the truth table of a Boolean function from $\{0,1\}^n$ to $\{0,1\}^m$ with maximum possible circuit complexity $M_{n,m}$, and runs in time polynomial in the length of its output, then there is a $(1 - 1/r^3)$-hitting set generator for circuits $C : \{0,1\}^r \to \{0,1\}$ of size r, running in time $r^{O(1)}$, and hence* **P=BPP**.

Proof Let $f_{n,m}$ be the function generated on input n, m. Fix r. Let I_i with $0 \leq i \leq r$ be the subset of $\{0,1\}^r$ obtained by discarding the last i bits of each of the elements of $\{f_{2 \log r, r+i}(x) | x \in \{0,1\}^{2 \log r}\}$. We shall argue that at least one of the sets I_i is a $(1 - 1/r^3)$-hitting set for circuits of size r. Thus, the union of these sets is a hitting set, and as their union can be generated in polynomial time, we are done.

Assume, to the contrary, that no set I_i is such a hitting set. In particular this is true for $i = 0$, i.e., for the image I of the function $f = f_{2 \log r, r}$. Thus, there is a circuit of size r mapping r bits to 1 bits, so that if we let $Z = \{y \in \{0,1\}^r | C(y) = 0\}$, we have $|Z| \leq 2^r/r^3$ and $I \subseteq Z$.

Now let $\mathcal{H}$ be the universal class of hash functions of Lemma 7, mapping $\{0,1\}^r$ to $\{0,1\}^{r-1}$. Consider picking an $h \in \mathcal{H}$ at random. We have

$$\Pr_h[\exists x \in I, y \in Z : x \neq y \wedge h(x) = h(y)] \leq r^2 \cdot 2^r/r^3 \cdot 2^{r-1} < 1$$

Thus, there exists $h \in \mathcal{H}$, so that $\forall x \in I, y \in Z, x \neq y \Rightarrow h(x) \neq h(y)$. Fix such h. Let $f' : \{0,1\}^{2 \log r} \to \{0,1\}^{r-1}$ be the map given by $f'(x) = h(f(x))$.

Now suppose we are given a circuit for f'. We can compute $f(x)$ by computing $f'(x)$, computing $\{y_1, y_2\} = h^{-1}(f'(x))$ and then returning whichever y_i has $C(y_i) = 0$. The right value (i.e., $f(x)$) has this property (as $I \subseteq Z$) and since h does not collide I and Z, the other value does not have this property.

As the circuit complexity of f is $M_{2\log r,r}$ by assumption, and the circuit complexity of f' is at most $M_{2\log r,r-1}$, we have

$$M_{2\log r,r} \leq M_{2\log r,r-1} + O(r)$$

Arguing similarly for each function $f_{2\log r,r+i}, i = 1..n$, we get

$$M_{2\log r,r+i} \leq M_{2\log r,r+i-1} + O(r),$$

i.e.

$$M_{2\log r,2r} \leq M_{2\log r,r} + O(r^2),$$

contradicting Theorem 1.   ∎

The assumption that a function with maximum circuit complexity can be generated efficiently is very strong. To see how we can get the proof to work with an assumption more like the assumption in the Impagliazzo-Wigderson theorem, it is useful to visualize the set I in the proof above as an $r^2 \times r$ 0-1 matrix. The proof works by *compressing* each row of the matrix by 1 bit in a way that makes it possible to retrieve each row again efficiently from the compressed image. This is inconsistent with the function corresponding to I having maximum circuit complexity.

Now, to get an assumption involving smaller circuits, we need to be able to compress each row by much more than one bit. The current compression is possible if the set I is not a $(1 - 1/r^3)$-hitting set. To do really significant compression, we need a stronger assumption about the non-hitting property of the set I: We need to assume that it is not a $(1 - \epsilon(r))$-hitting set for $\epsilon(r)$ much closer to 0 than r^{-3}. But this is not a problem: Lemma 30 tells us that to derandomize **BPP**, it is enough to construct a hitting set which is $(1 - 2^{-r+r^\epsilon})$-hitting for circuits of size r^q for a constant q, depending on ϵ. This enables us to compress each row from r bits all the way down to r^ϵ bits. However, a more serious problem arises: If we compress by, say, t bits, the circuit we end up constructing for the function f needs to do an exhaustive search through 2^t different possibilities for the uncompressed value it is trying to compute. Thus, if we compress each row from r bits to r^ϵ bits, $t = r - r^\epsilon$, and we would have to search through 2^{r-r^ϵ} possibilities. This would make us lose all we had gained from the bigger compression.

A way out of this dilemma is to switch to a hardness assumption involving a more powerful kind a circuits, namely *nondeterministic* circuits. If the circuit we are trying to construct may be nondeterministic,

918

we can easily perform the exhaustive search above by a simply *guessing* the decompressed value and verifying its correctness.

The above discussion leads to the following theorem, also proved by Andreev et al., 1997, though phrased somewhat differently there.

Theorem 19 *There is a $\delta > 0$ so that the following holds. If there is a language L in* **E**, *so that for all n, the nondeterministic circuit complexity of $L \cap \{0,1\}^n$ is at least $2^{(1-\delta)n}$, then* **BPP** = **P**.

Proof By Lemma 30, it is enough to show that there is a generator which on input r runs in time polynomial in r and outputs a set $I \subseteq \{0,1\}^r$ which is a $(1 - 2^{-r+r^{1/3}})$-hitting set for circuit of size r^q, for a certain constant q. We can assume that $q \geq 2$. The value of δ for which we will show the theorem will be $\delta = \frac{1}{2q}$.

The assumption allows us, on input l, to generate the truth table of a Boolean function $f : \{0,1\}^l \to \{0,1\}$ with nondeterministic circuit complexity at least $2^{(1-\delta)l}$, in time polynomial in the size of the table.

Given the parameter r, we let $l = \log(r^{q+1})$ and generate the truth table of a Boolean function $f : \{0,1\}^l \to \{0,1\}$. We arrange the 2^l bits of the truth table in an $r^q \times r$ matrix. Let I be the set of rows of this matrix. We claim that I is the desired hitting set, if the parameter r is sufficiently large.

Suppose not. Then there is a circuit C of size r^q mapping r bits to 1 bit, so that if we let $Z = \{y \in \{0,1\}^r | C(y) = 0\}$, we have $|Z| \leq 2^{r^{1/3}}$ and $I \subseteq Z$.

Now let $\mathcal{H}$ be the universal class of hash functions of lemma 6, mapping $\{0,1\}^r$ to $\{0,1\}^{2r^{1/3}}$. Consider picking an $h \in H$ at random. The probability that h introduces any collisions on the set Z is smaller than 1, so we can pick an h so that $\forall x, y \in Z, x \neq y \Rightarrow h(x) \neq h(y)$.

Let $f' : \{1, 2, \ldots, r^q\} \to \{0,1\}^{2r^{1/3}}$ be the map which, given the index of a row in I, returns the hashed value of this row, using the hash function h.

Now suppose we are given a table for f'. We can construct a nondeterministic circuit for the original Boolean function f as follows. Given an input y, we first compute the index of the row of A where we can find the value $f(y)$. Then we look up the hashed value of the row in the table for f'. We then *guess* the value $z \in \{0,1\}^r$ of the original row and verify that our guess is correct by checking that $C(z) = 0$ and that $h(z)$ has the right value. We then read off the relevant bit of z.

The nondeterministic circuit for f we have just described contains C, a circuit for h, and a table for f'. Beyond that, only straightforward computation is performed. Thus, we can make its size $O(r^{q+o(1)} + r^q r^{1/5}) = O(r^{q+1/5})$. This is smaller than $2^{(1-\delta)l}$ for sufficiently large r. ∎

Note that the assumptions of Theorem 18 and Theorem 19 are incomparable - and both much stronger than the Impagliazzo-Wigderson assumption. However, the assumption of Theorem 19 seems somewhat closer in spirit to the Impagliazzo-Wigderson assumption than the assumption of Theorem 18. Next, we will show that we can make it even closer, by replacing $2^{(1-\delta)n}$, for *some* small constant δ, with $2^{\epsilon n}$, for *any* desired constant $\epsilon > 0$.

To do this, we need to compress our matrix even more. We have already compressed each row more or less as much as we can, as the disperser we used was close to optimal. To compress it even more, we are going to compress it twice: *First we will compress each row, and then we will compress each column of the compressed matrix.* Now, there is no longer any reason for having the matrix having more columns than rows, so in the following discussion, we should have a square $r \times r$ matrix I in mind. To compress the matrix twice, the set we are going to assume is not hitting will be the set of rows *and* columns of this matrix. The proof should then proceed by first finding a hash function and use it to compress the rows of the matrix. This will be possible under the assumption that the set of rows is not a hitting set. Next, we will take our compressed matrix and now, under the assumption that the set of columns is not a hitting set, we will find a hash function and use it to compress the columns of the compressed matrix.

However, here a problem arises: *The columns of the row-compressed matrix have no relationship to the columns of the original matrix.* Indeed, applying a traditional hash function to a piece of data destroys all structure in the data completely. Thus, we *cannot* compress the columns of the row-compressed matrix under the assumption that the set of columns of the original matrix is not a hitting set, and we seem to be stuck.

To proceed, we need to be able hash the rows of a matrix, *without destroying the columns.* Fortunately, the family of hash function we use need only be weakly universal. Thus, Theorem 15 gives us a solution to the problem: If the rows of the matrix have been encoded in an error correcting code, we can use a projection to a small set of indices as our hash function. Applying such a hash function to all rows of a matrix yields a compressed matrix whose set of columns is just some subset of the columns of the original matrix. Thus, under the assumption that the set of columns of the original matrix is not a hitting set, we *can* compress the columns of the matrix, after having compressed the rows.

To get a hardness assumption involving circuits of size $2^{\epsilon n}$ for an arbitrary small constant $\epsilon > 0$, we need to carry out the above program, not on a square $n \times n$ matrix, but on a k-dimensional matrix of bits, for a constant value k, depending on ϵ and compress not twice, but k times, once along each dimension of the matrix. Thus, we need to encode our original data as a k-dimensional matrix of bits, so that each 1-dimensional slide of the encoding is a code word of an error correcting code e with good minimum distance. A simple way of doing this is to take the *tensor product* of our code e of Lemma 12 with itself k times. We omit the technical details of this and only state the following lemma.

Lemma 42 *There is an efficiently computable function taking as input a k-dimensional bit matrix $f : \{1, 2, \dots, n\}^k \to \{0, 1\}$ and outputting a k-dimensional bit matrix $\tilde{f} : \{1, 2, \dots, n^{10}\}^k \to \{0, 1\}$, so that*

- $\tilde{f}[i_1, \dots, i_k] = f[i_1, \dots, i_k]$, *for* $i_1, \dots, i_k \leq n$, *and*

- *each one-dimensional* slice *of the matrix* $\tilde{f}$, *i.e. each vector* $v = (\tilde{f}[i_1, i_2, \dots, i_{j-1}, x, i_{j+1} \dots, i_k])_{x=1..n^{10}}$, *for any setting of j and $i_1, i_2, \dots, i_{j-1}, i_{j+1}, \dots, i_k$, is a code word of e.*

Now we are ready for a more formal version of the proof sketch carried out above. The proof is due to Miltersen and Vinodchandran, 1999.

Theorem 20 *If there is an $\epsilon > 0$ and a language L in **E**, so that for all n, the nondeterministic circuit complexity of $L \cap \{0, 1\}^n$ is at least $2^{\epsilon n}$, then* **BPP** $=$ **P**.

Proof The assumption allows us, on input n, to generate the truth table of a Boolean function $f : \{0, 1\}^n \to \{0, 1\}$ with nondeterministic circuit complexity at least $2^{\epsilon n}$, in time polynomial in the size of the table.

We show that such a procedure allows us to construct a $1 - 2^{-r+r^{\epsilon'}}$-hitting set generator for circuits $C : \{0, 1\}^r \to \{0, 1\}$ of size r^q, where $\epsilon' = \epsilon/100$ and where q is *any* fixed integer. By Lemma 30, we'll be done.

Given r, we use our procedure to generate the truth table of a function $f : \{0, 1\}^m \to \{0, 1\}$, where $m = \frac{k}{10} \log r$, and k is a constant to be decided later, and arrange the bits of the truth table in a k-dimensional matrix. More precisely, we identify f with a function $f : (\{0, 1\}^{\frac{\log r}{10}})^k \to \{0, 1\}$ in some canonical way. Using Lemma 42, we encode f as $\tilde{f} : \{1, 2, \dots, r\}^k \to \{0, 1\}$.

Now let H be the set of all one-dimensional slices of $\tilde{f}$, i.e., the members of H are the vectors $\tilde{f}[i_1, i_2, \dots, i_{j-1}, x, i_{j+1} \dots, i_k])_{x=1..r}$ for any setting of j and $i_1, i_2, \dots, i_{j-1}, i_{j+1}, \dots, i_k$.

We now claim that H is a $(1 - 2^{-r+r^{\epsilon'}})$-hitting set for circuits of size r^q, if we set k appropriately.

Assume not. Let C be a circuit establishing this, i.e., C maps $\{0,1\}^r$ to $\{0,1\}$, it has size r^q, and if we by Z denote $\{x \in \{0,1\}^r | C(x) = 0\}$, then $|Z| \le 2^{r^{\epsilon'}}$ and $H \subseteq Z$.

By Lemma 15, there is a projection $\pi_S : \{0,1\}^r \to \{0,1\}^{r^{2\epsilon'}}$ to a set of indices $S \subseteq \{1, 2, \dots, r\}$, so that π_S is 1-1 on $Z \cap W$, where W is the set of code words of e.

We will now construct a small nondeterministic circuit for $\tilde{f}$ and hence f. Into the circuit we build as advice the circuit C, the set S, and a table of the restriction of $\tilde{f}$ to S^k.

To compute $\tilde{f}(a_1, a_2, \dots, a_k)$ the procedure does as follows.

For every element $u \in S^{k-1}$ it *guesses* the vector $v = (\tilde{f}(j, u))_{j \in \{1,\dots,r\}}$. It checks that its guess is correct by checking that v is a code word of e, checking that $C(v) = 0$, and checking that the entries of v corresponding to indices in S are correct, by consulting the table for $\tilde{f}$, restricted to S^k. By construction, the value $v = (\tilde{f}(j, u))_{j \in \{1,\dots,r\}}$ is the only value with these properties. After having ensured that v is the correct value, it keeps the value $\tilde{f}(a_1, u)$ and throws away the rest of the vector v. Having done this for every possible u, the procedure has now built a table of $\tilde{f}$, restricted to $\{a_1\} \times S^{k-1}$.

Now, for every element $u \in S^{k-2}$ the procedure guesses the vector $v = (\tilde{f}(a_1, j, u))_{j \in \{1,\dots,r\}}$. It checks that each guess is correct by checking that v is a code word, checking that $C(v) = 0$ and checking that the entries of v corresponding to indices in S are correct, by consulting the table for $\tilde{f}$, restricted to $\{a_1\} \times S^{k-1}$. After having ensured that v is the correct value, it keeps the value $\tilde{f}(a_1, a_2, u)$ and throws away the rest of the vector v. Having done this for every possible u, the procedure has now built a table of $\tilde{f}$, restricted to $\{a_1\} \times \{a_2\} \times S^{k-2}$.

Now it goes through a similar loop for every element $u \in S^{k-3}$, building a table of $\tilde{f}$, restricted to $\{a_1\} \times \{a_2\} \times \{a_3\} \times S^{k-3}$, and so on, until in the end it has the value of $\tilde{f}(a_1, a_2, \dots, a_k)$ which was the value we wanted.

Converting the procedure into a nondeterministic circuit, building in the advice, we get a circuit of size at most $O(k(r^{3\epsilon'})^k r^q)$.

By fixing k to an appropriate constant, this is smaller than $2^{\epsilon m}$ for sufficiently large m. As a circuit for $\tilde{f}$ can also be used as a circuit for f, this contradicts the assumption on f. ∎

Though the compression proof technique falls just short of proving the Impagliazzo-Wigderson theorem, it can be used to prove a derandomization result for the class **AM** we do not at present know how to prove using the techniques based on locally decodable codes and Nisan-Wigderson designs. This theorem is described in the next section.

9. OTHER HARDNESS BASED DETERMINISTIC SIMULATION RESULTS

In this section, we discuss hardness based deterministic simulation results for other randomized complexity classes, besides **BPP** and **RP**.

In general, the derandomization theorems we consider will have the form "If **A** contains a language not in **B** then **C** is contained in **D**", where **C** is the randomized class we are considering, **D** is the deterministic, uniform class in which we want to do the deterministic simulation, **A** is a uniform complexity class, and **B** is the class of problems infinitely often having a certain complexity in a non-uniform model.

The proofs we are considering proceed by adapting the proof techniques of Sections 6, 7 and 8. Thus, we first identify a appropriate non-uniform class of computational objects so that making pseudorandom or hitting set generators for objects of this class derandomizes **C**. This step is usually routine. For instance, for **RP** and **BPP**, the relevant objects are *circuits*, as shown in Lemma 2. For **BP$_\mathbf{H}$L**, they are *finite automata*, as shown in Lemma 3.

Now, given a class **A**, we consider constructing a pseudorandom or hitting set generator using a language of **A** by the same transformations as in Section 6, 7 or 8, or using one of alternative strategies indicated there. The complexity of a deterministic machine, going through all outputs of such a generator and using them as the random bits in a computation of type **C**, defines the class **D**. Getting the class **D** to be the desired one may require some fine-tuning of the construction of the generator.

To see which class **B** can be obtained, we assume that the pseudorandom generator does not work, i.e. that there is a non-uniform computational object of appropriate type on which it fails. Then we redo the sequence of lemmas of the last sections (or a similar sequence following one of the alternative strategies) that proves the correctness of the generator, with these objects replacing circuits, and get some kind of non-uniform simulation of **A**. This defines the class **B**.

We present a sample of results obtained this way. For several other variations, we refer to the papers cited below.

We have only worried about the time-efficiency of the pseudorandom generator. To derandomize space bounded computation, one needs to check that the computation of the pseudorandom generator can be done space-efficiently. This was done by Allender et al., 1999. This leads to the following theorem, a slightly weaker version of which is stated in Klivans and van Melkebeek, 1999.

Theorem 21 *If there is a language L in* **LINSPACE** *and an $\epsilon > 0$, so that for all n, no branching program of size $2^{\epsilon n}$ decides $L \cap \{0,1\}^n$, then* $\mathbf{BP_H L = L}$

For definition and discussion of the branching program model, see, e.g., Boppana and Sipser, 1990. One could perhaps think that since we can actually prove that certain explicit functions are not decided by small automata, it might be possible to construct a hardness based generator using no unproven assumption. However, the class of functions computed by small automata lacks nice closure properties, so if we try to let **B** be the class of functions requiring big automata, we find that we cannot make the necessary lemmas go through. In particular, the analogy of Lemma 33 would seem to fail. The class of functions having small branching programs is, essentially, the closure of the class of functions having small automata under the operations necessary to carry out the steps of the various constructions.

Since the computation of the pseudorandom generators can be done space-efficiently, they can also be done by shallow circuits. The class **BPNC** is the class of problems decided by a uniform family of two-sided error randomized circuits of polynomial size and polylogarithmic depth. Nisan, 1991, obtains the following result.

Theorem 22 *If there is a language L in* **PSPACE** *and an $\epsilon > 0$, so that for all n, the circuit depth of $L \cap \{0,1\}^n$ is at least n^ϵ, then* **BPNC** *is contained in* **SPACE**(polylog n).

Furthermore, the following theorem is implicit in Impagliazzo and Wigderson, 1997.

Theorem 23 *If there is a language L in* **LINSPACE** *and an $\epsilon > 0$, so that for all n, the circuit size of $L \cap \{0,1\}^n$ is at least $2^{\epsilon n}$, then* **BPNC** $=$ **NC**.

Let $\mathbf{BPAC^0}$ be the class of languages decided by a uniform family of two-sided error randomized unbounded fan-in circuits of polynomial size and *constant* depth. To derandomize $\mathbf{BPAC^0}$ we need to construct pseudorandom generators for non-uniform $\mathbf{AC^0}$ circuits. This can be done using no unproven assumptions, as there are easily computed functions, such as the *parity* function, so for which all small constant depth circuits must err on almost half the inputs. Thus, a hardness based generator can be built along the lines of Section 6, and we can even skip the two first steps of the construction. Thus, Nisan, 1991, obtains the following *unconditional* derandomization of $\mathbf{BPAC^0}$.

Theorem 24 BPAC^0 *has uniform constant depth circuits of quasipoly-nomial size (i.e.,* **BPAC^0** *is contained in quasi-*AC^0.)

An interesting family of randomized complexity classes is the classes defined by *interactive proof systems*. In particular, **IP** is the class of languages decided by a randomized interactive proof system of polynomial complexity (we refer to Papadimitriou, 1994 for exact definitions). The natural goal would be to prove that randomized polynomial time interactive proof systems are not more powerful than deterministic polynomial time interactive proof systems, the latter systems capturing exactly **NP**. However, the classical result of Lund et al., 1992 and Shamir, 1992, shows that **IP=PSPACE**. Thus we cannot hope for such a derandomization of **IP** without proving a breakthrough result about **PSPACE**. More tractable are the classes **MA** and **AM** of *bounded round* interactive proof systems.

Along the lines of Impagliazzo and Wigderson, 1997, Goldreich and Zuckerman, 1997, proves[6]

Theorem 25 *If there is a language L in* **NE $\cap$ coNE** *and an $\epsilon > 0$, so that for all n, the circuit complexity of $L \cap \{0,1\}^n$ is at least $2^{\epsilon n}$, then* **MA $=$ NP.**

It is interesting to note that the assumption of Theorem 25 is slightly weaker than the assumption of Theorem 14, as we only need to find L in **NE $\cap$ coNE** rather than **E**. Thus, perhaps **MA** is easier to derandomize than **BPP**.

A *strong nondeterministic circuit* for a function $f : \{0,1\}^n \to \{0,1\}$ is a pair of nondeterministic circuits: One for f and one for $\neg f$. Along the lines of Nisan and Wigderson, 1994, Arvind and Köbler proves the following theorem.

Theorem 26 *If there is a language L in* **NE $\cap$ coNE** *and an $\epsilon > 0$, so that for all n, no strong nondeterministic circuit of size $2^{\epsilon n}$ decides $L \cap \{0,1\}^n$ correctly on a $\frac{1}{2} + 2^{-\epsilon n}$ fraction of the inputs, then* **AM $=$ NP.**

Klivans and van Melkebeek, 1999, observed that the steps outlined above leading to a derandomization result can sometimes be "automatized", using the fact that the Impagliazzo-Wigderson theorem *relativizes*. By plugging in the right oracle in the relativized version of the theorem(for instance, an oracle for SAT), a derandomization theorem pops out. They carry out this program for several randomized complexity classes and randomized constructions in complexity theory. In particular, for **AM** they prove

Theorem 27 *If there is a language L in* **NE** $\cap$ **coNE** *and an* $\epsilon > 0$, *so that for all n, no SAT-circuit of size* $2^{\epsilon n}$ *decides* $L \cap \{0, 1\}^n$, *then* **AM** = **NP**.

Miltersen and Vinodchandran, 1999, improved Theorem 26 as well as Theorem 27 by proving the following theorem.

Theorem 28 *If there is a language L in* **NE** $\cap$ **coNE** *and an* $\epsilon > 0$, *so that for all n, no strong nondeterministic circuit of size* $2^{\epsilon n}$ *decides* $L \cap \{0, 1\}^n$, *then* **AM** = **NP**.

Theorem 28 is proved using the compression technique along the lines of the proof of Theorem 20. It is not known if one can also obtain this result using locally decodable codes and Nisan-Wigderson designs. Thus, the proof of Theorem 28 is one of the few cases where the compression technique seems to be superior. The compression technique does not work very well for proving "low end results". Thus, it is at the time of writing an open problem if the statement "**NE** $\cap$ **coNE** requires superpolynomial strong nondeterministic circuits implies **AM** is in **SUBNEXP**" can be shown. The statement one obtains by replacing "strong nondeterministic circuits" with "SAT-circuits" was proved correct by Klivans and van Melkebeek, 1999.

All of the above results assume that a uniform class **A** is not in a small non-uniform class **B**. Impagliazzo and Wigderson, 1998 gave a highly interesting variation of the above scheme, where the non-uniform class **B** is replaced with a randomized uniform class. They prove the following theorem.

Theorem 29 *If* **BPP**$\neq$**EXP** *then* **BPP** $\subseteq$ *heur-***SUBEXP**$_{i.o.}$

Here, a language L is in *heur-***SUBEXP**$_{i.o.}$ if there, for any k, is a **SUBEXP**-machine M which, on infinitely many input lengths n, decides L correctly on at least a $1 - 1/n^k$ fraction of $\{0, 1\}^n$. The assumption **BPP**$\neq$**EXP** is the weakest possible assumption about the power of **BPP** we can make. Indeed, if **BPP**=**EXP**, no nontrivial deterministic simulation is possible, by the time hierarchy theorem. Yet from this assumption, we can still get a non-trivial derandomization of **BPP**. Impagliazzo and Wigderson call their theorem a "gap"-theorem: Either **BPP** is everything or it is "small". It can't be "almost everything".

The proof is an ingenious modification of the proof of Theorem 12 we presented in Section 6 (or actually, the corresponding proof for Theorem 13). We outline it below and refer to the original paper for details.

The proof of Theorem 12 in Section 6 is, essentially, a combination of Lemma 34, Lemma 35, Lemma 33, and Lemma 2. To make the proof of Section 6 a proof of Theorem 29 rather than Theorem 12, we would have to redo the lemmas replacing "circuits" with "efficient randomized algorithms". A global view of the new proof is as follows: we use the same sequence of transformation of a truth table, through other truth tables, into a pseudorandom generator. The old proof would then, under the assumption that the generator does not work, prove the existence of a small circuit for the original truth table. In the new proof we shall also find such a circuit, but we want the proof to be constructive in the sense that it can be carried out by an efficient randomized algorithm. In a first attempt, such a proof should proceed by using a new version of Lemma 2, Lemma 33, Lemma 35 and Lemma 34, in that order.

The first obstacle is found in the application of Lemma 2. If the generator does not work on some machine and some input x, we used that lemma to build x into the machine, obtaining a non-uniform circuit. Now we want to construct this circuit, *without knowing x*. The solution is to pick an input x to be built into the circuit at random. Since we are only trying to prove by contradiction that $\mathbf{BPP} \subseteq heur\text{-}\mathbf{SUBEXP}_{i.o}$ and not $\mathbf{BPP} \subseteq \mathbf{SUBEXP}_{i.o}$, we can assume that such a random input will work with non-negligible probability. Indeed, this is the only reason why we shall only obtain this weaker conclusion.

The rest of the proof is more tricky. A crucial observation is this: If $\mathbf{EXP} \nsubseteq \mathbf{P/poly}$, we already have the desired conclusion, by Theorem 13. Thus, we can assume without loss of generality that $\mathbf{EXP} \subseteq \mathbf{P/poly}$. Then, a theorem by Karp and Lipton, 1980, states that $\mathbf{EXP} = \Sigma_p^2$. A theorem of Toda, 1991, states that the permanent of a matrix is hard for Σ_p^2 (indeed, it is hard for the polynomial hierarchy). As the permanent of a matrix can be computed in $\mathbf{EXP}$, we conclude that we can without loss of generality assume that the permanent is complete for $\mathbf{EXP}$. Thus, if there is a language L in $\mathbf{EXP}\text{-}\mathbf{BPP}$, we can assume that it is the permanent and use special properties of the permanent below.

The next observation is that if we use the permanent language to construct the pseudorandom generator, the use of Lemma 34 is unnecessary: The permanent is *random self-reducible*. This means that if the assumption about f of Lemma 34 holds for f being the permanent, then the conclusion about f' holds for f' being the permanent as well.

Thus, we only need to look at the transformations of Lemma 35 and Lemma 33 and in the circuits constructed in their proofs. Inspecting the constructions, we find that the only parts that cannot be carried out by an efficient randomized algorithm is the tabulation of $f(y_{S_j})$ as a function of $y_{S_i \cap S_j}$ in the proof of Lemma 33 and the use of a free random source of $(y, f(y))$-pairs in the proof of Lemma 35. In both cases the function f is an XOR of several copies of the permanent function. Thus, with an *oracle* for the permanent, the construction could be carried out efficiently.

Summing up, if the generator does not work, there is a small circuit for the permanent and we can construct it by an efficient randomized algorithm with oracle access to the permanent. In short, the permanent is *learnable*. Inspecting the proof further, we see that to construct the circuit on instances of length n, we would need to query the oracle on instances of length at most n.

Without any assumptions, we know that the permanent is *downward self-reducible*: There is an efficient procedure for computing the permanent on inputs of length n, given an oracle for the permanent on all smaller input lengths.

Combining the learnability and the downward self-reducibility of the permanent, we can now construct a sequence of small circuits for the permanent on inputs of length $1, 2, \ldots, n$: Having already constructed circuits for inputs of length $1, 2, \ldots, i$, we can learn the circuit for inputs of length $i + 1$, as the circuit for inputs of length up to i, combined with the downward self-reducibility of the permanent, gives us an oracle for the permanent for all input lengths up to and including $i + 1$. This completes the proof.

10. UNCONDITIONAL DERANDOMIZATION OF SPACE-BOUNDED COMPUTATION

In this section, we present unconditional deterministic simulation results for randomized space-bounded computation. It may, at first, be surprising that such results are possible at all, as this is not the case for randomized time-bounded computation. However, by Lemma 3, to derandomize space-bounded computation, we only have to construct pseudorandom generators for *automata*, rather than *circuits*, and as we understand how automata compute much better than we understand how circuits compute, this is a much easier task. However, the ultimate goal, to prove $\mathbf{L} = \mathbf{BP_H L}$, is still open (though we already saw a hardness based generator giving this statement under an reasonable unproven assumption in Theorem 21).

10.1 THE NISAN-ZUCKERMAN GENERATOR

We prove the following theorem, due to Nisan and Zuckerman, 1996, implying Statement 8 of the introduction. The theorem improves an earlier theorem of Ajtai et al., 1987.

Theorem 30 *Let $s(n) \geq \log n$ be a proper space bound. If a language L is decided by a two-sided error randomized Turing machine using space $O(s(n))$ and making $s(n)^{O(1)}$ random moves, then L is decided by a deterministic Turing machine using space $O(s(n))$.*

As we aim for a derandomization when the number of random moves made is limited, we cannot directly use Lemma 3 in a proof of Theorem 30. However, inspecting the proof of that lemma, we see that to prove Theorem 30, it is enough to consider the following promise problem, s-LAYERED AUTOMATON ACCEPTANCE PROBABILITY ESTIMATION. As input, we are given a finite automaton A with n states. The automaton must be *layered*, meaning that for all states v in the automaton, all paths from the start state to v have the same length. In particular, the automaton must be acyclic. The set of nodes at distance i from the start state is called *layer i*. The maximum length of any path in the automaton (the *depth* of the automaton) must be at most $l = s(n)$. The automaton has a single accepting state. We are given as promise that *either* (a) $\Pr[A \text{ accepts } x] \leq \frac{1}{3}$ *or* (b) $\Pr[A \text{ accepts } x] \geq \frac{2}{3}$ for a randomly chosen $x \in \{0,1\}^l$, and we should tell which of (a) or (b) is the case.

To prove Theorem 30, it is enough to show that for all functions s with $s(n) = (\log n)^k$ for constant k, s-LAYERED AUTOMATON ACCEPTANCE PROBABILITY ESTIMATION is in **pBP$_H$L**. Using the terminology of Section 5, to show that this problem is in **pBP$_H$L**, it is enough to show that we can generate, on input n, using space $O(\log n)$, a pseudorandom set in $\{0,1\}^{(\log n)^k}$ for layered automata of size n and depth $(\log n)^k$.

In the proof, we shall actually make a generator using somewhat more space, namely $O((\log n)^{k-1/2})$ and then derive the stronger conclusion by "bootstrapping".

To define our generator G, we shall need an explicit extractor

$$E : \{0,1\}^{6\log n} \times \{0,1\}^{\sqrt{\log n}} \to \{0,1\}^{\log n}$$

with min-entropy threshold $3\log n$ and error $\epsilon < (\log n)^{-k-1}$. For instance, an extractor with the parameters of Theorem 6 or Theorem 7 (proved using either iterated universal hashing or codes and designs) will certainly do. The first extractor with the desired parameters was constructed for the application at hand, by Nisan and Zuckerman, 1996. For the application at hand, we need the map E to be computable in linear space (i.e., in space $O(\log n)$). This mild space constraint is satisfied by any of the constructions.

On input n, G goes through all strings $x \in \{0,1\}^{6 \log n + (\log n)^{k-1/2}}$ and outputs a string in $\{0,1\}^{(\log n)^k}$ for each of them. For each x, it divides x into a a block $x' \in \{0,1\}^{\log n}$ and $l = (\log n)^{k-1}$ small blocks, $y_1, y_2, \dots, y_l$, with $y_i \in \sqrt{\log n}$. Now, it outputs $E(x, y_1) \circ E(x, y_2) \circ \cdots \circ E(x, y_l)$.

Lemma 43 *G is a pseudorandom generator with error at most $\frac{1}{\log n}$ for layered automata of size n and depth $(\log n)^k$.*

Proof Given a layered automaton of size n and depth $(\log n)^k$, split its layers into $(\log n)^{k-1}$ blocks of depth $\log n$ each and collapse each block into a single layer, by changing the alphabet of the automaton from $\{0,1\}$ to $\{0,1\}^{\log n}$. Thus, we can view the automaton as a layered automaton of size n, alphabet size n and depth $l = (\log n)^{k-1}$.

Let U_i be the distribution on nodes in layer i induced by running the automaton on input uniformly chosen from $(\{0,1\}^{\log n})^i$ and let P_i be the distribution on nodes in layer i induced by running the automaton on a random member of the set generated by G, i.e., on input $E(x', y_1) \circ E(x', y_2) \cdots \circ E(x', y_i)$ for random uniformly chosen $x' \in \{0,1\}^{6 \log n}, y_1, y_2, \dots, y_i \in \{0,1\}^{\sqrt{\log n}}$. Note that $U_0 = P_0$. We show, by induction in i, that $\|U_i - P_i\|_1$ is small. The statement of the theorem will follow by taking $i = l$.

So suppose that $\|U_{i-1} - P_{i-1}\|_1$ is small. We shall bound $\|U_i - P_i\|_1$ by $\|U_i - M_i\|_1 + \|M_i - P_i\|_1$, where M_i is the distribution obtained by running the automaton on $E(x', y_1) \circ E(x', y_2) \circ \cdots \circ E(x', y_{i-1}) \circ v$ where $x', y_1, y_2, \dots, y_{i-1}$ are as before, and $v \in \{0,1\}^{\log n}$ is uniformly and independently chosen. Thus, M_i is defined by running the automata on a pseudorandom input for $i-1$ steps and with a real random symbol in the i'th step. Note that by Lemma 19, $\|U_i - M_i\|_1 \le \|U_{i-1} - P_{i-1}\|_1$, so we only have to bound $\|M_i - P_i\|_1$.

Let v be any state of the $i-1$'th layer of the automaton which receives probability at least $\frac{1}{n^2}$ according to distribution P_{i-1}.

Consider the distribution of x, conditioned by the event V that state v is the state visited in the $i-1$'st layer in the corresponding computation. Any outcome of x has conditional probability at most $n^2 2^{-6 \log n}$, so x has min-entropy is at least $(6-2) \log n > 3 \log n$, conditioned by V. Thus, conditioned by V, $E(x', y_i)$ is ϵ-close to uniform, where $\epsilon < \frac{1}{(\log n)^{k+1}}$. This means that the distributions P_i and M_i, conditioned by V are ϵ-close. Now let G be the event that *some* state v of P_{i-1}-probability more than $\frac{1}{n^2}$ is visited. The conditional distributions of P_i and M_i are ϵ-close, given G. The probability of $\neg G$, under P_{i-1}, is at most $\frac{n}{n^2} = \frac{1}{n}$. By Lemma 18, this means that P_i and M_i are $\epsilon + 2 \cdot \frac{1}{n}$ close.

Thus, U_l and P_l are $l\epsilon + 2l\frac{1}{n}$-close, in particular, they are $\frac{1}{\log n}$-close. ∎

As an immediate corollary of Lemma 43, we get that a space $s(n)$ computation using $s(n)^{\frac{3}{2}}$ random binary moves can be simulated deterministically in space $O(s(n))$. For a space $s(n)$ computation using $s(n)^k$ random bits for a bigger constant k, it seems we only get a deterministic simulation in space $O(s(n)^{k-1/2})$. However, using the pseudorandom generator to do deterministic simulation, by cycling through all the strings in the pseudorandom set is not the only way we can use it. Looking at the way the pseudorandom set is defined, we see that we can *sample* a random element from the set using $s(n)^{k-1/2}$ random moves and using space only $s(n)$. Thus, we get the following lemma.

Lemma 44 *Let $k \geq \frac{3}{2}$ be a rational number and let $s(n) \geq \log n$ be a proper space bound. If a language L is decided by a two-sided error randomized Turing machine using space $O(s(n))$ and making $s(n)^k$ random moves, then L is decided by a deterministic Turing machine using space $O(s(n))$ and making $O(s(n)^{k-1/2})$ random moves.*

Now, Theorem 30 follows, by applying Lemma 44 a sufficient number of times.

10.2 THE IMPAGLIAZZO-NISAN-WIGDERSON GENERATOR

The generator of Impagliazzo et al., 1994 uses space $O((\log n)^2)$ and produces, on input n, a pseudorandom set in $\{0, 1\}^n$ for automata of size n. Thus, it can be used to derandomize space-bounded computation without a bound on the number of random bits used, by Lemma 3.

The following explanation of the generator is due to Raz and Reingold, 1999.

Recall that the Nisan-Zuckerman generator generates a set $\{G(x, y_1, \ldots, y_k) | x, y_1, y_2, \ldots, y_k\}$ where $G(x) = E(x, y_1) \circ E(x, y_2) \circ \cdots \circ E(x, y_l)$ and E is an extractor of appropriate parameters. Similarly, the Impagliazzo-Nisan-Wigderson generator generates a set $\{G_l(x) | x \in \{0, 1\}^{O(\log^2 n)}\}$, where $l = O(\log n)$ and G_l is defined by the following recursion:

$G_0(x) = x$

$G_i(xy) = G_{i-1}(x) \circ G_{i-1}(E_i(x, y)),$

where in the last equation, $x \in \{0, 1\}^{O(i \log n)}$, $y \in \{0, 1\}^{O(\log n)}$ and E_i is an extractor of appropriate parameters. G_l expands $O((\log n)^2)$ bits to

n bits. By a proof similar to the proof of the correctness of the Nisan-Zuckerman generator, the set of outputs of G_l is a pseudorandom set for automata of size n.

Note that we want the input and output domain of E_i to be identical. Appropriate extractors with this property are the ones directly based on expanders, described in Section 4.1. Thus in the construction of Impagliazzo et al., 1994, $E_i(x, y)$ is the y'th neighbor of x in (a power of) an explicit expander.

Using the pseudorandom generator, we can deterministically simulate any randomized space $s(n)$ computation in deterministic space $O(s(n)^2)$. As we mentioned in Section 2.1, that this is possible was known since the early 80's and without using pseudorandom generators. To get new simulation results, one has to use a modified generator and explore its structure carefully. This is done in the next subsection.

10.3   NISAN'S GENERATOR

Nisan's pseudorandom generator of Nisan, 1990 is very similar to the Impagliazzo-Nisan-Wigderson generator, and though it historically precedes it, it is at this point of our presentation best explained in terms of that generator.

Given a parameter n, let $m = c \log n$ for a suitable constant c, and let $\mathcal{H}$ be a strongly universal family of hash functions mapping $\{0,1\}^m$ to $\{0,1\}^m$.

We now define maps $G_i : \{0,1\}^m \times \mathcal{H}^i \to \{0,1\}^{2^i m}$, $i = 1, \ldots, m$, by
$G_0(x) = x$,
$G_i(x, h_1, \ldots, h_i) = G_{i-1}(x, h_1, \ldots, h_{i-1}) \circ G_{i-1}(h_i(x), h_1, \ldots, h_{i-1})$.

The claim is that $\{G_i(x)|x\}$ for $i \leq \log n$ are pseudorandom sets with small error for automata of size n. If we think of the value $h_i(x)$ as $E(x, h_i)$ where E is an extractor, and compare the recursion with the recursion in the definition of the Impagliazzo-Nisan-Wigderson generator, we see that we have essentially replaced $E(xh_1h_2 \ldots h_i)$ with $E(x, h_i)h_1h_2 \ldots h_{i-1}$, i.e., we do not feed the values $h_1, h_2, \ldots, h_{i-1}$ through the extractor, but just reuse them. Such a change would not work for any extractor, but it does work for *strongly* universal hash functions. The reason is that a stronger statement that just pseudorandomness of the resulting set can be shown by induction.

Recall that we use the term "pseudorandom set in $\{0,1\}^i$ for automata of size n" as a short hand for a pseudorandom set for sets in $\{0,1\}^i$ decided by a finite automaton with n states. Now consider a *fixed* finite automaton A with n states, and consider all n^2 variations of A obtained by moving the start state and single accepting state. These n^2 automata define n^2 languages over $\{0,1\}^i$. Now use the term "pseudorandom set in $\{0,1\}^i$ for A" as a short hand for a pseudorandom set for these n^2 languages.

The extra property that Nisan's generator enjoys is this: Given a *fixed* automaton A with n states, if we pick $h_1, h_2, \ldots, h_k$ at random and freeze them, then, with high probability, the distribution resulting from *then* picking just x at random, is pseudorandom for A.

In particular, in the bottom of the recursion, if h_1 is picked at random, then with high probability, the distribution $x \circ h_1(x)$ induced by random x, is pseudorandom.

The truth of this statement at the bottom of the recursion is the *hash mixing lemma* of Nisan, 1990, which we state without proof (the proof is a straightforward application of Chebyshev's inequality).

Lemma 45 *Let $\mathcal{H}$ be a strongly universal family of hash functions mapping $\{0,1\}^m$ to $\{0,1\}^m$. Let A and B be subsets $\{0,1\}^m$. Then*

$$\Pr_{h \in \mathcal{H}} \left[| \Pr_{x \in A}[h(x) \in B] - \mu(B)| \geq \epsilon \right] \leq \frac{\mu(B)(1 - \mu(B))}{\mu(A) 2^m \epsilon^2}$$

Note that it follows from $\Pr_{h \in \mathcal{H}}[| \Pr_{x \in A}[h(x) \in B] - \mu(B)| \geq \epsilon] \leq \frac{\mu(B)(1-\mu(B))}{\mu(A) 2^m \epsilon^2}$ that $| \Pr_{h \in \mathcal{H}, x \in A}[h(x) \in B] - \mu(B)| \leq \epsilon + \frac{\mu(B)(1-\mu(B))}{\mu(A) 2^m \epsilon^2}$. Thus, thinking of B as a statistical test, we see the hash mixing lemma implies an extraction property of the family $\mathcal{H}$, but it is the stronger mixing property that is relevant to us.

Using an induction based on Lemma 45, Nisan now shows that for a fixed automaton A of size n and random $h_1, h_2, \ldots, h_i$, with very high probability the set $\{G_i(x, h_1, h_2, \ldots, h_i) \,|\, x\}$ is a pseudorandom set in $\{0,1\}^{2^i m}$ for A.

In particular, the set $\{G_m(x, h_1, h_2, \ldots, h_i) | x, h_1, h_2, \ldots, h_i\}$ is a pseudorandom set for all automata of size n, which we can use by Lemma 3 to derandomize $\mathbf{BP_H L}$, again obtaining a simulation in deterministic space $O((\log n)^2)$, as we did using the Impagliazzo-Nisan-Wigderson generator.

However, *more can be said*. Suppose we have already found $h_1, h_2, \ldots, h_{i-1}$ so that $\{G_i(x, h_1, h_2, \ldots, h_{i-1}) | x\}$ is a pseudorandom set in $\{0,1\}^{2^i m}$ for A. Now we can do an *exhaustive search* for an appropriate setting of h_{i+1} so that $\{G_{i+1}(x, h_1, h_2, \ldots, h_{i+1}) | x\}$ is a pseudorandom set in $\{0,1\}^{2^{i+1} m}$ for A: We can check if a particular setting of h_{i+1} is okay by checking, for each pair of states a and b in the automaton, that a randomly chosen pseudorandom input from $\{G_{i+1}(x, h_1, h_2, \ldots, h_{i+1}) | x\}$ takes us from state a to state b with approximately the same probability as the concatenation of two randomly chosen inputs from

$\{G_i(x, h_1, h_2, \ldots, h_i)|x\}$ takes us from a to b. With a bit of care, this procedure can be made space-efficient. Thus, we can iteratively find appropriate values of $h_1, h_2, \ldots, h_{O(\log n)}$, thus obtaining a pseudorandom set in $\{0, 1\}^n$ for A. This implies the following theorem of Nisan, 1991, appearing as Statement 9 of the introduction.

Theorem 31 *Any language in* **BP$_H$L** *can be decided by a polynomial time algorithm using space* $O((\log n)^2)$.

Before Nisan's result, it was known that any language in **BP$_H$L** could be simulated in polynomial time by one algorithm and in space $O((\log n)^2)$ by another, but not that these time and space bounds could be obtained simultaneously.

10.4 SAKS AND ZHOU'S MATRIX EXPONENTIATION ALGORITHM

In this section, we sketch the proof of the following result of Saks and Zhou, 1999, yielding Statement 10 of the introduction.

Theorem 32 **BP$_H$L** *is a subset of* **SPACE**$((\log n)^{3/2})$

By lemma 4, we only have to show that APPROXIMATE STOCHASTIC MATRIX EXPONENTIATION can be solved using deterministic space $O((\log n)^{3/2})$.

APPROXIMATE STOCHASTIC MATRIX EXPONENTIATION can easily be solved using space $O((\log n)^2)$ by $O(\log n)$ repeated squarings, rounding off all entries in intermediate results to $O(\log n)$ bits. This yields Simon's result that **BP$_H$L$\subseteq$ SPACE**$((\log n)^2)$.

Alternatively, we can obtain the same space bound by using Lemma 4 to convert the problem into a question about the acceptance probability of a finite automaton, and then use Nisan's generator to generate, using space $O((\log n)^2)$, a pseudorandom set for the automaton and simulate the automaton on all elements of the set to answer the question.

Now consider the following *hybrid* between these two extremes. Let M be the original matrix. Then, $\sqrt{\log n}$ times, replace M with an approximation of M^l with $l = 2^{\sqrt{\log n}}$, the latter matrix obtained by converting all questions about entries in M^l to questions about the acceptance probability of a finite automaton with n states on an input of length $2^{\sqrt{\log n}}$, using the technique of Lemma 4, and then using Nisan's generator G_l to generate a pseudorandom set for this automaton and simulate the automaton on the set to answer the question.

If the latter hybrid technique is done as described, we would again use space $O((\log n)^2)$. However, recall that the pseudorandom set is indexed by $x, h_1, h_2, \ldots, h_l$, and for a fixed automaton A, if $h_1, h_2, \ldots, h_l$ are fixed to random values, with very high probability the distribution induced by picking x at random will be pseudorandom for A.

Now, if $h_1, h_2, \ldots, h_l$ are set to random values at the beginning of the computation and the above algorithm is carried out, *assuming* that the distribution induced by picking x at random will be a pseudorandom set for each of the automata occurring during the computation, we have a randomized algorithm using space $O((\log n)^{3/2})$ and using $O((\log n)^{3/2})$ random bits - and thus a deterministic algorithm using space $O((\log n)^{3/2})$, as desired.

However it is not clear that the assumption made will be correct with high probability, so the simulation may be incorrect. The reason is this. Picking random $h_1, h_2, \ldots, h_l$ at random works with very high probability for any *fixed* automaton. However, the automata occurring in the above simulation after the first approximate exponentiation has been done *depend on the $h_1, h_2, \ldots, h_l$ chosen*, as they depend on the first approximate matrix power obtained. If there was no such dependence, the above simulation would work.

Saks and Zhou solve this problem by modifying the simulation so that with very high probability over the choices of $h_1, h_2, \ldots, h_l$, the matrices (and hence the automata) occurring as subresults are actually a fixed sequence of matrices. That this is possible follows from the fact that though the matrices occurring as subresults of the computation depend on $h_1, h_2, \ldots, h_l$, they are, with high probability, a very good approximation of the original matrix to a certain power, say M^r. Thus, if we do a *rounding* of all entries in the matrix to a certain number of binary digits and do the same rounding of the entries in M^r, we are very likely to end up at the same matrix. Note that this may *not* hold if the entry in one matrix, is, say, 0.1010001, and the corresponding entry in the other is 0.1001111. Then, rounding down both entries to three digits would make the matrices different. Similar cases could occur for all other deterministic rounding schemes. Thus, Saks and Zhou use a *randomized rounding* scheme, essentially by picking a random offset δ, subtracting δ from each matrix entry, rounding down the entry, and adding δ again. Doing this for every matrix obtained as a subresult in the simulation ensures that, with high probability, the sequence of matrices occurring depends on δ and the original matrix M only. Thus, the simulation gives the right result with high probability. The extra randomness needed for the offset δ is negligible, so we can derandomize the choice of δ by running through all possible values.

10.5 FURTHER RESULTS

Armoni, 1998, presents a hybrid between the generators of Nisan and Zuckerman, Impagliazzo, Nisan and Wigderson, and Nisan. For certain setting of the parameters (number of random bits vs. space) the generator obtained gives a better deterministic simulation result than is obtainable using either of the 3 generators.

The Impagliazzo-Nisan-Wigderson generator extracts remaining randomness in the seed x after having finished half the computation. Raz and Reingold, 1999, observes that some randomness may be present in the *state* of the computation at this point and that this randomness could be extracted, making certain assumptions about the computation. Thus, they obtain improved derandomization results for restricted classes of space bounded computation.

For the special case of UNDIRECTED GRAPH CONNECTIVITY, probably the most natural example of a problem in $\mathbf{BP_H L}$ not known to be in $\mathbf{L}$, Armoni et al., 1997, present an $O((\log n)^{4/3})$ space algorithm, based on refining the Saks-Zhou technique described above.

Notes

1. The chapter was essentially completed in February 2000 and only results known to the author at that time have been incorporated.

2. The fact that this answer is either "yes" or "no" is less important. Thus, most of the results can be adapted to deal with algorithms seeking a particular numerical value, such as Monte Carlo integration, but we shall not do so.

3. One can try, however, to get the expected number of random bits used to be close to the entropy of the desired output distribution. This problem was considered by Impagliazzo and Zuckerman, 1989.

4. Though they shall not be covered explicitly, both the technique of conditional probabilities and the technique of small sample spaces play a role for generic derandomization as well, as will be apparent.

5. The notation is not quite standard. Buhrman and Fortnow, 1999, use **promiseC**, but this is a bit too awkward if one really wants to use the notation as we do.

6. Goldreich and Zuckerman, 1997 only states the result for $L \in \mathbf{E}$, but the proof is easily seen to generalize.

References

Adleman, L. (1978). Two theorems on random polynomial time. In *19th Annual Symposium on Foundations of Computer Science*, pages 75–83.

Ajtai, M., Komlós, J., and Szemerédi, E. (1987). Deterministic simulation in LOGSPACE. In *Proceedings of the Nineteenth Annual ACM Symposium on Theory of Computing*, pages 132–140.

Aleliunas, R., Karp, R. M., Lipton, R. J., Lovász, L., and Rackoff, C. (1979). Random walks, universal traversal sequences, and the complexity of maze problems. In *20th Annual Symposium on Foundations of Computer Science*, pages 218–223.

Allender, E., Reinhardt, K., and Zhou, S. (1999). Isolation, matching, and counting uniform and nonuniform upper bounds. *J. Comput. System Sci.*, 59(2):164–181.

Alon, N. (1986). Eigenvalues and expanders. *Combinatorica*, 6(2):83–96.

Alon, N. and Milman, V. D. (1985). λ_1, isoperimetric inequalities for graphs, and superconcentrators. *J. Combin. Theory Ser. B*, 38(1):73–88.

Alon, N. and Spencer, J. H. (1992). *The probabilistic method.* John Wiley & Sons Inc., New York.

Andreev, A. E., Clementi, A. E. F., and Rolim, J. D. P. (1997). Worst-case hardness suffices for derandomization: A new method for hardness-randomness trade-offs. In *Automata, Languages and Programming, 24th International Colloquium*, volume 1256 of *Lecture Notes in Computer Science*, pages 177–187. Springer-Verlag.

Andreev, A. E., Clementi, A. E. F., and Rolim, J. D. P. (1998). A new general derandomization method. *Journal of the Association for Computing Machinery*, 45(1):179–213.

Andreev, A. E., Clementi, A. E. F., Rolim, J. D. P., and Trevisan, L. (1999). Weak random sources, hitting sets, and BPP simulations. *SIAM J. Comput.*, 28(6):2103–2116.

Armoni, R. (1998). On the de-randomization of space-bounded computations. In *Randomization and Approximation Techniques in Computer Science, Second International Workshop*, volume 1518 of *Lecture Notes in Computer Science*, pages 47–59.

Armoni, R., Ta-Shma, A., Wigderson, A., and Zhou, S. (1997). $SL \subseteq L^{\frac{4}{3}}$. In *Proceedings of the Twenty-Ninth Annual ACM Symposium on Theory of Computing*, pages 230–239.

Babai, L., Fortnow, L., Levin, L. A., and Szegedy, M. (1991a). Checking computations in polylogarithmic time. In *Proceedings of the 23rd Annual ACM Symposium on the Theory of Computing*, pages 21–31.

Babai, L., Fortnow, L., and Lund, C. (1991b). Nondeterministic exponential time has two-prover interactive protocols. *Computational Complexity*, 1(1):3–40.

Bellare, M., Goldreich, O., and Sudan, M. (1998). Free bits, PCPs, and non-approximability - towards tight results. *SIAM J. Comput.*, 27:804–915.

Berlekamp, E. and Welch, L. (1986). Error correction of algebraic block codes. US Patent Number 4,633,470.

Blum, M. and Micali, S. (1984). How to generate cryptographically strong sequences of pseudo-random bits. *SIAM Journal on Computing*, 13(4):850–864.

Boppana, R. B. and Sipser, M. (1990). The complexity of finite functions. In *Handbook of theoretical computer science, Vol. A*, pages 757–804. Elsevier, Amsterdam.

Buhrman, H. and Fortnow, L. (1999). One-sided versus two-sided error in probabilistic computation. In *Theoretical Aspects of Computer Science, 16th Annual Symposium*, volume 1563 of *Lecture Notes in Computer Science*, pages 100–109. Springer-Verlag.

Carter, J. L. and Wegman, M. N. (1979). Universal classes of hash functions. *J. Comput. System Sci.*, 18(2):143–154.

Chor, B. and Goldreich, O. (1989). On the power of two-point based sampling. *Journal of Complexity*, 5:96–106.

Clementi, A. E., Rolim, J. D., and Trevisan, L. (1998). Recent advances towards proving P = BPP. *Bulletin of the European Association for Theoretical Computer Science*, 64:96–103.

Cohen, A. and Wigderson, A. (1989). Dispersers, deterministic amplification, and weak random sources (extended abstract). In *30th Annual Symposium on Foundations of Computer Science*, pages 14–19.

Dietzfelbinger, M. (1996). Universal hashing and k-wise independent random variables via integer arithmetic without primes. In *13th Annual Symposium on Theoretical Aspects of Computer Science*, volume 1046 of *lncs*, pages 569–580. Springer.

Gabber, O. and Galil, Z. (1981). Explicit construction of linear sized superconcentrators. *Journal of Computer and System Sciences*, 22:407.

Goldreich, O. (1999). *Modern cryptography, probabilistic proofs and pseudorandomness*. Springer-Verlag, Berlin.

Goldreich, O. and Levin, L. A. (1989). A hard-core predicate for all one-way functions. In *Proc. 21st Annual ACM Symposium on Theory of Computing*, pages 25–32.

Goldreich, O., Nisan, N., and Wigderson, A. (1995). On Yao's XOR-lemma. Technical Report TR95-050, Electronic Colloquium on Computational Complexity.

Goldreich, O. and Wigderson, A. (1994). Tiny families of functions with random properties: A quality-size trade-off for hashing (preliminary version). In *Proceedings of the Twenty-Sixth Annual ACM Symposium on the Theory of Computing*, pages 574–583.

Goldreich, O. and Wigderson, A. (1999). Improved derandomization of BPP using a hitting set generator. In *Randomization and Approximation Techniques in Computer Science, Third International Workshop.*, volume 1671 of *Lecture Notes in Computer Science*, pages 131–139.

938

Goldreich, O. and Zuckerman, D. (1997). Another proof that BPP sub-seteq PH (and more). Technical Report TR97-045, Electronic Collo-quium on Computational Complexity.

Impagliazzo, R. (1995). Hard-core distributions for somewhat hard prob-lems. In *Proc. 36th Annual IEEE Symposium on Foundations of Com-puter Science*, pages 538–547.

Impagliazzo, R., Levin, L. A., and Luby, M. (1989). Pseudo-random generation from one-way functions. In *Proceedings of the Twenty First Annual ACM Symposium on Theory of Computing*, pages 12–24.

Impagliazzo, R., Nisan, N., and Wigderson, A. (1994). Pseudorandom-ness for network algorithms. In *Proceedings of the Twenty-Sixth An-nual ACM Symposium on the Theory of Computing*, pages 356–364.

Impagliazzo, R., Shaltiel, R., and Wigderson, A. (1999). Near-optimal conversion of hardness into pseudo-randomness. In *40th Annual Sym-posium on Foundations of Computer Science*, pages 181–190.

Impagliazzo, R., Shaltiel, R., and Wigderson, A. (2000). Extractors and pseudo-random generators with optimal seed length. In *Proceedings of the Thirty-Second Annual ACM Symposium on Theory of Computing*, pages 1–10.

Impagliazzo, R. and Wigderson, A. (1997). P=BPP if E requires expo-nential circuits: Derandomizing the XOR lemma. In *Proc. 29th Annual ACM Symposium on Theory of Computing*, pages 220–229.

Impagliazzo, R. and Wigderson, A. (1998). Randomness vs. time: De-randomization under a uniform assumption. In *Proceedings of the 39th Annual Symposium on Foundations of Computer Science*, pages 734–743.

Impagliazzo, R. and Zuckerman, D. (1989). How to recycle random bits. In *30th Annual Symposium on Foundations of Computer Science*, pages 248–253.

Karp, R., Pippenger, N., and Sipser, M. (1986). Expanders, randomness, or time versus space. In *First Annual Conference on Structure in Complexity Theory*, pages 325–329.

Karp, R. M. and Lipton, R. J. (1980). Some connections between nonuni-form and uniform complexity classes. In *Conference Proceedings of the Twelfth Annual ACM Symposium on Theory of Computing*, pages 302–309, Los Angeles, California.

Klivans, A. and Servedio, R. (1999). Boosting and hard-core sets. In *40th Annual Symposium on Foundations of Computer Science*, pages 624–633.

Klivans, A. R. and van Melkebeek, D. (1999). Graph nonisomorhism has subexponential size proofs unless the polynomial-time hierarchy collapses. In *Proc. 31st ACM Symposium on Theory of Computing*, pages 659–667.

Knuth, D. E. (1981). *The art of computer programming. Vol. 2.* Addison-Wesley Publishing Co., Reading, Mass., second edition. Seminumerical algorithms, Addison-Wesley Series in Computer Science and Information Processing.

Lautemann, C. (1983). BPP and the polynomial hierarchy. *Information Processing Letters*, 17(4):215–217.

Levin, L. A. (1987). One way functions and pseudorandom generators. *Combinatorica*, 7(4):357–363.

Lubotzky, A., Phillips, R., and Sarnak, P. (1986). Explicit expanders and the Ramanujan conjectures. In *Proceedings of the Eighteenth Annual ACM Symposium on Theory of Computing*, pages 240–246, Berkeley, California.

Lund, C., Fortnow, L., Karloff, H., and Nisan, N. (1992). Algebraic methods for interactive proof systems. *Journal of the ACM*, 39(4):859–868.

Lupanov, O. (1965). A method of synthesis of control system - the principle of local coding. *Problemy Kibernet.*, 10:31–110.

Mansour, Y., Nisan, N., and Tiwari, P. (1993). The computational complexity of universal hashing. *Theoretical Computer Science*, 107(1):121–133.

Margulis, G. A. (1973). Explicit constructions of expanders. *Problemy Peredači Informacii*, 9(4):71–80.

Miltersen, P. B. (1998). Error correcting codes, perfect hashing circuits, and deterministic dynamic dictionaries. In *Proc. 9th Annual ACM-SIAM Symposium on Discrete Algorithms*, pages 556–563.

Miltersen, P. B. and Vinodchandran, N. (1999). Derandomizing Arthur-Merlin games using hitting sets. In *40th Annual Symposium on Foundations of Computer Science*, pages 71–80.

Motwani, R. and Raghavan, P. (1995). *Randomized algorithms.* Cambridge University Press, Cambridge.

Nisan, N. (1990). Psuedorandom generators for space-bounded computation. In *Proceedings of the Twenty Second Annual ACM Symposium on Theory of Computing*, pages 204–212.

Nisan, N. (1991). Pseudorandom bits for constant depth circuits. *Combinatorica*, 1:63–70.

Nisan, N. (1994). RL ⊆ SC. *Computational Complexity*, 4(1):1–11.

Nisan, N. and Ta-Shma, A. (1999). Extracting randomness: a survey and new constructions. *J. Comput. System Sci.*, 58(1, part 2):148–173.

Nisan, N. and Wigderson, A. (1994). Hardness vs randomness. *Journal of Computer and System Sciences*, 49(2):149–167.

940

Nisan, N. and Zuckerman, D. (1996). Randomness is linear in space. *Journal of Computer and System Sciences*, 52(1):43–52.

Papadimitriou, C. (1994). *Computational Complexity*. Addison-Wesley.

Rabin, M. O. (1976). Probabilistic algorithms. In *Algorithms and complexity (Proc. Sympos., Carnegie-Mellon Univ., Pittsburgh, Pa., 1976)*, pages 21–39. Academic Press, New York.

Radhakrishnan, J. and Ta-Shma, A. (1997). Tight bounds for depth-two superconcentrators. In *38th Annual Symposium on Foundations of Computer Science*, pages 585–594.

Raz, R. and Reingold, O. (1999). On recycling the randomness of states in space bounded computation. In *Proceedings of the Thirty-First Annual ACM Symposium on Theory of Computing*, pages 168–178.

Raz, R., Reingold, O., and Vadhan, S. (1999a). Error reduction for extractors. In *40th Annual Symposium on Foundations of Computer Science*.

Raz, R., Reingold, O., and Vadhan, S. (1999b). Extracting all the randomness and reducing the error in Trevisan's extractor. In *Proceedings of the Thirty-First Annual ACM Symposium on Theory of Computing*, pages 149–158.

Saks, M. (1996). Randomization and derandomization in space-bounded computation. In *Proceedings, Eleventh Annual IEEE Conference on Computational Complexity*, pages 128–149, Philadelphia, Pennsylvania. IEEE Computer Society Press.

Saks, M., Srinivasan, A., and Zhou, S. (1998). Explicit OR-dispersers with polylogarithmic degree. *Journal of the Association for Computing Machinery*, 41(1):123–154.

Saks, M. and Zhou, S. (1999). $BP_H SPACE(S) \subseteq DSPACE(S^{3/2})$. *J. Comput. System Sci.*, 58(2):376–403.

Santha, M. and Vazirani, U. V. (1986). Generating quasirandom sequences from semirandom sources. *J. Comput. System Sci.*, 33(1):75–87.

Schönhage, A. and Strassen, V. (1971). Schnelle Multiplikation grosser Zahlen. *Computing (Arch. Elektron. Rechnen)*, 7:281–292.

Shamir, A. (1992). IP=PSPACE. *Journal of the ACM*, 39:869–877.

Shannon, C. E. (1949). The synthesis of two-terminal switching circuits. *Bell. Syst. Tech. J.*, 28:59–98.

Shoup, V. (1994). Fast construction of irreducible polynomials over finite fields. *J. Symbolic Comput.*, 17(5):371–391.

Simon, J. (1981). On tape-bounded probabilistic Turing machine acceptors. *Theoretical Computer Science*, 16(1):75–91.

Sipser, M. (1983). A complexity theoretic approach to randomness. In *Proceedings of the Fifteenth Annual ACM Symposium on Theory of Computing*, pages 330–335.

Sipser, M. (1988). Expanders, randomness, or time versus space. *Journal of Computer and System Sciences*, 36:379–383.

Srinivasan, A. and Zuckerman, D. (1999). Computing with very weak random sources. *SIAM J. Comput.*, 28(4):1433–1459.

Sudan, M., Trevisan, L., and Vadhan, S. (1999). Pseudorandom generators without the XOR lemma. In *Proc. 31st Annual ACM Symposium on Theory of Computing*, pages 537–546.

Ta-Shma, A. (1996). On extracting randomness from weak random sources (extended abstract). In *Proceedings of the Twenty-Eighth Annual ACM Symposium on the Theory of Computing*, pages 276–285.

Ta-Shma, A. (1998). Almost optimal dispersers. In *Proc. 30th Annual ACM Symposium on the Theorem of Computing*, pages 196–202.

Tanner, R. M. (1984). Explicit concentrators from generalized N-gons. *SIAM J. Algebraic Discrete Methods*, 5(3):287–293.

Toda, S. (1991). PP is as hard as the polynomial-time hierarchy. *SIAM Journal on Computing*, 20(5):865–877.

Trevisan, L. (1999). Constructions of near-optimal extractros using pseudorandom generators. In *Proc. 31st ACM Symposium on Theory of Computing*, pages 141–148.

Wigderson, A. and Zuckerman, D. (1999). Expanders that beat the eigenvalue bound: explicit construction and applications. *Combinatorica*, 19(1):125–138.

Yao, A. C. (1982). Theory and application of trapdoor functions. In *Proc. 23rd Annual IEEE Symposium on Foundations of Computer Science*, pages 80–91.

Zuckerman, D. (1990). General weak random sources. In *31st Annual Symposium on Foundations of Computer Science*, volume II, pages 534–543.

Zuckerman, D. (1991). Simulating BPP using a general weak random source. In *32nd Annual Symposium on Foundations of Computer Science*, pages 79–89.

Zuckerman, D. (1996). Randomness-optimal sampling, extractors, and constructive leader election. In *Proceedings of the Twenty-Eighth Annual ACM Symposium on the Theory of Computing*, pages 286–295.

INDEX

Combinatorial Optimization

1. E. Çela: *The Quadratic Assignment Problem*. Theory and Algorithms. 1998
    ISBN 0-7923-4878-8
2. M.Sh. Levin: *Combinatorial Engineering of Decomposable Systems*. 1998
    ISBN 0-7923-4950-4
3. A.I. Barros: *Discrete and Fractional Programming Techniques for Location Models*.
    1998   ISBN 0-7923-5002-2
4. V. Boltyanski, H. Martini and V. Soltan: *Geometric Methods and Optimization Problems*. 1999   ISBN 0-7923-5454-0
5. P.M. Pardalos and S. Rajasekaran (eds.): *Advances in Randomized Parallel Computing*. 1999   ISBN 0-7923-5714-0
6. D.-Z. Du, J.M. Smith and J.H. Rubinstein (eds.): *Advances in Steiner Trees*. 2000
    ISBN 0-7923-6110-5
7. P.M. Pardalos and L.S. Pitsoulis (eds.): *Nonlinear Assignment Problems*. Algorithms
    and Applications. 2000   ISBN 0-7923-6646-8
8. C. Wynants: *Network Synthesis Problems*. 2000   ISBN 0-7923-6689-1
9. S. Rajasekaran, P.M. Pardalos, J.H. Reif and J. Rolim (eds.): *Handbook of Randomized
    Computing*. Volume I. 2001   ISBN 0-7923-6957-2; Set: ISBN 0-7923-6959-9
    S. Rajasekaran, P.M. Pardalos, J.H. Reif and J. Rolim (eds.): *Handbook of Randomized
    Computing*. Volume II. 2001   ISBN 0-7923-6958-0; Set: ISBN 0-7923-6959-9

KLUWER ACADEMIC PUBLISHERS – DORDRECHT / BOSTON / LONDON